W9-DFH-571

Nuts & Seeds in Health and Disease Prevention

Nuts & Seeds in Health and Disease Prevention

Edited by

Victor R. Preedy

Department of Nutrition and Dietetics,
Nutritional Sciences Division, School of Biomedical & Health Sciences,
King's College London, Franklin-Wilkins Building, London, UK

Ronald Ross Watson

University of Arizona,
Division of Health Promotion Sciences,
Mel and Enid Zuckerman College of Public Health, and School of Medicine,
Arizona Health Sciences Center, Tucson, AZ, USA

Vinood B. Patel

Department of Biomedical Sciences, School of Biosciences,
University of Westminster, London, UK

AMSTERDAM • BOSTON • HEIDELBERG • LONDON • NEW YORK • OXFORD
PARIS • SAN DIEGO • SAN FRANCISCO • SINGAPORE • SYDNEY • TOKYO

Academic Press is an imprint of Elsevier

Academic Press is an imprint of Elsevier
32 Jamestown Road, London NW1 7BY, UK
30 Corporate Drive, Suite 400, Burlington, MA 01803, USA
525 B Street, Suite 1800, San Diego, CA 92101-4495, USA

First edition 2011

Notice

No responsibility is assumed by the publisher for any injury and/or damage to persons or property as
a matter of products liability, negligence or otherwise, or from any use or operation of any methods,
products, instructions or ideas contained in the material herein. Because of rapid advances in the medical
sciences, in particular, independent verification of diagnoses and drug dosages should be made

British Library Cataloguing-in-Publication Data
A catalogue record for this book is available from the British Library

Library of Congress Cataloging-in-Publication Data
A catalog record for this book is available from the Library of Congress

ISBN: 978-0-12-375688-6

For information on all Academic Press publications
visit our website at elsevierdirect.com

Typeset by TNQ Books and Journals Pvt Ltd.
www.tnq.co.in

Printed and bound in United States of America

11 12 13 14 15   10 9 8 7 6 5 4 3 2 1

Working together to grow
libraries in developing countries

www.elsevier.com | www.bookaid.org | www.sabre.org

ELSEVIER   BOOK AID International   Sabre Foundation

CONTENTS

v

CONTENTS

vi

CONTENTS

viii

CONTENTS

CONTENTS

CONTENTS

xiv

Mansurah A. Abdulazeez
Department of Biochemistry, Faculty of Science, Ahmadu Bello University Zaria, Kaduna State, Nigeria

Mahinda Abeywardena
CSIRO Food and Nutritional Sciences, Adelaide, South Australia, Australia

Mujitaba S. Abubakar
Department of Pharmacognosy and Drug Development, Ahmadu Bello University, Zaria, Nigeria

G.O. Adegoke
Department of Food Technology, University of Ibadan, Ibadan, Oyo State, Nigeria

Oluyemisi Elizabeth Adelakun
Department of Food Science and Engineering, Ladoke Akintola University of Technology, Ogbomoso, Oyo State, Nigeria

Shyam S. Agrawal
HOD Department of Pharmacology, Delhi Institute of Pharmaceutical Sciences and Research, New Delhi, India

Francis Agyemang-Yeboah
Department of Molecular Medicine, School of Medical Science, Kwame Nkrumah University of Science and Technology, Kumasi, Ghana

Cecilia Aiello
Dipartimento di Genetica e Biologia Molecolare, Sapienza Università di Roma, Roma, Italy

Ibironke Adetolu Ajayi
Industrial Unit, Chemistry Department, Faculty of Science, University of Ibadan, Ibadan, Oyo State, Nigeria

Olubunmi Bolanle Ajayi
Department of Biochemistry, University of Ado-Ekiti, Nigeria

Teresa A. Akeng'a
Academic Division, Bondo University College (Constituent College of Maseno University), Bondo, Kenya

Aderemi Caleb Aladeokin
Department of Pharmacology and Therapeutics, University of Ibadan, Ibadan, Nigeria

Mohamed Ali Al-Farsi
Date Processing Research Laboratory, Ministry of Agriculture, Oman

Amna Ali
Beth Israel Deaconess Medical Center, Division of Cardiology, Boston, Massachusetts, USA

Niloufer Sultan Ali
Department of Family Medicine, The Aga Khan University, Karachi, Pakistan

Ryszard Amarowicz
Institute of Animal Reproduction and Food Research of the Polish Academy of Sciences, Olsztyn, Poland

Sunday J. Ameh
Department of Medicinal Chemistry and Quality Control, National Institute for Pharmaceutical Research and Development (NIPRD), Garki, Abuja, Nigeria

Larissa Lovatto Amorin
Universidade Federal de Ouro Preto, Departamento de Ciências Biológicas, Ouro Preto, Minas Gerais, Brasil

B. Andallu
Department of Home Science, Sri Sathya Sai University, Anantapur, Andhra Pradesh, India

Paula B. Andrade
REQUIMTE/Department of Pharmacognosy, Faculty of Pharmacy, Porto University, Porto, Portugal

Sirajudheen Anwar
Department of Pharmacology, Prin.K.M. Kundanai College of Pharmacy, Cuffe Parade, Mumbai, India

Harry Archimède
INRA, UR143 Unité de Recherches Zootechniques, Centre INRA-Antilles-Guyane, Domaine de Duclos, Petit-Bourg, Guadeloupe (French West Indies)

Ignasius Radix Astadi
Department of Food Technology, Widya Mandala Surabaya Catholic University, Surabaya, Indonesia

Everaldo Attard
Institute of Earth Systems, Division of Rural and Food Systems, University of Malta, Msida MSD, Malta

Marcos Aurélio de Santana
Universidade Federal dos Vales do Jequitinhonha e Mucuri, Departamento de Farmácia, Diamantina, Minas Gerais, Brasil

Cecile Badet
Laboratoire Odontologique de Recherche, UFR d'Odontologie, Université Victor Segalen, Bordeaux Cedex, France

Sachin L. Badole
Department of Pharmacology, Poona College of Pharmacy, Bharati Vidyapeeth Deemed University, Maharashtra, India

Daniel J. Ballhorn
Department of Plant Biology, University of Minnesota, St Paul, Minnesota, USA

Monica S. Banach
Clinical Nutrition & Risk Factor Modification Center, St Michael's Hospital, Toronto, Ontario, Canada; Department of Pharmaceutical Sciences, Faculty of Medicine, University of Toronto, Toronto, Ontario, Canada

Douglas C. Baxter
ALS Scandinavia AB, Luleå, Sweden

Dorothea Bedigian
Missouri Botanical Garden, St Louis, Missouri, USA

Mikhail A. Belozersky
A.N. Belozersky Institute of Physico-Chemical Biology, Moscow State University, Moscow, Russia

Valerio Berardi
Dipartimento di Genetica e Biologia Molecolare, Sapienza Università di Roma, Roma, Italy;
On leave at SUNY Downstate Medical Center, Department of Physiology and Pharmacology,
Brooklyn, New York, USA

Trust Beta
Department of Food Science, University of Manitoba, Winnipeg, Manitoba, Canada

Aruna Bhatia
Immunology and Immunotechnology Laboratory, Department of Biotechnology, Punjabi
University, Punjab, India

Sanjib Bhattacharya
Bengal School of Technology (A College of Pharmacy), Sugandha, Hooghly, West Bengal,
India

Jitendra D. Bhosale
National Research Institute of Basic Ayurvedic Sciences, CCRAS, Nehru Garden, Kothrud,
Pune, India

Sujit K. Bhutia
Department of Biotechnology, Indian Institute of Technology, Kharagpur, India

Hans Konrad Biesalski
Institute for Biological Chemistry and Nutrition, University of Hohenheim, Stuttgart,
Germany

Abdullah bin Habeeballah bin Abdullah Juma
Department of Orthopaedics & Trauma, Faculty of Medicine, King Abdul Aziz University,
Jeddah, Saudi Arabia

Rune Blomhoff
Department of Nutrition, Institute of Basic Medical Sciences, Faculty of Medicine, University
of Oslo, Norway

Judith Boateng
Alabama A & M University, Normal, Alabama, USA

Subhash L. Bodhankar
Department of Pharmacology, Poona College of Pharmacy, Bharati Vidyapeeth University,
Maharashtra, India

Anders Borgen
Agrologica, Mariager, Denmark

Tomasz Brzozowski
Department of Physiology, Jagiellonian University Medical College, Cracow, Poland

H.N. Büyükkartal
Ankara University, Faculty of Science, Department of Biology, Tandoğan, Ankara, Turkey

Deon V. Canyon
Disaster Health and Crisis Management Unit, Anton Breinl Centre for Public Health
and Tropical Medicine, James Cook University, Townsville Qld, Australia

Monica H. Carlsen
Department of Nutrition, Institute of Basic Medical Sciences, Faculty of Medicine,
University of Oslo, Norway

Benildo S. Cavada
BioMol-Lab, Federal University of Ceará, Fortaleza/Ceará, Brazil

Pierre Champy
Laboratoire de Pharmacognosie, UMR CNRS 8076 BioCIS, Faculté de Pharmacie, Université Paris-Sud 11, France

Cheng-Jie Chen
Department of Naval Medicine, Second Military Medical University, Shanghai, China

Jianwei Chen
Department of Pharmaceutical Chemistry, School of Pharmacy, Nanjing University of Chinese Medicine, Nanjing, Jiangsu, China

Yong Chen
Department of Pharmaceutical Chemistry, School of Pharmacy, Nanjing University of Chinese Medicine, Nanjing, Jiangsu, China

Eliton Chivandi
School of Physiology, Faculty of Health Sciences, University of the Witwatersrand, Johannesburg, Republic of South Africa

Chanya Chaicharoenpong
Institute of Biotechnology and Genetic Engineering, Chulalongkorn University, Bangkok, Thailand

Yau-Sang Chan
School of Biomedical Sciences, Faculty of Medicine, The Chinese University of Hong Kong, Hong Kong, China

Randy C.F. Cheung
School of Biomedical Sciences, Faculty of Medicine, The Chinese University of Hong Kong, Hong Kong, China

Adriana Chicco
University of Litoral, Department of Biochemistry, School of Biochemistry, Santa Fe, Argentina

Pei-Yi Chu
Department of Environmental and Occupational Health, National Cheng Kung University Medical College, Tainan, Taiwan

M. Carmen Cid
Instituto Canario de Investigaciones Agrarias, La Laguna, Spain

William de Castro Borges
Universidade Federal de Ouro Preto, Departamento de Ciências Biológicas, Ouro Preto, Minas Gerais, Brasil

Luis Cisneros-Zevallos
Department of Horticultural Sciences, Fruit and Vegetable Improvement Center, Texas A&M University, College Station, Texas, USA

Wayne Coates
The University of Arizona, Sonoita, Arizona, USA

Marie M. Cochard
Adult & Child Allergy Unit, University Hospitals of Geneva, Geneva, Switzerland

Marina Cocchi
Department of Chemistry, University of Modena and Reggio Emilia, Modena, Italy

Hatice Çölgeçen
Zonguldak Karaelmas University, Faculty of Arts and Science, Department of Biology, İncivez, Zonguldak, Turkey

Marina Contini
Dipartimento di Scienze e Tecnologie Agroalimentari, Tuscia University, Viterbo, Italy

Rajesh Dabur
National Research Institute of Basic Ayurvedic Sciences, CCRAS, Nehru Garden, Kothrud, Pune, India

Patrick A. Dakia
Unité de Formation et de Recherche des Sciences et Technologies des Aliments (UFR-STA), Université d'Abobo-Adjamé (UAA), Abidjan, Côte d'Ivoire

B. Daramola
Department of Food Technology, Federal Polytechnic, Ado-Ekiti, Ekiti State, Nigeria

Teresa Delgado
CIMO/Escola Superior Agrária, Instituto Politécnico de Bragança, Bragança, Portugal

Sandhya Desai
Department of Pharmacology, Prin.K.M. Kundanai College of Pharmacy, Cuffe Parade, Mumbai, India

Saikat Dewanjee
Advanced Pharmacognosy Research Laboratory, Department of Pharmaceutical technology, Jadavpur University, Kolkata, India

Giuseppa Di Bella
Dipartimento di Scienze degli Alimenti e dell'Ambiente "G. Stagno d'Alcontres", Università di Messina, Messina, Italy

Irene Dini
Dipartimento di Chimica delle Sostanze Naturali, Università di Napoli "Federico II", Naples, Italy

Danuta Drozdowicz
Department of Physiology, Jagiellonian University Medical College, Cracow, Poland

Noélia Duarte
iMed.UL, Faculty of Pharmacy, University of Lisbon, Lisbon, Portugal

Anuradha Dube
Division of Parasitology, Central Drug Research Institute, Lucknow, India

Giacomo Dugo
Dipartimento di Scienze degli Alimenti e dell'Ambiente "G. Stagno d'Alcontres", Università di Messina, Messina, Italy

Yakov E. Dunaevsky
A.N. Belozersky Institute of Physico-Chemical Biology, Moscow State University, Moscow, Russia

María del Carmen Durán-de-Bazúa
Chemical Engineering Department, Facultad de Química, Universidad Nacional Autónoma de México, México

Caterina Durante
Department of Chemistry, University of Modena and Reggio Emilia, Modena, Italy

Rafael Cypriano Dutra
Department of Food and Toxicology, Faculty of Pharmacy and Biochemistry, Federal University of Juiz de Fora, Juiz de Fora, Brazil

George A. Dyer
The Macaulay Land Use Research Institute, Craigiebuckler, Aberdeen, UK

Tsezi A. Egorov
Shemiakin and Ovchinnikov Institute of Bioorganic Chemistry, Russian Academy of Sciences, Moscow, Russia

Akram Eidi
Department of Biology, Science and Research Branch, Islamic Azad University, Tehran, Iran

Maryam Eidi
Department of Biology, Varamin Branch, Islamic Azad University, Varamin, Iran

Philippe A. Eigenmann
Child Allergy Unit, University Hospitals of Geneva, Geneva, Switzerland

Mohamed Elleuch
Unité de Valorisation des Résultats Scientifiques, Centre de Biotechnologie de Sfax, Sfax, Tunisia

Emma Engström
ALS Scandinavia AB, Luleå, Sweden

Kennedy H. Erlwanger
School of Physiology, Faculty of Health Sciences, University of the Witwatersrand, Johannesburg, Republic of South Africa

Amin Esfahani
Clinical Nutrition & Risk Factor Modification Center, St Michael's Hospital, Toronto, Ontario, Canada; Department of Nutritional Sciences, Faculty of Medicine, University of Toronto, Toronto, Ontario, Canada

Letícia Estevinho
CIMO/Escola Superior Agrária, Instituto Politécnico de Bragança, Bragança, Portugal

Mohd Fadzelly Abu Bakar
Laboratory of Natural Products, Institute for Tropical Biology and Conservation, Universiti Malaysia Sabah, Kota Kinabalu, Sabah, Malaysia

Evandro F. Fang
School of Biomedical Sciences, Faculty of Medicine, The Chinese University of Hong Kong, Hong Kong, China

Fan-Fu Fang
Department of Traditional Chinese Medicine, Changhai Hospital, Shanghai, China

E. Olatunde Farombi
Drug Metabolism and Toxicology Research Laboratories, Department of Biochemistry, College of Medicine, University of Ibadan, Nigeria

Afrânio G. Fernandes
Center of Sciences and Technology, State University of Ceará, Fortaleza/CE, Brazil

Maria-José U. Ferreira
iMed.UL, Faculty of Pharmacy, University of Lisbon, Lisbon, Portugal

Federico Ferreres
Research Group on Quality, Safety and Bioactivity of Plant Foods, Department of Food Science and Technology, CEBAS (CSIC), Espinardo, Murcia, Spain

Swaran J.S. Flora
Division of Pharmacology and Toxicology, Defence Research and Development Establishment, Gwalior, India

Giorgia Foca
Department of Agricultural and Food Sciences, University of Modena and Reggio Emilia, Reggio Emilia, Italy

Octavio Luiz Franco
Centro de Analises Proteomicas e Bioquímicas, Universidade Católica de Brasília, Brasília-DF, Brazil

Maria Teresa Frangipane
Dipartimento di Scienze e Tecnologie Agroalimentari, Tuscia University, Viterbo, Italy

Jeffrey R. Fry
School of Biomedical Sciences, University of Nottingham Medical School, Queen's Medical Centre, Nottingham, UK

Shin Yee Fung
Department of Molecular Medicine, Faculty of Medicine, University of Malaya, Kuala Lumpur, Malaysia

Magaji Garba
Department of Pharmaceutical and Medicinal Chemistry, Ahmadu Bello University, Zaria, Nigeria

Manohar L. Garg
Nutraceuticals Research Group, School of Biomedical Sciences, The University of Newcastle, Callaghan, NSW, Australia

Mahavir H. Ghante
Department of Chemistry, J. L Chaturvedi College of Pharmacy, MIDC, Nagpur, India

Hasanah Mohd Ghazali
Department of Food Science, Faculty of Food Science and Technology, Universiti Putra Malaysia, Serdang, Selangor D.E, Malaysia

Kesturu Subbaiah Girish
Department of Studies in Biochemistry, University of Mysore, Manasagangothri, Mysore, India

Daniele Giuffrida
Dipartimento di Scienze degli Alimenti e dell'Ambiente "G. Stagno d'Alcontres", Università di Messina, Messina, Italy

Alexandre Gonçalves Santos
Universidade Federal de Ouro Preto, Departamento de Ciências Biológicas, Ouro Preto, Minas Gerais, Brasil

Mónica González
Instituto Canario de Investigaciones Agrarias, La Laguna, Spain

Milton Hércules Guerra de Andrade
Universidade Federal de Ouro Preto, Departamento de Ciências Biológicas, Ouro Preto, Minas Gerais, Brasil

Asheesh Gupta
Department of Biochemical Pharmacology, Defence Institute of Physiology and Allied Sciences, Timarpur, Delhi, India

Ashish Deep Gupta
Mangalayatan University, Institute of Biomedical Education & Research, Department of Biotechnology, Uttar Pradesh, India

Ella H. Haddad
Department of Nutrition, School of Public Health, Loma Linda University, Loma Linda, California, USA

Bente Lise Halvorsen
Department of Nutrition, Institute of Basic Medical Sciences, Faculty of Medicine, University of Oslo, Norway

Hans-Peter Hanssen
University of Hamburg, Institute of Pharmaceutical Biology and Microbiology, Hamburg, Germany

Ken-ichi Hatano
Department of Chemistry and Chemical Biology, Faculty of Engineering, Gunma University, Kiryu, Gunma, Japan

Mahadevappa Hemshekhar
Department of Studies in Biochemistry, University of Mysore, Manasagangothri, Mysore, India

Blanca Hernández-Ledesma
Department of Nutritional Sciences and Toxicology, University of California at Berkeley, Berkeley, California, USA; Instituto de Fermentaciones Industriales. Consejo Superior de Investigaciones Científicas (CSIC) Madrid, Spain

María del Rosario Hernández-Medel
Instituto de Ciencias Básicas, Universidad Veracruzana, Xalapa, Ver., México

Elisa Yoko Hirooka
Department of Food Science and Technology, State University of Londrina, Londrina, Paraná, Brazil

Michal Holčapek
University of Pardubice, Faculty of Chemical Technology, Department of Analytical Chemistry, Pardubice, Czech Republic

Chia-Chien Hsieh
Department of Nutritional Sciences and Toxicology, University of California at Berkeley, Berkeley, California, USA

Dur-Zong Hsu
Department of Environmental and Occupational Health, National Cheng Kung University Medical College, Tainan, Taiwan

Xiaojie Hu
Department of Food Science and Nutrition, and Institute of Agrobiology and Environmental Science, Zhejiang University, Hangzhou, China; and Center of Nutrition & Food Safety, Asia Pacific Clinical Nutrition Society

Wen-Chuan Huang
Graduate Institute of Biomedical Sciences, College of Medicine, Chang Gung University, Tao-Yuan, Taiwan, China; Present address: TCM Biotech International Corp. 7F, Neihu District, Taipei, Taiwan, Republic of China

Uford S. Inyang
Department of Medicinal Chemistry and Quality Control, National Institute for Pharmaceutical Research and Development (NIPRD), Garki, Abuja, Nigeria

Zafar Iqbal
Department of Parasitology, University of Agriculture, Faisalabad, Pakistan

Gerhard Jahreis
Friedrich Schiller University, Institute of Nutrition, Jena, Germany

Anna Jaromin
Faculty of Biotechnology, University of Wroclaw, Department of Lipids and Liposomes, Wrocław, Poland

David J.A. Jenkins
Clinical Nutrition & Risk Factor Modification Center, Department of Nutritional Sciences, St. Michael's Hospital, Toronto, Ontario, Canada; Department of Medicine, Division of

Endocrinology and Metabolism, St. Michael's Hospital, Toronto, Ontario, Canada;
Departments of Nutritional Sciences and of Medicine, Faculty of Medicine, University
of Toronto, Toronto, Ontario, Canada

Sunny Jhamnani
Beth Israel Deaconess Medical Center, Division of Cardiology, Boston, Massachusetts, USA

Yueming Jiang
South China Botanical Garden, Chinese Academy of Sciences, Guangzhou, China

Rod Jones
Knoxfield, Victoria, Australia

Rachid Kacem
Department of Biology, Faculty of Sciences, Ferhat Abbas University, Setif, Algeria

Monica Kachroo
Al-Ameen College of Pharmacy, Bangalore, India

Daein Kang
Korean Pharmacopuncture Institute (KPI), Yeoksam-Dong, Gangnam-Gu, Seoul,
Republic of Korea

Dayanand M. Kannur
Department of Pharmacognosy, Shree Chanakya Education Society's Indira College
of Pharmacy, Pune, India

Kempaiah Kemparaju
Department of Studies in Biochemistry, University of Mysore, Manasagangothri, Mysore,
India

Cyril C.W. Kendall
Clinical Nutrition & Risk Factor Modification Center, St. Michael's Hospital, Toronto,
Ontario, Canada; Department of Nutritional Sciences, Faculty of Medicine, University of
Toronto, Toronto, Ontario, Canada; College of Pharmacy and Nutrition, University of
Saskatchewan, Saskatoon, Saskatchewan, Canada.

Natalya V. Khadeeva
N.I. Vavilov Institute of General Genetics, Russian Academy of Sciences, Moscow, Russia

Tanvir Khaliq
Division of Medicinal & Process Chemistry, Central Drug Research Institute, Lucknow, India

Ali Khan Khuwaja
Department of Family Medicine, The Aga Khan University, Karachi, Pakistan

David H. Kinder
College of Pharmacy, Ohio Northern University, Ada, Ohio, USA

Jiyoung Kim
Research Institute for Agriculture and Life Sciences, Seoul National University, Seoul, Republic
of Korea

Soressa M. Kitessa
CSIRO Food and Nutritional Sciences, Adelaide, South Australia, Australia

Merih Kivanc
Department of Biology, Faculty of Science, Anadolu University, Eskisehir, Turkey

Kathryn T. Knecht
School of Pharmacy, Loma Linda University, Loma Linda, California, USA

Kazunori Koba
Department of Nutritional Science, University of Nagasaki, Siebold, Nagasaki, Japan

U. Koca
Gazi University, Faculty of Pharmacy, Department of Pharmacognosy, Etiler, Ankara, Turkey

Mongomaké Koné
Laboratoire de Biologie et Amélioration des Productions Végétales, UFR des Sciences de la Nature, Université d'Abobo-Adjamé, Côte d'Ivoire

Stanislaw J. Konturek
Department of Physiology, Jagiellonian University Medical College, Cracow, Poland

Mariola Korycińska
Faculty of Biotechnology, University of Wroclaw, Department of Lipids and Liposomes, Wrocław, Poland

Arkadiusz Kozubek
Faculty of Biotechnology, University of Wroclaw, Department of Lipids and Liposomes, Wrocław, Poland

Vera Krimer-Malešević
Public Health Institute, Sanitary Chemistry, Subotica, Serbia

Nikhil Kumar
Betel Vine Biotechnology Laboratory, National Botanical Research Institute, Lucknow, India

A.V. Vinay Kumar
Institute of Environmental and Human Health, Texas Tech University, Lubbock, Texas, USA

Slawomir Kwiecien
Department of Physiology, Jagiellonian University Medical College, Cracow, Poland

Sze-Kwan Lam
School of Biomedical Sciences, Faculty of Medicine, The Chinese University of Hong Kong, Hong Kong, China

Karen Lapsley
Almond Board of California, Modesto, California, USA

Carl J. Lavie
Department of Cardiovascular Diseases, John Ochsner Heart and Vascular Institute, Ochsner Clinical School, The University of Queensland School of Medicine, New Orleans, Louisiana, USA

Sonaly Cristine Leal
Universidade Federal de Ouro Preto, Departamento de Ciências Biológicas, Ouro Preto, Minas Gerais, Brasil

Chang Young Lee
Department of Food Science and Technology, Cornell University, Geneva, NY, USA

John H. Lee
Department of Cardiovascular Diseases, John Ochsner Heart and Vascular Institute, Ochsner Clinical School, The University of Queensland School of Medicine, New Orleans, Louisiana, USA

Hyong Joo Lee
Department of Agricultural Biotechnology, College of Agriculture and Life Sciences, Seoul National University, Seoul, Republic of Korea

Ki Won Lee
Department of Bioscience and Biotechnology, Konkuk University, Seoul, Republic of Korea

Yang Deok Lee
Department of Internal Medicine, Eulji University School of Medicine, Daejeon, Korea

Yann-Lii Leu
Graduate Institute of Natural Products, College of Medicine, Chang Gung University, Tao-Yuan, Taiwan, Republic of China

Delfín Rodríguez Leyva
Cardiovascular Research Division, V.I. Lenin University Hospital, Holguin, Cuba

Duo Li
Department of Food Science and Nutrition, and Institute of Agrobiology and Environmental Science, Zhejiang University, Hangzhou, China; and Center of Nutrition & Food Safety, Asia Pacific Clinical Nutrition Society

Min Li
Department of Naval Medicine, Second Military Medical University, Shanghai, China

Xiang Li
Department of Pharmaceutical Chemistry, School of Pharmacy, Nanjing University of Chinese Medicine, Nanjing, Jiangsu, China

Ying Liang
Department of Environmental Engineering, Guilin University of Electronic Technology, Guilin, China

Chang-Quan Ling
Department of Traditional Chinese Medicine, Changhai Hospital, Shanghai, China

Miroslav Lísa
University of Pardubice, Faculty of Chemical Technology, Department of Analytical Chemistry, Pardubice, Czech Republic

Ming-Yie Liu
Department of Environmental and Occupational Health, National Cheng Kung University Medical College, Tainan, Taiwan

Qin Liu
School of Food Science and Engineering, Nanjing University of Finance and Economics, Nanjing, Jiangsu, China

Xianxian Liu
Department of Chemistry and Engineering Technology, Guilin Normal College, Guilin, China

Mario Li Vigni
Department of Chemistry, University of Modena and Reggio Emilia, Modena, Italy

M. Gloria Lobo
Instituto Canario de Investigaciones Agrarias, La Laguna, Spain

Tirupur Venkatachalam Logaraj
Department of Plant Biology & Biotechnology, Government College of Arts and Science, Coimbatore, Tamil Nadu State, India

Leonardo Lombardini
Department of Horticultural Sciences, Fruit and Vegetable Improvement Center, Texas A&M University, College Station, Texas, USA

Yolanda B. Lombardo
University of Litoral, Department of Biochemistry, School of Biochemistry, Santa Fe, Argentina

Giovanna Loredana La Torre
Dipartimento di Scienze degli Alimenti e dell'Ambiente "G. Stagno d'Alcontres", Università di Messina, Messina, Italy

Ben O. de Lumen
Department of Nutritional Sciences and Toxicology, University of California at Berkeley, Berkeley, California, USA

Xingming Ma
Department of Immunology, School of Basic Medical sciences, Lanzhou University, China

Senka Mađarev-Popović
University of Novi Sad, Faculty of Technology, Department of Applied and Engineering Chemistry, Novi Sad, Serbia

Maurice Mahieu
INRA, UR143 Unité de Recherches Zootechniques, Centre INRA-Antilles-Guyane, Domaine de Duclos, Petit-Bourg, Guadeloupe (French West Indies)

Anup Maiti
Pharmacy College Ipaura, Chandeswar, Azamgarh, Uttar Pradesh, India

Tapas K. Maiti
Department of Life science, National Institute of Technology, Rourkela, Orissa, India

Manisha Deb Mandal
Department of Physiology and Biophysics, KPC Medical College and Hospital, Jadavpur, Kolkata, India

Shyamapada Mandal
Department of Microbiology, Bacteriology and Serology Unit, Calcutta School of Tropical Medicine, Kolkata, India; Present address Department of Zoology, Gurudas College, Narkeldanga, Kolkata-700054, India

Tusharkanti Mandal
National Research Institute of Basic Ayurvedic Sciences, CCRAS, Nehru Garden, Kothrud, Pune, India

Andrea Marchetti
Department of Chemistry, University of Modena and Reggio Emilia, Modena, Italy

Carine Marie-Magdeleine
INRA, UR143 Unité de Recherches Zootechniques, Centre INRA-Antilles-Guyane, Domaine de Duclos, Petit-Bourg, Guadeloupe (French West Indies)

Colin R. Martin
Faculty of Education, Health and Social Sciences, University of the West of Scotland, Ayr, UK

Sarfaraz Khan Marwat
University WENSAM College, Gomal University, Dera Ismail Khan, KPK, Pakistan

Riccardo Massantini
Dipartimento di Scienze e Tecnologie Agroalimentari, Tuscia University, Viterbo, Italy

Richard D. Mattes
Department of Foods and Nutrition, Purdue University, West Lafayette, Indiana, USA

Richelle S. McCullough
Institute of Cardiovascular Sciences, St Boniface Hospital Research Centre, and Department of Physiology, University of Manitoba, Winnipeg, Manitoba, Canada

Krystyna Michalak
Wrocław Medical University, Department of Biophysics, Wrocław, Poland

Richard Milani
Department of Cardiovascular Diseases, John Ochsner Heart and Vascular Institute, Ochsner Clinical School, The University of Queensland School of Medicine, New Orleans, Louisiana, USA

Arash Mirrahimi
Clinical Nutrition & Risk Factor Modification Center, St Michael's Hospital, Toronto, Ontario, Canada; Department of Nutritional Sciences, Faculty of Medicine, University of Toronto, Toronto, Ontario, Canada

Tulika Mishra
Immunology and Immunotechnology Laboratory, Department of Biotechnology, Punjabi University, Punjab, India

Pragya Misra
Division of Parasitology, Central Drug Research Institute, Lucknow, India

Takuya Miyakawa
Department of Applied Biological Chemistry, Graduate School of Agricultural and Life Sciences, The University of Tokyo, Bunkyo-ku, Tokyo, Japan

Abdulkarim Sabo Mohammed
Department of Food Science, Faculty of Food Science and Technology, Universiti Putra Malaysia, Serdang, Selangor D.E, Malaysia

Maryati Mohamed
Laboratory of Natural Products, Institute for Tropical Biology and Conservation, Universiti Malaysia Sabah, Kota Kinabalu, Sabah, Malaysia

Joseph Molnár
Department of Medical Microbiology, University of Szeged, Szeged, Hungary

Alisa Mori
Department of Foods and Nutrition, Purdue University, West Lafayette, Indiana, USA

Siham Mostafa Ali El-Shenawy
Department of Pharmacology, National Research Centre, Dokki, Cairo, Egypt

Michael Murkovic
Institute of Biochemistry, Graz University of Technology, Graz, Austria

Subban Nagarajan
Plantation Products, Spices and Flavour Technology Department, Central Food Technological Research Institute, Mysore, India

Marcelo H. Napimoga
Laboratory of Biopathology and Molecular Biology, University of Uberaba, Uberaba/MG; Laboratory of Immunology and Molecular Biology, São Leopoldo Mandic Institute and Research Center, Campinas, SP, Brazil

Tadigoppula Narender
Division of Medicinal & Process Chemistry, Central Drug Research Institute, Lucknow, India

Tzi-Bun Ng
School of Biomedical Sciences, Faculty of Medicine, The Chinese University of Hong Kong, Hong Kong, China

Patrick H.K. Ngai
School of Life Sciences, Faculty of Science, The Chinese University of Hong Kong, Hong Kong, China

Nguyen Nguyen
Beth Israel Deaconess Medical Center, Division of Cardiology, Boston, Massachusetts, USA

Peter D. Nichols
CSIRO Marine and Atmospheric Research, Hobart, Tasmania, Australia

Julius Enyong Oben
Laboratory of Nutrition and Nutritional Biochemistry, Faculty of Science, University of Yaounde I, Yaounde, Cameroon

Obiageli O. Obodozie
Department of Medicinal Chemistry and Quality Control, National Institute for Pharmaceutical Research and Development (NIPRD), Garki, Abuja, Nigeria

Steve Ogbonnia
Department of Pharmacognosy, University of Lagos, Lagos, Nigeria

James H. O'Keefe
Mid America Heart Institute, University of Missouri-Kansas City, Missouri, USA

María Eugenia Oliva
University of Litoral, Department of Biochemistry, School of Biochemistry, Santa Fe, Argentina

Elisabete Yurie Sataque Ono
Department of Biochemistry and Biotechnology, State University of Londrina, Londrina, Paraná, Brazil

Mario Augusto Ono
Department of Pathological Sciences, State University of Londrina, Londrina, Paraná, Brazil

Ana G. Ortiz-Quezada
Department of Horticultural Sciences, Fruit and Vegetable Improvement Center, Texas A&M University, College Station, Texas, USA

Olusegun James Oyelade
Department of Food Science and Engineering, Ladoke Akintola University of Technology, Ogbomoso, Oyo State, Nigeria

Folake Lucy Oyetayo
Department of Biochemistry, University of Ado-Ekiti, Nigeria

Yingming Pan
School of Chemistry and Chemical Engineering, Guangxi Normal University, Guilin, China

Vidhu Pachauri
Division of Pharmacology and Toxicology, Defence Research and Development Establishment, Gwalior, India

Madan Mohan Padhi
National Research Institute of Basic Ayurvedic Sciences, CCRAS, Nehru Garden, Kothrud, Pune, India

Robert Pajdo
Department of Physiology, Jagiellonian University Medical College, Cracow, Poland

Alistair G. Paice
Department of Nutrition and Dietetics, King's College London, London, UK

Dilipkumar Pal
College of Pharmacy, Institute of Foreign Trade & Management, Lodhipur Rajput, Moradabad, Uttar Pradesh, India

Yingming Pan
School of Chemistry & Chemical Engineering of Guangxi Normal University, Guilin, China

Michal Pawlik
Department of Physiology, Jagiellonian University Medical College, Cracow, Poland

Wieslaw W. Pawlik
Department of Physiology, Jagiellonian University Medical College, Cracow, Poland

Patrícia Barbosa Pelegrini
Centro de Analises Proteomicas e Bioquímicas, Universidade Católica de Brasília, Brasília-DF, Brazil

Draginja Peričin
University of Novi Sad, Faculty of Technology, Department of Applied and Engineering Chemistry, Novi Sad, Serbia

David M. Pereira
REQUIMTE/Department of Pharmacognosy, Faculty of Pharmacy, Porto University, Porto, Portugal

José Alberto Pereira
CIMO / Escola Superior Agrária, Instituto Politécnico de Bragança, Bragança, Portugal

Amorn Petsom
Department of Chemistry, Faculty of Science, Chulalongkorn University, Bangkok, Thailand

Grant N. Pierce
St Boniface General Hospital, and Department of Physiology, University of Manitoba, Winnipeg, MB Canada

Alexander M. Popov
Pacific Institute of Bioorganic Chemistry, Far Eastern Branch of the Russian Academy of Science, Vladivostok, Russia

Y. Pramodini
Government Dental hospital, Hyderabad, Andra Pradesh, India

Nagendra Prasad
South China Botanical Garden, Chinese Academy of Sciences, Guangzhou, China

Lu-Ping Qin
Department of Pharmacognosy, School of Pharmacy, Second Military Medical University, Shanghai, China

Yang Qiu
Department of Food Science, University of Manitoba, Winnipeg, Manitoba, Canada

K. Vijaya Rachel
Department of Biochemistry, GITAM University, Visakhapatnam, Andra Pradesh, India

Ljiljana Radulović
University of Novi Sad, Faculty of Technology, Department of Applied and Engineering Chemistry, Novi Sad, Serbia

Asmah Rahmat
Department of Nutrition and Dietetics, Faculty of Medicine and Health Sciences, Universiti Putra Malaysia, UPM, Serdang, Selangor, Malaysia

P.S. Rajini
Food Protectants & Infestation Control Department, Central Food Technological Research Institute, Mysore, India

Deepak Rajpurohit
College of Horticulture and Forestry, Department of Biotechnology, Rajasthan, India

C.U. Rajeshwari
Department of Home Science, Sri Sathya Sai University, Anantapur, Andhra Pradesh, India

Elsa Ramalhosa
CIMO / Escola Superior Agrária, Instituto Politécnico de Bragança, Bragança, Portugal

Gajendra Rao
National Research Institute of Basic Ayurvedic Sciences, CCRAS, Nehru Garden, Kothrud, Pune, India

Lingamallu Jagan Mohan Rao
Plantation Products, Spices and Flavour Technology Department, Central Food Technological Research Institute, Mysore, India

Nádia Rezende Barbosa Raposo
Department of Food and Toxicology, Faculty of Pharmacy and Biochemistry, Federal University of Juiz de Fora, Juiz de Fora, Brazil

Francesca Ricci
Dipartimento di Genetica e Biologia Molecolare, Sapienza Università di Roma, Roma, Italy; Currently at: Dipartimento di Biologia Molecolare Cellulare e Animale, Università di Camerino, Macerata, Italy

Gianfranco Risuleo
Dipartimento di Genetica e Biologia Molecolare, Sapienza Università di Roma, Roma, Italy

Simone J. Rochfort
Bundoora, Victoria, Australia

Ilia Rodushkin
Division of Applied Geology, Luleå University of Technology, Luleå, Sweden; ALS Scandinavia AB, Luleå, Sweden

Carolina Nachi Rossi
Department of Biochemistry and Biotechnology, State University of Londrina, Londrina, Paraná, Brazil

Hafiz Abubaker Saddiqi
Department of Parasitology, University of Agriculture, Faisalabad, Pakistan

Kanagal Sahana
Plantation Products, Spices and Flavour Technology Department, Central Food Technological Research Institute, Mysore, India

Marcello Saitta
Dipartimento di Scienze degli Alimenti e dell'Ambiente "G. Stagno d'Alcontres", Università di Messina, Messina, Italy

Ibrahim Sani
Department of Biochemistry, Faculty of Science, Ahmadu Bello University Zaria, Kaduna State, Nigeria

Nalan Yilmaz Sariozlu
Department of Biology, Faculty of Science, Anadolu University, Eskisehir, Turkey

Yoriko Sawano
Department of Applied Biological Chemistry, Graduate School of Agricultural and Life Sciences, The University of Tokyo, Bunkyo-ku, Tokyo, Japan

Ulrich Schäfer
Friedrich Schiller University, Institute of Nutrition, Jena, Germany

Markus Schmitz-Hübsch
Rosen-Apotheke, Hamburg, Germany

Dante Selenscig
University of Litoral, Department of Biochemistry, School of Biochemistry, Santa Fe, Argentina

S. Sengottuvelu
Department of Pharmacology, Nandha College of Pharmacy & Research Institute, Tamil Nadu, India

Essam Shaalan
Zoology Department, Aswan Faculty of Science, South Valley University, Aswan 81528, Egypt

John L. Sievenpiper
Clinical Nutrition & Risk Factor Modification Center, St Michael's Hospital, Toronto, Ontario, Canada; Department of Pathology and Molecular Medicine, Faculty of Health Sciences, McMaster University, Hamilton, Ontario, Canada

Simona Sighinolfi
Department of Chemistry, University of Modena and Reggio Emilia, Modena, Italy

Si Mui Sim
Department of Pharmacology, Faculty of Medicine, University of Malaya, Kuala Lumpur, Malaysia

Aline Siqueira Ferreira
Department of Food and Toxicology, Faculty of Pharmacy and Biochemistry, Federal University of Juiz de Fora, Juiz de Fora, Brazil

Julio A. Solís-Fuentes
Instituto de Ciencias Básicas, Universidad Veracruzana, Xalapa, Ver., México

Korbua Srichaikul
Clinical Nutrition & Risk Factor Modification Center, St Michael's Hospital, Toronto, Ontario, Canada; Department of Nutritional Sciences, Faculty of Medicine, University of Toronto, Toronto, Ontario, Canada

Welma Stonehouse
Institute of Food, Nutrition and Human Health, Massey University, North Shore City, Auckland, New Zealand

M.T. Ravi Subbiah
Department of Internal Medicine, University of Cincinnati Medical Center, Cincinnati, Ohio, USA

Katarzyna Suchoszek-Łukaniuk
Oleofarm, Producer of Pharmaceutical & Cosmetic Raw Materials, Dietary Supplements and Healthy Food, Pietrzykowice, Poland

Reka Szollosi
Department of Plant Biology, University of Szeged, Hungary

Nget Hong Tan
Department of Molecular Medicine, Faculty of Medicine, University of Malaya, Kuala Lumpur, Malaysia

Masaru Tanokura
Department of Applied Biological Chemistry, Graduate School of Agricultural and Life Sciences, The University of Tokyo, Bunkyo-ku, Tokyo, Japan

Lorenzo Tassi
Department of Chemistry, University of Modena and Reggio Emilia, Modena, Italy

Edson H. Teixeira
Faculty of Medical School of Sobral, Federal University of Ceara, Sobral/CE, Brazil

Christine D. Thomson
Department of Human Nutrition, University of Otago, Dunedin, New Zealand

Sriniwas Tiwari
Division of Medicinal & Process Chemistry, Central Drug Research Institute, Lucknow, India

Yaya Touré
Laboratoire de Biologie et Amélioration des Productions Végétales, UFR des Sciences de la Nature, Université d'Abobo-Adjamé, Côte d'Ivoire

Alessandro Ulrici
Department of Agricultural and Food Sciences, University of Modena and Reggio Emilia, Reggio Emilia, Italy

A. Umasankar
Department of Biochemistry, Faculty of Medicine, Al Tahadi University, Sirt, Libya

Solomon Umukoro
Department of Pharmacology and Therapeutics, University of Ibadan, Ibadan, Nigeria

Nitin K. Upadhyay
Department of Biochemical Pharmacology, Defence Institute of Physiology and Allied Sciences, Timarpur, Delhi, India

Fazal ur Rehman
Department of Pharmacognosy, Faculty of Pharmacy, Gomal University, Dera Ismail Khan, NWFP, Pakistan

Vellingiri Vadivel
Institute for Biological Chemistry and Nutrition, University of Hohenheim, Stuttgart, Germany

Patrícia Valentão
REQUIMTE/Department of Pharmacognosy, Faculty of Pharmacy, Porto University, Porto, Portugal

Andras Varga
Department of Molecular Parasitology, Humboldt University, Berlin, Germany

Žužana Vaštag
University of Novi Sad, Faculty of Technology, Department of Applied and Engineering Chemistry, Novi Sad, Serbia

Martha Verghese
Alabama A&M University, Normal, Alabama, USA

Y. Vimala
Department of Microbiology, GITAM University, Visakhapatnam, Andra Pradesh, India

Lloyd T. Walker
Alabama A&M University, Normal, Alabama, USA

He-xiang Wang
State Key Laboratory of Agrobiotechnology, Department of Microbiology, China Agricultural University, Beijing, China

Hengshan Wang
School of Chemistry and Chemical Engineering, Guangxi Normal University, Guilin, China

Kai Wang
Department of Chemistry and Engineering Technology, Guilin Normal College, Guilin, China

Xiao-li Wang
Department of Naval Medicine, Second Military Medical University, Shanghai, China

Cornelius W. Wanjala
Department of Physical Sciences, South Eastern University College (Constituent College of the University of Nairobi), Kitui, Kenya

Francine K. Welty
Beth Israel Deaconess Medical Center, Division of Cardiology, Boston, Massachusetts, USA

Jack H. Wong
School of Biomedical Sciences, Faculty of Medicine, The Chinese University of Hong Kong, Hong Kong, China

Lisa G. Wood
Centre for Asthma and Respiratory Diseases, Level 3, Hunter Medical Research Institute, John Hunter Hospital, Newcastle, NSW, Australia

Teruyoshi Yanagita
Department of Applied Biochemistry and Food Science, Saga University, Saga, Japan

Bao Yang
South China Botanical Garden, Chinese Academy of Sciences, Guangzhou, China

Xiujuan Ye
School of Biomedical Sciences, Faculty of Medicine, The Chinese University of Hong Kong, Hong Kong, China

Xianghui Yi
Department of Chemistry and Engineering Technology, Guilin Normal College, Guilin, China

Jau-Song Yu
Graduate Institute of Biomedical Sciences, College of Medicine, Chang Gung University, Tao-Yuan, Taiwan, Republic of China; Department of Cell and Molecular Biology, College of Medicine, Chang Gung University, Tao-Yuan, Taiwan, Republic of China

Alam Zeb
Department of Biotechnology, University of Malakand, Chakdara, Pakistan

Ye Zhang
Department of Chemistry and Engineering Technology, Guilin Normal College, Guilin, China

Cheng-Jian Zheng
Department of Pharmacognosy, School of Pharmacy, Second Military Medical University, Shanghai, China

Adel Zitoun
Unité de Valorisation des Résultats Scientifiques, Centre de Biotechnologie de Sfax, Sfax, Tunisia

The objective of this book is to bring together scientific material relating to the health benefits, and, where appropriate, adverse effects of nuts and seeds. In general, nuts and seeds are important not only from a nutritional point of view, but also in terms of their putative medicinal or pharmacological properties. This book aims to describe these properties in a comprehensive way. However, at the same time it is recognized that harmful effects also arise. Some "nuts" and "seeds," for example, are poisonous when ingested in large quantities, but extracts have putative effects on tissues that may offer some therapeutic potential. Many of the nuts and seeds described in this book are components of traditional remedies without any present day evidence to support their claims; their properties await rigorous elucidation and scientific investigation. Thus, the book embraces nuts and seeds in an unbiased way. The Editors also recognize that there is a wide interpretation of the terms nuts and seeds, and indeed some authorities have claimed that there are at least 12 seed types. The Editors have largely excluded cereals (grains) and other staple food crops unless there was cause to include them, such as with buckwheat seeds. They have also selected some specific legumes, where there is some therapeutic potential in their extracts or interesting properties.

The book *Nuts and Seeds in Health and Disease Prevention* is divided into two parts. Part I, General Aspects and Overviews, contains holistic information, with sections on Overviews, Composition, Effects on Health, and Adverse Aspects. In Part II, Effects of Specific Nuts and Seeds, coverage is more specific. Each chapter in Part II contains sections entitled Botanical description, Historical cultivation and usage, Present day cultivation and usage, Applications to health promotion and disease prevention (the main article), and, finally, Adverse effects and reactions. The Editors were faced with a difficult choice in organizing the chapters in Part II, and this was done using the simplest method available. Thus, in Part II the nuts and seeds are listed alphabetically in terms of their common names, although each chapter contains full botanical terminology. We realize this is not perfect — for example, there are numerous types of cabbage seeds, and some nuts and seeds may have as many as 20 common names depending on the country where they are grown — but navigation and the retrieval of specific information is aided by a comprehensive index system.

This book is designed for health scientists, including nutritionists and dietitians, pharmacologists, public health scientists, those in agricultural departments and colleges, epidemiologists, health workers and practitioners, agriculturists, botanists, healthcare professionals of various disciplines, policy-makers, and marketing and economic strategists. It is designed for teachers and lecturers, undergraduates and graduates.

The Editors

General Aspects and Overviews

SECTION A

Overviews

CHAPTER 1

A Primer on Seed and Nut Biology, Improvement, and Use

George A. Dyer
The Macaulay Land Use Research Institute, Craigiebuckler, Aberdeen, UK

5

You know that the seed is inside the horse-chestnut tree;
and inside the seed there are the blossoms of the tree, and the chestnuts, and the shade.
So inside the human body there is the seed, and inside the seed there is the human body again.

Kabir

INTRODUCTION

Seeds are the centerpiece of plants' sexual reproductive strategies and a critical stage in their life cycle. Seed traits play a direct role in plants' fitness. The diversity of these traits across species reflects the evolutionary pressures operating on plants. Adaptation to new environments is greatly facilitated by the genetic diversity and distinctiveness of every new seed cohort. Seed-producing plants (i.e., spermatophytes) invest significant resources in seeds, which is why seeds are an invaluable resource for innumerable animal species, including humans. Seeds have played a vital role in human diet and health since prehistory, yet they have numerous other roles in modern society. Science and technology are continuously expanding the range of potential uses for seeds. Substantial contributions to human wellbeing and health could derive from promising applications in emerging fields such as biopharmaceuticals. However, realizing this potential requires that we acknowledge the complexity of the issues involved — a task that entails recourse to the natural sciences as well as the humanities. In this chapter, we

Nuts & Seeds in Health and Disease Prevention. DOI: 10.1016/B978-0-12-375688-6.10001-5

consider some of these issues. We devote the first half of the chapter to definitions and concise descriptions of seed anatomy, development, and ecology. In the second half we turn to the social sciences for a brief outline of the historical interaction of seed plants and humans, and some of the consequences that this has had on both groups.

SEEDS, FRUITS, AND NUTS
Definition and Classification

Seeds can be defined in various ways. In science, they are defined from a developmental perspective: a seed is a mature, fertilized ovule (Copeland & McDonald, 2001). In everyday language, by contrast, seeds are defined by function — i.e., the term "seed" ordinarily refers to various structures involved in plant propagation (Table 1.1). Clearly, common usage of the term is much broader than scientific usage: fruits (containing one or more seeds) and tubers (which do not involve seed at all) often are referred to as "seed." Scientists sometimes speak of "true seed" when referring to structures derived from an ovule in order to distinguish between the scientific and colloquial connotations of the word. Scientific and popular usage of the term "fruit" also differs. From a developmental perspective, a fruit is a mature ovary that contains one or more seeds (Copeland & McDonald, 2001). In this case, therefore, the scientific definition is broader than the colloquial: apples, oranges, and peaches are fruit, as are legume pods, peppers, and cereal grains. False fruits are fruit-like structures not derived from an ovary (Table 1.1). Strictly speaking, a nut is a dry, one-seeded fruit with an extremely hard pericarp.[1] However, the term is commonly used in reference to various edible, oily seeds, often (but not necessarily) derived from a true nut (Copeland & McDonald, 2001).

Seeds and fruits exhibit remarkable diversity in shape, size, color, and chemical composition. There is no pre-eminent way to classify this diversity. Classifications can be based on a functional, morphological, or developmental perspective (Dickie & Stuppy, 2003). Anatomical classifications have the disadvantage that similar structures might have different origins in separate taxa. A classification based on developmental homology is more likely to generate monophyletic groups exhibiting anatomical and physiological similarities, but a rigorous developmental classification can be exceedingly complex. The most useful and widespread

TABLE 1.1 Definitions of Seeds, Fruits, and Nuts[*]		
	Definition	**Examples**
True seed	A mature, fertilized ovule	Beans, cashew nuts
False seed	Any structure that works as a propagule	Cereal grains, potatoes and other tubers
True fruit	A mature ovary	Apples, legume pods, cereal grains, chestnuts
False fruit	Any fleshy, sweet-tasting, structure; structures involved in zoochorous seed dispersal	Cashew apples (marañón); arils, ovuliferous scales and other appendages of gymnosperm seed
True nut	A dry, one-seeded fruit with an extremely hard pericarp	Acorns, hazels and chestnuts
False nut	Any oily, edible seed, including those derived from true nuts	Peanuts, cashew nuts

[*]In science, seeds, nuts, and fruits are specific structures associated with sexual reproduction — i.e., the mature ovule and ovary. In everyday language, use of these terms is not as restrictive but applied to a wider set of structures involved in plant propagation.

[1] The pericarp, mesocarp, and endocarp are, respectively, the outer, middle, and inner layers of the ovary wall.

TABLE 1.2 Classification of Seeds

According to:	Types of Seed	Description
Ovule morphology	Anotropous	Ovules and seeds with a raphe (a continuation of the funicle that ends at the chalaza)
	Orthotropous	Ovules in which funicle, chalaza, nucellus, and micropyle are in a direct line
	Campylotropous	Ovules with curved embryo sacs
Embryo morphology	Basal	Small embryo restricted to the lower half of the seed
	Peripheral	Large, elongate, often curved embryo contiguous to the testa
	Axile (or axial)	Small to total embryo, central, straight, curved, coiled, bent, or folded
Origin of the mechanical layer of the seed coat	Exotegmic or exotestal	Inner (tegmen) or outer (testa) derived from outer epidermis
	Mesotegmic or mesotestal	Inner (tegmen) or outer (testa) derived from middle layer
	Endotegmic or endotestal	Inner (tegmen) or outer (testa) derived from inner epidermis
Endosperm	Exalbumious	Seeds with little or no endosperm.
	Albuminous	Seeds with a well-developed endosperm or perisperm

Scientific classifications of seeds are often based on internal morphology (Dickie & Stuppy, 2003).

TABLE 1.3 Classification of Fruits

	Mesocarp Dehiscence	Examples
Dry fruits	Dehiscent	Legume (e.g., bean, soybean), follicle (e.g., milkweed), capsule (e.g., tulip), or silicle
	Indehiscent	Achene (e.g., sunflower), utricle (e.g., Russian thistle), caryopsis (e.g., grasses), samara (e.g., maple), nut (e.g., acorn, hazel, filbert, chestnut), nutlet (e.g., mint family), or a schizocarp (e.g., carrot family)
Fleshy fruits	Dehiscent	Pomegranate, nutmeg
	Indehiscent	Berry (e.g., tomato), pepo (e.g., squash), pome (e.g., apple), drupe (e.g., cherry, coconut, walnut), or hesperidia (e.g., oranges)

Fruits are classified into dry or fleshy, dehiscent or indehiscent types according to the atrophy and dehiscence of the mesocarp (Copeland & McDonald, 2001; Dickie & Stuppy, 2003).

classifications tend to be eclectic (Copeland & McDonald, 2001; Dickie & Stuppy, 2003). Informal seed classifications are often based on external morphology (e.g., winged seeds); scientific classifications tend to rely on internal morphology (Table 1.2). As regards fruit, classifications commonly focus on characteristics of the mesocarp and pericarp (Table 1.3).

Structure and Development

A typical seed consists of three elements: the embryo (or young sporophyte), endosperm, and seed coat (Copeland & McDonald, 2001). Each element has a particular chemical composition (and consequently nutritious value) that depends as much on its genetic identity as on its role in development. Seed development can be divided into three distinct stages circumscribed by four defining events: ovule fertilization, seed abscission, physiological maturity, and germination. Despite some commonalities, seed development differs considerably across the two groups of spermatophytes (Dickie & Stuppy, 2003). In angiosperms, ovules derive from meristematic tissue on the ovary wall — i.e., they are enclosed within the ovary, where fertilization and development occur. Gymnosperms' ovules, by contrast, are not borne within ovaries, but are naked, or protected by cones or scales.

The embryo, endosperm, and seed coat develop from distinct parts of the ovule. Seed development begins with an archesporial cell in the ovary wall giving rise to a haploid (1N) megaspore through meiosis (Copeland & McDonald, 2001; Werker, 1996). After a series of mitotic divisions, the megaspore is transformed into the mature female gametophyte or embryo sac, which contains a haploid egg and a diploid (2N) nucleus. The embryo sac is enclosed within specialized tissue called the nucellus, and both of them are enveloped by one or more layers of tissue (i.e., the integuments), leaving only a small opening at the apex, the micropyle. The embryo sac, nucellus, and integuments constitute the archetypical angiosperm ovule, which is connected to the ovary wall by the funiculus.

After landing on the flower's stigma, the pollen tube descends through the style and into the embryo sac through the micropyle, delivering two sperm cells (Werker, 1996). One cell fuses with the polar nucleus, forming a triploid (3N) endosperm nucleus; the other fuses with the egg cell to form the diploid fertilized egg (or zygote) that later will become the embryo. This is the characteristic double fertilization of angiosperms. Embryo and endosperm thus have distinct genetic identities because they derive from separate fertilization events. The embryo is diploid: one half maternal, one half paternal. The endosperm is triploid, derived from the fusion of a haploid sperm cell with the diploid polar nucleus.[2] Since the seed coat derives from the integuments of the mature ovule, it is of strictly maternal origin and haploid in nature.

At every stage in the seed's development, nutrients are drawn from several sources and used in the growth process or stored as reserve materials for later use (Werker, 1996). In some species, the nucellus functions as a temporary source of nutrients for the megaspore and then disappears. In other species, it continues to transport and store nutrients for the developing seed and seedling in the form of the pericarp. The embryo derives its nutrients from the nucellus and integuments through the embryo sac, via the endosperm, or directly from the ovary wall through the funiculus or specialized outgrowths called haustoria. In albuminous seeds, which include monocotyledons, endosperm and perisperm act as the main reserve tissue. In exalbuminous seeds, including most dycotyledons, the endosperm is completely consumed during development and absent in the mature seed, so reserve materials accumulate in the seed coat or the embryo itself, particularly the cotyledons. In addition to nutrients for the developing embryo (i.e., carbohydrates, fats, and proteins), seeds contain chemical compounds involved in growth control (e.g., vitamins and plant hormones), as well as reserves to be used by the quiescent embryo after abscission from the parent plant and upon germination (Copeland & McDonald, 2001). Seeds also contain antimetabolites and other compounds whose role requires that we address species-to-species interactions and other issues within the scope of plant ecology.

Ecology and Evolution

Seed traits are best explained from an evolutionary perspective — i.e., in terms of fitness advantages (Fenner & Thompson, 2005; Moles et al., 2005). In many species, successful regeneration depends on traits that improve the sporophyte's survival through unfavorable growing conditions. Alkaloids, glycosides, and a thick seed coat help deter seed predators,

---

[2] In gymnosperms, the endosperm develops from the female gametophyte and is usually haploid.

while a large seed size increases the sporophyte's survival after germination. Other traits associated with seed dispersal and dormancy improve the sporophyte's chances of reaching a more favorable environment — i.e., a "safe site" (Fenner & Thompson, 2005). Some species depend on wind or water as dispersal vectors; others entice animal dispersers by providing a reward in the form of fruit or nutritious tissue on the seed's exterior, known as the elaiosome. In most cases, a sturdy coat protects seeds as they pass through animals' guts to end in feces and the soil, but sometimes significant numbers of seeds are digested. Sacrificing some seeds so that others might reach a safe site seems to blur the line between dispersion and predation. In fact, some traits play complex roles that defy easy understanding; for example, tannins provide resistance to infections but also influence the uptake of nutrients by predators. Although the advantage of individual traits might seem readily apparent, it still is unclear how different traits combine into a successful reproductive strategy (Moles *et al.*, 2005).[3]

Allocation of resources to seed and other sexual organs necessarily compromises plants' own growth and survival (Fenner & Thompson, 2005). Ovules and seed have a higher nutrient content than other tissues, yet few survive into the next stage in the life cycle. Allocation of resources to different seed traits presents similar compromises — for example, between producing numerous small, easily-dispersed seeds or larger albeit fewer seeds with a greater chance of establishment. According to the principle of allocation, plants distribute resources optimally in order to maximize their fitness given prevalent environmental conditions (Fenner & Thompson, 2005; Moles *et al.*, 2005). Vegetative propagation gives plants a foothold in undisturbed, highly competitive environments where seedling establishment is difficult; seed banks, by contrast, are most advantageous in unstable environments. Explaining a particular reproductive strategy is a more elusive task. Twenty years ago, it was suggested that seed size, dispersal, and dormancy co-evolve to maximize the sporophyte's survival up to the seedling stage. After innumerable studies, it is becoming clear that this explanation is too narrow. Seed traits might not be explicable without accounting for a species' complete life cycle (Moles *et al.*, 2005). In the case of crop species, explaining many seed traits requires addressing their interactions with humans.

A COMPLEX RELATIONSHIP

Ecological interactions between plants and humans going back thousands of years have been unique in many ways. For instance, substances that inhibit other predators — such as caffeine and nicotine — have been readily consumed by humans for their pharmacological effects. However, the relationship between these two groups has been significant in a more profound way, particularly since it developed into a symbiotic relationship where crop plants became a dietary staple for humans but lost the ability to survive in the wild. In this process, humans hijacked crop reproductive strategies, radically altering their ecology and evolution. The consequences for both groups are more readily apparent from a historical vantage point.

Historical perspective

Continuous innovations in dental and masticatory structures (documented in the fossil record) bespeak concomitant changes in the food resources exploited by human ancestors. Anatomical novelties allowed succeeding hominin species to consume previously inaccessible resources, and this had crucial implications. Access to hard nuts and roots as fallback foods opened up new habitats to colonization. Simultaneous increases in body size, robusticity, and encephalization had similar consequences. Increasing brain size presumably facilitated exploitation of an ever-wider resource base, leading to new rounds of changes that ultimately gave rise to modern humans. It has been suggested that the superior intelligence, long lifespan, cooperative

9

---

[3] A reproductive strategy is a complex of traits encompassing the spatial and temporal allocation of energy to reproduction.

behavior, and other attributes that characterize humans co-evolved after a dietary shift to high-quality, difficult-to-access food resources. At the same time, our ancestors must have begun cooking food early on, realizing that cooked foods are easier to assimilate — presumably leading to digestive alterations that prevent us from processing raw foods efficiently. A precise account of the forces that shaped humans still might be a long time coming, but there is no doubt that an increasing ability to manipulate food resources had significant consequences on human diet and health. An important milestone in this process was domestication.

Around 10,000 years ago, numerous plant species under cultivation experienced dramatic morphological and physiological changes (Gepts, 2005). A breakdown of seed dispersal and dormancy and other changes affecting the size, shape, and physiology of seeds and fruits (i.e., the so-called domestication syndrome) arose almost simultaneously, yet independently, around the globe. Early scholars described domestication as a natural response to human contact — a biological reaction to an unintentional shift in environmental conditions. More recent discovery of single-gene mutations with large-step effects of agronomic value seems to suggest a conscious effort (on the part of humans) to improve crop qualities. In either view, selection is considered the main force in crop evolution, and diversity a result of crops' adaptive radiation across environments and cultures (Gepts, 2005). Yet such a perspective neglects the larger social context of crop domestication and evolution. Although increases in agronomic value were an undoubtedly important goal for early farmers, since its beginning the relationship between crops and humans has been equally imbued with social, political, and cultural values. Such values constitute powerful forces with important implications for both seed plants and humans.

We often conceive of the rise of civilization as a period of constant increases in the amount, quality, and diversity of resources available to society, entailing a continuous improvement in human nutrition and health. Scholars argue, nevertheless, that the transition from hunting-gathering to farming and contemporary society has been marked by the upsurge of chronic health problems (Cordain et al., 2005). Unable to adapt to continuous changes in diet, our digestive system still resembles that of ancestors whose diet was much richer in vegetable protein and fiber. Unlike pre-agricultural foragers, whose diverse diet included a variety of wild seeds and nuts, agricultural societies came to depend increasingly on domesticated grains. A cereal-based, carbohydrate-rich diet resulted in a sharp increase in dental decay, metabolic acidosis, and other afflictions among farmers. Dietary breadth was further reduced as humans transited from a rural to an urban lifestyle, gradually leading to a deficiency of essential vitamins, minerals, and other nutrients. The result was a long decline in human health extending up to the modern age, particularly in the western hemisphere (Steckel & Rose, 2002). Of course, this was not a result of crops' nutritional qualities alone, but ultimately the product of an inequitable society. Civilization was founded as much on agriculture as on a hierarchical social order where resources are distributed through complex rules (Newman, 1990). In recent years, malnutrition has come to be defined not only by undernourishment but also by overeating. While millions of people in the developing world remain under-nourished, seemingly unlimited availability (and consumption) of energy-rich food in developed countries has led to the spread of high blood pressure, heart disease, insulin resistance, and other "diseases of affluence" (Cordain et al., 2005).

Crop improvement today

Continuous advances in crop improvement since the early 20th century have fostered a steady growth of food supplies in industrialized countries (Jauhar, 2006). Introduced to the developing world later in the century, heterosis breeding and cytogenetics have achieved impressive results around the world; however, modern breeding has not reached everyone. Many farmers continue to grow landraces — i.e., low-yielding, traditional crop varieties. Ironically, the genetic diversity embodied in landrace seeds constitutes an invaluable resource for modern breeders, who continuously pore over seed stocks for traits that might help sustain yields in the future. In the short term, nevertheless, diversity can be an obstacle to breeding. Isolating highly desirable traits from unfavorable ones is a painstaking task. In out-crossing species, including most perennial fruit and nut trees, crosses between attractive parents can yield economically worthless progeny. Since trees must be cultivated for several years before seed and fruit qualities are expressed, cross-pollination can be a serious setback. Early farmers solved this problem by simply cloning individuals of exceptional quality, avoiding sexual reproduction altogether; however, this practice greatly limited the genetic diversity of tree crops. Added to their long juvenile period, asexual reproduction has largely left these crops outside the purview of classical crop breeding. Substituting new clones for old ones has been the main way to improve stocks, but, given the noticeably low rates of turnover, most fruit and nut cultivars are not markedly different from their wild progenitors (Gepts, 2005).

In the past few years, nevertheless, crop improvement has been transformed profoundly by biotechnology.[4] Application of marker-assisted selection, for instance, has shortened the time requirements of conventional breeding in various crops, including tomatoes (Jauhar, 2006). Although advances in perennial fruit and nut crops have been relatively modest, many expect genetic engineering to have its most profound impact on these and other asexually reproduced crops (Litz, 2005). Genetic engineering constitutes a novel approach to breeding because it allows the transfer of genes (and potentially useful traits) across taxonomic groups (Jauhar, 2006). Although four annual species represent the bulk of its commercial applications to date, biotechnology has been (or soon will be) applied to numerous perennial crops, including some profiled in other chapters in this volume — for example, walnuts, olives, cashews, pistachios, chestnuts, almonds, coconuts, and cocoa (Litz, 2005).

It was widely anticipated that genetically modified (GM) crops would lower the price of food and improve the quality of crops, among other benefits, but controversy has notoriously hindered the industry's development (Qaim, 2009). The advent of nutritionally enhanced crops has been further delayed by technical and institutional problems. Most GM varieties released to date exhibit some type of agronomic advantage — for example, herbicide tolerance or insect resistance. Advocates claim that these "first generation" crops have already raised agricultural outputs while reducing environmental impacts, but debate persists on both counts (Gurian-Sherman, 2009; Qaim, 2009). Widespread adoption of GM crops in developed countries has resulted mostly in changes in pest-management strategies, with few yield increases.[5] In developing countries, where biotechnology has the potential to increase yields because pest control is lacking, evidence of actual gains is still limited (Smale *et al.*, 2006). The implications of GM crops for the environment are also controversial. Although insect-resistant and herbicide-tolerant varieties have been linked to multiple environmental gains, they have also raised numerous concerns (Cedreira & Duke, 2006; Wolfenbarger *et al.*, 2008).[6] A major unresolved controversy involves the risk of transgene flow into non-GM cultivars and their wild relatives (Gepts & Papa, 2003). Widely underplayed until recently, the likelihood of transgenes spreading into landraces in centres of crop diversity is significant, while the implications for the (generally poor) farmers who grow them are completely unknown (Dyer *et al.*, 2009).

A new generation of GM plants under development — so-called pharma crops — has exacerbated the complexity of these issues. Intended as a vehicle for the production of

---

[4] Biotechnology consists of an assortment of molecular methods, including genetic engineering, marker-assisted selection, somatic embryogenesis, and *in vitro* mutagenesis.

[5] Current GM crops have the same intrinsic yields (i.e., maximum yields) as their non-GM isolines. Yields under field conditions (i.e., operational yields) might be higher in GM crops or not, depending on which pest-management practices are in place.

[6] Adoption of GM crops has resulted in a reduction of toxic agrochemicals and promoted soil-conserving practices, but it has also favored herbicide tolerance in weeds, and its effects on non-target organisms have been mixed.

pharmaceutical and industrial proteins, pharma crops could have substantial advantages over alternative systems, which some believe would make medicines affordable to the world's poor (Ma *et al.*, 2005).[7] Scientists working in this field advocate the use of staple crops as hosts, highlighting the potential benefits of producing, storing, distributing, and even delivering pharmaceuticals within seeds; however, the possibility of admixtures with food and feed supplies has raised concerns. Although the likelihood of commingling could be reduced through various containment strategies, all measures are susceptible to human error, particularly wherever institutional oversight is lacking (Ma *et al.*, 2005; Spok *et al.*, 2008). Containment strategies have been designed and tested under conditions prevalent in developed countries, but a substantial gap exists in our knowledge of conditions in developing countries, where seeds are saved across cycles and the risks of containment failures could be much higher (Dyer *et al.*, 2009).

As documented throughout this volume, seed and nut cultivars possess countless nutritional, medical, and pharmacological attributes that have, or could have, application in different areas of human endeavor. Converting potential applications into actual improvements in wellbeing and health, particularly for those in greatest need, will require deliberate effort. Advances in science and technology rarely translate directly or immediately into improvements for most of the human population, particularly when the market is the main means of achieving it. If human wellbeing is the ultimate purpose of science, new approaches that bridge the gap between scientific research and downstream applications are needed. Such approaches will require that we earnestly reassess the role of public institutions in society. Acknowledging the complex social processes that separate scientists' laboratories from the realities of the human condition is a step forward.

References

Cedreira, A. L., & Duke, S. O. (2006). The current status and environmental impacts of glyphosate-resistant crops: a review. *Journal of Environmental Quality, 35*, 1633–1658.

Copeland, L. O., & McDonald, M. B. (2001). *Principles of seed science and technology* (4th ed.). Boston, MA: Kluwer Academic Publishers.

Cordain, L., Eaton, S. B., Sebastian, A., Mann, N., Lindeberg, S., Watkins, B. A., et al. (2005). Origins and evolution of the Western diet: health implications for the 21st century. *American Journal of Clinical Nutrition, 81*, 341–354.

Dickie, J. B., & Stuppy, W. H. (2003). Seed and fruit structure: significance in seed conservation operations. In R. D. Smith, J. B. Dickie, S. H. Linington, H. W. Pritchard, & R. J. Probert (Eds.), *Seed conservation: Turning science into practice* (pp. 254–279). Kew, UK: Royal Botanic Gardens.

Dyer, G., Serratos, A., Piñeyro, A., Perales, H., Gepts, P., Chávez, A., et al. (2009). Escape and dispersal of transgenes through maize seed systems in Mexico. *PLoS ONE, 4*(5), e5734, doi:10.1371/journal.pone.0005734.

Fenner, M., & Thompson, K. (2005). *The ecology of seeds.* Cambridge, UK: Cambridge University Press.

Gepts, P. (2005). Plant and animal domestication as human-made evolution. In J. Cracraft, & R. W. Bybee (Eds.), *Evolutionary science and society: Educating a new generation* (pp. 180–186). Colorado Springs, CO: Biological Sciences Curriculum Study.

Gepts, P., & Papa, R. (2003). Possible effects of (trans)gene flow from crops on the genetic diversity from landraces and wild relatives. *Environmental Biosafety Research, 2*, 89–103.

Gurian-Sherman, D. (2009). *Failure to yield: Evaluating the performance of genetically engineered crops.* Cambridge, MA: Union of Concerned Scientists Publications.

Jauhar, P. P. (2006). Modern biotechnology as an integral supplement to conventional plant breeding: the prospects and challenges. *Crop Science, 46*, 1841–1859.

Litz, R. E. (Ed.). (2005). Biotechnology of fruit and nut crops. *Biotechnology in Agriculture Series, No. 29.* Wallingford, UK: CABI Publishing.

Ma, J. K.-C., Chikwamba, R., Sparrow, P., Fischer, R., Mahoney, R., & Twyman, R. M. (2005). Plant-derived pharmaceuticals — the road forward. *Trends in Plant Science, 10*, 580–585.

[7] These proteins include monoclonal antibodies, enzymes, blood proteins, hormones, growth regulators, and vaccines.

Moles, A. T., Ackerly, D. D., Webb, C. O., Tweddle, J. C., Dickie, J. B., Pitman, A. J., et al. (2005). Factors that shape seed mass evolution. *Proceedings of the National Academy of Sciences of the United States of America, 102,* 10540−10544.

Newman, L. F. (Ed.). (1990). *Hunger in history: Food shortage, poverty and deprivation.* Cambridge, UK: Basil Blackwell.

Qaim, M. (2009). The economics of genetically modified crops. *Annual Review Resource Economics, 1,* 665−693.

Smale, M., Zambrano, P., Falck-Zepeda, J., & Gruere, G. (2006). Parables: Applied economics literature about the impact of genetically engineered crop varieties in developing countries. *EPT Discussion Paper 158.* Washington, DC: International Food Policy Research Institute.

Spok, A., Twyman, R. M., Fischer, R., Ma, J. K.-C., & Sparrow, P. A. C. (2008). Evolution of a regulatory framework for pharmaceuticals derived from genetically modified plants. *Trends in Biotechnology, 26,* 506−517.

Steckel, R., & Rose, J. (Eds.). (2002). *The backbone of history: Health and nutrition in the western hemisphere.* Cambridge, UK: Cambridge University Press.

Wolfenbarger, L. L., Narnajo, S. E., Lundgren, J. G., Bitzer, R. J., & Wagtrud, L. S. (2008). *Bt* crop effects on functional guilds of non-target arthropods: a meta-analysis. *PloS ONE, 3*(5), e2118, doi:10.1371/journalpone.0002118.

Werker, E. (1996). Seed anatomy, *Vol. 10, Part 3 of* Encyclopedia of plant anatomy. Berlin, Germany: Borntraeger.

CHAPTER 2

Seeds as Herbal Drugs

Sanjib Bhattacharya
Bengal School of Technology (A College of Pharmacy), Sugandha, Hooghly,
West Bengal, India

CHAPTER OUTLINE

INTRODUCTION

The seed is the transportation stage in the plant's life. Seed is derived from the fertilized ovule; it contains an embryo, and is constructed so as to facilitate its transportation and germination in favorable conditions. Seed represents a condensed form of life, and is a characteristic of phenograms. Care must be taken to distinguish seeds from fruits or parts of fruits containing a single seed (for example, cereals and mericarps of the umbelliferous plants).

The use of plants and plant-based products for treatment of diseases is as old as mankind. Likewise, other plant parts as well as seeds have been used as natural drugs for the management of several ailments as advocated by traditional healers since time immemorial. Seeds are still widely used as herbal drugs, either in their crude form or as preparations thereof, or as sources of medicinally active natural products to be used in traditional or contemporary medicine. In the past few decades there has been an exponential growth in the field of herbal medicine. It is becoming popular in developed and developing countries as people consider herbal remedies to be a better option for their healthcare needs, owing to their natural origin and reduced side effects. Apart from dietary (for edible oils) and industrial (mainly as lubricants) purposes, seeds serve as important herbal drugs intended for several therapeutic indications. Numerous types of seeds have traditionally been used for medicinal purposes in different parts of the world, and among these are certain seeds that are still being used today, with renewed interest in their medicinal properties. The present chapter provides a broad overview of medicinally important seeds of contemporary interest.

MEDICINAL CONSTITUENTS OF SEEDS

The chemical constituents of seeds are responsible for their medicinal properties. Most seeds have a certain general chemical composition. All seeds contain reserve foods for the

Nuts & Seeds in Health and Disease Prevention. DOI: 10.1016/B978-0-12-375688-6.10002-7

nourishment of the embryo during germination. These foods may be present in the endo-sperm or the perisperm, or in both, or they may be present in the embryo itself either in the cotyledons or in the axis. Sometimes, as in linseed, the food is found in both endosperm and cotyledons. The reserve foods present are usually of two types: carbohydrates and fixed oils, which supply the elements carbon, hydrogen, and oxygen; and proteins, which supply nitrogen and phosphorus in addition to the former three elements. The commonest carbohydrate is starch; others are cellulose, hemicellulose, and sugar. Starches, sugars, fixed oils, and proteins are stored in the cell cavities, whereas celluloses are present as heavily thickened cell walls. These constituents give rise to starchy or farinaceous seeds, such as in the calabar bean; oily seeds, such as in linseed and the umbelliferous seeds; and very hard horny seeds, such as in nux vomica and the ignatius bean. Endosperm, which is present in many seeds, is composed of a cellulose-walled parenchyma containing food reserves. The walls are usually thin, but in some seeds (such as nux vomica and ignatius bean) the walls become very thick, being largely composed of hemi-cellulose. Seeds also contain protein reserves (aleurone grains), sclereids, mucilage, pigments, enzymes, several other nutrients, etc. (Wallis, 1985).

Apart from the general constituents, comprising both primary and secondary plant metabo-lites, the seeds contain a spectrum of specific secondary metabolites, most of which are very limited in distribution and some of which have complex chemical structures. Examples include alkaloids in nux vomica, calabar bean, and colchicum seeds; essential oils (terpenoids) in cardamom and nutmeg seeds; cardenolides in thevitia and stropanthus seeds; cyanogenetic glycosides in bitter almond seed; isothiocyanate glycosides in mustard seed; flavonolignans in milk thistle seed; bitter principles in neem and karela seeds; saponins in fenugreek seed; resinous matter in croton seed; gum in guar gum seed; liquid wax in jojoba seed, etc. It has been found that all of these secondary metabolites are principally responsible for the medicinal activity of seeds. Different specific secondary metabolites and the general seed constituents like fixed oils, carbohydrates, proteins, mucilage, etc., together contribute to the biological activity of seeds used as herbal drugs (Evans, 2002).

FACTORS INFLUENCING MEDICINAL PROPERTIES OF SEEDS

The seeds that reach the herbal market pass through various stages, all of which influence the nature and quantity of chemical constituents, and hence their therapeutic activity.

1. *Cultivation.* Environmental factors, including temperature, rainfall, length of daylight, radiation characteristics, altitude, soil and fertilizer/pesticide use, etc., may influence the chemical constituents.
2. *Collection.* To ensure maximum therapeutic potential, the seeds must be collected when perfectly ripe. Cleaning or washing is an important step, as the seeds must be free from fruit pulp before drying. Seeds which are obtained from mucilaginous fruits, such as nux vomica and cocoa, should be washed well to ensure that they are free from pulp. Some seeds, such as linseed, are separated from their husks by threshing and winnowing the fruits. Other factors include the nature of the plant (wild or cultivated, and its age), the season and time of collection, the personnel involved, etc.
3. *Drying.* Immediately after collection the seeds should be dried in the open air; generally, they should not be kiln-dried. Moist seeds are liable to develop microbial growth if not dried immediately. In some instances, however, drying by artificial heat becomes necessary, especially in tropical countries. Cocoa seeds are fermented by slow drying at a moderate temperature prior to sun drying. Some seeds, such as those of guarana, cola, and nutmeg, are decorticated during or after drying and their kernels are removed. These seeds require special drying techniques.
4. *Storage.* The dry seeds are normally stored in a cool, dry place. Microbial contamination must be avoided. Properly stored seeds retain their medicinal values for several years.

SEEDS AS A SOURCE OF MEDICINALLY IMPORTANT FIXED OILS

As agricultural crops, seeds used for the extraction of fixed oils are rated second in importance to cereals. Over the past six decades the production of oils for the food industry has increased enormously, whereas consumption by industrial and other users has remained relatively static although there have been interesting developments in the pharmaceutical industry.

Fixed oils or fatty oils are the reserve food materials of plant and animals. Those that are liquid at 15.5–16.5°C are termed fixed oils, while those that are solid or semi-solid at that temperature are termed fats. Fixed oils derived from plant sources generally occur in seeds (Kokate *et al.*, 2008).

Chemically fixed oils are esters of glycerol and various straight-chain monocarboxylic acids, known as fatty acids, which may be saturated, monounsaturated, or polyunsaturated. They contain various minor components (i.e., vitamins, sterols, antioxidants, phospholipids, pigments, and traces of hydrocarbons and ketones), some of which contribute to the medicinal properties.

Physiologically fixed oils are emollients and demulcents, and have nutritional value. The unsaturated fatty acids, namely linoleic, linolenic, and arachidonic acids (present in several seed fixed oils), are termed essential fatty acids, as they are not produced in the human body and must be provided in the diet. Fixed oils containing essential fatty acids (such as evening primrose oil) serve as nutritive and dietary supplements (nutraceuticals). Polyunsaturated fatty acids including omega-3 fatty acids are present in various seed fixed oils, namely safflower, flaxseed, mustard, and soya oils, etc. They help to reduce cholesterol formation and/or deposition, and hence to decrease the risk of atherosclerosis and ischemic heart disease. Fixed oils containing mainly unsaturated fatty acids are used as nutraceuticals in the prophylaxis of hypercholesteremia and atherosclerosis. Phospholipids present in fixed oils are hepatoprotective (Awang, 2009). Apart from these specific therapeutic uses, fixed oils are also utilized in the preparation of ointment, suppository, and cream bases; in dermatological preparations; as oily vehicles for oil-soluble injectable drugs; and as demulcents, emollients, and lubricants, etc. (Duke *et al.*, 2002). Table 2.1 provides a comprehensive list of the names, sources, constituents, and pharmacological (or therapeutic) activities and uses of medicinally important fixed seed oils.

Fixed oils are generally obtained from seeds by expression, centrifugation, and/or solvent extraction, depending on the product. Some fixed oils, such as arachis or linseed oil, are obtained by hot expression (or pressing); others, such as castor oil and chaulmoogra oil, are obtained by cold pressing. In some cases — for example, chaulmoogra oil — the seeds are decorticated and the kernels are pressed. Some oils, such as theobroma oil, are obtained from roasted seeds. The crude oil obtained by pressing then requires appropriate refining. Cold-pressed oils usually require nothing further than filtration, although sometimes the crude oil obtained is treated to get rid of unwanted or toxic constituents — for example, crude castor oil is steamed at 80°C to destroy the enzyme lipase and ricin (a toxic protein) to render it suitable for medicinal use, and linseed oil is left to settle so as to precipitate mucilage and coloring matter, before being treated with alkali to remove free fatty acids. Some of the fixed oils are bleached (for example, linseed oil and arachis oil), while others may be refined industrially according to their specific medicinal purposes (for example, as a vehicle for intramuscular oily injections) (DerMarderosian, 2001).

SEEDS AS HERBAL DRUGS AND A SOURCE OF MEDICINALLY ACTIVE COMPOUNDS

This section deals with the seeds used, either in crude form or as preparations, or as a source of medicinally active natural products other than fixed oils and fatty acids, in traditional and contemporary medicine. These seeds show diverse medicinal actions due mainly to the

TABLE 2.1 Seeds as a Source of Medicinally Important Fixed Oils

SI No.	Common Name	Biological Source	Chemical Constituents of Seeds	Actions and Uses
1	Arachis or earthnut or groundnut or peanut oil	*Arachis hypogaea* (Leguminosae)	Glycerides of fatty acids, *viz.* oleic, linoleic, palmitic, stearic, arachidic and other acids	Vehicle for intramuscular injections, preparation of liniments, plasters, soap, margarine.
2	Castor or ricinus oil	*Ricinus communis* (Euphorbiaceae)	Glycerides of ricinoleic, isoricinoleic, linoleic, and other acids	Purgative, lubricant
3	Coconut oil	*Cocos nucifera* (Palmae)	Glycerides of mainly lauric, myristic, and other acids	Non-aqueous medium for oral administration of drugs
4	Hydnocarpus or chaulmoogra oil	*Hydnocarpus wightiana*, *Taraktogenous kurzii* (Flacourtiaceae)	Glycerides of hydnocarpic, chaulmoogric, and palmitic acids	Antibacterial, antileprotic, antitubercular, in psoriasis and rheumatism
5	Linseed (Flax seed) oil	*Linum usitatissimum* (Linaceae)	Glycerides of fatty acids, sterols, tocopherol, pectin, mucilage, phenylpropanoids, flavonoids, linamarin (glycoside)	Demulcent, as poultice, lotions, and liniments in skin diseases
6	Palm kernel oil	*Elaeis guineensis* (Palmae)	Glycerides of mainly lauric and other acids	As a suppository base
7	Sesame oil (Gingelly oil, Teel oil)	*Sesamum indicum* (Pedaliaceae)	Glycerides of higher fatty acids, olein, sesamol, lignans (sesamin, sesamolin)	Nutritive, laxative, demulcent, emollient; in preparation of liniments and ointments; vehicle for intramuscular oily injections
8	Mangosteen oil (Kokum oil)	*Garcinia indica* (Guttiferae)	Glycerides of stearic, oleic, hydroxycapric, palmitic, and linoleic acids	Nutritive, demulcent, astringent, emollient; as ointment and suppository bases
9	Safflower oil	*Carthamus tinctorius* (Compositae)	Glycerides of palmitic, stearic, arachidic, oleic, linoleic, and linolenic acids	In preparation of oleomargarine, as dietary supplement in hypercholesteremia and atherosclerosis
10	Black mustard oil	*Brassica nigra*, or *Brassica juncea* (Brassicaceae)	Glycerides of fatty acids, sinigrin (glycoside), myrosin (enzyme)	Local irritant, externally as rubefacient and vesicant; internally as emetic
11	White mustard oil	*Brassica alba* (Brassicaceae)	Glycerides of fatty acids, mucilage, sinalbin (glycoside), myrosin (enzyme)	Used externally as rubefacient and vesicant

18

Continued

TABLE 2.1   Seeds as a Source of Medicinally Important Fixed Oils—continued

Sl No.	Common Name	Biological Source	Chemical Constituents of Seeds	Actions and Uses
12	Poppy seed oil or poppy oil	*Papaver somniferum* (Papaveraceae)	Glycerides of mainly linoleic, palmitic, arachidic, and oleic acids; no narcotic principles	In preparation of iodized oils, soaps
13	Karanja oil	*Pongamia glabra* (Papilionaceae)	Glycerides of different fatty acids, β-sitosterol, karanjin, pongapin, pongamol	In scabies, herpes, leukoderma, and other cutaneous diseases; in rheumatism
14	Neem oil	*Azadirachta indica* (Meliaceae)	Glycerides of mainly oleic and stearic acids, sulfur-containing bitters, nimbidin, nimbin, nimbinin, nimbidiol, and nimbosterol	Antiviral, insect repellent, spermicide, pesticide; in rheumatism.
15	Soya oil	*Glycine max* or *G. soja* (Leguminosae)	Glycerides of linoleic, oleic, palmitic, linolenic, and stearic acids, phosphatidylcholine, proteins, isoflavones	Nutritive, hepatoprotective, lipotropic agent, hypolipidemic, antitumor
16	Croton oil	*Croton tinglum* (Euphorbiaceae)	Glycerides of different fatty acids, croton resin, protein (crotin), phorbol esters	Local irritant, counter-irritant, internally drastic purgative
17	Cotton seed oil	*Gossypium herbaceum* (Malvaceae)	Glycerides of mainly palmitic, oleic, and linoleic acids	Pediculicide, acaricide, and laxative in veterinary medicine
18	Theobroma oil	*Theobroma caco* (Sterculiaceae)	Glycerides of mainly stearic, palmitic, and oleic acids	Base for suppositories, ointments, creams
19	Jojoba oil	*Simmondsia chinensis, S. californica* (Buxaceae)	Liquid wax, i.e. esters of monounsaturated acids and alcohols, tocopherols, and phytosterols	Cosmetics, vehicle of dermatological preparations, lubricant
20	Evening primrose oil	*Oenothera biennis* (Onagraceae)	Glycerides of mainly linoleic acid and γ-linolenic acid	Dietary supplement for essential fatty acids, antithrombotic; in premenopausal syndrome, atopic eczema, rheumatic arthritis, diabetic neuropathy; hepatoprotective
21	Rapeseed oil	*Brassica napus, B. campestris* (Brassicaceae)	Glycerides of mainly omega-3 fatty acids	Nutritive, antiplatelet; as dietary supplement in hypercholesteremia
22	Sweet almond oil	*Prunus communis* (Rosaceae)	Glycerides of different fatty acids, proteins, emulsion (enzyme),	Demulcent, nutritive

19

Continued

TABLE 2.1   Seeds as a Source of Medicinally Important Fixed Oils—continued

Sl No.	Common Name	Biological Source	Chemical Constituents of Seeds	Actions and Uses
23	Tung oil	*Aleurites moluccna* (Euphorbiaceae)	gum, sucrose, asparagin Glycerides of fatty acids, *viz.* eleostearic, linolenic, linoleic, and oleic acids, proteins	Wood preservative; in varnishes, paints; toxic to humans
24	Sunflower oil	*Helianthus annus* (Asteraceae)	Glycerides of unsaturated fatty acids, stigmasterol, β-sitosterol, tocopherol	Dietary supplement

presence of different secondary plant metabolites as the pharmacologically active chemical constituents, thus demonstrating therapeutic activities for a multitude of diseases. They may also contain fixed oils. Some of them are toxic in nature, especially when consumed in appreciable amounts (Barceloux, 2008). Most of their chemical constituents have been isolated and pharmacologically evaluated, and some of the active constituents are commercially extracted and isolated from the seeds for prophylactic and therapeutic uses in medicine. Table 2.2 provides a detailed list of the names, sources, active constituents, and pharmacological (or therapeutic) activities and uses of such important seeds. However, numerous references to the multiple biological activities of various seeds and their constituents can be found in the literature, and new reports of pharmacological screening continually appear.

TABLE 2.2 Seeds as Herbal Drugs and a Source of Medicinally Active Compounds

Sl No.	Common Name	Biological Source	Chemical Constituents of Seeds	Actions and Uses
1	Nux vomica	*Strychnos nuxvomica* (Loganiaceae)	Alkaloids, *viz.* strychnine, brucine, vomicine, α-colubrine, pseudostrychnine, chlorogenic acid	Bitter stomachic, tonic, central nervous system stimulant; in cardiac failure, as rodenticide
2	Ignatius beans	*Strychnos ignatti* (Loganiaceae)	Alkaloids, *viz.* strychnine, brucine	Similar to those of nux vomica
3	Physostigma or Calabar bean	*Physostigma venenosum* (Leguminosae)	Alkaloids, *viz.* physostigmine, eseramine, physovenine, 8-norphysostigmine, isophysostigmine	Parasympathomimetic, anticholinesterase; as miotic for contraction of eye pupils, antidote for belladonna poisoning
4	Belladonna	*Atropa belladonna* (Solanaceae)	Alkaloids, *viz.* atropine, L-hyoscyamine, hyoscine, belladonine, scopoletin	Parasympatholytic; as antisecretory in peptic ulcer, mydriatic (pupil dilatation), antispasmodic, antidote of chloral hydrate poisoning
5	Datura	*Datura metel* (Solanaceae)	Alkaloids, *viz.* hyoscine, atropine, l-hyoscyamine	Parasympatholytic, CNS depressant; in bronchial asthma, cough, cerebral excitement, preoperative medication

Continued

TABLE 2.2   Seeds as Herbal Drugs and a Source of Medicinally Active Compounds—continued

Sl No.	Common Name	Biological Source	Chemical Constituents of Seeds	Actions and Uses
6	Stramonium	*Datura stramonium* (Solanaceae)	Alkaloids, *viz.* L-hyoscyamine, hyoscine, proteins, fixed oil	Similar to those of belladonna; in bronchial asthma, parkinsonism
7	Lobelia, Asthma weed	*Lobelia nicotianaefolia* (Campanulaceae)	Alkaloids, *viz.* lobeline, lobelidine, lobelanine, isolobelanine	In bronchial asthma, chronic bronchitis, as respiratory stimulant
8	Areca	*Areca catechu* (Palmae)	Alkaloids, *viz.* arecoline, arecaidine, guvacine, guvacoline	Parasympathomimetic, sialogogue; as anthelmintic (vermicide and taenifuge) in veterinary practice
9	Coffee	*Coffea arabica* (Rubiaceae)	Caffeine (alkaloid), tannin, fixed oil, protein, chlorogenic acid, sugars	Stimulant, diuretic; to antagonize toxic effects of CNS-depressant drugs
10	Cocoa	*Theobroma cocoa* (Sterculiaceae)	Alkaloids, mainly theobromine, caffeine, polyphenols	Stimulant, diuretic, nutritive
11	Kola (Bissy or Gooroo)	*Cola nitida* (Sterculiaceae)	Alkaloids, *viz.* caffeine, theobromine, kolacatechin (tannin)	Stimulant, diuretic; in aerated beverages
12	Colchicum	*Colchicum leutum, C. autumnale.*	Alkaloids, *viz.* colchicines, demecolcine	Antitumor; in gout and rheumatism; colchicine causes polyploidy
13	Cardamom	*Elettaria cardamomum* (Zingiberaceae)	Essential oil comprising cineole, terpineol, borneol, terpinene; fixed oil, protein	Aromatic, carminative, stimulant, flavoring agent, condiment
14	Carrot	*Dacuus carota* (Umbelliferae)	Essential oil comprising α-pinene, β-pinene, limonene, carotol, daucol, β-elemene, geraniol, etc.	Carminative, aromatic, diuretic, smooth-muscle relaxant, vasodilator, antidysenteric, aphrodisiac
15	Nutmeg	*Myristica fragrans* (Myristicaceae)	Essential oil comprising myristicin, elimicin, saffrole; fixed oil comprising glycerides of mainly myristic and other fatty acids	Aromatic, carminative, stimulant, flavoring agent, condiment; poisonous and narcotic in large amounts
16	Annatto	*Bixa orellana* (Bixaceae)	Annatto oleo-resin, bixin (carotenoid pigment)	Coloring agent for food, cosmetics and pharmaceuticals
17	Guar gum	*Cyamposis tetragonolobus* (Leguminosae)	Gum comprising guaran (hydrocolloid polysaccharide), sugars, proteins	Laxative, anoretic, anti-ulcer, hypocholesterolemic, oral hypoglycemic, emulsifying agent
18	Isapgol	*Plantago ovata* (Plantaginaceae)	Mucilage, fixed oils, protein	Demulcent, laxative, emollient, antidysenteric
19	Psyllium, Flea seed	*Plantago psyllium* (Plantaginaceae)	Mucilage, aromatic. carboxylic acids, alkaloids, amino acids, flavonoids, fats, iridoids, sugars	Laxative, demulcent, antihyperlipidemic, antitumor
20	Thevetia	*Thevetia peruviana* (Apocynaceae)	Cardenolides, *viz.* thevetin A and B, peruvoside, neriifolin, thevenarin, peruvosidic acid	Cardiotonic, emetic febrifuge, abortifacient, purgative; in rheumatism and dropsy; poisonous in large amounts

Continued

TABLE 2.2 Seeds as Herbal Drugs and a Source of Medicinally Active Compounds—continued

Sl No.	Common Name	Biological Source	Chemical Constituents of Seeds	Actions and Uses
21	Stropanthus	*Stropanthus kombe* (Apocynaceae)	Cardenolides, *viz.* K-stropanthoside, K-stropanthoside β-cymarin, cymarol; mucilage, resin, trigonelline, fixed oil	Cardiotonic in congestive heart failure; poisonous in large amounts
22	Stropanthus	*Stropanthus gratus* (Apocynaceae)	Cardenolide: G-stropanthidin (ouabain)	Emergency cardiotonic in acute cardiac failure; poisonous in large amounts
23	Bitter almond	*Prunus amygdalus* (Rosaceae)	Fixed oil, proteins, enzymes, amygdalin (cyanogenetic glycoside)	Sedative, demulcent, flavoring agent; poisonous in large amounts
24	Milk-thistle	*Silybum marianum* (Asteraceae)	Silymarin (a mixture of three isomeric flavonolignans), and other flavonolignans, betaine, silybonol	Hepatoprotective in hepatitis, cirrhosis and fatty liver; antihepatotoxic; as bitter tonic
25	Tonka-bean	*Dipteryx odorata* (Leguminosae)	Coumarin, fats	Flavoring agent, fixative in perfumery
26	Visnaga	*Ammi visnaga* (Umbelliferae)	Furanocoumarins, *viz.* khellin, visnagin, khelloside, fixed oil comprising samidine, dihydrosamidine visnadine, etc.	Smooth-muscle relaxant, coronary vasodilator in angina pectoris; in uterine and renal colic, bronchial asthma, and whooping cough
27	Ammi	*Ammi majas* (Umbelliferae)	Furanocoumarins, *viz.* xanthotoxin, bergapten, isopimpilin, imperatorin	Melanizing agent, in idiopathic vitiligo, leukoderma
28	Psoralea	*Psoralea corylifolia* (Leguminosae)	Essential oil, fixed oil, resin, pigments; furanocoumarins, *viz.* psoralen, psoralidin, isopsoralidin	In leukoderma, leprosy, psoriasis, skin inflammations
29	Ambrette	*Abelmoschus moschatus* (Malvaceae)	β-sitosterol, cholesterol, stigmasterol, ergosterol, campesterol; fixed oil comprising higher fatty acids; ambrettolide (lactone)	Aphrodisiac, antispasmodic, carminative, demulcent, diuretic, stomachic, tonic, flavoring agent; dermal irritant in large amounts
30	Asparagus	*Asparagus officinale* (Liliaceae)	Polysaccharides, proteins	As coffee substitute, diuretic, laxative, analgesic, antirheumatic
31	Bitter melon, Karela	*Momordica charantia* (Cucurbitaceae)	Fixed oil comprising triglycerides of stearic, linoleic, oleic acids; vicine (nucleoside), α and β-momorcharin (glycoproteins), lectins, amino acids	Oral hypoglycemic, antimicrobial; not recommended in pregnancy
32	Celery	*Apium graveolens* (Umbelliferae)	d-limonene, selinene, 3-*n*-butylphthalide, sedanenolide, sedanonic anhydride	Sedative, anti-arthritic, antirheumatic, analgesic, diuretic, nervine tonic, hypoglycemic, antifungal
33	Corn cockle	*Agrostema githago* (Caryophyllaceae)	Aromatic amino acids, L (+) citrullin, fixed oil, sugars, starch, saponins	Diuretic, expectorant, vermifuge, emmenagogue; in jaundice, gastritis; fatally poisonous in large amounts
34	Cucurbita	*Cucurbita pepo* (Cucurbitaceae)	Fixed oil, carotenoids, cucurbitin	Anthelmintic; in prostate disorders

Continued

22

TABLE 2.2 Seeds as Herbal Drugs and a Source of Medicinally Active Compounds—continued

SI No.	Common Name	Biological Source	Chemical Constituents of Seeds	Actions and Uses
			(carboxypyrrolidine), flavonols, sterols, amino acids, tocopherols	
35	Guarana	*Paullinia cupana* (Sapindaceae)	Caffeine, guaranatin, catechutannic acid, fat, starch	Nervine stimulant, diuretic, anoretic
36	Quince seeds	*Cydonia oblongata* (Rosaceae)	Mucilage, fixed oil, proteins, amygdalin	Demulcent, suspending agent, emulsifying agent
37	Stavesacre	*Delphinium stapisagria* (Ranunculaceae)	Alkaloids, *viz.* delphinine, delphisine, delphinoidine, staphisagroine; fixed oil	Antiparasitic; in neuralgia; extremely poisonous in large amounts
38	Cevadilla or Sabadilla	*Schoenocaulon officinale* (Liliaceae)	Alkaloids, *viz.* cevadine, verarine, sebadilline, sebadine, sebadinine; fats	Antiparasitic, insecticide, counter-irritant, local anesthetic; in neuralgia
39	Grains of paradise	*Aframomum melegueta* (Zingiberaceae)	Essential oil, paradol (pungent principle)	Stimulant, analgesic, condiment
40	Fenugreek	*Trigonella foenum-graecum* (Leguminosae)	Coumarin derivatives, alkaloids (trigonelline, gentianine, carpaine), fixed oil, saponins (smilagenin, sarasapogenin, yuccagenin), mucilage	Oral hypoglycemic, hypocholesterolemic, anti-inflammatory, diuretic, condiment
41	Grape seed	*Vitis vinifera* (Vitaceae)	Essential fatty acids, tocopherols, oligostilbenes, procyanidines, and other polyphenols	Nutritive, hepatoprotective, antitumor
42	Horse chestnut	*Aesculus sp.* (Sapindaceae)	Fixed oil comprising glycerides of mainly oleic acids; proteins, sugars, triterpene, coumarin and saponin glycosides	Circulatory tonic in chronic venous insufficiency, anti-inflammatory, antipyretic; poisonous in large amounts
43	Black cumin	*Nigella sativa* (Ranunculaceae)	Fixed oils, proteins, nigellone, thymoquinone, essential oil, saponins, alkaloids, ascorbic acid	Anti-inflammatory, antimicrobial, anthelmintic; in respiratory diseases (cough, bronchitis, asthma, flu); not recommended in pregnancy
44	Custard apple, Pawpaw	*Asimia triloba* (Annonaceae)	Acetogenins	Antitumor, antimicrobial, pesticide
45	Jequirty seed	*Abrus precatorius* (Fabaceae)	Alkaloids, *viz.* abrine, hyaphorine, precatorine, lectins; proteins, *viz.* abrin A, B and C	Analgesic, neuromuscular blocker; fatally poisonous, especially in children
46	Kavach, Cowharge	*Mucuna pruriens or M. prurita* (Fabaceae)	Amino acids, mainly L 3,4-dihydroxy phenyl alanine (L-dopa); lecithin, alkaloids, fat	Antiparkinsonian; in manufacture of antiparkinsonian drug levodopa

23

SUMMARY POINTS

- Seeds have been traditionally used as herbal drugs, and are still being used in this way.
- Seeds contain several constituents, including certain secondary plant metabolites in specific seeds which are mainly responsible for their medicinal values.
- Environmental and pre- and post-harvesting factors can influence the medicinal properties of seeds.

- Seeds serve as sources of different medicinally important fixed oils.
- Seeds serve as herbal drugs for various therapeutic indications, and as source of medicinally active compounds, most of which are putative secondary metabolites.

References

Awang, D. V. C. (2009). *Tyler's herbs of choice: The therapeutic uses of phytomedicinals* (3rd ed.). Boca Raton, FL: CRC Press.

Barceloux, D. G. (2008). *Medical toxicology of natural substances.* Hoboken, NJ: John Wiley & Sons, Inc.

DerMarderosian, A. (2001). *The reviews of natural products* (1st ed.). St Louis, MO: Facts and Comparisons.

Duke, J. A., Bogenschuz-Godwin, M. J., duCellier, J., & Duke, P. K. (2002). *Handbook of Medicinal Plants* (2nd ed.). Boca Raton, FL: CRC Press.

Evans, W. C. (2002). *Trease and Evans pharmacognosy* (15th ed.). New Delhi, India: Reed Elsevier India Pvt. Ltd.

Kokate, C. K., Purohit, A. P., & Gokhale, S. B. (2008). *Pharmacognosy* (36th ed.). Pune, India: Nirali Prakashan.

Wallis, T. E. (1985). *Textbook of pharmacognosy* (5th ed.). New Delhi, India: CBS Publishers and Distributors.

CHAPTER 3

Seeds, Nuts, and Vector-Borne Diseases

Essam Abdel-Salam Shaalan[1], Deon V. Canyon[2]
[1] Zoology Department, Aswan Faculty of Science, South Valley University, Aswan 81528, Egypt
[2] Disaster Health and Crisis Management Unit, Anton Breinl Centre for Public Health and Tropical Medicine, James Cook University, Townsville, Queensland, Australia

25

LIST OF ABBREVIATIONS

DDT, dichlorodiphenyltrichloroethane
DEET, N,N-diethyl-*meta*-toluamide
IGR, insect growth regulator
LC_{50}, lethal concentration for 50% of the target population
LD_{50}, lethal dose for 50% of the target population

INTRODUCTION

Many arthropods, such as mosquitoes, sand flies, black flies, fleas, lice, and ticks, can vector serious human diseases. Vector control has relied upon synthetic insecticides since the discovery of DDT. However, owing to their adverse effects on humans and the environment, interest in other methods has increased. Chemicals of botanical origin, called phytochemicals, have displayed a range of acute and chronic insecticidal effects against a variety of insects, and are promising candidates as future synthetic insecticides. Almost all research on the activity of seed extracts against disease vectors has focused on mosquitoes. This is because mosquitoes carry vector diseases such as malaria, filariasis, and dengue, and cause most arthropod-borne disease. The World Health Organization has indicated that malaria infects 500 million people annually, and kills more than 1.5 million annually — particularly African children. Thus, the focus on other arthropod disease vectors has been minimal. While mosquito control relies extensively on synthetic insecticides, mosquitoes have also been the primary focus for seed-based phytochemical and bioinsecticide investigations.

Nuts & Seeds in Health and Disease Prevention. DOI: 10.1016/B978-0-12-375688-6.10003-9

Research has focused on testing plants that have previously demonstrated some degree of activity. Since there are thousands of plants, and bioinsecticide research is underfunded, a broadside biodiscovery approach has never been possible. Thus, prior plant activity as determined through non-scientific anecdotal "herb lore" has tended to provide the first clues. Not surprisingly, active phytochemical components have been extracted from a broad selection of plant types and species. Phytochemicals have been extracted from medicinal plants, citrus plants, some leguminous plants, and some marine weeds. They also come from different parts of plants, such as the leaves, stems, roots, tubers, fruits, seeds and seeds kernels, and shoot systems (Shaalan *et al.*, 2005). They have been found to be toxic against different developmental mosquito stages, such as eggs (ovicide), larvae (larvicide), pupae (pupicide), and adults (adulticides and repellents), and some have shown growth-regulating activity (insect growth regulators, IGR). Furthermore, some have displayed considerable synergistic activity with current synthetic chemical insecticides, which is an advantage because it may prolong the usability of some synthetic insecticides when resistance builds up in pest populations. In fact, in original articles and a few reviews, the literature shows hundreds of plant species to have demonstrated mosquitocidal activity, but none of the work specifically reviewed seeds and nuts. Hence, this chapter highlights the mosquitocidal activity that can be ascribed to phytochemicals of seed and nut origin.

MOSQUITOCIDAL ACTIVITY OF SEEDS

The seeds of over 40 plant species have been found to display some mosquitocidal activity (Table 3.1). Seed extracts and some of their constituent compounds have been observed to possess larvicidal, growth regulating, and repellent activity against different mosquito vector species. Since adulticides often only act to reduce the adult mosquito population temporarily, most mosquito control programs are based on the judicious and evidence-based use of larvicides. Larvicides are considered to be more effective, since they provide longer-term control of larval stages in their breeding habitats. However, identifying plants with these qualities can be difficult, because results can be unexpected. When looking at neem-producing trees, for instance, an acetone seed extract of *Melia volkensii* was equally toxic to both larvae and pupae with a LD_{50} of 30 μg/ml, while an extract of a closely related species, *M. azedarach*, was exclusively larvicidal with a LD_{50} of 40 μg/ml and had no inhibitory effect on the pupal stage (Al-Sharook *et al.*, 2009). Similarly, when looking at some spices, the array of effects can be broad but very selective. For instance, a seed extract of anise (*Pimpinella anisum*) was better as a larvicide and an ovicide than in other activities; a seed extract of cumin (*Cuminum cyminum*) proved to be more effective as an adulticide, an oviposition deterrent, and a repellent than as a larvicide and an ovicide; and a seed extract of black cumin (*Nigella sativa*) was more effective as an oviposition deterrent and larvicide against *Aedes aegypti*, *Anopheles stephensi*, and *Culex quinquefasciatus* (Prajapati *et al.*, 2005).

Mosquitocidal activity, whether ovicidal, larvicidal, growth regulating, adulticidal, oviposition deterrent, or repellent, is basically a toxic effect that is governed by the familiar pharmaceutical dose–response curve. Thus, different effects can be produced at different concentrations of the same extract. For instance, a petroleum ether extract of *Argemone mexicana* seeds exhibited larvicidal and growth inhibiting activity against the second instar *Ae. aegypti* larvae at concentrations between 25 and 200 ppm, while chemosterilant activity (reduction in blood meal utilization, reduction in fecundity, adult mortality, and sterility of first-generation eggs) occurred at a lower concentration of 10 ppm (Sakthivadivel & Thilagavathy, 2003).

It may be assumed that seeds are the most potent part of a plant; however, the mosquitocidal activity of seed extracts is only occasionally better than that of other plant parts. For instance, a seed acetone extract from *Tribulus terrestris* produced 100% larval mortality against four species of mosquito larvae (*An. Culicifacies*, *An. stephensi*, *Cx. quinquefasciatus*, and *Ae. aegypti*) at 100 ppm, while a leaf acetone extract only killed all larvae at 200 ppm (Singh *et al.*, 2008).

TABLE 3.1 Mosquitocidal Activity of Seeds

Scientific Name	Activity Type	Mosquito Species	Reference
Agave americana	Larvicide	Aedes aegypti Anopheles stephensi Culex quinquefasciatus	Dharmshaktu et al., 1987
Annona crassiflora A. glabra A. squamosa Pterodon polygalaeflorus Senna occidentalis Dorstenia sp.	Larvicide	Ae. aegypti	De Omena et al., 2007
Ammi visnaga	Larvicide	Cx. quinquefasciatus	Pavela, 2008
Annona squamosa	Larvicide, adulticide, IGR	An. stephensi	Senthilkumar et al., 2009
Apium graveolens	Larvicide, adulticide, repellent	Ae. aegypti	Choochote et al., 2007
Apium graveolens Carum carvi	Adulticide	Ae. aegypti	Chaiyasit et al., 2006
Carum carvi	Repellent	Ae. aegypti	Choochote et al., 2007
Apium graveolens Pimpinella anisum	Larvicide	Ae. aegypti	Shaalan & Canyon 2008
Argemone mexicana	Larvicide, adulticide, IGR, chemosterilant	Ae. aegypti	Sakthivadivel & Thilagavathy 2003
Azadiracta indica	Repellent	Ae. aegypti	Hati et al., 1995
Azadiracta indica (neem seed kernel)	Larvicide, IGR	Ae. aegypti Ae. togoi An. stephensi Cx. quinquefasciatus	Zebitz, 2009
Azadiracta indica Melia azedarach	Larvicide	Ae. aegypti	Wandscheer et al., 2004
Azadiracta indica Momordica charantia Ricinus communis	Larvicide	An. stephensi	Batabyal et al., 2007
Azadiracta indica Pongamia glabra	Larvicide, IGR	Cx. quinquefasciatus	Sagar & Sehgal 1996
Carica papaya	Larvicide	Cx. quinquefasciatus	Rawani et al., 2009
Chenopodium spp.	Repellent	Ae. albopictus	Chio & Yang 2008
Cinnamomum camphora	Larvicide	Cx. pipines pallens	Zhou et al., 2000
Citrus reticulate	Larvicide	Ae. aegypti Cx. quinquefasciatus	Sumroiphon et al., 2006
Clitoria ternatea	Larvicide	Ae. aegypti, An. stephensi Cx. quinquefasciatus	Mathew et al., 2009
Coriander sativum Petroselinum crispum Pimpinella anisum	Larvicide	Ochlerotatus caspius	Knio et al., 2008
Cuminum cyminum Nigella sativa Pimpinella anisum	Adulticide, larvicide, ovicide, oviposition deterrent, repellent	Ae. aegypti, An. stephensi Cx. quinquefasciatus	Prajapati et al., 2005
Daucus carota Khaya senegalensis	Larvicide	Cx. annulirostris	Shaalan et al., 2006
Delonix regia Raphia vinifera	Larvicide	An. gambiae	Aina et al., 2009

Continued

TABLE 3.1 Mosquitocidal Activity of Seeds—continued

Scientific Name	Activity Type	Mosquito Species	Reference
Eucalyptus globulus	Larvicide	Cx. pipiens	Elbanna, 2006
Gloriosa superba	Larvicide	An. subpictus	Abduz Zahir et al.,
Solannum trilobatum		Cx. tritaeniorhynchus	2009
Jatropha curcas	Larvicide	Och. triseriatus	Georges et al., 2008
Datura innoxia			
Litsea cubeba	Repellent, oviposition deterrent	Ae. aegypti	Tawatsin et al., 2006
		Ae. albopictus	
		An. dirus	
		Cx. quinquefasciatus	
		Ae. aegypti	
Melia volkensii	Larvicide, IGR	Cx. pipiens molestus	Al-Sharook et al., 2009
M. azedarach			
Millettia dura	Larvicide	Ae. aegypti	Yenesew et al., 2003
Ocimum basilicum	Larvicide, repellent	Anopheles sp.	Nour et al., 2009
Pimpinella anisum	Repellent	Cx. pipiens	Erler et al., 2006
Piper guineense	Larvicide	Ae. aegypti	Oke et al., 2001
Piper guineense	Larvicide	An. gambiae	Aina et al., 2009
Jatropha curcas			
Polylophium involvucratum	Larvicide	An. stephensi	Reza & Abbas 2007
		Cx. pipiens	
Pongamia glabra	Larvicide, IGR	Ae. aegypti	Sagar et al., 1999
		Cx. quinquefasciatus	
Rumex obtusifolius	Larvicide	Ae. aegypti	Kim et al., 2002
		Cx. pipiens pallens	
Solanum torvum	Larvicide	An. subpictus	Kamaraj et al., 2009
		Cx. tritaeniorhynchus	
Sterculia guttata	Larvicide	Ae. aegypti	Katade et al., 2006
		Cx. quinquefasciatus	
Tribulus terrestris	Larvicide, repellent	Ae. aegypti	Singh et al., 2008
		An. culicifacies	
		An. stephensi	
		Cx. quinquefasciatus	
Trachyspermum ammi	Larvicide, repellent, oviposition deterrent, vapor toxic	An. stephensi	Pandey et al., 2009

See Shaalan and Canyon (2010) for references cited in this Table.

Likewise, a seed extract from *Eucalyptus globulus* was more effective as a larvicide than was a leaf extract against *Cx. pipiens* (Elbanna, 2006). The lowest LC$_{50}$ values discovered for a seed extract, thus indicating the most effective toxicity, were 2.6 mg/l recorded for *Annona squamosa* (custard apple) seeds against *An. stephensi* (Senthilkumar et al., 2009), and 0.06 mg/l recorded against *Ae. aegypti* (De Omena et al., 2007). The abnormal movements observed in larvae exposed to *Annona graveolens* indicate that the active phytochemicals most likely affected the nervous system (Choochote et al., 2007).

Earlier research on *An. squamosa* seeds found that they were effective against head lice. While head lice are not vectors of disease, they are a disease in themselves — pediculosis. The earliest study on the pediculicidal efficacy of *An. squamosa* seed extracts was carried out in 1980 in Thailand (Intaranongpai et al., 2006). This clinical head-lice study used a seed extract diluted in coconut oil (1:2), which was found to cause 98% mortality of head lice within 2 hours of application to the heads of participants. A leaf extract showed less potency. Two other studies on petroleum ether extract of custard apple seeds that used oil diluents reported 90–93% mortality in *in vitro* tests after 1–3 hours (Intaranongpai et al., 2006). In a clinical study on schoolgirls, Gritsanapan *et al.* (1996) found that 20 g of a 20% cream caused $95 \pm 9\%$ mortality of head lice 3 hours after application.

A recent study focused on the separation and identification of the active compounds from a hexane extract of the seeds (Intaranongpai *et al.*, 2006). Two primary compounds were identified: oleic acid, and a triglyceride with one oleate ester. In *in vitro* tests against head lice where all test substances were diluted with coconut oil (1 : 1), crude hexane extract killed all the lice after 31 minutes, oleic acid caused 100% mortality after 50 minutes, and the triglyceride with one oleate ester was 100% effective after only 11 minutes. Since head-lice preparations are usually applied to the head for up to 20 minutes, this finding has commercial potential.

Although there have only been a few field studies on the evaluation of seed extracts for mosquito control, they have shown excellent results. Essential oils derived from caraway and celery seeds were tested as adulticides in the laboratory and then in natural field conditions against laboratory and wild strains of the dengue vector, *Ae. aegypti*, in Chiang Mai Province, Thailand (Chaiyasit *et al.*, 2006). It was suggested that these bioinsecticides had potential for field control and eradication of mosquito vectors.

An African village study evaluated the efficacy of neem seed extract for sustainable malaria vector control (Gianotti *et al.*, 2008). The study was conducted in Banizoumbou village, western Niger. Neem seeds, collected around the village, were dried, ground into a coarse powder, then sprinkled onto known *Anopheles* breeding habitats twice weekly during the 2007 rainy season. Weekly adult mosquito captures were compared to those from 2005 and 2006 during the same seasonal period. Adult mosquitoes were captured in a nearby village, Zindarou, as a control, and were compared to those from Banizoumbou. Results revealed that twice-weekly applications of the powder to breeding habitats of *Anopheles* larvae in 2007 resulted in 49% fewer adult female *An. gambiae* s.l. mosquitoes in Banizoumbou, compared with previous captures under similar environmental conditions in 2005 and 2006. Results of this study suggest that larval control using neem seed powder offers a sustainable additional tool for vector control that can be employed by local people at minimal cost.

Some seed extracts have demonstrated strong repellent activity against mosquito vectors. For instance, a 10% seed acetone extract of *Tribulus terrestris* was 100% effective in repelling adult *An. culicifacies* species A and *An. stephensi* mosquitoes for up to 6 hours, and *Cx. quinquefasciatus* mosquitoes for up to 4 hours (Singh *et al.*, 2008). A survey of the knowledge and usage of traditional insect mosquito-repellent plants in Addis Zemen Town, South Gonder, North Western Ethiopia, revealed that less than 1% (0.17%) of local residents used seeds of plants as repellents against mosquitoes and other insects compared to the other parts of plants, including leaves, flowers, roots, stems, and bark (Karunamoorthi *et al.*, 2009). These parts are underutilized either due to processing issues, or because they are less effective.

In another repellent study, a gel formulation of anise seed hexane extract was found to provide remarkable repellency with a median protection time of 4.5 hours (4.5− 5 h) (Tuetun *et al.*, 2008). This was greater than that of ethanolic DEET solution (25% DEET, 3.5 h), and comparable to that of the best commercial repellent, Insect Block 28 (28.5% DEET, 4.5 h). This laboratory investigation against female *Ae. aegypti* found that the gel formulation caused no irritation in volunteers. Such promising results, including high repellent activity and no skin irritation, are significant, because they provide proof that plant-based alternatives for synthetic mosquito repellents are a reality.

The formulation of commercial products may take many forms. In one example in Thailand, the dried root powder of a local plant, *Rhinacanthus nasutus*, was packaged as 10% tablets. These were tested against *Ae. aegypti* and *Cx. Quinquefasciatus*, for which the LC_{50}s were 14.2 mg/l and 17.3 mg/l, respectively (Rongsriyam *et al.*, 2006). Furthermore, acute-toxicity bioassays with fish (*Poecilia reticulata*) showed that these prepared tablets could be used to

TABLE 3.2 Mosquitocidal Activity of Nuts

Scientific Name	Activity Type	Mosquito Species	References[*]
Anacardium occidentale (shell)	Larvicide	Ae. aegypti	Laurens et al., 1997
Anacardium occidentale	Ovicide, larvicide pupicide	Ae. aegypti	Farias et al., 2009
Atriplex canescens (saltbrush nut)	Larvicide, adulticide, ovicide, oviposition detterent	Cx. quinquefasciatus	Ouda et al., 1998
Myristica fragrans	Larvicide, adulticide, IGR	An. stephensi	Senthilkumar et al., 2009

[*]See Shaalan and Canyon (2010) for these citation references.

TABLE 3.3 Safety of Seed and Nut Extracts

Scientific Name	Effect on Non-Target Organisms	References[*]
Anacardium occidentale	Safe for mice at high dose (0.3 g/kg).	Farias et al., 2009
Apium graveolens	No adverse effects on human skin observed.	Choochote et al., 2004
Azadiracta indica	More toxic against Aphanius dispar (killifish) than An. stephensi larvae	Khan et al., 2000
Citrus reticulate	Safe for Oreochromis niloticus Nile Tilapia fishes at 2.3 ppm	Sumroiphon et al., 2006
Carica papaya	Safe for the bug Diplonychus annulatum and the midge Chironomus circumdatus. No change in behavior and survival noticed at LC_{50}	Rawani et al., 2009

[*]See Shaalan and Canyon (2010) for these citation references.

control mosquito vectors, and should be considered for inclusion in mosquito control programs. The remaining obstacle is how to make production acceptable, efficient, and affordable in order for local affected populations to prevent mosquito bites and interrupt mosquito-borne disease transmission.

MOSQUITOCIDAL ACTIVITY OF NUTS

By comparison, only three plants — cashew, saltbrush, and nutmeg — have nuts that contain mosquitocidal phytochemicals (Table 3.2). The lowest LC_{50} published was 2.22 ppm, recorded for a *Myristica fragrans* nut extract against *An. stephensi* larvae (Senthilkumar et al., 2009). This is an excellent result, and merits further research.

SAFETY

There have been only a few safety studies, but these provide an indication of the safety of essential oils extracted from seeds, for aquatic organisms, predaceous insects, mammals, and humans, when applied topically to the skin for protection against mosquito bites (Table 3.3). Despite the eco-friendly advantages of seed extracts, research is required into the non-target effects of inert surfactants used in extract formulation for those extracts that are destined to be used as insecticides (Kumar et al., 2000).

CONCLUSIONS

In comparison to other plant parts, relatively few seeds and nuts have been evaluated and used in the control of arthropod disease vectors. A moderate number of seeds display some mosquitocidal activity, while very few nuts have been tested or show any promise (Table 3.1). A wide variety of larvicidal, adulticidal, repellent, and oviposition deterrent effects were observed against serious mosquito vectors. However, these effects are not easily transferred into practical application because the mosquitocidal efficacy of extracts from seeds or nuts is influenced by the mosquito species, plant species, and extractive solvent (Shaalan *et al.*, 2005). The thermostability and water solubility of some extracts, such as bioactive compounds from *M. volkensii*, give them another advantage over synthetic insecticides (Al-Sharook *et al.*, 2009). Better formulation technology is needed for topical repellents. The provision of more effective fixation for the essential oil content and incorporation of strategies for controlled release of essential oil vapors, whilst also providing solutions for the problem of potential dermal irritancy, would all be beneficial.

SUMMARY POINTS

- Seeds and nuts of different plant species have shown different mosquitocidal activity against both nuisance and mosquito vector-borne diseases.
- The number of types of seeds that produce such activity is larger than that of nuts (40 plant species and 3 plant species, respectively).
- Such mosquitocidal activities are not only against mosquitoes, adult stage, but also against immature stages (eggs, larvae and pupae).
- Experiments and observations have indicated that seed and nut extracts are safe for humans, fish, predaceous insects, and mammals.
- Capacity, eco-friendliness, and other physical factors, such as thermostability and water solubility, will give seeds and nuts advantages over synthetic insecticides for both commercialization and mosquito control.

31

References

Al-Sharook, Z., Balan, K., Jiang, Y., & Rembold, H. (2009). Insect growth inhibitors from two tropical Meliaceae: effect of crude seed extracts on mosquito larvae. *Journal of Applied Entomology, 111*, 425–430.

Chaiyasit, D., Choochote, W., Rattanachanpichai, E., Chaithong, U., Chaiwong, P., Jitpakdi, A., et al. (2006). Essential oils as potential adulticides against two populations of *Aedes aegypti*, the laboratory and natural field strains, in Chiang Mai province, northern Thailand. *Parasitology Research, 99*, 715–721.

Choochote, W., Chaithong, U., Kamsuk, K., Jitpakdi, A., Tippawangkosol, P., Tuetun, B., et al. (2007). Repellent activity of selected essential oils against *Aedes aegypti*. *Fitoterapia, 78*, 359–364.

De Omena, C., Navarro, F., de Paula, E., Luna, S., Ferreira de Lima, R., & Sant'Ana, G. (2007). Larvicidal activities against *Aedes aegypti* of some Brazilian medicinal plants. *Bioresource Technology, 98*, 2549–2556.

Elbanna, M. (2006). Larvaecidal effects of eucalyptus extract on the larvae of *Culex pipiens* mosquito. *International Journal of Agriculture and Biology, 8*, 896–897.

Gianotti, L., Bomblies, A., Dafalla, M., Issa-Arzika, I., Duchemin, J., & Eltahir, B. (2008). Efficacy of local neem extracts for sustainable malaria vector control in an African village. *Malaria J, 7*, 138.

Gritsanapan, W., Somanabandhu, A., Titirungruang, C., & Lertchaiporn, M. (1996). A study on the antiparasitic activities and chemical constituents of extracts from the leaves and seeds of custard apple (*Annona squamosa* Linn.). *Proceedings of Third NRCTJSPS Joint Seminar*, 209–215.

Intaranongpai, J., Chavasiri, W., & Gritsanapan, W. (2006). Anti-head lice effect of *Annona squamosa* seeds. *Southeast Asian Journal of Tropical Medicine and Public Health, 37*, 532–535.

Karunamoorthi, K., Mulelam, A., & Wassie, F. (2009). Assessment of knowledge and usage custom of traditional insect/mosquito repellent plants in Addis Zemen Town, South Gonder, Northwestern Ethiopia. *Journal of Ethnopharmacology, 121*, 49–53.

Kumar, A., Dunkel, V., Broughton, J., & Sriharan, S. (2000). Effect of root extracts of Mexican Marigold, *Tagetes minuta* (Asterales: Asteraceae), on six nontarget aquatic macroinvertebrates. *Environmental Entomology., 29*, 140–149.

Prajapati, V., Tripathi, A., Aggarwal, K., & Khanuja, S. (2005). Insecticidal, repellent and oviposition-deterrent activity of selected essential oils against *Anopheles stephensi*, *Aedes aegypti* and *Culex quinquefasciatus*. *Biores. Technol., 96*, 1749–1757.

Rongsriyam, Y., Trongtokit, Y., Komalamisra, N., Sinchaipanich, N., Apiwathnasorn, C., & Mitrejet, A. (2006). Formulation of tablets from the crude extract of *Rhinacanthus nasutus* (Thai local plant) against *Aedes aegypti* and *Culex quinquefasciatus* larvae: a preliminary study. *Southeast Asian Journal of Tropical Medicine and Public Health, 37*, 265–271.

Sakthivadivel, M., & Thilagavathy, D. (2003). Larvicidal and chemosterilant activity of the acetone fraction of petroleum ether extract from *Argemone mexicana* L. seed. *Bioresource Technology, 89*, 213–216.

Senthilkumar, N., Varma, P., & Gurusubramanian, G. (2009). Larvicidal and adulticidal activities of some medicinal plants against the malarial vector. *Anopheles stephensi (Liston). Parasitology Research, 104*, 237–244.

Shaalan, E. A., & Canyon, D. V. (2010). *A bibliography of recent references from 1987 to 2009 that relate to mosquitocidal activity caused by phytochemicals of seed and nut origin.* Perth: West Wind Publications.

Shaalan, E. A., Canyon, D. V., Younes, M., Abdel-Wahab, H., & Mansour, A. (2005). A review of botanical phytochemicals with mosquitocidal potential. *Environment International, 31*, 1149–1166.

Singh, P., Raghavendra, K., Singh, K., Mohanty, S., & Dash, P. (2008). Evaluation of *Tribulus terrestris* Linn (Zygophyllaceae) acetone extract for larvicidal and repellence activity against mosquito vectors. *Journal of Communication Disorders, 40*, 255–261.

Tuetun, B., Choochote, W., Pongpaibul, Y., Junkum, A., Kanjanapothi, D., Chaithong, U., et al. (2008). Celery-based topical repellents as a potential natural alternative for personal protection against mosquitoes. *Parasitology Research, 104*, 107–115.

Composition

CHAPTER 4

Fatty Acid Content of Commonly Available Nuts and Seeds

Duo Li, Xiaojie Hu
Department of Food Science and Nutrition, and Institute of Agrobiology and Environmental Science, Zhejiang University, Hangzhou, China
Center of Nutrition & Food Safety, Asia Pacific Clinical Nutrition Society

CHAPTER OUTLINE

LIST OF ABBREVIATIONS

α-ESA, α-eleostearic acid 18: 3(9*c*,11*t*,13*c*)
β-ESA, β-eleostearic acid 18: 3(9*t*,11*t*,13*t*)
CA, catalpic acid 18: 3(9*t*,11*t*,13*c*)
CLNA, conjugated linolenic acid
DAG, diacylglycerol
FFA, free fatty acid
MUFA, monounsaturated fatty acid
PA, punicic acid 18: 3(9*c*,11*t*,13*t*)
PUFA, polyunsaturated fatty acid
S, sterol
SE, sterol esters
SFA, saturated fatty acid
TAG, triacylglycerol

INTRODUCTION

Nuts and seeds are good dietary sources of unsaturated fatty acids. Linoleic acid (18:2n-6) and alpha-linolenic acid (18:3n-3) are two essential fatty acids in humans and are precursors of C20 and C22 polyunsaturated fatty acids (PUFAs). Other uncommon fatty acids, such as

Nuts & Seeds in Health and Disease Prevention. DOI: 10.1016/B978-0-12-375688-6.10004-0

stearidonic acid (18:4n-3), conjugated linolenic acids (CLNAs), and ximenynic acid (a triple bond fatty acid), are also presented in certain seeds and nuts. The substantial epidemiological evidence shows that fatty acids from seeds and nuts are associated with different health effects (Li *et al.*, 2002). The aim of this chapter is to summarize the composition and content of lipids and fatty acids in commonly available nuts and seeds worldwide.

FATTY ACIDS IN VARIOUS NUTS AND SEEDS

The predominant lipid in all nuts and seeds investigated was triacylglycerol (TAG), which was found at levels above 90%, reaching 98.4% in macadamia nuts. The total lipid concentration in the samples ranged from 2.2 g/100 g in *ginkgo biloba* to 75.4 g/100 g in walnut. Apart from peanut and *T. kirilowii* Maxim. seed, most of the analyzed samples contained phytosterols; pistachio contained the highest amount, at 5.0%. Some seeds contained diacylglycerol and free fatty acids. For instance, diacyglycerol comprised 4.8% of the total lipids in *Cannabis sativa*, and free fatty acids comprised 1.7% of the total lipids in *Ginkgo biloba*. Phytosterol ester ranged from 0.2% of total lipids in peanut seed to 7.1% in grand torreya seed; however, it was not detected in walnut, pistachio, almond, and black melon seed (Table 4.1).

The primary saturated fatty acids (SFAs) identified in the 20 nuts and seeds were palmitic acid (16:0) and stearic acid (18:0), with a particularly high content of the former in the fig-leaf gourd seed, and of the latter in the Brazil nut (15.4% and 11.8% of the total, respectively). The proportion of total SFAs ranged from 6.34% to 26.21%, with the Brazil nut yielding the greatest percentage and the pecan nut the least. The most predominant SFA was 16:0, ranging from 4.28% in pecan nut to 15.4% in fig-leaf gourd (Tables 4.2—4.4). The levels of total unsaturated fatty acids ranged from 73% in the Brazil nut to 93% in the pecan. The total proportion of PUFAs was the highest in black melon seed, at 75.8%, and the lowest in macadamia, at 2.8%. Furthermore, PUFAs were predominant over monounsaturated fatty acids (MUFAs) in all nut and seed samples except the pistachio, filbert, almond, macadamia, pumpkin, and cashew nuts, which were found to contain MUFAs ranging from 42.7% in pumpkin to 82.6% in macadamia (Figure 4.1).

There were three main fatty acids in all nuts and seeds: 18:2n-6, 18:1n-9, and 16:0. The former, 18:1n-9, was present in high levels in the macadamia and filbert, at 57.1% and 74.7% of total fatty acids respectively, and black melon seed had the lowest level, at 8.63%. The proportion of

TABLE 4.1 Lipid Content (g/100 g) and Lipid Compositions (% of Total Lipid) of Commonly Consumed Nuts and Seeds

	Lipid Content (g/100g)	Lipid Composition (% of Total Lipid)				
		SE	TAG	DAG	FFA	S
Grand torreya seed	51.2 ± 1.0	7.1 ± 4.5	92.1 ± 4.5	nd	nd	0.8 ± 0.3
Walnut	75.4 ± 2.5	nd	97.0 ± 0.2	nd	nd	3.0 ± 0.4
Pistachio	53.9 ± 0.5	nd	95.0 ± 14.2	nd	nd	5.0 ± 1.0
Filbert	60.0 ± 2.5	3.2 ± 0.5	95.1 ± 0.8	nd	nd	1.7 ± 0.2
Pine nut	66.6 ± 1.2	1.4 ± 0.2	95.9 ± 2.1	nd	nd	2.7 ± 0.3
Almond	53.5 ± 0.8	nd	95.9 ± 0.4	2.0 ± 0.4	nd	2.1 ± 0.5
Macadamia	70.1 ± 0.6	0.7 ± 0.1	98.4 ± 0.3	nd	nd	0.9 ± 0.2
Cannabis sativa	49.5 ± 3.0	1.4 ± 0.8	91.1 ± 4.2	4.8 ± 1.3	nd	2.7 ± 1.2
Peanut	35.4	0.2	97.3	0.6	0.2	nd
Ginkgo biloba	2.2	3.7	85.5	3.6	1.7	1.6
T. kirilowii Maxim. seed	49.4 ± 1.0	1.1 ± 0.8	98.1 ± 0.8	0.2 ± 0.2	0.4 ± 0.2	nd
Black melon seed	34.6	nd	95.5	nd	1.4	3.1

Mean $\pm$ SD, $n = 3$, nd = not detected.
SE, sterol esters; TAG, triacylglycerols; DAG, diacylglycerols; FFA, free fatty acids; S, sterol.
Sources: Li *et al.* (2006); Yoshida *et al.* (2005).

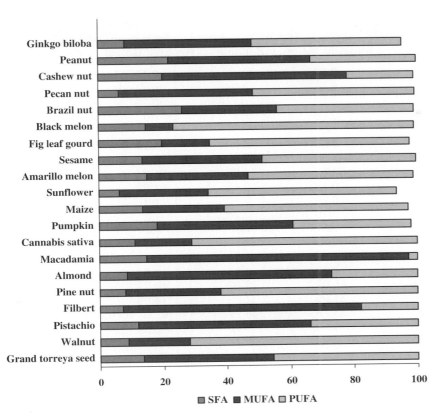

FIGURE 4.1

Proportion of saturated fatty acids (SFA), monounsaturated fatty acids (MUFA), and polyunsaturated fatty acids (PUFA) in oil from 20 edible nuts and seeds. *Sources: Hierro et al., 1996; Li et al., 2006; Ryan et al., 2006; Were et al., 2006; Sabudak, 2007; Bernardo-Gil et al., 2009.*

16:1 in analyzed nuts was very low, at less than 1% of the total fatty acids, except in macadamia (25.5%), ginkgo nut (3.43%), and almond (1.1%). The fatty acid 18:2n-6 was the most abundant PUFA, with lesser amounts of 18:3n-3. The content of 18:2n-6 varies remarkably, ranging from 2.5% in macadamia to 75.7% in black melon seed, but most nuts and seeds are rich in this fatty acid — especially walnut "pine nut", *Cannabis sativa*, maize, sunflower seed, "Amarillo melon", fig-leaf gourd seeds, and pecan nut, where the proportion is approximately 50—60%. Similarly, the content of 18:3n-3 also varies greatly, ranging from 0.01% in sunflower seed to 15.2% in *Cannabis sativa*. *Ginkgo biloba* nut contained 20:2n-6 and 20:3n-6 (Tables 4.2—4.4).

The total content of CLNA amounted to 36.9% in *T. kirilowii* Maxim seed and 83.4—87.9% in pomegranate seed. The content of punicic acid, the predominant isomer of CLNA,

37

TABLE 4.2 Fatty Acid Composition (% of Total Fatty Acid) of Common Consumed Nuts

Fatty Acids	Grand Torreya	Walnut	Pistachio	Filbert	Pine Nut	Almond	Macadamia	Cannabis Sativa	*P*-Value
16:0	9.6 ± 0.3	6.7 ± 0.8	10.5 ± 3.5	4.8 ± 0.2	5.4±0.3	7.1 ± 0.3	10.3±0.0	7.7 ± 0.2	< 0.001
16:1n-7	0.2 ± 0.1	0.2 ± 0.1	0.8 ± 0.3	0.1±0.0	0.2 ± 0.2	1.1±0.4	25.5 ± 0.2	0.3±0.1	< 0.001
17:0	0.1 ± 0.0	nd	0.1 ± 0.0	nd	0.2 ± 0.2	nd	nd	0.1 ± 0.0	
18:0	3.9 ± 0.2	2.1 ± 0.2	1.4 ± 0.1	2.5 ± 0.1	2.6 ± 0.3	1.5 ± 0.1	4.3 ± 0.3	3.4 ± 0.0	< 0.001
18:1	40.7 ± 1.4	19.3 ± 6.8	53.4 ± 1.8	74.7 ± 1.0	29.6 ± 1.0	63.2 ± 1.3	57.1 ± 0.4	17.7 ± 0.5	< 0.001
18:2n-6	44.7 ± 1.4	60.3 ± 6.7	33.1 ± 1.8	16.8 ± 0.1	61.6 ± 1.8	26.9 ± 1.3	2.5 ± 0.2	55.7 ± 0.5	< 0.001
18:3n-3	0.8 ± 0.2	11.4 ± 1.4	0.7 ± 0.1	1.0 ± 0.8	0.5 ± 0.1	0.2 ± 0.2	0.3 ± 0.1	15.2 ± 0.3	< 0.001
SFA	13.7 ± 0.4	8.8 ± 0.7	12.0 ± 3.6	7.3 ± 0.2	8.2 ± 0.7	8.6 ± 0.4	14.7 ± 0.3	11.1 ± 0.2	< 0.001
MUFA	40.9 ± 1.4	19.4 ± 6.8	54.2 ± 2.0	74.8 ± 1.0	61.6 ± 1.7	29.7 ± 1.1	64.2 ± 1.5	82.6 ± 0.2	70.9 ± 0.7	< 0.001
PUFA	45.4 ± 1.5	71.7 ± 7.5	33.8 ± 1.8	17.8 ± 0.9	62.1 ± 1.7	27.1 ± 1.2	2.8 ± 0.3	70.9 ± 0.7	< 0.001

Mean ± SD, *n* = 3, nd = not detected.
Source: Reproduced from Li *et al.* (2006), with permission.

TABLE 4.3 Fatty Acid Composition (% of Total Fatty Acid) of Commonly Consumed Seeds

Fatty Acids	Pumpkin	Maize	Sunflower	Amarillo Melon	Sesame	Fig Leaf Gourd	Black Melon
14:0	0.12	0.09	0.05	0.07	nd	0.16	nd
16:0	12.26	11.03	4.66	8.51	8.24	15.4	9.31
16:1n-7	trace	0.06	0.02	0.08	nd	0.29	nd
17:0	trace	0.06	0.06	0.08	nd	nd	nd
17:1	0.05	nd	0.05	nd	nd	nd	nd
18:0	5.22	1.7	0.4	6.09	4.89	4.21	5.43
18:1	42.49	25.52	27.73	31.5	37.64	14.7	8.63
18:2n-6	36.99	56.9	59.22	51.6	47.82	61.0	75.70
18:3n-3	0.11	1.04	0.01	0.19	0.45	1.92	0.13
20:0	0.37	0.37	0.26	0.29	0.50	nd	0.16
20:1n-9	0.1	0.05	0.01	0.16	0.24	nd	nd
22:0	0.11	0.19	0.74	nd	nd	nd	nd
22:1	nd	nd	0.01	0.25	nd	nd	nd
24:0	0.07	0.22	0.29	nd	nd	nd	nd
24:1	0.1	nd	0.01	nd	nd	nd	nd
SFA	18.15	13.66	6.46	15.04	13.63	19.77	14.9
MUFA	42.74	25.63	27.83	31.99	37.88	14.99	8.63
PUFA	37.1	57.94	59.23	51.79	48.27	62.9	75.83
Others	2.01	2.77	6.48	1.18	0.22	2.32	0.64

nd = not detected.
Sources: Were et al. (2006); Sabudak (2007); Bernardo-Gil et al. (2009).

TABLE 4.4 Fatty Acid Composition (% of Total Fatty Acid) of Commonly Consumed Nuts

Fatty Acids	Brazil Nut	Pecan Nut	Cashew Nut	Peanut	Ginkgo Biloba
14:0	0.06	nd	0.07	nd	trace
16:0	13.50	4.28	9.93	11.49	6.62
16:1n-9	nd	nd	nd	nd	0.13
16:1n-7	0.33	0.09	0.36	nd	3.30
17:0	0.22	0.10	0.14	nd	nd
18:0	11.77	1.80	8.70	3.97	0.99
18:1n-9	29.09	40.63	57.24	43.72	13.76
18:1n-7	nd	nd	nd	nd	21.53
18:2n-6	42.80	50.31	20.80	33.30	38.99
18:3n-3	0.20	0.65	0.23	nd	nd
18:3n-6	nd	nd	nd	nd	1.61
20:0	0.54	Tr.	0.97	1.90	0.37
20:1n-9	0.21	1.21	0.25	0.89	0.44
20:1n-7	nd	nd	nd	nd	0.66
20:2n-6	nd	nd	nd	nd	0.90
20:3n-6	nd	nd	nd	nd	5.70
22:0	0.12	0.16	0.39	3.46	0.40
22:1	0.34	0.25	0.28	nd	nd
24:0	nd	nd	nd	1.26	nd
SFA	26.21	6.34	20.20	22.08	8.38
MUFA	29.97	42.18	58.13	44.61	39.82
PUFA	43.00	50.96	21.03	33.30	47.2
Others	0.82	0.52	0.64	0.01	4.6

nd = not detected.
Sources: Hierro et al. (1996); Ryan et al. (2006).

reached 32.6% in *T. kirilowii* Maxim seed, while it rose to 73.4−77.5% in pomegranate seed (Table 4.5).

All *Santalum* species contained significant amounts of ximenynic acid, especially *S. obtusifolium*, *S. insulare*, and *S. album*; the latter had the greatest quantity, at 82.8% (Table 4.6).

Regarding the five families of *Ribes* berries, Boraginaceae, Scrophulariaceae, Onagraceae, and Ranunculaceae, 18:3n-3 was detected in all the samples, and variable contents of γ-linolenic acid (18:3n-6) and 18:4n-3 were found in seeds of the *Ribes* berries and Boraginaceae species (Table 4.7).

CONCLUSIONS

All the nuts, seeds, currants, and *Santalum* kernels contained a low level of saturated fatty acids and a high level of unsaturated fatty acids. A high content of 18:3n-3 was found in *sativa*, walnut, *Ribes* berries, and Ranunculaceae; variable levels of 18:3n-3 were also detected in the Scrophulariaceae, Onagraceae, and Boraginaceae plant families; and 18:4n-3 was detected in *Ribes* berries and Boraginaceae species. Meanwhile, *T. kirilowii* Maxim and pomegranate seeds were rich in CLNA, while *Santalum* kernels contained ximenynic acid.

TABLE 4.5 Fatty Acid Composition (% of Total Fatty Acids) of Some Seeds from China

Fatty Acids	*T. Kirilowii* Maxim	Pomegranate		
		San bai yu	Qing pi ruan zi	Tian lv zi
16:0	5.8 ± 0.3	2.2 ± 0.2	2.3 ± 0.0	2.4 ± 0.0
17:0	nd	trace	trace	trace
18:0	2.4 ± 0.2	1.3 ± 0.1	1.8 ± 0.0	1.6 ± 0.3
18:1n-9	22.6 ± 1.1	2.5 ± 0.1a	5.2 ± 0.0b	4.4 ± 0.9b
18:2(iso)	nd	0.3 ± 0.0	0.2 ± 0.0	0.2 ± 0.0
18:2n-6	32.6 ± 0.6	5.4 ± 0.2	6.4 ± 0.0	5.6 ± 0.9
18:3n-3	nd	0.1 ± 0.0	trace	trace
20:0	nd	0.3 ± 0.0	0.2 ± 0.0	0.3 ± 0.0
21:1n-9	nd	0.5 ± 0.0	0.5 ± 0.2	0.6 ± 0.1
PA	32.6 ± 0.8	77.5 ± 1.3a	73.4 ± 1.4b	75.5 ± 1.6ab
α-ESA	3.0 ± 0.0	nd	nd	nd
CA	0.9 ± 0.1	8.4 ± 0.8	7.8 ± 0.7	8.3 ± 0.5
β-ESA	nd	1.4 ± 0.2	2.2 ± 0.5	1.1 ± 0.1

Mean ± SD, *n* = 3, nd = not detected.
PA, punicic acid 18:3(9*c*,11*t*,13*t*); CA, catalpic acid 18:3(9*t*,11*t*,13*c*); α-ESA, α-eleostearic acid 18:3(9*c*,11*t*,13*c*); β-ESA, β-eleostearic acid 18:3(9*t*,11*t*,13*t*).
Sources: Yuan, *et al.* (2009); Zhou (2010).

TABLE 4.6 Fatty Acid Composition (% of Total Fatty Acid) of Various *Santalum* Species Kernel Oils

Species	Album	Acuminatum	Murrayanum	Obtusifolium	Spicatum	Insulare
16:0	0.8	2-2.9	2.4	0.6	3.5	1.0
16:1n-7	0.5	0.3−2.7	0.3	0.4	0.7	0.6
18:0	1	1.1−2.3	2.1	1.2	1.9	1.0
18:1n-9	12.3	43.8−57.7	54.8	14.3	54.4	18.1
18:2n-6	nd	0.3−1.4	1.4	0.7	0.6	0.5
18:3n-3	0.8	0−2.5	2.3	3.2	nd	1.0
Ximenynic	82.8	32.2−46.2	35.5	71.5	33.4	74.5
Others	1.8	2.3	1.2	8.1	5.5	3.3

nd = not detected,
Sources: Liu *et al.* (1996); Butaud *et al.* (2008).

TABLE 4.7 Fatty Acid Composition (% of Total Fatty Acid) of Seed Oils from Five Plant Families

Species	14:0	16:0	16:1n-7	18:0	18:1n-9	18:1n-7	18:2n-6	18:3n-6	18:3n-3	18:4n-3	20:0	20:1n-9	20:2n-6	22:0	22:1n-9	24:0	24:1n-9
Ribes																	
nigrum	0.1(2.7)	5.2(0.9)	—	1.8(0.2)	10.3(0.2)	0.5(1.0)	48.2(0.3)	11.3(0.3)	17.5(0.1)	3.0(0.2)	0.2(7.6)	0.9(1.0)	0	0	0	0	0
spicatum	—	3.8(7.4)	—	1.2(0.8)	13.5(5.2)	0.5(4.2)	38.5(1.5)	16.1(3.5)	19.1(7.0)	6.0(4.5)	0.2(2.4)	0.1(6.2)	0	0	0	0	0
alpinum	0.1(44.2)	5.2(0.9)	0.1(1.8)	1.1(1.0)	18.6(0.1)	0.8(1.1)	36.1(0.6)	13.3(1.6)	18.2(1.0)	5.1(0.4)	0.1(1.2)	0.1(5.0)	0	0	0	0	0
Boraginaceae																	
Anchusa azurea	0.09	8.63	0.35	2.19	24.1	0.43	41.78	11.11	0.43	0.08	0.23	3.57	0.18	0.37	0	0	0
Anchusa undulata	0.16	8.76	0.54	2.15	24.4	0	25.37	8.35	17.99	3.55	0.24	4.21	0.15	0.32	0	0	0
Asperugo procumbens	0.12	8.07	0.16	1.85	15.48	0.62	15.2	5.35	36.46	11.75	0.19	2.03	0.12	0.24	0	0	0
Borago officinalis	0.10	9.57	0.18	6.18	20.92	0.46	33.21	19.2	1.00	0.49	0.43	3.95	0.15	0.27	0.05	0	0
Buglosoides arvensis	0.23	9.41	0.15	2.81	6.83	0.61	14.8	6.44	39.68	14.08	1.63	0.98	0	0.28	0	0	0
Cynoglossum cheirifolium	1.56	17.34	0	3.41	7.93	0	13.81	1.52	39.35	3.24	2.03	0.54	0.02	1.27	0.08	0.05	0
Cynoglossum creticum	1.55	16.25	0.35	2.92	8.57	0.35	18.27	0.66	44.75	1.16	1.88	0.42	1.27	1.93	0	0	0
Cynoglossum nebrodense	0.19	6.01	0.11	1.61	46.42	0	6.79	1.42	16.53	2.83	0.68	5.31	0.08	0.68	0	0	0
Cynoglossum officinale	0.21	7.01	0.15	1.4	42.6	0	9.02	1.68	16.14	2.48	0.68	5.14	0.34	0.73	0	0	0
Echium asperrimum	0.08	7.7	0.13	2.77	14.68	0	16.33	9.62	35.3	21.06	0.09	0.98	0.05	0.08	0.34	0	0.07
Echium boissieri	0	5.48	0.09	2.28	14.7	0	8.64	5.52	47.14	14.31	0.1	0.74	0.04	0.06	0.06	0	0
Echium creticum	0.05	5.58	0.06	2.98	8.18	0.35	14.31	9.70	42.68	14.73	0.11	0.58	0.06	0.06	0.06	0	0
Echium flavum	0	6.29	0.07	2.12	21.05	0.52	24.16	8.38	32.23	3.14	0.09	1.01	0.06	0.07	0.07	0	0
Echium humile	6.69	7.28	0.43	3.95	17.18	0.45	24.43	7.95	31.21	5.88	0.44	0.52	0.22	0	0	0	0
Echium sabulicola	0.06	5.51	0.08	2.42	8.03	0.36	16.31	10.94	40.39	14.72	0.08	0.65	0.10	0.06	0.06	0	0
Echium vulgare	0.14	7.38	0.06	2.52	11.14	0.44	21.18	11.74	34.14	9.68	0.1	0.74	0.07	0.08	0	0	0.04
Myosotis alpina	0.04	8.12	0.02	2.3	24.92	0.45	27.02	4.38	18.03	8.38	0.45	3.65	0.14	0.33	1.03	0	0.05
Myosotis nemorosa	0.16	13.15	0.35	3.89	20.79	0	30.76	20.25	4.69	1.56	0.27	2.57	0.10	0.16	1.23	0	0.02
Myosotis secunda	0.08	8.22	0.16	4.08	25.22	0	23.07	12.17	15.62	4.29	0.36	3.44	0.13	0.29	2.89	0	0
Nonea vesicaria	0.21	9.49	0.17	2.57	26.23	0.4	26.52	9.39	13.88	4.90	0.29	3.40	0.13	0.23	0.05	0	0
Ranunculaceae																	
Delphinium gracile	0.40	22.43	0.14	2.25	11.18	0.91	38.44	0	19.87	0	0.67	0.01	0.01	0	0	0	0
Ranunculus repens	0.37	10.46	0.20	2.07	7.36	0.73	36.55	0	39.71	0	0.27	0.13	0.31	0	0	0	0
Ranunculus peltatus	0.37	13.35	1.69	1.50	8.49	0.94	28.44	0	37.89	0	0.41	0.01	0.38	0	0	0	0

Onagraceae

Taxon																	
Epilobium hirsutum	0.04	10.93	0.12	3.84	9.25	0.49	71.55	0.02	1.84	0	0.66	0.10	0.08	0.17	0	0	0
Epilobium lanceolatum	0.05	10.56	0.14	2.60	6.47	0.21	76.75	0.08	1.71	0	0.56	0.13	0.11	0.18	0	0	0

Scrophulariaceae

Taxon																	
Antirrhinum barrelieri	0.04	5.13	0.17	2.58	18.04	0.81	71.77	0	0.34	0	0.13	0.11	0.04	0.02	0	0	0
Antirrhinum charidemi	0.04	6.40	0.21	1.54	15.67	0.87	73.20	0	0.53	0	0.12	0.12	0	0.08	0	0	0
Antirrhinum hispanicum	0.04	5.48	0.15	1.8	16.15	1.79	72.73	0.07	0.42	0	0.15	0.13	0.06	0.14	0	0	0
Antirrhinum majus	2.15	16.14	0.78	7.61	3.90	1.04	31.41	0	25.89	0	3.09	2.63	0.3	1.48	0	0	0
Antirrhinum molle	0.12	6.23	0.29	1.89	16.76	1.94	70.43	0.06	0.59	0	0.21	0.13	0.11	0.18	0.02	0	0
Bellardia trixago	0.05	8.71	0.17	1.76	18.37	1.06	43.56	0	24.8	0	0.36	0.16	0.16	0	0	0	0
Chaenorhinum macropodum	0.15	7.55	0	1.83	10.84	0.98	75.73	0	2.22	0	0.01	0.01	0.01	0	0	0	0
Chaenorhinum origanifolium	0.10	6.00	0.22	2.07	12.75	1.12	73.34	0	3.42	0	0.17	0.09	0.11	0.10	0.02	0	0
Chanaenorhinum villosum	0.23	5.71	0	3.87	2.03	0	73.14	0.53	2.02	0	0.21	0.12	0	0.23	0.19	0.18	0.2
Cymbalaria muralis	0.14	5.12	0	2.56	22.66	0	63.08	1.33	3.05	0	0.14	0.38	0.42	0.09	0.15	0.02	0
Digitalis obscura	0.15	8.04	0.15	2.69	18.97	2.11	60.57	0.06	4.44	0	0.42	0.07	0.04	0.21	0	0	0
Lafuentea rotundifolia	1.16	14.85	0.46	2.94	7.32	0.39	24.04	0	19.89	0	1.51	0.01	0.64	0.07	0.03	0	0
Linaria aeruginea	0.12	26.4	0	9.35	6.66	0	26.38	0	11.23	0	0.69	0.32	0.12	0.21	0.11	0.29	0
Linaria amoi	0.08	6.29	0.12	1.96	19.49	0.12	67.43	0.46	2.04	0	0.18	0.14	0.01	0	0	0	0
Misopates orontium	0.04	6.90	0.11	2.34	16.38	0.54	71.45	0	0.38	0	0.37	0.12	0.19	0.02	0	0	0
Odontites longiflora	0.09	8.12	0.84	2.07	32.38	2.97	11.86	0	40.49	0	0.16	0.09	0.05	0.04	0	0	0.12
Parentucela viscosa	0.21	8.93	0.33	2.63	21.65	0.12	37.28	0.21	28.56	0	0.28	0.16	0.12	0.15	0.18	0	0.12
Scrophularia auriculata	0.06	12.01	0.15	2.48	14.89	2.63	63.45	2.66	0.36	0	0.39	0.13	0.03	0.13	0	0.12	0.23
Scrophularia nodosa	0.12	26.4	0	3.6	13.92	0	60.33	2.26	4.39	0	0.12	0.89	0.11	0	0	0	0
Scrophularia sciophila	0.06	11.05	0.10	2.12	10.82	0.71	63.57	10.17	0.38	0	0.34	0.10	0.05	0.17	0	0	0
Verbascum phlomoides	0.19	6.39	0.17	3.11	17.03	0.59	69.58	0	1.26	0	0.53	0.09	0.05	0.01	0	0	0
Verbascum thapsus	0.04	6.4	0.17	2.61	16.75	0.59	70.84	0	1.03	0	0.47	0.15	0.05	0.01	0.08	0	0
Veronica anagalloides	0.35	10.81	0.25	2.21	19.12	0.42	52.12	0	12.24	0	0.36	0.16	0.43	0	0	0	0
Veronica persica	1.24	15.82	0.68	2.43	28.87	0.54	28.39	0	19.24	0	0.59	0.41	0.4	0	0	0	0[a]

Sources: Horrobin (1992); Johansson (1997); Guerrero et al. (2001).

SUMMARY POINTS

1. Nuts, seeds, currants and *Santalum* kernels contained a high proportion of unsaturated fatty acid, of which 18:2n-6 was the most abundant PUFA, with lesser amounts of 18:3n-3.
2. The Boraginaceae, Scrophulariaceae, Onagraceae, Ranunculaceae, and *Ribes* berries families contained relatively high levels of 18:3n-6.
3. The *Ribes* berries and Boraginaceae families contained 18:4n-3.
4. *T. kirilowii* Maxim and pomegranate seed contained relatively high CLNA.
5. *Santalum* kernels contained ximenynic acid.

References

Bernardo-Gil, M. G., Casquilho, M., Esquível, M. M., & Ribeiro, M. A. (2009). Supercritical fluid extraction of fig leaf gourd seeds oil: fatty acids composition and extraction kinetics. *Journal of Supercritical Fluids, 49,* 32–36.

Butaud, J. F., Raharivelomanana, P., Bianchini, J. P., & Gaydou, E. M. (2008). *Santalum insulare* acetylenic fatty acid seed oils: comparison within the *Santalum* genus. *Journal of American Oil Chemists' Society, 85,* 353–356.

Guerrero, J. L. G., Maroto, F. F. G., & Giménez, A. G. (2001). Fatty acid profiles from forty-nine plant species that are potential new sources of γ-linolenic acid. *Journal of American Oil Chemists' Society, 78,* 677–684.

Hierro, M. T. G., Robertson, G., Christie, W. W., & Joh, Y. G. (1996). The fatty acid composition of the seeds of *Ginkgo biloba. Journal of American Oil Chemists' Society, 73,* 575–579.

Horrobin, D. F. (1992). Nutritional and medical importance of gamma-linoleic acid. *Progress in Lipid Research, 31,* 163–194.

Johansson, A. (1997). Characterization of seed oils of wild, edible Finnish berries. *Zeitschrift für Lebensmittel-luntersuchung und -Forschung A, 204,* 300–307.

Li, D., Bode, O., Drummond, H., & Sinclair, A. J. (2002). Omega-3 (n-3) fatty acids. In Frank D. Gunstone (Ed.), *Lipids for functional foods and nutraceuticals* (pp. 225–262). Bridgewater, UK: Oily Press.

Li, D., Yao, T., & Siriamornpun, S. (2006). Alpha-linolenic acid content of commonly available nuts in Hangzhou. *International Journal for Vitamin and Nutrition Research, 76,* 18–21.

Liu, Y. D., Longmore, R. B., & Fox, J. E. D. (1996). Separation and identification of ximenynic acid isomers in the seed oil of *Santalum spicatum* RBr as their 4,4-dimethyloxazoline derivatives. *Journal of American Oil Chemists' Society, 73,* 1729–1731.

Ryan, E., Galvin, K., O'Connor, T. P., Maguire, A. R., & O'Brien, N. M. (2006). Fatty acid profile, tocopherol, squalene and phytosterol content of Brazil, pecan, pine, pistachio and cashew nuts. *International Journal of Food Sciences and Nutrition, 57,* 219–228.

Sabudak, T. (2007). Fatty acid composition of seed and leaf oils of pumpkin, walnut, almond, maize, sunflower and melon. *Chemistry of Natural Compounds, 43,* 465–467.

Were, B. A., Onkware, A. O., Gudu, S., Welander, M., & Carlsson, A. S. (2006). Seed oil content and fatty acid composition in East African sesame (*Sesamum indicum* L.) accessions evaluated over 3 years. *Field Crop Research, 97,* 254–260.

Yoshida, H., Hirakawa, Y., Tomiyama, Y., Nagamizu, T., & Mizushina, Y. (2005). Fatty acid distributions of triacylglycerols and phospholipids in peanut seeds (*Arachis hypogaea* L.) following microwave treatment. *Journal of Food Composition and Analysis, 18,* 3–14.

Yuan, G. F., Sinclair, A. J., Xu, C. J., & Li, D. (2009). Incorporation and metabolism of punicic acid in healthy young humans. *Molecular Nutrition and Food Research, 53,* 1336–1342.

Zhou, C. Q. (2010). Analysis of nutritional compositions and preparation of O/W microemulsion of edible *Trichosanthes kirilowii* Maxim. seed kernels. Master's Dissertation.

CHAPTER 5

Triacylglycerols in Nut and Seed Oils
Detailed Characterization Using High-performance Liquid Chromatography/ Mass Spectrometry

Miroslav Lísa, Michal Holčapek
University of Pardubice, Faculty of Chemical Technology, Department of Analytical Chemistry, Pardubice, Czech Republic

43

LIST OF ABBREVIATIONS

APCI, atmospheric pressure chemical ionization
CN, carbon number
DB, double bond
ECN, equivalent carbon number
FA, fatty acid
HPLC, high-performance liquid chromatography
NARP, non-aqueous reversed-phase
RF, response factor
TG, triacylglycerol

Fatty Acid Abbreviations
Cy, caprylic (CN:DB, C8:0)
C, capric (C10:0)

Nuts & Seeds in Health and Disease Prevention. DOI: 10.1016/B978-0-12-375688-6.10005-2

La, lauric (C12:0)

M, myristic (C14:0)

C15: 0, pentadecanoic

P, palmitic (C16:0)

Po, palmitoleic (Δ9-C16:1)

Ma, margaric (C17:0)

Mo, margaroleic (Δ9-C17:1)

S, stearic (C18:0)

O, oleic (Δ9-C18:1)

L, linoleic (Δ9,12-C18:2)

Ln, *alpha*-linolenic (Δ9,12,15-C18:3)

γLn, *gamma*-linolenic (Δ6,9,12-C18:3)

St, stearidonic (Δ6,9,12,15-C18:4)

C19: 0, nonadecanoic (C19:0)

A, arachidic (C20:0)

G, gadoleic (Δ9-C20:1)

C20: 2, eicosadienoic (Δ11,14-C20:2)

C21: 0, heneicosanoic (C21:0)

B, behenic (C22:0)

C22: 1, erucic (Δ13-C22:1)

C23: 0, tricosanoic (C23:0)

24: 1, nervonic (Δ15-C24:1)

Lg, lignoceric (C24:0)

C25: 0, pentacosanoic (C25:0)

C26: 0, hexacosanoic (C26:0)

INTRODUCTION

Triacylglycerols (TGs) form an important part of the human diet, serve as a source of energy stored in fat tissues, provide a thermal and mechanical protective layer surrounding important organs, and are a source of essential FAs (linoleic and linolenic acids), fat-soluble vitamins and other non-polar compounds. The main sources of TGs in the human diet are oil plants, and, especially, the oils prepared from them (Gunstone, 2006; Leray, 2009). The final use of plant oils depends on their composition, and comprehensive triacylglycerol profiling provides valuable information in this respect. Individual TGs can differ in the number of carbon atoms (CNs), the number of double bonds (DBs), and the *cis/trans* configuration of the double bonds. Also, different stereochemical positions *sn*-1, -2, or -3 of FAs on the glycerol backbone (regioisomers), or *R/S* optical configuration of TGs esterified in *sn*-1 and *sn*-3 positions by two different FAs (optical isomers), lead to enormous complexity. Two techniques of high-performance liquid chromatography (HPLC) are widely used in the analysis of TG mixtures: silver ion normal-phase HPLC (Ag-HPLC), and non-aqueous reversed-phase HPLC (NARP-HPLC). Ag-HPLC is widely used for the separation of lipids according to the number, position, and *cis/trans* configuration of double bonds. TG regioisomers can be also separated under carefully optimized chromatographic conditions (Holčapek *et al.*, 2009; Lísa *et al.*, 2009a). In NARP-HPLC, TGs are separated according to acyl chain lengths and the number of double bonds. The retention of TGs is governed by the equivalent carbon number (ECN), which is defined as ECN = CN − 2DB. The separation of most TGs within one ECN group is feasible under optimized chromatographic conditions (Lísa & Holčapek, 2008; Lísa *et al.*, 2009b). The separation of *cis/trans* isomers or double-bond positional isomers is also possible in NARP-HPLC (Lísa *et al.*, 2007; Holčapek *et al.*, 2009). Mass spectrometry (MS) coupled to HPLC is the most powerful tool for the identification of lipids. Atmospheric pressure chemical ionization (APCI) provides the best results for TGs (Holčapek *et al.*, 2005; Lísa *et al.*, 2009c) because of the full compatibility with common NARP conditions, easy ionization of non-polar TGs, and the presence of both protonated molecules $[M+H]^+$ and fragment ions $[M+H-R_iCOOH]^+$. The APCI quantitation approach is based on response factors (RFs) of 23 single-acid TG standards calculated from calibration parameters of these TGs related to triolein, as one of the most widespread TGs in nature (Holčapek *et al.*, 2005). RFs of mixed-acid TGs are calculated as the arithmetic mean of RFs of individual FAs present in each TG.

METHOD OF TRIACYLGLYCEROL ANALYSIS IN NUTS AND SEEDS

Full details of triacylglycerol analysis can be found in Holčapek *et al.* (2005) and Lísa and Holčapek (2008). Briefly, the process includes the following:

- a liquid chromatograph, Waters 616 (Milford, MA, USA)
- two chromatographic columns, Nova-Pak C_{18}, a total length 45 cm
- an acetonitrile-2-propanol mobile phase gradient
- an Esquire 3000 ion trap mass analyzer (Bruker Daltonics, Germany)
- positive-ion atmospheric pressure chemical ionization (APCI)
- UV detection at 205 nm.

Chromatographic behavior of triacylglycerols

Most of TGs within one ECN group were separated (Figures 5.1–5.3), including the partial separation of critical pairs of TGs — i.e., SLO ($t_R = 85.1$ min) and OOP (85.4) with ECN = 48. The retention behavior of TGs in one ECN group was strongly influenced by the FA composition in individual TGs, mainly by the unsaturation degree and acyl chain lengths. For example, the group of OOO ($t_R = 84.0$ min), OOP (85.4), POP (87.0), and PPP (88.7) with ECN = 48 was well resolved. The retention of TGs within one ECN group increased with decreasing double-bond number in acyl chains — for example, with replacement of oleic acid by palmitic acid, or linoleic acid by palmitoleic acid; i.e., pairs LLL ($t_R = 65.3$ min) and LLPo (65.7) with ECN = 42, OLL (71.8) and OLPo (72.2) with ECN = 44, etc.

Retention times of identified TGs in the wide range of analyzed samples were used for the identification of TGs containing FAs with unusual positions of double bonds (double-bond positional isomers). TG double-bond positional isomers shifted retention times in comparison to common TGs containing FAs with the same number of carbon atoms and the same number of double bonds. Characteristic shifts in their retention factors (Δk) were used for the identification of TGs containing unusual gamma-linolenic acid ($\Delta 6,9,12$-18:3, γLn) in blackcurrant and redcurrant (Figure 5.2C) oils. In NARP-HPLC systems, TGs containing γLn had a higher retention in comparison to TGs containing only Ln — for example, pairs of double-bond positional isomers LnLnLn ($t_R = 48.3$ min) and LnLnγLn (49.0), LnLSt (49.4) and γLnLSt (50.1), etc. The differences in retention factors of TGs containing one γLn and TGs containing only Ln were constant, with an average value $\Delta k = 0.22$. Differences in retention factors for TGs containing two and three γLn acids corresponded approximately two times and slightly more than three times with the Δk value of one γLn — i.e., $\Delta k = 0.44$ and $\Delta k = 0.72$, respectively. TGs containing only γLn without any Ln were identified in borage (Figure 5.2D) and evening primrose (Figure 5.3B) oils.

APCI-MS PROFILING OF TRIACYLGLYCEROL COMPOSITION IN NUTS AND SEEDS

TGs were identified using positive-ion APCI-MS based on both protonated molecules $[M+H]^+$ and fragment ions $[M+H-R_iCOOH]^+$. The standard notation of TGs using initials of FA trivial names (Table 5.1) arranged according to their *sn-1*, *sn-2*, and *sn-3* positions was used. FAs in *sn-1* and *sn-3* positions were not resolved using NARP-HPLC/APCI-MS, and they were considered as equivalent. FAs in these positions were arranged according to their decreasing molecular weights. Unlike in the *sn-1* and *sn-3* positions, the FA in the *sn-2* position was identified

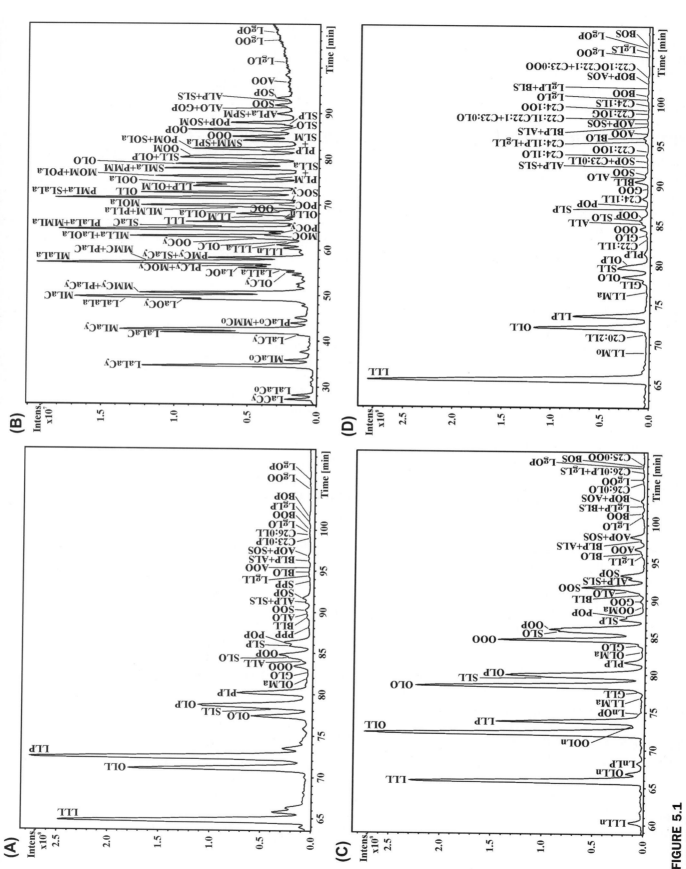

FIGURE 5.1

NARP-HPLC/APCI-MS analysis of plant oils: (A) cottonseed (*Gossypium hirsutum*), (B) coconut (*Cocos nucifera*), (C) sesame (*Sesamum indicum*), and (D) safflower (*Carthamus tinctorius*).

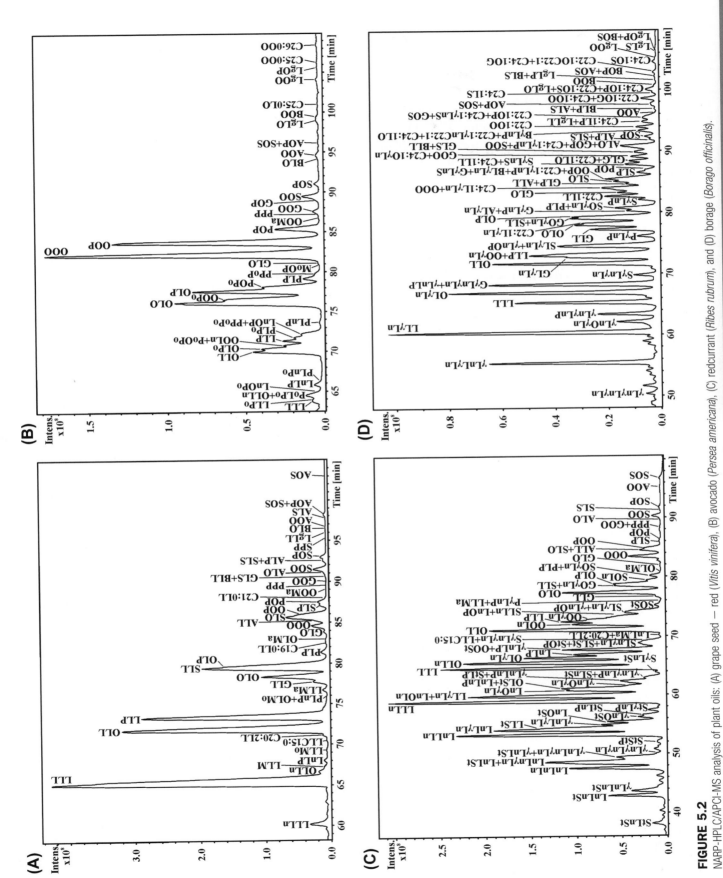

FIGURE 5.2

NARP-HPLC/APCI-MS analysis of plant oils: (A) grape seed — red (*Vitis vinifera*), (B) avocado (*Persea americana*), (C) redcurrant (*Ribes rubrum*), and (D) borage (*Borago officinalis*).

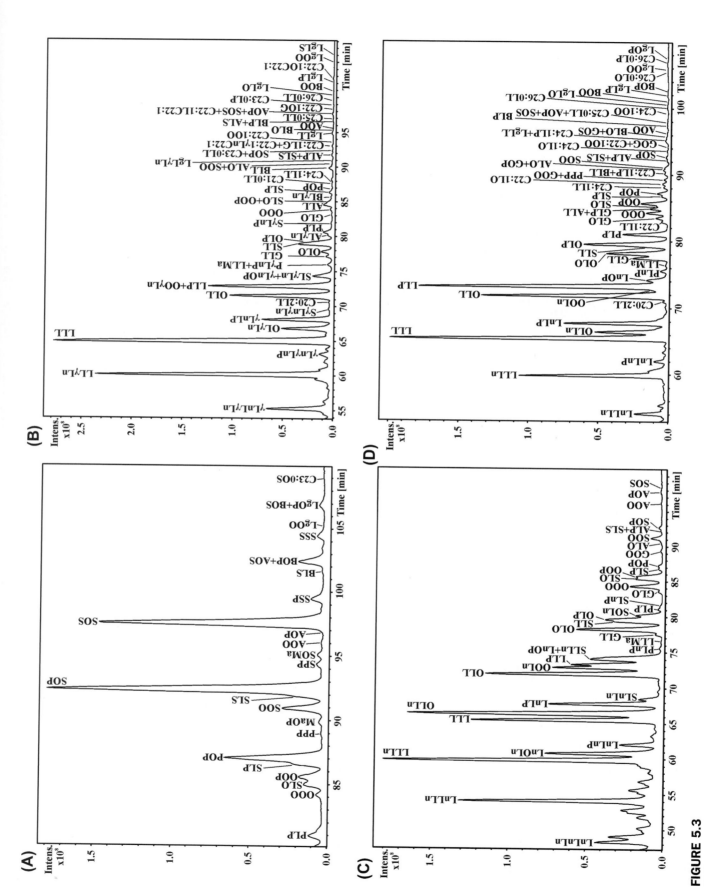

FIGURE 5.3
NARP-HPLC/APCI-MS analysis of plant oils: (A) cocoa butter (*Theobroma cacao*), (B) evening primrose (*Oenothera biennis*), (C) kukui (*Aleurites moluccana*), and (D) wheat germ (*Triticum vulgare*).

TABLE 5.1 Relative Concentrations [%] of Individual Fatty Acids in Analyzed Plant Oils Calculated From NARP-HPLC/APCI-MS of Triacylglycerols With Their Response Factors (RF)

Fatty Acid	Symbol	CN:DB	RF	Palm	Rape	Soybean	Sunflower	Peanut	Cotton	Coconut	Maize	Olive	Sesame	Almond	Safflower	Grape Seed – White	Grape Seed – Red	Hazelnut
Caproic	Co	6:0	134.76							1.37								
Caprylic	Cy	8:0	74.44							15.60								
Capric	C	10:0	17.62							3.70								
Lauric	La	12:0	6.04							37.05								
Myristic	M	14:0	2.77	2.36			0.18			18.80						0.08	0.13	
	C15:0	15:0	1.75													0.03	0.03	
Palmitoleic	Po	Δ9-16:1	1.33		0.13							1.10						0.20
Palmitic	P	16:0	1.32	40.57	6.51	11.66	7.69	9.47	22.12	7.33	11.95	11.75	10.86	9.47	6.60	9.40	10.50	10.60
Margaroleic	Mo	Δ9-17:1	0.81		0.09	0.12	0.06	0.12			0.04	0.08	0.05	0.23	0.02	0.05	0.05	0.18
Margaric	Ma	17:0	0.81	0.03		0.13	0.11	0.08	0.02		0.04			0.09	0.03	0.12	0.10	0.10
Stearidonic	St	Δ6,9,12,15-18:4	0.23															
alpha-Linolenic	Ln	Δ9,12,15-18:3	0.40	0.23	12.87	12.52	0.14	0.27		0.01	1.73	0.77	0.63			0.60	0.61	0.19
gamma-Linolenic	γLn	Δ6,9,12-18:3	0.29															
Linoleic	L	Δ9,12-C18:2	0.57	10.26	19.21	51.76	61.52	35.63	57.27	2.20	55.99	8.53	41.52	27.03	73.96	63.20	65.07	17.61
Oleic	O	Δ9-C18:1	1.00	41.36	57.67	19.18	22.94	43.50	18.15	10.73	27.41	73.85	40.91	61.65	15.15	22.21	19.36	67.83
Stearic	S	18:0	0.61	4.59	1.46	3.41	5.15	1.94	1.93	3.18	1.44	2.57	4.95	1.43	1.86	3.69	3.63	2.91
	C19:0	19:0	0.49				<0.01									0.01	0.01	
	C20:2	Δ11,14-20:2	0.36			0.05	0.03	0.04			0.02			0.04	0.04	0.04	0.05	
Gadoleic	G	Δ9-20:1	0.36	0.05	0.97	0.16	0.15	1.90	0.01	<0.01	0.34	0.34	0.16	0.04	0.25	0.28	0.24	0.18
Arachidic	A	20:0	0.40	0.38	0.45	0.32	0.49	0.92	0.25	0.03	0.65	0.49	0.59	0.06	0.40	0.25	0.21	0.20
	C21:0	21:0	0.39					0.02				0.01			0.12	0.01	0.01	
Erucic	C22:1	Δ13-22:1	0.42		0.40	0.42	1.14	0.22				0.28						
Behenic	B	22:0	0.46	0.07	0.02	0.08	0.06	3.49	0.15		0.16	0.04	0.19	0.19	0.95	0.02		
	C23:0	23:0	0.40		0.08			0.03	0.01						0.05	<0.01		
Nervonic	C24:1	Δ15-24:1	0.40								0.01	0.01	<0.01		0.33			
	C25:0	25:0	0.39								0.02	0.02	0.01					
Cerotic	C26:0	26:0	0.39			0.01	0.01	0.38	0.01									

Continued

49

TABLE 5.1 Relative Concentrations [%] of Individual Fatty Acids in Analyzed Plant Oils Calculated From NARP-HPLC/APCI-MS of Triacylglycerols With Their Response Factors (RF)—continued

Fatty acid	Symbol	CN:DB	RF	Linseed	Poppy Seed	Walnut	Avocado Pear	Blackcurrant	Redcurrant	Borage	Cocoa Butter	Evening Primrose	Kukui Oil	Wheat Germ
Caproic	Co	6:0	134.76											
Caprylic	Cy	8:0	74.44											
Capric	C	10:0	17.62											
Lauric	La	12:0	6.04											
Myristic	M	14:0	2.77											
	C15:0	15:0	1.75	0.08					0.03					
Palmitoleic	Po	Δ9-16:1	1.33				6.52							
Palmitic	P	16:0	1.32	6.90	10.95	8.67	16.75	9.23	5.76	10.97	27.03	9.02	7.29	16.52
Margaroleic	Mo	Δ9-17:1	0.81	0.04	0.16	0.04	0.01							
Margaric	Ma	17:0	0.81	0.04	0.12	0.07	0.01	0.03			0.32	0.06	0.01	0.02
Stearidonic	St	Δ6,9,12,15-18:4	0.23					3.54	5.32					
alpha-Linolenic	Ln	Δ9,12,15-18:3	0.40	52.32	1.68	16.59	1.32	15.80	20.10				25.56	8.06
gamma-Linolenic	γLn	Δ6,9,12-18:3	0.29					13.73	9.20	18.41		13.04		
Linoleic	L	Δ9,12-18:2	0.57	15.89	66.05	52.69	13.48	37.86	36.34	35.42	1.89	67.49	39.36	54.85
Oleic	O	Δ9-18:1	1.00	20.82	18.58	19.34	60.98	17.29	21.20	22.88	34.58	7.66	24.95	17.81
Stearic	S	18:0	0.61	3.65	2.10	2.19	0.46	1.52	1.86	4.18	34.51	1.71	2.66	0.76
	C19:0	19:0	0.49											
	C20:2	Δ11,14-20:2	0.36		0.04	0.02		0.07	0.02			0.04		0.05
Gadoleic	G	Δ9-20:1	0.36	0.11	0.12	0.15	0.20	0.73	0.09	3.37		0.17	0.13	1.11
Arachidic	A	20:0	0.40	0.06	0.17	0.24	0.07	0.16	0.05	0.28	1.05	0.36	0.04	0.21
	C21:0	21:0	0.39									0.01		
Erucic	C22:1	Δ13-22:1	0.42							2.53		0.15		0.21
Behenic	B	22:0	0.46	0.09	0.01	<0.01	0.06	0.03		0.35	0.45	0.16		0.12
	C23:0	23:0	0.40	0.01				<0.01			0.01	0.02		
Nervonic	C24:1	Δ15-24:1	0.40							1.49		0.01		0.13
	C25:0	25:0	0.39	0.03	0.02		0.10	0.01		0.12	0.16	0.09		0.10
Cerotic	C26:0	26:0	0.39	<0.01			0.02					<0.01		0.01
							0.02					0.01		0.04

according to the ratio of fragment ions $[M+H-R_iCOOH]^+$, because the neutral loss of FA from the middle *sn*-2 position was less favored in comparison to *sn*-1 and *sn*-3 positions, and therefore it provided the fragment ion with lower relative abundance than statistically expected. In nature, regioisomers were present in mixtures, and relative abundances of fragment ions were composed from fragment ions of all isomers. Relative abundances of $[M+H-R_iCOOH]^+$ fragment ions also depended on the number of double bonds and carbon atoms; thus, the precise determination of the FA in the *sn*-2 position was achieved only by using regioisomeric standards, but not all of them were commercially available. Therefore, only predominant FAs in the *sn*-2 position were designated. In agreement with previously published data, the *sn*-2 position in plant oils was preferentially occupied by unsaturated FAs, mainly linoleic acid.

APCI-MS QUANTITATION OF TRIACYLGLYCEROLS

The previously developed HPLC/APCI-MS method of quantitative analysis of TGs using RFs (Holčapek *et al.*, 2005) was applied for the quantitation of TGs in natural samples. Briefly, the RFs of individual FAs (Table 5.1) were calculated from calibration parameters of 23 commercially available single-acid TG standards of $R_1R_1R_1$ type (i.e., single-acid saturated TGs from C7:0 to C22:0 and unsaturated single-acid TGs C16:1, C18:1, C18:2, C18:3, γC18:3, C20:1, and C22:1) using the ratios of individual TG calibration slopes to the calibration slope of triolein — for example, $RF(LLL) = a_{LLL}/a_{OOO}$. RFs of mixed-acid TGs were calculated as the arithmetic mean of RFs of FAs present in TGs; for example, $RF(OLP) = (RF(OOO) + RF(LLL) + RF(PPP))/3$. Single-acid TG standards from 9 identified trace FAs were not available, and therefore their RFs were approximated by the following way. RFs of saturated FAs — i.e., hexanoic (C6:0), tricosanoic (C23:0), tetracosanoic (C24:0), pentacosanoic (C25:0), and hexacosanoic (C26:0) acids — were calculated from the equation $y = 2959 \, H \exp(-0.5134x) + 0.3824$ ($R^2 = 0.999$) corresponding to the dependence of RFs of 16 saturated TGs on their carbon atoms, where y was the RF and x was the number of carbon atoms. The RF of stearidonic acid (C18:4) was calculated from the equation corresponding to the dependence of RFs of unsaturated TGs with 18 carbon atoms and 1 (OOO), 2 (LLL), and 3 (LnLnLn) double bonds on the number of double bonds, i.e., $y = 1.615 \exp(-0.4904x)$ ($R^2 = 0.988$), where y was the RF and x was the number of double bonds. The RF values of monounsaturated palmitoleic (C16:1, RF = 1.33) and saturated palmitic (C16:0, 1.32) acids with the same number of carbon atoms, gadoleic (C20:1, 0.36) and arachidic (C20:0, 0.40) acids, or erucic (C22:1, 0.42) and behenic (C22:0, 0.46) acids, were very similar, therefore RFs of margaroleic (C17:1) and nervonic (C24:1) acids were considered the same as their saturated analogs, i.e. RF(C17:1) = F(C17:0) and RF(C24:1) = RF(C24:0). Similarly, the RF of diunsaturated C20:2 acid was considered the same as its monounsaturated analog, i.e. RF(C20:2) = RF(C20:1). TGs containing these FAs were present in analyzed samples only at trace levels, at a maximum of a few tenths of a percent, with the exception of 1.49% nervonic acid in borage oil (Table 5.1). In all cases, FAs with approximated RFs formed only one third of mixed-acid TGs.

TRIACYLGLYCEROL AND FATTY ACID COMPOSITION IN NUTS AND SEEDS

In total, 264 TG species were identified and quantified using HPLC/APCI-MS in 26 plant oils important in different branches of the nutrition and cosmetic industries. Identified TGs were composed of 28 FAs with 6–26 carbon atoms and 0–4 double bonds (Table 5.1). ECNs of all identified TGs ranged from 32 to 58. Only six TGs (PLP, OOO, OOP, POP, SOO and SOP) were detected in all samples. The number of TGs (Table 5.2) ranged from simple almond oil and cacao butter containing only 25 TGs to very complex blackcurrant oil with 77 TGs, redcurrant oil with 78 TGs, and borage oil with 88 TGs. Generally, about five to six TGs formed the main constituents in each sample at a relative concentration level > 5% (Table 5.2).

TABLE 5.2 Number of Identified Triacylglycerols and Fatty Acids, Average Parameters, and Relative Concentration of Essential Fatty Acids [%] (Linoleic and Linolenic Acids) and of Saturated (Sat) and of Monounsaturated (Mono), and Polyunsaturated (Poly) Fatty Acids [%] in All Analyzed Oils, Calculated from NARP-HPLC/APCI-MS results

Plant	Number of TGs/FAs	Major TGs (> 5%)	aECN	aCN	aDB	Essential FAs [%]	Sat [%]	Mono [%]	Poly [%]
Oil palm	41/11	POP,OOP,OLP,PLP,SOP,OOO,PPP	15.85	17.07	0.61	10.49	48.10	41.41	10.49
Rape	55/13	OOO,OLO,OOLn,OLLn,OLL,OOP	15.20	17.91	1.36	32.08	9.07	58.85	32.08
Soybean	66/14	LLL,LLP,OLL,LLLn,LnLP,OLP,OLO	14.59	17.79	1.60	64.28	16.21	19.46	64.33
Sunflower	50/16	LLL,OLL,LLP,OLO,SLL,OLP	14.97	17.90	1.46	61.66	15.16	23.15	61.69
Peanut	60/16	OLL,OLO,OOO,OLP,LLP,OOP	15.69	18.05	1.18	35.90	18.32	45.74	35.94
Cotton	38/11	LLP,LLL,OLL,OLP,PLP	14.93	17.55	1.31	57.27	24.57	18.16	57.27
Coconut palm	85/13	MLaCy,LaLaCy,PLaCy,LaOCy	12.10	12.36	0.13	2.21	87.06	10.73	2.21
Maize	46/12	OLL,LLL,LLP,OLO,OLP	14.90	17.78	1.44	57.72	14.4	27.79	57.74
Olive	37/15	OOO,OOP,OLO,SOO,OLP	15.88	17.76	0.94	9.30	15.33	75.37	9.30
Sesame	49/12	OLL,OLO,LLL,OOO,OLP,LLP,OOP	15.29	17.79	1.25	42.15	16.78	41.07	42.15
Almond	25/8	OOO,OLO,OLL,OLP,OOP	15.49	17.80	1.15	27.03	11.05	61.92	27.03
Safflower	55/14	LLL,OLL,LLP	14.67	17.94	1.63	73.96	10.13	15.87	74.00
Grape vine white	46 /17	LLL,OLL,LLP,OLP,OLO,SLL	14.81	17.81	1.50	63.80	13.62	22.54	63.84
Grape vine red	46/16	LLL,OLL,LLP,OLP,OLO,SLL	14.77	17.78	1.51	65.68	14.64	19.63	65.73
Hazel	30/10	OOO,OOP,OLO,OIL,OLP,SOO	15.70	17.77	1.04	17.80	13.81	68.39	17.80
Linseed (Flax)	63/13	LnOLn,LnLnLn,LnLLn,LnLnP,OLLn,OOLn	13.68	17.86	2.09	68.21	10.86	20.93	68.21
Opium poppy	33 /12	LLL,LLP,OLL,OLO,OLP	14.67	17.77	1.55	67.73	13.37	18.86	67.77
Walnut	43/11	LLL,OLL,LLLn,LLP,OLLn,LnLP,OLO	14.34	17.82	1.74	69.28	11.17	19.53	69.30
Avocado pear	44/14	OOO,OOP,OLO,OLP,OOPo	15.57	17.52	0.97	14.80	17.49	67.71	14.80
Blackcurrant	77/14	OLL,LLL,LLLn,OLLn,LLP, γLnLP	13.91	17.82	1.94	53.66	10.98	18.02	71.00
Redcurrant	78/12	LLLn,OLLn,OLL,LLL	13.82	17.88	2.00	56.44	7.73	21.29	70.98
Borage	88/11	OLγLn,LLγLn,OLL, γLnLP	14.90	17.98	1.56	35.42	15.90	30.27	53.83
Cacao	25/9	SOP,SOS,POP	16.72	17.47	0.37	1.89	63.53	34.58	1.89
Evening primrose	61/17	LLL,LLγLn,LLP,OLL,γLnLP	14.21	17.82	1.81	67.49	11.44	7.99	80.57
Kukui nut tree	38/8	OLLn,LLLn,OLL,LLL,LnLLn,LnLP, OLO,OOLn,LnOLn	14.25	17.85	1.80	64.92	10.00	25.08	64.92
Wheat germ	61/15	LLP,LLL,OLL,OLP,LnLP,LLLn,OLO	14.67	17.70	1.51	62.91	17.78	19.26	62.96

FA composition calculated from TG composition using NARP-HPLC/APCI-MS was in good agreement with the FA composition obtained from validated GC/FID analysis of fatty acid methyl esters after the transesterification of TGs by sodium methanoate (Holčapek *et al.*, 2005). Therefore, the FA composition of analyzed plant oils was calculated directly from TG composition using HPLC/APCI-MS (Table 5.1) to compare their FA profiles in individual samples. Analyzed samples were composed almost exclusively of FAs with 16 (palmitic and palmitoleic acids) and 18 (stearic, oleic, linoleic, linolenic, gamma-linolenic, and stearidonic acids) carbon atoms, with the total concentration in samples ranging from 97.01% in palm oil to 99.82% in kukui oil. A lower concentration of C16 and C18 FAs was observed only in borage (91.86%) and peanut (90.81%) oils, with a higher concentration of FAs with long acyl chains (C20 and longer), and in coconut oil (23.45%), with a well-known high content of short-chain FAs (C6:0 to C14:0). The most abundant FAs present in all analyzed samples were palmitic acid (C16:0), with the concentration ranging from 5.76% in redcurrant oil to 40.57% in palm oil; stearic acid (C18:0), with the concentration ranging from 0.46% in avocado oil to 34.51% in cocoa butter; oleic acid (C18:1), with the concentration ranging from 7.66% in evening primrose oil to 73.85% in olive oil; linoleic acid (C18:2), with the concentration ranging from 1.89% in cocoa butter to 73.96% in safflower oil; and arachidic acid (C20:0), with the concentration ranging from 0.03% in coconut oil to 1.05% in cocoa butter.

NUTRITIONAL PARAMETERS OF PLANT OILS FROM NUTS AND SEEDS

The physical and nutritional properties of plant oils were governed by the FA composition in TGs. Various contents of saturated and unsaturated FAs in TG mixtures resulted in their different melting points at room temperature (oils vs fats), oxidation stability, digestion, or relation to the harmful low-density lipoprotein cholesterol. The average composition of FAs was expressed by average parameters calculated from the TG composition in individual samples (Table 5.2). Values of calculated average parameters were compared with values of the relevant parameters of FAs — i.e., ECN = 16, CN = 18, and DB = 1 for oleic acid (C18:1), etc. For example, average ECN (aECN) = 15.88, aCN = 17.76, and aDB = 0.94 for olive oil corresponded to the high content of oleic acid with 18 carbon atoms and 1 double bond; in fact, with the total content of oleic acid (73.85%). As an example, the aECN = 14.21, aCN = 17.82, and aDB = 1.81 for evening primrose oil corresponded to the high content of linoleic acid with 18 carbon atoms and 2 double bonds; in fact, with the total content of linoleic acid (67.49%). Average ECN values of analyzed samples ranged from 14 to 16, with several exceptions. Coconut oil (aECN = 12.10) was typical, with its high content of short-chain FAs with a low ECN value. Lower aECN values for linseed (aECN = 13.68), redcurrant (13.82), and blackcurrant (13.91) oils were caused by a high content of linolenic acid (ECN = 12), or linolenic and gamma-linolenic (ECN = 12) acids in the case of blackcurrant and redcurrant oils. In contrast, the higher value for cocoa butter (aECN = 16.72) resulted from the high content of stearic acid with ECN = 18. Plant oils were composed predominantly of FAs with 18 carbon atoms, which also corresponded to an average number of carbon atoms ranging from 17.40 to 18.00 in plant oils. Exceptions were found for coconut oil (aCN = 12.10), with its high content of short-chain FAs; palm oil (aCN = 17.07), with its high content of palmitic acid; and peanut oil (aCN = 18.05), with its higher content of long-chain FAs from C20 to C24. The content of saturated and unsaturated FAs in samples was a valuable nutritional parameter in the human diet. In analyzed samples, aDB ranged from 0.9 to 1.9, apart from in the highly saturated coconut (aDB = 0.13), cacao butter (0.37), and palm (0.61) oils, or the highly unsaturated blackcurrant (1.94), redcurrant (2.00), and linseed (2.09) oils.

Other important nutritional parameters of plant oils were expressed by the sums of essential, saturated, monounsaturated, and polyunsaturated FAs (Table 5.2). FAs with double bonds at positions $\Delta 12$ and $\Delta 15$ ($\omega 3$ and $\omega 6$ FAs) are essential for humans, and have to be obtained by food; therefore, their content in plant oils correlates with the nutritional value of these oils.

53

The sum of essential FAs in most samples was found to range from 10 to 70%, apart from in cacao butter (1.89%), coconut oil (2.21%), olive oil (9.3%), and highly essential safflower oil (73.96%). The sum of saturated FAs in analyzed plant oils ranged from 10 to 25% in common plant oils, apart from redcurrant (7.73%) and rapeseed (9.07%) oils, and the highly saturated oils (48.10% of saturated FAs in palm oil, 63.53% in cocoa butter, and 87.06% in coconut oil). In analyzed samples, the sum of monounsaturated FAs ranged from 15 to 65%, except for evening primrose oil (7.99%), coconut oil (10.73%), avocado oil (67.71%), hazelnut oil (68.39%), and olive oil (75.37%). The sum of polyunsaturated FAs ranged from 10 to 70% in samples, except for cocoa butter (1.89%), coconut oil (2.21%), olive oil (9.30%), redcurrant oil (70.98%), blackcurrant oil (71.00%), safflower oil (74.00%), and evening primrose oil (80.57%).

SUMMARY POINTS

- TGs are an important part of the human diet.
- NARP-HPLC allows separation of triacylglycerols according to their fatty acid composition.
- APCI mass spectrometry is used for the identification of triacylglycerols.
- The NARP-HPLC/APCI-MS response factor approach is used in triacylglycerol quantitation.
- Average parameters calculated from triacylglycerol composition characterize the nutritional properties of plant oils.

ACKNOWLEDGEMENTS

This work was supported by the Grant Project MSM0021627502, sponsored by the Ministry of Education, Youth and Sports of the Czech Republic, and Project Nos. 203/09/P249 and 203/09/0139, sponsored by the Czech Science Foundation.

References

Gunstone, F. D. (Ed.). (2006). *Modifying lipids for use in food*. Cambridge, UK: Woodhead Publishing.

Holčapek, M., Lísa, M., Jandera, P., & Kabátová, N. (2005). Quantitation of triacylglycerols in plant oils using HPLC with APCI-MS, evaporative light-scattering, and UV detection. *Journal of Separation Science, 28*, 1315–1333.

Holčapek, M., Velínská, H., Lísa, M., & Česla, P. (2009). Orthogonality of silver-ion and non-aqueous reversed-phase HPLC/MS in the analysis of complex natural mixtures of triacylglycerols. *Journal of Separation Science, 32*, 3672–3680.

Leray, C. (2009). Cyberlipid Center, Paris, 2009, http://www.cyberlipid.org.

Lísa, M., & Holčapek, M. (2008). Triacylglycerols profiling in plant oils important in the food industry, dietetics and cosmetics using high-performance liquid chromatography–atmospheric pressure chemical ionization mass spectrometry. *Journal of Chromatography A, 1198*, 115–130.

Lísa, M., Holčapek, M., Řezanka, T., & Kabátová, N. (2007). HPLC/APCI-MS and GC/FID characterization of Δ5-polyenoic fatty acids in triacylglycerols from conifer seed oils. *Journal of Chromatography A, 1146*, 67–77.

Lísa, M., Velínská, H., & Holčapek, M. (2009a). Regioisomeric characterization of triacylglycerols using silver-ion HPLC/MS and randomization synthesis of standards. *Analytical Chemistry, 81*, 3903–3910.

Lísa, M., Holčapek, M., & Sovová, H. (2009b). Comparison of various types of stationary phases in non-aqueous reversed-phase high-performance liquid chromatography–mass spectrometry of glycerolipids in blackcurrant oil and its enzymatic hydrolysis mixture. *Journal of Chromatography A, 1216*, 8371–8378.

Lísa, M., Holčapek, M., & Boháč, M. (2009c). Statistical evaluation of triacylglycerol composition in plant oils based on high-performance liquid chromatography–atmospheric pressure chemical ionization mass spectrometry data. *Journal of Agricultural and Food Chemistry, 57*, 6888–6898.

CHAPTER 6

Antioxidants in Nuts and Seeds

Monica H. Carlsen, Bente Lise Halvorsen, Rune Blomhoff
Department of Nutrition, Institute of Basic Medical Sciences, Faculty of Medicine, University of Oslo, Norway

55

LIST OF ABBREVIATIONS

FRAP, ferric reducing ability of plasma
ORAC, oxygen radical absorbance capacity
RNS, reactive nitrogen species
ROS, reactive oxygen species
TEAC, the 6-hydroxy-2,5,7,8-tetramethylchroman-2-carboxylic acid equivalent antioxidant capacity

INTRODUCTION

Nuts and seeds are not easily defined in a manner which would be compatible with popular usage yet acceptable to botanists. For example, the groundnut or peanut is a legume, and the chufa nut, which is common in south Europe and Africa, is a tuber. Some languages, such as French, even lack an umbrella word equivalent to nuts. "Noix" in French looks like one, but just means walnuts. In this chapter we will include not only botanically defined nuts, but also data on some foods which have traditionally been defined as nuts (Davidson, 1999).

Nuts & Seeds in Health and Disease Prevention. DOI: 10.1016/B978-0-12-375688-6.10006-4

Nuts and seeds are highly nutritious, and are of prime importance for people in several regions in Asia and Africa. Most nuts and seeds contain a lot of fat (e.g., pecan 70%, macadamia nut 66%, Brazil nut 65%, walnut 60%, almonds 55%, and peanut butter 55%). Most have a good protein content (in the range of 10–30%), and only a few have a very high starch content (Davidson, 1999). Recently, many nuts and seeds have also been identified as being especially rich in antioxidants (Carlsen *et al.*, 2010). Nuts and seeds therefore constitute some of the most nutritionally concentrated kinds of food available. Most nuts, left in the shell, have a remarkably long shelf-life, and can conveniently be stored for winter use.

Nuts and seeds contain numerous different antioxidants which may contribute to their nutritious value. For example, in almonds there is an array of flavonoids, including catechins, flavonols, and flavonones in their aglycone and glycoside forms. Peanuts and pistachios contain several flavonoids, and are enriched in resveratrol, while walnuts contain a variety of polyphenols and tocopherols. Polyphenols in walnuts are typically non-flavonoid ellagi-tannins. In cashews, alkyl phenols are found in abundance. These antioxidants, in cooperation with many other major and minor antioxidants in nuts and seeds, may potentially reduce oxidative stress and risk of related diseases (Blomhoff *et al.*, 2006).

The aim of this chapter is to discuss the potential role of antioxidants in nuts and seeds in oxidative stress and related diseases. We begin with an introduction to oxidative stress and antioxidant defense, and continue with a presentation of total antioxidants in foods, with a special focus on nuts and seeds.

OXIDATIVE STRESS

Free radicals and other reactive oxygen and nitrogen species (ROS and RNS) are formed as a result of normal cellular oxidative metabolic reactions. Such molecules are also formed as a consequence of diseases (e.g., inflammations), and from tobacco smoke, environmental pollutants, natural food constituents, drugs, ethanol, and radiation. If not quenched by antioxidants, these highly reactive compounds will react non-enzymatically with, and potentially alter, the structure and function of several cellular or extracellular components, such as cell membranes, lipoproteins, proteins, carbohydrates, RNA, and DNA (Halliwell, 1996; Gutteridge & Halliwell, 2000).

When the critical balance between generation of free radicals and other ROS or RNS and the antioxidant defenses is unfavorable, oxidative damage can accumulate. Oxidative stress is defined as "a condition that is characterized by accumulation of non-enzymatic oxidative damage to molecules that threaten the normal function of the cell or the organism" (Blomhoff, 2005). Compelling evidence has emerged in the past two decades demonstrating that oxidative stress is intimately involved in the pathophysiology of many seemingly unre-lated types of disease. Thus, oxidative stress is now thought to make a significant contribution to all inflammatory diseases (arthritis, vasculitis, glomerulonephritis, lupus erythematosus, adult respiratory distress syndrome), ischemic diseases (heart disease, stroke, intestinal ischemia), cancer, hemochromatosis, acquired immunodeficiency syndrome (AIDS), emphysema, organ transplantation, gastric ulcers, hypertension and pre-eclampsia, neurologic diseases (multiple sclerosis, Alzheimer's disease, Parkinson disease, amyotrophic lateral scle-rosis, muscular dystrophy), alcoholism, and smoking-related diseases (Halliwell, 1996; Gutteridge & Halliwell, 2000; McCord, 2000).

ANTIOXIDANT DEFENSE

The term "antioxidant" cannot be defined purely chemically; it is always related to the cellular or organismal context, and to oxidative stress. Furthermore, every molecule can be both an oxidant and a reductant; this is determined by the reduction potential of the molecule with which it reacts. An antioxidant is therefore defined as "a redox active compound that limits oxidative stress by reacting non-enzymatically with a reactive oxidant," while an antioxidant enzyme is "a protein that limits oxidative stress by catalysing a redox reaction with a reactive oxidant" (Blomhoff, 2005).

A complex endogenous antioxidant defense system has been developed to counteract oxidative damage and oxidative stress. Such an antioxidant defense is essential for all aerobic cells. The endogenous antioxidant defense has both enzymatic and non-enzymatic components that prevent radical formation, remove radicals before damage can occur, repair oxidative damage, and eliminate damaged molecules (Halliwell, 1996; Gutteridge & Halliwell, 2000; Lindsay & Astley, 2002). The endogenous antioxidant defense, which is produced by cells themselves, consists of components such as glutathione, thioredoxin, and various antioxidant enzymes. Mutations in genes coding for these peptides or proteins often lead to increased incidence of oxidative stress-related diseases, as well as premature death.

In addition to the endogenous antioxidant defense, it has been hypothesized that dietary components also may contribute to the antioxidant defense either by providing redox active compounds that can directly scavenge or neutralize free radicals or other ROS and RNS, or by providing compounds that can induce gene expression of the endogenous antioxidants (Blomhoff, 2005).

DIETARY COMPOUNDS WITH THE ABILITY TO INDUCE PRODUCTION OF ENDOGENOUS ANTIOXIDANTS

An important antioxidant defense mechanism involves detoxification enzymes such as members of the glutathione S-transferase family, γ-glutamyl cysteine synthetase, and NAD(P)H:quinone reductase (quinone reductase) (Talalay, 2000). These enzymes are generally referred to as phase 2 enzymes because they catalyze conversion of xenobiotics, mutagenic metabolites, or their precursors to compounds that are more readily excreted. It is believed that if benign or non-damaging plant compounds induce the phase 2 enzymes, cells are more readily able to neutralize toxic agents such as free radicals and other toxic electrophiles when they appear.

The major plant compounds believed to be able to support the antioxidant defense via this mechanism includes the glucosinolates and several other sulfur-containing plant compounds. Glucosinolates are widespread plant constituents, and it is believed that glucosinolate breakdown products (such as the isothiocyanate sulforaphane) induce phase 2 enzymes and are therefore responsible for the protective effects shown by Brassica vegetables. Allium vegetables contain a number of other sulfur-containing compounds that also may induce phase 2 enzymes. These compounds include the cysteine sulfoxides and the dithiolthiones. Like the glucosinolates, the active principles found in allium vegetables result from enzymatic degradation of the plant compounds (Talalay, 2000; Lindsay & Astley, 2002).

Dietary plants rich in compounds that induce phase 2 detoxification enzymes include the vegetables broccoli, Brussels sprouts, cabbage, kale, cauliflower, carrots, onions, tomatoes, spinach and garlic. The evidence for phase 2 enzyme inductions at ordinary intake levels of plant foods in humans is, however, limited, and the importance of this defense mechanism in the overall protection against oxidative damage is still uncertain.

DIETARY COMPOUNDS WITH THE ABILITY TO DIRECTLY SCAVENGE OR NEUTRALIZE REACTIVE OXIDANTS

In addition to the well-known antioxidants, vitamin C, vitamin E, and selenium, there are numerous other antioxidants in dietary plants. Carotenoids are ubiquitous in the plant kingdom, and as many as 1000 naturally occurring variants have been identified. At least 60 carotenoids occur in fruit and vegetables commonly consumed by humans (Lindsay & Astley,

57

2002). Besides the provitamin A carotenoids, α- and β-carotene, and β-cryptoxanthin, lycopene and the hydroxy-carotenoids (xanthophylls) lutein and zeaxanthin are the main carotenoids present in the diet. Their major role in plants is related to light harvesting as auxiliary components, and quenching of excited molecules that might be formed during photosynthesis.

Phenolic compounds are also ubiquitous in dietary plants (Lindsay & Astley, 2002). They are synthesized in large varieties belonging to several molecular families, such as benzoic acid derivatives, flavonoids, proanthocyanidins, stilbenes, coumarins, lignans, and lignins. Over 8000 plant phenols have been isolated. Plant phenols are antioxidants by virtue of the hydrogen-donating properties of the phenolic hydroxyl groups.

Glutathione, thioredoxin, and many antioxidant enzymes are present in abundant amounts in the diet, but they are not absorbed as such from the diet. They are broken down to their constituent amino acids by digestion. The dietary availability of sulfur amino acids can, however, modulate the cellular glutathione production (Blomhoff, 2005).

MEASUREMENTS OF TOTAL ANTIOXIDANT CONTENTS IN FOODS — METHODOLOGICAL CONSIDERATIONS

Several assays have been used to assess the total antioxidant content of foods. The 6-hydroxy-2,5,7,8-tetramethylchroman-2-carboxylic acid (Trolox) equivalent antioxidant capacity (TEAC) assay (Miller & Rice-Evans, 1996), the ferric reducing ability of plasma (FRAP) assay (Benzie & Strain, 1996), and the oxygen radical absorbance capacity (ORAC) assay (DeLange & Glazer, 1989) are among the methods most often used. Different chemical reaction mechanisms are involved in the various assays. The assays can mainly be divided into two groups: inhibition assays and reduction assays. The inhibition assays are based on antioxidants' ability to react with or neutralize free radicals generated in the assay system (e.g., ORAC, TEAC), while the reduction assays are based on the ability of antioxidants to reduce an oxidant which also functions as a probe that changes color when it is reduced (e.g., FRAP).

Based on careful considerations (Blomhoff, 2005), we chose to use a modified version of the FRAP assay by for total antioxidant analysis (Halvorsen *et al.*, 2002, 2006; Dragland *et al.*, 2003; Carlsen *et al.*, 2010). Most importantly, the modified FRAP assay is a simple, fast, and inexpensive assay with little selectivity. Assay conditions, such as extraction solvents, were optimized regarding detection of both lipophilic and hydrophilic antioxidants. The FRAP assay directly measures antioxidants with a reduction potential below the reduction potential of the Fe^{3+}/Fe^{2+} couple. Thus, the FRAP assay does not measure glutathione. Most other assays have higher reduction potentials, and measure glutathione and other thiols. This may be an advantage when using the FRAP assay, because glutathione is found in high concentrations in foods but is degraded in the intestine and poorly absorbed by humans. A disadvantage of the FRAP assay is its inability to detect other small molecular weight thiols and sulfur-containing molecules of, for example, garlic.

Results obtained by the various assays correlate reasonably well, implying that even though the assays produce different values of antioxidant content, the ranking of the products is often similar (Pellegrini *et al.*, 2003, 2006).

TOTAL AMOUNTS OF ANTIOXIDANTS IN FOODS

We have recently performed a systematic measurement of the total antioxidant content of more than 3100 foods. This novel antioxidant food table enables us to calculate the total antioxidant content of complex diets, to identify and rank potentially good sources of antioxidants, and to provide the research community with comparable data on the relative antioxidant capacity of a wide range of foods.

There is not necessarily a direct relationship between the antioxidant content of a food sample consumed and the subsequent antioxidant activity in the target cell. Factors influencing the bioavailability of phytochemical antioxidants include the food matrix, absorption, and metabolism. Also, the methods measuring total antioxidant capacity do not identify single antioxidant compounds, and they are therefore of limited use when investigating the mechanisms involved. This is, however, not the scope of this initial study. With the present study, food samples with high antioxidant content are identified, but further investigation into each individual food and phytochemical antioxidant compound is needed to identify those which may have biological relevance, and the mechanisms involved.

The aim of the present study was to screen foods to identify the total antioxidant capacity of fruits, vegetables, beverages, spices, and herbs, in addition to common everyday foods. In nutritional epidemiologic and intervention studies, the antioxidant food database may be utilized to identify and rank diets and subjects with regard to antioxidant intake, and as a tool in planning dietary antioxidant interventions.

The results show large variations both between as well as within each food category; all of the food categories contain products almost devoid of antioxidants (Table 6.1). Readers are

TABLE 6.1 Characteristics of the Antioxidant Food Table

		Antioxidant Content (mmol/100 g)						
	n	mean	median	min.	max.	25th Percentile	75th Percentile	90th Percentile
Plant-based foods[1]	1943	11.57	0.88	0.00	2897.11	0.27	4.11	24.30
Animal-based foods[2]	211	0.18	0.10	0.00	1.00	0.05	0.21	0.46
Mixed foods[3]	854	0.91	0.31	0.00	18.52	0.14	0.68	1.50
Categories								
1. Berry products	119	9.86	3.34	0.06	261.53	1.90	6.31	37.08
2. Beverages	283	8.30	0.60	0.00	1347.83	0.15	2.37	3.64
3. Breakfast cereals	90	1.09	0.89	0.16	4.84	0.53	1.24	1.95
4. Chocolates/sweets	80	4.93	2.33	0.05	14.98	0.82	8.98	13.23
5. Dairy products	86	0.14	0.06	0.00	0.78	0.04	0.14	0.44
6. Desserts and cakes	134	0.45	0.20	0.00	4.10	0.09	0.52	1.04
7. Egg	12	0.04	0.04	0.00	0.16	0.01	0.06	0.14
8. Fats and oils	38	0.51	0.39	0.19	1.66	0.30	0.50	1.40
9. Fish and seafood	32	0.11	0.08	0.03	0.65	0.07	0.12	0.21
10. Fruit and fruit juices	278	1.25	0.69	0.03	55.52	0.31	1.21	2.36
11. Grain products	227	0.34	0.18	0.00	3.31	0.06	0.38	0.73
12. Herbal medicine	59	91.72	14.18	0.28	2897.11	5.66	39.67	120.18
13. Infant foods	52	0.77	0.12	0.02	18.52	0.06	0.43	1.17
14. Legumes	69	0.48	0.27	0.00	1.97	0.12	0.78	1.18
15. Meat products	31	0.31	0.32	0.00	0.85	0.11	0.46	0.57
16. Miscellaneous	44	0.77	0.15	0.00	15.54	0.03	0.41	1.70
17. Mixed food entrees	189	0.19	0.16	0.03	0.73	0.11	0.23	0.38
18. Nuts and seeds	90	4.57	0.76	0.03	33.29	0.44	5.08	15.83
19. Poultry products	50	0.23	0.15	0.05	1.00	0.12	0.23	0.59
20. Snacks, biscuits	66	0.58	0.61	0.00	1.17	0.36	0.77	0.97
21. Soups, sauces, etc.	251	0.63	0.41	0.00	4.67	0.25	0.68	1.27
22. Spices and herbs	425	29.02	11.30	0.08	465.32	4.16	35.25	74.97
23. Vegetable products	303	0.80	0.31	0.00	48.07	0.17	0.68	1.50
24. Supplements	131	98.58	3.27	0.00	1052.44	0.62	62.16	316.93

Source: Carlsen et al., 2010.
[1]Categories 1, 2, 3, 10, 11, 12, 14, 18, 22, 23.
[2]Categories 5, 7, 9, 15, 19.
[3]Categories 4, 6, 8, 13, 16, 17, 20, 21.

referred to the antioxidant food table published as an electronic supplement to the paper by Carlsen *et al.* (2010) for the FRAP results on all 3139 products analyzed. The updated database is available online at www.blomhoff.no (link to "Scientific Online Material"). The categories "Spices and herbs," "Herbal/traditional plant medicine," and "Vitamin and dietary supplements" include the most antioxidant-rich products analyzed in the study. The categories "Berries and berry products," "Fruit and fruit juices," "Nuts and seeds," "Breakfast cereals," "Chocolate and sweets," "Beverages," and "Vegetables and vegetable products" include most of the common foods and beverages which have medium to high antioxidant values. We find that plant-based foods are generally higher in antioxidant content than animal-based and mixed food products, with median antioxidant values of 0.88, 0.10, and 0.31 mmol/100 g, respectively. Furthermore, the 75th percentile of plant-based foods is 4.11 mmol/100 g, compared to 0.21 and 0.68 mmol/100 g for animal-based and mixed foods, respectively. The high mean value of plant-based food is due to a minority of products with very high antioxidant values, found among the plant medicines, spices, and herbs. Table 6.1 summarize the results from the 24 categories.

TOTAL AMOUNTS OF ANTIOXIDANTS IN NUTS AND SEEDS

Our analyses (Table 6.2) demonstrate that walnuts contain massive amounts of antioxidants — about 20 mmol antioxidants per 100 g. Pecans, chestnuts, peanuts, pistachios, and sunflower

TABLE 6.2 Antioxidants in Nuts and Seeds

Product	Manufacturer/Product Label/Country of Origin	Procured in	Antioxidant Content (mmol/100 g)
Almonds, with pellicle	Kjøkkensjefens	Norway	0.23
Almonds, with pellicle	Eldorado	Norway	0.37
Almonds, with pellicle	Coop Chef's	Norway	0.28
Almonds, with pellicle		USA	0.53
Almonds, with pellicle, sliced	Blue Diamond, USA	Norway	0.26
Almonds, without pellicle (scalded using hot water)	Eldorado	Norway	0.13
Almonds, without pellicle (scalded using hot water)	Coop Chef's	Norway	0.22
Almonds, without pellicle, sliced	Freia	Norway	0.20
Brazil nuts		USA	0.47
Brazil nuts, with pellicle (partly)	Den Lille Nøttefabrikken	Norway	0.50
Cashews, without pellicle		USA	0.66
Cashews, without pellicle, roasted	Den Lille Nøttefabrikken, Norway	Norway	0.40
Chest nuts, with pellicle (purchased with shell)		Italy	4.67
Chest nuts, without pellicle (purchased with shell)		Italy	0.75
Flaxseed	Peru	Peru	0.64
Flaxseed, ground		USA	1.13
Flaxseed, whole brown		USA	0.80
Hazelnuts, roasted with salt and spices, with pellicle	Iran	Iran	0.46
Hazelnuts, with pellicle	Den Lille Nøttefabrikken, Norway	Norway	0.49
Hazelnuts, with pellicle		Norway	0.50
Hazelnuts, with pellicle	Sunport	Norway	0.69
Hazelnuts, with pellicle		USA	0.94
Hazelnuts, without pellicle	Sunrise Food	Norway	0.08
Hazelnuts, without pellicle	Sunport	Norway	0.16

Continued

TABLE 6.2 Antioxidants in Nuts and Seeds—continued

Product	Manufacturer/Product Label/ Country of Origin	Procured in	Antioxidant Content (mmol/100 g)
Kernel from watermelon, roasted with salt and spices	Iran	Iran	3.27
Kernels from pumpkin, roasted with salt and spices	Iran	Iran	0.40
Macadamia nuts, without pellicle	Den Lille Nøttefabrikken, Norway	Norway	0.55
Macadamia nuts, without pellicle		USA	0.44
Peanut butter, coarse type	Mills, Norway	Norway	0.47
Peanut butter, creamy	Skippy	USA	0.66
Peanut butter, creamy	Jif	USA	0.57
Peanut butter, crunchy	Store Brand	USA	0.51
Peanut butter, crunchy	Skippy	USA	0.55
Peanuts, Malawi nuts, Traditional African Roasted Peanuts	Rab Processors Ltd, Malawi	Malawi	0.89
Peanuts, Polly, roasted, with salt, without pellicle	KiMs, Norway	Norway	0.62
Peanuts, roasted, with pellicle (purchased with shell)	Food Man	Norway	1.97
Peanuts, without pellicle		USA	0.35
Pecans, with pellicle	Den Lille Nøttefabrikken, Norway	Norway	8.24
Pecans, with pellicle	San Lázara	Mexico	10.62
Pecans, with pellicle	Sunport	Norway	7.31
Pecans, with pellicle	The Green Valley	Norway	9.24
Pecans, with pellicle		Norway	6.32
Pecans, with pellicle	La Pasiega, Mexico	Mexico	7.91
Pecans, with pellicle		USA	9.67
Pine nuts	Den Lille Nøttefabrikken, Norway	Norway	0.52
Pine nuts		USA	0.71
Pine nuts	Davy's, Holland	Norway	0.10
Pine nuts, ecologically grown	Urtekram, Denmark	Norway	0.07
Pistachios	India	Norway	4.98
Pistachios	Den Lille Nøttefabrikken, Norway	Norway	0.78
Pistachios	Sunport	Norway	1.08
Pistachios	Iran	Iran	1.16
Pistachios		Norway	1.00
Pistachios		USA	1.43
Pistachios (purchased with shell)	Mexico	Mexico	1.41
Pistachios, roasted with salt and spices	Iran	Iran	1.38
Poppy seeds	Spice Cargo	Mexico	0.44
Poppy seeds		USA	0.03
Poppy seeds, dried	Black Boy, Rieber og søn	Norway	0.31
Poppy seeds, dried	TRS Wholesale CO, England	Norway	0.16
Sesame seeds	Nutana, Denmark	Norway	1.21
Sesame seeds	Spice Cargo	Mexico	0.95
Sesame seeds (ajonjoli)	La Surtidora	Mexico	1.32
Sesame seeds, black	India	India	0.26
Sesame seeds, hulled		USA	0.06
Sesame seeds, white	India	India	0.49
Sesame seeds, with shell	Risenta, Finland	Norway	1.16
Sesame seeds, without shell	Hakon, Norway	Norway	0.30
Sesame seeds, without shell	NatuVit, Denmark	Norway	0.43
Sunflower seeds	Hakon, Norway	Norway	7.50
Sunflower seeds	NatuVit, Denmark	Norway	5.39
Walnuts, with pellicle	USA	Norway	15.16

Continued

TABLE 6.2 Antioxidants in Nuts and Seeds—continued

Product	Manufacturer/Product Label/ Country of Origin	Procured in	Antioxidant Content (mmol/100 g)
Walnuts, with pellicle	Den Lille Nøttefabrikken, Norway	Norway	14.29
Walnuts, with pellicle		Norway	25.41
Walnuts, with pellicle	Diamond	Norway	16.02
Walnuts, with pellicle		USA	13.13
Walnuts, with pellicle	India	India	15.84
Walnuts, with pellicle	Demanter, Bio'noix, France	Norway	19.75
Walnuts, with pellicle (purchased with shell and cupule)		Italy	18.67
Walnuts, with pellicle (purchased with shell)	Shells BI	Norway	33.09
Walnuts, with pellicle (purchased with shell)		Italy	33.04
Walnuts, with pellicle (purchased with shell)	India	India	15.76
Walnuts, with pellicle (purchased with shell)	Natural	Norway	31.38
Walnuts, with pellicle (purchased with shell)		Norway	33.29
Walnuts, without pellicle		Norway	1.81
Walnuts, without pellicle (purchased with shell and cupule)		Italy	0.46
Walnuts, without pellicle (purchased with shell)	Shells	Norway	0.74
Walnuts, without pellicle (purchased with shell)		Italy	1.04
Walnuts, without pellicle (purchased with shell)		Norway	0.79
Walnuts, without pellicle (purchased with shell)	Natural	Norway	1.27

Source: Carlsen *et al.* (2010).

seeds are also very rich in total antioxidants, with about 8, 5, 1, 1 and 6 mmol/100 g, respectively. Hazelnuts, almonds, Brazil nuts, macadamias, pine kernels, cashew nuts, flax seeds, poppy seeds, and sesame seeds also contain significant amounts of total antioxidants (about 0.5–1.0 mmol/100 g). Peanut butter, a common form of nut consumption, contains only about 50% of the total antioxidants of peanuts; though not in the top ranks of antioxidant-containing foods, it is still an important contributor.

Remarkably, in the walnuts most of the antioxidants are located in the pellicle, since less than 10% is retained in the walnuts when pellicle is removed. In most other cases, nuts without a pellicle contained less than 50% of the total antioxidants compared to nuts with a pellicle. Also, Brazil nuts and pistachios purchased without the outer shell contained far less antioxidants than did nuts purchased with the shell. Thus, a significant portion of nut antioxidants is located in the pellicle, and nuts stored with outer shell tend to contain more antioxidants than nuts stored without the outer shell intact. The total antioxidant content of most nuts is much higher than total antioxidant content values of food of animal origin, such as meat, cheese, and milk, which contain 0.0–0.1 mmol/100 g (Carlsen *et al.*, 2010).

As demonstrated in the present study, the variation in the antioxidant values of otherwise comparable products is large. Like the content of any food component, antioxidant values will differ for a wide array of reasons, such as growing conditions, seasonal changes, genetically different cultivars, storage conditions, and differences in manufacturing procedures and processing.

POTENTIAL HEALTH EFFECTS OF DIETARY ANTIOXIDANTS

The highly reactive and bioactive phytochemical antioxidants are postulated to in part explain the protective effect of plant foods. An optimal mixture of different antioxidants with complementary mechanisms of action and different redox potentials is postulated to work in synergistic interactions. Still, it is not likely that all antioxidant-rich food items are good sources, or that all antioxidants provided in the diet are bioactive. Bioavailability differs greatly from one phytochemical to another, so the most antioxidant-rich food items in our diet are not necessarily those leading to the highest concentrations of active metabolites in target tissues. The antioxidants obtained from foods include many different molecular compounds and families with different chemical and biological properties that may affect absorption, transport, and excretion, cellular uptake and metabolism, and, eventually, their effects on oxidative stress in various cellular compartments. Biochemically active phytochemicals found in plant-based foods also have many powerful biological properties which are not necessarily correlated with their antioxidant capacity, including acting as inducers of antioxidant defense mechanisms *in vivo*, or as gene expression modulators. Thus, a food low in antioxidant content may have beneficial health effects due to other food components or phytochemicals executing bioactivity through other mechanisms.

Understanding the complex role of diet in chronic diseases is challenging because a typical diet provides more than 25,000 bioactive food constituents, many of which may modify a multitude of processes that are related to these diseases. Because of the complexity of this relationship, it is likely that a comprehensive understanding of the role of these bioactive food components is needed to assess the role of dietary plants in human health and disease development. We suggest that both their numerous individual functions as well as their combined additive or synergistic effects are crucial to their health beneficial effects; thus, a food-based research approach is likely to elucidate more health effects than those derived from each individual nutrient. Most bioactive food constituents are derived from plants; those so derived are collectively called phytochemicals. The large majority of these phytochemicals are redox active molecules, and therefore defined as antioxidants. It is hypothesized that antioxidants originating from foods may work as antioxidants in their own right *in vivo*, as well as bringing about beneficial health effects through other mechanisms, including acting as inducers of mechanisms related to antioxidant defense, longevity, cell maintenance, and DNA repair.

The antioxidant food table is a valuable research contribution, expanding the research evidence base for plant-based nutritional research, and may be utilized in epidemiological studies where reported food intakes can be assigned antioxidant values. It can also be used to test antioxidant effects and synergy in experimental animal and cell studies, or in human clinical trials. The ultimate goal of this research is to combine these strategies in order to understand the role of dietary phytochemical antioxidants in the prevention of cancer, cardiovascular diseases, diabetes, and other chronic diseases related to oxidative stress.

SUMMARY POINTS

- The total content of antioxidants has been assessed in more than 3100 foods.
- The results show large variations both between as well as within each food category.
- Nuts and seeds are among the food categories that include the most antioxidant-rich food items.
- Walnuts contain massive amounts of antioxidants.
- Pecans, chestnuts, peanuts, pistachios, and sunflower seeds are very rich in total antioxidants.
- Hazelnuts, almonds, Brazil nuts, macadamias, pine kernels, cashew nuts, flax seeds, poppy seeds, and sesame seeds contain significant amounts of total antioxidants.
- A significant portion of nut antioxidants is located in the pellicle.

References

Benzie, I. F. F., & Strain, J. J. (1996). The ferric reducing ability of plasma (FRAP) as a measure of "antioxidant power": the FRAP assay. *Analytical Biochemistry, 239,* 70—76.

Blomhoff, R. (2005). Dietary antioxidants and cardiovascular disease. *Current Opinion in Lipidology, 16,* 47—54.

Blomhoff, R., Carlsen, M. H., Andersen, L. F., & Jacobs, D. R., Jr. (2006). Health benefits of nuts: potential role of antioxidants. *British Journal of Nutrition, 96*(Suppl. 2), S52—S60.

Carlsen, M. H., Halvorsen, B. L., Holte, K., Bohn, S. K., Dragland, S., Sampson, L., et al. (2010). The total antioxidant content of more than 3100 foods, beverages, spices, herbs and supplements used worldwide. *Nutrition Journal, 9,* 3.

Davidson, A. (1999). *The Oxford companion to food.* Oxford, UK: Oxford University Press.

DeLange, R. J., & Glazer, A. N. (1989). Phycoerythrin fluorescence-based assay for peroxyl radicals: a screen for biologically relevant protective agents. *Analytical Biochemistry, 177,* 300—306.

Dragland, S., Senoo, H., Wake, K., Holte, K., & Blomhoff, R. (2003). Several culinary and medicinal herbs are important sources of dietary antioxidants. *Journal of Nutrition, 133,* 1286—1290.

Gutteridge, J. M., & Halliwell, B. (2000). Free radicals and antioxidants in the year 2000. A historical look to the future. *Annals of the New York Academy of Sciences, 899,* 136—147.

Halliwell, B. (1996). Antioxidants in human health and disease. *Annual Review of Nutrition, 16,* 33—50.

Halvorsen, B. L., Holte, K., Myhrstad, M. C. W., Barikmo, I., Hvattum, E., Remberg, S. F., et al. (2002). A systematic screening of total antioxidants in dietary plants. *Journal of Nutrition, 132,* 461—471.

Halvorsen, B. L., Carlsen, M. H., Phillips, K. M., Bohn, S. K., Holte, K., Jacobs, D. R., Jr., et al. (2006). Content of redox-active compounds (i.e., antioxidants) in foods consumed in the United States. *American Journal of Clinical Nutrition, 84,* 95—135.

Lindsay, D. G., & Astley, S. B. (2002). European research on the functional effects of dietary antioxidants — EUROFEDA. *Molecular Aspects of Medicine, 23,* 1—38.

McCord, J. M. (2000). The evolution of free radicals and oxidative stress. *American Journal of Medicine, 108,* 652—659.

Miller, N. J., & Rice-Evans, C. A. (1996). Spectrophotometric determination of antioxidant activity. *Redox Report, 2,* 161—168.

Pellegrini, N., Serafini, M., Colombi, B., Del Rio, D., Salvatore, S., Bianchi, M., et al. (2003). Total antioxidant capacity of plant foods, beverages and oils consumed in Italy assessed by three different *in vitro* assays. *Journal of Nutrition, 133,* 2812—2819.

Pellegrini, N., Serafini, M., Salvatore, S., Del Rio, D., Bianchi, M., & Brighenti, F. (2006). Total antioxidant capacity of spices, dried fruits, nuts, pulses, cereals and sweets consumed in Italy assessed by three different *in vitro* assays. *Molecular Nutrition and Food Research, 50,* 1030—1038.

Talalay, P. (2000). Chemoprotection against cancer by induction of phase 2 enzymes. *Biofactors, 12,* 5—11.

CHAPTER 7

Inorganic Constituents of Nuts and Seeds

Ilia Rodushkin[1,2], Douglas C. Baxter[2], Emma Engström[1,2]
[1] Division of Applied Geology, Luleå University of Technology, Luleå, Sweden
[2] ALS Scandinavia AB, Luleå, Sweden

65

LIST OF ABBREVIATIONS

AAS, atomic absorption spectrometry
EFSA, European Food Safety Authority
FNIC, Food and Nutrition Information Center
ICP, inductively coupled plasma
ICP-MS, inductively coupled plasma mass spectrometry
ICP-OES, inductively coupled plasma optical emission spectrometry
PTWI, provisional tolerable weekly intake
RDA, recommended dietary allowance
RDI, reference (recommended) daily intake
RfD, reference dose
RSD, relative standard deviation (standard deviation divided by arithmetic mean)
TDI, tolerable daily intake
TWI, tolerable weekly intake
UL, tolerable upper intake level

INTRODUCTION

Nuts and seeds are rich in essential elements (Mann & Truswell, 1998; Yang, 2009), consisting of both macro elements (i.e., calcium, magnesium, sodium, potassium, and phosphorus) and trace elements or "micronutrients" with requirements of no more than few mg/day (i.e., iron, zinc, copper, manganese, selenium, and iodine). According to Segura *et al.* (2006), nuts have an optimal nutritional density with respect to calcium, magnesium, and potassium, while the sodium content of nuts is very low. Thus nut consumption is associated with protection against bone demineralization, arterial hypertension, insulin resistance, and overall cardiovascular risk.

Nuts & Seeds in Health and Disease Prevention. DOI: 10.1016/B978-0-12-375688-6.10007-6

As for other foods, these inorganic constituents of nuts and seeds are often collectively referred to as "minerals." The mineral richness of nuts probably contributes to the prevention of diabetes and coronary heart disease observed in epidemiological studies of populations with frequent nut consumption. Dietary selenium, with Brazil nuts being probably the richest food source, is associated with protection against tumor development in laboratory animal studies (Yang, 2009). Although high concentrations do not necessary result in high daily intakes — for example, as a result of infrequent consumption — nuts are nevertheless referred to as the best dietary source of manganese and a good source of boron, especially for vegetarians (Rainey et al., 1999).

Even essential elements may become harmful when their ingestion rates are excessive, resulting in adverse physiological effects. Other elements have no physiological benefits at all, and are recognized as being toxic for humans (for example, mercury, cadmium, lead, and arsenic). Therefore, deficiencies, excesses, or imbalances in the supply of inorganic constituents from dietary sources can have important deleterious effects on human health (WHO, 1996). To evaluate the contribution of a particular food group to the dietary intake of elements, and thus make informed estimates of the nutritional importance from the perspectives of beneficial or adverse health effects, the following background information is necessary:

- Average daily consumption
- Typical element concentrations (or concentration ranges) for food products forming the group
- RDA/RDI/RfD and UL values for essential elements, and PTWI, TWI, TDI or toxic levels for non-essential elements.

The aim of the present work is to provide such an evaluation of the nutritional and toxicological significance of hazelnuts (*Corylus maxima*), walnuts (*Juglans regia*), almonds (*Prunus amygdalus v. dulcis*), pecans (*Carya illinionensis*), cashews (*Anacardium occidentale*), Brazil nuts (*Bertholletia excelsa*), pistachios (*Pistacia vera*), pine nuts (*Pinus pinea*), peanuts (*Arachis hypogaea*), coconuts (*Cocos nucifera*), pumpkin seeds (*Cucurbita pepo*), and sunflower seeds (*Helianthus annuus*) in the human diet. All data presented herein correspond to raw, unsalted, unprocessed nuts and seeds, the terms being used in this chapter with their trivial meaning, not in the botanical sense.

CONSUMPTION OF NUTS AND SEEDS

Nuts and seeds are relatively seldom included in modern mixed diet or nutritional studies, making an accurate assessment of consumption challenging. In a study of the minerals and trace elements in total diets in The Netherlands (Van Dokkum et al., 1989), nut consumption of 11 g/day was reported for 18-year-old males — a group selected on the basis of their presumably highest food consumption. The total intake of pure nut and seed products (composed of at least 90% nuts and/or seeds) for the European population as determined from dietary questionnaires was provided in a study on the association of nut and seed intake with colorectal cancer risk (Jenab et al., 2004). This study included 478,040 subjects. The average daily intake of nuts and seeds was 4.6 g for men (range 0—300 g/day) and 4.1 g (range 0—266 g/day) for women, with a range of < 1 g/day (Swedish women) to 11 g/day for Dutch men, thus confirming the previous estimate by Van Dokkum et al. (1989). The average intake of nuts in the USA is 3—7 g/day (Allen, 2008), with the tendency being to consume nuts as snacks as opposed to meal-time foods. In a study by Frasier et al. (1992), about 32% of the nuts eaten in the USA were peanuts, 29% almonds, 16% walnuts, and 23% of all other varieties. However, it should be pointed out that all of these intake figures may underestimate actual consumption, as they neglect possible contributions from food products with a relatively high nut and seed content, including salads, sauces, bread, desserts, and sweets.

Serving portions suggested on nuts packages sold as snacks vary from two tablespoons (approximately 10 g for pine nuts) to 100 g (peanuts). About 30 g of nuts per day is typically recommended in public health guidelines, and some studies have suggested an optimal intake level of 40 g or more to gain the maximum benefits.

INORGANIC COMPOSITION

Table 7.1 summarizes measured concentrations of 27 elements in 10 nut and 2 seed varieties. This selection covers essential, potentially essential, and toxic elements. For potassium, phosphorus, magnesium, calcium, iron, and zinc, national nutritional tables and textbooks on nutrition were used as primary sources (i.e., FNIC, 2009; Institute of Medicine, 2009). When concentration data provided by different tables were entirely identical for all elements and all nut and seed varieties, only one set was used in the calculations. For the remaining elements, relevant publications on the inorganic constituents of nuts and seeds provided the bulk of the data (Waheed *et al.*, 2007; Rodushkin *et al.*, 2008 and references therein; Welna *et al.*, 2008 and references therein; Yang, 2009).

The number of collected data sets varies greatly, both between different nuts/seeds and between analytes, with the general trend of the least data being available for trace elements and less common specimens — specifically, pecans, pistachios, pine nuts, and pumpkin seeds. These trends reflect differences in either nutritional or toxicological significance, as well as analytical challenges related to measurement at trace levels. In total, data from more than 60 independent data sources were used for this compilation. When three or more independent values were available for a given element in a given nut or seed variety, and the ratio between maximum and minimum concentrations was below 10, the median concentration is presented together with the RSD (%). For two available sets of data, or when the range of reported concentrations for a given element exceeded an order of magnitude, the concentration range is used instead. In some cases, only a single value was available.

As a rule, RSD values below 30% are typical for the majority of essential element concentrations (potassium, phosphorus, magnesium, calcium, iron, zinc, copper, and manganese), confirming the low variability between nut or seed compositions harvested in different locations (Rodushkin *et al.*, 2008) and the fact that the accurate determination of macro elements in these matrices has been possible for decades (Mitchell & Beadles, 1937). However, given that even changes in the degree of dehydration during storage can provide concentration variations in excess of 5% RSD for nuts or seeds from the same location, the praxis of providing concentration information in some nutrition tables with three or even four significant figures seems questionable. The spread in reported values for trace elements is considerably greater than 30% RSD, reaching several orders of magnitude in some cases (e.g., mercury in Brazil nuts). There are several potential reasons for these variations:

- differences in soil composition, climate, and agricultural practices (e.g., use of fertilizers). These factors were shown to be responsible for variable concentrations of selenium, barium, and radon in Brazil nuts (Parekh *et al.*, 2008), cadmium in sunflower seeds (Andersen & Hansen, 1984), and aluminum in pecans (Rodushkin *et al.*, 2008)
- contamination during shelling, peeling, handling and packaging — for example, contact with unprotected skin may significantly alter sodium and chlorine concentrations in almonds, hazelnuts and walnuts
- analytical artefacts, including incomplete recovery of elements from the samples during work-up, inadequate control of and correction for laboratory contamination, and failure to recognize and account for non-specificity of the instrumental analysis. The potential for such sources of bias increases for analytes present at concentrations close to or at the detection limits for such techniques as flame AAS or ICP-OES. In spite of developments in analytical instrumentation, providing reliable data at low or sub-µg/g concentrations still represents a significant challenge.

TABLE 7.1 Concentration of Elements in Pine Nuts

Elements (No. of Sources)	Hazelnuts	Walnuts	Almonds	Pecans	Cashews	Brazil Nuts	Pistachios	Pine nuts	Peanuts	Coconuts	Pumpkin Seeds	Sunflower Seeds
Essential Elements												
K(6–14)	7100(10)	4800(20)	7300(5)	4700(20)	6600(20)	6300(10)	10000(5)	6700(10)	6700(10)	6000(30)	8100(5)	5900(30)
P(6–15)	3100(20)	3900(10)	5300(20)	3100(20)	5600(10)	6900(10)	4900(10)	6300(10)	4300(20)	1800(20)	11000(10)	7700(20)
Mg(6–13)	1600(10)	1600(10)	2800(5)	1300(10)	2800(5)	4100(10)	1300(5)	2500(10)	2000(10)	900(40)	5300(5)	4000(10)
C(6–11)	1400(10)	1000(20)	2500(5)	720(>50)	410(10)	1700(20)	1100(20)	130(30)	540(20)	200(50)	430(5)	1000(20)
Na(4–7)	30(>50)	10(>50)	60(>50)	5(>50)	110(30)	30(>50)	30(40)	30(>50)	20(>50)	370(10)	180(40)	30(50)
Cl(1–3)	60	70	100(5)	100	200(>50)	100	200(>50)	600(>50)	250(>50)	50	20	30
Fe(5–14)	39(20)	33(20)	41(20)	25(10)	64(5)	27(20)	50(30)	55(10)	24(20)	31(5)	130(30)	53(30)
Zn(5–15)	23(20)	29(20)	33(10)	52(20)	57(10)	42(10)	22(5)	65(10)	32(10)	17(30)	75(5)	52(5)
Cu(5–18)	15(20)	15(20)	11(10)	10(10)	21(20)	18(20)	12(10)	15(30)	8(10)	7(10)	13(10)	20(40)
Mn(3–7)	60(5)	34(5)	20(30)	37(10)	23(30)	13(5)	12(10)	91(5)	15(30)	29(5)	40(30)	22(20)
Se(4–13)	0.02(>50)	0.12(>50)	0.03(30)	0.04(>50)	0.12(50)	1–50	0.05(40)	0.01(50)	0.05(40)	0.01(>50)	0.06(>50)	0.6(40)
Cr(1–4)	0.003–0.4	0.001–0.4	0.001–0.4	0.003	0.01–0.3	0.02–8	0.01–0.4	0.003–0.3	0.01–0.4	0.006	0.005	0.006–0.7
Ni(1–3)	0.5(20)	0.5(30)	0.6(30)	2	3(>50)	7(30)	0.5(30)	1(>50)	0.4(30)	0.2	3(20)	3(>50)
Mo(1–2)	0.08	0.3	0.4(20)	0.07	0.3(>50)	0.02	0.2(10)	0.2(20)	2(>50)	0.04	0.7	0.4
I(2–3)	0.04(>50)	0.03(>50)	0.04(30)	0.04(>50)	0.06(>50)	0.05(>50)	0.03(>50)	0.03(>50)	0.03(>50)	0.02(>50)	0.02(>50)	0.02(>50)
Co(1–2)	0.14	0.04	0.05(10)	0.18	0.03(5)	0.7	0.05(50)	0.05(5)	0.05(30)	0.003	0.04	0.06
B(1)	30	20	30	10	30	40	10	20	20	2	10	10
V(1)	0.0005	0.0004	0.0003	0.001	0.001	0.0004	0.0008	0.001	0.001	0.0003	0.0003	0.001
Toxic Elements												
Cd(2–3)	0.006–0.01	0.0004–0.006	0.0005–0.008	0.04	0.0005	0.0005	0.005	0.05–0.2	0.01–0.09	0.005	0.01	0.01–1
Pb(2–4)	0.003–0.4	0.0003–0.5	0.0005–0.3	0.0007	0.002–0.3	0.001	0.003–0.3	0.006–0.4	0.004–0.3	0.001	0.003	0.002–0.3
Hg(1–2)	0.002	0.0007	0.0005–0.06	0.003	0.0005–0.07	0.0001–0.4	0.0008–0.03	0.004–0.07	0.0001–0.12	0.0003	0.0002	0.0001
As(1–3)	0.006–0.3	0.005–0.8	0.005	0.005	0.003	0.001–0.02	0.01–0.6	0.006–0.4	0.005–0.2	0.001	0.002	0.006–0.02
Al(1–4)	0.3–12	0.3–5	0.2–20	10	0.8–13	0.2–5	0.4–11	4–40	0.8–14	0.1	0.2	0.2–10
Sb(1)	0.0004	0.0002	0.0005	0.0004	0.0003	0.0003	0.0003	0.0004	0.0004	0.001	0.0004	0.0004
Sn(1)	0.01	0.0003	0.004	0.005	0.001	0.0006	0.007	0.05	0.004	0.005	0.004	0.003
U(1)	<0.0001	<0.0001	0.0004	0.0001	<0.0001	<0.0001	0.0005	<0.0001	0.0007	0.0004	0.002	0.001
Ba(1–4)	10	2	2	10	0.1	70–2000	1	0.03	3	0.01	10	0.1
Tl(1–2)	0.007	0.002	0.003	0.009	0.0002	0.002	0.0003	0.0008	0.0002	0.002	0.0004	0.002–0.01

Concentrations of elements in nuts, expressed as median or range values (µg/kg), with RSD values (%) in parentheses.

NUTRITIONAL SIGNIFICANCE

For the general population, the daily intake of elements from nuts and seeds can be estimated using relevant consumption data and concentration information (Table 7.1). The largest uncertainty in this approach is likely to be associated with defining the exact composition of the food basket fragment vaguely defined as nuts and seeds — i.e., the proportions of each of the different varieties included. As an alternative approach, Table 7.2 presents the calculated consumption of each nut and seed variety that would be sufficient to provide 10% of the daily intake level of an element (nutrient) considered to be sufficient to meet the requirements of nearly all healthy individuals (approach taken from RDI definition). It is, of course, completely arbitrary to set the importance of a particular nutrition source to 10% of needs, but the values in Table 7.2 can easily be recalculated to any desired threshold. The consumption thus obtained can then be compared to the 1.9 kg of food the average person eats daily, to provide information on nutritional density.

Data from Table 7.2 reaffirm that nuts and seeds are very rich nutrition sources for magnesium, copper, manganese, and nickel, and important ones for iron, zinc, and potassium. For these elements, pumpkin seeds have the highest nutrition potential, followed by sunflower seeds, whereas coconuts occupy the opposite end of the scale. In spite of the significant variability of selenium concentrations in Brazil nuts (Table 7.1), consumption of even a single nut with the lowest published selenium content will contribute significantly to the daily intake of this element. Even sunflower seeds, cashews, and walnuts are relatively rich selenium sources. On the other hand, with the exception almonds, nuts and seeds can hardly be considered an important dietary source of calcium.

Since published chromium concentrations span almost two orders of magnitude for some nuts and seeds (Table 7.1), statements on the nutritional value will vary from insignificant (assuming Cr concentrations at the low ng/g level) to high (0.2—0.3 µg/g level). Peanuts, pumpkin seeds, and sunflower seeds are important sources of molybdenum, while pecans, cashews, Brazil nuts, and coconuts are not. Regarding boron, Brazil nuts, hazelnuts, almonds, and cashews have significantly greater nutritional significance than coconuts. At the recommended nuts and seed consumption rate of 40 g/day, the intakes of sodium, chlorine, iodine, and vanadium will be negligible.

ADVERSE EFFECTS

For essential elements that can be harmful in large amounts, comparison with UL data can be made to access risks for excessive intake. Of course, such risks can only occur when nuts and seeds are proven to be a rich nutritional source for the element in question (see above). Daily nut and seed consumption levels well above 300 g are needed to reach respective UL values for phosphorus, manganese, copper, nickel, zinc, and iron. However, 10% of the UL threshold will be exceeded by an average 10—50 g daily consumption of pumpkin seeds (for phosphorus, iron, zinc, nickel, and manganese), sunflower seeds (for nickel and copper), cashew (for nickel and manganese), hazelnuts and pine nuts (for manganese), or Brazil nuts (for selenium and nickel). Brazil nuts grown on selenium-rich soil may provide more than 0.1 mg selenium in one nut (Chang *et al.*, 1995), and thus the daily consumption of just four nuts will exceed the UL for this element.

For elements with known toxicity to humans, comparison of average intakes via nut or seed consumption with tolerable intakes (PTWI, TWI, or TDI) can be useful. Where PTWI or TDI data are unavailable, toxic intake levels for mammals can be used as rough estimates. As a rule, toxic elements are present in very low concentrations in raw unprocessed nuts and seeds (Waheed *et al.*, 2007; Rodushkin *et al.*, 2008 and references therein). Consequently, even daily nut or seed consumption of several hundred grams will contribute less than a few percent of the unsafe amounts of silver, aluminum, antimony, tin, thallium, tellurium, thorium, or uranium.

TABLE 7.2 Daily Consumption (g) Corresponding to 10% of Recommended Intake (Essential Elements), or Tolerable Intake for 70-kg Adult (Toxic Elements)

	Hazelnuts	Walnuts	Almonds	Pecans	Cashews	Brazil Nuts	Pistachios	Pine Nuts	Peanuts	Coconuts	Pumpkin Seeds	Sunflower Seeds
Essential Elements												
K	50	80	50	80	60	60	40	60	60	60	50	60
P	30	30	20	30	20	10	20	20	20	60	10	10
Mg	20	20	10	30	10	10	30	20	20	40	10	10
Ca	70	100	40	140	240	60	90	>300	190	>300	230	100
Na	>300	>300	>300	>300	>300	>300	>300	>300	>300	>300	>300	>300
Cl	>300	>300	>300	>300	>300	>300	>300	>300	>300	>300	>300	>300
Fe	30	40	30	50	20	40	30	20	50	40	10	20
Zn	70	50	50	30	30	40	70	20	50	90	20	30
Cu	10	10	20	20	10	10	20	10	20	30	10	10
Mn	10	20	30	10	20	40	40	5	30	20	10	20
Se	>300	60	230	180	40	0.2	90	>300	140	>300	120	10
Cr	30	30	30	>300	40	2	30	40	30	>300	>300	20
Ni	30	30	20	7	5	3	30	10	40	70	5	5
Mo	190	50	40	210	>300	>300	80	80	10	>300	20	40
I	>300	>300	>300	>300	>300	>300	>300	>300	>300	>300	>300	>300
Co	100	>300	280	80	250	20	280	280	280	>300	>300	230
B	30	50	30	100	30	20	100	50	50	>300	100	100
Toxic Elements												
Cd	250	>300	60	>300	>300	>300	>300	20	30	>300	250	3
Pb	60	50	80	>300	80	>300	80	60	80	>300	>300	80
Hg	>300	>300	60	>300	50	8	110	30	>300	>300	>300	>300
As	50	20	>300	>300	>300	>300	30	40	80	>300	>300	>300
Ba	>300	>300	>300	>300	>300	10	>300	>300	>300	>300	>300	>300

Daily consumption (g) corresponding to 10% of recommended intake (essential elements) or tolerable intake for 70-kg adult (toxic elements). The following intakes were used for calculations: 3800 mg K (RDA); 1000 mg P (RDI); 400 mg Mg (RDI); 1000 mg Ca (RDI); 600 mg Na (RDI); 2300 mg Cl (RDI); 12 mg Fe (RDI); 15 mg Zn (RDA); 2 mg Cu (RDI); 5 mg Mn (RDI); 70 μg Se (RDI); 120 μg Cr (RDI); 140 μg Ni (RDI); 150 μg Mo (RDI); 150 μg I (RDI); 140 μg Co (RfD); 10 mg B (average intake of healthy adult in USA); 25 μg Cd (recalculated from TWI); 250 μg Pb (recalculated from PTWI); 33 μg Hg (recalculated from PTWI); 150 μg As (recalculated from PTWI); 200 mg Ba (toxic level).

The notable exception is cadmium — a heavy metal that is primarily toxic to the kidney, but can also cause bone demineralization, and has been classified as carcinogenic to humans by the International Agency for Research on Cancer (EFSA, 2009). Foodstuffs are the main source of cadmium exposure for the non-smoking population, and nuts and seeds have been acknowledged as being among significant contributors to human dietary exposure (EFSA, 2009). High levels of cadmium in sunflower seeds intended for human consumption were reported by Andersen and Hansen (1984). Based on the latter study, Danish consumers were advised not to eat large amounts of sunflower seeds over extended periods of time. Later studies indicated that individuals who ingest large amounts of sunflower seeds, with corresponding cadmium intake up to 100 µg per day, do not have levels of cadmium in the blood or urine that are higher than in individuals with far lower intake levels (Reeves & Vanderpool, 1997). However, the EFSA panel on contaminants in the food chain has recently reduced the TWI for cadmium to 2.5 µg/kg body weight per week, corresponding to 25 µg daily intake for a 70-kg adult (EFSA, 2009). Therefore, consumption of sunflower seeds, as well as of almonds, pine nuts, and peanuts (Table 7.1), cannot be neglected in the assessment of dietary exposure to this element. In fact, estimation based on the highest reported cadmium concentration in sunflower seeds shows that the new TWI will be exceeded by consuming 30 g daily.

For some nut and seed varieties, grossly different arsenic, lead, and mercury concentrations have been reported in studies conducted during the past decade. Assuming the lower range of reported values provides the most reliable and accurate estimates, daily consumption of even > 100 g nuts and seeds will correspond to less than 1% of tolerable intakes. On the other hand, for high extremes of published concentrations, the recommended (40 g/day) consumption of walnuts, pistachios, and pine nuts will contribute > 10% of safe levels for arsenic, while Brazil nut and pine nut consumption has to be considered in calculating dietary mercury intake. However, the distinct probability that these elevated concentrations have been caused by analytical artefacts (see above) should be considered before drawing far-reaching conclusions of toxicological relevance (Rodushkin *et al.*, 2008). As a result of very high barium concentrations in Brazil nuts (Table 7.1), daily consumption in excess of 50 g will result in an intake exceeding the safe level of 200 mg/day.

SUMMARY POINTS

- In spite of the relatively modest average daily consumption, nuts and seeds are an important nutritional supply of many essential elements (magnesium, copper, manganese, phosphorus, nickel, iron, zinc, potassium, and, to a lesser extent, boron and molybdenum). Sunflower seeds, and especially Brazil nuts, are rich sources of dietary selenium, though extremely high concentrations reported for Brazil nuts from some geographical areas raise overconsumption concerns.
- Nuts and seeds represent a moderate nutritional source of calcium, but a negligible source of sodium, chlorine, iodine, and vanadium.
- Nuts and seeds contain very low concentrations of most toxic and potentially toxic elements, such as silver, aluminum, antimony, tin, tellurium, thorium, and uranium. Relatively high arsenic, lead, and mercury concentrations reported in some studies for selected nuts and seeds are likely the result of contamination or analytical errors.
- High levels of cadmium present in almonds, pine nuts, peanuts, and especially sunflower seeds have to be taken into account when estimating dietary exposure. The same is true for barium from Brazil nuts, where intake may be significant even with very moderate consumption.

References

Allen, L. H. (2008). Priority areas for research on the intake, composition, and health effects of tree nuts and peanuts. *Journal of Nutrition, 138*(9), 1763S–1765S.

Andersen, A., & Hansen, M. (1984). High cadmium and nickel contents in sunflower kernels. *Zeitschrift für Lebensmitteluntersuchung und -Forschung A, 179*, 399–400.

Chang, J. C., Gutenmann, W. H., Reid, C. M., & Lisk, D. J. (1995). Selenium content of Brazil nuts from two geographic locations in Brazil. *Chemosphere, 30*(4), 801–802.

European Food Safety Authority (EFSA) (2009) http://www.efsa.europa.eu/EFSA/efsa_locale-1178620753812_1211902396126.htm (accessed January 22, 2010).

Food and Nutrition Information Center (FNIC) (2009). http://fnic.nal.usda.gov/nal_display/index.php?info_center=4&tax_level=3&tax_subject=256&topic_id=1342&level3_id=5140 (accessed January 22, 2010).

Frasier, G. E., Sabete, J., Beeson, W. L., & Strachan, T. M. (1992). A possible protective effect of nut consumption on risk of coronary heart disease. The Adventist Health Study. *Archives of Internal Medicine, 152*, 1416–1424.

Institute of Medicine (2009). http://www.iom.edu/Global/News%20Announcements/~/media/48FAAA2FD9E74D95BBDA2236E7387B49.ashx (accessed January 22, 2010).

Jenab, M., Ferrari, P., Slimani, N., Norat, T., Casagrande, C., Overad, K., et al. (2004). Association of nut and seed intake with colorectal cancer risk in the European prospective investigation into cancer and nutrition. *Cancer Epidimiology Biomarkers and Prevention, 13*, 1595–1603.

Mann, J., & Truswell, A. S. (1998). *Essentials of human nutrition.* Oxford, UK: Oxford University Press.

Mitchell, H. H., & Beadles, J. R. (1937). The nutritive value of the proteins of nuts in comparison with the nutritive value of beef proteins. *Journal of Nutrition, 14*(6), 597–606.

Parekh, P. P., Khan, A. R., Torres, M. A., & Kitto, M. E. (2008). Concentrations of selenium, barium, and radium in Brazil nuts. *Journal of Food Composition and Analysis, 21*, 332–335.

Rainey, C. J., Nyquist, L. A., Christensen, R. E., Strong, P. L., Culver, B. D., & Coughlin, J. R. (1999). Daily boron intake from the American diet. *Journal of American Dietetic Association, 99*, 335–340.

Reeves, P. G., & Vanderpool, R. A. (1997). Cadmium burden of men and women who report regular consumption of confectionery sunflower kernels containing a natural abundance of cadmium. *Environmental Health Perspectives, 105*, 1098–1104.

Rodushkin, I., Engström, E., Sörlin, D., & Baxter, D. (2008). Levels of inorganic constituents in raw nuts and seeds on the Swedish market. *Science of the Total Environment, 392*(2–3), 290–304.

Segura, R., Javierre, C., Lizarraga, M. A., & Ros, E. (2006). Other relevant components of nuts: phytosterols, folate and minerals. *British Journal of Nutrition, 96*, 36–44.

Van Dokkum, W., De Vos, R. H., Muys, T. H., & Wesstra, J. A. (1989). Minerals and trace elements in total diets in The Netherlands. *British Journal of Nutrition, 61*, 7–15.

Waheed, S., Siddique, N., & Rahman, A. (2007). Trace element intake and dietary status of nuts consumed in Pakistan: study using INAA. *Radiochimica Acta, 95*, 239–244.

Welna, M., Klimpel, M., & Zyrnicki, W. (2008). Investigation of major and trace elements and their distributions between lipid and non-lipid fractions in Brazil nuts by inductively coupled plasma atomic optical spectrometry. *Food Chemistry, 111*, 1012–1015.

World Health Organization. (1996). *Trace elements in human nutrition and health.* Geneva, Switzerland: WHO.

Yang, J. (2009). Brazil nuts and associated health benefits: a review. *LWT Food Science and Technology, 42*, 1573–1580.

Effects on Health

CHAPTER 8

Nuts and Seeds in Cardiovascular Health

John H. Lee[1], Carl J. Lavie[1], James H. O'Keefe[2], Richard Milani[1]
[1] Department of Cardiovascular Diseases, John Ochsner Heart and Vascular Institute, Ochsner Clinical School, The University of Queensland School of Medicine, New Orleans, Louisiana, USA
[2] Mid-America Heart Institute, University of Missouri-Kansas City, Missouri, USA

CHAPTER OUTLINE

75

LIST OF ABBREVIATIONS

ALA, α-linoleic acid
CHD, coronary heart disease
CRP, C-reactive protein
CV, cardiovascular
DHA, docosahexaenoic acid
EPA, eicosapentaenoic acid
HDL-C, high-density lipoprotein cholesterol
LDL-C, low-density lipoprotein cholesterol
MI, myocardial infarction
MS, metabolic syndrome
MUFA, monounsaturated fatty acid
PUFA, polyunsaturated fatty acids
SCD, sudden cardiac death
VCAM-1, vascular cell adhesion molecules-1

INTRODUCTION

Coronary heart disease (CHD) has been consistently reported as the leading cause of death in developed countries around the world. Although many novel pharmaceutical agents, devices,

Nuts & Seeds in Health and Disease Prevention. DOI: 10.1016/B978-0-12-375688-6.10008-8

and procedures have been developed in an attempt to curb progression of this disease, an abundance of data has accumulated suggesting that dietary and lifestyle changes can also greatly reduce the disease process. One dietary pattern that has received substantial study is the "Mediterranean diet," which includes a relatively large amount of nut consumption that is high in both monounsaturated fatty acids (MUFAs) and polyunsaturated fatty acids (PUFAs). Although increased saturated fatty acid consumption has been associated with increased CHD events, MUFAs and PUFAs, on the other hand, have been observed to be inversely related to CHD events (Kromhout *et al.*, 1995). In this chapter we will review the epidemiologic and randomized clinical trial data that support the consumption of nuts and seeds in improving cardiovascular (CV) outcomes, and then discuss some of the potential mechanisms by which nut consumption confers its cardioprotective effects.

COMPOSITION OF NUTS

Nuts are an energy-rich food that can provide up to 30 kJ per gram of nut. They are predominantly composed of fat, which can make up to 76% (in the case of macadamias) of this plant-based food (Table 8.1). Peanuts (also included in Table 8.1) are actually considered a legume but are consumed interchangeably with nuts in the American diet, and any reference to nuts in this review will include peanuts. The nut varieties that differ significantly from others are chestnuts and peanuts. Peanuts have higher protein content than other varieties, while chestnuts have a higher carbohydrate to fatty acid ratio. Most other nuts differ simply in their PUFA to MUFA ratio, with walnuts having the highest PUFA content. Although fat concentration in nuts is quite high, nut consumption is not harmful to health because the saturated fatty acid content is relatively low. Nuts are made up of mostly MUFAs and, in the case of walnuts, PUFAs. The PUFA content in walnuts is a combination of linoleic acid and α-linoleic acid (ALA), which, like its related compounds docosahexaenoic acid (DHA) and eicosapentaenoic acid (EPA), have been shown to reduce CV risk (Lee *et al.*, 2008). Aside from the high content of MUFAs and PUFAs, nuts also contain an assortment of important bioactive nutrients such as protein, L-arginine (amino acid precursor to nitric oxide), folate, antioxidant vitamins and compounds, calcium, magnesium, and potassium, which all may be linked to the reduction of CV disease (Ros, 2009).

EPIDEMIOLOGIC DATA

There is an accumulating body of epidemiologic evidence that suggests the benefit of nut consumption in decreasing the incidence of CHD. Kris-Etherton and colleagues reported a pooled analysis of four major US studies assessing nut consumption and CHD incidence, and found that subjects in the highest nut consumption group had an approximately 35%

TABLE 8.1 Nutritional Content of Commonly Consumed Nuts

Nut Type	MUFAs	PUFAs	Sat	Fiber	Mg	K	ALA
Almond	31	12	4	12	268	705	
Peanut	24	16	7	9	168	8	
Walnut	9	47	6	7	158	441	9
Hazelnut	46	8	5	10	163	680	
Pistachio	23	14	5	10	121	1025	
Pecan	40	22	6	9	121	410	
Macadamia	59	2	12	9	130	368	
Cashew	24	8	8	3	292	660	
Chestnut	1	Low	Low		84	447	

Measurements in g/100 g; Sat, saturated fats; Mg, magnesium; K, potassium.
Adapted from USDA website http://www.nal.usda.gov/fnic/foodcomp/search/

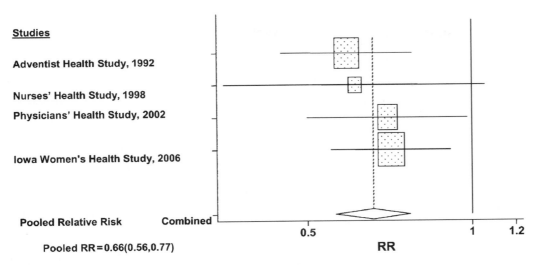

Studies

Adventist Health Study, 1992

Nurses' Health Study, 1998

Physicians' Health Study, 2002

Iowa Women's Health Study, 2006

Pooled Relative Risk   Combined

0.5   1   1.2

RR

Pooled RR=0.66(0.56,0.77)

FIGURE 8.1
Pooled analysis of epidemiologic studies on nut consumption and CHD risk (Kris-Etherton *et al.*, 2008).

decreased risk of CHD incidence (Fraser *et al.*, 1992; Hu *et al.*, 1998; Ellsworth *et al.*, 2001; Albert *et al.*, 2002; Kris-Etherton *et al.*, 2008) (Figure 8.1) The Adventist Health Study of over 30,000 subjects found that individuals who consumed nuts more than four times a week had significantly ($P < 0.05$) fewer fatal CHD events (RR 0.52; 95% CI 0.36—0.76) and non-fatal myocardial infarctions (MI) (RR 0.49; 95% CI 0.28—0.89) when compared to those who consumed nuts less than once a week (Fraser *et al.*, 1992). In the Nurses' Health Study of over 80,000 women, there was also a reported decrease in CHD events (RR 0.66; 95% CI 0.47—0.93, $P = 0.005$) in women who consumed nuts (more than five times a week) when compared to those who did not, after adjustments for baseline characteristics and other dietary habits. In the Physicians' Health Study of over 21,000 men, dietary nut intake (more than twice a week) was associated with a significantly reduced risk of sudden cardiac death (SCD) (RR 0.53; 95% CI 0.30—0.92, $P = 0.01$) that was noted before and after multivariate adjustments (Albert *et al.*, 2002). On the other hand, and in contrast to prior studies, nut consumption was not associated with decreased non-fatal MI and non-SCD. This disparity could be attributed to the lower consumption of nuts observed in the Physicians' Health Study when compared to other studies. In the Iowa Women's Health Study, 34,111 postmenopausal women with no known history of CHD were questioned on their frequency of nut consumption and followed for a mean of 12 years. There was an inverse association of frequent nut consumption (greater than twice a week) and death from CHD, but this association did not meet statistical significance (Ellsworth *et al.*, 2001), possibly owing to the lower nut consumption levels in this particular cohort of subjects.

The reported disparities between the above epidemiologic studies would suggest a dose-dependent effect of nut consumption. The studies that investigated individuals with a higher intake of nuts, such as the Adventist and Nurses' Health Studies (nut consumption more than four times per week) showed a clear benefit of decreased CHD events when compared to low or non-nut consuming individuals. The two studies with lower levels of consumption (greater than twice a week), i.e. the Physicians' Health Study and the Iowa Women's Health Study, showed a less robust association between nut consumption and CV endpoints. Nonetheless, a recent systematic review conducted by Mente and colleagues demonstrated a significant CV risk reduction with nut consumption (RR 0.70, 95% CI 0.57—0.82) as well as with MUFAs (RR 0.80, 95% CI 0.67—0.93), which are a main component of most forms of nuts (Mente *et al.*, 2009).

77

MECHANISMS OF RISK REDUCTION

There are several proposed mechanisms in which nut consumption confers its cardioprotective benefits: lipid lowering, anti-inflammatory, improvements in vascular reactivity, anti-arrhythmic properties, and antioxidant effects.

One of the first clinical trials that investigated the role of nut consumption in CV disease was performed by Sabate and colleagues, and evaluated the effects of implementing walnuts into a standard cholesterol-lowering diet. It found that subjects who consumed nuts in their diet had significantly lower total cholesterol and more favorable lipoprotein profiles (Sabate *et al.*, 1993). Individuals randomized to a nut consumption diet had a 16% lower level of low-density lipoprotein cholesterol (LDL-C) and a 12% lower LDL-C to high density lipoprotein cholesterol (HDL-C) ratio (Sabate *et al.*, 1993). Subsequent to Sabate and colleagues' study, many more published studies added to the evidence that nut consumption had beneficial effects on lipid profiles. In a study conducted by Jenkins and colleagues, patients were randomized to a low saturated fat diet, to the same diet with lovastatin, or to a diet with cholesterol-lowering foods that included plant sterols and nuts (Jenkins *et al.*, 2003). These investigators found that the diet alone only yielded an 8% reduction of LDL-C, while the diet plus lovastatin and cholesterol-lowering food diet arms showed similar LDL-C reductions at 31 and 29%, respectively (Jenkins *et al.*, 2003) (Figure 8.2). In a recent meta-analysis of 13 studies of walnut consumption, Banel and Hu found that diets supplemented with walnuts resulted in significantly greater decreases in total and LDL-C when compared to individuals on the control diets, observing a 5% greater decrease in total cholesterol and a 7% decrease in LDL-C ($P < 0.001$) (Banel & Hu, 2009) (Figure 8.3).

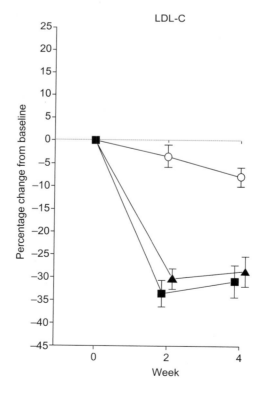

FIGURE 8.2

Changes from baseline in LDL cholesterol. Circle = control, square = statin, triangle = diet including nuts (Jenkins *et al.*, 2003). *Reproduced with permission from the American Medical Association, Copyright © (2003). All rights reserved.*

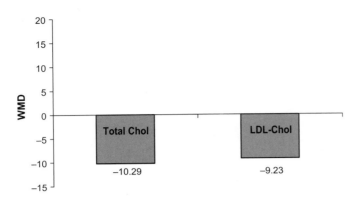

FIGURE 8.3

Meta-analysis of 12 trials of walnut consumption showing significant reductions in total cholesterol and LDL cholesterol with combined weighted mean differences (WMD) as illustrated (Banel & Hu, 2009).

Nut consumption has also been observed to have anti-inflammatory properties. Several studies have illustrated this point by demonstrating the significant reduction in C-reactive protein (CRP) in individuals who consume nuts. The abovementioned study by Jenkins and colleagues showed reductions in CRP in individuals who consumed nuts: CRP was reduced by 33% in the lovastatin arm, 38% in the nut consumption arm, and only 10% in the control group (Jenkins *et al.*, 2003). Jiang and colleagues studied the effects of nut and seed consumption on markers of inflammation in over 6000 individuals, and found that there was a significant inverse relationship between the amount of nut consumption and several markers of inflammation: CRP, interleukin-6, and fibrinogen (Jiang *et al.*, 2006) (Figure 8.4). A similar relationship between nut consumption and reduction in CRP was also observed in three other studies that included supplementation with almonds (Jenkins *et al.*, 2003, 2005) and walnuts (Zhao *et al.*, 2004). However, no significant reductions in CRP were seen in Ros and colleagues' study of walnut supplementation in hypercholesterolemia patients (Ros *et al.*, 2004), or Estruch and colleagues' study of subjects on a Mediterranean diet enriched with nuts (Estruch *et al.*, 2006). The Estruch study did, however, show a decrease in interleukin-6, which is also a marker of inflammation. Although there are some conflicting data regarding CRP reduction and nut consumption, the balance seems to weigh in favor of beneficial effects, in reducing systemic inflammation, from nut consumption.

The improvements in endothelial function with nut consumption were predominantly seen in studies involving walnuts. A study that investigated endothelial function was performed by Ros and colleagues; this study randomized patients to either a cholesterol-lowering Mediterranean diet, or a similar diet with 32% of energy from MUFAs replaced with walnuts (Ros *et al.*, 2004). This study showed that individuals randomized to the walnut diet had improved endothelium-dependent vasodilatation by ultrasound, as well as reduced levels of vascular cell adhesion molecules-1 (VCAM-1). In a study by Cortés and colleagues of individuals who were fed high-fat meals with the addition of either walnuts or olive oil, the walnut-fed individuals had significantly better flow-mediated dilatation when compared to the olive-oil group (Cortés *et al.*, 2006).

The potential anti-arrhythmic effect of nut consumption is also an area that has been studied by several investigators. Albert and colleagues showed a 47% risk reduction of SCD in individuals who consumed nuts two or more times a week (RR 0.53, 95% CI 0.30—0.92) (Albert *et al.*, 2002). The anti-arrhythmic property of nuts may be associated with the PUFA content (particularly ALA), which has the potential to be converted to EPA — a compound that has been shown in many epidemiologic and randomized trials to be beneficial in reducing CV risk (Lee *et al.*, 2008). ALA, which is only present in significant amounts in walnuts, is an 18-carbon, 3-double bond (C18:3n-3) compound that is a precursor to EPA (C20:5n-3) and DHA

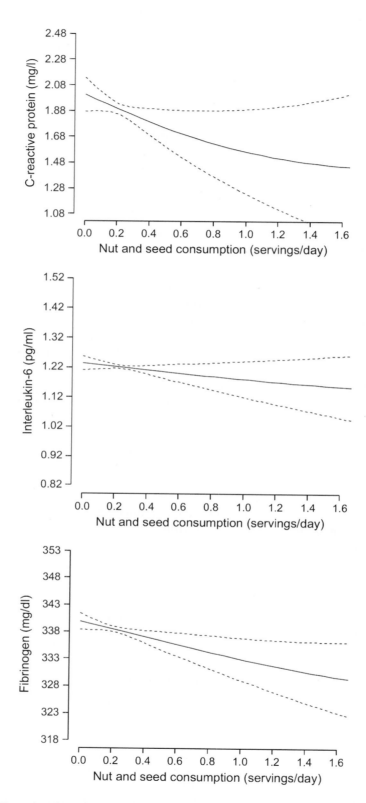

FIGURE 8.4

Inverse relationship of nut consumption and markers of inflammation from the Multi-Ethnic Study of Atherosclerosis after adjustment for age, gender, race/ethnicity, site, educational attainment, family income, smoking status, pack years of smoking, alcohol consumption, total physical activity, use of fish oil supplements, dietary intake of fruits, vegetables, trans-fat, fish and total energy with 95% CI (Jiang *et al.*, 2006)

(C22:6n-3). Aside from walnuts, ALA is most notably found in flaxseed oil, which is the major plant-based source of this PUFA. It is unknown exactly how much of the ALA can be converted to EPA endogenously, but it has been reported that there can be up to 10% conversion, depending on the exact method used to quantify this transformation (Harris, 2005). ALA is purported to provide its cardioprotective effect by decreasing inflammation, improving lipid profiles, and decreasing blood pressure (Harris, 2005).

In addition to the abovementioned cardioprotective properties of nuts, another potential mechanism for reducing CV risk is by decreasing oxidative stress. Nuts are a good source of both tocopherols and phenolic antioxidants, which may play a hand in reducing LDL oxidation. Melatonin, present in walnuts, has also been shown to be an active antioxidant in rat models (Kris-Etherton et al., 2008). Despite the higher fat content of nuts, the predominant MUFA composition of nuts lends it resistance to oxidation. The PUFA component, which is more susceptible to oxidation, may be somewhat protected by the other antioxidant components of nuts (Kris-Etherton et al., 2008). The antioxidant components of nuts have been reported to reside largely in the outer soft shell, and more than half are purported to be lost when the skin is removed.

Nut consumption may also act in reducing some risk factors of CV disease, such as metabolic syndrome (MS). Salas-Salvado and colleagues compared individuals on a Mediterranean diet with olive oil or mixed nuts, and those with just the recommendation for a low-fat diet. These investigators found that, at 1 year, the prevalence of MS was significantly reduced in those consuming a Mediterranean diet with olive oil (7% reduction) or with mixed nuts (14% reduction) when compared to the control diet (2% reduction), with the Mediterranean and mixed nuts diet exhibiting the greatest reduction (Salas-Salvado et al., 2008).

One potential criticism regarding the recommendation of daily nut consumption for CV health is that the high calorie content of nuts may promote weight gain and obesity, which is negatively associated with CHD. However, there is substantial evidence (Banel & Hu, 2009) that suggests that there is no association between increased nut consumption and weight gain. The lack of weight gain despite the higher caloric content of nuts may be secondary to the increased satiety associated with nut consumption. Nuts should be substituted for less healthy snacks, rather than simply added to the diet.

CONCLUSION

There is an abundance of evidence to illustrate the cardioprotective benefits of implement nut consumption in a healthy diet. The CV benefits of nut consumption act on multiple pathways that include decreasing cholesterol, improving vasoreactivity, and acting as an anti-inflammatory, antioxidant, and anti-arrhythmic agent. In conclusion, nut consumption should be encouraged in all individuals at risk for CV disease.

SUMMARY POINTS

- Nut consumption is associated with improved cardiovascular outcomes.
- Nuts are predominantly composed of monounsaturated and polyunsaturated fatty acids.
- Nuts also contain other beneficial bioactive compounds, such as protein, L-arginine (amino acid precursor to nitric oxide), folate, antioxidant vitamins and compounds, calcium, magnesium, and potassium.
- Nuts may confer cardiovascular benefit through their ability to decrease LDL-C and improve vasoreactivity, and their ability to act as an anti-inflammatory, antioxidant, and anti-arrhythmic agent.

References

Albert, C. M., Gaziano, J. M., Willett, W. C., & Manson, J. E. (2002). Nut consumption and decreased risk of sudden cardiac death in the Physicians' Health Study. *Archives of Internal Medicine, 62*, 1382–1387.

Banel, D. K., & Hu, F. B. (2009). Effects of walnut consumption on blood lipids and other cardiovascular risk factors: a meta-analysis and systematic review. *American Journal of Clinical Nutrition, 90*, 56–63.

Cortés, B., Nunez, I., Cofan, M., Gilabert, R., Perez-Heras, A., Casals, E., et al. (2006). Acute effects of high-fat meals enriched with walnuts or olive oil on postprandial endothelial function. *Journal of the American College of Cardiology, 48*, 1666–1671.

Ellsworth, J. L., Kushi, L. H., & Folsom, A. R. (2001). Frequent nut intake and risk of death from coronary heart disease and all causes in postmenopausal women: the Iowa Women's Health Study. *Nutrition Metabolism and Cardiovascular Diseases, 11*, 372–377.

Estruch, R., Martinez-Gonzalez, M. A., Corella, D., Salas-Salvado, J., Ruiz-Gutierrez, V., Covas, M. I., et al. (2006). Effects of a Mediterranean-style diet on cardiovascular risk factors: a randomized trial. *Annals of Internal Medicine, 145*, 1–11.

Fraser, G. E., Sabate, J., Beeson, W. L., & Strahan, T. M. (1992). A possible protective effect of nut consumption on risk of coronary heart disease. The Adventist Health Study. *Archives of Internal Medicine, 152*, 1416–1424.

Harris, W. S. (2005). Alpha-linolenic acid: a gift from the land? *Circulation, 111*, 2872–2874.

Hu, F. B., Stampfer, M. J., Manson, J. E., Rimm, E. B., Colditz, G. A., Rosner, B. A., et al. (1998). Frequent nut consumption and risk of coronary heart disease in women: prospective cohort study. *British Medical Journal, 317*, 1341–1345.

Jenkins, D. J., Kendall, C. W., Marchie, A., Faulkner, D. A., Wong, J. M., de Souza, R., et al. (2003). Effects of a dietary portfolio of cholesterol-lowering foods vs lovastatin on serum lipids and C-reactive protein. *Journal of the American Medical Association, 290*, 502–510.

Jenkins, D. J., Kendall, C. W., Marchie, A., Faulkner, D. A., Josse, A. R., Wong, J. M., et al. (2005). Direct comparison of dietary portfolio vs statin on C-reactive protein. *European Journal of Clinical Nutrition, 59*, 851–860.

Jiang, R., Jacobs, D. R., Jr., Mayer-Davis, E., Szklo, M., Herrington, D., Jenny, N. S., et al. (2006). Nut and seed consumption and inflammatory markers in the multi-ethnic study of atherosclerosis. *American Journal of Epidemiology, 163*, 222–231.

Kris-Etherton, P. M., Hu, F. B., Ros, E., & Sabate, J. (2008). The role of tree nuts and peanuts in the prevention of coronary heart disease: multiple potential mechanisms. *Journal of Nutrition, 138*, 1746S–1751S.

Kromhout, D., Menotti, A., Bloemberg, B., Aravanis, C., Blackburn, H., Buzina, R., et al. (1995). Dietary saturated and trans fatty acids and cholesterol and 25-year mortality from coronary heart disease: the Seven Countries Study. *Preventive Medicine, 24*, 308–315.

Lee, J. H., O'Keefe, J. H., Lavie, C. J., Marchioli, R., & Harris, W. S. (2008). Omega-3 fatty acids for cardioprotection. *Mayo Clinic Proceedings, 83*, 324–332.

Mente, A., de Koning, L., Shannon, H. S., & Anand, S. S. (2009). A systematic review of the evidence supporting a causal link between dietary factors and coronary heart disease. *Archives of Internal Medicine, 169*, 659–669.

Ros, E. (2009). Nuts and novel biomarkers of cardiovascular disease. *American Journal of Clinical Nutrition, 89*, 1649S–1656S.

Ros, E., Nunez, I., Perez-Heras, A., Serra, M., Gilabert, R., Casals, E., et al. (2004). A walnut diet improves endothelial function in hypercholesterolemic subjects: a randomized crossover trial. *Circulation, 109*, 1609–1614.

Sabate, J., Fraser, G. E., Burke, K., Knutsen, S. F., Bennett, H., & Lindsted, K. D. (1993). Effects of walnuts on serum lipid levels and blood pressure in normal men. *New England Journal of Medicine, 328*, 603–607.

Salas-Salvado, J., Fernandez-Ballart, J., Ros, E., Martinez-Gonzalez, M. A., Fito, M., Estruch, R., et al. (2008). Effect of a Mediterranean diet supplemented with nuts on metabolic syndrome status: one-year results of the PREDIMED randomized trial. *Archives of Internal Medicine, 168*, 2449–2458.

Zhao, G., Etherton, T. D., Martin, K. R., West, S. G., Gillies, P. J., & Kris-Etherton, P. M. (2004). Dietary alpha-linolenic acid reduces inflammatory and lipid cardiovascular risk factors in hypercholesterolemic men and women. *Journal of Nutrition, 134*, 2991–2997.

CHAPTER 9

Brassica Seeds: Metabolomics and Biological Potential

David M. Pereira[1], Federico Ferreres[2], Patrícia Valentão[1], Paula B. Andrade[1]
[1] REQUIMTE/Department of Pharmacognosy, Faculty of Pharmacy, Porto University, Porto, Portugal
[2] Research Group on Quality, Safety and Bioactivity of Plant Foods, Department of Food Science and Technology, CEBAS (CSIC), Espinardo, Murcia, Spain

CHAPTER OUTLINE

LIST OF ABBREVIATIONS

AChE, acetylcholinesterase
AD, Alzheimer's disease
DPPH·, 1,1′-Diphenyl-2-picrylhydrazyl free radical
EDTA, ethylenediaminetetraacetic acid
HOCl, hypochlorous acid
HPLC-PAD, high-pressure liquid chromatography-photoarray detection
HPLC-PAD-MS/MS-ESI, high-pressure liquid chromatography-photoarray detection-mass spectrometry-electrospray ionization
X/XO, xanthine/xanthine oxidase

Nuts & Seeds in Health and Disease Prevention. DOI: 10.1016/B978-0-12-375688-6.10009-X

INTRODUCTION

Brassicacea includes a large number of species that are widely used in the human diet, being highly valuable from a nutritional point of view. These species possess a very rich chemistry, which reflects on their biological activities. Regarding *Brassica oleracea* varieties, studies are available concerning the seeds' chemical composition, mainly phenolics (flavonoids and phenolic acids), organic acids, and glucosinolates, as well as the assessment of this material's antioxidant potential and acetylcholinesterase inhibition.

BOTANICAL DESCRIPTION
Tronchuda cabbage

The *Brassica oleracea* L. var. *costata* DC plant resembles a thick-stemmed collard with large floppy leaves. The leaves are close together, round, smooth, and slightly notched at the margins, and can be eaten raw or cooked. The internal leaves are pale yellow, tender, and sweeter than the external leaves, which are dark green (Rosa & Heaney, 1996; Ferreres *et al.*, 2005).

Kale

Brassica oleracea var. *acephala* L. can be either green or purple, and the central leaves do not form a head. It is considered to be closer to wild cabbage than most domesticated forms.

HISTORICAL CULTIVATION AND USAGE

Brassica oleracea grows along the coasts in Europe and north Africa. The earliest records of *B. oleracea* being cultivated for food come from the Greeks, at around 600 BC.

PRESENT-DAY CULTIVATION AND USAGE
Tronchuda cabbage

Tronchuda cabbage is especially popular in some Mediterranean countries, having a determinant role in the diet and agricultural systems. It is a hardy crop that is high yielding, less susceptible to pests and diseases, and well-adapted to a wide range of climates, and is generally grown with little or no agrochemical input.

Kale

The most important growing areas lie in central and northern Europe and North America; it is rarer in tropical regions because it prefers cooler climates, and here it often comes in exotic colours. Kale is the most robust cabbage type — indeed, its hardiness is unmatched by any other vegetable. Kale also tolerates nearly all soils, provided that drainage is satisfactory. Another advantage is that kale rarely suffers from the pests and diseases of other members of the cabbage family.

APPLICATIONS TO HEALTH PROMOTION AND DISEASE PREVENTION
Tronchuda cabbage
PHENOLICS

Among phenolic compounds, flavonoids are of exceptional interest, as they represent one of the most ubiquitous and widely spread groups of natural products. The aqueous lyophilized extract of tronchuda cabbage seeds was analyzed by HPLC-PAD-MS/MS-ESI, revealing a rather complex profile (Figure 9.1A), characterized by the existence of several phenolics (Ferreres *et al.*, 2009a). These compounds have been considered as UV screens in young seedlings, and have been associated with seedling vigor, height, and weight (Randhir & Shetty, 2005).

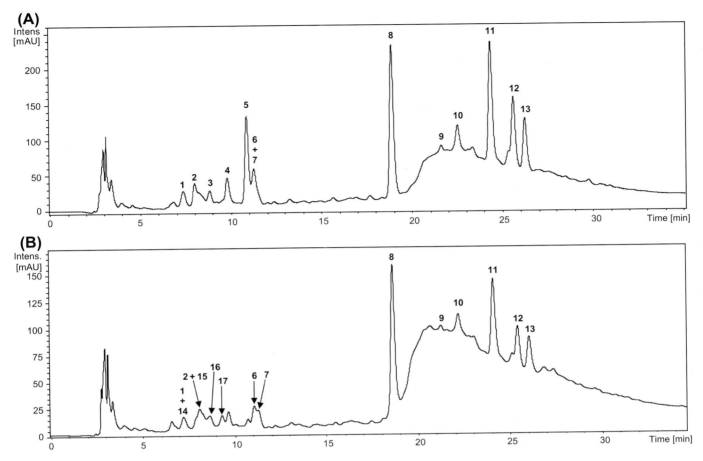

FIGURE 9.1

HPLC-PAD-MSn-ESI phenolic profile of (A) tronchuda cabbage seeds and (B) kale seeds. Peaks: (1) sinapoylgentiobioside; (2) 1-sinapoylglucoside isomer; (3) sinapoylgentiobioside isomer; (4) 1-sinapoylglucoside isomer; (5) 1-sinapoylglucoside; (6) kaempferol-3-*O*-(sinapoyl)triglucoside-7-*O*-glucoside; (7) kaempferol-3-*O*-(sinapoyl)diglucoside-7-*O*-glucoside; (8) sinapoylcholine; (9) 1,2-disinapoylgentiobioside isomer; (10) 1,2-disinapoylgentiobioside isomer; (11) 1,2-disinapoylgentiobioside; (12) 1,2,2′-trisinapoylgentiobioside; (13) 1,2-disinapoylglucoside; (14) quercetin-3-*O*-diglucoside-7-*O*-glucoside; (15) kaempferol-3-*O*-triglucoside-7-*O*-glucoside; (16) kaempferol-3-*O*-diglucoside-7-*O*-glucoside; (17) isorhamnetin-3-*O*-diglucoside-7-*O*-glucoside. Detection at 330 nm. *Adapted from Ferreres* et al. *(2009), J. Agric. Food Chem., 57, 8884–8892, with permission. Copyright © American Chemical Society (2009).*

MSn ion-trap electrospray ionization fragmentation studies indicated that some compounds are sinapic acid derivatives: 1-sinapoylglucoside (5), 1,2-disinapoylgentiobioside (11), 1,2-disinapoylgentiobioside isomers (9 and 10), 1,2,2′-trisinapoylgentiobioside (12), and 1,2-disinapoylglucoside (13). In addition, two 1-sinapoylglucoside isomers (compounds 2 and 4) and two sinapoylgentiobiosides (compounds 1 and 3), with an MSn fragmentation distinct from that of the sinapate esters described above, have been found.

The total phenolic content of tronchuda cabbage seed extract was *ca.* 24.0 g/kg. Sinapoylcholine (8) was the compound present in higher amounts, representing *ca.* 28% of total compounds. Quercetin derivatives were found in trace amounts, while isorhamnetin derivatives were absent.

1,2-Disinapoylgentiobioside isomer was a minor compound, accounting for 2% of the total phenolics. In addition, tronchuda cabbage seeds contained sinapoylgentiobioside isomer (3), sinapoylglucoside isomer (4), and sinapoylglucoside (5), representing 2, 4, and 11%, respectively, of the total compounds in this variety.

ORGANIC ACIDS

Organic acids are a group of primary metabolites that exert a number of roles in living organisms. Ascorbic, shikimic, and quinic acids are quite widespread through a number of species. Shikimic and quinic acids, two cyclohexane carboxylic acids, are very important, as they are precursors of aromatic compounds in plants.

Regarding organic acids, tronchuda cabbage seeds presented a profile composed of eight identified compounds: oxalic, aconitic, citric, pyruvic, malic, fumaric, quinic, and shikimic acids.

The total organic acid content of tronchuda cabbage seeds (42.9 g/kg) (Ferreres *et al.*, 2009a) was much higher than that previously found in the leaves (Ferreres *et al.*, 2006; Vrchovska *et al.*, 2006). In addition, seeds exhibited a distinct profile in which citric acid was the main compound, representing 59.2% of total identified organic acids, followed by malic plus quinic acids (27.1% of compounds). Oxalic, pyruvic, and fumaric acids accounted for 1.2, 1.8, and 0.5%, respectively. Shikimic acid was present in the least amount (0.2%).

ANTIOXIDANT ACTIVITY

Many of the beneficial effects in human health resulting from the consumption of vegetables are strongly associated with their content of antioxidants, which can protect against a number of diseases, ranging from atherosclerosis to cancer.

Given the high content of phenolics in tronchuda cabbage seeds, their antioxidant potential was assessed against an array of reactive species, including both oxygen and nitrogen reactive ones, as well as DPPH· radicals (Ferreres *et al.*, 2007).

The lyophilized aqueous extract of tronchuda cabbage seeds displayed a strong concentration-dependent antioxidant potential against DPPH·, with an IC_{25} of 64 µg/ml (Table 9.1).

An aqueous extract of tronchuda cabbage seeds was evaluated for its ability to scavenge superoxide radicals (Ferreres *et al.*, 2007) generated by the enzymatic xanthine/xanthine oxidase (X/XO) system, and a concentration-dependent effect, with and IC_{25} of 197 µg/ml, was found. The capacity of the lyophilized extract to scavenge superoxide radicals, in a concentration-dependent manner, was confirmed when this radical was generated by a chemical system which indicated an IC_{25} of 118 µg/ml (Figure 9.2, Table 9.1).

The lyophilized extract was also revealed to be a potent scavenger of hydroxyl radicals (Ferreres *et al.*, 2007) — one of the reactive species with the highest deleterious effects in human health — generated in a Fenton system ($IC_{25} = 4$ µg/ml; Figure 9.3, Table 9.1). When ascorbic acid was omitted in the assay, in order to check the pro-oxidant capacity of the extract, the lyophilized extract of tronchuda cabbage seeds proved to be an effective substitute for this acid at

TABLE 9.1 Inhibitory Concentration (IC) Values of *B. oleracea* var. *costata* and *B. oleracea* var. *acephala* Seeds Against Reactive Species

	DPPH·	$O_2^{-·a}$	·NO	·OH	HOCl
Brassica oleracea var. *costata*	64[b]	118[b]	356[d]	4[b]	87[c]
Brassica oleracea var. *acephala*	120[b]	19[b]	151[e]	nd	nd

ICs obtained against DPPH·, superoxide radical (O_2^-), nitric oxide (·NO), hydroxyl radical (·OH) and HOCl.
nd, no data available.
Data are expressed as µg/ml lyophilized extract.
[a] O_2^- generated in a chemical system.
[b] 25% inhibition.
[c] 10% inhibition.
[d] 50% inhibition.
[e] 20% inhibition.

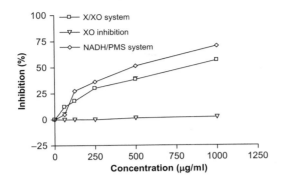

FIGURE 9.2

Effect of tronchuda cabbage seeds against superoxide radicals generated in an enzymatic (X/XO) and a chemical (NADH/PMS) system, and on XO activity. Tronchuda cabbage seeds displayed a concentration-dependent effect against superoxide radical and no activity on XO. Values show mean $\pm$ SE from three experiments performed in triplicate. *Adapted from Ferreres* et al. *(2009), J. Agric. Food Chem., 57, 8884–8892, with permission. Copyright © American Chemical Society (2009).*

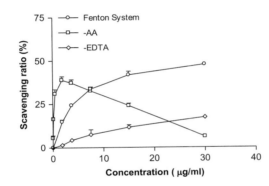

FIGURE 9.3

Tronchuda cabbage seeds' non-specific hydroxyl radical scavenging activity, pro-oxidant activity (-AA) and specific hydroxyl radical scavenging (-EDTA). Tronchuda cabbage seeds displayed a concentration-dependent effect against hydroxyl radical and ability to chelate iron ions (-EDTA), although a pro-oxidant activity was noticed for concentrations above 1.9 µg/ml (-AA). Values show mean $\pm$ SE from three experiments performed in triplicate. *Adapted from Ferreres* et al. *(2009), J. Agric. Food Chem., 57, 8884–8892, with permission. Copyright © American Chemical Society (2009).*

concentrations above 1.9 µg/ml (Figure 9.3). It seems that, at the tested concentrations, tronchuda cabbage seeds have both antioxidant and pro-oxidant effects, with the former being more pronounced than the latter. Inhibition of iron-dependent deoxyribose degradation in the absence of EDTA depends not only on the ability of a scavenger to react with hydroxyl radical, but also on its ability to form complexes with iron ions (Halliwell *et al.*, 1987). In the assay performed in the absence of EDTA, the lyophilized extract of tronchuda cabbage seeds displayed a concentration-dependent ability to chelate iron ions, with an IC_{10} of 12 µg/ml (Figure 9.3).

The lyophilized extract of tronchuda cabbage seeds also exhibited a concentration-dependent protective activity against HOCl damage ($IC_{10} = 87$ µg/ml; Figure 9.4, Table 9.1) (Ferreres *et al.*, 2007).

Regarding reactive nitrogen species, IC_{50} values of 356 and 302 µg/ml were found against nitric oxide and peroxynitrite, respectively (Sousa *et al.*, 2008). The occurrence of high amounts of phenolic compounds and organic acids in tronchuda cabbage seeds suggests that these compounds protect storage lipids from oxidation, as observed with tocopherols, contributing to the viability of seeds and their rapid germination when oxygen demand during germination is high (Sattler *et al.*, 2004).

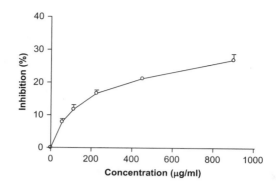

FIGURE 9.4

Effect of tronchuda cabbage seeds against HOCl. Tronchuda cabbage seeds displayed a concentration-dependent effect against HOCl. Values show mean ± SE from three experiments performed in triplicate. *Adapted from Ferreres* et al. *(2009), J. Agric. Food Chem., 57, 8884–8892, with permission. Copyright © American Chemical Society (2009).*

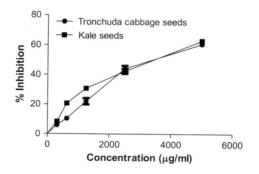

FIGURE 9.5

Effect of tronchuda cabbage and kale seeds on AChE. Tronchuda cabbage and kale seeds displayed a concentration-dependent AChE inhibitory activity. Values show mean ± SE from three experiments performed in triplicate. *Adapted from Ferreres* et al. *(2009), J. Agric. Food Chem., 57, 8884–8892, with permission. Copyright © American Chemical Society (2009).*

ACETYLCHOLINESTERASE INHIBITION

Alzheimer's disease (AD) is a neurodegenerative disorder of the central nervous system that is characterized by profound memory impairment, emotional disturbance, and, in late stages, personality changes. Neurochemical and neuroanatomical studies led to the cholinergic hypothesis, which associates AD symptoms with cholinergic deficiency. Therefore, symptomatic treatment for AD has focused upon augmenting brain cholinergic neurotransmission. An effective way to increase acetylcholine levels is to inhibit acetylcholinesterase (AChE), the enzyme responsible for degrading acetylcholine in the synaptic cleft (Greenblatt *et al.* 2004).

Tronchuda cabbage seeds exhibited a concentration-dependent AChE inhibitory capacity (Figure 9.5). Under assay conditions, the IC_{50} found for tronchuda cabbage seed extract corresponded to 3399 µg/ml (Figure 9.5), containing 22.8 µg/ml of sinapine (Ferreres *et al.*, 2009a).

Kale

PHENOLICS

HPLC-PAD-MSn-ESI analysis of kale seeds allowed the identification of 14 phenolic compounds (Figure 9.1B). In order to obtain a better characterization of the aqueous lyophilized extract, phenolic compounds were quantified by HPLC-PAD.

The total phenolic content of kale seeds was *ca.* 12.2 g/kg (Ferreres *et al.*, 2009a), being similar to that previously reported for its leaves (*ca.* 11.1 g/kg) (Ferreres *et al.*, 2009b). However, the seeds were richer in phenolic acids, while the leaves were mainly characterized by the presence of flavonols (Ferreres *et al.*, 2009a).

Kale seeds contained less phenolics than tronchuda cabbage seeds. Sinapoylcholine (8) was the major compound, representing 42% of the total compounds. 1,2-Disinapoylgentiobioside (11) was also an important compound, representing 15% of the total phenolics in kale seeds.

Kaempferol-3-*O*-diglucoside-7-*O*-glucoside (16) and isorhamnetin-3-*O*-diglucoside-7-*O*-glucoside (17) were the compounds present in lower levels in kale seeds, each accounting for 2% of the total phenolics in this matrix.

Thus, the differences found between the seeds may be used to distinguish these two varieties.

ORGANIC ACIDS

The analysis of kale seeds revealed a chemical profile comprising six organic acids: oxalic, aconitic, citric, pyruvic, malic, and fumaric acids.

From a quantitative point of view, the total organic acid content of kale seeds (*ca.* 41.4 g/kg) was similar to that found in tronchuda cabbage seeds (*ca.* 42.9 g/kg) (Ferreres *et al.*, 2009a), and almost three times lower than that previously found in kale leaves (*ca.* 112.3 g/kg) (Ferreres *et al.*, 2009b). This can be justified by plants' primary metabolism, which is much more active in leaves than in seeds due to their quiescent state (Eastmond & Graham, 2001).

The acids present in higher amounts were malic (50.4%) and citric (42%) acids. Oxalic, pyruvic, and fumaric acids were minor components, accounting for *ca.* 1.3, 0.8, and 0.4%, respectively, of the total acids in kale seeds.

ANTIOXIDANT ACTIVITY

The antioxidant ability of the aqueous lyophilized extract of kale seeds was screened by the DPPH· assay. In this assay, kale seeds exhibited a strong concentration-dependent antioxidant potential ($IC_{25} = 120$ µg/ml) (Ferreres *et al.*, 2009a). Given this value, kale seeds proved to be less effective in DPPH· scavenging than tronchuda cabbage seeds (Table 9.1).

Kale seeds also provided concentration-dependent protection against nitric oxide, with an IC_{20} of 151 µg/ml (Ferreres *et al.*, 2009a). It is not possible to compare this value directly with that found for tronchuda cabbage seeds, as only an IC_{50} (of 356 µg/ml) is recorded in the literature (Table 9.1) (Sousa *et al.*, 2008).

Regarding superoxide radicals, kale seeds displayed a potent protective effect, with an IC_{25} of 19 µg/ml (Ferreres *et al.*, 2009a), thus proving to be much more potent than tronchuda cabbage seeds (Table 9.1).

In general, comparing the two varieties, tronchuda cabbage seeds exhibited a higher protective effect than kale seeds. The observed differences can be explained, at least partially, by higher amounts of phenolic compounds in the former.

ACETYLCHOLINESTERASE INHIBITION

An aqueous extract of kale seeds exhibited a concentration-dependent AChE inhibitory capacity (Figure 9.5). Under assay conditions the IC_{50} found for kale seed extract was 3438 µg/ml of lyophilized extract (Ferreres *et al.*, 2009a).

Some flavonoids, such as quercetrin, quercetin, or 3-methoxy-quercetin, have also been described in the literature as AChE inhibitors. However, despite the presence of flavonoids in the matrix (namely quercetin derivatives), none of the above described compounds were found. Clearly, regardless of the displayed activity against AChE, further studies are required,

89

particularly concerning bioavailability, before these species' seeds can be employed as functional foods for the treatment of detrimental health conditions.

Glucosinolates

The glucosinolates are a group of compounds that are quite widespread in Brassicacea. They consist of glycosides (α-thioglucose moiety, a sulfonated oxime moiety, and a variable side chain derived from an amino acid) which are enzymatically hydrolyzed in damaged plant tissues, giving rise to both potentially toxic and health-promoting compounds. Hydrolysis of the glucoside bond via myrosinase (a thioglucosidase enzyme) leads to a thiohydroximate sulfonate. This compound usually yields an isothiocyanate, prompted by the sulfate group leaving. Under certain conditions, dependent on pH, or the presence of metal ions or other enzymes, related compounds, such as thiocyanates (RSCN) or nitriles (RCN), may be formed from glucosinolates.

Glucosinolates and their breakdown products have been shown to have a chemoprotective effect against certain cancers in humans (Fahey *et al.*, 2001), via a mechanism involving the induction of phase II enzymes. However, as presented below, they may also constitute a potential source of toxicity in humans.

ADVERSE EFFECTS AND REACTIONS (ALLERGIES AND TOXICITY)

There is evidence that consumption of the hydrolysis products from glucosinolates in food crops may induce goiter, an enlargement of the thyroid gland. The goitrogenic effects of glucosinolates cannot be alleviated merely by the administration of iodine. Other deleterious effects are also associated with high glucosinolates intake, such as hepatotoxicity, tumor promotion, and neurological effects (Anilakumar *et al.*, 2006).

Research regarding the glucosinolate composition of tronchuda cabbage and kale seeds is scarce (Velasco & Becker, 2000), although glucosinolates are well characterized in these species' leaves (Fahey *et al.*, 2001). This indicates the importance of studying the effect of these compounds in human health in more detail.

SUMMARY POINTS

- *Brassica oleracea* varieties, in particular *B. oleracea* var. *costata* and *B. oleracea* var. *acephala*, are extremely rich in phenolics.
- These species present both diversity and high contents of those compounds.
- Flavonoids and phenolic acids exert potent antioxidant activity, which can explain some of the health-promoting properties of these crops, namely in inflammation and some types of cancer.
- These seeds also exhibit acetylcholinesterase inhibitory capacity, which may be explored for the treatment of Alzheimer's disease.
- Other compounds present, namely glucosinolates, can be associated with some pathologies, like goiter.
- For safer and healthier exploitation of these matrices, identification and quantification of these compounds in seeds used for human consumption is needed.

References

Anilakumar, K. R., Khanum, F., & Bawa, A. S. (2006). Dietary role of glucosinolate derivatives: a review. *Journal of Food Science and Technology, 43,* 8–17.

Eastmond, P. J., & Graham, I. A. (2001). Re-examining the role of the glyoxylate cycle in oilseeds. *Trends in Plant Science, 6,* 72–77.

Fahey, J. W., Zalcmann, A. T., & Talalay, P. (2001). The chemical diversity and distribution of glucosinolates and isothiocyanates among plants. *Phytochemistry, 56,* 5–51.

Ferreres, F., Valentão, P., Llorach, R., Pinheiro, C., Cardoso, L., Pereira, J. A., et al. (2005). Phenolic compounds in external leaves of tronchuda cabbage (*Brassica oleracea* L. var. *costata* DC). *Journal of Agricultural and Food Chemistry, 53,* 2901–2907.

Ferreres, F., Sousa, C., Vrchovska, V., Valentão, P., Pereira, J. A., Seabra, R. M., et al. (2006). Chemical composition and antioxidant activity of tronchuda cabbage internal leaves. *European Food Research and Technology, 222,* 88–98.

Ferreres, F., Sousa, C., Valentão, P., Seabra, R. M., Pereira, J. A., et al. (2007). Tronchuda cabbage (*Brassica oleracea* L. var. *costata* DC) seeds: phytochemical characterization and antioxidant potential. *Food Chemistry, 101,* 549–558.

Ferreres, F., Fernandes, F., Sousa, C., Valentão, P., Pereira, J. A., & Andrade, P. B. (2009a). Metabolic and bioactivity insights into *Brassica oleracea var acephala. Journal of Agricultural and Food Chemistry, 57,* 8884–8892.

Ferreres, F., Fernandes, F., Oliveira, J. M., Valentão, P., Pereira, J. A., & Andrade, P. B. (2009b). Metabolic profiling and biological capacity of *Pieris brassicae* fed with kale (*Brassica oleracea* L. var. *acephala*). *Food and Chemical Toxicology, 47,* 1209–1220.

Greenblatt, H. M., Guillou, C., Guenard, D., Argaman, A., Botti, S., Badet, B., et al. (2004). The complex of a bivalent derivative of galanthamine with *Torpedo* acetylcholinesterase displays drastic deformation of the active-site gorge: implications for structure-based drug design. *Journal of the American Chemical Society, 126,* 15405–15411.

Halliwell, B., Gutteridge, M. C., & Aruoma, O. I. (1987). The deoxyribose method: a simple test-tube assay for determination of rate constants for reactions of hydroxyl radicals. *Analytical Biochemistry, 165,* 215–219.

Randhir, R., & Shetty, K. (2005). Developmental stimulation of total phenolics and related antioxidant activity in light- and dark-germinated corn by natural elicitors. *Process Biochemistry, 40,* 1721–1732.

Rosa, E., & Heaney, R. (1996). Seasonal variation in protein, mineral and glucosinolate composition of Portuguese cabbages and kale. *Animal Feed Science and Technology, 57,* 111–127.

Sattler, S. E., Gilliland, L. U., Magallanes-Lundback, M., Pollard, M., & Della Penna, D. (2004). Vitamin E is essential for seed longevity and for preventing lipid peroxidation during germination. *Plant Cell, 16,* 1419–1432.

Sousa, C., Valentão, P., Ferreres, F., Seabra, R. M., & Andrade, P. B. (2008). Tronchuda cabbage (*Brassica oleracea* L. var. *costata* DC): scavenger of reactive nitrogen species. *Journal of Agricultural and Food Chemistry, 56,* 4205–4211.

Velasco, L., & Becker, H. (2000). Variability for seed glucosinolates in a germplasm collection of the genus *Brassica*. *Genetic Resources and Crop Evolution, 47,* 231–238.

Vrchovska, V., Sousa, C., Valentão, P., Ferreres, F., Pereira, J. A., Seabra, R. M., et al. (2006). Antioxidative properties of tronchuda cabbage (*Brassica oleracea* L. var. *costata* DC) external leaves against DPPH, superoxide radical, hydroxyl radical and hypochlorous acid. *Food Chemistry, 98,* 416–425.

Nuts and Seeds Used in Health and Disease in Pakistan

Zafar Iqbal, Hafiz Abubaker Saddiqi
Department of Parasitology, University of Agriculture, Faisalabad, Pakistan

93

INTRODUCTION

Pakistan is an agricultural country, and more than 70% of the population resides in rural areas. Those who have migrated into cities in search of a better life also have a rural background. Although the prevalence of common diseases (blood pressure, hypertension, heart attack, and obesity) is negligible, the farming community faces many bacterial, viral, and parasitic diseases and has had its own indigenous practices/remedies to combat these problems since prehistoric times. The most commonly used household seeds of medicinal plants include those of *Tachyspermum ammi* L., *Linum usitatissimum* L., *Lepidium sativum* L., *Foeniculum vulgare* Mill., and *Elettaria cardamom* Maton. Poor people in rural areas try to treat common ailments by the use of seeds available in their own homes, from the fields, or from local herbal shops. Nuts and seeds are easily available from traditional healers in villages, and from traditional pharmacies in towns. Commonly used nuts and seeds in health and diseases in Pakistan, along with their botanical and English/common names, are summarized in Table 10.1. Both urban and rural communities in Pakistan consult traditional healers for their health, who prescribe medicinal formulations (*Majunes*) containing nuts and seeds. Currently, the use of nuts and seeds for treatment and prevention of different ailments is being validated by scientists. Ethnopharmacology has therefore emerged as an important area of focus for those researching new drugs.

HISTORICAL ASPECTS OF THE USE OF NUTS AND SEEDS FOR HEALTH IN PAKISTAN

History relates the similarity in the cultural heritage of Pakistan and India. The Materia medica of the Greco-Arab (*Unani* system) that is utilized in the subcontinent contains an inventory of

Nuts & Seeds in Health and Disease Prevention. DOI: 10.1016/B978-0-12-375688-6.10010-6

TABLE 10.1 Commonly Used Nuts and Seeds in Health and Diseases in Pakistan

Botanical Name	English/ Common Name	Nuts/Seeds	Indications/Uses/Activities
Abutilon indicum L.	Indian mallow	Seeds	Mucilaginous, and laxative (Hayat *et al.*, 2008)
Amomum sabulatum Roxb.	Black cardamom	Fruit/nuts	Aromatic, and tonic to digestive system (Qureshi *et al.*, 2009)
Anemone nemorosa L.	Vulgaris	Seeds	Chronic headache, excretion of nasal secretion, brain tonic, and analgesic (Kabirudin & Khan, 2003)
Anethum graveolens L.or *Anethum sowa* Roxb.	Dill	Seeds	Gastric disorders, facial palsy, and paralysis (Hayat *et al.*, 2008)
Apium graveolens L.	Karafs	Seeds	Expectorant, antifebrile, lumbago, ascitis, dropsy, kidney, and bladder stones (Usmanghani *et al.*, 1997; Rizvi *et al.*, 2007)
Areca catechu L.	Betelnut	Nuts/fruit	Antidiarrheal, bleeding gums, and ophthalmia (Qureshi *et al.*, 2009)
Artemisia parviflora Roxb.		Seeds	Abdominal pain, stomach ache, high blood pressure, diabetes, and anthelmentic (Ahmad *et al.*, 2006; Hussain *et al.*, 2007).
Avena fatua L.	Oat	Seed	Nerve tonic, stimulant, and laxative (Islam *et al.*, 2006)
Azadirachta indica A. Juss.	Neem, Indian lilac	Seeds	Blood purifier, anti-lice, vermicide, hemorrhoids, skin diseases such as scabies, wounds, leprosy, earache, and liver disorders (Sabeen & Ahmad, 2009)
Bergenia stracheyi H & T. Engl.	Bergenia	Seeds	Pimples, eczema, toothache, and bleeding gums (Ahmad *et al.*, 2006)
Brassica compestris L.	Mustard plant	Seeds	Aphrodisiac, and skin infections (Hussain *et al.*, 2007; Hayat *et al.*, 2008)
Brassica juncea L.	Brown mustard, Indian mustard	Seeds	Anodyne, rubefacient, stimulant, arthritis, footache, lumbago, rheumatism, skin eruptions, and ulcers (Sabeen & Ahmad, 2009)
Butea monosperma (Lam.) Taub	Bastard teak	Seeds	Rubefacient (Rizvi *et al.*, 2007), anthelmintic (Anonymous, 1987)
Cannabis sativa L.	Cannabis	Seeds	Blood purifier, scabies, sedative, appetizer, anodyne, earache, malaria, anthrax, sore throat, and hemorrhoids (Ahmad *et al.*, 2006; Sabeen & Ahmad, 2009)
Cardiospermum halicacabum L.	Heart pea	Seeds	Aphrodisiac (Kabirudin & Khan, 2003)
Carthamus oxycantha Bieb.	Carthamus	Seeds	Dressing of ulcers, and itching (Shinwari & Khan., 2000; Hayat *et al.*, 2008)
Carthamus tinctorious L.	Bastards saffron, safflower	Seeds	Chicken pox (Ahmad *et al.*, 2006), Respiratory tract ailments, aphrodisiac, and spermatopoetic (Rizvi *et al.*, 2007; Hussain *et al.*, 2007)
Carum carvi L.	Caraway	Seeds	Nausea, and stomach ache (Ahmad *et al.*, 2006)
Carum copticum L.	Omum	Seeds	Antifebrile, stomachic, gastrointestinal disorders, molluscicidal, menstruation problems, expectorant, antibacterial, and antidote to opium (Ali *et al.*, 2003; Sabeen & Ahmad, 2009)
Carum petroselinum (BENTH.)	Parsley	Seeds	Gastritis, antifebrile, constipation, and diabetes mellitus (Kabirudin & Khan, 2003)
Cassia fistula L.	Golden shower	Seeds	Asthma, cough, wounds, jaundice, bronchitis, menstrual problems, for the removal of umbilical cord, and to ease labor (Rizvi *et al.*, 2007; Hayat *et al.*, 2008)
Chenodium album L.	Goose foot, lambs quarter, peg weed	Seeds	Antifebrile, jaundice, and hepatitis (Kabirudin & Khan, 2003), Gastrointestinal disorders, anthelmintic, and purgative (Shinwari & Khan, 2000; Qureshi *et al.*, 2009)

Continued

TABLE 10.1 Commonly Used Nuts and Seeds in Health and Diseases in Pakistan—continued

Botanical Name	English/ Common Name	Nuts/Seeds	Indications/Uses/Activities
Cichorium intybus L.	Common chicory	Seeds	Liver complaints like jaundice, typhoid, constipation, and nausea (Anonymous, 1987; Ahmad *et al.*, 2006)
Citrullus lanatus (Thunb)	Water melon	Seeds	Stomachic, immunomodulator, jaundice, urinary blockage, and kidney stones (Hayat *et al.*, 2008: 20)
Cocos nucifera L.	Coconut	Fruit/kernel	Anthelmintic (Kabirudin & Khan, 2003)
Cordia obliqua Willd.	Geiger tree	Seeds	Cough, and stomach ulcers (Hayat *et al.*, 2008)
Coriandrum sativum L.	Coriander	Seeds	Antiperoxidative (Rizvi *et al.*, 2007)
Cotoneaster nummularia Fisch & Mey.		Seeds	Blood purifier (Hussain *et al.*, 2007)
Cucumis melo L.	Melons	Seeds	Liver disorders, skin infections, flexion, diuretic, and to treat kidney and bladder stones (Hayat *et al.*, 2008)
Cucumis melo var. utilissimus Roxb.	Cucumber, gourd, melon	Seeds	Urethritis, spleen diseases, and jaundice (Hayat *et al.*, 2008)
Cucumis sativus L.	Cucumber	Seeds	Diuretic (Marwat *et al.*, 2009)
Cucurbita maxima Duch. ex Lam.	Turban squash, Turk's cap	Seeds	Cough (Hussain *et al.*, 2007)
Cuminum cyminum L.	Cumin/carum	Seeds	Expectorants, squamous cell carcinoma, digestive disorders, carminative (Shinwari & Khan, 2000; Kabirudin & Khan, 2003), ear ache, cough, toothache, chest infection, pneumonia, antidiabetic, allergy, acne, dysmenorrheal, and anticancer (Sabeen & Ahmad, 2009)
Cuscuta reflexa Roxb.	Air creeper	Seeds	Hepatitis, gastritis, constipation, antifebrile, carminative, and anthelmintic (Shinwari & Khan, 2000)
Cydonia oblonga Mill.	Quince	Seeds	Antifebrile, respiratory ailments, cardiac problems, dysentery, and diarrhea (Marwat *et al.*, 2009)
Dacus carota L.	Carrot	Seeds	Regulates menstruation, and high doses are abortificant (Sabeen & Ahmad, 2009)
Datura inoxia Mill.	Thorn apple	Seeds	Aches, antifebrile, asthma, narcotic, anodyne, and antiseptic (Shinwari & Khan, 2000; Sabeen & Ahmad, 2009)
Dolichos sinensis L.	Chinese doli kas	Seeds	Cough, menstrual cycle, and aphrodisiac (Kabirudin & Khan, 2003)
Eruca sativa Mill.	Rocket	Seeds	Dandruff (Hayat *et al.*, 2008)
Erysimum cheiri L.	Wall flower	Seeds	Spermatopoieric, glactagouge, and respiratory ailments (Kabirudin & Khan, 2003)
Euphorbia helioscopia L.	Sun spurge	Seeds	Cholera (Sabeen & Ahmad, 2009)
Foeniculum vulgare Mill	Fennel fruit	Seeds	Expectorant, stomachic, amenorrhoea, lactogogue, weak eyesight, toothache, and pneumonia (Ahmad *et al.*, 2006; Hayat *et al.*, 2008; Sabeen & Ahmad, 2009)
Gossypium herbaceum L.	Cotton	Seeds	Aphrodisiac, spermatopoetic, lactagogue, expectorant, and liniment (Kabirudin & Khan, 2003)
Helianthus annuus L.	Sunflower	Seeds	Seed decoction used to treat cough and fever (Hayat *et al.*, 2008)
Holarrhena antidysenterica Wall.	Kurchi bark	Seeds	Chronic diarrhea (Kabirudin & Khan, 2003)
Hordeum vulgare L.	Barley	Seeds	Headache, indigestion, liver complaints, antifebrile, antitussive, antidiarrheal, and jaundice (Hayat *et al.*, 2008)
Hygrophila auriculata (Schum.) Heine	Schumach	Seeds	Jaundice, hepatic obstruction, rheumatism, inflammation, urinary infections, sexual vigor, gout, and asthma (Ali *et al.*, 2003)

Continued

TABLE 10.1 Commonly Used Nuts and Seeds in Health and Diseases in Pakistan—continued

Botanical Name	English/Common Name	Nuts/Seeds	Indications/Uses/Activities
Hyoscyamus niger L.	Henbean	Seeds	Expectorant, analgesic, for burnps, toothache, cough, hemoptysis (Usmanghani *et al.*, 1997), antispasmodic, anticolic, for irritable bladder, and antitumor (Rizvi *et al.*, 2007)
Ipomia nil L.	Kaladana	Seeds	Vermicide, blood purifier, warts, eczema, and other skin diseases (Kabirudin & Khan, 2003)
Ipomoea pentaphylla L.	Greater yam, Yam	Seeds	Jaundice, intestinal pain, & worms (Qureshi *et al.*, 2009)
Juglans regia L.	Walnut	Nuts	Brain tonic, aphrodisiac, cold, and cough (Islam *et al.*, 2006)
Juniperus excelsa M.B.	Greek juniper	Fruit	Anthelmintic (Hussain *et al.*, 2007)
Lactuca scariola L.	Salad lettuce	Seeds	Aphrodisiac, diuretic, tonic, stimulant, placental retention, anthelmintic, galactogogue, cough, scabies, headache, spermatorrhoea, sedative, and jaundice (Kabirudin &Khan, 2003)
Lallemantia royleana Benth	Lallemantia	Seeds	Aphrodisiac, antiarrhythmic, mental diseases, insanity, griping, gingivitis, gleets, bronchitis, thirst, diuretic, and in urinary troubles (Rizvi *et al.*, 2007)
Lathyrus aphaca L.	Yellow-flowered pea	Seeds	Narcotic, and resolvent (Shinwari & Khan, 2000)
Linum usitatissimum L.	Common flax/linseed	Seeds	Wound healing, burns, gonorrhea, dermatitis, enema, and lumbago (Ahmad *et al.*, 2006; Hussain *et al.*, 2007)
Malva neglecta L.	Common mallow	Seeds	Nephritis, cough, bronchitis, ulceration of bladder, skin diseases, and in hemorrhoids (Shinwari & Khan, 2000)
Malva parviflora L.	Cheesewood/ Egyptian mallow	Seeds	Purgative, anthelmintic, cough, influenza, and fever (Hayat *et al.*, 2008; Qureshi *et al.*, 2009)
Mangifera indica L.	Mango	Seeds	Chronic dysentery and diarrhea (Kabirudin & Khan, 2003)
Melilotus indicus Linn.	Sweet clover	Seeds	Infantile diarrhea and bowel complaints (Shinwari & Khan, 2000)
Mirabilis jalapa L.	Four 'o'clock flower or marvel of Peru	Seeds	Astringent, leukorrhea, gonorrhea, and menstrual problems (Hayat *et al.*, 2008)
Nepeta cataria L.	Catmint	Seeds	Injuries, and backache (Ahmad et al., 2006)
Nigella sativa L.	Black seeds/ black cumin	Seeds	Expectorant, diuretic, carminative, cleansing, deobstruent in renal stones, amenorrhoea, respiratory ailments, ulcerative scabies, vitiligo, antiphlegmatic attenuant, anthelmintic, antidyspeptic, local anesthetic, hypotensive, antibacterial, anticancer, aids, and antiparasitic (Rizvi *et al.*, 2007)
Ocium sanctum L.	Basil	Seeds	Bronchitis, rheumatism, pyrexia, urogenital diseases such as gonorrhea and syphilis, otitis, cardiotonic, stomachic, and analgesic (Kabirudin & Khan, 2003)
Oxalis corniculata L.	Creeping lady's sorrel, creeping oxalis, Creeping wood sorrel	Seeds	Skin diseases, dysentery, and sensitive teeth (Qureshi *et al.*, 2009)
Papaver somniferum L.	Opium	Seeds	Headache, fatigue, lumbago, toothache, eye ache, brain weakness, cold, analgesic, hypnotic, sedative,

96

Continued

TABLE 10.1 Commonly Used Nuts and Seeds in Health and Diseases in Pakistan—continued

Botanical Name	English/ Common Name	Nuts/Seeds	Indications/Uses/Activities
Peganum harmala L.	Syrian rue	Seeds	spasmodic, antiperiodic, antipyretic, and antiabortive (Hayat *et al.*, 2008) Aphrodisiac, expectorant, asthma, gastric problems, mental diseases, paralysis, facial palsy, vaginal complaints, and toothache (Kabirudin & Khan, 2003; Hayat *et al.*, 2008)
Phyllanthus emblica L.	Emblic myrobalan	Seeds	Asthma, bronchitis, and biliousness (Kabirudin & Khan, 2003)
Phyllanthus maderaspatensis L.	Kanocha	Seeds	Carminative, laxative, tonic to the liver, diuretic, and useful in bronchitis, earache, headache, griping, ophthalmia, and ascites (Ali *et al.*, 2003)
Pinus roxburghii Sargent	Chir pine	Seeds	Stimulant, antispasmodic, astringent, diuretic, and antibacterial (Sabeen & Ahmad, 2009)
Pistacia vera L.	Pistachio	Culinary nut	Tonic for brain and heart, aphrodisiac, cough, phlegm, and kidneys (Kabirudin & Khan, 2003)
Plantago lanceolata L.	Narrow leaf plantain	Seeds	Gastrointestinal disorders such as constipation, loose motions, and diarrhea (Ahmad *et al.*, 2006)
Plantago major L.	Greater plantain	Seeds	Antidiarrheal, abdominal pain, antibleeding, phthiasis, bronchitis, and toothache (Ahmad *et al.*, 2006; Hussain *et al.*, 2007)
Plantago ovata Forssk	Spoqal	Seeds	Antidysenteric, both laxative and as an antidiarrheal, colic disorders, decreases serum cholesterol level, spermatopoetic, antitussive, and carminative (Usmanghani *et al.*, 1997; Rizvi *et al.*, 2007; Hayat *et al.*, 2008)
Podophyllm emodi Wall ex. Royle	Royle	Seeds	Liver and stomach disorders (Sabeen & Ahmad, 2009)
Portulaca oleracea L.	Purslane	Seeds	Demulcent, diuretic, and vermifuge (Islam *et al.*, 2006)
Portulaca oleraceae L.	Purslane	Seeds	Vermifuge, jaundice, typhoid, and skin allergy (Hayat *et al.*, 2008; Sabeen & Ahmad, 2009)
Prunus armeniaca L.	Apricot	Seeds	Stomachic, and health tonic (Hussain *et al.*, 2007)
Prunus dulcis L.	Almond sweet	Seeds	Brain tonic, aphrodisiac, spermatopoietic, cardiotonic, immunostimulant, expectorant, skin conditions such as psoriasis and eczema, hemorrhoids, otosclerosis, and purgative (Ahmad *et al.*, 2006)
Prunus persica L.	Peach	Seeds	Skin infections (Ahmad *et al.*, 2006; Hussain *et al.*, 2007)
Psophocarpus tetragonolobus L.	Winged bean	Seeds	Expectorant, asthma, and cough (Kabirudin & Khan, 2003)
Psoralea corylifolia L.	Babchi	Seeds	Antiparasitic, blood purifier, scabies, constipation, hemorrhoids, diuretic and diaphoretic, and skin problems such as leukoderma, leprosy, and psoriasis (Kabirudin & Khan, 2003)
Punica granatum L.	Pomegranate	Seeds	Vermicide, headache, cold, influenza, menstruation problems, and jaundice (Sabeen & Ahmad, 2009)
Punica granatum Linn.	Pomegranate	Seeds	Jaundice (Sabeen & Ahmad, 2009)
Ricinus communis L.	Caster	Seed oil	Paralysis, facial palsy, expectorants, dermatitis, eczema, purgative, anthelmintic, and mastitis (Usmanghani *et al.*, 1997; Hayat *et al.*, 2008)
Rosa indica L.	Rose	Seeds	Heart disease (Hayat *et al.*, 2008)
Rumex acetosa L.	Common sorrel	Seeds	Gastrointestinal problems, and as an antidote for scorpions' sting (Kabirudin & Khan, 2003)
Rumex vesicarius L./*Spinacea oleracea*	Spinach	Seeds	Tuberculosis, analgesic, and inflammation of urinary tract (Sabeen & Ahmad, 2009)

97

Continued

TABLE 10.1 Commonly Used Nuts and Seeds in Health and Diseases in Pakistan—continued

Botanical Name	English/ Common Name	Nuts/Seeds	Indications/Uses/Activities
Salvadora oleoides Decne.	Toothbrush tree	Seeds	Rheumatism, postparturient infections (Hayat *et al.*, 2008)
Salvia moorcroftiana Wall.	Wild sage	Seeds	Dysentery (Shinwari & Khan, 2000)
Sapindus mukorossi Gaertn.	Chinese soap berry	Seeds	Hemorrhoids (Sabeen & Ahmad, 2009)
Secale cereale L.	Rye	Seeds	Eye diseases, and in dissolving the opacity of the cornea (Shinwari & Khan, 2000)
Sida cordata (Burm. f.) Boiss.	Bhumibala	Seed	Aphrodisiac, laxative, demulcent, gonorrhoea cystitis, colic, and hemorrhoids (Shinwari & Khan, 2000)
Silene conoidea L.	Large sand catchfly	Seeds	For pimples, and backache (Ahmad *et al.*, 2006)
Silybum marianum L.	Milk thistle	Seeds	Demulcent, used in hemorrhage and liver diseases (Shinwari & Khan, 2000)
Sisymbrium irio L.	London rocket	Seeds	Expectorant, stimulant, and restorative (Shinwari & Khan, 2000), bronchitis (Ahmad *et al.*, 2006), and dropsy (Qureshi *et al.*, 2009)
Solanum miniatum Benth.	Black nightshade	Seeds	Asthma, bronchitis, and biliousness (Kabirudin & Khan, 2003)
Strychnos nux vomica L.	Nux vomica	Seeds	Paralysis, facial palsy, lumbago, stomachic, brain tonic, purgative, syphilis, decreases excessive urination in old age, and to alleviate urinary bladder weakness in children (Kabirudin & Khan, 2003)
Syzigium cumini L. Skeels	Blackberry	Seed kernels	Antidiabetic (Hayat *et al.*, 2008; Sabeen & Ahmad, 2009)
Syzygium aromaticum L.	Clove	Seeds	Brain tonic, cardiotonic, and aphrodisiac (Kabirudin & Khan, 2003)
Thymus serphyllum L.	Wild thyme	Seeds	Used as a lotion to massage a woman's back after delivery (Ahmad *et al.*, 2006)
Tinosporia cordifolia (Willd.) Miers	Heart-leaved moon seed	Seeds	Tuberculosis (Hayat *et al.*, 2008)
Tribulus terrestris L.	Calotrops seeds, puncture vine	Seeds	Kidney and bladder stones, aphrodisiac, and backache (Hayat *et al.*, 2008)
Trigonalla foenum-graceum L.	Fenugreek	Seeds	Antidiarrheal, and emmenegogue (Ahmad *et al.*, 2006)
Trigonella foenum graecum L.	Fenugreek	Seeds	Antidiabetic, hypocholesterolemic, asthma, backache, and nerve weakness (Hayat *et al.*, 2008)
Urtica dioica L.	Stinging nettle	Seed	Diuretic, astringent, and tonic (Sabeen & Ahmad, 2009)
Verbascum thapsus Linn.	Great or common mullein	Seeds	Aphrodisiac, narcotic, and used as a fish poison (Shinwari & Khan, 2000)
Vicis vinifera L.	Raisin	Seeds	Stomachic, and antidiarrheal (Kabirudin & Khan, 2003)
Viola serpens Wall. Roxb		Seed	Purgative, and diuretic (Shinwari & Khan, 2000; Sabeen & Ahmad, 2009)
Vitex negundo L.	Five-leaved chaste tree	Seeds	Catarrah, allergy, tympany, galactogogue, flatulence, and cholera (Hayat *et al.*, 2008)
Withania coagulens DUNAL.	Vegetable rennet	Seeds	Aphrodisiac, stomachic, and tympany (Kabirudin & Khan, 2003)
Zanthoxylum armatum D.C.	Japanese pepper	Seed	Carminative, and stomachic (Sabeen & Ahmad, 2009)
Zyzyphus jujuba Mill.	Jujube fruit	Fruit/seeds	Influenza, cough, allergy, wounds, scabies, and whelm (Kabirudin & Khan, 2003)

about 30,000 plants (Kabirudin & Mahmood, 1937). Pakistan is blessed with a wide diversity of plants that can be grown in one or other seasons of the year, and have been cultivated since ancient civilizations. History reveals that Arabs brought a revolution in food items, leading to the inclusion of nuts and raisins in different dishes native to India. Valuable work on the pharmacology and phytochemistry of commonly used medicinal plants has been carried out in Pakistan (Ikram & Hussain, 1978; Usmanghani *et al.*, 1997).

Most of the nuts and seeds used in health and disease in Pakistan are cultivated in the country, although some are partially imported from different regions — for example, almonds from Iran. Some plants grow wild, such as *Tribulus terrestris* L. (*Bhakhda*), and grasses along the banks of canals, rivers and in deserts. A number of plants are obtained as byproducts of different crops — for example, *Hordeum vulgare* L. (*Jo*) and *Cichorium intybus* L. (*Kasni*) from wheat crops. Nuts are seeds that are covered with a hard shell, and both nuts and seeds are a vital source of nutrients for human beings. Some fruit and vegetable seeds (melon, water melon, pumpkin, etc.) are stored by local people and used for the maintenance of health. Coconut, sunflower, safflower, rapeseed, corn, or sesame seed oils are used to treat skin problems, and are most often used as carrier oils.

THE PRESENT-DAY USE OF NUTS AND SEEDS FOR HEALTH PROMOTION AND DISEASE PREVENTION IN PAKISTAN

In Pakistan, the majority of the rural population is catered for by *Unani* herbal practitioners, who prescribe single-herb or compound formulations to alleviate disease or health problems. About three-fourths of the population depend upon folklore and the tradition system of medicine, due to both the high costs and the side effects of allopathic medicines, which have encouraged the manufacture of Greco-Arab and Ayurvedic medicines (Shinwari & Khan, 2000). Most of the people in rural areas seek the advice of their grandparents or of traditional healers for frequently occurring problems such as influenza, constipation, diarrhea, aches and pains, and skin diseases before going to allopathic doctors, and are successfully cured by the local products of plants. Different medicinal plants and their products can be purchased from local herbal drug stores.

SUMMARY POINTS

- About 75% of the Pakistani population are dependent on alternative medicine based on the nuts and seeds of different plants.
- Nuts and seeds have been used for the promotion of health and prevention of disease for centuries, and there is an inventory of hundreds of plants in local languages.
- Most of the nuts and seeds used for health are indigenous to Pakistan and are available throughout the year, as the climate of one or other regions of the country supports their cultivation.
- Nuts and culinary nuts are added to different sweet dishes, and their oils also have medical significance.
- Local people are well aware of the beneficial and adverse effects (if any) of nuts and seeds.

References

Ahmad, S., Ali, A., Beg, H., Dasti, A. A., & Shinwari, Z. K. (2006). Ethnobotanical studies on some medicinal plants of Booni Valley, district Chitral Pakistan. *Pakistan Journal of Weed Science Research, 12*, 183—190.

Ali, S., Qazi, A. H., & Khan, M. R. (2003). Protease activity in seeds commonly used as herbal medicine. *Pakistan Journal of Medical Research, 42*, 2.

Anonymous. (1987). *Standardization of single drugs of Unani medicine. CCRUM, Part 1, Publication No. 25.* New Delhi, India: Central Council for Research in Unani Medicine. pp. 252—257.

Hayat, M. Q., Khan, M. A., Ahmad, M., Shaheen, N., Yasmin, G., & Akhter, S. (2008). Ethnotaxonomical approach in the identification of useful medicinal flora of Tehsil Pindigheb (District Attock) Pakistan. *Ethnobotany Research and Applications, 6*, 35—62.

Hussain, F., Shah, S. M., & Sher, H. (2007). Traditional resource evaluation of some plants of Mastuj, district Chitral. *Pakistan. Pakistan J Bot, 39*, 339–354.

Ikram, M., & Hussain, S. F. (1978). *Compendium of medicinal plants. National products and fine chemical division.* Peshawar, Pakistan: Pakistan Council of Scientific and Industrial Research (PCSIR) Laboratories. p. 167.

Islam, M., Ahmad, H., Rashid, A., Razzaq, A., Akhtar, N., & Khan, I. (2006). Weeds and medicinal plants of Shawar valley, District Swat. *Pakistan Journal of Weed Science Research, 12*, 83–88.

Kabirudin, M. H., & Khan, H. M. A. (2003). *Kitab-ul-Mufridat.* Lahore, Pakistan: Maktaba Danyal. [in Urdu].

Kabirudin, H., & Mahmood, H. (1937). *Lughat-e-Kabeer [in Urdu], Vol. II, Lughat al Adviya* (3rd ed.). Delhi, India: Daftar al-Maseeh. p. 473.

Marwat, S. K., Khan, M. A., Ahmad, M., Zafar, M., Fazal-ur-Rehman, & Sultana, S. (2009). Fruit plant species mentioned in the Holy Qura'n and Ahadith and their ethnomedicinal importance. *American-Eurasian Journal of Agricultural and Environmental Science, 5*, 284–295.

Qureshi, R., Waheed, A., Arshad, M., & Umbreen, T. (2009). Medico-ethnobotanical inventory of Tehsil Chakwal, Pakistan. *Pakistan Journal of Botany, 41*, 529–538.

Rizvi, M. A., Saeed, A., & Zubairy, H. N. (2007). *Medicinal plants: History, cultivation and uses.* Karachi, Pakistan: Hamdard Institute of Advanced Studies and Research.

Sabeen, M., & Ahmad, S. S. (2009). Exploring the folk medicinal flora of Abbotabad City, Pakistan. *Ethnobotanical Leaflets, 13*, 810–833.

Shinwari, M. I., & Khan, M. A. (2000). Folk use of medicinal herbs in Marghalla Hills National Park Islamabad. *Journal of Ethnopharmacology, 69*, 45–46.

Usmanghani, K., Saeed, A., & Alam, M. T. (1997). *Indusyunic medicine, traditional medicine of herbal, animal and mineral origin in Pakistan.* Karachi, Pakistan: University of Karachi BBC & T Press. pp. 314–315.

CHAPTER 11

Seed Components in Cancer Prevention

Blanca Hernández-Ledesma[1,2], **Chia-Chien Hsieh**[1], **Ben O. de Lumen**[1]
[1] Department of Nutritional Sciences and Toxicology, University of California at Berkeley, Berkeley, California, USA
[2] Food Research Institute (CIAL, CSIC-UAM), Nicolás Cabrera, 9, Campus de la Universidad, Autónoma de Madrid, 28049 Madrid, Spain

101

LIST OF ABBREVIATIONS

AICR, American Institute for Cancer Research
BBI, Bowman-Birk protease inhibitor
CLA, conjugated linolenic acid
HIV-1, human immunodeficiency virus-1
RAR, retinoic acid receptor
Rb, retinoblastoma protein
RXR, retinoid X receptor

INTRODUCTION

Cancer has been defined by the American Cancer Society (ACS) as "a group of diseases characterized by uncontrolled growth and spread of cells." The three critical phases in this process are initiation, promotion, and progression involving sequential generations of cells

Nuts & Seeds in Health and Disease Prevention. DOI: 10.1016/B978-0-12-375688-6.10011-8

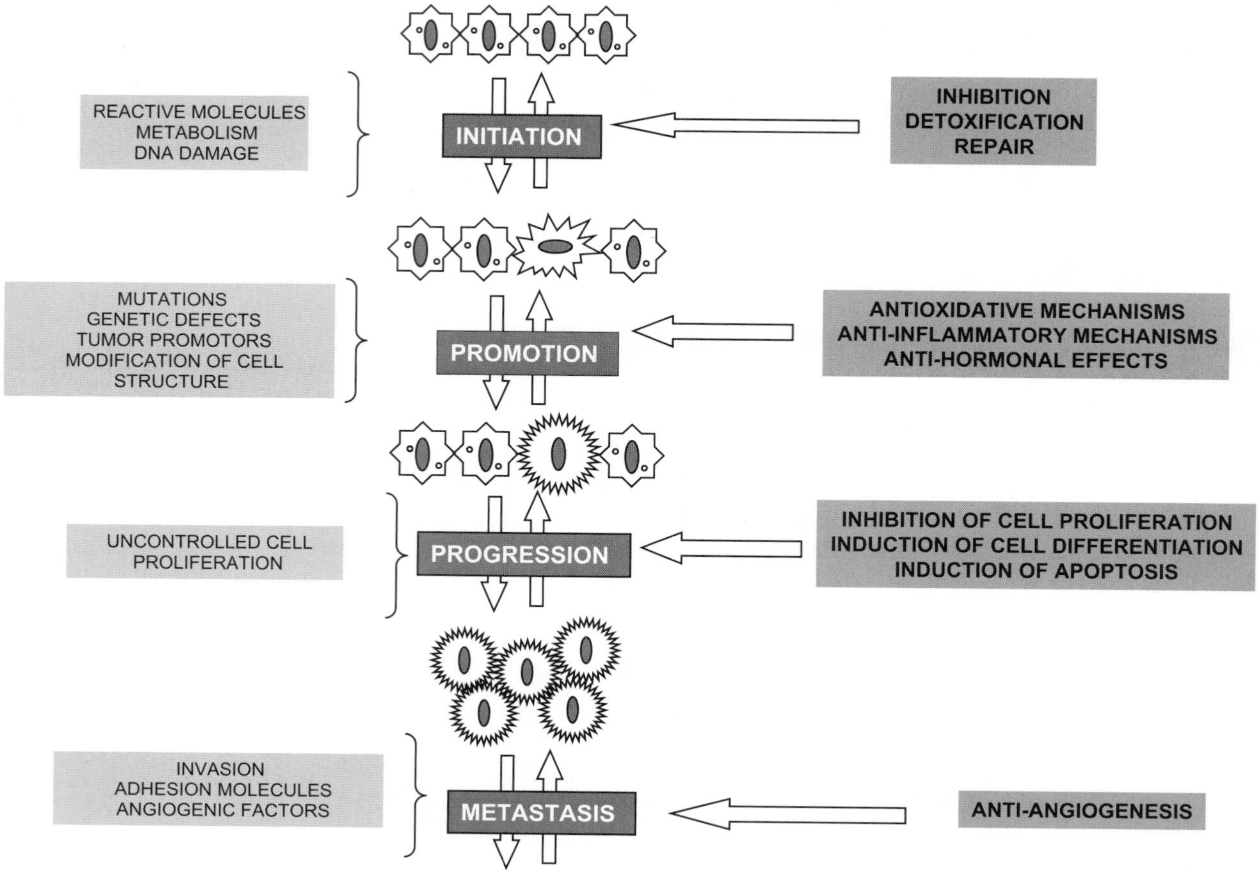

FIGURE 11.1

Initiation, promotion, and progression steps of carcinogenesis process. The repairing mechanisms of chemopreventive phytochemicals. Phytochemicals act on the initiation, promotion, and progression phases of carcinogenesis through different mechanisms of action.

that exhibit continuous disturbance of cellular and molecular signal cascades (Vincent & Gatenby, 2008) (Figure 11.1).

It has been estimated that 30–40% of all cancers can be prevented by appropriate diet, physical activity, and maintenance of appropriate weight (Rigas *et al.*, 2008). Epidemiological evidence, cell culture, and animal tumor model studies have demonstrated that a large number of natural compounds present in the diet could lower cancer risk and even sensitize tumor cells in anticancer therapies (de Kok *et al.*, 2008). Figure 11.2 shows the main molecular targets of bioactive phytochemicals.

SEEDS: BOTANICAL DESCRIPTION

A seed is a small embryonic plant enclosed in a covering called the seed coat, usually with some stored foods. Seeds serve several functions for the plant. Keys among these functions are nourishment of the embryo, dispersal to a new location, and dormancy during unfavorable conditions. Seeds are classified into beans, also called legumes or pulses; cereals (including pseudocereals); and some nuts. "Bean" is the common name for a large plant seed of several genera of the family *Fabaceae* used for human or animal food. The main edible seeds are summarized in Table 11.1. Pulse grains contain a large number of bioactive compounds which have a metabolic benefit when consumed on a regular basis (Rochfort & Panozzo, 2007).

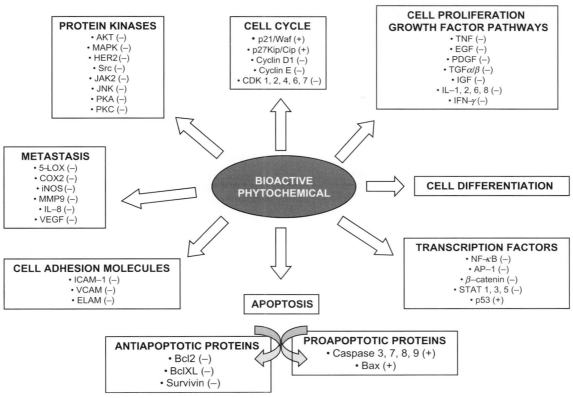

FIGURE 11.2
Molecular targets of bioactive phytochemicals. Phytochemicals act on the cell cycle, cell proliferation and differentiation, apoptosis, metastasis, protein kinases, cell adhesion molecules, transcription factors, and growth factor pathways.

Cereals, grains, or cereal grains are the edible seeds or grains of the grass family *Graminae*. The main cereals are summarized in Table 11.2. Based on epidemiological studies and biologically plausible mechanisms, scientific evidence shows that regular consumption of whole-grain foods may help mitigate oxidative-stress related disease conditions, cardiovascular disorders, and certain cancers (Aisbitt *et al.*, 2008). "Nut" is a general term for a large, dry, oily seed or fruit of some plants. Nuts are produced by plants from the families *Fagaceae* and *Betulaceae*. The main seed nuts are summarized in Table 11.3.

This chapter summarizes the main components contained in seeds for which cancer-preventive properties have been reported.

SEED COMPONENTS AND THEIR ROLE IN CANCER PREVENTION
Phenolic compounds

Phenolic compounds or polyphenols constitute one of the most numerous and widely distributed groups of substances in the plant kingdom. They are formed to protect the plant from photosynthetic stress, reactive oxygen species, wounds, and herbivores. In legumes and cereals, the main polyphenols are flavonoids, phenolic acids, and tannins.

FLAVONOIDS

Flavonoids are important in contributing flavor, color, and taste, and are also central to the normal growth, development, and defense of a plant. Several *in vitro* and *in vivo* studies have demonstrated the effect of flavonoids in suppressing carcinogenesis through different mechanisms of action (Garcia-Lafuente *et al.*, 2009).

TABLE 11.1 Main Edible Legumes (Latin and Common Names), Origin, and Main Uses

Latin Name	Common Name(s)	Origin	Uses
Arachis hypogaea	Peanut or groundnut	South America, Mexico, Central America	Culinary, medicinal
Cajanus cajan	Pigeon pea	Asia	Culinary, medicinal
Cicer arietinum	Chick pea, garbanzo bean, Indian pea	Middle East	Culinary
Gycine max	Soybean	East Asia	Culinary, medicinal
Lablab vulgaris	Hyacinth bean	Africa, India, Indonesia	Culinary, medicinal
Lens culinaris	Lentil, daal, dal	Near East	Culinary
Lupinus albus	White lupin	Mediterranean area	Culinary
Moringa oleifera	Moringa	Middle East, Africa	Culinary, medicinal
Mucuna pruriens	Velvet bean	Tropical Africa, India, the Caribbean	Culinary, medicinal
Pachyrhizum	Yam bean	South America	Culinary
Phaseolus vulgaris	Common bean	Mesoamerica, the Andes	Culinary
Pisum sativum	Pea	Mediterranean area, Near East	Culinary, bioplastics
Psophocarpus tetragonolobus	Winged bean	Papua New Guinea	Culinary
Sterculia	Tropical chestnuts	India	Culinary
Vicia faba	Broad bean, fava bean	North Africa, Southwest Asia	Culinary, medicinal, religious
Vigna subterranean	Bambara groundnut	West Africa	Culinary
Vigna unguiculata	Cowpea	Semi-arid tropics of Asia, Africa, Southern Europe, Central and South America	Culinary

TABLE 11.2 Main Cereals (Latin and Common Names), Origin, and Main Uses

Latin Name	Common Name(s)	Origin	Uses
Zea mays L. ssp. *mays*	Maize, corn	Mesoamerica	Culinary, medicinal, biofuel, ornamental
Oryza sativa	Rice	East, South, Southeast Asia, Middle East, Latin America, West Indies	Culinary, biofuel
Triticum spp.	Wheat	Near East	Culinary, biofuel
Hordeum vulgare	Barley	Near East	Culinary, animal feed, measurement
Sorghum spp.	Sorghum	Africa, Central America, South Asia	Culinary
Avena sativa	Oat	Near East	Culinary, animal feed, cosmetics
Secale cereale	Rye	Central and Eastern Turkey	Culinary, medicinal
Triticosecale spp.	Triticale	Laboratory hybrid of wheat and rye	Culinary
Digitaria exilis	Fonio	West Africa, India	Culinary
Chenopodium quinoa	Quinoa	Andean region of South America	Culinary
Eragrostis tef	Teff	North Ethiopia	Culinary
Zizania spp.	Wild rice	North America, Asia	Culinary, religious, ornamental
Amaranthus spp.	Amaranth, pigweed	Aztec nation, Mexico	Culinary, seed oil production

TABLE 11.3 Main Seed Nuts (Latin and Common Names), Origin, and Main Uses

Latin Name	Common Name(s)	Origin	Uses
Prunus dulcis	Almond	Middle East	Culinary, medicinal, oil production, religious
Juglans spp.	Walnut	Worldwide	Culinary, oil production, wood, medicinal, cosmetics
Bertholletia excelsa	Brazil nut	South America	Culinary, cosmetics, lubricant
Aleurites moluccana	Candle nut, candleberry, Indian walnut	Tropical areas of New and Old Worlds	Culinary, oil production, medicinal, religious
Pinus spp.	Pine nut	Europe, Asia, North America	Culinary, medicinal, oil production
Pistacia vera	Pistachio	Greece, Syria, Iran, Turkey, Afghanistan	Culinary

Isoflavone genistein and soybean

Soybean (*Glycine max*) is an ancient legume consumed worldwide, but most commonly in Asian countries. A number of epidemiological studies have demonstrated an association between the consumption of soybean and improved health, particularly as a reduced risk for cardiovascular diseases and cancer. Genistein has been identified as the predominant isoflavone contained in soybean. Extensive experiments have concluded that genistein functions as a promising carcinogenesis inhibitor through different molecular mechanisms of action (Banerjee *et al.*, 2008). Genistein is structurally similar to 17β-estradiol. It interacts with estrogen receptors, blocking the binding of more potent estrogens and affecting estrogen metabolism, thereby exerting a potential favorable role in the prevention of hormone-related cancers such as breast and prostate cancer. Moreover, genistein possesses free-radical scavenging activity, inhibiting the expression of stress-response related genes. It also inhibits the growth of several cancer cell lines through the modulation of genes intimately related to the regulation of cell cycle and apoptosis. Genistein intervenes too in several cellular transduction signaling pathways, and is involved in the regulation of gene activity by modulating epigenetic events such as DNA methylation and/or histone acetylation, either directly or through the estrogen-receptor dependent process (Banerjee *et al.* 2008). Genistein, ingested through soybean or other food sources, exerts anticarcinogenic effects through pleiotropic molecular mechanisms of action on the cell cycle, apoptosis, angiogenesis, invasion, and metastasis.

Proanthocyanidins and grape seeds

Vitis vinifera, known as the grapevine, is indigenous to southern Europe and western Asia, and is today cultivated worldwide. Grape seeds are byproducts of grapes separated during the industrial production of grape juice and wine. They are a potent source of proanthocyanidins, which are also found in almonds, cashews, hazelnuts, pecans, pistachios, peanuts, and walnuts. The *in vitro* and *in vivo* experimental data support the concept that proanthocyanidins exert chemoprotective properties against free radicals and oxidative stress, anti-inflammatory activity, and anticarcinogenic properties. Potential cancer chemopreventive activities include cell proliferation inhibition, apoptosis induction, and cell-cycle arrest in tumor cells. They also modulate the expression and activity of NF-κB and its targeted genes, including the invasion and metastasis-specific molecular targets (Nandakumar *et al.*, 2008).

PHENOLIC ACIDS

Phenolic acids are present in many fruits and vegetables, and they are especially abundant in the coffee bean. The phenolic acids exert a direct antiproliferative action in human breast cancer cells and oral cancer cells at low concentrations, comparable with those found in biological fluids after the ingestion of foods rich in phenolic acids (Kampa *et al.*, 2004).

STILBENES: RESVERATROL

Resveratrol is the most abundant member of the stilbenes family in grapes and wines. It exerts anticarcinogenic properties through different mechanisms, including antioxidant and anti-inflammatory activity, apoptosis inducing properties, cell-cycle arrest, suppression of NF-κB, AP-1 activation, inhibition of protein kinases, and suppression of angiogenesis (Goswami & Das, 2009).

TANNINS

Tannins are mainly present in grains, such as sorghum, barley, dry beans, faba beans, and other legumes. Many biological effects have been attributed to tannins, such as antimicrobial, antimutagenic and anticarcinogenic effects, which may be related to their antioxidative properties in protecting cellular components from oxidative damages, including DNA single-strand breakage and lipid peroxidation (Chung et al., 1998). They have also been reported to exert other effects, such as reducing blood pressure, immunological effects, and protection against caries.

LIGNANS

Lignans are typically isolated from the woody portions of plants, the seed coat of seeds, and the bran layer in grains. Lignans possess cancer-preventive properties through a variety of mechanisms (Webb & McCullogh, 2005). Due to its structural similarity to 17β-estradiol, lignans may influence cancer via estrogen-mediated pathways through direct or indirect interactions with estrogen receptors. Additionally, lignans may act through the mediation of other growth hormones, such as insulin-like growth factor and vascular endothelial growth factor. Lignans have also shown free-radical scavenging activity, inhibition of oxidation, and no pro-oxidant activity. Moreover, lignans modulate apoptotic pathways, which may contribute to their antitumorigenic and antimetastasic properties.

Terpenoids

Terpenoids, also referred as terpenes, constitute the largest group of natural compounds playing a variety of roles in many different plants, serving a wide range of important physiological functions (Rabi & Bishayee, 2009).

DITERPENES: RETINOIDS

Retinoids exert chemopreventive properties in a variety of rodent mammary gland, prostate, bladder, skin, and liver tumor models. Experimental evidence suggests that these compounds act by modulating multiple signal transduction pathways, resulting in direct and indirect effects on gene expression. Binding of retinoids to the nuclear receptors, namely retinoic acid receptor (RAR) and retinoid X receptor (RXR), which are ligand-activated transcription factors, leads to regulation of several cellular processes, including growth, differentiation, and apoptosis. Growth inhibition has been associated with induction of the expression of RAR, which may act as a tumor suppressor and appears to inhibit the proliferation of cancer cells (Rabi & Bishayee, 2009).

TRITERPENOID: LIMONIN

Limonin has been identified in citrus seeds. It has been found to inhibit fore-stomach neoplasia in a mouse model, and to inhibit p-24 antigen activity, protease activity, and replication of HIV-I. Limonoids also inhibit colon cancer and ovarian cancer in animal models; human neuroblastoma cells in in vitro studies; growth in estrogen receptor-negative and receptor-positive human breast cancer cells; and the development of oral tumors in a hamster cheek pouch model (Patil et al., 2009).

TETRATERPENOID: CAROTENOIDS

Carotenoids include α-carotene, β-carotene, lycopene, lutein, astaxanthin, cryptoxanthin, and zeaxanthin. Lycopene is the most efficient quencher of singlet oxygen species, whereas lutein and zeaxanthin are scavengers of radical oxygen species. In addition to antioxidant properties, lycopene has demonstrated an array of biological effects, including cardioprotective, anti-inflammatory, antimutagenic, and anticarcinogenic activities. Many studies have shown that lycopene could be promising as a chemopreventive agent for mammary and other types of tumors. Lycopene is known to induce apoptosis and cell-cycle arrest, and to inhibit cell invasion and metastasis. Other possible cancer-preventive mechanisms of action of this compound include the modulation of redox signaling, prevention of oxidative DNA damage, modulation of cell-cycle regulatory protein expression, upregulation of tumor-suppressor proteins, and modulation of carcinogen-metabolizing enzymes (Patil *et al.*, 2009). Lutein has been demonstrated to exhibit chemopreventive activity in animal breast-cancer models, also enhancing cell-mediated immune responses, and, consequently, resistance to tumor formation (Rabi & Bishayee, 2009).

Seed oil

Many studies using *in vivo* and *in vitro* models have shown that conjugated linolenic acid (CLA) suppresses the development of multistage carcinogenesis at different sites. Seed oils from pomegranate and bitter melon contain rich 9*cis*, 11*trans*, 13*cis*-CLA, and catalpa contains a high amount of 9*trans*, 11*trans*, 13*cis*-CLA, which inhibited the occurrence of colonic aberrant crypt foci induced by azoxymethane in F344 rats; the inhibition is associated in part with increased PPARγ protein and inhibited COX-2 mRNA in the colonic mucosa (Suzuki *et al.*, 2006). In addition, flax-seed oil and meal are good sources of omega-3 fatty acids, which have been shown to reduce the incidence of crypt foci in colon-cancer animal models.

Vitamin E is primarily available in plant-oil rich foods such as vegetable oils, seeds, and nuts. Vitamin E may help to prevent cancer by decreasing the formation of mutagens arising from the oxidation of fecal lipid, and by molecular mechanisms that influence cell death, cell cycle, and transcriptional events (Kline *et al.*, 2007).

Bioactive proteins and peptides

The seeds of legumes and cereals produce a variety of biologically active proteins that play a role in defense and protection against pathogenic micro-organisms and predatory insects. In addition, they may exhibit potentially exploitable activities, including antiproliferative, anti-tumor, and immunomodulatory activities (Ho & Ng, 2008). Also, bioactive peptides like lunasin have been found to exert chemopreventive properties.

BOWMAN-BIRK PROTEASE INHIBITOR

Soybean Bowman-Birk protease inhibitor (BBI) is a 71 amino-acid polypeptide with demonstrated cancer chemopreventive activity in both *in vitro* and *in vivo* assays (Losso, 2008). BBI can keep free radicals from being produced in cells, and thereby decrease the amount of oxidative damage, which is also related to the potent anti-inflammatory activity of this protease inhibitor that may be its major cancer-preventive mechanism of action. Other mechanisms are the direct inhibition of the major proteases involved in inflammation processes, and suppression of carcinogenesis by affecting the amount of certain types of proteolytic activities or the expression of certain proto-oncogenes, both of which are thought to play important roles in carcinogenesis.

LUNASIN

Lunasin is a 43 amino-acid peptide identified in soybean, wheat, barley, and other seeds. It has been found to have chemopreventive properties by *in vitro* and *in vivo* assays. Lunasin has been demonstrated to inhibit post-translational modifications of histones, such as acetylation, which neutralizes the positive charge of these histones and disrupts the electrostatic interactions between them and DNA. This promotes chromatin unfolding, which has been associated with transcription and gene expression. Moreover, lunasin has demonstrated its capacity to inhibit retinoblastoma protein (Rb) phosphorylation induced by cyclin D1 affecting the cell cycle (Hernández-Ledesma et al., 2009). Recently, new mechanisms of action of this peptide have been identified, such as arrest of cell cycle, induction of apoptosis, and modification of signaling transduction genes expression. Moreover, antioxidant and anti-inflammatory activities of lunasin can contribute to its chemopreventive properties.

SUMMARY POINTS

- Chemoprevention is a new and promising strategy to prevent cancer by using natural and/ or synthetic substances that block, reverse, or retard the process of carcinogenesis.
- An ideal chemopreventive agent should: (1) have little or no toxicity (2) have high efficacy in multiple sites (3) have the capacity for oral consumption, (4) have known mechanisms of action, (5) be of low cost, and (6) be acceptable to humans.
- Seeds are classified as beans, cereals, nuts, and other seeds. They contain a large number of bioactive phytochemicals with potent cancer-preventive activity.
- Different mechanisms of action have been reported to be responsible for the anticancer properties of these bioactive compounds. They are important due to their antioxidant and anti-inflammatory activities. Moreover, many of them exert their chemopreventive properties by interfering with multiple cell-signaling pathways, cell proliferation, and apoptosis.

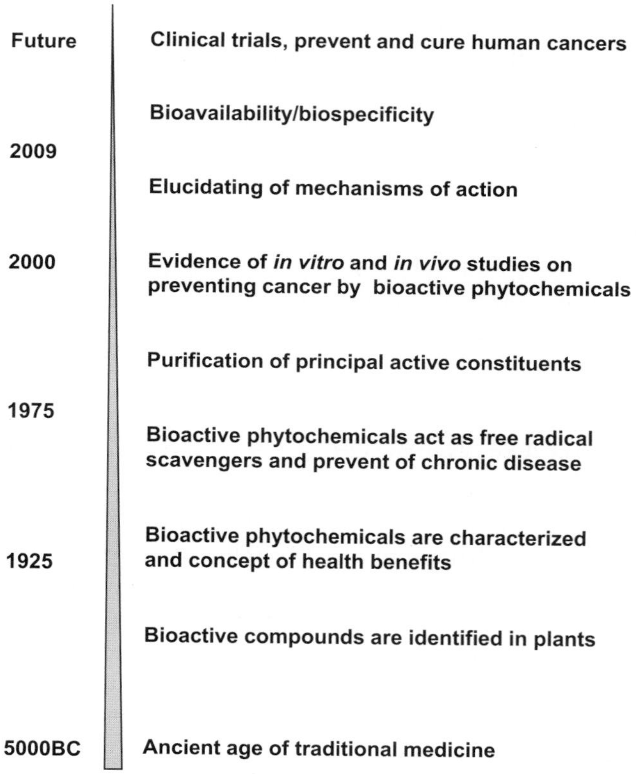

Future	Clinical trials, prevent and cure human cancers
	Bioavailability/biospecificity
2009	
	Elucidating of mechanisms of action
2000	Evidence of *in vitro* and *in vivo* studies on preventing cancer by bioactive phytochemicals
	Purification of principal active constituents
1975	
	Bioactive phytochemicals act as free radical scavengers and prevent of chronic disease
1925	Bioactive phytochemicals are characterized and concept of health benefits
	Bioactive compounds are identified in plants
5000BC	Ancient age of traditional medicine

FIGURE 11.3

The journey of knowledge on bioactive phytochemicals, from ancient ages to current research. Studies of the effect of phytochemicals on chronic diseases began thousands of years ago. The most important advances are shown and expected during the 20th and 21st centuries.

- Understanding of the role of bioactive phytochemicals in cancer prevention is currently emerging (Figure 11.3).
- To successfully convert a potent food bioactive to a clinically viable drug will require studies that should investigate: (1) the *in vivo* pharmacokinetics of food bioactives in human body, (2) the anticarcinogenic activity in each target organ and tissue, (3) the biological mechanisms of action involved in the anticancer properties, and (4) the application of these bioactives in medicine and industry.

ACKNOWLEDGEMENTS

The authors would like to acknowledge the American Institute for Cancer Research (AICR) for research funding, and the European Commission and the Spanish National Research Council for the Marie Curie post-doctoral fellowship of Blanca Hernández-Ledesma.

References

Aisbitt, B., Caswell, H., & Lunn, J. (2008). Cereals — current and emerging nutritional issues. *Nutrition Bulletin, 33,* 169–185.

Banerjee, S., Li, Y., Wang, Z., & Sarkar, F. H. (2008). Muti-targeted therapy of cancer by genistein. *Cancer Letters, 269,* 226–242.

Chung, K.-T., Wong, T.-Y., Wei, C.-I., Huang, Y.-W., & Lin, Y. (1998). Tannins and human health: a review. *Critical Reviews in Food Science and Nutrition, 38,* 421–464.

de Kok, T. M., van Breda, S. G., & Manson, M. M. (2008). Mechanisms of combined action of different chemopreventive dietary compounds. *European Journal of Nutrition, 47,* 51–59.

García-Lafuente, A., Guillamon, E., Villares, A., Rostagno, M. A., & Martínez, J. A. (2009). Flavonoids as anti-inflammatory agents: implications in cancer and cardiovascular disease. *Inflammation Research, 58,* 537–552.

Goswami, S. K., & Das, D. K. (2009). Resveratrol and chemoprevention. *Cancer Letters, 284,* 1–6.

Hernández-Ledesma, B., Hsieh, C.-C., & de Lumen, B. O. (2009). Lunasin: a novel seed peptide for cancer prevention. *Peptides, 30,* 426–430.

Ho, V. S. M., & Ng, T. B. (2008). A Bowman-Birk trypsin inhibitor with antiproliferative activity from Hokkaido large black soybeans. *Journal of Peptide Science, 14,* 278–282.

Kampa, M., Alexaki, V.-I., Notas, G., Nifli, A.-P., Nistikaki, A., Hatzoglou, A., et al. (2004). Antiproliferative and apoptotic effects of selective phenolic acids on T47D human breast cancer cells: potential mechanisms of action. *Breast Cancer Research, 6,* R63–R74.

Kline, K., Lawson, K. A., Yu, W., & Sanders, B. G. (2007). Vitamin E and cancer. *Vitamins and Hormones, 76,* 435–461.

Losso, J. N. (2008). The biochemical and functional food properties of the Bowman-Birk Inhibitor. *Critical Reviews in Food Science and Nutrition, 48,* 94–118.

Nandakumar, V., Singh, T., & Katiyar, S. K. (2008). Multi-targeted prevention and therapy of cancer by proantho-cyanidins. *Cancer Letters, 269,* 378–387.

Patil, B. S., Jayaprakasha, G. K., Chidambara Murthy, K. N., & Vikran, A. (2009). Bioactive compounds: historical perspectives, opportunities, and challenges. *Journal of Agricultural and Food Chemistry, 57,* 8142–8160.

Rabi, T., & Bishayee, A. (2009). Terpenoids and breast cancer chemoprevention. *Breast Cancer Research and Treatment, 115,* 223–239.

Rigas, B., Rao, C. V., Cooney, R., & Singh, S. (2008). Food components, alternative medicine, and cancer: progress and promise. *Nutr. Cancer, 60*(S1), 1.

Rochfort, S., & Panozzo, J. (2007). Phytochemicals for health, the role of pulses. *J. Agric. Food Chem., 55,* 7981–7994.

Suzuki, R., Yasui, Y., Kohno, H., Miyamoto, S., Hosokawa, M., Miyashita, K., et al. (2006). Catalpa seed oil rich in 9t, 11t, 13c-conjugated linolenic acid suppresses the development of colonic aberrant crypt foci induced by azoxymethane in rats. *Oncology Reports, 16,* 989–996.

Vincent, T. L., & Gatenby, R. A. (2008). An evolutionary model for initiation, promotion, and progression in carcinogenesis. *International Journal of Oncology, 32,* 729–737.

Webb, A. L., & McCullough, M. L. (2005). Dietary lignans: potential role in cancer prevention. *Nutr. Cancer, 51,* 117–131.

CHAPTER 12

Nuts, Seeds, and Oral Health

Cecile Badet
Laboratoire Odontologique de Recherche, UFR d'Odontologie, Université Victor Segalen, Bordeaux Cedex, France

111

LIST OF ABBREVIATIONS

CBH, cacao bean husk
GSE, grape seed extract
MIC, minimum inhibitory concentration

INTRODUCTION

A major factor of the oral ecosystem is dental plaque, which develops naturally on the hard and soft tissues of the mouth. Pathologies such as dental caries or periodontitis may arise when an imbalance occurs among the indigenous bacteria. Some bacterial species are strongly linked to oral diseases. *Streptococcus mutans* and *Streptococcus sobrinus* play a major role in the development of dental caries, but other species, such as *Lactobacillus sp.* and *Actinomyces sp.*, are also implicated. *Porphyromonas gingivalis* and *Fusobacterium nucleatum* are closely associated with

Nuts & Seeds in Health and Disease Prevention. DOI: 10.1016/B978-0-12-375688-6.10012-X

various forms of periodontal diseases. While tooth-brushing is an effective method to remove plaque mechanically, it is not always sufficient, and chemical antiplaque agents could be helpful. At present, the search continues for active ingredients that could prevent dental plaque formation without affecting the biological equilibrium within the oral cavity. Natural products including nut and seed extracts have already shown various biological effects that could be useful in the preservation of oral health.

NUTMEG
Description and general usage

Myristica fragrans Houtt is commonly named nutmeg or mace. It is an aromatic evergreen tree with spreading branches and a yellow fleshy fruit similar in appearance to an apricot or peach.

Nutmeg was probably imported into Europe during the 12th century, by Arab merchants. It has been used in folk medicine for multiple purposes (antithrombotic, antitumor, and anti-inflammatory) (Chung *et al.*, 2006).

Applications to oral health promotion and disease prevention

Nutmeg contains 25—30% fixed oils and 5—15% volatile oils (camphrene, eugenol, etc.), and also other molecules such as myristic acid, myristicin, and lignin compounds. Among these molecules, eugenol is widely used in dentistry, and is very effective in antibacterial activity against oral bacteria. However, only a few studies have been conducted using nutmeg extracts against oral bacteria.

Chung and colleagues identified macelignan as an active compound. It showed antibacterial effect against *Streptococcus mutans*, *Streptococcus sanguis*, *Streptococcus salivarius*, and *Lactobacillus casei*. On the contrary, it showed only weak activity against *Actinomyces viscosus*, *Porphyromonas gingivalis*, and *Staphylococcus aureus* (Chung *et al.*, 2006).

CACAO (*THEOBROMA CACAO*)
Description and general usage

The cacao tree is a handsome evergreen tree with pale-colored wood. The bark is brown, and the leaves are lanceolate, bright green, and full. The flowers are small and reddish. The fruits are yellowy-red and smooth. The seeds are fleshy and mucilaginous, and have either white or purple cotyledons.

Theobromine, the alkaloid contained in the beans, resembles caffeine in its action, but its effect on the central nervous system is less powerful. Its action on the muscles, kidneys, and heart is more pronounced. It is used principally for its diuretic effect due to stimulation of the renal epithelium. It is also employed for high blood pressure, as it dilates the blood-vessels.

Applications to oral health promotion and disease prevention

Recent studies have focused on the biological activity of various extracts of cacao, such as from cacao bean husk, which is known to contain high concentrations of polyphenols.

Cacao bean husk extracts (CBH) are prepared from ground cacao bean husks by treatment with cellulose, and then extraction with 30% ethanol. These extracts have proven effective at reducing the growth rates and the cariogenicity of *Streptococcus mutans* and *Streptococcus sobrinus*. They also reduced the rate of acid production by these species, and inhibited *in vitro* biofilm formation (Figure 12.1, Table 12.1) (Ooshima, 2004; Matsumoto *et al.*, 2004; Percival *et al.*, 2006).

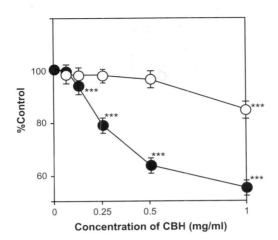

FIGURE 12.1

CBH inhibits insoluble glucan formation by either cell-associated glucosyltransferase of *S. mutans* or crude glucosyltransferase of *S. sobrinus*. Inhibitory effect of CBH extract on insoluble glucan synthesis by crude glucosyltransferases of mutans streptococci: statistically significant difference between control (no CBH) and CBH (0.06—1 mg/ml) (***$P < 0.001$). Symbols: —○— *Strep. mutans* MT8148R; —●— *Strep. sobrinus* 6715. *Reprinted from Ooshima et al. (2000), Arch. Oral Biol., 45, 639—645, with permission.*

TABLE 12.1 Viable Counts of *Streptococcus sanguinis* and *Streptococcus mutans* in Biofilms, 4 and 24 h After Surface Pretreatment with 35-μM Cocoa Polyphenol Pentamer

Bacterium	Exposure for 15 min			Exposure for 60 min		
	Log₁₀ CFU/ml ($n = 6$)		Percentage Increase (+)/ Decrease (−)	Log₁₀ CFU/ml ($n = 6$)		Percentage Increase (+)/ Decrease (−)
	Control	Treated		Control	Treated	
S. sanguinis (− sucrose)	4.20 ± 0.25	4.23 ± 0.15	+6	5.12 ± 0.21	5.01 ± 0.34	−23
S. sanguinis (+ sucrose)	3.85 ± 0.16	3.97 ± 0.37	+46	3.14 ± 0.18	3.00 ± 0.32	−19
S. mutans (− sucrose)	4.92 ± 0.37	5.29 ± 0.45	+62	4.32 ± 0.03	$5.33 \pm 0.26^{**}$	+92
S. mutans (+ sucrose)	4.82 ± 0.34	$5.21 \pm 0.20^{*}$	+52	4.76 ± 0.50	4.56 ± 0.47	V27

S. mutans and *S. sanguinis* form biofilm in the presence or absence of sucrose. Generally, *S. sanguinis* forms less biofilm in terms of CFU/ml.
Cells were grown in the presence (+) or absence (−) of 1% (w/v) sucrose.
(+), viable counts increased after treatment, compared with control; (−), viable counts decreased after treatment, compared with control; CFU, colony-forming units.
Reprinted from Percival *et al.* (2006), *European Journal of Oral Sciences, 114*, 343—348, with permission.
*$P < 0.05$.
**$P < 0.001$.

Cacao bean husk extract showed a caries-inhibitory effect in specific pathogen-free rats. A significant reduction in the plaque index was found at concentrations of more than 0.5 mg/ml, while the caries score was reduced at concentrations of more than 1 mg/ml (Ooshima *et al.*, 2000). Some authors isolated the cariostatic substances in CBH, and identified them as epicatechin and epicatechin-4α-benzyl thioether (Osawa *et al.*, 2001).

Studies carried out on human volunteers showed that mouth rinses with an alcoholic solution of CBH (1 mg/ml) reduced dental plaque deposit (Matsumoto *et al.*, 2004). Likewise, a study carried out on children using CBH-extract mouth rinses, showed a 20.9% decrease in *Streptococcus mutans* counts and a 49.6% decrease in plaque score in the CBH-extract group as compared to the placebo group (Srikanth *et al.*, 2008).

ARECA CATECHU SEED
Description and general usage

Areca catechu (LINN.), also named betel palm, is a handsome tree cultivated throughout the warmer parts of Asia for its yellowish-red fruits, which contain conical-shaped seeds with a flattened base. They are brownish in color externally, and internally are mottled like a nutmeg. *Areca catechu* seeds, also known as betel nuts, have been widely used in traditional Chinese medicine as an antiparasitic agent for thousands of years.

Some constituents of betel nut have antihypertensive properties. Both leaves and nuts are used for the treatment of diarrhea, dropsy, sunstroke, beri beri, throat inflammations, edema, lumbago, bronchial catarrh, and urinary disorders.

Applications to oral health promotion and disease prevention

According to some studies, the regular mastication of betel nut has a cariostatic effect (Möller *et al.*, 1977; Nigam & Srivastava, 1990). Hada and colleagues showed that the antibacterial principals of betel nut against *Streptococcus mutans* were a mixture of fatty acids (myristic acid, and oleic acids). Moreover, they tested betel nut extracts against glucosyltransferase (GTF) activity. This enzyme catalyzes the formation of glucanes, and is implicated in the adhesion of *S. mutans*. These authors found that the anti-GTF molecules were procyanidins (Hada *et al.*, 1989).

Later, Iwamoto and colleagues isolated new 5′-nucleotidase inhibitors from the seeds of *Areca catechu*, which have an inhibitory effect on the growth of *S. mutans* and on glucan formation. These inhibitors were found to be polyphenolic substances (Iwamoto *et al.*, 1991).

de Miranda and colleagues confirmed previous works on *S. mutans*, and showed antibacterial activity against *Fusobacterium nucleatum*. They also found that baked and boiled nuts extracts were more effective than raw nuts extracts (de Miranda *et al.*, 1996).

Adverse effects and reactions

The pathogenic effect of *Areca catechu* is linked to the chewing of betel quid habits. Epidemiological studies have shown that this is associated with an increased risk of oral cancer, and promotes subgingival infection with periodontal pathogens which may lead to greater severity of periodontal diseases (Ling *et al.*, 2001).

PERILLA SEED
Description and general usage

Perilla (*P. frutescens* Britton var. *japonica* Hara) is a green or purple annual herb, with a distinctive aroma and taste, which has been cultivated for centuries in Korea. Its leaves and extracted essential oil are often used as flavoring ingredients for soups, stews, and roasting. Perilla is also grown in many other countries, including Japan, China, Vietnam, Turkey, and the United States. It has long been used in traditional Japanese medicine.

Applications to oral health promotion and disease prevention

The ethanolic extract of defatted perilla seed weakly inhibited the growth of the oral pathogenic strains *S. mutans*, *S. sobrinus*, and *P. gingivalis*. The ethyl acetate extract strongly inhibited the growth of *S. mutans* and *P. gingivalis*. This effect could be due to luteolin, its phenolic component (Yamamoto & Ogawa, 2002).

GRAPE (*VITIS VINIFERA*) SEED
Description and general usage

Grapes are native to Asia near the Caspian Sea, but were brought to North America and Europe around the 1600s. This plant's climbing vine has large, jagged leaves, and its stem bark tends to peel. The grapes may be green, red, or purple.

Grape seeds are considered to be rich sources of polyphenolic compounds, mainly mono-meric catechin and epicatechin, gallic acid, and polymeric and oligomeric procyanidins. They have showed various biological effects, including antioxidant or antimicrobial (Baydar et al., 2006).

Applications to oral health promotion and disease prevention

There has been limited research on the action of grape seed extracts (GSE) in oral diseases.

Smullen and colleagues found that red GSEs were bacteriostatic, and prevented acid production when added at the MIC to culture of S. mutans (Smullen et al., 2007). Later, other authors showed that GSEs had a bacteriostatic effect on Porphyromonas gingivalis and Fusobacterium nucleatum. At the concentration of 2000 μg/ml, it significantly decreased the formation of dental biofilm (Figure 12.2) (Furiga et al., 2009).

PSORALEA CORYLIFOLIA
Description and general usage

Psorolea corylifolia belongs to the Papilionaceae family. It is an erect herbaceous plant that possesses broad elliptic leaves. Its flowers are yellow or bluish purple in color. The plant bears pods containing a single seed. It is widely found in India.

According to Ayurveda, P. corylifolia root is useful in dental caries, whereas the leaves are good for diarrhea. The fruit is diuretic, and causes biliousness. The seeds are refrigerant, alternative, laxative, antipyretic, anthelmintic, and useful for heart problems. The seed oil is applied externally to treat skin-related troubles.

Bakuchiol is isolated from the seeds and leaves of Psoralea corylifolia Linn. It is a phenolic isoprenoid, and exhibits antimutagenic and antibacterial properties.

115

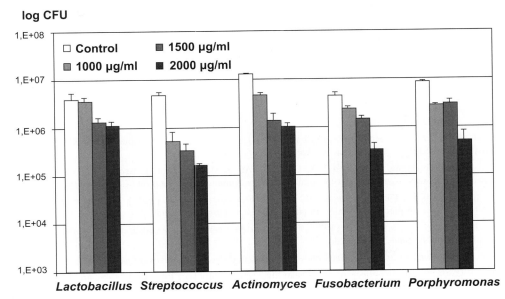

FIGURE 12.2

Effect of different concentrations of GSE on bacteria composing a multi-species biofilm. Results are expressed as means and standard deviations of triplicate experiments. The inhibitory effect of GSE on oral biofilm is dose-dependent. It shows its greatest activity at a concentration of 2000 μg/ml, even if antibacterial effects are different among the bacteria composing the biofilm. *Reprinted from Furiga et al., (2009), Food Chem. 113, 1037–1040, with permission.*

Applications to oral health promotion and disease prevention

Katsura and colleagues evaluated, *in vitro*, the antimicrobial activity of bakuchiol against oral pathogens. They found that the bacteriostatic effect required concentrations from 1 to 4 µg/ml, and that the bactericidal effect required concentrations from 5 to 20 µg/ml. The antibacterial activity was effective for both Gram-positive and Gram-negative bacteria. Moreover, bakuchiol at the concentration of 100 µg/ml inhibited glucan formation by *S. mutans* (Katsura *et al.*, 2001).

SUMMARY POINTS

- Macelignan shows a strong antibacterial effect against *Streptococcus mutans*, *Streptococcus sanguis*, *Streptococcus salivarius*, and *Lactobacillus casei*, and weak activity against *Actinomyces viscosus*, *Porphyromonas gingivalis*, and *Staphylococcus aureus*.
- Cacao bean husk reduces the growth rates and the cariogenicity of *Streptococcus mutans* and *Streptococcus sobrinus*. The cariostatic substances in cacao bean husk are epicatechin and epicatechin-4α-benzyl thioether.
- Betel nuts have an inhibitory effect on the growth of *Streptococcus mutans* and *Fusobacterium nucleatum*, and on glucan formation.
- Grape seed extract has a bacteriostatic effect on *Streptococcus mutans*, *Porphyromonas gingivalis*, and *Fusobacterium nucleatum*. It significantly decreases the formation of dental biofilm.
- Bakuchiol, which is isolated from the seeds and leaves of *Psoralea corylifolia* Linn, shows antimicrobial activity against oral pathogens.

References

Baydar, N. G., Sagdic, O., Ozkan, G., & Cetin, S. (2006). Determination of antibacterial effects and total phenolic contents of grape (*vitis vinifera*) seed extracts. *International Journal of Food Science and Technology, 41*, 799–804.

Chung, J. Y., Choo, J. H., Lee, M. H., & Hwang, J. K. (2006). Anticariogenic activity of macelignan isolated from *Myristica fragrans* (nutmeg) against *Streptococcus mutans*. *Phytomedicine, 13*, 261–266.

de Miranda, C. M., van Wyk, C. W., van der Biji, P., & Basson, N. J. (1996). The effect of areca nut on salivary and selected oral microorganisms. *International Dental Journal, 46*, 350–356.

Furiga, A., Lonvaud-Funel, A., & Badet, C. (2009). *In vitro* study of antioxidant capacity and antibacterial activity on oral anaerobes of a grape (*Vitis vinifera*) seed extract. *Food Chemistry, 113*, 1037–1040.

Hada, S., Kakiuchi, N., Hattori, M., & Namba, N. (1989). Identification of antibacterial principles against *Streptococcus mutans* and inhibitory principles against glucosyltransferase from the seed of *Areca catechu* L. *Phytotherapy Research, 3*, 140–144.

Iwamoto, M., Uchino, K., Toukairin, T., Kawaguchi, K., Tatebayashi, T., Ogawara, H., et al. (1991). The growth inhibition of *Streptococcus mutans* by 5′-nucleotidase inhibitors from *Areca catechu* L. *Chemical & Pharmaceutical Bulletin, 39*, 1323–1324.

Katsura, H., Tsukiyama, R. I., Suzuki, A., & Kobayashi, M. (2001). *In vitro* antimicrobial activities of bakuchiol against oral microorganisms. *Antimicrob. Agents Chemother, 45*, 3009–3013.

Ling, L. J., Hung, S. L., Tseng, S. C., Chen, Y. T., Chi, L. Y., Wu, K. M., et al. (2001). Association between betel quid chewing, periodontal status and periodontal pathogens. *Antimicrobial Agents and Chemotherapy, 16*, 364–369.

Matsumoto, M., Tsuji, M., Okuda, J., Sasaki, H., Nakano, K., Osawa, K., et al. (2004). Inhibitory effects of cacao bean husk extract on plaque formation *in vitro* and *in vivo*. *European Journal of Oral Sciences, 112*, 249–252.

Möller, I. J., Pindborg, J. J., & Effendi, I. (1977). The relation between betel chewing and dental caries. *Scandinavian Journal of Dental Research, 85*, 64–70.

Nigam, P., & Srivastava, A. B. (1990). Betel chewing and dental decay. *Federation of Operative Dentistry, 1*, 36–38.

Ooshima, T. (2004). Inhibitory effects of cacao bean husk extract on plaque formation *in vitro* and *in vivo*. *European Journal of Oral Sciences, 112*, 249–252.

Ooshima, T., Osaka, Y., Sasaki, H., Osawa, K., Yasuda, H., Matsumura, M., et al. (2000). Caries inhibitory activity of cacao bean husk extract in *in vitro* and animal experiments. *Archives of Oral Biology, 45*, 639–645.

Osawa, K., Miyazaki, K., Shimura, S., Okuda, J., Matsumoto, M., & Ooshima, T. (2001). Identification of cariostatic substances in the cacao bean husk: their anti-glucosyltransferase and antibacterial activities. *Journal of Dental Research, 80*, 2000–2004.

Percival, R. S., Devine, D. A., Duggal, M. S., Chartron, S., & Marsh, P. D. (2006). The effect of cocoa polyphenols on the growth, metabolism, and biofilm formation by Streptococcus mutans and Streptococcus sanguinis. *European Journal of Oral Sciences, 114,* 343–348.

Smullen, J., Koutsou, G. A., Foster, H. A., Zumbé, A., & Storey, D. M. (2007). The antibacterial activity of plant extracts containing polyphenols against *Streptococcus mutans. Caries Research, 41,* 342–349.

Srikanth, R. K., Shashikiran, N. D., & Subba Reddy, V. V. (2008). Chocolate mouth rinse: effect on plaque accumulation and *mutans streptococci* counts when used by children. *Journal of Indian Society of Pedodontics and Preventive Dentistry, 26,* 67–70.

Yamamoto, H., & Ogawa, T. (2002). Antimicrobial activity of perilla seed polyphenols against oral pathogenic bacteria. *Bioscience Biotechnology and Biochemistry, 66,* 921–924.

Adverse Aspects

CHAPTER 13

Mycotoxins in Seeds and Nuts

Elisabete Yurie Sataque Ono[1], Elisa Yoko Hirooka[2], Carolina Nachi Rossi[1], Mario Augusto Ono[3]

[1] Department of Biochemistry and Biotechnology
[2] Department of Food Science and Technology
[3] Department of Pathological Sciences, State University of Londrina, Londrina, Paraná, Brazil

121

LIST OF ABBREVIATIONS

AFB_1, aflatoxin B_1
AFB_2, aflatoxin B_2
AFG_1, aflatoxin G_1
AFG_2, aflatoxin G_2
AFM_1, aflatoxin M_1
AFM_2, aflatoxin M_2
a_w, water activity
CAST, Council for Agricultural Science and Technology
DNA, deoxyribonucleic acid
FAO, Food and Agriculture Organization of the United Nations
HCC, hepatocellular carcinoma
HIV, human immunodeficiency virus
HPRT, hypoxanthine guanine phosphoribosyltransferase
IARC, International Agency for Research on Cancer

INTRODUCTION

Mycotoxins are toxic secondary metabolites produced by fungi which contaminate agricultural staples in the field, and during harvest, transport, and storage. These compounds have different chemical structures and are produced by field and storage fungi, including many species of

Nuts & Seeds in Health and Disease Prevention. DOI: 10.1016/B978-0-12-375688-6.10013-1

Fusarium, Aspergillus, and *Penicillium.* The Food and Agriculture Organization (FAO) has estimated the worldwide mycotoxin contaminated crop at 25%.

In addition to serious economic losses, the natural occurrence of mycotoxins in agricultural staples can cause acute and/or chronic intoxication symptoms in both humans and animals at low concentration levels (mg/kg to µg/kg range) (CAST, 2003).

The human health hazard associated with the natural occurrence of harmful mycotoxins has compelled several countries to adopt regulatory guidelines, but the maximum tolerated levels for mycotoxins vary widely between countries (FAO, 2004), tending to be higher in tropical and subtropical producing countries, and lower in importing countries with a temperate climate.

Although other mycotoxins (zearalenone, cyclopiazonic acid, ochratoxin A) can occur, the most reported mycotoxins in nuts and seeds are aflatoxins, a group of related bisfuranocoumarin compounds produced mainly by *Aspergillus flavus* and *A. parasiticus.* Twenty different analogs have been identified; however, the major naturally occurring toxins are aflatoxin B_1 (AFB_1), AFB_2, AFG_1, and AFG_2 (Figure 13.1). They have been shown to cause mutagenic, teratogenic, and hepatocarcinogenic effects, and the liver is the primary target (CAST, 2003). The International Agency for Research on Cancer (IARC) has classified naturally occurring mixtures of aflatoxins as carcinogenic to humans (Group 1). The toxicity order is $AFB_1 > AFG_1 > AFB_2 > AFG_2$ (IARC, 1993).

AFM_1 and AFM_2, the hydroxylated derivatives of AFB_1 and AFB_2, are detected in the milk of dairy cattle or lactating mothers exposed to aflatoxins. AFM_1 is less mutagenic and carcinogenic than AFB_1, but shows acute toxicity similar to other aflatoxins. The IARC (1993) classified it as a Group 2B carcinogen (possibly carcinogenic to humans).

Among aflatoxins, AFB_1 has triggered the most research efforts due to its high toxicity and worldwide occurrence in staple foods and feeds. At least 99 countries involved in international trade have established regulatory limits for AFB_1, or the sum of AFB_1, AFB_2, AFG_1, and AFG_2 (Table 13.1), in foods and/or feeds (FAO, 2004). Table 13.2 shows the maximum tolerated levels of aflatoxins in foodstuffs in some countries in 2003. Harmonization of tolerance levels is taking place in some free trade zones (European Union, MERCOSUR, Australia/New Zealand).

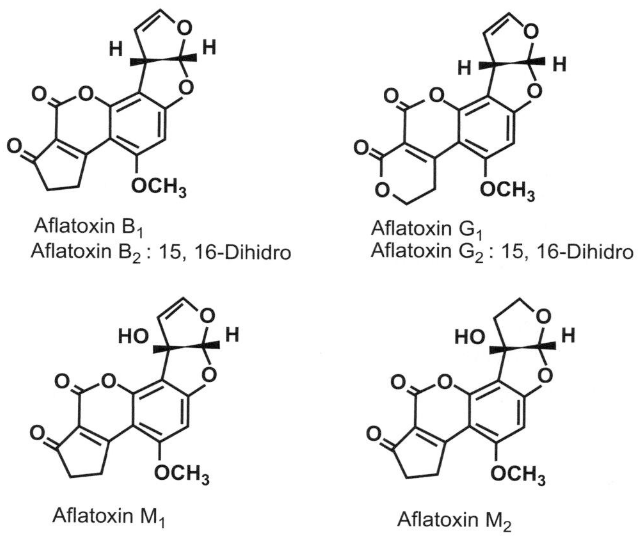

Aflatoxin B_1
Aflatoxin B_2 : 15, 16-Dihidro

Aflatoxin G_1
Aflatoxin G_2 : 15, 16-Dihidro

Aflatoxin M_1

Aflatoxin M_2

FIGURE 13.1
Chemical structures of aflatoxins B_1, B_2, G_1, G_2, M_1 and M_2.

TABLE 13.1 Medians and Ranges of Maximum Tolerated Levels (μg/kg) for Some (Groups of) Aflatoxins in 2003, and Numbers of Countries Known to have Relevant Regulations

Aflatoxin/Matrix Combination	Median (μg/kg)	Range (μg/kg)	No. of Countries
AFB$_1$ in foodstuffs	5	1–20	61
AFB$_1$ + AFB$_2$ + AFG$_1$ + AFG$_2$ in foodstuffs	10	0–35	76
AFM$_1$ in milk	0.05	0.05–15	60

Adapted from FAO (2004), with permission.

TABLE 13.2 Maximum Tolerated Levels of Aflatoxins in Foodstuffs in Some Countries in 2003

Countries	Aflatoxin(s) (μg/kg)	Limit (μg/kg)
European Union (Austria, Belgium, Denmark, Finland, France, Germany, Greece, Ireland, Italy, Luxembourg, The Netherlands, Portugal, Spain, Sweden, United Kingdom)		
Groundnuts, nuts and dried fruit and processed products thereof, intended for direct human consumption or as an ingredient in foodstuffs	AFB$_1$ AFB$_1$ + AFB$_2$ + AFG$_1$ + AFG$_2$	2 4
Groundnuts to be subjected to sorting, or other physical treatment, before human consumption or use as an ingredient in foodstuffs	AFB$_1$ AFB$_1$ + AFB$_2$ + AFG$_1$ + AFG$_2$	8 15
Nuts and dried fruit to be subjected to sorting, or other physical treatment, before human consumption or use as an ingredient in foodstuffs	AFB$_1$ AFB$_1$ + AFB$_2$ + AFG$_1$ + AFG$_2$	5 10
United States of America All foods except milk	AFB$_1$ + AFB$_2$ + AFG$_1$ + AFG$_2$	20
Japan All foods	AFB$_1$	10
MERCOSUR (Argentina, Brazil, Paraguay, Uruguay) Peanuts, maize and products thereof	AFB$_1$ + AFB$_2$ + AFG$_1$ + AFG$_2$	20

Adapted from FAO (2004), with permission.

NATURAL OCCURRENCE

Fungal growth and aflatoxin production depend on biological (susceptible crop) and environmental factors, with the emphasis on regional climatic conditions during plant development and crop harvest. Aflatoxin contamination is also favored by stress or damage to the crop due to drought before harvest, insect infestation, and inadequate drying of the crop before storage (CAST, 2003).

Aflatoxins occur worldwide, but they are more frequent in subtropical and warm temperate climates. It has been shown that water availability (water activity, a_w) and temperature play an important role in determining the extent of aflatoxin production. The ideal temperature for growth and mycotoxin production ranges from 25 to 35°C for *A. flavus* strains, while a_w values for aflatoxin production by *A. flavus* range from 0.95 to 0.99, with a minimum a_w value of 0.82 (ICMSF, 1996).

In addition to corn and spices, peanuts, cottonseed, Brazil nuts, pistachios, and copra are included in the group of commodities with a high risk of aflatoxin contamination, whereas almonds, pecans, and walnuts are in the lower risk group (CAST, 2003).

A number of survey and monitoring programs on aflatoxins have been carried out in several countries. Table 13.3 shows selected examples of natural occurrence of aflatoxins in seeds and nuts.

Abdulkadar and colleagues (2000) analyzed 81 nut samples imported into Qatar in 1997 for aflatoxins, and detected contamination in 19 samples with total aflatoxin levels ranging from

TABLE 13.3 Selected Examples of Natural Occurrence of Aflatoxins in Seeds and Nuts

Commodity	Country	Samples Analyzed	Positive Samples	Range (μg/kg)	Reference
Pistachios, peanuts, mixed nuts	Imported in Qatar	61	19	0.53–289.0[a]	Abdulkadar et al. (2000)
Other nuts (almonds, cashew nuts, walnuts and hazel nuts)		20	—	nd	
Peanut kernels	Brazil	60	20	7.0–116.0[b]	Nakai et al. (2008)
Melon seeds	Nigeria	137	37	2.3–47.7[c]	Bankole et al. (2006)

nd, not detected.
[a]Total aflatoxins ($AFB_1 + AFB_2 + AFG_1 + AFG_2$).
[b]Mean AFB_1 levels.
[c]AFB_1.

0.53 to 289 μg/kg. Aflatoxin contamination was detected in pistachios and peanuts, but not in other nuts such as almonds, cashew nuts, walnuts, and hazel nuts. Nakai et al. (2008) analyzed 60 stored peanut samples from São Paulo State, Brazil, and showed that in kernels 20 samples were contaminated with AFB_1 (mean levels of 7.0–116 μg/kg).

EFFECT OF ROASTING

Due to health hazards to both humans and animals, great efforts have been made to minimize the aflatoxin levels in foods. Although prevention is the most effective approach, chemical, biological and physical methods to reduce aflatoxin levels in foodstuffs have been investigated. Aflatoxins are highly heat stable, but heat inactivation in contaminated food has been attempted. AFB_1 levels have been reduced by 30–44% in peanuts by roasting them for 30 minutes at 150°C (Pluyer et al., 1987). Aflatoxin degradation was found to be time- and temperature-dependent. Roasting naturally-contaminated whole pistachio kernels with 144 μg/kg and 235 μg/kg AFB_1 at 150°C for 30 minutes reduced the levels of AFB_1 to 63% and to 19% of the initial levels, respectively. Roasting at 150°C for 120 minutes degraded more than 95% of AFB_1, but caused alterations in the appearance and taste of the pistachio nuts (Yazdanpanah et al., 2005).

TOXICOLOGICAL EFFECTS IN HUMANS

Nowadays there is increasing concern about aflatoxin involvement in human diseases, based on epidemiological evidence and several studies on animals. The most prominent means of humans' exposure to aflatoxins is through ingestion of contaminated food, but other routes, such as inhalation and contact, should be considered. Exposure by inhalation can result from handling contaminated materials, or from airborne fungal components. The disease resulting from exposure to aflatoxins, called aflatoxicosis, can be classified as acute or chronic (Table 13.4), and ranges from rapid death to tumor formation. Acute aflatoxicosis is more likely to occur following exposure to high levels, whereas the chronic form is more likely from prolonged exposure to small quantities of toxin (CAST, 2003).

Acute aflatoxicosis is characterized by severe, acute hepatotoxicity with early symptoms including anorexia, malaise, and low-grade fever. Moreover, acute high-level exposure can lead to death by hepatitis, with vomiting, abdominal pain, jaundice, and fulminate hepatic failure (Strosnider et al., 2006).

Several outbreaks of acute aflatoxin poisoning caused by contamination of inadequately stored staple foods have been reported in developing countries. Exposure to aflatoxins is generally higher in developing countries due to food insufficiency, lack of diversified diet, less emphasis on legislating maximum tolerated levels, and frequent lack of enforcement of such legislation.

TABLE 13.4 Toxicological Effects of Aflatoxins in Humans

Aflatoxicosis	Reference
Acute	
Anorexia	Strosnider *et al.*, 2006
Malaise	
Fever	
Vomiting	
Abdominal pain	
Jaundice	
Hepatic failure	
Chronic	
Hepatocellular carcinoma	Wang *et al.*, 1996
Impairment of chemotaxis and phagocytosis[*]	Ubagai *et al.*, 2008
	Cusumano *et al.*, 1996
Lower levels of activated T and B cells and lower levels of cytotoxic T cells with perforin and granzyme	Jiang *et al.*, 2005
Lower counts of T regulatory and cytotoxic T cells with perforin in HIV patients	Jiang *et al.*, 2008

[*]In vitro *studies.*

In April 2004, an outbreak of jaundice caused by ingestion of aflatoxin-contaminated homegrown maize resulted in 317 cases and 125 deaths in Kenya. Maize collected from the affected areas showed high AFB_1 levels, with 55% of samples being contaminated above the Kenyan legal limit (20 µg/kg), 35% having levels above 100 µg/kg, and 7% having levels above 1000 µg/kg. Another outbreak occurred in 2005 in Eastern Kenya, which led to 75 cases and 32 deaths (Strosnider *et al.*, 2006).

Epidemiological studies in humans have shown that exposure to aflatoxins is one of the major risk factors in the multifactorial etiology of hepatocellular carcinoma (HCC) in Africa, the Philippines, and China. The contribution of AFB_1 to HCC is affected by several factors, such as infection by hepatitis B or hepatitis C virus, the nutritional state and age of the individual, as well as the exposure level to AFB_1 (Wang *et al.*, 1996).

It is believed that the toxicity and carcinogenicity of AFB_1 is due to its enzymatic activation by a microsomal cytochrome P450 (CYP450) dependent epoxidation of the terminal furan ring of AFB_1 forming a highly reactive AFB_1 8,9-epoxide. The AFB_1 8,9-epoxide can bind to cellular macromolecules such as DNA and proteins, particularly to serum albumin, forming stable adducts (Wang & Groopman, 1999). The covalent binding to DNA is considered the critical step in hepatocarcinogenesis.

The measurement of aflatoxin—albumin adducts has been used as an exposure marker in several studies, and has been shown to correlate with aflatoxin intake, but the major limitation of this marker is that it only represents recent past exposure (2—3 months) (Wild *et al.*, 1990).

Specific mutations in the hypoxanthine guanine phosphoribosyltransferase (HPRT) gene in human lymphocytes after *in vitro* exposure (Cariello *et al.*, 1994) and in the p53 gene in the human liver following exposure to aflatoxin (Hsu *et al.*, 1991) have been reported, and can provide longer-term markers for aflatoxin exposure. A study of patients with HCC from regions with low and high AFB_1 intake in the diet showed higher mutation in the p53 tumor suppressor gene in patients living in an area with high AFB_1 contamination levels in foodstuffs (Pineau *et al.*, 2008).

The immunosuppressive effect of aflatoxins has been described in several studies carried out mainly in animal models. The immunotoxic effects in humans have been inferred by means of *in vitro* experiments, and studies in populations naturally exposed to aflatoxins.

In vitro effects of AFB_1 on the immune system have been described in human polymorphonuclear and mononuclear cells. The chemotaxis of human polymorphonuclear cells was reduced by high concentration of AFB_1 (Ubagai *et al.*, 2008). Phagocytosis and intracellular killing of *Candida albicans* were impaired in mononuclear cells treated with low doses of AFB_1 (0.1 pg/ml), although the pretreatment of mononuclear cells with AFB_1 did not modify the antiviral activity against herpes simplex virus or superoxide production (Cusumano *et al.*, 1996).

The evaluation of the immune status in humans exposed to high levels of AFB_1 showed lower activated T and B cells, and the percentages of cytotoxic T cells with pore-forming protein (perforin) and serine proteases (granzyme A) were significantly lower when compared to the individuals exposed to low doses of AFB_1. These proteins are required for the cytolytic activity of effector cytotoxic T cells against host infected cells. The impairment of different immunological mechanisms by aflatoxins could result in increased susceptibility to several infectious diseases, taking into account that T and B cell responses are protective against intracellular and extracellular pathogens, respectively (Jiang *et al.*, 2005).

HIV (Human Immunodeficiency Virus)-positive individuals with high exposure to aflatoxin, evaluated by the AFB_1—albumin adduct, showed lower perforin expression on cytotoxic T cells, and lower percentages of regulatory T cells, compared to HIV-positive individuals with low exposure to aflatoxin, suggesting that these immunological changes may affect the HIV disease progression (Jiang *et al.*, 2008).

SUMMARY POINTS

- Although other mycotoxins (zearalenone, cyclopiazonic acid, ochratoxin A) can occur, the most reported mycotoxins in nuts and seeds are aflatoxins, a group of toxic secondary metabolites produced mainly by *Aspergillus flavus* and *A. parasiticus*.
- Aflatoxins are heat stable, and they are not completely destroyed by industrial processing.
- The health impact of aflatoxins is a complex issue, as multiple mycotoxins can coexist in the same commodity and individuals may be simultaneously exposed to several mycotoxins.
- Chronic aflatoxin exposure resulting in HCC, generally in association with hepatitis B virus, has been well documented, but additional effects have not been widely studied. There is some evidence suggesting an interaction between chronic aflatoxin exposure and malnutrition, impaired growth, immunosuppression, and, consequently, susceptibility to infectious diseases; therefore, further investigations on the impact of this association on health are required.
- Taking into account that more and more stricter guidelines on aflatoxin contamination have been imposed by the importing countries, constant monitoring throughout the food producing chain is essential to minimize health risks and comply with trade requirements.

ACKNOWLEDGEMENTS

The authors thank the CNPq (the Brazilian Government Organization for grant aid and fellowship to Brazilian researchers), CNPq/Ministry of Agriculture, the Araucária Foundation, PPSUS/Brazilian Ministry of Health, Paraná Fund/SETI and CAPES (Co-ordination for Formation of High Level Professionals) for the financial support for projects on mycotoxins. The authors also thank the CNPq for the productivity fellowships granted to E.Y.S. Ono, E.Y. Hirooka, and M.A. Ono. A special thanks to CAST (the Council for Agricultural Science and Technology) and the FAO (Food and Agriculture Organization of the United Nations) for permission to use information from *Mycotoxins — Risks in Plant, Animal and Human Systems*, and tables from the FAO Food and Nutrition Paper, respectively.

References

Abdulkadar, A. H. W., Al-Ali, A., & Al-Jedah, J. (2000). Aflatoxin contamination in edible nuts imported in Qatar. *Food Control, 11,* 157—160.

Bankole, S. A., Ogunsanwo, B. M., Osho, A., & Adewuyi, G. O. (2006). Fungal contamination and aflatoxin B_1 of "egusi" melon seeds in Nigeria. *Food Control, 17,* 814—818.

Cariello, N. F., Cui, L., & Skopek, T. R. (1994). *In vitro* mutational spectrum of aflatoxin B, in the human hypoxanthine guanine phosphoribosyltransferase gene. *Cancer Research, 54,* 4436—4441.

Council for Agricultural Science and Technology (CAST). (2003). Mycotoxins — risks in plant, animal and human systems. *Task Force Report 139.* Ames, IO: CAST.

Cusumano, V., Rossano, F., Merendino, R. A., Arena, A., Costa, G. B., Mancuso, G., et al. (1996). Immunobiological activities of mould products: functional impairment of human monocytes exposed to aflatoxin B_1. *Research in Microbiology, 147,* 385—391.

Food and Agriculture Organization of the United Nations (FAO). (2004). Worldwide regulations for mycotoxins in food and feed in 2003. In *Food and Nutrition Paper, Vol. 81.* (p. 9), Rome, Italy: FAO.

Hsu, I. C., Metcalf, R. A., Sun, T., Welsh, J. A., Wang, N. J., & Harris, C. C. (1991). Mutational hotspot in the p53 gene in human hepatocellular carcinomas. *Nature, 350,* 427—428.

International Agency for Research on Cancer (IARC). (1993). Some naturally occurring substances: food items and constituents. Heterocyclic aromatic amines and mycotoxins.. In *IARC Monograph on the evaluation of carcinogenic risk to humans, Vol. 56* Lyon, France: IARC.

International Commission on Microbiological Specifications for Foods (ICMSF). (1996). Toxigenic fungi: *Aspergillus.* In *Microorganisms in Foods, Vol. 5, Characteristics of Food Pathogens* (pp. 347—381). London, UK: Academic Press.

Jiang, Y., Jolly, P. E., Ellis, W. O., Wang, J. S., Phillips, T. D., & Williams, J. H. (2005). Aflatoxin B_1-albumin adduct levels and cellular immune status in Ghanaians. *International Immunology, 17,* 807—814.

Jiang, Y., Jolly, P. E., Preko, P., Wang, J. S., Ellis, W. O., Phillips, T. D., et al. (2008). *Aflatoxin-related immune dysfunction in health and in human immunodeficiency virus disease.* Clinical and Developmental Immunology, *ID* 790309, 12 pages, doi:1155/2008/790309.

Nakai, V. K., Rocha, L. O., Gonçalez, E., Fonseca, H., Ortega, E. M. M., & Corrêa, B. (2008). Distribution of fungi and aflatoxins in a stored peanut variety. *Food Chemistry, 106,* 285—290.

Pineau, P., Marchio, A., Battiston, C., Cordina, E., Russo, A., Terris, B., et al. (2008). Chromosome instability in human hepatocellular carcinoma depends on p53 status and aflatoxin exposure. *Mutation Research, 653,* 6—13.

Pluyer, H. R., Ahmed, E. M., & Wei, C. I. (1987). Destruction of aflatoxin on peanut by oven- and microwave-roasting. *Journal of Food Protection, 50,* 504—508.

Strosnider, H., Azziz-Baumgartner, E., Banziger, M., Bhat, R. V., Breiman, R., Brune, M. N., et al. (2006). Workgroup report: public health strategies for reducing aflatoxin exposure in developing countries. *Environmental Health Perspectives, 114,* 1898—1903.

Ubagai, T., Tansho, S., Ito, T., & Ono, Y. (2008). Influences of aflatoxin B_1 on reactive oxygen species generation and chemotaxis of human polymorphonuclear leukocytes. *Toxicoogy. In Vitro, 22,* 1115—1120.

Wang, J.-S., & Groopman, J. D. (1999). DNA damage by mycotoxins. *Mutation Research, 424,* 167—181.

Wang, L. Y., Hatch, M., Chen, C.-J., Levin, B., You, S.-L., Lu, S.-N., et al. (1996). Aflatoxin exposure and risk of hepatocellular carcinoma in Taiwan. *International Journal of Cancer, 67,* 620—625.

Wild, C. P., Jiang, Y.-Z., Allen, S. J., Jansen, L. A. M., Hall, A. J., & Montesano, R. (1990). Aflatoxin-albumin adducts in human sera from different regions of the world. *Carcinogenesis, 11,* 2271—2274.

Yazdanpanah, H., Mohammadi, T., Abouhossain, G., & Cheraghali, A. M. (2005). Effect of roasting on degradation of aflatoxins in contaminated pistachio nuts. *Food and Chem. Toxicology, 43,* 1135—1139.

CHAPTER 14

Cyanogenic Glycosides in Nuts and Seeds

Daniel J. Ballhorn
Department of Plant Biology, University of Minnesota, St Paul, Minnesota, USA

LIST OF ABBREVIATIONS

BMAA, β-methylamino-L-alanine
HCN, hydrogen cyanide
HDL, high density lipoprotein cholesterin
INC, International Nut and Dried Fruit Foundation
LDL, low density lipoprotein cholesterin
USDA, United States Department of Agriculture
WP ALS-PD, Western Pacific Amyotrophic Lateral Sclerosis/Parkinsonism-Dementia Complex

INTRODUCTION

Cyanide production by plants is widespread. While all higher plants probably form low levels of hydrogen cyanide (HCN) as a coproduct of ethylene biosynthesis in leaves and fruits, many plant species actively synthesize and accumulate cyanide-containing compounds. These cyanogenic plants have the ability to release HCN up to toxic thresholds in response to cell disintegration — for example, when damaged by herbivores (Ballhorn *et al.*, 2009). Over 3000 species, representing more than 550 genera and 130 families, produce cyanide-containing

Nuts & Seeds in Health and Disease Prevention. DOI: 10.1016/B978-0-12-375688-6.10014-3

compounds, including many economically important food plants (Jones, 1998). The enzymatically accelerated release of HCN from these inactive precursors is called cyanogenesis. While in certain sapindaceous seeds HCN may arise during cyanolipid hydrolysis, HCN production in plants frequently results from the catabolism of cyanogenic glycosides. The approximately 75 documented cyanogenic glycosides are all O-β-glycosidic derivatives of α-hydroxynitriles (Poulton, 1990). Most cyanogenic glycosides are derived from the five hydrophobic protein amino acids, tyrosine, phenylalanine, valine, leucine, and isoleucine (Figure 14.1). In the course of the biosynthetic pathway the specific amino acids are hydroxylated to N-hydroxylamino acids, aldoximes, and then to nitriles serving as intermediates. In addition to cyanogenic monosaccharides, in which the unstable cyanohydrin moiety is stabilized by glycosidic linkage to a single sugar residue (e.g., prunasin, linamarin, dhurrin; Figure 14.1), cyanogenic disaccharids (e.g., amygdalin and linustatin; Figure 14.1) or trisaccharides (e.g., xeranthin) occur. Because HCN generally is toxic to all eukaryotes — including the plant itself — the accumulation of this toxin as inactive precursors is essential to prevent autotoxicity. To avoid uncontrolled release of HCN, in the intact plant cyanogenic glycosides are separated on cellular or tissue level from specific β-glucosidases. In case of cell disentegration, β-glucosidases are brought into contact with the cyanogenic glycosides. By hydrolysis of the glycosides, α-hydroxynitriles are formed that are relatively unstable, and dissociate either spontaneously or are enzymatically accelerated by α-hydroxynitrile lyases into HCN and an aldehyde or a ketone (Figure 14.2). Thus, the production of HCN from cyanogenic plants depends both on the biosynthesis of cyanogenic glycosides and the coexistence of one or more degrading enzymes.

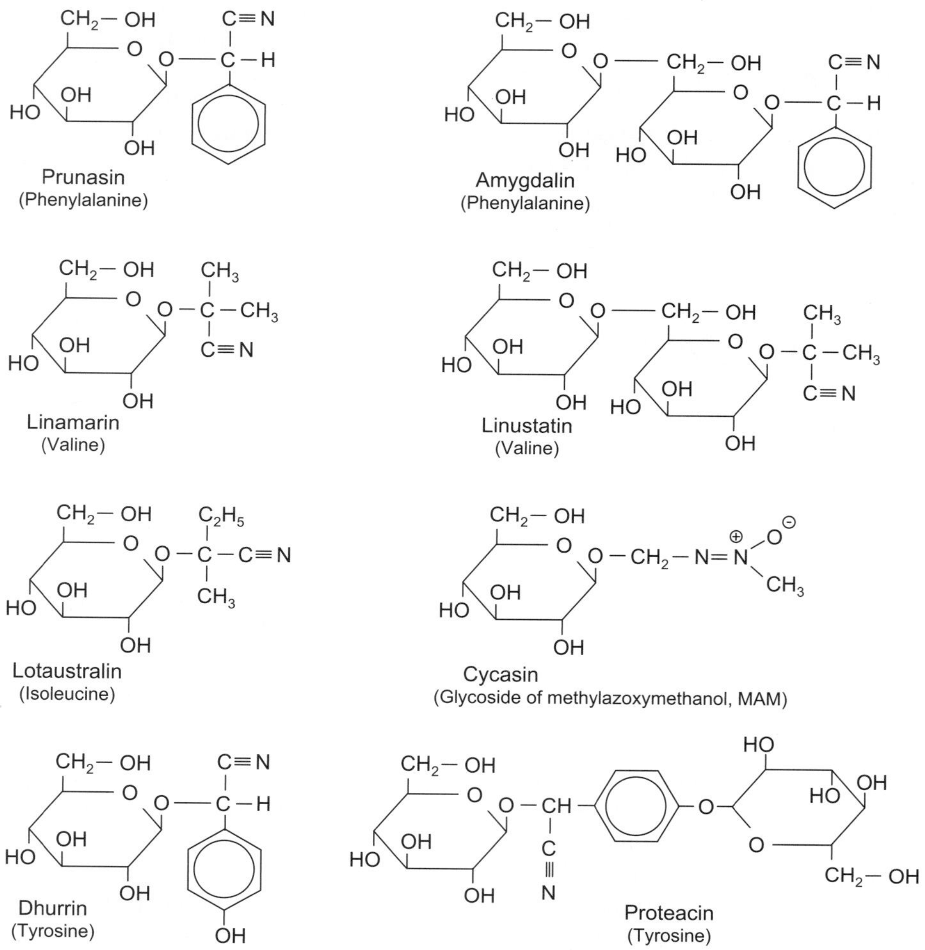

FIGURE 14.1

Important cyanogenic glycosides in food plants. Prunasin and amygdalin are central cyanogenic glycosides in rosaceous plants such as almonds, apricots, peaches, cherries, plums, and apples. Linamarin, linustatin, and lotaustralin occur in a wide range of different families (such as Fabaceae, Linaceae, Euphorbiaceae), for example in lima bean, flaxseed, and cassava. Cycasin (which is not a true cyanogenic glycoside since it does not contain a cyano group) is found in cycad seeds, whereas dhurrin occurs in *Sorghum* sp. and — together with proteacin — in macadamia nuts. The amino acid precursors of the respective cyanogenic glycosides are given in parentheses.

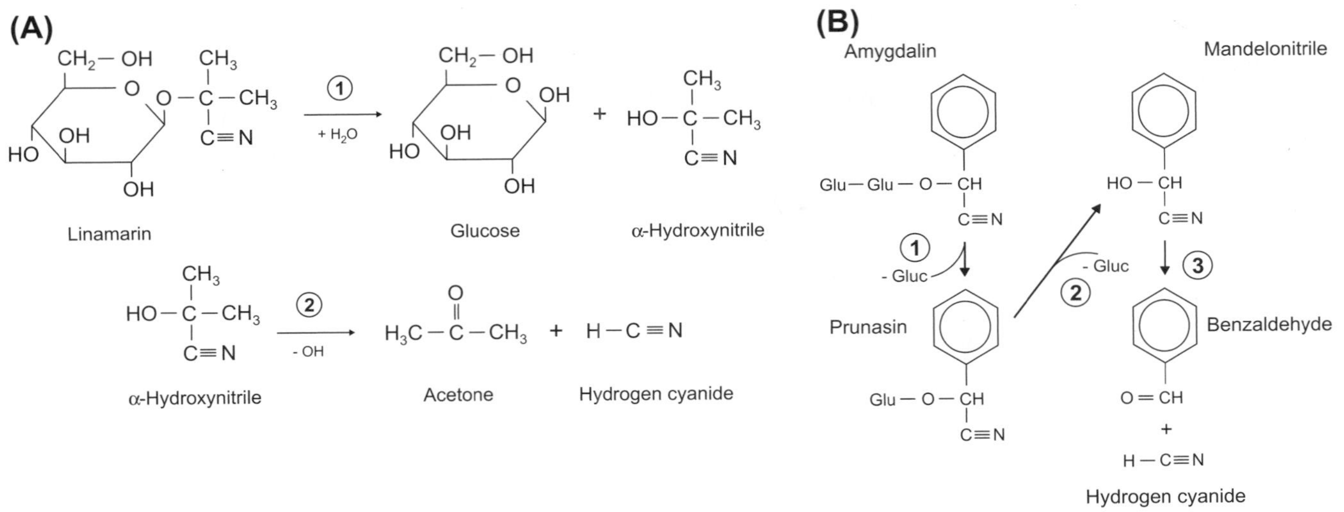

FIGURE 14.2
Release of HCN from cyanide-containing precursors. In response to cell damage, HCN is released by a two-step (A) or three-step (B) mechanism, depending on the type of the cyanogenic glycoside. In the case of cyanogenic monoglycosides such as linamarin (A), by the activity of β-glucosidases (1) unstable α-hydroxynitriles are formed that dissociate either spontaneously or are enzymatically accelerated by α-hydroxynitrile lyases (2) into HCN and a corresponding carbonyl compound. In the case of the cyanogenic diglycoside amygdalin (B), amygdalin hydrolase (1) and prunasin hydrolase (2) are β-glycosidases that catalyze the hydrolysis of the compound in a stepwise reaction by removing two glucose residues and producing mandelonitrile. This decomposes, enzymatically accelerated by a mandelonitrile lyase (3) into benzaldehyde and HCN.

While cyanide production in leaves can doubtlessly be attributed to herbivore defense (Ballhorn *et al.*, 2009), the ecological function of cyanogenic precursors in seeds is less understood. It appears reasonable that during seed germination, the defensive cyanide-containing compounds are transferred to the growing seedling for its defense. However, in addition to defense-associated functions, cyanogenic glycosides may serve as storage compounds for nitrogen that is required for seedling growth, or they may act as germination inhibitors.

Many plants store cyanogenic precursors in their seeds. Prominent and economically important examples presented in this chapter are various species belonging to the Rosaceae, for example *Prunus amygdalus* (almond), *P. armeniaca* (apricot), and *P. persica* (peach); Linaceae, for example *Linum usitatissimum* (flax); Fabaceae, for example *Phaseolus lunatus* (lima bean); and Proteaceae, such as *Macadamia integrifolia* and *M. tetraphylla* (macadamia nuts). In addition to these world-economically important plants, regional species such as *Cycas* sp. (e.g., *Cycas revoluta*) are known to produce cyanogenic seeds.

BOTANICAL DESCRIPTION, HISTORICAL AND PRESENT-DAY CULTIVATION

Almonds, Apricots, and Peaches

These rosaceous plants have long been known to contain cyanogenic compounds in their seeds. Almonds (*Prunus amygdalus*) originated in the Middle East and have been cultivated for 4000 years, mostly in temperate and subtropical regions, with an annual worldwide production in 2007 of $2.1*10^6$ tonnes (with shell) (FAOSTAT, 2009). The almond plant is a small deciduous tree, growing to 4–10 meters in height, with a trunk of up to 30 centimeters in diameter. In botanical terms, an almond is a stone fruit (not a nut). In contrast to other species of the genus *Prunus*, such as the peach, apricot, plum, and cherry, which show fleshy mesocarp, the mesocarp of almonds is a thick leathery greyish to greenish coat with a hairy epidermis. The endocarp is sclerotisized, and covers the edible seed. Like almonds, apricots (*Prunus armeniaca*) and peaches (*Prunus persica*) originated from Asia and have been cultivated

131

for 3000–4000 years. Both species are small trees that require a temperate continental to subtropical climate. Although mainly produced for their fruits, the seeds (kernels), especially of apricots, are also used for human consumption.

Flaxseed

Flax (Linaceae: *Linum usitatissimum*) is an annual herbaceous plant of 30–120 cm in height that is native to the region extending from the eastern Mediterranean to India. Today flax is an important crop mainly in Europe and the US. The worldwide production of flaxseed in 2007 amounted $1.9*10^6$ tonnes (FAOSTAT, 2009).

Lima bean

The lima bean *Phaseolus lunatus* (Fabaceae) is a herbaceous bush, 30–90 cm in height, or a twining vine 2–4 m long with trifoliate leaves, white or violet flowers, and pods of 5–12 cm containing two to four seeds. The lima bean is a grain legume of Andean and Mesoamerican origin. Two separate domestication events are believed to have occurred; one in the Andes around 2000 BC, and one in Mesoamerica most likely around AD 800. The South American genotypes produce large seeds (Lima type), while the Mesoamerican varieties produce small seeds (Sieva type). By 1301 cultivation had spread to North America, and in the 16th century the plant began to be cultivated in the Eastern Hemisphere. Today, lima beans are cultivated in the tropics and subtropics around the world.

Macadamia nuts

Macadamia nut trees (Proteaceae) are small to medium-sized trees growing 2–12 m in height. They are native to tropical rainforests in eastern Australia (seven species), New Caledonia (one species, *M. neurophylla*), and Sulawesi in Indonesia (one species, *M. hildebrandii*). *Macadamia* sp. prefer fertile, well-drained soils, a rainfall of 1000–2000 mm, and temperatures not falling below 10°C, with an optimum temperature of 25°C. Only two species, *Macadamia integrifolia* and *Macadamia tetraphylla*, are of commercial importance, and these are in fact the only native Australian food plants of economic importance. According to the International Nut and Dried Fruit Foundation (http://www.nutfruit.org/), the annual production of macadamia nuts in 2008 was about 26,000 tonnes worldwide.

Cycad seeds

Cycas plants belong to the order Cycadales (family Cycadaceae), which consists of 11 genera of tropical and subtropical plants that produce terminal oblong cones containing orange-yellow seeds. Within the Cycadales, *C. revoluta* is the most cultivated species. It is sometimes named the "Sago palm" or "King Sago palm," which is misleading, since the common Sago palm, *Metroxylon sagu*, belongs to the Arecaceae. *Cycas revoluta* is a slow-growing plant that reaches 2–5 m in height. It is native to southern Japan, and has been naturalized throughout temperate and tropical habitats.

APPLICATIONS TO HEALTH PROMOTION AND DISEASE PREVENTION – ADVERSE EFFECTS AND REACTIONS

Almonds, Apricots, and Peaches

Seeds from sweet almonds are a valuable food source. They contain very low levels of carbohydrates, and may therefore be made into flour for low-carbohydrate diets or for patients suffering from diabetes mellitus or any other form of glycosuria. Almond flour is gluten-free, and is therefore a popular ingredient in cookery in place of wheat flour for gluten-sensitive people, and people with wheat allergies and celiac disease. In addition, almonds are a rich source of riboflavin, magnesium, manganese, and, especially, vitamin E (alpha tocopherol), containing 24 mg/100 g (USDA, 2008). They are also rich in monounsaturated fatty acids, and almonds in the daily diet reduced LDL cholesterol by as much as 9.4%, reduced the LDL : HDL ratio by 12.0%, and increased HDL cholesterol by 4.6% (Jenkins *et al.*, 2002). Claimed health benefits of almonds furthermore include improved complexion, improved transition of food through the colon, and even the prevention of cancer. Recent research associates the inclusion of almonds in the diet with elevating the blood levels of beneficial high density lipoproteins, and lowering the levels of low density lipoproteins. High concentrations of phenolics and flavonoids in the testa provide for antioxidative efficacy (Wijeratne *et al.*, 2006). Bitter almonds (*Prunus amygdalus* var. *amara*) in particular accumulate substantial amounts of the cyanogenic di-glycoside amygdalin (D-mandelonitrile-β-D-gentiobioside; Figure 14.1) in their seeds. The seeds can contain up to 5% amygdalin (∼1 mg hydrogen cyanide per seed), and 10–15 seeds are considered lethal for children while 50–60 seeds represent a critical amount for adults (for lethal dose of cyanide, see below). Seeds of sweet almonds (*Prunus amygdalus* var. *dulcis*) contain much lower levels of cyanide; however, up to 2% of sweet almond seeds are bitter – i.e., contain amounts of amygdalin comparable to *P. amygdalus* var. *amara*.

Like almonds, both *Prunus armeniaca* (apricot) and *P. persica* (peach) accumulate amygdalin in their seeds. The amount of cyanide in seeds of apricots ranges from 0.05 to about 4 mg/g, with an average amount of 0.5 mg of cyanide, but varies widely depending on a variety of poorly defined factors, including cultivation practices, variety, and moisture content. In general, the HCN content in peach kernels is lower than in bitter almond or apricot kernels, and ranges from 0.4 to 2.6 mg/g. In addition to amygdalin, apricot and peach seeds also contain prunasin (the corresponding monoglycoside) and several minor cyanogenic glycosides, including amygdalinic acid, mandelic acid β-D-glucopyranoside, benzyl β-gentiobioside, and benzyl β-D-glucopyranoside (Fukuda *et al.*, 2003). Like bitter almonds, apricot kernels can sometimes be strongly bitter tasting. Consumed excessively, they can produce severe symptoms of cyanide poisoning or even death.

Flaxseed

Flaxseed (or linseed) is a valuable oil seed that is very low in cholesterol and sodium (USDA, 2008). It is also a good source of magnesium, phosphorus, and copper, and an excellent source of dietary fiber, thiamin, and manganese. At the same time, flaxseed can contain considerable amounts of cyanogenic diglycosides, primarily linustatin and neolinustatin, together with the corresponding monoglycosides linamarin and, at lower concentrations, lotaustralin (Figure 14.1). The cyanide content of different cultivars of flaxseed can range from 124 to 196 µg/g (Chadha *et al.*, 1995). The seeds contain about 35–45% oil, and are a desirable food product because of the high content of α-linolenic acid and lignans. Cyanogenic compounds are generally not detectable in processed flaxseed oil.

Lima bean

Dry, mature lima bean seeds are very low in saturated fat, cholesterol, and sodium (USDA, 2008). They are also a good source of protein, iron, magnesium, phosphorus, potassium, copper, dietary fiber, folate, and manganese (USDA, 2008). Depending on genotype, lima beans contain varying concentrations of cyanogenic glycosides, primarily linamarin (Figure 14.1), ranging from 2.1 (white seeds from Burma) to 3.1 mg HCN/g in seeds from Puerto Rico (black). The modern high-yielding and generally white varieties contain much less cyanide. Clinical toxicity in humans following the ingestion of lima beans is not well documented.

Macadamia nuts

These seeds are a valuable food due to their very low cholesterol and sodium content. Furthermore, they are rich in thiamin, and an excellent source of manganese (USDA, 2008). All *Macadamia* spp. acumulate cyanogenic glycosides (proteacin and dhurrin, Figure 14.1) in their seeds. Cyanide concentrations in the commercially used seeds are low (∼4.05 µg HCN/g

133

fresh weight), while *Macadamia whelanii* and *M. ternifolia* accumulate considerably higher amounts of cyanogenic glycosides in their seeds (260 μg/g in *M. ternifolia*) and are considered inedible (Dahler *et al.*, 1995). Indigenous people in Australia, however, process these seeds to reduce the cyanide content so they can use these species as well.

Cycad seeds

The cycad plants have a long history of use as food and medicine. In traditional Chinese medicine, cycad seeds are used to treat hypertension, musculoskeletal disorders, gastrointestinal distress, cough, and amenorrhea. Although the seeds are also frequently consumed as a source of carbohydrates, detailed data on nutritional facts are not available. Cycad seeds contain the toxic compounds cycasin (0.2–0.3%; Figure 14.1) and neocycasin (methylazoxymethanol β-D-glycosides), which are unique toxins present in cycad species (DeLuca *et al.*, 1980). These azoxyglucosides are glycosides of the same aglycone, methylazoxymethanol. The glycosides do not contain cyano groups, and thus are considered "pseudo-cyanogens," but can decompose to yield HCN. However, the liberation of cyanide from this process is a minor pathway compared to the formation of nitrogen gas, formaldehyde, and methanol. While acute poisoning has been observed arising after consumption of cycad seeds, the role of HCN as the agent determining toxicity is not fully resolved. Patients suffering from acute poisoning after consumption of cycad seeds and tested for blood cyanide concentrations had elevated blood cyanide levels, but the blood cyanide concentrations were below the values (0.5–1 mg/l) causing serious toxicity. In addition to acute poisoning, neuro-degenerative diseases (WP ALS-PD; Western Pacific Amyotrophic Lateral Sclerosis/Parkinsonism-Dementia Complex) associated with consumption of cycad seeds have been reported (Cox & Sacks, 2002). However, the non-protein amino acid β-methyl-amino-L-alanine (BMAA) is the most likely neurotoxic compound, rather than a cyanogenic compound.

TOXICITY AND DETOXIFICATION OF CYANIDE

In general, cyanide-containing compounds in nuts and seeds are detrimental rather than beneficial agents. However, since many cyanogenic nuts and seeds are of high nutritive value, these edibles have been used for a long time and in many cases food safety has been substantially improved by breeding low-cyanogenic plant varieties. Today, extremely low-cyanogenic cultivars of naturally sometimes high-cyanogenic plants, such as *Prunus* sp., lima bean, and *Macadamia* spp., exist. Cyanogenic nuts and seeds used for human consumption, or the products derived from them, generally — if properly processed — are below critical cyanide thresholds.

When consumed in low (i.e., sub-lethal) doses, hydrogen cyanide in mammals can be efficiently metabolized. The major defense of the organism to counter the toxic effects of cyanide is its conversion to thiocyanate, mediated by the enzyme rhodanese (sulfur transferase) located in mitochondria. The enzymatic detoxification requires sulfur donors, which are mostly provided from the dietary sulfur amino acids cysteine and methionine. Rhodanese catalyzes *in vitro* the formation of thiocyanate and sulfite from cyanide and thiosulfate or other suitable sulfur donors (Figure 14.3), while *in vivo* the enzyme is multifunctional.

$$\text{HCN} + \text{S}_2\text{O}_3^{2-} \xrightarrow{\text{Rhodanese}} \text{SO}_3^{2-} + \text{SCN}^-$$

Hydrogen cyanide   Thiosulfate   Sulfite   Thiocyanate

FIGURE 14.3

Detoxification of cyanide. Rhodanese is a mitochondrial enzyme which detoxifies cyanide by converting it to thiocyanate. The enzyme contains an active disulfide group, which reacts with the thiosulfate and cyanide. This detoxification requires sulfur donors, which are provided by dietary sulfur-containing amino acids. Thiocyanate is excreted in the urine.

Although most cyanogenic foods in industrialized countries are safe, and lower doses of hydrogen cyanide can be efficiently metabolized by humans, HCN released from cyanide-containing precursors is a strong respiratory poison, and the potential toxicity of cyanogenic food plants should be kept in mind. The toxicity of cyanide to vertebrates is mainly based on inhibition of the mitochondrial respiration pathway. Hydrogen cyanide inactivates the enzyme cytochrome oxidase in the mitochondria of cells by binding to the Fe^{3+}/Fe^{2+} ion contained in the enzyme. This causes a decrease in the utilization of oxygen in the tissues. Cyanide causes an increase in blood glucose and lactic acid levels, and a decrease in the ATP/ADP ratio, indicating a shift from aerobic to anaerobic metabolism. Cyanide activates glycogenolysis and shunts glucose to the pentose phosphate pathway, decreasing the rate of glycolysis and inhibiting the tricarboxylic acid cycle. Hydrogen cyanide ingestion reduces the energy availability in all cells, but its effect is most immediate on the respiratory system and the heart. This effect is further supported by the occupation of the oxygen-binding site in hemoglobin by cyanide, thus reducing oxygen transport capacities in blood. Beyond this, cyanide can inhibit several other metalloenzymes, most of which contain iron, copper, or molybdenum (e.g., alkaline phosphatase), as well as enzymes containing Schiff base intermediates (e.g., 2-keto-4-hydroxyglutarate aldolase). Cyanide that is released from cyanogenic glycosides in the gastrointestinal tract is readily absorbed by most animals. Oral lethal doses of HCN are 2.0 mg/kg body weight for a cat, 2.0 mg/kg for a sheep, 3.7 mg/kg for a mouse, 4.0 mg/kg for a dog, and 10 mg/kg for a rat. The oral lethal dose of HCN for humans is 0.5–3.5 mg/kg body weight, or about 50–250 mg for a typical adult human (Ballhorn *et al.*, 2009, and references therein).

In humans, the clinical features associated with the ingestion of cyanogenic glycosides mimic cyanide poisoning, and the severity of the intoxication correlates to the dose of active cyanogenic compounds. The onset of vomiting, abdominal pain, weakness, dyspnea, and diaphoresis, followed by convulsions, stupor, disorientation, hypotension, metabolic acidosis, coma, respiratory failure, and cardiovascular collapse, develops after exposure to high doses of cyanogenic glycosides (Akintonwa & Tunwashe, 1992). In addition to acute cyanide poisoning, chronic sub-lethal dietary cyanide frequently cause neuronal diseases, such as tropical ataxic polyneuropathy, Konzo, or WP ALS-PD, as well as reproductive effects including lower birth rates and increased numbers of neonatal deaths (Cox & Sacks, 2002).

Despite their potential to release highly toxic HCN, cyanogenic glycosides have been used as nutritional supplements and in cancer treatment. However, amygdalin, also sometimes falsely referred to as vitamin B17 or laetrile, is neither a vitamin nor an efficient pharmaceutical. Amygdalin and laetrile generally are different chemical compounds. Laetrile, which was patented in the US, is a semi-synthetic molecule sharing part of the amygdalin structure, while the "laetrile" made in Mexico is usually amygdalin — the natural product obtained from crushed apricot pits, or neoamygdalin. Clinical trials of amygdalin revealed no beneficial effect in cancer treatment, and considerable numbers of patients included in the studies suffered from cyanide toxicity (Milazzo *et al.*, 2006).

SUMMARY POINTS

- Many cyanogenic nuts and seeds have a long history of use as a food source or medicine, and nowadays some species are crop plants of world-economic importance.
- The concentrations of cyanide-containing compounds in nuts and seeds vary widely; however, in some species, such as *Prunus amygdalus* (almonds), *P. armeniaca* (apricots), and *Phaseolus lunatus* (lima bean), they can reach toxic thresholds.
- Although potentially toxic, cyanogenic nuts and seeds are often characterized by exceptionally high nutritive values. All of the economically most important cyanogenic nuts and seeds (i.e., almonds, macadamia nuts, lima beans, and flaxseed) contain health-promoting compositions of fatty acids, minerals, vitamins, and carbohydrates.

- Taking into consideration the possible health risk arising from consumption of inappropiately processed plant materials or the selection of highly cyanogenic crop plant genotypes, cyanogenic nuts and seeds represent excellent food sources.

References

Akintonwa, A., & Tunwashe, O. L. (1992). Fatal cyanide poisoning from cassava-based meal. *Human & Experimental Toxicology, 11,* 47–49.

Ballhorn, D. J., Kautz, S., & Rakotoarivelo, F. P. (2009). Quantitative variability of cyanogenesis in *Cathariostachys madagascariensis* – the main food plant of bamboo lemurs in southeastern Madagascar. *American Journal of Primatology, 71,* 305–315.

Chadha, R. K., Lawrence, J. F., & Ratnayake, W. M. (1995). Ion chromatographic determination of cyanide released from flaxseed under autohydrolysis conditions. *Food Additives and Contaminants, 12,* 527–533.

Cox, P. A., & Sacks, O. W. (2002). Cycad neurotoxins, consumption of flying foxes, and ALS-PDC disease in Guam. *Neurology, 58,* 956–959.

Dahler, J. M., McConchie, C. A., & Turnbull, C. G. N. (1995). Quantification of cyanogenic glycosides in seedlings of three *Macadamia* (Proteaceae) species. *Australian Journal of Botany, 43,* 619–628.

De Luca, P., Moretti, A., Sabato, S., & Siniscalco Gigliano, G. (1980). The ubiquity of cycasin in cycads. *Phytochemistry, 19,* 2230–2231.

FAOSTAT (2009). http://faostat.fao.org/site/567/default.aspx#ancor

Fukuda, T., Ito, H., Mukainaka, T., Tokuda, H., Nishino, H., & Yoshida, T. (2003). Anti-tumor promoting effect of glycosides from *Prunus persica* seeds. *Biological & Pharmaceutical Bulletin, 26,* 271–273.

Jenkins, D. J. A., Kendall, C. W. C., Marchie, A., Parker, T. L., Conelly, P. W., Qian, W., et al. (2002). Dose response of almonds on coronary heart disease risk factors: blood lipids, oxidized low-density lipoproteins, lipoprotein(a), homocysteine, and pulmonary nitric oxide: a randomized, controlled, crossover trial. *Circulation, 106,* 1327–1332.

Jones, D. A. (1998). Why are so many food plants cyanogenic? *Phytochemistry, 47,* 155–162.

Milazzo, S., Lejeune, S., & Ernst, E. (2006). Laetrile for cancer: a systematic review of the clinical evidence. *Support. Care Cancer, 15,* 583–595.

Poulton, J. E. (1990). Cyanogenesis in plants. *Plant Physiology, 94,* 401–405.

USDA National Nutrient Database for Standard Reference Release 21 (2008). http://www.nal.usda.gov/fnic/foodcomp/search/

Wijeratne, S. S. K., Abou-Zaid, M. M., & Shahidi, F. (2006). Antioxidant polyphenols in almond and its coproducts. *Journal of Agricultural and Food Chemistry, 54,* 312–318.

CHAPTER 15

Allergies to Nuts and Seeds

Marie M. Cochard, Philippe A. Eigenmann
Child Allergy Unit, University Hospitals of Geneva, Geneva, Switzerland

137

LIST OF ABBREVIATIONS

IgE, immunoglobulin E
IL, interleukin
OAS, oral allergy syndrome
FCεRI, high-affinity IgE receptor

INTRODUCTION

In recent decades the number of patients with different allergic diseases has grown rapidly in the industrialized world. The prevalence of food allergies is following a similar trend, with increasing numbers. In this context, allergy to peanuts, for example — one of the most severe allergies — doubled over a period of 5 years in North America (Sicherer *et al.*, 2003).

Besides the usual food allergens (e.g., milk and egg white), "new" allergies to certain food proteins have emerged; among these are several to nuts and seeds. This chapter reviews the major features of allergy, with a specific focus on nuts and seeds.

Nuts & Seeds in Health and Disease Prevention. DOI: 10.1016/B978-0-12-375688-6.10015-5

PREVALENCE OF ALLERGY

Allergies, and in particular food-related allergies, are a growing entity in the industrialized population. The true prevalence and risk factors are still not well known. Prevalence has different patterns in different age groups, and recent meta-analysis reported occurrences varying from 3 to 35% (Rona *et al.*, 2007). Of note, the number of correctly diagnosed food allergies largely differs from the number of perceived reactions to any food in population-based studies (Zuidmeer *et al.*, 2008).

Since 2005, a large prospective birth cohort study (EuroPrevall) has been running in nine countries across Europe. This study will provide helpful results regarding the prevalence, patterns, and associated factors of food challenge-diagnosed food allergies in European children (Keil *et al.*, 2009).

A study in the United States has shown that 5.3% of adults reported a doctor-diagnosed food allergy and 9.1% reported a self-perceived food allergy. The prevalence of food allergy to the eight most common allergens (peanut, tree nuts, egg, milk, wheat, soybeans, fish, and crustacean shellfish) was reported as 2.7% (Vierk *et al.*, 2007).

WHAT IS AN ALLERGIC REACTION?
Pathophysiology

An allergy is an abnormal, inappropriate, exaggerated reaction of the immune system subsequent to contact with a foreign protein. These foreign proteins, usually well tolerated by the healthy population, are known as allergens. The term "hypersensitivity" is often used as a synonym for allergy.

As an example, immune reactions related to foods require an initial encounter with the antigen. The initial sensitization procedure to a food allergen occurs mostly via the gut mucosa, or by skin contact. A primary sensitization via aerosolized foods has also been described (Eigenmann, 2009). There are two types of immunological reaction, according to the mechanism involved; IgE mediated, or non-IgE mediated. In the IgE mediated type, initial exposure to the antigen leads to a priming of naïve T cells via the antigen presenting cells. This stimulation of T cells induces the production of interleukins (IL 4 and 13), which initiate food-specific IgE production by plasmocytes.

IgE antibodies present in the serum will then bind to FCεRI receptors on mast cells present in potential target organs (e.g., the skin and the respiratory tract). Upon re-exposure to the food, antigens will activate FCεRI-bound specific IgE and initiate degranulation of mast cells. These cells will release inflammatory mediators (interleukins, cytokines, etc.), activating cell-mediated inflammation and the various symptoms recognized as being an allergy (Figure 15.1).

Symptoms

Food-induced allergy may present with various symptoms. In 90% of reactions the skin is involved, with either acute urticaria (hives) or angioedema. These symptoms usually appear within minutes of ingesting the triggering food. Conjunctival and nasal symptoms are also common in acute systemic food reactions, but occur infrequently as isolated manifestations. Gastrointestinal symptoms are also very common, usually starting within minutes to 2 hours after the food has been ingested. Symptoms are various, and include nausea, vomiting, abdominal pain or cramping, and diarrhea. Respiratory reactions manifest as acute wheezing, respiratory distress, dysphonia, and dysphagia.

The symptoms of food allergy can range from very mild, to life-threatening reactions with multisystemic severe shock. A reaction involving several organs is known as "anaphylaxis," which is generally graded in four stages. The degree of reaction determines the emergency

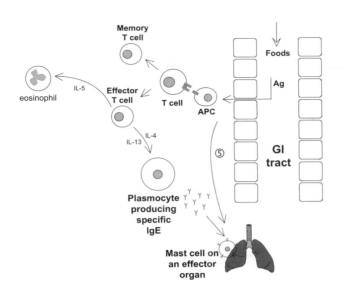

FIGURE 15.1

Mechanism of food IgE-mediated hypersensitivity. (1) Antigen (AG) passes the gut mucosa; (2) Antigen is taken up by antigen presenting cell (APC) and presented to T cell; (3) T cell develops to become a memory T cell to this antigen, or an effector T cell which stimulates plasmocytes to produce antibodies (AB) against the AG or eosinophil; (4) Antibodies coat mast cells in effector organs like the lung, the nasal mucosa, or the skin; (5) On the next exposure to the antigen, mast cells will be activated directly.

treatment, as well as the prognosis and long-term treatment. One helpful rating scale often used is the Mueller's Anaphylaxis Scale (Table 15.1).

Pollen allergy is very common. Up to 25% of the northern European population suffers from allergic rhinitis and conjunctivitis. Pollen-related food allergy — i.e., oral allergy syndrome (OAS) — is reported by more than half of the patients with seasonal rhinitis. Symptoms are usually restricted to the oral cavity, with itching and light mucosal swelling, but swelling of the laryngeal mucosa may also happen and cause significant distress. Symptoms of OAS are usually self-limited, and disappear rapidly.

Treatment and clinical management

An allergic reaction is an emergency, and needs rapid and efficient management to avoid a potentially fatal issue. The Mueller classification of reactions (Table 15.1) indicates which medication is needed. Table 15.2 shows a management plan according to the degree of severity.

TABLE 15.1 Grading of the Anaphylactic Reaction

Grade	Reactions
I	Generalized urticaria, pruritis, malaise, anxiety
II	One or more symptoms of grade I with a minimum of two of the following: angiodema (if isolated also grade II), abdominal pain, nausea, vomiting, diarrhoea, vertigo, chest pain
III	One or more symptoms of grade I or II with a minimum of two of the following: stridor, wheezing, respiratory distress, dysphagia, dysarthria, confusion, death anxiety
IV	One or more symptoms of grades I—III with a minimum of two of the following: cyanosis, arterial hypotension, collapse, loss of consciousness, incontinence (urinary, fecal)
Other (late onset reaction or unusual)	Serum sickness, generalized vasculitis, glomerulonephritis, nephrotic syndrome, neurological/hematological syndromes

Anaphylactic reactions can be graded from I, localized and mild symptoms, to IV, involving potentially life-threatening reactions.
Reproduced from: Wassenberg, J., Lauener, R., Kuenzli, M., Eigenmann, P., Eng, PI, Hofer, M. (2006). Recommandations pour la prise en charge de l'allergie aux venins d'hyménoptères chez l'enfant. *Paediatrica*, 17(3).

TABLE 15.2 Treatment of an Anaphylactic Reaction	
Grade	**Emergency Treatment**
I	Oral antihistaminic
II	Oral/i.v. antihistaminic
	Oral/i.v. steroid
III	Oxygen
	Epinephrine 0.01 mg/kg i.m.
	Antihistaminic i.v.
	Steroid i.v.
	If asthma: salbutamol aerosol
	If laryngeal edema, epinephrine aerosol
IV	ABC and same as grade III

Treatments for anaphylaxis have to be adapted to the severity of symptoms — i.e., oral antihistaminic for mild and epinephrine for severe symptoms.

All patients with an anaphylactic or allergic reaction should have a subsequent allergy work-up by a trained specialist. This assessment relies on a good history, skin prick tests, serum IgE measurement, and, in selected cases, oral food challenges. The patient should be advised on how to avoid the allergen, and how to handle an accidental reaction. There is no specific treatment for food allergy, but numerous novel strategies for efficient therapy are being studied, such as sublingual/oral immunotherapy, injection of anti-IgE antibodies, cytokine/anticytokine therapies, Chinese herbal therapies, and immunotherapies with engineered proteins and strategic immunomodulators (Sicherer & Sampson, 2009).

ECONOMIC IMPACT OF FOOD ALLERGY

Very little is known regarding the economic impact of food allergy. However, it is evident that food allergy induces a cost related to individual health expenses, as well as indirect costs to families and the community. A recent study by a British and Dutch team reported on a questionnaire to measure the cost of food allergy in Europe. They compared the answers given in questionnaires from households with or without a food-allergic member. Annual direct costs were €1000 higher, indirect costs €2500 higher, and health costs €274 higher for households with a food-allergic member (Fox *et al.*, 2009).

ALLERGIES TO NUTS AND SEEDS
Peanut allergy

The peanut is the seed of an annual plant from the Leguminosae family. More than 2.4 billion pounds of peanuts are consumed each year in the USA. They are eaten fresh, roasted, or boiled. Peanuts account, with other nuts, for the majority of severe allergic reactions. Peanuts have been implicated in the majority (59%) of deaths due to anaphylaxis in individuals with fatal reactions; data documented by the Food Anaphylaxis Fatality Registry in the United States (Sicherer *et al.*, 2003). These life-threatening reactions can be elicited with only a small amount of protein allergen. Peanut allergy tends to manifest itself early in life, and only some affected patients will outgrow their allergy.

The prevalence of peanut allergy is increasing rapidly. In 2002, 1.1% (i.e., 3 million) of the population of the United States reported hypersensitivity to peanut and tree nuts, while the proportion in 1997 was only 0.6%. The cause of this increasing prevalence is unknown, but may be related to the overall increase of atopic diseases, the increase in peanut consumption, or early prenatal sensitization through maternal exposure. The food matrix also plays an important part. Roasting enhances allergenicity by increasing the stability of major peanut allergens. The peanut is a legume, and its protein shares homology with other legumes such as soy, lupine, and peas, leading to cross-reactive positive diagnostic tests (< 95%). Most importantly, this cross-reactivity is without clinical consequences — there is an absence of clinical reactions to other legumes in most peanut allergic reactions (Pansare & Kamat, 2009).

Peanut allergy has an impact on socialization, and affects the quality of life not only of allergic patients but also of their families, who live in constant fear of a reaction.

Hazelnut allergy

Hazelnuts are commonly used in the food industry, in particular in chocolate and pastries. Worldwide production is estimated at 600,000 tonnes of shelled hazelnuts per year (www.fao. org). Hazelnut allergy can range from mild symptoms to a severe anaphylactic reaction. The milder form is related to the "oral allergy syndrome," a food-induced local reaction of the oral and laryngeal mucosa after primary respiratory sensitization to a cross-reacting pollen (e.g., birch tree pollen). Hazelnuts can also elicit more severe symptoms, and a full-blown anaphylactic reaction. Together with peanuts, hazelnuts account for 90% of fatalities due to food allergy. This type of severe allergic reaction occurs in subjects with a primary sensitization directly to the food (Flinterman *et al.*, 2008).

Other Tree Nut Allergies (Almond, Walnut, Pecan, Macadamia, Brazil, Cashew, Pistachio, etc.)

The list of edible tree nuts is large: almonds, walnuts, pecans, cashews, pistachios, Brazils, macadamias, pine nuts, chestnuts, black walnuts, and coconuts. Most of them are shelled and used in products such as cookies, ice cream and confectionary, and as snacks. Walnuts, almonds, cashews, and Brazils are most frequently cited as causative allergy triggers, whereas pistachios, macadamias, and pine nuts are less commonly responsible for allergic reactions. Chestnut allergy is most often seen as a cross-reactive allergy with natural rubber latex. Coconut rarely causes an allergy, and has been seldom reported in the literature.

Tree nut allergy severity and frequency of sensitization vary between the five continents, as well as within some countries. Onset of tree nut allergy is generally observed later than peanut allergy; however, it most commonly appears during childhood. There is a potential cross-reactivity between tree nuts and peanuts, and cross-reactivity between different tree nuts has also been reported. However, separate tree nut allergies exist without cross-reactivity. In fact, it is prudent to advise avoidance of all tree nuts even when allergy to only one type of nut is diagnosed, because of the high risk of contamination during processing (Teuber *et al.*, 2003).

Sesame allergy

Sesame, a member of the Pedaliaceae family, has been cultivated for human use since 2450 BC, when it was found in Babylon. It is still widely cultivated in many countries, especially in India and the Middle East, where it is eaten in paste or pastries. Recently, sesame has become a very common ingredient in Western food industries, where it is used as a garnish and in many bakery products. Sesame seeds contain 50—60% oil; this oil is used for consumption, but also in pharmaceutical and cosmetic products. Prevalence of sesame allergy is unknown globally, but anaphylactic reactions have been reported in the literature since 1950 and are constantly increasing. In Australian children, the prevalence of sesame allergy has been estimated at 0.42%, placing it fourth, behind egg, milk, and peanut allergies. In Israeli children it is the third most common food allergy, with a prevalence of 1.7% — higher even than that of peanut allergy. Sesame allergy occurs in individuals of all ages, from infancy to adulthood. Symptoms vary from subject to subject and include all allergic patterns, although there is a predominance of IgE mediated symptoms. Contact dermatitis is also seen, but this is generally attributed to the use of sesame oil. Sensitization occurs by contact with food containing sesame proteins, but also by occupation (in a bakery or the food industry) or use in therapeutics (intramuscular injections of progesterone, ophthalmic preparations, etc.) and cosmetics (lipsticks, ointments, soaps, plasters, etc.). It appears that sesame allergy is rarely outgrown (Gangur *et al.*, 2005).

Lupine allergy

Lupine, a member of the legumes family, was legally licensed for use in industrially prepared foods in the mid-1990s. It may be used as an additive or a substitute for wheat flour, and added to breads, pasta, jams, sauces, and meat products. Since then, allergic reactions have been described in adult patients known to have a peanut allergy — for example, after inhalation of lupine flour in employees working in the food industry, or as a non-cross-reactive food allergy (Wassenberg & Hofer, 2007).

OTHER SEED ALLERGIES (SUNFLOWER, MUSTARD, QUINOA, BUCKWHEAT, ETC.)

In theory, each individual seed could be responsible for an allergic reaction, but in practice few cases have been published. Recently, two cases of sunflower seed allergy have been reported in children with positive oral challenge. The first patient responded positively to the food challenge at a low dose (2 g), but the second ingested 14 g and developed a severe systemic reaction, requiring treatment with epinephrine (Caubet *et al.*, 2010). The same report outlined a case of an anaphylactic reaction with positive oral challenge to pumpkin seeds. Seeds were eaten as snacks in each of these cases.

Mustard plants are members of the Brassicaceae family, and seeds are used as a condiment, mainly in sauces and as seasoning in dishes, or as an ingredient in salad dressings. France is the largest European producer and consumer of mustard, and in this country mustard is the fourth most important allergen for children, after eggs, peanuts, and cows' milk (Rance, 2003). A study has suggested an association between mustard hypersensitivity and mugwort pollen sensitization (Figueroa *et al.*, 2005).

Quinoa, which has long been a staple food in the Andean regions, has now reached Europe, and is largely available in grocery shops. In 2009 the first case of food allergy to quinoa was described in France (Astier *et al.*, 2009), and more can be expected as this food is increasingly consumed in Europe.

Allergy to buckwheat has been recorded since 1961, and was first described in Japan. Buckwheat is a common seed in Asia and Russia, and its consumption is now increasing in Europe and the USA, where it is used in pasta, sandwiches, and pancakes. It has good nutritional values, and is used as a substitute for wheat flour for celiacs, as it does not contain gluten. Buckwheat is known to induce a respiratory allergic reaction in some patients, with rhinitis and asthma. However, these symptoms are rare, and might be seen in patients using a buckwheat pillow, or in those exposed to the seed when working in pancake restaurants or noodle factories.

In summary, allergies to nuts and seeds are among the most common allergies seen in the Western world. They are characterized by symptoms with a high potential for severity (IgE mediated symptoms), which merit an appropriate diagnostic work-up and require measures to prevent subsequent accidental reactions. Allergies to nuts and seeds are usually long-lasting, and are currently incurable. Continuous introduction of new foods into the Western diet will inevitably lead to new seed and nut allergies.

SUMMARY POINTS

- Allergy is a common symptom of food hypersensitivity causing issues ranging from light to severe with life-threatening symptoms.
- Potentially, all proteins may induce an allergic reaction.

- Peanuts and tree nuts are responsible for severe food-induced reactions.
- Nuts and seeds are first-line allergens.
- No definite treatment is yet available for nut and seed allergies.

References

Astier, C., Moneret-Vautrin, D. A., Puillandre, E., & Bihain, B. E. (2009). First case report of anaphylaxis to quinoa, a novel food in France. *Allergy, 64,* 819—820.

Caubet, J. C., Hofer, M. F., Eigenmann, P. A., & Wassenberg, J. (2010). Snack seeds allergy in children. *Allergy, 65,* 136—137.

Eigenmann, P. A. (2009). Mechanisms of food allergy. *Pediatric Allergy and Immunology, 20,* 5—11.

Figueroa, J., Blanco, C., Dumpierrez, A. G., Almeida, L., Ortega, N., Castillo, R., et al. (2005). Mustard allergy confirmed by double-blind placebo-controlled food challenges: clinical features and cross-reactivity with mugwort pollen and plant-derived foods. *Allergy, 60,* 48—55.

Flinterman, A. E., Akkerdaas, J. H., Knulst, A. C., van Ree, R., & Pasmans, S. G. (2008). Hazelnut allergy: from pollen-associated mild allergy to severe anaphylactic reactions. *Current Opinion in Allergy Clinical Immunology, 8,* 261—265.

Fox, M., Voordouw, J., Mugford, M., Cornelisse, J., Antonides, G., & Frewer, L. (2009). Social and economic costs of food allergies in Europe: development of a questionnaire to measure costs and health utility. *Health Services Research, 44,* 1662—1678.

Gangur, V., Kelly, C., & Navuluri, L. (2005). Sesame allergy: a growing food allergy of global proportions? *Annals of Allergy, Asthma and Immunology, 95,* 4—11.

Keil, T., McBride, D., Grimshaw, K., Niggemann, B., Xepapadaki, P., Zannikos, K., et al. (2009). The multinational birth cohort of EuroPrevall: background, aims and methods. *Allergy, 65,* 482—490.

Pansare, M., & Kamat, D. (2009). Peanut allergies in children — a review. *Clinical Pediatrics (Phila), 48,* 709—714.

Rance, F. (2003). Mustard allergy as a new food allergy. *Allergy, 58,* 287—288.

Rona, R. J., Keil, T., Summers, C., Gislason, D., Zuidmeer, L., Sodergren, E., et al. (2007). The prevalence of food allergy: a meta-analysis. *Journal of Allergy and Clinical Immunology, 120,* 638—646.

Sicherer, S. H., Munoz-Furlong, A., & Sampson, H. A. (2003). Prevalence of peanut and tree nut allergy in the United States determined by means of a random digit dial telephone survey: a 5-year follow-up study. *Journal of Allergy and Clinical Immunology, 112,* 1203—1207.

Sicherer, S. H., & Sampson, H. A. (2009). Food allergy: recent advances in pathophysiology and treatment. *Annual Review of Medicine, 60,* 261—277.

Teuber, S. S., Comstock, S. S., Sathe, S. K., & Roux, K. H. (2003). Tree nut allergy. *Current Allergy and Asthma Reports, 3,* 54—61.

Vierk, K. A., Koehler, K. M., Fein, S. B., & Street, D. A. (2007). Prevalence of self-reported food allergy in American adults and use of food labels. *Journal of Allergy and Clinical Immunology, 119,* 1504—1510.

Wassenberg, J., & Hofer, M. (2007). Lupine-induced anaphylaxis in a child without known food allergy. *Annals of Allergy, Asthma and Immunology, 98,* 589—590.

Zuidmeer, L., Goldhahn, K., Rona, R. J., Gislason, D., Madsen, C., Summers, C., et al. (2008). The prevalence of plant food allergies: a systematic review. *Journal of Allergy and Clinical Immunology, 121,* 1210—1218.

PART 2

Effects of Specific Nuts and Seeds

Potential Usage of African Ebony (*Diospyros mespiliformis*) Seeds in Human Health

Eliton Chivandi, Kennedy H. Erlwanger
School of Physiology, Faculty of Health Sciences, University of the Witwatersrand, Johannesburg, Republic of South Africa

147

LIST OF ABBREVIATIONS

ALA, α-linolenic acid
ANF, antinutritional factor
DHA, docosahexaenoic acid
DSM, *Diospyros mespiliformis*
EFA, essential fatty acid
LA, linoleic acid
PUFA, polyunsaturated fatty acid

INTRODUCTION

The search for, development, and domestication of fruit trees with potential as complements to staple food crops is increasing. *Diospyros mespiliformis* (DSM), also known as the African ebony tree (family Ebenaceae), is a protected species in South Africa (Janick & Paull, 2008). Its distribution stretches from Southern Africa to the Sudan (Dalziel, 1937), and it grows on

Nuts & Seeds in Health and Disease Prevention. DOI: 10.1016/B978-0-12-375688-6.10016-7

TABLE 16.1 Characteristics of the Seeds of *Diospyros mespiliformis*

Seed Characteristic	
Number per fruit	4—6
Shape	bean-shaped
Colour	brown

Adapted from Janick and Paull (2008).

fertile riverine and anthill soils (Janick & Paull, 2008). Subsistence farmers recognizing its multiple uses normally leave it standing in crop fields (Palgrave, 1981). Research on the potential uses of DSM has focused largely on the ethno-medical potential of the tree's leaf, bark, and root extracts (Adzu *et al.*, 2002). This chapter profiles the potential usage of DSM fruit seeds in human nutrition and health.

BOTANICAL DESCRIPTION

Diospyros mespiliformis is a densely crowned evergreen tree, attaining heights of 15—50 m. The stem is rounded, buttressed, and has grey-black or black bark (Palgrave, 1981). The bark is smooth in young trees, but rough with regular scales in older trees. The leaves are alternate, oblong to elliptical in shape, 4—7 cm long and 1.5—5.5 cm wide, shiny above, paler beneath, and hairy when young (Janick & Paull, 2008). The white flowers are fragrant. Dioecious in nature, pollination is by bees. In Southern Africa, the tree flowers during the rainy season (October to November) and fruit matures during the April to September dry season (Palgrave, 1981; Janick & Paull, 2008). Its globose, fleshy fruit is about 3 cm in diameter, greenish and pubescent when young, yellowish to orange when ripe, and with an enlarged calyx. The physical characteristics of *Diospyros mespiliformis* seeds are summarized in Table 16.1. The tree is easily propagated from seed (their dormancy is broken by soaking the seeds in hot water for a few minutes), coppice cuttings, seedlings, and root suckers (Janick & Paull, 2008).

HISTORICAL CULTIVATION AND USAGE

Historically, the tree was not cultivated. Nevertheless, due to its importance as a fruit tree, it was one of the trees that was spared when clearing land for subsistence agriculture (Palgrave, 1981). Its contribution to communal household food security, together with other indigenous fruits trees, was more manifest in agriculturally lean years. Various components of the tree were used in ethno-medicine.

PRESENT-DAY CULTIVATION AND USAGE

Currently, although the tree is still not routinely cultivated as is the case with conventional crop plants, recognition of its contribution to the ecosystem has seen increased efforts to propagate and conserve it. Usage of the tree and its by-products is broad. The bark, root, and leaf extracts are widely employed in ethno-medicine (Mohammed *et al.*, 2009). A leaf decoction is reportedly used in the treatment of malaria and headaches, and as an antihelminthic (Etkin, 1997). The bark extract reportedly provides good relief for coughing (Nwude & Ebong, 1980). Extracts from the plant have also been shown to have analgesic and antipyretic effects (Adzu *et al.*, 2002). The fruit is eaten fresh, or is dried and stored for later use. It is also used as an ingredient in the preparation of other foods and beverages, such as brandy (Janick & Paull, 2008). The woody component of the tree makes fine sculptures (Palgrave, 1981).

APPLICATIONS TO HEALTH PROMOTION AND DISEASE PREVENTION

Generally, the seeds are discarded after harvesting the fruit and use of the fruit pulp. Not much chemical characterization of the seed has been done. DSM fruit pulp largely consists of

TABLE 16.2 Proximate Composition, Soluble Sugars, and Starch Content of *Diospyros mespiliformis* Seed

Proximate Component	Fresh Mass (%)
Moisture	8.99
Crude protein (N × 6.25)*	5.44–5.46
Fat	5.46
Ash	2.89
Total carbohydrates	77.89
Soluble sugars & starch	
Total soluble sugars	14.61
Total starch	37.81

Modified from Ezeagu *et al.* (1996) and Petzke *et al.* (1997).
*Crude protein was estimated by multiplying the nitrogen content (N) by a factor of 6.25.

carbohydrates; it has little protein, but contains appreciable amounts of minerals (calcium, iron, magnesium, and phosporus) as well as vitamin C. DSM fruit pulp also has low concentrations of oxalate, phytate, saponin, and tannin (Umaru *et al.*, 2007).

DSM seeds: Potential in human nutrition and health

Generally, the nutrient and antinutritional factor (ANF) composition of seeds plays a crucial role in defining the beneficial and potentially beneficial effects of the given plant seed in human nutrition and health. Little has been published regarding the nutrient and ANF composition of DSM seeds. The proximate composition, soluble sugars, and starch contents of the seeds are summarized in Table 16.2. The seed has a high carbohydrate concentration but low protein content. As a human food resource, the soluble sugars and starch (both highly metabolizable energy sources) make DSM seeds a potential source of dietary energy. In drought years when conventional crops (cereals) fail to thrive, the highly energy dense seeds could be used to replace calories that would otherwise be derived from cereal grain. However, from a nutritional quality standpoint, the low protein content in the seed means that if there is a dependence on the seed as the "staple" food/calorie source then supplementation with protein concentrates becomes essential in order to guard against protein deficiency and subsequent malnutrition. The amino acid profile of the DSM seed further confirms it as a poor source of protein (Table 16.3). Over and above the low crude protein content in comparison to protein-rich plant seeds such as soya bean, ground nut, and sesame, dietary limiting amino acids such as lysine and methionine are in very low concentrations in DSM seeds. Protein malnutrition, especially in growing, recently weaned children, weakens their bodies, making them more susceptible to disease. Nutritional disorders such as kwashiorkor set in, negatively affecting both physical and mental development of children. This means that if DSM seeds are used to ameliorate food (energy) shortages when conventional crops fail, there is dire need to complement its use with some protein rich dietary supplement.

DSM seed oil: Potential applications in human nutrition and health

Although the oil yield from the DSM seed is low at 0.70 ± 0.17 % on a dry matter basis (Chivandi *et al.*, 2009), the proportion of various oil fractions (saturated, monounsaturated, and polyunsaturated fatty acids) opens up tremendous potential opportunities for its utilization in the improvement of human nutrition and health (Table 16.4). The diversity of the fatty acid profile lends the oil to potential exploitation in human nutrition and health, chiefly as a source of essential fatty acids (EFAs) and as a raw material for synthesis (*de novo*) of other critical lipids vital for supporting key physiological process in the body.

Linoleic acid (LA) constitutes more than 95% of the polyunsaturated fatty acid (PUFA) content of DSM seed oil, thus making the oil a potential source of LA, which has vast potential

TABLE 16.3 Amino Acid Composition of *Diospyros mespiliformis* Seeds

Amino Acid	Concentration (mg/gN)
Aspartic acid	507
Threonine	231
Serine	291
Glutamic acid	1002
Proline	324
Glycine	336
Alanine	295
Valine	287
Isoleucine	228
Leucine	398
Tyrosine	191
Phenylalanine	243
Histidine	126
Lysine	301
Arginine	501
Cysteine	151
Methionine	113
Tryptophan	92

Modified from Petzke *et al.* (1997).
The concentrations of 18 amino acids in *Diospyros mespiliformis* seeds as analyzed by ion-exchange chromatograph and expressed milligrams per gram nitrogen (mg/gN).

TABLE 16.4 Fatty Acid Profile of *Diospyros mespiliformis* Seed Oil

Fatty Acid	Fatty Acid Content (%)
C14:0 (myristic acid)	0.62 ± 0.04
C16:0 (palmitic acid)	30.06 ± 0.61
C18:0 (stearic acid)	7.74 ± 0.44
C20:0 (arachidic acid)	1.12 ± 0.02
Total saturated fatty acids	39.54
C14:1n7 (myristoleic acid)	0.05 ± 0.00
C16:1n7 (palmitoleic acid)	29.37 ± 0.38
Total monounsaturated fatty acids	29.42
C18:2n6 (linoleic acid — n6PUFA)	28.71 ± 1.79
C18:3n3 (α-linolenic acid — n3PUFA)	0.95 ± 0.61
Total polyunsaturated fatty acids	29.66
TPUFA : TSFA	0.75 : 1
n6PUFA : n3PUFA	30.22 : 1

Modified from Chivandi *et al.* (2009).
Lipid profile of oil extracted from *Diospyros mespiliformis* seeds expressed as mean $\pm$ SD.
TPUFA, total polyunsaturated fatty acids; TSFA, total saturated fatty acids.

physiological applications, and can be used as a raw material for *in vitro* and *de novo* synthesis of eicosanoids (Chivandi *et al.*, 2009). LA and α-linolenic acid (ALA) are EFAs that are vital in the maintenance of some of the key physiological functions of the human body. LA is essential for maintaining integrity of the skin, cell membranes, immune system and for the synthesis of eicosanoids (Dupont *et al.*, 1990). Eicosanoids, derived *de novo* from LA, are vital for reproductive, cardiovascular, renal, and gastrointestinal functions, and resistance to disease. Dietary supplementation with DSM seed oil could be useful in providing raw material for the synthesis of eicosanoids in the body.

ALA, which constitutes 1% of the DSM seed oil, is the precursor molecule of the omega-3 essential fatty acid family. It is converted in the body to docosahexaenoic acid (DHA) and eicosapentaenoic acid (EPA). DHA and EPA are two other members of the omega-3 group family. Research indicates that omega-3 fatty acids attenuate inflammation and help reduce the risk factors associated with chronic pathological conditions such as cardiovascular disease, cancer, and arthritis (Simopoulos, 1999). Omega-3s are highly concentrated in the brain, and appear to be particularly important for cognitive (brain memory and performance) and behavioral functions (Simopoulos, 1999). Studies have shown that infants who do not get enough omega-3 fatty acids from their mothers during pregnancy are at risk of developing vision and nerve problems (Simopoulos, 1999). Clinical signs and symptoms of omega-3 fatty acid deficiency include extreme fatigue, poor memory, dry skin, heart problems, mood swings or depression, and poor circulation (Simopoulos, 1999). Decreased tissue concentration of DHA has been implicated in the etiology of non-puerperal and postpartum depression (Levant *et al.*, 2008). Dietary deficiencies of DHA are associated with decreased hippocampal brain-derived neurotrophic factor gene expression and increased relative corticosterone response to an intense stressor (Levant *et al.*, 2008). DSM seed oil has the potential to be exploited as a source of metabolic raw material for both DHA and EPA in the form of ALA. In the DSM seed oil, ALA could be used as a dietary supplement to arrest the development and progression of DHA-deficiency triggered depression and other physiological conditions associated with decreased concentration of omega-3s in the body.

ADVERSE EFFECTS AND REACTIONS (ALLERGIES AND TOXICITY)

There is no published literature on toxicity directly attributable to DSM seeds. While Umaru and colleagues (2007) reported the fruit pulp of DSM as having very low levels of ANFs, Utsunomiya *et al.* (1998) reported fruits of other *Diospyros* species as being used to poison fish — an indication of the presence of toxic compounds in the fruit and, possibly, the seed. Contact with the leaves, bark, fruit, root, or wood dust of some members of the Ebenaceae family has been reported to cause dermatitis and irritation of the respiratory tract. Hausen (1981) contends that naphthoquinones occurring in some *Diospyros* species could be responsible for the development of allergic contact dermatitis. Such reports indicate that care should be taken when harvesting the fruit for extraction of the seeds, and also form a premise for the need to characterize the ANF content of the DSM seed. The bark, root and leaf extracts of DSM are known to contain a number of secondary plant metabolites, such as triterpenes and naphthoquinones, that are key to the ethno-pharmacological applications of the plant (Fallas & Thompson, 1968; Lajubutu *et al.*, 1995). Such metabolites are likely to be found as constituents of the DSM seed. Some plant metabolites interfere with protein digestion and absorption (tannins), and nutrient absorption from the gastrointestinal tract (lectins, phytates, oxalates and tannins), and some are toxic (saponins) (Umaru *et al.*, 2007). It is therefore important to fully characterize the ANF content of the DSM seed before wholesale recommendation is given for its use as a food resource.

SUMMARY POINTS

- *Diospyros mespiliformis* has a wide ecological distribution, and is recognized as an important fruit tree with medicinal properties.
- *Diospyros mespiliformis* seeds have high concentrations of soluble sugars and starch, making them possible energy sources in human diets.
- *Diospyros mespiliformis* seed oil is rich in the essential fatty acids linoleic acid and α-linolenic acid.
- *Diospyros mespiliformis* seed oil could be exploited as raw material for synthesis of omega-3s.
- Omega-3s are essential in the reduction of inflammation and chronic diseases.
- There is need to fully characterize the antinutritional factor content of *Diospyros mespiliformis* seed, and breed plant varieties with higher seed oil yields but maintaining the current fatty acid profile.

References

Adzu, B., Amos, S., Dzarma, S., Muazzam, I., & Gamaniel, K. S. (2002). Pharmacological evidence favouring folkloric use of diospyros mespiliformis in the relief of pain and fever. *Journal of Ethnopharmacology, 82,* 191–195.

Chivandi, E., Erlwanger, K. H., & Davidson, B. C. (2009). Lipid content and fatty acid profile of the fruit seeds of *Diospyros mespiliformis. International Journal of Integrative Biology, 5,* 121–124.

Dalziel, J. M. (1937). The useful plants of West Tropical Africa. *Crown Overseas Agent for Colonies and Administration.* London, UK: ISSN (ASIN) B000850VJ8. pp. 347–348.

Dupont, J., White, P. J., Carpenter, M. P., Schaefer, E. J., Meyidani, S. N., Elson, E. C., et al. (1990). Food uses and health effects of corn oil. *Journal of the American College of Nutrition, 9,* 438–470.

Etkin, N. L. (1997). Antimalarial plants used by Hausa in northern Nigeria. *Tropical Doctor, 27,* 12–16.

Ezeagu, I. E., Metges, C. C., Proll, J., Petzke, K. J., & Okinsoyinu, A. O. (1996). Chemical composition of some wild-gathered tropical plant seeds. *Food and Nutrition Bulletin, 17,* 275–285.

Fallas, A. L., & Thompson, R. H. (1968). Ebenaceae extractives – 111 – binaphthoquinones from *Diospyros* species. *Journal of the Chemical Society, 18,* 2279–2282.

Hausen, B. M. (1981). *Woods injurious to human health. A manual.* Berlin, Germany: Walter de Gruyter & Co.

Janick, J., & Paull, R. E. (2008). Diospyros mespiliformis. In J. Janick, & R. E. Paull (Eds.), *The encyclopedia of fruit and nuts* (pp. 337). Wallingford, UK: CABI.

Lajubutu, B. A., Pinney, R. J., Robert, M. F., Odeloa, H. A., & Oso, B. A. (1995). Antibacterial activity of diosquinone and plumbagin from the root of *D. mespiliformis* (Hostch) (Ebenaceae). *Phytotherapy Research, 9,* 346–350.

Levant, B. O., Ozias, M. K., Davis, P. F., Winter, M., Russell, K. L., Carlson, S. E., et al. (2008). Decreased brain docohexaenoic acid content produces neurobiological effects associated with depression: interactions with reproductive status in female rats. *Psychoneuroendocrinology, 33,* 1279–1292.

Mohammed, I. E., El Nur El Bushra, E., Choudhary, M. I., & Khan, S. N. (2009). Bioactive natural products from two Sudanese medicinal plants *Diospyros mespiliformis* and *Croton zambesicus. Records Natural Products, 3,* 198–203.

Nwude, N., & Ebong, O. O. (1980). Some plants used in the treatment of leprosy in Africa. *Leprosy review, 51,* 11–18.

Palgrave, K. C. (1981). *Trees of Southern Africa.* Cape Town: (p. 748). Republic of South Africa: Struik Publishers.

Petzke, K. J., Ezeagu, I. E., Proll, J., Akinsoyinu, A. O., & Metges, C. C. (1997). Amino acid composition and available lysine content and *in vitro* protein digestibility of selected tropical crop seeds. *Plant Foods for Human Nutrition (Dordrecht, Netherlands), 50,* 151–162.

Simopoulos, A. P. (1999). Essential fatty acids in health and chronic disease. *The American Journal of Clinical Nutrition, 70,* 560–569.

Umaru, H. A., Adam, R., Dahiru, D., & Nadro, M. S. (2007). Levels of antinutritional factors in some wild edible fruits of Northern Nigeria. *African Journal of Biotechnology, 6,* 1935–1938.

Utsunomiya, N., Subhadrqbandhu, S., Yonemori, K., Oshida, M., Kanzaki, S., Nakatsubo, F., et al. (1998). *Diospyros* species in Thailand: their distribution, fruit morphology and uses. *Economic Botany, 52,* 343–351.

CHAPTER 17

Therapeutic Potential of Ajwain (*Tracyspermum ammi* L.) Seeds

C.U. Rajeshwari[1], A.V. Vinay Kumar[2], B. Andallu[1]
[1] Department of Home Science, Sri Sathya Sai University, Anantapur, Andhra Pradesh, India
[2] The Institute of Environmental and Human Health, Texas Tech University, Lubbock, Texas, USA

153

LIST OF ABBREVIATIONS

AF, *Aspergillus flavus*
AN, *Aspergillus niger*
AO, *Aspergillus oryzae*
AO′, *Aspergillus ochraceus*
BHA, butylated hydroxy anisole
BHT, butylated hydroxy toluene
CL, *Curvularia lunata*
FID, flame ionization detection
FG, *Fusarium graminearum*
FM, *Fusarium monoliform*
PC, *Penicillium citrium*
PM, *Penicillium madriti*
PV, *Penicillium viridicatum*
TBA, thiobarbituric acid

Nuts & Seeds in Health and Disease Prevention. DOI: 10.1016/B978-0-12-375688-6.10017-9

INTRODUCTION

Currently there is an increased interest globally in identifying antioxidant compounds that are pharmacologically potent and have low or no side effects, for use in preventive medicine and in the food industry. As several metabolic diseases and age-related degenerative disorders are closely associated with oxidative processes in the body, the use of herbs and spices as a source of natural antioxidants warrants further attention. The antioxidant activity of herbs may be attributable to active components of the essential oil, and also due to synergistic interactions between them. Therefore, supplementing a balanced diet with herbs may have beneficial health effects (Lee & Shibamoto, 2002). At present, recommendations are warranted to support the consumption of foods rich in bioactive components, such as herbs and spices. With time, a greater body of scientific evidence supporting the benefits of herbs and spices in the overall maintenance of health and protection from disease will be seen. Immediate studies should focus on validating the antioxidant capacity of herbs and spices after harvesting, as well as testing their effects on markers of oxidation (Ali *et al.*, 2008).

BOTANICAL DESCRIPTION

Ajwain (*Trachyspermum ammi* L. Sprague) is an annual aromatic and herbaceous plant of the family Apiaceae. It is an erect annual herb with a striate stem and originated in the eastern regions of Persia and India. Its fruits are small, and grayish-brown in color.

Ajwain has several other common names in English, including carom, Ethiopian cumin, wild parsley, and bishop's weed, as well as numerous names in other languages — *netch* (white) *azmad* (Amharic), *ajwan, kamun al-mulaki, taleb el koub* (Arabic), *joni-gutti* (Assamese), *jowan, yamani* (Bengali), *yan-jhon-wuihheung* (Cantonese), *nanava* (Farsi), *ajowan* (Dutch, French, German, Italian), *ayamo, yavan* (Gujarati), *ajwain, carom omum* (Hindi), *ajamoda, oma* (Kannada), *ayowan* (Korean), *ajowan* (Japanese, Spanish), *ayamodakam* (Malay), *yin-dou-zeng-hui-xiang* (Mandarin), *javano* (Nepali), *oregano-semente, ajowan* (Portugese), *ajavain* (Punjabi), *assamodum* (Singhalese), *omam* (Tamil), *omu* (Telegu), *chilan* (Thai), and *misir anason* (Turkish).

HISTORICAL CULTIVATION AND USAGE

Ajwain originated in the Middle East, possibly in Egypt, before spreading to the Mediterranean region and south-west Asian countries. Ajwain seeds have long been used in Indian, Greek, Unani, and Egyptian medicine. The seeds contain variable amounts of nutrients and crude fiber (Table 17.1), and important chemical constituents include β-pinene, para-cymene, α-pinene, limonene, and γ- and β-terpinenes.

TABLE 17.1 Nutrient Composition of Ajwain Seeds	
Parameter	**g%**
Moisture	8.9
Protein	15.4
Fat	18.1
Fiber	11.9
Carbohydrates	38.6
Minerals	7.1
Calcium	1.42
Phosphorus	0.30
Iron	14.6 mg%

Source: The Wealth of India (1976).

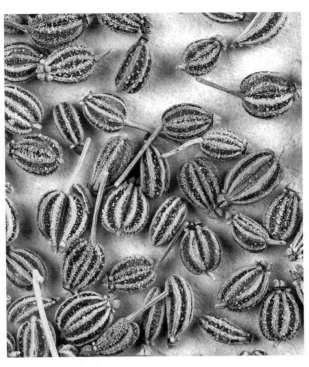

FIGURE 17.1
Ajwain (*Trachyspermum ammi*) seeds. *Source: www. kitchenwonders.blogsspot.com.*

PRESENT-DAY CULTIVATION AND USAGE

Ajwain, although primarily grown and used in the Indian subcontinent, is also found in Iran, Egypt, and Afghanistan. It is cultivated extensively as a cold-season crop in the plains, and as a summer crop in the hills. It requires a long warm and frost-free growing season. It can be grown in any loamy or sandy-loam soil, and even in black soils. In India, it is grown using both rain-fed and irrigated cultivation systems (Aggarwal *et al.*, 2009).

Ajwain is sometimes used as an ingredient in berbere, a spice mixture favored in Eritrea and Ethiopia. In India, the major ajwain-producing states are Rajasthan and Gujarat; Rajasthan alone produces about 90% of India's total production. The most utilized part of ajwain is the small caraway-like fruit, which is particularly popular in Indian savory recipes, savory pastries, and snacks (http://en.wikipedia.org/wiki/ajwain). The chief sensory constituent of the essential oil is thymol, which is used for flavoring, culinary, household, and cosmetic purposes (The Wealth of India, 1976). The grayish-brown seeds and fruits of ajwain are also used as a spice in cooking (http://www.webindia123.com/spices) (Figure 17.1).

APPLICATIONS TO HEALTH PROMOTION AND DISEASE PREVENTION

Ajwain is highly valued in India as a gastrointestinal medicine and as an antiseptic. It is combined with salt and hot water, and taken after meals to relieve bowel pain or colic, and to improve digestion. Ajwain is also a traditional remedy for cholera and fainting spells. Westerners generally use it for coughs and throat issues (Nadkarni, 1976).

Ajwain is also an ingredient in mouthwashes and toothpastes because of its antiseptic properties, and is valued for the following problems: flatulence, indigestion, polyuria, asthma, bronchitis, common cold, toothache, various other gastrointestinal disorders, earache, pain in the throat, arthritic and rheumatic pains, and migraine. It is also used as an aphrodisiac, to enhance virility and prevent premature ejaculation (Figure 17.2).

Chemical composition

The volatile oil present in the seeds of ajwain is one of the principal constituents responsible for providing a typical flavor, owing to the presence of thymol. It also contains a cumene and terpene — "thymene." The fruits of ajwain yield 2—4% essential oil, containing thymol. The remainder of the oil consists of p-cymene, β-pinene, α-pinene, and carvacrol (Pruthi, 1980).

Investigation of the aromatic compounds of the essential oils, and acetone extract of dried fruits of ajwain, was carried out by means of GC and GC-MS, to identify the target component responsible for the characteristic odor of valuable spice and food flavoring products. The analysis of the oil and acetone extract of ajwain was undertaken by using previously stated HP-5 series. On analysis, the volatile oil showed the presence of 26 identified components that accounted for 96.3% of the total amount. Thymol (39.1%) was found as a major component, along with p-cymene (30.8%), γ-terpinene (23.3%), α-pinene (1.7%), terpinene-4-ol (0.8%), and several other components in minor quantities. The acetone extract showed the presence of 18 identified components, which accounted for 68.8% of the total amount. The major component was thymol (39.1%), followed by oleic acid (10.4%), linoleic acid (9.6%), γ-terpinene (2.6%), p-cymene (1.6%), palmitic acid (1.6%), and 4-hydroxy-4-methylpenta-2-one (1.1%). It is interesting to note that thymol content in the oil and in the acetone extract was almost the same (Singh *et al.*, 2004). Chialva and colleagues (1993) had already reported that thymol is a major component of this oil, and there is a compositional variation between a steam-distilled oil and oleoresin.

Chemical constituents and the biological activities of *Trachyspermum ammi* L. are given in Table 17.2 (www.ars-grin.gov/duke).

There are many reports in the literature regarding the chemical composition and antimicrobial studies (Singh *et al.*, 2002) of the volatile oil of ajwain. Ajwain oil and thymol are known to

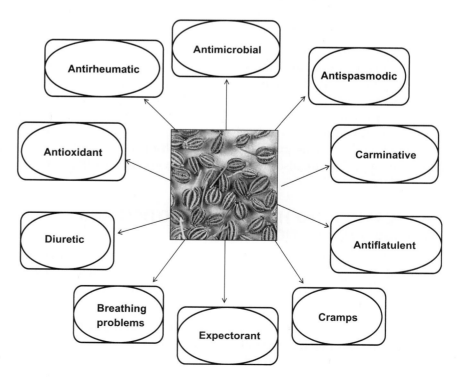

FIGURE 17.2
Traditional uses of Ajwain (*Trachyspermum ammi* L.).

TABLE 17.2 Chemicals in *Trachyspermum ammi* L. (ajwain) Seeds and their Biological Activities

Sl.No	Chemical Constituents	Part of Plant	Antitumor	Anti-Inflammation	Anti-aging	Anticancer	Antioxidant	Antidiabetic	Antimicrobial	Anti-ulcer
1	α-Phellandrene	Plant							✓	
2	α-Pinene	Fruit			✓				✓	
3	β-Pinene	Plant			✓					
4	Carvacrol	Fruit			✓		✓	✓		
5	Copper	Fruit			✓			✓		
6	δ-3-Carena	Fruit			✓				✓	
7	Dipentene	Fruit							✓	
8	Fiber	Fruit	✓					✓		✓
9	γ-Terpinene	Fruit					✓			
10	Iodine	Fruit							✓	
11	Limonene	Fruit	✓	✓	✓				✓	
12	Manganese	Fruit						✓		
13	Myrcene	Fruit					✓		✓	
14	p-Cymene	Fruit							✓	
15	Thymol	Fruit		✓			✓		✓	
16	Zinc	Fruit						✓		✓

Source: www.ars_grin.gov/duke.

157

possess a number of functional properties (Meera & Sethi, 1994); namely, antimicrobial, antiflatulent, antispasmodic, antirheumatic, diuretic, stimulant, carminative, and expectorant (Singh *et al.*, 2004).

Antioxidative activity

The methanolic extracts of ajwain seeds possess natural antioxidant properties. However, the acetone extract showed better antioxidative activity for linseed oil compared to synthetic antioxidants such as butylated hydroxy toluene and butylated hydroxy anisole (Frag *et al.*, 1990). These results correlated well with those obtained previously using the peroxide value and thiobarbituric acid (TBA) methods. It is known that the effectiveness of added antioxidants varies depending on the food and on the processing and storage conditions (Yanishlieva *et al.*, 2002) — for example, BHT is very effective in animal fat, but less effective in vegetable oils. Volatile oil and BHT may also be lost during heating, because of their volatility. BHA is known to be a very effective antioxidant for vegetable oils, and it is more stable at high temperatures than BHT (Gordon & Kourimska, 1995).

It is most likely that acetone extracts are more effective than BHA and BHT at high temperatures. In terms of retarding the formation of primary and secondary oxidation products, the effectiveness of samples at a concentration of 200 ppm can be put into the following order: acetone extract > oil > BHA > BHT > control. Several factors can induce changes in the antioxidant activity during the distillation of volatile constituents from spices. The volatile oil and acetone extract both contain thymol (39.1%) as a major component, which indicates in general, the antioxidative properties of these extracts. In acetone extract, oleic acid (10.4%) and linoleic acid (9.6%) do not possess antioxidant activity. Other substances, which could be present in acetone extract and not identified by gas chromatography (GC), may contribute to improving the antioxidant activity of acetone extract; this needs further investigation. Heat- and water-induced chemical reactions can also change the activity of a complex extract system consisting of numerous compounds with different chemical and physical properties (Singh *et al.*, 2004).

Fungicidal activity

Ajwain oil exhibited a broad spectrum of fungitoxic behavior against all tested fungi, including *Aspergillus niger* (AN), *Aspergillus flavus* (AF), *Aspergillus oryzae* (AO), *Aspergillus ochraceus* (AO′), *Fusarium monoliforme* (FM), *Fusarium graminearum* (FG), *Pencillium citrium* (PC), *Penicillium viridicatum* (PV), *Pencillium madriti* (PM), and *Curvularia lunata* (CL), as absolute mycelial zone inhibition was obtained at a 6-μl dose of the oil. Using the Inverted Petriplate technique, the oil was found to be 100% antifungal against all the tested fungi at a 6-μl dose. It was highly effective in controlling mycelial growth of tested *Penicillium* species even at a 2-μl dose of the oil. The oil was also found to be active against AO′, PC, and CL at a 2-μl dose, as more than 80% mycelial zone inhibition was obtained; however, when using the same method, acetone extract was found to be less effective than the volatile oil, as only more than 50% mycelial zone inhibition of AF, AO′, FG, FM, and CL was obtained even at a 6-μl dose. Moreover, using the food poisoning-testing technique, the oil was found to be highly effective against AN, AF, AO, AO′, FM, and PV, as absolute mycelial zone inhibition was obtained at 6 μl. It is very interesting to note that 100% activity of the fungi AN, FM, and PV was obtained even at a 2-μl dose of the oil; whereas, using the same method, acetone extract was found to be comparatively less effective than the volatile oil, as only more than 70% mycelial zone inhibition was obtained for FG, FM, PC, and PM. Various researchers have reported that this oil exhibits a broad fungitoxic spectrum, inhibiting the mycelial growth of a number of fungi at 100, 200, and 300 ppm (Diwivedi & Singh, 1999). The oil was thermostable and more efficacious than various synthetic fungicides, and the data were found to be highly significant ($P < 0.05$) (Meera & Sethi, 1994).

ADVERSE EFFECTS AND REACTIONS

No adverse effects and reactions or toxicities have been reported for ajwain seeds.

SUMMARY POINTS

- Ajwain seed oil and thymol are known to possess a number of functional properties, namely antimicrobial, antiflatulent, antispasmodic, antirheumatic, diuretic, stimulant, carminative, and expectorant.
- Ajwain is also valued for the following properties: flatulence, indigestion, polyuria, asthma, bronchitis, common cold, toothache, various other gastrointestinal disorders, cardialgia, earache, pain in the throat, arthritic and rheumatic pains, and migraine. It is also used as an aphrodisiac.
- Ajwain is highly valued in India as a gastrointestinal medicine and as an antiseptic.
- Ajwain is widely used to enhance virility and premature ejaculation; it is also a traditional remedy for cholera and fainting spells.
- Ajwain seed oil exhibited a broad spectrum of fungitoxic behavior by inhibiting the mycelial growth of all tested fungi.
- The methanolic extracts of ajwain seeds possess natural antioxidant properties.
- The volatile oil and acetone extract of ajwain seeds have been proven to be alternative sources of natural antioxidants, and more efficacious than various synthetic antioxidants such as BHA and BHT.

References

Aggarwal, B. B., van Kuiken, Michelle E., Iyer Laxmi, H., Harikumar Kuzhuvelil, B., & Sung, B. (2009). Molecular targets of nutraceuticals derived from dietary spices: potential role in suppression of inflammation and tumorigenesis. *Experimental Biology and Medicine, 234,* 1–35.

Ali, S. S., Kasoju, N., Luthra, A., Sing, A., Sharanabasava, H., Sahu, A., et al. (2008). Indian medicinal herbs as source of antioxidants. *Food Research International, 41,* 1–260.

Chialva, F., Monguzzi, F., Manitto, P., & Akgul. (1993). Essential oil constituents of *Trachyspermum capticum* (L.) Link fruits. *Journal of Essential Oil Research, 5,* 105–106.

Diwivedi, S. K., & Singh, K. P. (1999). Fungitoxicity of some higher plant products against *Macrophomina phaseolina* (Tassi) Goid. *Flavour and Fragrance Journal, 14,* 315–318.

Frag, R. S., Ali, M. N., & Taha, S. H. (1990). Use of some essential oils as natural preservatives for butter. *Journal of the American Oil Chemists' Society, 68,* 188–191.

Gordon, H. H., & Kourimska, L. (1995). The effect of antioxidants on changes in oils during heating and deep frying. *Journal of the Science of Food and Agriculture, 68,* 347–353.

Lee, K. G., & Shibamoto, T. (2002). Determination of antioxidant potential of volatile extracts isolated from various herbs and spices. *Journal of Agricultural and Food Chemistry, 50,* 4947–4952.

Meera, M. R., & Sethi, V. (1994). Antimicrobial activity of essential oils from spices. *Journal of Food Science & Technology, 31,* 68–70.

Nadkarni, K. M. (1976). *Indian Materia Medica, Vol 1* (3rd ed.). India: Popular Prakashan. pp. 11–1243.

Pruthi, J. S. (1980). *Spices and condiments: chemistry, microbiology, technology.* New York, NY: Academic Press. p. 449.

Singh, G., Kapoor, I. P., Pandey, S. K., Singh, S. K., Singh, U. K., & Singh, R. K. (2002). Studies on essential oils, Part 10: Antibacterial activity of volatile oils of some spices. *Phytotherapy Research, 16,* 680–682.

Singh, G., Maurya, S., Catalan, C., & De Lampasona, M. P. (2004). Chemical constituents, antifungal and antioxidative effects of ajwain essential oil and its acetone extract. *Journal of Agricultural and Food Chemistry, 52,* 3292–3296.

The Wealth of India. (1976). A dictionary of Indian raw materials and industrial products. *Vol. X. Council of Scientific and Industrial Research.* New Delhi, India: NISCAIR. pp. 267–273.

Yanishlieva, N., Kamal-Eldin, A., Marinova, E. M., & Toheva, A. S. (2002). Kinetics of antioxidant action of γ- and α-tocopherols in sunflower and soybean triacylglycerols. *European Journal of Lipid Science and Technology, 104,* 262–270.

CHAPTER 18

Almond (*Prunus dulcis*) Seeds and Oxidative Stress

Arash Mirrahimi[1,3], **Korbua Srichaikul**[1,3], **Amin Esfahani**[1,3], **Monica S. Banach**[1,5], **John L Sievenpiper**[1,6], **Cyril W.C. Kendall**[1,3,7], **David J.A. Jenkins**[1,2,3,4]

[1] Clinical Nutrition & Risk Factor Modification Center,
[2] Department of Medicine, Division of Endocrinology and Metabolism, St Michael's Hospital, Toronto, Ontario, Canada
[3] Department of Nutritional Sciences,
[4] Department of Medicine,
[5] Department of Pharmaceutical Sciences, Faculty of Medicine, University of Toronto, Toronto, Ontario, Canada
[6] Department of Pathology and Molecular Medicine, Faculty of Health Sciences, McMaster University, Hamilton, Ontario, Canada
[7] College of Pharmacy and Nutrition, University of Saskatchewan, Saskatoon, Saskatchewan, Canada

161

LIST OF ABBREVIATIONS

BMI, body mass index
CHD, coronary heart disease
CVD, cardiovascular disease
LDL, low-density lipoprotein
MUFA, monounsaturated fatty acid
ROS, reactive oxidative species

INTRODUCTION

The word almond, sometimes used to define the shape of objects, comes from the Greek *amygdala*, which literally refers to the well known seed. Though the commonly used term "nut"

Nuts & Seeds in Health and Disease Prevention. DOI: 10.1016/B978-0-12-375688-6.10018-0

refers to a wide range of seeds, based on botanical definitions almonds are not actually true nuts (Rushforth, 1999). While hazelnuts meet the botanical definition of nuts, almonds (along with pistachios and walnuts) are seeds of drupe fruits. Despite this inconsistency, these variable seeds have been clustered together under the collective term "tree nuts." Almonds have historically and culturally been associated with good health, in India with good brain health, and in China with female beauty. Years of research have come to validate some of the age-old beliefs about these seeds. The various bioactive components and the unique macronutrient profile of nuts have been linked to lowering oxidative stress and inflammation, improving lipid profiles (Sabate *et al.*, 2003), endothelial function (Ros, 2009), and blood pressure (Estruch *et al.*, 2006). The ability of these seeds to improve the blood lipid profile and assist in diabetes management (Kris-Etherton *et al.*, 2001; Lovejoy *et al.*, 2002; Josse *et al.*, 2007; Kendall *et al.*, 2010), as well as reducing coronary heart disease (CHD) risk, has progressively become more established and considered as therapeutic (Ternus *et al.*, 2006). Furthermore, with increasing rates of chronic diseases (including diabetes and cancer), known for causing greater levels of oxidative stress, the antioxidant effects of almonds are, now more than ever, of particular interest.

BOTANICAL DESCRIPTION

The almond is the seed of *Prunus Dulcis*, a small deciduous species of tree (4–10 m tall). The *Prunus Dulcis* drupe consists of an outer hull (exocarp) and a hard shell (endocarp) with the seed inside. Almond trees are not immediately productive, and only bear fruit after 5 years. It is in autumn, 7 to 8 months after flowering, that the fruit becomes mature (Griffiths & Anthony, 1992; Rushforth, 1999).

HISTORICAL CULTIVATION AND USAGE

Prunus Dulcis is native to regions of the Middle East with Mediterranean climates. The wild form of domesticated almonds grows in parts of the Levant, suggesting that almonds must first have been cultivated in this region. The fruit of the wild form contains the glycoside amygdalin, which is transformed into hydrogen cyanide upon infliction of injury to the seed. Domesticated almonds are not toxic due to a common genetic mutation resulting in the absence of glycoside amygdalin, and this mutant was grown by early farmers. Almonds are believed to have been one of the earliest domesticated fruit trees (Zohary & Maria, 2000). Domesticated almonds appear to have been cultivated as early as the Early Bronze Age (3000–2000 BC) of the Near East. An archaeological example of almond consumption, likely imported from the Levant, is the fruit found in Tutankhamun's tomb in Egypt (*c.* 1325 BC) (Zohary & Maria, 2000). The almond has since been spread by humans along the Mediterranean shores into northern Africa, southern Europe, and, more recently, other parts of the world, including California (Rieger, 2006). Almonds have been an integral part of the diet in many of the Levant region cultures, being frequently used in pastries and other foods. For centuries, almonds have also been utilized therapeutically — both the wild variant (bitter) and the domesticated variant (sweet). For example, sweet almond oil, obtained from the dried kernel of sweet almonds, has traditionally been used for massage therapy, and medicinally ingested in the Greco-Persian system of medicine.

PRESENT-DAY CULTIVATION AND USAGE

The global production of almonds is around 1.7 million tonnes annually, with the USA (41%), Spain (13%), Syria (7%), Italy (6%), Iran (6%), and Morocco (5%) being the major producers, according to Food and Agriculture Organization (FAO). In the United States, production is mainly in California, with almonds being California's third leading agricultural product. India imports over 94% of its almond consumption, and is the largest global and US market for in-shell almonds. Interestingly, the pollination of California's almonds is the largest annually managed pollination event in the world; approximately half of all USA beehives

(1 million beehives) are trucked in to the almond groves in February for this task. Pollination of almonds has been heavily impacted by the colony collapse disorder affecting the bee population globally. In 2006, because of two cases of Salmonella traced to almonds, the Almond Board of California proposed pasteurization of almonds available to the public; the USDA has approved this as part of the almond distribution process.

Almonds continue to be used in various foods, but they have now become of particular interest in nutraceuticals as a means of therapy. They have been linked to disease prevention and management in numerous studies. An example of almond incorporation into present-day diets is the use of almond milk. Almond milk, a dairy milk substitute, is processed from almonds, and makes an efficient and well-liked soy-free choice analog for lactose-intolerant people and for vegans. More innovative methods of incorporating almonds in diets as healthy alternatives continue to be explored and developed. Recently, a low carbohydrate nut-bread recipe was conceived at the St Michael's Hospital Clinical Nutrition and Risk Factor Modification Center in Toronto, and this, due to its high palatability and beneficial nutrient profile, was used in the successful "EcoAtkins" weight loss and cholesterol lowering trial (see Table 18.1 for the recipe) (Jenkins *et al.*, 2009).

APPLICATIONS TO HEALTH PROMOTION AND DISEASE PREVENTION

At one time, nuts (including almonds) were considered to be unhealthy foods in western societies due to their high fat content. This perception has since changed, with the determination of the fatty acid profile of almonds and other nuts, as well as the association of their consumption with reduced BMI and their inclusion in weight-maintaining diets (Bes-Rastrollo *et al.*, 2009). The metabolic benefits of almonds stem from their low saturated fat and high monounsaturated fatty acid (MUFA, regarded as a "healthy" fat) content, as well as their vegetable protein, fiber, phytosterols, polyphenols, vitamins, and minerals (Table 18.2). It has been proposed that the almond bioactive compounds may help lower the risk factors for cardiovascular disease (CVD) by improving endothelial function (Ros, 2009), blood pressure (Estruch *et al.*, 2006), and the serum lipid profile (Kris-Etherton *et al.*, 2001; Lovejoy *et al.*, 2002; Sabate *et al.*, 2003; Kendall *et al.*, 2010), in addition to lowering oxidative stress (Kendall *et al.*, 2010) and inflammation (Ros, 2009; Kendall *et al.*, 2010).

163

TABLE 18.1 Recipe for Low Carbohydrate Bread

- Mix all of the ingredients below together in a large bowl, using a wooden spoon or your hands
- Let it rest for 10 minutes
- Continue to knead on a flat surface, using up all remaining nuts and gluten, before transferring dough to a loaf pan (or shape into two round loaves, and place on a baking sheet)
- Let dough rise for about 20 minutes or so
- Bake for 30 minutes at 300°F.
- Keep an eye on the loaves, baking them until they are lightly browned on top.

Ingredients:
180 g ground almonds (1¾ cups)
180 g ground, toasted hazelnuts (1¾ cups)
220 g gluten flour (1⅔ cups)
10 g sugar (2½ tsp)
8 g salt (1 tsp)
16 g instant yeast (1½ tbsp)
20 g psyllium (¼ cup)

Add:
400 g warm water (1⅔ cups)

Ideally, all ingredients should be weighed, but if you don't have a scale, approximate volumes are provided.

TABLE 18.2 Constituents of Almond Seeds	
Nutrients*	Almond (*Prunus Dulcis*)
Energy (kcal)	575
Protein (g)	21.2
Fat (g)	49.4
SFAs (g)	3.7
MUFAs (g)	30.89
PUFAs (g)	12.1
Carbohydrate (g)	21.7
Fiber (g)	12.2
Sugars (g)	3.89
α-tocopherol (mg)	26.22
β-tocopherol (mg)	0.29
γ-tocopherol (mg)	0.65
Δ-tocopherol (mg)	0.05
Selenium (μg)	2.5
Vitamin C (mg)	0
Thiamin (mg)	0.21
Folate (μg)	50
Arginine (g)	2.45
Flavanoids and/or polyphenols	EGC, cyanidin, GCG, EC, Clsorhamnetin 3-O-glucoside, isorhamnetin 3-O-rutinoside, C, kaempferol 3-o-rutinoside, EC, quercetin 3-O-galactoside, isorhamnetin 3-O-galactoside

*Estimates are based on United States Department of Agriculture (USDA) National Nutrient Database standard references; values are given per 100 grams. All approximations are based on different numbers of data points.

Almonds have also been linked to the risk of developing type 2 diabetes. According to the American Diabetes Association, glycemic control is of crucial importance for the prevention and management of diabetes (American Diabetes Association, 2007). Evidence suggests that when eaten alone or in combination with mixed meals, the low available carbohydrate content of nuts can help lower the postprandial glucose and insulin responses, and, as such, play a beneficial role in glycemic control (Josse *et al.*, 2007; Kendall *et al.*, 2010). Moreover, it has been suggested that high MUFA consumption can also improve glycemic control (Kendall *et al.*, 2010), possibly by displacing the carbohydrates and effectively decreasing the glycemic load. Since hyperglycemia is a major cause of oxidative stress in diabetic patients (Brownlee, 2001), these benefits of almond consumption may have considerable potential in reducing reactive oxidative species (ROS) through reducing the glycemic excursions of foods consumed with almonds (Monnier *et al.*, 2006; Kendall *et al.*, 2010). ROS have been associated with multiple disease states, including CVD and cancer (Mercuri *et al.*, 2000).

Almonds and other nuts contain a multitude of antioxidants (Table 18.2) that may prove to reduce the extent of oxidative damage incurred to cells through metabolism (Chen *et al.*, 2005). A recent study examined the acute effects of nut consumption in healthy human subjects. The investigators found that both almonds and walnuts (in the form of smoothies) led to significant increases in the plasma polyphenol concentration following the nut meals (Torabian *et al.*, 2009). The same study showed that the susceptibility of plasma lipids to peroxidation decreased after 90 minutes on both nut treatments, but not on the control. Similar results have been noted in a trial of 60 male smokers, where almonds were shown to decrease lipid peroxidation and reduce oxidative damage to DNA (Li *et al.*, 2007). Moreover, flavonoids found in the skin of almonds may further contribute to oxidative protective benefits. Flavonoids are considered to possess a variety of biological activities, including antioxidant and anti-inflammatory capability. The antioxidant properties of almond flavonoid were recently shown to be highly biologically available and work synergistically with other antioxidants (Vitamins C and E) to protect against LDL oxidation *in vitro*, as well as enhancing resistance to Cu^{2+} induced oxidation of LDL *ex vivo* (Chen *et al.*, 2005). These data suggest that almonds are a rich source of bioactive compounds and antioxidants that can prevent lipid, DNA, and protein peroxidation.

In conclusion, through the many macro- and micronutrients that almond seeds contain (Table 18.2), they may contribute to lowering oxidative damage in both healthy individuals and patients with chronic disease. These effects may be achieved either indirectly, via lowering the overall glycemic load and the glycemic index of meals consumed concomitantly (Brownlee, 2001; Monnier *et al.*, 2006; Kendall *et al.*, 2010), or more directly through their intrinsic polyphenols and flavonoids. Therefore, it is not surprising that almond seeds and other nuts have been so prevalently associated with significant reductions in risk factors for developing CHD, CVD, diabetes, and other chronic diseases (Ros, 2009; Kendall *et al.*, 2010).

ADVERSE EFFECTS AND REACTIONS (ALLERGIES AND TOXICITY)

Almonds are ranked third (15% reactive), behind walnut and cashew nuts, as a tree nut allergen (Roux *et al.*, 2003). The total annual number of anaphylactic mortalities in the United States due to all food allergens is 100, most of which are due to peanuts, but also include shellfish, eggs and other tree nuts, including almonds (Matasar & Neugut, 2003).

The wild bitter variants of almonds contain trace amounts of cyanide (found in most seeds). However, the sweet or domesticated variant of almonds does not contain this toxin. Hence, cyanide toxicity resulting from almond seed consumption has been rarely reported (Roux *et al.*, 2003).

Another consideration when harvesting and storing almonds is the possibility of the presence of aflatoxins. The UK Food Standards Agency, in a survey of nut products available in the market, tested 154 samples of tree nuts (including almonds) for aflatoxins. The findings of this survey suggested that of the almond samples tested, only one had aflatoxin levels higher than the acceptable limit of 4 μg/kg (The Food Standards Agency, 2004).

SUMMARY POINTS

- The high levels of MUFAs in almond (*Prunus Dulcis*) have been shown to improve glycemic control; hence the management of type 2 diabetes.
- The low carbohydrate content of nuts results in a low glycemic index and a moderate postprandial serum glucose excursion.
- The highly bioavailable flavonoids and polyphenols that almonds contain help to reduce lipid, protein, and DNA peroxidation.
- Through reduction of postprandial glycemia, almonds have been shown to reduce oxidative stress, and therefore to decrease the risks of CHD and cancer.
- Nut consumption has been indicated for weight loss, and as a part of a weight-maintaining diet.

References

American Diabetes Association. (2007). Nutrition Recommendations and Interventions for Diabetes: a position statement of the American Diabetes Association. *Diabetes Care, 30*(Suppl. 1), S48–S65.

Bes-Rastrollo, M., Wedick, N. M., Martinez-Gonzalez, M. A., Li, T. Y., Sampson, L., & Hu, F. B. (2009). Prospective study of nut consumption, long-term weight change, and obesity risk in women. *The American Journal of Clinical Nutrition, 89*, 1913–1919.

Brownlee, M. (2001). Biochemistry and molecular cell biology of diabetic complications. *Nature, 414*, 813–820.

Chen, C. Y., Milbury, P. E., Lapsley, K., & Blumberg, J. B. (2005). Flavonoids from almond skins are bioavailable and act synergistically with vitamins C and E to enhance hamster and human LDL resistance to oxidation. *Journal of Nutrition, 135*, 1366–1373.

Estruch, R., Martinez-Gonzalez, M. A., Corella, D., Salas-Salvado, J., Ruiz-Gutierrez, V., Covas, M. I., et al. (2006). Effects of a Mediterranean-style diet on cardiovascular risk factors: a randomized trial. *Annals of Internal Medicine, 145*, 1—11.

Griffiths, M. D., & Anthony, J. H. (1992). *The New Royal Horticultural Society dictionary of gardening.* London, UK: Macmillan Press.

Jenkins, D. J., Wong, J. M., Kendall, C. W., Esfahani, A., Ng, V. W., Leong, T. C., et al. (2009). The effect of a plant-based low-carbohydrate ("Eco-Atkins") diet on body weight and blood lipid concentrations in hyperlipidemic subjects. *Archives of Internal Medicine, 169*, 1046—1054.

Josse, A. R., Kendall, C. W., Augustin, L. S., Ellis, P. R., & Jenkins, D. J. (2007). Almonds and postprandial glycemia — a dose—response study. *Metabolism, 56*, 400—404.

Kendall, C. W., Esfahani, A., Truan, J., Srichaikul, K., & Jenkins, D. J. (2010). Health benefits of nuts in prevention and management of diabetes. *Asia Pacific Journal of Clinical Nutrition, 19*, 110—116.

Kris-Etherton, P. M., Zhao, G., Binkoski, A. E., Coval, S. M., & Etherton, T. D. (2001). The effects of nuts on coronary heart disease risk. *Nutrition Reviews, 59*, 103—111.

Li, N., Jia, X., Chen, C. Y., Blumberg, J. B., Song, Y., Zhang, W., et al. (2007). Almond consumption reduces oxidative DNA damage and lipid peroxidation in male smokers. *Journal of Nutrition, 137*, 2717—2722.

Lovejoy, J. C., Most, M. M., Lefevre, M., Greenway, F. L., & Rood, J. C. (2002). Effect of diets enriched in almonds on insulin action and serum lipids in adults with normal glucose tolerance or type 2 diabetes. *The American Journal of Clinical Nutrition, 76*, 1000—1006.

Matasar, M. J., & Neugut, A. I. (2003). Epidemiology of anaphylaxis in the United States. *Current Allergy and Asthma Reports, 3*, 30—35.

Mercuri, F., Quagliaro, L., & Ceriello, A. (2000). Oxidative stress evaluation in diabetes. *Diabetes Technology & Therapeutics, 2*, 589—600.

Monnier, L., Mas, E., Ginet, C., Michel, F., Villon, L., Cristol, J. P., et al. (2006). Activation of oxidative stress by acute glucose fluctuations compared with sustained chronic hyperglycemia in patients with type 2 diabetes. *The Journal of the American Medical Association, 295*, 1681—1687.

Rieger, M. (2006). *Introduction to fruit crops.* New York, NY: Food Products Press.

Ros, E. (2009). Nuts and novel biomarkers of cardiovascular disease. *The American Journal of Clinical Nutrition, 89*, 1649S—1656S.

Roux, K. H., Teuber, S. S., & Sathe, S. K. (2003). Tree nut allergens. *International Archives of Allergy and Immunology, 131*, 234—244.

Rushforth, K. (1999). *Collins Wildlife Trust guide to trees: a photographic guide to the trees of Britain and Europe.* London, UK: Harper Collins.

Sabate, J., Haddad, E., Tanzman, J. S., Jambazian, P., & Rajaram, S. (2003). Serum lipid response to the graduated enrichment of a Step I diet with almonds: a randomized feeding trial. *The American Journal of Clinical Nutrition, 77*, 1379—1384.

Ternus, M., McMahon, K., Lapsley, K., & Johnson, G. (2006). Qualified health claim for nuts and heart disease prevention: development of consumer-friendly language. *Nutrition Today, 41*, 62—66.

The Food Standards Agency. (2010). *The Food Standards Agency Aflatoxins in Nuts Survey.* London, UK: FSA.

Torabian, S., Haddad, E., Rajaram, S., Banta, J., & Sabate, J. (2009). Acute effect of nut consumption on plasma total polyphenols, antioxidant capacity and lipid peroxidation. *Journal of Human Nutrition and Dietetics, 22*, 64—71.

Zohary, D., & Maria, H. (2000). *Domestication of plants in the old world: the origin and spread of cultivated plants in West Asia, Europe, and the Nile Valley.* Oxford, UK: Oxford University Press.

CHAPTER 19

Almonds (*Prunus dulcis*): Post-Ingestive Hormonal Response

Alisa Mori[1], Karen Lapsley[2], Richard D. Mattes[1]
[1] Department of Foods and Nutrition, Purdue University, West Lafayette, Indiana, USA
[2] Almond Board of California, Modesto, California, USA

LIST OF ABBREVIATIONS

CCK, cholecystokinin
GLP-1, glucagon-like peptide-1
HOMA-B, homeostasis model assessment of β cell function
HOMA-IR, homeostasis model assessment of insulin resistance
iAUC, incremental area under the curve
MUFA, monounsaturated fatty acid

INTRODUCTION

There is increasing evidence that nut consumption may reduce the risk for a variety of chronic diseases, such as cardiovascular disease and diabetes. Further, despite their high energy density, findings from epidemiological studies and clinical trials consistently indicate that nut

Nuts & Seeds in Health and Disease Prevention. DOI: 10.1016/B978-0-12-375688-6.10019-2

consumption is not associated with weight gain. Thus, while adding desirable sensory properties to the diet, there are also recommendations to increase consumption for health promotion. Almonds are an especially nutrient-dense nut, and there is growing scientific substantiation of its health benefits as reviewed below.

BOTANICAL DESCRIPTION

The almond fruit is classed botanically as a drupe, with a pubescent exocarp (skin), a fleshy but thin mesocarp (hull), and a distinct hardened endocarp (shell). The mesocarp undergoes only limited enlargement during development, becoming dry and leathery, and dehiscing at maturity. The tree, while relatively slow growing, can survive for 100 years or more, reaching heights exceeding 20 m. Almond's outlier status within the *Prunus* species has confounded its botanical classification. Presently, the most widely accepted scientific name, *Prunus dulcis*, acknowledges its taxonomic affinities with other *Prunus* based on similar morphology, molecular-genetic relatedness, and reported hybridization with peach, apricot, and some plums. Because it was the first to be proposed in the literature, *Prunus dulcis* has superseded the scientific name *Prunus amygdalus* still commonly found in the European literature (Gradziel, 2009). In its Central Asian center of origin and diversity, the taxonomic experts most familiar with almond species in their native ecosystems have preferred to classify them in a separate genus, *Amygdalus communis* (Ladizinsky, 1999), arguing that their evolution of specialized botanical structures and development patterns in these often extreme environments justify a separate genus.

HISTORICAL CULTIVATION AND USAGE

Almonds have played an important role in human diets since pre-agricultural times, being native to central and southwest Asia in a region historically described as the Fertile Crescent, which stretches across Israel, Palestine, and Lebanon, up through Syria and eastern Turkey and across to Iraq. Almonds were collected there, in the wild, 10,000 years ago, and they were among the first fruit trees to be domesticated in Old World agriculture, around the 3rd millennium BC (Ladizinsky, 1999). Almonds are consistently found in archaeological sites, and were even present in the famous tomb of Tutankhamen in Egypt, 1352 BC (Zohary & Hopf, 1993).

The oldest and most extensive medical system that first recorded the health uses for almonds derives from ancient Greece, with Hippocrates, the father of Western Medicine. The classical Western world held almonds to be a heating and purgative food, and its uses were deduced from these basic properties. Another key figure who summarized these ideas was Galen of Pergamum, a Greek physician serving several Roman Emperors in succession at the end of the 2nd century AD (Albala, 2009).

Since almonds thrive in hot weather but cannot tolerate frost or high humidity, it is not surprising that almond cultivation spread in a narrow horizontal band westward through the Mediterranean to Spain, with the successive Greek, Roman, and Arab invasions (Ladizinsky, 1999). The Romans introduced almonds to England. From medieval times to the 18th century, nuts were a source of substitute "milk." The first French cookbook, written in about 1300, gave great attention to almond milk as one of the two basic sauce elements in medieval cooking, since it was used as a thickener before starch was "discovered" (Albala, 2009).

PRESENT-DAY CULTIVATION AND USAGE

Horticulturally, almonds are classified as a nut, with the commercial product being the edible seed or kernel. It is the earliest deciduous fruit and nut tree to bloom in spring, due to its relatively low winter chilling requirement and quick response to warm growing temperatures. The almond growth cycle is well adapted to a Mediterranean climate, where plants are dormant during winter precipitation and associated low temperatures. The early flowering makes almond trees very susceptible to spring frosts in more temperate growing regions, and limits plantings to more moderate, almost subtropical, but dry climates.

Globally, the US is the largest almond producer, accounting for 81% of the 2008 global production, while Spain and Australia contributed 9 and 3%, respectively. California produces 100% of the US output, and exports over 70% of the crop to over 90 countries. The European Union, China, and India are currently the largest export destinations. Valued for their nutritional quality and sensory properties, almonds are eaten as snacks, and are used extensively in confectionary and baked goods.

APPLICATIONS TO HEALTH PROMOTION AND DISEASE PREVENTION

There is growing interest in dietary modifications to curb the global incidence and severity of complications related to obesity and type 2 diabetes. Recent studies have explored the acute effects of almond consumption on glucose, insulin, and gut hormone concentrations, with potential implications for chronic use. Additional work is also emerging on the bioaccessibility of almond components, and the fractions most responsible for their health effects.

Effect of acute almond consumption on insulin sensitivity

Acute almond consumption favorably influences metabolic factors implicated in the progression of type 2 diabetes. Reductions in glycemia and insulinemia are observed after healthy subjects have consumed meals including 60 g of almonds (e.g., ~46 almonds) (Jenkins *et al.*, 2006). In one trial, consumption of almond oil with defatted almond flour, intended to mimic bioaccessible almond composition, significantly decreased the 3-hour blood glucose incremental area under the curve (iAUC) compared to when small almond particles were consumed. However, there was no difference in insulin iAUC (Berry *et al.*, 2008). In contrast, other work revealed that both whole almonds and almond oil decreased postprandial blood glucose iAUC compared to a high glycemic index vehicle (Mori, 2009). The discrepant blood glucose findings between consumption of whole almonds and large almond particles may be due to differences in almond particle size distribution (e.g., naturally masticated versus predefined), with resulting altered bioaccessibility of almond components. Similar to other findings, there was no difference in postprandial insulin iAUC, indicating that greater insulin output was not responsible for lower blood glucose concentrations. This is suggestive of improved insulin sensitivity, although the role of gastric emptying cannot be discounted, as it was not measured (Mori, 2009).

Effect of long-term almond consumption on insulin sensitivity

Longer-term studies (1–4 months) evaluating of the effects of almond consumption on glycemia and changes in insulin sensitivity have yielded varied conclusions. Improvements in fasting insulin, insulin sensitivity (HOMA-IR), and β-cell function (HOMA-B) were observed in a 16-week parallel-group trial in pre-diabetic individuals. Almonds comprised 20% of the energy intake in the intervention group, while the control group followed an almond-free diet (Wien, 2008). A subsequent study supplemented a similar percentage of energy (~22%) from almonds (73 ± 3 g/day) in full-dose, half-dose, and no-dose portions in a randomized crossover study, with each treatment lasting 4 weeks (Jenkins *et al.*, 2008). The hyperlipidemic subjects showed no improvements in fasting glucose, insulin, C-peptide, or insulin sensitivity. However, 24-hour urinary C-peptide was significantly decreased in both of the treatments containing almonds. Cumulative 24-hour insulin secretion was therefore improved with the consumption of almonds. In healthy and diabetic adults following a 4-week, almond-enriched diet (100 g of almonds/day in healthy participants and 57–113 g/d of almonds to equal 10% of fat intake in the diabetics), no effect on glycemia was noted (Lovejoy *et al.*, 2002). However, these negative findings may stem from contravening influences from incorporation of the almonds into high glycemic index foods (e.g., trail mix, muffins, and cookies). Within the context of the entire diet, prudent nut consumption may be a useful dietary management tool to decrease the incidence and severity of a number of chronic disease risk factors. However, more long-term studies are needed in various populations to conclusively assess the specific role of almond consumption in decreasing risk for type 2 diabetes.

Inherent protective mechanisms of almond consumption on diabetes risk

Almonds are a rich source of fiber, highly digestible protein, unsaturated fat, and vitamin E (USDA, 2007) (Table 19.1). Moreover, they contain phytates and phenolics that inhibit the enzyme activity of amylase, and are thought to act synergistically to decrease the digestibility of starch (Thompson *et al.*, 1984) (Table 19.2). The decreased rate of carbohydrate digestion may explain reported increases in satiety and blunted blood glucose response with almond consumption, which defines them as a low-GI food. In addition, nuts are a low-carbohydrate food. Further, their high fiber content decreases the effective carbohydrate "dose" by decreasing the amount of available carbohydrate. Replacement of high-carbohydrate foods in the diet with almonds may therefore result in an attenuated postprandial blood glucose and insulin response. This would potentially translate to decreased long-term β-cell demand and glycolytic damage. Short-term alterations in gene expression and enzyme activities, and long-term changes in membrane fluidity, have been noted as a potential mechanism whereby nuts may be protective against development of type 2 diabetes (Risérus *et al.*, 2009).

TABLE 19.1 Almond Nutrient Composition	
Energy (kcal)	598
Carbohydrate (g)	19.3
Protein (g)	22.1
Total fat (g)	52.2
Saturated fat (g)	4.0
Monounsaturated fat (g)	33.6
Polyunsaturated fat (g)	12.7
Dietary fiber (g)	11.8
Calcium (mg)	266
Iron (mg)	4.5
Magnesium (mg)	287
Phosphorus (mg)	489
Potassium (mg)	746
Sodium (mg)	0
Zinc (mg)	3.5
Copper (mg)	1.2
Manganese (mg)	2.6
Selenium (μg)	2.8
Riboflavin (mg)	0.9
Niacin (mg)	3.8
Folate (μg)	32.9
Vitamin E (mg)	26.1

Based on 100 g dry roasted unsalted almond portion; values rounded to nearest tenth.

Almonds are a rich source of nutrients. Components most closely linked to health benefits include Vitamin E, unsaturated fat, fiber and protein.

Source: United States Department of Agriculture (USDA) data. Nutrient Database for Standard Reference, Release 20 (2007). Available at: www.ars.usda.gov/ba/bhnrc/ndl.

TABLE 19.2 Flavonoid Content (mg/100 g) and Antioxidant Capacity of Almonds	
Flavonoids	
Flavan-3-ols	4.47
Flavanones	0.38
Flavonols	7.93
Anthocyanins	2.46
Isoflavones	0.01
Antioxidant capacity	
ORAC (lipophilic + hydrophilic) μmol Trolox Equivalents/g	44.5
FRAP μmol Fe^{2+}/g	41.3
TRAP μmol Trolox Equivalents/g	6.3
TEAC μmol Trolox Equivalents/g	13.4

Almonds are a rich source of phytochemicals. The phytates and phenolics they contain may inhibit amylase activity and slow carbohydrate digestion, leading to beneficial effects on postprandial glycemia. Adapted from Bolling, B.W. *et al.* (2010). The phytochemical composition and antioxidant actions of tree nuts. *Asia Pac. J. Clin. Nutr.*, *19*, 117–123, with permission.

Modulators of almond influences on health outcomes: Mastication and lipid bioavailability

Evidence suggests that the lipid fraction of almonds may be the predominant contributor to alterations in insulin sensitivity and satiety, and that lipid bioaccessibility, determined by mastication efficiency or processing, is influential. Chewing mechanically disrupts the cell walls of nuts, increasing the amount of lipid released and available for digestion and absorption. Literature on the associations between mastication, hormonal responses, and appetitive sensations is limited. In one report on healthy adults, subjective hunger ratings were lower, fullness ratings were higher, and GLP-1 (glucagon-like peptide-1) concentrations were increased after chewing almonds 40 versus 25 times before swallowing (Cassady *et al.*, 2009). However, there were no clear relationships between the number of chews and changes in glucose and insulin concentrations. Further, no significant correlations were noted between the particle size distributions resulting from the different required number of chews, and satiety or hormonal responses. This suggests an independent contribution of orosensory stimulation on satiety, but lipid bioavailability may also be influential. In one trial, more bioavailable lipid loads (e.g., almond oil or safflower/corn oil compared to whole almonds) elicited greater and more sustained concentrations of cholecystokinin (CCK) and lower hunger without altering glucose and insulin concentrations (Burton-Freeman *et al.*, 2004). However, these findings were not confirmed in a second study, where whole almonds were a more effective satiety stimulus than almond oil (Mori, 2009). Although consumption of both whole almonds and almond oil produced a significantly lower postprandial blood glucose iAUC and a non-significantly greater and sustained GLP-1 response (Mori, 2009), that may have contributed to satiety.

Interventions not likely to be confounded by bioaccessibility issues yield clearer insights related to lipid influences on insulin sensitivity. Replacement of dietary saturated fat or dietary carbohydrate with monounsaturated fatty acids (MUFAs) results in improved insulin sensitivity, independent of weight loss (Rocca *et al.*, 2001). Insulin resistant individuals randomized to consume a diet high in MUFA for 28 days decreased fasting glucose concentrations. Postprandial glucose and insulin AUC was decreased and glucagon-like peptide-1 (GLP-1) was increased at the end of the study period (Paniagua *et al.*, 2007). Hypothesized mechanisms include changes in membrane fluidity, increased efficiency of insulin receptor signaling, and increased glycemic control through stimulation of the incretin and ileal-brake hormone, GLP-1 (Rocca *et al.*, 2001).

171

Energy and nutrient bioavailability are important considerations when developing recommendations for almond consumption. More available forms (e.g., butter) or components (e.g., oil), can be incorporated into the diet to enhance nutrient, fiber, or phytochemical intake, but may also increase energy consumption. Other forms (e.g., whole nuts) may enhance satiety and moderate energy intake, but could be less efficient for delivering desired nutrients. Maximal benefit may be derived by matching the form with the desired function.

Conclusions on almond consumption and chronic disease risk reduction

Although it is difficult to make comparisons across studies with different subject population characteristics, especially when dissimilar amounts of almonds were provided, the preponderance of evidence indicates that acute almond consumption has either neutral or beneficial effects on glycemia/insulinemia and satiety. Potential secondary effects related to changes in blood lipids and decreases in oxidative and glycolytic damage may lead to further positive metabolic responses that hold potential health benefits. The long-term effects of including nuts in the diet are not sufficiently established, although epidemiological evidence shows an inverse relationship between nut consumption and risk of developing type 2 diabetes (Jiang *et al.*, 2002). Additional research evaluating the component(s) of almonds responsible for metabolic effects, and more specific research in at-risk or diseased, rather than healthy, populations is needed. Information about the almond fraction(s) responsible for metabolic benefits can be used to formulate more healthful foods, as well as assist in making public health recommendations.

ADVERSE EFFECTS AND REACTIONS (ALLERGIES AND TOXICITY)

Our current understanding of almond allergy derives from what has been learned during the course of investigation of tree nuts as a collective group of allergenic foods. Prevalence of tree nut allergy in US and Europe is estimated at less than 1% of the adult population, but for those individuals affected the reaction can range from mild oral symptoms to severe anaphylaxis. Most research work on almond allergy has focused on the chemical identification of the allergenic proteins, and the effects of various food matrices on allergenicity (Tiwari *et al.*, 2010). From a European and US regulatory viewpoint, analytical detection techniques of these proteins are needed to assess their presence as an ingredient in processed foods. Detection levels between 1 and 100 mg/kg are of interest, and, slowly, analytical methods of these levels are being approached.

SUMMARY POINTS

- Almonds have a long history of use in culinary and health.
- Frequent consumption of almonds has been associated with reduced risk of a variety of chronic diseases, including obesity, heart disease, and diabetes.
- Inherent properties of almonds, such as their high fiber, unsaturated fat, antioxidant, and phytochemical content, likely explain their health protective effects.
- Acute almond consumption has either neutral or beneficial effects on glycemia/insulinemia and satiety, and further research of the effect on gut peptides is needed.
- The lipid fraction of almonds may be the predominant contributor to alterations in insulin sensitivity and satiety.
- Almond lipid bioaccessibility is largely determined by mastication efficiency and processing.
- There is little risk to consuming almonds unless a tree-nut allergy is present.

References

Albala, K. (2009). Almonds along the Silk Road: the exchange and adaptation of ideas from West to East. *Petits Propos Culinaires, 88,* 17–32.

Berry, S. E. E., Tydeman, E. A., Lewis, H. B., Phalora, R., Rosborough, J., Picout, D. R., et al. (2008). Manipulation of lipid bioaccessibility of almond seeds influences postprandial lipemia in healthy human subjects. *The American Journal of Clinical Nutrition, 88,* 922–929.

Bolling, B. W., McKay, D. L., & Blumberg, J. B. (2010). The phytochemical composition and antioxidant actions of tree nuts. *Asia Pacific Journal of Clinical Nutrition, 19,* 117—123.

Burton-Freeman, B., Davis, P. A., & Schneeman, B. O. (2004). Interaction of fat availability and sex on postprandial satiety and cholecystokinin after mixed-food meals. *The American Journal of Clinical Nutrition, 80,* 1207—1214.

Cassady, B., Hollis, J. H., Fulford, A. D., Considine, R. V., & Mattes, R. D. (2009). Mastication of almonds: effects of lipid bioaccessibility, appetite, and hormone response. *The American Journal of Clinical Nutrition, 89,* 794—800.

Gradziel, T. M. (2009). Almond *(Prunus dulcis)* breeding. In S. M. Jain, & M. Priyadarshan (Eds.), *Breeding of plantation tree crops* (pp. 1—31). New York, NY: Springer Science.

Jenkins, D. J. A., Kendall, C. W. C., Josse, A. R., Salvatore, S., Brighenti, F., Augustin, L. S. A., et al. (2006). Almonds decrease postprandial glycemia, insulinemia, and oxidative damage in healthy individuals. *Journal of Nutrition, 136,* 2987—2992.

Jenkins, D. J. A., Kendall, C. W. C., Marchie, A., Josse, A. R., Nguyen, T. H., Faulkner, D. A., et al. (2008). Effect of almonds on insulin secretion and insulin resistance in nondiabetic hyperlipidemic subjects: a randomized controlled crossover trial. *Metabolism Clinical and Experimental, 57,* 882—887.

Jiang, R., Manson, J. E., Stampfer, M. J., Liu, S., Willett, W. C., & Hu, F. B. (2002). Nut and peanut butter consumption and risk of type 2 diabetes in women. *The Journal of the American Medical Association, 288,* 2554—2560.

Ladizinsky, G. (1999). On the origin of almond. *Genetics Resources and Crop Evolution, 46,* 143—147.

Lovejoy, J. C., Most, M. M., Lefevre, M., Greenway, F. L., & Rood, J. C. (2002). Effect of diets enriched in almonds on insulin action and serum lipids in adults with normal glucose tolerance or type 2 diabetes. *The American Journal of Clinical Nutrition, 76,* 1000—1006.

Mori, A. (2009). *Acute post-ingestive and second-meal effects of almond form on diabetes risk factors.* MS Thesis, Purdue University, Department of Foods and Nutrition.

Paniagua, J. A., Gallego de la Sacristana, A., Sánchez, E., Romero, I., Vidal-Puig, A., Berral, F. J., et al. (2007). A MUFA-rich diet improves postprandial glucose, lipid and GLP-1 responses in insulin-resistant subjects. *Journal of American College of Nutrition, 26,* 434—444.

Risérus, U., Willett, W. C., & Hu, F. B. (2009). Dietary fats and prevention of type 2 diabetes. *Progress in Lipid Research, 48,* 44—51.

Rocca, A. S., LaGreca, J., Kalitsky, J., & Brubaker, P. L. (2001). Monounsaturated fatty acid diets improve glycemic tolerance through increased secretion of glucagon-like-peptide 1. *Endocrinology, 142,* 1148—1155.

Thompson, L. U., Yoon, J. H., Jenkins, D. J. A., Wolever, T. M. S., & Jenkins, A. L. (1984). Relationship between polyphenol intake and blood glucose response of normal and diabetic individuals. *The American Journal of Clinical Nutrition, 39,* 745—751.

Tiwari, R. S., Venkatachalam, M., Sharma, G. M., Su, M., Roux, K. H., & Sathe, S. K. (2010). Effect of food matrix on amandin, almond *(Prunus dulcis* L.) major protein, immunorecognition and recovery. *LWT-Food Science and Technology, 43,* 675—683.

United States Department of Agriculture (USDA) (2007). *Nutrient Database for Standard Reference,* Release 20(2007). Available at. www.ars.usda.gov/ba/bhnrc/ndl.

Wien, M. A. (2008). Almonds and pre-diabetes. Powerpoint presentation at *Experimental Biology.* April 4, 2008. San Diego, California.

Zohary, D., & Hopf, M. (1993). *Domestication of plants in the Old World* (3rd ed.). Oxford, UK: Oxford University Press. pp. 135—171.

CHAPTER 20

Aniseeds (*Pimpinella anisum* L.) in Health and Disease

B. Andallu, C.U. Rajeshwari
Department of Home Science, Sri Sathya Sai University, Anantapur, Andhra Pradesh, India

175

LIST OF ABBREVIATIONS

b.w., body weight
GRAS, generally recognized as safe
LD, lethal dose

INTRODUCTION

Centuries ago, Hippocrates remarked, "Let food be thy medicine and medicine be thy food." From the beginning of civilization, through trial and error, humans have experimented with higher plants in an attempt to find remedies against various illnesses and diseases, and found medicinal properties in the seeds, bark, and roots of certain plants. In spite of the advent of

Nuts & Seeds in Health and Disease Prevention. DOI: 10.1016/B978-0-12-375688-6.10020-9

modern high-throughput drug discovery and screening techniques, traditional knowledge systems have still given clues to the discovery of valuable drugs (Buenz *et al.*, 2004). Traditional medicinal plants are often cheaper, locally available, and easily consumable, either raw or as simple medicinal preparations. Considerable research on pharmacognosy, chemistry, pharmacology, and clinical therapeutics has been carried out on Ayurvedic medicinal plants in order to establish the scientific basis of their therapeutic potentials.

BOTANICAL DESCRIPTION

Pimpinella is a large genus of annual herbs, and *Pimpinella anisum* L. belongs to the celery family. Part of the Umbelliferae family, it is commonly known as aniseed. The seeds are the parts of the plant used; these contain anethole, anisaldehyde, & methyl-chavicol, and have carminative and stomachic activities.

Nutrient composition

Aniseeds contain nutrients and crude fiber (Table 20.1) and are considered to be a mild expectorant, stimulating, carminative, and diuretic. The chief active constituent of aniseed oil is anethole, which is used as an ingredient in cough lozenges in combination with liquorice.

Chemical composition

On steam distillation, aniseed yields a colorless or pale yellow liquid, with the characteristic odor and taste of the fruit, known as oil of aniseed. The oil also contains methyl chavicol, p-methoxy phenyl acetone, small amounts of terpenes, and sulfur-containing compounds (Parry, 1962). Furthermore, small amounts of estragol (0.5−5.0%), pseudoisoeugenyl-2 methylbutyrate, epoxypseudo isoeugenyl-2-butyrate, and anisaldehyde (0.1−1.4%) are present in the volatile oil. The *European Pharmacopoeia* specifies a minimum essential oil content of 20 ml/kg. The trans-anethole content of the volatile oil is in the range of 87−94% (Staesche *et al.*, 1994).

HISTORICAL CULTIVATION AND USAGE

Pimpinella anisum L. (Figure 20.1), an annual herb, is a native of the Eastern Mediterranean region, and is widely cultivated in southern and central Europe. In India, it is grown as a culinary herb, and a large quantity of aniseed is exported from India.

PRESENT-DAY CULTIVATION AND USAGE

Aniseed is widely grown for its fruits in tropical low lands, as it requires plenty of warmth and sunshine. The plant prefers a light, fertile, or moderately rich, well-drained sandy loam soil, and requires frequent and thorough cultivation throughout the growing season, with

TABLE 20.1 Nutrient Composition of Aniseeds	
Parameter	**g%**
Moisture	9–13
Protein	18
Fatty oil	8–23
Essential oil	2–7
Sugars	3–5
Starch	5
Nitrogen free extract	22–28
Crude fiber	12–25
Ash	6–10

FIGURE 20.1

Plant, flower, and seeds of *Pimpinella anisum* L.

occasional weeding. Aniseed is listed by the Council of Europe as a natural source of feed flavoring, and in the United States of America, it is listed as GRAS (generally recognized as safe) (Barnes *et al.*, 2002).

The fruits are extensively used as a spice for flavoring bakery products, and the essential oil is added to foodstuffs and liquors as a sensory and flavoring agent. Aniseed possesses a sweet, aromatic taste, and when crushed, emits a characteristic agreeable odor. The active component, anethole imparts a pleasant and characteristic flavor. The active ingredients are methyl chavicol, and some amounts of p-methoxyphenol, acetone, and terpenes. If aniseed is boiled for too long, it is liable to be divested of its essential components due to the heating/boiling process (www. diabetesmellitus-information.com).

Applications to health promotion and disease prevention

A concoction of seeds in hot water is used as a carminative, antiseptic, diuretic and digestive, and as a folk remedy for insomnia and constipation (Bisset, 1994). Several therapeutic effects, including for digestive disorders, gynecological problems and dyspnea, as well as anticon-vulsant and anti-asthma effects were described for the seeds of *Pimpinella anisum* L. in ancient medical books (Aboabrahim, 1970). Aniseeds possess expectorant, antispasmodic, carmina-tive, and parasiticidal properties. In traditional medicine, the drug is used internally for bronchial catarrh, pertussis, spasmodic cough, and flatulent colic, and externally for pedicu-losis and scabies. Furthermore, it is used as an estrogenic agent. It increases milk secretion, and promotes menstruation (Barnes *et al.*, 2002) (Figure 20.2).

A statistically significant bronchodilatory effect of the essential oil (0.02 ml), aqueous extract (0.6 ml equivalent to 1.5 g of aniseed) and ethanol extract (0.1 ml equivalent to 0.25 g of aniseed) was detected in guinea pigs (Boskabady & Ramazani-Assari, 2001).

Secretolytic and expectorant effects

The volume of respiratory secretion of anesthetized rabbits was increased dose-dependently from 19 to 82% following administration of aniseed oil by inhalation (in steam) in doses of 0.7–6.5 g/kg b.w. via a vaporizer, but signs of tissue damage and a mortality rate of 20% were observed at the highest dose level. Inhalation of anethole did not affect the volume, but produced a dose-dependent (1–9 mg/kg) decrease in the specific gravity of respiratory

177

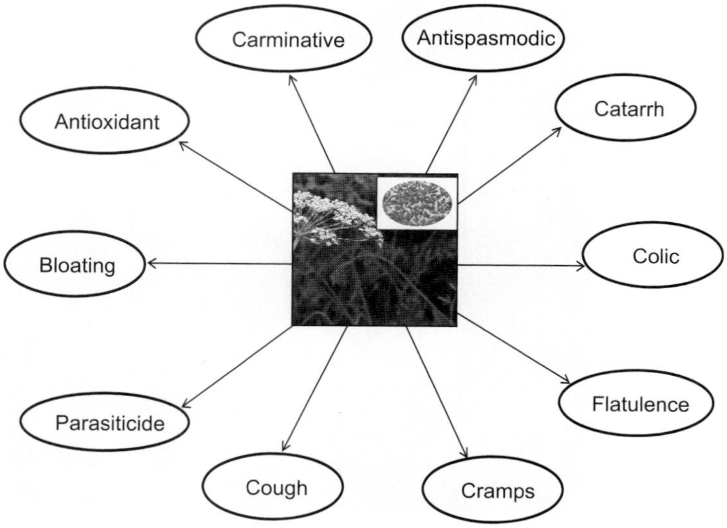

FIGURE 20.2
Traditional therapeutic uses of aniseeds (*Pimpinella anisum* L.).

tract fluid in urethanized rabbits in doses from 1 to 243 mg/kg b.w. (Boyd & Sheppard, 1968).

The effect of a mixture of herbal extract containing aniseeds was compared to that of the flavonoid quercetin. The clinical results showed significant reductions in sleep discomfort, cough frequency, and cough intensity in the herbal tea-using subjects compared to the placebo tea-using subjects (Haggag *et al.*, 2003).

Antibacterial and antifungal effects

The essential oil of aniseed is described as exerting fungicidal activity in a concentration-dependent manner (Soliman & Badeaa, 2002).

The aqueous decoction of aniseed exhibited maximum antibacterial activity against *Micrococcus roseus*. Singh and colleagues (2002) reported that the essential oil has strong antibacterial activity against eight human pathogenic bacteria. In addition, essential oil of aniseed possesses anticonvulsant activity in the mouse (Pourgholami *et al.*, 1999). An acetone extract of aniseed inhibited the growth of bacteria, including *Escherichia coli* and *Staphylococcus aureus*, and also exhibited antifungal activity against *Candida albicans* and other organisms (Maruzzella & Freundlich, 1959).

The aniseed and oil were found to have a high antibacterial activity against *Staphylococcus aureus* (responsible for boils, sepses, and skin infections), *Streptococus haemoliticus* (throat and nasal infections), *Bacillus subtilis* (infection in immuno-compromised patients), *Pseudomonas aeruginos* (hospital-acquired infection), and *Escherichia coli* (urogenital tract infections and diarrhea) (Singh *et al.*, 2002).

Local anesthetic activity

Trans-anethole concentration-dependently reduced electrically-evoked contractions of rat phrenic nerve hemidiaphragm by 10.3% at $10-3$ µg/ml, by 43.9% at $10-2$ µg/ml, by 79.7% at $10-1$ µg/ml, and by 100% at 1 µg/ml (Ghelardini *et al.*, 2001).

Estrogenic and anti-estrogenic effects

Trans-anethole, administered orally to immature female rats at 80 mg/kg b.w. for 3 days, significantly increased uterine weight to 2 g/kg compared to 0.5 g/kg in controls, and 3 g/kg in

TABLE 20.2 Chemical Compounds Found in *Pimpinella anisum* L. Seeds, and Their Biological Activities

Sl No.	Chemical Constituents	Part of the Plant	Antitumor	Anti-Inflammation	Anti-Aging	Anticancer	Antioxidant	Antidiabetic	Antimicrobial	Anti-Ulcer
1	α-Pinene	Fruit			✓					✓✓
2	α-Terpineol	Fruit		✓						
3	α-Zingiberene	Plant								
4	Ar-curcumene	Fruit			✓✓✓			✓		✓✓
5	Ascorbic acid	Fruit				✓	✓			
6	Bergapten	Fruit	✓							
7	β-Amyrin	Plant								
8	β-Bisabolene	Plant			✓					
9	β-Pinene	Fruit						✓✓		✓✓
10	Caffeic acid	Plant	✓✓	✓✓			✓			
11	Camphene	Fruit	✓✓✓					✓✓		
12	Chlorogenic acid	Plant			✓					
13	Copper	Seed	✓							
14	Eugenol	Plant					✓			
15	Fiber	Fruit			✓		✓			
16	Hydroquinone	Plant	✓							
17	Imperatorin	Leaf			✓					
18	Isoorientin	Plant			✓					
19	Limonene	Fruit	✓							
20	Magnesium	Fruit			✓✓✓		✓✓	✓✓✓		
21	Manganese	Fruit								
22	Mannitol	Root								
23	Myristicin	Plant	✓							✓
24	Rutin	Fruit			✓✓✓✓		✓✓			
25	Sabinene	Fruit								
26	Scoparone	Leaf								
27	Scopoletin	Fruit	✓				✓✓			
28	Squalene	Plant			✓					
29	Stigmasterol	Plant								
30	Umbelliferone	Fruit								
31	Zinc	Fruit		✓						

Source: www.ars-grin.gov/duke

animals given estradiol valerate subcutaneously at 0.1 μg/rat per day, confirming that transanethole has estrogenic activity (Dhar, 1995). Estrogenic activity of trans-anethole at high concentrations was determined by a sensitive and specific bioassay using recombinant yeast cells expressing the human estrogen receptor (Howes *et al.*, 2002).

Sedative effect

The pentobarbital-induced sleeping time of mice was increased by 93.5% after simultaneous intra-peritoneal administration of essential oil of aniseeds at 50 mg/kg b.w., and trans-anethole gave similar results (Marcus & Lichtenstein, 1982). An aqueous extract of aniseed exhibited weak *in vitro* cytotoxic activity against melanoma cells (Sathiyamoorthy *et al.*, 1999).

Chemical constituents and biological activities of *Pimpinella anisum* L. are given in Table 20.2.

ADVERSE EFFECTS AND REACTIONS (ALLERGIES AND TOXICITY)
Acute toxicity

The oral lethal dose of aniseed oil has been reported to be in the range of 50 to 500 mg/kg b.w. for human beings (Gosselin *et al.*, 1984). Intraperitoneal LD_{50} values for trans-anethole were determined as 0.65–1.41 g/kg in mice and 0.9–2.67 g/kg in rats (Lin, 1991).

Subchronic toxicity

In 90-day experiments in rats, 0.1% trans-anethole in the diet induced no toxic effects, whereas a dose-related edema of the liver was reported at levels of 0.3 and 3.0% concentration. Rats treated with 0.2, 0.5, 1.0, or 2% anethole in their diet for 12–22 months showed no effects on clinical chemistry, hematology, histopathology, or mortality, but a lower body weight and reduced fat storage were observed at 1.0 and 2.0% dose levels (Lin, 1991).

Reproductive toxicity

Trans-anethole exerted dose-dependent anti-implantation activity after oral administration to adult female rats on days 1–10 of pregnancy. When compared with control animals (all of which delivered normal offspring at completion of term), trans-anethole administered at 50, 70, and 80 mg/kg b.w. inhibited implantation by 33, 66, and 100%, respectively. When rats were administered trans-anethole on days 1–2 of pregnancy, normal implantation and delivery occurred. No gross malformations of offspring were observed (Dhar, 1995).

SUMMARY POINTS

- A concoction of aniseeds is a carminative, an antiseptic, a diuretic, and a digestive, and a folk remedy for insomnia and constipation.
- Aniseed oil increases pulmonary resistance in bronchopulmonary congestion.
- Anethole, the active component of aniseed oil, increases intestinal motility.
- The essential oil of aniseed exerts antifungicidal and antibacterial activities.
- An aqueous extract of aniseed exhibited a weak *in vitro* cytotoxic activity against melanoma cells.

References

Aboabrahim, Z. (1970). Zakhirah Kharazmshahi, Vol. 2, (p.141). National Works Publications, Teheran. In "Relaxant effect of *Pimpinella anisum* L. on isolated guinea pig tracheal chains and its possible mechanism(s)". Boskabady, M. H. & Ramazani-Assari, M. H. (2001), *Journal of Ethnopharmacology, 74*(1), 83–88.

Barnes, J., Anderson, L. A., & Phillipson, J. D. (2002). Aniseed. In *Herbal Medicines — A guide for healthcare professionals* (2nd edn). (pp. 51–54). London: Pharmaceutical Press.

Bisset, N. G. (Ed.). (1994). *Herbal Drugs and Phytopharmaceuticals. Medpharm* (pp. 73–75). Stuttgart: CRC Press.

Boskabady, M. H., & Ramazani-Assari, M. (2001). Relaxant effect of *Pimpinella anisum* L. on isolated guinea pig tracheal chains and its possible mechanism(s). *Journal of Ethnopharmacology, 74*(1), 83–88.

Boyd, E. M., & Sheppard, E. P. (1968). The effect of steam inhalation of volatile oils on the output and composition of respiratory tract fluid. *Journal of Pharmacology and Experimental Therapeutics, 163*(1), 250–256.

Buenz, E. J., Schnepple, D. J., Bauer, B. A., Elkin, P. L., Riddle, J. M., & Motley, T. J. (2004). Techniques: Bio-prospecting historical herbal texts by hunting for new leads in old tomes. *Trends in Pharmacological Sciences, 25*, 494–498.

Dhar, S. K. (1995). Anti-fertility activity and hormonal profile of trans-anethole in rats. *Indian Journal of Physiology and Pharmacology, 39*, 63–67.

Ghelardini, C., Galeotti, N., & Mazzanti, G. (2001). Local anaesthetic activity of monoterpenes and phenylpropanes of essential oils. *Planta Medica, 67*(6), 564–566.

Gosselin, R. E., Smith, R. P., & Hodge, H. C. (1984). *Clinical toxicology of commercial products, Vol. II* (5th ed., pp.11–230). Baltimore: Williams & Wilkins.

Haggag, E. G., Abou-Moustafa, M. A., Boucher, W., & Theoharides, T. C. (2003). The effect of a herbal water-extract on histamine release from mast cells and on allergic asthma. *Journal of Herbal Pharmacotherapy, 3*(4), 41–54.

Howes, M. J., Houghton, P. J., Barlow, D. J., Pocock, V. J., & Milligan, S. R. (2002). Assessment of estrogenic activity in some common essential oil constituents. *Journal of Pharmacy and Pharmacology, 54*(11), 1521–1528.

Lin, F. S. D. (1991). Trans-anethole. In *Joint FAO/WHO Expert Committee on Food Additives. Toxicological evaluation of certain food additives and contaminants, WHO Food Additives Series 28* (pp. 135–152). Geneva: World Health Organization.

Marcus, C., & Lichtenstein, E. P. (1982). Interactions of naturally occurring food plant components with insecticides and pentobarbital in rats and mice. *Journal of Agricultural and Food Chemistry, 30*, 563–568.

Maruzzella, J. C., & Freundlich, M. (1959). Antimicrobial substances from seeds. *Journal of the American Pharmacists Association, 48*, 356–358.

Parry J. W. (1962). *Spices: their morphology, histology and chemistry.* New York: Chemical Publishing Co. Inc. In: *The Wealth of India, Vol. XII* (pp. 60–64). Council of Scientific and Industrial Research New Delhi, India: NISCAIR.

Pourgholami, M. H., Majzoob, S., Javadi, M., Kamalinejad, M., Fanaee, G. H. R., & Sayyah, M. (1999). The fruit essential oil of *Pimpinella anisum* exerts anticonvulsant effects in mice. *Journal of Ethnopharmacology, 66*(2), 211–215.

Sathiyamoorthy, P., Lugasi Evgil, H., Schlesinger, P., Kedar, I., Gopas, J., Pollack, Y., et al. (1999). Screening for cytotoxic and antimalarial activities in desert plants of the Negev and Bedouin market products. *Pharmaceutical Biology, 37*(3), 188–195.

Singh, G., Kapoor, I. P. S., Pandey, S. K., Singh, U. K., & Singh, R. K. (2002). Studies on essential oil: Part 10; Antibacterial activity of volatile oils of some spices. *Phytotherapy Research, 16*, 680–682.

Soliman, K. M., & Badeaa, R. I. (2002). Effect of oil extracted from some medicinal plants on different mycotoxigenic fungi. *Food and Chemical Toxicology, 40*(11), 1669–1675.

Staesche, K., Schleinitz, H., & Burger, A. (1994). Pimpinella. In *Hager's Handbuch der Pharmazeutischen Praxis* (H. Hänsel, K. Keller, H. Rimpler & G. Schneider, Eds.), *Vol. 6* (pp. 135–156). Berlin: Springer Verlag.

CHAPTER 21

Bambangan (*Mangifera pajang*) Seed Kernel: Antioxidant Properties and Anti-cancer Effects

Mohd Fadzelly Abu Bakar[1], Maryati Mohamed[1], Asmah Rahmat[2], Jeffrey R. Fry[3]
[1] Laboratory of Natural Products, Institute for Tropical Biology and Conservation, Universiti Malaysia Sabah, Kota Kinabalu, Sabah, Malaysia
[2] Department of Nutrition and Dietetics, Faculty of Medicine and Health Sciences, Universiti Putra Malaysia, UPM, Serdang, Selangor, Malaysia
[3] School of Biomedical Sciences, University of Nottingham Medical School, Queen's Medical Centre, Nottingham, United Kingdom

LIST OF ABBREVIATIONS

Caov3, human ovarian carcinoma cell line
DPPH, 1,1-diphenyl-2-picrylhydrazyl
FRAP, ferric reducing/antioxidant power
GR, glutathione reductase
GSH, glutathione
HepG2, human Caucasian hepatoma carcinoma cell line
HPLC, high performance liquid chromatography
HT-29, human colon carcinoma cell line
MCF7, Hormone-dependent breast carcinoma cell line

Nuts & Seeds in Health and Disease Prevention. DOI: 10.1016/B978-0-12-375688-6.10021-0

MDA-MB-231, non-hormone-dependent breast carcinoma cell line
MTT, 3-(4,5-dimethylthiazo-2-yl)-2,5-diphenyltetrazolium bromide
t-BHP, tertiary-butyl hydroperoxide

INTRODUCTION

The *Mangifera* genus comprises about 40 species, of which at least 26 are known to produce edible fruits (Verheij & Coronel, 1991). The most common species in this genus is *Mangifera indica*, or the common mango, which is cultivated in many tropical regions and consumed worldwide. This fruit has been extensively utilized as a foodstuff, as well as an ingredient for nutraceuticals and functional foods. Other than common mango, one of the important species in this genus is *Mangifera pajang*, also known as bambangan (in Malay) or wild mango. This tree grows in the wild, and is endemic to Borneo Island. Nowadays, this tree is also cultivated by local people on Borneo Island.

BOTANICAL DESCRIPTION

Mangifera pajang (bambangan) is native to Borneo Island (Malaysia — Sabah and Sarawak; Brunei; and Indonesia — Kalimantan). The fruit of the bambangan weighs about 0.5—1kg or more, and is the largest known in the *Mangifera* genus. It has a rough, potato-brown skin which distinguishes it from the other *Mangifera* fruits. When ripe, the flesh is bright yellow in colour, sweet-sour, and juicy, although rather fibrous, with a somewhat strong turpentine aroma (Wong & Siew, 1994). The fruit of *M. pajang* is shown in Figure 21.1.

184

FIGURE 21.1
Fruit of *Mangifera pajang*. (A) Fruits on a tree; (B) Whole fruit; (C) Fruit with skin removed.

HISTORICAL CULTIVATION AND USAGE

The bambangan tree grows in the wild, and has also been cultivated by seed propagation by local people (i.e., the Dayak and Kadazan-dusun) on Borneo Island. The flesh, which represents 60–65% of the total weight, is used as a food, whilst the kernel (15–20% of the total weight) and peel (10–15% of the total weight) are usually discarded (Abu Bakar *et al.*, 2009a). The flesh is eaten fresh by indigenous people, and also made into pickle or cooked with fish, chicken, or meat for a distinctive "mangoey" and sour flavour. The flesh of bambangan is also sometimes cooked with onion and chilli, and served as side dishes or "sambal". The flesh of the fruit is used to make juice, and, together with the grated kernel, is used to make pickle, which is delicious eaten with the local staple food (rice).

PRESENT-DAY CULTIVATION AND USAGE

Currently, this tree is cultivated commercially in farms or planted locally in the backyards of people's homes. Bambangan fruits and products (i.e., pickle and juice) are sold commercially in markets on Borneo Island.

APPLICATIONS TO HEALTH PROMOTION AND DISEASE PREVENTION

The kernel of bambangan is considered to be a waste product, since it is not eaten fresh, and is only occasionally used to make pickle. Our recent study has shown that bambangan kernel contains diverse antioxidant phytochemicals, and displays numerous potential health-beneficial properties.

Phytochemical components

The kernel of *M. pajang* has been investigated in terms of its selected phytochemical components. Abu Bakar *et al.* (2009a) reported that an alcoholic extract of the kernel is a rich source of plant phenolics, with the total phenolic content representing approximately 10% of its total weight, although no anthocyanins were detected. Further analysis by HPLC was conducted to identify individual phenolic phytochemicals in the kernel, and the results are presented in Table 21.1. A variety of phenolic acids (gallic, coumaric, sinapic, caffeic, ferulic, chlorogenic) and flavonoids (naringin, hesperidin, rutin, diosmin) were identified, many of which are recognized as potent antioxidants with demonstrable health-beneficial properties (Abu Bakar *et al.*, 2010).

TABLE 21.1 Polyphenol Phytochemical Composition of Bambangan Kernel

Phytochemical	Concentration (µg/g Dry Weight)
Gallic acid	236.6 ± 28.3
p-Coumaric acid	301.1 ± 19.1
Sinapic acid	20.1 ± 1.0
Caffeic acid	150.0 ± 13.5
Ferulic acid	5334.0 ± 513.9
Chlorogenic acid	14.1 ± 0.9
Naringin	294.5 ± 98.2
Hesperidin	221.6 ± 24.9
Rutin	721.1 ± 34.1
Diosmin	2386.0 ± 325.1

Data are presented as mean $\pm$ SEM ($n = 3$)
Adapted from Abu Bakar *et al.* (2010).

Antioxidant activity

Our recent study showed that bambangan fruit displayed high antioxidant activity (Abu Bakar *et al.*, 2009a). The fruits were separated into flesh, peel, and kernel, and the antioxidant assessment of alcoholic extracts was conducted by a free radical scavenging assay (using DPPH) and a ferric reduction (FRAP) assay. The results showed that the kernel extract of the fruit displayed the highest DPPH scavenging effect, followed by the peel and flesh extracts (Table 21.2). The reducing ability of the tested extracts was in the order of kernel > peel > flesh; the same trend as shown in the DPPH free radical scavenging assay. The previously identified polyphenol phytochemicals (see above) are most likely responsible for the antioxidant activity of the kernel of bambangan.

Anticancer activity

Abu Bakar *et al.* (2010) investigated the cytotoxic effects of the kernel, peel, and flesh of bambangan against the proliferation *in vitro* of cancer cell lines derived from human liver (HepG2), colon (HT-29), ovary (Caov3), and breast (MCF-7 and MDA-MB-231), using MTT assay as the marker for proliferation. The results showed that bambangan kernel induced strong cytotoxic activity against all cancer cell lines tested, especially the breast cancer lines (IC$_{50}$ for MCF-7, 23.0 µg/ml, and MDA-MB-231, 30.5 µg/ml) (Table 21.3). Again, the inhibition of proliferation of the selected cancer cell lines can be explained, at least partially, by the presence of the phenolic phytochemicals described previously, many of which have well-characterized antitumor actions. These phenolic phytochemicals might act additively, synergistically, and/or antagonistically with other (unidentified) compounds to display the antiproliferative activity (Yang *et al.*, 2009).

Cytoprotective activity

The potential cellular antioxidant and cytoprotective effects of bambangan kernel, peel, and flesh extracts were tested against oxidative stress toxicity induced by t-BHP in the HepG2 cell

TABLE 21.2 Antioxidant Properties of Different Parts of Bambangan

Sample	DPPH Free Radical Scavenging[1]	FRAP[2]
Flesh	9.9 ± 0.2^d	150.0 ± 1.4^c
Peel	20.3 ± 0.1^b	343.2 ± 7.2^b
Kernel	23.2 ± 0.0^a	3130.0 ± 35.5^a

Values are presented as mean $\pm$ SEM ($n = 3$); different superscript letters indicate significant difference at $P < 0.05$ (ANOVA and Duncan's post-hoc test)
Adapted from Abu Bakar *et al.* (2009a).
[1]DPPH free radical scavenging activity was expressed as mg ascorbic acid equivalent antioxidant capacity (AEAC) in 1 g of dry sample
[2]FRAP was expressed as µM ferric reduction to ferrous in 1 g of dry sample.

TABLE 21.3 Inhibition of Proliferation of Cancer Cell Lines by Crude Extracts of Bambangan

Sample	HepG2	HT-29	Caov3	MCF-7	MDA-MB-231
Flesh	> 100	> 100	> 100	> 100	> 100
Peel	36.5	> 100	55.0	> 100	> 100
Kernel	34.5	63.0	92.0	23.0	30.5

Inhibition expressed as IC$_{50}$ values (µg/ml; concentration of extract inhibiting cell proliferation by 50%); data are presented as mean of three separate experiments.
Adapted from Abu Bakar *et al.* (2010).

line (Abu Bakar *et al.*, 2009b). The results showed that only the kernel extract displayed protective activity against oxidative damage caused by t-BHP in HepG2, with a cytoprotective index value (concentration producing 50% protection) of $1.2 \pm 0.1\,\mu g/ml$. Meanwhile, the positive control (quercetin) showed lower cytoprotective activity, as indicated by a higher cytoprotective index value of $5.3 \pm 0.7\,\mu g/ml$. In addition, both quercetin and *M. pajang* kernel extract significantly induced the expression of glutathione reductase (GR), indicating that both materials (quercetin and kernel extract) may exert their cytoprotective action by a combination of direct (radical scavenging) and indirect (protein upregulation) antioxidant activities. t-BHP has been reported to exert its cytotoxic activity via glutathione (GSH) depletion (Lima *et al.*, 2007); upregulation of GR serves to maintain the GSH status and thereby provide protection against oxidative stress.

ADVERSE EFFECTS AND REACTIONS (ALLERGIES AND TOXICITY)

No adverse effects or toxicity have been reported for bambangan kernel in humans or animals. In addition, no cytotoxic activity was observed in normal human (MRC-5) and rat fibroblast (3T3) cell lines after incubation with bambangan kernel extract at concentrations of up to $100\,\mu g/ml$ (unpublished data). The kernel of bambangan can be regarded as safe, since it is eaten with the flesh as a pickle by the local community. However, possible toxic effects of bambangan kernel extracts at high doses and/or with long-term exposure should be considered.

SUMMARY POINTS

- Bambangan kernel contains diverse phenolic phytochemicals, such as phenolic acids and flavonoids, which might be responsible for the potent health-beneficial properties.
- The kernel extract has been shown to display superior antioxidant activity compared to other parts of the fruit.
- Bambangan kernel extract has displayed a broad spectrum of anticancer properties, especially against breast cancer cells.
- Bambangan has also been shown to protect against t-BHP induced oxidative stress in mammalian cell lines, comparable to the well-known cytoprotective compound, quercetin.
- Preliminary data reveal that bambangan kernel extract has great potential as a chemopreventive and chemotherapeutic agent. Further study is needed to determine the efficacy of this extract *in vivo* and in human clinical trials.

187

References

Abu Bakar, M. F., Mohamed, M., Rahmat, A., & Fry, J. R. (2009a). Phytochemicals and antioxidant activity of different parts of bambangan (*Mangifera pajang*) and tarap (*Artocarpus odoratissimus*). *Food Chemistry, 113,* 479–483.

Abu Bakar, M. F., Mohamed, M., Rahmat, A., Burr, S. A., & Fry, J. R. (2009b). Cytoprotective activity of Bambangan (*Mangifera pajang*) extracts against t-BHP induced oxidative damage in hepatocyte (HepG2) cell line. Kuching, Sarawak, Malaysia: 4th Global Summit on Medicinal and Aromatic Plants. Four Points by Sheraton.

Abu Bakar, M. F., Mohamed, M., Rahmat, A., Burr, S. A., & Fry, J. R. (2010). Cytotoxicity and polyphenol diversity in selected parts of *Mangifera pajang* and *Artocarpus odoratissimus* fruits. *Nutrition and Food Science, 40,* 29–38.

Lima, C. F., Valentao, P. C. R., Andrade, P. B., Seabra, R. M., Fernandes-Ferreira, M., & Pereira-Wilson, C. (2007). Water and methanolic extracts of *Salvia officinalis* protect HepG2 cells from t-BHP induced oxidative damage. *Chemico-Biological Interactions, 167,* 107–115.

Verheij, E. W. M., & Coronel, R. E. (Eds.). (1991). *Plant resources of South East Asia, No. 2, Edible fruits and nuts.* Wageningen: Pudoc.

Wong, K. C., & Siew, S. S. (1994). Volatile components of the fruits of bambangan (*Mangifera pajang* Kostermans) and binjai (*Mangifera caesia* Jack). *Flavour and Fragrance Journal, 9,* 173–178.

Yang, J., Liu, R. H., & Halim, L. (2009). Antioxidant and antiproliferative activities of common edible nut seeds. *LWT-Food Science and Technology, 42,* 1–8.

CHAPTER 22

Bambara Groundnut [*Vigna subterranea* (L.) Verdc. (Fabaceae)] Usage in Human Health

Mongomaké Koné[1], Alistair G. Paice[2], Yaya Touré[1]
[1] Laboratoire de Biologie et Amélioration des Productions Végétales,
UFR des Sciences de la Nature, Université d'Abobo-Adjamé, Côte d'Ivoire
[2] Department of Nutrition and Dietetics, King's College London, London, UK

189

LIST OF ABBREVIATIONS

BG, bambara groundnut
CT, condensed tannins
FAO, Food and Agriculture Organization
L, Linneum
PEM, protein energy malnutrition
Verdc, Verdecourt

INTRODUCTION

Bambara groundnut (BG) is an indigenous grain legume mainly cultivated by subsistence farmers in sub-Saharan Africa. It is primarily grown for its seeds, which contain significant quantities of protein, carbohydrate, and fat (Rowland, 1993). The crop has the ability to support a wide range of agro-ecological conditions, making it popular among resource-poor farmers. Therefore, it plays an important role in alleviating hunger, and thus contributes to food security in Africa. Besides its nutritional qualities, the plant's medical properties have also

Nuts & Seeds in Health and Disease Prevention. DOI: 10.1016/B978-0-12-375688-6.10022-2

been exploited to overcome various diseases. In the present chapter, we report on bambara groundnut seed utilization in human health.

BOTANICAL DESCRIPTION

Bambara groundnut landraces differ in many aspects from each other, with a wide variety of seed and pod colors, and growth habits varying from bunch type, to semi bunch and spreading. The crop is a prostrate, herbaceous annual legume, with a well-developed compact tap-root system bearing abundant geotropic lateral roots. In association with *Rhizobium*, the roots form rounded and sometimes lobed nodules. The leaves are trifoliate, with a long, grooved and stiff petiole that is thickened at the base. Usually two papilionaceous flowers are attached to the peduncle. The flowers are yellow, and flowering is indeterminate. During pollination and fertilization, the peduncle elongates to bring the ovaries to the ground level. After fertilization, the pedicles penetrate the soil to form pods with either one or two seeds. Like the groundnut (*Arachis hypogaea*), BG forms pods on or just below the soil surface (Figure 22.1A—G).

FIGURE 22.1

Morphology and reproductive traits of Vigna subterranea (L.) cultivated on experimental field of University of Abobo-Adjamé (Côte d'Ivoire). (A) Bambara groundnut, bunchy type, with oval leaves (L) at the petiole (Pet) end and many lateral roots (IR) derived from a well-developed tap root (tR) ; (B) formation of yellow flowers (arrow) at the base of the plant, followed by elongation and penetration of peduncle (Pe) carrying ovaries in the soil. Ovaries developed in pods (P) containing one or two seeds, of which the size and coloration (C—G) vary with landraces. *Source: Mongomaké Koné.*

FIGURE 22.1   Cont'd

191

HISTORICAL CULTIVATION AND USAGE

Bambara groundnut, an indigenous legume, has been cultivated for centuries in the tropical regions south of the Sahara. It is also widely grown in Eastern Africa and Madagascar, and is even found in parts of South and Central America. Usually intercropped with sorghum, maize, and tuberous crops, BG gives the best yields on a deeply ploughed field with a fine seedbed. The seeds are used for human consumption, and to feed chicks (Oluyemi *et al.*, 1976). The haulm has been found to be palatable (Doku & Karikari, 1971), and is an important source of livestock feed.

PRESENT-DAY CULTIVATION AND USAGE

Knowledge about the bambara groundnut in the Côte d'Ivoire is limited, and few people know the role of the crop in society. BG is mainly cultivated in the savannah area, but it is also well known in western and central regions of the country. Small-scale farmers and a limited number of people in rural areas grow BG mainly as a subsistence crop. Farmers sow seeds of various landraces in different agro-ecological environments. There is little information available on yield levels. BG yields are highest when seeds are sown in July or August, especially after sufficient rainfall. Generally, seeds are boiled with or without salt and pepper (Figure 22.2). Seeds can also be ground, and the subsequent flour used to make cakes. BG cultivars that are resistant to foliar diseases have a dual role in providing pods for human use, and fodder for livestock feed.

FIGURE 22.2
Woman selling boiled seeds of Vigna subterranea (L.) in Adjamé Market, Abidjan, Côte d'Ivoire. Pods containing seeds are boiled with or without salt; this is one of the main ways of preparing and consuming bambara groundnut in Côte d'Ivoire. A handful of boiled pods costs 50 Frs CFA (0.08 euros, or 0.10 US dollars). *Source: Mongomaké Koné.*

APPLICATIONS IN HEALTH PROMOTION AND DISEASE PREVENTION

In Africa, bambara groundnut is generally cultivated by women on small plots. Production is primarily at subsistence level, and only the surplus is sold. There has been relatively little research on the BG crop, and this has mainly concentrated on characterization of landraces and their testing to assess the potential yield of components and yield-related characters (Adeniji *et al.*, 2008).

The contribution of BG crops to improving human health can be classified into two broad categories. First, the plant has a significant impact on human nutrition (Mkandawire, 2007) by improving the quality of nutrition and, consequently, the health of many rural communities in Africa. Thus, to meet food sufficiency, farmers are being encouraged to grow the crop. BG produces a food of exceptional nutritional quality, and thus plays an important role in building and maintaining a solid foundation for good health. The seeds have a good nutrient balance, and are a rare example of a complete food. They contain sufficient quantities of proteins, carbohydrates, and lipids to sustain human life (Table 22.1). This legume can therefore contribute to food security and help alleviate nutritional problems (Brough & Azam-Ali, 1992). In addition, seed proteins contain more of the essential amino acid methionine than almost any other bean (Table 22.2). Therefore, this crop could be used to overcome undernutrition in many parts of Africa.

A diet based only on cereal is unbalanced in the absence of animal products such as meat and milk. The inclusion of a legume in this diet would help overcome these deficiencies in the domestic diet. Blends of sorghum, BG, and sweet potatoes have good protein quality (Nnam, 2001). Fortification of maize meal (cereal) with BG (legume) can alleviate problems of protein energy malnutrition (PEM). Fortified food prepared with bambara nut and maize is nutritious, and conforms to specifications as recommended by the National Institute of Nutrition and Food and Agriculture Organization (FAO) to combat malnutrition, especially in low economic groups (Mbata *et al.*, 2009). Thus, BG plays an important role in the traditional diets of rural people in most countries of sub-Saharan Africa, and helps to overcome Kwashiorkor, the common protein deficiency disease in young children. Of note in pediatric nutrition, seeds with a cream-colored testa are recommended in

TABLE 22.1 Bambara Groundnut: Composition of Dry, Mature Seeds

Elements	Composition (per 100 g edible portion)
Food energy	367 calories
Moisture	10.3%; (38); 5.8—14.8
Protein	18.8 g; (38); 13.8—21.1
Lipid	6.2 g; (39); 4.5—9.2
Carbohydrate	61.3 g
Fiber	4.8 g; (33); 1.4—7.2
Ash	3.4 g; (49); 1.3—5.1
Calcium	62 mg; (40); 29—217
Phosphorus	276 mg; (40); 105—475
Iron	12.2 mg; (26); 1.0—16.0
β-carotene	10 µg; (1)
Thiamin	0.47 mg; (32); 0.22—0.62
Riboflavin	0.14 mg (20); 0.12—0.15
Niacine	1.8 mg; (20); 0.6—2.4
Ascorbic acid	Trace; (18); 0—8

Bambara groundnut is a high energy, balanced food, containing most of the nutrients essential for human health. The number of analyses is indicated in parentheses; these were not necessarily by the same laboratory. The range values are also given.
Reprinted from Leung, W.-T. W., Busson, F., & Jardin, C. (1968). *Food composition table for use in Africa*. Bethesda, MD: US Department of Health, Education, and Welfare and the Food and Agriculture Organization of the United Nations (FAO).

TABLE 22.2 Amino Acid Content of Bambara Groundnut

Amino Acid	mg/g Total Nitrogen
Tryptophane	68
Lysine	403 ± 5.92
Methionine	110 ± 5.39
Phenylalanine	350 ± 12
Threonine	218 ± 9.27
Valine	331 ± 9.07
Leucine	489 ± 6.93
Isoleucine	274 ± 8.54
Cystine	65 ± 5.29
Tyrosine	218 ± 9.27
Arginine	396 ± 16.31
Histidine	189 ± 9.38
Alanine	282 ± 8.49
Aspartic acid	735 ± 14.56
Glutamic acid	1047 ± 9.06
Glycine	251 ± 9.43
Proline	317 ± 13.38
Serine	349 ± 10.30

Analyses were carried out by the column chromatographic (CC) method. In the case of cystine, data were obtained following oxidization of cystine to cysteic acid before hydrolysis (Schram, Moore, & Bigwood, 1954; Moore & Stein, 1963).
Reprinted from FAO (1970), *Amino Acid Content of Foods and Biological Data on Proteins*, Nutritional Studies No. 24. Rome: FAO.

weaning formulae, as the condensed tannin (CT) content is acceptable for this age group. From a nutritional perspective, farmers and consumers can select landraces with a high nutritive value but low CT and trypsin inhibitor content for cultivation and human consumption.

The second broad category of health benefit exhibited by the plant is its use in traditional medicine therapies. To date, there is no information available on the specific pharmaceutical and chemical properties of bambara groundnut. All data reported are based on ethno-botanical information obtained from local communities. In many parts of Africa, bambara has recognized medicinal uses. Seeds are currently used in Burkino Faso to overcome numerous human diseases (Nacoulma-Ouedraogo, 1996). In north-eastern Nigeria, the seeds are not only consumed as food, but also used for medicinal purposes (Atiku, 2000). Seeds are used by the Igbos tribe in Nigeria to treat venereal diseases, and, according to the Luo tribe in Kenya, seeds are also used to overcome diarrhea. The consumption of roasted seeds is recommended in the treatment of polymenorrhea. The cooking water of BG seeds is used as drink, as a remedy for internal bruising; here, in order to speed up the resorption of hematomas, sufferers are advised to drink seed flour diluted in water. Crushed seeds mixed with water are also administered to treat cataracts.

Amongst the beans, bambara groundnut seeds have the highest concentration of soluble fiber — a non-nutrient believed to reduce the incidence of heart disease and to help prevent colon cancer. In addition, in Botswana, the black-seeded landraces have the reputation for being a treatment for impotence. The chewing and swallowing of immature fresh seeds is supposed to arrest nausea and vomiting; this remedy is often used to treat morning sickness in pregnant women. Bambara groundnut seeds have also been used to treat some malignancies and inflammatory disorders in Africa. However, the biochemical basis for these effects remains unclear (Hye-Kyung *et al.*, 2004). Recent research, though, has indicated that bambara groundnut seeds with condensed tannins in low concentrations have beneficial effects in human nutrition and health (Akindahunsi & Salawu, 2005).

In northern Côte d'Ivoire, we surveyed the local communities of more than 40 villages regarding the medical utilization of the bambara groundnut plant. Results indicated that farmers do not use the plant leaves, stem, and roots for therapy, but, rather, the seeds. According to these farmers, the seeds are used to treat cases of anemia. This leads to application for peri-partum women within 1 month of delivery. In numerous cases, bambara groundnut seeds, with and without various components of other plant mixtures, are used to cure several diseases. Thus, the juice obtained from boiling maize and BG seeds is used to treat diarrhea (Goli *et al.*, 1991). The flour obtained from seeds of BG is mixed with flour obtained from the fruit of *Puvpalia lappacea* (L.) (*Amaranthaceae*), the mixture is then dissolved in water and used as a hemostatic drink, to treat menorrhagia during pregnancy, and rectal bleeding. A decoction of seeds, associated with the leaves of *Terminalia laxiflora* (L.) (*Combretaceae*), is used as a drink to treat gonorrhea. According to one farmer (the chief of traditional hunters, commonly called the Dozo) in the village of Sinématiali, the black seeds, mixed with a plant whose identity has not yet been identified, can be used to treat ulcers. These black seeds, in addition to their therapeutic properties, are also used in the field of mysticism by traditional hunters.

ADVERSE EFFECTS AND REACTIONS (ALLERGIES AND TOXICITY)

As already mentioned, no dedicated pharmacological studies have been reported in the literature regarding the bambara groundnut. No cases of allergy and toxicity have been recorded from ethno-botanical studies carried out thus far.

The digestion, and subsequent bioavailability, of nutrients contained within the bambara seeds is limited by "antinutrients" such as trypsin inhibitors and condensed tannins (Apata & Ologhobo, 1997). Removal of these would be necessary to allow effective utilization of the protein, carbohydrate, and mineral content for human nutrition. The condensed tannin content in cream landraces is lower, and these may therefore be better as a source of nutrition. The most common processing method used for bambara groundnut preparation is boiling the seeds. Heat-labile phenolic compounds contained within the seeds are leached out into the water during this process (Tibe *et al.*, 2007). The dehulling, soaking, and boiling of the seeds, then discarding the cooking water, may lower the condensed tannin content of the foodstuff (Champ, 2002). There seems to be no adverse effect of trypsin inhibitor activity in the landraces if consumed cooked, but more research needs to be done in this field.

SUMMARY POINTS

- Bambara groundnut is a neglected and underutilized grain legume mainly cultivated by small-scale farmers (usually women) in many parts of Africa.
- As a well-balanced food, bambara groundnut is an excellent supplement in helping to achieve a balanced diet and overcome malnutrition, especially among children.
- The seeds have potential therapeutic properties that may contribute to the health of local communities.
- The seeds, with or without other plant mixtures, are used in traditional medicine.
- High temperature negates antinutrients and inhibitor factors contained within the seeds.
- Pharmacological studies needed to confirm these ethno-botanical reports remain to be performed.
- Currently, no detailed biochemical studies have been published on toxicity and potential adverse effects associated with the medicinal usage of bambara groundnut.

References

Adeniji, O. T., Peter, J. M., & Bake, I. (2008). Variation and interrelationships for pod and seed yield characters in bambara groundnut (*Vigna subterrenea*) in Adamawa State, Nigeria. *African Journal of Agriculture and Social Research, 3*, 617−621.

Akindahunsi, A. A., & Salawu, S. O. (2005). Phytochemical screening and nutrient/antinutrient composition of selected tropical green leafy vegetables. *African Journal of Biotech, 4*, 497−501.

Apata, D. F., & Ologhobo, D. (1997). Trypsin inhibitor and other antinutritional factors in tropical legume seeds. *Tropical Science, 37*, 52−59.

Atiku, A. A. (2000). *Bambara groundnut processing and storage practices in North Eastern Nigeria. Postgraduate seminar paper, Department of Agricultural Engineering*. Maiduguri, Nigeria: University of Maiduguri.

Brough, S. H., & Azam-Ali, S. N. (1992). The effect of soil moisture on proximate composition of bambara groundnut [*Vigna subterranean* (L.) Verdc.]. *Journal of the Science of Food and Agriculture, 60*, 197−203.

Champ, M. M. J. (2002). Non-nutrient bioactive substances of pulses. *British Journal of Nutrition, 88*, 307−319.

Doku, E. V., & Karikari, S. K. (1971). The role of ants in pollination and pod production in bambara groundnut (Vigna subterranea). *Economic Botany, 25*, 357−362.

Goli, A. E., Begemann, F., & Ng, N. Q. (1991). Germplasm diversity in bambara groundnut and prospects for crop improvement. In N.Q. Ng, P. Perrino, F. Attere, & H. Zedan. *Crop Genetic Resources of Africa, Vol. 2*, 195−202.

Hye-Kyung, Na, Kensese, S. M., Ji-Yoon, L., & Young-Joon, S. (2004). Inhibition of phorbol ester-induced COX-2 expression by some edible African plants. *BioFactors, 21*, 149−153.

Leung, W.-T. W., Busson, F., & Jardin, C. (1968). *Food composition table for use in Africa*.

Mbata, T. I., Ikenebomeh, M. J., & Alaneme, J. C. (2009). Studies on the microbiological, nutrient composition and antinutritional contents of fermented maize flour fortified with bambara groundnut (*Vigna subterranea L*). African. *Journal of Food Science and Biotechnology, 3*, 165−171.

Mkandawire, C. H. (2007). Review of Bambara groundnut (*Vigna subterranea* (L.) Verdc.) production in sub-Saharan Africa. *Agricultural Journal, 2*, 464−470.

Moore, S., & Stein, W. H. (1963). Chromatographic determination of amino acids by use of automatic recording equipment. *Methods Enzymol, 6*, 819.

Nacoulma-Ouedraogo, O. G. (1996). *Plantes médicinales et pratiques médicales traditionnelles au Burkina-Faso; cas du plateau central; Thèse de Doctorat*. Faculté des Sciences et Techniques de l'Université de Ouagadougou. p. 266.

Nnam, N. M. (2001). Comparison of the protein nutritional value of food blends based on sorghum, bambara groundnut and sweet potatoes. *International Journal of Food Sciences and Nutrition, 52*, 25−29.

Oluyemi, J. A., Fetuga, B. L., & Endeley, H. N. L. (1976). The metabolizable energy value of some feed ingredients for young chicks. *Poultry Science, 55*, 11−618.

Rowland, J. R. J. (1993). Bambara groundnut. In J. R. J. Rowland (Ed.), *Dry farming in Africa* (pp. 278–282). London, UK: MacMillan Limited.

Schram, E. O., Moore, S., & Bigwood, E. J. (1954). Chromatographic determination of cystine as cysteic acid. *Biochemical Journal, 57*, 33.

Tibe, O., Amartefio, J. O., & Njogu, R. M. (2007). Trypsin inhibitor activity and condensed tannin content in bambara groundnut (*Vigna Subterranea* (L.) Verdc) grown in Southern Africa. *Journal of Applied Sciences and Environmental Management, 11*, 159–164.

CHAPTER 23

Betel Nut (*Areca catechu*) Usage and Its Effects on Health

Niloufer Sultan Ali, Ali Khan Khuwaja
Department of Family Medicine, The Aga Khan University, Karachi, Pakistan

197

INTRODUCTION

Betel nut is a seed of the *Areca Catechu* tree, which is cultivated in the tropical Pacific Islands and parts of Asia and Africa. Its use is widely prevalent throughout the world. Medical use of betel nut is limited, while scientific evidence reveals its various health hazards. Research has proved that regular use of betel nut is associated with different types of cancers, systemic illnesses, and various other diseases. In addition, use of betel nut is also linked to health emergencies, toxicities, and drug interactions. This chapter describes important information about betel nut use and its effects on health.

BOTANICAL DESCRIPTION

The betel nut, also known as the *Areca* nut, is a seed of the *Areca catechu* tree, which is a species of palm tree that grows in parts of the tropical Pacific Islands, Asia, and Africa. Commonly known as the betel nut tree, it can grow to a height of 20—28 meters, and bears fruit throughout the year (Figure 23.1). *Areca catechu* is part of the Arecaceae family (referred to as the palm family), which contains over 200 genera and about 2600 species. Most members of the Arecaceae family only grow in tropical or subtropical climates. Leaves from the Piper Betle plant (commonly known as betel) are often chewed together with areca nut and edible lime (also called calcium hydroxide, limbux, or slaked lime), and, because of this association, the areca nut has come to be known as betel nut.

Nuts & Seeds in Health and Disease Prevention. DOI: 10.1016/B978-0-12-375688-6.10023-4

FIGURE 23.1
Areca catechu tree and Areca nut (betel nut). *Source: http://commons.wikimedia.org/wiki/File:Areca_catechu_Blanco2. 350.png.*

HISTORICAL CULTIVATION AND USAGE

Areca catechu is originally native to the Malaysian peninsula, but its use has resulted in a long history of cultivation and naturalization throughout South and South-east Asia. It requires a warm, humid, tropical climate to thrive. Seeds are the only means of propagation. Chewing betel nut is an ancient practice among Asians, where it is socially acceptable among all sections of society, including women and children. The nut may be used fresh or dried, fermented, or after boiling, baking, and roasting. It is also used wrapped in betel leaves, with or without slaked lime (Figures 23.2 and 23.3).

PRESENT-DAY CULTIVATION AND USAGE

Betel nut seeds usually germinate in a minimum time of 6 weeks, and 1- to 2-year-old seedlings are planted into their permanent sites at a density of 1000–1500 palms per hectare. The palms usually begin to flower and fruit after 7 years. It takes 10–15 years to achieve maximum fruit production, which continues for 45–75 years. Use of inorganic fertilizers has, however, increased the production of betel nut in recent years. Around 10–20% of world's population uses betel nut (Gupta & Warnakulasuriya, 2002) (Figure 23.4). Recently, the trend has changed to chewing processed betel products known as pan-masalas and gutka. These are available in small, attractive, and convenient sachets for individual use (Figure 23.5), which contain a mixture of betel nut, cardamom, fennel, and slaked lime, and may also contain tobacco. Table 23.1 summarizes the different methods of use of betel nut.

FIGURE 23.2
Preparation of betel nut, with betel quid. A betel quid (sulemani pan — without slaked lime) prepared in the traditional way using betel nuts and betel leaves. *Source: http://en.wikipedia.org/wiki/File:Paan_Making.jpg.*

FIGURE 23.3
Betel nut with betel quid, slaked lime, and tobacco. A betel quid (pan) prepared in the traditional way using betel nuts, betel leaves, slaked lime, and tobacco.

APPLICATIONS TO HEALTH PROMOTION AND DISEASE PREVENTION

The significance of betel nut usage in health promotion and disease prevention is limited. It was one of the earliest psychoactive substances used in the world, mainly in the South Pacific islands, South-east Asia, and South Asia. It is the fourth most widely used addictive substance in the world, after caffeine, nicotine, and alcohol. Its use is also becoming increasingly frequent among youth, where it may serve as a gateway to tobacco use (Chandra & Mulla, 2007). People chew it for stress reduction, for a feeling of well-being, and for heightened awareness. It contains several psychoactive compounds. Arecoline, the principal alkaloid in betel nut, acts as a stimulant of the nervous system, and increases the levels of noradrenaline and acetylcholine. This leads to subjective effects of increased well-being, alertness, and stamina. Arecaidine, another active ingredient, may have anxiolytic properties through inhibition of gamma-amino

FIGURE 23.4
Dried betel nuts.

FIGURE 23.5
Sachets of pan-masala and gutka. These sachets contain a mixture of betel nut, cardamom, fennel, and slaked lime, and may also contain tobacco.

butyric acid reuptake. These properties mean that there is a strong potential for the abuse of and dependence on the betel nut. The preferred route of intake is chewing, which leads to rapid absorption of these alkaloids through the buccal mucosa. The onset of effects starts within 5 minutes, and lasts for about 2—3 hours (Winstock, 2002).

It has been suggested that betel nut chewing may confer protection against dental caries. Researchers have reported a lower prevalence of dental caries among betel nut chewers than in non-chewers (Moller *et al.*, 1997). The cariostatic properties of betel nut are not well known, but it has been suggested that the betel stain, which often coats the surface of the teeth, may act as a protective varnish. *In vitro*, evidence has suggested that the tannin content of areca may have antimicrobial properties, and this may contribute to the cariostatic role of areca (De Miranda *et al.*, 1996). Moreover, the pH in the oral cavity might be increased by the process of chewing, leading to production of copious amounts of saliva in the mouth; this may

TABLE 23.1 Betel Nut: Means of Consumption, Contents, and Practice

Local Products	Contents	Practice
Pan (Sulemani)	Betel nut, with or without tobacco, wrapped in betel leaf	Chewed slowly; juices are either swallowed or spat out
Pan	Betel nut and slaked lime, with or without tobacco, wrapped in betel leaf	Chewed slowly; juices are either swallowed or spat out
Gutka	Betel nut, slaked lime, and powdered tobacco, with spices and flavoring agents	Powder is kept in the mouth, chewed slowly, and swallowed
Pan Masala	Powdered betel nut and slaked lime, with spices and flavoring agents like cardamom and fennel	Powder is kept in the mouth, chewed slowly, and swallowed
Mawa	Betel nut, slaked lime, and tobacco	After mixing the contents, they are kept in the mouth and chewed slowly
Maimpori	Small pieces of betel nut and tobacco	Contents are chewed slowly and then swallowed

act as a buffer against acid formed in plaque on teeth. However, some investigators have shown that there is no difference in the prevalence of dental caries between betel nut chewers and non-chewers in some Asian populations (Williams *et al.*, 1996). Hence, its alleged beneficial effects on dental caries need further evaluation.

Among South Asians, betel nut is thought to be good for health, and it is used as a traditional Ayurvedic medicine. It is used as a mouth freshener after meals, a taste enhancer, a purgative, and to help digestion. Some people use it for parasitic intestinal infections, impotence, and gynecological problems (Strickland, 2002). It is also used among the lower socio-economic classes, to avoid boredom and to suppress hunger (Croucher & Islam, 2002; Strickland, 2002). Many other uses for betel nut have been suggested, based on tradition or on scientific theories. However, these uses have not been thoroughly studied in humans, and there is limited scientific evidence regarding its safety and effectiveness. Some of these suggested uses are for conditions that are potentially very serious.

Some evidence suggests that betel nut extract may improve speech, bladder control, and muscle strength after cerebrovascular accidents or stroke. Literature also suggests improvement in symptoms of schizophrenia. Betel nut chewers believe that it improves digestion. Some research also reports that betel nut chewing may reduce the risk of anemia in pregnant women (Natural Standard and Harvard Medical School, 2009). It is also used to prevent morning sickness and against unpleasant odors during pregnancy (Senn *et al.*, 2009). However, it is not clear how these effects are associated with the use of betel nuts. Hence, additional research-based evidence is needed in this area before recommending the use of betel nut in cases of stroke, schizophrenia, and other indicated conditions.

Chewing betel nut on a habitual basis is known to be deleterious to human health, and has significant adverse effects. Use of betel nut in any form is not safe for health. Over the past 40 years, many large-scale epidemiological and experimental studies have shown that, even when consumed without tobacco or lime, betel nut may have potentially harmful effects on the oral cavity (Trivedy *et al.*, 2002). These effects can be divided into two broad categories: those affecting the dental hard tissues, which include the teeth, their supporting perio-dontium, and the temporomandibular joint; and the soft tissues, which make up the mucosa that lines the oral cavity. The regular chewing of betel nut has major effects on the teeth leading to severe wear and tear of incisal and occlusal tooth surfaces, mainly the enamel covering. The loss of enamel may also result in dentinal sensitivity, and root fractures have been demonstrated among regular betel nut chewers. These effects depend on various factors, including the consistency of the betel nut, frequency of chewing, and duration of the habit. Researchers have reported that loss of periodontal attachments and calculus formation is greater among betel nut chewers (Anerud *et al.*, 1991). However, it is difficult to establish the biological effects of betel nut on periodontal health, and further studies are suggested in this regard. There is enough evidence to suggest that betel nut products, even those without tobacco, are associated with an increased risk for the development of oral malignancy (oral squamous cell carcinoma) and its precursor leukoplakia. It is well established that chewing betel nut is the single most important etiological factor for developing oral submucous fibrosis, which is a chronic disorder characterized by fibrosis of the mucosal lining of the oral cavity, oro- and hypopharynx, and the upper third of the esophagus. The risk of developing these conditions is even greater among paan masala users (Trivedy *et al.*, 2002). A study from Pakistan, a country with a high prevalence of betel nut consumers, has reported that there is a higher relative risk of developing oral submucous fibrosis amongst chewers of betel nut alone as compared to users of betel leaf with or without tobacco (Maher *et al.*, 1994).

Betel nut chewing is linked not only to the development of cancers of the oral cavity and head and neck, but also with cancers of the esophagus, stomach, prostate, cervix, lungs (Zhang & Reichart, 2007; Wikipedia, 2009), and sweat glands (Natural Standard and Harvard Medical School, 2009). Betel nut chewing is also associated with cirrhosis and malignancy of the liver (Hsiao *et al.*, 2007). There may be a higher risk of many other chronic non-communicable diseases (Yen *et al.*, 2006; Natural Standard and Harvard Medical School, 2009), such as obesity, type 2 diabetes, hypertension, hyperlipidemia, metabolic syndrome, chronic kidney disease, vision abnormalities, and abnormal thyroid function, among people who chew betel nuts regularly. Furthermore, evidence has shown that betel nut chewing is independently associated with greater cardiovascular disease and all-cause mortality after control for possible confounders (Lin *et al.*, 2008). It has been found to be a common cause of airway obstruction in children, with potentially fatal complications (Tariq, 1999). Betel nut chewing also causes bronchoconstriction, and may aggravate asthma (Wikipedia, 2009). The effects of chronic betel nut usage in man are at least as diverse as those of smoking, and the habit increases the risk of ill health many times.

Betel nut use may lead to increased production of body secretions, including saliva, tears, and sweat. Some people may experience nausea, vomiting, diarrhea, abdominal cramps, urinary incontinence, fever, and flushing with the use of betel nut (Natural Standard and Harvard Medical School, 2009). Prolonged chewing of betel nut causes significant alteration in the functions of intestinal epithelial cells, and can lead to malabsorption of nutrients. Some observations have reported that intake of betel nut in the form of pan masala and gutka correlates with chronic urticaria.

Use of betel nut during pregnancy can increase the risk of birth defects, spontaneous abortions, and birth-weight reduction. Metabolic syndrome in babies has also been associated with betel nut use during pregnancy (Natural Standard and Harvard Medical School, 2009). Because of its adverse effects, its use is discouraged particularly during pregnancy and childhood.

Interaction of betel nut usage with drugs, herbs, and supplements is well reported. Side effects such as muscle stiffness, tremors, and an increase in blood pressure are reported when betel nut is taken with certain anti-emetic and antidepressant drugs. As betel nut may also alter blood sugar levels, careful blood sugar monitoring is imperative among people with diabetes who are on oral hypoglycemic drugs or using insulin. Betel nut should also be used with care with drugs such as digoxin, propranolol, and verapamil, as it may cause the heart rate to become very slow. Moreover, drugs for glaucoma, antibiotics, antidepressants, anti-inflammatories, antipsychotics, immunosuppressants, and vitamin D all have interactions with betel nut. It is also thought that betel nut may reduce the beneficial effects of thiamine (vitamin B1) (Natural Standard and Harvard Medical School, 2009).

ADVERSE EFFECTS AND REACTIONS (ALLERGIES AND TOXICITY)

Betel nut chewing can produce significant cholinergic, neurological, cardiovascular, and gastrointestinal manifestations. High doses of its usage can cause hypercalcemia, hypokalemia, and metabolic alkalosis. Betel nut-induced extrapyramidal syndrome has been reported (Wikipedia, 2009). People should avoid betel nut if they have a known allergy to it; signs of allergy include rash, itching, or difficulty in breathing. Betel nut can cause tremors, muscle stiffness, involuntary movements of the mouth and face, and seizures. Vision abnormalities can also occur. Betel nut usage has been associated with confusion, memory lapse, and anxiety. Acute effects of betel chewing include exacerbation of asthma, hypertension, and tachycardia (Wikipedia, 2009). Other effects, such as muscle stiffness and tremor, may be increased when betel nut is used with drugs such as prochlorperazine, and blood pressure may rise to dangerously high levels if betel nut is taken with phenelzine (Natural Standard and Harvard Medical School, 2009). The effects of betel nut chewing on the heart can result in heart attacks, irregular heart rhythms, rapid heartbeat, and altered blood pressure (Natural Standard and Harvard Medical School, 2009).

SUMMARY POINTS

- Betel nut is a seed of the *Areca catechu* tree, cultivated in tropical humid regions.
- Up to 20% of world's population uses betel nut; more so in South and South-east Asia.
- Medical use of betel nut is limited.
- Globally, betel nut is widely used as an addictive substance, mouth freshener, and to assist digestion.
- Regular use of betel nut is a strong risk factor for cancers of the mouth, pharynx, esophagus, and stomach, and also increases risk of prostate, cervix, and lung cancer.
- Prolonged use of betel nut is associated with diabetes, hypertension, and cardiovascular diseases, and is also associated with increased all-cause mortality.
- The use of betel nut during pregnancy is discouraged due to unwanted effects on the fetus.
- Betel nut has serious interactions with a large number of herbs, drugs, and medications.
- Due to its various ill effects on health, use of betel nut should be discouraged.

203

References

Anerud, A., Loe, H., & Boysen, H. (1991). The natural history and clinical course of calculus formation in man. *Journal of Clinical Periodontology, 18,* 160—170.

Chandra, P. S., & Mulla, U. (2007). Areca nut: the hidden Indian "gateway" to future tobacco use and oral cancers among youth. *Indian Journal of Medical Sciences, 61,* 319—321.

Croucher, R., & Islam, S. (2002). Socio-economic aspects of areca nut use. *Addiction Biology, 7,* 139—146.

De Miranda, C. M., Van Wyk, C. W., Van der Biji, P., & Basson, N. J. (1996). The effect of areca nut on salivary and selected organisms. *International Dental Journal, 46,* 350—356.

Gupta, P. C., & Warnakulasuriya, S. (2002). Global epidemiology of areca nut usage. *Addiction Biology, 7,* 77—83.

Hsiao, T. J., Liao, H. W., Hsieh, P. S., & Wong, R. H. (2007). Risk of betel quid chewing on the development of liver cirrhosis: a community-based case—control study. *Annals of Epidemiology, 17,* 479—485.

Lin, W. Y., Chiu, T. Y., Lee, L. T., Lin, L. C., Huang, C. Y., & Huang, K. C. (2008). Betel nut chewing is associated with increased risk of cardiovascular disease and all-cause mortality in Taiwanese men. *American Journal of Clinical Nutrition, 87,* 1204—1211.

Maher, R., Lee, A. J., Warnakulasuriya, K. A. A. S., Lewis, J. A., & Johnson, N. W. (1994). Role of areca nut in the causation of oral submucous fibrosis — a case control study in Pakistan. *Journal of Oral Pathology & Medicine, 23,* 65—69.

Moller, I. J., Pindborg, J. J., & Effendi, I. (1997). The relation between betel chewing and dental caries. *Scandinavian Journal of Dental Research, 85,* 64—70.

Natural Standard and Harvard Medical School. (2009). Betel nut. Available at: http://www.intelihealth.com/IH/ihtPrint/WSIHW000/8513/31402/351498 (accessed November 18, 2009).

Senn, M., Baiwog, F., Wimai, J., Mueller, I., Rogerson, S., & Senn, N. (2009). Betel nut chewing during pregnancy, Madang province, Papua New Guinea. *Drug Alcohol Depend.*, *105*, 126–131.

Strickland, S. S. (2002). Anthropological perspectives on use of the areca nut. *Addiction Biology, 7*, 85–97.

Tariq, P. (1999). Foreign body aspiration in children: a persistent problem. *Journal of Pakistan Medical Association, 49*, 33–36.

Trivedy, C. R., Craig, G., & Warnakulasuriya, S. (2002). The oral health consequences of chewing areca nut. *Addiction Biology, 7*, 115–125.

Wikipedia. (2009). Areca nut. http://en.wikipedia.org/wiki/Areca_nut Last updated 18 November, 2009.

Williams, S. A., Summers, R. M., Ahmed, I. H., & Prendergast, M. J. (1996). Caries experience, tooth loss and oral health related behaviours among Bangladeshi women resident in West Yorkshire, UK. *Community Dental Health, 13*, 150–156.

Winstock, A. (2002). Areca nut – abuse liability, dependence and public health. *Addiction Biology, 7*, 133–138.

Yen, A. M., Chiu, Y. H., Chen, L. S., Wu, H. M., Huang, C. C., Boucher, B. J., et al. (2006). A population-based study of the association between betel-quid chewing and the metabolic syndrome in men. *American Journal of Clinical Nutrition, 83*, 1153–1160.

Zhang, X., & Reichart, P. A. (2007). A review of betel quid chewing, oral cancer and precancer in Mainland China. *Oral Oncologyl, 43*, 424–430.

CHAPTER 24

Swietenine, Big Leaf Mahogany (*Swietenia macrophylla*) Seed Extract as a Hypoglycemic Agent

Saikat Dewanjee[1], Anup Maiti[2]
[1] Advanced Pharmacognosy Research Laboratory, Department of Pharmaceutical Technology, Jadavpur University, Kolkata, India
[2] Pharmacy College Ipaura, Chandeswar, Azamgarh, Uttar Pradesh, India

INTRODUCTION

The big-leaf mahogany, *Swietenia macrophylla* King (Meliaceae), is a lofty, large tree that is found in Central America, Mexico, and South America, and almost all tropical and subtropical regions of the world (Cámara-Cabrales & Kelty, 2009). The plant is popular for its solid timber, which is used for making fine furniture and cabinets. The leaves are used as a dyeing agent. During a study of folk usage of *Swietenia macrophylla* seeds, we investigated their effectiveness against diabetes. In the present chapter, we present our research approach to isolating a novel antidiabetic molecule from *Swietenia macrophylla* seeds.

BOTANICAL DESCRIPTIONS

Swietenia macrophylla King is an evergreen tree, which grows to 40–60 m in height and 3–4 m in girth. The trunk is straight and cylindrical with a buttressed base. The leaves are usually paripinnate, 20–45 cm long, and are made up of three to six pairs of lanceolate or ovate leaflets. The leaflets are asymmetric, 5–12 cm long, with a whole margin and an acute or acuminate apex, and are light green or reddish when young, becoming dark green when mature. The yellow-green flowers of both sexes are in the same inflorescence and are arranged

Nuts & Seeds in Health and Disease Prevention. DOI: 10.1016/B978-0-12-375688-6.10024-6

in panicles. The season of flowering and fruiting differs with the geographic location. The fruits are capsular, oblong, 10–40 cm long and 6–12 cm in diameter, and each fruit contains 20–70 samaroid seeds. The seeds are 7–12 cm long and 2–2.5 cm wide, including the wing. The seed coat is differentiated into the testa and tegmen.

HISTORICAL CULTIVATION AND USAGE

Swietenia macrophylla is an evergreen tree native to tropical America, Mexico, and South America. The trees regenerate naturally through seeds in their native countries, and grow luxuriantly under favorable conditions. The plant is well known for its fast growth and adaptability. This exotic was introduced into southern India and some other parts in 1872, using seeds obtained from Honduras, as an ornamental tree and for timber. The big-leaf mahogany has had long and successful history regarding the use of its woods and seeds. The wood is used as timber for many purposes; it was so extensively used in tropical America, and exported, that its trade ended by the 1950s. The seeds have long been known for their ethono-medicinal significance against a number of diseases, being used for the treatment of leish-maniasis and abortion by an Amazonian Bolivian ethnic group, and as a folk medicine in Indonesia for the treatment of hypertension, diabetes, and malaria.

PRESENT-DAY CULTIVATION AND USAGE

Nowadays, artificial regeneration via direct sowing, transplanting, and stump-planting is encouraged over the natural regeneration for the cultivation of big-leaf mahogany. Trans-planting gives the best result, and is most widely used in India. Direct sowing gives good results only on rich soil and under humid climatic conditions. In South-eastern Asia, ball planting of 4-month-old seedlings gave excellent success. Big-leaf mahogany is still very popular for its valuable, solid lumber, and is now used to produce veneers. Besides this, the medicinal attributions of big-leaf mahogany have attracted the interest of herbal industries in utilizing its therapeutic potential.

APPLICATIONS TO HEALTH PROMOTION AND DISEASE PREVENTION

The seeds of *Swietenia macrophylla* have been reported to possess hypotensive, anti-inflammatory, antimutagenic, antitumor and antibabesial (Guevera *et al.*, 1996; Falah *et al.*, 2008) activities. Petroleum ether extract of *Swietenia macrophylla* seeds has been reported to produce significant antidiarrheal activity (Maiti *et al.*, 2007). Aqueous extract of *Swietenia macrophylla* seeds has been reported to be effective against *Plasmodium falciparum* (Falah *et al.*, 2008). Methanol extract of *Swietenia macrophylla* seeds was found to be effective against *Escherichia coli*, *Bacillus* species, *Staphylococcus aureus*, *Vibrio cholerae*, *Aspergillus flavus*, and *Candida albicans* (Dewanjee *et al.*, 2007).

The ethno-medicinal literature reveals that the *Swietenia macrophylla* seeds were used as a folk medicine in Indonesia for the treatment of diabetes. In order to substantiate the folklore claim, preclinical pharmacological assay was performed at our laboratory in streptozotocin-induced diabetic rats. The seed extract was found to possess significant antidiabetic activity in experi-mental diabetic rats (Maiti *et al.*, 2009a). The methanol extract of *Swietenia macrophylla* seeds at dosages of 200 and 300 mg/kg p.o. was found to possess significant dose-dependent hypo-glycemic and hypolipidemic activities. The extract also significantly improved the liver glycogen content and body weight of diabetic rats. In an oral glucose tolerance test, a blood glucose lowering effect was observed 2 hours after administration of seed extract. This reflects the efficiency of the extract in controlling elevated blood glucose levels.

The observed oral hypoglycemic activity encouraged our research group to initiate a subse-quent study, employing a concept of reverse pharmacology and bioassay guided drug discovery, to identify novel antidiabetic lead from *Swietenia macrophylla* seeds. In search of the exact mechanism of action, it was found that the *Swietenia macrophylla* seeds improved peripheral glucose utilization (Maiti *et al.*, 2009b). Subsequent work was carried out to isolate the chemical constituent from seed, employing the bioassay guided fractionation method. *In vitro* peripheral glucose utilization assay, using the isolated rat hemidiaphragm method, was employed to isolate lead from *Swietenia macrophylla* seeds (Figure 24.1).

The seeds were dried under shade, pulverized into a coarse powder, and macerated with 70% ethanol for 48 hours. The resulting extract was subsequently fractionated successively with petroleum ether (60−80°C), chloroform, and ethyl acetate. Respective fractions were assayed for peripheral glucose utilization, employing the isolated rat hemidiaphragm method (Walaas & Walaas, 1952; Chattopadhyay *et al.*, 1992). The chloroform fraction exhibited significant

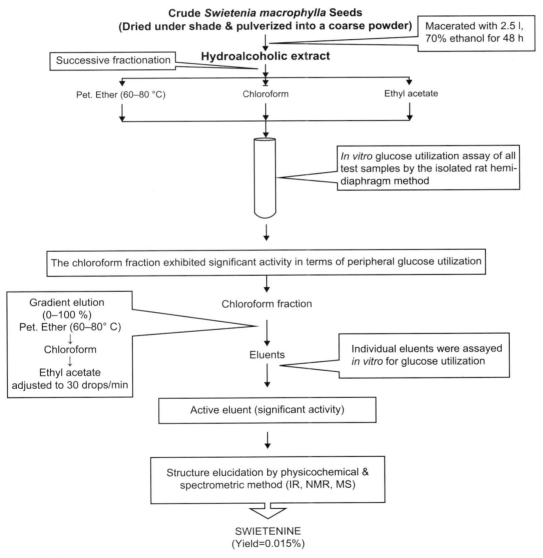

FIGURE 24.1

Schematic diagram representing bioassay-guided isolation of antidiabetic lead from *Swietenia macrophylla* seeds.
The hydroalcoholic extract was subjected to bioassay-guided isolation employing peripheral glucose utilization assay. The chloroform fraction of the parent extract was identified as the active fraction and subsequently subjected to chromatography coupled with *in vitro* bioassay to identify the lead, swietenine.

207

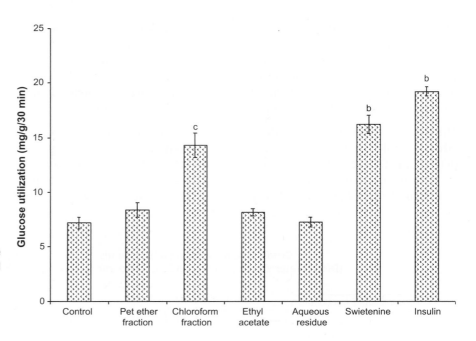

FIGURE 24.2

Effect of different fractions and isolated lead (swietenine) from *Swietenia macrophylla* seeds on peripheral glucose uptake by isolated rat hemidiaphragm test. The chloroform fraction exhibited the best activity in terms of peripheral glucose utilization, and the antidiabetic lead, swietenine, was identified in the chloroform fraction of *Swietenia macrophylla* seed extract. Values are expressed as mean ± SEM, $n = 6$. b, Values differ significantly from control ($P < 0.01$); c, values differ significantly from control ($P < 0.05$).

activity ($P < 0.05$) in terms of peripheral glucose utilization. The chloroform fraction was then subjected to bioassay guided isolation by classical silica gel column chromatographic separation employing the gradient elution technique with hexane and chloroform (0–100%). Individual eluents were assayed *in vitro* for glucose utilization to identify the lead molecule (Figure 24.2). One tetranortriterpenoid, swietenine (Figure 24.3), was identified as a lead

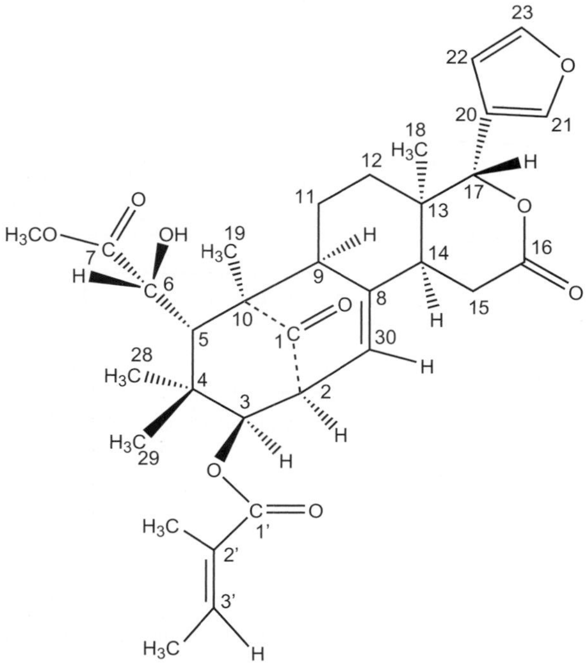

FIGURE 24.3
Molecular structure of swietenine.

molecule that exhibited significant peripheral glucose utilization ($P < 0.01$), and the effect was comparable to that of human insulin (Maiti *et al.*, 2009b).

The hypoglycemic effect of swietenine was substantiated with *in vivo* assay in streptozotocin-nicotinamide induced type 2 diabetic rats. Diabetes was induced in overnight fasted rats by a single i.p. injection of streptozotocin (65 mg/kg) in citrate buffer (pH 4.5), followed, 15 minutes later, by the administration of nicotinamide (110 mg/kg i.p.) (Dewanjee *et al.*, 2009a). The animals exhibiting fasting glucose levels of 140–200 mg/dl after 7 days were screened as type 2 diabetic rats, and used for the experiment. Fasting blood glucose levels were estimated on days 0, 1, 4, 7, 14, and 28 with the help of single touch glucometer (Ascensia Entrust®, Bayer Health Care, USA). Serum cholesterol (Demacker *et al.*, 1980) and triglyceride (Foster & Dunn, 1973) profiles were estimated on day 28 by using standard enzymatic colorimetric kits (Span Diagnostics Ltd., Surat, India). The liver glycogen level was estimated on day 28 (Caroll *et al.*, 1956).

Oral administration of swietenine at 25 and 50 mg/kg body weight per day to diabetic rats was found to produce significant dose-dependent hypoglycemic activity (Figure 24.4). A maximal 39.91% ($P < 0.01$) reduction in fasting blood glucose level was found with 50 mg/kg of swietenin p.o., on day 28. A significant ($P < 0.01$) reduction of serum lipid profiles was observed in diabetic rats, with maximal effects of 14.11% and 11.98% reductions in cholesterol and triglyceride levels, respectively, at 50 mg/kg of swietenin p.o., on day 28 (Figure 24.5). Treatment with swietenine also significantly improved liver glycogen levels when compared with diabetic control rats (Figure 24.6).

Maintenance of blood glucose levels in diabetic rats vindicated the effectiveness of swietenine as an oral hypoglycemic agent, which acted by increasing the peripheral utilization of glucose —i.e., insulinmimic action (Kar *et al.*, 2006). It is well known that insulin activates enzyme lipoprotein lipase, which hydrolyzes triglyceride under normal conditions (Dewanjee *et al.*, 2009b). The

209

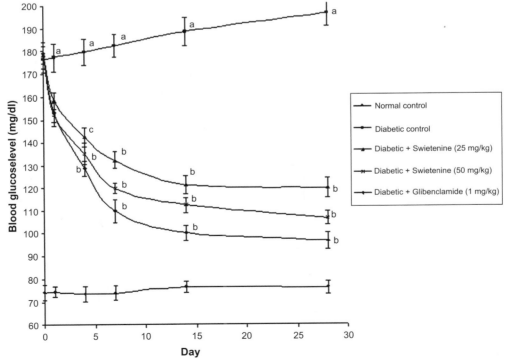

FIGURE 24.4

Effect of swietenine on fasting blood glucose level in type 2 diabetic rats. Swietenine exhibited a significant dose-dependent reduction of fasting blood glucose levels compared with diabetic control rats. Values are expressed as mean ± SEM, $n = 6$. a, Values differ significantly from normal control ($P < 0.01$); b, values differ significantly from diabetic control ($P < 0.01$); c, values differ significantly from diabetic control ($P < 0.05$).

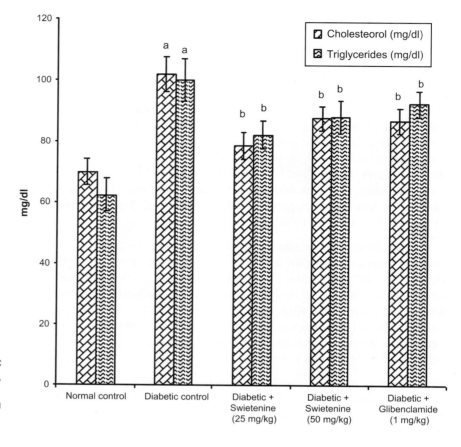

FIGURE 24.5

Effect of swietenine on serum cholesterol and triglyceride levels on day 28 in type 2 diabetic rats. Swietenine exhibited a significant dose-dependent reduction of serum cholesterol and triglyceride levels on day 28 compared with diabetic control rats. Values are expressed as mean ± SEM, $n = 6$. a, Values differ significantly from normal control ($P < 0.01$); b, values differ significantly from diabetic control ($P < 0.01$).

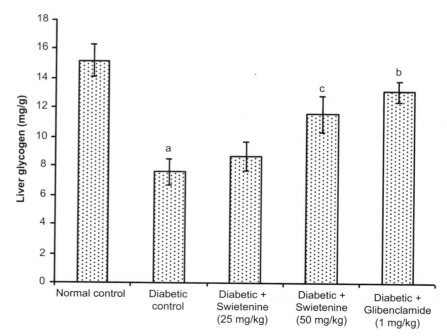

FIGURE 24.6

Effect of swietenine on liver glycogen level on day 28 in type 2 diabetic rats. Swietenine exhibited a significant dose-dependent reduction of the liver glycogen level in experimental animals on day 28. Values are expressed as mean ± SEM, $n = 6$. a, Values differ significantly from normal control ($P < 0.01$); b, values differ significantly from diabetic control ($P < 0.01$); c, values differ significantly from diabetic control ($P < 0.05$).

significant control of plasma lipid levels supported the insulinmimic activity of swietenine. Maintenance of the liver glycogen level in diabetic rats also supported the hypothesis that swietenine acts by improving peripheral glucose utilization *in vivo*.

ADVERSE EFFECTS AND REACTIONS (ALLERGIES AND TOXICITY)

Ethno-medicinal literature does not mention any adverse effects and reactions of *Swietenia macrophylla* seeds. However, it was worth studying the toxic profile of the isolated molecule before reporting the activity. Acute toxicity studies showed that the LD_{50} value of swietenine was 1.6 g/kg body weight p.o. in experimental mice. A subacute toxicity study was performed at the therapeutic dose of 50 mg/kg p.o. to experimental rats for 8 weeks. Swietenine did not produce any significant change in the hematological parameters or hepatic enzyme profile of experimental rats. No significant morphological and ultrastructural changes were observed in selected organs — namely, the liver, kidney, heart, and brain — after 8 weeks.

SUMMARY POINTS

- The ethno-medicinal literature reveals that *Swietenia macrophylla* seeds have long been used in Indonesia as a folk medicine treatment for diabetes.
- Preliminary pharmacological assay revealed that *Swietenia macrophylla* seeds possess significant hypoglycemic activity in experimental diabetic rats.
- *Swietenia macrophylla* seeds produce their hypoglycemic effect by improving peripheral glucose utilization.
- Swietenine, a novel hypoglycemic molecule, was identified employing bioassay guided isolation.
- *In vivo* antidiabetic bioassay of swietenine revealed that it exhibited significant hypoglycemic and hypolipidemic activity at the doses of 25 and 50 mg/kg p.o.
- Acute and subacute toxicity studies revealed that swietenine is non-toxic at its therapeutic dose in experimental animals.

References

Cámara-Cabrales, L., & Kelty, M. J. (2009). Seed dispersal of big-leaf mahogany (*Swietenia macrophylla*) and its role in natural forest management in the yucatán peninsula, Mexico. *Journal of Tropical Forest Science, 21*, 235–245.

Caroll, N. V., Longley, R. W., & Roe, J. H. (1956). The determination of glycogen in liver and muscle by use of anthron reagent. *Journal of Biological Chemistry, 220*, 583–593.

Chattopadhyay, R. R., Sarkar, S. K., Ganguly, S., Banerjee, R. N., & Basu, T. K. (1992). Effect of leaves of *Vinca rosea* Linn. on glucose utilization and glycogen deposition by isolated rat hemidiaphragm. *Indian Journal of Physiology and Pharmacology, 36*, 137–138.

Demacker, P. N., Hijmans, A. G., Vos-Janssen, H. E., Van't Laar, A., & Jansen, A. P. (1980). A study of the use of polyethylene glycol in estimating cholesterol in high density lipoproteins. *Clinical Chemistry, 26*, 1775–1779.

Dewanjee, S., Kundu, M., Maiti, A., Majumdar, R., Majumdar, A., & Mandal, S. C. (2007). *In vitro* evaluation of antimicrobial activity of crude extract from plants *Diospyros peregrina*, *Coccinia grandis* and *Swietenia macrophylla*. *Tropical Journal of Pharmaceutical Research, 6*, 773–778.

Dewanjee, S., Das, A. K., Sahu, R., & Gangopadhyay, M. (2009a). Antidiabetic activity of *Diospyros peregrina* fruit: effect on hyperglycemia, hyperlipidemia and augmented oxidative stress in experimental type 2 diabetes. *Food and Chemical Toxicology, 47*, 2679–2685.

Dewanjee, S., Maiti, A., Sahu, R., Mandal, V., & Mandal, S. C. (2009b). Antidiabetic and antioxidant activity of the methanol extract of *Diospyros peregrina* (Ebenaceae) fruit on type I diabetic rats. *Pharmaceutical Biology, 47*, 1149–1153.

Falah, S., Suzuki, T., & Katayama, T. (2008). Chemical constituents from *Swietenia macrophylla* bark and their antioxidant activity. *Pakistan Journal of Biological Sciences, 11*, 2007–2012.

Foster, J. B., & Dunn, R. T. (1973). Stable reagents for determination of serum triglyceride by colorimetric condensation method. *Clinica Chemica Acta, 19*, 338–340.

Guevera, A. P., Apilado, A., Sakarai, H., Kozuka, M., & Tokunda, H. (1996). Anti-inflammatory, antimutagenecity and antitumor activity of mahogony seeds *Swietenia macrophylla* (Meliaceae). *Philippine Journal of Science, 125*, 271–278.

Kar, D. M., Maharana, L., Pattnaik, S., & Dash, G. K. (2006). Studies on hypoglycaemic activity of *Solanum xanthocarpum* Schrad. & Wendl. fruit extract in rats. *Journal of Ethnopharmacology, 108,* 251−256.

Maiti, A., Dewanjee, S., & Mandal, S. C. (2007). *In vivo* evaluation of antidiarrhoeal activity of the seed of *Swietenia macrophylla* King. (Meliaceae). *Tropical Journal of Pharmaceutical Research, 6,* 711−716.

Maiti, A., Dewanjee, S., Kundu, M., & Mandal, S. C. (2009a). Evaluation of antidiabetic activity of methanol extract of the seeds of *Swietenia macrophylla* (Meliaceae) in diabetic rats. *Pharmaceutical Biology, 47,* 132−136.

Maiti, A., Dewanjee, S., & Sahu, R. (2009b). Isolation of hypoglycemic phytoconstituent from *Swietenia macrophylla* seed. *Phytotherapy Research, 23,* 1731−1733.

Walaas, E., & Walaas, O. (1952). Effect of insulin on rat diaphragm under anaerobic conditions. *Journal of Biological Chemistry, 195,* 367−373.

CHAPTER 25

Bitter Kola (*Garcinia kola*) Seeds and Health Management Potential

B. Daramola[1], G.O. Adegoke[2]
[1] Department of Food Technology, Federal Polytechnic, Ado-Ekiti, Ekiti State, Nigeria
[2] Department of Food Technology, University of Ibadan, Ibadan, Oyo State, Nigeria

LIST OF ABBREVIATIONS

GK, *Garcinia kola*
GLC, gas liquid chromatography
HPLC, high performance liquid chromatography
MS, mass spectrophotometer
NMR, nuclear magnetic resonance
TLC, thin-layer chromatography
VLC, vacuum liquid chromatography

INTRODUCTION

Food value is defined based on three roles: the primary function, which involves nutrients and metabolism; the secondary function, which deals with taste and satisfaction; and the tertiary function, which is concerned with food components as body modulators resulting

Nuts & Seeds in Health and Disease Prevention. DOI: 10.1016/B978-0-12-375688-6.10025-8

in the prevention, risk reduction, or management of diseases (Arai & Fujimaki, 2004). In addition to their primary applications, there is catalog of seeds and fruits indigenous to Africa, handed by tradition down the generations, which have also been found to be boosters of immunity against some infectious diseases and are known to be efficacious in the management of disease conditions. One such is GK seed, a lesser known plant food that is known to have existed for centuries and is important in ethno-botanical medicinal usage and as a traditional stimulant used in extending ceremonial activities (Nwokeke, 2004).

This chapter reports on applications of GK seed in the treatment and management of a wide spectrum of diseases. Putative scientific bases for biological activities, and the direction of future research on seeds of GK, are briefly discussed.

BOTANICAL DESCRIPTION

Bitter kola is a member of the genus *Garcinia* and belongs to the botanical family *Guttiferae*, a polygamous tree distributed throughout tropical Africa and Asia, with more than 180 species. Some other important members in the family are G. *afzelli*, G. *handburii*, G. *indica*, G. *mangostana*, G. *mannii*, and G. *Morella* (Sakariah *et al.*, 2002). Among the family *Guttiferae*, *Garcinia* is the genus most localized to West Africa. In West Africa, the tree can be found growing domestically as well as naturally in the forest, especially in western Nigeria. It grows to a height of about 12–14 m, and its fruits mature during the rainy season, between June and August. Selected physical characteristics of bitter kola fruits and seeds are presented in Table 25.1 and Figure 25.1.

TABLE 25.1 Selected Characteristics of Fruit and Seed of Bitter Kola (*Garcinia kola*)

Physical Characteristics	Fruit	Seed
Shape	Spherical	Ovoid
Color	Reddish-brown	Brown
Average weight	200.53 g	7.06 g
Number	One	2–4
Skin appearance	Smooth	Rough
Width	80–85 mm	20–21 mm
Length	80–85 mm	35–40 mm

Source: Adapted from Daramola (2005).

FIGURE 25.1
Bitter kola *(Garcinia kola)*: (A) fruit, and (B) seeds.

HISTORICAL CULTIVATION AND USAGE

Garcinia kola is an evergreen tree that grows wild in the wet and moist zones of Nigeria, Ghana, and the Republic of Congo. The tree is now planted in farms to serve as a shade tree in cocoa plantations; however, it has been difficult to obtain information regarding the history of GK cultivation, and this may be because the GK seed is a lesser known plant food in western parts of Nigeria despite having local commercial importance. It is traditionally used during ceremonial displays and as a gift to nobles, because it is believed to influence longevity, and has been used in the treatment of ailments since ancient times.

PRESENT-DAY CULTIVATION AND USAGE

Bitter kola seed is chewed principally for health maintenance or curative reasons, and its therapeutic efficacy has gained universal acceptance. The tree is still sparsely cultivated, because seedling growth is slow, and natural regeneration of the species is poor. This consequently results in low production of the seed during harvest. However, the extensive potential of GK seed against a spectrum of disease conditions should renew interest in the domestication of the species for effective utilization for public health management. In addition, the bitterness inherent to GK seed could be exploited as a substitute for hops.

APPLICATIONS TO HEALTH PROMOTION AND DISEASE PREVENTION

The ubiquitous use of the *Garcinia kola* (GK) seed in the western part of Nigeria in the treatment of ailments has existed since ancient times. The seed is used in herbal preparations that are important in traditional African medicine for the treatment of various respiratory diseases, notably the cough and asthma (Adeniji, 2003). It is also used as an antidote to food-borne disorders and snakebites, and has been found to be effective in increasing a low sperm count, as well as sometimes being used as an aphrodisiac. As a herbal medicine, it is involved in the management of dysentery and chest colds. GK seed chewing sticks are used as oral and dental medicaments for diseases such as toothache, mouth ulcers/sores, thrush, periodontitis, and gingivitis (Hollist, 2004).

Other uses as folk remedies are for the treatment of liver disorders (antihepatotoxic activity), hepatitis, laryngitis, bronchitis, and gonorrhoea (Okojie *et al.*, 2009). Iwu (1993) has also reported the antidiabetic potential of GK seed. Some studies have shown that GK seed extract exhibits a dilatory effect on the alveolar ducts and sacs, and alveoli, thus improving respiratory activity (see Okojie *et al.*, 2009, for a review). The extract also enhances the functionality of the gall bladder, indicating that it has detoxification and cleansing properties. Other therapeutic uses are in the management of tuberculosis and diarrhea, and the treatment of measles and mumps in children.

In a recent report, Okojie and colleagues (2009) suggested a physiological mechanism underlying the use of GK seed for the treatment of asthma: the authors hypothesized that the phytochemical components, xanthones and flavonoids, inhibit calcium influx and histamine release, respectively, stimulated by IgE-dependent ligands. As expressed by the authors, asthma is believed to be caused by immunological dysfunction. As shown by the immunological model, asthma is mediated by reaginic (IgE) antibodies bound to mast cells in the airway mucosa. It has been postulated that enzymes such as histamine, tryptase, and other neural proteases, leukotrienes, and some specific prostaglandins are the cause of muscular contraction and vascular leakage, which induces bronchoconstriction.

Antibiological activity

Some aspects of the pharmacological and antimicrobial activities of GK seed were reported by Sodipo *et al.* (1991). Ethanolic and diethyletheric crude and fractionated extracts of GK seed have been shown to exhibit antibiological activities against some micro-organisms that are of public health importance with respect to food poisoning and spoilage (Daramola & Adegoke, 2007).

Preliminary characterization of bioactive components

Using standard methods elicited in Trease and Evans (2002), a spectrum of bioactive components with phenolic bases were identified in GK seed, including biflavonoids (Trease & Evans, 2002), alkaloids, flavonoids in the form of flavonols, and glycosides; trace amounts of anthraquinones were also detected (Daramola *et al.*, 2009). The identification of the spectrum of phenolic bioactive components and their relationship with antioxidative activities serves as an impetus to studies on the antioxidative properties of GK seed bioactive components compatible with food systems; this will be discussed briefly later in the chapter.

Beside phenolics, other organic and inorganic compounds of biological importance have been identified in the GK seed. Physical and physiological characteristics of the phenolics and other organics and inorganics identified in GK seed extracts and that are known to exhibit biological activities are shown in Table 25.2.

As previously mentioned, exhibition of antioxidative activity is one typical example of the pharmacological consequences of all the bioactive components, most especially phenolics, identified in GK seed. Pro-oxidation or generation of radicals is inevitable in nature, because they are a by-product of the metabolic process in all living systems. Free radicals have been implicated in playing some role in more than 100 diseases, including cancer, atherosclerosis, rheumatoid arthritis, inflammatory bowel disease, and cataracts. Therefore, the identified bioactive components in the GK seed suggest that it could exhibit antioxidative activity and, consequently, a reduction of free radicals. Reduction of free radicals by antioxidants can lead to a decrease in vulnerability to the named degenerative diseases.

216

TABLE 25.2 Some Bioactive Groups Identified in Bitter Kola (*Garcinia kola*) seed

Components	Characteristics	Biological Activities/Comments	References
Alkaloids	Nitrogenous-based compounds that unite with acids to form salts	Although most alkaloids are toxic, their medicinal value can be harnessed if applied at a low dosage; however, alkaloids in bitter kola seed are unlikely to be toxic, because humans have consumed bitter kola seed since prehistoric times	Trease & Evans (2002)
Flavonoids	Constitutes the largest group of naturally occurring phenols; usually appear yellow (Latin *flavus*, yellow)	Antioxidative, anti inflammatory, enzyme-inhibitory, immunomodulatory, antitumor, antibacterial, and antifungal properties, among others	Trease & Evans (2002)
Peptides	Oligomers of amino acids usually with less than 10,000 molecular weight	Therapeutic properties of peptides include hypoglycemic, antibacterial, enzymatic, and hormonal modulations	Olaniyi *et al.* (1998)
Dietary acids	Water-soluble vitamin (e.g., ascorbic acid)	Have antioxidant and anti-scurvy activities	Okojie *et al.* (2009)
Anthraquinones/ xanthones	The two are structurally related, often orange-red compounds	Possess purgative properties	Trease & Evans (2002)
Dietary elements	Inorganic substances usually required in minute quantities for physiological activity	Structural constituent of both hard and soft components of all biological systems — bone, tissue, fluids (e.g, blood), enzymes, hormones	Olaniyi *et al.* (1998)

The principle underlying the activity of antioxidants can be put simply as follows. The hydroxyl moiety in phenolics interacts with electrons of the benzene rings, and the phenyl radicals generated from such interactions are stabilized by delocalization of the electrons. The phenyl radicals modify radical-mediated oxidation processes by inhibiting the further propagation of primary radicals prior to attacking biological cells. Moreover, many phenolics chelate the metal ions that are catalysts of degenerative conditions. The metal ions are continuously generated by normal metabolic processes, such as during the action of glycoprotein oxidases. Moreover, phenolics are good hydrogen donors to radicals, thereby quenching the radicals before exerting deleterious effects (Giese, 1996). They also form stable complexes with proteins; this is the basis of phenolics modifying enzymic activities that act as a major contributor to disease conditions.

In addition, hydrolytic products of phenolic salts or glycosides produce phenolic acids, which result in an alteration of the ionic concentration, with the effect of modifying the reactants' environment and consequently altering their reaction pathway or rate of activity. This exerts bactericidal or bacteriostatic effects on agents of health disorders, such as bacteria, viruses, and deleterious enzymes.

In our investigation using TLC and VLC techniques (Daramola *et al.*, 2009), separation of the ethanolic crude extract (ETH) of GK seed showed various colorimetric bands (dark brown, dirty red, and golden yellow spots). Both the crude and fractionated extracts showed prospects for antioxidative activity, when screened using chromogenic reagents, with respect to phenolics, reductone, free electrons, or unsaturation. Moreover, the antioxidant screen tested positive for the ethanolic and fractionated extracts of GK seed using chromogenic reagents, and the ultraviolet spectral characteristics of the samples showed strong absorption in the UV region — a diagnostic feature of unsaturation or unbound electrons in the bioactive components (Shriner *et al.*, 1979), which is a prerequiste for antioxidative activity. In addition, samples were examined using the infrared spectrophotometric method for functional groups. Functional groups were assigned to IR absorption peaks using a functional-group frequency chart. Peaks of diagnostic value were selected on the basis of functional-group antioxidant potentials. The selected antioxidant functional groups of prime interest were as follows: phenol hydroxyl group (~ 3390 cm^{-1}), methylene bond or unsaturation (1600–1670 cm^{-1}), and aromaticity (~ 800 cm^{-1}). Excerpts from the investigation are shown in Table 25.3.

Ordinarily, when GK seed is chewed it is bitter to taste — hence the name "bitter kola." A bitter taste is one of the primitive methods used in suspecting that a natural product possesses medicinal value. However, GK seed is characterized by a sweet aftertaste that was detected in some of the fractionated extracts, especially fractions without alkaloids. Regardless of the taste type, both the crude and fractionated extracts showed antiperoxidative effectiveness comparable to that of some synthetic and dietary antioxidants — namely, butylated hydroxyltoluene, and tocopherol — using the oil-storage stability test (Daramola *et al.*, 2009).

217

TABLE 25.3 IR Spectral Characteristics of Crude and Fractionated Extracts of Seed of Bitter Kola (*Garcinia kola*)

Sample	V max (KBr)/cm					
F$_1$	3423(b)	1368–1040(b)		736(s)	713(s)	
F$_2$	3356(b)			833(w)		
F$_3$	3296(b)	1368(s)		811(s)	730(s)	639(s)
F$_4$	3314(b)	1620(s)	1362(s)	811.69(sh)		
F$_5$	3529(v)	1368(sh)	1233(s)	907(v)		
F$_6$	3369(b)					
ETH	3399(b)	1626(sh)	1168(w)	967(w)	833(w)	

F, fraction; s, strong; w, weak; v, variable; b, broad; sh, sharp; ETH, ethanolic crude extract of GK seed.

Mechanism of antioxidant action

Antioxidants can be divided into three groups by their mechanism: (1) primary antioxidants, which function essentially as free radical terminators (scavengers); (2) secondary antioxidants, which are important preventive antioxidants that function by retarding chain initiation; and (3) tertiary antioxidants, which are concerned with the repair of damaged biomolecules. Further information on the functionality of antioxidants can be found in a review by Giese (1996). As a sequel to the information given here, a follow-up study is ongoing in our laboratory to designate the class of antioxidants in GK seed. Such investigation should assist in providing information regarding appropriate applications of the bioactive components of GK seed.

Some aspects of nutritional composition

In addition to the phytochemicals in GK seed, there is a plethora of both inorganic and organic components that occur in the form of primary nutrient foods and have been established to possess medicinal activity (Olaniyi et al., 1998). Some of these primary food nutrients of GK seed have been evaluated. Oyenuga (1968) determined the proximate composition; Essien and colleagues (1995) reported the lipid composition and fatty acid profile of GK seed; and Daramola and Adegoke (2007) reported the mineral composition and amino acid profile of GK seed. Considering the physiological roles of dietary elements and amino acids and peptides (Olaniyi et al., 1998), it could be speculated that the food constituents might have a contributory role to the medicinal value of GK seed.

Future prospects and directions for further studies

The future appears bright regarding the relevance of GK seed in the pharmacopeia of herbal remedies, for the following reasons. First, GK seed-derived medicines are natural products, and therefore relatively safer compared to synthetic analogs, that are capable of offering affordable and profound therapeutic benefits. Second, humans have consumed GK seed since ancient times, and consequently its application is not expected to be constrained by stiff legislation. Third, the fact that the source is renewable ensures the continuity of supply of the basic material.

Even with the currently limited scientific findings, it is clear that the active components in GK seed are polygamous. Therefore, there is a need for the application of advanced techniques for the separation and purification of bioactive components that could yield varieties of potent constituents to give assorted natural products with unique therapeutic values. Some of these could then serve as templates for the design and preparation (synthesis) of natural identical drugs. Although there are claims regarding the isolation of and absolute structural determination of bioactive components of GK seed with respect to flavonoids, there are other bioactive components in GK seed with yields larger than flavonoids — for instance, alkaloids. Therefore, the significance of the abovementioned assertion is that there is a need for extensive and comprehensive studies which should isolate, purify and elucidate the absolute chemical structure of all the bioactive components in GK seed, using techniques such as GLC, HPLC, MS, NMR, and X-ray crystallography.

The pharmacological activity of active components should be established in relation to present traditional applications using approved experimental methods, rather than relying on anecdotal evidence. It is also important to note that numerous research opportunities exist regarding the biochemistry underlying the physiology of bioactive components in GK seed. Other studies on GK seed should include alexipharmic characteristics, if any; the possibility of control of fungal and bacterial mutation; and the effect on cryptic natural toxins. Extensive studies should also focus on determining the mechanism of the pharmacological activity of each of the purified bioactive components in GK seed. Such therapeutic mechanistic investigations (Olaniyi, 2000) should span the action on some enzymes that are important in human health.

Put briefly, the GK seed has enormous potential for the treatment of a spectrum of diseases. The reported therapeutic activities of GK seed cover antioxidant, antibacterial, antiviral, antifungal, and anti-inflammatory properties. However, studies with respect to chemical and biological activities available in literature are not adequate enough to match the stated span of such therapeutic activities, and there is a need for extensive studies to match the pharmacological potential of GK seed with a view to making a significant contribution to health promotion and disease prevention.

ADVERSE EFFECTS AND REACTIONS (TOXICITY AND ALLERGIES)

Although GK seed is a natural product that has been consumed by humans since prehistoric times, the totality of the active components or individual constituents may possess some inherent toxicity or allergy, or induce adverse reactions, in some vulnerable individuals or groups — for example, pregnant women and lactating mothers. This therefore also requires investigation.

SUMMARY POINTS

- The seed of bitter kola, a member of the genus *Garcinia*, is a lesser known plant food that is chewed not necessarily for its food value but rather for its curative properties.
- Bitter kola seed is used in folk medicine in the treatment and management of a spectrum of disease conditions.
- Using standard procedures, pharmacological and antimicrobial activities of bitter kola seed have been established.
- The bioactive components of bitter kola seed exhibit antioxidative activity.
- A comparison of the traditional applications of bitter kola seed with the available scientific bases for activity showed a gross imbalance, which signals the need for extensive further investigation.

References

Adeniji, M. O. (2003). *Herbal treatment of human diseases.* Cynx International Ltd Nigeria. p. 24.

Arai, S., & Fujimaki, M. (2004). The role of biotechnology in functional foods. In R. J. Nesser, & J. B. German (Eds.), *Bioprocesses and Biotechnology for Functional Foods and Nutraceuticals* (pp. xvii–xxxii). New York, NY: Marcel Dekker Inc.

Daramola, B. (2005). *Nutrient profile, antioxidant and antimicrobial properties of the active components of Garcinia kola.* Nigeria: PhD thesis, Department of Food Technology, University of Ibadan.

Daramola, B., & Adegoke, G. O. (2007). Nutritional composition and antimicrobial activity of fractionated extracts of *Garcinia kola* Heckel. *Pakistan Journal of Scientific and Industrial Research, 50,* 104–108.

Daramola, B., Adegoke, G. O., & Osanyinlusi, S. A. (2009). Fractionation and assessment of antioxidant activities of active components of *Garcinia kola* seed. *Journal of Food Agriculture and Environment, 7,* 27–30.

Essien, E. U., Esenowo, G. J., & Akpananbiatu, M. I. (1995). Lipid composition of lesser-known tropical seeds. *Plant Foods for Human Nutrition, 48,* 135–140.

Giese, J. (1996). Antioxidants: tools for preventing lipid oxidation. *Food Technol, 50,* 73–81.

Hollist, N. O. (2004). *A collection of traditional Yoruba oral and dental medicaments.* Ibadan, Nigeria: Bookbuilders. p. 98.

Iwu, M. (1993). *Handbook of African medical plants.* Boca Raton, FL: CRC Press.

Nwokeke, C. (2004). Health benefits of bitter kola. *Daily Champion* (Newspaper), Lagos Nigeria: June 10.

Okojie, A. K., Ebomoyin, I., Ekhator, C. N., Emeri, O., Okosun, J., Onyesu, G., et al. (2009). Efficacy of bitter kola in amelioration of respiratory diseases. *Internet Journal of Pulmonary Medicine* 1.

Olaniyi, A.A. (2000). *Essential medicinal chemistry.* Ibadan, Nigeria: Shaneson C.J. Nigeria Ltd, p. 49.

Olaniyi, A.A., Ayim, J.S.K., Ogundaini, A.O., & Olugbade, T.A. (1998). *Essential inorganic and organic pharmaceutical chemistry.* Ibadan, Nigeria: Shaneson C.J. Nigeria Ltd, pp. 27–90, 546.

Oyenuga, V. A. (1968). *Nigerian's food and feeding stuffs: Chemistry and nutritive value.* Ibadan, Nigeria: Ibadan University Press.

Sakariah, K. K., Jena, B. S., Jayaprakasha, G. K., & Singh, R. P. (2002). Chemistry and biochemistry of (−)-hydroxycitric acid from Garcinia. *Journal of Agricultural and Food Chemistry, 50,* 10–22.

Shriner, R. L., Fuson, R. C., Curtin, D. Y., & Morrill, T. C. (1979). *The systematic identification of organic compounds, a laboratory manual* (6th ed.). New York, NY: John Wiley. pp. 416—430.

Sodipo, O. A., Akanji, M. A., Kolawole, F. B., & Odutuga, A. A. (1991). Saponin is the active antifungal principle in *Garcinia kola* Heckel seeds. *Biological Scientific Research Communication, 3*, 171.

Trease, G. E., & Evans, W. C. (2002). *Pharmacognosy* (15th ed.). Edinburgh, UK: Harcourt Publishers.

CHAPTER 26

Bitter Kola (*Garcinia kola*) Seeds and Hepatoprotection

E. Olatunde Farombi
Drug Metabolism and Toxicology Research Laboratories, Department of Biochemistry, College of Medicine, University of Ibadan, Nigeria

CHAPTER OUTLINE

LIST OF ABBREVIATIONS

AFB1, aflatoxin B1
AP-1, activator protein-1
AUC, area under the curve
CCl_4, carbon tetrachloride
COX-2, cyclooxygenase
DMN, dimethyl nitrosamine (DMN)
iNOS, inducible nitric oxide synthase
LDL, lipoprotein,
NF-κB, nuclear factor kappa B

Nuts & Seeds in Health and Disease Prevention. DOI: 10.1016/B978-0-12-375688-6.10026-X

INTRODUCTION

The risk of developing hepatic diseases is highest in developing countries, due to exposure to mycotoxins. Also, synergistic interaction of mycotoxins with hepatitis B virus has been implicated as a major risk factor in the etiology of hepatocellular carcinoma, which has one of the poorest 5-year survival rates and accounts for 15% of total cancer mortality (Premalatha & Sachdanandam, 2000). At present certain drugs are used to treat liver diseases, but alternative therapies for this purpose are desirable. Moreover, since the increase in the use of synthetic chemicals in cancer therapy has led to many toxic effects, there is a worldwide trend towards exploiting naturally occurring plant sources that are therapeutically effective, culturally acceptable, and economically within reach of the general population.

Identification of naturally occurring modulators for hepatocarcinogenesis that can be incorporated in the human diet at minimal cost would be relevant, because the economic limitations of the majority of the populace, especially in third world countries, may not allow for the purchase of prophylactic drugs. Coincidentally, the edible *Garcinia kola* nut occupies a prominent position in the social customs of the people in Nigeria and other parts of West Africa. In spite of its very bitter taste, this nut is consumed recreationally; it is commonly offered to guests and shared at social gatherings.

BOTANICAL DESCRIPTION

Garcinia kola is a tropical flowering plant found in western and central Africa, which produces large, orange fruits and brown, nut-like seeds embedded in an orange-colored pulp (Figure 26.1). It is well-branched, evergreen, and grows to a medium-sized tree, reaching 12 m in height in 12 years. It is found in humid forest regions throughout West and Central Africa. The leaves are broad, elongated, and leathery, with distinct resinous canals, and have 10 pairs of lateral veins running parallel to the margin. The tree bears male and female flowers separately, around December—March and May—August, respectively. The female flowers are yellow and fleshy, while male flowers are smaller, with more distinct stamens. Garcinia kola has a regular fruiting cycle, and the tree produces fruits every year.

HISTORICAL CULTIVATION AND USAGE

Historically, *Garcinia kola* has been cultivated as an ideal tree for shade around homesteads, and twigs are used as chewing sticks because of their bitter taste and the antibacterial activities

	R1	R2	R3	R4
GB1	OH	H	OH	H
GB2	OH	H	OH	OH
Kolaflavanone	OH	H	OCH$_3$	OH

Kolaviron

FIGURE 26.1

The plant *Garcinia kola* (Guttiferae), from which kolaviron (*Garcinia* biflavonoid) is derived, and its structure.

of the extracts (Taiwo *et al.*, 1999). In Nigeria, low populations of *G. kola* are found in domestic gardens, and few stands are found in the wild due to the rapid deforestation and heavy exploitation of the natural forests.

Garcinia kola seed plays a very important role in African ethno-medicine. The seed is employed as a general tonic, and is believed to cure impotence. Traditionally, the seeds are used in the treatment of inflammatory disorders and liver disease. Extracts of the seeds led to remarkable improvement of liver function in patients with chronic hepatitis and cholangitis after treatment for 14 days at a Nigerian herbal home (Iwu, 1982). Other medicinal uses include as a purgative, an antiparasitic, an antimicrobial, and an antiviral. The seeds are used in the treatment of bronchitis and throat infections, and to prevent and relieve colic, cure head or chest colds, and relieve coughs (Figure 26.2). Constituents include xanthones and benzophenones. The anti-microbial properties of this plant are attributed to the benzophenone flavanones, while the hepatoprotective properties have been linked to the biflavonoid constituents of the plant. Specifically, a de-fatted fraction of an alcoholic extract of *Garcinia kola* seeds, characterized as kolaviron (a biflavonoid complex containing GB1, GB2, and kolaflavanone) (Figure 26.1), has been demonstrated to possess antioxidant and hepatoprotective properties (Iwu, 1985).

PRESENT-DAY CULTIVATION AND USAGE

Garcinia kola has been successfully applied to treat patients suffering from osteoarthritis of the knee. It is known for its anti-inflammatory and antioxidant properties, and is used to prevent infections and viruses, especially of the immune system. *Garcinia kola* is used as a substitute for hops in brewing lager beer, and is especially useful in preventing beer spoilage. Laboratory brewing trials with *Garcinia kola* and with hops gave beers with similar chemical properties. Organoleptically, *Garcinia kola* beer was as acceptable to tasters as hopped beer, except that it had a greater bitterness. *Garcinia kola* and hop extracts exerted similar antimicrobial effects on two beer spoilage micro-organisms (*Lactobacillus delbruckii* and *Candida vini*).

Bitter kola seeds have been formulated into tablet form, and are also added to many herbal preparations, either singly or in combination with other plants. For instance, Hepa-Vital Tea, a blend of *Garcinia kola* and *Combretum micranttum*, and Hangover Tonic, comprising kolaviron and *Cola nitida*, are manufactured and marketed as phytomedicines. Hangover tonic is used traditionally in Africa, and is known for its anti-stress and refreshing activities.

223

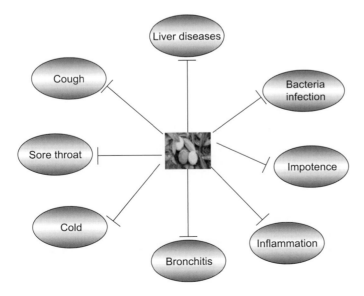

FIGURE 26.2
Traditional uses of *Garcinia kola* seed.

Owing to the presence of potent antioxidative biflavonoids in the seed, kolaviron is currently being considered as a candidate for clinical trials in patients with hepatocellular carcinoma.

APPLICATIONS TO HEATH PROMOTION AND DISEASE PREVENTION

Kolaviron and drug-induced hepatotoxicity

Garcinia kola seeds and seed extracts have shown to be of benefit to health, and relevant in the management and chemoprevention of life-threatening diseases. The effects range from anti-diabetic, immunomodulatory, antiviral, anti-inflammatory, and antioxidant activities to strong hepatoprotective properties. In this section, the focus will be on the experimental evidence and biochemical and molecular mechanisms underlying the hepatoprotective properties of *Garcinia kola*, and especially the bioflavonoid kolaviron isolated from it. Kolaviron has been demonstrated by many researchers to have protective effects against several hepatocarcinogens in both *in vitro* and *in vivo* model systems. Figure 26.3 shows the wide array of chemical compounds affected by kolaviron.

Kolaviron and other isolates of *Garcinia kola seeds* were shown to protect against CCl_4-induced liver injury, reduce significantly CCl_4-induced increases in serum transaminases and sorbitol dehydrogenases, and protect against CCl_4-mediated hepatotoxicity by modulating the cholesterol:phospholipid ratio as well as the toxic onslaught imposed on liver microsomal marker enzymes (Farombi, 2000).

Akintonwa and Essien (1990) showed that kolaviron, when administered at a dose of 100 mg/kg three times a day for 5 days, reduced paracetamol-induced lethality, with a concomitant reduction in serum enzymes and histologic scores. Intraperitoneal administration of kolaviron also protected against thioacetamide-induced increases in the activities of serum enzymes, as well as thiopental-induced sleep.

Kolaviron given orally to rats as a pretreatment prior to challenge with 2-acetyl aminofluorene reversed its effects on ornithine carbamyl transferase and other biochemical indices of hepatic damage (Farombi *et al.*, 2000). The protective effect of kolaviron was comparable to the effect of butylated hydroxyl anisole, indicating that kolaviron may act as chain-breaking *in vivo* antioxidant.

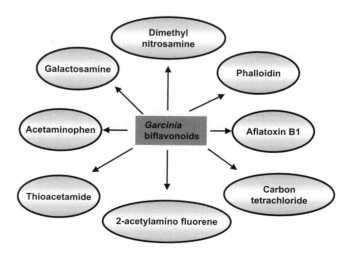

FIGURE 26.3
Garcinia biflavonoids and hepatotoxicants.

Of particular interest is the remarkable effect of kolaviron on AFB1-induced liver damage. We demonstrated the chemopreventive effect of kolaviron against the hepatic oxidative damage induced by aflatoxin in rats (Farombi *et al.*, 2005) by mechanisms involving induction of phase 2 antioxidant enzymes capable of detoxifying toxic aflatoxin metabolites.

Biochemical and molecular basis for the hepatoprotective action of kolaviron

Earlier hypotheses regarding the biochemical mechanisms of the hepatoprotective effect of kolaviron stated that it protected against carcinogen-induced liver injury by stabilizing the membrane, and by interfering with the distortion of the cellular ionic environment associated with xenobiotic treatment; also possibly by the interference of kolaviron with hepatic drug metabolism (Braide, 1991) These studies, however, did not elucidate the hepatoprotective mechanism of action of kolaviron.

EFFECT OF KOLAVIRON ON XENOBIOTIC METABOLIZING ENZYME COMPLEX SYSTEM

We therefore investigated the hepatoprotective action of kolaviron using CCl_4 as a model compound. Our data showed that kolaviron treated with CCl_4 in rats preserved the activities of certain phase 1 enzymes, especially aniline hydroxylase, a representative CYP 2E1 enzyme (Farombi, 2000), while administration of kolaviron alone did not alter the activity of these enzymes.

Following CCl_4 administration to rats, the cholesterol:phospholipid ratio was increased, suggesting a decrease in membrane fluidity with a resultant alteration in membrane function (Farombi, 2000). Co-administration of kolaviron with CCl_4 reversed the CCl_4-induced elevation in the cholesterol:phospholipid ratio and preserved some microsomal marker enzymes, suggesting that kolaviron may protect biomembranes against damage, maintain membrane fluidity (Farombi, 2000), and stabilize the drug metabolizing enzyme complex.

ANTIOXIDANT AND FREE RADICAL SCAVENGING ACTION OF KOLAVIRON

One of the mechanisms of chemoprevention is the antioxidative activity of a chemopreventive agent; thus, the ability of natural compounds to reverse or mitigate carcinogen-induced hepatotoxicity may be related to their intrinsic antioxidant properties. Chromatographic fractionation and spectroscopic analysis of the methanolic fraction of *Garcinia kola* seeds revealed the presence of four compounds — namely, *Garcinia* biflavonoids GB1 and GB2, garcinal, and garcinoic acid — that elicited strong antioxidant effects by the inhibition of nitric oxide production in lipopolysaccharide activated macrophage U937 cells (Okoko, 2009).

Using several *in vitro* chemical models involving the chemical generation of reactive oxygen species, we evaluated the antioxidant properties of kolaviron. Kolaviron showed significant reducing power, and a dose-dependent inhibition of oxidation of linoleic acid (Farombi *et al.*, 2002). Kolaviron inhibited H_2O_2, was more effective than butylated hydroxyanisole and β-carotene, and was comparable with α-tocopherol — a radical chain-breaking antioxidant. Kolaviron also significantly scavenged superoxides generated by phenazine methosulfate NADH. Furthermore, kolaviron scavenged hydroxyl radicals by inhibition of the oxidation of deoxyribose. The ability of kolaviron to scavenge hydroxyl radicals by inhibiting the oxidation of deoxyribose may relate directly to prevention of the propagation of the process of lipid peroxidation, and account for its hepatoprotective properties *in vivo* (Farombi, 2000; Farombi *et al.*, 2000).

In human-derived liver HepG2 cells with morphological characteristics of liver parenchyma cells, kolaviron dose-dependently inhibited the intracellular ROS induced by H_2O_2 (Nwankwo *et al.*, 2000). The report supports the role of kolaviron as an antioxidant, and a potential candidate for the chemoprevention of chemically induced DNA damage. Our

225

studies on the effect of *Garcinia* biflavonoids on DNA damage show that kolaviron concentration-dependently decreased H_2O_2-induced DNA strand breaks, and oxidized purine and pyrimidine bases in both human lymphocytes and rat liver cells (Farombi *et al.*, 2004).

Oxidation of low density lipoprotein (LDL) is generally believed to promote atherosclerosis primarily by leading to an increased uptake of oxidized LDL (ox-LDL) by macrophages, and subsequent foam-cell formation. Therefore, attempting to inhibit or reduce the process of LDL oxidation by antioxidants, which may slow the progression of atherosclerosis, appears to be a rational and pragmatic approach to preventing cardiovascular diseases. In rats treated with kolaviron for 7 days, lipoprotein resistance to copper-induced oxidation was highly improved, as shown by a significant increase in lag time, and a decrease in the area under the curve (AUC) and the slope of propagation (Farombi & Nwaokeafor, 2005). Markers of lipid oxidation significantly decreased in the kolaviron-treated rats, with an attendant significant increase in the total antioxidant activity compared to control. Additionally, kolaviron inhibited the Cu^{2+}-induced oxidation of rat serum lipoprotein in a concentration-dependent manner, and elicited a significant chelating effect on Fe^{2+}. Furthermore, kolaviron effectively prevented microsomal lipid peroxidation induced by iron/ascorbate in a concentration-dependent manner (Farombi & Nwaokeafor, 2005). In totality, the results demonstrate that kolaviron protected against the oxidation of lipoprotein by mechanisms involving metal chelation and antioxidant activity, and, as such, might be of importance in relation to the development of atherosclerosis.

KOLAVIRON AND MODULATION OF TRANSCRIPTION FACTORS AND STRESS RESPONSE PROTEINS

Transcription factors are nuclear proteins activated by the cell transduction pathways in response to a variety of stimuli. They bind to specific DNA sequences on the promoter of target genes, and translate short-term biochemical signals generated by signaling cascades into long-term changes in gene expression (Liu *et al.*, 2002). A causal link between the constitutive activation of NF-κB and hepatocarcinogenesis, via transcriptional regulation of genes involved in cellular transformation, proliferation, survival, invasion, and metastasis, has been defined by several investigations. Recent findings on the molecular mechanisms underlying the hepatoprotective action of kolaviron suggest that it can suppress certain pro-inflammatory genes whose expression has been shown to be regulated by transcription factors. Farombi and colleagues demonstrated that kolaviron abolished the expression of COX-2 and iNOS proteins in dimethyl nitrosamine (DMN)-treated rat liver, suggesting that kolaviron may be important not only in alleviating liver inflammation, but also probably in the prevention of liver cancer (Farombi *et al.*, 2009). NF-κB and AP-1 are key regulators of inflammatory proteins such as COX-2 and iNOS (Nanji *et al.*, 2003), and pretreatment of rats with kolaviron abrogated the DNA binding activity of NF-κB and AP-1 induced by DMN (Farombi *et al.*, 2009) (Figure 26.4). Therefore, the inhibition of DMN-mediated DNA binding of these transcription factors and expression of pro-inflammatory proteins by kolaviron partly explains the molecular basis of the hepatoprotective effect of kolaviron in drug-induced hepatotoxicity and possibly hepatocarcinogenesis, and may be an additional therapeutic application of the naturally occurring flavonoid. The results also confirm the previously reported anti-inflammatory activity of kolaviron and other isolates from *Garcinia kola*, as demonstrated in both acute and chronic models of inflammation, in comparison with phenyl butazone and hydroxycortisone.

ADVERSE EFFECT AND REACTIONS (ALLERGIES AND TOXICITIES)

Available data have shown that *Garcinia kola* is relatively safe, especially when used traditionally, and experiments with animals as well as *in vitro* studies have shown no toxicity of *Garcinia kola*. However, one study carried out by Braide (1990) revealed that a diet including 10% dry powdered seeds of *Garcinia kola* for 6 weeks caused vacuolation of duodenal villous epithelium, intracytoplasmic vacuoles in hepatocytes, and mild hydropic degeneration in the

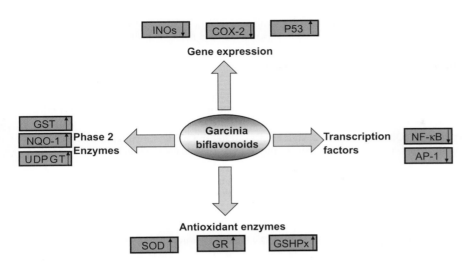

FIGURE 26.4

Molecular targets shown to be regulated by *Garcinia* biflavonoid.

renal proximal tubular epithelium. Also, *Garcinia kola* has antidiabetic and hypoglycemic effects, and therefore, caution is advised when using herbs or supplements that may also lower blood sugar levels. Blood glucose levels may require monitoring, and doses may need adjustment. In view of the importance of this seed in the management of various diseases, more animal and clinical studies are needed to define its therapeutic usage and application in the management of ailments.

SUMMARY POINTS

- *Garcinia kola*, a tropical flowering plant found in western and central Africa, produces large orange fruits and brown, nut-like seeds.
- The seed plays a very important role in African ethno-medicine, in the treatment of coughs, colds, hoarseness of voice, liver diseases, and as aphrodisiac.
- *Garcinia kola* is used as a substitute for hops in brewing lager beer, and in preventing beer spoilage.
- The seeds have been formulated into tablets and are also used in many herbal preparations, either singly or in combination with other plants. Hepa-Vital Tea, a blend of *G. kola* and *Combretum micranttum*, and Hangover Tonic, comprising kolaviron and *Cola nitida*, are manufactured and marketed as phytomedicines.
- The constituents isolated from the seed include xanthones, benzophenones, and kolaviron.
- The hepatoprotective properties of the seed have been associated with the isolated bioflavonoid complex kolaviron.
- Kolaviron protects against carcinogen-induced hepatotoxicity by free radical scavenging, metal chelation, upregulation of the detoxification system, inhibition of stress response proteins, and downregulation of NF-κB and AP-1.

References

Akintonwa, A., & Essien, A. R. (1990). Protective effects of *Garcinia kola* seed extract against paracetamol-induced hepatotoxicity in rats. *Journal of Ethnopharmacology, 29*, 207–211.

Braide, V. B. (1991). Antihepatotoxic biochemical effects of kolaviron, a biflavonoid of *Garcinia kola* seeds. *Phytotherapy Research, 5*, 35–37.

Farombi, E. O. (2000). Mechanisms for the hepatoprotective action of kolaviron: studies on hepatic enzymes, microsomal lipids and lipid peroxidation in carbon tetrachloride-treated rats. *Pharmacological Research, 42*, 75–80.

Farombi, E. O., & Nwaokeafor, I. A. (2005). Antioxidant mechanisms of action of kolaviron: studies on serum lipoprotein oxidation, metal chelation and oxidative microsomal membrane damage in rats. *Clinical and Experimental Pharmacology & Physiology, 32*, 667–674.

227

Farombi, E. O., Tahnteng, J. G., Agboola, O., Nwankwo, J. O., & Emerole, G. O. (2000). Chemoprevention of 2-acetyl aminofluorene-induced hepatotoxicity and lipid peroxidation in rats by kolaviron-A *Garcinia kola* seed extract. *Food and Chemical Toxicology, 38,* 535–541.

Farombi, E. O., Akanni, O. O., & Emerole, G. O. (2002). Antioxidative and scavenging activities of kolaviron *in vitro. Pharmaceutical Biology, 40,* 107–116.

Farombi, E. O., Moller, P., & Dragsted, L. O. (2004). *Ex-vivo* and *in vitro* protective effects of kolaviron against oxygen-derived radical-induced DNA damage and oxidative stress in human lymphocytes and rat liver cells. *Cell Biology and Toxicology, 20,* 71–82.

Farombi., E. O., Adepoju, B. F., Ola-Davies, O. E., & Emerole, G. O. (2005). Chemoprevention of aflatoxin B1-induced genotoxicity and hepatic oxidative damage in rats by Kolaviron, a natural biflavonoid of Garcinia kola seeds. *European Journal of Cancer Prevention, 14,* 207–214.

Farombi, E. O., Shrotriya, S., & Surh, Y. J. (2009). Kolaviron inhibits dimethyl nitrosamine-induced hepatotoxicity by suppressing COX-2 and iNOS expression: NF-κB and AP-1 as potential molecular targets. *Life Sciences, 84,* 149–155.

Iwu, M. M. (1982). Flavonoids of *Garcinia kola* seeds. *Journal of Natural Products, 45,* 650–651.

Iwu, M. M. (1985). Antihepatotoxic constituents of *Garcinia kola* seeds. *Experientia, 41,* 699–700.

Liu., J., Kadiiska, M. B., Corton, J. C., Qu, W., Waalkes, M. P., Mason, R. P., et al. (2002). Acute cadmium exposure induces stress-related gene expression in wild-type and metallothionein-I/II-null mice. *Free Radical Biology & Medicine, 32,* 525–535.

Nanji, A. A., Jokelainen, K., Tipoe, G. L., Rahemtulla, A., Thomas, P., & Dannenberg, A. J. (2003). Curcumin prevents alcohol-induced liver disease in rats by inhibiting the expression of NF-kappa B-dependent genes. *American Journal of Physiology. Gastrointestinal and Liver Physiology, 284,* G321–G327.

Nwankwo, J. O., Tahnteng, J. G., & Emerole, G. O. (2000). Inhibition of aflatoxin B_1 genotoxicity in human liver-derived HepG2 cells by kolaviron biflavonoids and molecular mechanisms of action. *European Journal of Cancer Prevention, 9,* 351–361.

Okoko, T. (2009). *In vitro* antioxidant and free radical scavenging activities of *Garcinia kola* seeds. *Food and Chemical Toxicology, 47,* 2620–2623.

Premalatha, B., & Sachdanandam, P. (2000). Modulating role of *Semecarpus anacardium* L. nut milk extract on aflatoxin B(1) biotransformation. *Pharmacological Research, 41,* 19–24.

Taiwo, O., Xu, H. X., & Lee, S. F. (1999). Antibacterial activities of extracts from Nigerian chewing sticks. *Phytotherapy Research, 13,* 675–679.

228

CHAPTER 27

Black Soybean (*Glycine max* L. Merril) Seeds' Antioxidant Capacity

Ignasius Radix Astadi[1], Alistair G. Paice[2]
[1] Department of Food Technology, Widya Mandala Surabaya Catholic University, Surabaya, Indonesia
[2] Department of Nutrition and Dietetics, King's College London, London, UK

229

CHAPTER OUTLINE

LIST OF ABBREVIATIONS

ACE, angiotensin-converting enzyme

ACF, aberrant crypt foci

COX-2, cyclooxygenase-2

HFD, high fat diet

iNOS, inducible nitric oxide synthase

IRF-1, interferon regulatory transcription factor-1

LDL, low density lipoprotein

PGE 2, prostaglandin E2

ROS, reactive oxygen species

SOD, super oxide dismutase

TBARS, thiobarbituric acid reactive substances

TNF-α, tumor necrosis factor-α

VCAM-1, vascular cell adhesion molecule-1

Nuts & Seeds in Health and Disease Prevention. DOI: 10.1016/B978-0-12-375688-6.10027-1

INTRODUCTION

Black soybean has been used as a health-food ingredient and herb in China for centuries. Recently, studies on the biological activities of the black soybean have increased significantly, especially in the investigation of its anthocyanin content. Anthocyanin is a bioactive compound found in the seed coat of the black soybean that is reported to inhibit several diseases, due at least in part, to its inherent antioxidant activity. Researchers have tried to elucidate the molecular mechanisms behind the inhibition of these diseases.

This chapter describes the black soybean as a plant and a source of food products, and illustrates the activity of its extract and subsequent food products in preventing diseases and promoting health.

BOTANICAL DESCRIPTION

The soybean (*Glycine max L. Merril*) is a constituent species of the Leguminoceae group. It is also widely known as *Glycine soja* or *soja max*. Soybeans come in different sizes, forms, and colors, depending on the variety, one of which is the black soybean. Black soybeans are characterized by a near-spherical shape, a black hull (seed coat), and yellow cotyledons (seed interior). The hull is easily removed to expose the cotyledon. Characteristically for soybeans, the hilum is longer and thinner than that of other beans. Black soybeans are a short-season crop, needing 3—5 months to grow and yield, and are cultivated within a wide range of temperatures. They are considered as rain-fed bush crops. On average, black soybean plants need 350—450 mm of rainfall for optimal growth. However, excess water can lead to impaired germination, increasing pathogenic activity and creating anaerobic conditions.

HISTORICAL CULTIVATION AND USAGE

Black soybeans were originally grown in Asia. It is believed that they were first cultivated in Northern China during the Shang Dynasty (1700—1100 BC), and were designated as one of the five "sacred" grains of the time. In ancient China, the black soybean was commonly used as a medicinal food and herb rather than as a staple food. By the 16th century, soybeans were being grown throughout Asia. In Korea, Japan, and Indonesia, black soybeans were used as an integral part of many traditional ceremonies. After the discovery of fermentation methods, black soybean-based foods also became popular. The utilization of black soybean for food products is illustrated in Table 27.1.

TABLE 27.1 Black Soybean-based Food Products

Products	Description	Countries of Production
in si, tau si	Dried by-product of the mash black soybean sauce fermented with *Aspergillus oryzae*	China, Phillipines
Tempeh	Traditional food from black or yellow soybean fermented with *Rhizopus oligosporus*	Indonesia
Tofu	Protein gel-like product from soybean	Asian countries
Soy sauce	Sauces fermented with *Aspergillus oryzae* and *Aspergillus soyae*, used as condiment	Asian countries
Natto	Traditional Japanese soybean product fermented with *Bacillus subtilis*	Japan
Chungkookjang	Steamed black soybeans fermented with *Bacillus* species	Korea

PRESENT-DAY CULTIVATION AND USAGE

Cultivation and utilization of the black soybean in Asia is still less popular than that of the yellow soybean. Black soybean is still used in traditional ceremonies, whilst modern industries have also developed large-scale black soybean-based food products such as tofu, tempeh, soy milk, soy sauce, natto, and miso. In Indonesia, cultivation of the black soybean has become part of community empowerment programs to help reduce poverty. The black soybean output from these programs is purchased by industry to produce soy sauce, which is a popular condiment amongst Indonesians. Industrial processes can also extract anthocyanin from the black soybean seed coat. Based upon the latest findings of the biological properties of black soybean anthocyanin, these commercially viable extracts could then be used in alternative medicines.

APPLICATIONS TO HEALTH PROMOTION AND DISEASE PREVENTION

Studies on the antioxidant properties of plant food are increasing rapidly. In food systems, antioxidants can help prevent undesirable flavors and nutritional degradation. In the human body, antioxidants are believed to play an important role in the inhibition of several degenerative diseases, such as cancer, coronary heart disease, atherosclerosis, and diabetes.

Studies into the antioxidant properties of soybeans have generally been conducted using the yellow seed-coat soybean. Isoflavons are commonly found in the yellow soybean cotyledons and hypocotyls. Antioxidants in the yellow seed coat have not been greatly investigated, as they were assumed to be unimportant. Black soybean seed coats, on the other hand, are rich in phenolic substances, in which secondary metabolites from plants are biologically active as antioxidants. It has been widely demonstrated that significant antioxidative activity in plant-based food is related to their polyphenol content. Polyphenols are divided into several categories, including flavonoids, isoflavons, flavonols, flavones, and anthocyanins. The black soybean seed coat contains anthocyanins. This class of polyphenol is a water-soluble pigment responsible for the black, red, and blue color of the plants, and well known as having considerable pharmaceutical activity. The chemical structure of anthocyanin is shown in Figure 27.1.

According to Choung *et al.* (2001), the first studies on anthocyanin in the black soybean were conducted by Nagai (1921). Other researchers identified several types of anthocyanins in black soybeans, especially in the seed coat. The studies on individual anthocyanins in black soybeans are presented in Table 27.2. As demonstrated in Table 27.2, cyanidin-3-glucoside is the major anthocyanin found in black soybean (Xu & Chang, 2008; Lee *et al.*, 2009). Petunidin-3-glucoside, which is considered as a new anthocyanin, has also been discovered in the black soybean seed coat (Choung *et al.*, 2001). The anthocyanin contents of the black soybean seed coat vary according to the species, but are relatively high compared to other sources of anthocyanin, such as berries, grapes, and sorghum. This is illustrated in Table 27.3.

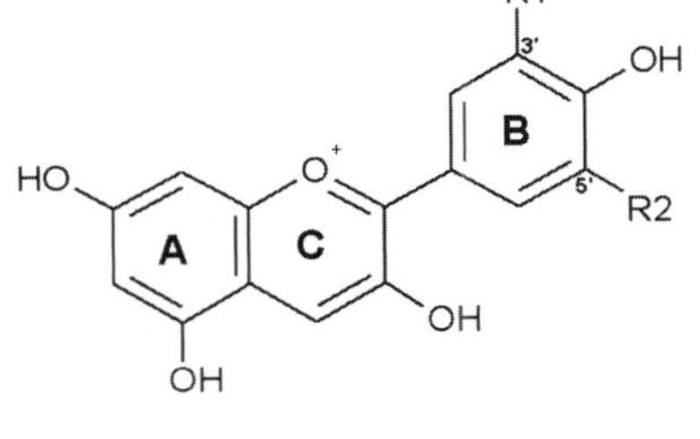

231

FIGURE 27.1
Structure of anthocyanin.

TABLE 27.2 Individual Anthocyanins of Black Soybean

Black Soybean Varieties/Sources	Individual Anthocyanin	References
NAFC Yeocheon	Cyanidin-3-glucoside, delphinidin-3-glucoside, petunidin-3-glucoside	Kim et al. (2008)
Hyoukei-kuro-3goh	Cyanidin-3-O glucoside	Takahashi et al. (2005)
Milyang 95	Cyanidin-3-glucoside, delphinidin-3-glucoside, petunidin-3-glucoside	Choung et al. (2001)
Casselton,ND	Cyanidin-3-glucoside, delphinidin-3-glucoside, peonidin-3-glucoside	Xu & Chang (2008)
Cheongja 3	Catechin-cyanidin-3-O-glucoside Delphinidin-3-O-galactoside Delphinidin-3-O-glucoside Cyanidin-3-O-galactoside Cyanidin-3-O-glucoside Petunidin-3-O-glucoside Pelargonidin-3-O-glucoside Peonidin-3-O-glucoside Cyanidin	Lee et al. (2009)

TABLE 27.3 Anthocyanin Content of Different Varieties of Black Soybean

Black Soybean Varieties/Sources	Anthocyanin Content (mg/g)	References
Milyang 95	9.83	Choung et al. (2001)
Geomjeongol	10.62	Choung et al.(2001)
IT 180220	18.81	Choung et al. (2001)
YJ100-1	20.18	Choung et al. (2001)
NAFC Yeocheon	0.59	Kim et al. (2008)
Casselton,ND	1.08	Xu & Chang (2008)
Mallika	13.63	Astadi et al. (2009)
Cikuray	14.68	Astadi et al. (2009)

A diet rich in anthocyanin is believed to have both anti-obesity and hypolipidemic effects. Rats fed on a high-fat diet had any weight gain moderated by the supplementation of black soybean anthocyanin into their food. The supplemented diet intermediately suppressed HFD-induced weight gain in the liver, and tended to decrease that of epididymal and perirenal fat pads. Black soybean meals also improved the lipid profile. They significantly reduced the levels of serum triglyceride and cholesterol, but increased the high-density lipoprotein-cholesterol concentrations (Park et al., 2007).

The beneficial properties of black soybean may be linked to its ability to prevent LDL oxidation. In the human body, LDL is a transport mechanism for cholesterol. Oxidized LDL is thought to play a central role in the development of atherosclerosis, leading to diseases such as coronary heart disease. Daily intake of black soybean proteins is associated with reduced cardiovascular disease risk (Messina, 1999). Several studies into the ability of black soybean to inhibit LDL oxidation have been conducted. For example, a study by Takahashi and colleagues (2005) showed that black soybean extract had a longer LDL oxidation "lag time" than that of yellow soybean, due to a higher total polyphenol content in its seed coat. Astadi and colleagues (2009) also demonstrated the antioxidative properties of black soybean seed coat anthocyanin on human LDL. The mechanism of action is probably through its ability to act as a free radical scavenger, preventing free radicals reacting with LDL. This is

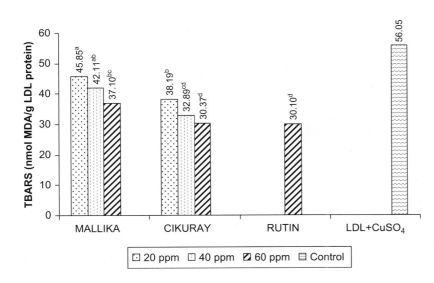

FIGURE 27.2

Inhibition of Cu^{2+}-induced human LDL oxidation *in vitro* by black soybean seed coat extract (Astadi *et al.*, 2009).

illustrated in Figure 27.2. Anthocyanin extracts inhibit LDL oxidation in a dose-dependent manner. Figure 27.2 demonstrates the dose–response relationship between these extracts, and the inhibition of "TBARS" formation. The hydroxyl group of anthocyanin donates hydrogen atoms to the radical compound, to produce a non-radical product. Anthocyanin also acts as a chelating agent.

Lacaille-Dubois and colleagues (2001) demonstrated that proantocyanidin has ACE inhibitory activity. ACE inhibitory activity may improve the cognitive performance of mice using the habituation test. In addition, bioactive compounds with ACE inhibitory activity are used extensively in the treatment of hypertension and congestive heart failure.

There are relatively consistent epidemiological findings showing that consumption of soybean foods that contain antioxidative compounds is strongly associated with a reduction in the risk of cancer. A case–control study conducted by Do *et al.* (2007) showed that black soybean consumption reduces the risk of breast cancer in Korean women. Anthocyanin is believed to play a crucial role in this. An *in vitro* and *in vivo* study on colonic inflammation by Kim *et al.* (2008) demonstrated that black soybean anthocyanin might have anti-inflammatory and antiproliferative effects. Inflammation has long been implicated in the development of cancer. In this research, cyanidin and delphinidin significantly inhibited cell growth and suppressed COX-2 and iNOS mRNA production in HT-29 human colon adenocarcinoma cells. A diet rich in black soybean seed coat may decrease ACF and plasma PGE2 levels in carcinogen-treated F344 rats.

The inhibitory effect of black soybean anthocyanin on inflammatory processes was also investigated by Nizamutdinova *et al.* (2009). In this study, it was postulated that anthocyanin could inhibit the antigen-induced TNF-α stimulation of VCAM-1 via regulation of GATAs and IRF-1. VCAM-1 is believed to be a target for highly metastatic human melanoma cells. These cells have a "high affinity" conformation at their cell surface, facilitating adherence and subsequent migration. The interferon regulatory transcription factor-1 (IRF-1) and transcription factor genes bind to DNA sequence GATA (GATAs) in the VCAM-1 gene promoter region. These metastatic cells have a pathological role in inflammatory processes that eventually lead to cancer and atherosclerosis. Stimulation of cells with TNF-α increases VCAM-1 expression. Nizamutdinova and colleagues reported that pretreatment of cells with anthocyanins inhibited VCAM-1 expression, and can reduce the nuclear levels of GATAs and IRF-1.

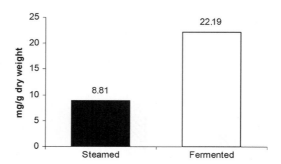

FIGURE 27.3
Polyphenol content of processed black soybean (Kwak *et al.*, 2007).

Anthocyanin may also have the ability to protect the skin from the effects of aging and the occurrence of skin cancer. Tsoyi *et al.* (2008) reported that anthocyanins from black soybeans protected keratinocytes from UVB-induced cytotoxicity and apoptosis, through inhibition of the caspase-3 pathway and reduction of the level of the pro-apoptotic Bax protein. Anthocyanin also prevented apoptotic cell death in mouse skin. These results are probably due to the ability of anthocyanin to modulate the UVB-mediated ROS production soon after UVB exposure. ROS are believed to have crucial roles in apoptosis pathway mediation, such as blocking caspase activation. Thus, cellular membrane lipid peroxidation and oxidative damage to DNA and cellular proteins could be prevented.

Studies have also been undertaken on food products derived from black soybean. Tofu is the most common black soybean-based product. The antioxidant activity of black bean tofu was determined using the thiocyanate method. Black bean tofu has higher antioxidative activity, lower potential for lipid peroxidation, and a longer shelf life than yellow soybean tofu (Shih *et al.*, 2002).

Kwak and colleagues (2007) investigated the properties of traditional fermented black soybean paste from Korea (chungkookjang). The antioxidant activity of chungkookjang, evaluated using the DPPH assay, was higher than that of unfermented black soybean. *In vivo* studies revealed that a diet rich in chungkookjang could increase SOD and catalase activity, and significantly reduce hepatic TBARS. Fermentation can increase the polyphenol content of the black soybean. The relevant data are shown in Figure 27.3. This increase in the polyphenol content is probably due to a partial cleavage of the glycosides, resulting from the increased glycosidase and glucuronidase activities, releasing potential antioxidant substances by the transformation of falconoid.

Black soybean has a biological activity which can help prevent human disease and promote health, due to its antioxidant activity. This is especially true of its seed coat. Importantly, processing does not significantly reduce the antioxidant activity of black soybean. Thus, consumption of processed foods containing black soybean should be recommended.

ADVERSE EFFECTS AND REACTIONS (ALLERGIES AND TOXICITY)

There are only limited reports of adverse effects following the consumption of black soybean and black soybean food products. As the black soybean has similar characteristics and nutritional content to yellow soybean, it is probably valid to use research conducted on yellow soybean allergies and toxicity as a substitute. Soybean contains a trypsin inhibitor and phytic acid. Trypsin inhibitors have the potential to derange protein digestion, leading to chronic deficiencies in amino acid uptake. Phytic acid can inhibit the uptake of essential minerals such as zinc, magnesium, calcium, copper, and iron.

The protein content of soybean could cause allergies. It contains several proteins that have been reported as major allergens, such as soy hydrophobic protein, soy hull protein, soy vacuolar protein, glycinin, and β-conglycinin. However, this potential may not be realized. Allergy secondary to soybean is relatively rare, and depends on the interplay between individual responses, the clinical and laboratory investigations used, and the populations that are being studied.

SUMMARY POINTS

- Black soybean, with a black seed coat, is extensively cultivated in China, Korea, Japan, and Indonesia.
- Black soybean is usually used to produce fermented products such as tempeh, natto, miso, and soy sauce, as well as non-fermented products such as tofu and soy milk.
- The black soybean seed coat contains anthocyanin, which can act as an antioxidant. The major anthocyanin found in black soybean seed coat is cyanidin-3-glucoside.
- Anthocyanin in black soybean has reported anti-inflammatory, anticarcinogenic, antitumor, and antimutagenic activities; it also enhances spatial memory and cognition, and inhibits LDL oxidation.
- Processing does not significantly decrease the antioxidative activity of black soybean-based food products.
- Despite its beneficial properties, black soybean contains potential allergenic and toxic components. However, adverse effects following soy consumption are rare, and depend on an individual's response.

References

Astadi, I. R., Astuti, M., Santoso, U., & Sih Nugraheni, P. (2009). *In vitro* antioxidant activity of anthocyanins of black soybean seed coat in human low density lipoprotein (LDL) oxidation. *Food Chemistry, 112*, 659–663.

Choung, M. G., Baek, I. Y., Kang, S. T., Han, W. Y., Shin, D. C., Moon, H. P., et al. (2001). Isolation and determination of anthocyanins in seed coats of black soybean (*Glycine max* (L.) Merr.). *Journal of Agricultural and Food Chemistry, 49*. 5848-5841.

Do, M. H., Lee, S. S., Jung, P. J., & Lee, M. H. (2007). Intake of fruits, vegetables, and soy foods in relation to breast cancer risk in Korean women: a case–control study. *Nutrition and Cancer, 57*, 20–27.

Kim, J. M., Kim, J. S., Yoo, H., Choung, M. G., & Sung, M. K. (2008). Effects of black soybean [*Glycine max* (L.) Merr.] seed coats and its anthocyanidins on colonic inflammation and cell proliferation *in vitro* and *in vivo*. *Journal of Agricultural and Food Chemistry, 56*, 8427–8433.

Kwak, C. S., Lee, M. S., & Park, S. C. (2007). Higher antioxidant properties of Chungkookjang, a fermented soybean paste, may be due to increased aglycone and malonylglycoside isoflavone during fermentation. *Nutrition Research, 27*, 719–727.

Lacaille-Dubois, M. A., Franck, U., & Wagner, H. (2001). Search for potential angiotensin converting enzyme (ACE)-inhibitors from plants. *Phytomedicine, 8*, 47–52.

Lee, J. H., Kang, N. S., Shin, S. O., Shin, S. H., Lim, S. G., Suh, D. Y., et al. (2009). Characterisation of anthocyanins in the black soybean (Glycine max L.) by HPLC-DAD-ESI/MS analysis. *Food Chemistry, 112*, 226–231.

Messina, M. J. (1999). Legumes and soybeans: overview of their nutritional profiles and health effects. *The American Journal of Clinical Nutrition, 70*, 439S–459S.

Nagai, I. (1921). A genetico-physiological study of the formation of anthocyanins and brown pigments in plant. *Journal of the College of Agriculture, Imperial University of Tokyo, 8*, 1.

Nizamutdinova, I. T., Kim, Y. M., Chung, J. I., Shin, S. C., Jeong, Y. K., Seo, H. G., et al. (2009). Anthocyanins from black soybean seed coats preferentially inhibit TNF-α-mediated induction of VCAM-1 over ICAM-1 through the regulation of GATAs and IRF-1. *Journal of Agricultural and Food Chemistry, 57*, 7324–7330.

Park, K. Y., Kwon, S. H., Ahn, I. S., Kim, S. O., Kong, C. S., Chung, H. Y., et al. (2007). Weight reduction and lipid lowering effects of black soybean anthocyanins in rats fed a high fat diet. *The FASEB Journal: Official Publication of the Federation of American Societies for Experimental Biology, 21*, 842–846.

Shih, M. C., Yang, K. T., & Kuo, S. J. (2002). Quality and antioxidative activity of black soybean tofu as affected by bean cultivar. *Journal of Food Science, 67*, 480–484.

Takahashi, R., Ohmori, R., Kiyose, C., Momiyama, Y., Ohsuzu, F., & Kondo, K. (2005). Antioxidant activities of black and yellow soybeans against low density lipoprotein oxidation. *Journal of Agricultural and Food Chemistry, 53*, 4578–4582.

235

Tsoyi, K., Park, H. B., Kim, Y. M., Chung, J. I., Shin, S. C., Shim, H. J., et al. (2008). Protective effect of anthocyanins from black soybean seed coats on UVB-induced apoptotic cell death *in vitro* and *in vivo*. *Journal of Agricultural and Food Chemistry, 56*, 10600–10605.

Xu, B., & Chang, S. K. C. (2008). Total phenolics, phenolic acids, isoflavones, and anthocyanins and antioxidant properties of yellow and black soybeans as affected by thermal processing. *Journal of Agricultural and Food Chemistry, 56*, 7165–7175.

CHAPTER 28

Antidiabetic and Antihyperlipidemic Activity of Bonducella (*Caesalpinia bonducella*) Seeds

Dayanand M. Kannur
Department of Pharmacognosy, Shree Chanakya Education Society's, Indira College of Pharmacy, Pune, India

CHAPTER OUTLINE

LIST OF ABBREVIATIONS

BGL, blood glucose level
BUN, blood urea nitrogen
HDL, high density lipoprotein
KEE, seed kernel ethanolic extract
LDL, low density lipoprotein
LD_{50}, lethal dose
PEE, seed kernel petroleum ether extract
SC, serum creatinine
SEE, ethanolic extracts of the seed coat
TC, total cholesterol

Nuts & Seeds in Health and Disease Prevention. DOI: 10.1016/B978-0-12-375688-6.10028-3

INTRODUCTION

Caesalpinia bonducella (synonym *Caesalpinia bonduc)* is a plant belonging to the Caesalpiniaceae family. It is one of the important herbs found in folklore medicine, and its medicinal uses have been quoted in various Ayurvedic texts and scriptures. Traditionally, it is used to cleanse the uterus in the postpartum period, and also acts as a uterine stimulant. It is known to alleviate fever, edema, and abdominal pain during this period. *Caesalpinia bonducella* is a prickly shrub and has numerous synonyms, including Duhsparsa ("difficult to touch"), Kantaki karanja ("having prickles"), Vajra bijaka ("having hard seeds"); Kanta phala ("fruits covered with prickles"), etc. It has many other common names, including fever nut, Indian filbert, and nickar (English); Karanja, Kanthekaranja, Karanjuvu, Karanju, and Lata Karanju (Hindi); Gajaga, Sagargota, Kanchaki, and Karbath (Marathi); and Akitmakit and Bunduk (Urdu).

BOTANICAL DESCRIPTION

Caesalpinia bonducella is a large, straggling, very thorny shrub, and is an extensive climber. The branches are armed with hooks and are straight, hard yellow prickles (Kirtikar & Basu, 1995).

238

FIGURE 28.1
Caesalpinia bonducella. (A) plant; (B) Inflorescence; (C) Pod; (D) Seeds; (E) Kernels; (F) Pods and seeds.

The leaves (see Figure 28.1A) are compound, 30—60 cm long, with prickly petioles. The stipules have a pair of round pinnae at the base of the leaf, each furnished with a long mucronate point. There are six to eight pairs of pinnae, 5—7.5 cm long, with a pair of hooked stipulary spines at the base, and six to nine pairs of leaflets, 2—3.8 × 1.3—2.2. Leaflets are membranous, elliptic—oblong, obtuse, strongly mucronate, glabrous on the upper surface and more or less puberlous beneath; the petilolules are very short (Sutaria, 1969).

The flowers are pale yellow in color, dense (usually spicate), with a long peduncled terminal and supra-axillary racemes at the top. Racemes droop downwards and are 15—25 cm long; the pedicles are very short in bud, elongating to 5 mm in flowers and 8 mm in fruit; they are brown, downy and oblanceolate. Filaments are declinate, flattened at the base, and clothed with long white silky hairs (Figure 28.1B). The flowering season starts in June, and the shrub bears fruits by November.

The fruits are inflated oblongate pods, 5—7.5 × 4.5 cm, covered with prickles, containing one or two seeds per pod. The pods have short stalks. The seeds are 1—2 cm in size, globular, hard, bluish-gray in color, and have a smooth, shiny surface (Sutaria, 1969) (Figure 28.1C—F).

HISTORICAL CULTIVATION AND USAGE

Caesalpinia bonducella is a large, prickly shrub that grows naturally throughout the hotter parts of India, Myanmar, Sri Lanka, and the West Indies, along the sea coasts and at heights of up to 800 meters on the hills. It is commonly found in the southern part of India.

Caesalpinia bonducella seeds have long been used in traditional medicine in treating symptoms and ailments, including abdominal pain, colic, leprosy, fever, edema, and malaria. It is also used as a uterine stimulant, and to cleanse the uterus during the post-partum period.

PRESENT-DAY CULTIVATION AND USAGE

Caesalpinia bonducella is propagated by seeds. These are sown at the start of the rainy season; they are first soaked overnight, and then generally sown at intervals of 50 cm to form a hedge. It is essential to irrigate immediately after sowing the seeds. Sandy loam soil gives optimum growth. After 3—4 weeks the plants sprout, growing to their maximum height in 2 to 2½ years. In the initial stages it is very important to irrigate, and plants should be pruned every fortnight.

Pharmacognostical analysis

Caesalpinia bonducella seed kernels (Figure 28.1E) are oblong to ovoid in shape, brittle, pale yellow to cream in color, with a characteristic odor and bitter taste. The seed coat is brownish-gray in colour, with a characterisic odor and an astringent taste (Figure 28.1D).

The seeds contain furanoditerpenes (Peter *et al.*, 1997) α-caesalpin, β-caesalpin, γ-caesalpin, δ-caesalpin, ε-caesalpin, and caesalpinia-F (Keitho *et al.*, 1986); fatty acids (palmitic, stearic, octadeca-4-enoic and octadeco-2-4-dienoic, lignocenic, oleic and linoleic); phytosterinin; β-sitosterol; a homoisoflavone, bonducellin (Purushothaman *et al.*, 1982); amino acids (aspartic, arginine, and citrulline); carbohydrates (starch, sucrose); β-carotene; a glycoside, bonducin; and gums and resins (Williamson, 2002).

Caesalpinia bonducella fruits contain D(+) pinitol, and the leaves contain the glycosidal compounds Brazilin and bonducin. The roots contain cassane furan diterpene caesalpinin; cassane diterpenes caesaldekarins C, F, and G; bonducellpins A, B, C, D (Peter & Tinto, 1997); and the steroidal saponin diosgenin (Figure 28.2).

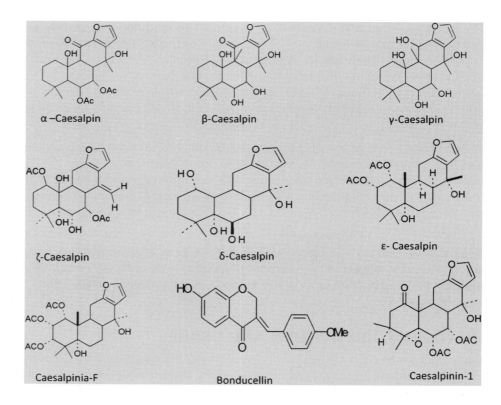

FIGURE 28.2
Chemical constituents of *Caesalpinia bonducella*.

Extraction process of the bonducella seed

Extraction is carried out by breaking the seed coat; the kernels are then crushed to a coarse powder. This powder is defatted with petroleum ether (60/80). The petroleum ether is distilled off on a rotary flash evaporator to leave a yellow, oily extract (22% w/w). The defatted marc is then dried and further extracted with 95% ethanol, which yields a sticky ethanolic reddish brown extract (16% w/w).

The physical constants of *Caesalpinia bonducella* are shown in Table 28.1; results of the analysis of the fixed oil obtained from the seed kernels are shown in Table 28.2.

TABLE 28.1 Determination of Physical Constants of *Caesalpinia bonducella* Seeds

SL No.	Quantitative Standards	Seed Kernels (% w/w)	Seed Coat (% w/w)
1	Loss on drying	10.22	10.06
2	*Extractives:*		
	a) Water soluble	52.00	36.20
	b) Alcohol soluble	21.44	14.64
	c) Pet. ether	44.80	03.36
3	*Ash values:*		
	a) Total ash	2.60	3.40
	b) Acid-insoluble ash	1.56	1.94
	c) Water-soluble ash	0.87	1.07

TABLE 28.2 Analytical Values of Oil of *Caesalpinia bonducella* Seeds

Acid value	4.71
Saponification value	144.46
Ester value	139.75

APPLICATIONS TO HEALTH PROMOTION AND DISEASE PREVENTION

Caesalpinia bonducella seeds have been used to cure various diseases and disorders. The seeds are used as antidiabetics, and various scientific studies have proved their claim to activity. They have been proven to be anti-inflammatory, anthelminitic, and antimalarial, and are also effective as stomachic and digestives. They are used as a liver tonic in the treatment of jaundice and various liver disorders. The seeds are also considered to be an aphrodisiac, and a general tonic helping in the rejuvenation of the body (Shrikantha Murthy, 2000a). The roasted seed powder is used as an antileprotic, antiperiodic, antipyretic, etc. The topical application of seed oil helps in rheumatic disorders and arthritis.

In Indian traditional medicine, *Caesalpinia bonducella* is used in a vast range of diseases. It is the best cure for abdominal pain due to flatulence. The roasted seed powder relieves pain when administered in an oily formulation. A mixture of roasted seed powder, asafoetida, ghee, and a small amount of salt eliminates abdominal pain during the postpartum period (Shrikantha Murthy, 2000b). It is the best medication for malaria, where a combination of the roasted seed powder with *Piper longum* (1:1) is given with honey, approximately 0.5 g, three times a day for 3—4 days.

Antidiabetic potential of *Caesalpinia bonducella* seeds

The antidiabetic potential of *Caesalpinia bonducella* seeds has been investigated since 1987 by Simon and colleagues, who showed that the aqueous extract possesses antihyperglycemic action (Simon *et al.*, 1987). Rao and colleagues found that the powder of *Caesalpinia bonducella* exhibited antidiabetic potential, and they attributed this property to reduced absorption from the gut (Rao *et al.*, 1994). Sharma and colleagues (1997) suggested that the aqueous and 50% ethanolic extracts of the seeds exhibited significant antihyperglycemic as well as hypoglycemic action. They proposed that the action may be similar to that of Glibenclamide and stimulate the β cells, enhancing the secretion of insulin.

Chakrabarti and colleagues strongly claim that the extracts of seeds of *Caesalpinia bonducella* significantly lower the serum glucose level in type 2 diabetes (Chakrabarti *et al.*, 2003); they also investigated the insulin secretion stimulating potential of *Caesalpinia bonducella* seeds and concluded that the antihyperglycemic activity exhibited is due to insulin secretagogue activity (Chakrabarti *et al.*, 2005).

Kannur and colleagues stated that the seed kernel ethanolic extract (KEE) and seed kernel petroleum ether extract (PEE) as well as the ethanolic extracts of the seed coat (SEE) of *C. bonducella* exhibited a remarkable blood glucose lowering effect in the glucose tolerance test (Figure 28.3). These extracts also significantly lower the elevated blood glucose level in alloxan induced hyperglycemia (Table 28.3; Figure 28.4) (Kannur *et al.*, 2006), as well as elevated blood urea nitrogen (Figure 28.5). All these studies prove that the seeds possess a strong capacity to reduce an elevated blood glucose level, and thus can be used in treating diabetes mellitus.

Antihyperlipidemic activity of *Caesalpinia bonducella* seeds in hyperglycemic conditions

In most of the studies to date, the seeds of *Caesalpinia bonducella* have shown significant antihyperlipidemic actions. The effect of the extracts on diabetes-induced hyperlipidemia has also been studied. It has been observed that due to diabetes, there is an increase in the total cholesterol level as well as the triglyceride level (Kannur *et al.*, 2006). In the diabetic animals, the HDL was reduced and the LDL level was increased significantly. All extracts of *Caesalpinia bonducella* showed a significant decrease in the total cholesterol level and the triglyceride level (Table 28.4). In particular, the kernel ethanolic extract (KEE) exhibited significant action. It also increased the HDL and was successful in suppressing the LDL level as compared to the standard drug (Figure 28.6).

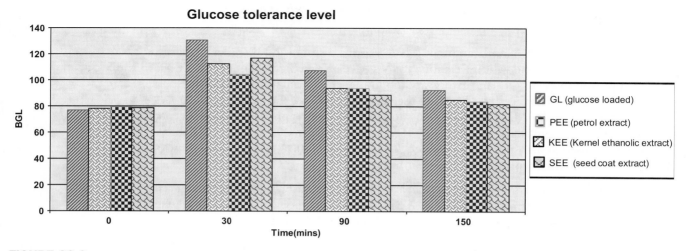

FIGURE 28.3

Effect of *Caesalpinia bonducella* seed extracts in the glucose tolerance test.

TABLE 28.3 Effect of *C. bonducella* Extracts on the Blood Glucose Level (BGL), Blood Urea Nitrogen (BUN), and Serum Creatinine (SC) Levels in Alloxan-induced Hyperglycemia in Rats[a]

Samples (mg%)	Normal	Diabetes Induced	Diabetes + Rosiglitazone 0.01 mg/kg	Diabetes + KEE	Diabetes + PEE	Diabetes + SEE
BGL	83.0 ± 2.42	578.0 ± 74.95	$87.0^c \pm 2.27$	$324.5^b \pm 47.97$	$329.0^b \pm 23.84$	$196.0^c \pm 40.86$
BUN	27.85 ± 1.08	39.20 ± 3.11	$28.65^b \pm 2.85$	$22.22^c \pm 0.83$	$33.35^{NS} \pm 0.76$	$28.55^c \pm 2.33$
SC	0.5 ± 0.04	0.7 ± 0.5	$0.63^{NS} \pm 0.3$	$0.63^{NS} \pm 0.03$	$0.65^{NS} \pm 0.03$	$0.65^{NS} \pm 0.03$

[a]Given by oral route at dose of 300 mg/kg; values are mean $\pm$ S.E.M.
[b]$P < 0.05$ when compared with diabetic control;
[c]$P < 0.001$ when compared with diabetic control.
[NS]Not significant.

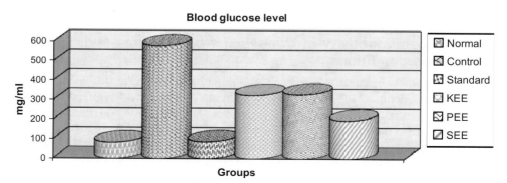

FIGURE 28.4

Effect of *Caesalpinia bonducella* seed extracts on elevated blood glucose levels in diabetic rats.

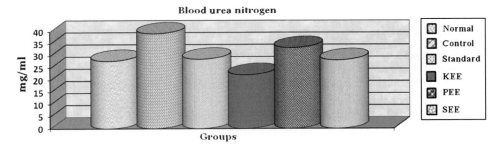

FIGURE 28.5
Effect of *Caesalpinia bonducella* seed extracts on blood urea nitrogen levels in diabetic rats.

TABLE 28.4 Effect of *C. bonducella* Extracts on Total Cholesterol (TC), Triglyceride (TG), HDL, LDL, and VLDL Levels in Rats[a]

Test (mg%)	Normal	Diabetes Induced	Standard	KEE	PEE	SEE
TC	137.25 ± 1.89	177.00 ± 4.02	145.75 ± 7.09^b	134.00 ± 7.62^c	142.50 ± 2.63^c	140.25 ± 2.10^c
TG	88.75 ± 1.81	107.75 ± 5.75	93.25 ± 9.99^b	88.50 ± 1.09^c	94.00 ± 6.01^b	86.00 ± 8.90^c
HDL	44.25 ± 2.78	33.75 ± 1.43	42.00 ± 1.47^c	49.25 ± 1.25^c	41.25 ± 5.21^c	37.75 ± 0.63^b
LDL	75.75 ± 1.11	127.25 ± 2.16	85.75 ± 8.20^c	67.50 ± 7.53^c	83.00 ± 6.22^c	85.75 ± 1.80^c

[a]Given by oral route at dose of 300 mg/kg; values are mean $\pm$ S.E.M.
[b]$P < 0.05$ when compared with diabetic control;
[c]$P < 0.001$ when compared with diabetic control.

243

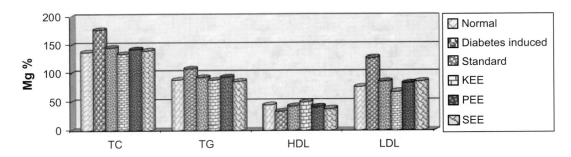

FIGURE 28.6
Antihyperlipidemic action of Caesalpinia bonducella seed extracts. TC, total cholesterol; TG, triglyceride level; HDL, high density lipoprotein; LDL, low density lipoprotein.

The presence of various sterols, and especially β-sitosterol, in the seeds may be responsible for their antihyperlipidemic action.

ADVERSE EFFECTS AND REACTIONS (ALLERGIES AND TOXICITY)

No serious adverse effects have been reported for *Caesalpinia bonducella* medicines, but overdosing can lead to nausea and vomiting. The LD_{50} has been found to be 3000 mg/kg body weight (Kannur *et al.*, 2006).

SUMMARY POINTS

- *Caesalpinia bonducella* seeds are a rich source of fixed oil. The seed kernels as well as the oil obtained from them show significant pharmacological actions. Seed oil can be used as a rubefacient as well as an anti-arthritic agent.
- *Caesalpinia bonducella* has been extensively investigated for various medicinal attributes; it has been validated scientifically, and can be used to treat inflammation and hyperlipidemia, as well as hyperglycemia. It can also be used as a stomachic. It is a proven immunomodulator (Shukla *et al.*, 2009a) as well as an adaptogenic agent. The seeds have also been proven to be antioxidant (Shukla *et al.*, 2009b) in nature.
- The various studies carried out to date support the claims of *Caesalpinia bonducella* seeds as a promising herbal antidiabetic drug. The extracts of *Caesalpinia bonducella* seeds may stimulate the secretion of insulin by the beta cells, and thus show a blood-glucose lowering action.
- The properties possessed by *Caesalpinia bonducella* have significant commercial value and use. It is essential that investigations are carried out in more depth in order to validate other claims as well as to find out the constituents responsible for its antidiabetic potential.

References

Chakrabarti, S., Biswas, T. K., Rokeya, B., Ali, L., Mosihuzzaman, M., Nahar, N., et al. (2003). Advanced studies on the hypoglycemic effect of *Caesalpinia bonducella* F. in type 1 and 2 diabetes in Long Evans rats. *Journal of Ethnopharmacology, 84*, 41–46.

Chakrabarti, S., Biswas, T. K., Seal, T., Rokeyac, B., Ali, L., Azad Khan, A. K., et al. (2005). Antidiabetic activity of *Caesalpinia bonducella* F. in chronic type 2 diabetic model in Long-Evans rats and evaluation of insulin secretagogue property of its fractions on isolated islets. *Journal of Ethnopharmacology, 97*, 117–122.

Kannur, D. M., Hukkeri, V. I., & Akki, K. S. (2006). Antidiabetic activity of *Caesalpinia bonducella* seed extracts in rats. *Fitoterapia, 77*, 546–549.

Keitho, P., Burke, A., & Chan, W. R. (1986). Caesalpin-F, a new furanoditerpene from *Caesalpinia bonducella*. *Journal of Natural Products, 49*, 913–915.

Kirtikar, K. R., & Basu, B. D. (1995). *Indian Medicinal Plants, 2*, 842–845.

Peter, S. R., & Tinto, W. F. (1997). Bonducellpins A–D, new Cassane furanoditerpene of *Caesalpinia bonducella*. *Journal of Natural Products, 60*, 1219–1221.

Peter, S. R., Tinto, W. F., McLean, S., Reynolds, W. F., & Tay, L.-L. (1997). Caesalpinin, a rearranged cassane furanoditerpene of *Caesalpinia bonducella*. *Tetrahedron Letters, 38*, 5767–5770.

Purushothaman, K. K., Kalyani, K., & Subramanian, K. (1982). Structure of boducellin, a new homoisoflavone from *Caesalpinia bonducella*. *Indian Journal of Chemistry, 21*, 383.

Rao, V. V., Dwivedi, S. K., & Swarup, D. (1994). Hypoglycemic effect of *Caesalpinia bonducella* in rabbits. *Fitoterapia, LXV* 245–247.

Sharma, S. R., Dwivedi, S. K., & Swarup, D. (1997). Hypoglycemic, antihyperglycemic and hypolipidemic activities of *Caesalpinia bonducella*. *Journal of Ethnopharmacology, 58*, 39–44.

Shrikantha Murthy, K. R. (2000a). Bhavaprakasa of Bhavamisra, Vol. 1, p. 246; Vol. 2, p. 627. Varanasi, India: Krishnadas Academy.

Shrikantha Murthy, K. R. (2000b). Vagabhata's Ashtang Hridayam, *Vol. 2, p. 386*. Varanasi, India: Krishnadas Academy.

Shukla, S., Mehta, A., John, J., Mehta, P., Vyas, S. P., & Shukla, S. (2009a). Immunomodulatory activities of the ethanolic extract of *Caesalpinia bonducella* seeds. *Journal of Ethnopharmacology, 125*, 252–256.

Shukla, S., Mehta, A., John, J., Singh, S., Mehta, P., & Vyas, S. P. (2009b). Antioxidant activity and total phenolic content of ethanolic extract of *Caesalpinia bonducella* seeds. *Food and Chemical Toxicology, 47*, 1848–1851.

Simon, O. R., Singh, N., Smith, K., & Smith, J. (1987). Effect of an aqueous extract of Nichol (*Caesalpinia bonduc*) in blood glucose concentration: evidence of an antidiabetic action presented. *Journal Council for Scientific and Industrial Research* (Australia), *6*, 25–32.

Sutaria, R. N. (1969). *A textbook of systematic botany, Vol. IV*. Ahmedabad, India: Khadayata Book Depot. 177–179.

Williamson, E. M. (2002). *Major herbs of Ayurveda*. Sahibabad Ghaziabad, India: The Dabur Research Foundation & Dabur Ayurvet Ltd.

CHAPTER 29

Brazil Nuts (*Bertholletia excelsa*): Improved Selenium Status and Other Health Benefits

Christine D. Thomson
Department of Human Nutrition, University of Otago, Dunedin, New Zealand

LIST OF ABBREVIATIONS

CVD, cardiovascular disease
GPx, glutathione peroxidase
MUFA, monounsaturated fatty acid
PUFA, polyunsaturated fatty acid
SFA, saturated fatty acid

INTRODUCTION

Brazil nuts are not the most widely consumed of all nuts; however, they are unique in several respects. Brazil nut trees grow mainly in the Amazon regions of South America, where nuts are sourced almost entirely from wild trees. The collection and export of Brazil nuts is an

Nuts & Seeds in Health and Disease Prevention. DOI: 10.1016/B978-0-12-375688-6.10029-5

important industry economically for countries in this region. Brazil and Bolivia are the main exporters, followed by Cote d'Ivoire and Peru (Mori, 1992).

Brazil nuts, like other nuts, are good sources of many nutrients, including vitamins, minerals, and other bioactive compounds, which may contribute to health-promoting effects. They are unique as the highest known food source of the essential trace element, selenium.

BOTANICAL DESCRIPTION

Brazil nuts are obtained from the giant tree, *Bertholletia excelsa*. This tree belongs to the Lecythidaceae family, and is native to Brazil, the Guianas, Venezuela, eastern Colombia, eastern Peru, and eastern Bolivia, growing as scattered trees in large forests on the banks of the Amazon, Rio Negro, and Orinoco rivers (Mori, 1992).

The tree is amongst the largest in Amazon rainforests, reaching 30–45 m in height with a 1–2 m diameter trunk. The fruit is a large capsule 10–15 cm in diameter, weighing up to 2 kg, with a hard, woody shell 8–12 mm thick containing 8–24 triangular seeds 4–5 cm long. It is not a true nut in the botanical sense, but the seed of a woody capsule fruit (Wikipedia, 2009).

PROPAGATION OF THE SPECIES

Almost all Brazil nuts consumed worldwide come from wild trees rather than plantations (Mori, 1992). The tree produces fruit only in virgin forests undisturbed by human activity. Propagation is an example of an intricate ecosystem to which plants, animals, and insects contribute. Reproduction depends on the orchid *Coryanthes vasquezii*; this produces a scent that attracts the long-tongued orchid bee (*Euglossa* spp.), which pollinates the Brazil nut tree-flower (Mori, 1992; Wikipedia, 2009).

The capsule is tough but contains a small hole at one end, enabling large rodents such as the agouti to gnaw it open. They eat some nuts, and bury others in a cache for subsequent use; some forgotten seeds may germinate 12 to 18 months later (Mori, 1992; Wikipedia, 2009). The trees grow slowly, taking up to 10–30 years before producing nuts. The tree, bee, and agouti are all dependent on one another for survival.

PRESENT-DAY CULTIVATION AND USAGE

The Brazil nut is an important economic plant for Brazil, Bolivia, Peru, and Cote d'Ivoire. Most seeds are consumed in England, France, the USA, and Germany (Mori, 1992). World production of shelled Brazil nuts in 2006/2007 was 20,100 megatonnes, compared with 683,286 for almonds, 394,632 for cashews, and 28,030 for macadamia nuts (Alsalvar & Shahidi, 2009).

Brazil nuts are harvested almost entirely from wild trees during the rainy season, gathered by migrant workers — *castanheiros* (Mori, 1992). However, Brazil nut gathering alone does not provide a living; the castanheiros depend on other activities such as timber extraction and agriculture to supplement their income, both of which are damaging to the forest (Silverton, 2004). As a result, production has declined in recent years, and there is some uncertainty about the future supply of Brazil nuts (Mori, 1992).

Because a specific bee is required to pollinate the flowers, and because of the slow growth of the trees, Brazil nuts may be unsuitable and unprofitable for plantation cultivation. Attempts at cultivation have been made in parts of the Amazon (Mori, 1992), but production is low, and to date there are no examples of economically successful plantations (Mori, 1992; Silverton, 2004).

APPLICATIONS TO HEALTH PROMOTION AND DISEASE PREVENTION

Nutrient composition

Like other nuts, Brazil nuts are good sources of nutrients, including protein, fiber, minerals, vitamins, and other bioactive compounds (Alsalvar & Shahidi, 2009).

MACRONUTRIENT CONTENT

Brazil nuts are relatively high in palmitic acid (16:0), resulting in the highest saturated fat content of nuts (except coconut) (Table 29.1). Of the mono-unsaturated fatty acids (MUFAs) and polyunsaturated fatty acids (PUFAs), Brazil nut oil contains mainly oleic (18:1,n-9), linoleic (18:2,n-6) and α-linolenic acids (18:3,n-3) (Shahidi & Tan, 2009).

The protein in Brazil nuts is high in the sulfur-containing amino acids cysteine and methionine, and also rich in glutamic acid and arginine.

SELENIUM CONTENT

Brazil nuts are the richest known source of selenium, with mean concentrations reported as being between 2.35 and 43.9 µg/g (Table 29.2) (Chang et al., 1995; Pacheco & Scussel, 2007; Thomson et al., 2008). Concentrations in individual nuts vary considerably (0.03−512 µg/g) depending mainly on soil selenium concentrations. Initial reports that nuts sourced with shells had higher selenium concentrations than those without appear now to relate to the area of source (Chang et al., 1995). Brazil nuts consumed in the USA come from two regions; those sourced with the shell removed come from the western areas of Brazil bordered by Bolivia and Peru (Rondonia and Acre), while those sourced with the shell come from the eastern tributaries of the Amazon (Manaus and Belem). This was confirmed in a recent study showing that concentrations in Brazil nuts from eastern regions of the Amazon basin were higher than those from western areas (Table 29.3), corresponding to higher soil selenium in eastern (5.2−9.4 mg/kg) compared with western regions (1.0−2.4 mg/kg) (Pacheco & Scussel, 2007).

TABLE 29.1 Concentrations of Macronutrients, Fatty Acids, and Amino Acids in Brazil Nuts

Nutrient	Concentration (g/100 g Nuts)	
	Brazil Nuts	Other Nuts
Macronutrients		
Protein	14.2	4.5−20.2
Carbohydrate	12.3	0.0−40.8
Total lipid	66.6	2.6−71.0
Total SFA	15.1	0.2−29.7
Total MUFAs	24.6	0.4−58.7
Total PUFAs	20.6	0.5−47.2
Fatty acids		
16:0	9.1	0.2−6.0
18:1,n9	24.2	0.4−45.7
18:2,n6	20.5	0.40−38.1
Amino acids		
Arginine	2.2	0.2−4.7
Glutamic acid	3.2	0.2−5.2
Methionine	1.0	0.0−0.4
Cysteine	0.4	0.0−0.4
Dietary fiber	7.5	2.4−33.5

Sources: Sasthe, S.K., Monaghan, E.K., Kshirsagar, H.H., & Venkatachalam, M. (2009). Chemical composition of edible nut seeds and its implications in human health. In A. Alasalvar & F. Shahidi (eds), *Tree nuts: Composition, phytochemicals and health effects* (pp. 11−36). Boca Raton, FL: CRC Press.

TABLE 29.2 Concentrations of Vitamins, Minerals, and Other Bioactive Compounds in Brazil Nuts

Nutrient	Concentration (mg/100 g nuts)	
	Brazil nuts	**Other nuts**
α-tocopherol	5.7	0–25.9
γ-tocopherol	7.9	0–24.24
Calcium	160	1–248
Copper	1.7	0.02–2.2
Iron	2.4	0.04–5.5
Magnesium	376	0–376
Phosphorous	725	0–725
Potassium	659	356–1017
Selenium	0.08–4.39	0.001–0.02
Zinc	4.1	0.4–6.5
Squalene	1377.8	39.5–151.7
β-sitosterol	65.5	

Sources: Sasthe, S.K., Monaghan, E.K., Kshirsagar, H.H., & Venkatachalam, M. (2009). Chemical composition of edible nut seeds and its implications in human health. In A. Alasalvar & F. Shahidi (eds), *Tree nuts: Composition, phytochemicals and health effects* (pp. 11–36). Boca Raton, FL: CRC Press.

Selenium distribution in Brazil nuts depends on the absorption ability of the tree, according to factors such as soil selenium content, the chemical form of selenium in soils, presence of heavy metals, rain intensity, tree maturity, and root systems. Selenium content may also vary with the position of the nut on the tree, which may relate to variation in efficiency of the root system (Pacheco & Scussel, 2007; Shahidi & Tan, 2009).

One possible reason for the high content is that selenium is chemically similar to sulfur, an essential constituent of methionine and cysteine. Sulfur is often deficient in Amazon soils and plants may use selenium instead of sulfur, thus enhancing selenium absorption. Speciation of selenium in Brazil nuts indicates that selenomethionine is the principal species, with some selenocysteine also present (Shahidi & Tan, 2009).

OTHER MICRONUTRIENTS

Like other nuts, Brazil nuts are good sources of phenolics (mainly tocopherol) (Table 29.2), many of which have antioxidant properties, and phytosterols (β-sitosterol), which may

TABLE 29.3 Selenium Concentration (µg/g) in Brazil Nuts from Eastern and Western Amazon Regions

	West Amazon: Acre-Rondonia	*East Amazon: Menaus-Belem*
Sourced without shell[c]	3.06 ± 4.01[a] (0.03–31.7)	
Sourced in shell[c]		36.0 ± 50.0[a] (1.25–512)
Sourced without shell[d]	21.8–25.3[b] (8.5–35.1)	43.7–3.9[b] (20.7–69.7)
Sourced in shell[d]	11.9–13.5[b] (9.2–18.5)	20.5–29.2[b] (11.1–38.6)

[a]Values are mean (± SD); range in parenthesis;
[b]Values are means of two areas; range in parenthesis;
[c]Chang et al. (1995);
[d]Pacheco and Scussel (2007).

interfere with the absorption of cholesterol (Shahidi & Tan, 2009; Yang, 2009). Brazil nuts have the highest content of squalene, a precursor of steroids and an essential building block of hormones.

Brazil nuts are also excellent sources of minerals and vitamins (Table 29.2) (Shahidi & Tan, 2009; Yang, 2009). In addition to essential minerals, Brazil nuts contain relatively high levels of radium (Parekh *et al.*, 2008), because the extensive tree root system extracts radium from the soil. Barium levels are also high (~0.1–0.3%), but are variable and likely reflect the amount of barium in regional soils (Parekh *et al.*, 2008).

Source of selenium in the diet

There is increasing interest in the health benefits of adequate selenium intake. Growing evidence that such intake may protect against cancer and other chronic conditions provides a strong argument for increasing selenium consumption. Brazil nuts provide a convenient source of dietary selenium.

A recent comparison showed that selenium in Brazil nuts was equally as bioavailable as a similar amount of selenomethionine in raising plasma selenium concentrations (Figure 29.1), and the selenoenzyme glutathione peroxidase (GPx) in plasma (Figure 29.2) and whole blood (Figure 29.3) (Thomson *et al.*, 2008). A simple public health recommendation to include one or two Brazil nuts daily in the diet would avoid the need for fortification of foods or expensive supplements, and would provide a cost-effective way of improving or maintaining adequate selenium status.

Selenium, present in 25 selenoproteins in humans, has several important functions as an antioxidant (e.g., GPxs), in thyroid hormone metabolism (iodo-thyronine deiodinases), in redox reactions (thioredoxin reductases), and in other undefined functions (Reeves & Hoffmann, 2009). As such, selenium is implicated in optimal functioning of the immune and thyroid systems, and protection against oxidative stress. In turn, these functions may afford protection against some cancers, viral infection, and progression of HIV, and possible protection against CVD (Boosalis, 2008), although the evidence is sometimes conflicting and controversial.

In rats, Brazil nuts as a source of selenium were as effective as a similar amount of selenium as selenite in maintaining selenoenzyme activity and preventing mammary cancer (Ip & Lisk, 1994). Increasing evidence suggests that adequate selenium intake is also correlated with

249

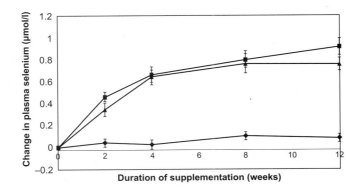

FIGURE 29.1
Changes in plasma selenium concentrations during 12 weeks' daily supplementation with two Brazil nuts (▲), 100 µg selenium as selenomethionine (■), or placebo (●). Values adjusted for age, sex, and BMI. Time-by-treatment interaction significant, $P < 0.0001$. By week 12, differences between Brazil nut and placebo groups, and selenomethionine and placebo groups, were significant, $P < 0.0001$. There was no significant difference between Brazil nut and selenomethionine groups, $P = 0.840$. Error bars represent SEs. *Reproduced from Thomson* et al. *(2008), with permission.*

FIGURE 29.2

Changes in plasma glutathione peroxidase (GPx) activities during twelve weeks' daily supplementation with two Brazil nuts (▲), 100 µg selenium as selenomethionine (■), or placebo (●). Time-by-treatment interaction significant, $P < 0.0001$. By week 12, differences between Brazil nut and placebo groups, $P < 0.0001$, and selenomethionine and placebo groups, $P < 0.014$, were significant. There was a tendency for change in plasma GPx activity to be greater in Brazil nut than selenomethionine group, $P = 0.065$. Error bars represent SEs. *Reproduced from Thomson* et al. *(2008), with permission.*

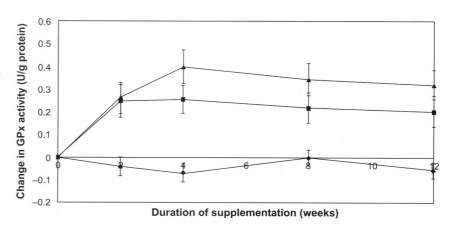

FIGURE 29.3

Changes in whole blood glutathione peroxidase (GPx) activities during twelve weeks' daily supplementation with two Brazil nuts (▲), 100 µg selenium as selenomethionine (■), or placebo (●). Time-by-treatment interaction significant, $P < 0.005$. By week 12, there was a significant difference between Brazil nut and placebo groups, $P < 0.0001$. The change was greater in the Brazil nut than selenomethionine group, $P = 0.032$. There was no significant difference between the selenomethionine and placebo groups, $P = 0.395$. Error bars represent SEs. *Reproduced from Thomson* et al. *(2008), with permission.*

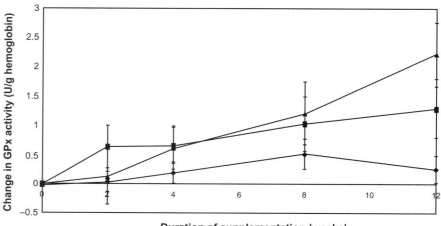

a reduced risk of some cancers in humans, which has led health commentators and nutritionists to recommend consumption of Brazil nuts as a protective measure (Thomson *et al.*, 2008; Yang, 2009). However, there is some controversy about the role of selenium in chronic disease protection; recent investigations of selenium supplementation and prostate cancer in subjects with adequate selenium status were negative (Hatfield & Gladyshev, 2009), and in fact appeared to increase the risk of diabetes (Boosalis, 2008). This, however, does not rule out a possible protective effect of enhanced selenium intake in subjects with inadequate selenium status. It is likely that there is a threshold effect of selenium status, below which there is an increased risk of chronic disease but above which there are no further benefits, and, possibly, adverse effects of higher than recommended intakes.

Prevention of chronic disease

Apart from their contribution in maintaining adequate selenium status, health benefits of Brazil nuts alone have not been studied to any extent. While the Brazil nut has its own attributes, it is tree nuts as a group that has the potential for prevention of CVD and other chronic diseases. Epidemiological and clinical studies consistently show positive effects of nuts in reducing the risk of CVD and related risk factors (Ternus *et al.*, 2009). Recent research on nuts and satiety, body weight and diabetes are also promising, and support recommendations to include a few nuts every day as part of a balanced diet (Ternus *et al.*, 2009).

There are two main mechanisms by which inclusion of Brazil nuts in the diet might contribute to chronic disease prevention. The first is as an excellent source of selenium, such that inclusion of a few nuts per day provides the recommended intake. Secondly, like other tree nuts, Brazil nuts have favorable lipid and fatty acid profiles, high protein content, and several bioactive nutrients such as oils, fiber, antioxidants, and other micronutrients (Shahidi & Tan, 2009). Nuts contribute to a diet protective of CVD through lipid-lowering effects of MUFAs, PUFAs, and fiber (Ros & Mataix, 2006). Another probable mechanism is the antioxidant effect of vitamin E, minerals, and flavenoids in protection against oxidative stress and LDL oxidation (López-Uriarte *et al.*, 2009). Therefore, Brazil nuts, when included as part of a diet high in nuts in general, may contribute in the prevention of CVD (Ternus *et al.*, 2009).

Epidemiological studies on the effects of nuts on the risk of cancer in humans are limited, and most include nuts with seeds and legumes in the analysis. World Cancer Research Fund and American Institute for Cancer Research concluded, "While there are theoretical reasons to believe that diets high in nuts and seeds might protect against some cancers, the evidence is currently lacking" (AICR & WCRF, 1997). A protective effect on colon and rectum cancer is plausible because of a high fiber, micronutrient, and antioxidant content, but evidence is inconclusive (González & Salas-Salvadó, 2006). Some studies suggest a protective effect on prostate cancer. Potential mechanisms for reducing cancer risks include antioxidant activity, regulation of cell differentiation and proliferation, reduction of tumor initiation or promotion, repair of DNA damage, regulation of immune function and inflammatory response, and the supply of fiber and MUFAs (González & Salas-Salvadó, 2006). The high selenium content of Brazil nuts may enhance many of these mechanistic pathways.

Weight control

Because of their high lipid content, nuts are often perceived as being associated with weight gain. However, current evidence indicates that nut consumption leads either to no change or to a decrease in body weight, and is therefore beneficial if consumed as part of a balanced diet (Ternus *et al.*, 2009). Brazil nuts have not been studied specifically in weight control, but they could contribute as part of a high nut diet. This effect may be due to satiating properties of nut-seed fat, fiber, or high protein content. Incomplete digestion of nuts and release of fatty acids in nuts may also contribute (Ternus *et al.*, 2009).

ADVERSE EFFECTS AND REACTIONS (ALLERGIES AND TOXICITY)

In spite of the previously discussed benefits of including Brazil nuts as part of a healthy diet, daily consumption should be limited. Because of extremely high and variable concentrations of selenium as selenomethionine, an accumulation in tissues may occur with regular high consumption. Therefore, it is recommended that daily consumption be limited to one to three nuts, if consumed on a regular basis (Thomson *et al.*, 2008).

Furthermore, as noted earlier, Brazil nuts contain unusually high and variable concentrations of barium and radium (Parekh *et al.*, 2008). Studies in rats and mice have not shown carcinogenic effects of barium from Brazil nuts, but there is no reference dose for barium in humans, and therefore uncertainty about biological effects. Although the amount of radium is small, most of which the body does not retain, levels are 1000 times higher than in other foods. Consequences of high intakes of barium and radium are unknown.

Because of these concerns and the recommended limit in intake, the benefits of Brazil nuts in reducing risk of CVD and other potential effects are likely to be limited.

SUMMARY POINTS

- Brazil nuts are an excellent source of the essential trace element selenium, and daily consumption of a few nuts will provide the recommended dietary intake.

- There is an increasing interest in the health benefits of adequate selenium intake, and growing evidence that selenium may protect against cancer and other chronic conditions provides a strong argument for increasing selenium intake.
- The favorable lipid profile and content of other bioactive compounds make Brazil nuts a desirable constituent of a high nut diet, which has been shown to be protective against CVD and possibly cancer.
- Because of possible adverse effects of high intakes of selenium, barium, and radium from Brazil nuts, intake should be limited to inclusion in the diet with other nuts, as part of the recommended 30 g of nuts per day.

References

AICR & WCRF. (1997). *Food, nutrition and the prevention of cancer: A global perspective.* Menasha, WI: BANTA Book Group.

Alsalvar, C., & Shahidi, F. (2009). Tree nuts: composition, phytochemicals and health effects. In A. Alasalvar & F. Shahidi (Eds.), *Tree nuts: Composition, phytochemicals and health effects* (pp. 1–10). Boca Raton, FL: CRC Press.

Boosalis, M. G. (2008). The role of selenium in chronic disease. *Nutrition in Clinical Practice, 23,* 152–160.

Chang, J., Gutenmann, W., Reid, C., & Lisk, D. (1995). Selenium content of Brazil nuts from two geographic locations in Brazil. *Chemosphere, 30,* 801–802.

González, C. A., & Salas-Salvadó, J. (2006). The potential of nuts in the prevention of cancer. *The British Journal of Nutrition, 96,* S87–S94.

Hatfield, D., & Gladyshev, V. N. (2009). The outcome of Selenium and Vitamin E Cancer Prevention Trial (SELECT) reveals the need for better understanding of selenium biology. *Molecular Interventions, 9,* 18–21.

Ip, C., & Lisk, D. (1994). Bioactivity of selenium from Brazil nut for cancer prevention and selenoenzyme maintenance. *Nutrition and Cancer, 21,* 203–212.

López-Uriarte, P., Bulló, M., Casa-Agustench, P., Babio, N., & Salas-Salvadó, J. (2009). Nuts and oxidation: a systematic review. *Nutrition Reviews, 67,* 497–508.

Mori, S. A. (1992). The Brazil nut industry — past, present and future. In M. J. Plotkin & L. M. Famolare (Eds.), *Sustainable harvest and marketing of rain forest products* (pp. 241–251). Washington, DC: Island Press.

Pacheco, A. M., & Scussel, V. M. (2007). Selenium and aflotoxin levels in raw Brazil nuts from the Amazon basin. *Journal of Agricultural and Food Chemistry, 55,* 11087–11092.

Parekh, P. P., Khan, A. R., Torres, M. A., & Kitto, M. E. (2008). Concentrations of selenium, barium, and radium in Brazil nuts. *Journal of Food Composition and Analysis, 21,* 332–335.

Reeves, M. A., & Hoffmann, P. R. (2009). The human selenoproteome: recent insights into functions and regulation. *Cellular and Molecular Life Sciences, 66,* 2457–2478.

Ros, E., & Mataix, J. (2006). Fatty acid composition of nuts — implications for cardiovascular health. *The British Journal of Nutrition, 96,* S29–S35.

Sasthe, S. K., Monaghan, E. K., Kshirsagar, H. H., & Venkatachalam, M. (2009). Chemical composition of edible nut seeds and its implications in human health. In A. Alasalvar & F. Shahidi (Eds.), *Tree nuts: Composition, phytochemicals and health effects* (pp. 11–36). Boca Raton, FL: CRC Press.

Shahidi, F., & Tan, Z. (2009). Bioactives and health benefits of Brazil nut. In A. Alasalvar & F. Shahidi (Eds.), *Tree nuts: Composition, phytochemicals and health effects* (pp. 143–156). Boca Raton, FL: CRC Press.

Silverton, J. (2004). Sustainability in a nut shell. *Trends in Ecology and Evolution, 19,* 276–278.

Ternus, M. E., Lapsley, K., & Geiger, C. J. (2009). Health benefits of tree nuts. In A. Alasalvar & F. Shahidi (Eds.), *Tree nuts: Composition, phytochemicals and health effects* (pp. 37–64). Boca Raton, FL: CRC Press.

Thomson, C. D., Chisholm, A., McLachlan, S. K., & Campbell, J. M. (2008). Brazil nuts: an effective way to improve selenium status. *The American Journal of Clinical Nutrition, 87,* 379–384.

Wikipedia (2009). Brazil nut. http://en.wikipedia.org/wiki/Brazil_nut (last updated 9 November 2009).

Yang, J. (2009). Brazil nuts and associated health benefits: a review. *LWT-Food Science and Technology, 42,* 1573–1580.

CHAPTER 30

Glucosinolate Phytochemicals from Broccoli (*Brassica oleracea* L. var. *botrytis* L.) Seeds and Their Potential Health Effects

Simone J. Rochfort[1], Rod Jones[2]
[1] Bundoora, Victoria, Australia
[2] Knoxfield, Victoria, Australia 3180

253

LIST OF ABBREVIATIONS

DAD, diode array detector
EGF, epidermal growth factor
GST, glutathione S-transferase
GSTM1, glutathione S-transferase mu 1
HPLC, high performance liquid chromatography
IL1β, interleukin 1 beta
Nrf2, nuclear factor (erythroid-derived 2)-like 2, also known as NFE2L2

Nuts & Seeds in Health and Disease Prevention. DOI: 10.1016/B978-0-12-375688-6.10030-1

TGF-β, transforming growth factor beta
TNF-α, tumor necrosis factor alpha

INTRODUCTION

Broccoli and its botanical relatives have been consumed for thousands of years. The leaves, flowering heads, and, more recently, the sprouted seeds are recognized as a good source of vitamins and minerals. In addition, there has been increasing focus on the brassicas as a source of glucosinolates — health-promoting phytochemicals. This chapter discusses the origins of broccoli, and explores the evidence for the health-promoting effects of glucosinolates.

BOTANICAL DESCRIPTION

Brassica oleracea L. belongs to the Brassicaceae family (previously called the Cruciferae) (Figure 30.1). Broccoli, cabbage, cauliflower, and Brussels sprouts are well known vegetables of the *Brassica* genus, which also includes mustards and the oil-seed plants canola (*B. napus* L. and *Brassica rapa* L). The word *brassica*, meaning to cut off the head, is probably derived from Latin, whilst the word broccoli, meaning bucktooth, was probably colloquial Latin applied to any projecting shoots of the cabbage family. The terms asparagus broccoli, sprouting broccoli, and Christmas Calabrian broccoli appear to be synonymous for the Bruttinm broccoli (Buck, 1956).

Broccoli flowers are borne on afaceted floral shoots so that the inflorescence terminates each axis of the plant. The general morphology of the plant is described in Figure 30.2. Each plant contains one primary and several secondary inflorescences; the primary inflorescence is commonly called a head, and consists of many functional floral buds. At the time of harvesting the primary inflorescence bears a large number of immature flower buds, known as beads. If left unharvested, each inflorescence produces flowers that will produce seed pods which, once dry, can be harvested for broccoli seed. Seed size varies between varieties, but in general the

Kingdom: Plantae – plants

Subkingdom: Tracheobionta – vascular plants

Superdivision: Spermatophyta – seed plants

Division: Magnoliophyta – flowering plants

Class: Magnoliopsida – dicotyledons

Order: Capparales

Family: Brassicaceae – mustard family

Genus: *Brassica* L. – mustard

Species *Brassica oleracea* L. – cabbage

Varieties

Brassica oleracea L. var. *acephala* – kale

Brassica oleracea L. var. *botrytis* L. – cauliflower

Brassica oleracea L. var. *capitata* – cabbage

Brassica oleracea L. var. *costata* DC. – tronchuda cabbage

Brassica oleracea L. var. *gemmifera* DC. – Brussels sprouts

Brassica oleracea L. var. *gongylodes* L. – kohlrabi

Brassica oleracea L. var. *italica* Plenck – sprouting broccoli

FIGURE 30.1

Taxonomy of broccoli. The vegetables broccoli, cabbage, cauliflower, etc., are closely related varieties of the species *Brassica oleracea. Adapted from http://plants.usda.gov and http://www.ncbi.nlm.nih.gov/Taxonomy/Browser.*

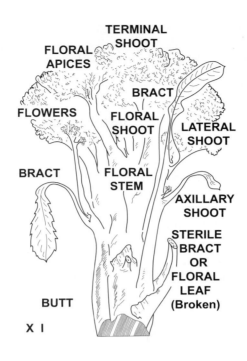

FIGURE 30.2
Morphology of the floral shoot of sprouting broccoli. Broccoli plants of today produce a floral head that can be consumed fresh, or left to go to seed for the production of glucosinolate-rich seeds. *Reproduced from* Economic Botany *(1956), with permission.*

seeds can be described as circular, approximately 1 mm in diameter, with approximately 300 seeds per gram.

HISTORICAL CULTIVATION AND USAGE

The brassicas have a long history of cultivation, being grown for their leaves, edible inflorescence, fleshy stems, roots, or oils (which are extracted from the seeds). Broccoli was developed from leafy Brassica forms, commonly known as "Calabrese broccoli," found in the northeastern Mediterranean and southern Europe. The probable progenitor of modern broccoli has been traced to coastal Europe, where it still grows wild today. The modern varieties have been developed in the past 300 years, and there are now three modern forms: typical broccoli heads; long-stemmed small-headed forms, sometimes called "Calebrese;" and the angular, fractal heads called "Romanesco" (Buck, 1956).

PRESENT-DAY CULTIVATION AND USAGE

Today, brassica plants are widely grown as a vegetable food or oil source. Spain, France, and China were the top three exporters of cauliflower and broccoli in 2006, together producing more than half a million tonnes of the vegetables (ERS, 2009). In the 1980s the potential human health benefits of glucosinolates was realized, with studies demonstrating the brassicas to be an excellent source of these phytochemicals. Since then broccoli seed has become important as a source of sprouts for direct human consumption, and as a source of glucosinolates to add to other nutritional products.

Glucosinolate content in broccoli seed

Compared to leaf and inflorescence tissues, broccoli seed has an exceptionally high content of glucosinolates and glucoraphanin in particular, with levels of glucoraphanin of up to 105 µmol/g fresh weight in some cultivars. Broccoli seeds are the best source of glucoraphanin,

TABLE 30.1 Glucosinolate Profile of Broccoli cv "Marathon" Seeds (after West *et al.*, 2004)

Glucosinolate	Mean Content (µmol/g)	% Content
Glucoraphanin	59.81	46.9
Glucoiberin	26.40	20.7
Glucoerucin	21.31	16.7
4OH-Glucobrassicin	15.10	11.8
Glucoibervirin	4.84	3.9

Broccoli seed is particularly high in glucosinolates, with glucoraphanin (precursor to sulforaphane) the most abundant.

but there is considerable variation amongst cultivars (West *et al.*, 2004). A typical broccoli seed glucosinolate profile is shown in Table 30.1.

APPLICATIONS TO HEALTH PROMOTION AND DISEASE PREVENTION

Glucosinolates are widely distributed in the genus *Brassica*, and have generated significant interest with respect to their proposed role in cancer prevention. There are approximately 120 different glucosinolates, with the general structure shown in Figure 30.3. They are defined by variation in the side chain, R, with two broad classes, the aliphatic glucosinolates and the aromatic glucosinolates.

Recent interest has focused primarily on glucoraphanin and glucobrassicin. Glucoraphanin, abundant in seed, is the precursor of sulforaphane, the isothiocyanate formed by the hydrolysis of glucoraphanin by myrosinase (Figure 30.4). The enzyme myrosinase is naturally present in the plant, and is released from plant vacuoles after mechanical stress — for example, during cutting or chewing. Although glucoraphanin is the best studied molecule, other glucosinolates also break down in the presence of myrosinase. Glucobrassicin, an indole glucosinolate, forms indole-3-carbinol after myrosinase action, which also has potent anti-cancer action.

Epidemiological studies

Epidemiological data, supported by many experimental studies with cell and animal models, and, more recently, small-scale human intervention trials, suggest that glucosinolates and their degradation products may have a wide range of beneficial effects on health. The multiple actions of sulforaphane in humans have been widely studied since 1992, when Talalay and associates discovered its action as an inducer of detoxifying enzyme systems (Zhang *et al.*, 1992). Sulforaphane stimulates one of the body's natural defence systems (phase II enzymes), which means the body is better able to detoxify and remove potentially harmful chemicals (such as carcinogens). Sulforaphane has also been shown to help prevent cancer cell growth,

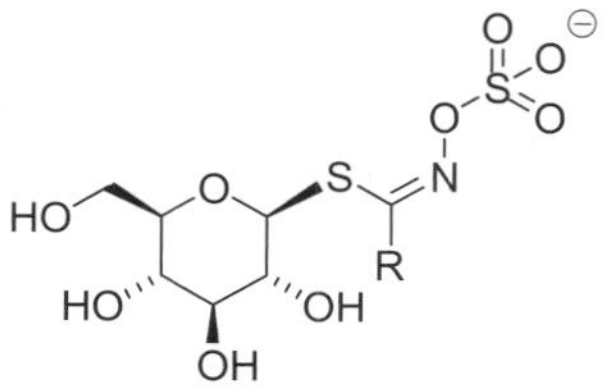

FIGURE 30.3

Chemical structure of glucosinolates. Glucosinolates consist of a sugar moiety, an isothiocyanate, and a sulfate group, as well as a variable group, R, which may be aromatic or aliphatic in nature.

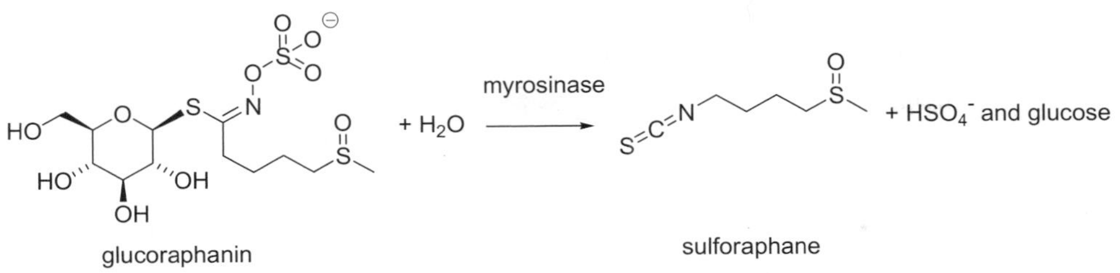

FIGURE 30.4
Conversion of a glucosinolate (glucoraphanin) to the corresponding isothiocyanate (sulforaphane). Glucosinolates are hydrolyzed by the enzyme myrosinase (also present in broccoli) to form isothiocyanates. The best studied for biological activity is sulforaphane, which is derived from glucoraphanin.

cancer cell-to-cell signaling, angiogenesis, and several other processes in the cancer cascade. Sulforaphane has been shown to reduce inflammation that can lead to heart disease (Juge et al., 2007).

There have been several epidemiological studies published that indicate people who consume more than two or more servings of brassica vegetables per week have a lower risk of disease, particularly prostate, intestinal, pancreatic, and breast cancers. A meta-review on Brassica consumption and cancer risk found 67% of all studies reported an inverse association between consumption of total Brassica vegetable intake and risk of cancer (Verhoeven et al., 1996). In the more recent EPIC-Heidelberg cohort study, which involved 11,405 men, the association between intake of glucosinolates and the risk of prostate cancer was assessed. It was found that there was significant association between glucosinolate intake and reduced risk of prostate cancer, with the aliphatic glucosinolates (including glucoraphanin) having the strongest effect (Steinbrecher et al., 2009).

Bioavailability

Bioavailability studies indicate that glucoraphanin can be absorbed intact, but when eaten within broccoli it is primarily converted to sulforaphane during chewing, which is then absorbed in the upper intestine. Intact glucoraphanin can also be converted to sulforaphane by gut microflora (Holst & Williamson, 2004). A study of rats indicated that glucoraphanin purified from broccoli seed is recovered in the urine, with no glucoraphanin or metabolites found in feces. Urinary products of glucoraphanin were studied, with both oral and intra-peritoneal delivery, with 20% and 45% of the dose recovered in urine, respectively. The study indicated that if glucoraphanin is absorbed intact, it undergoes enterohepatic circulation and is hydrolyzed in the gut (Bheemreddy & Jeffery, 2007). This absorption of glucoraphanin, rather than the biologically active derivative sulforaphane, is important, since cooking can destroy myrosinase, thereby preventing pre-absorption metabolism of glucoraphanin.

Diet—Genome interaction

Epidemiological studies also suggest that genetic polymorphisms in glutathione S-transferases (GSTs) may mediate the benefit of glucosinolate consumption. There are three GST families that are expressed in the cytosol, mitochondria, and microsome, with GSTM1 and GSTP1 having the highest activity towards sulforaphane. GSTM1 has a high level of expression in the liver. Several studies have shown a positive correlation between high broccoli consumption and reduced risk of cancer when consumed by individuals with the GSTM1 gene. Gasper and co-workers demonstrated that GSTM1-positive genotypes gain more protection from sulfor-aphane ingestion, as approximately 30% is retained in the body, whereas GSTM1-null geno-types excrete all sulforaphane within 24 hours (Gasper et al., 2005). Of most interest was the finding that when GSTM1-null genotypes consumed "super-broccoli," containing sulfor-aphane levels three times that of standard broccoli, they too retained approximately 30%, indicating that consuming high-sulforaphane broccoli, such as Booster™ broccoli recently released in Australia, may confer greater health benefits to both genotypes.

A study into the effect of cruciferous vegetables on prostate cancer investigated the effect of 12 months of consumption of a broccoli rich diet (Traka et al., 2008). In this study the authors also addressed the potential interaction between individuals with the GSTM1 genotype. The authors propose that glutathione S-transferase catalyzes the release of sulforaphane in the plasma, where it is then free to react with plasma proteins, including signaling molecules such as TGF-β, EGF, and insulin (Figure 30.5). The conjugation of sulforaphane to TGF-β1, EGF, and insulin peptides to form thioureas, and enhanced TGF-β1/Smad-mediated transcription, has been demonstrated (Traka et al., 2008). This study suggested that consuming broccoli interacts with the GSTM1 genotype, inducing changes to signaling pathways associated with inflammation and carcinogenesis in the prostate.

Interestingly, there is evidence that consumption of broccoli or sprouts may also reduce the incidence of gastric cancer through mechanisms associated with direct reduction of colonization by Helicobacter pylori. Infection by H. pylori is one causative agent of gastric ulcers, and is highly associated with gastric cancer. In a recent study, mice were placed on a high salt diet,

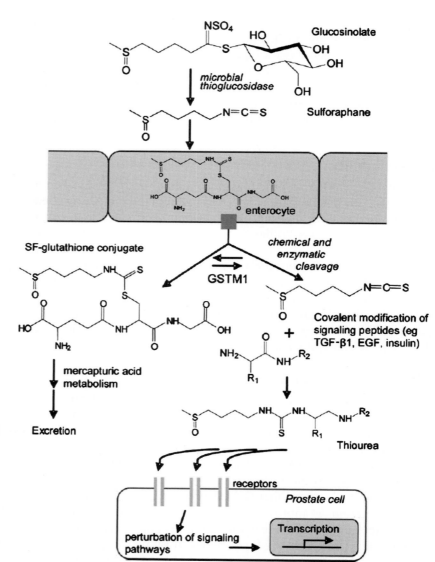

FIGURE 30.5

Metabolism of 4-methylsulfinylbutyl glucosinolate and sulforaphane. Upon entry into enterocytes, sulforaphane is rapidly conjugated to glutathione, exported into the systemic circulation, and metabolized through the mercapturic acid pathway. Within the low glutathione environment of the plasma the SF-glutathione conjugate may be cleaved, possibly mediated by GSTM1, leading to circulation of free sulforaphane in the plasma. This free sulforaphane can modify plasma proteins including signaling molecules, such as TGF-β, EGF, and insulin. *Reproduced from PLoS One under Creative Commons Attribution License (CCAL) (Traka* et al., *2008)*

infected with *H. pylori*, and subsequently fed broccoli sprouts. They showed reduced colonization and a prevention of high salt-induced gastric corpus atrophy (Yanaka *et al.*, 2009). These positive effects were absent when the Nrf2 gene was deleted. Following on from these animal studies, a human intervention study involved 48 *H. pylori*-infected patients being randomly assigned to consuming broccoli sprouts (70 g/d, containing 420 μmol of glucoraphanin) for 8 weeks or to consumption of an equal weight of alfalfa sprouts as placebo. Biomarkers of *H. pylori* infection decreased in the broccoli group, as did biomarkers for gastic inflammation. However, levels of these biomarkers had returned to normal levels 2 months after the treatment was discontinued. This indicates that, for protective effect, glucosinolates must be incorporated into the diet in an ongoing manner, and as such broccoli seed can play a key role in health.

While the mode of action of isothiocyanates is becoming clearer, additional research is required to fully elucidate the nutrigenomic interactions that lead to positive outcomes for people of certain genotypes.

ADVERSE EFFECTS AND REACTIONS (ALLERGIES AND TOXICITY)

Brassica vegetables have been safely consumed for thousands of years, and there is little evidence of adverse effects or allergy in humans. Historically, there has been an emphasis in canola breeding on decreasing the total seed glucosinolate concentration to less than 20 μmol/g, and to less than 30 μmol/g in the oil-extracted meal, as the presence of glucosinolates in the meal reduces the feed quality for livestock. In humans, no adverse effects have been identified at doses of glucosinolates close to normal daily consumption, or even with concentrated broccoli sprout extracts (Shapiro *et al.*, 2006).

The other phytochemical of health concern present in brassicas is erucic acid, which, in the 1970s, was thought to have a growth-retarding effect in animals and potentially adverse effects on the heart and liver. Broccoli seed has a high level of erucic acid (12.1 g/100 g). According to USA and Canada guidelines for erucic acid limits (canola oil), a relatively small amount of seed (about 35 g/wk) equaled calculated exposure limits (West *et al.*, 2002).

Glucoraphanin itself appears to be very well tolerated. A randomized, placebo-controlled chemoprevention trial of 200 individuals in Qidong (China) also suggests glucoraphanin is generally safe and well tolerated. In the study, 200 healthy adults drank infusions containing either 400 or < 3 μmol glucoraphanin nightly for 2 weeks with no adverse effects observed (Kensler *et al.*, 2005). Interestingly, the authors did note great inter-individual differences in bioavailability.

Further research is needed to assess both the benefits and potential adverse effects of glucosinolates on people with different genotypes and microflora, since bioavailability and response to consumption may differ considerably. Despite this, the majority of evidence suggests there is a strong health benefit in the consumption of glucoraphanin.

Methods of analysis and purification

Due to the importance of glucosinolates as molecules for human health, there has been considerable effort devoted to the development of rapid and accurate methods for quantitation for glucosinolates. Many of these methods for quantitation initially focused on HPLC with UV-Vis or DAD detection. The majority of the HPLC-based methods involved a desulfation step, since the removal of this polar moiety made the metabolites more tractable for analysis by reverse phase chromatography. More recently LC-MS has been employed, as it is not only more sensitive but also provides molecular weight and fragmentation information to confirm glucosinolate identity even in a complex mixture, and desulfation is not required. More rapid MS infusion methods have also been utilized as a screening tool to assess glucosinolate content. Both infusion and LC-MS methods are sensitive (part per trillion or low nmol/g), and have been demonstrated in a number of matrices (Rochfort *et al.*, 2008).

Obtaining pure glucosinolates is important to provide standards for analysis and to provide material for research into biological effects. Purified glucosinolates may also be incorporated in functional foods — for example, glucoraphanin has been incorporated into health-enhancing teas. Glucoraphanin is heat stable, water soluble, has few or no negative flavor attributes, and is a direct precursor to sulforaphane, and is therefore considered an ideal candidate for inclusion in functional foods. Most purification methods start with an extraction of plant material, generally seed, in hot water, which denatures myrosinase, allowing gluco-sinolates to be extracted intact. The aqueous mixture may be semi-purified using solid phase extraction (SPE) techniques (Rochfort *et al.*, 2006). This allows the removal of uncharged species, resulting in a much cleaner extract for final purification. Most methods rely on preparative HPLC, and this provides good quantities of pure metabolites, but alternate tech-niques such as counter current chromatography (CCC) have also been utilized. Both methods can quickly provide grams of purified glucosinolates.

SUMMARY POINTS

- Brassica vegetables have long been a part of the human diet, and, aside from being a good source of vitamins and minerals, they also contain phytochemicals that are beneficial for human health.
- Broccoli contains the highest level of glucoraphanin amongst commonly consumed Brassica vegetables, with broccoli seed and sprouts containing the highest levels during the life cycle.
- There is strong evidence that sulforaphane derived from glucoraphanin is protective against cancer for many individuals.
- Diet—genotype interactions still need to be better explored to determine how this benefit is mediated.
- Glucoraphanin is well tolerated in both animal and human studies at normal dietary levels, with no serious adverse reactions reported.

References

Bheemreddy, R. M., & Jeffery, E. H. (2007). The metabolic fate of purified glucoraphanin in F344 rats. *Journal of Agricultural and Food Chemistry, 55,* 2861–2866.

Buck, P. A. (1956). Origin and taxonomy of broccoli. *Economic Botany, 10,* 250.

ERS. (2009). *US Broccoli Statistics.* USDA: Economics, Statistics and Market Information System.

Gasper, A. V., Al-Janobi, A., Smith, J. A., Bacon, J. R., Fortun, P., Atherton, C., et al. (2005). Glutathione S-transferase M1 polymorphism and metabolism of sulforaphane from standard and high-glucosinolate broccoli. *The American Journal of Clinical Nutrition, 82,* 1283–1291.

Holst, B., & Williamson, G. (2004). A critical review of the bioavailability of glucosinolates and related compounds. *Natural Product Reports, 21,* 425–447.

Juge, N., Mithen, R. F., & Traka, M. (2007). Molecular basis for chemoprevention by sulforaphane: a comprehensive review. *Cellular and Molecular Life Sciences, 64,* 1105–1127.

Kensler, T. W., Chen, J. G., Egner, P. A., Fahey, J. W., Jacobson, L. P., Stephenson, K. K., et al. (2005). Effects of glucosinolate-rich broccoli sprouts on urinary levels of aflatoxin—DNA adducts and phenanthrene tetraols in a randomized clinical trial in He Zuo township, Qidong, People's Republic of China. *Cancer Epidemiology, Biomarkers & Prevention, 14,* 2605–2613.

Rochfort, S., Caridi, D., Stinton, M., Trenerry, V. C., & Jones, R. (2006). The isolation and purification of glucor-aphanin from broccoli seeds by solid phase extraction and preparative high performance liquid chromatog-raphy. *Journal of Chromatography A, 1120,* 205–210.

Rochfort, S. J., Trenerry, V. C., Imsic, M., Panozzo, J., & Jones, R. (2008). Class targeted metabolomics: ESI ion trap screening methods for glucosinolates based on MSn fragmentation. *Phytochemistry, 69,* 1671–1679.

Shapiro, T., Fahey, J., Dinkova-Kostova, A., Holtzclaw, W., Stephenson, K., Wade, K., et al. (2006). Safety, tolerance, and metabolism of broccoli sprout glucosinolates and isothiocyanates: a clinical phase I study. *Nutrition and Cancer, 55,* 53–62.

Steinbrecher, A., Nimptsch, K., Husing, A., Rohrmann, S., & Linseisen, J. (2009). Dietary glucosinolate intake and risk of prostate cancer in the EPIC-Heidelberg cohort study. *International Journal of Cancer, 125,* 2179–2186.

Traka, M., Gasper, A. V., Melchini, A., Bacon, J. R., Needs, P. W., Frost, V., et al. (2008). Broccoli consumption interacts with GSTM1 to perturb oncogenic signalling pathways in the prostate. *PLoS One. 3*, e2568.

Verhoeven, D. T., Goldbohm, R. A., van Poppel, G., Verhagen, H., & van den Brandt, P. A. (1996). Epidemiological studies on brassica vegetables and cancer risk. *Cancer Epidemiology, Biomarkers & Prevention, 5*, 733–748.

West, L., Tsui, I., Balch, B., Meyer, K., & Huth, P. J. (2002). Determination and health implication of the erucic acid content of broccoli florets, sprouts, and seeds. *Journal of Food Science, 67*, 2641–2643.

West, L. G., Meyer, K. A., Balch, B. A., Rossi, F. J., Schultz, M. R., & Haas, G. W. (2004). Glucoraphanin and 4-hydroxyglucobrassicin contents in seeds of 59 cultivars of broccoli, raab, kohlrabi, radish, cauliflower, Brussels sprouts, kale, and cabbage. *Journal of Agricultural and Food Chemistry, 52*, 916–926.

Yanaka, A., Fahey, J. W., Fukumoto, A., Nakayama, M., Inoue, S., Zhang, S., et al. (2009). Dietary sulforaphane-rich broccoli sprouts reduce colonization and attenuate gastritis in *Helicobacter pylori*-infected mice and humans. *Cancer Prevention Research, 2*, 353–360.

Zhang, Y., Talalay, P., Cho, C. G., & Posner, G. H. (1992). A major inducer of anticarcinogenic protective enzymes from broccoli: isolation and elucidation of structure. *Proceedings of the National Academy of Sciences of the United States of America, 89*, 2399–2403.

CHAPTER 31

Proteinase Inhibitors from Buckwheat (*Fagopyrum esculentum* Moench) Seeds

Yakov E. Dunaevsky[1], Natalya V. Khadeeva[2], Tsezi A. Egorov[3], Mikhail A. Belozersky[1]
[1] A.N. Belozersky Institute of Physico-Chemical Biology, Moscow State University, Moscow, Russia
[2] N.I. Vavilov Institute of General Genetics, Russian Academy of Sciences, Moscow, Russia
[3] Shemiakin and Ovchinnikov Institute of Bioorganic Chemistry, Russian Academy of Sciences, Moscow, Russia

263

INTRODUCTION

Protein proteinase inhibitors are widely distributed in plants and obtained from many sources. Although the biological role of protein proteinase inhibitors is not yet clear, it is proposed that they may exert three main functions: to serve as storage proteins, to regulate the activity of endogenous proteinases, and to protect plants against insects and pathogenic microflora (Ryan, 1981). In the present chapter we summarize data on identification and characterization of a group of protease inhibitors from buckwheat seeds. An effort was also made to find out what function these inhibitors perform in the seeds.

BOTANICAL DESCRIPTION

Buckwheat (*Fagopyrum esculentum* Moench) is a broadleaf plant native to northern Asia. Seeds are pointed, broad at the base, and triangular to nearly round in cross-section. They vary in size, according to variety, from about 4 mm at the maximum width and 6 mm long to 2 mm

Nuts & Seeds in Health and Disease Prevention. DOI: 10.1016/B978-0-12-375688-6.10031-3

wide and 4 mm long. The seed consists of an outer layer or hull, an inner layer, the seed coat proper, and within this a starchy endosperm and the germ.

HISTORY OF CULTIVATION AND USAGE

Common buckwheat was domesticated and first cultivated in southeast Asia, possibly around 6000 BC, and from there spread to Europe, Central Asia, and Tibet. Domestication most likely took place in the western Yunnan region of China. Buckwheat was one of the earliest crops introduced by Europeans to North America. Dispersal around the globe was complete by 2006, when a variety developed in Canada was widely planted in China.

PRESENT-DAY CULTIVATION AND USAGE

Common buckwheat is by far the most important buckwheat species, economically, accounting for over 90% of the world's buckwheat production. Today, Russia and China are the world's top producers of buckwheat. Japan, Poland, Canada, Brazil, South Africa, and Australia also grow significant quantities.

APPLICATIONS TO HEALTH PROMOTION AND DISEASE PREVENTION

Buckwheat is considered to have beneficial health effects on the intestinal tract, to favor cerebral circulation, to bind cholesterol, and to support vision. It is also used for weakness of the pancreas and thyroid glands, as well as nephritis. Buckwheat contains rutin, which strengthens capillary walls, reducing hemorrhaging in people with high blood pressure and increasing microcirculation in people with chronic venous insufficiency. Buckwheat substitutes for potato and bread in diabetic diets. The high nutritional and medicinal value of buckwheat is due to a high content of iron, calcium, potassium, phosphorus, iodine, fluoride, molybdenum, cobalt, vitamins B1, B2, B9 (folic acid), PP, and E, as well as easily assimilable proteins similar in physiological value to egg and milk proteins.

Among buckwheat proteins are a considerable amount of proteinase inhibitors (BWI) that play an important role in buckwheat metabolism, and in defending it from pathogenic microorganisms and insect pests. The first homogeneous preparation of proteinase inhibitor from buckwheat seeds was obtained in 1978 (Pokrovsky *et al.*, 1978). Soon afterwards, a few proteinase inhibitors from buckwheat seeds were purified and characterized by Japanese researchers (Ikeda & Kusano, 1983; Kiyohara & Iwasaki, 1985). By the end of the 1980s our laboratory had started to study proteinase inhibitors from buckwheat seeds systematically (Voskoboynikova *et al.*, 1990; Belozersky *et al.*, 1995, 2000; Dunaevsky *et al.*, 1996; Dunaevsky *et al.*, 1997; Tsybina *et al.*, 2004; Khadeeva *et al.*, 2009).

In the course of research on storage protein proteolysis in buckwheat seeds, we found an inhibitor of endogenous metalloproteinase carrying out initial degradation of the main storage protein, 13S globulin (Voskoboynikova *et al.*, 1990). The inhibitor (BWI-m) was obtained in a highly purified form and was described (Table 31.1). Besides metalloproteinase from buckwheat seeds, BWI-m inhibitor also suppressed activity of carboxypeptidase T from *Thermoactinomyces vulgaris*, and metalloproteinase from *Legionella pneumophilla*, and affected neither collagenase from *Clostridium histolyticum* nor carboxypeptidase A from calf pancreas.

On the other hand, dry buckwheat seeds contain a group of inhibitors active towards exogenous animal proteinases (trypsin and chymotrypsin). They also suppress activities of proteases secreted by filamentous fungi able to affect buckwheat seeds. These inhibitors in buckwheat seeds are separated into two subgroups — anionic and cationic inhibitors — according to their behavior on ion-exchangers. Three anionic (BWI-1a, BWI-2a, and BWI-4a) and four cationic (BWI-1c, BWI-2c, BWI-3c and BWI-4c) inhibitors were purified to

TABLE 31.1 Main Properties of Proteinase Inhibitors from Buckwheat Seeds

Inhibitor	Molecular Mass, Da	Isoelectric Point	S—S Bonds	Reactive Center	Specificity
BWI-m	12000	4.8	n.d.	n.d.	Metalloproteinases from buckwheat and *L. pneumophilla*, carboxypeptidase T
BWI-1a	7744	5.8	1	Arg-Asp	Trypsin, chymotrypsin
BWI-2a	7632	5.6	1	Arg—Asp	Trypsin, chymotrypsin
BWI-4a	7478	5.2	1	Arg—Asp	Trypsin
BWI-1c	5203	8.2	2	Arg(Lys)—X	Trypsin, chymotrypsin
BWI-2c	5347	8.3	2	Arg—X	Trypsin, chymotrypsin
BWI-3c	7760	8.5	1	Lys—X	Trypsin, chymotrypsin, subtilisin
BWI-4c	6031	8.1	1	Lys—X	Trypsin, chymotrypsin, subtilisin

Dry buckwheat seeds contain a group of inhibitors of exogenous proteinases (trypsin, chymotrypsin, subtilisin) divided into two subgroups — anionic (BWI-1a, BWI-2a, BWI-4a) and cationic (BWI-1c, BWI-2c, BWI-3c, BWI-4c) inhibitors.
n.d., not determined.

homogeneity and characterized. Molecular masses of anionic inhibitors were in the range 7.7—9.2 kDa, and of cationic inhibitors were 6.0 kDa (Table 31.1).

All inhibitors were highly pH and temperature stable. The data obtained are indicative of high pH stability of buckwheat seed protease inhibitors, and of relatively high temperature stability at acid pH. The combination of high temperature and neutral or basic pH apparently caused denaturation of inhibitors, with a loss of activity. The high stability of inhibitors is likely to be due to the rigid globular structure of their molecules (Dunaevsky *et al.* 1996; Tsybina *et al.*, 2004).

Compared with anionic inhibitors, which are rich in glutamic acid/glutamine and valine, and poor in aromatic and sulfur-containing amino acids, cationic inhibitor BWI-2c, with a high content of glutamic acid/glutamine and relatively rich in sulfur-containing (cysteine and methionine) amino acids, contained less valine, alanine, and proline. BWI-4c contained less valine, and glutamic acid/glutamine, but had more threonine, and contained tyrosine — in contrast to all other inhibitors studied.

The N-terminal sequences of the BWI-3c and BWI-4c inhibitors have a remarkable homology with those of the buckwheat anionic protease inhibitors (Belozersky *et al.*, 1995, 2000). In the N-terminal region, all of these inhibitors show a high degree of sequence homology with proteins of the potato proteinase inhibitor I family (Figure 31.1). The degree of homology with the N-terminal sequence of inhibitor AmTI from amaranth seeds is 55.5% for inhibitor BWI-3c, and 45% for inhibitor BWI-4c (Valdes-Rodriguez *et al.*, 1993). In nearly all inhibitors of the mentioned family, residues are conserved at positions 7(Lys), 11(Pro)-12(Glu)-13(Leu), and 15(Gly). Lys7 and Trp10 residues, present in all inhibitors except BWI-4c, are replaced in the BWI-4c molecule by Leu and Glu residues, respectively. The amino acid sequence of BTI-1 buckwheat trypsin inhibitor was established by Pandya and colleagues (1996) (Figure 31.1). Inhibitor BTI-2, studied by the same authors, was identical to BWI-1a.

The N-terminal sequences of BWI-1c and BWI-2c inhibitors are highly homologous to each other, and to trypsin inhibitor BWI-2b (Park *et al.*, 1997). According to their sequence data, these inhibitors cannot be assigned to any of the presently known inhibitor families of proteolytic enzymes. However, inhibitors BWI-1c and BWI-2c have 30—40% sequence homology with parts of the N-terminal region of the first exon of the storage protein vicilin, from cotton seeds *Gossypium hirsutum* (Chlan *et al.*, 1986), with the antimicrobial polypeptide from *Macadamia integrifolia* (Marcus *et al.*, 1999), and with the arginine/glutamate-rich polypeptide (6.5 k-AGRP) from the seeds of luffa *Luffa cylindrica* (Kimura *et al.*, 1997) — the functions of which are not yet understood. All of these proteins contain the same structural motif: Cys—$(Xaa)_3$—Cys—$(Xaa)_{10-12}$—Cys—$(Xaa)_3$— Cys.

(A)

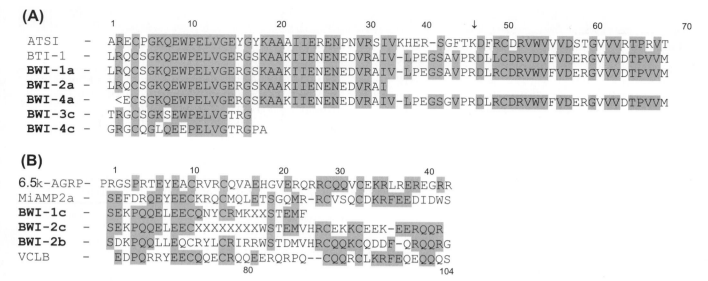

```
                1          10             20             30             40         ↓     50             60          70
ATSI    –  ARECPGKQEWPELVGEYGYKAAAIIERENPNVRSIVKHER-SGFTKDFRCDRVWVVVDSTGVVVRTPRVT
BTI-1   –  LRQCSGKQEWPELVGERGSKAAKIIENENEDVRAIV-LPEGSAVPRDLLCDRVDVFVDERGVVVDTPVVM
BWI-1a  –  LRQCSGKQEWPELVGERGSKAAKIIENENEDVRAIV-LPEGSAVPRDLRCDRVWVFVDERGVVVDTPVVM
BWI-2a  –  LRQCSGKQEWPELVGERGSKAAKIIENENEDVRAI
BWI-4a  –   <ECSGKQEWPELVGERGSKAAKIIENENEDVRAIV-LPEGSGVPRDLRCDRVWVFVDERGVVVDTPVVM
BWI-3c  –  TRGCSGKSEWPELVGTRG
BWI-4c  –  GRGCQGLQEEPELVGTRGPA
```

(B)

```
               1          10           20          30           40
6.5k-AGRP–  PRGSPRTEYEACRVRCQVAEHGVERQRRCQQVCEKRLREREGRR
MiAMP2a  –  SEFDRQEYEECKRQCMQLETSGQMR-RCVSQCDKRFEEDIDWS
BWI-1c   –  SEKPQQELEECQNYCRMKXXSTEMF
BWI-2c   –  SEKPQQELEECXXXXXXXXWSTEMVHRCEKKCEEK-EERQQR
BWI-2b   –  SDKPQQLLEQCRYLCRIRRWSTDMVHRCQQKCQDDF-QRQQRG
VCLB     –  EDPQRRYEECQQECRQQEERQRPQ--CQQRCLKRFEQEQQQS
                       80                       104
```

FIGURE 31.1

Amino acid sequences of proteinase inhibitors from buckwheat seeds. (A) This figure indicates that anionic proteinase inhibitors from buckwheat seeds as well as cationic inhibitors BWI-3c and BWI-4c belong to the potato proteinase inhibitor I family. Comparison of amino acid sequences of anionic proteinase inhibitors BWI-1a, BWI-2a, BWI-4a and cationic proteinase inhibitors BWI-3c and BWI-4c from buckwheat seeds with inhibitors of the potato proteinase inhibitor I family (Belozersky *et al.*, 1995, 2000). BTI-1, trypsin inhibitor from buckwheat seeds (Pandya *et al.*, 1996); AmTI, trypsin inhibitor from amaranth *Amaranthus hypochondriacus* seeds (Valdes-Rodriguez *et al.*, 1993). (B) BWI-1c and BWI-2c inhibitors together with BWI-2b inhibitor cannot be assigned to any of the presently known inhibitor families of proteolytic enzymes. Comparison of amino acid sequences of cationic proteinase inhibitors BWI-1c and BWI-2c with homologous proteins. BWI-2b, trypsin inhibitor from buckwheat seeds (Park *et al.*, 1997); VCLB, vicilin from cotton *Gossypium hirsutum* seeds (Chlan *et al.*, 1986); MiAMP2a, antimicrobial polypeptide from *Macadamia integrifolia* (Marcus *et al.*, 1999); 6.5k-AGRP, arginine/glutamate-rich polypeptide from the seeds of luffa *Luffa cylindrica* (Kimura *et al.*, 1997).

The reactive sites of BWI-3c and BWI-4c inhibitors contain a lysine residue, and those of inhibitors BWI-1a, BWI-2a, BWI-4a and BWI-2c an arginine residue (Table 31.1). Loss of the activity of inhibitor BWI-1c after modification of both lysine and arginine residues may be due to a modified residue that disturbs the interaction of the inhibitor with the target enzyme, though not being present in the inhibitor's reactive site.

It is noteworthy that the studied anionic and cationic inhibitors did not interact with endogenous seed proteinases, but efficiently suppressed the activity of animal proteolytic enzymes as well as of extracellular enzymes of filamentous fungi and bacterial subtilisins. The inhibition constants for trypsin, chymotrypsin, and subtilisin-72 of the inhibitors from buckwheat seeds are presented in Table 31.2. The data indicate that the investigated proteins are effective inhibitors of these enzymes. Remarkably, inhibitor BWI-3c is highly active toward all three enzymes.

TABLE 31.2 Inhibition Constants of Proteinase Inhibitors from Buckwheat Seeds

	K_i, M		
Inhibitor	**Trypsin**	**Chymotrypsin**	**Subtilisin-72**
BWI-1a	1.1×10^{-9}	6.7×10^{-7}	no inh.
BWI-1c	3.8×10^{-10}	2.0×10^{-9}	no inh.
BWI-2c	4.7×10^{-10}	4.8×10^{-8}	no inh.
BWI-3c	4.1×10^{-10}	8.4×10^{-9}	6.5×10^{-9}

Proteinase inhibitors from buckwheat seeds are effective inhibitors of exogenous proteinases — trypsin, chymotrypsin, and subtilisin-72.

Taking into consideration that isolated inhibitors suppressed the activity of extracellular proteinases of micromycetes, capable of infecting buckwheat seeds, and that there is enough indirect evidence that one of the possible functions of proteinase inhibitors is participation in the defense of plants against pathogens and pests, the direct effect of individual preparations of the studied inhibitors from buckwheat seeds on the growth and development of filamentous fungi *Alternaria alternata* and *Fusarium oxysporum* has been studied. An investigation into the effects of the inhibitors on the growth of these fungi revealed that the growth of their mycelium was suppressed around the wells with the inhibitors.

Cationic inhibitors were also able to suppress mycelium growth of fungi, but in higher concentrations than anionic inhibitors. On the other hand, the ability of cationic inhibitors to suppress the activity of bacterial proteinases may point to their possible participation in plant defense against bacterial infection.

It is therefore concluded that buckwheat seeds contain both endogenous proteinase inhibitors involved in the regulation of storage protein metabolism, and a group of exogenous protease inhibitors capable of suppressing the growth and development of pathogenic microflora. The latter group of inhibitors appear to form a part of the barrier protecting a host plant from pathogens.

Recently there has been increased interest in buckwheat proteinase inhibitors, due to the results of their application in biomedical experiments, indicating that these inhibitors specifically inhibited proliferation of human B lymphoblastoid cells (IM-9 from patients with multiple myeloma), human solid tumor cells (EC9706, HepG2, HeLa), as well as T-acute lymphoblastic leukemia cells (T-ALL) (Park & Ohta, 2004; Zhang *et al.*, 2007; Li *et al.*, 2009). In all cases, the studied buckwheat seed inhibitors induced apoptosis in tumor cells; this indicates that they may have potential for use in antitumor therapy.

Since one of the possible functions of seed proteinase inhibitors is the defense of plants against insects and pathogenic microflora, it is also attractive to use the genes of buckwheat seed proteinase inhibitors in gene-engineering work to obtain plants with increased resistance to a wide range of crop pests.

An attempt was made to use one of the trypsin inhibitors from buckwheat seeds (BWI-1a) in obtaining transgenic plants with increased resistance to pathogenic microflora (Khadeeva *et al.*, 2009). This inhibitor was present quantitatively in buckwheat seeds, and was able to inhibit the growth of hyphae and spore germination of fungi of the phytopathogenic species *Alternaria alternata* and *Fusarium oxysporum*, grown in Petri dishes or under "hanging drop" conditions (Dunaevsky *et al.*, 1997).

The structure of the target gene was deduced from the amino acid sequence of buckwheat seed anionic proteinase inhibitor BWI-1a. The 213 bp inhibitor gene (*ISP*) was directly assembled from a number of synthetic polynucleotides by PCR technique.

The *Agrobacterium tumefaciens* A281 strain with pBI 101 vector, containing the *ISP* gene under the 35S promoter of cauliflower mosaic virus, was used for transformation of aseptic tobacco cv. Samsun (*Nicotiana tabacum* L.) and potato cv. Reserv (*Solanum tuberosum* L.) plants.

Analysis of the presence of the *ISP* gene in the cells of transformed potato and tobacco lines indicated that all transformed lines contained a common fragment of 0.6×10^3 bp corresponding to the fragment of gene construction. Its expression was confirmed by Northern blot analysis.

Antibacterial activity of transgenic lines was determined using dish tests. Tissues of non-transformed control tobacco plant (C_{nt}) and that transformed by the *ISP* gene-free vector (C_O) hardly inhibited growth of any used bacterium, whereas tissues of all analyzed transgenic plants exhibited antibacterial activity (Table 31.3). Evidently, transformants synthesized a target functional protein exhibiting an inhibitory effect on bacterial growth. However,

TABLE 31.3 Antibacterial Activity of Tissues of Primary Tobacco Transformants

Line	Mean Diameter of the Zone of Bacteria Growth Inhibition Including Diameter of the Well (0.5 cm)	
	Pseudomonas syringae	*Clavibacter michiganensis*
C_{nt}	0.5*	0.5*
C_0	0.5*	0.5*
C1	2.27 ± 0.24	1.28 ± 0.08
C2	1.63 ± 0.13	1.23 ± 0.11
C7	1.82 ± 0.21	1.20 ± 0.07
C8	2.09 ± 0.18	1.25 ± 0.13
C10	1.92 ± 0.17	1.26 ± 0.11
C11	1.98 ± 0.19	1.24 ± 0.09
C12	0.5*	0.5*
C15	1.81 ± 0.22	1.32 ± 0.13
C16	1.84 ± 0.12	1.21 ± 0.07
C18	1.77 ± 0.16	1.56 ± 0.21
C19	1.78 ± 0.26	1.18 ± 0.08
C20	1.92 ± 0.15	1.43 ± 0.16
C22	2.18 ± 0.19	1.63 ± 0.11
C37	1.80 ± 0.18	1.15 ± 0.08
C41	1.76 ± 0.14	1.31 ± 0.14
C44	1.92 ± 0.23	1.52 ± 0.13
C47	2.01 ± 0.19	1.42 ± 0.09

Data presented indicate that tissues of non-transformed control tobacco plants (C_{nt}) practically did not inhibit growth of any used bacterium, whereas tissues of all transgenic plants demonstrated antibacterial activity.
Reprinted from Khadeeva *et al.* (2009), *Biochemistry (Moscow)*, 74, 260–267, with permission.

significant differences were detected in the extent of inhibition of different bacterial species. It should be noted that selection of bacterial strains *Clavibacter michiganensis* and *Pseudomonas syringae* is explained by the fact that they secrete serine proteinases during their vital activities. Transgenic clones inhibited growth of *P. syringae* and, to a lesser extent, growth of *C. michiganensis*.

Exposure of tobacco plants to glasshouse whiteflies (*Trialeurodes vaporariorum*), observed in the greenhouse, showed that transgenic plants were resistant to the effect of these insects, whereas non-transformed ones were strongly affected by them. Numerous eggs were laid by the whitefly on the control plant leaves, and full-value insects emerged from these eggs (Figure 31.2B). Only single ovipositions of one or two eggs were observed on transgenic plants, and no progeny appeared in this case (Figure 31.2A). It should be noted that, unlike a number of different insects, serine proteinases are prevalent in the whitefly alimentary canal. Thus, introduction of just a single gene of serine proteinase inhibitor into the plants of heterologous groups has shown the possibility of obtaining protective effects against insects and phyto-pathogenic bacteria.

The results indicate that, unlike control plants, those grown by us are able to synthesize certain functional proteins that exhibit a protective effect and inhibit the development of bacteria. It appears that the presence of just a single gene of serine proteinase inhibitor provides for sufficient protection against at least two phytopathogenic bacteria.

Thus, the possibility of involvement of a recombinant plant inhibitor of proteolytic enzymes in the protection of various plants against pathogenic microorganisms and insects is shown in this work. In connection with recent achievements of biotechnology in the creation of genetically modified plants characterized by increased resistance to different unfavorable

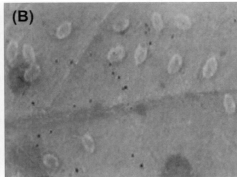

FIGURE 31.2
Effect of whiteflies on control tobacco plant (B) and the absence of egg laying on transgenic tobacco plant, containing buckwheat seed BWI-1a inhibitor gene (A), in the greenhouse. The figure shows that transgenic tobacco plants were resistant to the effect of whiteflies (*Trialeurodes vaporariorum*) observed in the green house. *Reprinted from Khadeeva* et al. *(2009),* Biochemistry (Moscow), 74, *260—267, with permission.*

effects, such an approach is becoming more and more real because it not only enables the increase in productivity of cultured plants, but also contributes to improvement of ecological conditions due to reduced usage of highly toxic protective chemicals.

ADVERSE EFFECTS AND REACTIONS (ALLERGIES AND TOXICITY)

Weak allergenic activity of buckwheat seed proteinase inhibitors has been reported (Park *et al.*, 1997).

SUMMARY POINTS

- Buckwheat seeds contain both inhibitors of endogenous seed proteinases and inhibitors of exogenous proteinases.
- Inhibitors of exogenous proteinases are capable of suppressing the growth and development of pathogenic microflora.
- Buckwheat proteinase inhibitors suppressing exogenous proteinases are subdivided into two subgroups — anionic and cationic inhibitors.
- Molecular masses of anionic inhibitors are 7.7—9.2 kDa and of cationic inhibitors are 6.0 kDa. Both anionic and cationic inhibitors are highly pH and temperature stable.
- Buckwheat inhibitors efficiently suppressed activity of animal and bacterial proteinases, as well as extracellular proteinases of filamentous fungi.
- All buckwheat seed anionic proteinase inhibitors and two of the cationic inhibitors belong to the potato proteinase inhibitor 1 family.
- The possibility of obtaining transgenic plants resistant to bacterial infections and insects by the introduction of the gene of serine proteinase inhibitor BWI-1a from buckwheat seeds has been demonstrated.

ACKNOWLEDGEMENTS

This work was supported by the Russian Foundation for Basic Research (Grants 09-04-00955, 10-04-00739) and International Scientific and Technology Center (Grant 3455).

References

Belozersky, M. A., Dunaevsky, Y. E., Musolyamov, A. Kh., & Egorov, T. A. (1995). Complete amino acid sequence of the protease inhibitor from buckwheat seeds. *FEBS Letters, 371,* 264—266.

Belozersky, M. A., Dunaevsky, Y. E., Musolyamov, A. Kh., & Egorov, T. A. (2000). Amino acid sequence of the protease inhibitor BWI-4a from buckwheat seeds. *IUBMB Life, 49,* 273—276.

Chlan, C. A., Pyle, J. B., Legocki, A. B., & Dure, L., III (1986). Developmental biochemistry of cottonseed embryogenesis and germination. XVIII. cDNA and amino acid sequences of the members of the storage protein families. *Plant Molecular Biology, 7,* 475—489.

Dunaevsky, Y. E., Pavlukova, E. B., & Belozersky, M. A. (1996). Isolation and properties of anionic protease inhibitors from buckwheat seeds. *Biochemistry and Molecular Biology International, 40,* 199—208.

Dunaevsky, Y. E., Gladysheva, I. P., Pavlukova, E. B., Beliakova, G. A., Gladyshev, O. P., Papisova, A. I., et al. (1997). The anionic protease inhibitor BWI-1 from buckwheat seeds: kinetic properties and possible biological role. *Physiologia Plantarum, 101,* 483—488.

Ikeda, K., & Kusano, T. (1983). Purification and properties of the trypsin inhibitors from buckwheat seed. *Agricultural and Biological Chemistry, 47,* 1481—1486.

Khadeeva, N. V., Kochieva, E. Z., Tcherednitchenko, M. Yu., Yakovleva, E. Yu., et al. (2009). The use of buckwheat seed protease inhibitor gene for improvement of tobacco and potato plant resistance to biotic stress. *Biochemistry (Moscow), 74,* 260—267.

Kiyohara, T., & Iwasaki, T. (1985). Purification and properties of trypsin inhibitors from buckwheat seeds. *Agricultural and Biological Chemistry, 49,* 581—588.

Kimura, M., Park, S. S., Yamasaki, N., & Funatsu, G. (1997). Primary structure of 6.5k- arginine/glutamate rich polypeptide (6.5k-AGRP) from the seeds of sponge guard *(Luffa cylindri*ca). *Bioscience. Biosci. Biotechnology, and Biochemistry, 61,* 984—988.

Li, Y. Y., Zhang, Z. H., Wang, H. W., Zhang, L., & Zhu, L. (2009). rBTI induces apoptosis in human solid tumor cell lines by loss in mitochondrial transmembrane potential and caspase activation. *Toxicology Letters, 189,* 166—175.

Marcus, J. P., Green, J. L., Goulter, K. C., & Manners, J. M. (1999). A family of antimicrobial peptides is produced by processing of a 7S globulin protein in *Macadamia integrifolia* kernels. *The Plant Journal, 19,* 699—710.

Pandya, M. J., Smith, D. A., Yarwood, A., Gilroy, J., & Richardson, M. (1996). Complete amino acid sequence of two trypsin inhibitors from buckwheat seed. *Phytochemistry, 43,* 327—331.

Park, S. S., & Ohta, H. (2004). Suppressive activity of protease inhibitors from buckwheat seeds against human T-acute lymphoblastic leukemia cell lines. *Applied Biochemistry and Biotechnology, 117,* 65—74.

Park, S. S., Abe, K., Kimura, M., Urisu, A., & Yamasaki, N. (1997). Primary structure and allergenic activity of trypsin inhibitors from the seeds of buckwheat *(Fagopyrum esculentum* Moench). *FEBS Letters, 400,* 103—107.

Pokrovsky, S. N., Labazina, N. Y., & Belozersky, M. A. (1978). Protein inhibitor of chymotrypsin from buckwheat seeds. *Russian Journal of Bioorganic Chemistry, 4,* 269—275.

Ryan, C.A. (1981). Proteinase inhibitors. In A. Marcus (Ed.), *The biochemistry of plants, Vol. 6* (pp. 351—370). New York, NY: Academic Press.

Tsybina, T., Dunaevsky, Y., Musolyamov, A., Egorov, T., Larionova, N., Popykina, N., et al. (2004). New protease inhibitors from buckwheat seeds: properties, partial amino acid sequences and possible biological role. *Biological Chemistry, 385,* 429—434.

Valdes-Rodriguez, S., Segura-Nieto, M., Chagolla-Lopez, A., Vardas-Cortina, A., Martinez-Gallardo, N., & Blanco-Labra, A. (1993). Purification, characterization and complete amino acid sequence of a trypsin inhibitor from amaranth *(Amaranthus hypochondriacus)* seeds. *Plant Physiology, 103,* 1407—1412.

Voskoboynikova, N. E., Dunaevsky, Y. E., & Belozersky, M. A. (1990). A metalloproteinase inhibitor from dry buckwheat seeds. *Biochemistry (Moscow), 55,* 839—847.

Zhang, Z. H., Li, Y., Li, C., Yuan, J., & Wang, Z. (2007). Expression of a buckwheat trypsin inhibitor gene in *Escherichia coli* and its effect on multiple myeloma IM-9 cell proliferation. *Acta Biochimica et Biophysica Sinica, 39,* 701—707.

CHAPTER 32

Seed Extract of the West African Bush Mango (*Irvingia Gabonensis*) and its Use in Health

Julius Enyong Oben
Laboratory of Nutrition and Nutritional Biochemistry, Faculty of Science, University of Yaounde I, Yaounde, Cameroon

LIST OF ABBREVIATIONS

IG, Irvingia gabonensis
IGOB131, Irvingia gabonensis seed extract

INTRODUCTION

Irvingia is a genus of the family Irvingiaceae (Harris, 1996), with two species: *Irvingia gabonensis* and *Irvingia wombulu*. Though these two species differ in the taste of the succulent fruit (sweet vs bitter) as well as in their rheological properties, both species are generally known by the common names wild mango, African mango, or bush mango. They bear edible mango-like fruits, and are especially valued for their fat- and protein-rich nuts, known as *ogbono*, *etima*, *odika*, *oro*, or *dika* nuts. In the Central African sub-region, *Irvingia gabonensis* is one of the most

Nuts & Seeds in Health and Disease Prevention. DOI: 10.1016/B978-0-12-375688-6.10032-5

important non-timber forestry products, contributing to the economy of both the rural and urban populations.

BOTANICAL DESCRIPTION

The *Irvingia gabonensis* tree grows to a height of 15—40 m. The bark of the tree is gray and very slightly scaly. The leaves are leathery, dark green, and glossy on the upper surface; they are 5—15 × 2.5—6 cm in size, elliptical to slightly obovate, and one margin is often a little more rounded than the other. They have five to ten pairs of irregular lateral veins, the lower ones running out nearly to the margin. The flowers are yellowish to greenish-white, and occur in slender, clustered racemes or small panicles on the branchlets. The fruits are either green or yellow when ripe (Figure 32.1), and are of variable sizes, depending on the variety. The skin of the fruit covers a succulent fibrous pulp which surrounds the hard nut containing the edible *Irvingia* kernel; this is presently the most important part of the plant.

HISTORICAL CULTIVATION AND USAGE

Irvingia gabonensis seeds comprise a main staple for many tribes in Nigeria and Cameroon, with the frequency of consumption in parts of Nigeria being between six and seven meals per week. Despite this very high consumption, IG has historically been obtained mainly from random harvest of the mature fruit from the wild. With the massive deforestation that has been taking place in the past 20 years in the tropical forests of the Congo Basin, there has been a drop in the availability of IG and thus the need to start cultivation and domestication of this plant. There is, however, a report by Okundaye (2009) that makes reference to the cultivation and existence of *Irvingia* plantations in Benin, Nigeria, as far back as 1816.

PRESENT-DAY CULTIVATION AND USAGE

Present day cultivation of IG has been driven by its high demand as a food component, but more recently by its demand as a nutritional supplement. The *Irvingia* kernels form an important part of the West and Central African diet, providing carbohydrate and protein (Onyeike *et al.*, 1995). Agbor (1994) reports that the kernels may be roasted to enhance their flavoring effect, and that crushed pieces of the roasted kernels may be used in frying vegetables. The ground kernel of IG is used as a soup, sauce, or stew additive, for improving the flavor and consistency, as well as for thickening (Agbor, 1994; Leakey & Newton, 1994; Vivien & Faure, 1996). At present, it is difficult to ascertain how much IG seed is produced in the Congo Basin area. The majority of the produced IG seed is destined for cooking in households and restaurants, with *Ogbono* soup featuring on the menus of most restaurants in Nigeria. The demand in Southern Nigeria alone was estimated at 80,000 tonnes per year in 1997 (Ndoye *et al.*, 1998).

FIGURE 32.1
Photographs of *Irvingia gabonensis* fruits (A and B) and dried seed (C).

Since the publication by Ngondi and colleagues (2005), awareness has been raised regarding the health benefits of IG. This has been paralleled by an increasing demand and the need for cultivation. In Cameroon, the cultivation of IG was further hastened by the fall in prices of the key cash crops coffee and cocoa; this pushed farmers of these crops to switch to the cultivation of IG, which has witnessed a steady price increase in the local markets over the past 4 years. The fact that the *Irvingia* trees usually reach maturity and begin flowering at 10–15 years of age (Moss 1995; Ladipo *et al.*, 1996) implies that there will only be a noticeable boost in IG production and availability from 2015 onwards. However, much earlier fruiting has been reported, and Ladipo *et al.* (1996) have described trees that produced fruit at 6 years of age. Improved horticultural techniques, like the use of cuttings for propagation, have also resulted in varieties of IG that fruit after much shorter periods.

APPLICATIONS TO HEALTH PROMOTION AND DISEASE PREVENTION

Different parts of the IG plant have been used in the treatment of many ailments. Both animal and human studies have shown the ability of tree bark extracts to treat myringitis (www.mamaherb.com/myringitis-home-remedy-using-irvingia-gabonensis), and to act as an anti-diarrheal, anti-ulcer agent (Raji *et al.*, 2001), and analgesic (Okolo *et al.*, 1995).

The seeds of *Irvingia gabonensis* have, however, received the most attention in terms of therapeutic applications. These applications are linked to the chemical composition of IG, which is rich in fats, proteins, and fiber. These seeds have found applications in the management of the various components of metabolic syndrome.

Obesity

273

Obesity is a multifaceted problem that is on the rise worldwide. Once thought to be limited to developed countries, obesity-related complications are a major cause of mortality in developing countries. Initial interest in IG at the University of Yaoundé 1, Cameroon, was from epidemiological evidence collected from tribes in Southern Cameroon and Nigeria, where the consumption of Irvingia soup (*Ogbono* soup) was high. The prevalence of overweight and obesity in people from these regions who ate *Ogbono* soup 7–10 times a week was found to be very low, compared to the general population (Fezeu *et al.*, 2008).

The weight-loss properties of IG were initially linked to its high content of soluble dietary fiber (Ngondi *et al.*, 2005). Compared to the placebo group, participants in their pilot study who received 3.15 g of IG per day showed decreases in weight of 2.91 and 5.6% after 2 and 4 weeks, respectively. Results from subsequent testing with other fibere-containing material, as well as analysis of IG, suggested that the earlier observation of Ngondi *et al.* (2005) could not be accounted for by fiber content alone (Ngondi *et al.*, 2006).

Ngondi and colleagues (2009), using a proprietary extract of *Irvingia gabonensis* (IGOB131) administered at 300 mg/day over a 10-week period to overweight and obese participants, observed a 12 kg decrease in weight (Figure 32.2). This weight loss was accompanied by an 11 cm decrease in waist circumference and a 4% decrease in body fat compared to the placebo group. These anthropometric changes could be a result of significant inhibition of adipogenesis in adipocytes, an effect that appears to be mediated through the down-regulated expression of adipogenic transcription factor (PPARγ) and adipocyte specific protein (leptin) (Figure 32.3), and the up-regulation of adiponectin (Figure 32.4) (Oben *et al.*, 2008). These two hormones produced by the fat cells are at the center of current mechanisms controlling fat metabolism and obesity. The decrease in leptin levels of obese people on IGOB131 parallels the decrease in body fat stores resulting from a possible reversal of leptin insensitivity and increased satiety. The

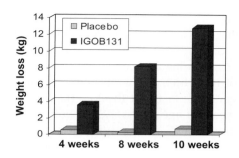

FIGURE 32.2
Percentage change in weight of overweight and obese participants after 10 weeks' treatment on 300 mg per day of IGOB131.

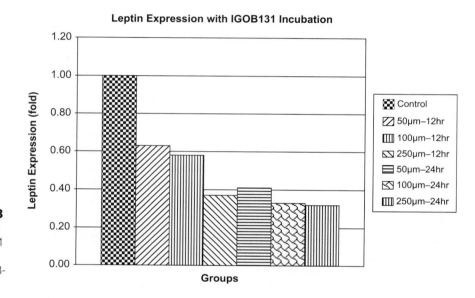

FIGURE 32.3
Effect of different doses of IGOB131 on leptin gene expression in 3T3-L1 adipocytes.

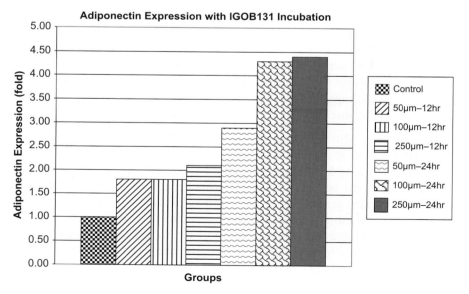

FIGURE 32.4
The effect of different concentrations of IGOB131 on adiponectin gene expression in 3T3-L1 adipocytes.

increased gene expression of the remaining fat cells could mean a more efficient production of adiponectin, which will imply the use of fat stores in metabolism, resulting in weight loss.

Diabetes and hyperglycemia

Diabetes, a condition where a patient has higher than normal blood glucose levels, is correlated with the increase in the incidence of obesity. Researchers theorize that the more fat tissue a person has, the less sensitive that person is to insulin — the hormone that controls the the uptake and use of glucose. Effective management of diabetes should therefore take into account the different aspects of diabetes: weight, and elevated blood glucose levels.

Irvingia gabonensis seeds and seed extracts have been shown to have hypoglycemic properties as well as antihyperglycemic properties (Ngondi *et al.*, 2006). Studies on the decreased α-amylase (Figure 32.5), pyruvate kinase and lactate dehydrogenase activities, provide a mechanistic hypothesis for the numerous studies supporting IG's antidiabetic potential via its ability to reduce fasting blood glucose levels (Ozolua *et al.*, 2006).

Feeding diabetics for 4 weeks on IG brought their blood glucose levels to normal (Adamson *et al.*, 1986). Ngondi *et al.* (2009) reported a 22.5% decrease in fasting blood glucose levels of overweight people who received 300 mg of IG extract (IGOB131) per day over a 10-week period. This suggests the potential use of *Irvingia gabonensis* in the management of diabetes, as well as conditions of hyperglycemia.

Dyslipidemia

Dyslipidemia is a disorder of lipoprotein metabolism, but generally refers to conditions where there is an elevation of some lipid fractions in the blood. It is closely linked to obesity, with increased triglycerides, total cholesterol, and decreased HDL cholesterol being present. The dyslipidemia associated with obesity plays a major role in the development of atherosclerosis and cardiovascular disease (Howard *et al.*, 2003). The consumption of ground IG over a 4-week period brought about a 39.21% reduction of total cholesterol, a 44.9% decrease in triglycerides, and a 45.58% reduction in LDL cholesterol. These changes were accompanied by a 46.85% increase in HDL cholesterol, resulting in a decrease in the atherogenicity index of overweight individuals (Ngondi *et al.*, 2005).

Similar findings were reported by Ngondi *et al.* (2009) for the use of the IGOB131 in overweight and obese humans over a 10-week period, in which the plasma total cholesterol levels dropped significantly (Figure 32.6). This was accompanied by a significant drop in LDL cholesterol levels.

275

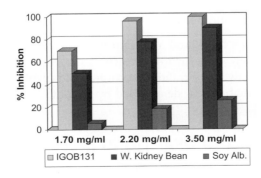

FIGURE 32.5

Comparative anti-α-amylase activity of IGOB131 compared with soy albumin and white kidney bean extract.

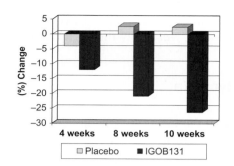

FIGURE 32.6

Changes in total plasma cholesterol levels in overweight and obese participants on IGOB131 for 10 weeks.

TABLE 32.1 Changes in Serum CRP Concentrations of Overweight and Obese Participants on 300 mg/day of IGOB131 for 10 Weeks				
	C-reactive Protein (mg/dl)			
	0	**4 Weeks**	**8 Weeks**	**10 Weeks**
Placebo	14.62 ± 4.50	14.55 ± 4.11	14.45 ± 3.20	14.45 ± 3.96
IGOB131	14.90 ± 4.12	9.11 ± 4.92	7.12 ± 4.44	7.15 ± 4.49

Inflammation

Inflammation is part of the body's reaction to injury or infection. In lower respiratory tract, Chlamydia and schistosomiasis infections, increases in the levels of inflammatory markers have been noted (Albazzaz *et al.*, 1994). A chronic sub-acute state of inflammation often accompanies the accumulation of excess lipids in adipose tissue and liver, which can be determined by increases in biochemical markers of inflammation like C-reactive protein (CRP). The consumption of *Irvingia gabonensis* extract (IGOB131) by overweight and obese patients significantly reduced the plasma levels of the inflammatory marker CRP (Ngondi *et al.*, 2009) (Table 32.1).

Oxidative stress

Oxidative stress seems to be present in many disease conditions. This generally results from an overproduction of pro-oxidants, or the body's inability to appropriately neutralize these oxidants. Oxidative stress, which is a consequence of disease, is known to aggravate the disease state. This is seen in oxidative stress in diabetes, where free radicals are formed dispropor-tionately by glucose oxidation, non-enzymatic glycation of proteins, and the subsequent oxidative degradation of glycated proteins, thereby resulting in insulin resistance and a worsening of the diabetic condition.

Recently, antioxidant therapy has found application in the management of diseases. Plant-based antioxidants are widely used due to their rich content of various bioactive components. Matsinkou (2010) has effectively used the high content of polyphenols in the pulp of IG to reduce oxidative stress in diabetes, as determined by a significant reduction of plasma malondialdehyde (MDA) levels.

ADVERSE EFFECTS AND REACTIONS (ALLERGIES AND TOXICITY)

Irvingia gabonensis is part of the staple diet, and is consumed in very large quantities, in various Nigerian and Cameroonian tribes. It is available and sold worldwide in supermarkets and restaurants as *Ogbono* and *Ogbono soup*, respectively. Unlike certain nuts and seeds, which form the most common food allergens, there are no reports in the literature of IG consumption resulting in an allergic reaction. Toxicity studies by Matsinkou (2010) failed to establish any level of toxicity of the ground IG seeds.

In clinical trials reported by Ngondi *et al.* (2009), a small number of participants reported difficulty sleeping, headaches, and intestinal flatulence after consumption of IGOB131. These conditions were, however, also reported in participants on placebo, indicating that the effect may not necessarily be due to IG.

SUMMARY POINTS

- *Irvingia gabonensis* is widely used as a food in many countries in Central Africa.
- The cultivation of IG has only occurred in recent years, as a result of deforestation and an increase in the demand for this non-timber forestry product.
- The seeds of IG are the most important and sought-after part of the plant.
- The seeds and the seed extracts of IG have many health benefits, including the management of the different components of metabolic syndrome and oxidative stress.
- The beneficial health properties of IG are a result of its rich content of various bioactive phytochemicals.

References

Adamson, I., Okafor, C., & Abu-Bakare, A. (1986). Erythrocyte membrane ATPases in diabetes: effect of dikanut (*Irvingia gabonensis*). *Enzyme, 36*(3), 212–215.

Agbor, L. O. N. (1994). *Marketing trends and potentials for* Irvingia gabonensis *products in Nigeria*. ICRAF-IITA Conference on Irvingia gabonensis, *Ibadan, Nigeria, May 1994*.

Albazzaz, M. K., Pal, C., Berman, P., & Shale, D. J. (1994). Inflammatory markers of lower respiratory tract infection in elderly people. *Age Aging, 23*, 299–302.

Fezeu, L. K., Assah, F. K., Balkau, B., Mbanya, D. S., Kengne, A. P., Awah, P., et al. (2008). Ten-year changes in central obesity and BMI in rural and urban Cameroon. *Obesity, 16*(5), 1144–1147.

Harris, D. J. (1996). A revision of the Irvingiaceae in Africa. *Bulletin du Jardin Botanique de Belgique, 65*(1–2), 143–196.

Howard, B. V., Ruotolo, G., & Robbins, D. C. (2003). Obesity and dyslipidemia. *Endocrinology and Metabolism Clinics of North America, 32*(4), 855–867.

Ladipo, D. O., Foudoun, J. M., & Ganga, N. (1996). Domestication of the bush mango (*Irvingia* spp.): some exploitable intraspecific variations in West and Central Africa. From: Domestication and commercialization of non-timber forest products in agroforestry systems. Proceedings of an international conference held in Nairobi, Kenya 19–23 February 1996. *Non Wood Forest Products, 9*, 193–205.

Leakey, R., & Newton, A. (1994). Domestication of tropical trees for timber and non-timber products. *Man and the Biosphere Digest, 17*, 67–68.

Matsinkou, R. (2010). Etude des proprietes biologiques d'Irvingia wombolu: Activites antioxydante et anti-diabetique. *PhD Thesis*. Cameroon: University of Yaounde I.

Moss, R. (1995). Underexploited tree crops: components of productive and more sustainable farming systems. *Journal for Farming Systems Research Extension, 5*(1), 107–117.

Ndoye, O., Ruiz-Perez, M., & Eyebe, A. (1998). NTFPs markets and potential forest resources degradation in Central Africa: the role of research for a balance between welfare improvement and forest conservation. Presented at the International Export Workshop on Non-wood Forest Products (NWFPs) for Central Africa. *Limbe Botanic Gargen*, 10–15, May.

Ngondi, J. L., Oben, J. E., & Minka, S. (2005). The effect of *Irvingia gabonensis* seeds on body weight and blood lipids of obese subjects in Cameroon. *Lipids in Health and Disease, 4*, 12.

Ngondi, J. L., Fossouo, Z., Djiotsa, J., & Oben, J. (2006). Glycaemic variations after administration of *Irvingia gabonensis* seed fractions in normoglycaemic rats. *African Journal of Traditional, Complementary and Alternative Medicines, 4*, 94–100.

Ngondi, J. L., Etoundi, B. C., Nyangono, C. B., Mbofung, C. M. F., & Oben, J. E. (2009). IGOB131, a novel seed extract of the West African plant *Irvingia gabonensis*, significantly reduces body weight and improves metabolic parameters in overweight humans in a randomized double-blind placebo controlled investigation. *Lipids in Health and Disease, 8*, 9.

Oben, J. E., Ngondi, J. L., & Blum, K. (2008). Inhibition of *Irvingia gabonensis* seed extract (IGOB131) on adipogenesis as mediated via down-regulation of the PPAR gamma and leptin genes and up-regulation of the adiponectin gene. *Lipids in Health and Disease, 7*, 44.

Okolo, C. O., Johnson, P. B., Abdurahman, E. M., Abdu Aguye, I., & Hussaini, I. M. (1995). Analgesic effect of *Irvingia gabonensis* stem bark extract. *Journal of Ethnopharmacology, 45*(2), 125—129.

Okundaye, P. (2009). Living slim with *Irvingia gabonensis*. Available at: http://www.articlesbase.com/health-articles/living-slim-and-healthy-with-irvingia-gabonensis-1355008.html.

Onyeike, E. N., Olungwe, T., & Uwakwe, A. A. (1995). Effect of heat-treatment and defatting on the proximate composition of some Nigerian local soup thickeners. *Food Chemistry, 53*(2), 173—175.

Ozolua, R. I., Eriyamremu, G. E., Okene, E. O., & Ochei, U. (2006). Hypoglycaemic effects of viscous preparation of *Irvingia gabonensis* (Dikanut) seeds in streptozotocin-induced diabetic Wistar rats. *Journal of Herbs, Spices & Medicinal Plants, 12*(4), 1—9.

Raji, Y., Ogunwande, I. A., Adesola, J. M., & Bolarinwa, A. F (2001). Anti-diarrhegenic and anti-ulcer properties of *Irvingia gabonensis* in rats. *Pharmaceutical Biology, 39*(5), 340—345.

Vivien, J., & Faure, J. J. (1996). Ministère Français de la Coopération. *Fruitiers Sauvages d'Afrique*. Paris: France.

CHAPTER 33

Therapeutic Use of Caper (*Capparis spinosa*) Seeds

Tzi-Bun Ng[1], Sze-Kwan Lam[1], Randy C.F. Cheung[1], Jack H. Wong[1], He-Xiang Wang[2], Patrick H.K. Ngai[3], Xiujuan Ye[1], Yau-Sang Chan[1], Evandro F. Fang[1]
[1] School of Biomedical Sciences, Faculty of Medicine, The Chinese University of Hong Kong, Hong Kong, China
[2] State Key Laboratory of Agrobiotechnology, Department of Microbiology, China Agricultural University, Beijing, China
[3] School of Life Sciences, Faculty of Science, The Chinese University of Hong Kong, Hong Kong, China

INTRODUCTION

Capers (*Capparis spinosa*) are frequently used in Mediterranean cuisine, meat dishes, pizzas, salads, and sauces. Caper buds and fruits, when salted and pickled, serve as a seasoning or garnish. Caper leaves are used in fish dishes, in salads, and as a replacement for rennet in cheese-making. When the buds are picked, an intense flavor attributed to mustard oil (glucocapparin) is emitted. This reaction also results in the appearance of rutin, which is manifested as white spots on the bud surface.

BOTANICAL DESCRIPTION

The caper (*Capparis spinosa*) is a perennial deciduous thorny bush indigenous to the Mediterranean region. It develops numerous branches, and bears round, glossy alternate leaves, and sweet-smelling white or pink flowers with numerous purple stamens and a long stigma. It is found on walls or in rocky coasts in the vicinity of other *Capparis* species such as *C. aegyptia*, *C. orientalis*, and *C. sicula*. Capers are classified and marketed in accordance with their size, the smallest being the most favored. The various sizes include non-pareil (up to 7 mm), surfines (7–8 mm), capucines (8–9 mm), capotes (9–11 mm), fines (11–13 mm), and grusas (14+ mm).

Nuts & Seeds in Health and Disease Prevention. DOI: 10.1016/B978-0-12-375688-6.10033-7

HISTORICAL CULTIVATION AND USAGE

It is believed that capers are indigenous to the Mediterranean basin, but their probable origin is the dry areas in western and central Asia. The ancient Greeks applied capers as a carminative, included them as an ingredient in cooking, and used the roots and leaves for medicinal purposes. Caper seeds and dried roots have been found in tombs in China, Egypt, India, Iran, and Turkey. Ancient Greeks and Romans used capers for reducing flatulence, to aid slimming, and as an aphrodisiac.

PRESENT-DAY CULTIVATION AND USAGE

Capers are found growing wild in the Mediterranean region, and are cultivated in Algeria, Morocco, France, Italy, and Spain. Seedlings develop from fresh seeds in 3—4 weeks. However, seeds which have been left in storage for quite some time remain dormant and need to undergo cold stratification before germination. Cuttings taken from shoots in the autumn may be employed for asexual reproduction. Watering is needed periodically in the summer, and much less often during the winter.

In response to an abrupt rise in humidity, capers produce wart-like pock marks on the leaf surfaces. Capers start to flower between the second and third years, and live for two to three decades. Pruning results in a good yield, since flower buds only appear on 1-year-old branches. Regular harvesting is conducted throughout the growing season — for example, every 8—12 days.

APPLICATIONS TO HEALTH PROMOTION AND DISEASE PREVENTION

In Greek ethnic medicine, a herbal tea composed of caper roots and young shoots has allegedly antirheumatic activity. Sprouts, leaves, roots, and seeds are used for treating inflammation. In ancient Xinjiang, in China, root bark of caper was consumed for the treatment of asthma, cough, dermatologic disease, paralysis, scrofula, splenic disease, and toothache. In Turpan, almost ripe caper fruits are collected by indigenous (Uigur) pharmacologists for treating rheumatism. Traditionally, *Capparis spinosa* has been used for the treatment of flatulence, rheumatism, liver function, arteriosclerosis, and kidney infection, and as an antihelminthic and tonic. Its root bark extract has been employed for anemia, arthritis, edema, and gout. In Israel, capers have also been used for treating hypoglycemia. Rutin is a non-toxic antioxidant bioflavonoid formed in capers, and utilized as a dietary supplement for capillary fragility. Capers have an extremely high abundance of the powerful antioxidant, quercetin. Caper seeds have a high content of protein, oil, and fiber, and have a potential value as food (Akgul & Ozcan, 1999). They can be used as a condiment due to their peppery flavor. In Egypt, the seeds are added to wine to prevent deterioration. Caper seeds also have a medicinal value because of the presence of ferulic acid and sinapic acid. Seeds that have been boiled in vinegar may alleviate toothache.

The caper plant exerts diuretic and antihypertensive actions. Flavonoids (kaempferol and quercetin derivatives) and hydrocinammic acids with antioxidant and anti-inflammatory actions are present in the plant extract. The methanolic extract of flowering buds is able to counteract the adverse effects induced by the proinflammatory cytokine IL-1β on human chondrocyte cultures, as judged by the generation of molecules released during chronic inflammatory events (nitric oxide, glycosaminoglycans, prostaglandins, and reactive oxygen species). This chondroprotective potency outstrips that of indomethacin, employed for the treatment of joint diseases. Hence, it might be utilized in the management of inflammatory cartilage damage (Panico *et al.*, 2005).

Liv-52® (Himalayan Co. India) is composed of ferric oxide and *C. spinosa* as one of its seven herbal ingredients, and has been claimed to be potentially efficacious for treating cirrhosis.

Although *C. spinosa* is abundant in the antioxidant flavonoid, rutin, *C. spinosa* may have the following applications in humans: anti-inflammatory, antimicrobial, antioxidant, hepato-protective, and photoprotective.

Linoleic acid, oleic acid and its isomer, γ-tocopherol, δ-tocopherol, α-tocopherol, sitosterol, campesterol, stigmasterol, delta-5-avenasterol, and glucosinolates of *C. ovata* and *C. spinosa* have been found in caper seed oils (Matthäus & Ozcan, 2005).

A chromatographic procedure involving the use of DEAE-cellulose, SP-Sepharose, CIM-Q, and Superdex 75 was used to isolate two proteins with therapeutic effects from caper seeds. Anion exchange chromatography of the caper seed extract on DEAE-cellulose yielded a very large unadsorbed fraction (D1) and two adsorbed fractions (D2 eluted with 0.1 M NaCl, and D3 eluted with 1 M NaCl) of similar size. Fraction D3, with antiproliferative and hemagglutinating activities, was resolved on SP-Sepharose into a large unadsorbed fraction (P1) and a smaller adsorbed fraction (P2). Fraction P2 with hemagglutinating activity was resolved on CIM-Q into a small unadsorbed fraction (Q1) and several large adsorbed fractions (Q2–Q5). Hemag-glutinating activity was restricted to adsorbed fraction Q4 eluted near the second half of the 0–0.3 M NaCl gradient. This fraction was then purified on Superdex 75 to form two fractions, S1 and S2. Fraction (S1), with hemagglutinating activity, was eluted from Superdex 75 as a single 62-kDa peak. This lectin appeared as a single 31-kDa band in SDS-PAGE. The subunit molecular mass (31 kDa) and dimeric nature of *C. spinosa* lectin resembled those of many plant lectins (Ye *et al.*, 2001). The N-terminal sequence of the lectin closely resembled a partial sequence of ribosomal subunit interface protein from *Roseobacter* sp., but was different from reported lectin sequences. The hemagglutinating activity of *C. spinosa* lectin can be inhibited by a multiplicity of simple sugars; by D(+)galactose, α-lactose, raffinose, and rhamnose at 1 mM concentration, by 25-mM L(+)arabinose, and by 100 mM D(+)glucosamine. For comparison, French bean lectin is inhibited by simple sugars (Leung *et al.*, 2008), Emperor banana lectin by mannose and glucose (Wong & Ng, 2006), *Canavalia gladiata* lectin by mannose, glucose, and rhamnose (Wong & Ng, 2005), and *Pleurotus citrinopileatus* lectin by o/p-nitrophenyl-β-D-glucuronide, o/p-nitrophenyl-β-D-galactopyranoside and maltose, and a polysaccharide, inulin (Li *et al.*, 2008). *C. spinosa* lectin was characterized by pH stability from pH 1–12 and moderate thermostability up to 40°C, similar to most of the lectins (Leung *et al.*, 2008). It inhibited mycelial growth in *V. mali* with an IC_{50} of 18 μM, but there was no effect on *M. arachidicola*, *F. oxysporum*, *H. maydis*, and *R. solani*. To date, only a few lectins have been reported with antifungal activity — for instance, red kidney bean lectin inhibits *F. oxysporum*, *Coprinus comatus, and R. solani* (Ye *et al.*, 2001). It requires a higher concentration of the lectin (1 μM) than ConA (150 nM) to elicit maximal mitogenic response. The maximal response evoked by the lectin was 75% of the maximal response to ConA. It displayed potent antiproliferative activity against hepatoma and breast cancer cells, in line with similar observations for other lectins (Wong & Ng, 2006). It could induce apoptosis in both HepG2 and MCF-7 tumor cells, with an IC_{50} of approximately 2 μM. Only galectin-9 (Yamauchi *et al.*, 2006) and mistletoe lectin (Pae *et al.*, 2000) are known to induce apoptosis in MCF-7 cells, while a large number of reports show that lectins can induce apoptosis in HepG2 cells, e.g., *Pouteria torta* lectin (Boleti *et al.*, 2008). *C. spinosa* lectin potently inhibits HIV-1 reverse transcriptase with an IC_{50} of 0.28 μM. An IC_{50} of 1–35 μM is demonstrated by some other lectins.

The unique features of *C. spinosa* lectin isolated in this study comprise: (1) a novel N-terminal sequence; (2) pH stability of hemagglutinating activity; (3) inhibition of hemagglutinating activity by a multiplicity of six sugars; and (4) a greater diversity of biological activities than other lectins, comprising mitogenic activity, antifungal activity, highly potent HIV-1 reverse transcriptase activity, and antiproliferative activity due to induction of apoptosis (Lam *et al.*, 2009).

Fraction P1 with antiproliferative activity was separated by gel filtration on Superdex 75 to yield four fractions. Antiproliferative activity was confined to the second fraction, S2. This fraction exhibited a molecular mass of 38 kDa in SDS-PAGE and gel filtration on the Superdex 75. The N-terminal sequence of the protein resembled a partial sequence of imidazoleglycerol phosphate synthase. The protein exhibited antiproliferative activity toward HepG2, HT29, and MCF-7 tumor cells, with IC_{50} values of 1, 40, and 60 μM, respectively. Its HIV-1 reverse transcriptase inhibitory activity ($IC_{50} = 0.23$ μM) was much more potent than many other anti-HIV-1 natural products (Lin *et al.*, 2007). The mechanism of inhibition is likely protein–protein interaction, similar to the inhibition of HIV-1 reverse transcriptase by the homologous retroviral protease. The protein exhibited specific antifungal activity toward *Valsa mali*, but there was no activity toward *M. arachidicola*, *F. oxysporum*, *H. maydis*, and *R. solani*. Thus, it resembles asparagus DNase and shallot antifungal protein, which inhibit only one fungal species out of the several tested. It was devoid of hemagglutinating, mitogenic, ribonuclease, or protease inhibitory activity.

The antiproliferative potency of the isolated protein toward HepG2 cells ($IC_{50} = 1$ μM) was relatively high, but its potency toward MCF-7 cells ($IC_{50} = 60$ μM) was much lower compared with other plant proteins. In comparison, the IC_{50} values of Bowman-Birk trypsin inhibitor from Hokkaido large black soybean (Ho & Ng, 2008), French bean hemagglutinin (Leung *et al.*, 2008), and lipid transfer protein from *Brassica campestris* (Lin *et al.*, 2007) toward hepatoma HepG2 cells and breast cancer MCF-7 cells were, respectively, 140 and 35 μM, 13 and 6.6 μM, and 5.8 and 35 μM. Differences in the antiproliferative potency of a given protein toward different tumor cell lines have been reported; for example, choriocarcinoma cells are much more sensitive than hepatoma cells to the ribosome inactivating proteins trichosanthin and α-momorcharin (Tsao *et al.*, 1990).

Some lectins (Wong & Ng, 2006), ribonucleases (Lam & Ng, 2001), protease inhibitors (Ng *et al.*, 2003), and antifungal proteins (Wang & Ng, 2006) manifest antiproliferative activity toward tumor cells. Yet the protein with antiproliferative activity from *C. spinosa* seeds lacks hemagglutinating, mitogenic, and ribonuclease and protease inhibitory activities (Lam & Ng, 2009).

A comparison of the biochemical characteristics and biological activities of antifungal protein and lectin from caper seeds is presented in Table 33.1.

ADVERSE EFFECTS AND REACTIONS (ALLERGIES AND TOXICITY)

Pantescal™ (Bionap, Italy), present in food supplements, is a nutraceutical derived from a conglomerate of plants, including *Capparis spinosa*, *Olea europaea*, *Panax ginseng*, and *Ribes nigrum*. In a randomized, double-blind, placebo-controlled study, 60 patients with an allergy to common aeroallergens were recruited and assigned to two groups; a Pantescal-treated group, and a placebo group. Two *in vitro* tests were conducted on blood samples collected from patients prior to treatment, and also at 2 hours, 2 days, 3 days, and 10 days following treatment. Tests for allergic biomarkers, including: (i) the cellular antigen stimulation test (CAST) to assess sulphidoleukotriene (SLT) production; and (ii) the flow-cytometric antigen stimulation test (FAST) to ascertain expression of basophil degranulation marker (CD63) after 2 and 3 days, revealed a much more prominent inhibition of CD63 expression than suppression of SLT production. The results indicate that cell membrane stabilization constitutes the major mechanism of the prophylactic actions of Pantescal (Caruso *et al.*, 2008).

SUMMARY POINTS

- Caper seeds exhibit various activities which can be useful for therapeutic purposes – for example, for treatment of inflammation.
- Caper seeds produce two proteins; one is a lectin with hemagglutinating activity, whereas the other is a non-lectin protein without hemagglutinating activity.

TABLE 33.1 Characteristics of Proteins with Medicinal Applications in Caper Seeds (Lam & Ng, 2009; Lam *et al.*, 2009)

	Antifungal Protein Without Hemagglutinating Activity	Lectin with Antifungal Activity
Molecular mass	38 kDa (monomeric)	62 kDa (dimeric)
N-terminal sequence	SYDTQAEAAL	ATETYSGFDA
Chromatographic behavior on:		
i) DEAE-cellulose	Adsorbed	Adsorbed
ii) SP-Sepharose	Unadsorbed	Adsorbed
iii) CIM-Q	Not used	Adsorbed
iv) Superdex75		
Antifungal activity against:		
i) *Valsa mali*	Present	$IC_{50} = 0.28\,\mu M$
ii) *Mycosphaerella arachidicola*	No inhibition	No inhibition
iii) *Fusarium oxysporum*	No inhibition	No inhibition
iv) *Helminthosporium maydis*	No inhibition	No inhibition
v) *Rhizoctonia solani*	No inhibition	No inhibition
Antiproliferative activity against:		
i) HepG2 cells	$IC_{50} = 1\,\mu M$	$IC_{50} = 2\,\mu M$
ii) MCF-7 cells	$IC_{50} = 60\,\mu M$	$IC_{50} = 2\,\mu M$
HIV-1 reverse transcriptase inhibitory activity	$IC_{50} = 0.23\,\mu M$	$IC_{50} = 0.28\,\mu M$
Mitogenic activity	Absent	Present
Hemagglutinating activity	Absent	Inhibited by multiplicity of sugars, pH stability from pH 1—12, thermostability up to 40°C

- The non-lectin protein displays antiproliferative activity toward a variety of cancer cells, including hepatoma HepG2 cells, colon cancer HT29 cells, and breast cancer MCF-7 cells, with IC_{50} values of approximately 1, 40, and 60 μM, respectively. The lectin inhibits MCF-7 cells and HepG cells with an IC_{50} of about 2 μM.
- Both proteins exert antifungal activity toward *Valsa mali*.
- The lectin, but not the non-lectin protein, exerts mitogenic activity on mouse spleen cells.

References

Akgul, A., & Ozcan, M. (1999). Some compositional characteristics of capers (*Capparis* spp.) seed and oil. *Grasas Aceites, 50*, 49—52.

Boleti, A. P., Ventura, C. A., Justo, G. Z., Silva, R. A., de Sousa, A. C., Ferreira, C. V., et al. (2008). Pouterin, a novel potential cytotoxic lectin-like protein with apoptosis-inducing activity in tumorigenic mammalian cells. *Toxicon, 51*, 1321—1330.

Caruso, M., Frasca, G., Di Giuseppe, P. L., Pennisi, A., Tringali, G., & Bonina, F. P. (2008). Effects of a new nutraceutical ingredient on allergen-induced sulphidoleukotrienes production and CD63 expression in allergic subjects. *International Immunopharmacology, 8*, 1781—1786.

Ho, V. S., & Ng, T. B. (2008). A Bowman-Birk trypsin inhibitor with antiproliferative activity from *Hokkaido* large black soybeans. *Journal of Peptide Science, 14*, 278—282.

Lam, S. K., & Ng, T. B. (2001). Isolation of a novel thermolabile heterodimeric ribonuclease with antifungal and antiproliferative activities from roots of the sanchi ginseng. *Panax notoginseng, Biochemical and Biophysical Research Communication, 285*, 419—423.

Lam, S. K., & Ng, T. B. (2009). A protein with antiproliferative, antifungal and HIV-1 reverse transcriptase inhibitory activities from caper (*Capparis spinosa*) seeds. *Phytomedicine, 16*, 444—450.

Lam, S. K., Han, Q. F., & Ng, T. B. (2009). Isolation and characterization of a lectin with potentially exploitable activities from caper (*Capparis spinosa*) seeds. *Bioscience Reports, 29*, 293—299.

Leung, E. H., Wong, J. H., & Ng, T. B. (2008). Concurrent purification of two defense proteins from French bean seeds, a defensin-like antifungal peptide and a hemagglutinin. *Journal of Peptide Science, 14*, 349—353.

Li, Y. R., Liu, Q. H., Wang, H. X., & Ng, T. B. (2008). A novel lectin with potent antitumor, mitogenic and HIV-1 reverse transcriptase inhibitory activities from the edible mushroom. *Pleurotus citrinopileatus. Biochimica Biophysica Acta, 1780,* 51–57.

Lin, P., Xia, X., Wong, J. H., Ng, T. B., Ye, X. Y., Wang, S., & Shi, X. (2007). Lipid transfer proteins from *Brassica campestris* and mung bean surpass mung bean chitinase in exploitability. *Journal of Peptide Science, 18,* 642–648.

Matthäus, B., & Ozcan, M. J. (2005). Glucosinolates and fatty acid, sterol, and tocopherol composition of seed oils from *Capparis spinosa* var. spinosa and *Capparis ovata* Desf. var. canescens (Coss.) Heywood. *Agricultural and Food Chemistry, 53,* 7136–7141.

Ng, T. B., Lam, S. K., & Fong, W. P. (2003). A homodimeric sporamin-type trypsin inhibitor with antiproliferative, HIV reverse transcriptase-inhibitory and antifungal activities from wampee (*Clausena lansium*) seeds. *Biological Chemistry, 384,* 289–293.

Pae, H. O., Seo, W. G., Oh, G. S., Shin, M. K., Lee, H. S., Lee, H. S., et al. (2000). Potentiation of tumor necrosis factor-alpha-induced apoptosis by mistletoe lectin. *Immunopharmacology and Immunotoxicology, 22,* 697–709.

Panico, A. M., Cardile, V., Garufi, F., Puglia, C., Bonina, F., & Ronsisvalle, G. (2005). Protective effect of *Capparis spinosa* on chondrocytes. *Life Sciences, 77,* 2479–2488.

Tsao, S. W., Ng, T. B., & Yeung, H. W. (1990). Toxicities of trichosanthin and alpha-momorcharin, abortifacient proteins from Chinese medicinal plants, on cultured tumor cell lines. *Toxicon, 28,* 1183–1192.

Wang, H. X., & Ng, T. B. (2006). An antifungal peptide from baby lima bean. *Applied Microbiology and Biotechnology, 73,* 576–581.

Wong, J. H., & Ng, T. B. (2005). Isolation and characterization of a glucose/mannose/rhamnose-specific lectin from the knife bean *Canavalia gladiata. Archives of Biochemistry and Biophysics, 439,* 91–98.

Wong, J. H., & Ng, T. B. (2006). Isolation and characterization of a glucose/mannose-specific lectin with stimulatory effect on nitric oxide production by macrophages from the emperor banana. *International Journal of Biochemistry & Cell Biology, 38,* 234–243.

Yamauchi, A., Kontani, K., Kihara, M., Nishi, N., Yokomise, H., & Hirashima, M. (2006). Galectin-9, a novel prognostic factor with antimetastatic potential in breast cancer. *Journal of Breast, 12,* S196–200.

Ye, X. Y., Ng, T. B., Tsang, P. W., & Wang, J. (2001). Isolation of a homodimeric lectin with antifungal and antiviral activities from red kidney bean (*Phaseolus vulgaris*) seeds. *Journal of Protein Chemistry, 20,* 367–375.

CHAPTER 34

Cardamom (*Elettaria cardamomum* Linn. Maton) Seeds in Health

Singaravel Sengottuvelu
Department of Pharmacology, Nandha College of Pharmacy & Research Institute, Tamil Nadu, India

INTRODUCTION

Historically known as the "Queen of Spices," cardamom has been used in India since ancient times. As long ago as the Vedic period (3000 BC), there was a mention of the use of this spice in wedding ceremonies. It has been implemented as a digestive aid and as a fat reducer, according to Ayurvedic (traditional Indian medicine) texts which date back to 1400 BC. Ancient Egyptians used cardamom as a mouth freshener. The botanical name of the plant is *Elettaria cardamomum*, and it belongs to the family *Scitaminaceae* (*Zingiberceae*). The plant shows much diversity in the size and percentage of the different chemical constituents of the volatile oil in the seeds, and in the geographical area of cultivation. The ethno-botanical uses of *Elettaria cardamomum* are shown in Table 34.1. As a spice, cardamom is used widely in Indian cuisine. The edible part of the plant is the fruit. The seeds and the essential oil are used as flavoring components in a variety of foods, including beverages, desserts, candy, baked goods, condiments, gravies, meat, and meat products. The extract of the seeds also finds applications in many pharmaceuticals and neutraceuticals, as well as in cosmetic formulations.

Nuts & Seeds in Health and Disease Prevention. DOI: 10.1016/B978-0-12-375688-6.10034-9

TABLE 34.1 Ethnobotanical Uses of *Elettaria cardamomum*		
Plant	**Ethnobotanical Uses**	**Reference**
Elettaria cardamomum	As an abortifacient, laxative, diuretic, stomachic, and carminative In asthma, bronchitis, hemorrhoids, strangury, scabies, pruritis; diseases of the bladder, kidney, rectum and throat; inflammation, headache, earache, snake bite and scorpion sting	Kiritikar & Basu, 1999
	In indigestion, colds, bronchitis, and asthma As an antiflatulent and to stimulate the appetite	Miriam & Christopher, 1992

World production of cardamom is estimated at 30,000 MT. Presently, the major producer is Guatemala, recording an average annual production of 18,000—20,000 MT. India is the second largest producer, with an average production of 11,000—12,000 MT. Indian cardamom is considered to be of superior quality, but in the international markets India has always been out-priced by Guatemala, which has negligible domestic consumption, leading to low pricing. Despite its numerous applications in the culinary methods of Sri Lanka, India, and Iran, 60% of the world production is exported to Arab (South-west Asia, North Africa) countries, where the majority is used to prepare coffee. Cardamom-flavored coffee is symbolic of Arab hospitality.

BOTANICAL DESCRIPTION

Cardamom consists of seeds of the dried fruits of *Elettaria cardamomum* (Linn.) Maton and its varieties (Fam. Zingiberaceae), a stout large perennial herb that grows naturally in the moist forests of Western Ghats up to a height of 1500 m. It is also cultivated in many other parts of south India, at elevations of 750—1500 m. Fruits are 1—2 cm long, ovoid or oblongate, and more or less three-sided with rounded angles, and greenish to pale-buff or yellowish in color. Their base is rounded, or has the remains of a pedicle, and the apex is shortly beaked; the surface of the fruit is almost smooth or has slight longitudinal striations. The fruits are small and trilocular, each containing about 15—20 seeds in a double row, adhering together to form a compact mass (Figure 34.1). The cardamom seed is dark brown to black, about 4 mm long

FIGURE 34.1
Elettaria cardamonmum Linn.

and 3 mm broad, irregularly angular, transversely wrinkled but not pitted, with a longitudinal channel containing raphe, and enclosed in a colorless and membranous aril. It has a characteristic odor and a strongly aromatic taste.

HISTORICAL CULTIVATION AND USAGE

Cardamomum is said to be the Greek name for an Indian spice, *kardamon*. In ancient Greek and Roman times the meaning of the word "cardamom" seems never to have been consistent, as merchants (and thus everyone else at that time) used it for one spice or another, or a group of spices that could vary in content from one market to another. Over time, in Europe the name has settled on *Elettaria cardamomum*. However, there must be sympathy for modern historians who have to try to unravel its early historical details from highly confusing records. Facts, such as its progress into the Mediterranean and Europe, are uncertain, and the following information is offered on that understanding.

Before the early centuries AD, cardamom was a common spice throughout India, used particularly in hotly-spiced sauces. It was no longer limited to its native habitat in the south of the country and neighboring areas further south still, but had already penetrated beyond India's northern borders. Ancient Egyptians chewed the seeds to whiten teeth, and cardamom was being imported by the ancient Greeks by the 4th century BC; Pedanius Dioscorides, the 1st century Greek physician, refers to it in his *De materia medica*. It was also being imported by the Romans by the 1st century AD. Both the Greeks and the Romans filled shells with a mixture of cardamom and wax to use as a perfume worn in their hair or on their clothing. This was a fashion that persisted for several centuries, not least among some of the Christian clergy, who were berated for it in the 4th century by the Italian scholar and Latin Church father, St Jerome (*c.* 342–420). In addition, the use of cardamom for dental hygiene was not confined to the Egyptians, as the Europeans also found the taste of garlic (*Allium sativum*) was neutralized if the cardamom seeds were chewed after a meal.

By the Middle Ages, cardamom, then often called Amomum, had penetrated more northern areas of Europe. It was to be found in some of the kitchens of the wealthy, where the cooks used it for flavoring; cardamom, along with other foreign spices, was a sign of prestige. Apparently it was far more popular in England than in France, and records show that while Jean le Bon (1319–1364) of France, who had been taken prisoner at Poitiers in 1356 by Edward the Black Prince (1330–1376), was held at his cousin's home in England awaiting payment of the requisite ransom, cardamom was one of the spices included in his household expenses. At the turn of the 20th and 21st centuries, it is now the Swedes who consume the most cardamom in Europe. They are said to account for one-fourth of India's output.

More recently, in the Middle East, it has been used as a flavoring in coffee (in Iran particularly) and in mulled wine. In Arabian countries, especially where cardamon coffee is offered as a symbol of hospitality, cardamom is believed to enhance digestion, and, for good measure, to act as an aphrodisiac. In India, cardamom has long had a reputation for being an aphrodisiac, and the sugar-coated seeds feature in Hindu festivals and ceremonies.

Ground cardamom is primarily used as a spice, and is a significant ingredient in Eastern as well as Scandinavian cookery. In Malaysia, it is chewed with betel nut, *Areca catechu*. These Middle Eastern and Eastern flavoring practices are beginning to gain popularity in other continents, including North America and Europe. Here, not least in parts of South America, cardamom can now be found added to coffee, tea, and chocolate, on occasion.

Cardamom can be used as a flavoring to disguise the taste of medicine. In classical times, it was one of at least 36 ingredients used by Mithridates (c.132–63 BC), the 1st century King of Pontus (northern Turkey), in a poison antidote (known as *Antidotum Mithradaticum* or *Theriac*) which he took daily to acquire an overall immunity — an important consideration, remembering that he gained his position of power by poisoning his opposition!

287

PRESENT-DAY CULTIVATION AND USAGE
Cultivation practices

Cardamom is grown commercially in plantations under the shade of tall forest trees, and is a very labor-intensive crop. The fruits are picked individually by hand before they are fully ripe, over a period of several months.

In India, cardamom grows under natural conditions in the evergreen forests in the Western Ghats. It thrives best in tropical forests at altitudes ranging from 600–1500 m, where there is a well-distributed rainfall of over 150 cm and a temperature of 10–35°C. Its optimum growth and development is observed in warm and humid places, under the canopy of lofty evergreen forest trees. It is highly sensitive to wind and drought, and therefore areas liable to be affected by these conditions are unsuitable. The crop does not survive in waterlogged ground or excessive moisture. The ideal site is sloping, with good drainage.

Fruits mature about 3–4 months after flowering. They are small trilocular capsules, each containing about 15–20 seeds. On maturity, fruits turn pale green. Under favorable environmental conditions, a healthy adult plant will produce about 200 capsules annually, with a green weight of about 900 g; on processing, this yields about 200 g of dry capsules (Anonymous, 2009).

Harvesting period

Cardamom plants normally start bearing 2 years after planting. Throughout the cropping season of cardamom (i.e., from August to March), approximately six pickings are carried out at 45-day intervals. In most areas, the peak period of harvest is during October–November. Ripe capsules are harvested in order to achieve the optimum green coloration during curing.

The output of cardamom is greatly influenced by climatic conditions. The cardamom plant requires a continuous spell of rain, interspersed with periods of good sunshine. The plant is very susceptible to attack by pests and diseases. Moreover, the cardamom-growing tracts in the country have faced severe ecological degradation over the past two decades due to diminishing forest cover, leaving the region open to devastation by floods and droughts. Cardamom requires tropical forest conditions for growth, and a lack of such areas makes very few states in the country suitable for its cultivation.

Climate and soil

The natural habitat of cardamom is the evergreen forests of Western Ghats. It is found to grow within an altitude ranging between 600 and 1200 m above MSL. Though considerable variations in both the total rainfall pattern and its distribution are noticed in the cardamom tracts (900–4000 mm), a well-distributed rainfall of 1500–2500 mm with not less than 200 mm in summer showers and a mean temperature of 15–25°C provide ideal conditions. Cardamom generally grows well in loamy soils that are acidic in nature, preferably at a pH of 5.5–6.5. Cardamom soils have high levels of organic matter and nitrogen, low to medium levels of available phosphorus, and medium to high levels of available potassium.

Nursery management

In order to raise a cardamom plantation, suckers or seedlings of high-yielding varieties that are suited to the location are used. If virus-free production of planting material can be ensured, vegetative propagation through suckers is the best method. However, vegetative propagation has the inherent disadvantage of reducing the genetic base of cardamom. Sucker propagation is the accepted practice in Kerala and Tamilnadu. Traditionally, cardamom plantations have been raised from seeds, and this is still the common and advisable practice in Karnataka, mainly because of the rampant viral disease infestation.

Applications

Today the essential oil is a commercial ingredient used by the perfumery industry, particularly in some eau de colognes. It is also used by the food industry as a flavoring in cakes, gingerbread, sausages, pickles, and curry powders. The drinks industry uses the oil as a flavoring in cordials, bitters, and liqueurs, and it also provides commercial flavorings for the pharmaceutical industry in proprietary medicines.

Medicinally, cardamom has been used locally in India and some other Asian countries to treat depression, some heart disorders, dysentery, and diarrhea. It has also been used to counter vomiting and nausea. Where cardamom is easily accessible, it is believed by some that its overuse could cause impotency. On the other hand, a daily dose is said to maintain general health and improve eyesight.

As a spice, cardamom is used for culinary purposes in curry, coffee, cakes, and bread, and for flavoring sweet dishes and drinks. The seeds and the essential oil are used as a flavoring component in a variety of foods, including alcoholic and non-alcoholic beverages, frozen desserts, candies, baked goods, puddings, condiments, relishes, gravies, meat, and meat products. It is used as a spice in Moroccan tajines, and is generally used widely in cooking, including that of meats. Cardamom is also utilized in traditional Chinese and Indian medicine as a digestive aid, and for the treatment of flatulence. It is added to massage oils and lotions, as well as soaps, detergents, and perfumes, because of its soothing properties (Ravindran, 2002).

APPLICATIONS TO HEALTH PROMOTION AND DISEASE PREVENTION

The seed, which is where the essential oils are mainly found, has potential applications as an antimicrobial, antibacterial, and antioxidant, and is also reported to act as an efficient skin-permeation enhancer for certain drugs. Cardamom is officially listed in the Pharmacopoeia of India, the UK, and the USA.

Skin-penetration enhancing activity

The essential oils from the seeds of *Elettaria cardamomum* have been reported to show good skin permeation activity for certain drugs. The oils from *Elettaria cardamomum* interact with the lipids of the horny layer of the skin, resulting in the destruction of the structural order of the skin and thus increasing the diffusion capacity of the active components by the lipid intercellular pathway. An *in vitro* study on the permeation of estradiol through hairless mouse skin revealed that complex terpenes are responsible for the enhancement of transdermal permeation for moderately lipophilic drugs like estradiol (Monti *et al.*, 2002). The *in vivo* and *in vitro* studies on the permeation of indomethacin showed that permeation was significantly enhanced after pretreatment with cardamom oil. This increased permeation was mainly due to the presence of cyclic monoterpenes from *Elettaria cardamomum* (Huang *et al.*, 1999).

Anticarcinogenic activity

The oils from *Elettaria cardamomum* seeds exhibited *in vitro* anticarcinogenic activity by inhibiting the formation of DNA adducts by aflatoxin B1 in a microsomal enzyme-mediated reaction (Hashim *et al.*, 1994). This enzymatic modulation may be due to the chemical constituents of the oils, which form the basis for their potential anticarcinogenic roles. *Elettaria cardamomum* seed extract also showed a protective effect on platelets against aggregation and lipid peroxidation (Suneetha & Krishnakantha, 2005). An aqueous suspension of cardamom has shown protective effects on experimentally induced colon carcinogenesis (Sengupta *et al.*, 2005).

Anti-ulcerogenic activity

The petroleum ether soluble extract from *Elettaria cardamomum* seeds was screened for aspirin-induced anti-ulcerogenic activity in rats. The petroleum ether soluble extract inhibited lesions by nearly 100% at 12.5 mg/kg (Jamal *et al.*, 2006).

Antimicrobial activity

The extracts from *Elettaria cardamomum* exhibited antimicrobial activity against oral microbes. The essential oils from *Elettaria cardamomum* showed marked inhibitory effects to select pathogenic and spoilage microorganisms. The alcoholic extracts of cardamom seeds were found to possess antibacterial activity against the human pathogenic strain of *Salmonella typhi* (Singh *et al.*, 2008). *Elettaria cardamomum* is one of the ingredients of a herbal syrup that has been found to have antimicrobial action against *E.coli, B.proteus, Klebsiella*, and *Pseudomonas*.

Miscellaneous

Elettaria cardamomum seed extract is one of the ingredients of the polyherbal formulation for treating the dementia of Alzheimer's disease (Aaishwarya *et al.*, 2005).The extract is also used in herbal combinations utilized in the treatment of anxiety, tension, and insomnia (Prema-latha & Rajgopal, 2005). An *in vivo* study of an ayurvedic formulation containing cardamom as one of the ingredients shows that the formulation has CNS-depressant and anticonvulsant activity in mice (Achliya *et al.*, 2004). A multi-ingredient herbal formulation with *Elettaria cardamomum* as one of the ingredients is found to be useful in the treatment of sore throats (Prakash, 2001). Cardamom is one of the ingredients of a Tibetan herbal formulation that was found to inhibit cell proliferation accompanied by the accumulation of CEM-C7H2 cells in subG1 phase, fragmentation of poly (ADP-ribose) polymerase (PARP), and nuclear body formation (Jenny *et al.*, 2005). The volatile oils from *Elettaria cardamomum* are also reported to reduce edema by increasing the permeation of ion-paired DS across viable skin (Sapra *et al.*, 2000). Eugenol inhibited tobacco-induced mutagenicity at concentrations of 0.5 and 1 mg/plate, and eugenol and the plant extracts also inhibited the nitrosation of methyl urea in a dose-dependent manner (Sukumaran & Kuttan, 1995). The antispasmodic activity was studied in a rabbit intestine preparation, using acetylcholine as agonist. The results from this study showed that cardamom seed oil exerts its antispasmodic action through muscarinic receptor blockage (Al-Zuhair *et al.*, 1996). The extract from the seeds of cardamom also exhibited diuretic properties (Gilani *et al.*, 2008)

ADVERSE EFFECTS AND REACTIONS (ALLERGIES AND TOXICITY)

Traditionally, cardamom seeds are regarded as safe, owing to their long-term use as a home remedy and common spice in various food items and beverages. No adverse effect or toxicity is reported in the literature. However, there are reports that cardamom seeds can trigger gallstone colic, and hence they not recommended for self-medication in patients with gallstones.

SUMMARY POINTS

- Cardamom, admired as the "Queen of Spices," belongs to the ginger family, Zingiberaceae. It is one of the most exotic and highly priced spices across the globe, after vanilla and saffron.
- Oil from the seeds is used in processed food, tonics, liquors, and perfumes. The fruit also finds significant usage in Ayurvedic medicine, as it has healing effects in dental infections, digestive disorders, etc.
- Cardamom is generally produced in the tropical regions of the world. The total world production of this species is around 35,000 tonnes per annum, and the largest producing country is Guatemala, followed by India. The major consuming countries are the Middle Eastern countries, India, Pakistan, European countries, the US, and Japan.

- Studies have revealed its use as an effective skin penetration enhancer for certain activities, and as an anticarcinogenic, anti-ulcerogenic, antimicrobial, and anticonvulsant.
- The extract also finds wide application as an ingredient in formulations for treating dementia of Alzheimer's disease, anxiety, tension, insomnia, and sore throats. The seed extract finds applications in some of the herbal-based cosmetic formulations, which include skin-whitening, anti-dandruff, hair-shine, and hair-growth preparations.
- India and Guatemala are committed to finding more productive, drought-tolerant, and disease-resistant cardamom varieties through genetic improvement. In the past 15 years, global production and consumption have increased almost 2.5 times, and it appears that cardamom has a bright future, facing a steady increase in demand and supply.
- Due to the extensive appliance of cardamom, its production needs to be increased by adopting modern technology.

References

Aaishwarya, B. D., Natvarlal, J. P., & Rashwin, J. P. (2005). An Ayurvedic strategy for treatment of dementia of Alzheimer's disease. *Pharmacognosy Magazine, 1,* 144–151.

Achliya, G. S., Wadodkar, S. G., & Dorle, A. K. (2004). Evaluation of sedative and anticonvulsant activities of Unmadnashak Ghrita. *Journal of Ethnopharmacology, 94,* 77–83.

Al-Zuhair, H., El-Sayeh, B., Ameen, H. A., & Al-Shoora, H. (1996). Pharmacological studies of cardamom oil in animals. *Pharmacological Research, 34,* 79–82.

Anonymous. (2009). Cultivation practices for cardamom *Elettaria cardamomum* Maton species. Kochi, India: Ministry of Commerce and Industry.

Gilani, A. H., Jabeen, Q., Khan, A., & Shah, A. J. (2008). Gut modulatory, blood pressure lowering, diuretic and sedative activities of cardamom. *Journal of Ethnopharmacology, 115,* 463–472.

Hashim, S., Aboobaker, V. S., Madhubala, R., Bhattacharya, R. K., & Rao, A. R. (1994). Modulatory effects of essential oils from spices on the formation of DNA adduct by aflatoxin B1 *in vitro*. *Nutrition and Cancer, 21,* 169–175.

Huang, Y. B., Fang, J. Y., Hung, C. H., Wu, P. C., & Tsai, Y. H. (1999). Cyclic monoterpene extract from cardamom oil as a skin permeation enhancer for indomethacin: *in vitro* and *in vivo* studies. *Biological & Pharmaceutical Bulletin, 22,* 642–646.

Jamal, A., Kalim Javed, A., Aslam, M., & Jafri, M. A. (2006). Gastroprotective effect of cardamom, *Elettaria cardamomum* Maton, fruits in rats. *Journal of Ethnopharmacology, 103,* 149–153.

Jenny, M., Schwaiger, W., Bernhard, D., Wrulich, O. A., Cosaceanu, D., Fuchs, D., et al. (2005). Apoptosis induced by the Tibetan herbal remedy PADMA 28 in the T cell-derived lymphocytic leukaemia cell line CEM-C7H2. *Journal of Carcinogenesis, 4,* 15–21.

Kiritikar, K. R., & Basu, B. D. (1999). Indian medicinal plants (2nd ed.). New Delhi, India: International Book Distributors.

Miriam, P., & Christopher, R. (1992). *The natural pharmacy, an encyclopedic illustrated guide to medicines from nature.* New York, NY: Dorling Kindersley.

Monti, D., Chetoni, P., Burgalassi., S., Najarro, M., Saettone, M. F., & Boldrini, E. (2002). Effect of different terpene-containing essential oils on permeation of estradiol through hairless mouse skin. *Journal of Pharmaceutics, 26,* 209–214.

Prakash, T. (2001). Koflet lozenges in the treatment of sore throat. *Antiseptic, 4,* 124–127.

Premalatha, B., & Rajgopal, G. (2005). Cancer – an ayurvedic perspective. *Pharmacological Research, 51,* 19–30.

Ravindran, M. K. (Ed.). (2002). Cardamom: The genus *Elettaria*. New York, NY: Taylor & Francis.

Sapra, B., Gupta, S., & Tiwary, A. K. (2000). Role of volatile oil pretreatment and skin cholesterol on permeation of ion-paired diclofenac sodium. *Indian Journal of Experimental Biology, 38,* 895–900.

Sengupta, A., Ghosh, S., & Bhattacharjee, S. (2005). Dietary cardamom inhibits the formation of azoxymethane induced aberrant crypt foci in mice and reduces COX-2 and iNOS expression in the colon. *Asian Pacific Journal of Cancer Prevention, 6,* 118–122.

Singh, G., Shashi, K., Palanisamy, M., Valery, I., & Vera, V. (2008). Antioxidant and antimicrobial activities of essential oil and various oleoresins of *Elettaria cardamomum* (seeds and pods). *Journal of the Science of Food and Agriculture, 88,* 280–289.

Sukumaran, K., & Kuttan, R. (1995). Inhibition of tobacco-induced mutagenesis by eugenol and plant extracts. *Mutation Research, 343,* 25–30.

Suneetha, W. J., & Krishnakantha, T. P. (2005). Cardamom extract as inhibitor of human platelet aggregation. *Phytotherapy Research, 19,* 437–440.

CHAPTER 35

Carob (*Ceratonia siliqua* L.) Seeds, Endosperm and Germ Composition, and Application to Health

Patrick A. Dakia
Unité de Formation et de Recherche des Sciences et Technologies des Aliments (UFR-STA), Université d'Abobo-Adjamé (UAA), Abidjan, Côte d'Ivoire

CHAPTER OUTLINE

LIST OF ABBREVIATIONS

AA, amino acids
Ala, alanine
ALA, α-linolenic acid
Arg, arginine
Asp, aspartic acid
Cys, cysteine
EAA, essential amino acids
FAO, Food and Agriculture Organization of the United Nations
Glu, glutamic acid
Gly, glycine
Ile, isoleucine
LA, linoleic acid
LBG, locust bean gum
Leu, leucine
Lys, lysine

Nuts & Seeds in Health and Disease Prevention. DOI: 10.1016/B978-0-12-375688-6.10035-0

Met, methionine
Phe, phenylalanine
Thr, threonine
Trp, tryptophan
Val, valine

INTRODUCTION

World production of commercial carob seeds is about 32,000 tonnes per year (Batlle & Tous, 1997). The seed is composed of the husk, the endosperm, which is the source of carob gum, and the germ. One benefit of carob seed gum (also called locust bean gum, LBG) is its ability to enhance the texture of foods and non-food products. Carob seed germ, obtained as a by-product of the carob-producing industries, is exploited as a dietetic or as animal feed due to its protein content ($> 50\%$) (Dakia *et al.*, 2007). In addition to discussing the carob seed's technologic and nutritive values, this chapter aims to describe its medicinal or pharmacological properties in a comprehensive way.

BOTANICAL DESCRIPTION

The carob tree, *Ceratonia siliqua* (also called algarroba) is a polygamous, termophilous and typical evergreen species of the leguminous tree. It is a member of the pea family, Fabaceae, which grows throughout the Mediterranean basin, mainly in Spain, Italy, Portugal, Morocco, and Turkey. Carob trees grow best in calcareous soil, preferably near the sea. They are drought resistant, but tolerate only light frost. They have been introduced into many other countries with warm climates, mainly the USA (Florida and California), Australia, and Argentina. Carob trees are handsome trees with pinnately compound leaves that have two to six pairs of oval leaflets; the trees can grow to a height of 15 m. Most carob trees are dioecious (Batlle & Tous, 1997).

HISTORICAL CULTIVATION AND USAGE

The Carob tree is exploited for its edible fruit pod, which contains sweet pulp. Some people believe that in Biblical times, Saint John's wilderness diet of "locusts and wild honey" actually included carob pods, which resemble locusts. Hence, carob is also known today as the locust bean, or Saint John's bread (Matthew 3:4). There is also a belief that Jesus referred to the fruit of the carob tree in his parable of the prodigal son, who, after squandering his inheritance, was reduced to eating the husks given to animals for fodder (Luke 15:16). Each pod can contain up to 12 uniform seeds, similar to watermelon seeds. The term "carat," the unit by which diamond weight is measured, is also derived from the Greek word *kerátion*, alluding to an ancient practice of weighing gold and gemstones against the seeds of the carob tree by people in the Middle East. The system was eventually standardized, and one carat was fixed at 0.2 g. The seeds contain a white and translucent endosperm, known as carob bean gum or locust bean gum. Ancient Egyptians used the gummy properties of carob seed as an adhesive in binding mummies (Batlle & Tous, 1997; Dakia *et al.*, 2008).

PRESENT-DAY CULTIVATION AND USAGE

Today, carob trees are grown for reforestation in the poor soils and dry environments of the coastal areas of some Mediterranean regions. The traditional uses of the carob bean (pod) until the 1960s were of the pulp (90% of the fruit) for animal feeding, or as a chocolate or cocoa substitute. Unlike cocoa, carob does not contain caffeine or theobromine. People who are allergic to chocolate can generally eat carob without a problem. It is also low in sodium and high in potassium, making it desirable for diets used to treat congestive heart failure and hypertension (Yousif & Alghzawi, 2000). In Europe, the part of the fruit that currently has the highest value is the seed (kernel or garrofín) from which a gum (LBG or E411) is extracted for use as a stabilizer and thickener in food and non-food industries. The seed is composed of the husk (30–33%), the endosperm (42–46%), which is the source of gum, and the germ (23–25%) (Dakia et al., 2008). Carob seed germ is used as a protein source in food and in animal feed (Batlle & Tous, 1997; Dakia et al., 2007, 2008).

APPLICATIONS TO HEALTH PROMOTION AND DISEASE PREVENTION

Carob seed endosperm

The composition of carob seed endosperm flour (crude locust bean gum) is given in Table 35.1. Galactomannan, comprising more than 90%, is the major compound in carob seed gum. It is is a linear polysaccharide based on a β-(1→4)-mannane backbone to which single D-galactopyranosyl residues are attached via α-(1→6) linkages (da Silva & Gonçalves, 1990). As stated above, LBG is widely used as an additive in the food and non-food industries, owing to its abilities to provide high viscosity at low concentrations (0.1–1%) and to function as a water binder. LBG solutions are only slightly affected by pH, added ions, and heat processing. It is used in ice cream preparation, in paper and textile manufacturing (as a strengthening agent), for rheology (flow) control and as a thickener in latex paints, in oil well fracturing and drilling, and in gels used in blasting agents. Control of the rheological properties of the aqueous phase confers the desired product properties, such as stability, texture, and controlled release of activity in various pharmaceutical and cosmetic products (Sittikijyothin et al., 2005).

Being non-digestible, locust bean gum is considered to be a dietary fiber in foods (Dakia et al., 2008). It increases the dietary fiber in a food product without increasing the calories, and because of this it is useful in developing reduced calorie food. It is used it as a fat substitute in mayonnaise, and in many dairy products, and is considered to be an excellent alternative to gluten because it is non-allergenic and safe. The presence of this substance in foods also increases the swelling of the food once in the stomach, which encourages a feeling of fullness. It is considered to be a natural appetite suppressant.

Locust bean gum can be used in the treatment or control of hyperlipidaemia (high cholesterol in plasma). It can also be made into sugar-free, starch-free flour for diabetics (Wenzl et al., 2003).

Another medicinal value of locust bean gum is its ability to reduce gastrointestinal conditions, especially diarrhea in infants. Utilization of carob gum as an adjunct to oral rehydrating solution showed promising results for the treatment of infantile diarrhea (Cruz, 1955). Carob gum is able to act as a thickener to absorb water and help bind together watery stools.

295

TABLE 35.1 Composition of Carob Seed Endosperm Flour (% on the Basis of Dry Matter)

	Proximate Analysis (%)
Moisture	5.9
Ashes	0.7
Proteins	5.2
Lipids	1.3
Galactomannan (as carbohydrates)	92.8

The higher content of galactomannan in carob seed endosperm shows that locust bean gum is a galactomannan polysaccharide.
Reprinted from from Dakia et al. (2008), Food Hydrocolloids, 22(5), 807–818, with permission.

In addition, infants with recurrent reflux and regurgitation disease improve when fed a formula thickened with carob bean gum. Recurrent reflux and regurgitation in infants (aged 6–12 months) can inflame and damage the esophagus, thus affecting the nutritional status and resulting in poor growth. Locust bean gum as dietary fiber may make food more viscous in the stomach, and thus interfere with acid reflux into the esophagus (Wenzl *et al.*, 2003).

Carob seed germ

Carob seed germ is a by-product of the carob seed gum-producing industries. The nutritional composition of carob germ meal is shown in Table 35.2. Carob seed germ represents an important source of protein for food and feed because it has a high protein content (54.7%) and contains all the known amino acids. The germ can therefore be considered a "complete protein." According the the FAO/WHO (1991) standard, all essential amino acids (EAA) are

TABLE 35.2 Overall Composition of Carob Seed Germ Meal (% on the Basis of Dry Matter)

Proximate Analysis (%)		
Moisture	8.3	
Ash	6.5	
Protein	54.7	
Lipids	6.6	
Carbohydrates (nitrogenous free extract including fiber)	23.9	
Crude energy (kJ/g)	17.5	

Amino acid content	(% protein)	(% of FAO/WHO for EAA)
Asp + Asn	7.86	
Thr	3.29	(94%)
Ser	4.57	
Glu + Gln	31.44	
Pro	3.29	
Gly	4.57	
Ala	4.02	
Val	3.83	(111%)
Met	0.91	
Cys − Cys	1.09	
(Met + Cys)	2.3	(92%)
Ile	3.10	(128%)
Leu	6.03	(91%)
Phe	3.10	
Tyr	2.92	
(Phe + Tyr)	6.0	(95%)
His	2.56	(137%)
Lys	5.48	(98%)
Arg	12.25	
Trp	0.73	(73%)

Fatty acid content	(% oil)	
Palmitic (C16)	16.2	
Stearic (C18)	3.4	
Oleic (C18:1)	34.4	
Linoleic (C18:2)	44.5	
Linolenic (C18:3)	0.7	

These results show that carob seed germ has a high content of protein, amino acids and unsaturated fatty acids.
Reprinted from Dakia *et al.* (2007), *Food Chemistry, 102(4)*, 1368–1374, with permission.

present in significant amounts. The proportion of all EAA is about 33.32%. Trp is the first limiting EAA (73%), followed by Met + Cys (92%) (Yust *et al.*, 2004; Dakia *et al.*, 2007). Carob germ flour can also yield an isolate (by using an alkaline extraction technique followed by isoelectric precipitation of proteins) with a protein content of over 95%. This protein isolate, again with a well-balanced amino acid composition, may be an attractive ingredient for use in the production of human dietetic foods, as it has reduced levels of undesirable components (Bengoechea *et al.*, 2008).

The amino acid profile is dominated by glutamic acid (31.44%), arginine (12.25%), and aspartic acid (7.86%). The amino acid glutamine has many important functions in the body — for example, it acts as the primary vehicle for the transfer of amino nitrogen from skeletal muscle to visceral organs, as a fuel for the rapidly dividing cells of the gastrointestinal tract and immune system, and as a substrate that permits the kidneys to excrete acid loads and protect the body against acidosis. Glutamine is considered to be a conditionally essential amino acid for critically ill and other stressed patients, because during periods of illness the metabolic rate of glutamine increases and the body is not able to synthesize sufficient glutamine to meet its needs. This is particularly true during episodes of stress, such as sepsis, injury, burns, inflammation, diarrhea, and surgery. The administration of glutamine-supplemented diets to preterm babies, or during periods of stress, or to athletes, has resulted in improvement of the person's condition. For example, glutamine-supplemented diets have been shown to regenerate muco-proteins and the intestinal epithelium, support gut barrier function, shorten hospital stays, improve immune function, and enhance patient survival (Lacey & Wilmore, 1990; Gianotti *et al.*, 1995).

Arginine is also classified as a semi-essential or conditionally essential amino acid, depending on the developmental stage and health status of the individual. Arginine plays an important role in cell division, the healing of wounds, removing ammonia from the body, immune function, and the release of hormones (Stechmiller *et al.*, 2005).

The lipid content of carob seed germ is about 6%. Due to the relatively low fat content, the seed germ has low caloric value, which makes it an interesting healthy food. Carob seed germ oil contains approximately 19.6% saturated fatty acids (palmitic and stearic), 34.4% mono-unsaturated fats (mainly oleic, 18:1n-9), and 44.5% polyunsaturated fats (mainly linoleic, 18:2n-6, omega-6). Unfortunately, the amount of n-3 fatty acid (α-linolenic acid, omega-3) is too low, at 0.7% of the total fats, for carob germ oil to be considered as a high nutritional quality oil (omega-3 fatty acids are compounds with potential cardiovascular benefit). However, carob seed oil is a good source of n-6 fatty acid (linoleic acid), which is defined as an "essential" fatty acid, like α-linolenic acid, since it is not synthesized in the human body and is mostly obtained from the diet. Deficits in n-6 "essential" fatty acids were correlated with the severity of atopic dermatitis, in affecting skin barrier function and cutaneous inflammation (Russo, 2009). In general, essential fatty acids, and their longer-chain more-unsaturated derivatives in particular, are essential for normal growth and cognitive development during childhood (Vlaardingerbroek *et al.*, 2006).

ADVERSE EFFECTS AND REACTIONS (ALLERGIES AND TOXICITY)

Although previous studies have been inconclusive about LBG toxicity, some authors have suggested that fibers, like LBG, used to thicken infant formulas can decrease the availability of minerals such as calcium and iron (Bosscher, 2001). Regarding carob seed germ, some searchers argue that extracts of carob seed germ were found to decrease trypsin activity (Filioglou & Alexis, 1987) and protein digestibility (Filioglou & Alexis, 1989). Fortunately, the antinutritive substances, such as trypsin inhibitors, generally contained in leguminosae seeds (Weder, 1986) can be deactivated by heating (for example, at 121°C for 25 minutes, according to Martinez-Herrera *et al.*, 2006) to make carob seed germ flour suitable for food and animal feed products. Carob seed germ also contains tannins which could create palatability problems due to their astringent taste, and could also reduce feed digestibility. It has been suggested that protein digestibility is reduced by tannins either by direct binding to certain parts of the protein molecule, or through non-competitive inhibition of the digestive enzymes (Filiglou & Alexis, 1989).

SUMMARY POINTS

- Being non-digestible, carob seed gum is considered as a dietary fiber in foods, and is used as a fat substitute in mayonnaise.
- As a water binder, carob seed gum is also used to treat infants with diarrhea and recurrent regurgitation.
- Carob seed germ represents an important source of protein, which contains all essential amino acids, in the preparation of dietary products.
- Carob germ protein is a good source of amino acid glutamine for patients suffering from gastro-intestinal injury or diseases.
- Carob seed germ oil is a good source of polyunsaturated fatty acids (n-6), which can used in ameliorating pathological conditions such as growth retardation.
- Carob seed endosperm flour presents no known allergie or toxicity, although carob seed germ flour contains trypsin inhibitors (whicih may depress growth). Fortunately, such antinutritive substances (generally contained in leguminosae seeds) can be deactivated by the heating process.

References

Batlle, I., & Tous, J. (1997). *Properties, agronomy, processing. In Carob tree Ceratonia siliqua L.* Rome, Italy: IPGRI. CGIAR. 61–6224–29.

Bengoechea, C., Romero, A., Villanueva, A., Moreno, G., Alaiz, M., Millan, F., et al. (2008). Composition and structure of carob (*Ceratonia siliqua* L.) germ proteins. *Food Chemistry, 107*, 675–683.

Bosscher, D., Van Caillie-Bertrand, M., & Deelstra, H. (2001). Effect of thickening agents, based on soluble dietary fiber, on the availability of calcium, iron, and zinc from infant formulas. *Nutrition, 17*(7–8), 614–618.

Cruz, M. (1955). Treatment of infant diarrhea with carob flour. *Acta Pediatrica España, 13*(151), 469–488.

Dakia, P. A., Wathelet, B., & Paquot, M. (2007). Isolation and chemical evaluation of the carob (*Ceratonia siliqua* L.) seeds germ. *Food Chemistry, 102*(4), 1368–1374.

Dakia, P. A., Blecker, C., Robert, C., Wathelet, B., & Paquot, M. (2008). Composition and physicochemical properties of locust bean gum extracted from whole seeds by acid or water dehulling pre-treatment. *Food Hydrocolloids, 22*(5), 807–818.

da Silva, J. A. L., & Gonçalves, M. P. (1990). Studies on a purification method for locust bean gum by precipitation with isopropanol. *Food Hydrocolloids, 4*, 277–287.

FAO/WHO. (1991). *Protein quality evaluation. Report of a joint FAO–WHO expert consultation. Food and Nutrition Paper No. 51.* Rome: FAO/WHO.

Filioglou, M. D., & Alexis, M. N. (1987). Use of the carob products in trout nutrition: the effects of growth inhibitors of the carob seed germ meal on digestion and studies on deactivation methods. *Proceedings of the Second Panhellenic Symposium of Oceanography and Fisheries* 618–624.

Filioglou, M. D., & Alexis, M. N. (1989). Protein digestibility and enzyme activity in the digestive tract or rainbow trout fed diets containing increasing levels of carob seed germ meal. In N. De Pauw, E. Jaspers, H. Ackefors, & N. Wilkins (Eds.), *Aquaculture. A biotechnology in progress* (pp. 839–843). Bredene, Belgium: European Aquaculture Society.

Gianotti, L., Alexander, J. W., Pyles, T., Gennari, R., Pyles, T., & Babcock, G. F. (1995). Oral glutamine decreases bacterial translocation and improves survival in experimental gut-origin sepsis. *Journal of Parenteral and Enteral Nutrition, 14*, 69–74.

Lacey, J. M., & Wilmore, D. W. (1990). Is glutamine a conditionally essential amino acid? *Nutrition Review, 48*(8), 297–309.

Martínez-Herrera, J., Siddhuraju, P., Francis, G., Dávila-Ortíz, G., & Becker, K. (2006). Chemical composition, toxic/antimetabolic constituents, and effects of different treatments on their levels, in four provenances of *Jatropha curcas* L. from Mexico. *Food Chemistry, 96*, 80–89.

Russo, G. L. (2009). Dietary n-6 and n-3 polyunsaturated fatty acids: from biochemistry to clinical implications in cardiovascular prevention. *Biochemical Pharmacology, 77*, 937–946.

Sittikijyothin, W., Torres, D., & Gonçalves, M. P. (2005). Modelling the rheological behaviour of galactomannan aqueous solutions. *Carbohydrate Polymers, 59*(3), 339–350.

Stechmiller, J. K., Childress, B., & Cowan, L. (2005). Arginine supplementation and wound healing. *Nutrition in Clinical Practice, 20*(13), 52–61.

Vlaardingerbroek, H., Hornstra, G., De Koning, T. J., Smeitink, J. A. M., Bakker, H. D., de Klerk, H. B. C., et al. (2006). Essential polyunsaturated fatty acids in plasma and erythrocytes of children with inborn errors of amino acid metabolism. *Molecular Genetics and Metabolism, 88*(2), 159–165.

Weder, J. K. P. (1986). Nutritional and toxicological significance of enzyme inhibitors in foods. In M. Friedman (Ed.), *Advances in experimental medicine and biology, Vol. 199* (pp. 239–279). New York, NY: Plenum Press.

Wenzl, T. G., Schneider, S., Scheele, F., Silny, J., Heimann, G., & Skopnik, H. (2003). Effects of thickened feeding on gastroesophageal reflux in infants: a placebo-controlled crossover study using intraluminal impedance. *Pediatrics, 111*, 355–359.

Yousif, A. K., & Alghzawi, H. M. (2000). Processing and characterization of carob powder. *Food Chemistry, 69*, 283–287.

Yust, M. M., Pedroche, J., Girón-Calle, J., Vioque, J., Millán, F., & Alaiz, M. (2004). Determination of tryptophan by high-performance liquid chromatography of alkaline hydrolysates with spectrophotometric detection. *Food Chemistry, 85*, 317–320.

CHAPTER 36

Cashew Nut (*Anacardium occidentale* L.) Skin Extract as a Free Radical Scavenger

Padmanabhan S. Rajini
Food Protectants & Infestation Control Department, Central Food Technological Research Institute (CSIR Lab), Mysore, India

301

LIST OF ABBREVIATIONS

ABTS, 2,2′-azino-bis (3-ethylbenzthiazoline-6-sulphonic acid
CNSL, cashew nut shell liquid
EDTA, ethylenediamine tetra-acetic acid
HPLC, high performance liquid chromatography
LPO, lipid peroxidation
OPI, organophosphorus insecticides
TBARS, thiobarbituric acid reactive species

INTRODUCTION

Free radical scavengers have attracted attention because they can protect the human body from free radicals, which are known to influence the pathogenesis of many human diseases, including cancer, and also lead to the aging process. Although human beings have antioxidant defenses against oxidative damage, these antioxidants can be inefficient. Further, several

Nuts & Seeds in Health and Disease Prevention. DOI: 10.1016/B978-0-12-375688-6.10036-2

synthetic antioxidants are commonly used in processed foods, and there has been growing concern about their safety and toxicity in long-term use. Hence, the research on natural antioxidant sources has attracted much attention. There have been reports on the extraction and identification of antioxidant components from peels, hulls, and seed coats, which are usually discarded as the by-products of agro-industries (Amin & Mukhrizah, 2006). Phenolic compounds, especially tannins, present in these fractions are known to render them anti-oxidative. The tannins located in the seed coats (hulls) are reported to play an important role in the defense system of the seeds, which are exposed to oxidative damage (Troszynska & Balasinska, 2002). Several studies have been conducted to determine and suggest uses of "tannin-rich seed coats," such as those from pea (Duenas *et al.*, 2006), soybean (Takahata *et al.*, 2001), peanut (Yen *et al.*, 2005), and almonds (Chen *et al.*, 2005), as a source of antioxidants. Cashew kernel testa (skin) is reported to contain huge amounts of tannin, and the tannin extracted from cashew kernel testa is used in the leather industry (Nayudamma & Rao, 1967). The testa has also been used as a wholesome poultry feed (Subramanian & Nair, 1969). However, its commercial use has been limited. The objective of this chapter is to provide information on the free radical scavenging activity of cashew kernel testa.

BOTANICAL DESCRIPTION

The cashew (*Anacardium occidentale*, Linn.) is a member of the *Anacardium* genus of the Anacardiaceae family. It is a small tree, with leaves that are alternate, simple, entire, obtuse, and borne on short leaf stalks. The flowers are abundant, small, and fragrant, and are produced in terminal, loose panicles. The enlarged juicy peduncle that bears the nut is known as the "cashew apple." When ripe, it is of a golden-yellow color, obovate in shape, has a pleasant, acid flavor, and is somewhat astringent. The cashew nut hangs from the end of the cashew apple, and is kidney-shaped and about 2.5 cm long. It consists of an edible kernel, surrounded by two shells. The outer shell is smooth and of a bright brown color. Between the two shells, there is a very caustic oily substance. The cashew kernel is considered to be of high nutritive quality, and is covered with a thin reddish brown skin or testa.

HISTORICAL CULTIVATION AND USAGE

Very little effort has been made to collect historical evidence of cashew cultivation. The French naturalist, Thevet, gave the first illustrative description of cashew, in 1558 AD. The Portuguese introduced the cashew tree, which was originally native to Brazil, to Mozambique and then India in the 16th century, as a means of controlling coastal erosion. It was not until the 19th century that plantations were developed, and the tree then spread to a number of other countries in Africa, Asia, and Latin America. The cashew is now distributed throughout the tropics and in parts of the warm subtropics. Cashew processing, using manual techniques, began in India in the first half of the 20th century, when cashews were exported to the wealthy western markets — particularly the United States. In the 1960s, some of the producing countries in East Africa began to process nuts domestically rather than sending them to India for processing. This allowed them to benefit from the sale of both processed nuts and the extracted cashew nut shell liquid.

PRESENT-DAY CULTIVATION AND USAGE

In the early 1970s the majority of global cashew production (68% of the total) took place in African countries, in particular in Mozambique and Tanzania. Over the following 30 years production trends shifted, with Asian countries emerging as the world leaders in cashew production. Today, India commands about 40% of the international market in cashew production. Other Asian countries, particularly Vietnam and Indonesia, are beginning to expand their production capacities. Currently, the four main cashew-producing regions are India, Brazil, Nigeria, and Tanzania (Azam-Ali & Judge, 2001).

Cashew is grown throughout the tropics and in parts of the warm subtropics, but mostly between the Tropic of Cancer (20°N) and the Tropic of Capricorn (24°S). It is now widely cultivated for its nuts and other products in the coastal regions of South Africa, Madagascar, and Tanzania; in South Asia, from Sri Lanka to the Philippines; and also in Australia, in a limited area. In most countries, smallholders carried out cashew planting individually. In India, the forest corporations and Cashew Development Corporations started large-scale cashew plantations in virgin forestlands. Apart from the organized plantations, cashew has semi-wild growth, and is found growing naturally in most of the cashew-growing regions. The cashew tree is easily cultivated, with a minimum of attention. It is usually found from sea level to an altitude of 1000 m (3000 ft), in regions with annual rainfall as low as 500 mm (20 inches) and as high as 3750 mm (150 inches). For maximum productivity, good soil and adequate moisture are essential. Optimal conditions include an annual rainfall of at least 890 mm (35 inches) and not more than 3050 mm (120 inches). The tree has an extensive root system, which helps it to tolerate a wide range of moisture levels and soil types, but commercial production is only advisable in well-drained, sandy loam, or red soils. The cashew tree can flourish in the sand of open beaches, but it grows poorly in heavy clay or limestone (Azam-Ali & Judge, 2001).

All parts of the cashew tree are of value. The three main cashew products that are traded on the international market are the raw nuts, cashew kernels, and cashew nut shell liquid (CNSL). A fourth product, the cashew apple, is generally processed and consumed locally. The fruit of the cashew tree, which surrounds the kernel, can be made into a juice with a high vitamin C content, and fermented to give a high percentage proof spirit. The raw cashew nut is the main commercial product of the cashew tree, though the yield of cashew apples is eight to ten times the weight of the raw nuts. Nuts are either exported raw, or processed prior to export. Processing of the raw nuts releases the by-product CNSL, which has industrial and medicinal applications. The skin of the nut is high in tannins (40%), which can be recovered and used in the tanning of hides.

APPLICATIONS TO HEALTH PROMOTION AND DISEASE PREVENTION

Cashew kernel testa (skin) is an important by-product obtained during the processing of cashew. With an annual global consumption of over 1,000,000 tonnes of cashew kernel, the resultant testa is a serious option for commercial exploitation. Although the various biochemical constituents of cashew kernels have been characterized (Nagaraja & Nampoothiri, 1986; Nagaraja, 1987a, 1987b, 1989), literature pertaining to the composition of cashew kernel testa is minimal.

Cashew kernel testa has been reported to be a good source of hydrolyzable tannins (Pillai *et al.*, 1963). The tannin content of cashew kernel testa from various varieties of cashew samples has been quantified, and reported to be higher than that of almond skin testa. Cashew kernel testa from different varieties and industrial samples has also been analyzed for protein, sugar, starch, and phenols, and significant variation in the composition of testa among different released varieties and industrial samples has been reported. Chromatographic studies show that the main constituents are catechins (Subramanian & Nair, 1969) especially (+) catechin and (−) epicatechin, which account for about 6 and 7.5%, respectively, of the dried cashew kernel testa. Together, they represent more than 40% of the total polyphenols. In cashew kernel testa, as in most other plant products, the polymeric proanthocyanidins account for a little less than 40%. The monomeric proanthocyanidins are represented by two leukocyanidins and two leukopelargonidins. Leukocyanidin was found to be the predominant leukoanthocyanidin (Mathew & Parpia, 1970). However, leukopelargonidin was found to occur in larger amounts than leukodelphinidin. Earlier, Kantamoni (1965) had reported the presence of gallic, caffeic, and quinic acids, as well as a catechin and leukocyanidin. Our

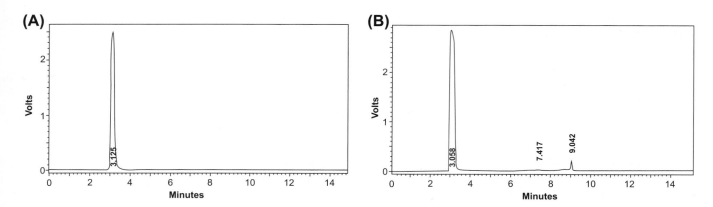

FIGURE 36.1
HPLC chromatogram of (A) standard epicatechin, and (B) antioxidant fraction of cashew kernel testa extract showing presence of epicatechin. The HPLC chromatogram (at 280 nm) of the active spot obtained by TLC showed the presence of epicatechin as major component, along with other polyphenols. The identities of the peaks were confirmed by comparing the relative retention time, and by spiking with standard epicatechin.

analysis of the ethanolic extract of cashew kernel testa showed epicatechin to be one of the major components (Figure 36.1).

Various extracts of cashew kernel testa have also been analyzed for protein, carbohydrate, and fiber content. Methanol-extracted cashew kernel testa was reported to contain higher protein, carbohydrate, and crude fiber contents, compared to cashew kernel testa. Methanol extraction of cashew kernel testa also resulted in a considerable decrease in tannin content, while it improved the water absorption capacity, and the *in vitro* digestibility of protein decreased considerably. The protein and carbohydrate contents of methanol-extracted cashew kernel testa were slightly higher than the reported values (Nagaraja, 2000). Crude fiber from cashew kernel testa had lower protein, sugar, and *in vitro* digestibility of carbohydrates, compared to methanol-extracted cashew kernel testa. Elemental analysis of cashew kernel testa has shown that it is also rich in phosphorus, potassium, and calcium. Hence, cashew kernel testa, after suitable treatment, could be a good source for developing mineral-, protein-, and crude fiber-rich foods/animal feed formulations.

Free radical scavenging activity of cashew kernel testa (skin) extract

In view of the chemical composition of the cashew kernel testa, we undertook investigations in our laboratory to establish the free radical scavenging activity of ethanolic extract of cashew kernel testa (Kamath & Rajini, 2007). Dried cashew kernel testa obtained from a local cashew nut processing industry was sun-dried, pulverized in a multi-mill, and passed through a 0.5-mm sieve to obtain a fine powder. The testa powder was mixed with five parts of ethanol, and kept in a rotatory shaker for 3 hours at 37°C. The extracts were separated by centrifugation, and used for quantitation of total phenolic compounds and assessment of free radical scavenging activities, employing a battery of *in vitro* assay systems.

The cashew kernel testa powder yielded large amounts of extract (0.45 g/g powder). The total phenolic content of cashew kernel testa extract determined using Folin-Ciocalteu reagent revealed the presence of 245 mg gallic acid equivalent of phenolic compounds/g powder. The antioxidant rich extract of cashew kernel testa was resolved by silica thin layer chromatography, and an antioxidant spot was visualized by spraying β-carotene-linoleate solution. HPLC characterization of the antioxidant spot yielded a peak similar to that of standard epicatechin (Figure 36.1), suggesting its predominant role in the antioxidative potential of the extract.

We assessed the free radical scavenging ability of cashew kernel testa extract by employing ABTS (2,2′-azino-bis(3-ethylbenzthiazoline-6-sulfonic acid)) radical and superoxide anion radical scavenging, deoxyribose oxidation, the β-carotene-linoleate model, and lipid peroxidation assay systems. Significant scavenging of superoxide anion radicals by cashew kernel testa extract was evident in a concentration-dependent manner, suggesting that cashew kernel testa extract possesses the potential to neutralize superoxide anion radicals. Cashew kernel testa extract also exhibited concentration-dependent scavenging of hydroxyl radicals — a property which could confer its potential to inhibit lipid peroxidation (Figure 36.2). Potent antioxidant activity of cashew kernel testa extract was also evident, as cashew kernel testa extract prevented linoleate-radical mediated discoloration of β-carotene (Figure 36.3).

We employed a rat brain homogenate system subjected to Fe^{2+}-H_2O_2 induced lipid peroxidation as a test to assess the ability of cashew kernel testa extract to prevent lipid peroxidation. Our results clearly demonstrate that cashew kernel testa extract significantly inhibited the formation of thiobarbituric acid reactive species (TBARS) in rat brain homogenate (Figure 36.4), suggesting its potential to inhibit lipid peroxidation. Employing the deoxyribose system, we successfully demonstrated the hydroxyl radical scavenging activity of cashew kernel testa extract. Therefore, inhibition of induced lipid peroxidation in rat brain homogenate by cashew kernel testa extract may be attributable to its ability to scavenge hydroxyl radicals. Since

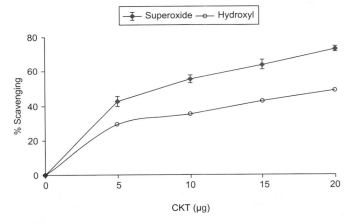

FIGURE 36.2

Superoxide and hydroxyl radical scavenging effect of cashew kernel testa extract. The percent radical scavenging versus concentration of cashew kernel testa in the extract revealed a concentration-related response in both the assay systems. Results are expressed as mean ± SEM of three determinations each.

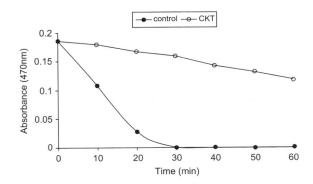

FIGURE 36.3

Antioxidant activity of cashew kernel testa extract in β-carotene-linoleic acid system. β-carotene undergoes rapid discoloration in the absence of an antioxidant (control). The presence of cashew kernel testa extract (20 μg) hindered the β-carotene destruction.

FIGURE 36.4

Inhibition of lipid peroxidation by cashew kernel testa extract in rat brain homogenate.

Peroxidation was induced in rat brain homogenate by ferrous chloride ($FeCl_2$)—hydrogen peroxide (H_2O_2), and the extent of protection offered by cashew kernel testa extract against induced lipid peroxidation (LPO) was monitored by measuring the formation of TBARS at 535 nm. A concentration-dependent protection was offered by cashew kernel testa extract against LPO. Values are expressed as mean $\pm$ SEM of three determinations each.

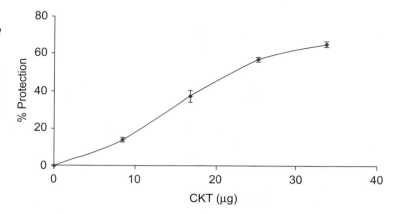

FIGURE 36.5

Ferrous ion chelating effect of cashew kernel testa extract.

Cashew kernel testa extract exhibited a concentration-dependent ferrous ion chelating effect although at higher concentrations. Values are expressed as mean $\pm$ SEM of three determinations each.

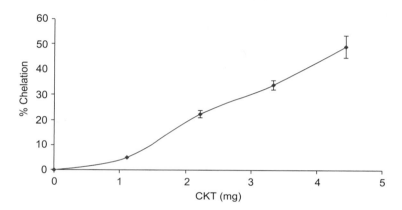

TABLE 36.1 EC_{50} **of Cashew Kernel Testa Extract in Various Antioxidant Assay Systems**

Antioxidant Assay	EC_{50}
ABTS scavenging[a]	1.30 ± 0.02
Superoxide scavenging[a]	10.69 ± 1.13
Deoxyribose oxidation[a]	17.70 ± 0.05
LPO in rat brain homogenate[a]	24.66 ± 0.32
Iron chelation[b]	6.00 ± 0.24

The EC_{50} value in each assay was determined by regression analysis of activity as a function of concentration of cashew kernel testa in the extract. Values are mean $\pm$ SEM of three determinations.
[a]$\mu g/ml$ extract
[b]mg/ml extract.

metal chelation is an important antioxidant property, we evaluated the ability of cashew kernel testa extract to compete with ferrozine for iron (II). The study revealed that cashew kernel testa extract exhibited relatively lower iron-binding capacity (Figure 36.5). Table 36.1 shows data from various *in vitro* assays depicted as EC_{50} values. The order of effectiveness of cashew kernel testa extract in antioxidant assays was as follows.

ABTS > superoxide > deoxyribose > inhibition of lipid peroxidation > iron chelation

Therefore, the antioxidant property of cashew kernel testa extract may be attributable to its radical scavenging activity due to the presence of free radical scavenging polyphenols.

With establishment of the radical scavenging potential of cashew kernel testa extract *in vitro*, it was of interest for us to investigate whether the antioxidant potential of cashew kernel testa could be exploited to abrogate oxidative stress *in vivo*. Organophosphorus insecticides (OPI) are neurotoxicants that primarily act by inhibiting acetylcholinesterase enzyme. Alterations in glucose homeostasis are another aspect of OPI toxicity, and oxidative stress is believed to be one of the mechanisms of OPI-induced alterations in glucose homeostasis. We investigated the possible role of oxidative stress in alterations in glucose homeostasis induced by dimethoate (OPI). The strategy of providing cashew kernel testa extracts supplements to dimethoate-treated rats confirmed involvement of oxidative stress dimethoate-induced alterations in glucose homeostasis, as cashew kernel testa extract offered protection against dimethoate-induced oxidative stress in rat pancreas as well as alterations in glucose homeostasis (Kamath *et al.*, 2008). The above two lines of investigation clearly establish the usefulness of cashew skin extract as an antioxidant, the activity of which is predominantly attributable to the presence of polyphenols that confer on cashew kernel testa an ability to scavenge free radicals. Cashew kernel testa, with its chemical composition of potent bioactive phenolic compounds, could be of interest to food and pharmaceutical industries, where it might be used as an additive or a source of natural antioxidants.

ADVERSE EFFECTS AND REACTIONS (ALLERGIES AND TOXICITY)

Plant tannins are a large, diverse group of polyphenolic compounds found throughout several species in the plant kingdom. Tannins have a protective function in the bark of the roots and stems, and any outer layers of plants. They are astringent in nature, due to their high polyphenol content. This attribute confers the ability to form strong complexes with proteins, starches, and other macromolecules. Although controversy still surrounds the use of hydrolyzed tannins at large doses or for prolonged periods because they can bind essential minerals in the human diet, condensed tannins have a relatively safe profile. While research on condensed tannins is limited, people have been consuming them for centuries without reported adverse affects. Oral consumption of condensed tannins can be assumed to be safe in humans (Clinton, 2009). Based on the above data, cashew kernel testa, being tannin-rich, is anticipated to cause no allergies or toxicity to humans on consumption.

SUMMARY POINTS

- Cashew kernel testa, a by-product of the cashew industry, is rich in tannins.
- Cashew kernel testa exhibits significant antioxidant activity due to the presence of radical scavenging polyphenols.
- The antioxidant potential of cashew kernel testa has also been demonstrated *in vivo*.
- The testa could be exploited as a source of natural antioxidants by the food and pharmaceutical industries.
- Tannin-stripped cashew kernel testa has protein, starch, and sugar content comparable to that of other seed testa, and hence may be viewed as a probable food and animal feed additive.

References

Amin, I., & Mukhrizah, O. (2006). Antioxidant capacity of methanolic and water extracts prepared from food-processing by-products. *Journal of the Science of Food and Agriculture, 86*, 778–784.

Azam-Ali., & Judge, E. C. (2001). Small-scale cashew nut processing. Available at: http://www.fao.org/ag/ags/agsi/Cashew/Cashew.html.

Clinton, C. (2009). Plant tannins: a novel approach to the treatment of ulcerative colitis. *Journal of Natural Medicines* 1–4.

Chen, C.-Y., Milbury, P. E., Lapsley, K., & Blumberg, J. B. (2005). Flavonoids from almond skins are bioavailable and act synergistically with vitamins C and E to enhance hamster and human LDL resistance to oxidation. *Journal of Nutrition, 135*, 1366–1373.

Duenas, M., Hernandez, T., & Estrella, I. (2006). Assessment of *in vitro* antioxidant capacity of the seed coat and cotyledons of legumes in relation to their phenolic contents. *Food Chemistry, 98*, 95–103.

Kamath, V., & Rajini, P. S. (2007). The efficacy of cashew nut (*Anacardium occidentale* L.) skin extract as a free radical scavenger. *Food Chemistry, 103*, 428–433.

Kamath, V., Joshi, A. K. R., & Rajini, P. S. (2008). Dimethoate-induced biochemical perturbations in rat pancreas and its attenuation by cashew nut skin extract. *Pesticide Biochemistry and Physiology, 90*, 58–65.

Kantamoni, S. (1965). Isolation and characterization of the constituents of cashew kernel testa. *Leather Science, 12*, 396–398.

Mathew, A. G., & Parpia, H. A. B. (1970). Polyphenols of cashew kernel testa. *Journal of Food Sciences, 35*, 140–143.

Nagaraja, K. V. (1987a). Proteins of high yielding varieties of cashew (*Anacardium occidentale* L). *Plant Foods for Human Nutrition, 37*, 69–75.

Nagaraja, K. V. (1987b). Lipids of high yielding varieties of cashew (*Anacardium occidentale* L). *Plant Foods for Human Nutrition, 37*, 307–311.

Nagaraja, K. V. (1989). *In vitro* digestibility of cashew kernel proteins. *Current Science, 58*, 769–771.

Nagaraja, K. V. (2000). Biochemical composition of cashew (*Anacardium occidentale* L.) kernel testa. *Journal of Food Science and Technology, 37*, 554–556.

Nagaraja, K. V., & Nampoothiri, V. M. K. (1986). Chemical characterization of high yielding varieties of cashew. *Plant Foods for Human Nutrition, 36*, 201–206.

Nayudamma, Y., & Rao, K. (1967). Cashew testa. Its use in leather industry. *Indian Cashew Journal, 4*, 12–13.

Pillai, M. K. S., Kedlaya, K. J., & Selvarangan, R. (1963). Cashew seed skin as a tanning material. *Leather Sciences, 10*, 317–319.

Subramanian, S. S., & Nair, A. G. R. (1969). Catechins from cashew nut testa. *Current Sciences, 39*, 494–495.

Takahata, Y., Ohnishi-Kameyama, M., Furuta, S., Takahashi, M., & Suda, I. (2001). Highly polymerized procyanidins in brown soybean seed coat with a high radical scavenging activity. *Journal of Agricultural and Food Chemistry, 49*, 5843–5847.

Troszynska, A., & Balasinska, B. (2002). Antioxidant activity of crude tannins of pea (*Pisum sativum* L.) seed coat and their hypocholesterolemic effect in rats. *Polish Journal of Nutrition Sciences, 11*, 33–38.

Yen, W.-J., Chang, L.-W., & Duh, P.-D. (2005). Antioxidant activity of peanut seed testa and its antioxidative component, ethyl protocatechuate. *LWT-Food Science and Technology, 38*, 193–200.

CHAPTER 37

Whole and Ground Chia (*Salvia hispanica* L.) Seeds, Chia Oil – Effects on Plasma Lipids and Fatty Acids

Wayne Coates
The University of Arizona, Tucson, Arizona, USA

CHAPTER OUTLINE

LIST OF ABBREVIATIONS

ALA, α-linolenic acid
BV, biological value
CHD, coronary heart disease
CRP, C-reactive protein
DHA, docosahexaenoic acid
ELISA, enzyme-linked immunosorbant assay
EPA, eicosapentaenoic acid
FDA, US Department of Food Administration
HDL, high density lipoprotein

Nuts & Seeds in Health and Disease Prevention. DOI: 10.1016/B978-0-12-375688-6.10037-4

IgE, immunoglobulin E
LDL, low density lipoprotein
NEFA, non-esterified fatty acid
PUFA, polyunsaturated fatty acid
TAG, triacylglycerol
RNPR, relative net protein ration

INTRODUCTION

In pre-Columbian times, chia (*Salvia hispanica* L.) was one of the four basic foods of Central American civilizations. It was less important than corn and beans, but more important than amaranth. Chia seed was not just a food, but was also used for medical purposes, and as an offering to the Aztec gods.

The use of chia in Aztec religious ceremonies led the Spanish conquistadors to try to eliminate it, replacing it with species brought from the Old World. They came close to eradicating chia as a crop, and it was pushed into obscurity for 500 years, only being grown in small patches in scattered mountain areas of southern Mexico and northern Guatemala.

This was the situation until the Northwestern Argentina Regional Project began researching chia in 1991. At that time, growers, commercial entities, and scientific personnel began collaborating on the production of chia. The idea was to provide growers with an alternative crop, and to improve human health by reintroducing chia into Western diets as a source of omega-3 fatty acids, antioxidants, and fiber.

BOTANICAL DESCRIPTION

Chia belongs to the *Labiatae* or *Lamiaceae* family (mint family). It is an annual herb, normally about 1 m high, having branched quadrangular stems, and opposite leaves with a 40-mm petiole. The leaves are oval to oblong, almost smooth, with few pubescens that are whitish and very short. The leaves have serrated edges, and are 80−100 mm long by 40−60 mm wide. The flowers are produced in terminal or axillary spikes; the seeds are present in groups of four, and are a smooth, shiny, grayish-black color with irregular spots measuring 2 mm by 1.5 mm.

HISTORICAL CULTIVATION AND USAGE

There is evidence that chia seed was first used as food as early as 3500 BC, and was a cash crop in the center of Mexico between 1500 and 900 BC. Chia was cultivated in the Central Valley of Mexico between 2000 and 2600 BC by the Teotihuacan and Toltecs, and was one of the four main components of the Aztec diet. Interestingly, the main area of natural dispersion of chia coincides with the region that is considered the probable origin of the Aztecs.

The southern Mexican State of Chiapas, which is located within what was ancient Mayan territory, derives its name from the Nahuatl words *chia* and *apan* (meaning in or on the water). Hence, Chiapan means chia river or chia water. Chiapas City was founded on the banks of the Grijalba River, by people that were called *chiapas* or *chiapanecas*.

Chia was used by the Aztecs as food, mixed in water and consumed as a beverage, ground into flour, included in medicines, fed to birds, and pressed for oil for use as a base for face and body paints and to protect religious statues and paintings from the elements. Chia seed and flour could be stored for many years, and was considered to be a highly energetic grain; hence, it was an essential food when conducting military exercises. The priest Francisco Javier Clavijero (1780) refers to this use in the *Ancient History of Mexico*.

Friar Bernardino de Sahagun In 1579, described bars made of almonds, chia, and syrup being sold in markets. Chia was also used to prepare a refreshing and nourishing drink (Sahagun,

310

1989; Torquemada, 1723). It is quite probable that the drink *chia fresca*, consumed today in Mexico, Arizona, and California, comes from this.

Chia was also highly regarded for medicinal use. Friar Bernardino de Sahagun described (in the *Codex Florentino*) the use of chia for treatment of different illnesses, either consumed alone or along with other herbs (Sahagun, 1989). In the *Nueva Farmacopea de Mexico*, published in 1874, chia is listed as having pharmaceutical uses, and as an emollient which, when introduced into the eye, was used for extracting extraneous bodies.

PRESENT-DAY CULTIVATION AND USAGE
Chia production

Historically, chia has been cultivated in tropical and subtropical environments, from frost-free areas to regions where frosts occur every year, and from sea level up to 2500 m. Today, chia is commercially grown in Argentina, Ecuador, Peru, Paraguay, Australia, Mexico, Nicaragua, Guatemala, and Bolivia. In Argentina and Mexico it is a summer—autumn crop, similar to corn, soybeans, and beans. In Bolivia, chia is an autumn—winter crop, sown following the harvest of other crops, and competes with winter wheat and sunflower. In Ecuador the crop can be grown all year round, with three to four harvests possible, depending upon location.

Agronomic aspects

Chia is not frost-tolerant. It develops best in sandy loam soils; however, it can be grown in clay loam soils with adequate drainage. Generally, chia grows well in soils containing a range of nutrients; however, a low nitrogen content appears to be a barrier to good seed yield.

Chia is sown at a rate of 5—6 kg/ha, with row spacings of 0.7 or 0.8 m. It requires wet soil to germinate, but once seedlings are established chia does well with limited water. It can be grown either dryland or irrigated.

The first 45 days of growth are critical, as chia initially grows very slowly, and weeds can compete for light, nutrients, and water. As no herbicide has been found satisfactory for chia, weed control is of utmost importance when the plants are small. Once established, chia can be mechanically weeded until the canopy closes; weeds then become a minor problem under most conditions. Pests and diseases are essentially non-existent.

The seed used in commercial plantings is sufficiently uniform to allow harvest with a combine. Commercial seed yields generally are 500—600 kg/ha; however, some growers have obtained up to 1200 kg/ha. Experimental plots in Argentina have yielded 2500 kg/ha with irrigation and nitrogen fertilizer. Variations in yield indicate the need for germplasm that is adapted to a production zone, as well as good management practices to maximize commercial yields.

Planting date influences production. Biomass and seed yields have been significantly higher for plots planted earlier, rather than those planted later, even though different planting dates flowered at the same time. The difference in yield is probably due to the larger plants that developed because of a longer vegetative growth period.

Oil content and fatty acid composition are affected by location. A negative correlation between α-linolenic fatty acid content and mean temperatures has been found with chia (Ayerza & Coates, 2009).

Variations in chia life cycle

As chia is sensitive to day length, the growing season depends on the latitude where it is planted. One chia selection sown in Columbia was harvested in 90 days; however, when sown in Argentina it needed 150 days to mature. In higher latitudes, like Choele-Choel (39°11′ south) in Argentina, or Tucson (32°14′ north) in Arizona, chia does not produce seeds, because the plant is killed by frost before the seed matures.

311

Seed that originated in western Guatemala was compared with chia commercially grown in Argentina. Under typical northwestern Argentina conditions, it needed 30 days more to meet daylight requirements for flowering to begin, than did the commercial chia.

APPLICATIONS TO HEALTH PROMOTION AND DISEASE PREVENTION

Composition of chia

The lipid, protein, fiber, and antioxidant contents of chia are significantly higher than in many other crops. Although chia seed serves mainly as a source of omega-3 fatty acids, it also contains a number of other components that are important for human nutrition. The oil content of chia seed ranges between 30 and 34%, and it contains one of the highest percentages of α-linolenic acid known (62–64%). Tests by independent laboratories of seed coming from numerous sources (Coates, 2009, unpublished) showed total fatty acid contents ranging from a low of 22.9% to a high of 31.7%, with α-linolenic acid contents ranging from 14.1 to 20.47 g/100 g (wet basis). Regarding percentage of fatty acid, values ranged from 55 to 65.8%.

Chia seed contains 20–23% protein. This is higher than wheat (14.7%) and corn (14%). The lysine content is quite high, and methionine plus cystine compares favorably with other oil seeds; consequently, chia has no limiting factors in the adult diet from an amino acid standpoint (Weber *et al.*, 1991). This means chia can be incorporated into human diets, as a balanced protein source.

Pallaro and colleagues (2004) found, with Wistar rats, RNPR and BV were $75.1 \pm 3.4\%$ and $76.1 \pm 7.9\%$, respectively, with a digestibility of 74.81%. The methionine + cystine component was found to be the limiting factor, at 73.2 . Lysine was the next lowest, at 76.5 of preschool children's requirement.

Chia seed has 2.3, 2.6, 8.3, and 9.8 times more fiber per 100 g of edible portion than wheat, oats, corn, and rice, respectively. Reyes-Caudillo *et al.* (2008) found the total digestible fiber of chia seed to range from 36 to 41 g/100 g, depending upon the test method employed. No significant differences among methods were found in terms of soluble or insoluble fractions, with soluble fiber being slightly more than 6 g/100 g. Coates (unpublished, 2009) analyzed seed from various countries and locations within them, and found insoluble fiber to vary between 23 and 46%, and soluble fiber between 2.5 and 7.1%.

Water and methanol extracts of chia seed meal remains, following pressing to remove the oil, have demonstrated strong antioxidant activity (Taga *et al.*, 1984). These antioxidants make chia a stable source of omega-3 fatty acids, explaining why the Aztecs were able to store chia seed and flour for extended periods of time without them becoming rancid.

The main antioxidants in chia are chlorogenic and caffeic acids, as well as myricetin, quercetin, and kaempferol flavonols (Taga *et al.*, 1984; Castro-Martinez *et al.*, 1986). Caffeic and chlorogenic acids have been shown to inhibit lipid peroxidation, and are significantly stronger than common antioxidants such as vitamin C and vitamin E (Kweon *et al.*, 2001). Quercetin can prevent oxidation of lipids and proteins, and its antioxidant properties are significantly more effective than those of some flavonol compounds (Makris & Rossiter, 2001).

Reyes-Caudillo and colleagues (2008) looked at the antioxidant activity of chia seeds from two Mexican sources. They found quercetin and kaempferol to be the major components, with caffeic and chlorogenic acids present in low concentrations.

Ayerza and Coates (2009) analyzed four selections to determine whether chia seed composition varies when grown under the same conditions. Variations were found to exist, with Iztac II having the highest protein content (24.43%); however, this was only significantly ($P < 0.05$)

different from Iztac I. All selections showed a similar relationship regarding antioxidant composition, with caffeic acid > chlorogenic acid >> quercetin > kaempherol.

The most common form of chia found in the marketplace today is composed of approximately 95% black seed and 5% white, although there are some retailers selling pure black seed and others selling pure white seed. The seed most readily available is commonly referred to as black chia. Comparisons of pure white and pure black seed, when grown side by side, show relatively few differences (Ayerza & Coates, 2006, unpublished) There is evidence that black seed contains greater amounts of antioxidants than does white seed (Coates, 2008, unpublished).

Using chia to produce healthier foods

A number of studies (Ayerza & Coates, 2001; Azcona *et al.*, 2008) have fed chia to chickens, to incorporate omega-3 into either the eggs or the meat. These trials have shown good incorporation rates, often with a reduction in saturated fat content as well. When compared to flaxseed, incorporation rates have been higher. When fed to pigs (Coates & Ayerza, 2009), similar results have been found. Hence, feeding chia to domestic animals can lead to the production of healthier foods.

Effects of chia consumption on rat plasma

Several studies have fed chia to rats to assess its effect on plasma composition. In one (Ayerza & Coates, 2005), 24 male Wistar rats were fed *ad libitum* diets that contained equal energy levels derived from corn oil (T_1), chia seed (T_2), or chia oil (T_3), for 4 weeks. Rats fed chia showed a significant ($P < 0.05$) decrease in serum triglycerides. There was a significant ($P < 0.05$) increase in serum HDL content, with 22 and 51% increases for T_2 and T_3, respectively, with the difference between chia diets also being significant ($P < 0.05$). Total cholesterol was significantly ($P < 0.05$) lower for T_2 than T_1. Significantly higher ($P < 0.05$) α-linolenic fatty acid contents were found for T_2 and T_3, compared with T_1.

313

In a second study (Ayerza & Coates, 2007), 32 male Wistar rats were fed *ad libitum* isocaloric diets for 30 days, with the energy derived from corn oil (T_1), whole chia seed (T_2), ground chia seed (T_3), or chia oil (T_4). Chia decreased serum triglyceride and increased HDL contents. Only with the T_2 diet was triglyceride significantly ($P < 0.05$) lower, and only with the T_3 diet was HDL significantly ($P < 0.05$) higher, than the control diet. Chia significantly ($P < 0.05$) increased the 18:3n-3, 20:5n-3, and 22:6n-3 plasma contents compared to the control diet, with no significant ($P < 0.05$) difference among chia diets detected.

Chicco and colleagues (2009) fed chia seed to Wistar rats to examine its effect on adiposity, hypertriacylglycerolemia, and insulin resistance. The two test protocols fed either a sucrose-rich diet, or a sucrose-rich diet along with chia. Addition of chia led to lower weight gain and fat-pad weights, although differences were not statistically significant. Cholesterol was significantly lower in the chia group, with chia preventing development of peripheral insulin resistance and decreasing hyperinsulinemia in the rats fed the sucrose diet. An increase in long-chain omega-3 PUFA serum levels was found with chia seed.

Fernandez and colleagues (2008) assessed the effect of chia on the immune system of Wistar rats divided into three groups: 150 g/kg of diet as ground chia seed; 50 g/kg of chia oil; and no chia (control). No significant differences were observed in IgE levels; nor were symptoms of dermatitis, diarrhea, animal growth, or behavior observed. Hence, chia did not produce any of the problems found with other omega-3 fatty acid sources, such as flaxseed or marine products, like fishy flavor (feed rejection).

Effects of chia consumption on human plasma

A 28-day preliminary trial was conducted with 16 subjects (Coates & Ayerza, 2002, unpublished), in which half received 28 g of chia seed each day, the others receiving a placebo. Serum cholesterol, HDL, LDL, and triglyceride levels were measured the day before the trial began, and at the end of the trial. Results were inconclusive. Significant differences in cholesterol, HDL, LDL, and triglyceride levels between groups were not detected; however, an analysis of covariance showed HDL and triglyceride levels to differ between groups, with the difference favoring consumption of chia.

Vuksan and colleagues (2007) divided 20 adults with type 2 diabetes into two groups, providing one with 37 g of chia daily and the other with wheat bran, for 12 weeks. Chia significantly reduced ($P < 0.05$) systolic blood pressure, diastolic blood pressure, high-sensitivity C-reactive protein, and vonWillebrand factor. Total LDL and HDL cholesterol levels all decreased, but not significantly. Triglycerides increased, but not significantly.

Nieman and colleagues (2009) divided 90 subjects into two groups, one ingesting 25 g of chia seed twice daily, the other a placebo. At the end of 12 weeks, body mass and composition showed no differences between groups. Although plasma ALA increased significantly ($P < 0.05$) in the chia group, no significant differences in EPA and DHA were detected. No significant differences in disease risk factors, including serum CRP, plasma cytokines, blood lipoproteins, and blood pressure, were detected.

ADVERSE EFFECTS AND REACTIONS (ALLERGIES AND TOXICITY)

One aspect of any food that must be considered is allergies. A study conducted in the United Kingdom (Atkinson, 2003) found no evidence that chia exhibited any allergic response. This was the case even with individuals having peanut and tree nut allergies. In another study, Ayerza and Coates (2006, unpublished) used the enzyme-linked immunosorbent assay (ELISA) to compare the serum of rats fed chia seed (whole and ground) and chia oil to a control diet. As the ELISA test has found applications in the food industry in detecting potential food allergens such as milk, peanuts, walnuts, almonds, and eggs, it was thought that this might be useful for chia. Although the results of the trials were not statistically significant, the overall trend of the data did not support a finding that allergenicity was an issue.

SUMMARY POINTS

- Chia was one of the most important foods of the Aztecs.
- Chia is an old food that is rapidly gaining popularity as people learn not only of its existence, but also of its many benefits in human and animal nutrition.
- In 2005 the FDA concluded that chia was considered a food, since it has been used as such for hundreds of years, and that there have been no known issues related to its consumption.
- Chia has been shown to be an excellent source of omega-3 fatty acid, fiber (soluble and insoluble), protein, and antioxidants.
- When fed to chickens and pigs, chia improved the composition of the foods produced, making them healthier for human consumption.
- When fed to rats, chia decreased serum triglyceride levels and increased HDL-cholesterol and omega-3 fatty acid contents.
- Human trials have been less conclusive, with two of the three studies showing improved plasma composition.
- These findings suggest that chia appears to be an alternative omega-3 fatty acid source for vegetarians and people allergic to fish and fish products.

References

Atkinson, H. (2003). *Studies to assess the allergenic potential of chia* (Salvia hispanica L.) *BIBRA Report 4095/4194.* Carshalton, Surrey, UK: BIBRA International Ltd.

Ayerza, R., & Coates, W. (2001). The omega-3 enriched eggs: the influence of dietary alpha-linolenic fatty acid source combination on egg production and composition. *Canadian Journal of Animal Science, 81,* 355–362.

Ayerza, R., & Coates, W. (2005). Ground chia seed and chia oil effects on plasma lipids and fatty acids in the rat. *Nutrition Research, 25,* 995–1003.

Ayerza, R., & Coates, W. (2007). Effect of dietary α-linolenic fatty acid derived from chia when fed as ground seed, whole seed and oil on lipid content and fatty acid composition of rat plasma. *Nutrition & Metabolism, 51,* 27–34.

Ayerza, R., & Coates, W. (2009). Influence of environment on growing period and yield: protein, oil and alpha-linolenic content of three chia (*Salvia hispanica* L.) selections. *Industrial Crops and Products, 30,* 321–324.

Azcona, J. O., Schang, M. J., Garcia, P. T., Gallinger, C., Ayerza, R., & Coates, W. (2008). Omega 3 enriched broiler meat: The influence of α-linolenic-ω-3 fatty acid sources on growth, performance and meat fatty acid composition. *Canadian Journal of Animal Science, 88,* 257–269.

Castro-Martinez, R., Pratt, D. E., & Miller, E. E. (1986). Natural antioxidants of chia seeds. In American Oil Chemists' Society. (Ed.), *Proceedings of the World Conference on Emerging Technologies in the Fats and Oils Industry* (pp. 392–396). Champaign, IL: American Oil Chemists' Society Press.

Chicco, A. G., D'Alessandro, M. E., Hein, G. J., Oliva, M. E., & Lombardo, Y. B. (2009). Dietary chia seed (*Salvia hispanica* L.) rich in α-linolenic acid improves adiposity and normalizes hypertriacylglycerolaemia, an insulin resistance in dyslipaemic rats. *British Journal of Nutrition, 101,* 41–50.

Clavijero, F. J. (1780). *Storia antica del Messico avanta da miglioria storici spagnoli e da'manos critti, e dalla pitture antiche degli' indiani divisa in dieci libri, e corredata di corte geografiche, e di varce figure e dissertazioni sulla terra, sugli animali e sugli abitatori del Messino.* Bologna, Italy: Gregorio Biasini.

Coates, W., & Ayerza, R. (2009). Chia (*Salvia hispanica* L.) seed as an ω-3 fatty acid source for finishing pigs: effects on fatty acid composition and fat stability of the meat and internal fat, growth performance, and meat sensory characteristics. *Journal of Animal Sciences, 87,* 3798–3804, doi:10.2527/jas.2009-1987.

Fernandez, S., Vidueiros, M., Ayerza, R., Coates, W., & Pallaro, A. (2008). Impact of chia (*Salvia hispanica* L) on the immune system: preliminary study. *Proceedings of the Nutrition Society, 67.* Issue OCE, E12.

Kweon, M. H., Hwand, H. J., & Sung, H. C. (2001). Identification and antioxidant activity of novel chlorogenic acid derivatives from bamboo (*Phyllostachys edulis*). *Journal of Agricultural and Food Chemistry, 49,* 4646–4655.

Makris, D. P., & Rossiter, J. T. (2001). Comparison of quercetin and a non-orthohydroxy flavonol as antioxidant by competing *in vitro* oxidation reactions. *Journal of Agricultural and Food Chemistry, 49,* 3370–3377.

Nieman, D. C., Cayea, E. J., Austin, M. D., Henson, D. A., McAnulty, S. R., & Jin, F. (2009). Chia seed does not promote weight loss or alter disease risk factors in overweight adults. *Nutrition Research, 29,* 414–418.

Pallaro, A. N., Feliú, M. S., Vidueiros, S. M., Slobodiank, N., Ayerza, R., Coates, W., et al. (2004). Study of a non-traditional source of protein. Congreso Internacional de Ciencia y Tecnología de los Alimentos. p. 1.

Reyes-Caudillo, E., Tecante, A., & Valdivia-Lopez, M. A. (2008). Dietary fiber and antioxidant activity of phenolic compounds present in Mexican chia (*Salvia hispanica* L.) seeds. *Food Chemistry, 107,* 656–663.

Sahagun, B. (1989). Historia general de las cosas de Nueva Espana (*Codex Florentino*). In A. M. Garibay (Ed.). Mexico City: Editorial Porrua. 1579.

Taga, M. S., Miller, E. E., & Pratt, D. E. (1984). Chia seeds as a source of natural lipid antioxidants. *Journal of the American Oil Chemists' Society, 61,* 928–931.

Torquemada, J. (1723). *Veinte I un libros riturales I monarquia Indiana con el origen y guerras de lost indios ocidentales de sus poblaciones descubrimiento, conquista, conversion y otras cosas maravillosas de la mesma tierra distybuidos en tres tomos. Pt. 2.* Madrid, Spain: Office and Acosta by Nicolas Rodriguez Franco. 1615.

Vuksan, V., Whitham, E., Sievenpiper, J. L., Jenkins, A. L., Rogovik, A. L., Bazinet, R. P., et al. (2007). Supplementation of conventional therapy with the novel grain Salba (*Salvia hispanica* L.) improves major and emerging cardiovascular risk factors in type 2 diabetes. *Diabetes Care, 30.* 2804-2011.

Weber, C. W., Gentry, H. S., Kohlhepp, E. A., & McCrohan, P. R. (1991). The nutritional and chemical evaluation of chia seeds. *Ecology of Food & Nutrition, 26,* 119–125.

315

Antiproliferative Activities of Chinese Cabbage (*Brassica parachinensis*) Seeds

Tzi-Bun Ng[1], Jack H. Wong[1], Randy C.F. Cheung[1], Patrick H.K. Ngai[2], He-Xiang Wang[3], Xiujuan Ye[1], Sze-Kwan Lam[1], Yau-Sang Chan[1], Evandro F. Fang[1]
[1] School of Biomedical Sciences, Faculty of Medicine, The Chinese University of Hong Kong, Hong Kong, China
[2] School of Life Sciences, Faculty of Science, The Chinese University of Hong Kong, Hong Kong, China
[3] State Key Laboratory of Agrobiotechnology, Department of Microbiology, China Agricultural University, Beijing, China

317

INTRODUCTION

The name "Chinese cabbage" is applied to many types and varieties, including *Brassica campestris* and *Brassica parachinensis*.

Napins and napin-like polypeptides have been isolated from the seeds of *Brassica* species (Ngai & Ng, 2004). There are also reports about the antifungal activity of *Brassica* seeds such as *B. juncea* and *B. alboglabra* (Lin & Ng, 2008; Ye & Ng, 2009). The isolation and characterization of a lipid transfer protein with remarkably stable antifungal and antiproliferative activities from *B. campestris* seeds, a napin-like protein with antiproliferative and nitric oxide inducing activities, and an antifungal protein from *B. parachinensis* seeds have been described (Ng & Ngai, 2004; Lin *et al.*, 2007).

Nuts & Seeds in Health and Disease Prevention. DOI: 10.1016/B978-0-12-375688-6.10038-6

BOTANICAL DESCRIPTION

Brassica parachinensis, or "choy sum" (literally means the heart or the center of a vegetable), is a small, branching, green-stemmed, yellow-flowered plant, similar to but not the same as *B. chinensis*. This brassica refers to a small, delicate version of *B. chinensis*. It more closely resembles rapini or broccoli rabe in appearance than the typical *B. chinensis*. In English, it is also known as the "flowering Chinese cabbage" because of its little yellow flowers.

Brassica campestris, commonly known as field mustard or turnip mustard, is a plant widely cultivated as a leaf vegetable, a root vegetable, and an oilseed.

HISTORICAL CULTIVATION AND USAGE

Chinese cabbage originated in China, where it was discovered around the 5th century AD. No wild cabbage has ever been found, and it is probably a cross between the southern *B. chinensis* and the northern turnip. Contemporary varieties are primarily Japanese hybrids; however, the vegetable did not arrive in Japan until the 1860s, and there was no breeding until the 1920s. *B. parachinensis* has elliptical green foliage and small yellow flowers. Either the shoot tips with the edible flowers are harvested, or the entire plant is picked.

PRESENT-DAY CULTIVATION AND USAGE

Chinese cabbage has been cultivated in Asia since the 5th century AD, but cultivation began in the United States only about a century ago. It is now grown for sale mainly in California, New Jersey, Hawaii, and Florida. In Florida, the main production sites lie in the organic soils of the Everglades and Central Florida, although some production takes place in other counties on sandy soils — for example, in Martin County. Home gardeners around the state grow Chinese cabbage in the fall, and also in the winter.

APPLICATIONS TO HEALTH PROMOTION AND DISEASE PREVENTION

Napins are storage proteins that are 1,1 disulfide-linked complexes of a 4.5-kDa small subunit and a 10-kDa large subunit (Gehrig & Biemann, 1996). Napins from the oilseed rape *Brassica napus* are the most extensively studied (Ye & Ng, 2001). The nitrogen storage function of napin is reflected in its rich content of amides and arginine residues (Muntz, 1998). Napins exhibit trypsin-inhibiting activity, but the subunits have no such activity. However, both napin and its subunits serve as calmodulin antagonists and as substrates for plant calcium-dependent protein kinases, because calmodulin and its small subunit manifest similar α-helix—hinge—α-helix motifs (Neumann *et al.*, 1996a, 1996b). Napins can reduce the activity of calmodulin-dependent myosin light-chain kinase (Gehrig & Biemann, 1996; Neumann *et al.*, 1996a, 1996b). Napins may also exhibit antifungal action (Barciszewki *et al.*, 2000).

A napin-like polypeptide has been isolated from *Brassica parachinensis* seeds. The seeds were soaked and homogenized, and the supernatant was loaded on a DEAE-cellulose column. The unbound fraction with translation-inhibitory activity and eluted with 10 mM Tris-HCl buffer (pH 7.4) was chromatographed on an Affi-gel blue gel column. Following elution of un-absorbed proteins without translation-inhibitory activity, adsorbed proteins were desorbed with a linear NaCl concentration (0—2 M) gradient in 10 mM Tris-HCl buffer (pH 7.4). The second adsorbed peak with translation-inhibitory activity was then dialyzed before cation exchange chromatography on a Mono S column using fast protein liquid chromatography (FPLC). After elution of unadsorbed proteins without translation-inhibitory activity, adsorbed proteins were eluted with two consecutive linear NaCl concentration gradients (0—0.3 M and 0.3—1 M) in 10 mM NH$_4$OAc (pH 5.5). The peak eluted with the first NaCl gradient, which exhibited translation-inhibitory activity, was resolved by FPLC-gel filtration on a Superdex 75

HR column in 20 mM NH_4HCO_3 buffer (pH 9.4). The first eluted peak constituted purified napin-like polypeptide.

The napin-like polypeptide from Chinese cabbage (*Brassica parachinensis*) seeds had a molecular weight of 13.8 kDa and is heterodimeric. The *N*-terminal sequence of the 8.8-kDa subunit of the polypeptide (PQGPQQRPPKLLQQQTNEEHE) demonstrated marked similarity to napins, albumins, and trypsin inhibitors, but exhibited little resemblance to the 5-kDa subunit. The 5-kDa subunit of *B. parachinensis* napin-like polypeptide was highly homologous in *N* terminal sequence to napins isolated from other species (Ngai & Ng, 2003). The inhibitory translation activity of *Brassica parachinensis* napin-like polypeptide in a cell-free rabbit reticulocyte lysate system (Ngai & Ng, 2003) was relatively stable at pH 6–11 and at 10–50°C (Ngai & Ng, 2003). The polypeptide enhanced nitrite production by mouse peritoneal macrophages. The viability of L1210 leukemia cells was reduced dose-dependently in response to the napin-like polypeptide. At a dose of 300 μg, the polypeptide did not exert antifungal activity on fungi. It did not exhibit ribonuclease activity when tested at 100 μg. Its trypsin inhibitory activity was more potent than its chymotrypsin inhibitory activity, but it lacked ribonuclease and antifungal activities (Ng & Ngai, 2004).

The 8.8-kDa subunit of *B. parachinensis* napin-like polypeptide displayed little sequence resemblance to the 5-kDa subunit. However, there was a pronounced homology to napins and to trypsin inhibitor. The napin-like polypeptide from *B. parachinensis* dose-dependently inhibited trypsin and chymotrypsin when the molar ratio of napin-like polypeptide to protease was raised. It enhanced nitric oxide production from murine peritoneal macrophages in a concentration-dependent manner. The antiproliferative activity of *B. parachinensis* napin-like polypeptide toward a leukaemia cell line is reminiscent of findings on some antifungal proteins (Lam *et al.*, 2000), ribosome inactivating proteins (Lam & Ng, 2001), lectins (Wong & Ng, 2003), and trypsin inhibitors (Ng *et al.*, 2003).

A lipid transfer protein (LTP) was prepared from *B. campestris* seeds. The crude seed extract was applied to anion exchange chromatography on a column of Q-Sepharose previously equilibrated and then eluted with 10 mM Tris-HCl buffer (pH 7.8). After the unadsorbed proteins (fraction Q1) had been eluted, the column was eluted with 10 mM Tris-HCl buffer (pH 7.8) containing 1 M NaCl to yield fraction Q2. Fraction Q1 was then loaded on an Affi-gel blue gel column in 10 mM Tris-HCl buffer (pH 7.8). Unadsorbed proteins (fraction B1) were eluted with the same buffer. Adsorbed proteins (fraction B2) were desorbed with 10 mM Tris-HCl buffer (pH 7.8) containing 1 M NaCl. Fraction B2 was chromatographed on a Mono S column in 10 mM NH_4OAc buffer (pH 4.5). Following removal of the unadsorbed proteins, the column was eluted with a linear 0–0.1 M NaCl gradient in the starting buffer to desorb the first adsorbed fraction S2, which was then purified on a Superdex Peptide column. The main peak constituted the purified antifungal peptide.

The purified peptide displayed a molecular mass below 14.4 kDa in SDS-PAGE, 9.4 kDa in gel filtration in Superdex peptide, and 9414 kDa in mass spectrometry. Its *N*-terminal sequence closely resembled those of nsLTPs from seeds and leaves of other species, including that of *P. mungo*. It inhibited mycelial growth in *Fusarium oxysporum*, *Helminthosporium staivum*, *Mycosphaerella arachidicola*, *Sclerotinia sclerotiorum, and Verticivium albotarum*. The IC_{50} values of its antifungal activity toward *F. oxysporum* and *M. arachidicola* were 8.3 μM and 4.5 μM, respectively. The antifungal activity was stable at various temperatures between 20 and 100°C; at various pH values between 0 and 4, and between 9 and 14; and after treatment with either trypsin or pepsin for 1–4 hours.

B. campestris LTP exerted antiproliferative activity toward hepatoma Hep G2 cells and breast cancer MCF 7 cells with IC_{50} values of 5.8 and 1.6 μM, respectively, and inhibited the activity of HIV-1 reverse transcriptase with an IC_{50} of 4 μM. *Brassica* LTP had no antibacterial activity, and was devoid of mitogenic activity toward splenocytes (Lin *et al.*, 2007).

From the seeds of *Brassica parachinensis*, a 5716-Da antifungal peptide designated as brassiparin was purified by using a procedure entailing ion exchange chromatography on SP-Sepharose, affinity chromatography on Affi-gel blue gel, and gel filtration on Superdex Peptide. It inhibited mycelial growth in *Fusarium oxysporum*, *Helminthosporium maydis*, *Mycosphaerella arachidicola*, and *Valsa mali* with IC_{50} values of 3.9, 4.7, 2.6, and 0.22 µM, respectively. The antifungal activity of brassiparin toward *M. arachidicola* displayed marked thermostability (40–100°C) and pH stability (pH 1–3 and 10–13). It inhibited proliferation of hepatoma (HepG2) and breast cancer (MCF7) cells and the activity of HIV-1 reverse transcriptase with IC_{50} values of 19.2, 4.8, and 17.8 µM, respectively. Its N-terminal sequence differed from antifungal proteins reported to date. Brassiparin represents one of the few antifungal proteins reported to date from *Brassica* species. Its antifungal activity has pronounced pH stability and thermostability. Brassiparin exhibits other exploitable activities, such as antiproliferative

TABLE 38.1 Characteristics of Napin-like Polypeptide with Antiproliferative Activity from Chinese Cabbage Seeds (Ngai & Ng, 2003)

	Napin-like Polypeptide from *Brassica parachinensis* Seeds
Molecular mass	Heterodimeric (one 8.8-kDa subunit and one 5-kDa subunit)
N-terminal sequence	8.8-kDa subunit PQGPQQRPPKLLQQQTNEEHE 5-kDa subunit, PAGPFRIPKKRKKEE
Chromatographic behavior on	
i) DEAE-cellulose	Unadsorbed
ii) Affi-gel blue gel	Adsorbed
iii) Mono S	Adsorbed
iv) Superdex75	
Antiproliferative activity toward	
i) L1210 cells	$IC_{50} = 3$ µM
Antifungal activity	Absent
Cell-free translation inhibitory activity	$IC_{50} = 62$ nM, stable at pH 6–11 and 10–50°C
Macrophage nitric oxide stimulatory activity	0.1–1 µM
Trypsin and chymotrypsin inhibitory activity	Present

TABLE 38.2 Characteristics of Lipid Transfer Protein with Antiproliferative Activity from Chinese Cabbage Seeds (Lin *et al.*, 2007)

	Lipid Transfer Protein from *Brassica campestris* Seeds
Molecular mass	Monomeric (9 kDa)
N-terminal sequence	ALSCGTVSGNLAACAGYV
Chromatographic behavior on	
i) Affi-gel blue gel	Adsorbed
ii) Mono S	Adsorbed
iii) SP-Sepharose	Adsorbed
iv) Superdex Peptide	
Antiproliferative activity toward	
i) HepG2 cells	$IC_{50} = 5.8$ µM
ii) MCF-7 cells	$IC_{50} = 1.6$ µM
HIV-1 reverse transcriptase inhibitory activity	$IC_{50} = 4$ µM
Antifungal activity	Thermostable and pH-stable activity against various fungal spp.
Trypsin and chymotrypsin inhibitory activity	Absent

TABLE 38.3 Characteristics of Antifungal Peptide with Antiproliferative Activity from Chinese Cabbage Seeds (Lin & Ng, 2009)

	Antifungal Peptide from *Brassica parachinensis* Seeds
Molecular mass	Monomeric (5.7 kDa)
N-terminal sequence	DQFPQEYPGDVQFSFNALHIYPSPQVVV
Chromatographic behavior on	
v) Affi-gel blue gel	Adsorbed
vi) Mono S	Adsorbed
vii) Q-Sepharose	Adsorbed
viii) Superdex Peptide	
Antiproliferative activity toward	
iii) HepG2 cells	$IC_{50} = 19.2\ \mu M$
iv) MCF-7 cells	$IC_{50} = 4.8\ \mu M$
HIV-1 reverse transcriptase inhibitory activity	$IC_{50} = 17.8\ \mu M$
Antifungal activity	Thermostable and pH-stable activity against various fungal spp.
Trypsin and chymotrypsin inhibitory activity	Absent

activity toward hepatoma and breast cancer cells, and inhibitory activity toward HIV-reverse transcriptase (Lin & Ng, 2009).

Characteristics of a napin-like protein, an antifungal peptide from *B. parachinensis* seeds, and a lipid transfer protein from *B. campestris* seeds, all with antiproliferative activity on tumor cells, are summarized in Tables 38.1–38.3.

ADVERSE EFFECTS AND REACTIONS (ALLERGIES AND TOXICITY)

Cabbage may be goitrogenic due to interference with iodine uptake, organification in thyroid cells, and inhibition of the production of thyroxine and triiodothyronine. The enhanced secretion of thyroid stimulating hormone due to low negative feedback of thyroid hormones causes hypertrophy of the thyroid (goiter). Thus, patients with goiter have to avoid eating cabbage.

Glucosinolate, sulfur-containing organic anions bonded to glucose, is present in plants, including cabbages. Glucosinolate is hydrolyzed by myrosinase, and various by-products are generated. One such product, thiocyanate, exerts harmful effects on the thyroidal metabolism owing to competition with iodine. The Korean diet contains large amounts of cabbage and radishes, which may be correlated with a high incidence of thyroid dysfunction. Whether the average daily intake of thiocyanate through Brassicaceae vegetables in Korea has any effect on health remains to be seen (Han & Kwon, 2009).

Allergy to cabbage is rare. A case of facial and throat swelling in an atopic female after eating cabbage has been reported. The patient presented 4+ reactions to *Brassica* plants by skin testing. A RAST employing cabbage extract was positive for specific IgE antibody. Thus, IgE sensitivity to *Brassica* plants can occur (Blaiss *et al.*, 1987).

Some people experience digestive discomfort because of gas originating from cabbage. This can be avoided by blanching the entire or chopped cabbage for 5 minutes, replacing the water with fresh water, and then cooking the cabbage.

Toxic amounts of S-methyl-L-cysteine sulfoxide, an alpha-amino acid that can induce hemolytic anemia in livestock, are produced by all *Brassica* vegetables. However, these vegetables are usually safe for human consumption. Poisoning occurs only when they are used almost exclusively as animal fodder (Benevenga *et al.*, 1989).

SUMMARY POINTS

- The viability of leukemia L1210 cells cultured *in vitro* is reduced in the presence of a 13.8-kDa heterodimeric napin-like polypeptide from Chinese cabbage (*Brassica parachinensis*) seeds.
- The polypeptide also stimulates mouse peritoneal macrophages to produce nitric oxide. It inhibits trypsin with a higher potency than it inhibits chymotrypsin; however, it does not exhibit antifungal activity.
- The viability of hepatoma HepG2 cells and breast cancer MCF-7 cells is reduced by a 9-kDa lipid transfer protein from *Brassica campestris* seeds with IC_{50} values of 5.8 and 1.6 μM, respectively.
- The lipid transfer protein also displays thermostable, pH-stable, and protease-stable antifungal activity. It has no mitogenic activity toward mouse splenocytes; nevertheless, it inhibits HIV-1 reverse transcriptase with an IC_{50} of 4 μM.
- A 5716-Da antifungal peptide from *B. parachinensis* seeds inhibits proliferation of HepG2 and MCF7 cells with IC_{50} values of 19.2 and 4.8 μM, respectively. It also inhibits HIV-1 reverse transcriptase.

References

Barciszewski, J., Szymanski, M., & Haertle, T. (2000). Minireview. Analysis of rape seed napin structure and potential roles of the storage protein. *Journal of Protein Chemistry, 19,* 249−254.

Benevenga, N. J., Case, G. L., & Steele, R. D. (1989). Occurrence and metabolism of s-methyl-l-cysteine and s-methyl-l-cysteine sulfoxide plants and their toxicity and metabolism in animals. In P. R. Cheeke (Ed.), *Toxicants of plant origin, Vol. III, Proteins and amino acids* (pp. 203−228). Boca Raton, FL: CRC Press.

Blaiss, M. S., McCants, M. L., & Lehrer, S. B. (1987). Anaphylaxis to cabbage, detection of allergens. *Annals of Allergy, 58,* 248−250.

Gehrig, P. M., & Biemann, K. (1996). Assignment of the disulfide bonds in napin, a seed storage protein from *Brassica napus,* using matrix-assisted laser description ionization mass spectrometry. *Peptide Research, 9,* 308−314.

Han, H., & Kwon, H. (2009). Estimated dietary intake of thiocyanate from Brassicaceae family in Korean diet. *Journal of Toxicology and Environmental Health, 72,* 1380−1387.

Lam, S. K., & Ng, T. B. (2001). Hypsin, a novel thermostable ribosome-inactivating protein with antifungal and antiproliferative activities from fruiting bodies of the edible mushroom. *Hypsizigus marmoreus. Biochemical and Biophysical Research Communications, 285,* 1071−1075.

Lam, Y. W., Wang, H. X., & Ng, T. B. (2000). A robust cysteine-deficient chitinase-like antifungal protein from inner shoots of the edible chive. *Allium tuberosum. Biochemical and Biophysical Research Communications, 279,* 74−80.

Lin, P., & Ng, T. B. (2008). A novel and exploitable antifungal peptide from kale (*Brassica alboglabra*) seeds. *Peptides, 29,* 1664−1671.

Lin, P., & Ng, T. B. (2009). Brassiparin, an antifungal peptide from *Brassica parachinensis* seeds. *Journal of Applied Microbiology, 106,* 554−563.

Lin, P., Xia, L., Wong, J. H., Ng, T. B., Ye, X., Wang, S., et al. (2007). Lipid transfer proteins from *Brassica campestris* and mung bean surpass mung bean chitinase in exploitability. *Journal of Peptide Science, 13,* 642−648.

Muntz, K. (1998). Deposition of storage proteins. *Plant Molecular Biology, 38,* 77−79.

Neumann, G. M., Condron, R., Thomas, I., & Polya, G. M. (1996a). Purification and sequencing of multiple forms of *Brassica napus* seed napin large chains that are calmodulin antagonists and substrates for plant calcium dependent protein kinases. *Biochimica Biophysica Acta, 1295,* 23−33.

Neumann, G. M., Condron, R., Thomas, I., & Polya, G. M. (1996b). Purification and sequencing of multiple forms of *Brassica napus* seed napin small chains that are calmodulin antagonists and substrates for plant calcium dependent protein kinases. *Biochimica Biophysica Acta, 1295,* 34−43.

Ng, T. B., & Ngai, P. H. (2004). The trypsin-inhibitory, immunostimulatory and antiproliferative activities of a napin-like polypeptide from Chinese cabbage seeds. *Journal of Peptide Science, 10,* 103−108.

Ng, T. B., Lam, S. K., & Fong, W. P. (2003). A homodimeric sporamin-type trypsin inhibitor with antiproliferative, HIV reverse transcriptase-inhibitory and antifungal activities from wampee (*Clausena lansium*) seeds. *Biological Chemistry, 384,* 289−293.

Ngai, P. H., & Ng, T. B. (2003). Isolation of a napin-like polypeptide with potent translation-inhibitory activity from Chinese cabbage (*Brassica parachinensis* cv green-stalked) seeds. *Journal of Peptide Science, 9,* 442–449.

Ngai, P. H., & Ng, T. B. (2004). A napin-like polypeptide with translation-inhibitory, trypsin-inhibitory, anti-proliferative and antibacterial activities from kale seeds. *Journal of Peptide Research, 64,* 202–208.

Wong, J. H., & Ng, T. B. (2003). Purification of a trypsin-stable lectin with antiproliferative and HIV-1 reverse transcriptase inhibitory activity. *Biochemical and Biophysical Research Communications, 301,* 545–550.

Ye, X. Y., & Ng, T. B. (2001). Peptides from pinto bean and red bean with sequence homology to cowpea 10-kDa protein precursor exhibit antifungal, mitogenic and HIV-1 reverse transcriptase inhibitory activities. *Biochemical and Biophysical Research Communications, 285,* 424–429.

Ye, X., & Ng, T. B. (2009). Isolation and characterization of juncin, an antifungal protein from seeds of Japanese takana (*Brassica juncea* var. integrifolia). *Journal of Agricultural and Food Chemistry, 57,* 4366–4371.

CHAPTER 39

Cancer and Treatment with Seeds of Chinese Fan Palm (*Livistona chinensis* R. Brown)

Wen-Chuan Huang[1], Yann-Lii Leu[2], Jau-Song Yu[1,3]
[1] Graduate Institute of Biomedical Sciences, College of Medicine, Chang Gung University, Tao-Yuan, Taiwan, Republic of China
[2] Graduate Institute of Natural Products, College of Medicine, Chang Gung University, Tao-Yuan, Taiwan, Republic of China
[3] Department of Cell and Molecular Biology, College of Medicine, Chang Gung University, Tao-Yuan, Taiwan, Republic of China

325

LIST OF ABBREVIATIONS

CK2, casein kinase 2
EGF, epidermal growth factor
EGFR, epidermal growth factor receptor
GSK-3α, glycogen synthase kinase 3α
JNK1, c-Jun N-terminal kinase 1
LC, *Livistona chinensis* R. Brown
MALDI-TOF, matrix-assisted laser desorption/ionization-time of flight
MAPK, mitogen-activated protein kinase
NPC, nasopharyngeal carcinoma
PAK2, p21-activated kinase 2
PARP, poly (ADP-ribose) polymerase
PKA, cAMP-dependent protein kinase
PKC, protein kinase C

Nuts & Seeds in Health and Disease Prevention. DOI: 10.1016/B978-0-12-375688-6.10039-8

INTRODUCTION

The seed of *Livistona chinesis* R. Brown (LC) is a traditional Chinese medical herb used in Southern China for treating various tumors, including nasopharyngeal carcinoma (NPC). The ethanolic and/or water extracts of the seeds of LC have been reported to possess antitumor, anti-angiogenic, antioxidant, and antibacterial activity in both *in vitro* and *in vivo* studies. This chapter aims briefly to review these studies and discuss the most recent progress in the possible underlying mechanisms through which the active components in LC extracts can exert their antitumor activity.

BOTANICAL DESCRIPTION

LC is one of the 36 species of the *Livistona* genus, Arecaceae family (palms), and is distributed throughout Southern China, Southern Japan, and Taiwan. It is an evergreen tree with a straight and non-branching trunk that can grow up to 10–15 m high and 30 cm in diameter. It has a rough taupe stipe, and the large, bright-green fan-shaped leaves are 1.2–1.5 m in length and 1.5–1.8 m wide. The leaves droop downward to give a graceful fountain-like aspect, inspiring its alternative common name of the Chinese fountain palm. The petiole is 1–2 m in length, triangular in shape, and has a jagged thorny edge. The flower stalk comes from the base of the leaf, and is 1.8 m long; yellow flowers are produced in late spring (March to April). When the oval or subglobose seeds are ripe, the seed color changes from dark blue to blue-gray.

HISTORICAL CULTIVATION AND USAGE

There are no historical references regarding the cultivation of LC. It has traditionally been used as an anticancer drug in Chinese folk medicine; usage involves decocting the dried seed of LC (37.5 g) with lean pork in boiling water for about 1–2 hrs (Quan Guo Zhong Cao Yao Hui Bian, 1996).

PRESENT-DAY CULTIVATION AND USAGE

LC grows well in moist, organically-rich soils, and thrives at 20–30°C in sheltered and sunny situations. The stones of LC can be preserved under dry, ventilated conditions for 2 years after removed from the pericarp. Seed should be sown in the spring to avoid frost damage. The usage of LC in treating various cancers is listed below (information obtained from the Tak Tai Ginseng Firm Ltd, Hong Kong):

- Esophageal or nasopharyngeal cancer: put 37.5 g LC seed and 150 g lean pork in simmering water to make a decoction
- Nasopharyngeal cancer: put 75 g seed and 75 g root of LC in simmering water for nasopharyngeal cancer.
- Choriocarcinoma: put 75 g LC seed and 75 g *Scutellaria barbata* Don over simmering water
- Early leukemia: put LC seed, roots of *Plumbago zeylanica*, *Hedyotis diffusa* Wilid, *Verbena officinalis* Linn (each 37.5 g), and *Prunella vulgaris* L (18.75 g) in simmering water for early leukemia.

APPLICATIONS TO HEALTH PROMOTION AND DISEASE PREVENTION

The seeds of LC have long been documented in several Chinese medical books as being a useful herb for treating various cancers, including nasopharyngeal carcinoma (NPC), esophageal cancer, and leukemia. To the best of our knowledge, the first English literature reporting the antitumor activity of extracts of seeds of LC (hereafter designated as LC extracts) appeared in 1987; this showed that the hot water extracts of LC significantly inhibited the growth of murine B16-F10 melanoma cells *in vitro*, as well as the tumor nodules metastatic to lung in C57BL/6 mice when these mice were given LC extracts orally every other day for

3 weeks (Liu *et al.*, 1987). Following that, Sartippour and colleagues (2001) reported the antiproliferative activity of hot water LC extracts against human umbilical vein endothelial cells (HUVEC), and cancer cell lines (mouse fibrosarcoma FSA, human breast cancer MCF-7, and human colon cancer HT-29) *in vitro*, as well as the subcutaneous growth of fibrosarcoma tumors in mice. In addition, they found that antitumor activity existed in the shell of LC seeds but not in the seed kernels.

In 2005, investigations by Cheung and Tai showed that both ethanolic extract and hot water extract of LC inhibited the growth of the human premyelomonocytic cell line HL60, and that the ethanolic extract, but not the water extract, of LC could significantly induce HL60 cell differentiation into monocyte/macrophage lineage (Cheung & Tai, 2005). They used cell-free Trolox equivalent antioxidant capacity hydroxyl radical scavenging assay to demonstrate that both ethanolic extract and hot water extract of LC displayed similar antioxidant activity. However, they did not detect any effects of either extract on the nitric oxide production in non-stimulated or in lipopolysaccharide (LPS)-stimulated mouse RAW 264.7 cells. Moreover, they found that although neither extract affected the mRNA expression of tumor necrosis factor-α and inducible nitric oxide synthase in LPS-stimulated cells, both extracts could reduce interleukin-1β mRNA expression in LPS-stimulated cells, and only the ethanolic extract had the ability to inhibit cyclooxygenase-2 mRNA expression in LPS-stimulated cells (Cheung & Tai, 2005). Collectively, these experimental results partly confirm the past observations regarding use of the seeds of LC as a medicinal herbal integrant for cancer treatment. These results also indicate the existence of common active component(s) in both ethanolic and hot water LC extracts, as well as unique active component(s) in the ethanolic LC extract.

In addition to the antiproliferative activity against tumor cells, results from some recent studies also indicate the anti-angiogenic potential of the LC extract. As mentioned above, Sartippour *et al.* (2001) reported the antiproliferative activity of hot water LC extracts against HUVEC, a simple *in vitro* assay to preliminarily assess the anti-angiogenic potential of a substance. The water extract of LC seeds has been treated as a natural health product with potential direct and indirect anti-angiogenic activity (Sagar *et al.*, 2006). Strong evidence for the LC extract being potentially useful anti-angiogenic agent came from a recent study by Wang and colleagues (2008). They showed that both secretion of vascular endothelial growth factor (VEGF) protein by four human tumor cell lines (chronic myelogenous leukemia K562, ovarian neoplasm SKOV3, colon carcinoma HT-29, and bladder cancer T24) and VEGF-induced expression of fetal liver kinase (Flk-1, also known as VEGF receptor-2) protein and mRNA in HUVEC could be significantly inhibited by the ethyl acetate part of an alcohol extract from seeds of LC. As both secretion of VEGF by tumor cells and the response of HUVEC to VEGF are critical events for tumor growth and angiogenesis (Kerbel, 2008), these findings suggest that direct interference in the VEGF-mediated signaling pathway may partly explain the antitumor effect of the LC extracts.

Although several studies have shown the antitumor activity of LC extracts *in vitro* and *in vivo*, little is known about the molecular mechanism responsible for these observed effects. In 1995, a preliminary study performed by Huang and colleagues showed that the ethanolic extract of LC seeds contains inhibitory activity towards protein kinase C (PKC), a family of protein kinases that play important roles in modulating diverse cell functions and pathological settings (Huang *et al.*, 1995; Cohen, 2002; Mackay & Twelves, 2007). However, it is unclear whether the protein kinase inhibitory activity in LC extracts contributes to the antitumor property of this herb. To address this issue, we carried out a series of experiments to partly isolate the active components with potent protein kinase inhibitory activity in the ethanolic LC extracts, and characterize its effects on cell cycle progression and the intracellular signaling pathway in human tumor cell lines (Huang *et al.*, 2007). We traced both protein kinase inhibitory activity (using the catalytic fragment of PAK2 as the kinase source) and anti-proliferative activity (using the human NPC cell line NPC-TW02 as the model cell) during the purification of active components by a serial of column chromatography, and found that both activities were well co-eluted in all the purification steps. This finding suggests that the anti-proliferative activity of LC extracts could be mainly attributed to the protein kinase inhibitory activity. Characterization of a highly purified fraction, LC-X, with the most potent inhibitory activity obtained after the last chromatography step revealed that LC-X can inhibit various protein kinases *in vitro*, including PAK2, PKA, PKC, GSK-3α, CK2, MAPK, and JNK1. Among them, MAPK appears to be the most sensitive ($IC_{50} \sim 1$ µg/ml); the IC50 values for the other five kinases (PAK2, PKC, GSK-3α, CK2, and JNK1) are approximately 10 µg/ml, whereas the value for PKA is about 40 µg/ml. Regarding the antiproliferative activity against human tumor cell lines, LC-X displayed stronger activity towards NPC (NPC-TW02 and -TW04) and breast (MCF-7) cancer cells than to epidermoid carcinoma A431 and cervical cancer HeLa cells. We found that LC-X at a dose > 50 µg/ml could significantly induce cell cycle arrest at the G_2/M phase and DNA fragmentation (a hallmark of apoptosis) of NPC-TW02 cells, indicating the perturbation of cell cycle progression as one of the mechanisms for LC extracts to inhibit tumor cell growth.

Like many solid tumors, NPC is of epithelial origin with overexpression of EGFR, and tyrosine kinase inhibitors of EGFR could suppress NPC cell growth *in vitro* and in animal models (Sun *et al.*, 1999; Zhu *et al.*, 2001). We have examined the effects of LC-X on the EGF-triggered signal events in NPC-TW02 cells, and found that LC-X treatment could dramatically inhibit the EGF-induced tyrosine phosphorylation of EGFR and MAPK activation, as well as decrease the EGFR protein levels in NPC-TW02 cells in a dose-dependent manner (Figure 39.1A). Unlike

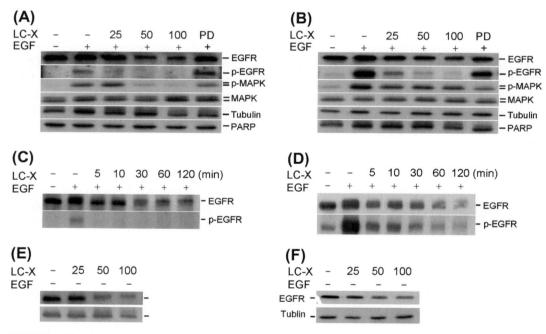

FIGURE 39.1

LC-X treatment decreases the protein levels of EGFR and inhibits the EGF-mediated MAPK activation in NPC-TW02 and A431 cells. **(A, B)** NPC-TW02 (A) or A431 (B) cells were serum-starved for 3 h, and then incubated with various concentrations of LC-X (0—100 µg/ml) or PD98059 (PD, 50 µM) for 2 h. After treatment, cells were stimulated with or without EGF (50 ng/ml) for 10 min. Cell extracts (60 µg) were then analyzed by Western blotting using antibodies against EGFR, phosphotyrosine, phospho-MAPK (p-MAPK), MAPK, tubulin, and PARP, respectively. PD98059-treated, EGF-stimulated cells were used as positive controls to show that PD98059 could block EGF-induced MAPK activation, but did not alter the protein level of EGFR. **(C, D)** NPC-TW02 (C) or A431 (D) cells were serum-starved for 3 h, and then incubated with LC-X (100 µg/ml) for various time intervals as indicated. After treatment, cells were stimulated with EGF and processed for Western blotting as described above. **(E, F)** NPC-TW02 (E) or A431 (F) cells were serum-starved for 3 h, and then incubated with various concentrations of LC-X (0—100 µg/ml) for 2 h. After treatment, cells were processed for Western blotting as described above. *Reprinted from Huang et al. (2007) Cancer Lett., 248, 137—146, with permission.*

the reduction of EGFR protein levels, LC-X treatment did not affect the protein levels of MAPK and other cytosolic proteins, such as tubulin and PARP. Similar results were observed when these experiments were performed on A431 cells, an epidermoid carcinoma cell line known to have high-level EGFR expression (Figure 39.1B). The LC-X-elicited decrease of EGFR protein level was time-dependent, and detectable in both EGF-stimulated and unstimulated cells (Figures 39.1C—F), indicating that both activated and non-activated EGFR are susceptible to LC-X treatment. The decrease of EGFR protein in LC-X-treated A431 cells occurred mainly in the plasma membrane, but not the cytosolic fraction (Figure 39.2). Interestingly, LC-X treatment resulted in a selective depletion of a 170-kDa protein in the plasma membrane fraction of A431 cells, as evidenced by silver staining of the SDS-gel (Figure 39.2B), which was further identified as EGFR by MALDI-TOF mass spectrometry (Huang *et al.*, 2007). Further study using purified EGFR protein showed that incubation of purified EGFR protein with LC-X prior to SDS-PAGE caused the loss of the 170-kDa EGFR protein at the predicted position, but generated concomitantly a dense staining band on the top of the gel (Figure 39.2C, asterisk). Although the exact reason for this observation is unknown, we speculate that LC-X may chemically modify the EGFR protein to alter its mobility in SDS-gel *in vitro*, and selectively modify EGFR in the plasma membrane of tumor cells. More recently, several studies have reported various biological activities of phenolic compounds extracted from the seeds of LC, including hemolytic activity, antibacterial and membrane-damaging activity, and cell protective activity against H_2O_2-induced cell damage (Kaur & Singh, 2008; Singh & Kaur, 2008; Yuan *et al.*, 2009). As it has been reported that phenolics can strongly interact with proteins via different mechanisms (Xu & Diosady, 2000; Kaur & Singh, 2008), whether the isolated LC-X

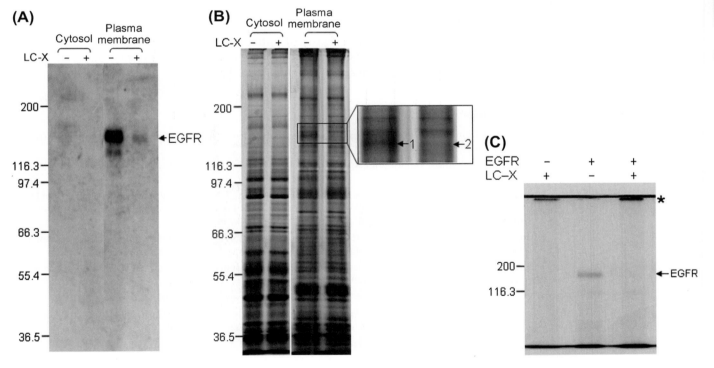

FIGURE 39.2
LC-X treatment selectively modifies EGFR protein in the plasma membrane of A431 cells. (A, B) The serum-starved A431 cells were treated with or without LC-X (100 μg/ml) for 2 h, and the plasma membrane and cytosolic fractions were isolated by differential centrifugation. Proteins (15 μg) from both fractions were resolved by 8—16% gradient SDS-PAGE, followed by Western blot analysis using anti-EGFR antibody (A), or by silver staining (B). Protein bands 1 and 2, indicated by arrows, were excised from the stained gel, digested by trypsin, and analyzed by MALDI-TOF mass spectrometry. EGFR protein was identified in band 1 but not in band 2. **(C)** Purified EGFR (100 ng) was incubated with or without LC-X (0.3 μg) in a 30-μl reaction mixture at 30°C for 15 min. The reaction products were resolved in 10% SDS-PAGE and then stained with silver nitrate. The position of EGFR is denoted by an arrow, and the band detected at the top of the gel is indicated by an asterisk. *Reprinted from Huang* et al. *(2007) Cancer Lett., 248, 137—146, with permission.*

fraction contains phenolic compounds, and whether phenolic compounds isolated from LC extracts possess protein kinase inhibitory activity, and EGFR-modifying activity, obviously represent intriguing issues that warrant further investigation.

ADVERSE EFFECTS AND REACTIONS (ALLERGIES AND TOXICITY)

Although the use of LC extracts has been documented in Chinese folk medicine, it has not been evaluated in clinical trials in an evidence-based, rigorous way. The research reports about LC are focused solely on the cell culture and mice models. The toxicity of LC extracts in mice was not found to be significantly different when using the micronucleus assay at the doses of 1.375, 2.75, 5.5, and 11 g/kg, respectively (Zhong *et al.*, 2009). Some research has indicated that LC extracts can be used as a dietary supplement without the need for establishing toxicity data in humans. For rigorous qualitative and quantitative control in modern medicine, it is apparent that we have to clarify the CMC (chemical, manufacture, and control) and toxicity data of LC extracts before the official use as a botanical drug.

SUMMARY POINTS

- The water and ethanolic extracts of the seeds of LC possess antitumor, anti-angiogenic, antioxidant, and antibacterial activity.
- The ethanolic extracts of the seeds of LC contain protein kinase inhibitory activity, which is well co-purified with antitumor activity in various chromatography steps.
- A highly purified fraction with protein kinase inhibitory activity, LC-X, can inhibit various protein kinases *in vitro*.
- LC-X can induce cell cycle arrest at the G_2/M phase, and apoptosis of tumor cell lines.
- LC-X can block the EGFR-mediated MAPK signaling pathway in various tumor cell lines.
- LC-X can selectively modify EGFR protein in the plasma membrane of tumor cells.
- Blockage of the functions of EGFR and intracellular kinase activities may account for the antitumor activity of the active components in LC-X.

References

Cheung, S., & Tai, J. (2005). *In vitro* studies of the dry fruit of Chinese fan palm *Livistona chinensis*. *Oncology Reports, 14*, 1331–1336.

Cohen, P. (2002). Protein kinases – the major drug targets of the twenty-first century. *Nature Review Drug Discovery, 1*, 309–315.

Huang, W. C., Hsu, R. M., Chi, L. M., Leu, Y. L., Chang, Y. S., & Yu, J. S. (2007). Selective downregulation of EGF receptor and downstream MAPK pathway in human cancer cell lines by active components partially purified from the seeds of *Livistona chinesis* R. Brown. *Cancer Letters, 248*, 137–146.

Huang, C., Qin, Y. M., & Liang, N. M. (1995). Effects of Dayecai (*Selaginella doederleinii*) and Chinese Livistona (*Livistona chinensis*) on the activity of protein kinase C [in Chinese]. *Zhongcaoyao, 26*, 414–415.

Kaur, G., & Singh, R. P. (2008). Antibacterial and membrane damaging activity of *Livistona chinensis* fruit extract. *Food and Chemical Toxicology, 46*, 2429–2434.

Kerbel, R. S. (2008). Tumor angiogenesis. *New England Journal of Medicine, 358*, 2039–2049.

Liu, S. Y., Hsu, C. C., & Ho, I. C. (1987). Anti-tumor metastatic components in *Trapa taiwanensis* Nakai and *Livistona chinensis* R. Br. *Bulletin of the Institute of Zoology, Academia Sinica, 26*, 143–150.

Mackay, H. J., & Twelves, C. J. (2007). Targeting the protein kinase C family: are we there yet? *Nature Reviews Cancer, 7*, 554–562

Quan Guo Zhong Cao Yao Hui Bian. (1996). *Quan Guo Zhong Cao Yao Hui Bian (Whole China Herb Conglomeration Edition)*, (2nd ed.). Beijing, China: People's Hygiene Press. p. 901.

Sagar, S. M., Yance, D., & Wong, R. K. (2006). Natural health products that inhibit angiogenesis: a potential source for investigational new agents to treat cancer – Part 1. *Current Oncology, 13*, 14–26.

Sartippour, M. R., Liu, C., Shao, Z. M., Go, V. L., Heber, D., & Nguyen, M. (2001). *Livistona* extract inhibits angiogenesis and cancer growth. *Oncology Reports, 8*, 1355–1357.

Singh, R. P., & Kaur, G. (2008). Hemolytic activity of aqueous extract of *Livistona chinensis* fruits. *Food and Chemical Toxicology, 46*, 553–556.

Sun, Y., Fry, D. W., Vincent, P., Nelson, J. M., Elliott, W., & Leopold, W. R. (1999). Growth inhibition of naso-pharyngeal carcinoma cells by EGF receptor tyrosine kinase inhibitors. *Anticancer Research, 19*, 919–924.

Wang, H., Li, A., Dong, X. P., & Xu, X. Y. (2008). Screening of anti-tumor parts from the seeds of *Livistona chinensis* and its anti-angiogenesis effect [in Chinese]. *Zhong Yao Cai, 31*, 718–722.

Xu, L., & Diosady, L. L. (2000). Interactions between canola proteins and phenolic compounds in aqueous media. *Food Research International, 33*, 725–731.

Yuan, T., Yang, S. P., Zhang, H. Y., Liao, S. G., Wang, W., Wu, Y., et al. (2009). Phenolic compounds with cell protective activity from the fruits of. *Livistona chinensis. Journal of Asian Natural Products Research, 11*, 243–249.

Zhong, Z. G., Luo, P., Huang, Y., Pan, L. N., Huang, J. L., Li, P., et al. (2009). Micronucleus experiment on the extracts of the seeds of *Livistona chinensis* in mice. *Lishizhen Med. Materia Medica Resolution, 20*, 775–776.

Zhu, X. F., Liu, Z. C., Xie, B. F., Li, Z. M., Feng, G. K., Yang, D., et al. (2001). EGFR tyrosine kinase inhibitor AG1478 inhibits cell proliferation and arrests cell cycle in nasopharyngeal carcinoma cells. *Cancer Letters, 169*, 27–32.

CHAPTER 40

Climbing Black Pepper (*Piper guineense*) Seeds as an Antisickling Remedy

Sunday J. Ameh[1], Obiageri O. Obodozie[1], Uford S. Inyang[1], Mujitaba S. Abubakar[2], Magaji Garba[3]

[1] Department of Medicinal Chemistry and Quality Control, National Institute for Pharmaceutical Research and Development (NIPRD), Garki, Abuja, Nigeria
[2] Department of Pharmacognosy and Drug Development, Ahmadu Bello University, Zaria, Nigeria
[3] Department of Pharmaceutical and Medicinal Chemistry, Ahmadu Bello University, Zaria, Nigeria

INTRODUCTION

Climbing black pepper, *Piper guineense*, is also called West African black pepper, Ashanti pepper, Guinea cubeb, and Benin pepper. It is native to the tropical rain forest of Africa, where it thrives mainly in the wild. It is, however, partly cultivated in Southern Nigeria, where the leaves and seeds are used to flavor soup. The culinary and other uses, first documented by Dalziel in 1937, are today well known. All the initial reports on the use of *P. guineense* in sickle cell disorder (SCD) emanated from Nigeria during the late 1990s. In those reports the seeds of *P. guineense* were used with parts of *Pterocarpus osun*, *Eugenia caryophyllata*, and *Sorghum bicolor* to produce Niprisan — a phytomedicine for SCD (Wambebe *et al.*, 2001). SCD crisis involves polymerization of hemoglobin, blockade of capillaries, extreme pain, and widespread

Nuts & Seeds in Health and Disease Prevention. DOI: 10.1016/B978-0-12-375688-6.10040-4

necrosis. In this chapter we assemble evidence linking the known constituents of *P. guineense* with one or other of the aspects of SCD crisis, and conclude that piperine, capsaicin, cubebin, and caryophyllene or closely related compounds are the likely agents that palliate the crisis.

BOTANICAL DESCRIPTION

Piper guineense is a perennial vine, with a hairless stem, growing to over 20 m in tropical Africa. The leaves are ovate, hairless, and occur alternately on the stem; the flowers are dioecious, yellow-green spikes, 2–3 cm long, and arise opposite terminal or lateral leaves. The male flowers have two or three stamens, and the female flowers three to five stigmas. The fruits, called berries or peppercorns, are prolate-elliptical, 3–8 mm in diameter when fresh, and vary in color from red to reddish brown, orange to buff, or green to brown. Since the genus *Piper* has over a thousand species that produce seeds similar to those of *P. guineense*, a large number of vines qualify as "climbing black pepper." However, two species — *Piper nigrum* and *Piper cubeba* — deserve special mention: first, because of their particularly close affinity with *P. guineense*, there is often a crisis of identity; and second, because these two are more widely known, the term "black pepper" more commonly refers to them. The key differences between the three are as follows. *P. guineense* is mainly African, while *P. nigrum* and *P. cubeba* are Asian. *P. guineense* berries tend to be less spherical than those of *P. nigrum* and *P. cubeba*. The *P. guineense* berries usually retain their stalks, distinctly curved in a way that makes the berries look like tadpoles. By contrast, the stalks of *P. nigrum* are usually absent, while those of *P. cubeba* are more often retained, appearing like tails; hence, the berries are also called "tailed cubeb." In all three, however, the berries are clustered into an elongated bunch that looks like ears of guinea corn; hence the berries are also called peppercorns. Some features and the antisickling effect of *P. guineense* seeds are shown in Table 40.1.

334

TABLE 40.1 Some Features and Antisickling Effects of *Piper guineense* Seeds and Other Components of Niprisan — the Antisickling Drug

Feature %w/w (mean ± SD)	*P. guineense* Seeds	*E. caryophyllata* Flower Buds	*P. osun* Stem Parts	*S. bicolor* Leaf Stalk
Loss on drying	7.84 ± 0.66 (n = 20)	8.68 ± 0.59 (n = 20)	5.74 ± 0.54 (n = 28)	7.85 ± 0.47 (n = 7)
Total ash	14.22 ± 3.09 (n = 4)	5.47 ± 0.63 (n = 20)	1.48 ± 0.34 (n = 4)	6.91 ± 0.87 (n = 19)
Water extractability	14.45 ± 0.94 (n = 4)	25.56 ± 2.22 (n = 28)	2.84 ± 0.40 (n = 4)	7.54 ± 1.44 (n = 20)
Hexane extractability	13 ± 0.8 (n = 4)	18 ± 2 (n = 6)	1.6 ± 0.2 (n = 4)	3.4 ± 0.6 (n = 4)
Thin layer chromatography				

1. *Eugenia caryophyllata*
2. *Piper guineense.*
3. *Pterocarpus osun.*
4. *Sorghum bicolor.*

Above is the diagrammatized TLC of the ethanolic extracts of Niprisan components. Stationary phase: K5 Silica. Mobile phase: hexane—ethylacetate (70 : 30).

TABLE 40.1 Some Features and Antisickling Effects of *Piper guineense* Seeds and Other Components of Niprisan — the Antisickling Drug—continued

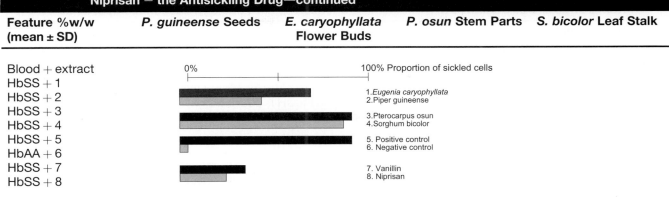

Feature %w/w (mean ± SD)	*P. guineense* Seeds	*E. caryophyllata* Flower Buds	*P. osun* Stem Parts	*S. bicolor* Leaf Stalk

The features were determined as per WHO (1998). The proportion of sickled cells was determined by the sickle cell slide test, as per Chessbrough (2000), using sodium metabisulfite ($Na_2S_2O_5$) as the reducing reagent. The blood, drug, and $Na_2S_2O_5$ were introduced in that order to the slide, each as a drop; covered with a cover slip; and observed every 5 minutes for 1 hour at room temperature. The numbers of sickled cells (SC) and normal cells (NC) within the field of view were determined microscopically. The percentage of sickled cells is calculated as $(SC \times 100)/(SC + NC)$. Sickle cell blood (HbSS) was drawn from a confirmed sickler attending the Institute's Sickle Cell Clinic. Normal blood (HbAA) was drawn from a healthy volunteer. The blood samples were drawn into EDTA bottles and used immediately. The herbs were used as 1 mg of ethanol extract per ml of 50% glycerin. Vanillin and Niprisan were used as solutions containing 1 mg of each drug per ml of 50% glycerin. The control samples received 50% glycerin without the drugs. The positive control utilized sickle cell blood (HbSS), while the negative utilized normal blood (HbAA). Glycerin was used to retard dehydration of the samples during the experiment. It is noted that *P. guineense* and *E. caryophyllata* possess demonstrable antisickling effects, and are known to share some key attributes — for example, *P. guineense* contains the vanilloids piperine and capsaicin, while *E. caryophyllata* contains vanillin and methyl salicylate. These four compounds are by-products of shikimic acid, and possess antisickling effects. The two herbs also contain the cannabinoid, β-caryophyllene.

Chemical constituents

P. guineense contains about 5–8% of piperine as the main pungent principle. It also contains capsaicin, myristicine, elemicin, safrole, and dillapiol, accounting for about 10% (Celtnet.org, 2009). The oil, obtained by steam distillation, contains some 62 principles, including monoterpenes (e.g., limonene and δ-carene), sesquiterpenes (e.g., β-caryophyllene, germacrene D, α-copaene, β-elemene, and α-caryophyllene), and other terpenes (e.g., aromadendrene, α- and β-selinene, β-cubebene, allo-aromadendrene, α-amorphenes, viridiflorene, α-fernesene, g-cadinene and E-calamenene (Onyenekwe, 2009). *P. cubeba* is found to contain similar compounds including cubebol (Katzer, 1998). Tables 40.2 and 40.3 show some of the more notable constituents of *P. guineense* seeds.

HISTORICAL CULTIVATION AND USAGE

P. guineense is native to the tropical rain forests of Africa, and although it was well traded in Europe during the Middle Ages, today its use is limited to Africa, where it is still the "King of Spices."

In the Old World, *Piper* species, including *P. Guineense*, have been in use for centuries in various official and anecdotal remedies. These include use in mouthwash and dental diseases, halitosis, loss of voice and sore throat, fever, and cough, and as a counter-irritant (Schmidt, 2009). In traditional Chinese medicine, *Piper* is used for its alleged warming effect. In Tibetan medicine, *Piper* is one of the six herbs claimed to benefit specific organs, being assigned to the spleen. Sir Richard Burton's book, *The Book of One Thousand and One Nights*, mentioned cubeb as the main ingredient of an aphrodisiac remedy for infertility. Similarly, the 1827 edition of the *London Dispensatorie* informed that cubeb "stir[s] up venery... very profitable for cold grief of the womb" (Katzer, 1998). Furthermore, in England, a small amount of *Piper* was often included in lozenges designed to alleviate bronchitis, owing to its antiseptic and expectorant properties.

TABLE 40.2 A Profile of the Alkaloidal Constituents of *Piper guineense*

Name/Structure of Constituent	Chemistry/Pharmacology/Economic Importance
Piperine and chavicine $CH:CHCH:CHCHNC_5H_{10}CO$ Piperine - a vanilloid	Piperine, mp 130°C, and chavicine are geometric isomers responsible for the pungency of *P. guineense*, of which they constitute ~5–8%, and are used ethnomedicines and insecticides. Piperine is tasteless at first but has a burning after taste; it is soluble in water (40 mg/l). Its UV_{max} in MeOH is 333 nm. Chavicine converts to piperine on storage; its UV_{max} in MeOH is 318 nm.
Capsaicin $CH_2NHCO(CH_2)_4CHCHCH_2Me_2$ Capsaicin - a vanilloid	Capsaicin, mp 65°C, is (8-methyl-*N*-vanillyl-6-nonenamide), a pungent constituent that produces a sensation of burning in all tissues. Capsaicin and related compounds, called capsaicinoids or vanilloids, are phytochemicals that deter herbivores and fungi. Pure capsaicin is hydrophobic, colorless, and odorless.
Dillapiole $CH_2CH:CH_2$ MeO MeO dillapiole - a vanilloid	Dillapiole, mainly extracted from dill, is a calminative for treating colic, especially in children. It has a warm taste, and is the active ingredient of gripe water.
Myristicine $CH_2CH:CH_2$ MeO myristicin - a vanilliod	Myristicine is more commonly associated with nutmeg, parsley, and dill. It is insecticidal, acaricidal, and partly responsible for the psychoactivity of nutmeg. It is a weak inhibitor of monoamine oxidase.
Cubebin cubebin	Cubebin, mp 132°C, is tetrahydrodiperonyl-2-furanol. The furanyl and piperonyl (vanilyl) groups draw attention to the palliative roles of furan and vanilloids in SCD crisis. Cubebine is the French designation for ether extract of P. cubeba.

TABLE 40.3 A Profile of Some Terpenic Constituents of *Piper guineense*

Name/Structure of Constituent	Chemistry/Pharmacology/Economic Importance
Limonene 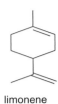 limonene	Limonene, bp 176°C, has lemon aroma, and is a wetting agent, with a warm effect on the skin. It is found in the rind of citrus fruits. Its slight anesthetic and rubefacient effects contribute to pain relief and improved microcirculation in the crisis condition.
Carene 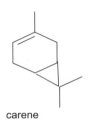 carene	Carene, bp 169⁰ has a pungent, sweet and characteristic aroma, and a natural constituent of turpentine, sometimes as high as 40%. It is a bicyclic monoterpene with slight anesthetic and rubefacient effects. It seems of interest that turpentine is a key ingredient of liniments used for controlling pain in nerves, bones and muscles.
Copaene 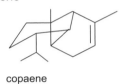 copaene	Copaene is an odorless, oily tricyclic sesquiterpine, bp 246–251°C. Its established economic importance is in agriculture, in that it strongly attracts *Ceratitis capitata*, a Mediterranean fruit fly that is critical to the citrus agribusiness. The structure of copaene appears interesting, but is not known for any medicinal effect.
Caryophyllene 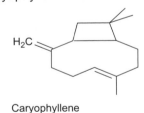 Caryophyllene	Caryophyllene, (−)-β-caryophyllene, is one of the warm constituents of *P. guineense* that is also found in cloves, Indian hemp, and rosemary. It occurs as a mixture with α-caryophyllene, and it is notable for having a cyclobutane ring — a rarity in nature. It has been found to bind selectively to cannabinoid receptor type 2. This is a significant finding, given the role of this receptor in endogenous pain control.

337

PRESENT-DAY CULTIVATION AND USAGE

P. guineense is native to the tropical rain forests of Africa, where it thrives mainly in the wild, although it is partly cultivated in Southern Nigeria.

Black and white peppers from *P. guineense* (Omafuvbe & Kolawole, 2004) are produced in the same way as their counterparts from *P. nigrum* (Schmidt, 2009) and *P. cubeba* (Katzer, 1998). The terms "black" and "white" pepper stem partly from their color, and partly from the process of producing them. To produce black pepper, the fully developed berries are picked before they are fully ripe. They are allowed to dry, shrinking and darkening in the process, and are then ground to yield "black pepper." To produce white pepper, the berries are allowed to ripen fully before harvesting, and are then treated with water to free the seeds from epidermal tissues.

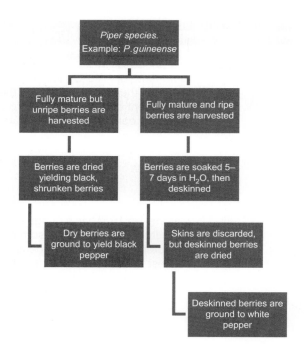

FIGURE 40.1

The processes of production of black pepper and white pepper from *Piper* species. Sun-drying the berries to a moisture content of 8—10% takes 7—10 days. Best-quality black pepper results if the berries are first dipped in hot water before drying. The yield of white pepper is about 25% that of black pepper. White pepper can be produced from *P. guineense*, but this is seldom the case in Africa, where *P. guineense* is used to produce black pepper. Most commercial black and white peppers are produced from *P. nigrum*, in Vietnam and India. Next in importance are the peppers produced from *P. cubeba*, especially in Java and Sumatra.

Subsequently, the naked seeds are dried and ground to yield the white pepper. These processes are illustrated in Figure 40.1.

Apart from the traditional uses of pepper described above, another important therapeutic application of the drug is in treating gonorrhea (Katzer, 1998). Similarly, there are reports that *P. guineense* accentuates mating behavior in male rats by elevating the level of circulating testosterone (Mbongue *et al.*, 2005).

Initial reports on the use of *P. guineense* in sickle cell disorder (SCD) emanated from Nigeria during the late 1990s. In those reports, the seeds of *P. guineense* were used with parts of *Pterocarpus osun*, *Eugenia caryophyllata*, and *Sorghum bicolor* to produce Niprisan — a phytomedicine for SCD (Wambebe *et al.*, 2001). The use of *P. guineense* in SCD conditions is described in the next section.

APPLICATIONS TO HEALTH PROMOTION AND DISEASE PREVENTION

Biochemistry and histopathology of sickle cell crisis

Sickle cell crisis refers to the symptoms experienced by people who are homozygous for the gene that codes for sickle hemoglobin (HbS). The disease is hereditary, and destroys red blood cells (RBC) by causing them to assume a rigid "sickle" shape. HbS occurs mainly in descendants from Africa, the Middle East, the Mediterranean, and India, and is a defective variant of the normal hemoglobin (HbA), the protein in red blood cells that transports oxygen to the tissues. While the β-chain of HbA has glutamic acid in the sixth position, the corresponding amino acid in HbS is valine. This subtle difference accounts for the high sensitivity of HbS to oxygen deficiency. When RBCs release their oxygen to the tissues, the HbS, in contrast to the HbA, tends to stack up within the cells as filaments, twisted into helical rods. These rods cluster into bundles that distort and elongate the cells, causing them to assume a rigid, sickle shape. This is the sickling phenomenon, and is partly reversible, if the RBCs are reoxygenated; however, repeated sickling leads to irreversible distortion of the cells. Such sickled cells eventually clog the capillaries, causing obstruction to microcirculation, and leading to severe pain and tissue necrosis.

Features of hemoglobin in relation to antisickling agents

There are two quaternary states of hemoglobin (Hb) — the deoxygenated conformation (called the Tense or T state), and the oxygenated conformation (called the Relaxed or R state). In HbS, the presence of the βVal6 mutation leads to hydrophobic interactions between the mutation region of one HbS molecule and a region defined by βPhe85 and βLeu88 in the heme pocket of another HbS. These interactions occur only in the T state, leading the HbS molecules to polymerize into fibers. Thus, a key approach to the crisis lies in finding a means of inhibiting the polymerization of T state HbS, or of causing it to revert to the R state. Safo and colleagues (2004) have shown that both HbA and HbS possess allosteric sites with which suitable chemical entities can interact to shift the equilibrium in favor of the R state, and have identified several such agents, called allosteric regulators. These regulators, in the case of HbS, act as antisickling agents, defined as entities that can prevent or reverse the pathological events leading to the sickling of RBCs. Compounds known to possess this effect include the "alternative aspirins," such as acetyl-3,5-dibromosalicylic acid (Walder *et al.*, 1977); furfural derivatives (Safo *et al.*, 2004); and a variety of compounds called capsaicinoids or vanilloids that possess the vanilyl functional group, or an approximation of it, as in vanillin or related compounds (Abraham *et al.* 1991). These vanilloids include some substituted benzaldehydes (Nnamani *et al.*, 2008), and several shikimic acid by-products. The structures of some of these antisickling entities, including the "alternative aspirins" and "vanilloids," are shown in Figure 40.2.

Mechanisms of *P. guineense* involvement in the management of pain

Physical pain is an unpleasant sensation associated with actual or potential tissue damage. It falls into two categories: nociceptive pain, defined in relation to the peripheral nervous system, which arises from excitations caused by chemicals, heat, electricity, or mechanical events; and neurogenic pain, defined in relation to the totality of the nervous system, which arises from damage to any part of the system. Physical pain is essential in an organism's defense and coordination system — for example, nociceptive pain prompts an organism to disengage from the source of the noxious stimulus. Since pain tends to persist beyond its immediate purpose, organisms are equipped with endogenous systems for controlling or alleviating pain. Such systems are orchestrated by a complex interplay of nerve cells, ion channels, and receptors.

ION CHANNELS

Ion channels are pore-forming proteins that act to establish and control the small voltage gradient across the plasma membrane of all living cells by allowing the flow of ions across their electrochemical gradient (Venkatachalam & Montell, 2007). Some channels permit the passage of ions based solely on charge. Some have pores that are only one or two atoms wide; such channels conduct a specific species of ions. In some ion channels, passage through the pore is governed by a "gate," which may be opened or closed by a neurotransmitter or chemical ligand (ligand-gated), a change in temperature, a mechanical force, or a voltage difference (voltage-gated). A special group consists of the transient receptor potential channels (or TRP channels), which have features common to both voltage- and ligand-gated channels. The

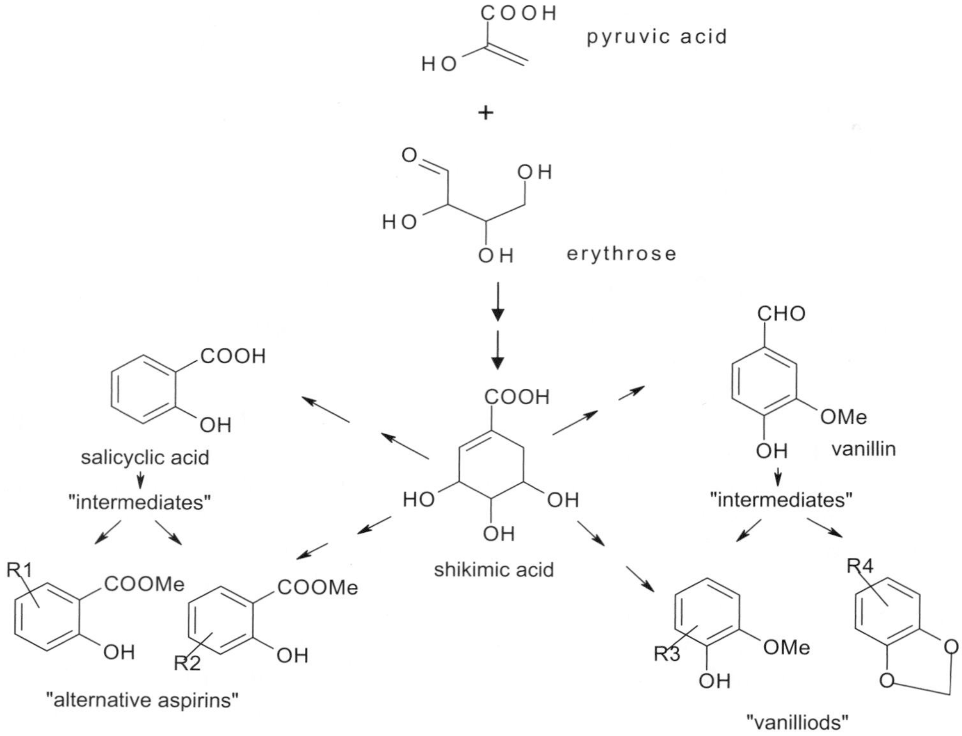

FIGURE 40.2

Biosynthesis and relationship of shikimic acid and its byproducts/intermediates. Piperine, capsaicin, and cubebin, as by-products of shikimic acid, are the likely antisickling agents in *P. guineense*. The shikimic acid pathway is a key biosynthetic pathway for several phytochemicals (secondary metabolites) known for their medicinal attributes. The figure illustrates the biosynthesis of shikimic acid from pyruvic acid and erythrose (primary metabolites), and the relationship between the acid and its by-products and intermediates, some of which possess aspirin-like effects like analgesia, and desickling of sickled RBCs. Such by-products/intermediates include salicyclic acid derivatives, vanillin, piperine, capsaicin, and cubebin.

group has 28 members that differ from one another in the way they are activated. Some are constitutively open, while others are gated by voltage, pH, redox state, osmolarity, heat, or mechanical stretch (Camerino *et al.*, 2007).

Outstanding among the TRP superfamily of ion channels is the family called the Transient Receptor Potential Vanilloids (TRPVs), which has six subfamilies, designated TRPV1 to TRPV6 (Vennekens *et al.*, 2008). They are selective for calcium and magnesium over sodium ions, and, like other TRPs, they can be activated by several mechanisms (Venkatachalam & Montell, 2007). In vertebrates, the TRPVs are so sensitive to temperature that they are regarded as molecular thermometers. TRPV1 is activated at 43°C, and is activated by noxious heat, acidic pH, allicin in garlic, the vanilloids (such as piperine and capsaicin), and the endocannabinoids (such as anandamide and *N*-arachidonoyl-dopamine) (Szallasi *et al.*, 2007). To illustrate how the vanilloids act, it has been shown that capsaicin selectively binds to TRPV1 on the membranes of pain- or heat-sensing neurons. As a heat activated calcium channel, TRPV1 normally opens at 37–45°C; however, when capsaicin binds to TRPV1 it causes the channel to open below 37°C (body temperature), which is why capsaicin is linked to the sensation of heat. Prolonged activation of these neurons by capsaicin leads to a depletion of presynaptic substance P — a neurotransmitter for pain and heat. Neurons lacking TRPV1 are unaffected by capsaicin (Caterina *et al.*, 1997; Szallasi *et al.*, 2007). Table 40.4 presents a summary of the functions and properties of the different TRPVs (Venkatachalam & Montell, 2007).

TABLE 40.4 Functions, Distribution, and Properties of Transient Receptor Potential Vanilloids (TRPVs) as Ion Channels

Ion Channel	Function/Activating Stimulus	Main Organ Distribution	Ca^{2+}/Na^+ Selectivity	Heteromeric Association	Other Associated Proteins
TRPV1	Vanilloid (capsaicin) receptor Noxious stimuli Thermosensor − 43°C	CNS PNS	9 : 1	TRPV2 TRPV3	Calmoduli PI3 kinase
TRPV2	Osmolarity Noxious stimuli Thermosensor − 52°C	CNS Spleen Lung	3 : 1	TRPV1	—
TRPV3	Warmth sensor channel (33−39°C)	Skin CNS PNS	12 : 1	TRPV1	—
TRPV4	Osmolarity Warmth sensor channel: 27−34°C	CNS Internal organs	6 : 1		Aquaporin 5 Calmodulin Pacsin 3
TRPV5	Calcium-selective TRP channel	Intestine Kidney Placenta	100 : 1	TRPV6	Annexin II/ S100A10 Calmodulin
TRPV6	Calcium-selective TRP channel	Kidney Intestine	130 : 1	TRPV5	Annexin II/ S100A10 Calmodulin

CANNABINOID RECEPTORS IN PAIN CONTROL

The cannabinoid receptors are a class of cell membrane receptors that are activated by lipids called cannabinoids (Graham *et al.*, 2009). Some, the endocannabinoids, are endogenous, while others, such as the psychoactive constituents of *Cannabis sativa* and the spicy phytochemical, caryophyllene, are exogenous. At least two subtypes, cannabinoid types 1 (CB_1) and 2 (CB_2), are well known. CB_1 is expressed mainly in the CNS, lungs, liver, and kidneys, while CB_2 is expressed in the immune system, hematopoietic cells, and peripheral nerve terminals, where they play a role in the perception of pain.

Caryophyllene or β-caryophyllene is a constituent of many oils, such as those from *Cannabis sativa* and *Eugenia caryophyllata* — a component of Niprisan. Incidentally, and significantly, caryophyllene is also one of the constituents of *P. guineense*. Caryophyllene has been found to bind selectively to CB_2, and to exert significant cannabimimetic anti-inflammatory effects in mice (Gertsch *et al.*, 2008). It thus seems likely that caryophyllene can relieve pain in humans, and thus be of benefit to SCD patients.

ADVERSE EFFECTS AND REACTIONS (ALLERGIES AND TOXICITY)

Aside from minor gastric upsets in some subjects, attributable to *Piper guineense* and clove in Niprisan, there have been no reports of any untoward effects from normal consumption (Wambebe *et al.*, 2001). However, a number of studies from West Africa have demonstrated the androgenizing effects of *Piper guineense* seeds. For example, it is reported from the Cameroons (Mbongue *et al.*, 2005) that acute treatment of male rats with aqueous extracts at 245 mg/kg body weight for 8 days significantly raised the level of testicular and serum testosterone and cholesterol; epididymal α-glucosidase activity; and fructose in the seminal fluid. By contrast, chronic treatment (122.5 mg/kg body weight for 55 days), while increasing testicular and serum cholesterol, lowered both epididymal α-glucosidase activity and the level of fructose in seminal fluid by over 20%. Although both acute and chronic treatments increased sexual activity, the chronic treatment reduced male fertility by 20%. From Nigeria, Ekanem and colleagues (2010) report that ethanolic extracts of the herb had a contraceptive effect in mice at doses of 10−40 mg/kg body weight. Thus, while the well reported aphrodisiac effect of the herb in man is corroborated by animal studies, a net negative effect on human reproduction has not been reported. Indeed, an anecdotal claim for promoting fertility in women (Katzer, 1998), mentioned earlier, suggests that the reverse might be the case. Furthermore, although the herb may raise serum cholesterol in humans, neither this nor any consequences thereof have been reported. Moreover the doses used in these studies far exceed those encountered in human use.

SUMMARY POINTS

- The application of *P. guineense* seeds in sickle cell disorder probably relies on their content of piperine, capsaicin, cubebin, and caryophyllene, or closely related compounds.
- Piperine and capsaicin are vanilloids that interact with receptors/ion channels associated with pain; both possess aspirin-like effects such as allosteric binding to hemoglobin — a property shared with several other by-products of shikimic acid.
- Cubebin, a tetrahydrodiperonyl-2-furanol, is also a vanilloid that contains furan — an allosteric regulator of hemoglobin possessing an antisickling effect.
- β-Caryophyllene is an agonist of cannabinoid receptor type 2, which functions in endogenous pain control.
- Remarkably, vanillin, methyl salicylate, and β-caryophyllene are also present in *E. caryophyllata*, which, like *P. guineense*, is one of the components of Niprisan — an antisickling drug developed in Nigeria.

References

Abraham, D. J., Mehanna, A. S., Wireko, F. C., Whitney, J., Thomas, R. P., & Orringer, E. P. (1991). Vanillin, a potential agent for the treatment of sickle cell anemia. *Blood, 77*, 1334—1341.

Camerino, D. C., Tricarico, D., & Desaphy, J. F. (2007). Ion channel pharmacology. *Neurotherapeutics, 4*, 184—198.

Caterina, M. J., Schumacher, M. A., Tominaga, M., Rosen, T. A., Levine, J. D., et al. (1997). The capsaicin receptor: a heat-activated ion channel in the pain pathway. *Nature, 389*, 816—824.

Celtnet.org (2009). Ashanti pepper. Available at: http://www.celtnet.org.uk/recipes/spice-entry.hph?term=Ashanti%20Pepper (accessed October 10, 2009).

Chessbrough, M. (2000). Investigation of sickle cell disease. *District laboratory practice in tropical countries.* Cambridge, UK: Cambridge University Press. Part 2 pp. 334—340.

Ekanem, A. P., Udoha, F. V., & Oku, E. E. (2010). Effects of ethanol extract of *Piper guineense* seeds (Schum. and Thonn) on the conception of mice (*Mus Musculus*). *African Journal of Pharmacy and Pharmacology, 4*(6), 362—367.

Gertsch, J., Leonti, M., Raduner, S., Racz, I., Chen, J., Xie, X., et al. (2008). Beta-caryophyllene is a dietary cannabinoid. *Proceedings of the National Academy of Sciences of the United States of America, 105*, 9099—1104.

Graham, E. S., Ashton, J. C., & Glass, M. (2009). Cannabinoid receptors: a brief history and "what's hot". *Frontiers in Bioscience, 14*, 944—957.

Katzer, G. (1998). Cubeb pepper (Cubebs, *Piper cubeba*). *Gernot Katzer's Spice Pages.* Available at: http://www.uni-graz.at/~katzer/engl/Pipe_cub.html (accessed October 10, 2009).

Mbongue, F. G. Y., Kamtchouing, P., Essame, O. J. L., Yewah, P. M., Dimo, T., & Lontsi, D. (2005). Effect of the aqueous extract of dry fruits of *Piper guineense* on the reproductive function of adult male rats. *Indian journal of pharmacology, 37*, 30—32.

Nnamani, I. N., Joshi, G. S., Danso-Danquah, R., Abdulmalik, O., Toshio Asakura, T., Abraham, D. J., et al. (2008). Pyridyl derivatives of benzaldehyde as potential antisickling sgents. *Chemistry and Biodiversity, 5*, 1762—1769.

Omafuvbe, B. O., & Kolawole, D. O. (2004). Quality assurance of stored pepper (*Piper guineense*) using controlled processing methods. *Pakistan Journal of Nutrition, 3*(4), 244—249.

Onyenekwe, P. C. (2009). The essential oil of *Piper guineense* berries. Available at: http://TheessentialoilofiPiper-guineense-i-berries.htm

Safo, M. K., Abdulmalik, O., Danso-Danquah, R., Burnett, J. C., Nokuri, S., Gajanan, S., et al. (2004). Structural basis for the potent antisickling effect of a novel class of 5-membered heterocyclic aldehydic compounds. *Journal of medicinal chemistry, 47*, 4665—4676.

Schmidt, R. J. (2009). BoDD (Botanical Dermatology Data Base) — PIPERACEAE. Available at: http://bodd.cf.ac.uk/BotDermFolder/PIPE.html (accessed October 10, 2009).

Szallasi, A., Cortright, D. N., Blum, C. A., & Eid, S. R. (2007). The vanilloid receptor TRPV1: 10 years from channel cloning to antagonist proof-of-concept. *Nature reviews. Drug discovery, 6,* 357–372.

Venkatachalam, K., & Montell, C. (2007). TRP channels. *Annu. Rev. Biochem., 76,* 387–417.

Vennekens, R., Owsianik, G., & Nilius, B. (2008). Vanilloid transient receptor potential cation channels: an overview. *Current pharmaceutical design, 14,* 18–31.

Walder, J. A., Zaugg, R. H., Iwaoka, R. S., Watkin, W. G., & Klotz, I. M. (1977). Alternative aspirins as antisickling agents: acetyl-3,5-dibromosalicylic acid. *Proceedings of the National Academy of Sciences of the United States of America, 74,* 5499–5503.

Wambebe, C., Khamofu, H., Momoh, J. A., Ekpeyong, M., Audu, B. S., Njoku, S. O., et al. (2001). Double-blind, placebo-controlled, randomized cross-over clinical trial of NIPRISAN in patients with sickle cell disorder. *Phytomedicine, 8,* 252–261.

WHO. (1998). *Quality control methods for medicinal plant materials.* Geneva, Switzerland: World Health Organization. pp. 1–128.

CHAPTER 41

Antifungal and Mitogenic Activities of Cluster Pepper (*Capsicum frutescens*) Seeds

Tzi-Bun Ng[1], Patrick H.K. Ngai[2], Randy C.F. Cheung[1], Jack H. Wong[1], Sze-Kwan Lam[1], He-Xiang Wang[3], Xiujuan Ye[1], Yau-Sang Chan[1], Evandro F. Fang[1]
[1] School of Biomedical Sciences, Faculty of Medicine, The Chinese University of Hong Kong, Hong Kong, China
[2] School of Life Sciences, Faculty of Science, The Chinese University of Hong Kong, Hong Kong, China
[3] State Key Laboratory of Agrobiotechnology, Department of Microbiology, China Agricultural University, Beijing, China

345

INTRODUCTION

Pepper is a favorite vegetable both in the West and in the Orient. Capsaicin (which has analgesic activity), the pigment capsanthin, and the bitter poisonous alkaloid solanine (which has narcotic and insecticidal properties) have been isolated from pepper. However, to date, very few proteinaceous constituents have been reported. Red cluster pepper lectin possesses antifungal and mitogenic activities (Ngai & Ng, 2007). Lectins are known to be present in different living organisms, such as microbes, plants, fungi, invertebrates, and vertebrates (Knibbs *et al.*, 1991; Pusztai, 1991; Wang *et al.*, 1996; Tateno *et al.*, 1998; Inamori & Saito, 1999; Leteux & Chai, 2000).

BOTANICAL DESCRIPTION

Capsicum frutescens plants have smooth, medium-sized, elliptical leaves, and slender branches which are 30−120 cm long. The plants attain their full size in a hot climate. A chili plant can

Nuts & Seeds in Health and Disease Prevention. DOI: 10.1016/B978-0-12-375688-6.10041-6

produce about 120 pods of intermediate pungency in a season, with Scoville ratings of between 30,000 and 60,000 units. Pods are erect, and exhibit less variation in color, shape, and size than do *Capsicum annuum*, *C. chinense*, and *C. baccatum*. The phenomenon is probably due to lack of enthusiasm for breeding *Capsicum frutescens* varieties.

HISTORICAL CULTIVATION AND USAGE

Capsicum has been known for many centuries. *C. frutescens* was first cultivated in Panama, later spreading to Mexico and the Caribbean, and then to India and the Far East. It was part of the human diet as early as 7500 BC. Native Americans cultivated chili plants between 5200 and 3400 BC.

PRESENT-DAY CULTIVATION AND USAGE

In South and Central America, the five domesticated *Capsicum* species are *C. annuum*, *C. baccatum*, *C. chinense*, *C. frutescens*, and *C. pubescens*. The smaller wild *C. annuum* var. *aviculare* is also harvested and sold in the marketplace.

C. frutescens has many cultivars, including Piri piri (African birdseye, African devil), Brazilian Malagueta pepper, Tabasco pepper, and Thai pepper (Chili Padi or Siling labuyo).

The most popular generic name for members of *C. frutescens* is "bird peppers." Well known examples include the Brazilian Malaguete and African Zimbabwe varieties. Fermented Tabasco chili sauce, produced by the McIlhenny family in Louisiana, is world renowned. The species probably originated from the Amazon Basin in Brazil, where wild Malagueta can be found to this date.

Many *C. frutescens* varieties are used to make hot sauces, both fresh and fermented, and are used in cuisine in the dried form when fresh supplies are unavailable.

APPLICATIONS TO HEALTH PROMOTION AND DISEASE PREVENTION

The pepper is a source of dietary fiber, xanthophylls, carotenoids, and phenolics, and of vitamins A, C, and E. Capsaicin has hypoglycemic, anti-inflammatory, and anticancer actions. The saponin CAY-1 from pepper has antifungal activity against *Aspergillus* spp., *Candida albicans*, *Trichophyton mentagrophytes*, *T. rubrum*, *T. tonsurans*, and *Microsporum canis*.

Lectins are proteins that selectively agglutinate blood cells of a particular blood group in the ABO blood group system. A monomeric 29.5-kDa mannose/glucose-binding lectin has been isolated from the seeds of red cluster pepper *Capsicum frutescens* L. var. *fasciculatum* with a procedure that entails anion exchange chromatography (AIEC) on diethylaminoethyl (DEAE)-cellulose and Q-Sepharose, and fast protein liquid chromatography-AIEC on Mono Q (Ngai & Ng, 2007). Ion exchange chromatography of pepper seed extract on DEAE-cellulose yielded five absorbance peaks. Hemagglutinating activity was confined to D2, eluted with 200-mM NH_4HCO_3 (pH 9.2). D2 was subsequently resolved on Q-Sepharose, into three adsorbed peaks: Q1, Q2, and Q3. Only Q2 exhibited hemagglutinating activity. Q2 was separated into two adsorbed peaks — a large peak, M1, and a small peak, M2 — upon FPLC on MonoS. Hemagglutinating activity resided in M1, which represented purified pepper seed lectin. The lectin displayed a molecular mass of 29.5 kDa in SDS-PAGE, and also in gel filtration. Its N-terminal sequence was GQRELKL, which exhibited resemblance to C-terminal fragments of viral RNA polymerases, and the N-terminal sequence of rabbit lectin-like oxidized low-density lipoprotein receptor.

In accordance with their carbohydrate-binding specificities, lectins can be classified into various groups, comprising mannose-binding, mannose- and glucose-binding, galactose binding, fructose-binding, N-acetylgalactosamine-binding, N-acetylglucosamine-binding, and sialic acid-binding lectins. Examples of the mannose- and glucose-binding group include concanavalin A (Agrawal & Goldstein, 1967) and chestnut lectin (Nomura *et al.*, 1998). Mannose-binding lectins include those from the monocot families, such as Alliaceae (Lam & Ng, 2001), Amaryllidaceae (Ooi *et al.*, 2000), Liliaceae (Singh Bains *et al.*, 2005), and Orchidaceae (Van Damme *et al.*, 1994). A preferentially mannose-binding lectin has been isolated from the seeds of the pepper. D-(+)-mannose was the most potent and D-(+)-glucose the second most potent in inhibiting hemagglutination induced by the lectin. The minimum sugar concentrations for inhibiting three hemagglutinating units of the lectin were 31 mM for D-(+)-mannose; 125 mM for D-(+)-glucose; 250 mM for L-(−)-mannose, D-(+)-galactose, and α-L(−)-fucose; and > 1000 mM for D-(+)-glucosamine, D(+)-galactosamine, D-(+)-mannosamine, D-(+)-raffinose, α-D(+)-melibiose, α-(+)-lactose, and L-rhamnose.

The hemagglutinating activity of pepper seed lectin was heat labile and pH sensitive. A minor reduction in the hemagglutinating activity of the lectin was noted when the incubation temperature was raised to 30°C (10% decline) and then to 40°C (15% decline). At 50°C, 40% of the activity was lost; at 60°C, about 90% of the hemagglutinating activity had disappeared; and above 60°C, the activity was destroyed. Maximal hemagglutinating activity was noted at pH 7 and 8, and full or almost full activity was seen between pH 8 and 9. At pH 6 and 10, about 80% and 60% of the activity was left, respectively. At pH 5 and 11, only about 30% of the activity was retained. When the pH was further reduced to pH 4 or elevated to pH 12, no activity could be detected. NH_4Cl, $NaCl$, KCl, $MgCl_2$, and $FeCl_3$ had no effect, whereas $CaCl_2$ and $MnCl_2$ potentiated the activity of the lectin two-fold. Augmentation of the hemagglutinating activity of pepper seed lectin by Ca^{2+} and Mn^{2+} ions is in line with similar findings on Con A, and the common practice of addition of the two aforementioned cations in buffer used to keep Con A-Sepharose, or in buffer for affinity chromatography on Con A-Sepharose.

Lectins exhibit interesting activities, some of which are exploitable, including antiproliferative, antitumor, immunomodulatory (Wang *et al.*, 1996, 2000; Yu *et al.*, 2001; Singh Bains *et al.*, 2005), antifungal (Ye *et al.*, 2001), anti-insect (Machuka & Oladapo, 2000), and antiviral (Ooi *et al.*, 2000) activities. The pepper seed lectin potently inhibited growth and spore germination in the fungi *F. moniliforme* and *A. flavus*. However, it had minimal effects on other fungi such as *Botrytis cinerea*, *Fusarium graminearum*, *Fusarium solani*, and *Physalospora piricola*. At a lectin concentration of 33 μM, (65 ± 5)% inhibition of hyphal growth in *F. moniliforme* and (80 ± 5)% inhibition in *A. flavus* resulted. At 3.3 μM lectin, the widths of inhibition zone in the spore germination test in the two fungi were 4 ± 0.4 and 5 ± 0.4 mm (mean $\pm$ SD, $n = 3$), respectively. Thus, pepper seed lectin is similar to lectins from red kidney bean, potato tuber, and *Urtica diocia* (Gozia *et al.*, 1993; Broekaert *et al.*, 1998; Ye *et al.*, 2001) in the manifestation of antifungal activity. Its antifungal activity shows species-specificity. The lectin exhibited mitogenic activity toward mouse splenocytes; however, the activity was weaker in comparison with Con A. Maximal response was elicited by pepper lectin at 0.27-μM concentration.

Pepper seed lectin has some distinctive features, and a summary of the biochemical characteristics and biological activities is presented in Table 41.1. It exhibits similarity to the N-terminal sequence in rabbit lectin-like oxidized low-density lipoprotein receptor-1, and the C-terminal sequences of some viral polymerases. Pepper seed lectin has potentially exploitable antifungal and mitogenic activities. Its sugar-binding characteristics are unique; a number of simple sugars, such as galactose and fucose, can inhibit, although less potently than mannose, its hemagglutinating activity. Pepper seed lectin is thus an interesting lectin in some aspects. It deserves mention that lectin activity is not found in the seeds of the bell pepper and long pepper, which are varieties closely related to the red cluster pepper. Pepper lectin constitutes an addition to the list of compounds produced by the pepper, including capsaicin — well known for its extensive application in neuroscience research (Szolcsanyi & Bartho, 2001); capsanthin; and solanine. In traditional Chinese medicine, pepper is used for the treatment of anorexia, indigestion, gastrointestinal distension, rheumatism, loin pain, frost bite, and furunculosis.

TABLE 41.1 Characteristics of Lectin with Antifungal and Mitogenic Activities from Cluster Pepper Seeds

Molecular mass	29.5 kDa
N-terminal sequence	GQREKL
Chromatographic behavior on	
i) DEAE-cellulose	Adsorbed
ii) Q-Sepharose	Adsorbed
iii) Mono S	Adsorbed
Antifungal activity against	
i) *Aspergillus flavus*	Spore germination and hyphal growth inhibited
ii) *Fusarium moniliforme*	Spore germination and hyphal growth inhibited
iii) *Botrytis cinerea*	No inhibition
iv) *F. graminearum*	No inhibition
v) *F. solani*	No inhibition
vi) *Physalospora piricola*	No inhibition
Mitogenic activity	Maximum mitogenic response induced by 0.27 μM lectin, potency weaker than ConA
Hemagglutinating activity	Potentiated by Ca^{2+} and Mn^{2+} ions, specific to mannose and glucose, sensitive to changes in ambient pH and temperature

ADVERSE EFFECTS AND REACTIONS (ALLERGIES AND TOXICITY)

The fruits and leaves of the cluster pepper contain capsaicin, and can irritate the eye, mouth, and skin as described below. Eating excessive quantities of pepper can also produce gastrointestinal symptoms.

Symptoms of hot pepper poisoning include the following:

- a burning or stinging sensation
- severe pain in the mouth
- a burning sensation of the lip after the mouth has been in contact with pepper
- a burning sensation in the throat
- diarrhea, nausea, and vomiting after ingestion of pepper
- a burning sensation of the eyes, and severe eye pain and tears after the eyes have been exposed to pepper
- blistering of the skin after exposure of the skin to pepper.

SUMMARY POINTS

- Cluster pepper seeds produce a lectin with both antifungal and mitogenic activities.
- The lectin inhibits spore germination and hyphal growth in two fungal species: *Aspergillus flavus* and *Fusarium moniliforme*. However there is no inhibitory activity on other fungi, including *Botrytis cinerea*, *F. graminearum*, *F. solani*, and *Physalospora piricola*.
- The lectin elicits a mitogenic response from murine splenocytes *in vitro*, with a reduced potency compared with that of the jackbean lectin Concanavalin A.
- The lectin has a molecular mass of 29.5 kDa, and demonstrates specificity toward glucose and mannose.
- The lectin can be isolated by using anion exchange chromatography on DEAE-cellulose and Q-Sepharose, and fast protein liquid chromatography on MonoQ. It is adsorbed on all three ion exchangers.
- The lectin activity is enhanced by Ca^{2+} and Mn^{2+} ions. However, the activity is affected by changes in ambient temperature and pH.

References

Agrawal, B. B., & Goldstein, I. J. (1967). Physical and chemical characterization of Concanavalin A, the hemagglutinin from jack bean (*Canavalia ensiformis*). *Biochimica et Biophysica Acta, 33*, 376–379.

Broekaert, W. F., Van Parijs, J., Leyn, F., Joos, H., & Peumans, W. (1998). A chitin-binding lectin from stinging nettle rhizomes with antifungal properties. *Science, 245*, 1100–1102.

Gozia, O., Ciopraga, J., Bentia, T., Lunga, M., Zamfirescu, I., Tudor, R., et al. (1993). Antifungal properties of lectin and new chitinases from potato tubers. *FEBS letters, 370*, 245–249.

Inamori, K., & Saito, T. (1999). A newly identified horseshoe crab lectin with specificity for blood group A antigen recognizes specific O-antigens of bacterial lipopolysaccharides. *Journal of Biological Chemistry, 274*, 3272–3278.

Knibbs, R. N., Goldstein, I. J., Patcliffe, R. M., & Shibuya, N. (1991). Characterization of the carbohydrate binding specificity of the leukoagglutinating lectin from *Maakia amurensis*. Comparison with other sialic acid-specific lectins. *Journal of Biological Chemistry, 266*, 83–88.

Lam, Y. W., & Ng, T. B. (2001). A monomeric mannose-binding lectin from inner shoots of the edible chive (*Allium tuberosum*). *Journal of Protein Chemistry, 20*, 361–366.

Leteux, C., & Chai, W. (2000). The cysteine-rich domain of the macrophage mannose receptor is a multispecific lectin that recognizes chondroitin sulfates A and B and sulfated oligosaccharides of blood group Lewis(a) and Lewis(x) types in addition to the sulfated N-glycans of lutropin. *Journal of Experimental Medicine, 191*, 1117–1126.

Machuka, J. S., & Oladapo, G. (2000). The African yam bean seed lectin affects the development of the cowpea weevil but does not affect the development of larvae of the legume pod borer. *Phytochemistry, 53*, 667–674.

Ngai, P. H., & Ng, T. B. (2007). A lectin with antifungal and mitogenic activities from red cluster pepper (*Capsicum frutescens*) seeds. *Applied Microbiology and Biotechnology, 74*, 366–371.

Nomura, K., Ashida, H., Uemura, N., Kushibe, S., Ozaki, T., & Yoshida, M. (1998). Purification and characterization of a mannose/glucose-specific lectin from *Castanea crenata*. *Phytochemistry, 49*, 667–673.

Ooi, L. S., Ng, T. B., Geng, Y., & Ooi, V. E. (2000). Lectins from bulbs of the Chinese daffodil *Narcissus tazetta* (family Amaryllidaceae). *Biochemistry and Cell Biology, 78*, 463–468.

Pusztai, A. P. (1991). *Chemistry and pharmacology of natural products*. In: *Plant lectins, Vol 1*. Cambridge, UK: Cambridge University Press.

Singh Bains, J., Singh, J., Kamboj, S. S., Nijar, K. K., Agrewale, J. N., Kumar, V., et al. (2005). Mitogenic and antiproliferative activity of a lectin from the tubers of Voodoo lily (*Sauromatum venosum*). *Biochimica et biophysica acta, 1723*, 163–174.

Szolcsanyi, J., & Bartho, L. (2001). Capsaicin-sensitive afferents and their role in gastroprotection: an update. *Journal of physiology Paris, 95*, 181–188.

Tateno, H., Saneyoshi, A., Ogawa, T., Muramoto, K., Kamiya, H., & Saneyoshi, M. (1998). Isolation and characterization of rhamnose-binding lectins from eggs of steelhead trout (*Oncorhynchus mykiss*) homologous to low density lipoprotein receptor superfamily. *Journal of Biological Chemistry, 273*, 19190–19197.

Van Damme, J. M., Smeets, K., Torrekens, S., Van Leuven, F., & Peumans, W. J. (1994). Characterization and molecular cloning of mannose-binding lectins from the Orchidaceae species *Listera ovata, Epipactis helleborine* and *Cymbidium hybrid*. *European Journal of Biochemistry, 221*, 769–777.

Wang, H., Gao, J., & Ng, T. B. (2000). A new lectin with highly potent antihepatoma and antisarcoma activities from the oyster mushroom. *Pleurotus ostreatus. Biochemical and Biophysical Research Communications, 275*, 810–816.

Wang, H. X., Liu, W. K., Ng, T. B., Ooi, V. E., & Chang, S. T. (1996). The immunomodulatory and antitumor activities of lectins from the mushroom. *Tricholoma mongolicum Immunopharmacology, 31*, 205–211.

Ye, X. Y., Ng, T. B., Tsang, P. W., & Wang, J. (2001). Isolation of a homodimeric lectin with antifungal and antiviral activities from red kidney bean (*Phaseolus vulgaris*) seeds. *Journal of Protein Chemistry, 20*, 367–375.

Yu, L. G., Milton, J. D., & Fernig, D. G. (2001). Opposite effects on human colon cancer cell proliferation of two dietary Thomsen–Friedenreich antigen-binding lectins. *Journal of Cellular Physiology, 186*, 282–287.

CHAPTER 42

Cocoa (*Theobroma cacao*) Seeds and Phytochemicals in Human Health

Jiyoung Kim[1,2], Ki Won Lee[2,3], Hyong Joo Lee[1,2]
[1] Research Institute for Agriculture and Life Sciences, Seoul National University, Seoul, Republic of Korea
[2] Department of Agricultural Biotechnology, college of Agriculture and Life Sciences, Seoul University, Seoul, Republic of Korea
[3] Department of Bioscience and Biotechnology, Konkuk University, Seoul, Republic of Korea

LIST OF ABBREVIATIONS

eNOS, endothelial nitric oxide synthase
LDL, low density lipoproteins
NF-κB, nuclear factor-κB
NO, nitric oxide
TNF-α, tumor necrosis factor-α
UV, ultraviolet

INTRODUCTION

It has been about 15 years since the first indication in a medical journal that cocoa is a potential source of antioxidants. Cocoa has been found to have more phenolic

Nuts & Seeds in Health and Disease Prevention. DOI: 10.1016/B978-0-12-375688-6.10042-8

phytochemicals per serving than teas and red wine (Lee *et al.*, 2003). Cocoa phytochemicals have strong antioxidant and anti-inflammatory activity. This chapter summarizes recent interesting and relevant findings regarding cocoa and its phytochemicals in human health.

BOTANICAL DESCRIPTION

The cocoa plant was first given its botanical name by the Swedish natural scientist Carl Linnaeus, who called it *Theobroma* ("food of the gods") *cacao* (Knight, 1999). Cocoa, also called cacao, is the seed of *Theobroma cacao*, an evergreen tree that reaches a height of 6–12 m and grows in a limited geographical zone from approximately 20° North to 20° South of the Equator (Figure 42.1A). *Theobroma cacao* requires an annual rainfall of 1–5 $1/m^2$ for adequate moisture. *Theobroma cacao* belongs to the genus *Theobroma* and the family Sterculiaceae, alternatively Malvaceae (Knight, 1999). The cocoa seed is borne in a berry-like fruit cocoa pod, which is filled with sweet and mucilaginous pulp enclosing 25–75 large seeds (Figure 42.1B). Cocoa seeds are usually white, but become violet or reddish brown during the drying process (Figure 42.1C).

HISTORICAL CULTIVATION AND USAGE

Theobroma cacao was first cultivated in 250–900 AD by the ancient civilizations of the Mayas and Aztecs in the Mesoamerican region (Knight, 1999). From its divine origins, cocoa was

FIGURE 42.1

Cocoa tree, cocoa pod, and cocoa seed. (A) Cocoa tree; (B) Cocoa beans exposed from the white cocoa butter that surrounds them inside the cocoa pod; (C) Cocoa seed. *Reproduced with permission from The Beautiful Store.*

introduced to the Spanish Royal Court in the mid-1550s (Dillinger *et al.*, 2000). Cocoa had many uses, ranging from medicine to currency, in early Mesoamerican cultures (Dillinger *et al.*, 2000). More than 100 medicinal uses of cocoa have been documented in Europe and New Spain (Dillinger *et al.*, 2000).

PRESENT-DAY CULTIVATION AND USAGE

Cocoa is cultivated on lands covering over 70,000 km^2 worldwide. About 70% of the world's cocoa production takes places in the equatorial region of West Africa, and the rest in the equatorial regions of Central and South America, the West Indies, and tropical areas of Asia (Dillinger *et al.*, 2000). The global production of cocoa was 1,556,484 tonnes in 1974, and 3,607,052 tonnes in 2004, an increase of 131.7% in 30 years. The use of cocoa has evolved to what we now know as chocolate, made in solid or liquid form. Significant quantities of cocoa are used in the manufacture of numerous foods and beverages.

APPLICATIONS TO HEALTH PROMOTION AND DISEASE PREVENTION

In 1519, Aztec Emperor Montezuma called cocoa "a divine drink, which builds up resistance and fights fatigue. A cup of this precious drink permits a man to walk for a whole day without food" (documented by Hernán Cortés, 1485—1547). From over 3000 years ago to the present, cocoa has been used to treat anemia, mental fatigue, tuberculosis, fever, gout, kidney stones, and even poor sexual appetite (Dillinger *et al.*, 2000). Several health effects, including improved heart function, relief of angina pectoris, stimulation of the nervous system, facilitated digestion, and improved kidney and bowel function have been considered for its medicinal use (Dillinger *et al.*, 2000). Recent investigation on Panama's Kuna Indian population revealed that Kuna Indians living on the islands, and heavy consumers of cocoa, had significantly lower rates of heart disease, cancer, and diabetes mellitus compared to those on the mainland, who do not consume cocoa (Corti *et al.*, 2009), suggesting that cocoa might be useful in the prevention of heart disease, cancer, and diabetes mellitus.

Polyphenolic phytochemicals in cocoa

Cocoa is unusually rich in phytochemicals. In dried and unfermented cocoa beans, pigment cells make up about 11—13% of the tissue. These cells are rich in polyphenolic compounds, in particular catechins, although the characteristic purple color of cocoa beans is due to anthocyanins. The main phytochemicals contained in cocoa are listed in Table 42.1, together with estimates of their concentration where available (Knight, 1999). Published information may vary between unfermented beans, fermented beans, beans or cocoa mass after roasting, and cocoa powder or chocolate.

The phytochemicals in cocoa fall into a number of types. The polyphenolics are quantitatively the largest group, comprising around 13.5% of the dried unfermented cocoa beans (Table 42.1). Most of these are flavanols and flavonols (Figure 42.2). The former mostly comprise catechins and procyanidins, and, of these, by far the greatest single component after roasting and conching is (−)-epicatechin. The catechins are colorless, while the anthocyanins, present in somewhat lower amounts, give rise to the color (cocoa purple) in unfermented cocoa beans. As a result of fermentation, much of the anthocyanin is converted to quinonic compounds, giving the beans their characteristic brown color. Quercetin (a flavonol) and two derivatives (a glucoside and an arabinoside) in cocoa are structurally similar to the simple flavanols (Figure 42.2).

The procyanidins (also termed leukocyanidins) are complex polymeric forms of flavanols (Table 42.1; Figure 42.3). They are produced during fermentation, resulting from simple flavonols undergoing condensation reactions. The most common building block is

TABLE 42.1 Polyphenolic Phytochemicals in Cocoa

Compound	Dry Seeds (g/100 g)	After Roasting and Conching (g/100 g)	In Milk Chocolate (mg/100 g)
Flavanols:			
Catechins	3.0		
(+)-Catechin	1.6—2.75	0.03—0.08	0.02
(−)-Epicatechin			
(+)-Gallocatechin	0.25—0.45	0.3—0.5	—
(+)-Epigallocatechin			
Procyanidins (Leucocyanidins):			
L_1—L_4	2.7	L_1:0.08—0.17	—
Polymeric procyanidins	2.1—5.4	—	—
Anthocyanins		0.01	
3-α-L-arabinosyl cyanidin	0.3	—	—
3-β-D-arabinosyl cyanidin	0.1	—	—
Flavonols:			
Quercetin	—	—	—
Quercetin-3-glucoside	—	—	—
Quercetin-3-arabinoside	—	—	—
Total phenolics	13.5		

The phytochemicals in cocoa are flavanols and flavonols. The flavanols mostly comprise catechins and procyanidins. The flavonols are quercetin and two derivatives (a glucoside and an arabinoside).
Modified from Knight (1999), with permission.

epicatechin. Such oligomeric molecules are often termed tannins. Procyanidins are the source of astringency in cocoa products. A major set of reviews on phytochemicals present in cocoa appears in Knight (1999).

Antioxidant effects

Consuming chocolate has been reported to increase the total antioxidant capacity in human blood plasma. Cocoa polyphenols exhibit strong free radical scavenging activity. Monomeric epicatechins and oligomeric procyanidins are effective in inhibiting both dioxygenase and 5-lipoxygenase, and preventing oxidation (Schewe et al., 2002). An extract of cocoa seeds rich in polyphenols counteracted the increased level of lipid peroxide in rats fed a vitamin-E deficient diet (Yamagishi et al., 2002). Due to the natural antioxidant flavonoids contained in cocoa, chocolate was found to be resistant to spoilage.

Inflammation and immune function

Data from numerous studies suggest that cocoa can effectively modify the inflammatory process. Cocoa monomer catechins and oligomeric procyanidins have a significant anti-inflammatory effect, inhibit neutrophil oxidative burst, and reduce the expression of adhesion molecules. High doses of cocoa flavanols suppress the level of plasma leukotriens, which contribute to inflammation in asthma and bronchitis. Cocoa-derived flavanols can modulate the synthesis and effects of eicosanoids, and mediate acute inflammatory cascade. Comprehensive experimental evidence has demonstrated that cocoa-derived flavanols reduce nuclear factor-κB (NF-κB) activation, resulting in reduced production of tumor necrosis factor-α (TNF-α) (Selmi et al., 2006). The impact of cocoa on blocking NF-κB activation may constitute

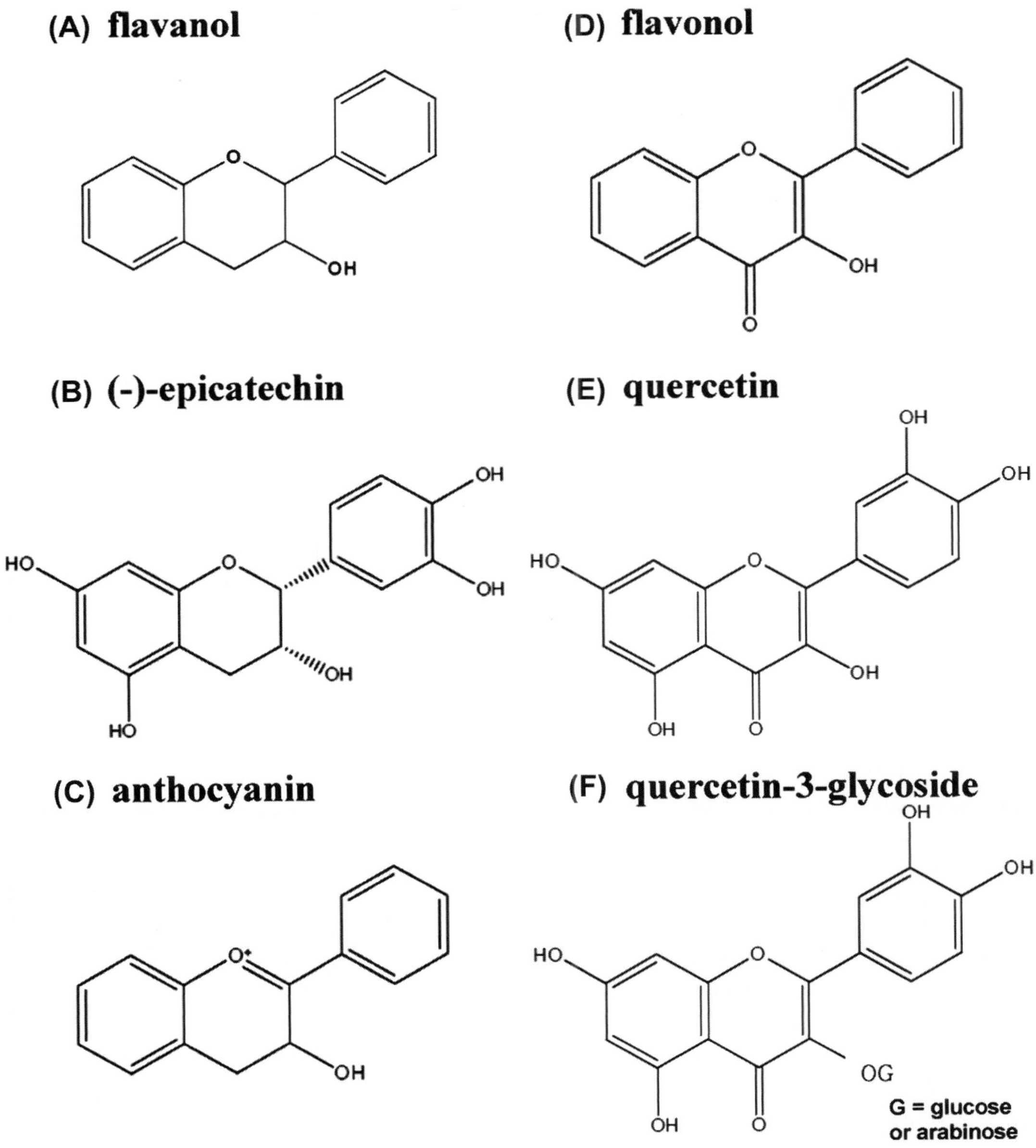

(A) flavanol

(D) flavonol

(B) (-)-epicatechin

(E) quercetin

(C) anthocyanin

(F) quercetin-3-glycoside

G = glucose
or arabinose

FIGURE 42.2

Structures of common flavanols and flavonols found in cocoa. (A) Common structure of flavanol; (B) (—)-Epicatechin, a representative flavanol found in cocoa seeds; (C) Anthocyanin, a flavanol found in cocoa seeds; (D) Common structure of flavonol; (E) Quercetin, a flavonol found in cocoa seeds; (F) Quercetin-3-glycoside, a flavonol found in cocoa seeds.

a common mechanism for all anti-inflammatory effects of cocoa. On the other hand, small molecular weight flavanols (monomer through pentamer) in cocoa can enhance the secretion of anti-inflammatory cytokine interleukin-5 (IL-5), which might protect against chronic infection (Selmi *et al.*, 2006).

Cardiovascular health and disease

Recent research demonstrates that cocoa consumption is beneficial to vascular and platelet function and blood pressure (Corti *et al.*, 2009). A single dose of a cocoa beverage reverses the endothelial dysfunction apparent in hyperlipidemia, hypertension, and coronary artery disease. Potential mechanisms through which cocoa might exert its beneficial effects on

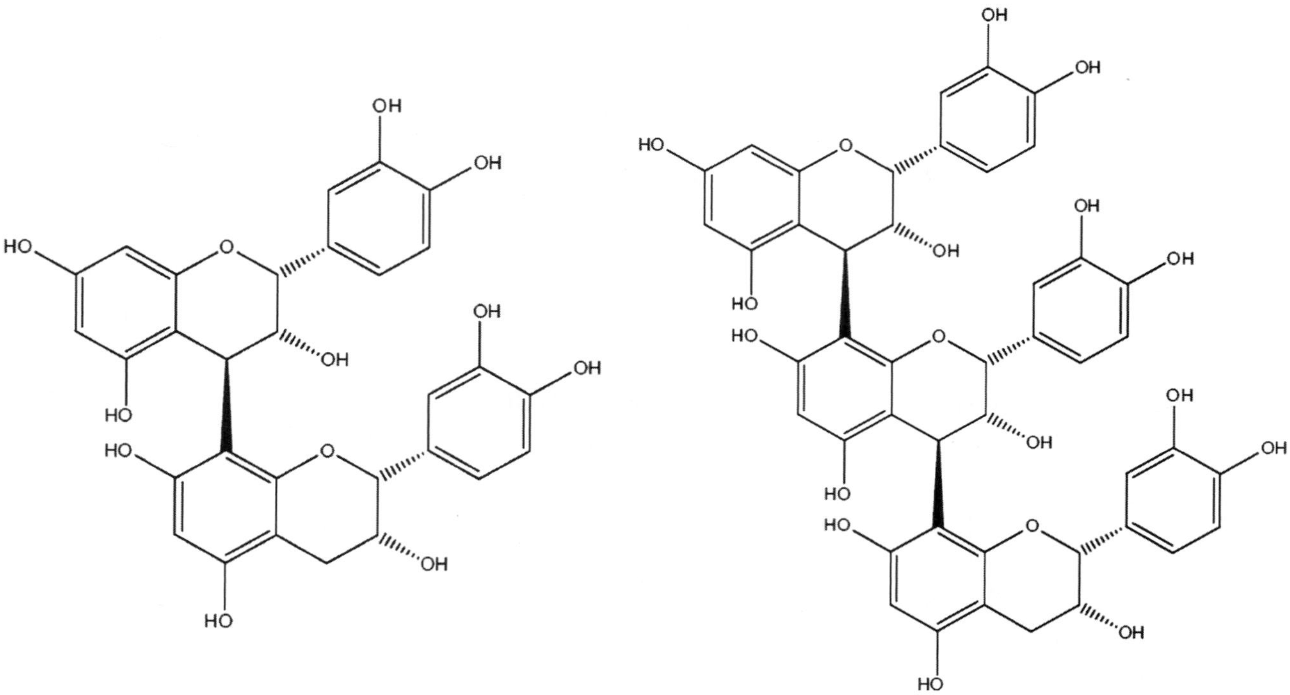

FIGURE 42.3
Structures of common procyanidins found in cocoa. Procyanidins found in cocoa are based on epicatechins.

356

cardiovascular health and disease have been proposed. The first mechanism proposes that cocoa mediates an increase in the bioavailability of nitric oxide (NO) in endothelium (Figure 42.4; Corti *et al.*, 2009). Cocoa lowers vascular arginase activity in endothelial cells, which augments the local levels of L-arginine required to synthesize NO by endothelial NO synthase (eNOS) (Corti *et al.*, 2009). NO from endothelium induces a relaxation of vascular smooth muscle cells, leading to vasodilation (Corti *et al.*, 2009). NO also prevents leukocyte adhesion and migration, smooth muscle cell proliferation, and platelet adhesion and aggregation (Corti *et al.*, 2009). In humans, a cocoa drink high in flavanol content (176–185 mg) rapidly enhances the circulating pool of bioactive NO by more than one third, and in turn augments vasodilation (Figure 42.5; Heiss *et al.*, 2005).

The second potential mechanism of cocoa on cardiovascular health is through its antioxidant effects (Corti *et al.*, 2009). Oxidative stress due to excess production of free radicals or reactive oxygen species is associated with a number of cardiovascular risk factors, such as dyslipidemias and hypertension. Oxidative modification of low density lipoproteins (LDL) is believed to be a major contributing factor in atherosclerosis. Cocoa, with its high quantity and quality of antioxidant flavonoids, can decrease the oxidation of LDL and reduce the cardiovascular risk factors (Corti *et al.*, 2009).

The third proposed mechanism of cocoa on cardiovascular health is through its anti-inflammatory effect (Selmi *et al.*, 2006; Corti *et al.*, 2009). Because inflammation promotes endothelial dysfunction and atherogenesis, it is now widely accepted that atherosclerosis is a chronic inflammatory disease. Cocoa-derived products rich in flavanols potentially modulate the inflammatory status that characterizes several chronic and acute cardiovascular diseases in a positive way (Selmi *et al.*, 2006).

Consumption of dark chocolate improves glucose metabolism, as well as insulin resistance and sensitivity. Experimental evidence in obese diabetic mice suggested that cocoa prevents hyperglycemia (Corti *et al.*, 2009).

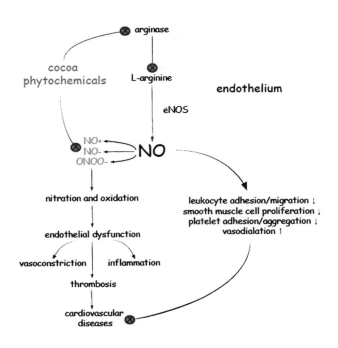

FIGURE 42.4
Cocoa-mediated nitric oxide (NO) production, and its intervention in cardiovascular diseases. In the endothelial cells, NO is synthesized from L-arginine by the endothelial nitric oxide synthase (eNOS). NO then diffuses into the vascular smooth muscle cells, resulting in vasodilation. Flavanols in cocoa may signal the release of NO and inhibit the generation of reactive nitrogen species nitrosonium (NO^+), nitroxyl anion (NO^-), and peroxynitrite ($ONOO^-$), linked to endothelial dysfunction, including vasoconstriction, inflammation, and thrombosis, and, finally, cardiovascular disease.

Cancer

357

Cancer chemopreventive activity of cocoa has been reported. For example, cocoa liquor procyanidins significantly reduced the incidence and multiplicity of lung carcinomas and also decreased thyroid adenomas developed in male rats (Yamagishi *et al.*, 2003). Selected procyanidins present in cocoa inhibit tumorigenesis, tumor growth, and angiogenesis (Kenny *et al.*, 2004). Cocoa liquor procyanidins inhibited mammary and pancreatic tumorigenesis in female Sprague-Dawley rats (Figure 42.6; Yamagishi *et al.*, 2002). Cocoa liquor polyphenols have been shown to have antimutagenic and anticlastogenic activity *in vitro* (Yamagishi *et al.*, 2002). Consumption of cocoa or dark chocolate can also decrease the burden and efficacy of epigenetic carcinogens (Kang *et al.*, 2008). Procyanidin-enriched cocoa seed extracts caused G2/M cell cycle arrest and 70% growth inhibition in Caco-2 colon cancer cells (Carnesecchi *et al.*, 2002). On the other hand, cocoa-derived pentameric procyanidin caused G0/G1 cell cycle arrest and selective growth inhibition in human breast cancer cells (Ramljak *et al.*, 2005). It has been also found that cocoa procyanidins reduce vascular endothelial growth factor activity and angiogenic activity associated with tumor pathology (Kenny *et al.*, 2004).

Fundamental mechanisms of carcinogenesis are associated with ROS, and population studies have demonstrated that people with a regular intake of foods containing antioxidants, such as vegetables, fruits, tea, or soy products, display a lower incidence of various types of cancer (Lee *et al.*, 2003; Kang *et al.*, 2008). It can be postulated, therefore, that consumption of cocoa or chocolate, which have high antioxidant activity, could be beneficial in decreasing the damage from genotoxic and epigenetic carcinogens, and inhibiting the complex processes leading to cancer.

Brain

Results of a pilot study showed that an acute dose of flavanol-rich cocoa can increase cerebral blood flow (Dinges, 2006). A number of vascular risk factors are likewise risk factors for

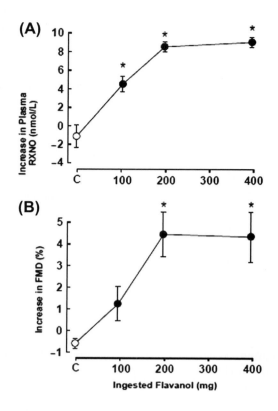

FIGURE 42.5

The level of circulating nitric oxide (NO) and vasodilation after cocoa consumption in humans. Flavanol-rich cocoa dose-dependently increased the circulating NO pool (RXNO) in plasma (A) and flow-mediated dilation (FMD) (B) at 2 hours (filled circles) in humans. RXNO and FMD did not significantly change after water control (open circles). Data represent means $\pm$ SE ($n = 4$); *$P < 0.05$. *Reprinted from Heiss et al. (2005)*, J. Am. Coll. Cardiol., 46, *1276–1283, with permission.*

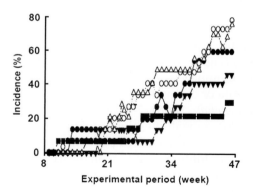

FIGURE 42.6

Incidences of palpable mammary tumors in rats treated with tumor inducer. ●, Tumor inducer, 2-amino-1-methyl-6-phenylimidazo[4,5-*b*]pyridine (PhIP) alone; Δ, 0.025% Cocoa liquor procyanidins together with PhIP treatment; ▼, 0.25% Cocoa liquor procyanidins together with PhIP treatment; ○, 0.025% Cocoa liquor procyanidins after PhIP treatment; ■, 0.25% Cocoa liquor procyanidins after PhIP treatment. *Reprinted from Yamagishi et al. (2002), Cancer Lett., 185:123–130, with permission.*

dementia, and many measures that boost endothelial NO function are linked to a decrease in risk for dementia (Dinges, 2006). The fact that flavanol-rich cocoa increases both the bioavailability of NO and the mean arterial flow velocity in the middle cerebral artery suggests cocoa as a possible treatment for dementia (Dinges, 2006). Regular consumption of cocoa flavanols might play a role in the retention of cognitive function as humans age.

Skin

Cocoa polyphenols induces a rapid improvement in skin tonus and elasticity. Consumption of high-flavanol cocoa improves skin hydration, and decreases transepidermal water loss (Heinrich *et al.*, 2006). A single dose of flavanol-rich cocoa leads to increase blood flow to cutaneous and subcutaneous tissues within 2 hours after ingestion (Gasser *et al.*, 2008). Oral or topical procyanidin also inhibits the ultraviolet (UV) radiation-induced erythema response (Heinrich *et al.*, 2006). Cocoa beverages rich in flavanols decrease the sensitivity of human skin to UV light (Heinrich *et al.*, 2006). The influence of cocoa polyphenols on parameters of connective tissues such as glycosaminoglycans and collagen is comparable or better than a commercially available anti-aging cream (Gasser *et al.*, 2008). When cocoa polyphenols are applied in conjunction with cocoa butter, there is an enhancing effect on the general skin morphology (Gasser *et al.*, 2008).

ADVERSE EFFECTS AND REACTIONS (ALLERGIES AND TOXICITY)

Because of cocoa's long-term use with no reported adverse effects in humans, there has not been great concern over its safety. In humans, long-term ingestion of extremely large amounts of cocoa products (daily equivalent of 100 g cocoa powder or 1.5 g methylxanthines) results in sweating, trembling, and severe headaches (Tarka, 1982).

SUMMARY POINTS

- Cocoa contains high concentrations of phytochemicals, including momomeric (epicatechin and catechin) and oligomeric (procyanidin) flavanols. Recent research has been identified the health-promoting effects of flavanols in cocoa.
- Cocoa and its phytochemicals have strong antioxidant and anti-inflammatory activity.
- Recent research demonstrates that cocoa consumption is beneficial to cardiovascular diseases and diabetes.
- Selected procyanidins present in cocoa prevent cancer by inhibiting tumorigenesis, tumor growth, and angiogenesis.
- Flavanol-rich cocoa increases cerebral blood flow, which suggests cocoa as a possible treatment for dementia.
- Cocoa has moisturizing properties in skin, improves skin appearance and texture, and confers substantial photoprotection from UV light.

References

Carnesecchi, S., Schneider, Y., Lazarus, S. A., Coehlo, D., Gosse, F., & Raul, F. (2002). Flavanols and procyanidins of cocoa and chocolate inhibit growth and polyamine biosynthesis of human colonic cancer cells. *Cancer letters*, *175*, 147–155.

Corti, R., Flammer, A. J., Hollenberg, N. K., & Luscher, T. F. (2009). Cocoa and cardiovascular health. *Circulation*, *119*, 1433–1441.

Dillinger, T. L., Barriga, P., Escarcega, S., Jimenez, M., Salazar Lowe, D., & Grivetti, L. E. (2000). Food of the gods: cure for humanity? A cultural history of the medicinal and ritual use of chocolate. *Journal of Nutrition, 130*, 2057S–2072S.

Dinges, D. (2006). Cocoa flavanols, cerebral blood flow, cognition, and health: going forward. *Journal of Cardiovascular Pharmacology, 47*, S221–S223.

Gasser, P., Lati, E., Peno-Mazzarino, L., Bouzoud, D., Allegaert, L., & Bernaert, H. (2008). Cocoa polyphenols and their influence on parameters involved in *ex vivo* skin restructuring. *International Journal of Cosmetic Science, 30*, 339–345.

Heinrich, U., Neukam, K., Tronnier, H., Sies, H., & Stahl, W. (2006). Long-term ingestion of high flavanol cocoa provides photoprotection against UV-induced erythema and improves skin condition in women. *Journal of Nutrition, 136*, 1565.

Heiss, C., Kleinbongard, P., Dejam, A., Perre, S., Schroeter, H., Sies, H., et al. (2005). Acute consumption of flavanol-rich cocoa and the reversal of endothelial dysfunction in smokers. *Journal of the American College of Cardiology, 46*, 1276–1283.

Kang, N. J., Lee, K. W., Lee, D. E., Rogozin, E. A., Bode, A. M., Lee, H. J., et al. (2008). Cocoa procyanidins suppress transformation by inhibiting mitogen-activated protein kinase. *Journal of Biological Chemistry, 283,* 20664–20673.

Kenny, T., Keen, C., Jones, P., Kung, H., Schmitz, H., & Gershwin, M. (2004). Pentameric procyanidins isolated from Theobroma cacao seeds selectively downregulate ErbB2 in human aortic endothelial cells. *Experimental Biology and Medicine, 229,* 255.

Knight, I. (1999). *Chocolate and cocoa: Health and nutrition* (1st ed.). Oxford, UK: Blackwell Science.

Lee, K., Kim, Y., Lee, H., & Lee, C. (2003). Cocoa has more phenolic phytochemicals and a higher antioxidant capacity than teas and red wine. *Journal of Agricultural and Food Chemistry, 51,* 7292–7295.

Ramljak, D., Romanczyk, L. J., Metheny-Barlow, L. J., Thompson, N., Knezevic, V., Galperin, M., et al. (2005). Pentameric procyanidin from *Theobroma cacao* selectively inhibits growth of human breast cancer cells. *Molecular Cancer Therapeutics, 4,* 537–546.

Schewe, T., Kuhn, H., & Sies, H. (2002). Flavonoids of cocoa inhibit recombinant human 5-lipoxygenase. *Journal of Nutrition, 132,* 1825–1829.

Selmi, C., Mao, T., Keen, C., Schmitz, H., & Eric Gershwin, M. (2006). The anti-inflammatory properties of cocoa flavanols. *Journal of Cardiovascular Pharmacology, 47,* S163.

Tarka, S. M., Jr. (1982). The toxicology of cocoa and methylxanthines: a review of the literature. *Critical Reviews in Toxicology, 9,* 275–312.

Yamagishi, M., Natsume, M., Osakabe, N., Nakamura, H., Furukawa, F., Imazawa, T., et al. (2002). Effects of cacao liquor proanthocyanidins on PhIP-induced mutagenesis *in vitro,* and *in vivo* mammary and pancreatic tumorigenesis in female Sprague-Dawley rats. *Cancer letters, 185,* 123–130.

Yamagishi, M., Natsume, M., Osakabe, N., Okazaki, K., Furukawa, F., Imazawa, T., et al. (2003). Chemoprevention of lung carcinogenesis by cacao liquor proanthocyanidins in a male rat multi-organ carcinogenesis model. *Cancer letters, 191,* 49–57.

CHAPTER 43

Health Benefits of Coconut (*Cocos nucifera* Linn.) Seeds and Coconut Consumption

Francis Agyemang-Yeboah
Department of Molecular Medicine, School of Medical Science, Kwame Nkrumah University of Science and Technology, Kumasi, Ghana

CHAPTER OUTLINE

LIST OF ABBREVIATION

CNO, coconut oil
IBS, irritable bowel syndrome
LDL, low density lipoprotein
MCFA, medium chain fatty acid
VCO, virgin coconut oil

INTRODUCTION

The coconut palm is widely recognized as one of nature's most useful and beautiful plants. The coconut has provided humans with food, shelter, and transport for millennia throughout the tropical regions of the world. Coconut fruit in the wild is light, buoyant, and highly water resistant, and evolved to disperse significant distances via marine currents (Werth, 1933; Foale, 2003).

In modern times, coconut products in the form of oils, fiber, and even charcoal are ubiquitous in consumer products such as soap, cosmetics, foods, and medicines worldwide. Coconut palms require warm conditions for successful growth, and are intolerant of cold

Nuts & Seeds in Health and Disease Prevention. DOI: 10.1016/B978-0-12-375688-6.10043-X

weather. Optimum growth is with a mean annual temperature of 27°C, and growth is reduced below 21°C.

The coconut palm is perhaps the most widely grown palm in the world (Chang & Elevitch, 2006). Coconuts feature as one of the main sources of income for producing countries. Coconuts are used as whole fruits or by their parts: mesocarp fibers, milk, kernel (or flesh), and husk. Copra (the dried "meat" of the coconut seed), from which oil is extracted, is a significant cash crop throughout the tropics. Coir, the fiber from the fruit, is used in manufacturing. The fruits, or coconuts, yield several food products at different stages of development (Tables 43.1, 43.2), and the leaves are used for thatch or are woven into baskets, mats, and clothing. Various studies carried out on the health benefits or otherwise of coconut/oil have shown that it has extensive usage and effects on various diseases. Even though some of these claims have not been fully elucidated, a few convincing studies on its ketogenic, glycemic, and atherogenic properties are available (Kritchevsky *et al.*, 1976; Young & Renner, 1977; Dayrit, 2006). Moreover, there have been various studies on its antiviral and antimicrobial benefits. It has been speculated by various researchers that some of the benefits of coconut oil can be attributed to the presence of lauric acid, capric acid, and caprylic acid (Goldberg & Enig, 1993; Ravnskov *et al.*, 2002).

BOTANICAL DESCRIPTION

The coconut palm (*Cocos nucifera*) (Figure 43.1) is an important member of the family Arecaceae (palm family) (Harries, 1978). The coconut palm can have a long life-span of up to

TABLE 43.1 Nutritional Composition of Coconut Fruit

	Nutritional Value per 100 g
Energy	1481 kJ (354 kcal)
Carbohydrates	15.23 g
Sugars	6.23 g
Dietary fiber	9.0 g
Fat	33.49 g
— saturated	29.70 g
— monounsaturated	1.43 g
polyunsaturated	0.37 g
Protein	3 g

Source: USDA Nutrient Database.

TABLE 43.2 Proximate Vitamin and Mineral Content of Coconut Fruit

	Nutritional Value per 100 g
Thiamine (Vit. B_1)	0.066 mg (5%)
Riboflavin (Vit. B_2)	0.02 mg (1%)
Niacin (Vit. B_3)	0.54 mg (4%)
Pantothenic acid (B_5)	0.300 mg (6%)
Vitamin B_6	0.054 mg (4%)
Folate (Vit. B_9)	26 μg (7%)
Vitamin C	3.3 mg (6%)
Calcium	14 mg (1%)
Iron	2.43 mg (19%)
Magnesium	32 mg (9%)
Phosphorus	113 mg (16%)
Potassium	356 mg (8%)
Zinc	1.1 mg (11%)

FIGURE 43.1
Cocos nucifera (coconut); Arecaceae (palm family): mature coconut palm with fruits. *Source: Species Profiles for Pacific Island Agroforestry (www.traditionaltree.org).*

100 years. It has a single trunk, 20—30 m tall, with smooth, gray bark marked by ringed scars left by fallen leafbases. The leaves are 4—6 m long, pinnate, and consist of linear-lanceolate, more or less recurved, rigid, bright-green leaflets. The inflorescences, arising at leaf axils and enveloped by a carinate spathe, are unbranched spadices. Female flowers are borne basally, male flowers at the apex. Flowers bear lanceolate petals, six stamens, and an ovary consisting of three connate carpels. Coconut is mostly propagated by cross-pollination, either anemophilous or entomophilous. The flowers of the coconut palm are polygamomonoecious (i.e., of a specied population containing plants that are polygamous and plants that are monoecious), with both male and female flowers in the same inflorescence. Flowering occurs continuously. Botanically, the coconut fruit is a drupe, not a true nut (Pearsall, 1999). Like other fruits, it has three layers: the exocarp, mesocarp, and endocarp (Figure 43.2). The exocarp and mesocarp

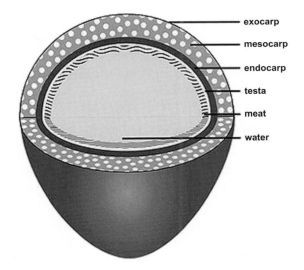

FIGURE 43.2
Mature coconut sectioned to display layers. As a fibrous drupe, the coconut has three layers: the exocarp, mesocarp, and endocarp. The exocarp and mesocarp make up the husk of the coconut. *Source: Own work by Kerina yin (2008).*

make up the husk of the coconut. Its fruit, as big as a person's head and 1–2 kg in weight, has a thin, smooth, grayish-brown epicarp, a fibrous, 4–8 cm thick mesocarp, and a woody endocarp; as it is rather light, it can be carried for considerable distances by water while maintaining its germinability for a long time. Inside, it contains one seed, rich in reserve substances located in the endosperm, which is partly liquid (coconut milk) and partly solid (flesh). The mature kernel of the coconut is the edible endosperm, located on the inner surface of the shell. Inside the endosperm layer, the coconut contains an edible clear liquid that is sweet, salty, or both (Pearsall, 1999).

The shell has three germination pores (stoma) or eyes that are clearly visible on its outside surface once the husk is removed. Within the shell is a single seed. When the seed germinates, the root (radicle) of its embryo pushes out through one of the eyes of the shell. The outermost layer of the seed, the testa, adheres to the inside of the shell. In a mature coconut, a thick albuminous endosperm adheres to the inside of the testa. Although coconut kernel contains less fat than many oilseeds and nuts such as almonds, it is noted for its high amount of medium-chain saturated fats (Enig, 2001). About 90% of the fat found in a coconut kernel is saturated — a proportion exceeding that of foods such as lard, butter, and tallow. It is relatively high in minerals such as iron, phosphorus, and zinc (Deosthale, 1981).

HISTORICAL CULTIVATION AND USAGE

The Indian state of Kerala is known as the Land of Coconuts. The name derives from *Kera* (the coconut tree) and *Alam* (place, or earth). Kerala has beaches fringed by coconut trees, and a dense network of waterways, flanked by green palm groves and cultivated fields. Coconuts form part of the daily diet, the oil is used for cooking, and coir is used for furnishing, decorating, etc.

Coconuts received their name from Portuguese explorers, the sailors of Vasco da Gama in India, who first brought them to Europe. The brown and hairy surface of coconuts reminded them of a ghost or witch called "Coco." Previously, it was called *nux indica* — a name given by Marco Polo in 1280. When coconuts arrived in England, they retained the coco name, and nut was added. Most authorities claim that it is native to South Asia (particularly the Ganges Delta), while some claim its origin is in north-western South America (Chang & Elevitch, 2006).

Fossil records from New Zealand indicate that small, coconut-like plants grew there as long as 15 million years ago. Even older fossils have been uncovered in other parts of the world, the oldest of which is the one uncovered in Khulna, Bangladesh. For many years, tropical communities have used coconut oil in key areas of their lives, such as cooking, to cure and prevent disease, and for hair- and skin-care, etc.

PRESENT-DAY CULTIVATION AND USAGE

Coconut palms are grown in more than 80 countries of the world, with a total production of 61 million tonnes per year. The palm is native to tropical eastern regions. Today, it is grown both on the Asian continent (India, Ceylon, and Indonesia) and in Central and South America (Mexico, Brazil) (Chang & Elevitch, 2006).

In Africa, the largest producing countries are Mozambique, Tanzania, and Ghana. The coconut palm is perhaps the most widely grown palm in the world, and features as one of the main sources of income for producing countries. However, other parts of this plant are used too; notably its leaves for making baskets, roofing thatch, etc.; the apical buds of adult plants are an excellent palm-cabbage; and an indigenous alcoholic drink is extracted from its sugar sap, tapped from the inflorescences by means of apposite cuttings.

APPLICATIONS TO HEALTH PROMOTION AND DISEASE PREVENTION

The health benefits of coconut have extensively been studied by various researchers (Kritchevsky *et al.*, 1976; Young & Renner, 1977). Even though the applications of coconut to health promotion and disease prevention have received wide publicity, some of these claims have not been fully elucidated. However, the ketogenic, glycemic, and atherogenic properties of coconut have been studied. In a study conducted to compare the rate and levels of blood glucose and ketones in chicks fed with "carbohydrate-free" diets derived from soya oil and coconut oil, results showed that chicks fed diets containing soybean oil grew faster, and had similar levels of blood glucose and blood ketone bodies but lower levels of liver glycogen than chicks fed diets containing coconut oil. The study indicated that when fed in triglyceride form, coconut oil and soybean oil did not differ in ketogenicity for the chick: however, when fed in unesterified form, coconut fatty acids were found to be more ketogenic than soybean fatty acids (Young & Renner, 1977). In another study in which the atherogenicity of various "cholesterol-free" diets including coconut oil (CNO) was compared, the results indicated that microscopic evaluation of the severity of aortic lesions occurring in experimental animals fed with coconut oil, peanut oil, or butter oil had the most frequent and severe lesions, and that the most severe gross atherosclerosis was observed in the CNO group. The most extensive lesions observed in the experimental animals were characterized by intimal proliferation spread over the area of lipid deposition (Kritchevsky *et al.*, 1976). However, other compelling evidence that coconut oil and its biologic actions as a medium-chain fatty acid are not atherogenic has also been put forward (Dayrit, 2006). In a related epidemiological study, it was observed that coconut oil and mustard oil, though saturated fats, do not show a wide omega-6 to omega-3 fatty acid ratio, which is quite high in polyunsaturated oils. The desirable ratio is less than 10:1. Increasing prevalence of diabetes and other related diseases are found to be correlated with an increasing omega-6 to omega-3 ratio. On the other hand, consumption of coconut oil, which is deficient in polyunsaturated fatty acids, has been found to enhance the secretion of insulin and utilization of blood glucose. It is beneficial to consume traditional edible fats such as coconut oil along with polyunsaturated fats to reduce the omega-6 intake and maintain an optimum omega-6 to omega-3 ratio in the diet (Sadikot, 2005).

Results from various studies have indicated that the lauric acid in coconut oil has antibacterial action, and that the medium-chain fats in coconut oil are similar to fats in human milk and have similar nutritional effects (Kabara, 2000; Enig, 2001) According to these studies, the monoglyceride monolaurin is the substance that keeps infants from getting viral or bacterial or protozoal infections. It should be noted that the saturated fatty acids contained in coconut oil do not comprise of a single family of fats, but of three subgroups: short (C2–C6), medium (C8–C12), and long (C14–C24)-chain fatty acids. The medium-chain fats are found exclusively in lauric oils. Indeed, further studies have confirmed that the health benefits of coconut oil can be attributed to the presence of lauric acid, capric acid, and caprylic acid (Goldberg & Enig, 1993; Ravnskov *et al.*, 2002). Preliminary clinical studies in Jakarta have shown some promising positive effects of virgin coconut oil (VCO) on immune responses among HIV-positive patients. The study concluded that the macronutrient intake, mostly in terms of energy, fats, and protein, was significantly improved among the VCO-supplemented group. In addition, the weight and nutritional status of the subjects, especially among the VCO-supplemented group, were maintained well throughout the study.

Fresh coconut water has been shown to be a natural electrolyte solution for intravenous replacement therapy (De La Cruz *et al.*, 1975). In a clinical trial, coconut water was given intravenously to 75 normal subjects and 25 patients with varying degrees of dehydration and electrolyte imbalance. The trial showed that intravenous coconut water causes a slight rise in serum and urine electrolytes. Coconut water, as extracted fresh and collected aseptically from green immature coconuts (*Cocos nucifera*), has higher potassium, calcium, magnesium, phosphate, and glucose content than human plasma, but lower content of sodium, chloride, and protein (Tables 43.1, 43.2). Coconut water is slightly hypotonic and more acid than the plasma. The study further indicated that coconut water produces marked diuresis in normal subjects, and relatively less marked diuresis in dehydrated patients. When more than 1 liter of coconut water is given to dehydrated patients, there is a positive fluid balance and a significant rise in serum potassium. It also supplies valuable glucose (Olurin *et al.*, 1972).

ADVERSE EFFECTS AND REACTIONS (ALLERGIES AND TOXICITY)

Coconut can be a food allergen (Fries & Fries, 1983). It is a top-five food allergy in India, where coconut is a common food source (Teuber & Peterson, 1999). On the other hand, food allergies to coconut are considered rare in Australia, the UK, and the USA. As a result, commercial extracts of coconut are not currently available for skin-prick testing in Australia or New Zealand. Coconut-derived products can cause contact dermatitis (de Groot *et al.*, 1987). They can be present in cosmetics, including some shampoos, moisturizers, soaps, cleansers, and handwashing liquids (Pinola *et al.*, 1993).

SUMMARY POINTS

- The coconut palm (*Cocos nucifera*) is an important member of the family Arecaceae (palm family).
- The palm is native to tropical eastern regions.
- Coconut fruit in the wild is light, buoyant, and highly water resistant, and evolved to disperse significant distances via marine currents.
- The coconut palm is grown throughout the tropics for decoration, as well as for its many culinary and non-culinary uses.
- For many years, tropical communities have used coconut oil in key areas of their lives, such as cooking, and for medicinal and cosmetic purposes.
- The benefits of coconut oil for health are countless and unparalleled.

References

Chang, E., & Elevitch, C. R. (2006). Species for Pacific island agroforestry. Available at www.traditionaltree.org.

Dayrit, C. S. (2006). Coconut products and virgin coconut oil (VCO) for health and nutrition — a strategy for making coconut globally competitive. In Proceedings of the XLII Cocotech Meeting, 21–25 August, 2006, Manila, The Philippines.

de Groot, A. C., de Wit, F. S., Bos, J. D., & Weyland, J. W. (1987). Contact allergy to cocamide DEA and lauramide DEA in shampoos. *Contact Dermatitis, 16*, 117–118.

De La Cruz, F., Velasco, R., Recio, P., & Jacalne, A. (1975). Fresh coconut water: a natural electrolyte solution for intravenous replacement therapy. *Journal of Medical Primatology, 4*, 348.

Deosthale, Y. G. (1981). Trace element composition of common oil seeds. *Journal of the American Oil Chemists Society, 58*, 988–990.

Enig, G. M. (2001). Health and nutritional benefits from coconut oil: an important functional food for the 21st century. *Coconuts Today*, Special Edition for the 13th Asian Pacific Congress of Cardiology.

Foale, M. (2003). *The Coconut Odyssey: the bounteous possibilities of the tree of life*. Canberra Australia: Australian Centre for International Agricultural Research.

Fries, J. H., & Fries, M. W. (1983). Coconut: a review of its uses as they relate to the allergic individual. *Annals of Allergy, 51*, 472–481.

Goldberg, M. L., & Enig, M. G. (1993). Palmitic and lauric acids and serum cholesterol. *American Journal of Clinical Nutrition, 58*, 244.

Harries, H. C. (1978). The evolution, dissemination and classification of *Cocos nucifera* L. *The Botanical Review, 44*, 3.

Kabara, J. J. (2000). Nutritional and health aspects of coconut oil In: Proceedings of the XXXVII COCOTECH Meeting/ICC 200, 24–28 July 2000, Chennai, India, pp. 101–109.

Kritchevsky, D., Tepper, S. A., & Kim, H. K. (1976). Experimental atherosclerosis in rabbits fed cholesterol free diets. 5. Comparison of peanut, corn, butter, and coconut oils. *Experimental and Molecular Pathology, 24*, 375–391.

Olurin, E. O., Durowoju, J. E. O., & Bassir, O. (1972). Intravenous coconut water therapy in surgical practice. *The West African Medical Journal, 21*, 124–131.

Pearsall, J. (Ed.). (1999). *"Cocoanut". Concise Oxford Dictionary* (10th ed.). Oxford, UK: Clarendon Press.

Pinola, A., Estlander, T., Jolanki, R., Tarvainen, T., & Kanerva, L. (1993). Occupational allergic contact dermatitis due to coconut diethanolamide (cocamide DEA). *Contact Dermatitis, 29*, 262–265.

Ravnskov, U., Allan, C., Atrens, D., Enig, M. G., Groves, B., Kaufman, J., et al. (2002). Studies of dietary fat and heart disease. *Science, 295*, 1464–1465.

Sadikot, S. M. (2005). *Coconut for health nutrition*. Jakarta Indonesia: APCC. 6.

Teuber, S. S., & Peterson, W. R. (1999). Systemic allergic reaction to coconut (*Cocos nucifera*) in 2 subjects with hypersensitivity to tree nut and demonstration of cross-reactivity to legumin-like seed storage proteins: new coconut and walnut food allergens. *The Journal of Allergy and Clinical Immunology, 103*, 1180–1185.

Werth, E. (1933). Distribution, origin and cultivation of the coconut palm (trans. R. Child). *Berichte der Deutschen Botanischen Gesellschaft, 51*, 301–304.

Young, S. J., & Renner, R. (1977). Ketogenicity of soybean oil, coconut oil and their respective fatty acids for the chick. *Journal of Nutrition, 107*, 2206–2212.

CHAPTER 44

Common Vetch (*Vicia sativum*) Seeds as a Source of Bioactive Compounds

Ryszard Amarowicz
Institute of Animal Reproduction and Food Research of the Polish Academy of Sciences, Olsztyn, Poland

CHAPTER OUTLINE

LIST OF ABBREVIATIONS

ABAP, 2,2′-azobis(2-amidopropane) hydrochloride
ABTS, 2,2′-azinobis(3-ethylbenzothiazoline-6-sulfonic acid)
ALP, alkaline phosphatase
ALT, alanine aminotransferase
AST, aspartate aminotransferase
BCA, β-cyanoalanine
CAT, catalase (CAT)
d.m., dry matter
DPPH, 2,2-diphenyl-2-picrylhydrazyl
FRAP, ferric-reducing antioxidant power
γ-gluBCA, γ-glutamyl-β-cyanoalanine
G6pdh, glucose-6-phosphate dehydrogenase
GSH, glutathione
GSH-Px, glutathione peroxidase
HPLC, high performance liquid chromatography
LDH, lactate dehydrogenase
MDA, malondialdehyde

Nuts & Seeds in Health and Disease Prevention. DOI: 10.1016/B978-0-12-375688-6.10044-1

PRTC, peroxyl radical-trapping capacity
SOD, superoxide dismutase
TTA, total antioxidant activity

INTRODUCTION

Mounting evidence from epidemiological studies indicates that the consumption of food of plant origin offers health benefits to man, and reduces the risk of developing chronic and neurodegenerative disorders such as cardiovascular diseases, cancers, and Alzheimer's disease. Vetch seeds contain numerous phytochemicals; these bioactive compounds possess antioxidant, radical scavenging, anti-inflammatory, and anticarcinogenic properties, which contribute to the chemopreventative potential of vetch.

BOTANICAL DESCRIPTION

The genera *Vicia* comprise 166 species of annual and perennial herbaceous legumes that are distributed throughout temperate and subtropical regions. Taxonomically, *Vicia* belong to the family Vicieae. An overview of the classification of genera *Vicia*, with a list of all species, can be found in Smartt (1990). Table 44.1 lists the main cultivated *Vicia* species.

HISTORICAL CULTIVATION AND USAGE

The historical background of the *Vicia* species as grain legumes dates back to antiquity. The proportion of vetch crops used for fodder increased during the 19th century due to the intensification of animal production (Enneking, 1994). Fodder vetches have been cultivated as grain and forage crops (in Germany, Bulgaria, Austria, Lithuania, and Spain), or only as green forage (in the UK, Hungary, Yugoslavia, and Sweden). *Vicia sativa*, *V. narbonensis* L., *V. villosa* Roth, *V. monantha*, and *V. ervilia* have long been used as feedstuff for farm animals. There was a suggestion that vetch could be used in a similar way to peas, but, because of their bitter taste, in lesser amounts — not more than 12.5% of other concentrated feeds (Oldershaw, 1931).

TABLE 44.1 Cultivated *Vicia* Species, According to Enneking (1994)

Species	Synonym English Name	English Name
Vicia articulata Hornem	*V. monanthos* (L.) Desf.	One-flowered vetch
V. benghalensis L.	*V. atropurpurea* Desf.	Purple vetch
V. monantha Retz	*V. calcarata* Desf. (Demehi)	Bard vetch
V. ciliatula Lipsky	*V. ciliata* Lipsky	Tufted vetch
V. cracca L.	*V. tenuifolia* Gren & Godron	Bitter vetch
ssp. *tenuifolia* (Roth) Gaudin	*V. tenuifolia* Roth	Scarlet vetch
V. ervilia Wild	*Ervum ervilia* L.	Hairy tare
V. fulgens Battand	*V. selloi* Vogel	Narbonne vetch
V. graminea Smith	*Orobus lathyroides* L.	Hungarian vetch
V. hirsute (L.) Gray	ssp. *dasycarpa* (Ten) Cavill	Broad pod vetch
V. johannis Taman		Pale flowered vetch
V. narbonensis L.		Common vetch
V. pannonica Crantz		Subterranean vetch
V. peregrina		Smooth vetch
V. pisiformis L.		Two leaved vetch
V. sativa L. ssp. *sativa*		Hairy vetch
ssp. *amphicarpa*		Wooly pod vetch
V. tetrasperma (L.)		
V. unijuga A.Br.		
V. villosa Roth		
ssp. *varia* (Host) Corbiere		

PRESENT-DAY CULTIVATION AND USAGE

Worldwide utilization of *Vicia* species as forage and green manure crops is well established. *Vicia* species are highly nutritious grains for sustainable agriculture. Common vetch (*Vicia sativa*) is grown extensively in certain regions of the world, and is used not only as cattle feed, but also as a cheap substitute for lentils, for human consumption (Islam *et al.*, 2003). The usage of *Vicia* species for human consumption was summarized by Enneking (1994) as follows:

- Young leaves of *V. amoena*, *V. amurensis*, *V. heptajuga*, *V. hirticalycina*, and *V. venosa* are used as culinary herbs in East Asia, Manchuria, Korea, and Siberia.
- The seeds or flour of *V. ervilia*, *V. sativa*, and *V. monantha* are used in soups and bread in Mediterranean countries.
- Seeds of *V. articulate* and *V. pisiformis* are used instead of lentils in Mediterranean countries.
- Seeds of *V. hirsuta* are cooked or roasted in Eurasia and North Africa.

According to Francis and colleagues (1999), vetches already occupy special niches in farming systems because of their wide adaptation and ability to grow where many food legumes cannot. These include in areas of extreme winter cold, and very dry conditions.

APPLICATIONS TO HEALTH PROMOTION AND DISEASE

Seeds, leaves, and other anatomical parts of vetch are a rich source of biologically active phenolic compounds. HPLC analysis of a crude extract of vetch (*Vicia sativa*) seeds revealed the presence of the phenolic acids such as vanillic, caffeic, *p*-coumaric, ferulic, and sinapic acids (after basic hydrolysis), as well as quercetin and kaempferol (after acid hydrolysis) (Amarowicz *et al.*, 2008).

A new flavonol glycoside, quercetin 3-*O*-galactosyl-(1→6)-glucoside, has been isolated from the above-ground parts of narrowleaf vetch (*Vicia angustifolia*) by Waage and Hedin (1985). Gamal-Elden and colleagues (2004) isolated nine flavonoids from the aerial parts of *Vicia sativa* growing in North Sinai (Egypt): apigenin 4-*O*-β-D-glucopyranoside, apigenin 6,8-di-C-β-D-glucopyranoside, apigenin 6-C-α-L-arabinopyranoside-8-C-β-D-glucopyranoside, kaempferol-3-*O*-β-L-dirhamnopyranosyl (1→2,1→6)-β-D-glucopyranoside, kaempferol 3-*O*-(4-β-D-xylopyranosyl)-α-L-rhamnopyranoside-7-*O*-L-rhamnopyranoside, kaempferol-3-*O*-α-L-rhamnopyranosyl(1→6)-β-D-glucopyranoside, luteolin-7-*O*-β-D-glucopyranoside, quercetin 3-*O*-β-D-glucopyranoside, and quercetin 3,7-di-*O*-glucopyranoside. Seven flavonoid aglycones were identified in leaves of 62 species, belonging to 18 sections of the genus *Vicia* (Webb & Harborne, 1991). In order of frequency, they were: luteolin (45%), kaempferol (45%), quercetin (44%), apigenin (24%), diosmetin (13%), myricetin (3%), and vitexin (3%). In general, the distribution patterns follow subgeneric and sectional classification. The patterns were more diverse in two sections, *Vicilla* and *Cracca*.

Two new acylated flavonol glycosides, named amurenosides A and B, together with quercetin 3-(2,6-di-*O*-α-rhamnopyranosyl-β-galactopyranoside), have been isolated from the whole plant of *Vicia amurensis* by Kang *et al.* (2000). Their chemical structures were elucidated as quercetin 3-*O*-α-L-(3-feruloylrhamnopyranosyl)(1→6)-[α-L-rhamnopyranosyl(1→2)]-β-D-galactopyranoside, and quercetin 3-*O*-α-L-(2-feruloylrhamnopyranosyl) (1→6)-[α-L-rhamnopyranosyl(1→6)]-β-D-galactopyranoside.

Regarding the extract from the seeds of *Vicia calcarata* Desf., Singab and colleagues (2005) confirmed the presence of two flavonol glycosides that rarely occur in the plant kingdom; identified as quercetin-3,5-di-*O*-β-D-diglucoside and kaempferol-3,5-di-*O*-β-D-diglucoside, in addition to the three known compounds identified as quercetin-3-*O*-α-L-rhamnosyl-(1→6)-β-D-glucoside (rutin), quercetin-3-*O*-β-D-glucoside (isoquercitrin), and kaempferol-3-*O*-β-D-glucoside (astragalin). Moreover, the spectrophotometric estimation of the flavonoid content revealed that the aerial parts of the plant contain an appreciable amount of flavonoids

(0.89%), analyzed as rutin. The content of total phenolics in extract of *Vicia sativa* seeds was assessed to be 97.8 mg /g (Orhan *et al.*, 2009), and 66 mg/g (Amarowicz *et al.*, 2008).

The phenolic compounds present in vetch extracts exhibit antioxidative properties. In the study by Amarowicz and colleagues (2008), a crude extract of vetch seeds and its fractions I (comprising low molecular-weight phenolic compounds) and II (tannins) were screened for their antioxidant and antiradical properties using a β-carotene-linoleate model system, a Randox® total antioxidant activity (TAA) kit, the 2,2-diphenyl-2-picrylhydrazyl (DPPH) radical scavenging activity assay, and a reducing power test. Results of the assays indicated that the antioxidant activity and radical scavenging capacity of the tannin fraction were greatest – for example, the TAA of the tannin fraction was 2.48 μmol Trolox/mg, whereas the crude extract from vetch seeds and fraction I showed activity of only 0.30 and 0.22 μmol Trolox/mg, respectively. The content of total phenolics in fraction II was the greatest, at 113 mg/g. The results of Zieliński (2002) showed that germinated vetch (*Vicia sativa*) seeds exhibited lower peroxyl radical-trapping capacity (PRTC) than the native seeds. Vetch seeds were more effective than germinated soybean when their peroxyl radical-trapping capacity was compared. A simple method of determining the peroxyl radical-trapping capacity, based on the use of 2,2′-azobis (2-amidopropane) hydrochloride (ABAP) decomposition as a free radical source and the use of 2,2′-azinobis(3-ethylbenzothiazoline-6-sulfonic acid) (ABTS) oxidation as the reaction indicator, was employed in the study. The values of PRTC dependent on the solvent system were used for the extraction. Application of 0.1 M phosphate buffer (pH 7.4) was more active (23.46 μmol of peroxyl radicals/g d.m.) than 80% methanol (7.28 μmol of peroxyl radicals/g d.m.).

The antioxidant activity of flavonoids isolated from *Vicia sativa* seeds was assessed by Gamal-Elden *et al.* (2004) using four different assays, including DPPH assay, quenching of ROO radicals, respiratory burst in rat neutrophils, and lipid peroxidation. Kaempferol-3-*O*-α-L-rhamnopyranosyl (1→6)-β-D-glucopyranoside showed effective inhibitory activity of both peroxyl radicals; apigenin 4-*O*-β-D-glucopyranoside, apigenin 6,8-di-*C*-β-D-glucopyranoside, apigenin 6-*C*-α-L-arabinopyranoside-8-*C*-β-D-glucopyranoside, kaempferol-3-*O*-α-L-rhamnopyranosyl (1→6)-β-D-glucopyranoside quercetin 3-*O*-β-D-glucopyranoside [8], and quercetin 3,7-di-*O*-glucopyranoside decreased both superoxide anion radical production in PMA-stimulated neutrophils; and compounds apigenin 6,8-di-*C*-β-D-glucopyranoside, apigenin 6-*C*-α-L-arabinopyranoside-8-*C*-β-D-glucopyranoside; kaempferol 3-*O*-(4-β-D-xylopyranosyl)- α-L-rhamnopyranoside-7-*O*-L-rhamnopyranoside, and luteolin-7-*O*-β-D-glucopyranoside were dramatically effective in inhibition of CCl_4-induced lipid peroxidation.

The extract of *Vicia sativa* seeds was screened by DPPH radical scavenging and ferric-reducing antioxidant power (FRAP) assays at 0.5-, 1.0-, and 2.0-mg/ml concentrations (Orhan *et al.*, 2009). The percentage of DPPH radicals scavenged ranged from 8.7% (0.5 mg/ml) to 39.9% (2.0 mg/ml). The absorbance at 700 nm showing FRAP increased from 0.135 (0.5 mg/ml) to 0.455 (2.0 mg/ml).

The antioxidant activity of the flavonol fraction (rutin, isoquercetin, astragalin, quercetin-3, 5-di-*O*-β-D-diglucoside, and kaempferol-3,5-di-*O*-β-D-diglucoside) obtained from aerial parts of *Vicis calcarata* Desf. markedly ameliorated the antioxidant parameters of rats intoxicated by CCl_4, including glutathione (GSH) content, glutathione peroxidase (GSH-Px), superoxide dismutase (SOD), plasma catalase (CAT), and packed erythrocytes glucose-6-phosphate dehydrogenase (G6PDH) activity, so they were comparable with normal control levels. In addition, it normalized liver malondialdehyde (MDA) levels and creatinine concentration (Singab *et al.*, 2005).

The antibacterial activity of quercetin 3-*O*-galactosyl-(1→6)-glucoside isolated from the above-ground parts of *Vicia angustifolia* against *Pseudomonas maltophilia* and *Enterobacter cloacae* was reported by Waage and Hedin (1985), and compared with that of several other flavonol glycosides.

Hepatoprotective action was observed for diosmine analog originating from *Vicia trunctula* seeds administered in a dose of 100 mg/kg to white male rats with a model of acute CCl_4 hepatitis (Dorkina, 2004). The effect was evaluated by the ability of tested substances to normalize the biochemical characteristics of the functional state of the liver, in comparison to the reference drug Carsil. Less active was diosmine from *Vicia tanuifolia* Roth; this was probably related to the higher bioaccessibility of diosmine analog ensured by a carbohydrate moiety of this compound, in combination with the C2—C3 double bond present in diosmetin (diosmin aglycon). In the toxicological experiments of Singab *et al.* (2005), treatment of the rats (25 mg/kg, b.w.) with the fraction of flavonol glycosides during the induction of hepatic damage by CCl_4 significantly reduced the indices of liver injuries. The hepatoprotective effects of flavonols significantly reduced the elevated levels of the following serum enzymes: alanine aminotransferase (ALT), aspartate aminotransferase (AST), alkaline phosphatase (ALP), and lactate dehydrogenase (LDH).

Purified flavonoids from *Vicia sativa* seeds showed low anti-inflammatory and antino-ciceptive activities in a range of 0.9—18.3% of inhibition, whereas the compounds apigenin 4-O-β-D-glucopyranoside, apigenin 6-C-α-L-arabinopyranoside-8-C-β-D-glucopyranoside, kaempferol-3-O-β-L-dirhamnopyranosyl $(1 \rightarrow 2, 1 \rightarrow 6)$-β-D-glucopyranoside, and quercetin 3-O-β-D-glucopyranoside were the most effective flavonoids (Gamal-Elden *et al.*, 2004). The anti-inflammatory and antinociceptive effects of *Vicia sativa* and its flavonoids were assessed by carragenan-induced inflammation in mice, acetic acid-induced writhing, and the formalin-licking test.

ADVERSE EFFECTS AND REACTIONS (ALLERGIES AND TOXICITY)

The seeds of the common vetch (*Vicia sativa*) contain a neurotoxin that is present as dipeptide γ-glutamyl-β-cyanoalanine (γ-gluBCA), as the free amino acid β-cyanoalanine (BCA), and as the cyanogenic glycosides vicianin and prunasin (Ressler & Tatake, 2001). In a study by Ressler *et al.* (1997), three specimens of common vetch seeds were found to contain 0.42—0.74% γ-gluBCA and 0.01—0.03 BCA. In the feeding experiments by Ressler *et al.* (1969), using chickens, acute toxicity of vetch seeds was correlated with the total content of BCA. Ressler and Tatake (2001) reported that vicinin, prunasin, and BCA were identified as principal dietary sources of the urinary thiocyanate in the rats fed a diet containing common vetch. The seeds contained 0.69 μmol vicianin/g and 0.16 μmol prunasin/g. Ressler and colleagues (1997) analyzed the stability of the vetch neurotoxins in cooking. When heated in water (100°C for 3 hours), γ-gluBCA cyclized to pyroglutamic acid and BCA. A modified cooking procedure, replacing the broth during cooking with fresh water and washing the seeds well, yielded cooked seeds without detectable neurotoxins.

Canavanine, which acts as an analog of arginine, is present in many *Vicia* species. The highest amounts were found in the species *stenophylla* Velen (5.2 g/kg), *V. disperma* DC (3.2—5.1 g/kg), *V. villosa* Roth (2.5—7.8 g/kg), and *V. benghalensis* L. (1.9—3.4 g/kg) (Enneking, 1994). According to Prete (1986), canavanine induced an autoimmune response in mice when administered as 0.725% of the diet. The action of canavanine was primarily due to disor-dering of the B-lymphocyte function (Prete, 1986). Canavanine also acted on suppressor-induced T cells to regulate antibody synthesis (Morimoto *et al.*, 1990). It was established that canavanine acts mainly on SD8(−L)Leu(+)cells, and that the lymphocyte response to cana-vanine depended primarily on which functional types of the cells were present.

Vicine and convicine present in *Vicia sativa* seeds are β-glycosides of the pirimidines and isoramil, respectively. Pyrimidine glycosides have been implicated in favism. After hydrolysis by β-glucosidase, the products of this reaction form radicals, which can cause a depletion of reduced glutathione (GSH) in G6pdh-deficient red blood cells, resulting in a hemolytic crisis (Marquardt, 1989).

SUMMARY POINTS

- Vetch species are rich sources of phenolic compounds.
- Flavonoids, phenolic acids, and condensed tannins present in seeds of the *Vicia* species exhibit strong antioxidant properties as measured using different chemical methods.
- These activities may help in the prevention and/or treatment of pathological disorders in which free radical oxidation, inflammation, and cellular activation are suggested to play a fundamental pathogenic role.
- Seeds of common vetch contain the neurotoxins γ-glutamyl-β-cyanoalanine, β-cyanoalanine, vicianin, and prunasin.
- The cooking procedure and technological process can reduce the content of the nerotoxins in seeds of *Vicia sativa*.
- Canavanin and pyrimidine glycosides are present in *Vicia sativa* seeds.

References

Amarowicz, R., Troszyńska, A., & Peg, R. B. (2008). Antioxidative and radical scavenging effects of phenolics from *Vicia sativum*. *Fitoterapia, 79*, 121–122.

Dorkina, E. G. (2004). Investigation of the hepatoprotector action of natural flavonoids. *Eksperimental'naia i klinicheskaia farmakologiia, 67*, 41–44.

Enneking, D. (1994). *The toxicity of* Vicia *species and their utilisation as grain legumes*. Australia: PhD dissertation, University of Adelaide.

Francis, C. M., Enneking, D., & El Moneim, A. (1999). When and where will vetches have an impact as grain legumes? Linking research and marketing opportunities for pulses in the 21st century. In R. Knight (Ed.), *Proceedings of the Third International Food Legume Research Conference* (pp. 671–683). Adelaide, Australia.

Gamal-Elden, A. M., Kawashty, S. A., Ibrahim, L. F., Shabana, M. M., & El-Negoumy, S. I. (2004). Evaluation of antioxidant, anti-inflammatory, and antinociceptive properties of aerial parts of *Vicia sativa* and its flavonoids. *Journal of Natural Remedies, 4*, 81–96.

Islam, M. R., Bari, A. S. M., Rahman, M. H., & Rahman, M. M. (2003). Evaluation of nutritional status of common vetch (*Vicia sativa*) on growing rats. *Bangladesh Journal of Veterinary Medicine, 1*, 57–61.

Kang, S. S., Chang, Y. S., & Kim, J. S. (2000). Two new acylated flavonol glycosides from *Vicia amurensis*. *Chemical & Pharmaceutical Bulletin, 48*, 1242–1245.

Marquardt, R. R. (1989). Vicine, convicine, and their aglycones — divicine and isoramil. In P. R. Cheeke (Ed.), *Toxicants of plant origin, Vol. II, Glycosides* (pp. 161–200). Boca Raton, FL: CRC Press.

Morimoto, I., Shiozawa, S., Tanaka, Y., & Fujita, T. (1990). L-Canavaninine acts on suppressor-inducer T cells to regulate antibody synthesis: lymphocytes of systemic lupus erythematosus patients are specifically unresponsive to L-canavanine. *Clinical Immunology and Immunopathology, 55*, 97–108.

Oldershaw, A. (1931). Vetch or tares. In H. Hunter (Ed.), *Bailliéres encyclopedia of scientific agriculture* (pp. 1234–1238). London, UK: Baillieres, Tindall, and Cox.

Orhan, I., Kartal, M., Abu-Asker, M., Sezer Senol, F., Yilmaz, G., & Sener, B. (2009). Free radical scavenging properties and phenolic characterization of some edible plants. *Food Chemistry, 114*, 276–281.

Prete, P. A. (1986). Membrane surface properties of lymphocytes of normal (DBA/2) and autoimmune (NZB/NZW) F1 mice: effects of L-canavanine and a proposed mechanism for diet induced autoimmune disease. *Canadian Journal of Physiology and Pharmacology, 64*, 1189–1196.

Ressler, C., & Tatake, J. (2001). Vicianin, prunasin, and β-cyanoalanine in common vetch seed as sources of urinary thiocyanate in the rat. *Journal of Agricultural and Food Chemistry, 49*, 5057–5080.

Ressler, C., Nigam, S. N., & Giza, Y.-H. (1969). Toxic principle in vetch. Isolation and identification of γ-glutamyl-β-cyanoalanine from common vetch seeds. Distribution in some legumes. *Journal of the American Chemical Society, 91*, 2758–2765.

Ressler, C., Tatake, J., Kaizer, E., & Putnam, D. H. (1997). Neurotoxins in vetch food: stability to cooking and removal of γ-glutamyl-β-cyanoalanine and β-cyanoalanine and acute toxicity from common vetch (*Vicia sativa* L.) legumes. *Journal of Agricultural and Food Chemistry, 45*, 189–194.

Singab, A. N., Youssef, D. T., Noaman, E., & Kotb, S. (2005). Hepatoprotective effect of flavonol glycosides rich fraction from Egyptian *Vicia calcarata* Desf. against CCl4-induced liver damage in rats. *Archives of Pharmacological Research, 28*, 791–798.

Smartt, J. (1990). Pulse of the classical world. In J. Smartt (Ed.), *Grain legumes. Evolution and genetic resources* (pp. 176–244). Cambridge, UK: Cambridge University Press.

Waage, S. K., & Hedin, P. A. (1985). Quercetin 3-O-galactosyl-(1→6)-glucoside, a compound from narrow vetch witch antibacterial activity. *Phytochemistry, 24,* 243–245.

Webb, M. E., & Harborne, J. B. (1991). Leaf flavonoid aglycone patterns and sectional classification in the genus *Vicia* (Leguminosae). *Biochemical Systematics and Ecology, 19,* 81–86.

Zieliński, H. (2002). Peroxyl radical-trapping capacity of germinated legume seeds. *Nahrung, 46,* 100–104.

CHAPTER 45

Usage of Coral Bean (*Erythrina latissima* E. Meyer) Seeds in Human Health

Cornelius W. Wanjala[1], Teresa A. Akeng'a[2]
[1] Department of Physical Sciences, South Eastern University College (Constituent College of the University of Nairobi), Kitui, Kenya
[2] Academic Division, Bondo University College (Constituent College of Maseno University), Bondo, Kenya

377

LIST OF ABBREVIATIONS

$CDCl_3$, deuterated chloroform
CD_3OD_3, deuterated methanol
$CHCl_3$, chloroform
DPPH, 2, 2-diphenyl-1-picrylhydrazyl
ECI, erythrina chymotrypsin inhibitor

Nuts & Seeds in Health and Disease Prevention. DOI: 10.1016/B978-0-12-375688-6.10045-3

EI-MS, electron ionization mass spectrum
EtOH, ethanol
GABA, gamma-amino butyric acid
HRMS, high resolution mass spectrum
IR, infrared
Me_2CO, acetone
MeOH, methanol
mp, melting point
NMR, nuclear magnetic resonance
Pet. ether, petroleum ether
TNF, tumor necrosis factor
TRAIL, TNF-related apoptosis-inducing ligand
UV, ultraviolet

INTRODUCTION

The genus *Erythrina* comprises over 170 species showing cosmopolitan distribution (Hemsley & Ferguson, 1985), and is known to produce C-prenylated flavanones, isoflavones, isoflavanones, pterocarpans (Chacha *et al.*, 2005), erythrinaline alkaloids from the seeds (Amer *et al.*, 1991; Juma & Majinda, 2004) and seed pods (Wanjala *et al.*, 2002). The genus *Erythrina* is popularly known for its ornamental and medicinal properties (Hennessy, 1972). *Erythrina latissima* E. Meyer (Fabaceae) is a tropical and subtropical flowering tree. The name *Erythrina latissima* describes its red flowers (erythro) and wide leaves (latissima); it is a decorative tree, commonly called the broad-leaved coral tree (Johnson & Johnson, 2002).

BOTANICAL DESCRIPTION

Erythrina latissima is an attractive, small to medium-sized tree, 5–10 m tall. The trunk and branches are woolly, and have prominent thorns and thick, corky bark. Unlike the other Erythrina spp., the leaves of *E. latissima* are large, soft, and velvety, becoming gray-green and leathery as they mature (Johnson & Johnson, 2002). The flowers are scarlet, with a gray, woolly calyx. The seed pod, up to 300 mm long, has constrictions between the bright red seeds. Each seed is marked with a black dot (Wanjala & Majinda, 2000). The tree is fairly slow-growing, taking 20–30 years to form a reasonable canopy, and will live for over 100 years. Older trees, however, do have a tendency to fall over (Johnson & Johnson, 2002).

HISTORICAL CULTIVATION AND USAGE

The orange-red seeds of *E. latissima*, commonly known as lucky beans or coral beans, are threaded onto string to make decorative necklaces, which some believe will ward off evil spirits. The bark is burned and then pounded into a powder which is used to dress wounds. Truncheons (made from large branches) of the larger Erythrinas are used as fence poles, which in time take root, creating a living fence (Johnson & Johnson, 2002).

E. latissima trees are used widely in the tropics and subtropics as street and park trees, especially in drier areas. The flowers are also widely used as floral emblems in some communities. The flowers, seeds, and seed pods of *E. latissima* contain potent Erythrina alkaloids, but some indigenous people use the seeds for medicinal and other purposes (Johnson & Johnson, 2002). The Erythrina alkaloids are toxic to some degree, and can cause fatal poisoning. The major active compounds in the seeds are the Erythrina alkaloids (Amer *et al.*, 1991) and hypaphorine (Ozawa *et al.*, 2008).

PRESENT-DAY CULTIVATION AND USAGE

Erythrina latissima easily grows from large truncheons or seed. Truncheons are made from branches which are more than 2 years old and should not be thinner than 100 mm; the bigger the truncheon, the more easily it roots. The trimmed branches are left to dry for 2 or 3 days before planting. A layer of river sand in the base of the hole enhances rooting. When the truncheons are not over-watered, they root in 2—3 months (Wanjala & Majinda, 2000).

Soaking the seeds in hot water overnight softens the hard seed coat in preparation for sowing. The soaked seeds are then planted in seed trays containing river sand or composted bark, and watered; germination begins in 5—8 weeks. When the seedlings have made a swollen base, they are transplanted into pots or plastic bags. Once established, the young plants should put on between 200—600 mm growth per year (Johnson & Johnson, 2002).

Stem borer is the major problem when the plants are young, and regular spraying with a systemic insecticide is essential. As the tree grows and matures, it becomes less of a problem. *E. latissima* provides a fast-food outlet for nectar-loving birds (Johnson & Johnson, 2002).

APPLICATIONS TO HEALTH PROMOTION AND DISEASE PREVENTION

Structure of erythrina alkaloids from the seeds of *E. latissima*

Erythrina alkaloids are tetrahydroisoquinoline types of alkaloids, known to be dextrorotatory with *3R-5S* absolute stereochemistry. The possible catabolism of the erythrina-type alkaloids is provided by lactonic rings like (+)-α-erythroidine, (+)-β-erythroidine, (+)-8-oxo-α-erythroidine, and (+)-8-oxo-β-erythroidine (Amer *et al.*, 1991), which are mostly products of *in vivo* oxidation of the aromatic ring **D** of the classical skeleton (Figure 45.1).

The numbering system for the Erythrina-type alkaloids is shown in Figure 45.1. The nomenclature of the Erythrina alkaloids is such that the prefix eryso- usually denotes the presence of phenolic function. The prefix erythroi- indicates that ring **D** is lactonic, while the prefix erythra- points to the classical skeleton (Figure 45.1). The dienoid alkaloids possess one carbon—carbon double bond in ring **A** and another in ring **B**, but alkenoids incorporate only one double bond, usually in ring **A** (Amer *et al.*, 1991). The compounds isolated from the seeds of *E. latissima* are glycosidic dienoid Erythrina alkaloids (Table 45.1), non-glycosidic dienoid Erythrina alkaloids (Table 45.2), and non-glycosidic alkenoid alkaloids (Table 45.3); and hypaphorine, the betaine of tryptophan, is the major compound which accumulates in seeds of most *Erythrina* spp. (Table 45.4). Spectroscopic data and physical properties are included in these tables for ease of reference.

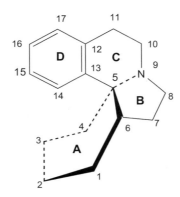

FIGURE 45.1
General structure of the Erythrina alkaloid.

TABLE 45.1 Glycosidic Dienoid Erythrina Alkaloids

S/No.	Name	Chemical Structure	Properties
1	(+)-15β-D-Glucoerysopine	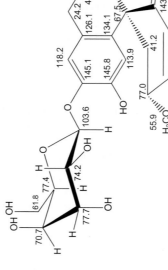	HRMS: $C_{23}H_{29}NO_8$: 447.1893. Mp: 150–152°C (*MeOH*) $[\alpha]_D^{+25} = +67.5°$ (c = 0.0057 MeOH) EI-MS: 447 [M]⁺, 299, 284, 268, 251 UV (MeOH): λ_{max} (log ε): 279 (3.74), 222 (4.40), 206 (4.38) nm IR: ν_{max}^{KBr} cm⁻¹ 3415, 2920, 1507 NMR: ¹³C (75.5 MHz, *CD₃OD*) (Wanjala & Majinda, 2000; Wanjala et al., 2009)
2	(+)-16β-D-Glucoerysopine		HRMS: $C_{23}H_{29}NO_8$: 447.1893. Mp: 158–160°C (*MeOH*) $[\alpha]_D^{+25} = +76.5°$ (c = 0.0057 MeOH) EI-MS: 447 [M]⁺, 299, 284, 268, 251 UV (MeOH): λ_{max} (log ε): 279 (3.74), 222 (4.40), 206 (4.38) nm IR: ν_{max}^{KBr} cm⁻¹ 3415, 2920, 1507. NMR: ¹³C (75.5 MHz, *CD₃OD*) (Wanjala & Majinda, 2000; Wanjala et al., 2009)
3	(+)-β-D-Glucoerysodine	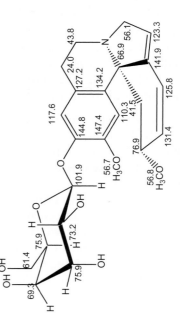	HRMS: $C_{24}H_{31}NO_8$: 461.2049. oil $[\alpha]_D^{+25} = +100.92°$ (c = 1.56 *CHCl₃*) EI-MS: 461 [M]⁺, 430, 299, 284, 268. UV (MeOH): λ_{max} (log ε): 281 (3.39), 225 (4.03) nm IR: ν_{max}^{KBr} cm⁻¹ 3415, 2920. NMR: ¹³C (75.5 MHz, *CD₃OD*) (Amer et al., 1991; Wanjala & Majinda, 2000)

TABLE 45.2 Non-glycosidic Dienoid Erythrina Alkaloids

S/No.	Name	Chemical Structure	Properties
4	(+)-Erysodine	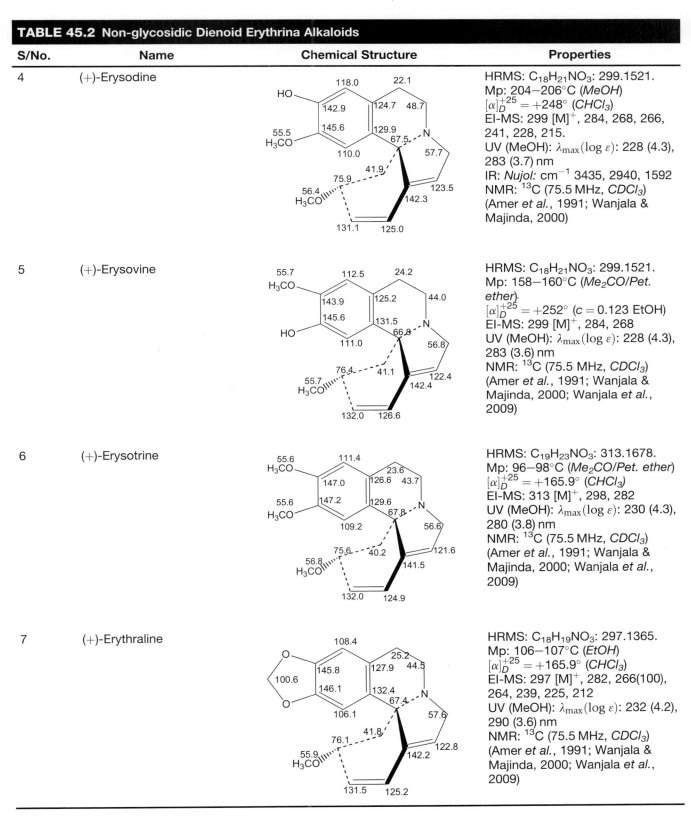	HRMS: $C_{18}H_{21}NO_3$: 299.1521. Mp: 204–206°C (*MeOH*) $[\alpha]_D^{+25} = +248°$ (*CHCl_3*) EI-MS: 299 [M]$^+$, 284, 268, 266, 241, 228, 215. UV (MeOH): $\lambda_{max}(\log \varepsilon)$: 228 (4.3), 283 (3.7) nm IR: *Nujol*: cm^{-1} 3435, 2940, 1592 NMR: ^{13}C (75.5 MHz, *CDCl_3*) (Amer *et al.*, 1991; Wanjala & Majinda, 2000)
5	(+)-Erysovine		HRMS: $C_{18}H_{21}NO_3$: 299.1521. Mp: 158–160°C (*Me_2CO/Pet. ether*) $[\alpha]_D^{+25} = +252°$ (c = 0.123 EtOH) EI-MS: 299 [M]$^+$, 284, 268 UV (MeOH): $\lambda_{max}(\log \varepsilon)$: 228 (4.3), 283 (3.6) nm NMR: ^{13}C (75.5 MHz, *CDCl_3*) (Amer *et al.*, 1991; Wanjala & Majinda, 2000; Wanjala *et al.*, 2009)
6	(+)-Erysotrine		HRMS: $C_{19}H_{23}NO_3$: 313.1678. Mp: 96–98°C (*Me_2CO/Pet. ether*) $[\alpha]_D^{+25} = +165.9°$ (*CHCl_3*) EI-MS: 313 [M]$^+$, 298, 282 UV (MeOH): $\lambda_{max}(\log \varepsilon)$: 230 (4.3), 280 (3.8) nm NMR: ^{13}C (75.5 MHz, *CDCl_3*) (Amer *et al.*, 1991; Wanjala & Majinda, 2000; Wanjala *et al.*, 2009)
7	(+)-Erythraline		HRMS: $C_{18}H_{19}NO_3$: 297.1365. Mp: 106–107°C (*EtOH*) $[\alpha]_D^{+25} = +165.9°$ (*CHCl_3*) EI-MS: 297 [M]$^+$, 282, 266(100), 264, 239, 225, 212 UV (MeOH): $\lambda_{max}(\log \varepsilon)$: 232 (4.2), 290 (3.6) nm NMR: ^{13}C (75.5 MHz, *CDCl_3*) (Amer *et al.*, 1991; Wanjala & Majinda, 2000; Wanjala *et al.*, 2009)

381

Continued

TABLE 45.2 Non-glycosidic Dienoid Erythrina Alkaloids—continued

S/No.	Name	Chemical Structure	Properties
8	(+)-8-Oxoerythraline		HRMS: $C_{18}H_{17}NO_4$: 311.1157. oil EI-MS: 311 [M]$^+$, 296, 280, 278 UV (MeOH): $\lambda_{max}(\log \varepsilon)$: 254 (3.44) nm IR: ν_{max}^{KBr} cm^{-1} 1665 NMR: ^{13}C (75.5 MHz, $CDCl_3$) (Amer et al., 1991; Wanjala & Majinda, 2000; Wanjala et al., 2009)
9	(+)-Erysotramidine		HRMS: $C_{19}H_{21}NO_4$: 327.1470. oil $[\alpha]_D^{+25} = +121°$ (c = 1.0 CHCl$_3$) EI-MS: 327 [M]$^+$, 312, 296, 294 UV (MeOH): $\lambda_{max}(\log \varepsilon)$: 316 (3.15 sh), 257 (3.7), 236 (4.07), 212 (4.02) nm IR: ν_{max}^{KBr} cm^{-1} 1665 NMR: ^{13}C (75.5 MHz, $CDCl_3$) (Amer et al., 1991; Wanjala & Majinda, 2000)
10	(+)-Erythravine		HRMS: $C_{18}H_{21}NO_3$: 299.1521. oil $[\alpha]_D^{+25} = +109°$ (c = 0.54 CHCl$_3$) EI-MS: 299 [M]$^+$, 282, 280, 266 UV (MeOH): $\lambda_{max}(\log \varepsilon)$: 282 (3.49) nm NMR: ^{13}C (75.5 MHz, $CDCl_3$) (Amer et al., 1991; Wanjala & Majinda, 2000)
11	(+)-11β-Hydroxyerysodine		HRMS: $C_{18}H_{21}NO_4$: 315.1470. NMR: ^{13}C (60.6 MHz) (Amer et al., 1991)

TABLE 45.3 Non-glycosidic Alkenoid Erythrina Alkaloids

S/No.	Name	Chemical Structure	Properties
12	(+)-Erythratidine		HRMS: $C_{19}H_{25}NO_4$: 331.1783. Mp: 120–120.5°C (*EtOAC/Pet.ether*) $[\alpha]_D^{+25} = +273°$ (*c* = 0.109 EtOH) EI-MS: 331 [M]$^+$, 300, 273, 257(100), 244 UV (MeOH): $\lambda_{max}(\log \varepsilon)$: 232 (3.76), 284 (3.41) nm IR: ν_{max}^{KBr} cm^{-1} 3509, 3387 NMR: ^{13}C (90.6 MHz) (Amer *et al.*, 1991)
13	(+)-Erysosalvine		HRMS: $C_{18}H_{23}NO_4$: 317.1627. EI-MS: 317 [M]$^+$, 286, 259, 243(100), 242 NMR: ^{1}H (60 MHz) (Amer *et al.*, 1991)

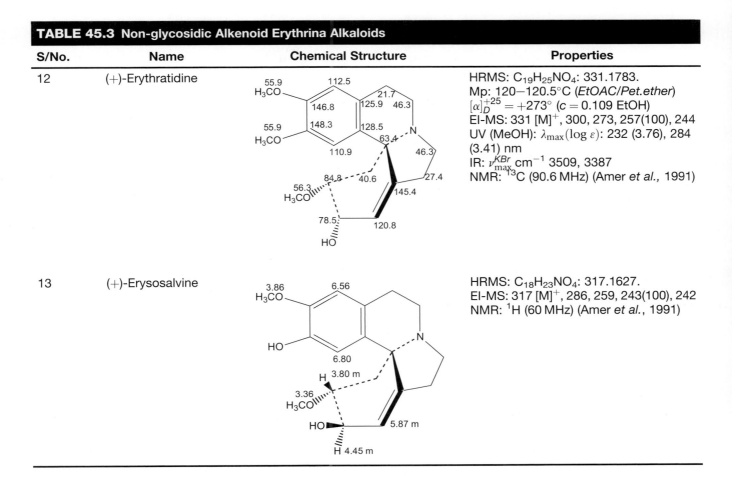

TABLE 45.4 Non-erythrina Alkaloid from the Seeds

S/No.	Name	Chemical Structure	Properties
14	Hypaphorine		HRMS: $C_{14}H_{18}N_2O_2$: 246.3049. Water-soluble salt (Ozawa *et al.*, 2008)

Activity of erythrina alkaloids from seeds of *E. latissima*

Iwanaga and colleagues (1999) reported that the strong intramolecular interactions between the primary binding loops that scaffold ECI play an important role in the strong inhibitory activity towards chymotrypsin. Surface plasmon resonance analysis revealed that the side chains on the primary binding loop of ECI contribute to both an increase in the association rate constant and a decrease in the dissociation rate constant for ECI—chymotrypsin interaction, whereas the backbone structure of the primary binding loop mainly contributes to a decrease in the dissociation rate constant.

The pharmacological activities of the individual isolated alkaloids reported from the seeds of *E. latissima* are discussed below, although it is worth noting that little research on pharmacological activity and clinical trials has been reported regarding the compounds from the seeds of this species compared to other *Erythrina* spp.

GLYCOSIDES

Three glycosidic alkaloids have been reported from the seeds of *E. latissima*, and so far only cytotoxicity activity of these compounds has been reported (Wanjala & Majinda, 2000). Although the antimicrobial activity of these glycosides, (+)-15β-D-glucoerysopine (**1**), (+)-16β-D-glucoerysopine (**2**), and (+)-β-D-glucoerysodine (**3**), against the gram negative bacteria *Escherichia coli* and *Pseudomonas aeruginosa*, gram positive *Staphylococcus aureus* and *Bacillus subtilis*, and single-celled yeast *Candida albicans* was negative (Wanjala & Majinda 2000), the aglycosides of these compounds are known to exhibit strong neuroparalytic activity. Compounds with reported neuroparalytic activity from *Erythrina* spp. include erysodine, erysopine, erysovine, and erythraline (Decker *et al.*, 1995). Erysodine, erysovine, and erythraline are some of the major compounds from the seeds of *E. latissima* (Wanjala & Majinda, 2000; Wanjala *et al.*, 2009). Erysopine, which can be obtained by hydrolysis of compounds **1** and **2**, is reported to exhibit anticonvulsant, CNS-depressant, hydrocholerectic, and myorelaxant activities (Decker *et al.*, 1995).

(+)-ERYSODINE (4)

Erysodine is a major and frequently isolated compound from the seeds of *Erythrina* spp. This compound has been documented as having neuromuscular effects characteristic of curare arrow poisons, and reported to have neuroparalytic activity. Studies also indicate that it can be useful as an antinicotine drug, it has demonstrated actions as a competitive antagonist blocking nicotine receptors. Both of these studies were published by major (and competing) pharmaceutical companies (Decker *et al.*, 1995). Erysodine also shows competitive enzyme inhibition on Rb efflux stimulated nicotine from KXα3β4R2 cells (Decker *et al.*, 1995; Majinda *et al.*, 2001), has been reported to have strong ($IC_{50} = 1.0\ \mu g/ml$) radical scavenging properties against stable DPPH radical (Juma & Majinda, 2004), and shows enhanced activity when combined with tumor necrosis factor (TNF)-related apoptosis-inducing ligand (TRAIL) (Ozawa *et al.*, 2009).

ERYSOVINE (5)

Iwu (1993) reported that erysovine, which is also found in large amounts in *E. beteroana*, *E. crista-galli, and E. fusca*, has anticonvulsant, CNS-depressant, hydrocholerectic, myorelaxant, and neuroparalytic activities. Erysovine has also been reported to show enhanced activity when combined with TRAIL (Ozawa *et al.*, 2009).

(+)-ERYSOTRINE (6)

Ozawa and colleagues (2009) reported that erysotrine does not show cytotoxic activity by itself, but exhibits significant cytotoxicity when combined with TRAIL. This compound also shows competitive enzyme inhibition on Rb efflux stimulated nicotine from KXα3β4R2 cells, as reported by Decker *et al.* (1995) and Majinda *et al.* (2001).

(+)-ERYTHRALINE (7)

(+)-Erythraline, one of the major compounds from the seeds of *E. latissima*, has been reported to enhance activity when combined with (TRAIL) (Ozawa *et al.*, 2009). (+)-Erythraline has also been reported to show competitive enzyme inhibition on Rb efflux-stimulated nicotine from KXα3β4R2 cells (Decker *et al.*, 1995; Majinda *et al.*, 2001). Iwu (1993) reported that (+)-erythraline also isolated from *E. crista-galli* and *E. fusca* has neuroparalytic activity.

(+)-8-OXOERYTHRALINE (8)

(+)-8-Oxoerythraline is the compound isolated from a wide range of *Erythrina* spp., including *E. crista-galli*, *E. brucei*, *E. chiriquensis*, *E. tabitensis*, *E. lysistemon*, and *E. abyssinica*. Despite the abundance of this compound in different *Erythrina* spp., only positive cytotoxity activity has been reported (Majinda *et al.*, 2001).

(+)-ERYSOTRAMIDINE (9)

(+)-Erysotramidine has strong antioxidant activity on peroxynitrite, a powerful oxidant that can damage a wide array of molecules in cells, including DNA and proteins (Tanaka *et al.*, 2008). (+)-Erysotramidine has also been reported to have moderate ($IC_{50} = 50 \, \mu g/ml$) radical scavenging properties against stable DPPH radicals, according to Juma and Majinda (2004).

(+)-ERYTHRAVINE (10)

In animal models (+)-erythravine has anxiolytic effects of anxiety in mice, and pharmacological activity in cholinergic serotonergic and/or GABAergic system models as reported by Flausino *et al.* (2007). Among the principal pharmacological actions of (+)-erythravine is its peripherical activity on the cholinergic system, which has been compared to the effects of d-tubocurarine (Flausino *et al.*, 2007). Erythravine also possesses sedative and anticonvulsant effects only in the strychnine-induced seizure model, suggesting its possible action in the glycine system and potentiation of pentobarbital sleeping time, and depressant action in the central nervous system, according to Vasconcelos *et al.* (2007).

(+)-11β-HYDROXYERYSODINE (11)

(+)-11β-Hydroxyerysodine (**11**) has also been previously isolated from other *Erythrina* spp., including *E. lysistemon*, *E. senegalensis*, and *E. livingstoniana* (Amer *et al.*, 1991), but no activity has been documented for this compound.

(+)-ERYTHRATIDINE (12)

(+)-Erythratidine (**12**) has also been previously reported from other *Erythrina* spp., including *E. crista-galli*, *E. senegalensis*, and *E. livingstoniana* (Amer *et al.*, 1991), but no activity has been reported for this compound.

(+)-ERYSOSALVINE (13)

(+)-Erysosalvine (**13**) has also been isolated from *E. arborescens*, *E. livingstoniana*, *E. tabitensis*, *E. burana*, *E. oliviae*, *E. salviiflora*, and *E. melanacantha* (Amer *et al.*, 1991), but no activity has been reported for this compound so far.

HYPAPHORINE (14)

The betaine of tryptophan is a quaternary ammonium salt, and is the major compound that accumulates in the seeds of most *Erythrina* species. Hypaphorine has been reported by Ozawa and colleagues (2008) to have sleep-promoting effects in mice, which showed significantly increased non-rapid eye movement (NREM) sleep time during the first hour after its administration. The NREM sleep time was enhanced by 33% in the experimental mice when compared to that of the controls. Apart from its sleep-promoting property, hypaphorine has also been reported to possess convulsant and tetamic properties.

ADVERSE EFFECTS AND REACTIONS (ALLERGIES AND TOXICITY)

Erythrina alkaloids are known to possess paralysis potency and curare-like neuromuscular blocking activities; thus, they cause collapse due to neuromuscular paralysis but without loss of consciousness (Reimann, 2007).

Kittler and colleagues (2002) suggested that the alkaloids in *Erythrina* "may alter GABAergic neurotransmission." GABA acts as a neurotransmitter in the brain, and abnormalities in its function are implicated in a number of diseases, including epilepsy, anxiety, and depression.

Neuroparalytic activity is mostly associated with erysodine, erysovine, and erythraline, all of which are dienoid Erythrina alkaloids (Iwu, 1993).

SUMMARY POINTS

- The genus *Erythrina* is popularly known for its ornamental and medicinal properties.
- *E. latissima* trees are widely distributed in the tropics and subtropics, taking 20 to 30 years to form a reasonable canopy, and living for over 100 years.
- The seeds of *E. latissima* contain the tetrahydroisoquinoline type of alkaloids and the betaine of tryptophan (an ionic compound) as the major compounds.
- The Erythrina alkaloids and hypaphorine are responsible for therapeutic properties of *E. latissima* seeds.
- Some Erythrina alkaloids from the seeds of *E. latissima* possess paralysis potency and curare-like neuromuscular blocking activities.

References

Amer, M. E., Shamma, M., & Freyer, A. J. (1991). The tetracyclic Erythrina alkaloids. *Journal of Natural Products, 54*, 329–363.

Chacha, M., Bojase-Moleta, G., & Majinda, R. R. T. (2005). Antimicrobial and radical scavenging flavonoids from the stem wood of *Erythrina latissima*. *Phytochemistry, 66*, 99–104.

Decker, M. W., Anderson, D. J., Brioni, J. D., Donnelly-Roberts, D. L., Kang, C. H., O'Neill, A. B., et al. (1995). Erysodine, a competitive antagonist at neuronal nicotinic acetylcholine receptors. *European Journal of Pharmacology, 280*, 79–89.

Flausino, O., Jr., Santos, Lde A., Verli, H., Pereira, A. M., Bolzani, Vda S., & Nunes-de-Souza, R. L. (2007). Anxiolytic effects of erythrinian alkaloids from *Erythrina mulungu*. *Journal of Natural Products, 70*, 48–53.

Hemsley, A. J., & Ferguson, I. K. (1985). Pollen morphology of the genus *Erythrina* (Leguminosae: Papilionoideae) in relation to floral structure and pollinators. *Annals of the Missouri Botanical Garden, 72*, 570–590.

Hennessy, E. (1972). *South African erythrinas*. Durban, South Africa: The Natal Branch of the Wildlife Protection and Conservation Society of South Africa.

Iwanaga, S., Nagata, R., Miyamoto, A., Kouzuma, Y., Yamasaki, N., & Kimura, M. (1999). Conformation of the primary binding loop folded through an intramolecular interaction contributes to the strong chymotrypsin inhibitory activity of the chymotrypsin inhibitor from *Erythrina Variegata* seeds. *Journal of Biochemistry, 126*, 162–167.

Iwu, M. M. (1993). *Handbook of African medicinal plants*. Boca Raton, FL: CRC Press. 435.

Johnson, D., & Johnson, S. (2002). In M. Whitaker (Ed.), *Down to earth: Gardening with indigenous trees* (p. 47). Capetown, South Africa: Struik.

Juma, B. F., & Majinda, R. R. T. (2004). Erythrinaline alkaloids from the flowers and pods of *Erythrina lysistemon* and their DPPH radical scavenging properties. *Phytochemistry, 65*, 1397–1404.

Kittler, J. T., McAinsh, K., & Moss, S. J. (2002). Mechanisms of GABA receptor assembly and trafficking: implications for the modulation of inhibitory neurotransmission. *Molecular Neurobiology, 26*, 251–268.

Majinda, R. R. T., Abegaz, B. M., Bezabih, M., Ngadjui, B., Wanjala, C. C. W. L., Mdee, K., et al. (2001). Recent results from natural products research at the University of Botswana. *Pure and Applied Chemistry, 73*, 1197–1208.

Ozawa, M., Kazuki, H., Izumi, N., Akio, K., & Ayumi, O. (2008). Hypaphorine, an indole alkaloid from Erythrina velutina, induced sleep on normal mice. *Bioorganic & Medicinal Chemistry Letters, 18*, 3992–3994.

Ozawa, M., Etoh, T., Masahiko Hayashi, M., Komiyama, K., Akio Kishida, A., & Ohsaki, A. (2009). TRAIL-enhancing activity of Erythrinan alkaloids from *Erythrina velutina*. *Bioorganic & Medicinal Chemistry Letters, 19*, 234–236.

Reimann, E. (2007). In W. Herz, H. Falk & G. W. Kirby (Eds.), Fortschritte der Chemie organischer Naturstoffe [*Progress in the chemistry of organic natural products*], Vol. 88 (pp. 1–62). Vienna, Austria: Springer.

Tanaka, H., Hattori, H., Tanaka, T., Sakai, E., Tanaka, N., Kulkarni, A., et al. (2008). A new Erythrina alkaloid from *Erythrina herbacea*. *Journal of Natural Medicines, 62*, 228–231.

Vasconcelos, S. M., Lima, N. M., Sales, G. T., Cunha, G. M., Aguiar, L. M., Silveira, E. R., et al. (2007). Anticonvulsant activity of hydroalcoholic extracts from *Erythrina velutina* and *Erythrina mulungu*. *Journal of Ethnopharmacology, 110*, 271–274.

Wanjala, C. C. W., & Majinda, R. R. T. (2000). Two novel glucodienoid alkaloids from *Erythrina latissima* seeds. *Journal of Natural Products, 63*, 871–873.

Wanjala, C. C. W., Akeng'a, T., Obiero, G. O., & Lutta, K. P. (2009). Antifeedant activities of the erythrinaline alkaloids from *Erythrina latissima* against *Spodoptera littoralis* (Lepidoptera Noctuidae). *Records Natural Products, 3*, 96–103.

Wanjala, C. C. W., Juma, B., Bojase, G., Gashe, B. A., & Majinda, R. R. T. (2002). Erythrina alkaloid and anti-microbial flavonoids from *Erythrina latissima*. *Planta Medica, 68*, 640–642.

CHAPTER 46

Coralwood (*Adenanthera pavonina* L.) Seeds and Their Protective Effect

Anna Jaromin, Mariola Korycińska, Arkadiusz Kozubek
Faculty of Biotechnology, University of Wroclaw, Department of Lipids and Liposomes, Wroclaw, Poland

INTRODUCTION

Adenanthera pavonina is a tropical tree, also known by many common names, such as the red sandalwood, coralwood, coral bean tree, or red bead tree. This plant is endemic to South-east China and India, and has been introduced throughout the humid tropics. It has been widely naturalized in Malaysia, Western and Eastern Africa, and most islands of the Pacific and Caribbean regions. This species is extensively cultivated as a valuable agroforestry species; there are also historical accounts of its uses in traditional medicine. *A. pavonina* is also known as a "food tree," because its seeds are often eaten by people.

BOTANICAL DESCRIPTION

Adenanthera pavonina (L.) (family Leguminosae, subfamily Mimosoideae) is a medium to large-sized deciduous tree, 6–15 m tall. Leaves are bipinnate, with two to six opposite pairs of pinnae. Flowers are small, creamy yellow, and fragrant. The curved pods contain the seeds, which are exposed when these pods curl back. Ripened pods remain on the tree for long periods. The seeds are hard, dark or bright red, about 8 mm wide, and there are 10–12 per pod (Figure 46.1).

The tree grows on a variety of soils in moist and seasonally moist tropical climates, preferring neutral to slightly acidic soils. Initial seedling growth is slow, but rapid height and diameter increments occur from the second year onward. The tree is susceptible to breakage in high winds, with the majority of damage occurring in the crown. This species is common throughout the lowland tropics up to 300–400 m.

Nuts & Seeds in Health and Disease Prevention. DOI: 10.1016/B978-0-12-375688-6.10046-5

FIGURE 46.1
Adenanthera pavonina with the red seeds. *Photograph by Pratheep (http://en.wikipedia.org/wiki/File:Adenanthera1.jpg).*

HISTORICAL CULTIVATION AND USAGE

The seeds have been used to weigh gold, silver, and diamonds, because they have a narrow range in weight. It is interesting that they are so very similar in weight, four seeds making up about 1 gram.

The bark contains saponins, and has been used to wash hair and clothing. The red dye from the tree has been used for dyeing clothes and for cosmetic purposes. Brahmins used this dye to mark religious symbols on their foreheads.

In ancient Indian medicine, ground seeds were used to treat boils and inflammation. A decoction of young leaves was used to treat gout and rheumatism.

PRESENT-DAY CULTIVATION AND USAGE

A. pavonina is planted in domestic gardens, often as shade trees for field vegetables; in coffee and cocoa plantations; and locally along field borders. It is also extensively cultivated as an ornamental tree for planting along roadsides and in common areas, especially for its red, glossy seeds.

In some countries, such as Java and Indonesia, the seeds are roasted, shelled, and then eaten with rice, whereas young leaves are eaten as a vegetable. In the Pacific Islands the wood is used as fuel, producing significant heat. The wood, which is very hard and strong, is also used for bridge and household construction, for flooring, furniture, and cabinet work. The red seeds are used as beads in bracelets or necklaces, and in toys.

APPLICATIONS TO HEALTH PROMOTION AND DISEASE PREVENTION

The seeds of *A. pavonina*, which are rich in pale yellowish oil, are a valuable food source. Detailed analysis showed that the seeds of this tree contain appreciable amounts of proteins (29.44 g/100 g), crude fat (17.99 g/100 g), and minerals, comparable to commonly consumed staples. Subsequent examination revealed a low total sugar content (8.2 g/100 g), with starch (41.95 g/100 g) constituting the major carbohydrate. Moreover, low levels of antinutrients have been detected. Free fatty acid levels were relatively high, but peroxide and saponification values of 29.6 mEq/kg and 164.1 mg KOH/g, respectively, point to a resemblance to oils processed for food (Ezeagu *et al.*, 2004). It should also be noted that chemicals possessing hemolytic properties against red blood cells (such as polyamines and 5-*n*-alkylresorcinols) were not detected in the oil obtained from the seeds of *A. pavonina*, (Zarnowski *et al.*, 2004). From this point of view, *A. pavonina* seeds represent a potential source of oil and protein that could alleviate shortages.

The anti-inflammatory and analgesic effects of methanol extract of the seeds of *A. pavonina* were evaluated in animal models by Olajide and colleagues (2004). As observed by the authors, the extract (50–200 mg/kg) produced statistically significant ($P < 0.05$) inhibition of carrageenan-induced paw edema in the rat (Table 46.1), as well as acetic acid-induced vascular permeability in mice. At doses of 100 and 200 mg/kg, inhibition of pleurisy induced with carrageenan was also detected. Moreover, the extract (50–200 mg/kg) exhibited dose-dependent and significant ($P < 0.05$) analgesic activity in acetic acid-induced writhing in mice (Table 46.2). Both early and late phases of formalin-induced paw-licking in mice were inhibited by the extract. All the obtained observations suggest that this seed extract has an analgesic effect through both inflammatory mechanisms, and a direct effect on nociceptors (Olajide *et al.*, 2004). A recent report by Oni and colleagues (2009) provides information about the central nervous system activities of the methanol extract of the seed. This extract (50–200 mg/kg) protected mice against picrotoxin- and pentylenetetrazole-induced seizures, and prolonged phenobarbitone-induced sleeping time dose-dependently. The authors suggest that this indicates anticonvulsant and depressant activities of the extract, as evidenced in the experimental animal model (Oni *et al.*, 2009).

TABLE 46.1 Effect of *A. pavonina* Methanol Extract on Carrageenan-induced Rat Paw Edema

Treatment	Paw Size at Third Hour[a] (cm)	Inhibition of Edema (%)
Control (10 ml/kg saline)	3.24 ± 0.07	—
Control *A. pavonina* (50 mg/kg)	2.84 ± 0.08[b]	34.4
A. pavonina (100 mg/kg)	2.62 ± 0.07[b]	47.5
A. pavonina (200 mg/kg)	2.56 ± 0.05[b]	49.2
Indomethacin (5 mg/kg)	2.10 ± 0.04[b]	90.2

Reproduced from Olajide *et al.* (2004), *Inflammopharmacology*, *12*, 196–202, with permission.
[a]Values are expressed as mean $\pm$ SEM for six independent observations (n = 6).
[b]Values are statistically significant at P < 0.05, compared with Tween-80 treatment, Student's t-test.

TABLE 46.2 Effect of *A. pavonina* Methanol Extract on Acetic Acid-induced Writhing in Mice

Treatment	No. of Writhings	Inhibition of Writhing (%)
Control (10 ml/kg saline)	43.0 ± 3.9	—
A. pavonina (50 mg/kg)	29.2 ± 2.0[a]	32.1
A. pavonina (100 mg/kg)	23.8 ± 1.7[a]	44.7
A. pavonina (200 mg/kg)	15.2 ± 2.6[a]	64.7
Indomethacin (5 mg/kg)	9.2 ± 0.9[a]	78.6

Values are expressed as mean $\pm$ SEM for six independent observations (*n* = 6).
Reproduced from Olajide *et al.* (2004), *Inflammopharmacology*, *12*, 196–202, with permission.
[a]Values are statistically significant at P < 0.05, compared with Tween-80 treatment, Student's t-test.

TABLE 46.3 Antihypertensive Effects of Methanol Extract of the Seeds of *A. pavonina* (*n* = 4)

Group	Mean Arterial Blood Pressure (mmHg)
A. Control	60 ± 3.4
B. Propanolol (1 mg/kg)	23 ± 4.2
C. Extract (200 mg/kg dose)	30 ± 3.8

Reproduced from Adedapo *et al.* (2009), *Records of Natural Products*, 3, 82–89, with permission.

The effect of methanol *A. pavonina* seed extract on the blood pressure of normotensive rats was investigated, as reported by Adedapo and colleagues (2009). Twelve adult male Wistar rats, divided into three groups, were treated orally with normal saline (control group), propanolol (positive control; 1 mg/kg), and 200 mg/kg of seed extract, respectively, over a 4-week period. The mean arterial blood pressure of the normal saline-treated animals was 60 mmHg, whereas those of the propanolol- and extract-treated animals was decreased (Table 46.3). Moreover, histopathological changes due to the methanol seed extract were investigated. It was observed that the extract did not cause any significant lesion changes in the liver, kidney, and even the testes, indicating that is relatively safe. During treatment with the seed extract, a significant decrease in the level of liver enzymes was observed. This may suggest that the extract has hepatoprotective effects (Adedapo *et al.*, 2009).

Leaf extracts from this tree were also investigated for potential anti-inflammatory activity. Ethanolic extracts from the leaves were tested, at doses of 250 and 500 mg/kg, for anti-inflammatory effects, using both acute and chronic inflammatory models. It was found that the doses possessed inhibitory effects on the acute phase of inflammation, as seen in carrageenan-induced hind paw edema, as well as in a subacute study of cotton pellet-induced granuloma formation. Antidiarrheal activity was also demonstrated, as evidenced by the delay in the formation of wet feces (Table 46.4) (Mayuren & Ilavarasan, 2009). The anti-inflammatory activity caused by the leaf extracts may be due to the presence of the active compounds such as glucosides of beta-sitosterol and stigmasterol as identified earlier (Nigam *et al.*, 1973; Yadav *et al.*, 1976).

Methanolic extracts of the root, bark, leaves, and fruits/seed kernels of *A. pavonina* plant were studied in the range for antifungal, antioxidant, and cytotoxic activities. Extracts of *A. pavonina* displayed expressed marginal antifungal activity, but stem-bark extracts demonstrated significant antioxidant activity compared to the reference, α-tocopherol, whereas root extracts showed moderate activity. Only the root and stem-bark extracts of *A. pavonina* showed cytotoxicity (Rodrigo *et al.*, 2007). There has also been a report that extracts of this tree exhibits

TABLE 46.4 Effect of Ethanolic Extract of *A. pavonina* Leaves on Castor Oil-induced Diarrhea

Treatment	Dose (mg/kg)	Time of First Appearance of Wet Feces (min)	Total Wet Feces at 4th Hour (mg)
Vehicle	—	36.33 ± 1.68	30.16 ± 2.24
Standard	20	$73.83 \pm 1.66a^*$	$12.5 \pm 1.25a^*$
Ethanolic extract of *A. Pavonina*	250	$50.66 \pm 2.52a^*$	$22.5 \pm 1.82a^*$
Ethanolic extract of *A. Pavonina*	500	$63.5 \pm 2.23a^*$	$20.66 \pm 1.62a$

Values are mean $\pm$ SEM, $n = 6$ animals in each group; 'a' represents comparisons between Group 2, 3, 4 *vs* Group 1; statistical evaluation by one-way ANOVA followed by Dunnett's *t*-test.
Reproduced from Mayuren and Ilavarasan (2009), *J. Young Pharmacist*, 1, 125–128, with permission.
*Represent, statistical significance P < 0.001.

antibacterial activity against methicillin-sensitive and resistant strains of *Staphylococcus aureus*, *Enterococcus faecalis*, and *Pseudomonas aeruginosa* (Jayasinghe *et al.*, 2006).

Seed powder of *A. pavonina* was analyzed regarding the control of root rot diseases caused by *Macrophomina phaseolina* (Tassi) Goid, *Rhizoctonia solani* Kühn, and *Fusarium* spp. in plants like the mung bean (*Vigna radiata* L.) and chick pea (*Cicer arietinum* L.). Natural treatments for the control of plant diseases are highly desirable, as fungicidal applications cause serious hazards to human health and increase environmental pollution. The results obtained by the authors from screen house application of *A. pavonina*, at 0.1 and 1% w/w showed significant control of tested root rot fungi. Another interesting finding is the fact that there was complete suppression of *Fusarium* spp. and *R. solani* infection in mung beans when soil was modified with seed powder of *A. pavonina* at 1% w/w. In the case of chick peas, this effect was observed for *M. phaseolina* for both 0.1 and 1% w/w seed powder (Ahmed *et al.*, 2009).

Another interesting aspect of the application of seeds of *A. pavonina* is the suitability of their galactomannans as edible coatings for different tropical fruits, in response to consumers' demands for food without chemical preservatives. Polysaccharide-based coatings can be used to increase the shelf-life of fruits or vegetables, to avoid dehydration, and to reduce the darkening of the surface. As reported by Cerqueira and colleagues (2009), novel galacto-mannan extracted from *A. pavonina* can be successfully applied to several fruits (acerola, cajá, mango, pitanga, and seriguela), based on their surface properties. The positive effects of the application of this coating to fruits, such as shelf-life extension with improved sensory quality, are expected to be of interest in the near future.

The potential usefulness of the seed oil in oil-in-water (o/w) emulsions — for example, as carriers for active compounds — has also been demonstrated (Zarnowski *et al.*, 2004; Jaromin *et al.*, 2006). The o/w emulsions developed from seed oil, stabilized with soybean lecithin (SPC) and an additional co-emulsifier, Tween 80, were very stable. These studies suggested that emulsions containing the edible oil of *A. pavonina* may be useful as an alternative formulation matrix for pharmaceutical or cosmetic applications (Jaromin *et al.*, 2006).

As can be seen, extracts from different parts of *A. pavonina* are under investigation in numerous laboratories, and many important activities and properties have already been proven.

ADVERSE EFFECTS AND REACTIONS (ALLERGIES AND TOXICITY)

The raw seeds of *A. pavonina* are poisonous, and boiling is required to neutralize their toxicity. An acute toxicity study on the extract of *A. pavonina* seeds, performed by Olajide and colleagues (2004), revealed that the extract produced a dose-dependent reduction of motor activity, with 800, 1600, and 3200 mg/kg of the extract producing loss of the righting reflex in mice. The LD_{50} value was found to be 1360 mg/kg. This clearly shows that the plant extract is relatively non-toxic, but can produce marked depressant effects on the central nervous system at low doses (Olajide *et al.*, 2004).

On the other hand, acute toxicity studies in rats revealed that the leaf extract is non-toxic at doses of up to 5000 mg/kg body weight, demonstrating the safety profile of this type of extract (Mayuren & Ilavarasan, 2009).

SUMMARY POINTS

- The seeds of *A. pavonina* are rich in proteins, crude fat, and minerals, with low levels of antinutrients detected.
- The methanol extract of the seeds of *A. pavonina* has anti-inflammatory and analgesic effects.
- *A. pavonina* seed extract has potential blood pressure lowering effects.

- The ethanolic leaf extract of this tropical tree possesses significant anti-inflammatory activity.
- The stem-bark extracts demonstrated significant antioxidant activity.
- The seed powder (at 0.1% and 1% w/w) reduced the infection of root-infecting fungi.
- Galactomannans of *A. pavonina* can be used as edible coatings for different tropical fruits.
- The seed oil can be applied in oil-in-water (o/w) emulsions.

References

Adedapo, A. D. A., Osude, Y. O., Adedapo, A. A., Moody, J. O., Adeagbo, A. S., Olajide, O. A., et al. (2009). Blood pressure lowering effect of *Adenanthera pavonina* seed extract on normotensive rats. *Records of Natural Products, 3,* 82–89.

Ahmed, Z. M., Dawar, S., & Tariq, M. (2009). Fungicidal potential of some local tree seeds for controlling root rot disease. *Pakistan Journal of Botany, 41,* 1439–1444.

Cerqueira, M. A., Lima, A. M., Teixeira, J. A., Moreira, R. A., & Vicente, A. A. (2009). Suitability of novel galactomannans as edible coatings for tropical fruits. *Journal of Food Engineering, 94,* 372–378.

Ezeagu, I. E., Gopal Krishna, A. G., Khatoon, S., & Gowda, L. R. (2004). Physico-chemical characterization of seed oil and nutrient assessment of *Adenanthera Pavonina,* L.: an underutilized tropical legume. *Ecology of Food & Nutrition, 43,* 295–305.

Jaromin, A., Zarnowski, R., & Kozubek, A. (2006). Emulsions of oil from *Adenanthera pavonina L. seeds* and their protective effect. *Cellular and Molecular Biology Letters, 11,* 438–448.

Jayasinghe, P. K. I. D. E., Bandara, B. M. R., Ekanayaka, E. W. M. A., & Thevanesam, V. (2006). Screening for antimicrobial activity of *Acronychia pedunculata* (Ankenda) and *Adenanthera pavonina* (Madatiya) against bacteria causing skin and wound infections in humans. *Proceedings of the Peradeniya University Research Sessions, Sri Lanka, 11,* 105.

Mayuren, C., & Ilavarasan, R. (2009). Anti-inflammatory activity of ethanolic leaf extracts from *Adenanthera pavonina* (L) in rats. *Journal of Young Pharmacists, 1,* 125–128.

Nigam, S. K., Misra, G., & Mitra, C. R. (1973). Stigmasterol glucoside a constituent of Adenanthera pavonina seed and leaf. *Plant Medicine, 23,* 145–148.

Olajide, O. A., Echianu, C. A., Adedapo, A. D., & Makinde, J. M. (2004). Anti-inflammatory studies on *Adenanthera pavonina* seed extract. *Inflammopharmacology, 12,* 196–202.

Oni, J. O., Awe, O. E., Olajide, A. O., & Makinde, M. J. (2009). Anticonvulsant and depressant activities of the seed extracts of *Adenanthera pavonina. Journal of Natural Products, 2,* 74–80.

Rodrigo, S. K., Jayasinghe, U. L. B., & Bandara, B. M. R. (2007). Antifungal, antioxidant and cytotoxic activity of *Acronychia pedunculata* and *Adenanthera pavonina. Proceedings of the Peradeniya University Research Sessions, Sri Lanka, 12,* 94–95.

Yadav, N., Misra, G., & Nigam, S. K. (1976). Triterpenoids of *Adenanthera pavonina* bark. *Plant Medicine, 29,* 176–178.

Zarnowski, R., Jaromin, A., Certik, M., Czabany, T., Fontaine, J., Jakubik, T., et al. (2004). The oil of *Adenanthera pavonina L.* seeds and its emulsions. *Z. Naturforsch, 59c,* 321–326.

CHAPTER 47

Effect of Coriander (*Coriandrum sativum* L.) Seed Ethanol Extract in Experimental Diabetes

Maryam Eidi[1], Akram Eidi[2]
[1] Department of Biology, Varamin Branch, Islamic Azad University, Varamin, Iran
[2] Department of Biology, Science and Research Branch, Islamic Azad University, Tehran, Iran

395

LIST OF ABBREVIATIONS

i.p., intraperitoneally
STZ, streptozotocin

INTRODUCTION

Many spices are used as a source of vitamins and minerals, and to treat human disorders (Chopra *et al.*, 1956; Nadkarni & Nadkarni 1976). Coriander (*Coriandrum sativum* L.), an annual herb belonging to the carrot family, is grown primarily as a spice crop for use in various cuisines all over the world (Gupta *et al.*, 1986). Extracts and preparations from spices such as coriander seeds have been reported to improve glucose tolerance (Pruthi, 1993). Recent studies have also demonstrated a hypoglycemic effect of coriander seed on carbohydrate metabolism (Gray & Flatt, 1999; Chithra & Leelamma, 2000). Apart from these, no demonstration of the effect of this spice on the insulin-releasing effects of pancreatic beta cell has been carried out. The present study evaluated the insulin-releasing activity of coriander seeds in STZ-induced diabetic rats.

Nuts & Seeds in Health and Disease Prevention. DOI: 10.1016/B978-0-12-375688-6.10047-7

TABLE 47.1 Classification of the Coriander Plant	
Kingdom	**Plantae (plants)**
Subkingdom	Tracheobionta (vascular plants)
Superdivision	Spermatophyta (seed plants)
Division	Magnoliophyta (flowering plants)
Class	Magnoliopsida (dicotyledons)
Subclass	Rosidae
Order	Apiales
Family	Apiaceae (the carrot family)
Genus	*Coriandrum L.*
Species	*Coriandrum sativum L.*

BOTANICAL DESCRIPTION

Coriander (*Coriandrum sativum* L.), an annual herb belonging to the carrot family, is grown primarily as a spice crop for culinary use throughout the world (Gupta *et al.*, 1986). *C. sativum* is an annual, herbaceous plant, originally from the Mediterranean and Middle Eastern regions. It grows 25—60 cm (9—24 inches) in height, and has thin, spindle-shaped roots, an erect stalk, alternate leaves, and small, pinkish-white flowers. The plant flowers from June to July, and yields round fruits consisting of two pericarps. The plant is cultivated for its aromatic leaves and seeds. The phylogenetic classification of *C. sativum* is provided in Table 47.1.

HISTORICAL CULTIVATION AND USAGE

Coriander seed has a long history of use. It is mentioned in Sanskrit literature as far back as 5000 BC, and in the Greek Eber Papyrus as early as 1550 BC. Coriander seed was used in traditional Greek medicine by Hippocrates (*ca.* 460—377 BC). The seeds of coriander were found in the ancient Egyptian tomb of Ramses II. The Egyptians called this herb the "spice of happiness," probably because it was considered to be an aphrodisiac. It was used for cooking, and for children's digestive upsets and diarrhea. The Greeks and Romans also used the seeds to flavor wine and as a medicine. Demand by the Romans for coriander seeds was so great, it was imported from as far away as Egypt. Subsequently, it was introduced into Great Britain by the Romans. The use of coriander seeds to accelerate childbirth has been cited in manuscript illustrations, from the early 13th century, on medieval midwifery. Thus, the seeds (dried) have been in use for almost 7000 years. The oil has been used as a food and fragrance ingredient since the 1900s. The etymology of coriander started with the Greek word *korannon*, a combination of koris and annon (a fragrant anise), referring to the ripe fruit. The Roman naturalist, Pliny the Elder, first used the genus name *Coriandrum*, derived from koris (a stinking bug), in reference to the fetid smell of the leaves and unripe fruit (Burdock & Carabin, 2009).

PRESENT-DAY CULTIVATION AND USAGE

Coriander is used as a spice all over the world (Gupta *et al.*, 1986). Also, coriander seed oil methyl esters are prepared and used as an alternative biodiesel fuel, and contain an unusual fatty acid hitherto unreported as the principle component in biodiesel fuels (Moser & Vaughn, 2010).

APPLICATIONS TO HEALTH PROMOTION AND DISEASE PREVENTION

Coriander seed is known for its wide range of healing properties. In folk medicine, the seeds find use against intestinal parasites. Different studies have proven its efficacy as a larvicidal, bactericidal, and fungicidal.

It is generally used in gastrointestinal complaints such as anorexia, dyspepsia, flatulence, griping pain, vomiting, sub-acid gastritis, and diarrhea, and in disorders resulting from delayed gastrointestinal transit, including anorexia and indigestion (Tatsuta & Iishi, 1993). Coriander seeds are used as stomachic, spasmolytic, and carminative, and for their digestive stimulation and antibilious effect (Usmanghani *et al.*, 1997). Coriander has been shown to stimulate gastric acid secretion by a cholinergic mechanism.

Coriander seed is also used as an anti-edemic, anti-inflammatory, antiseptic, emmenagogue, antihypertensive, and myorelaxant. Its use has been suggested with caution, because of potential allergic reactions from furanocoumarins (Brinker, 1998).

Coriander seeds inhibit the electrically-evoked contractions of spiral strips and tubular segments of isolated central ear artery from rabbit. Coriander seeds had hypotensive effects in rats. The results of this study suggest that coriander seeds may protect various tissues by preventing the formation of free radicals (Medhin *et al.*, 1986). In another study, it is reported that feeding coriander seed protected against 1,2-dimethylhydrazine-induced colon and intestine tumors in male rats (Chithra & Leelamma, 2000).

Coriander, as a major constituent of a spice mix added to a diet fed to female rats for 8 weeks, "favorably enhanced" the activities of pancreatic lipase, chymotrypsin, and amylase. Additionally, feeding the diet containing the spice mix significantly stimulated bile flow and bile acid secretion (Platel *et al.*, 2002). In a series of experiments on spice principles as antioxidants in the inhibition of lipid peroxidation of rat liver microsomes, it is reported that linalool, the principle component of coriander seed oil, had no significant effects on ascorbate/Fe^{2+}-induced lipid peroxidation of rat liver microsomes (Reddy & Lokesh, 1992). In a series of studies, Weber and colleagues observed the effects of coriander seed oil on lipid metabolism (Weber *et al.*, 2003). Feeding of "coriander seed oil" containing high proportions of a positional isomer of oleic acid (i.e., petroselinic (cis-6-octadecenoic) acid) to rats for 10 weeks resulted in a significant decrease in proportions of arachidonic acid in the cellular lipids of rats.

Different studies have proven coriander's efficacy as antidiabetic, hypolipidemic, and anti-oxidant. In these studies, Chithra and Leelamma (1999) investigated changes in lipid metabolism in rats fed a high-fat diet containing coriander seed powder for 75 days. The levels of total cholesterol and triglycerides were decreased significantly in serum, liver, and heart. The serum levels of very low and low-density lipoprotein cholesterol were decreased, while high-density lipoprotein cholesterol increased. In a further study, they examined the changes in levels of lipid peroxides and activity of antioxidant enzymes in female rats maintained on a high-fat diet containing 10% coriander seed powder for 90 days. Feeding a diet containing coriander seed powder resulted in a significant decrease in the levels of lipid peroxides, as determined by malondialdehyde, hydroperoxides and conjugated dienes in liver and heart. The levels of free fatty acids in serum, liver, and heart of the treated animals were significantly decreased. Antioxidant-related enzymes, such as superoxide dismutase, catalase, glutathione peroxidase, glutathione-S-transferase, glucose 6-phosphate dehydrogenase, and glutathione reductase were significantly increased in the liver and heart of the treated animals.

Administration of coriander ethanolic seed extract at doses of 200 and 250 mg/kg, i.p., produced a significant hypoglycemic effect, lasting for 1−5 h, in healthy fasted animals (Figure 47.1).

The administration of coriander seed ethanol extract at doses of 100, 200, and 250 mg/kg, i.p., significantly decreased serum glucose in STZ-induced diabetic fasted animals. Administration of glibenclamide, a standard hypoglycemic agent (600 µg/kg, i.p.), also decreased serum glucose in STZ-induced diabetic fasted animals (Figure 47.2).

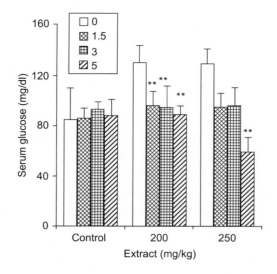

FIGURE 47.1

Administration of coriander ethanolic seed extract at doses of 200 and 250 mg/kg, i.p. produced a significant hypoglycaemic effect in healthy fasted animals lasting for 1–5 h after administration. Each column represents mean ± SD for six rats. Serum glucose levels at zero time or before administrations were considered as control values. The control group was administered with water as a vehicle. $^{**}P < 0.01$ different from before administration. *Reprinted from Eidi et al., 2008; Phytotherapy Research; 23: 404-406, with permission.*

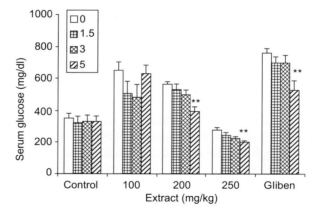

FIGURE 47.2

Effect of intraperitoneally administered coriander seeds ethanol extract at doses of 100, 200 and 250 mg/kg and glibenclamide (Gliben) at a dose of 600 μg/kg body weight on serum glucose in STZ-induced diabetic fasted rats for 1.5, 3 and 5 h after administration. Each column represents mean ± SD for six rats. Serum glucose levels at zero time or before administrations were considered as control values. The control group was administered with water as a vehicle. $^{**}P < 0.01$ different from before administration. *Reprinted from Eidi et al., Phytotherapy Research 2008; 23: 404-406, with permission.*

Administration of STZ decreased the number of beta cells with insulin-releasing activity in comparison with intact rats. However, treatment with coriander seed extract (200 mg/kg) significantly increased active beta cells in comparison with diabetic control rats (Figure 47.3).

Extracts and preparations from spices such as coriander seed have been reported to improve glucose tolerance. Recent studies have also demonstrated a hypoglycemic effect of coriander seed on carbohydrate metabolism (Chithra & Leelamma, 2000). They previously reported that the hypoglycemic action of coriander seeds may be due to the increased utilization of glucose in liver glycogen synthesis, and the decreased degradation of glycogen. It may also be due to an increased rate of glycolysis and decreased rate of gluconeogenesis. The present results indicate

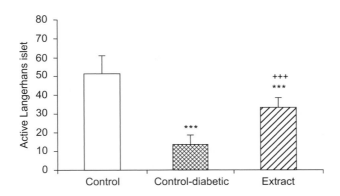

FIGURE 47.3

Effect of intraperitoneally administrered coriander seeds ethanol extract at a dose of 250 mg/kg body weight on the number of active Langerhan's islets in STZ-induced diabetic fasted rats 3 h after administration. Each column represents mean $\pm$ SD for six rats. The control group was administered with water as a vehicle. [***]P < 0.01 different from control intact rats. [+++]P < 0.001 different from control diabetic rats. *Reprinted from Eidi et al., 2008; Phytotherapy Research 23: 404-406, with permission.*

that coriander seeds may also decrease serum glucose by increasing insulin release from the pancreas.

ADVERSE EFFECTS AND REACTIONS (ALLERGIES AND TOXICITY)

The acute oral LD_{50} of the major constituent of coriander seed oil, linalool, in Osborne-Mendel rats was reported to be greater than 2.79 g/kg. The authors noted ataxia "soon" after treatment, and time of death to be 4–18 h (Jenner *et al.*, 1964). As seed oil contains approximately 70% linalool, the oral LD_{50} studies suggest that the lethality caused by the oil is from its constituent, linalool. In another report, coriander seed oil was classified as very mildly irritating to the skin (Tisserand & Balacs, 1995). In a 48-hour closed-patch test, seed oil at a concentration of 6% in petrolatum produced no irritation in 25 human subjects. It is reported that a 20% solution of linalool (major component of coriander seed oil) in petrolatum was non-irritating in a 48-hour patch test in humans (Kligman, 1970). Collectively, these studies indicate that seed oil and its major constituent, linalool (60–80%), are of slight acute toxic potential (Derelanko & Hollinger, 1995).

SUMMARY POINTS

- Many spices are used as a source of vitamins and minerals, and to treat human disorders.
- Coriander, an annual herb belonging to the carrot family, is grown primarily as a spice crop for use in various cuisines all over the world.
- Coriander seed is known for its wide range of healing properties.
- Coriander seeds decrease serum glucose by increasing insulin release from the pancreas.
- Coriander seed oil and its major constituent, linalool, are of slight acute toxic potential.

References

Brinker, F. (1998). Coriander. In *Herb contraindications and drug interactions* (2nd ed.). (p. 146) Sandy, OR: Eclectic Medical Publications.

Burdock, G. A., & Carabin, I. G. (2009). Safety assessment of coriander (*Coriandrum sativum* L.) essential oil as a food ingredient. *Food and Chemical Toxicology, 47*, 22–34.

Chithra, V., & Leelamma, S. (1999). *Coriandrum sativum* changes the levels of lipid peroxides and activity of anti-oxidant enzymes in experimental animals. *Journal of Biochemistry & Biophysics, 36*, 59–61.

Chithra, V., & Leelamma, S. (2000). *Coriandrum sativum* – effect on lipid metabolism in 1,2-dimethyl hydrazine induced colon cancer. *Journal of Ethnopharmacology, 71*, 457–463.

Chopra, R. N., Nayar, S. L., & Chopra, I. C. (1956). *Glossary of Indian medicinal plants*. New Delhi, India: CSIR. 470.

Derelanko, M. J., & Hollinger, M. A. (1995). Toxicity classifications. In *CRC handbook of toxicology* (p. 657). Boca Raton, FL: CRC Press.

Gray, A. M., & Flatt, P. R. (1999). Insulin-releasing and insulin-like activity of the traditional anti-diabetic plant *Coriandrum sativum* (coriander). *British Journal of Nutrition, 81*, 203–209.

Gupta, K., Thakral, K. K., Arora, S. K., & Wagle, D. S. (1986). Studies on growth, structural carbohydrates and phytate in coriander (*Coriandrum sativum*) during seed development. *Journal of the Science of Food and Agriculture, 54*, 43–46.

Jenner, P. M., Hagan, E. C., Taylor, J. M., Cook, E. L., & Fitzhugh, O. G. (1964). Food flavorings and compounds of related structure. I. Acute oral toxicity. *Food and Cosmetics Toxicology, 2*, 327–343.

Kligman, A. M. (1970). Report to the Research Institute for Fragrance Materials (RIFM), 7 October (cited in Opdyke, 1975).

Medhin, D. G., Bakos, P., & Hadhazy, P. (1986). Inhibitory effects of extracts of *Lupinus termis* and *Coriandrum sativum* on electrically induced contraction of the rabbit ear artery. *Acta Pharmaceutica Hungarica, 56*, 109–113.

Moser, B. R., & Vaughn, S. F. (2010). Coriander seed oil methyl esters as biodiesel fuel: unique fatty acid composition and excellent oxidative stability. *Biomass Bioenergy, 34*, 550–558.

Nadkarni, K. M., & Nadkarni, A. K. (1976). *Indian Materia Medica*. Bombay, India: Popular Prakashan. p. 414.

Platel, K., Rao, A., Saraswathi, G., & Srinivasan, K. (2002). Digestive stimulant action of three Indian spices mixes in experimental rats. *Nahrung, 46*, 394–398.

Pruthi, L. S. (1993). *Major spices of India: Crop management post harvest technology*. New Delhi, India: ICAR. p. 387.

Reddy, A. C., & Lokesh, B. R. (1992). Studies on spice principles as antioxidants in the inhibition of lipid peroxidation of rat liver microsomes. *Molecular and Cellular Biochemistry, 111*, 117–124.

Tatsuta, M., & Iishi, H. (1993). Effect of treatment with Liu-Jun-Zi-Tang (TJ-43) on gastric emptying and gastrointestinal symptoms in dyspeptic patients. *Alimentary Pharmacology & Therapeutics, 7*, 459–462.

Tisserand, R., & Balacs, T. (1995). Coriander seed. In *Essential oil safety* (p. 205). New York, NY: Churchill Livingstone.

Usmanghani, K., Saeed, A., & Alam, M. T. (1997). *Indusyunic medicine: Traditional medicine of herbal, animal and mineral origin in Pakistan*. Karachi, Pakistan: BCC and T Press, University of Karachi. 184–185.

Weber, N., Klein, E., & Mukherjee, K. D. (2003). Stereospecific incorporation of palmitoyl, oleoyl and linoleoyl moieties into adipose tissue triacylglycerols of rats results in constant sn-1:sn-2:sn-3 in rats fed rapeseed, olive, conventional or high oleic sunflower oils, but not in those fed coriander oil. *Journal of Nutrition, 133*, 435–441.

CHAPTER 48

Extracts of Cowhage (*Mucuna pruriens*) Seeds and Anti-Snake Venom Effects

Nget Hong Tan[1], Shin Yee Fung[1], Si Mui Sim[2]
[1] Department of Molecular Medicine
[2] Department of Pharmacology, Faculty of Medicine, University of Malaya, Kuala Lumpur, Malaysia

LIST OF ABBREVIATIONS

i.p., intraperitoneally
MPE, *Mucuna pruriens* seed extract

INTRODUCTION

Cowhage *(Mucuna pruriens* Linn.*)* is also known as velvet bean or cowitch. The seeds have been used widely in traditional medicine (Sathiyanarayanan & Arulmozhi, 2007), including the treatment of Parkinson's disease, edema, and impotence. In Nigeria, it is also used as an oral

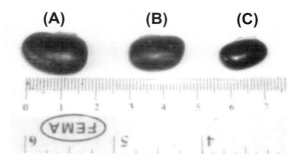

FIGURE 48.1
Cowhage seeds from India (A), South Africa (B), and Nigeria (C). Note that the size of the seed from India is almost double that from Nigeria.

prophylactic for snakebite. It has been claimed that when the *Mucuna pruriens* seeds are swallowed intact, the individual is protected for 1 year against snake venom poisoning (quoted by Aguiyi *et al.*, 1999).

BOTANICAL DESCRIPTION

Mucuna pruriens belongs to the Fabaceae family. The plant is a long, slender climber, with fleshy fruits containing three to six seeds. Its seeds are fawnish-brown, ellipsoid, and small (Figure 48.1).

HISTORICAL AND PRESENT-DAY CULTIVATION AND USAGE

Cowhage is indigenous to tropical regions, especially India and Africa, and certain parts of China. The plant is an important cover crop, and has been used widely in traditional medicine (Sathiyanarayanan & Arulmozhi, 2007).

Anti-snake venom activity of *Mucuna pruriens* seed

Aguiyi and colleagues (1999) were the first to demonstrate in the laboratory that immunization of mice with *Mucuna pruriens* seed extract (MPE) elicited production of immunoglobulins that could neutralize cobra venom. The seed extract was prepared by soaking dried *Mucuna pruriens* (from Nigeria) seed meal (50 g) in distilled water (100 ml) for 24 hours at 4°C. The extract was centrifuged at 10,000 g for 20 minutes, and the supernatant termed *Mucuna pruriens* seed extract (MPE). The antibody was produced in rabbits by subcutaneous injection with MPE (18 mg/kg) followed by a repeated immunization with a 9-mg/kg dose of MPE on day 14, and the animals were bled on day 30. Both *in vitro* and *in vivo* tests showed that the immunoglobins produced were effective in the neutralization of cobra venom.

Subsequently, Guerranti and colleagues (1999) demonstrated that injection of MPE into mice conferred protection against the lethal action of *Echis carinatus* venom. MPE (21 mg/kg) was injected intraperitoneally into the mice, and challenged with i.p. injection of the minimum lethal dose of the venom, 24 hours and 14 days after MPE pretreatment, respectively. More than 60% of the animals survived. However, pretreatment of mice with protein fraction (4.2 mg/kg, i.p.) isolated from MPE only conferred protection when the said snake venoms were injected 14 days after the treatment — presumably via an immunological mechanism. The results suggest that more than one active principle in the MPE is responsible for the protection against snake venoms: the active principle in the non-protein fraction that confers protection 24 hours after the pretreatment; and the active principle in the protein fraction, which presumably acts via an immunological mechanism. These conclusions were supported by further experiments by Guerranti *et al.* (2002), who again confirmed that the non-protein fraction (which contain L-DOPA, free amino acids, and fatty acids) confers short-term protection, whereas the protein fraction confers long-term protection against *Echis carinatus* venom (Table 48.1).

TABLE 48.1 Time Course of Protective Effects of *Mucuna pruriens* Seed Extract, Protein and Non-protein Fractions against *Echis carinatus* Venom

	Survival/total		
	1 day	1 week	3 weeks
MPE	6/8	5/8	8/8
Protein fraction	2/8	3/8	8/8
Non-protein fraction	5/8	3/8	2/8
Untreated (Control)	0/8	0/8	0/8

Mice were injected (i.p.) with the indicated fractions (at dose of 21 mg/kg MPE or equivalent), and treated with a minimum lethal dose of *Echis carinatus* venom (2 mg/kg) 24 h, 1 week, and 3 weeks later.
Source: Reproduced from Guerranti *et al* (2004), *Proteomics*, 8, 402–412, with permission.

While almost all the studies on the protective action of *Mucuna pruriens* seeds against snake venom poisoning were conducted using Nigerian seeds, preliminary studies indicated that seeds from India and South Africa also possess similar anti-cobra venom activity (unpublished observations). Morphologically the seeds look similar, but they differ in size. (Figure 47.1).

Mechanism of the protective effect of *Mucuna pruriens* seed extract pretreatment against *Echis carinatus* venom

Echis carinatus (the saw-scaled or carpet viper, also known as *Echis osellatus*) is the most significant cause of snakebite mortality in Western Africa. Victims of severe *Echis carinatus* venom poisoning develop incoagulable blood and spontaneous systemic bleeding, with evidence of disseminated intravascular coagulation (Warrell *et al.*, 1977). The major toxins include prothrombin activator, hemorrhagins, and myotoxic phospholipase A_2 (Warrell *et al.*, 1977; Yamada *et al.*, 1996; Zhou *et al.*, 2008).

Guerranti *et al.* (2001) reported that the protective effect of *Mucuna pruriens* seed against *Echis carinatus* venom poisoning was only evident when the animals were pretreated with MPE 24 hours before *Echis carinatus* venom challenge, and that there was no protection when the mice were injected with a pre-incubated mixture of *Echis carinatus* venom and the seed extract. This showed that there is no direct neutralization of the venom components by seed extract.

Aguiyi and colleagues (Aguiyi *et al.*, 2001; Guerranti *et al.*, 2001) also investigated the effect of MPE pretreatment on venom-induced elevation of the serum enzymes lactate dehydrogenase, glutamate pyruvic transaminase, and creatine kinase, as well as increased prothrombin time and levels of D-Dimer (a parameter associated with disseminated intravascular coagulation). They concluded that pretreatment with MPE (21 mg/kg, one injection 24 hours before venom challenge; or three weekly injections) was able to significantly inhibit the venom-induced myotoxic and cytotoxic effects and the elevation of D-Dimer levels. The pretreatment also decreased the clotting time of mice injected with the venom. The data could be partially explained by inhibition of the venom phospholipase A_2 following MPE pretreatment, as phospholipase A_2 is known to exhibit myotoxic (Zhou *et al.*, 2008) and anticoagulant activities.

Guerranti *et al.* (2002) further investigated the immunological mechanism of the protective action of MPE against *Echis carinatus* venom. Western blot analysis indicated that two proteins of *Echis carinatus* venom, with molecular weights of 16 kDa and 25 kDa, respectively, cross-reacted with anti-MPE IgG. Electrospray ionization-mass spectrometry analysis of the immunoreactive proteins identified the 16-kDa protein to be a phospholipase A_2.

Thus, the results of these studies suggest that the protective action of MPE pretreatment involves immunological neutralization of venom phospholipase A_2 by antibodies formed as

404

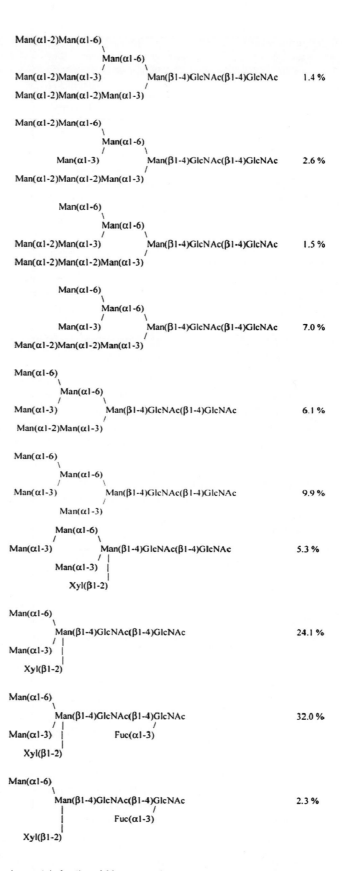

FIGURE 48.2

N-Glycans obtained from glycoprotein fraction of *Mucuna pruriens* seed extract and their relative amounts. *Source: Reproduced from Di Patrizi, L. (2006),* Glycoconjugate J., 23, *599—609, with permission.*

a result of the pretreatment. However, the nature of the 25-kDa cross-reacting protein was not established, and therefore the immunological protective action of MPE may involve neutralization of venom lethal toxins other than the phospholipase A_2.

An interesting result of this study is the demonstration that *Echis carinatus* venom proteins and the plant proteins in MPE possess common epitopes. Similar interesting phenomena of common epitopes shared by plant and snake venom proteins have also been reported (Kieliszewski *et al.*, 1994).

Further work by Guerranti *et al.* (2004) demonstrated that the immunogen in MPE that shares common epitopes with the 16-kDa protein of *Echis carinatus* venom proteins is a multiform glycoprotein, termed gpMuc. The glycoprotein represents about 16% of the total *Mucuna pruriens* extract proteins, which in turn constitute 20—28% of the total weight of the MPE. The glycoprotein was separated into seven isoforms using 2-D gel electrophoresis. The molecular weights of the isoforms ranged from 20.3 to 28.7 kDa, with pI values ranging from 4.8 to 6.5. N-terminal sequencing indicated that all seven isoforms contained the consensus sequence DDREPV-DT found in soybean Kunitz-type trypsin inhibitor. Complete removal of all glycans from gpMuc left only three protein spots in the 2-D gel, suggesting that there are only three different sequence types. The protein spots have pI values of 5.1—5.6 and molecular weight of 22 kDa. Immunoblotting studies showed that these three completely deglycosylated proteins lose their immunoreactivity with anti-gpMuc IgG, indicating that the anti-gpMuc epitopes do not reside in the protein component of the glycoprotein, but instead are found in its glycan chains.

Di Patrizi (2006) investigated the structure of the glycans of the gpMuc. Analysis of the oligosaccharides released from gpMuc by PNGase F indicated that they comprised a mixture of oligomannose-type structures and xylosylated structures; whereas the PNGase A-sensitive N-glycans contained ($\alpha 1,3$)-linked fucose. The authors concluded that the immunoreactivity of the glycoprotein is due to the presence of core ($\beta 1,2$)-linked xylose, and core ($\alpha 1,3$)-linked fucose-modified N-glycan chains (Figure 48.2).

Recently, Guerranti *et al.* (2008) reported proteomic analysis of the pathophysiological process involved in the protective effect of *Mucuna pruriens* seed pretreatment against *Echis carintus* venom in mice. Injection of the venom increased the levels of albumin, haptoglobin, fibrinogen, serum amyloid A, and serum amyloid P. These venom-induced changes are due to the venom's effects on coagulation, inflammation, and hemolysis. MPE pretreatment prevented the venom-induced elevation in haptoglobin. Overall, the results of the proteomic studies suggested that the mechanism of MPE protection involves the activation of counterbalancing processes to compensate for the toxic effects caused by *Echis carinatus* venom.

Protective action of *Mucuna pruriens* seeds against other snake venoms

Using indirect enzyme-linked immunosorbent assay, Tan *et al.* (2009) showed that there are high levels of cross-reactions between rabbit anti-MPE IgG and venoms from many elapids (*Ophiophagus hannah, Naja sputatrix, Naja nigricollis, Bungarus candidus, Notechis scutatus*, and *Pseudechis australis*) and vipers/pit vipers (*Calloselasma rhodostoma, Cryptelytrops purpureomaculatus, Bothrops asper, Agkistrodon piscivorus*, and *Daboia russelii russelii*). Pretreatment of rats with the seed extract (three weekly i.p. injections of 21 mg MPE/kg) conferred effective protection against *N.sputatrix* (Malayan cobra) venom and moderate protection against *C.rhodostoma* (Malayan pit viper) venom, but failed to protect against the minimum lethal dose of other venoms tested (Table 48.2). The pretreatment, however, substantially prolonged the survival time of the treated animals. Thus, *Mucuna pruriens* seed extract seems to possess anti-snake venom activity against a wide variety of snake venoms when injected into rats.

405

TABLE 48.2 Survival Ratio of Rats Pretreated with *Mucuna pruriens* Seed Extract and Challenged with Various Snake Venoms

Venom	Survival Ratio	
	Untreated	MPE Treated
Naja sputatrix	0/6	5/6
Calloselasma rhodostoma	0/6	2/6
Bungarus candidus	0/6	0/6[a]
Daboia russelli russelli	0/6	0/6[a]
Ophiophagus hannah	0/6	0/6[a]

After pretreatment with MPE (once a week for three subsequent weeks, at 21 mg/kg), the animals were challenged with 1.5 LD$_{50}$ of the various venoms and then monitored for 24 h.
Source: Fung (2008).
[a]*prolonged survival time.*

Protective action of *Mucuna pruriens* seed extract against cobra venom

Cobra (*Naja*) venom poisoning is characterized by neurotoxicity and cardiotoxicity. The main venom toxins include polypeptide neurotoxins, cardiotoxins, and phospholipases A$_2$ (Tan, 1991).

Tan *et al.* (2009) demonstrated that while MPE pretreatment in rats conferred effective protection against the lethality of *Naja sputatrix* venom after three weekly treatments, shorter periods of MPE pretreatment (at 24 hour, 1 week, or 2 weeks, i.p.) failed to confer effective protection against the venom. This is in contrast to the reported protective effects of MPE against *Echis carinatus* venom. Thus, the protective mechanism of MPE against snake venoms may differ from species to species.

Pre-incubation of rabbit anti-MPE serum with *Naja sputatrix* venom effectively neutralized the lethality of the venom (Tan *et al*, 2009). This is in agreement with the early report by Aguiyi *et al.* (1999), indicating that the protective action of *Mucuna pruriens* against cobra venom indeed also involves an immunological mechanism. The potency of the anti-MPE was determined to be 0.68 mg venom/ml, which is comparable to the potency of some commercial cobra antivenoms (Tan, 1983). The anti-MPE was also moderately effective in the neutralization of other *Naja* venoms (*Naja kaouthia*, *Naja nivea*). Thus, anti-MPE may be used as an alternative to cobra antivenom in treating snake venom poisoning. Western blotting analysis indicated that rabbit anti-MPE IgG cross-reacts not only with the venom phospholipases A$_2$, but also with the polypeptide neurotoxin (Fung, 2008). The result suggested that cobra venom neurotoxin also shares common epitopes with protein(s) from MPE. The nature of the cross-reactive MPE protein(s), however, has not been established.

Pharmacological studies using anesthetized rats and isolated rat atria indicated that the three injections of MPE pretreatment protected the animals against respiratory and cardiovascular (and, to a lesser extent, neuromuscular) depressant effects of *Naja sputatrix* venom (Fung, 2008). The pretreatment also significantly prevented the venom-induced drop in heart contractility and heart rate in the isolated atria. The ability of MPE pretreatment to attenuate the cardiovascular, respiratory, and neuromuscular depressant effects of the cobra venom is presumably related to the neutralization of the venom phospholipases A$_2$ and neurotoxins by the anti-MPE antibodies. Results of the isolated atria studies (where there were no circulating antibodies), however, indicated that the cardio-protective action of MPE pretreatment may be mediated by a direct action on the heart in addition to the immunological mechanism.

In a separate study, Fung *et al.* (2009) reported that the same MPE pretreatment in rats prevented the *Naja sputatrix* venom-induced histopathological changes in the heart.

Light-microscopic examination revealed that the injection of 1.5 LD_{50} of the venom (i.v.) induced disruption in the striations of the heart muscles, but the histological damage was completely prevented by the three injections of MPE pretreatment. The histopathological changes in the heart were probably caused by the combined action of venom cardiotoxins and phospholipases A_2. The venom did not cause observable histopathological damages to the other organs (brain, lung, spleen, and kidney) examined, except for a mild effect on the liver blood vessels, which was also prevented by MPE pretreatment. This study indicated that the ability of MPE pretreatment to counter the venom-induced toxic effects on heart plays an important role in the protective effects of MPE pretreatment against cobra venom poisoning, and the results are in agreement with the above-mentioned pharmacological studies. However, it is not clear from this study whether the protective effect is due to immunological neutralization of phospholipase A_2 alone, or includes the involvement of a direct, non-immunological cardio-protective action by MPE pretreatment, as previous studies showed that MPE pretreatment in rabbits did not elicit formation of cardiotoxin cross-reacting antibodies. Preliminary results from gene expression studies indicated that in MPE-pretreated rats, 49 genes in the heart were significantly up-regulated (Fung, 2008). Of these up-regulated genes, 20 were related to immune responses, and 6 were related to energy production and metabolism, and these may play a key role in maintaining the viability of the heart. Certain genes that are involved in the maintenance of homeostasis of the heart were also significantly up-regulated. These results suggested that MPE pretreatment is likely to have a direct cardio-protective effect.

Concluding remarks on the anti-snake venom properties of cowhage seed

It has been claimed that swallowing *Mucuna pruriens* seeds confers prophylactic protection against snake venom poisoning. Although the findings of the above-mentioned studies cannot be used as proof for the claimed anti-snakebite prophylactic protective effect of *Mucuna pruriens* seeds, because all the studies reported thus far involved pretreatment of the animals with intraperitoneal injection, and not oral ingestion of the seed extract, these studies nevertheless provide a possible scientific basis for the claim. Since, so far, the main established mechanism of protection is immunological, it is not certain how the oral mode of entry could trigger adequate antibody formation against the seed proteins. It would be interesting to analyze the sera samples of those who consume the seed orally for the presence of anti-*Mucuna pruriens* antibodies. Furthermore, to establish the oral prophylactic anti-snake venom effect of *Mucuna pruriens* seeds, it is necessary to investigate the protective effect of the seeds in animals fed orally with the seed meal.

407

ADVERSE EFFECTS AND REACTIONS (ALLERGIES AND TOXICITY)

The use of raw *Mucuna pruriens* seeds is often accompanied by toxic symptoms, including neurotoxicity, behavioral changes, and severe vomiting (Sathiyanarayanan & Arulmozhi, 2007).

SUMMARY POINTS

- Pretreatment of the animals by intraperitoneal injection of *Mucuna pruriens* seed extract did effectively protect the animals against the lethalities of *Echis carinatus* and *Naja sputatrix* venoms. The same treatment was moderately effective against lethalities of some other venoms.
- The protective mechanism involved an immunological mechanism. Injection of the seed extract elicited formation of antibodies that could cross-react with snake venom phospholipases A_2, polypeptide neurotoxins, and some other venom toxins. The antibodies may be used as an alternative to commercial antivenoms.

- The immunogen in *Mucuna pruriens* seeds that shares common epitopes with snake venom phospholipases A_2 is a multiform glycoprotein, and the epitopes were found to reside on the glycan chains.
- Immunological neutralization is not the only mechanism involved in the anti-snake venom properties of *Mucuna pruriens* seed extract. Survival studies, pharmacological experiments, and gene expression studies all suggest the involvement of a direct protective mechanism; in particular, cardio-protective action in the case of anti-cobra venom action.

ACKNOWLEDGEMENTS

Part of this work was supported by grants RG022/09HTM, RG084/09HTM, and RG 088/09 HTM from the University of Malaya.

References

Aguiyi, J. C., Igweh, A. C., Egesie, U. G., & Leoncini, R. (1999). Studies on possible protection against snake venom using *Mucuna pruriens* protein immunization. *Fitoterapia, 70,* 21—24.

Aguiyi, J. C., Guerranti, R., Pagani, R., & Marinello, E. (2001). Blood chemistry of rats pretreated with *Mucuna pruriens* seed aqueous extract MP101UJ after *Echis carinatus* venom challenge. *Phytotherapy Research, 15,* 712—714.

Di Patrizi, L. (2006). Structural characterization of the N-glycans of gpMuc from *Mucuna pruriens* seeds. *Glycoconjugate Journal, 23,* 599—609.

Fung, S. Y. (2008). Biomedical studies on the protection of *Mucuna pruriens* seed extract against the toxic actions of snake venom. *PhD thesis.* Kuala Lumpur, Malaysia: University of Malaya.

Fung, S. Y., Tan, N. H., Liew, S. H., Sim, S. M., & Aguiyi, J. C. (2009). The protective effects of *Mucuna pruriens* seed extract against histopathological changes induced by Malayan cobra (*Naja sputatrix*) venom in rats. *Tropical Biomedicine, 26,* 80—84.

Guerranti, R., Aguiyi, J. C., Leoncini, R., Pagani, R., Cinci, G., & Marinello, E. (1999). Characterization of the factor responsible for the antisnake activity of *Mucuna pruriens* seeds. *Journal of Preventive Medicine and Hygiene, 40,* 25—28.

Guerranti, R., Aguiyi, J. C., Errico, E., Pagani, R., & Marinello, E. (2001). Effects of *Mucuna pruriens* extract on activation of prothrombin by *Echis carinatus* venom. *Journal of Ethnopharmacology, 75,* 175—180.

Guerranti, R., Aguiyi, J. C., Neri, S., Leoncini, R., Pagani, R., & Marinello, E. (2002). Proteins from *Mucuna pruriens* and enzymes from *Echis carinatus* venom. Characterization and cross-reactions. *Journal of Biological Chemistry, 277,* 17072—17078.

Guerranti, R., Aguiyi, J. C., Ogueli, I. G., Onorati, G., Neri, S., Rosati, F., et al. (2004). Protection of *Mucuna pruriens* seeds against *Echis carinatus* venom is exerted through a multiform glycoprotein whose oligosaccharide chains are functional in this role. *Biochemical and Biophysical Research Communications, 323,* 484—490.

Guerranti, R., Ogueli, I. G., Bertocci, E., Muzzi, C., Aguiyi, J. C., Cianti, R., et al. (2008). Proteomic analysis of the pathophysiological process involved in the anti-snake venom effect of *Mucuna pruriens* extract. *Proteomics, 8,* 402—412.

Kieliszewski, M. J., Showalter, A. M., & Leykam, J. F. (1994). A modulator protein sharing sequence similarities with the extension family, the hevein lectin family, and snake venom disintegrins (platelet aggregation inhibitors). *Plant Journal, 5,* 849—861, Potato lectin.

Patrizi, L. D., Rosati, F., Guerranti, R., Pagani, R., Gerwig, G. J., & Kamerling, J. P. (2006). Structural characterization of the N-glycans of gpMuc from *Mucuna pruriens* seeds. *Glycoconjugate Journal, 23,* 599—609.

Sathiyanarayanan, L., & Arulmozhi, S. (2007). *Mucuna pruriens* Linn. — A comprehensive review. *Pharmacognosy Review, 1,* 157—162.

Tan, N. H. (1983). Improvement of Malayan cobra antivenom. *Toxicon, 21,* 75—79.

Tan, N. H. (1991). The biochemistry of venoms of some venomous snakes of Malaysia — a review. *Tropical Biomedicine, 8,* 91—103.

Tan, N. H., Fung, S. Y., Sim, S. M., Marinello, E., Guerranti, R., & Aguiyi, A. (2009). The protective effect of *Mucuna pruriens* seeds against snake venom poisoning. *Journal of Ethnopharmacology, 123,* 356—358.

Warrell, D. A., Davidson, N., Greenwood, B. M., Ormeror, L. D., Pope, H. M., Watkins, B. J., et al. (1977). Poisoning by bites of the saw-scaled or carpet viper (*Echis carinatus*) in Nigeria. *Quarterly Journal of Medicine, 46,* 33—62.

Yamada, D., Sekiya, F., & Morita, T. (1996). Isolation and characterization of carinactivase, a novel prothrombin activator in *Echis carinatus* venom with a unique catalytic mechanism. *Journal of Biological Chemistry, 271,* 5200—5207.

Zhou, X., Tan, T. C., Valiyaveettil, S., Go, M. L., Kini, R. M., Velazquez-Campo, A., et al. (2008). Structural characterization of myotoxic ecarpholin S from *Echis carinatus* venom. *Biophysical Journal, 95,* 3366—3380.

CHAPTER 49

Crabs Eye (*Abrus precatorius*) Seed and Its Immunomodulatory and Antitumor Properties

Sujit K. Bhutia[1], Tapas K. Maiti[2]
[1] Department of Life Science, National Institute of Technology, Rourkela, Orissa, India
[2] Department of Biotechnology, Indian Institute of Technology, Kharagpur, India

LIST OF ABBREVIATIONS

ABP, abrin-derived peptide

ABR, abrin

ADCC, antibody dependent cellular cytotoxicity

AGG, agglutinin

AGP, agglutinin derived peptide

DL, Dalton's lymphoma

EAC, Ehrlich's ascites carcinoma

ER, endoplasmic reticulum

IL-2, interleukin-2

IFN-γ, interferon-γ

NK, natural killer

RIP-II, ribosome inactivating proteins-II

TAM, tumor associated macrophage

TDA, tryptic digest of Abrus agglutinin

TNF-α, tumor necrosis factor-α

Nuts & Seeds in Health and Disease Prevention. DOI: 10.1016/B978-0-12-375688-6.10049-0

INTRODUCTION

Seeds of the Jequiriti bean (*Abrus precatorius* L.) have long been known for their medicinal use in Unani and Ayurvedic medicine. *Abrus precatorius* is a vine native to India and other tropical and subtropical areas of the world, and is known by a variety of names, including jequirty bean, rosary pea, crab's eye, and love bean. Owing to their spectacular appearance, the seeds are often used for jewelry, beadwork, and ornaments (Chopra *et al.*, 1956). While all parts of the plant are toxic, the highest concentrations of cytotoxic compounds are found in the seeds. Along with smaller concentrations of glycyrrhizin, aric acid, and *N*-methyltryptophan, *Abrus* seeds contain the toxic lectins, namely abrin (ABR) and the relatively less toxic agglutinin commonly known as *Abrus* agglutinin (AGG). Abrin is a 63-kDa heterodimeric glycoprotein, whereas agglutinin is a heterotetrameric glycoprotein with a molecular weight of 134 kDa. Both of these lectins belong to the ribosome inactivating proteins-II (RIP-II) family, and consist of a toxic subunit A chain (molecular weight 30 kDa) and a galactose-binding B subunit (molecular weight 31 kDa) connected by a single disulfide bond (Olsnes *et al.*, 1974).

BOTANICAL DESCRIPTION

As a member of the leguminosae, *Abrus precatorius* is a small, high-climbing perennial vine with alternately compound leaves, 5—12 cm long, with 5—15 pairs of oblong leaflets. A key characteristic in identifying the rosary pea is the lack of a terminal leaflet on the compound leaves. The flowers are small, pale, and violet to pink, clustered in leaf axils. The fruit is characteristic of a legume. The pod is oblong, flat, and truncate, roughly 3—5 cm long, and silky-textured. Each pod generally contains three to five oval-shaped seeds. This seed pod curls back when it opens, revealing the seeds. Seeds are usually bright scarlet, and have a jet-black spot surrounding the hilum, which is the point of attachment. The seed coat, or testa, is smooth and glossy, and becomes hard when the seed matures (Figure 49.1).

410

FIGURE 49.1

Seeds of *Abrus precatorius*.

HISTORICAL CULTIVATION AND USAGE

Abrus precatorius plants are seen growing wild throughout all tropical forests, and are propagated through seeds. Traditionally the seeds were used for decorative and gold-weighing purposes. The plant is used in traditional herbal formulations to treat many ailments, mainly scratches, sores, and wounds caused by dogs, cats, and mice. In addition, it is also used to treat leucoderma, tetanus, and rabies (Chopra *et al.*, 1956). The dry seeds are powdered and taken one teaspoonful once a day for 2 days to cure worm infections. Various African and Indian tribes use the powdered seeds as oral contraceptives. They have also been used against chronic eye diseases, and particularly against trachoma (Acharya, 2004). As well as the seeds, the leaves and roots of the plants are also used for medicinal purposes.

PRESENT-DAY CULTIVATION AND USAGE

Wild *Abrus precatorious* plants are naturalized and cultivated in different parts of the world, including northern Australia, South-east Asia, tropical Asia, and the Western Pacific. Presently, it is widely cultivated as a garden ornamental because of its attractive and decorative bright red and black colored seeds.

The safety and effectiveness of the traditional uses of *Abrus precatorious*-derived products have not yet been well explored scientifically. Investigations are ongoing to fully characterize the anti-inflammatory, antiallergic, immunomodulatory, and anticancer properties of *Abrus* seed derived products.

APPLICATIONS TO HEALTH PROMOTION AND DISEASE PREVENTION

The *in vitro* growth inhibition potential of *Abrus* lectins has been studied, and results show that the lectins inhibit the major protein synthesis pathway, which leads to cell death. Lectins (AGG and ABR) bind to the terminal galactose residue of cell surface receptors through the B subunit, enter the cells by receptor mediated endocytosis, and are finally transported to the endoplasmic reticulum (ER) by the retrograde pathway. The reduction of the A—B inter subunit disulfide bond, essential for the cytotoxicity that takes place in the ER, followed by translocation of the A chain to the cytosol A chain cleaves a specific adenine residue from the α-sarcin—ricin loop of the 28S rRNA in the cytosol of eukaryotes. The depurination prevents the binding of ribosome to elongation factors, thereby arresting the protein synthesis that causes cell death. The protein synthesis inhibitory activity of AGG ($IC_{50} = 3.5$ nM) is weaker than that of ABR (0.05 nM). As a consequence ABR is found to be extremely toxic, with an LD_{50} of 20 µg/kg of body weight, whereas the LD_{50} of the AGG is comparatively higher (5 mg/kg of body weight), implying lesser toxicity. Both the lectins trigger apoptosis by caspase-3 activation, and follow the intrinsic mitochondrial pathway involving potential damage to the the mitochondrial membrane, and reactive oxygen species production (Narayanan *et al.*, 2004). Moreover, *in vitro* growth inhibitory properties of *Abrus* lectins have been evaluated in several tumor cells. For instance, ABR could bring about cell death of Dalton's lymphoma (DL), as is evident from typical morphological changes associated with apoptosis. However, necrotic cell death is dominant when a higher dose of ABR is used. ABR induces apoptosis by stimulating the expression of pro-apoptotic caspase-3, at the same time blocking the expression of Bcl-2 (Ramnath *et al.*, 2007). Interestingly, AGG in a heat-denatured condition (prepared by maintaining it at 50°C for 30 minutes and then immediately at 100°C for 2 minutes in a water bath) exerts similar types of cellular growth inhibition potential as demonstrated in DL cells (Ghosh & Maiti, 2007).

Antitumor properties

An agent that can selectively induce cell death in tumor cells without affecting the normal cell population is considered to be an ideal candidate for cancer therapy. The selective tumor-targeting nature of *Abrus* lectins was investigated in normal and tumor cells, and it was observed that ABR showed greater cytoagglutination against human cultured cell lines derived from acute lymphoblastic leukemia and adult T cell leukemia, and weak agglutination against normal lymphocytes (Kaufman & McPherson, 1975). Likewise, AGG-mediated human lymphoproliferative activity *in vitro* was also observed (Closs *et al.*, 1975). Though the exact mechanisms are not known, this tumor-specificity of *Abrus* lectins may be investigated to develop a novel anticancer agent (Figure 49.2).

The first reports on the anticancer properties of *Abrus* lectins appeared before much was known about their mechanism of action. In the late 1960s and early 1980s, the effect of *Abrus* protein extract on growth inhibition of Yoshida sarcoma and Ehrlich ascites tumors in mice were investigated (Reddy & Sirsi, 1969; Lin *et al.*, 1982). In the following years, the anticancer effect of the ABR was investigated in different experimental models, including human tumors grown in nude mice (Fodstad *et al.*, 1977). Similarly, in another study, ABR showed a potent anti-tumor potential in a sarcoma mouse model, which is thought to be mediated by inhibition of DNA biosynthesis. These results were sufficiently promising to initiate a phase I clinical study of ABR. Though clinical study was found to be promising, unexpected observation in a few patients became an obstacle for further clinical investigation.

For the next couple of years, with improved protein separation techniques, a few papers appeared on *Abrus* lectin purification and crystallization, and the mechanism of intoxication. After the discovery of the detailed mechanism of cell growth inhibition, lectins again attracted the attention of the scientific community. Meanwhile, investigators came up with the idea that the toxins might be useful in treating cancer. ABR exhibits antitumor activity in mice at a sub-lethal dose and is able to decrease solid tumor mass development, as is evident from the study related to murine DL and the Ehrlich's ascites carcinoma (EAC) model (Ramnath *et al.*, 2002). ABR with interperitonial administration increased the lifespan of ascites tumor-bearing mice, and showed maximum antitumor effect when used simultaneously with tumor cells. Such a study showed that prophylactic administration of ABR was ineffective. Similarly, the antitumor activity of AGG was evaluated in a murine DL ascites tumorogenic model. It was observed that both pre- and post-treatment of agglutinin at

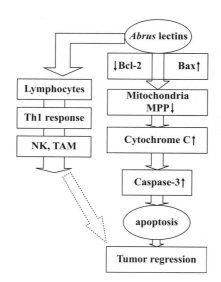

FIGURE 49.2

Mechanism of action of Abrus lectins on tumor regression. *Abrus* lectins contribute to tumor inhibition by direct killing of tumor cells through mitochondrial-dependent pathways. Further, lectins are able to stimulate tumor associated macrophages and proliferation of splenocytes, leading to Th1 response and NK cell activation in tumor-bearing mice, which involves an antitumor response.

a non-toxic dose significantly decreased the lymphoma cell numbers in DL-bearing mice. Moreover, heat-denatured AGG, like native AGG, was able to decrease the DL tumor cell number *in vivo*, and significantly increased the median survival time of DL bearing mice (Ghosh & Maiti, 2007). The *in vivo* tumoricidal property of lectins against DL and EAC-induced tumors was thought to be mediated through apoptosis (Ghosh & Maiti, 2007; Ramnath *et al.*, 2007).

Immunostimulatory properties

Along with direct antitumor potential, these lectins have a functional role in tumor defense by immunomodulation. The mitogenic properties of ABR and AGG were studied in human as well as mice immune cells, and both lectins showed significant adjuvant activities in oil emulsion and aqueous solution in animal models. It has been demonstrated that heat-denatured AGG in oil emulsion produces a humoral immune response comparable with that of traditional complete Freund's adjuvant against ovalbumin in a rat model (Suryakala *et al.*, 2000). Immunoadjuvant activity of AGG and ABR were investigated in aqueous solution with ovalbumin, diphtheria toxoid, and lysozyme as test antigens in mouse model, and induction of humoral immunity by native and heat-denatured AGG were comparable with that of Freund's adjuvant (Tripathi & Maiti, 2003). All these findings indicate that *Abrus* lectins may be used as immunoadjuvants in aqueous solution for raising a high antibody titer with a high avidity for weak antigens, thus potentiating the systemic immune response.

Lectins are known as polyclonal activators of lymphocytes, and work through the induction of a battery of cytokines. Treatment of AGG increases the expression of activation markers (CD25, CD71) in B and T cells, which implies that the mitogenic stimulus of agglutinin influences not only proliferation but also the activation status of splenocytes (Ghosh *et al.*, 2009). Both native and heat-denatured AGG activate splenocytes and induce the production of cytokines like interleukin-2 (IL-2), interferon-γ (IFN-γ), and tumor necrosis factor-α (TNF-α) indicating a Th1 type of immune response (Tripathi & Maiti, 2005). The natural killer (NK) cells, innate effector cells, are also activated in the presence of AGG. Furthermore, AGG stimulates the innate effector arms, like macrophages, by up-regulating pro-inflammatory cytokine expression. Increased production of nitric oxide and hydrogen peroxide, and a high phagocytic and bactericidal activity of macrophages, are potentiated by native as well as heat-denatured AGG. These results suggest that both forms of AGG can act as immunostimulants *in vitro*. *In vivo* administration of native and heat-denatured AGG showed macrophage and NK cell activation (Ghosh & Maiti, 2007). Similarly, ABR also augments the humoral and cell-mediated immune response of the host. A non-toxic dose of ABR, (1.25 µg/kg body weight) for 5 consecutive days of treatment in normal mice stimulates specific humoral responses, and a significant increase is observed in the total leukocyte count, lymphocytosis, the weights of the spleen and thymus, circulating antibody titer, antibody forming cells, bone marrow cellularity, and α-esterase positive bone marrow cell.

The host response to cancer therapies might be potentiated by the simultaneous administration of these type-1 immunoadjuvant *Abrus* lectins, along with cancer drugs. Splenocytes from native and heat-denatured AGG-treated DL-bearing mice exhibited the Th1 type of cytokine (IL-2 and IFN-γ) response, which could generate a robust antitumor immune response (Ghosh & Maiti, 2007). AGG, in its native and heat-denatured forms, modulates tumor associated macrophage (TAM) by increasing *in vitro* cytotoxicity towards tumor cells, and production of nitric oxide. Similarly, ABR augments antitumor immunity in tumor-bearing hosts by boosting the cell-mediated immune effector arms, such as NK cells of tumor-bearing hosts. Antibody-dependent cellular cytotoxicity (ADCC), which links humoral and cell-mediated immunity, was enhanced in the ABR-treated tumor-bearing mice. Early antibody-dependent complement-mediated cytotoxicity was also observed in the ABR-treated mice. Immunization with ABR-treated Meth-A tumor cells induces a strong antitumor immunity in

413

syngeneic BALB/c mice, and such an immunizing effect is stronger than that produced by an irradiated Meth-A tumor cell vaccine (Shionoya *et al.*, 1982). Studies on the adjuvant effects of ABR in a tumor condition indicate its potential use as an immunoadjuvant.

Many proteins are degraded or processed into smaller peptide products through natural mechanisms, giving rise to a newly recognized natural "peptidome," which potentially contains a valuable source of anticancer agents. *Abrus* lectin derived peptides represent a prime example of anticancer peptide segments encrypted within lectins. In an initial study, tryptic digestion of *Abrus* agglutinin stimulates macrophages, as is evident from the increasing phagocytic and bactericidal activity, as well as hydrogen peroxide production. It also proliferates splenocytes, leading to Th1 response and NK cell activation. Interestingly, tryptic-digested AGG and ABR derived peptides, named AGP and ABP respectively (obtained from 10-kDa molecular weight cut-off membrane permeate), molecular weight in the range of 500 Da to 1500 Da, induce mitochondria-dependent cell death. Sub-lethal doses of peptide fractions showed significant growth inhibitory properties in an *in vivo* DL model, which was found to be mediated through apoptosis (Bhutia *et al.*, 2009a). Along with direct antitumor potential, these peptides activate *in vitro* immune cells such as macrophages, NK cells, and B and T cells in the tumor microenvironment, which, in a concerted way, inhibits tumor progression (Bhutia *et al.*, 2009b).

ADVERSE EFFECTS AND REACTIONS (ALLERGIES AND TOXICITY)

At the cellular level, both lectins inhibit protein synthesis, thereby causing cell death. Oral ingestion of whole seeds often does not produce serious illness, since the shell protects the lectins from digestion. If the seeds are crushed and then ingested, more serious gastrointestinal toxicity, including death, can occur (Reedman *et al.*, 2008). *Abrus*-seed poisoning causes endothelial cell damage, and an increase in capillary permeability, with consequent fluid and protein leakage and tissue edema. Although the extract of jequirity bean has been used therapeutically, it also causes severe inflammation of the eye, including permanent damage to the cornea, and occasionally blindness. ABR is highly toxic by all routes of exposure (Dickers *et al.*, 2003), whereas the adverse effect of agglutinin has not yet been reported.

Management of *Abrus* lectin poisoning is symptomatic and supportive. Investigations showed that vaccination with abrin toxoid or administration of ascorbate may offer some protection against a subsequent abrin challenge (Dickers *et al.*, 2003). The sub-lethal doses of lectins showed potent antitumor and immunostumulatory responses, and the therapeutic effect of *Abrus* lectins can be obtained by masking the undesired toxic effect of lectins. Clinical trials showed that the human minimum lethal dose of ABR by intravenous injection is 0.3 µg/kg as an immunotoxin for cancer treatment. The report showed that patients tolerated a dose of 0.3 µg/kg without serious adverse effects (Gill, 1982).

SUMMARY POINTS

- The seeds of the Jequiriti bean (*Abrus precatorius* L.) have long been known for their medicinal use in Unani and Ayurvedic medicine.
- The selective tumor-targeting nature of *Abrus* lectins ensures its place as a potential anticancer agent, and both lectins (agglutinin and abrin) inhibit the growth of tumors in experimental animals through apoptosis induction.
- The immunoadjuvant phenomena of *Abrus* agglutinin and abrin provide a challenge for potentiating the systemic immune response.
- Native and heat-denatured *Abrus* agglutinin stimulates the Th1 type immune response by up-regulating IL-2, IFN-γ, and TNF-α production.
- The cell-mediated immune responses of agglutinin and abrin are highly relevant in the tumor cell destruction in the host body.

- *Abrus* lectin (agglutinin and abrin)-derived peptides showed potent *in vitro* and *in vivo* antitumor and immunostimulatory properties, which may be used further as a potential source of tumor-targeting therapeutic peptides.
- A detailed mechanistic study of *Abrus* lectins mediated cancer regression in advanced malignant disease condition is required to evaluate the potential of *Abrus* lectins as alternative chemotherapeutic agents.

References

Acharya, D. (2004). Medicinal plants for curing common ailments in India. *Positive Health, 102*, 28–30.

Bhutia, S. K., Mallick, S. K., Maiti, S., & Maiti, T. K. (2009a). Inhibitory effect of *Abrus* abrin-derived peptide fraction against Dalton's lymphoma ascites model. *Phytomedicine, 16*, 377–385.

Bhutia, S. K., Mallick, S. K., & Maiti, T. K. (2009b). *In vitro* immunostimulatory properties of *Abrus* lectins derived peptides in tumor bearing mice. *Phytomedicine, 16*, 776–782.

Chopra, R. N., Nayar, S. L., & Chopra, I. C. (1956). *Glossary of Indian medicinal plants*. New Delhi, India: CSIR.

Closs, O., Saltvedt, E., & Olsnes, S. (1975). Stimulation of human lymphocytes by galactose-specific *Abrus* and *ricinus* lectins. *Journal of Immunology, 115*, 1045–1048.

Dickers, K. J., Bradberry, S. M., Rice, P., Griffiths, G. D., & Vale, J. A. (2003). Abrin poisoning. *Toxicology Review, 22*, 137–142.

Fodstad, O., Olsnes, S., & Pihl, A. (1977). Inhibitory effect of abrin and ricin on the growth of transplantable murine tumors and of abrin on human cancers in nude mice. *Cancer Research, 37*, 4559–4567.

Ghosh, D., & Maiti, T. K. (2007). Immunomodulatory and anti-tumor activities of native and heat denatured *Abrus* agglutinin. *Immunobiology, 212*, 589–599.

Ghosh, D., Bhutia, S. K., Mallick, S. K., Banerjee, I., & Maiti, T. K. (2009). Stimulation of murine B and T lymphocytes by native and heat-denatured *Abrus* agglutinin. *Immunobiology, 214*, 227–234.

Gill, D. M. (1982). Bacterial toxins: a table of lethal amounts. *Microbiological Reviews, 46*, 86–94.

Kaufman, S. J., & McPherson, A. (1975). Abrin and hurin: two new lymphocyte mitogens. *Cell, 4*, 263–268.

Lin, J. Y., Lee, T. C., & Tung, T. C. (1982). Inhibitory effects of four isoabrins on the growth of sarcoma 180 cells. *Cancer Research, 42*, 276–279.

Narayanan, S., Surolia, A., & Karande, A. A. (2004). Ribosome-inactivating protein and apoptosis: abrin causes cell death via mitochondrial pathway in Jurkat cells. *Biochemical Journal, 377*, 233–240.

Olsnes, S., Saltvedt, E., & Pihl, A. (1974). Isolation and comparison of galactose-binding lectins from *Abrus precatorius* and *Ricinus communis*. *Journal of Biological Chemistry, 249*, 803–810.

Ramnath, V., Kuttan, G., & Kuttan, R. (2002). Antitumour effect of abrin on transplanted tumours in mice. *Indian Journal of Physiology and Pharmacology, 46*, 69–77.

Ramnath, V., Rekha, P. S., Kuttan, G., & Kuttan, R. (2007). Regulation of caspase-3 and Bcl-2 expression in Dalton's lymphoma ascites cells by abrin. *Evidence-based Complementary and Alternative Medicine, 4*, 1–6.

Reddy, V. V. S., & Sirsi, M. (1969). Effect of *abrus precatorius* L. on experimental tumors. *Cancer Research, 29*, 1447–1451.

Reedman, L., Shih, R. D., & Hung, O. (2008). Survival after an intentional ingestion of crushed *Abrus* seeds. *Western Journal of Emergency Medicine, 9*, 157–159.

Shionoya, H., Arai, H., Koyanagi, N., Ohtake, S., Kobayashi, H., Kodama, T., et al. (1982). Induction of antitumor immunity by tumor cells treated with abrin. *Cancer Research, 42*, 2872–2876.

Suryakala, S., Maiti, T. K., Sujatha, N., & Sashidhar, R. B. (2000). Identification of a novel protein adjuvant isolated from *Abrus precatorius*. *Food and Agricultural Immunology, 12*, 87–96.

Tripathi, S., & Maiti, T. K. (2003). Efficiency of heat denatured lectins from *Abrus precatorius* as immunoadjuvants. *Food and Agricultural Immunology, 15*, 279–287.

Tripathi, S., & Maiti, T. K. (2005). Immunomodulatory role of native and heat denatured agglutinin from *Abrus precatorius*. *International Journal of Biochemistry & Cell Biology, 37*, 451–462.

415

CHAPTER 50

Cumin (*Cuminum cyminum* L.) Seed Volatile Oil: Chemistry and Role in Health and Disease Prevention

Kanagal Sahana, Subban Nagarajan, Lingamallu Jagan Mohan Rao
Plantation Products, Spices and Flavour Technology Department, Central Food Technological Research Institute, Mysore, India

CHAPTER OUTLINE

LIST OF ABBREVIATIONS

ASTA, American Spice Trade Association
DPPH -1, 1diphenyl-2-picrylhydrazyl
GC, gas chromatography
GC-MS, gas chromatography and mass spectrometry
GRAS, generally regarded as safe
HD, hydrodistillation
MWE, microwave extraction
OS-SD, combination of organic solvent and steam distillation
SCFE, supercritical fluid extraction

Nuts & Seeds in Health and Disease Prevention. DOI: 10.1016/B978-0-12-375688-6.10050-7

SD, steam distillation
SFME, solvent-free microwave extraction
SIM, selective ion monitoring
SIRA, stable isotope ratio analysis
SWE, superheated water extraction
Sx Ex, Soxhlet extraction

INTRODUCTION

Spices are the products of agriculture, mainly used for flavoring, and basically contribute taste and aroma to foods. Processing of spices and the development of post-harvest technologies results in value-addition to raw materials, and better utilization, with enhanced bioactive potentials of end products. Spices produce biologically significant organic compounds, known as secondary metabolites, which include the volatile and non-volatile constituents. These constituents form the characteristic nature of the spice, and possess medicinal and pharmacological properties with a possible impact on human health. India is known as the "home" of spices, and is also a leading producer of major spices (Wealth of India, 2001). Among the spice-growing countries, such as India, Sri Lanka, Indonesia, and Malaysia, these are used extensively as natural food flavorings. In developed countries in the West, where processed foods are consumed in large quantities, 80–90% of spices are used by the industrial sector, with the exception of pepper. In addition to flavoring, the spices help in protecting food from oxidative deterioration, thereby increasing shelf-life, and also play a role in the body's defences against cardiovascular diseases, certain cancers, and conditions such as arthritis and asthma. The spice oils are made up of terpenes, sesquiterpenes, and their oxygenated derivatives, like alcohol, aldehydes, ketones, ethers, esters, etc. The volatile oil content of different spices varies widely; it may vary from 0.1 to 18%, on moisture-free basis (Parthasarathy et al., 2008).

418

The trade in spices

The global trade in spices has been reported to be 1.547 million tonnes, valued at US$3 billion, and is expected to increase with growing consumer demand in importing countries for more exotic ethnic tastes in food (Parthasarathy et al, 2008). India has about 63 spices as edible items, and is also a major exporter to the Western world. Extensive cultivation of these spices is undertaken for use in food and pharmaceutical applications.

BOTANICAL DESCRIPTION

Cumin (*Cuminum cyminum* L.) is an aromatic herb (Figure 50.1) of the Apiaceae family, and its dried seeds are used as a spice (Figure 50.2). In India it is commonly known as cumin or zeera, and is called kummel, comino, zirech-e sabz, cumino, kemon, zira, and kamun in various other parts of the world. It is native to India, Iran, the Mediterranean, and Egypt. Cumin is a mixture of united and separated mericarps, yellowish green/brown in color, elongated ovoid, and 3–6 mm in length. The surface has five primary ridges alternating with four less distinct secondary ridges bearing numerous short hairs. Some seeds have a short stalk (Peter, 2003).

HISTORICAL CULTIVATION AND USAGE

Cumin seed has been used as a spice since ancient times, and is mentioned in the Bible. It is native to the eastern Mediterranean, extending to East India. Cumin seeds excavated at a Syrian site have been dated to the second millennium BC, and seeds have also been reported from several New Kingdom levels of ancient Egyptian archaeological sites. Originally cultivated in Iran and the Mediterranean region, cumin was also known in ancient Greece and Rome. It was introduced to America by Spanish and Portuguese colonists.

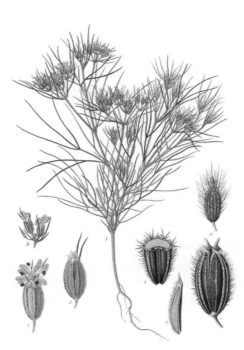

FIGURE 50.1
Aerial parts of *Cuminum cyminum* plant.

Cumin seeds have long been used for their unique aroma, and are popular in North African, Middle Eastern, Western, Chinese, Indian, and Mexican cuisine.

PRESENT-DAY CULTIVATION AND USAGE

Cumin is a very popular spice in Western to Central Asia, Central and South America, Burma, India, and Indonesia. Currently, it is grown in Iran, Uzbekistan, Tajikistan, Turkey, Morocco, Egypt, India, Syria, Mexico, and Chile. India is the largest producer and consumer of cumin, with an annual production ranging between 1.5 and 2.0 lakh tonnes.

Cumin is usually grown domestically, in the garden, from seed sown in the spring. It needs highly organic, well-drained, fertile, sandy and loamy soil. The crop needs well-regulated irrigation from the time of sowing to fruiting. The plants prefer a mild climate, and can be cultivated up to an elevation of 3800 m. The seeds are harvested by hand when they turn brown in color, indicating that the fruits are ripe. Several biotic stresses limit the production of cumin. Fungal toxins isolated from seeds of plants infected by Fusarium oxysporum lead to extensive damage. The crop is occasionally attacked by mildew (Oidiopsis taurica (Lev.)

FIGURE 50.2
Seed of cumin.

Salmon), which dries out the plants and blackens them. An effective control measure, such as sulfur dusting, is generally followed. Leaf-eating caterpillars and rootworms have occasionally been observed (Wealth of India, 2001).

Proximate composition and specification

Cumin seeds contain up to 14.5% lipids (Wealth of India, 2001). The ASTA Standard specification for cumin seeds indicates the contents to be ash 9.5%, acid-insoluble ash 1.5%, volatile oil 2.5%, and moisture 9%. The proximate composition shows: water 6.0 g, food energy 460 kcal, protein 18.0 g, fat 23.8 g, carbohydrates 44.6 g, ash 7.7 g, calcium 0.9 g, phosphorus 450 mg, sodium 160 mg, potassium 2100 mg, iron 47.8 mg, thiamine 0.730 mg, riboflavin 0.380 mg, niacin 2.5 mg, ascorbic acid 17 mg, and vitamin A activity 127 retinal equivalents, respectively, per 100 g (Donna & Antony, 1993).

Sensory quality and storage

Freshly ground cumin seeds have more aroma and flavor than pre-ground seeds. The flavor quality of volatile oil stored for 1 year deteriorated slightly. The quantity of cuminaldehyde decreased, while terpinene, linalyl acetate, cuminyl alcohol, β-pinene, and *p*-cymene content increased (Donna & Antony, 1993). Aluminum and polythene pouches packed under conditions of 37°C and 70% relative humidity are ideal for preservation (Peter, 2003).

APPLICATIONS TO HEALTH PROMOTION AND DISEASE PREVENTION

Cumin seeds possess a strong aromatic odor, warm and spicy, with a mild, bitter taste. Ground cumin is used in curries, chilli powder, to flavor soups and meats such as chicken, turkey, lamb and pork, and for seasoning bread and cakes. It is even added to olive oil when frying vegetables. Whole cumin is used in the preparation of pickles (Peter, 2003), in barbecue sauces and in snack foods, as well as in the food industry (Donna & Antony, 1993). Indian cumin is used extensively in foods, beverages, liquors, medicines, toiletries, and perfumery. In The Netherlands and France, cumin is appreciated for its culinary and therapeutic properties, and particularly for flavoring cheese.

Composition of volatile oil

The essential oil composition of cumin depends on many factors, such as the stage of maturity of the seeds from which oil is extracted, the method of extraction, type of cultivars, geographical origin, and storage conditions (Leopold *et al.*, 2005).

With steam distillation, cumin seeds yield 2.5–4.5% of volatile oil, which is colorless or pale yellow, turning dark during storage under ambient condition. The yield depends upon the quality and age of the seeds — the older seeds contain less oil. It has been reported that quantitative extraction of cumin oil using liquid carbon dioxide (4.5% for 2 hours at 58 bar pressure and 20°C) is faster than steam distillation (2.5% in 4 hours), without the loss of active flavor ingredients (Naik *et al.*, 1989).

The volatile oil from Mexico possesses 62.7% of cuminaldehyde, whereas samples from the Mediterranean and India showed 47.4 and 43%, respectively. Samples from Iran and Pakistan showed about 32.4 and 20% of cuminaldehyde, with significant differences in other physicochemical properties (Peter, 2003; Sushmita *et al.*, 2004). The characteristic odor of cumin is mainly attributed to the aldehyde present in the seeds — namely, cuminaldehyde (Figure 50.3), which forms nearly 20–40% of the oil. Other compounds (Figure 50.4) include γ-terpinene, sabinene, *p*-cymene, *a*-phellandrene, *a*-cadinene, myrcene, perilla alcohol, nerol, geraniol, citronellol, safranal, germacrene, β-bisabolene, *p*-mentha-1, 3-dien-7-al, and *p*-mentha-1,

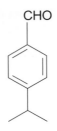

FIGURE 50.3
Cumin aldehyde (major constituent).

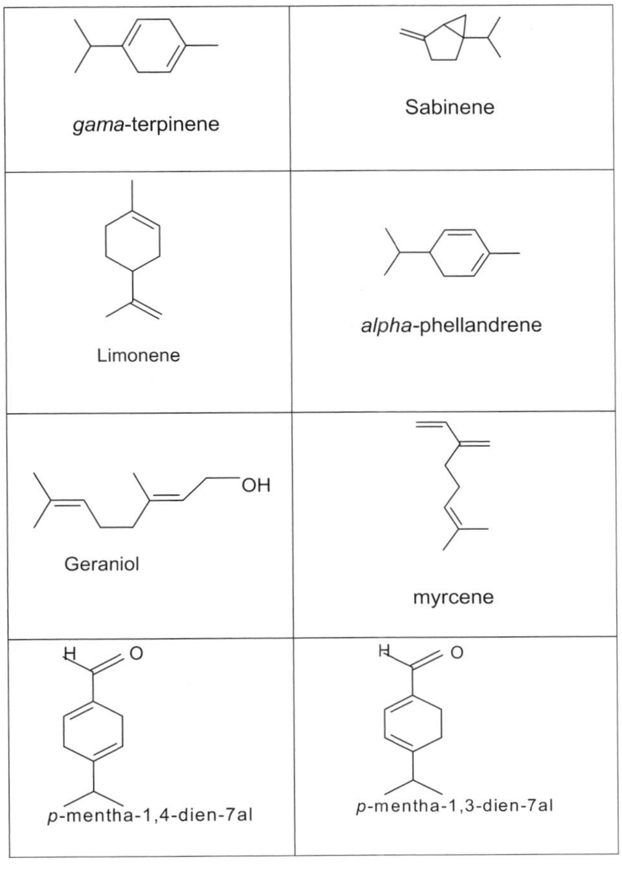

gama-terpinene	Sabinene
Limonene	*alpha*-phellandrene
Geraniol	myrcene
p-mentha-1,4-dien-7al	*p*-mentha-1,3-dien-7al

FIGURE 50.4
Other major compounds of cumin seed volatile oil.

4-dien-7-al. Prescribed specifications of the volatile oil are that it is a colorless or pale yellow oil, of specific gravity (25°C) 0.905–0.925, specific rotation (20°C) +3 to +8, refractive index 1.501–1.506, and cuminaldehyde content 40–52% (Parthasarathy *et al.*, 2008).

The chemical composition of the essential oil of cumin seeds has been isolated using various methods, such as SCFE, OS-SD, HD, SD, SWE, Sx Ex, and MWE (Table 50.1). Reports on volatile oil extraction by a combination technology of organic solvent with low boiling point (diethyl ether, dichloromethane and *n*-pentane) and steam distillation (OS-SD) was compared with the conventional method of hydrodistillation (HD) and supercritical fluid extraction (SCFE) using carbon dioxide. The results show that the three variables chosen, namely extraction time, temperature, and particle size, have a positive influence on the yield of oleoresin using the extraction technology by organic solvent with low boiling point. The

TABLE 50.1 The Chemical Composition of Essential Oil from *C. cyminum* Seed

S. No.	Compound	SCFE[a]	OS-SD[b]	HD[c]	SD[d]	SWE[e]	Sx Ex[f]	MWE[g]
1	2-Methyl-5-(1-methylethyl)-bicyclo [3.1.0] hex-2-ene	0.15	0.14	0.12	—	—	—	—
2	Cyclohexane, 1-methylethylideneyl	—	0.02	0.02	—	—	—	—
3	α-Pinene	0.40	0.34	0.42	29.1	—	—	0.09
4	β-Pinene	—	—	—	—	0.05	1.34	6.21
5	Camphene	0.25	0.24	0.21	—	—	—	—
6	Phellandrene	0.17	0.33	—	—	t	0.05	—
7	Sabinene	2.47	2.36	2.22	0.60	nd	0.06	0.28
8	Myrcene	0.27	—	—	0.20	t	0.12	—
9	3-Carene	0.09	0.04	0.07	0.20	—	—	—
10	1-Methyl-2-(1-methylethyl) benzene	3.05	2.98	2.69	—	—	—	—
11	Limonene	0.15	0.14	0.12	21.5	0.03	0.01	—
12	1,8-Cineole	0.25	0.25	0.43	—	—	—	—
13	1-Methyl-4-(1-methylethyl)-1,4-cyclohexadiene	6.12	5.93	5.87	—	—	—	—
14	γ-terpinene	17.13	17.12	17.07	—	—	—	3.98
15	β-Terpineol	0.36	0.45	0.33	—	—	—	—
16	γ-Terpinol	—	—	—	—	0.17	0.01	—
17	L-Fenchone	0.54	0.62	0.58	—	—	—	—
18	Linalool	—	—	0.11	10.40	—	—	—
19	6,6-Dimethyl-2-methylenebicyclo [2.2.1]heptan-3-one	0.06	0.12	0.09	—	—	—	—
20	Pinocarveol	0.55	0.61	0.59	0.070	—	—	—
21	Anethole	0.07	—	—	—	—	—	—
22	(5R)-5-Methyl-2-(1-methyl ethylidene) cyclohexanone	3.65	3.64	3.54	—	—	—	—
23	Ledene	—	0.29	0.23	—	—	—	—
24	γ-Elemene	0.57	—	0.56	—	—	—	—
25	Cuminal	31.73	30.63	29.84	—	8.66	1.40	32.86
26	4-(1-Methylethyl)-1-cyclohexene-1-carboxaldehyde	3.53	3.51	3.41	—	—	—	—
27	p-mentha-1,3-dien-7-al	—	—	—	—	—	—	12.44
28	p-mentha-1,4-dien-7-al	—	—	—	—	—	—	0.08
30	Cumin alcohol	—	—	—	—	—	—	0.07
31	2-Ethylidene-6-methyl-3,5-heptadienal	9.13	8.94	8.20	—	—	—	—
32	α-Proyl-benzenemethanol	8.05	7.87	7.84	—	—	—	—
33	4-(1-Methylethyl)-1,4-cyclohexadiene-1-methanol	0.37	0.44	0.34	—	—	—	—
34	6-Isopropylidene-1-methyl-bicyclo [3.1.0] hexane	0.06	0.02	0.05	—	—	—	—
35	Phenol, 2-methoxy-4-(1-propenyl)	—	0.04	0.09	—	—	—	—
36	Geraniol	0.10	—	—	1.10	—	—	—
37	Myrtenal	0.11	0.13	0.06	—	—	—	—

TABLE 50.1 The Chemical Composition of Essential Oil from *C. cyminum* Seed—continued

S. No.	Compound	SCFE[a]	OS-SD[b]	HD[c]	SD[d]	SWE[e]	Sx Ex[f]	MWE[g]
38	E-2-Nonenal	0.43	0.42	0.38	—	—	—	—
39	Rhodinol	0.79	0.72	0.69	—	—	—	—
40	7,11-Dimethyl-3-methylene-1,6,10-dodecatriene	0.26	0.28	0.24	—	—	—	—
41	2-Isopropyl-5-methyl-9-methylene bicyclo[4.4.0]dec-1-ene	0.08	—	0.06	—	—	—	—
42	p-Allylphen	0.42	0.42	0.40	—	—	—	—
43	E-Hexadecen-6-yne	0.04	0.04	0.01	—	—	—	—
44	Octahy-dro-3,8,8-trimethyl-6-methylene-1H-3a,7-methanoazulene	0.16	0.15	0.14	—	—	—	—
45	Thujopsene	0.42	0.36	0.40	—	—	—	—
46	1-(1,5-Dimethyl-4-hexenyl)-4-methylbenzene	—	—	0.02	—	—	—	—
47	5-(1,5-Dimethy-4-hexenyl)-2-methyl-1,3-cyclohexadiene	0.14	0.12	0.13	—	—	—	—
48	Copaene	0.08	0.09	0.08	—	—	—	—
49	β-Sesquiphellandrene	0.09	0.03	0.08	—	—	—	—
50	β-acoradiene	0.09	0.04	—	—	—	—	—
51	Carotol	0.12	0.15	0.09	—	—	—	—

nd, not detected.

t, traces.

[a]Supercritical fluid extraction.
[b]Combination of organic solvent and steam distillation.
[c]Hydrodistillation.
[d]Steam distillation.
[e]Superheated water extraction.
[f]Soxhlet extraction.
[g]Microwave extraction.

423

OS-SD method offered important advantages over HD and SCFE. Results were better in terms of rapidity (i.e., a shorter extraction time of 9 hours, against 12 hours for HD), cost (energy and equipment investment), yield (4.87% against 3.02% for HD), and higher contents of characteristic flavor-impacting compounds (Table 50.2), such as aldehydes (Li *et al.*, 2009). The volatile oils extracted by HD from cumin seeds were characterized by means of GC and GC-MS (Gachkar *et al.*, 2007). The SWE method resulted in a more valuable essential oil with respect to the oxygenated components (Eikani *et al.*, 2007).

Roasting of cumin seeds in the microwave at various power levels was compared with conventional roasting of cumin seeds at different temperatures. GC and GC-MS were used to

TABLE 50.2 Classification of Volatile Constituents of Cumin Oil Obtained by OS-SD Method

S. No.	Class of Compounds	%
1	Terpene hydrocarbons	30.53
2	Aldehydes/ketones	48.01
3	Alcohols	10.86
4	Oxygenated compounds	0.29

analyze the volatile oils distilled. The quantity of p-mentha-1,3-dien-7-al, β-myrcene increased, while that of p-mentha-1,4-dien-7-al, β-pinene, p-cymene decreased. Results also showed that there was better retention of the characteristic flavor-impact compounds (i.e., total aldehydes) in microwave samples than in samples roasted conventionally (Sushmita *et al.*, 2004).

Further improvements were made using microwave extraction, which is termed the "green technique;" namely, SFME, which is a combination of dry distillation and microwave heating without adding any solvent or water. Extraction time was reduced from 8 hours to 1 hour. This oil was characterized as having a higher concentration of oxygenated compounds and lower concentration of monoterpenes (Lucchesi *et al.*, 2004). Subsequently, an improved SFME method was reported in which a microwave absorption solid medium such as carbonyl iron powder was used for extraction without any pre-treatment; here, the extraction time reduced to 30 minutes with microwave energy of 85 W. However, there was not much difference in the essential oils obtained by these methods (Ziming *et al.*, 2006).

The essential oil composition of cumin seeds at different harvesting times was reported. The concentration of major constituents that varied at different harvesting times were cumin-aldehyde (19.9—23.6%), p-mentha-1,3-dien-7-al (11.4—17.5%), and p-mentha 1,4-dien-7-al (13.9—16.9%). GC and GC-MS analysis showed that fruits should be harvested at the ripe or mature stage for an ideal volatile oil yield and composition (Kan *et al.*, 2007). The yield of volatile oil was increased to 3.5—4.5% on flaking the cumin seeds; this is due to the cell rupturing, which allows the easy passage of steam (Sowbhagya *et al.*, 2008).

Cumin essential oil prevents butter from deterioration, thus showing an antihydrolytic effect, which has the advantage over conventional antioxidants such as butylated hydroxytoluene, known to be toxic and carcinogenic. Reports also indicate that it stimulates cytochrome P and b_5 as well as N-demethylase activity. Seeds can be used to prepare stable cyclodextrin inclusion compounds, which can be stored over extended periods. The seeds reportedly possess aphrodisiac properties. Volatile components of cumin possess antimicrobial activity (Wealth of India, 2001).

Essential oils extracted by HD from fruits of cumin demonstrated antibacterial activity against gram positive and gram negative bacteria. Antibacterial activity was particularly high against the genera *Clavibacter*, *Curtobacterium*, *Rhodococcus*, *Erwinia*, *Xanthomonas*, *Ralstonia*, and *Agrobacterium*, which causes plant and cultivated mushroom diseases. However, lower activity was observed against bacteria belonging to the genus *Pseudomonas*. Hence, a potential use for cumin volatile oils could be in the control of bacterial diseases in plants and mushrooms (Iacobellis *et al.*, 2005).

The antioxidant activity of extracts from cumin seed from Gorakpur (India) was evaluated by two different methods (namely, peroxide value and thiobarbituric acid value), and found to be effective in linseed oil. The volatile oil showed 100% antifungal activity against *Curvularia lunata* and *Fusarium monoliforme*. The acetone extract was found to be 85% effective in controlling the mycelia growth of *Aspergillus ochraceus*, *Aspergillus flavus*, and *Pencillium citrinum* (Singh *et al.*, 2006).

The essential oils extracted by HD from cumin exhibited stronger antimicrobial activity against *Escherchia coli*, *Staphylococcus aureus*, and *Listeria monocytogenes*. Radical scavenging and antioxidant properties were tested using 1,1-diphenyl-2-picrylhydrazyl (DPPH) assay and the β-carotene bleaching test, which demonstrates the potential of cumin volatile oil in food preservation (Gachkar *et al.*, 2007).

The spent cumin, after isolation of the essential oil, was studied for various enzymatic activities and antioxidant activity. Saline and hot aqueous extracts of spent cumin showed enzymatic activity similar to that of native cumin. The spent cumin had the following content: carbohydrate 23%, protein 19%, fat 10%, and soluble dietary fiber 5.5%. It also contained vitamins such as thiamine (0.05 mg/100 g), riboflavin (0.28 mg/100 g), and niacin (2.7 mg/100 g), and was a rich source of minerals, containing Fe^{2+} (6.0 mg/100 g) and Zn^{2+} (6.5 mg/100 g). The result indicates that the spent cumin can be potentially used in various health food formulations with improved digestibility and good nutrient composition (Muthamma *et al.*, 2008).

Medicinal values

Cumin seeds are a very good source of iron, which contributes to many vital roles in the body. Cumin as well as its distilled oil is used as a stimulant (i.e., it stimulates and maintains the digestive and the excretory system), an antispasmodic (in the treatment of spasms and associated troubles such as cramps, convulsions, non-stop coughs, and pains), a carminative (i.e., it expels gas from the intestine and prevents further formation of gases), an antiseptic (for application to external and internal cuts and wounds), a diuretic (i.e., it increases the frequency of urination, which is beneficial for the loss of excess fats, salts, water, and toxins from the body, and also reduces blood pressure and helps to clean the kidneys), and is an antimicrobial agent. Cumin is used widely in traditional medicine to treat flatulence, digestive disorders (it aids the discharge of bile and gastric juices, and also stimulates peristaltic motion of the intestines), and diarrhea (Peter, 2003). The aqueous extract of cumin seeds shows antifertility and abortifacient activity in female rats. The cumin oil shows fungitoxic activity, due to its cuminaldehyde content (Wealth of India, 2001). Rats infected with *Candida albicans* and exposed to cumin volatile oil vapors were found to undergo an immunostimulatory effect, depending upon the timing of exposure and the stage of infection (Lawrence, 1992).

Cumin in various forms is used to treat lung ailments, weakness of the liver, and indigestion. It is used in liquor production, and as an anticonvulsant (Ishikawa *et al.*, 2002). Seeds possess a cooling effect, and therefore form an ingredient of most prescriptions for gonorrhea, chronic diarrhea, and dyspepsia, and are also applied externally in the form of a poultice to allay pain and the irritation of abdominal worms. A mixture of powdered cumin seed, honey, salts, and butter is applied for scorpion bites (Shahnaz *et al.*, 2004). In addition to applications in traditional medicine, cumin an application in stimulating the appetite; this improves the secretion of pancreatic enzymes, which helps the digestive system. It is also helpful in reducing nausea during pregnancy. The effect of essential oil on the nervous system was demonstrated in induced epileptic activity by pentylentetrazol (PTZ) using intracellular technique (Parthasarathy *et al.*, 2008).

ADVERSE EFFECTS AND REACTIONS (ALLERGIES AND TOXICITY)

Cumin essential oil can be adulterated with synthetic cuminaldehyde, and this can be detected by modern analytical techniques such as SIRA, SIM, and measuring optical rotation (Peter, 2003).

Cumin seeds and their oil are regarded generally as safe (GRAS 2340 and GRAS 2343) under the regulatory system (Parthasarathy *et al.*, 2008). To date, no adverse effects have been reported from the use of whole cumin, processed cumin, or its essential oil. It is used as a medicine for aches, inflammation, worm infestation, diarrhea, skin diseases, fever, vomiting and nausea, and as an appetizer. Cumin is beneficial to the heart and uterus, and is administered to women post partum (Sowbhagya *et al.*, 2008). Cumin has a protective effect by decreasing the lipid level in alcohol, and thermally oxidized oil was found to induce hepatotoxicity.

SUMMARY POINTS

- Cumin is one of the important seed spices commonly used in food, beverages, liquors, medicines, etc.

- Cumin possesses a unique aroma and taste. It contains 2.3–5% of essential oil, which consists predominantly of aldehydes and ketones (50–70%), hydrocarbons (30–50%), alcohols (2–5%), and ethers < 1%, depending on the cultivar.
- Among the aldehydes, cuminaldehyde (20–40%) is the major constituent, with p-mentha,1,3-dien-7-al, p-mentha,1,4-dien-7-al in minor quantities.
- Steam distillation is largely employed for the extraction of essential oil by industries.
- Other extraction methods include supercritical fluid extraction (SCFE), a combination of organic solvent and steam distillation (OS-SD), hydrodistillation (HD), steam distillation (SD), superheated water extraction (SWE), soxhlet extraction (Sx Ex), microwave extraction (MWE), and solvent-free microwave extraction (SFME); the chemical composition of the extracted volatiles in each method has been reported.
- Applications of the cumin seed in terms of health have been established based on its nutrient content. The cumin seeds are a good source of iron, manganese, potassium, zinc, essential amino acids, proteins, and other unsaturated fatty acids.
- Cumin seeds possess potential therapeutic and medicinal value, such as antioxidant activity, and use as a remedy for digestive disorders, skin diseases, and selected disorders of the nervous system, etc. The essential oil also shows significant antimicrobial activity against many bacteria, possesses chemopreventive activities, and has potential medicinal value.

References

Donna, R. T., & Antony, T. G. (1993). *Spices & seasonings: A food technology hand book.* pp. 71–74. New York, NY: VCH Publishers.

Eikani, M. H., Golmohammad, F., Mirza, M., & Rowshanzamir, S. (2007). Extraction of volatile oil from cumin (*Cuminum cyminum* L.) with superheated water. *Journal of Food Process Engineering, 30,* 255–266.

Gachkar, L., Yadegari, D., Rezaei, M. B., Taghizadeh, M., Astanesh, S. A., & Rasooli, I. J. (2007). Chemical and biological characteristics of *Cuminum cyminum* and *Rosmarinus officinalis* essential oils. *Food Chemistry, 102,* 898–904.

Iacobellis, N. S., Cantore, P. I., Capasso, F., & Senatore, F. (2005). Antibacterial activity of *Cuminum cyminum* L. and *Carum carvi* L. essential oils. *Journal of Agricultural and Food Chemistry, 53,* 57–61.

Ishikawa, T., Takayanagi, T., & Kitajima, J. I. (2002). Water-soluble constituents of cumin: monoterpinoid glucosides. *Chemical & Pharmaceutical Bulletin, 50,* 1471–1478.

Kan, Y., Kartal, M., Ozek, T., Aslan, S., & Baser, K. H. C. (2007). Composition of essential oil of *Cuminum cyminum* L. according to harvesting times. *Turkish Journal of Pharmaceutical Sciences, 4,* 25–29.

Lawrence, B. (1992). Volatile oils in spices. *Perfume Flavour, 17,* 39–44.

Leopold, J., Gerhad, B., Albena, S. S., Evgenii, V. G., & Stanka, T. D. (2005). Composition, quality control and antimicrobial activity of the essential oil of cumin seeds from Bulgaria that had been stored for up to 36 years. *International Journal of Food Science & Technology, 40,* 305–310.

Li, X. M., Tian, S. L., Pang, Z. C., Shi, J. Y., Feng., Z. S., & Zhang, Y. M. (2009). Extraction of *Cuminum cyminum* essential oil by combination technology of organic solvent with low boiling point and steam distillation. *Food Chemistry, 115,* 1114–1119.

Lucchesi, M. E., Chemat, F., & Smadja, J. (2004). An original solvent free microwave extraction of essential oils from spices. *Flavour Fragrance Journal, 19,* 134–138.

Muthamma, M. K. S., Hemang, D., Purnima, K. T., & Prakash, V. (2008). Enhancement of digestive enzymatic activity by cumin (*Cuminum cyminum* L.) and role of spent cumin as a bionutrient. *Food Chemistry, 110,* 678–683.

Naik, S. N., Lentz, H., & Maheshwari, R. C. (1989). Extraction of perfumes and flavours from plant materials with liquid carbon-dioxide under liquid-vapour equilibrium conditions. *Fluid Phase Equilibria, 49,* 115–126.

Parthasarathy, V. A., Chempakam, B., & Zachariah, T. J. (2008). *Chemistry of spices.* p. 18. Oxford, UK: CABI Publishers.

Peter, K. V. (2003). *Handbook of herbs and spices, Vol. 1, pp.164–167.* Cambridge, UK: Woodhead Publishing Ltd.

Shahnaz, H., Hafsa, A., Bushra, K., & Khan, J. I. (2004). Lipid studies of *Cuminum cyminum* fixed oil. *Pakistan Journal of Botany, 36,* 395–401.

Singh, G., Marimuthu, P., de Lampasona, M. P., & Catalan, C. A. N. (2006). *Cuminum cyminum* L. chemical constituents, antioxidants and antifungal studies on volatile oil and acetone extract. *Indian Perfumer, 50,* 31–39.

Sowbhagya, H. B., Satyendra Rao, B. V., & Krishnamurthy, N. (2008). Evaluation of size reduction and expansion on yield and quality of cumin (Cuminum cyminum) seed oil. *Journal of Food Engineering,, 84*, 595–600.

Sushmita, B., Nagarajan, S., & Jagan Mohan Rao, L. (2004). Microwave heating and conventional roasting of cumin seeds (*Cuminum cyminum* L.) and effect on chemical composition of volatiles. *Food Chemistry, 87*, 25–29.

Wealth of India. (2001). *First supplement series.* Vol. 1, A–Ci, p. 257. New Delhi, India: National Institute of Science and Communication (NISCOM).

Ziming, W., Lan, D., Tiechun, L., Xin, Z., Lu, W., Hanqi, Z., et al. (2006). Improved solvent-free microwave extraction of essential oil from dried *Cuminum cyminum* L. and *Zanthoxylum bungeanum* Maxim. *Journal of Chromatography A, 1102*, 11–17.

CHAPTER 51

Acetogenins from the Seeds of the Custard Apple (*Annona squamosa* L.) and their Health Outcomes

Pierre Champy
Laboratoire de Pharmacognosie, UMR CNRS 8076 BioCIS, Faculté de Pharmacie, Université Paris-Sud 11, France

CHAPTER OUTLINE

LIST OF ABBREVIATIONS

ACG, annonaceous acetogenins
ATP, adenosine tri-phosphate
EC, effective concentration
FDA, Food and Drug Administration
HPLC–DAD, high performance liquid chromatography–diode array detection
IC, inhibitory concentration
NADH, nicotinamide adenine dinucleotide, reduced form
PSP, progressive supranuclear palsy
ROS, reactive oxygen species
SARs, structure activity relationships
THF, tetrahydrofuran
UQ, ubiquinone

Nuts & Seeds in Health and Disease Prevention. DOI: 10.1016/B978-0-12-375688-6.10051-9

INTRODUCTION

Annona squamosa L., a small tropical tree, is a famous cultivated Annonaceae. Its fruit is known as the custard apple, sugar apple, or *fruta do conde*. Its seeds are poisonous, and have multiple, mainly traditional, uses. They contain high amounts of annonaceous acetogenins (ACGs), for which a phytochemical update is proposed. This group of polyketides comprises the most potent inhibitors of mitochondrial complex I (Bermejo *et al.*, 2005). Recent biological outcomes are presented, in regard to antitumoral and pesticidal potential. ACGs are being proposed as environmental neurotoxins, toxicological data are summarized, along with concerns about seed uses.

BOTANICAL DESCRIPTION

The pseudosyncarpic fruits of *A. squamosa* are green, and display marked carpel protuberances. They are heart-shaped, measure approximately 7.5 cm in length, and weigh 100–400 g, depending on the cultivar and cultivation conditions. Their whitish, custard-like sweet pulp contains 35 to 50 black seeds of 1–1.5 cm in length and 0.5–0.8 cm in width, with a glossy cuticle (Figure 51.1).

HISTORICAL CULTIVATION AND USAGE

Originating from Central America, like most *Annona* species, the tree is believed to have spread to Mexico, South America, and the Caribbean in the 16th to 17th centuries, and is now commonly found in domestic gardens in tropical America. It was brought to India by the Portuguese during the same period, then to South-east Asia, and was also introduced into Africa and Oceania (Pinto *et al.*, 2005). Alimentary use of the fruit appears mostly to be a South-American and Asian habit. In tropical areas, various and convergent medicinal uses, mainly of bark and leaves, are reported.

430

PRESENT-DAY CULTIVATION AND USAGE

Custard apple grows at low altitudes (0–1500 m), and is widely cultivated in tropical to semi-arid regions, in orchards or on commercial farms. Considered to be a minor crop by the FAO, it is the third most commercially cultivated Annonaceae in South and Central America (behind *A. cherimolia* and *A. muricata*), Brazil being one of the main producers (cultivation > 1200 ha, and production > 11,000 tonnes, in 2000). The tree is also cultivated in India (44,000 ha in the 1980s), Sri Lanka, Malaysia, Viet Nam, the Philippines, and Taiwan (2500 tonnes per year).

FIGURE 51.1
A. squamosa fruits and seeds.

Smaller production areas are encountered in southern Florida, Australia (2 tonnes in 2003), tropical Africa, and Egypt (170 tonnes in 1997). Exportation to northern market is, however, limited. The ripe fruit was being sold at US$ 0.56/kg in 2004, for direct consumption or for industrial processing as juices or ice creams. For this use, prior peeling and removal of seeds is performed (Pinto *et al.*, 2005).

APPLICATIONS TO HEALTH PROMOTION AND DISEASE PREVENTION

Scarcity of traditional internal use of seeds, and convergence in topical treatment against external parasites with crushed seeds or oil, are remarkable. Seeds are also often reported as traditional pesticides, and, less frequently, as fish poison. Among various other bioactive secondary metabolites (i.e., isoquinolines, *ent*-kauranes, cyclopeptides), ACGs appear to support these uses. These white, waxy polyketides, specific to the Annonaceae family, have been encountered in all *Annona* species studied so far. Derived from long chain fatty acids, they constitute 35 or 37 carbon atoms, with an alkyl chain bearing a central oxygenated system (tetrahydrofuranic (THF) rings) and a terminal butyrolactone. Inner classification is based on the structure of these moieties (Figure 51.2).

Extensive chemical studies of the seeds of *A. squamosa* led to isolation of 74 ACGs, most bearing two adjacent THF rings (e.g., rolliniastatin-2 (**1**), squamocin (**2**)) or a single THF ring (e.g., annonacin (**3**); Figure 51.3). ACGs are also reported in the bark (Bermejo *et al.*, 2005) and fruit pulp (Champy *et al.*, 2008). Yang and colleagues, in a simultaneous HPLC–DAD determination of eight ACGs from a supercritical CO_2 extract of seeds from China, evidenced **1** and **2** as major representatives (0.58 and 0.37 mg/g; total ACGs, 2.29 mg/g)

431

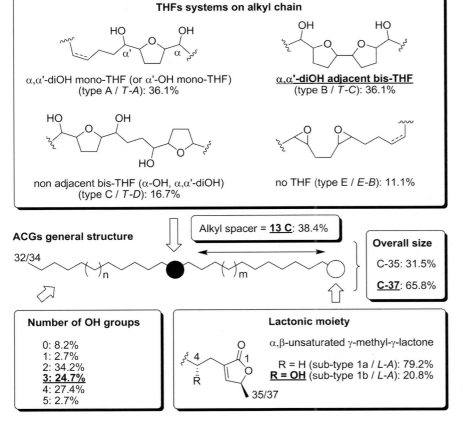

FIGURE 51.2
Structural features of ACGs from *A. squamosa* seeds. Structural characteristics that are the most favorable for complex I inhibition are underlined (compare with structure of (**1**), Figure 51.3). Types and sub-types are classified according to Cavé *et al.*, 1997, and Bermejo *et al.*, 2005 (in italics). Percentages are calculated with respect to the total number of ACGs isolated from seeds (Bermejo *et al.*, 2005; Souza *et al.*, 2008, and references cited in Figure 51.4). Length of 13 carbon atoms for alkyl spacer between lactone and hydroxylated THF system: ~50% of type A and type B ACGs. Note that a type D (three adjacent THF groups), a type E bearing three epoxides, a C-18, and a bis-lactonic C-22 representative were also obtained.

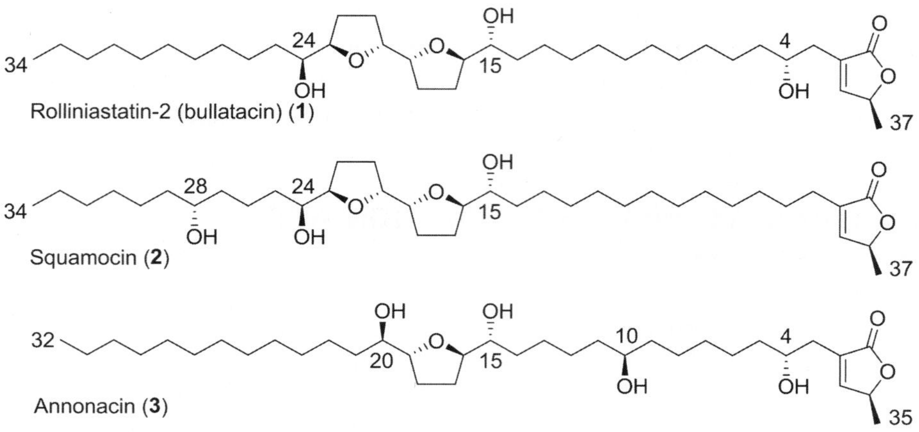

FIGURE 51.3
Prototypical ACGs from *A. squamosa* seeds.

(Yang *et al.*, 2009a). In our experience, with several batches from Brazil, **2** was the main ACG (~60% of total ACGs), with a similar yield.

Since the last review on ACGs (Bermejo *et al.*, 2005), eight articles have been published, describing isolation of 46 of these compounds from the seeds, of which 19 were obtained for the first time in the species (annotemoyins-1 and -2, bullatencin, *cis*-bullatencin, corepoxylone, diepomuricanins A and B, dieporeticenin, glabranin, glabrencin B, probably narumicin-II ("Compound 1" in Sousa *et al.*, 2008), reticulatains-1 and -2, solamin, *erythro*-solamin, tripoxyrollin, and uvariamicins I, II, and III; "isosquamocin" is also mentioned (Grover *et al.*, 2009)); 17 display original structures (note that homonymies exist for squamostatin-C and for squamocenin) (Figure 51.4).

ACGs are very strong inhibitors of mitochondrial complex I (NADH ubiquinone oxidoreductase). Most act as uncompetitive semiquinone antagonists, with the lactonic ring as a probable inhibitory pharmacophore, and THF system allowing positioning in mitochondrial internal membrane (Bermejo *et al.*, 2005; Barrachina *et al.*, 2007; see also Derbré *et al.*, 2005 and references cited in Kojima and Tanaka, 2009). Rolliniastatin-2 (**1**) is the most active representative, squamocin (**2**) displaying close potency (Table 51.1; see also structure—activity relationships (SARs) depicted in Figure 51.1).

ACGs show tremendous cytotoxicity, with IC$_{50}$ values ranging from $10\,\mu M$ to $10^{-4}\,nM$. However, striking discrepancies in SARs versus that for complex I appear in the literature. Differential intracellular distribution of these amphiphilic, apolar compounds might be implicated (Höllerhage *et al.*, 2009). Fluorescent analogs of **2** showed mitochondrial tropism in Jurkat cells (Derbré *et al.*, 2005). Alternative targets are also proposed but their relevance is unknown (Derbré *et al.*, 2008; Liaw *et al.*, 2008; Takahashi *et al.*, 2008). *In vitro* studies in various mammal cell lines or primary cultures reported death to be triggered by ROS production or ATP deprivation, depending on the *in vitro* paradigm. Apoptotic mechanisms consistent with a mitochondrial pathway were observed, notably with **2** as a pharmacological tool. The perspective of ACGs being anticancer agents has thus motivated most research on these metabolites during the past three decades, with promising milestones being achieved. Selectivity for cancerous cells in regard to normal ones was proposed, on the basis of discrepancies in ATP requirements, but this issue remains under discussion (Garcia-Aguirre *et al.*, 2008). ACGs also proved cytotoxic in multi-drug resistant cell lines expressing the ATP-dependent MDR efflux transporter (McLaughlin, 2008). Numerous semisynthesic analogs designed for activity enhancement or mechanistic studies were obtained, **2** being a lead

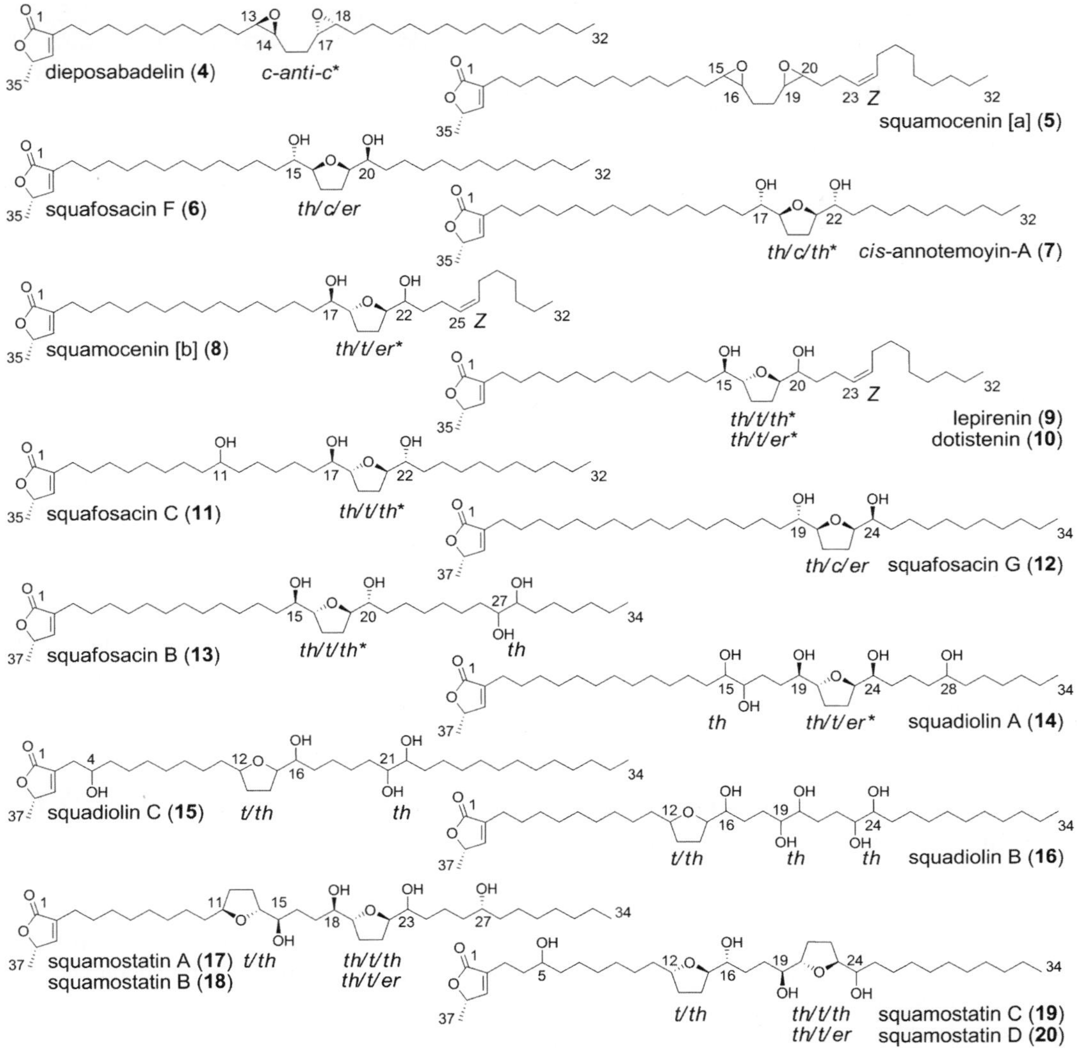

FIGURE 51.4

Original ACGs isolated from *A. squamosa* seeds between 2005 and 2009. Relative configurations: *er, erythro*; *th, threo*; *c, cis*; *t, trans*. *Undetermined absolute configurations. References: Compounds (**8**), Yu *et al.* (2005); (**6**,**7**, **11**–**16**), Liaw *et al.* (2008); (**4**,**5**,**9**,**10**), Bajin ba Ndob *et al.* (2009); (**17**, **18**), Yang *et al.* (2009b); (**19**, **20**), Yang *et al.* (2009c).

compound (Kojima & Tanaka, 2009). Among other ACGs, **1** underwent promising antitumor assays, and was reported as being well tolerated in several animal species (McLaughlin, 2008; see also Cuendet *et al.*, 2008). However, to our knowledge, no ACG passed preclinical evaluation, and no clinical studies were published. According to McLaughlin (2008), dietary supplements containing ACGs gave satisfactory results as oral adjuncts to chemotherapy in cancer patients. Indeed, poorly evaluated Annonaceae dietary supplements are sold for cancer treatment and prevention, on the Internet and in health stores. None contain *A. squamosa* seeds or seed extracts, but between 2007 and 2010 five patents related to the use of *A. squamosa* seeds ACGs in cancer were deposited in China.

ACGs also proved molluscidal and anthelmintic (antibacterial, antifungal, immunosuppressive, and antiparasitic activities have also been reported, Bermejo *et al.*, 2005). They display impressive acaricidal and insecticidal potency (McLaughlin, 2008): Among extracts containing ACGs, those of *A. squamosa* seeds were extensively studied (Grover *et al.*, 2009). Promising

TABLE 51.1 Complex I Inhibition Potential ACGs from *A. squamosa* Seeds

	NADH Oxidase IC$_{50}$ (nM)	NADH/DB Oxidoreductase IC$_{50}$ (nM)	K$_i$ (nM)
Rolliniastatin-2 (**1**)	0.85–1.2[a]		0.6[e]
Squamocin (**2**)	0.8[b]; 1.3[c]	2.0[b]	0.4[e]
Annonacin (**3**)	2.3 ± 0.3[d]	26.1 ± 3.2[d]	
Rotenone	30[c]; 5.1 ± 0.9[d]	28.8 ± 1.5[d]	4.0[e]

IC$_{50}$-Complex I, half-maximal concentration inhibiting NADH oxidase activity in the absence and presence of an exogenous ubiquinone analog (decyl-UQ; NADH-DB oxido-reductase activity), in bovine heart sub-mitochondrial particles; data from:
[a]Miyoshi et al.1998 and Fujita et al., 2005; cited in Kojima & Tanaka (2009);
[b]Duval et al. (2005); cited in Kojima & Tanaka (2009);
[c]Derbré et al., 2006, cited in Kojima & Tanaka (2009);
[d]Tormo et al. (2003); cited in Bermejo et al. (2005);
[e]Degli-Espoti et al. (1994), cited in Bermejo et al. (2005).

semisynthesic ACGs were also obtained, including β-amino-(**2**), a dual complex I/complex III inhibitor designed by Duval and colleagues (Kojima & Tanaka, 2009). A patent for an anti-head-lice shampoo containing a standardized extract of seeds of *A. squamosa* has been registered, alike *Asimina triloba* products containing similar ACGs (McLaughlin, 2008). Agrochemical valorization of *A. squamosa* seeds appears to be a potentially important outcome, with five publications in 2009 and three Indian patents on standardized apolar extracts since 2006. This is reminiscent of rotenoid-containing Fabaceae and of rotenone, a reference lipophilic complex I inhibitor sharing the enzyme binding domain of ACGs, with potency close to that of **3** (Table 51.1).

ADVERSE EFFECTS AND REACTIONS (ALLERGIES AND TOXICITY)

Seeds of *A. squamosa* are of notorious toxicity, and are thus barely used orally in traditional medicine (except as an abortive in India, where aqueous extracts are used). They are reported to cause irritation to the eye and mucosa. Oral ingestion provokes vomiting, related to the ACG content (McLaughlin, 2008). The plant is mentioned in the poisonous plants database of the FDA (American Food and Drug Administration), and the AFSSA (Agence Française de Sécurité Sanitaire des Aliments: Saisine 2007-SA-0231, 2007/12/21; pp 3, 5; Saisine 2008-SA-0171, 2010/04/28, 7 p.) has expressed safety concerns regarding its use in dietary supplements.

In relation to pesticide use, safety evaluation of a defatted seed extract (MeOH/CH$_2$Cl$_2$ 1:1) in female Wistar Rats was proposed by Grover *et al.* (2009). Mortality was observed at 2 g/kg p.o. At doses of 150 and 300 mg/kg, genotoxicity was evidenced in leukocytes and bone marrow, from 4 to 72 h after ingestion, possibly due to ACGs (Garcia-Aguirre *et al.*, 2008). Consistent with complex I inhibition, involvement of ROS was suggested by significantly enhanced lipid peroxidation, and decreased glutathione and glutathione S-transferase levels. However, an MeOH extract likely to contain ACGs did not increase oxidative markers in the livers of female Swiss mice (dosage 200 mg/kg, p.o., 10 days; Panda & Khar, 2007; see also Damasceno *et al.*, 2002; Pardhasaradhi *et al.*, 2005). Histological examination of liver and kidney revealed no lesions. Authors have expressed concern about the use of *A. squamosa* seed extract as a pesticide until more tests are carried out (Grover *et al.*, 2009).

Nevertheless, complex I dysfunction has been reported in Parkinson's disease (a movement disorder with progressive degeneration of dopaminergic neurons in *substantia nigra*), as well as in the tauopathy progressive supernuclear palsy (PSP), an atypical form of parkinsonism. Complex I inhibitors such as 1-methyl-4-phenylpyridinum, paraquat or rotenone are used to establish animal models of neurodegeneration, and are linked to the occurrence of parkinsonism (Gibson *et al.*, 2010). PSP-like syndromes were observed in genetically heterogeneous populations regularly consuming alimentary and medicinal Annonaceae products. Thus, in

TABLE 51.2 *In vitro* Neurotoxicity of ACGs (Striatal Primary Cultures)

	IC_{50}-Cpx I (nM)	EC_{50}-ATP (nM)	EC_{50}-ND (nM)	EC_5-Tau (nM)
Rolliniastatin-2 (**1**)	0.9	3.6[*]	1.1[*]	0.6[*]
Squamocin (**2**)	1.4	2.9	1.1	0.6
Annonacin (**3**)	54.8	134.0	60.8	44.1
Rotenone	6.8	7.3	8.1	7.2

IC_{50}-Cpx I, half-maximal concentration inhibiting complex I activity (brain homogenates); EC_{50}-ATP, half-maximal effective concentration inducing a decrease in ATP levels (cultures, 6 h); EC_{50}-ND, half-maximal effective concentration inducing neuronal cell death (cultures, 48 h); EC_5-Tau, concentration at which tau was redistributed in 5% of the neurons as a measure of minimum concentration inducing tauopathy (cultures, 48 h); see Escobar-Khondiker *et al.* (2007) and Höllerhage *et al.* (2009).

Guadeloupe (French West Indies), such patients account for two-thirds of all cases of parkinsonism, compared to approximately 30% of atypical forms in European countries. They display a combination of movement disorders and dementia, the disease being thoroughly characterized. Autopsies performed in three patients revealed accumulation of neuronal Tau-fibrils (see references cited in Camuzat *et al.*, 2008; Champy *et al.*, 2009). ACGs were identified as candidate toxins using PC12 cells (unpublished data), as confirmed for annonacin (**3**) in mesencephalic primary cultures. In striatal primary cultures, ACGs induced ATP loss, Tau hyperphosphorylation and redistribution, microtubular disruption, and cell death at low nanomolar concentrations (Table 51.2; Höllerhage *et al.*, 2009).

Subchronic systemic intoxication of Lewis rats with **3** (continuous i.v., 3.8; 7.6 mg/kg per day, 28 days) did not cause locomotor dysfunction or signs of illness. However, **3** crossed the blood—brain barrier, reduced cerebral ATP levels, and caused neuronal cell loss and gliosis in the brain stem and basal ganglia. These features are similar to those obtained with rotenone (Höglinger *et al.*, 2006), and are reminiscent of the human disease. ACGs are therefore proposed as etiological agents for cases of sporadic atypical parkinsonism and tauopathies worldwide, upon chronic exposure. However, pharmacokinetic parameters remain to be determined, and further epidemiological studies are needed before drawing firm conclusions. It is noteworthy that rotenone, widely used as an organic pesticide with low environmental reminiscence, was banned in 2008 in the European Union. In the absence of a defined benefit—risk balance, these facts challenge the various alternatives proposed for valorization of *A. squamosa* seeds.

SUMMARY POINTS

- *Annona squamosa* is a cultivated pantropical fruit tree, and its seeds are by-products.
- *Annona squamosa* seeds constitute a major source of Annonaceous acetogenins (ACGs), which are potent lipophilic complex I inhibitors.
- Sources of ACGs are proposed as antitumoral dietary supplements.
- The seeds have major potential as an organic pesticide, with patents applied.
- An extract of *A. squamosa* seeds was shown to be mildly genotoxic.
- An epidemiological link between Annonaceae and atypical parkinsonian syndromes was evidenced.
- ACGs are neurotoxic *in vitro* and *in vivo*.
- The benefit—risk balance of use of *A. squamosa* seeds remains undefined, and caution should therefore prevail.

References

Bajin ba Ndob, I., Champy, P., Gleye, C., Lewin, G., & Akendengué, B. (2009). Annonaceous acetogenins: Precursors from the seeds of Annona squamosa. *Phytochemistry Letters, 2,* 72—76.

Barrachina, I., Royo, I., Baldoni, H. A., Chahboune, N., Suvire, F., DePedro, N., et al. (2007). New antitumoral acetogenin "guanacone type" derivatives: Isolation and bioactivity. Molecular dynamics simulation of diacetyl-guanacone. *Bioorganic & Medicinal Chemistry, 15,* 4369—4381.

Bermejo, A., Figadère, B., Zafra-Polo, M.-C., Barrachina, I., Estornell, E., & Cortes, D. (2005). Acetogenins from Annonceae: Recent progress in isolation, synthesis and mechanisms of action. *Natural Product Reports, 22*, 269–303.

Camuzat, A., Romana, M., Dürr, A., Feingold, J., Brice, A., Ruberg, M., et al. (2008). The PSP-associated MAPT H1 subhaplotype in Guadeloupean atypical parkinsonism. *Movement Disorders, 23*, 2384–2391.

Cavé, A., Figadère, B., Laurens, A., & Cortes, D. (1997). Acetogenins from Annonaceae. In W. Herz, G. W. Kirby, R. E. Moore, W. Steglich, & Ch. Tamm. (Eds.), *In Progress in the chemistry of organic natural products, Vol. 70* (pp. 81–288). Vienna, Austria: Springer-Verlag.

Champy, P., Escobar-Khondiker, M., Bajin ba Ndob, I., Yamada, E., Lannuzel, A., Laprévote, O., et al. (2008). Atypical parkinsonism induced by Annonaceae: Where are we yet? Proceedings of the 7th Joint Meeting of the AFERP, ASP, GA, PSE & SIF, Athens, August 2008. *Planta Medica, 74*, 936–937.

Champy, P., Guérineau, V., & Laprévote, O. (2009). MALDI-TOF MS profiling of Annonaceous acetogenins in *Annona muricata* products of human consumption. *Molecules, 14*, 5235–5246.

Cuendet, M., Oteham, C. P., Moon, R. C., Keller, W. J., Peaden, P. A., & Pezzuto, J. M. (2008). Dietary administration of *Asimina triloba* (pawpaw) extract increases tumor latency in *N*-methyl-*N*-nitrosourea treated rats. *Pharmaceutical Biology, 46*, 3–7.

Damasceno, D. C., Volpato, G. T., Sartori, T. C. F., Rodrigues, P. F., Perin, E. A., Calderon, I. M. P., et al. (2002). Effects of *Annona squamosa* extract on early pregnancy in rats. *Phytomedicine, 9*, 667–672.

Derbré, S., Roué, G., Poupon, E., Susin, S.-A., & Hocquemiller, R. (2005). Annonaceous acetogenins: The hydroxyl groups and THF rings are crucial structural elements for targeting the mitochondria, demonstration with the synthesis of fluorescent squamocin analogues. *Chemistry and Biochemistry, 6*, 979–982.

Derbré, S., Gil, S., Taverna, M., Boursier, C., Nicolas, V., Demey-Thomas, E., et al. (2008). Highly cytotoxic and neurotoxic acetogenins of the Annonaceae: New putative biological targets of squamocin detected by activity-based protein profiling. *Bioorganic & Medicinal Chemistry Letters, 18*, 5741–5744.

Escobar-Khondiker, M., Höllerhage, M., Michel, P. P., Muriel, M.-P., Champy, P., Yagi, T., et al. (2007). Annonacin, a natural mitochondrial complex I inhibitor, causes Tau pathology in cultured neurons. *Journal of Neuroscience, 27*, 7827–7837.

Garcia-Aguirre, K. K., Zepeda-Vallejo, L. G., Ramon-Gallegos, E., Alvarez-Gonzalez, I., & Madrigal-Bujaidar, E. (2008). Genotoxic and cytotoxic effects produced by acetogenins obtained from *Annona cherimolia* Mill. *Biological & Pharmaceutical Bulletin, 31*, 2346–2349.

Gibson, G. E., Starkov, A., Blass, J. P., Ratan, R. R., & Beal, M. F. (2010). Cause and consequence: Mitochondrial dysfunction initiates and propagates neuronal dysfunction, neuronal death and behavioral abnormalities in age-associated neurodegenerative diseases. *Biochimica Biophysica Acta, 1802*, 122–134.

Grover, P., Singh, S. P., Prabhakar, P. V., Reddy, U. A., Balasubramanyam, A., Mahboob, M., et al. (2009). *In vivo* assessment of genotoxic effects of *Annona squamosa* seed extract in rats. *Food and Chemical Toxicology, 47*, 1964–1971.

Höglinger, G. U., Oertel, W. H., & Hirsch, E. C. (2006). The rotenone model of parkinsonism – the five year inspection. *Journal of Neural Transmission Supplementa, 70*, 269–272.

Höllerhage, M., Matusch, A., Champy, P., Lombès, A., Ruberg, M., Oertel, W. H., et al. (2009). Natural lipophilic inhibitors of mitochondrial complex I are candidate toxins for sporadic tau pathologies. *Experimental Neurology, 220*, 133–142.

Kojima, N., & Tanaka, T. (2009). Medicinal chemistry of Annonaceous acetogenins: Design, synthesis, and biological evaluation of novel analogues. *Molecules, 14*, 3621–3661.

Liaw, C.-C., Yang, Y.-L., Chen, M., Chang, F.-R., Chen, S.-L., Wu, S.-H., et al. (2008). Mono-tetrahydrofuran Annonaceous acetogenins from *Annona squamosa* as cytotoxic agents and calcium ion chelators. *Journal of Natural Products, 71*, 764–771.

McLaughlin, J. L. (2008). Paw paw and cancer: Annonaceous acetogenins from discovery to commercial products. *Journal of Natural Products, 7*, 1311–1321.

Panda, S., & Khar, A. (2007). *Annona squamosa* seed extract in the regulation of hyperthyroidism and lipid-peroxidation in mice: Possible involvement of quercetin. *Phytomedicine, 14*, 799–805.

Pardhasaradhi, B. V. V., Reddy, M., Ali, A. M., Kumari, A. L., & Khar, A. (2005). Differential cytotoxic effects of *Annona squamosa* seed extracts on human tumour cell lines: Role of reactive oxygen species and glutathione. *Journal of Bioscience, 30*, 237–244.

Pinto, A. C., de Q., Cordeiro, M. C. R., de Andrade, S. R. M., Ferreira, F. R., Figueiras, H. A., de, C., et al. (2005). *Annona* species *(pp. 1–268)*. Southampton, UK: International Center for Underutilised Crops, University of Southampton.

Souza, M. M. C., Bevilaqua, C. M. L., Morais, S. M., Costa, C. T. C., Silva, A. R. A., & Braz-Fhilo, R. (2008). Anthelmintic acetogenin from *Annona squamosa* L. seeds. *Anais Academia Brasileira Ciencias, 80*, 271–277.

Takahashi, S., Yonezawa, Y., Kubota, A., Ogawa, N., Maeda, K., Koshino, H., et al. (2008). Pyranicin, a non-classical annonaceous acetogenin, is a potent inhibitor of DNA polymerase, topoisomerase and human cancer cell growth. *International Journal of Oncology, 32*, 451–458.

Yang, H.-J., Li, X., Tang, Y., Zhang, N., Chen, J.-W., & Cai, B.-C. (2009a). Supercritical fluid CO_2 extraction and simultaneous determination of eight annonaceous acetogenins in *Annona* genus plant seeds by HPLC-DAD method. *Journal of Pharmaceutical and Biomedical Analysis, 49*, 140—144.

Yang, H.-J., Li, X., Zhang, N., He, L., Chen, J.-W., & Wang, M.-Y. (2009b). Two new cytotoxic acetogenins from *Annona squamosa*. *Journal of Asian Natural Products Research, 11*, 250—256.

Yang, H.-J., Zhang, N., Li, X., He, L., & Chen, J.-W. (2009c). New nonadjacent bis-THF ring acetogenins from the seeds of *Annona squamosa*. *Fitoterapia, 80*, 177—181.

Yu, J.-G., Luo, X.-Z., Sun, L., Li, D.-Y., Huang, W.-H., & Liu, C.-Y. (2005). Chemical constituents from the seeds of *Annona squamosa*. *Yao Xue Xue Bao, 40*, 153—158.

CHAPTER 52

Beneficial Aspects of Custard Apple (*Annona squamosa* L.) Seeds

Jianwei Chen, Yong Chen, Xiang Li
Department of Pharmaceutical Chemistry, School of Pharmacy, Nanjing University of Chinese Medicine, Nanjing, Jiangsu, China

CHAPTER OUTLINE

LIST OF ABBREVIATIONS

ATP, adenosine triphosphate
Bcl-2, B cell lymphoma/leukemia-2
HPLC, high performance liquid chromatography
IL-6, interleukin-6
LPO, lactoperoxidase
LPS, lipopolysaccharide
MTT, 3-(4,5)-dimethylthiahiazo(-z-y1)-3,5-di-phenytetrazoliumromide
Pam3Cys, (S)-(2,3-bis(palmitoyloxy)-(2RS)-propyl)-N-palmitoyl-(R)-Cys-(S)-Ser(S)-Lys4-OH, trihydrochloride
Ps, polysyndactyly
ROS, reactive oxygen species
TLR2/4, toll-like receptor2/4
TNF-α, tumor necrosis factor-α

Nuts & Seeds in Health and Disease Prevention. DOI: 10.1016/B978-0-12-375688-6.10052-0

INTRODUCTION

Custard apple (*Annona squamosa* L.) is an edible tropical fruit, and is also called sugar apple or sweetsop. Although the exact original home of custard apple is still being debated, experts agree that its native habitat extends from Southern Central America to tropical South America.

Custard apple seeds are well known for their interesting pharmacological activities, especially for antitumor activity. It was found from a number of *in vivo* experiments in animals that custard apple seed extract was effective in the treatment of leukemia, liver cancer, prostate cancer, pancreatic cancer, cervical cancer, etc. (Li & Fu, 2004; Liu *et al.*, 2007).

In recent years, phytochemical and pharmacological studies on the custard apple seeds have shown that the major bioactive compounds are annonaceous acetogenins (more than 100 compounds), which have a strong antitumor activity. In this chapter, the beneficial aspects of custard apple seeds are reviewed.

BOTANICAL DESCRIPTION

The custard apple is a small, semi-deciduous, well-branched shrub or small tree, 3—8 m in height, with thin, gray bark. The crown is spherical or a flattened ball (Figure 52.1A). The leaves are univalent alternate, lanceolate or oblong lanceolate, 6—17 cm long, sharp or blunt at the tip, and round or widely wedge-shaped at the base. The flowers are in clusters of two to four, and the length of each flower is about 2.5 cm. The outer petals are oblong, and green and purplish at the base. Inner petals reduce to minute scales, or are absent (Figure 52.1B). The fruit is round, heart-shaped, ovate or conical, 5—10 cm in diameter, with many round protuberances (Figure 52.1C), greenish-yellow when ripe, with a white, powdery bloom. The white pulp is edible, and has a sweetly aromatic taste. Each carpel holds an oblong, smooth, shiny, blackish or dark-brown seed, ranging in length ranged from 1.3 to 1.6 cm (Figure 52.1D). About 67% of the fruit is edible, and the average fresh weight of a single fruit is around 350 g (Jiang & Li, 1979). The major nutritional components in custard apple fruit were analyzed by Ding *et al.* (2006) (Table 52.1).

440

FIGURE 52.1

Photographs of custard apple (*Annona squamosa* L.). (A) whole plant; (B) flower; (C) fruit, (D) seeds. *The picture of the flower of* Annona squamosa *is reproduced from http://zh.wikipedia.org, with permission.*

TABLE 52.1 Major Nutrient Content in Fruit of Custard Apple (per 100 g Fresh Flesh)

Component	Content	Component	Content	Component	Content
Thiamine	0.11 mg	Vitamin C	12.74 mg	Protein	0.79 g
Riboflavin	0.13 mg	Vitamin E	0.01 mg	Carbohydrate	18.15 g
VitaminB6	0.78 mg	Flavonoid	35.8 mg	Crude fat	0.83 g
Vitamin K1	0.84 mg			Crude fiber	1.71 g

Reprinted from Ding *et al.* (2006), *Acta Nutrimenta Sinica*, *28*, 275–276, with permission.

HISTORICAL CULTIVATION AND USAGE

The custard apple is a native semi-deciduous tree in tropical America and the West Indies. It was planted in Puerto Rico as fruit trees in 1626, spreading from cultivated areas to roadsides and valleys. The Spaniards probably carried seeds from the New World to the Philippines, and the Portuguese are assumed to have introduced it to southern India before 1590 (Morton, 1987).

In addition to its edible fruits, custard seeds have traditionally been used in folk medicines. In India, the seeds were used to make a hair tonic to eliminate head lice. A decoction of the seeds was also used as an enema for children suffering from dyspepsia (Yang *et al.*, 2008). Ground seeds macerated in water were used as an insecticide, fish poison, powerful irritant of the conjunctiva, and abortifacient.

PRESENT-DAY CULTIVATION AND USAGE

Currently, the custard apple is widely cultivated as a fruit tree throughout tropical and subtropical areas, including southern Florida, South Asia, the South Pacific, some areas of central and western India, and in the Deccan Peninsula. Although Netherlanders introduced it to Taiwan in the 17th century, where it was first planted in the south, it was not popular because of its small fruits and low yield. However, in the 1980s cultivation began in Taitung county, in the southeast of Taiwan, where the climate and soil are suitable for the growth of the plants.

Phytochemical studies on custard apple were initiated in 1924. More than 400 compounds have been isolated from different tissues of the plant, and most of them can be classified into four major groups: alkaloids, annonaceous acetogenins, cyclic peptides, and *ent*-kaurane diterpenoids. Some of them have shown interesting pharmacological activities (Yang *et al.*, 2008). Annonaceous acetogenins, a new class of compounds isolated from the seeds, have been reported to have potent antibacterial, anti-ovulatory, anti-inflammatory, antithyroidal, and other activities. Interestingly, annonaceous acetogenins have shown antitumor activity (Pardhasaradhi *et al.*, 2005; Liu *et al.*, 2007).

APPLICATIONS TO HEALTH PROMOTION AND DISEASE PREVENTION

In traditional medicine, custard apple seeds were mainly used to treat various digestive disorders, and also as an insecticidal agent (Table 52.2). A simple seed extract, prepared by blending a handful of fresh custard apple seeds with water and filtering the mixture through several layers of fine cloth, showed promotion of abortion in rats (Damasceno *et al.*, 2002). The seed extract also showed a post-coital antifertility activity (Mishra *et al.*, 1979).

Pardhasaradhi and colleagues (2005) tested the organic and aqueous extracts from the defatted seeds of custard apple with different human tumor cell lines for antitumoral activity, and found that the extracts induced apoptosis in MCF-7 and K-562 cells. Treatment of MCF-7 and K-562 cells with organic and aqueous extracts resulted in nuclear condensation, DNA

TABLE 52.2 The Traditional Uses and Present Pharmacological Activity of Seeds of Custard Apple (*Annona squamosa* L.)

Traditional Uses	Present Phytochemicals and Pharmacological Active	
	Chemical Constituent/Extracts	Pharmacological Activity
Insecticide, enema for children with dyspepsia, hair tonic to eliminate head lice, and other usage as fish poison, powerful irritant of the conjunctiva, and abortifacient	Annonaceous acetogenins Cyclic peptides Cyclic octapeptides Total seed extracts	Antitumor Anti-inflammatory Hypotensive Antibacterial, anti-ovulatory, antifertility, antifungal, antiperoxidative, antithyroidal, etc.
	Extract of defatted seeds	Antitumor, central analgesic activity

Reprinted from Ding *et al.* (2006), *Acta Nutrimenta Sinica*, 28, 275–276, with permission.

fragmentation, induced reactive oxygen species (ROS) generation, and reduced intracellular glutathione levels. Down-regulation of Bcl-2 and PS externalization by Annexin-V staining suggested induction of apoptosis in MCF-7 and K-562 cells by both the extracts, through oxidative stress. However, COLO-205 cells showed only PS externalization, and no change in ROS and glutathione levels. These observations suggest that the induction of apoptosis by custard apple seed extracts may be selective for certain types of cancerous cells (Pardhasaradhi *et al.*, 2005). An application of annonareticin, a lactone compound from custard apple seeds, in the treatment of lung cancer or breast cancer, indicated that the antitumor activity of annonareticin on lung cancer or breast cancer cells was higher than that of 5-fluorouracil. Another *in vivo* pharmacological experiment with custard apple seed extract, which contains several different types of annonaceous acetogenins, showed that annonaceous acetogenin V has obvious inhibitory effects on lung cancer, breast cancer, and liver cancer, with antitumor activity being 100 times higher than that of 5-fluorouracil. The extracted annonaceous acetogenin V can be used as antitumor agent (Zhang, 2009). Yang and colleagues (2009a) found two new acetogenins in custard apple seeds, and their cytotoxicities were evaluated against the human tumor cell lines of human colon adenocarcinoma (HCT), human lung carcinoma (A-549), human breast carcinoma (MCF-7), and human prostate adenocarcinoma (PC-3), by the conventional MTT method (Yang *et al.*, 2009a). The results showed that both compounds exhibited significantly selective activity against all four human tumor cell lines in 2-day MTT tests (Gu *et al.*, 1995; Yang *et al.*, 2009a).

With the extensive investigations into the phytochemistry and pharmacology of custard apple seeds, a number of bioactive compounds have been isolated and identified (Table 52.2) (Gupta *et al.*, 2005). Yang *et al.* (2009b) developed a protocol for the simultaneous determination of eight annonaceous acetogenins from five different *Annona* plant seeds by HPLC–DAD (Figure 52.2). The total content of the eight compounds in the custard seeds was higher than that of four other *Annona* genus plant seeds.

Annonaceous acetogenins are a large family of fatty acid-derived natural products with unique structures. They display a broad spectrum of biological activities, especially in the treatment of leukemia, liver cancer, prostate cancer, pancreatic cancer, and cervical cancer. The strongest antitumor activity among the acetogenins is that of bullatacin; its antileukemia activity is about 300 times higher than that of paclitaxel. The relatively pure acetogenins exhibit a selective cytotoxicity for various human tumor cell lines (McLaughlin & Hopp, 1999). It is believed that acetogenins act directly at the terminal electron transfer step in complex I of mitochondria (Li & Fu, 2004; Liu *et al.*, 2007).

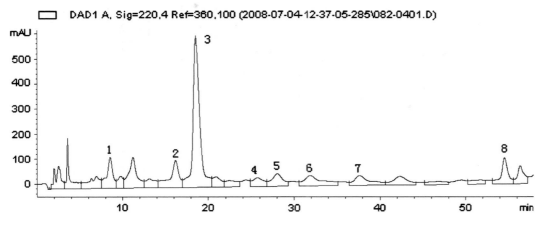

FIGURE 52.2
HPLC chromatograms of SFE extract of _Annona squamosa_ seeds. Peak identification: (1) 12, 15-cis-squamostatin-A; (2) squamostatin-A; (3) bullatacin; (4) squamostatin-D; (5) squamocin; (6) isodesacetyluvaricin; (7) asiminecin, and (8) desacetyluvaricin. _Reprinted from Yang et al. (2009b), J. Pharm. Biomed. Anal., 49, 140–144, with permission._

Yang _et al._ (2008) investigated the effects of major cyclic peptides in the custard seed on anti-inflammatory activity, using _in vitro_ model LPS- and Pam3Cys-stimulated macrophages and monitoring of the production of TNF-α and IL-6. It was found that the cyclic peptides inhibited TNF-α and IL-6 production by J774A.1 cells that were stimulated by LPS and Pam3Cys. It was suggested that the action target may be the TLR2/4 receptors (Yang _et al._, 2008). Morita and colleagues (2006) found that a cyclic octapeptide isolated from custard seeds showed a hypotensive effect on rat aorta. A possible mode of action was deduced to be inhibition of voltage-dependent calcium channels. The authors also reported that the custard apple seed extract showed remarkable antimicrobial activities, using the disc diffusion method and measuring cytotoxic activity by Brine shrimp bioassay.

Phytochemical analyses, including HPLC, identified quercetin from custard apple seed extract. It has been reported that quercetin may be involved in the regulation of hyperthyroidism and lipid-peroxidation in mice (Panda & Khar, 2007). The seed extract decreased hepatic LPO, suggesting that it is safe and has an antiperoxidative nature. Quercetin also decreased hepatic LPO with a higher efficacy as compared with propyl thiouracil (PTU), a standard antithyroidic drug.

Another study showed that an ethanol extract (ASE) of the defatted seeds of _Annona squamosa_, where phytochemical analyses revealed the presence of alkaloids, produced a depressant effect on the central nervous system, potentiated the hypnotic effect of pentobarbitone, at an oral dose of 250 mg/kg in rats, and showed anticonvulsant activity against electrically-induced convulsions and raised the pain threshold when tested by analgesiometer. Incubation with 0.3 mg/ml ASE stimulated rabbit and rat ileum, and the effect was not altered by atropine pretreatment. On frog rectus abdominis muscle, potentiation of acetylcholine action was observed (Saluja & Santani, 1994).

ADVERSE EFFECTS AND REACTIONS (ALLERGIES AND TOXICITY)

The seeds of custard apple must be kept away from the eyes due to a highly irritant effect which may lead to blindness. Londerhausen and colleagues (1991) first noted signs of the toxicity of acetogenins, such as reduction in movement and lethargy, in a number of animal experiments. Toxicity may be related to a lower level of ATP caused by respiratory depression (Londershausen _et al._, 1991). Custard apple seed extracts from four different organic solvents (i.e., petroleum ether, ether, chloroform, and ethanol) were tested for toxicity to the eyes and ear skin of rabbits by using diluted (1:10 w/v) dry extracts with propylene glycol as test substances. The results revealed that some extracts caused conjunctival redness, chemosis, a rugged cornea, skin erythema, and edema (Sookvanichsilp *et al.*, 1994).

SUMMARY POINTS

- The custard apple is a small, semi-deciduous fruit tree which is a native of tropical America and the West Indies, but is now widely cultivated throughout the tropics and subtropics.
- The seeds of custard apple have been used as folk medicines worldwide.
- The compounds and extracts of custard apple seeds show interesting pharmacological activities.
- Annonaceous acetogenins, a group of bioactive compounds in the custard apple seeds, show impressive antitumor activity.
- The seeds of custard apple may cause blindness, reduction in movement, and some other adverse effects.

ACKNOWLEDGEMENTS

The authors thank Dr Houman Fei and Dr Kevin Vessey for editing the English of this chapter.

References

Damasceno, D. C., Volpato, G. T., Sartori, T. C. F., Rodrigues, P. F., Perin, E. A., Calderon, I. M. P., et al. (2002). Effects of *Annona squamosa* extract on early pregnancy in rats. *Phytomedicine, 9*, 667–672.

Ding, L. J., Ren, N. L., Zhou, Y. F., Zhang, L. P., Xie, S. J., & Zhang, M. S. (2006). Analysis of the nutritional composition of custard apple flesh. *Acta Nutrimenta Sinica, 28*, 275–276.

Gu, Z. M., Zhao, G. X., Oberlies, N. H., Zeng, L., & McLaughlin, J. L. (1995). Phytochemistry of medicinal plants. In J. T. Arnason, R. Mata, & J. T. Romeo (Eds.), *Recent advances in phytochemistry: Annonaceous acetogenins Vol. 29* (pp. 249–310). New York, NY: Plenum Press.

Gupta, R. K., Kesari, A. N., Watal, G., Murthy, P. S., Chandra, R., & Tandon, R. (2005). Nutritional and hypoglycemic effect of fruit pulp of *Annona squamosa* in normal healthy and alloxan-induced diabetic rabbits. *Annals of Nutrition & Metabolism, 49*, 407–413.

Jiang, Y., & Li, B.-T. (1979). *Flora Reipublicae Popularis Sinica: Angiospermae dicotyledoneae Calycanthaceae Annonaceae Myristicaceae, 30(2)*. Beijing, China: Science Press. p. 171.

Li, Y. F., & Fu, L. W. (2004). Antitumoral effects of annonaceous acetogenins. *Chinese Pharmacology Bulletin, 20*(3), 245–247.

Liu, H. X., Huang, G. R., & Zhang, H. M. (2007). Annonaceous acetogenin mimics bearing a terminal lactam and their cytotoxicity against cancer cells. *Bioorganic & Medicinal Chemistry Letters, 17*, 3426–3430.

Londershausen, M., Leicht, W., Lieb, F., Moeschler, F., & Weiss, H. (1991). Molecular mode of action of annonins. *Pesticide Science, 3*, 427.

McLaughlin, J. L., & Hopp, D. C. (1999). *Bioactive acetogenins. US Patent 5955497.*

Mishra, A., Dogra, J. V. V., Singh, J. N., & Jha, O. P. (1979). Post-coital antifertility activity of *Annona squamosa* and *Ipomoea fistulosa*. *Planta Medica, 35*, 283–285.

Morita, H., Iizuka, T., Choo, C. Y., Chan, K. L., Takeya, K., & Kobayashi, J. (2006). Vasorelaxant activity of cyclic peptide, cyclosquamosin B, from *Annona squamosa*. *Bioorganic & Medicinal Chemistry Letters, 16*, 4609–4611.

Morton, J. F. (1987). Sugar apple (*Annona squamosa*. In F. Julia, & J. F. Morton (Eds.), *Fruits of warm climates* (pp. 69–72). Miami, FL: Ed, Media Inc.

Panda, S., & Khar, A. (2007). *Annona squamosa* seed extract in the regulation of hyperthyroidism and lipid-peroxidation in mice: possible involvement of quercetin. *Phytomedicine, 14*, 799–805.

Pardhasaradhi, B. V., Reddy, M., Ali, A. M., Kumari, A. L., & Khar, A. (2005). Differential cytotoxic effects of *Annona squamosa* seed extracts on human tumour cell lines: Role of reactive oxygen species and glutathione. *Journal of Biosciences, 30*, 237–244.

Saluja, A. K., & Santani, D. D. (1994). Pharmacological screening of an ethanol extract of defatted seeds of *Annona squamosa*. *International Journal of Pharmacognosy, 32*, 154–162.

Sookvanichsilp, N., Gritsanapan, W., Somanabandhu, A., Lekcharoen, K., & Tiankrop, P. (1994). Toxicity testing of organic solvent extracts from *Annona squamosa*: Effects on rabbit eyes and ear skin. *Phytotherapy Research, 8*, 365–368.

Yang, Y. L., Hua, K. F., Chuang, P. H., Wu, S. H., Wu, K. Y., Chang, F. R., et al. (2008). New cyclic peptides from the seeds of *Annona squamosa* L. and their anti-inflammatory activities. *Journal of Agricultural and Food Chemistry, 56,* 386—392.

Yang, H. J., Li, X., Zhang, N., Chen, J. W., & Wang, M. Y. (2009a). Two new acetogenins from *Annona squamosa. Journal of Asian Natural Products Research, 11,* 250—256.

Yang, H. J., Li, X., Tang, Y. P., Zhang, N., Chen, J. W., & Cai, B. C. (2009b). Supercritical fluid CO_2 extraction and simultaneous determination of eight annonaceous acetogenins in Annona genus plant seeds by HPLC-DAD method. *Journal of Pharmaceutical and Biomedical Analysis, 49,* 140—144.

Zhang, Y. H. (2009). *Annonaceous acetogenins compound V in the preparation of the treatment of anti-cancer drugs. CN Patent 101342202.*

CHAPTER 53

Usage of Date (*Phoenix dactylifera* L.) Seeds in Human Health and Animal Feed

Mohamed Ali Al-Farsi[1], Chang Young Lee[2]
[1] Date Processing Research Laboratory, Ministry of Agriculture, Oman
[2] Department of Food Science and Technology, Cornell University, Geneva, New York, USA

INTRODUCTION

Dates from the date palm tree are popular among the population of the Middle Eastern countries, providing a staple food for millions of people in arid and semi-arid regions of the world. The world production of dates has increased from about 4.6 million tonnes in 1994 to 7.2 million tonnes in 2009 (FAO, 2010). Date seeds, also called stones or pits, form part of the integral date fruit, which is composed of a fleshy pericarp and seed that constitutes between 10 and 15% of the date fruit's weight, depending on the variety and quality (Hussein *et al.*, 1998); thus, approximately 825,000 tons of date seeds are produced annually (FAO, 2010). As it is also known that date seeds contain valuable bioactive compounds, utilization of this by-product is highly desirable for the date industry.

BOTANICAL DESCRIPTION

The botanical name of the date palm is *Phoenix dactylifera* L., and it is an important member of the family Palmacea. There are over 2000 different date varieties, which vary in shape, size, and weight. Usually they are oblong in shape, although certain varieties may be almost round. The length ranges from 1.8 to 11.0 cm, and the width from 0.8 to 3.2 cm; the average weight per fruit is 2−60 g (Zaid, 2002). As with the fruit, the seed characteristics

Nuts & Seeds in Health and Disease Prevention. DOI: 10.1016/B978-0-12-375688-6.10053-2

also vary greatly according to variety, and environmental and growing conditions. The seed weight ranges from 0.5 g to 4 g, the length from 1.2 to 3.6 cm, and the width from 0.6 to 1.3 cm. The seed is usually oblong, ventrally grooved, with a small embryo, and with a hard endosperm made of a cellulose deposit on the inside of the cell walls (Zaid, 2002).

HISTORICAL CULTIVATION AND USAGE

Evidence of date palm cultivation goes as far back as 4000 BC in what is now southern Iraq (Zaid, 2002). The date palm is found in the Near East, North Africa, and the American continent, where dates are grown commercially in large quantities. Date palm trees are propagated by two techniques: seed and offshoot propagation. Seed propagation is the easiest and quickest method of propagation, but it is not an accurate propagation technique because no two seedlings will be alike. Because of its diversity, the seed approach can only be useful for breeding purposes. Date seeds are therefore discarded, or used as fodder for cattle, sheep, camels, and poultry. The use of date seed for animal feed in the traditional way is still likely the most common practice.

PRESENT-DAY CULTIVATION AND USAGE

The development of a tissue culture technique for the mass propagation of date palm plants has expanded the date palm industry. Substantial advantages are gained from this technique — a better production rate, greater strength, freedom from pests and diseases, and wider availability of valuable varieties. At present, seeds are still used mainly for animal feed. A coffee-like product is made from date seeds by drying, roasting, and grinding them in a similar way to coffee beans, to produce caffeine-free coffee. Date seed oil has been used to replace the portions of other vegetable oils in body creams, shampoos, and shaving soap formulations, and, in general, the quality of these cosmetic compositions is encouraging (Devshony *et al.*, 1992).

APPLICATIONS TO HEALTH PROMOTION AND DISEASE PREVENTION

Epidemiological studies have consistently shown that high fruit and vegetable consumption is associated with a reduced risk of several chronic diseases, such as coronary heart disease, cardiovascular disease, cancers, atherosclerosis, neurodegenerative diseases (such as Parkinson and Ahlzeimer), and inflammation, as well as aging. This is attributed to the fact that these foods may provide an optimal mixture of phytochemicals such as dietary fiber, natural anti-oxidants such as vitamins C, E, and beta-carotene, and phenolic compounds. Interestingly, the peel and seed fractions of some fruits possess higher antioxidant activity than the pulp fraction (Guo *et al.*, 2003). Date seeds appear to fit well into this category.

The moisture, protein, fat, ash, and carbohydrate contents of different date seed varieties from several studies are shown in Table 53.1. The reported composition of seeds varied as follows: 3.1—10.3% moisture, 2.3—6.4% protein, 5.0—13.2% fat, 0.9—1.8% ash, and 71.9—87.0% carbohydrates. Date seeds contain relatively high amounts of protein and fat compared to date flesh, where the levels were 1.5—3.0% and 0.1—1.4%, respectively (Al-Farsi *et al.*, 2007). Regarding the mineral content of date seeds, Ali-Mohamed and Khamis (2004) reported on six varieties: their values were as follows (mg/100 g): 459.8—542.2 potassium, 21.7—26.1 sodium, 6.5—11.3 calcium, 61.3—69.5 magnesium, 2.8—6.0 iron, 1.3—1.7 manganese, 1.0—1.4 zinc, and 0.4—0.6 copper.

Date seed protein contains the majority of essential amino acids; glutamic acid was the major amino acid in Deglet Nour and Allige date seeds, representing 17.8 and 16.8%, respectively (Bouaziz *et al.*, 2008). Al-Hooti and colleagues (1998) reported the fatty acid profile of five varieties of date seed; oleic acid was the predominant fatty acid (56.1%), followed by palmitic

TABLE 53.1 Date Seed Composition

Varieties	Moisture %	Protein %	Fat%	Ash %	Carbohydrate %	References
Mabseeli	3.1	3.9	5.0	1.0	87.0	Al-Farsi *et al.*, 2007
Um-sellah	4.4	5.4	5.9	1.2	83.1	Al-Farsi *et al.*, 2007
Shahal	5.2	2.3	5.1	0.9	86.5	Al-Farsi *et al.*, 2007
Fard	10.3	5.7	9.9	1.4	72.7	Hamada *et al.*, 2002
Khalas	7.1	6.0	13.2	1.8	71.9	Hamada *et al.*, 2002
Lulu	9.9	5.2	10.5	1.0	73.4	Hamada *et al.*, 2002
Deglet noor	9.4	5.0	9.2	1.0	75.4	Besbes *et al.*, 2004
Allig	8.6	4.7	11.6	1.0	74.1	Besbes *et al.*, 2004
Ruzeiz	5.4	6.4	9.7	1.0	77.5	Sawaya *et al.*, 1984
Sifri	4.5	5.9	10.0	1.1	78.5	Sawaya *et al.*, 1984
Average	6.8	5.1	9.0	1.1	78.0	

Data are expressed on wet weight basis.

acid (11.9%), linoleic acid (11.6%), lauric acid (8.3%), myristic acid (6.0%), and stearic acid (2.6%).

Table 53.2 shows the dietary fiber, phenolic, and antioxidant content of several date seed varieties. Date seeds are a very rich source of dietary fiber; the level varied between 64.5 and 80.15 g/100 g fresh weight. Insoluble dietary fiber (hemicellulose, cellulose, and lignin) is considered to be the major constituent of seed fiber (Al-Farsi & Lee, 2008). The high nutritional value of date seeds is based on their dietary fiber content, which makes them suitable for the preparation of fiber-based foods and dietary supplements. Dietary fiber has important therapeutic implications for certain conditions, such as diabetes, hyperlipidemia, and obesity, and may have a protective effect against hypertension, coronary heart disease, high cholesterol, colorectal and prostate cancers, and intestinal disorders (Tariq *et al.*, 2000).

Date seeds are also a rich source of phenolics and antioxidants, which ranges from 3102 to 4430 mg gallic acid equivalent/100 g, and 58,000 and 92900 μmol Trolox equivalent/100 g, respectively (Table 53.2). Al-Farsi and Lee (2008) reported the phenolic acids of date seeds; of the nine phenolic acids detected, *p*-hydroxybenzoic (9.89 mg/100 g), protocatechuic (8.84 mg/100 g), and *m*-coumaric (8.42 mg/100 g) acids were found to be among the highest. Since the dietary fiber (5.9–8.7 g/100 g), phenolic (172–246 mg gallic acid equivalent/100 g), and antioxidant (14,600–16,200 μmol Trolox equivalent/100 g) contents in date flesh are much lower than in date seeds, date seeds could potentially be utilized as a functional food ingredient (Al-Farsi *et al.*, 2007). Phenolic compounds of fruit seeds, such as phenolic acids and flavonoids, have been shown to possess many beneficial effects, including antioxidant, anticarcinogenic, antimicrobial, antimutagenic, and anti-inflammatory activities, and the

TABLE 53.2 Dietary Fiber, Phenolics, and Antioxidants of Date Seeds

Varieties	Fiber g/100 g	Phenolics mg/100 g	Antioxidants μmol/100 g	References
Mabseeli	79.84	4430	58000	Al-Farsi *et al.*, 2007
Um-sellah	80.15	4293	90300	Al-Farsi *et al.*, 2007
Shahal	77.75	3102	92900	Al-Farsi *et al.*, 2007
Fard	67.8	—	—	Hamada *et al.*, 2002
Khalas	64.5	—	—	Hamada *et al.*, 2002
Lulu	68.8	—	—	Hamada *et al.*, 2002
Average	73.1	3942	80400	

Data are expressed on wet weight basis.

reduction of cardiovascular disease (Shahidi & Naczk, 2004). Thus, it is important to increase the antioxidant intake in the human diet, and one way of achieving this is by enriching food with natural phenolics. As some synthetic antioxidants may exhibit toxicity, have high manufacturing costs, and have lower efficiency than natural antioxidants (Soong & Barlow, 2004), Al-Farsi and Lee (2008) have developed a technique that enriches the dietary fiber and phenolic content of date seeds to 93.5 g/100 g and 18.1 g ferulic acid equivalent/100 g, respectively.

Date seed oil is edible, but, due to the low extraction rate ($\sim$9%), it is not competitive with other oil crops. The average chemical characteristics of four varieties of date seed oil were: acid value 1.04, iodine value 49.5, saponification value 221.0, and unsaponifiable matter 0.8%. The major unsaturated fatty acid was oleic acid (42.3%), while the main saturated fatty acid was lauric (21.8%), followed by linoleic (13.7%), myristic (10.9%), and palmitic (9.6%) (Devshony et al., 1992). The percentage of un-saturation of Ruzeiz and of Sifri date seed oils was reported to be 52.49 and 54.52%, respectively, which is considered relatively low. Date seed oil has a lower degree of unsaturation and low content of linoleic acid compared with the commonly consumed vegetable oils, which may potentially be used in human and animal diets (Sawaya et al., 1984).

Seed oils of Deglet Nour and Allig cultivars were compared in terms of phenolic, toco-pherol, and sterol profiles (Besbes et al., 2004). The total phenols ranged from 22.0 to 52.1 mg caffeic acid equivalent/100 g, which is relatively high compared to most edible oils except for olive oil, which is considered to be a rich source of phenolic compounds in the Mediterranean diet (Besbes et al., 2004). For example, the total phenolic content in olive oil has been shown to range from 12.4 to 51.6 mg/100 g (Nissiotis & Tasioula-Margari, 2002). This may explain the fact that the oxidative stability of date seed oils was higher than that of most vegetable oils, and comparable to that of olive oil (Besbes et al., 2004). Date seed oil could also be considered as a potential source of natural phenolic compounds, in addition to their contribution to oxidative rancidity resistance and their participation in conferring a specific flavor to the oil (Caponio et al., 1999). Due to the high oxidative stability of date seed oil, it would be a good ingredient for cosmetic and pharmaceutical products such as sun-block creams that provide protection against both UV-A and UV-B, which are responsible for cellular damage (Besbes et al., 2004). α-Tocopherol was found to be the predominant tocopherol in date seed oils from Deglet Nour (24.97%) and Allig (38.85%) cultivars. The total sterol content of Deglet Nour and Allig cultivars was 350 and 300 mg/100 g, respectively; the sterol marker, β-sitosterol, accounted for 83.31 and 78.66% of the total sterols, respectively (Besbes et al., 2004). These minor components are not only very important for the functional properties of oils (oxidation resistance, taste, aroma, and color), but could also have many health benefits.

Several studies have reported the advantages of the incorporation of date seeds into animal diets (Elgasim et al., 1995; Hussein et al., 1998; Ali et al., 1999). Some of these advantages include increased weight gain, improved feed efficiency, and improved meat palatability. Adding date seed to the starter and finisher diets improved body weight gain, feed conversion, and growth performance, comparable to the corn–soybean meal diet of broiler chicks (Hussein et al., 1998). Results show that the date seed can be included at 10% in broiler diets to support and enhance growth performance (Hussein et al., 1998). Elgasim and colleagues (1995) found that the date seed was effective in increasing body weight gain and the depo-sition of back fat in sheep. Ali and colleagues (1999) found that a feeding treatment with normal date seeds (7–14%) significantly increased the testosterone in plasma, and the body weight of rats. The protein of date seeds has a higher concentration of lysine, which is often the limiting amino acid in diets based on cereals (Sawaya et al., 1984). Thus date seed can be used to replace some of the expensive vegetable proteins in livestock or poultry feed.

However, feed value is not only determined by composition, but also by the accessibility and digestibility of the components. The hard structure of date seeds is a real obstacle to optimizing the feed value, although it is also claimed that seeds are an excellent source of slow-release energy for camels during long desert journeys. Traditionally, date seeds were soaked in water before they were fed to ruminants. Date seeds submerged in water for 72 hours will gain 25% in weight, which may increase to 50% after a week (Zaid, 2002).

ADVERSE EFFECTS AND REACTIONS (ALLERGIES AND TOXICITY)

A few studies have reported allergy or hypersensitivity to date palm fruit and pollen (Waisel *et al.*, 1994; Kwaasi *et al.*, 1999). The high concentration of selenium detected in some date varieties, which is related to the selenium content of the soil, also gives some cause for concern (Al-Farsi *et al.*, 2005). However, no studies are available reporting adverse effects of date seeds.

SUMMARY POINTS

- Date seeds comprise up to 15% of the date fruit's weight.
- Date seeds have been used mainly for animal feed.
- The seeds contain a higher content of protein and fat compared to date flesh.
- The high content of dietary fiber and phenolics in date seeds makes them a good ingredient for functional foods.
- Date seed oil has high oxidation stability due to the high content of phenolics.
- Utilization of seed in animal feed will improve weight gain and feed efficiency.

References

Al-Farsi, M., & Lee, C. Y. (2008). Optimization of phenolics and dietary fibre extraction from date seeds. *Food Chemistry, 108*, 977—985.

Al-Farsi, M., Alasalvar, C., Morris, A., Baron, M., & Shahidi, F. (2005). Compositional and sensory characteristics of three native sun-dried date (*Phoenix dactylifera* L.) varieties grown in Oman. *Journal of Agricultural and Food Chemistry, 53*, 7586—7591.

Al-Farsi, M., Alasalvar, C., Al-Abid, M., Al-Shoaily, K., Al-Amry, M., & Al-Rawahy, F. (2007). Compositional and functional characteristics of dates, syrups, and their by-products. *Food Chemistry, 104*, 943—947.

Al-Hooti, S., Sidhu, J. S., & Qabazard, H. (1998). Chemical composition of seeds date fruit cultivars of United Arab Emirates. *Journal of Food Science and Technology, 35*, 44—46.

Ali, B. H., Bashir, A. K., & Alhadrami, G. (1999). Reproductive hormonal status of rats treated with date pits. *Food Chemistry, 66*, 437—441.

Ali-Mohamed, A. Y., & Khamis, A. S. H. (2004). Mineral ion content of seeds of six cultivars of Bahraini date palm (*Phoenix dactylifera*). *Journal of Agricultural and Food Chemistry, 52*, 6522—6525.

Besbes, S., Blecker, C., Deroanne, C., Bahloul, N., Lognay, G., Drira, N., et al. (2004). Phenolic, tocopherol and sterol profiles. *Journal of Food Lipids, 11*, 251—265.

Bouaziz, M. A., Besbes, S., Blecker, C., Wathelet, B., Deroanne, C., & Attia, H. (2008). Protein and amino acid profiles of Tunisian Deglet Nour and Allig date palm fruit seeds. *Fruits, 63*, 37—43.

Caponio, F., Alloggio, V., & Gomes, T. (1999). Phenolic compounds of virgin olive oil: Influence of paste preparation techniques. *Food Chemistry, 64*, 203—209.

Devshony, S., Eteshola, A., & Shani, A. (1992). Characterisation and some potential application of date palm (*Phoenix dactylifera* L.) seeds and seeds oil. *Journal of American Oil Chemists Society, 69*, 595—597.

Elgasim, E. A., Alyousef, Y. A., & Humeida, A. M. (1995). Possible hormonal activity of date pits and flesh fed to meat animals. *Food Chemistry, 52*, 149—152.

FAO (2010). Statistical Databases; http://faostat.fao.org, accessed October 24, 2010.

Guo, C., Yang, J., Wei, J., Li, Y., Xu, J., & Jiang, Y. (2003). Antioxidant activities of peel, pulp and seed fractions of common fruits as determined by FRAP assay. *Nutrition Research, 23*, 1719—1726.

Hamada, J. S., Hashim, I. B., & Sharif, A. F. (2002). Preliminary analysis and potential uses of date pits in foods. *Food Chemistry, 76*, 135—137.

Hussein, A. S., Alhadrami, G. A., & Khalil, Y. H. (1998). The use of dates and date pits in broiler starter and finisher diets. *Bioresource Technology, 66*, 219—223.

Kwaasi, A. A., Harfi, H. A., Parhar, R. S., Al-Sedairy, S. T., Colllison, K. S., Panzani, R. C., et al. (1999). Allergy to date fruits: characterization of antigens and allergens of fruits of date palm (*Phoenix dactylifera* L.). *Allergy, 54*(12), 1270–1277.

Nissiotis, M., & Tasioula-Margari, M. (2002). Changes in antioxidant concentration of virgin olive oil during thermal oxidation. *Food Chemistry, 77,* 371–376.

Sawaya, W. N., Khalil, J. K., & Safi, W. J. (1984). Chemical composition and nutritional quality of date seeds. *Journal of Food Science, 49,* 617–619.

Shahidi, F., & Naczk, M. (2004). *Phenolics in food and nutraceuticals.* Boca Raton, FL: CRC Press.

Soong, Y., & Barlow, P. J. (2004). Antioxidant activity and phenolic content of selected fruit seeds. *Food Chemistry, 88,* 411–417.

Tariq, N., Jenkins, D. J. A., Vidgen, E., Fleshner, N., Kendall, C. W. C., & Story, J. A. (2000). Effect of soluble and insoluble fibre diets on serum prostate specific antigen in men. *Journal of Urology, 163,* 114–118.

Waisel, Y., Keynan, N., Gil, T., Tayar, D., Bezerano, A., Goldberg, A., et al. (1994). Allergic responses to date palm and pecan pollen in Israel [in Hebrew]. *Harefuah, 126*(6), 305–310.

Zaid, A. (2002). Date palm cultivation. FAO Plant Production and Protection Paper 156, Rome, Italy.

CHAPTER 54

Piceatannol, an Antitumor Compound from *Euphorbia lagascae* Seeds

Maria-José U. Ferreira[1], Noélia Duarte[1], Krystyna Michalak[2], Andras Varga[3], Joseph Molnár[4]

[1] iMed.UL, Faculty of Pharmacy, University of Lisbon, Lisbon, Portugal
[2] Wrocław Medical University, Department of Biophysics, Wrocław, Poland
[3] Department of Molecular Parasitology, Humboldt University, Berlin, Germany
[4] Department of Medical Microbiology, University of Szeged, Szeged, Hungary

453

LIST OF ABBREVIATIONS

BCECF, 2′,7′-bis-(3-carboxypropyl)-5-(and-6)-carboxyfluorescein
FAR, fluorescence activity ratio value
MDR, multidrug resistance
MRP, Multidrug resistance associated protein
P-gp, P-glycoprotein

INTRODUCTION

Multidrug resistance (MDR) remains a major obstacle in cancer chemotherapy. The development of reversing agents has been considered one of the most promising strategies to overcome MDR. Since apoptosis plays a crucial role in MDR, another realistic approach is the discovery of new apoptosis inducers (Gottesman *et al.*, 2002; Szakács *et al.*, 2006). Therefore, in order to discover new, effective plant-derived compounds against drug-resistant cancer cells, we have evaluated, as MDR reversing and apoptosis inducing agents, the stilbene piceatannol, isolated from the seeds of *Euphorbia lagascae*, and some of its derivatives (Ferreira *et al.*, 2006; Wesolowska *et al.*, 2007). Their effects on the activity of voltage-gated potassium channels Kv1.3 were also investigated (Teisseyre *et al.*, 2009).

BOTANICAL DESCRIPTION

Euphorbia lagascae Spreng (Euphorbiaceae) is a species characteristic to the Iberic Peninsula and Corsica. It is an annual plant with smooth, light green stems, about 20—60 cm tall, which flowers in early spring and fruits in April and May, depending upon climate. The seeds have a grayish color, black-marked or speckled, and are ovate, truncate, and subcompressed. The Euphorbiaceae family consists of about 300 genera and 7500 species better developed in tropical and subtropical regions. The largest genus is *Euphorbia* L., with over 2000 species found in the tropical and subtropical regions of Africa and America, and also in temperate zones worldwide. *Euphorbia* species range from annual or perennial herbs, through woody shrubs and trees, to succulent plants, and are characterized by milky latex (Krewson & Scott, 1966; Castroviejo *et al.*, 1989).

HISTORICAL CULTIVATION AND USAGE

Euphorbia species have been widely used in traditional medicine all over the world, to treat several diseases such as skin ulcers and warts, as well as cancer tumors and intestinal parasites (Hartwell, 1969). The common name "spurge" derives from the French word *espurgier* (Latin *expurgare*), which means "to purge," due to the use of *Euphorbia* latex as a purgative. The botanical name *Euphorbia* derives from the Greek Euphorbus, in honor of the physician of Juba II, a Romanized king of Mauritania who was supposed to have used in his treatments a plant (*E. resinifera*) with milky latex and powerful medicinal properties (Appendino & Szallasi, 1997).

PRESENT-DAY CULTIVATION AND USAGE

These days, the main uses of *Euphorbia* species are horticultural. In fact, many *Euphorbia* species are cultivated for their brilliant, showy bracts, as well as for their frequently colorful foliage. This is the case of *E. pulcherrima* (poinsettia), which is cultivated for ornamental purposes as a popular Christmas decoration.

Euphorbia lagascae has been cultivated for the production of 12-epoxyoctadeca-*cis*-9-enoic acid (vernolic acid), which is found at high levels in its seeds. Vernolic acid is an unusual C_{18} epoxidated fatty acid with potential industrial value due to the unique chemical properties associated with the Δ^{12} epoxy group. Vernolic acid-enriched seed oils can be used as plasticizers of polyvinylchloride — a market that is currently served by petroleum-derived compounds such as phthalates. In addition, the ability of the epoxy group to crosslink makes vernolic acid-containing oils useful in adhesives and coating materials such as paint (Cahoon *et al.*, 2002).

APPLICATIONS TO HEALTH PROMOTION AND DISEASE PREVENTION

Euphorbia lagascae seeds are rich in piceatannol (*trans*-3,5,3′,4′-tetrahydroxystilbene, 1), a stilbene that was first identified in *Vouacapoua* species (King *et al.*, 1956), and is also present in rhubarb, sugar cane, berries, peanuts, red wine, and the skin of grapes (Roupel *et al.*, 2006).

Stilbenes are a class of plant polyphenols, biosynthesized via the phenylpropanoid pathway and characterized by the presence of a 1,2-diphenylethylene nucleus, mainly as *trans* stereoisomers. The majority of them belong to a class of defence molecules called phytoalexins that protect against stress, injury, ultraviolet irradiation, and fungal infection. The most well-known stilbene is resveratrol (*trans*-3,4′,5-trihydroxystilbene, **6**), which is produced by a number of unrelated plants, such as grapevines (*Vitis vinifera*), peanuts (*Arachis hypogaea*), mulberries (*Morus* species), and pines (*Pinus* species) (Roupel *et al.*, 2006). It has been reported to have numerous beneficial properties, such as potent anti-inflammatory, anticancer, antioxidant, anti-aging, and antiviral activities, as well as cardiovascular and neuroprotective effects. Furthermore, resveratrol has been identified as an effective candidate for cancer chemoprevention due its ability to block each step of the carcinogenesis process (Kundu & Surh, 2008). Nevertheless, recent research has shown that resveratrol undergoes metabolism by cytochrome P450 enzyme CYP1B1 to piceatannol (Potter *et al.*, 2002). This enzyme catalyzes aromatic hydroxylation reactions, and is overexpressed in a wide variety of human tumors. Importantly, it has been suggested that resveratrol may act as a pro-drug of piceatannol that would be the compound responsible for the biological effect on tumor cells. Structurally, piceatannol differs from resveratrol by having an additional aromatic hydroxyl group. Piceatannol was first identified as an antileukemic agent (Ferrigni *et al.*, 1984). However, most of the reported studies on piceatannol have focused on its potent protein-tyrosine kinase inhibitory properties. It was shown that piceatannol inhibits a variety of tyrosine kinases involved in cell proliferation, in the phosphorylation of DNA transcription factors, and several other tyrosine kinases which are overexpressed in different types of cancer (Geahlen & McLaughlin, 1989; Potter *et al.*, 2002). Moreover, it was found that piceatannol had an inhibitory effect on asexual maturation of *Plasmodium falsiparum* (Mishra *et al.*, 1999), and was also active against the extracellular and intracellular forms of *Leishmania* parasites (Duarte *et al.*, 2008).

The occurrence of multidrug resistance is considered to be the major obstacle to the successful chemotherapy of cancer. MDR can be defined as the intrinsic or acquired simultaneous resistance of cells to multiple classes of structurally unrelated drugs that do not have a common mechanism of action. The most widely studied mechanism of MDR is related to an increased drug efflux through overexpression of certain transport proteins, such a P-glycoprotein (P-gp) or the Multidrug Resistance Associated Proteins (MRP). These proteins belong to the ABC superfamily of transporters that act as ATP-dependent efflux pumps, and can prevent the accumulation of drugs by expelling them from the cell membrane before they are able to interact with their cellular targets (Szakács *et al.*, 2006). Nevertheless, MDR is a complex phenomenon resulting from several other biochemical mechanisms. One of these mechanisms is related to resistance to apoptosis, or programmed cell death, which is an essential physiologic process that plays a key role in tissue and organ development, immune system function, and tissue homeostasis. Deregulation of apoptosis has been implicated in several pathological conditions that include cancer. This might occur as a result of genetic mutations or changes in the cellular apoptotic pathways during exposure to the antineoplasic drugs (Gottesman *et al.*, 2002). Therefore, another realistic approach to overcome MDR is the discovery of new drugs able to modulate the expression of molecules involved in the apoptotic pathway and able to induce specific apoptosis for malignant tumor cells, providing new useful tools for the treatment of patients with drug resistant malignancies.

In our search for antitumor compounds from natural sources, and in particular from *Euphorbia* species, the stilbene piceatannol (**1**) was isolated in large quantities from *Euphorbia lagascae* seeds (Ferreira *et al.*, 2006). This compound was acetylated and methylated to afford the stilbenes *trans*-3,5,3′,4′-tetraacetoxystilbene (**2**), *trans*-3,5,3′,4′-tetramethoxystilbene (**3**), *trans*-3,5-dihydroxy-3′,4′-dimethoxystilbene (**4**), and *trans*-3,5,3′-trihydroxy-4′-methoxystilbene (**5**) (Ferreira *et al.*, 2006; Wesolowska *et al.*, 2007).

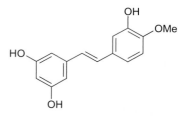

FIGURE 54.1

Chemical structures of stilbenes **1–6**.

Piceatannol (**1**)

Resveratrol (**6**)

trans-3,5,3',4'-tetracetoxystilbene (**2**)

trans-3,5,3',4'-tetramethoxystilbene (**3**)

trans-3,5-dihydroxy-3',4'-dimethoxystilbene (**4**)

trans-3,5,3'-trihydroxy-4'-methoxystilbene (**5**)

The stilbenes 1–5 (Figure 54.1) were evaluated on the reversion of multidrug resistance mediated by P-gp, by using the rhodamine-123 exclusion test, on human MDR1 gene-transfected mouse lymphoma cells. Moreover, stilbenes **1–3** were also assessed for their potential as inhibitors of MRP1, by measuring the efflux of the fluorescent MRP1 substrate BCECF, out of human erythrocytes (Wesolowska *et al.*, 2007).

Trans-3,5,3',4'-tetramethoxystilbene (**3**), was found to be a powerful modulator of P-gp activity in a dose-dependent manner (FAR = 56.0 at 4 μg/ml and FAR = 75.1 at 40 μg/ml; Table 54.1) and manifold activity when compared to that of the positive control verapamil (FAR = 18.7 at 10 μg/ml concentration). In contrast, piceatannol (**1**), and its derivatives **2**, **4**, and **5**, were found to be ineffective in this MDR reversal assay.

Regarding the MRP1 inhibitory assay, the most effective compound was piceatannol (**1**), which showed an IC_{50} value of *ca.* 42 μM (Table 54.2, Figure 54.2). On the other hand, its tetraacetoxy derivative (**2**) inhibited MRP1 transport to a lesser extent (IC_{50} *ca.* 52 μM), and 3,5,3',4'-tetramethoxystilbene (**3**) was not able to reduce BCECF efflux out of erythrocytes by more than 30% (Wesołowska *et al.*, 2007).

The five stilbenes tested have very similar structures, differing only in the substitution pattern of the benzene rings; piceatannol has four free hydroxyl groups, while in compounds **2** and **3** these groups are acetylated and methylated. In compounds **4** and **5**, two and one of the hydroxyl groups are replaced by a methoxyl group, respectively. It can be concluded that these structurally similar compounds can display considerably different MDR reversal effects mediated either by P-gp or MRP1. When compared to piceatannol (**1**), the existence of four methyl substituents in 3,5,3',4'-tetramethoxystilbene (**3**) significantly changes several molecular properties, such as the lipophilicity (logP values of 4.2 and 2.5, for compounds

TABLE 54.1 MDR Modulator Activities and Physico-chemical Properties of Stilbenes

Compound	R_1	R_2	R_3	R_4	N^OH Acc.	N^OH Don.	MW	log P	FAR (4 μg/ml)	FAR (40 μg/ml)
1	OH	OH	H	OH	4	4	244	2.5	0.9	0.8
2	Ac	Ac	Ac	Ac	8	0	412	2.2	1.0	2.9
3	Me	Me	Me	Me	0	4	300	4.2	56.0	75.1
4	OH	OH	Me	Me	4	2	272	3.1	1.4	—
5	OH	OH	OH	Me	4	3	258	2.8	1.1	1.4

Compounds with fluorescence activity ratio (FAR) values higher than 1 were considered active as P-gp modulators, and those with FAR values higher than 10 as strong MDR modulators. Compound **4** was tested at 2.7 μg/ml; physico-chemical parameters were determined by using the JME molecular editor (http://www.molinspiration.com/).
Acc., number of hydrogen acceptors; Don., number of hydrogen donors; MW, molecular weight; log P, octanol/water partition coefficient.
Data from Ferreira *et al.* (2006), with permission.

3 and **1**, respectively). These differences may explain the observed effects of stilbenes **1** and **3** in both efflux proteins, P-gp and MRP1. *Trans*-3,5,3′,4′-tetramethoxystilbene (**3**) is the most lipophilic and also the most active modulator in the rhodamine-123 assay. On the other hand, piceatannol (**1**), the least lipophilic stilbene, was shown to be the strongest MRP1 inhibitor. Thus, it is interesting to note that these results corroborate the fact that P-gp transports mainly lipophilic compounds, while MRP1 transports more hydrophilic molecules like organic anions, including glutathione and glucuronide conjugates (Szakács *et al.*, 2006).

Furthermore, the stilbenes **1** and **3**–**5** were also evaluated as apoptosis inducers in human MDR1 gene-transfected mouse lymphoma cells, using the annexin-V/propidium iodide assay. The most significant results were obtained for piceatannol (**1**), which showed strong activity (36.5 and 62.8% total apoptosis induction at 4 and 40 μg/ml, respectively; Table 54.3). Methylation of piceatannol decreased the apoptosis induction activity, showing the methylated derivatives **3**–**5** to have 24.3%, 23.6%, and 13.5% total apoptosis induction activity at the higher concentration tested (Ferreira *et al.*, 2006).

In recent years, an important role of ion channels in cancer has been revealed. Ion channels may affect different pathways regulating cell proliferation and apoptosis. It was discovered that potassium channels in particular participate in cancer development and apoptosis. For example, in human breast and colon cancer, or in gliomas, the expression of voltage-dependent Kv1.3 channels is up-regulated. The same channels are also involved in apoptosis of Jurkat T lymphocytes. K^+ channel blockers may inhibit cell proliferation, as this was shown in different cell lines. In human T lymphocytes, voltage-dependent Kv1.3 channels

457

TABLE 54.2 MRP1 Transport Activity Modulation by Stilbenes 1–3

Compound	Concentration (μM)	MRP1 Inhibition (%)
1	25	35.2
	75	61.3
2	25	17.1
	75	56.3
3	25	30.2
	75	28.5

Inhibition of MRP1 activity was calculated from the measurements of rate transport of BCECF (2′,7′-bis-(3-carboxypropyl)-5-(and-6)-carboxyfluorescein) in the presence of stilbene derivatives. Experiments were carried out on human erythrocytes, and BCECF was used as fluorescence substrate analog.
Data from Wesołowska *et al.* (2007), with permission.

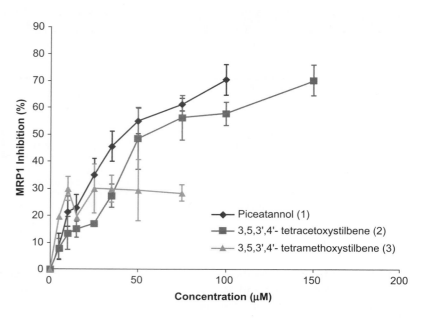

FIGURE 54.2

Inhibition of MRP1-mediated BCECF efflux from erythrocytes by piceatannol (**1**), *trans*-3,5,3′,4′-tetracetoxystilbene (**2**) and *trans*-3,5,3′,4′-tetramethoxystilbene (**3**). Fluorescence probe BCECF (2′,7′-bis-(3-carboxypropyl)-5-(and-6)-carboxyfluorescein) as substrate of multidrug resistance-associated protein 1 (MRP1) was extruded from erythrocytes and the rate of probe efflux was determined by fluorescence spectroscopy. The fluorescence of BCECF was measured in the supernatant. The percentage of transport inhibition in the presence of piceatannol derivatives was calculated for different concentrations of the compounds. Each point is the mean ± SD. *Data from Wesołowska* et al. *(2007), with permission.*

are expressed endogenously and are the main type of potassium channels in these cells. Therefore, human T lymphocytes were used as a model system in the patch-clamp experiments. Lymphocytes were isolated from peripheral blood of healthy donors. Patch-clamp recordings of Kv1.3 currents were performed in the whole-cell configuration, and the standard method of voltage stimuli depolarizing cell membrane from −100 mV to +40 mV was applied. The effect of piceatannol (**1**) and its two derivatives (**2**, **3**) on the activity of voltage-dependent potassium channels Kv1.3 was investigated (Teisseyre *et al.*, 2009). The influence of stilbenes **1**−**3** on Kv1.3 currents was compared with the previously studied effect of resveratrol (**6**). Voltage-dependent potassium channels were not inhibited by piceatannol, in contrast to the effect of resveratrol, which caused the concentration-dependent decrease of the potassium

TABLE 54.3 Effect of Stilbenes 1 and 3−5 on Apoptosis Induction in Human MDR1 Gene-transfected Mouse Lymphoma Cells

Compound	Concentration (μg/ml)	Early Apoptosis%	Total Apoptosis%	Necrosis%
Cell control without staining	−	0.02	0.04	0.10
Cell control annexin-V + PI	−	2.98	18.11	2.26
Cell control annexin-V − PI	−	0.08	18.31	0.01
M-627	50	0.02	98.0	1.8
1	4	6.8	36.5	0.8
	40	1.5	62.8	2.9
3	10	5.0	19.0	3.8
	40	1.6	24.3	2.5
4	10	2.2	8.5	4.7
	40	1.6	23.6	2.2
5	10	2.0	10.2	5.2
	40	1.4	13.5	1.4

Annexin-V assays are based on the observation that during apoptosis, phosphatidylserine (PS) is translocated from the inner to the outer leaflet of the plasma membrane. The fluorescent labeling of annexin-V enables the flow cytometric detection of externalized PS, and hence apoptotic cells. Disruption of membrane integrity during late apoptosis or necrosis makes the cell interior accessible to annexin-V, but they also stain positive for propidium iodide (PI), a DNA fluorescent binding dye that only stains non-viable cells. Thus, by using propidium iodide together with annexin-V, apoptotic cells could be identified and discriminated from necrotic cells.
M-627, positive control (12H-benzo(α)phenothiazine).
Data from Ferreira *et al.* (2006), with permission.

current amplitude (Teisseyre & Michalak, 2006). Taking into account that strong inhibition of membrane transporters and ion channels was often observed in cases of hydroxy-, methoxy-, acetoxy-, and prenyl-derivatives, compared to the effect of parent poliphenolic compounds, we have also determined the influence of acetoxy- and methoxy- derivatives of piceatannol on Kv1.3 currents in the same model system. Our studies revealed that only 3,5,3′,4′- tetramethoxystilbene (**3**) inhibits Kv 1.3 channels in T lymphocytes. In the presence of the methoxy-derivative at 30-µM concentration, the average peak current at +40 mV was 31 ± 4% of control value. The effect was fully reversible. In contrast, 3,5,3′,4′-tetracetoxystilbene (**2**) was ineffective as a Kv1.3 channel blocker.

Piceatannol and its derivatives exert differential biological effects by modifying multiple cellular targets. The influence of these stilbenoid compounds on membrane transport may be mediated, at least in part, by non-specific membrane effects similar to those observed in pure lipid model systems (Wesołowska *et al.*, 2009).

ADVERSE EFFECTS AND REACTIONS (ALLERGIES AND TOXICITY)

Despite its traditional medical applications, the use of *Euphorbia* species has been hampered by the occurrence of skin irritant latex that characterizes these plants.

SUMMARY POINTS

- Piceatannol, isolated in large quantities from *Euphorbia lagascae*, is an effective Multidrug Resistance-Associated Protein-1 inhibitor.
- Piceatannol was also shown to be a strong apoptosis inducer, although its methylated derivatives showed just moderate apoptosis induction activity.
- *Trans*-3,5,3′,4′-tetramethoxystilbene was found to be a powerful modulator of P-glycoprotein activity in a dose-dependent manner.
- Piceatannol and its derivatives 3,5,3′,4′-tetraacetoxystilbene, 3,5-dihydroxy-3′,4′-dimethoxystilbene, and 3,5,3′-trihydroxy-4′-methoxystilbene were found to be ineffective in the P-glycoprotein-mediated multidrug resistance reversal assay.
- Piceatannol and its derivatives exert differential biological effects by modifying multiple cellular targets.
- These data, together with previous work, point to stilbenes as a group of promising antitumor compounds.

References

Appendino, G., & Szallasi, A. (1997). Euphorbium: modern research on its active principle, resiniferatoxin, revives an ancient medicine. *Life Sciences, 60,* 681–696.

Cahoon, E., Ripp, K., Hall, S., & McGonigle, B. (2002). Transgenic production of epoxy fatty acids by expression of a cytochrome P450 enzyme from *Euphorbia lagascae* seed. *Plant Physiology, 128,* 615–624.

Castroviejo, S., Aedo, C., & Benedí, C. (1989). *Flora Iberica: plantas vasculares de la Península Iberica e Islas Baleares* (pp. 190–191, 225). Madrid Spain: Real Jardin Botánico.

Duarte, N., Kayser, O., Abreu, P., & Ferreira, M. J. (2008). Antileishmanial activity of piceatannol isolated from *Euphorbia lagascae* seeds. *Phytotherapy Research, 22,* 455–457.

Ferreira, M. J., Duarte, N., Gyémánt, N., Radics, R., Chepernev, G., Varga, A., & Molnár, J. (2006). Interaction between doxorubicine and a resistance modifier stilbene in the antiproliferative effect on multidrug resistant mouse lymphoma and human breast cancer cells. *Anticancer Research, 26,* 3541–3546.

Ferrigni, N. R., McLaughlin, J. L., Powell, R. G., & Smith, C. R., Jr. (1984). Use of potato disc and brine shrimp bioassays to detect activity and isolate piceatannol as the antileukemic principle from the seeds of *Euphorbia lagascae. Journal of Natural Products, 47,* 347–352.

Geahlen, R. L., & McLaughlin, J. (1989). Piceatannol (3,4,3′ 5′-tetrahydroxy-*trans*-stilbene) is a naturally occurring protein-tyrosine kinase inhibitor. *Biochemical and Biophysical Research Communications, 165,* 241–245.

Gottesman, M. M., Fojo, T., & Bates, S. (2002). Multidrug resistance in cancer: role of ATP-dependent transporters. *Nature Reviews Cancer, 2,* 48–58.

Hartwell, J. (1969). Plants used against cancer. A survey. *Lloydia, 32,* 153–205.

King, F. E., King, T. J., Godson, D., & Mannin, L. (1956). The chemistry of extractives from hardwoods. Part XXVIII. The occurrence of 3,4,3′,5′-tetrahydroxy- and 3,4,5,3′,5′-pentahydroxy-stilbene in *Vouacapoua* species. *Journal of Chemical Society* 4477–4480.

Krewson, C., & Scott, W. (1966). *Euphorbia lagascae* Spreng., an abundant source of epoxyoleic acid; seed extraction and oil composition. *Journal of the American Oil Chemists' Society, 43,* 171–174.

Kundu, J., & Surh, Y. (2008). Cancer chemopreventive and therapeutic potential of resveratrol: mechanistic perspectives. *Cancer Letters, 269,* 243–261.

Mishra, N., Sharma, M., & Sharma, A. (1999). Inhibitory effect of piceatannol, a protein tyrosine kinase inhibitor, on asexual maturation of *Plasmodium falciparum. Indian Journal of Experimental Biology, 37,* 418–420.

Potter, G. A., Patterson, L. H., Wanogho, E., Perry, P. J., Butler, P. C., Ijaz, T., et al. (2002). The cancer preventative agent resveratrol is converted to the anticancer agent piceatannol by the cytochrome P450 enzyme CYP1B1. *British Journal of Cancer, 86,* 774–778.

Roupel, K. A., Remsberg, C. M., Yáñez, J. A., & Davies, N. M. (2006). Pharmacometrics of stilbenes: Seguing towards the clinic. *Current Clinical Pharmacology, 1,* 81–101.

Szakács, G., Paterson, J., Ludwig, J., Booth-Genthe, C., & Gottesman, M. M. (2006). Targeting multidrug resistance in cancer. *Nature Reviews Drug Discovery, 5,* 219–234.

Teisseyre, A., & Michalak, K. (2006). Inhibition of the activity of lymphocyte Kv1.3 potassium channels by resveratrol. *Journal of Membrane Biology, 214,* 123–129.

Teisseyre, A., Duarte, N., Ferreira, M. J. U., & Michalak, K. (2009). Influence of the multidrug transporter inhibitors on the activity of KV1.3 voltage-gated potassium channels. *Journal of Physiology and Pharmacology, 60,* 69–76.

Wesołowska, O., Wiśniewski, J., Duarte, N., Ferreira, M.J., & Michalak, K. (2007). Inhibition of MRP1 transport activity by phenolic and terpenic compounds isolated from *Euphorbia* species. *Anticancer Research, 27,* 4127–4134.

Wesołowska, O., Kuźdżal, M., Štrancar, J., & Michalak, K. (2009). Interaction of the chemopreventive agent resveratrol and its metabolite, piceatannol, with model membranes. *Biochimica Biophysica Acta, 1788,* 1851–1860.

Usage and Significance of Fennel (*Foeniculum vulgare* Mill.) Seeds in Eastern Medicine

Hafiz Abubaker Saddiqi, Zafar Iqbal
Department of Parasitology, University of Agriculture, Faisalabad, Pakistan

461

INTRODUCTION

Fennel (*Foeniculum vulgare* Mill.) (Family: Apiacae) is an aromatic herb. Vernacular names of *Foeniculum vulgare* Mill. in international and some Asian languages are given in Table 55.1. Fennel has been used in Eastern medicine since prehistoric times for the treatment of different ailments, especially in China, India, Pakistan, and Middle East. All parts of the fennel plant are used in Eastern medicine. Fennel seeds, however, are most frequently used in digestive disorders, as a vermicide, and for the improvement of eyesight. Major constituents found in the essential oil of fennel seeds are anethole, fenchone, methyl chavicol, limonene, phyllandrene, camphene, pinine, anisic acid, and palmitic, oleic, linoleic and petroselenic acids (Duke, 1985). Though major usage of fennel seeds is based on empirical evidence of treatment, work has also been conducted on scientific validation (Rizvi *et al.*, 2007). Different aspects of *Foeniculum vulgare* Mill. are reviewed in the following subsections.

BOTANICAL DESCRIPTION

Fennel (*Foeniculum vulgare* Mill.) is a biennial or short lived perennial herb, 1.2—2 m in height, with a thick, spindle-shaped taproot (Figure 55.1A—E). The stem is pithy, smooth, or finely-fluted. Its leaves are three—four pinnatisect, and are often filiform. Flowers are

TABLE 55.1 Vernacular Names of Fennel (*Foeniculum vulgare* Mill.)

Language	Vernacular Name
Urdu/Punjabi/Hindi/Unani	Saunf
English	Fennel
Sanskrit	Satupusa
Tamil	Perunchirakam
Persian	Badyan
Arabic	Razyanaj
Chinese	Hui-hasinag
Japanese	Uiko
Bengali	Mauri
Nepalese	Marchyashupp
German	Garten feuchel
Marhati	Shopha
Gujrati	Variyali

in umbels, and hermaphrodite (Figure 55.1E). The fruit is 6−7 mm long, oblong, ellipsoid or cylindrical (Figure 55.1D). It is an ovoid, ribbed double achene that is bluish initially, then becoming brownish gray (Rizvi *et al.*, 2007) (Figure 55.1D). The leaves have fleshy sheaths at the base, and are finely divided. Vascular strands and vittae of this plant are more well-developed than those of common fennel (Kapoor, 2001). Fennel seeds are diverse in size, being small in the subcontinent and bigger in size in Europe. Florence fennel has a broad bulbous base (Figure 55.1B), and its leaves have an aniseed flavor.

462

FIGURE 55.1
Different parts and ripened seeds of fennel (*Foeniculum vulgare* Mill.) plant. (A) fennel flower head; (B) fennel bulb; (C) fennel in flower; (D) fennel seeds; (E) fennel plant with separate body parts.

CHAPTER 55

Usage and Significance of *Foeniculum vulgare* Mill. Seeds

HISTORICAL CULTIVATION AND USAGE

Fennel (*Foeniculum vulgare* Mill.) is native to the Mediterranean region, but is currently cultivated in many parts of the world. The name *Foeniculum* is derived from the Latin word *foenum*, meaning hay, as the smell of fennel resembles that of hay. The plants prefer sandy, medium (loamy), clayey and well-drained soil, which may be acidic, neutral or basic (alkaline). They cannot grow in the shade, and require dry or moist soil. The plant has capacity to tolerate drought and strong winds, but not maritime exposure. Fennel was described by Hippocrates (460−377 BC) and Discorides (60 AD) as a diuretic and emmenagogue. Ibn Sina (980−1037 AD) referred to it as a deobstruent and eye tonic. Fennel can easily be traced through American, British (1932, 1996), and French (1994) Pharmacopoeias. In the 1st century, Romans considered fennel seeds to be valuable; they used them in culinary and medicinal items.

PRESENT-DAY CULTIVATION AND USAGE

Fennel is now cultivated in most parts of the world, including Pakistan and India, preferably in sandy clay with sufficient lime and plenty of water in dry periods (Rizvi *et al.*, 2007). It is also cultivated in dry, stony, calcareous soils near the sea. Its cultivation is affected by many diseases, such as leaf spot, blight, bulb rot, root rot, and gray mold rot, caused by different pathogens. In Pakistan, it is grown in temperate and subtropical areas up to an altitude of 2000 m, and cultivated as an annual crop. It can also be cultivated on the foothills and plains (Ahmad *et al.*, 2008). Major insects that attack this plant are *Systole coriandri*, *S. albepennis*, and aphids. Production of fennel is also influenced by the sowing season, and spacing. Autumn sowing has been reported as better for production (Ahmad *et al.*, 2004). The principal fennel producing countries in the world are India, Argentina, China, Indonesia, Russia, Japan, and Pakistan. Different parts of this plant are used in Eastern medicine, including dried ripe fruit, dried roots, and steam-distilled fennel oil. Fennel seeds have been reported to be effective in relaxing smooth muscles, and in promoting the functions of the digestive system (Rizvi *et al.*, 2007).

APPLICATIONS TO HEALTH PROMOTION AND DISEASE PREVENTION

Fennel seeds and seed oil (*Foeniculum vulgare* Mill.) are used to promote health, in prevention of diseases, and as a flavoring agent in food items. The seeds are sweet, laxative, stomachic, and stimulant, and are used as an appetizer. They are also used to treat headache, madness, flu, eye problems, weakness of eyesight, as a brain tonic, and in deafness.

In Greco-Arab (Unani) medicine, fennel seeds are used to cure bloat, diarrhea, menstrual problems, and piles. For centuries, fennel seeds remained a traditional herbal medicine in Europe and China. Uses of fennel (*Foeniculum vulgare* Mill.) seeds in different forms, purposes, and cultures are given in Tables 55.2−55.4.

Fennel seeds have a long history of being a galactagogue. Fennel seeds boiled with barley are helpful for lactating women. A poultice prepared from fennel seeds is helpful to treat inflamed breasts. Tea prepared from fennel seeds contains vitamins (B and C) and vital minerals (potassium, magnesium, and calcium), and is used as a herbal remedy to treat colic, release spasms in the digestive tract, and for the treatment of gas and bloating (Ahmad *et al.*, 2004; Hussain *et al.*, 2008). Fennel-seed tea has also prescribed to expel hookworms, and to get rid of intestinal bacterial infections (Kapoor, 2001).

Fennel seeds are effective diuretics, and help to relieve hypertension. The seeds are ingested in Hindu and Chinese cultures to speed up the removal of snake and scorpion poisons (Duke, 1985). Fennel water is mixed with NaOH and syrup, and used as "gripe

TABLE 55.2 Health, Culinary, and Industrial uses of Fennel (*Foeniculum vulgare* Mill.) in Eastern Countries

Use/Activity/Indication	Recipe/Remedy/Preparation
Food and flavor	Seeds used as flavoring, digestive and refreshing agent in different foods and drinks, such as rice, meat, vegetables, salad, sauces, soups, pastries, bread, tea, alcoholic beverage, etc., and many herbal medicines and toothpastes (Yamini et al., 2002; Hussain et al., 2008).
Galactogogue	Seeds improve breast growth because of phytoestrogen content, and thus may increase milk production (Kabir-ud-Din & Mahmood, 1937; Yeung, 1985).
Gripe water and syrup	Seed extract is included in gripe water, traditionally used to treat abdominal pain in children, and in syrups used as liver tonic (Hussain *et al.*, 2008).
Spices	Dry and fresh umbels are added as spices in the subcontinent and China (Yeung, 1985; Hussain *et al.*, 2008).
Anti-inflammatory	Methanolic extract of fennel seeds has inhibitory effects against inflammatory diseases and type IV allergic reactions, and also exhibited a central analgesic effect (Eun & Jae, 2004).
Antioxidant/anticancer	Fennel seeds have been described to possess antioxidant activity (Cai *et al.*, 2004).
Anti-snake bite venom	In China, fennel seeds have been reported to be effective against snake venom (Duke, 1985).
Industry	Fennel essential oil is added to soaps, creams, condiments, perfumes, and liquor (Yamini *et al.*, 2002; Hussain *et al.*, 2008).

TABLE 55.3 Disease Prevention and Treatment by the Use of Fennel (*Foeniculum vulgare* Mill.) Seeds in Eastern Medicine

Disease	Part/Extract
Phlegm/cough	Aqueous extract of seeds is mixed with honey and advised for a cough, as an expectorant (Kabir-ud-Din & Mahmood, 1937).
Chest pain	Seed decoction is mixed with NaCl and used for the alleviation of chest pain (Chughtai, 1950).
Atrophic vaginitis	Phytoestrogens in the seed extracts are helpful in treating atrophic vaginitis due to lack of estrogens, and pain in the female genital tract (Jadhav & Bhutani, 2005).
Asthma	Fennel seeds are soaked in the milky white secretions of calotropis (*Calotropis gigentia* L.) and dried in the shade. This preparation is helpful in treating asthma, at extremely low doses in view of its toxicity. Essential oils have the potential to relax bronchial smooth muscles (Chughtai, 1950).
Tympany/bloat/colic	Seeds are boiled in water, and when the volume is reduced to one-fourth, a small amount of common salt is added; this mixture is administered to resolve tympany and bloat (Chughtai, 1950; Yeung, 1985).
Scalding of urine	Paste of the seeds is also utilized in the scalding of urine (Chughtai, 1950).
Diarrhea	The seed decoction has antidiarrheal efficacy (Chughtai, 1950; Yeung, 1985).
Stomachic	Seeds alone, or mixed with sweetener, are advised as a stomachic (Kabir-ud-Din & Mahmood, 1937; Yeung, 1985).
Fever	Seeds are used in drinks, having a cooling effect in fevers (Singh & Panda, 2005).
Piles	The seed decoction plus sweetener is effective for piles (Chughtai, 1950).
Gonorrhoea	Fennel seeds are helpful in treating gonorrhoea (Chughtai, 1950).
Amenorrhea/menstrual problems	A hot infusion of seeds is useful in amenorrhea, and when lacteal secretion is suppressed. It is also advised as a diaphoretic. Fennel has been found useful in painful menstrual cramps (Yeung, 1985; Singh & Panda, 2005).
Biliousness	Seeds are considered effective in treating biliousness (Kapoor, 2001).
Madness	A decoction of fennel seeds mixed with sweetener is found useful for madness (Chughtai, 1950).
Deafness	250 g fennel seeds in powder form plus equal amounts of sweetener and cow's milk are mixed and stored in clay pots. This is advised, with cow's milk, in the morning and evening, for deafness and as a brain tonic (Kabir-ud-Din & Mahmood, 1937; Chughtai, 1950).

CHAPTER 55

Usage and Significance of *Foeniculum vulgare* Mill. Seeds

TABLE 55.3 Disease Prevention and Treatment by the Use of Fennel (*Foeniculum vulgare* Mill.) Seeds in Eastern Medicine—continued

Disease	Part/Extract
Hookworms	Seed oil acts as an anthelmintic against hookworms (Kapoor, 2001).
Brain tonic	A spoon of seed powder is taken with water daily as a brain tonic, for constipation, and as a stomachic (Chughtai, 1950).
Restlessness/delirium	Fennel seeds are boiled in water, and when the water has reduced by a half then one-fourth of cow's milk is added. It is useful for drowsiness (Chughtai, 1950).
Sleepiness	A seed decoction taken with common salt helps in lessening sleepiness (Chughtai, 1950).
Headache/giddiness	Seed powder mixed with sweetener and used daily in the evening for some days is advised for headache/giddiness (Chughtai, 1950).
Flu and cold	Seeds are soaked in half a liter of water. After 3 hours, the mixture is boiled and sweetener added. This is effective in flu and colds (Chughtai, 1950).
Digestive/abdominal disorders, and as diuretic, expectorant, stimulant, antispasmodic, and stomachic	Fennel oil is mildly carminative, and is helpful in infantile colic and flatulence. It also relieves griping abdominal pain and distention (Yamini *et al.*, 2002; Ahmad *et al.*, 2004; Hussain *et al.*, 2008).
Eye ache/eyesight	Fennel seed aqueous extract is advised for eye problems. Seed powder is taken daily with cow's milk to improve eye sight. A seed infusion may be used in cases of conjunctivitis and blepharitis (Kabir-ud-Din & Mahmood, 1937; Chughtai, 1950; Singh & Panda, 2005; Hussain *et al.*, 2008).
Hoarseness	A semi-hot aqueous extract of fennel seeds mixed with sweetener is drunk to clear the voice. Fennel seeds are widely used in India as an after-dinner breath freshener (Chughtai, 1950).
Liver tonic	Seeds are helpful in opening obstructions of the liver, spleen, and gall bladder, and to relieve painful swellings and yellow jaundice. They also protect the liver from toxins. Fennel has been reported to enhance hepatic degeneration (Chughtai, 1950).
Digestive disorders in camels	200 g fennel seeds are brewed in water and drenched (Muhammad *et al.*, 2005).
Respiratory ailments in camels	250 g fennel seeds are used to treat respiratory problems ((Muhammad *et al.*, 2005).
Indigestion and halitosis in camels	125 g fennel is given on a weekly basis for prevention of indigestion and halitosis (Muhammad *et al.*, 2005).
Glaucoma	Fennel seed extract has been found to be effective in animal studies in the treatment of glaucoma (Agarwal *et al.*, 2008).
Scabies	*F. vulgare* is crushd along with *Coriandrum sativum* and the mixture is mixed with vanaspati and sugar, then taken orally, to treat scabies (Saikia *et al.*, 2006).

465

water" to relieve flatulence in infants (Hussain *et al.*, 2008). An infusion or decoction of dried fennel seeds is considered helpful for the digestion of fatty food. Other beneficial effects of fennel seeds include weight loss by enhancing the metabolism, soothing sore throats, and loosening/expelling phlegm from the respiratory tract. In the subcontinent, licorice-flavored fennel seeds are served after meals to cleanse the breath. Fennel seed powder is also added in carminative mixtures in Pakistan, and used as spices in Assam, Bengal, and Oriysa, India. Fennel seeds are added in different fish dishes, vegetable products, meat and teas in the subcontinent and China (Yamini *et al.*, 2002; Hussain *et al.*, 2008). They can also be used with purgatives to alleviate their side effects. A decoction of a fine powder of crushed seeds is used to wash the eyes to reduce irritation and eye strain.

Fennel seeds and their oil have common use in soap and confectionery items, as well as in liqueurs, sweet pickles, sweet rice, bread, meat sauces, and candies (Hussain *et al.*, 2008). They are also used as a flavoring agent in natural toothpastes. Fine fennel-seed powder is also used as a flea repellent in stables.

TABLE 55.4 Greco-Arab and Traditional Preparations of Fennel (*Foeniculum vulgare* Mill.) Seeds in Eastern Medicine

Preparation (Local Names)	Composition	Uses
Safoof akseer paichus	Fennel, coconut and sweetener in 1 : 1 : 2 ratio	Administered with fresh water for diarrhea and dysentery
Safoof	Fennel seeds and coriander (1 : 1) are ground, then cow's milk and sweetener are added.	Effective for allergy
Laooq	About 375 g fennel seeds are boiled in 1 l water. When it is reduced to one-third, 0.5 kg sweetener is added and the mixture is heated until it makes a paste.	Advised thrice a day as expectorant, and for throat problems
Ghota	Violet, black pepper, and coriander are dissolved in water and strained	Taken with sweetener for nose bleeds
Decoction	Fennel seeds, mint, clove, and rose petals are boiled in water. When volume is reduced to one-half, it is mixed thoroughly, strained, and offered to the patient in small quantities	Useful for cholera
Jildi Marham	Fennel seeds, sugar, tamarind bark and cloves (5 : 6 : 4 : 2) are mixed as a paste	For chronic skin diseases
Ethno-remedy for reproductive disorder in large ruminants	*Amomum subulatum* Roxb. (100 g) + *Foeniculum vulgare* Mill. (250 g) + *Trachyspermum ammi* (L.) (50 g) Sprague. ex Turrill. + *Papaver hybridum* L. (250 g) are mixed and administered orally for 3 consecutive days	Remedy is used to treat anestrous in cattle and buffaloes (Dilshad *et al.*, 2008).
Ethno-remedy for reproductive disorder in large ruminants	*Piper nigrum* L. + *Amomum subulatum* Roxb + *Foeniculum vulgare* Mill. + *Cinnamomum Zylanicum* Nees. 50 g of each component of preparation is mixed in jaggery and administered *orally* in four equal doses over 4 days.	Remedy is useful to treat rectal prolapse in cattle and buffaloes (Dilshad *et al.*, 2008).

ADVERSE EFFECTS AND REACTIONS (ALLERGIES AND TOXICITY)

Skin contact with the sap or essential oil of fennel seeds is said to cause photosensitivity and/or dermatitis in some people. Ingestion of the oil can cause vomiting, seizures, and pulmonary edema (Foster & Duke, 1990). Essential oil extracted from fully ripened and dried fennel seeds has medicinal valuem, but its use is discouraged in pregnant women. Ingestion of fennel-seed tea by a breastfeeding mother was reported to have caused neurotoxicity in newborn child (Rosti *et al.*, 1994). Breast cancer patients should also be careful about the use of fennel seeds in large quantities.

SUMMARY POINTS

- Fennel is a short-lived perennial herb that is native to the Mediterrarean region but is currently cultivated in many parts of the world.
- Fennel seeds are extensively used as a herbal medicine, in kitchen as a spice, and as a flavoring agent.
- Fennel seeds are very popular as effective household remedies for common ailments, particularly digestive disorders.
- Fennel seed oil may cause skin reactions, while ripened seeds may create problems in pregnant women and in infants.

References

Agarwal, R., Gupta, S. K., Agrawal, S. S., Srivastava, S., & Saxena, R. (2008). Oculohypotensive effects of *Foeniculum vulgare* in experimental models of glaucoma. *Indian Journal of Physiology and Pharmacology, 52,* 77–83.

CHAPTER 55

Usage and Significance of *Foeniculum vulgare* Mill. Seeds

Ahmad, M., Hussain, S. A., Zubair, M., & Rab, A. (2004). Effect of different sowing seasons and row spacing on seed production of fennel (*Foeniculum vulgare*). *Pakistan Journal of Biological Sciences, 7*, 1144–1147.

Ahmad, S., Koukab, S., Islam, M., Ahmad, K., Aslam, S., Aminullah, K., et al. (2008). Germplasm evaluation of medicinal and aromatic plants in highland Balochistan. *Pakistan Journal of Botany, 40*, 1473–1479.

Cai, Y., Luo, Q., Sun, M., & Corke, H. (2004). Antioxidant activity and phenolic compounds of 112 traditional Chinese medicinal plants associated with anticancer. *Life Sciences, 74*, 2157–2184.

Chughtai, M. H. (1950). *Rehnumai aqaqeer almaroof desi jadi bootian.* Gujarat, Pakistan: Shaukat Book Depot.

Dilshad, S. M. R., Najeeb-ur-Rehman, I. Z., Muhammad, G., Iqbal, A., & Ahmed, N. (2008). An inventory of the ethnoveterinary practices for reproductive disorders in cattle and buffaloes, Sargodha district of Pakistan. *Journal of Ethnopharmacology, 117*, 393–402.

Duke, J. A. (1985). *Handbook of medicinal herbs.* Boca Raton, FL: CRC Press. p. 677.

Eun, M. C., & Jae, K. H. (2004). Anti-inflammatory, analgesic and antioxidant activities of the fruit of *Foeniculum vulgare. Fitoterapia, 75*, 557–565.

Foster, S., & Duke, J. A. (1990). *A field guide to medicinal plants.* Eastern and Central N. America. Atlanta, GA: Houghton Mifflin.

Hussain, A., Khan, M. N., Iqbal, Z., & Sajid, M. S. (2008). An account of the botanical anthelmintics used in traditional veterinary practices in Sahiwal district of Punjab, Pakistan. *Journal of Ethnopharmacology, 119*, 185–190.

Jadhav, A. N., & Bhutani, K. K. (2005). Ayurveda and gynecological disorders. *Journal of Ethnopharmacology, 97*, 151–159.

Kabir-ud-Din, H., & Mahmood, H. (1937). In Lughat al Adviya. (Ed.), *Lughat-e-Kabeer Daftar al-Maseeh, Qurul Bagh, Dehli. Vol. II* (pp. 473), Vol. III.

Kapoor, L. D. (2001). *A handbook of Ayurvedic medicinal plants.* Boca Raton, FL: CRC Press.

Muhammad, G., Khan, M. Z., Hussain, M. H., Iqbal, Z., Iqbal, M., & Athar, M. (2005). Ethnoveterinary practices of owners of pneumatic-cart pulling camels in Faisalabad City (Pakistan). *Journal of Ethnopharmacology, 97*, 241–246.

Rizvi, M. A., Saeed, A., & Zubairy, H. N. (2007). *Medicinal Plants.* Karachi, Pakistan: Hamdard Institute of Advanced Studies and Research.

Rosti, L., Nardini, A., Bettinelli, M. E., & Rosti, D. (1994). Toxic effects of a herbal tea mixture in two newborns. *Acta Paediatrica, 83*, 683.

Saikia, A. P., Ryakala, V. K., Sharma, P., Goswami, P., & Bora, U. (2006). Ethnobotany of medicinal plants used by Assamese people for various skin ailments and cosmetics. *Journal of Ethnopharmacology, 106*, 149–157.

Singh, M. P., & Panda, H. (2005). *Medicinal herbs with their formulations.* New Delhi, India: Daya Publishers.

Yamini, Y., Sefidkon, F., & Pourmortazavi, S. M. (2002). Comparison of essential oil composition of Iranian fennel (*Foeniculum vulgare*) obtained by supercritical carbon dioxide extraction and hydrodistillation methods. *Flavour and Fragrance Journal, 17*, 345–348.

Yeung, H.-C. (1985). *Handbook of Chinese herbs and formulas.* Los Angeles, CA: Institute of Chinese Medicine.

CHAPTER 56

Antidiabetic Activities of Fenugreek (*Trigonella foenum-graecum*) Seeds

Sirajudheen Anwar[1], Sandhya Desai[1], Maryam Eidi[2], Akram Eidi[3]
[1] Department of Pharmacology, Prin. K.M. Kundanai College of Pharmacy, Cuffe Parade, Mumbai, India
[2] Department of Biology, Varamin Branch, Islamic Azad University, Varamin, Iran
[3] Department of Biology, Science and Research Branch, Islamic Azad University, Tehran, Iran

469

LIST OF ABBREVIATIONS

ALP, alkaline phosphatase
ALT, alanine aminotransferase
AST, aspartate aminotransferase
GL, galactomannan
MDA, malonyl dialdehyde

INTRODUCTION

Diabetes mellitus is currently growing at a fast rate throughout the world, and is the sixteenth leading cause of global mortality. Nature has been a source of medicinal treatments for

Nuts & Seeds in Health and Disease Prevention. DOI: 10.1016/B978-0-12-375688-6.10056-8

thousands of years, and plant-based systems continue to play an essential role in the primary health care of 80% of the world's underdeveloped and developing countries.

Various constituents of fenugreek (*Trigonella foenum-graecum*) seed are responsible for hypoglycemic activity, one of them being fenugreek galactomannan (soluble and insoluble). On the basis of our research, it can be concluded that fenugreek galactomannan is a potential blood glucose reducing agent, but its antidiabetic activity is not fully supported by different mechanisms like peripheral glucose uptake and antioxidant effect. Thus, the antioxidant effect of fenugreek galactomannan is not promising. Fenugreek galactomannan was also found to reduce elevated levels of diabetes-associated components, like glycosylated hemoglobin.

Further alloxan-induced reduction in liver glycogen was reversed in rats treated with galactomannan. Hence, it is possible that galactomannan may act through an extra-pancreatic route, such as inhibition of glycogenolysis by liver. Histopathological studies during long-term treatment have shown it to ameliorate alloxan-induced histological damage of the islets of Langerhans. Treatment with an alcoholic extract of fenugreek seeds significantly decreased serum total cholesterol, triacylglycerol, urea, uric acid, creatinine, AST, and ALT levels.

BOTANICAL DESCRIPTION

Fenugreek is an erect annual plant, 30—60 cm high. It has thin, 10- to 15-cm long sword-shaped pods, containing 10—20 seeds. The oblong, yellow-brown seeds (2—3 mm long) are hard, with a wrinkled surface. The plant grows in well-drained soil and a mild climate. The seed crop matures in 3—4 months, and yields 300—400 kg seeds per acre. In India, besides the seeds, green fenugreek leaves are used as a vegetable.

Dicotyledonous fenugreek seed consists of a wrinkled brown-yellow seed coat or husk, enclosing two whitish translucent endosperm halves mainly composed of soluble galactomannan polysaccharide. The seeds contain protein (25—30%), ether extract (7—9%), steroidal saponins (5—7%), galactomannan (25—30%), insoluble fiber (20—25%), and ash (3—4%). Between the endosperms is sandwiched the yellowish germ portion, which consists mainly of good quality edible proteins, along with ether extractables (7—9%, fatty acids and flavoring for essential oil) and alcohol extractables (5—7%, consisting of diosgenin, yellow pigments, trigonelline, free amino acids, vitamins of the B-complex group, and flavonoids). Fibrous material mainly contains insoluble cellulosic fiber, while the endosperm galactomannan is soluble (Reid & Meier, 1970).

HISTORICAL CULTIVATION AND USAGE

Fenugreek or Methi (*Trigonella-foenumgraceum* Linn.) is an old cultivated spice-bean crop, indigenous to India, the Middle East, and southern Europe.

PRESENT-DAY CULTIVATION AND USAGE

Currently, fenugreek is cultivated in the Middle East, North Africa, southern Europe, and, more recently, North America. India is the major exporting country, followed by France, Egypt, and Argentina (Zimmer, 1984).

APPLICATIONS TO HEALTH PROMOTION AND DISEASE PREVENTION
Fenugreek galactomannan in diabetes mellitus

Fenugreek seeds contain 40—45% galactomannan (soluble and insoluble), which was isolated from whole seeds and shows considerable hypoglycemic activity (Sirajudheen *et al.*, 2009).

TABLE 56.1 Effect of Fenugreek Galactomannan on Blood Glucose Level in Normal Glucose (1 g/kg)-loaded Rats

Treatment and Dose	Blood Glucose Level (mg/dl)			
Time Interval	−0.5 h	+0.5 h	+1 h	+2 h
Normal (CMC 1 ml/kg + glucose)	62.33 ± 0.95	85.5 ± 1.50	80.66 ± 1.94	75 ± 1.61
GL (150 mg/kg + glucose)	62.16 ± 1.30	71.83 ± 1.11[*]	67.66 ± 1.20[*]	60 ± 1.27[*]
GL (300 mg/kg + glucose)	63.33 ± 1.52	69.33 ± 0.49[*]	66.16 ± 2.27[*]	58.5 ± 1.84[*]
GL (600 mg/kg + glucose)	62 ± 1.29	62.66 ± 1.15[*]	60.33 ± 1.26[*]	52.16 ± 0.95[*]
Glibenclamide (4 mg/kg + glucose)	61.83 ± 1.82	60 ± 1.48[*]	52.33 ± 1.09[*]	50 ± 1.13[*]

% effect was calculated by % decrease in BGL at +2+h; GL, fenugreek galactomannan.
$n = 6$ in each group; values are mean ± SEM.
[*]$P < 0.01$ compared to normal group (ANOVA followed by Dunnett's).

Experimentation reveals that fenugreek galactomannan is an ideal antidiabetic agent; fenugreek galactomannan, like glibenclamide, showed significant ($P < 0.05$ and $P < 0.01$) dose-related manner hypoglycemic effects at 150, 300, and 600 mg/kg in glucose-loaded rats (Table 56.1). The results were more pronounced in glucose-loaded rats at all three doses used (i.e., 150 (25%), 300 (28.20%), and 600 mg/kg (43.78 %), which were comparable to glibenclamide 4 mg/kg (50%), the standard antidiabetic drug.

Experimentation in the alloxan model revealed that fenugreek galactomannan lowered blood glucose levels significantly, being almost comparable to glibenclamide. On repeated administration of either vehicle, GL or glibenclamide, for 15 days, a sustained and significant ($P < 0.01$) decrease in the BGL of the diabetic rats was observed with GL 300 mg/kg (29.11% effect), GL 600 mg/kg (34.48% effect), and glibenclamide (4 mg/kg) (38.69% effect), as compared to the vehicle-treated group (Table 56.2)

DIFFERENT MODES OF PHARMACOLOGICAL ACTION

Antioxidant activity

Diabetics and experimental animal models exhibit high oxidative stress due to persistent and chronic hyperglycemia, which depletes the activity of the antioxidative defense system, and thus promotes *de novo* generation of free radicals (Albers & Beal, 2000). The activity of antioxidant enzymes such as superoxide dismutase, catalase, and glutathione peroxidase, which is low in islet cells when compared to other tissues, becomes further worsened under diabetic conditions (Kawamura & Heinecke, 1994). Furthermore, the presence of higher glucose or glycated protein concentration enhances lipid peroxidation (Hicks & Delbridge, 1989), and lipid peroxides may increase the extent of advanced glycation end products (Tiedge & Lortz, 1997).

Lipid peroxidation may bring about protein damage and inactivation of membrane bound enzymes, either through direct attack by free radicals or through chemical modification by its end products, MDA and 4-hydroxy nonenal (Halliwell & Gutteridge, 1999). Increased lipid peroxidation under diabetic conditions may be attributed to increased oxidative stress in the cells as a result of the depletion of antioxidant systems. Diminished levels of both enzymatic antioxidants (CAT and GPx) and increased levels of MDA were observed in diabetic rats. It is observed that in fenugreek galactomannan-treated diabetic rats, there was slight decrease in lipid peroxidation (GL 300, 9.3%; GL 600, 18.09%; glibenclamide, 43.67%) and increase in the levels of antioxidant enzymes (CAT: GL 300, 70%; GL 600, 76.3%; glibenclamide, 88.2%; and GPx: GL 300, 79%; GL 600, 83.4%; glibenclamide, 94%) (Table 56.3).

TABLE 56.2 Effect of Fenugreek Galactomannan on BhjGL (Repeated Dose Study) in Alloxan-induced Diabetic Rats

Treatment and Dose (mg/kg) (p.o.)	Blood Glucose Level (mg/dl)									
	3rd day	% effect[a]	6th day	% effect[a]	9th day	% effect[a]	12th day	% effect[a]	16th day	% effect[a]
Group 1, Normal control	59.33 ± 1.05	—	61 ± 1.06	—	60.33 ± 0.72	—	63.16 ± 1.35	—	60.33 ± 1.12	—
Group 2, Diabetic control	311.83 ± 1.97[*]	—	304.83 ± 1.47[*]	—	301.33 ± 1.36[*]	—	276.5 ± 2.51[*]	—	261.5 ± 2.19[*]	—
Group 3, GL (300 mg/kg)	291.16 ± 6.22[**]	6.43	274 ± 5.61[***]	9.86	237.66 ± 8.58[***]	21.26	205.83 ± 8.77[***]	25.72	185.16 ± 5.52[***]	29.11
Group 4, GL (600 mg/kg)	280.33 ± 5.23[***]	9.96	260.83 ± 5.36[***]	14.47	219.16 ± 4.05[***]	27.24	189.5 ± 2.57[***]	31.52	171.16 ± 13.05[***]	34.48
Group 5, Glibenclamide (4 mg/kg)	278 ± 5.54[***]	10.6	252.66 ± 5.51[***]	17.10	220.16 ± 5.20[***]	26.91	185.66 ± 4.51[***]	32.97	160.83 ± 5.04[***]	38.69

$n = 6$ in each group, values are mean ± SEM.

[a]% reduction in BLG at 3 hours as compared to diabetic control group; GL, fenugreek galactomannan.

[*]$P < 0.01$ compared to normal group (ANOVA followed by Dunnett's);

[**]$P < 0.05$ compared to diabetic group (ANOVA followed by Dunnett's);

[***]$P < 0.01$ compared to diabetic group (ANOVA followed by Dunnett's).

TABLE 56.3 Antioxidant Effects of Fenugreek Galactomannan (Catalase Activity, Glutathione Peroxidase and Lipid Peroxidation Activity)

Treatment and Dose	CAT (U/mg Prot)	% effect[a]	GPx (U/mg Prot)	% effect[a]	% LP	% effect[b]
Normal (CMC 1 ml/kg)	3.23 ± 0.09	—	2.68 ± 0.13	—	—	—
Normal (CMC 1 ml/kg) $\pm$ Alloxan	$1.981 \pm 0.08^{*}$	61	$1.721 \pm 0.08^{*}$	64.2	100	0
GL (300 mg/kg) $\pm$ Alloxan	$2.261 \pm 0.09^{**}$	70	$2.121 \pm 0.16^{**}$	79	$91.44 \pm 1.27^{*}$	9.3
GL (600 mg/kg) + Alloxan;	$2.465 \pm 0.05^{***}$	76.3	$2.235 \pm 0.06^{***}$	83.4	$84.68 \pm 1.66^{***}$	18.09
Glibenclamide (4 mg/kg) $\pm$ Alloxan	$2.851 \pm 0.07^{***}$	88.2	$2.53 \pm 0.022^{***}$	94	$69.60 \pm 3.31^{***}$	43.67

$n = 6$ in each group; values are mean $\pm$ SEM; one-way ANOVA is applied for statistical analysis.
[a]% increase in CAT & GPx;
[b]% decrease in lipid peroxidation; GL, fenugreek galactomannan; CAT, catalase; GPx, glutathione peroxidase.
*$P < 0.01$ compared to normal group;
**$P < 0.05$ compared to diabetic group;
***$P < 0.01$ compared to diabetic group.

Antioxidant activity of fenugreek galactomannan was significant when compared to the diabetic untreated group, but it was much less when compared to the glibenclamide-treated group. This shows that fenugreek galactomannan does not possess very effective antioxidant activity in comparison with the standard drug.

GLUCOSE UPTAKE BY PERIPHERAL CELLS AND TISSUES

It is well known that insulin and antidiabetic drugs promote glucose uptake by peripheral cells and tissues (Melander, 1996), so this is one of the possible modes of action for lowering the blood glucose level. The hemidiaphragms taken from rats treated with fenugreek galactomannan and glibenclamide showed significant ($P < 0.01$) enhancement in the glucose uptake process as compared to diabetic untreated rats (26.31%) (Table 56.4). However, glucose uptake with fenugreek galactomannan (GL 300, 38.49%; GL 600, 51.86%) -treated hemidiaphragms was much less than that with glibenclamide (82.48%).

EFFECT OF FENUGREEK GALACTOMANNAN ON BIOCHEMICALS
Glycosylated hemoglobin

In diabetics, elevated glucose leads to increased glycation of HbA within red blood cells. Since red blood cells circulate for 120 days, measurement of HbA1c in diabetic patients serves as an

473

TABLE 56.4 Effect of Fenugreek Galactomannan on Glucose Uptake by Hemidiaphragm, Glycosylated Hemoglobin and Liver Glycogen levels in Alloxan (150 mg/kg × 1 i.p.) -induced Diabetic Rats

Treatment and Dose	Glucose Uptake (mg/g)	% effect[a]	% GHbA1c	Liver Glycogen (mg/100 g)
Normal (CMC 1 ml/kg)	11.825 ± 0.66	—	4.61 ± 0.08	2130.83 ± 25.83
Normal (CMC 1 ml/kg) + Alloxan	$3.11 \pm 0.41^{*}$	26.31	$7.14 \pm 0.06^{*}$	$807 \pm 6.21^{*}$
GL (300 mg/kg) $\pm$ Alloxan	$4.55 \pm 0.18^{**}$	38.49	$6.54 \pm 0.20^{***}$	$1479.83 \pm 6.46^{***}$
GL (600 mg/kg) + Alloxan	$6.13 \pm 0.13^{**}$	51.86	$5.64 \pm 0.15^{***}$	$1652.5 \pm 10.48^{***}$
Glibenclamide (4 mg/kg) $\pm$ Alloxan	$9.75 \pm 0.32^{**}$	82.48	$5.41 \pm 0.07^{***}$	$1933.33 \pm 12.13^{***}$

$n = 6$ in each group; values are mean $\pm$ SEM.
[a]% glucose uptake by hemidiaphragm is calculated.
*$P < 0.01$ compared to normal group (ANOVA followed by Dunnett's);
**$P < 0.05$ compared to diabetic group;
***$P < 0.01$ compared to diabetic group (ANOVA followed by Dunnett's).

index of glycemic control over the preceding 2—3 months (Stephen *et al.*, 2000). Fenugreek galactomannan lowered %GHbA$_{1c}$ significantly as compared to the diabetic control group. The results of glycoslyated hemoglobin tend to suggest that fenugreek galactomannan can help to control blood glucose levels on a long-term basis, and help to prevent diabetic complications (Table 56.4).

Effect of fenugreek galactomannan on glycogen content

The glycogen content in the liver was significantly increased following treatment with glibenclamide (4 mg/kg). An increase in the liver glycogen content after fenugreek galactomannan treatment can be brought about by an increase in glycogenesis and/or a decrease in glycogenolysis. Hence, it is likely that fenugreek galactomannan either stimulates glycogenesis and/or inhibits glycogenolysis in the liver of diabetic rats. The rise in the glycogen content of the liver after 15 days of treatment can therefore be correlated with the decline in blood glucose levels (Table 56.4)

EFFECT OF FENUGREEK GALACTOMANNAN ON HISTOPATHOLOGICAL FINDINGS OF THE PANCREAS

Histopathological findings of the pancreas of the diabetic rats showed necrosis, atrophy, and fibrotic changes in beta cells of the islets of Langerhans. However, pancreas from rats treated with fenugreek galactomannan and glibenclamide showed a protective effect on the pancreatic islets (Figure 56.1). These results tend to suggest that fenugreek galactomannan is a promising agent for the management of diabetes mellitus. Further research in diabetic human patients is warranted in exploring its potential as an antidiabetic agent

Effect of alcoholic extracts of fenugreek seeds on diabetes mellitus

In the present study, the antidiabetic effect of alcoholic extract of fenugreek seeds in normal and STZ-induced rats was evaluated and also compared with glibenclamide as a reference standard antidiabetic drug (Eidi *et al.*, 2007). The data indicated that there were significant elevations in serum total cholesterol, triacylglycerol, urea, uric acid, creatinine, AST, and ALT levels in diabetic rats. Administration of fenugreek seed extract (0.1, 0.25, and 0.5 g/kg body weight) and glibenclamide (600 µg/kg) significantly decreased serum triacylglycerol and total

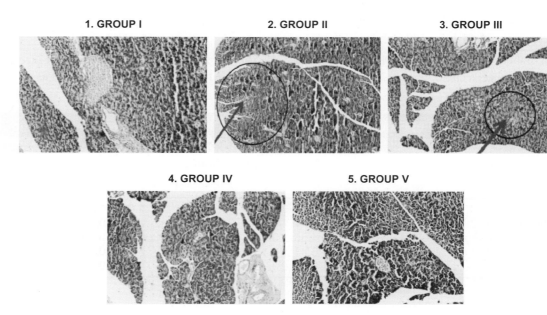

FIGURE 56.1
Histopathological section of pancreas.

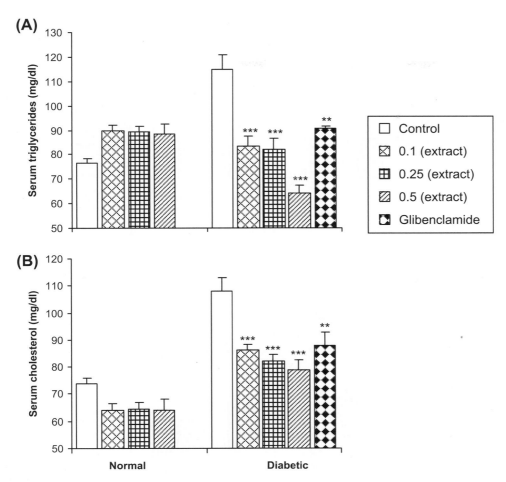

FIGURE 56.2

Effect of oral administration of fenugreek seed alcoholic extract on serum triglycerides and cholesterol in normal and diabetic rats. Effect of oral administration of fenugreek seeds alcoholic extract at doses of 0.1, 0.25, and 0.5 g/kg body weight on (A) serum triglyceride and (B) total cholesterol levels in normal and diabetic rats. Glibenclamide (600 µg/kg) was given only to diabetic rats. Each column represents mean ± SEM for eight rats. The control group had water as the vehicle. Bars with asterisks indicate differences from the diabetic control group. **$P < 0.01$; ***$P < 0.001$. *Reprinted from Eidi et al. (2007), Nutrition Research, 27, 728–733, with permission.*

cholesterol in diabetic but not in healthy rats (Figure 56.2). The results showed that serum urea, uric acid, and creatinine increased in diabetic rats. Administration of the fenugreek seed extract (0.5 g/kg body weight) and glibenclamide (600 µg/kg) significantly decreased serum urea, uric acid, and creatinine in diabetic but not in healthy rats (Figure 56.3). The results showed that serum AST and ALT levels increased in diabetic rats. The administration of the fenugreek seeds extract (0.25 and 0.5 g/kg body weight) and glibenclamide (600 µg/kg) similarly decreased AST and ALT levels in diabetic but not in healthy rats (Figure 56.4).

The hypolipidemic action of the soluble dietary fiber fraction could be the result of retardation of carbohydrate and fat absorption due to the presence of bioactive fiber in the agent. Our data showed that serum uric acid, urea, and creatinine levels were increased in diabetic rats. This may be due to metabolic disturbance in diabetes, reflected in high activities of xanthine oxidase, lipid peroxidation, and increased triacylglycerol and cholesterol levels. Moreover, protein glycation in diabetes may lead to muscle wasting and increased release of purine, the main source of uric acid, as well as increased activity of xanthine oxidase. Our data showed that fenugreek seed extract decreased the serum uric acid, urea, and creatinine levels

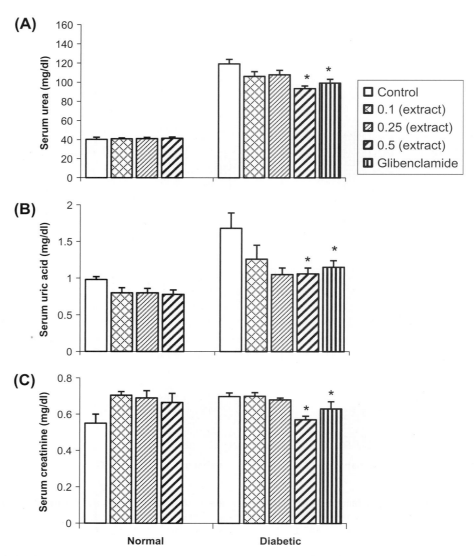

FIGURE 56.3
Effect of oral administration of fenugreek seed alcoholic extract on serum urea, uric acid, and creatinine in normal and diabetic rats. Effect of oral administration of fenugreek seeds alcoholic extract at doses of 0.1, 0.25, and 0.5 g/kg body weight on (A) serum urea, (B) uric acid, and (C) creatinine (down) levels in normal and diabetic rats. Glibenclamide (600 μg/kg) was given only to diabetic rats. Each column represents mean ± SEM for eight rats. The control group had water as the vehicle. Bars with asterisks indicate differences from the diabetic control group. *$P < 0.05$. *Reprinted from Eidi et al. (2007), Nutrition Research, 27, 728–733, with permission.*

in diabetic rats. Elevation of the serum urea and creatinine, as significant renal function markers, is related to renal dysfunction in diabetic hyperglycemia. Serum enzymes, including AST and ALT, are used in the evaluation of hepatic disorders. An increase in these enzyme activities reflects active liver damage. Inflammatory hepatocellular disorders result in extremely elevated transaminase levels.

ADVERSE EFFECTS AND REACTIONS (ALLERGIES AND TOXICITY)

From the preliminary acute toxicity study, fenugreek galactomannan seems to be safe up to 8 g/kg because even at this high dose no toxic or deleterious effects were observed immediately, or during the 3 days of the observation period.

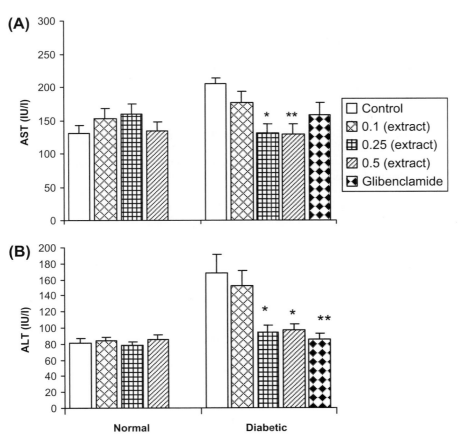

FIGURE 56.4

Effect of oral administration of fenugreek seed alcoholic extract on serum ALT and AST in normal and diabetic rats.
Effect of oral administration of fenugreek seed alcoholic extract at doses of 0.1, 0.25, and 0.5 g/kg body weight on serum
(A) AST and (B) ALT levels in normal and diabetic rats. Glibenclamide (600 μg/kg) was given only to diabetic rats. Each column
represents mean ± SEM for eight rats. The control group had water as the vehicle. Bars with asterisks indicate differences
from the diabetic control group. *$P < 0.05$; **$P < 0.01$. *Reprinted from Eidi* et al. *(2007)*, Nutrition Research, 27,
728—733, with permission.

SUMMARY POINTS

- Fenugreek galactomannan is one of the constituents of fenugreek seed, and shows promising antidiabetic properties.
- It shows little effect on glucose uptake by peripheral cells and antioxidant activity.
- Fenugreek galactomannan lowered %GHbA$_{1c}$, so it can be used in long-term therapy
- Increase in the liver glycogen content after fenugreek galactomannan treatment can be brought about by an increase in glycogenesis and/or a decrease in glycogenolysis; hence, it is likely that fenugreek galactomannan stimulates glycogenesis and/or inhibits glycogenolysis in the liver of diabetic rats.
- The alcoholic extracts of seeds have been shown to restore the activities of key enzymes of carbohydrate and lipid metabolism close to normal values.
- These results tend to suggest that fenugreek galactomannan and alcoholic extract of seeds are promising agents for the management of diabetes mellitus. Further research in diabetic human patients warrants exploration for its potential as an antidiabetic agent.

References

Albers, D. S., & Beal, M. F. (2000). Mitochondrial dysfunction and oxidative stress in ageing and neurodegerative disease. *Journal of Neural Transmission, 59*, 133–154.

Eidi, A., Eidi, M., & Sokhteh, M. (2007). Effect of fenugreek (*Trigonella foenum-graecum* L) seeds on serum parameters in normal and streptozotocin-induced diabetic rats. *Nutrition Research, 27*, 728–733.

Halliwell, B., & Gutteridge, J. M. C. (1999). *Free radicals in biology and medicines* (3rd ed.). Oxford, UK: Oxford University Press. pp. 20–37.

Hicks, M., & Delbridge, L. (1989). Increase in crosslinking of nonenzymatically glycosylated collagen induced by products of lipid peroxidation. *Archives of Biochemistry and Biophysics, 268*, 249–254.

Kawamura, M., & Heinecke, J. W. (1994). Pathophysiological concentrations of glucose promote oxidative modification of low density lipoprotein by a superoxide-dependent pathway. *Journal of Clinical Investigation, 94*, 771–778.

Melander, A. (1996). Glucose uptake by hemidiaphragm. *Diabetic Medicine, 13*. 143.

Reid, R. S. G., & Meier, H. Z. (1970). Composition of fenugreek galactomannan. *Phytochemistry, 9*, 513–516.

Sirajudheen, A., Rahul, M., & Sandhya, D. (2009). Exploring antidiabetic mechanisms of action of galactomannan: a carbohydrate isolated from fenugreek seeds. *Journal of Complementary and Integrative Medicine, 6*, 1–10.

Stephen, J. M., Vishwanath, R. L., & William, F. G. (2000). *An introduction to clinical medicine — Pathophysiology of disease* (3rd ed.). New York, NY: McGraw Hill. pp. 439–454.

Zimmer, R. C. (1984). Experiments on growth and maturation of fenugreek crop. *Canadian Plant Disease Survey, 64*, 33–35.

Tiedge, M., & Lortz, S. (1997). Relation between antioxidant enzyme gene expression and antioxidative defense status of insulin-producing cells. *Diabetes, 46*, 1733–1742.

CHAPTER 57

Five-Leaved Chaste Tree (*Vitex negundo*) Seeds and Antinociceptive Effects

Cheng-Jian Zheng, Lu-Ping Qin
Department of Pharmacognosy, School of Pharmacy, Second Military Medical University, Shanghai, China

479

LIST OF ABBREVIATIONS

EVNS, aqueous ethanol extract of *Vitex negundo* seeds
MeOH, methanol
NMR, nuclear magnetic resonance
s.c., subcutaneous

INTRODUCTION

The five-leaved chaste tree (*Vitex negundo* Linn.) is a medicinal shrub commonly found in tropical, subtropical, and warm temperate regions throughout the world, being a native of South Asia, China, Japan, Indonesia, East Africa, and South America. All parts of this plant have been used as folk medicine for many years in China and other Asian countries. VN seeds have been claimed to possess anti-inflammatory, analgesic, antioxidant, and anti-androgenic activity (Srinivas & Raju, 2002; Zheng *et al.*, 2009). Regarding the chemical constituents, the presence of lignans, flavonoids, and terpenoids has been previously reported (Ono *et al.*, 2004; Zheng *et al.*, 2009).

Nuts & Seeds in Health and Disease Prevention. DOI: 10.1016/B978-0-12-375688-6.10057-X

FIGURE 57.1

Plant and fruits of *Vitex negundo*. The five-leaved chaste tree is a small aromatic plant with typical five foliolate leaf pattern that flourishes abundantly in wastelands and is widely distributed in tropical to temperate regions.

BOTANICAL DESCRIPTION

Vitex negundo is an erect, branched, deciduous shrub or small tree, 2–5 m tall, with a slightly rough, pale reddish-brown bark that peels off in papery flakes. The leaves are opposite, digitately three to five foliolate (Figure 57.1); leaflets are lanceolate with an entire margin, 3–10 cm long, green above and gray pubescent beneath. Panicles are terminal, slightly hairy, many-flowered, and 10–20 cm long; additional axillary panicles are often present. The tree flowers from May to August. The scented flowers are hermaphrodite (have both male and female organs), and are pollinated by insects; they are blue to lavender in color and 5–7 mm long. Fruits are globose to broadly egg-shaped, 3–6 mm long and 2–3 mm in diameter, and black when mature (in October).

Vitex negundo is an easily grown plant that prefers a light well-drained loamy soil in a warm sunny position, but succeeds in poor dry soils. It is found in humid areas or along watercourses, in thickets, waste places, and mixed open forest. It is widely distributed throughout China (Figure 57.2), Korea, Japan, Indonesia, the Philippines, and other Asian countries.

HISTORICAL CULTIVATION AND USAGE

Vitex negundo is indigenous to Asia (China, Afghanistan, Japan, India, the Philippines, Sri Lanka, Bangladesh, Bhutan, Nepal, Pakistan, Cambodia, Myanmar, Vietnam, Indonesia, and Malaysia) and Africa (Kenya, Tanzania, Mozambique, and Madagascar). It can be propagated by use of vegetative cuttings, viable seeds, and root suckers, and has been cultivated for many years in China, India, Japan, the Philippines, and other Asian countries for various medicinal purposes.

Vitex negundo seed is a pepper substitute, and is occasionally used as a condiment for edible use. When washed to remove the bitterness, it can be ground into a powder and used as a flour, although it is very much a famine food used only when all else fails (Kunkel, 1984; Facciola, 1990). The seeds also make a cooling medicine for skin diseases and leprosy, and for analgesia, sedation, rheumatism, and joint inflammation (Ono *et al.*, 2004; Chawla *et al.*, 1992a).

PRESENT-DAY CULTIVATION AND USAGE

Vitex negundo is now widely cultivated in Europe, Asia, North America, and the West Indies, and is commonly found in tropical, subtropical, and warm temperate regions throughout the world. In addition to the above-mentioned conventional propagation methods, *in vitro* culture techniques have been established for micropropagation of *Vitex negundo* through high-frequency axillary shoot proliferation (Sahoo & Chand, 1998).

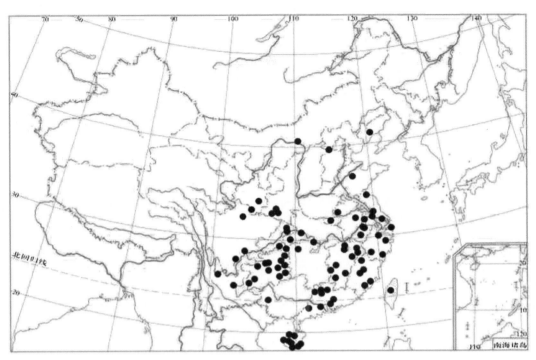

FIGURE 57.2

Geographical distribution of *Vitex negundo* in China. The five-leaved chaste tree is widely distributed in southern China, up to an altitude of 1500 m.

Vitex negundo seed has recently been proven to be an effective antineoplastic (Zhou *et al.*, 2009) and antioxidant (Zheng *et al.*, 1999), and has been considered as a botanical insecticide against various insect pest species (Yuan *et al.*, 2004).

APPLICATIONS TO HEALTH PROMOTION AND DISEASE PREVENTION

Vitex negundo seeds have been commonly used as folk medicine in southern China for the treatment of various pain disorders, such as stomach ache, hernia pain, dysmenorrhea, arthralgia, and piles. Based on previous investigations, our group studied the analgesic activity of an aqueous ethanol extract of *Vitex negundo* seeds (EVNS), and the antinociceptive activity of the petroleum ether, dichloromethane, acetoacetate, and *n*-butanol fractions from the aqueous ethanol extract, and found that the acetoacetate fraction had a powerful analgesic activity in preliminary experiments. We further examined the analgesic activity and the possible mechanism of the acetoacetate fraction, as well as elucidating its chemical components and providing a scientific basis for the clinical use of *Vitex negundo* seeds.

From the acetoacetate fraction, two phenylnaphthalene-type lignans (Figure 57.3) were obtained and identified as 6-hydroxy-4-(4-hydroxy-3-methoxy-phenyl)-3-hydroxymethyl-7-methoxy-3,4-dihydro-2-naphthaldehyde (**1**) and vitedoamine A (**2**), both of which have been previously reported and isolated from the seeds of *Vitex negundo* (Chawla *et al.*, 1992b; Ono *et al.*, 2004). The structures were elucidated unambiguously by spectroscopic methods, including 1D and 2D NMR analysis, and also by comparing experimental data with literature data.

In our study, the acetoacetate fraction and 6-hydroxy-4-(4-hydroxy-3-methoxy-phenyl)-3-hydroxymethyl-7-methoxy-3,4-dihydro-2-naphthaldehyde (**1**) of *Vitex negundo* seeds demonstrated dose-related and marked antinociception against two models of chemical nociception in mice; namely, acetic acid-induced abdominal constriction, and

481

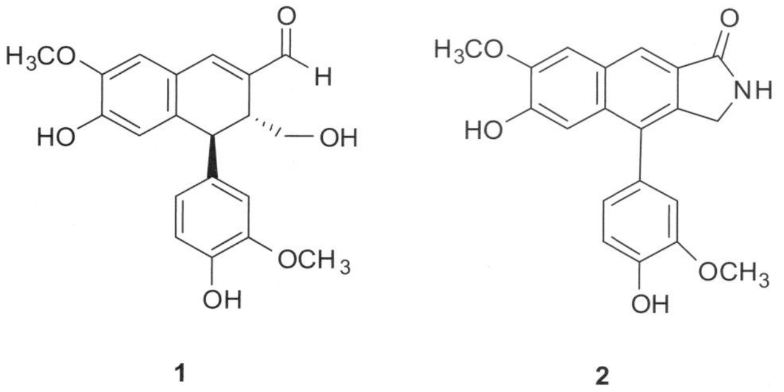

FIGURE 57.3

Chemical structures of two major phenylnaphthalene-type lignans isolated from *Vitex negundo* seeds. These two lignans were found to be the characteristic constituents isolated from *Vitex negundo* seeds.

formalin-induced licking response. Furthermore, they were also found to produce significant anti-inflammatory activities in the dimethyl benzene-induced ear edema test.

In the acetic acid-induced writhing test, all the extracts and fractions except for the aqueous fraction at the dose 5.0×10^3 mg/kg body wt caused a significant inhibition of the writhing responses induced by acetic acid as compared to the control, with values ranging from 29.52 to 55.72% protection (Table 57.1). At 2.5×10^3 mg/kg body wt, EVNS and the acetoacetate fraction still showed significant inhibition of the writhing responses, with inhibitions of 28.30

TABLE 57.1 Effects of *Vitex negundo* Seeds on Acetic Acid-induced Writhing Responses in Mice

Treatment	Dose (mg/kg Body wt)	Number of Writhings, Mean ± SEM	Inhibition (%)
Control	—	38.62 ± 4.47	
Ethanol extract of *Vitex negundo* seeds	5.0×10^3	21.75 ± 2.44^b	43.68
	2.5×10^3	27.69 ± 1.98^a	28.30
	1.2×10^3	32.11 ± 3.39	16.86
Petroleum ether fraction	5.0×10^3	24.50 ± 3.01^a	36.56
	2.5×10^3	28.49 ± 3.86	26.23
	1.2×10^3	34.38 ± 4.25	10.98
Dichlormethane fraction	5.0×10^3	22.82 ± 2.71^b	40.91
	2.5×10^3	35.41 ± 3.26	8.31
	1.2×10^3	37.53 ± 5.19	2.85
Acetoacetate fraction	5.0×10^3	17.10 ± 2.91^b	55.72
	2.5×10^3	23.32 ± 1.93^b	39.62
	1.2×10^3	24.21 ± 3.09^a	37.31
n-Butanol fraction	5.0×10^3	27.22 ± 2.18^a	29.52
	2.5×10^3	33.62 ± 4.29	12.95
	1.2×10^3	31.26 ± 3.07	11.06
Aqueous fraction	5.0×10^3	29.47 ± 2.76	23.69
	2.5×10^3	36.82 ± 4.22	4.66
	1.2×10^3	35.13 ± 3.14	9.04
Compound (1)	10.0	22.30 ± 3.91^a	42.26
	5.0	27.15 ± 1.82^a	29.70
	2.5	30.11 ± 2.89	22.04
Indomethacin	10.0	16.00 ± 2.46^b	58.57

Among the ethanol extract and some fractions with different polarity, the acetoacetate fraction showed the highest anti-nociceptive activity in the acetic acid-induced writhing test. Results are expressed as mean ± SEM, $n = 10$.
Reprinted from Zheng et al. (2009), *Phytomedicine*, *16*, 560–567, with permission.
^{a}P < 0.05.
^{b}P < 0.01 compared with control.

and 39.62%, respectively. In contrast, at a dose of 1.2×10^3 mg/kg body wt, only the acetoacetate fraction showed any significant analgesic effect. Interestingly, compound **1** also produced significant inhibition of writhing in a dose-dependent manner at the tested doses of 2.5, 5.0, and 10 mg/kg, with inhibitions of 22.04, 29.70, and 42.26%, respectively. The acetic acid-induced writhing method was widely used for the evaluation of peripheral antinociceptive activity, which was able to determine the antinociceptive effect of compounds or dose levels. Our results indicated that the acetoacetate fraction and compound **1** of *Vitex negundo* seeds could significantly reduce the amount of writhing in the animal model, showing powerful antinociceptive effects. However, the results of this writhing test alone didn't ascertain whether the antinociceptive effect was central or peripheral.

In order to confirm this, the formalin test was carried out. The advantage of the formalin model of nociception is that it can discriminate pain into its central and/or peripheral components. The test consists of two different phases which can be separated in time: the first one is generated in the periphery through the activation of nociceptive neurons by the direct action of formalin, and the second occurs through the activation of the ventral horn neurons at the spinal cord level (Tjolsen *et al.*, 1992). Central analgesic drugs, such as narcotics, inhibited equally in both phases, while peripherally acting drugs, such as steroids (hydrocortisone, dexamethasone) and NSAIDs (aspirin), suppressed mainly in the later phase (Trongsakul *et al.*, 2003). In the formalin test, the vehicle-treated animals showed mean licking times of 46.63 ± 6.64 s in the first phase and 28.13 ± 4.03 s in the second phase (Figure 57.4). Pretreatment with the acetoacetate fraction caused significant

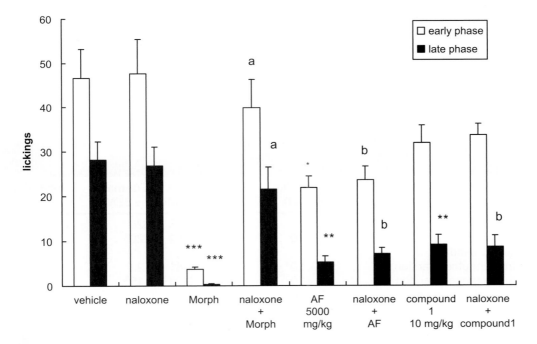

FIGURE 57.4

Effects of the acetoacetate fraction (AF) and compound 1 of *Vitex negundo* seeds and morphine on formalin-induced nociception in mice. The total time spent in licking the injected hind-paw was measured in the early phase (0–5 min, white column) and the late phase (20–25 min, black column). The vehicle (Control, 10 ml/kg), AF (acetoacetate fraction, 5.0×10^3 mg/kg), or compound **1** (10 mg/kg) were administered orally, and morphine (10 mg/kg) subcutaneously. AF and compound 1 were administered 1 h before and morphine 30 min before the test. Naloxone (Nalox, 1 mg/kg s.c.) was administered 15 min before AF, compound **1**, or morphine. Each column represented the mean $\pm$ SEM ($n = 10$). Asterisks indicated significant difference from control. $*P < 0.05$; $**P < 0.01$; $***P < 0.001$ vs control; a$P < 0.001$ vs Morph; b$P > 0.05$ vs AF or compound 1 (ANOVA followed by Dunnett's test). *Reprinted from Zheng* et al. *(2009), Phytomedicine, 16, 560–567, with permission.*

diminutions of both first-phase (21.83 ± 2.56 s) and second-phase (5.33 ± 1.35 s) pain responses, at the tested doses of 5.0×10^3 mg/kg. Compound **1** suppressed mainly in the later phase (first phase 31.83 ± 3.87 s, second phase 9.17 ± 2.17 s). Morphine (10 mg/kg), the reference drug, also significantly suppressed the formalin-response in both phases (first phase 3.67 ± 0.62, second phase 0.33 ± 0.16 s). When used alone, naloxone (1 mg/kg, s.c.) failed to modify the formalin-induced nociceptive responses in a significant manner (Figure 57.4) (naloxone: first phase 47.66 ± 7.81 s, second phase, 26.76 ± 4.25 s). Our data showed that the acetoacetate fraction of *Vitex negundo* seeds had the same anti-nociceptive activity as central analgesic drugs, while compound **1** acted peripherally and suppressed mainly in the second phase, similar to classical anti-inflammatory drugs. In the combination studies, naloxone notably antagonized the analgesic action of morphine but failed to reverse the antinociception produced by the acetoacetate fraction and compound **1**, which indicates that the observed analgesic effect is not mediated through opioid receptors.

As we know, prostaglandins play an important role in pain progress in chemical noci-ception models (Santos *et al.*, 1998), and are the target of action of commonly used anti-inflammatory drugs. Several other inflammatory mediators, such as sympathomimetic amines, tumor necrosis factor-α, interleukin-1β, and interleukin-8, are also involved in the nociceptive response to chemical stimulus in mice (Ferreira *et al.*, 1988, 1993; Santos *et al.*, 1998; Ribeiro *et al.*, 2000). Therefore, the dimethyl benzene-induced ear edema test was further employed to evaluate the anti-inflammatory activity of the acetoacetate fraction and compound **1**. In this test, the percentage inhibition of edema of EVNS was significant ($P < 0.05$) at the dose tested (31.54% at 5.0×10^3 mg/kg), in comparison with the control. The acetoacetate fraction exhibited significant dose-dependent anti-inflammatory activity at 1.2, 2.5, and 5.0×10^3 mg/kg. The dose–response effect was also observed consistently for compound **1** (Table 57.2). At a dose of 5 mg/kg, the anti-inflammatory activity of compound **1** was comparable to that of dexamethasone at a dose of 1 mg/kg (54.36% vs 45.97%). These results demonstrate that both the acetoacetate fraction and compound **1** showed significant anti-inflammatory activity, consistent with the indications revealed in

TABLE 57.2 Effects of *Vitex negundo* Seeds on Ear Edema Induced by Dimethyl Benzene in Mice

Treatment	Dose (mg/kg Body wt)	Weight of Edema (mg), Mean $\pm$ SEM	Inhibition (%)
Control	—	2.98 ± 0.28	
Ethanol extract of *Vitex negundo* seeds	5.0×10^3	2.04 ± 0.20^a	31.54
Petroleum ether fraction	5.0×10^3	1.81 ± 0.23^a	39.26
Dichlormethane fraction	5.0×10^3	2.03 ± 0.19^a	31.55
Acetoacetate fraction	5.0×10^3	1.16 ± 0.14^b	61.07
	2.5×10^3	1.53 ± 0.18^b	48.66
	1.2×10^3	1.98 ± 0.21^a	33.56
n-Butanol fraction	5.0×10^3	2.56 ± 0.30	14.09
Aqueous fraction	5.0×10^3	2.77 ± 0.25	7.05
Compound **(1)**	10.0	1.06 ± 0.09^b	64.42
	5.0	1.36 ± 0.10^b	54.36
	2.5	2.48 ± 0.22	16.78
Dexamethasone	1.0	1.61 ± 0.23^b	45.97

Both the acetoacetate fraction and compound (1) exhibited significant anti-inflammatory activities in the dimethyl benzene-induced ear edema test in a dose-dependent manner. Results are expressed as mean $\pm$ SEM, $n = 10$.
Reprinted from Zheng *et al.* (2009), *Phytomedicine*, 16, 560–567, with permission.
[a]$P < 0.05$
[b]$P < 0.01$ compared with control.

the second phase (inflammatory) of the formalin test, implying that the analgesic activity of the acetoacetate fraction and compound **1** may be mediated by its anti-inflammatory action.

In conclusion, our studies, taken together, gave pharmacological support to the validity of the significant analgesic activity of the acetoacetate fraction of *Vitex negundo* seeds in the test models of nociception induced by chemical stimuli. Bioassay-guided isolation from the acetoacetate fraction led to a productive compound, 6-hydroxy-4-(4-hydroxy-3-methoxy-phenyl)-3-hydroxymethyl-7-methoxy-3,4-dihydro-2-naphthaldehyde, with potent analgesic properties that could partially explain the analgesic effect of *Vitex negundo* seed extract. It is suggested that the observed antinociceptive activity might be related to its anti-inflammatory activity, which merits further studies regarding the precise site and the mechanism of action.

ADVERSE EFFECTS AND REACTIONS (ALLERGIES AND TOXICITY)

Anti-androgenic properties were reported from the flavonoid-rich fraction of the seeds (Bhargava, 1989; Das *et al.*, 2004), which was shown to reduce the weight of accessory sex organs, to decrease the quantity and quality of sperm in male Wistar rats, and to impair spermiogenesis at the late stage in adult male dogs.

SUMMARY POINTS

- Of the 80% ethanol extract and some fractions with different polarity, the acetoacetate fraction showed the highest antinociceptive activity.
- The analgesic bioguided isolation of the acetoacetate fraction yielded two major lignans: 6-hydroxy-4-(4-hydroxy-3-methoxy-phenyl)-3-hydroxymethyl-7-methoxy-3,4-dihydro-2-naphthaldehyde (**1**) and vitedoamine A (**2**).
- 6-hydroxy-4-(4-hydroxy-3-methoxy-phenyl)-3-hydroxymethyl-7-methoxy-3,4-dihydro-2-naphthaldehyde (**1**) exhibited significant antinociceptive activity at tested doses, and contributed to the analgesic properties of *Vitex negundo* seed extract.
- Co-administration of naloxone failed to antagonize the analgesic activity of compound (**1**) in the formalin test.
- Both the acetoacetate fraction and compound **1** showed significant anti-inflammatory activity, implying that the analgesic activity of the acetoacetate fraction and compound **1** may be mediated by their anti-inflammatory action.

References

Bhargava, S. K. (1989). Antiandrogenic effects of a flavonoid-rich fraction of *Vitex negundo* seeds: a histological and biochemical study in dogs. *Journal of Ethnopharmacology, 27*, 327–339.

Chawla, A. S., Sharma, A. K., & Handa, S. S. (1992a). Chemical investigation and anti-inflammatory activity of *Vitex negundo* seeds. *Journal of Natural Products, 55*, 163–167.

Chawla, A. S., Sharma, A. K., Handa, S. S., & Dhar, K. L. (1992b). A lignan from *Vitex negundo* seeds. *Phytochemistry, 31*, 4378–4379.

Das, S., Parveen, S., Kundra, C. P., & Pereira, B. M. J. (2004). Reproduction in male rats is vulnerable to treatment with the flavonoid-rich seed extracts of *Vitex negundo*. *Phytotherapy Research, 18*, 8–13.

Facciola, S. (1990). Cornucopia — A source book of edible plants. Vista, CA: Kampong Publications.

Ferreira, S. H., Lorenzetti, B. B., Bristow, A. F., & Poole, S. (1988). Interleukin-1β as a potent hyperalgesic agent antagonized by a tripeptide analogue. *Nature, 334*, 698–700.

Ferreira, S. H., Lorenzetti, B. B., & Poole, S. (1993). Bradykinin initiates cytokine mediated inflammatory hyperalgesia. *British Journal of Pharmacology, 110*, 1227–1231.

Kunkel, G. (1984). *Plants for human consumption.* Koenigstein, Germany: Koeltz Scientific Books.

Ono, M., Nishida, Y., Masuoka, C., Li, J. C., Okawa, M., Ikeda, T., et al. (2004). Lignan derivatives and a norditerpene from the seeds of *Vitex negundo*. *Journal of Natural Products, 67*, 2073–2075.

485

Ribeiro, R. A., Vale, M. L., Thomazzi, S. M., Paschoalato, A. B. P., Poole, S., Ferreira, S. H., et al. (2000). Involvement of resident macrophages and mast cells in the writhing nociceptive response induced by zymosan and acetic acid in mice. *European Journal of Pharmacology, 387*, 111–118.

Sahoo, Y., & Chand, P. K. (1998). Micropropagation of *Vitex negundo* L., a woody aromatic medicinal shrub, through high-frequency axillary shoot proliferation. *Plant Cell Reports, 18*, 301–307.

Santos, A. R., Vedana, E. M., & De Freitas, G. A. (1998). Antinociceptive effect of meloxicam in neurogenic and inflammatory nociceptive models in mice. *Inflammation Research, 47*, 302–307.

Srinivas, K., & Raju, M. B. V. (2002). Chemistry and pharmacology of *Vitex negundo*. *Asian Journal of Chemistry, 14*, 565–569.

Tjolsen, A., Berge, O. G., Hunskaar, S., Rosland, J. H., & Hole, K. (1992). The formalin test: an evaluation of the method. *Pain, 51*, 5–17.

Trongsakul, S., Panthong, A., Kanjanapothi, D., & Taesotikul, T. (2003). The analgesic, antipyretic and anti-inflammatory activity of *Diospyros variegata* Kruz. *Journal of Ethnopharmacology, 85*, 221–225.

Yuan, L., Xue, M., Xing, J., & Li, C. H. (2004). Toxicity of *Vitex negundo* extracts to several insect pests. *Chinese Journal of Pesticides, 43*, 70–72.

Zhou, Y. J., Liu, Y. E., Cao, J. G., Zeng, G. Y., Shen, C., Li, Y. L., et al. (2009). Vitexins, nature-derived lignan compounds, induce apoptosis and suppress tumor growth. *Clinical Cancer Research, 15*, 5161–5169.

Zheng, C. J., Huang, B. K., Han, T., Zhang, Q. Y., Zhang, H., Rahman, K., et al. (2009). Nitric oxide scavenging lignans from *Vitex negundo* seeds. *Journal of Natural Products, 72*, 1627–1630.

Zheng, G. M., Luo, Z. M., & Chen, D. M. (1999). Studies on the compositions of *Vitex negundo* L. seeds with antioxidant activity. *Guangdong Gongye Dexue Xuebao, 16*, 41.

CHAPTER 58

Flax Seed (*Linum usitatissimum*) Fatty Acids

Martha Verghese, Judith Boateng, Lloyd T. Walker
Alabama A&M University, Normal, Alabama, USA

CHAPTER OUTLINE

LIST OF ABBREVIATIONS

AA, arachidonic acid
ALA, alpha-linolenic acid
CVD, cardiovascular disease
COX, Cyclooxygenase
DHA, docosahexaenoic acid
EPA, eicosapentaenoic acid
FA, fatty acid
FFA, free fatty acid
IL-6, interleukin 6
LA, linoleic acid
LDL, low density lipoprotein
LDL-C, low density lipoprotein cholesterol
LOX, lipooxygenase
LXRα, liver X receptor α
PCNA, proliferating cell nuclear antigen

Nuts & Seeds in Health and Disease Prevention. DOI: 10.1016/B978-0-12-375688-6.10058-1

PPARγ, peroxisome proliferator activated receptor
PKC, protein kinase C
PUFA, polyunsaturated fatty acid
RXRα, retinoid X receptor α
SDG, secoisolariciresinol diglucoside
SREBP, sterol regulatory element-binding proteins
VCAM-1, vascular cell adhesion molecule-1
VLDL, very low density lipoprotein

INTRODUCTION

Flax seed is considered to be an important oil seed crop that is especially rich in essential fatty acids. The seed contains essential omega-6 and omega-3 fatty acids that our bodies require to maintain optimal health. It is the most abundant plant source of omega-3 fatty acid, specifically α-linolenic acid (ALA; 18:3n-3). Flax seed contains 41% oil by weight, of which 70% is polyunsaturated (PUFA), and more than half of the total fatty acid is α-linolenic acid (Bhatty, 1995, Jenkins *et al.*, 1999). Omega-6 fatty acid, such as linoleic acid (LA), is universally found in the vegetable kingdom, and is particularly rich in most (but not all) vegetable seeds and in the oils produced from the seeds (with coconut oil, cocoa butter, and palm oil being exceptions). ALA and LA are considered essential, from a nutritional standpoint. This is because neither can be synthesized by humans, and hence must be obtained from dietary sources. The essentiality of LA and ALA is due to the fact that they are precursors for some of the most important highly unsaturated fatty acids (AA, DHA, and EPA), which are required for the synthesis of the family of bioactive mediators known as the eicosanoids. Flax seed (or linseed, as it is also known) is among the richest source of omega-3 fatty acid α-linolenic acid (ALA), which accounts for more than 50% of its fatty acid content (Tables 58.1, 58.2). Other important vegetable sources are canola and soybean, with about 8—10% ALA.

Although flax seed is not commonly used as food oil, the oil from the seed is low in saturated fat (approximately 9%), contains moderate amounts of monounsaturated fat (approximately 18%), and is high in polyunsaturated fats (approximately 73%). Flax seed oil does not contain

TABLE 58.1 Comparison of Major Fatty Acids in Traditional and Modified Vegetable Oils

	Fatty Acids (% of total)					
	16:0	18:0	18:1	18:2	18:3	20:1
Canola						
Traditional	4.0	1.5	61.0	21.0	9.5	1.5
High-oleic	3.5	2.5	77.0	8.0	2.5	1.5
Low linolenic	4.0	1.0	61.0	27.0	2.0	1.0
Sunflower						
Traditional	6.0	4.0	18.5	71.0	0.5	—
High-oleic	4.0	4.0	78.5	11.5	—	—
Flax						
Traditional	6.5	4.5	19.5	16.5	53.0	—
Low-linolenic (Linola™)	6.0	4.0	16.0	71.5	2.0	—

TABLE 58.2 α-Linolenic Acid Content of Selected Vegetable Oils, Nuts, and Seeds

Food Sources	α-Linolenic Acid Content, g/5 ml
Olive oil	0.1
Walnuts, English	0.7
Soybean oil	0.9
Canola oil	1.3
Walnut oil	1.4
Flax seeds (linseed)	2.2
Flax seed (linseed) oil	8.5

trans fatty acids, which have been suggested to cause significant increases in coronary heart disease risk. Additionally, it contains both soluble and insoluble fiber, which accounts for about 28% of the weight of full-fat flax seeds. The major insoluble fiber fraction in flax seed consists of cellulose and, most importantly, lignin, which represents approximately 0.7–1.5% of the seed. Lignans are phytoestrogens, which are beneficial compounds that affect the metabolism of hormones such as estrogens in humans. The major lignan found in flax seed is secoisolariciresinol diglucoside, commonly referred to as SDG. SDG has been shown to be a potent antioxidant, and a known precursor of the mammalian lignans, enterolactone and enterodiol (Figures 58.1, 58.2).

Research over the past 30 years supports the many expected health benefits of consuming flax seed and cold-pressed flax seed oil. Omega-3 fatty acids are essential for growth and development, and have been associated with health and the prevention and treatment of heart disease, arthritis, inflammatory and autoimmune diseases, and cancer. Epidemiologic studies have shown that populations that consume high amounts of oils containing omega-3 fatty acids have lower rates of various types of cancers, including lung, breast, prostate, and colon.

BOTANICAL DESCRIPTION

Flax (*Linum usitatissimum*) belongs to the family Linaceae, which is composed of 22 genera. The genus contains more than 300 annual and perennial species. It is a self-pollinating diploid plant with a karyotype of 2n = 30. The plant has long, upright stems measuring approximately 1 m tall at maturity, and bears round seed capsules on its branched head (Bond & Hunter, 1987; Allaby *et al.*, 2005). The seeds of flax are tiny, smooth and flat, and range in color from light- to reddish-brown.

HISTORICAL CULTIVATION AND USAGE

Flax is one of the earliest cultivated crops, with evidence dating from the Stone Age. Early documents indicate that the Phoenicians were responsible for transporting flax into Europe from the fertile valleys of the Tigris and Euphrates Rivers in Mesopotamia (Stephens, 1997) during the period 2500–1200 BC. It is one of the earliest vegetable fibers domesticated by mankind, and as an oilseed (Dillman, 1938). It has been reported that flax was primarily utilized as a source of fiber for making linen cloth. However, there is evidence showing that flax was also consumed as a cereal and valued for its medicinal qualities; the oil, on the other hand, was used for frying food and as fuel for lamps (Vaisey-Genser & Morris, 2003). The earliest production of flax in the United States could be traced back to 1753. During the Colonial era, European colonists transported flax to North America, New Zealand, and Australia (Rosberg, 1996).

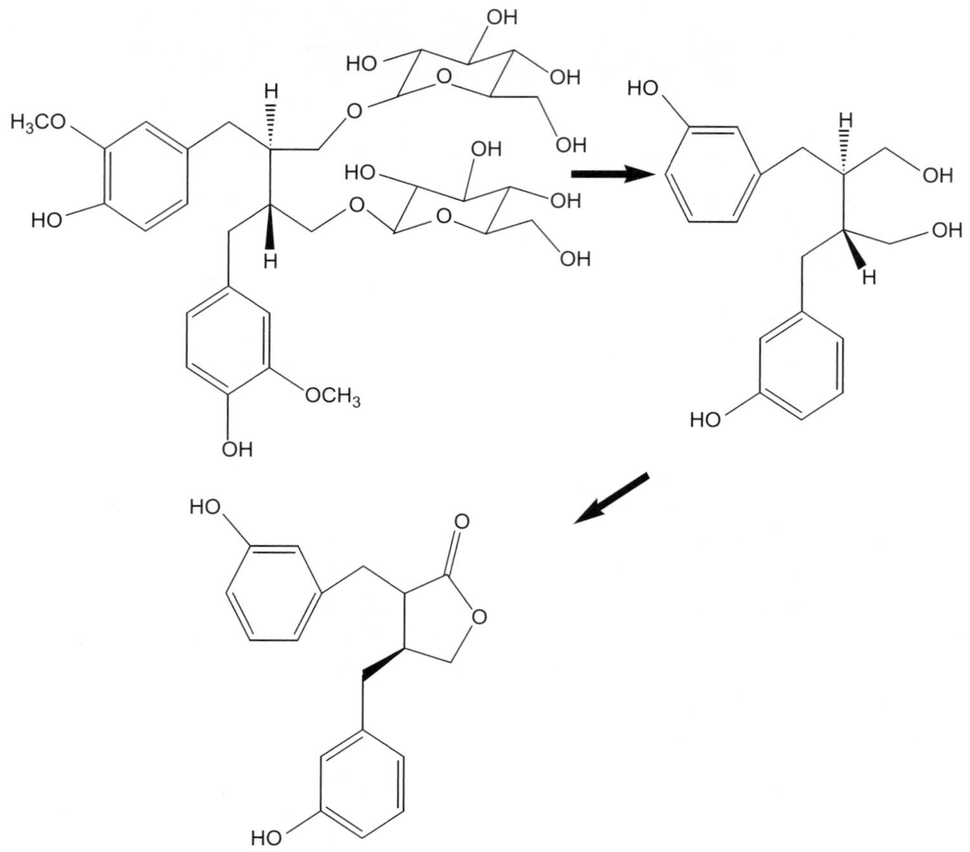

FIGURE 58.1
Chemical structures of the lignan secoisolariciresinol (Seco), diglucoside and its mammalian metabolites, enterodiol (ED), and enterolactone (EL).

PRESENT-DAY CULTIVATION AND USAGE

Cultivated flax has two varieties; one grown for oil and the other for fiber (Diederichsen & Richards, 2003). Flax is the sixth largest oilseed crop in the world; currently, flax seed is grown in nearly 47 countries, and in 2005 seed production was 1.903 million metric tonnes (Smith & Jimmerson, 2005). Canada, China, and the United States (Kansas, Minnesota, Montana, Nebraska, North Dakota, and Wisconsin) together are responsible for 64% of the total world flax seed output. Canada is currently the world's leader in the production and export of flax seed — a position it has held since 1994. Flax is presently produced for its rich oil, and as feed for livestock. In Europe, flax seed refers to the seeds grown for fiber or linen production, while linseed refers to those used for oil production (Vaisey-Genser & Morris, 2003). However, in North America both terms are used interchangeably to describe *Linum* species (Vaisey-Genser & Morris, 2003).

Metabolism of omega-3 fatty acids

The steps that transform LA and ALA to their higher unsaturated derivatives — the long chain eicosapentaenoic acid (EPA, 20:5n-3) and docosahexaenoic acid (DHA, 22:6n-3), the two types of omega-3 fatty acids more readily used by the body (Connor, 2000), and arachidonic acid (AA), the primary metabolite of LA — entail the activities of successive desaturation and elongation reactions (Makni *et al.* 2008, Russo, 2009). The n-6 PUFA

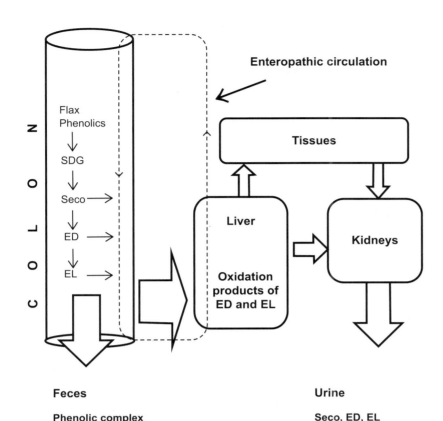

Enteropathic circulation

Tissues

Liver

Oxidation products of ED and EL

Kidneys

COLON

Flax Phenolics
↓
SDG
↓
Seco →
↓
ED →
↓
EL →

Feces

Phenolic complex
SDG, Seco, ED, EL
Conjugated ED and EL
with glucuronic acid

Urine

Seco, ED, EL
12 aromatic oxidation
products of ED and EL

FIGURE 58.2

Intestinal conversion of SDG from phenolic complex in flax seed, and absorption and excretion of secoisolariciresinol (Seco), enterodiol (ED), and enterolactone (EL).

arachidonic acid (AA) as well as the n-3 PUFAs, EPA and DHA, can be metabolized through a series of pathways. The metabolism of EPA and DHA produces an anti-inflammatory response, whereas AA is pro-inflammatory. The anti-inflammatory response produced by n-3 PUFAs is through a competitive inhibition mechanism that produces the 2-series eicosanoids, the less biologically active 3-series prostaglandins (PGE3 and PGF3α), and thromboxanes (TXA3) through the cyclooxygenase (COX) pathway, and the 5-series leukotrienes (LTB5, LTC5 and LTE5) through the lipooxygenase (LOX) pathway. Metabolism of AA is through 2-series eicosanoids, such as prostaglandins (PGE2 and PGF2α) and thromboxanes (TXA2), while metabolism through the LOX pathway yields the chemotactic 4-series leukotrienes (LTB4, LTC4, and LTE4) (Kang and Weylandt, 2008, Massaro *et al.*, 2008, and Micallef *et al.*, 2009) (Figure 58.3).

Delta-6-desaturase (Δ-6-d), which is one of the major enzymes involved in the metabolism of PUFA, recognizes and metabolizes LA and ALA, producing GLA and octadecatetraenoic (stearidonic) acids, respectively (Figure 58.3). However, the affinity of Δ-6-d for the PUFAs is different. The distinct functions of these two families make the balance between dietary n-6 and n-3 fatty acids an important consideration influencing cardiovascular health. This is mainly due to their competitive nature and their different biological roles, which are to ensure the conversion of ALA to EPA and DHA (Makni *et al.* 2008). As discussed by Makni and colleagues, the concentration of ALA required to inhibit GLA formation by 50% is about 10 times the concentration of the substrate (LA), suggesting that in the presence of higher concentration of LA (such as occurs in a living system) the pathway leading to AA is preferred (Russo, 2009). Because there is competition between the ratios of omega-3 to omega-6 fatty

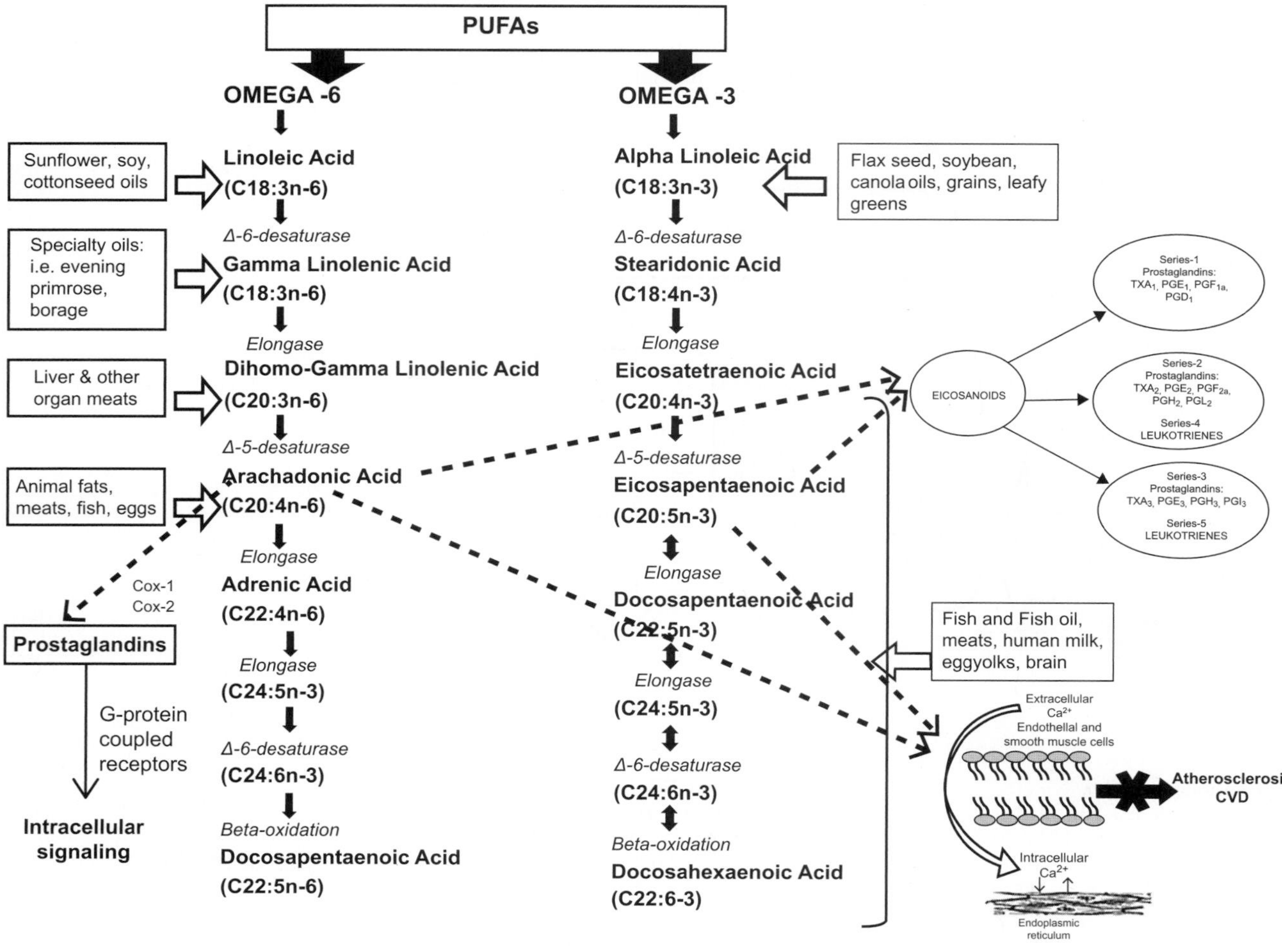

FIGURE 58.3

Metabolism and health-promoting properties of omega-6 and omega-3 polyunsaturated fatty acids. *Adapted from MacLean* et al. *(2005).*

acids for the desaturation enzymes, ALA is usually present at much lower levels in the diet and in the tissues of the body than is LA, the primary essential fatty acid of the n-6 family usually consumed in Western diet. As a result, diets rich in foods high in LA, such as corn, safflower, sunflower, and peanut oils, and significantly low in ALA, can lead to n-3 fatty acid deficiency, as such a high ratio of n-6 to n-3 fatty acids in the diet accentuates n-3 fatty acid deficiency.

APPLICATIONS TO HEALTH PROMOTION AND DISEASE PREVENTION

Metabolic diseases

The World Health Organization (WHO) predicts that in 15 years cardiovascular disease (CVD) will be the major killer worldwide, due to the accrual of several metabolic disorders, including obesity and diabetes (among others). Most would agree that diet is one of the most important factors in maintaining human health. Omega-3 fatty acids are described as pleiotropic molecules with a broad variety of biological actions, including hypotriglyceridemic, anti-aggregatory, anti-inflammatory, and anti-arrhythmic responses (Garg *et al.*, 2006, Micallef & Garg, 2009). They have been shown to reduce the risk of CVD, coronary heart disease, type 2 diabetes, and insulin resistance, among other conditions (Woods & Fearon, 2009). These effects are mediated by alterations in circulating plasma lipids, eicosanoids, and cytokines, and physicochemical properties in the phospholipid membrane.

The anti-atherogenic properties of n-3 PUFAs are perhaps their ability to modify serum and tissue lipid alterations, while the most consistent finding is a reduction in fasting and post-prandial serum triglycerides and free fatty acids (FFA) (Micallef & Garg, 2009). The displayed effects are decreased levels of very low density lipoprotein (VLDL) production by the liver, which occurs mostly through a reduction in the synthesis of triglycerides (Micallef & Garg, 2009), largely resulting from interference with most of the transcription factors that control the expression of enzymes responsible for both triglyceride assembly and FA oxidation. This leads to the decreased availability of FFAs released from adipose stores (Micallef & Garg, 2009). Omega-3 PUFAs have been shown to significantly reduce the expression of the sterol regulatory element-binding proteins (SREBP), which are transcription factors that regulate cholesterol, FA, and triglyceride-synthesizing enzymes (Micallef & Garg, 2009). This effect by EPA and DHA seems to induce inhibition of the liver X receptor α/retinoid X receptor α (LXRα/RXRα) heterodimer by binding to the promoter of the SREBP-1cgen e (Micallef & Garg, 2009), and the activation of PPAR.

Long-chain EPA and DHA are potent anti-arrhythmic agents that help to improve the vascular endothelial function and to lower blood pressure, platelet sensitivity, and serum triglyceride levels. Dietary ALA consumed as flax seed oil was shown to raise serum n-3 PUFAs, such as EPA and DHA levels (Mantzioris et al., 1994; Cunnane et al., 1995). In a recent report, intake of flax seed oil and milled flax seed for 4 weeks resulted in significant increases in plasma ALA levels (Dupasquier et al., 2006; Austria et al., 2008) and aortic tissue. Flax seed supplementation was also reported to reduce the n-6 to n-3 PUFA ratio, primarily as a result of the elevated levels of circulating n-3 PUFAs (Dupasquier et al., 2006).

Supplementation of ALA from linseed oil/flax seed oil significantly decreased inflammatory markers, including C-reactive protein, serum amyloid A, IL-6, and soluble VCAM-1, in dyslipidemic patients (Rallidis et al., 2003, 2004). Dupasquier et al. (2007) reported an anti-atherogenic action by dietary flax seed. In their study, the authors revealed for the first time an antiproliferative and anti-inflammatory action of flax seed. Feeding flax seed to LDL receptor-deficient (LDLrKO) mice (which closely mimic the human condition) suppressed the expression of inflammatory markers such as IL-6, mac-3, and VCAM-1, and the proliferative marker PCNA by reducing the infiltration of macrophages into the sub-endothelial space, and reduced the inflammatory and proliferative state of atherosclerotic lesions. The authors attributed these effects to the omega-3 fatty acid content of flax seed.

Cancer

Several molecular mechanisms whereby omega-3 PUFAs potentially affect carcinogenesis have been proposed (Larsson et al., 2004). These mechanisms include the following:

1. Suppression of arachidonic acid (AA, 20:4n-6)-derived eicosanoid biosynthesis, which results in altered immune response to cancer cells, and modulation of inflammation, cell proliferation, apoptosis, metastasis, and angiogenesis.
2. Influences on transcription factor activity, gene expression, and signal transduction, which lead to changes in metabolism, cell growth, and differentiation.
3. Alteration of estrogen metabolism, which leads to reduced estrogen-stimulated cell growth.
4. Increased or decreased production of free radicals and reactive oxygen species.
5. Mechanisms involving insulin sensitivity and membrane fluidity (Larsson et al., 2004).

Linoleic acid (LA) and arachidonic acid (AA) activate protein kinase C (PKC) and induce mitosis, but EPA and DHA appear to reverse the protein kinase C activity changes associated with colon carcinogenesis (McCarty, 1996; Rose & Connolly, 1999a, 1999b). The AA products of cyclooxygenase (COX) and lipoxygenase (LOX) enzymes increase mitosis; EPA and DHA

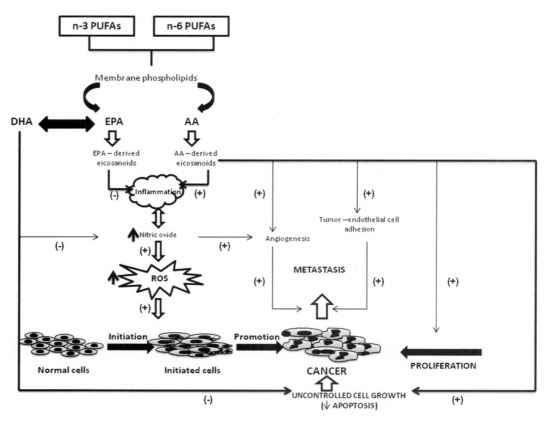

FIGURE 58.4

Proposed mechanisms by which omega-6 and omega-3 polyunsaturated fatty acids may promote and suppress carcinogenesis. *Adapted from Larsson* et al. *(2004).*

decrease mitosis and inhibit growth of breast and colon cancers (Rose *et al.*, 1995) (Figure 58.4). Flax seed oil was reported to reduce colon cancer (Williams *et al.*, 2007) and mammary tumor growth at a late stage of carcinogenesis in rodents (Thompson *et al.*, 1996a, 1996b); the authors attributed this effect to the high ALA content in the flax seed oil component, which was found to be more effective at the stage when tumors have already been established.

Further studies indicated a reduction in the growth and metastasis of human estrogen receptor negative (ER−) breast cancer in nude mice fed diets containing flax seed products (Wang *et al.*, 2005). The authors concluded that the tumor inhibitory effect of flax seed products was due to its oil content, specifically ALA. Some of the actions proposed were a greater reduction of Ki-67 LI, decreased cell proliferation, and induction of apoptosis. Flax seed is high in phytochemicals, and includes many antioxidants. In addition to lignans, flax contains phenolic acids, cinnamic acids, flavonoids, and lignins (Lambert *et al.*, 2005). Lignan metabolites from flax seed may decrease the risk of cancer by acting as antioxidants and free radical scavengers. Using human colonic cancer (SW480) cells, the lignan metabolites enterodiol and enterolactone were shown to inhibit cancer cell growth mediated by cytostatic and apoptotic mechanisms (Qu *et al.*, 2005). Animal studies have shown that dietary flax seed oil rich in the omega-3 polyunsaturated fatty acid ALA can decrease natural killer (NK) cell activity (Thies *et al.*, 2001). Many mechanisms seem to be operational in omega-3 fatty acid suppression of tumor growth. Most experimental studies have been conducted on primary cancers, but it seems reasonable that these same mechanisms might suppress the growth of metastatic cancer cells.

ADVERSE EFFECTS AND REACTIONS (ALLERGIES AND TOXICITY)

Flax seed is said to contain about 23–33 g/kg of phytic acid, depending on the cultivar, location, and year of harvest (Oomah *et al.*, 1996; Muir & Westcott, 2003). Phytic acid forms a complex with many important minerals, including iron, zinc, calcium, etc., thus making them less available. Antinutrients such as phytic acid are thought to have evolved as a defense mechanism so seeds such as flax can protect themselves from insect infestation. In any case, the phytic acid content in flax seed is lower or comparable to that of other oil seeds (Oomah *et al.*, 1996; Morris, 2003).

Flax seed contains a vitamin B-6 antagonist, linatine, a cyclic hydroxylamine derivative (Klosterman *et al.*, 1967). Linatine was found to adversely affect the growth of chickens fed linseed meal (Klosterman *et al.*, 1967; Klosterman, 1974). As a result, in animal studies diets containing high levels of flax meal are supplemented with extra vitamin B-6. However, Ratnayake *et al.* (1992) and Dieken (1992) found that flax seed meal, when fed 50 g to humans, did not affect vitamin B-6 levels or metabolism.

Other antinutrients include trypsin inhibitors, oxalates, and cyanogenic glycosides, to mention a few. Cyanogenic glycosides, which are also present in flax, can be found in a number of plants, including cassava — for which it is known. In flax, however, the release of hydrogen cyanide is reported to be below the toxic or lethal dose (Touré & Xueming, 2010). The levels of cyanogenic glycoside released from 1–2 tablespoons of flax seed meal is said to be approximately 5–10 mg of hydrogen cyanide, which, according to Roseling (1994), is significantly lower than the acute toxic dose for an adult, which is between 50 and 60 mg. In fact, Zimmerman (1988) reported that the levels detected in blood were significantly lower than those of somebody smoking tobacco.

Although rare, allergic reactions to flax are not well documented. A few cases have been reported by Muthiah and colleagues (1995), who reported an anaphylactic reaction to the carbohydrate component of flax. Meanwhile, Alonso *et al.* (1996) also recorded a case of an anaphylactic reaction after the intake of flax oil (Muir & Westcott, 2003). One of the better known reactions, known as byssonosis, has been reported to occur from exposure to flax dust during the harvesting and processing of flax. Individuals coming into contact with the dust particles report incidences of asthma, pneumonia, and bronchitis.

SUMMARY POINTS

- Flax is a self-pollinating diploid plant with a karyotype of 2n = 30.
- Cultivated flax has two varieties, one grown for oil and the other for fiber, and is the sixth largest oilseed crop in the world.
- Flax seed is the most abundant plant source of omega-3 fatty acids, specifically α-linolenic acid.
- Omega-3 fatty acids are essential for growth and development, and have been associated with health and the prevention and treatment of heart disease, arthritis, inflammatory and autoimmune diseases, and cancer.
- Dietary ALA in the form of flax seed oil and milled seed has been shown to raise serum n-3 PUFAs, to increase plasma ALA levels, and reduce the n-6 to n-3 PUFA ratio.
- Epidemiologic studies have shown that populations that consume high amounts of oils containing omega-3 fatty acids have lower rates of various types of cancers, including lung, breast, prostate, and colon.
- Few allergic reactions to flax have been documented, and anaphylactic reactions are rare. Byssonosis has been described following exposure to flax dust during harvesting and processing of flax, and there have been reports of asthma, pneumonia and bronchitis.
- Flax seed contains various antinutrients, including phytic acid, linatine, trypsin inhibitors, oxalates, and cyanogenic glycosides.

References

Allaby, R. G., Peterson, G. W., Merriwether, D. A., & Fu, Y. B. (2005). Evidence of the domestication history of flax (*Linum usitatissimum* L.) from genetic diversity of the sad2 locus. *Theoretical and Applied Genetics, 112*(1), 58—65.

Alonso, L., Marcos, M. L., Blanco, J. G., Navarro, J. A., Juste, S., del Mar Garcés, M., et al. (1996). Anaphylaxis caused by linseed (flaxseed) intake. *Journal of Allergy and Clinical Immunology, 98*(2), 469—470.

Austria, J. A., Richard, M. N., Chahine, M. N., Edel, A. L., Malcolmson, L. J., Dupasquier, C. M., et al. (2008). Bioavailability of alpha-linolenic acid in subjects after ingestion of three different forms of flaxseed. *Journal of the American College of Nutrition, 27*(2), 214—221.

Bhatty, R. S. (1995). Nutrient composition of whole flaxseed and flaxseed meal. In S. C. Cunnane, & L. U. Thompson (Eds.), *Flaxseed in human nutrition* (pp. 22—42). Champaign, IL: AOCS Press.

Bond, J. M., & Hunter, J. R. (1987). Flax growing in Orkney from the Norse period to the 18th century. *Proceedings of the Society of Antiquaries of Scotland, 117*, 175—181.

Connor, W. E. (2000). Importance of n-3 fatty acids in health and disease. *American Journal of Clinical Nutrition, 71* (Suppl.1), 1715—1755.

Cunnane, S. C., Hamadeh, M. J., Liede, A. C., Thompson, L. U., Wolever, T. M., & Jenkins, D. J. (1995). Nutritional attributes of traditional flaxseed in healthy young adults. *American Journal of Clinical Nutrition, 161*(1), 62—68.

Diederichsen, A., & Richards,, K. W. (2003). Cultivated flax and the genus Linum L. — taxonomy and germplasm conservation. In A. Muir, & N. Westcott (Eds.), *Flax, the Genus Linum* (pp. 22—54). *London, UK: Taylor & Francis.*

Dieken, H. A. (1992). Use of flaxseed as a source of omega-3 fatty acids in human nutrition. *Proceedings of the Flaxseed Institutional, 54*, 1—4.

Dillman, A. C. (1938). Natural crossing in flax. *Journal of American Society Agronomy, 30*, 279—286.

Dupasquier, C. M., Weber, A. M., Ander, B. P., Rampersad, P. P., Steigerwald, S., Wigle, J. T., et al. (2006). Effects of dietary flax seed on vascular contractile function and atherosclerosis during prolonged hypercholesterolemia in rabbits. *American Journal of Physiology Heart and Circulatory Physiology, 291*(6), H2987—2996.

Dupasquier, C. M., Dibrov, E., Kneesh, A. L., Cheung, P. K., Lee, K. G., Alexander, H. K., et al. (2007). Dietary flax seed inhibits atherosclerosis in the LDL receptor-deficient mouse in part through antiproliferative and anti-inflammatory actions. *American Journal of Physiology Heart and Circulatory Physiology, 293*(4), H2394—2402.

Garg, M. L., Leitch, J., Blake, R. J., & Garg, R. (2006). Long-chain n-3 polyunsaturated fatty acid incorporation into human atrium following fish oil supplementation. *Lipids, 41*(12), 1127—1132.

Jenkins, D. J., Kendall, C. W., Vidgen, E., Agarwal, S., Rao, A. V., Rosenberg, R. S., et al. (1999). Health aspects of partially defatted flaxseed, including effects on serum lipids, oxidative measures, and *ex vivo* androgen and progestin activity: a controlled crossover trial. *American Journal of Clinical Nutrition, 69*(3), 395—402.

Kang, J. X., & Weylandt, K. H. (2008). Modulation of inflammatory cytokines by omega-3 fatty acids. *Subcellular Biochemistry, 49*, 133—143.

Klosterman, H. J., Lamoureax, G. L., & Parsons, J. L. (1967). Isolation, characterization and synthesis of linatine. *Biochemistry, 6*, 170—175.

Klosterman, H. J. (1974). Vitamin B6 antagonists of natural origin. *Journal of Agricultural and Food Chemistry, 22*(1), 13—16.

Lambert, J. D., Hong, J., Guang-yu, Y., Liao, J., & Yang, C. (2005). Inhibition of carcinogenesis by polyphenols: evidence from laboratory investigations. *American Journal of Clinical Nutrition, 81*(Suppl.), 284S—291S.

Larsson, S. C., Kumlin, M., Ingelman-Sundberg, M., & Wolk, A. (2004). Dietary long-chain n-3 fatty acids for the prevention of cancer: a review of potential mechanisms. *American Journal of Clinical Nutrition, 79*(6), 935—945.

MacLean, C. H., Issa, A. M., Newberry, S. J., Mojica, W. A., Morton, S. C., Garland, R. H., et al. (2005). *Effects of omega-3 fatty acids on cognitive function with aging, dementia, and neurological diseases. Evidence Report/Technology Assessment No. 114 (Prepared by the Southern California Evidence-based Practice Center, under Contract No. 290-02-0003.) AHRQ Publication No. 05-E011-2. Rockville, MD: Agency for Healthcare Research and Quality.*

Makni, M., Fetoui, H., Gargouri, N. K., el Garoui, M., Jaber, H., Makni, J., et al. (2008). Hypolipidemic and hepatoprotective effects of flax and pumpkin seed mixture rich in omega-3 and omega-6 fatty acids in hypercholesterolemic rats. *Food and Chemical Toxicology, 46*(12), 3714—3720.

Mantzioris, E., James, M. J., Gibson, R. A., & Cleland, L. G. (1994). Dietary substitution with an alpha-linolenic acid-rich vegetable oil increases eicosapentaenoic acid concentrations in tissues. *American Journal of Clinical Nutrition, 59*(6), 1304—1309.

Massaro, M., Scoditti, E., Carluccio, M. A., Montinari, M. R., & De Caterina, R. (2008). Omega-3 fatty acids, inflammation and angiogenesis: nutrigenomic effects as an explanation for anti-atherogenic and anti-inflammatory effects of fish and fish oils. *Journal of Nutrigenetics and Nutrigenomics, 1*(1−2), 4−23.

McCarty, M. F. (1996). Fish oil may impede tumour angiogenesis and invasiveness by down-regulating protein kinase C and modulating eicosanoid production. *Medical Hypotheses, 46*(2), 107−115.

Micallef, M. A., & Garg, M. L. (2009). Anti-inflammatory and cardioprotective effects of n-3 polyunsaturated fatty acids and plant sterols in hyperlipidemic individuals. *Atherosclerosis, 204*(2), 476−482.

Micallef, M. A., Munro, I. A., & Garg, M. L. (2009). An inverse relationship between plasma n-3 fatty acids and C-reactive protein in healthy individuals. *European Journal of Clinical Nutrition, 63*(9), 1154−1156.

Morris, D. (2003). *Flax: a health and nutrition primer.* Winnipeg, Canada: Flax Council of Canada.

Muir, A., & Westcott, N. (Eds.). (2003). *Flax, the Genus Linum.* London, UK: Taylor & Francis.

Muthiah, R., Louthian, R., Knoll, M. T., & Kagen, S. (1995). Flaxseed-induced anaphylaxis (ANA): another cross-reactive carbohydrate food allergen. 51st Annual Meeting of the American Academy of Allergy and Immunology, New York. *Journal of Allergy Clinical Immunology, 95*, 370.

Oomah, B. D., Mazza, G., & Przybylski, R. (1996). Comparison of flaxseed meal lipids extracted with different solvents. *Lebensmittel-Wissenschaft and Technology, 29*(7), 654−658.

Qu, H., Madl, R. L., Takemoto, D. J., Baybutt, R. C., & Wang, W. (2005). Lignans are involved in the antitumor activity of wheat bran in colon cancer SW480 cells. *Journal of Nutrition, 135*(3), 598−602.

Rallidis, L. S., Paschos, G., Liakos, G. K., Velissaridou, A. H., Anastasiadis, G., & Zampelas, A. (2003). Dietary alpha-linolenic acid decreases C-reactive protein, serum amyloid A and interleukin-6 in dyslipidaemic patients. *Atherosclerosis, 167*(2), 237−242.

Rallidis, L. S., Paschos, G., Papaioannou, M. L., Liakos, G. K., Panagiotakos, D. B., Anastasiadis, G., et al. (2004). The effect of diet enriched with alpha-linolenic acid on soluble cellular adhesion molecules in dyslipidaemic patients. *Atherosclerosis, 174*(1), 127−132.

Ratnayake, W. M. N., Behrens, W. A., Fischer, P. W. F., L'Abbe, M. R., Mongeau, R., & Beare-Rogers, J. L. (1992). Chemical and nutritional studies of flaxseed (variety Linott) in rats. *Journal of Nutritional Biochemistry, 3*, 232−240.

Rosberg, R.J. (1996). Underexploited temperate, industrial and fiber crops. In *Proceedings of the Third National Symposium on New Crops: New Opportunities, New Technologies.* Indianapolis, Indiana. 22−25 October, 1995 (pp. 60−84). Alexandria, VA: ASHS Press.

Rose, D. P., & Connolly, J. M. (1999a). Omega-3 fatty acids as cancer chemopreventive agents. *Pharmacology & Therapeutics, 83*(3), 217−244.

Rose, D. P., & Connolly, J. M. (1999b). Antiangiogenicity of docosahexaenoic acid and its role in the suppression of breast cancer cell growth in nude mice. *International Journal of Oncology, 15*(5), 1011−1015.

Rose, D. P., Connolly, J. M., Rayburn, J., & Coleman, M. (1995). Influence of diets containing eicosapentaenoic or docosahexaenoic acid on growth and metastasis of breast cancer cells in nude mice. *Journal of National Cancer Institute, 87*(8), 587−592.

Roseling, H. (1994). Measuring effects in humans of dietary cyanide exposure to sublethal cyanogens from cassava in Africa. *Acta Horticulturae, 375*, 271−283.

Russo, G. L. (2009). Dietary n-6 and n-3 polyunsaturated fatty acids: from biochemistry to clinical implications in cardiovascular prevention. *Biochemical Pharmacology, 77*(6), 937−946.

Smith, H. V., & Jimmerson, J. (2005). *Briefing No. 56: Flaxseed.* Bozeman, MT: Agricultural Marketing Policy Center, Montana State University. http://www.ampc.montana.edu/briefings/briefing56.pdf Available at: (accessed April 5, 2010).

Stephens, G. R. (1997). *A manual for fiber flax production.* New Haven, CT: The Connecticut Agricultural Experimental Station.

Thies, F., Neve-Von-Caron, G., Powell, J., Yaqoob, P., Newsholme, E., & Calder, P. (2001). Dietary supplementation with eicosapentaenoic acid, but not with other long-chain n-3 or n-6 polyunsaturated fatty acids, decreases natural killer cell activity in healthy subjects aged more than 55 year. *American Journal of Clinical Nutrition, 73*, 539−548.

Thompson, L. U., Rickard, S. E., Orcheson, L. J., & Seidl, M. M. (1996a). Flax seed and its lignan and oil components reduce mammary tumor growth at a late stage of carcinogenesis. *Carcinogenesis, 17*(6), 1373−1376.

Thompson, L. U., Seidl, M. M., Rickard, S. E., Orcheson, L. J., & Fong, H. H. (1996b). Antitumorigenic effect of a mammalian lignan precursor from flaxseed. *Nutrition and Cancer, 26*(2), 159−165.

Touré, A., & Xueming, X. (2010). Flaxseed lignans: source, biosynthesis, metabolism, antioxidant activity, bio-active components, and health benefits. *Comprehensive Reviewsin Food Science and Food Safety, 9*, 261−269.

Vaisey-Genser, M., & Morris, D. H. (2003). History of the cultivation and uses of Flaxseed. *Flax: The genus Linum*. New York: Taylor & Francis Inc.

Wang, L., Chen, J., & Thompson, L. U. (2005). The inhibitory effect of flaxseed on the growth and metastasis of estrogen receptor negative human breast cancer xenograftsis attributed to both its lignan and oil components. *International Journal of Cancer, 116*(5), 793—798.

Williams, D., Verghese, M., Walker, L. T., Boateng, J., Shackelford, L., & Chawan, C. B. (2007). Flax seed oil and flax seed meal reduce the formation of aberrant crypt foci (ACF) in azoxymethane-induced colon cancer in Fisher 344 male rats. *Food Chemical Toxicology, 45*(1), 153—159.

Woods, V. B., & Fearon, A. M. (2009). Dietary sources of unsaturated fatty acids for animals and their transfer into meat, milk and eggs: A review. *Livestock Science, 126*(1—3), 1—20.

Zimmerman, D. C. (1988). Flax, linseed oil and human nutrition. *Proceedings of Flaxseed Institutional, 52*, 30.

CHAPTER 59

Fragrant Olive (*Osmanthus fragrans*) Seeds in Health

Yingming Pan[1], Ying Liang[2], Kai Wang [3], Hengshan Wang[1], Ye Zhang[3], Xianxian Liu[3]
[1] School of Chemistry and Chemical Engineering, Guangxi Normal University, Guilin, China
[2] Department of Environmental Engineering, Guilin University of Electronic Technology, Guilin, China
[3] Department of Chemistry and Engineering Technology, Guilin Normal College, Guilin, China

LIST OF ABBREVIATIONS

BHT, butylated hydroxytoluene
NMOF, natural melanin of *Osmanthus fragrans* seeds
OFS, *Osmanthus fragrans* seeds
RPOF, red pigment of *Osmanthus fragrans* seeds
IC_{50}, the concentration that scavenges 50% of the free radicals
SPF, sun protection factor

INTRODUCTION

Osmanthus fragrans is a traditional and famous ornamental plant in China. The flower is especially valued as an additive to tea and other beverages consumed in the Far East. The extracts

Nuts & Seeds in Health and Disease Prevention. DOI: 10.1016/B978-0-12-375688-6.10059-3

of the flower are of high value, and are only used in the most expensive perfumes (Wang *et al.*, 2006). *O. fragrans* (flower, seed, and root) is also used in folk medicine. Therefore, planting areas for *O. fragrans* trees in the south and middle of China have increased exponentially. In recent years, a number of studies have been conducted to explore the pharmacological potential and novel health benefits of *O. fragrans* seeds. This chapter describes the cultivation of *O. fragrans*, and reviews the pharmacological potential and health benefits of its seeds.

BOTANICAL DESCRIPTION

O. fragrans (also known as sweet osmanthus, *guìhuā* in China, *kinmokusei* in Japan, sweet olive, tea olive, or fragrant olive) belongs to the olive family (Oleaceae). It is an evergreen tree or shrub with a slow growth rate, native to China. It can grow to 3–12 m in height. The leaves are 7–15 cm long and 2.6–5.0 cm wide, with an entire or finely toothed margin. The flowers are small (1 cm long), with a four-lobed corolla 5 mm in diameter, and are produced in small clusters in the late summer and autumn. The flowers are usually hidden by foliage, and have a strong, apricot-like fragrance that can be smelled from a distance. The seed is found in a purple-black drupe, 10–15 mm long, that matures in the spring, about 6 months after flowering.

HISTORICAL CULTIVATION AND USAGE

O. fragrans is considered to be one of the four famous traditional flowers in China, and has been cultivated in China for over 2000 years. The book *Xijing Zaji* ("Miscellanies about West Capital") recorded sweet *O. fragrans* growing in Hanwu Emperor's Shanglin Gardens (about 11 BC). Similar records can also be found in other ancient Chinese books, such as *Sanfu Huangtu* and *Shanglin Fu*. Between the Han and South-and-North Dynasties in China, *O. fragrans* became a very famous flower. *O. fragrans* was introduced to Europe in the middle of the 19th century by the French botanist Jean Marie Delavay (Zang & Xiang, 2004). The flower is used in perfumery and foods, such as sweet Osmanthus wine, tea, sugar, juices, cakes and sauces (Wang *et al.*, 2009).

PRESENT-DAY CULTIVATION AND USAGE

Although there are increasingly cultivated areas of *O. fragrans* in other countries, China still has the largest. *O. fragrans* is widely distributed in the south and middle of China. Five cities (Guilin, Hangzhou, Suzhou, Chengdu, and Xianning) comprise the production center for *O. fragrans* (Zang *et al.*, 2003).

O. fragrans flowers (known as Gui Hua or Kwei Hwa) are widely used in Chinese medicine. There are many medicinal products made from sweet osmanthus buds, leaves, and bark. They are said to protect against coughs, used to flavor other medicines, and are added to cosmetics for hair and skin. A decoction of the stem bark is used in the treatment of boils and carbuncles, and a decoction of the lateral roots is used in the treatment of dysmenorrhoea, rheumatism, and bruises. The seeds are often used as an analgesic. An essential oil obtained from the flowers is used as an insect repellent for clothes. Products are also added to herbal medicines in order to disguise obnoxious flavors (Duke & Ayensu, 1984; Manandhar & Manandhar, 2002). The essential oil of *O. fragrans* is considered to be one of the best natural essences, and is only used in the most expensive perfumes and cosmetics (Jin *et al.*, 2006).

APPLICATIONS TO HEALTH PROMOTION AND DISEASE PREVENTION
Characterization and free radical scavenging activity of novel red pigment from OFS

Natural colorants have attracted widespread interest because of their general health and safety benefits, and numerous other important properties. Therefore, the development of food colorants from natural sources is receiving increasing attention (Es-Safi, 2004). Recently, a natural red pigment of *O. fragrans* (RPOF) has been isolated from OFS in our laboratory (Pan *et al.*, 2009). RPOF can be dissolved in alkaline or acidic water solutions, and common hydrophilic organic solvents. The color of a water solution of RPOF changes with pH (Table 59.1). RPOF is stable to heat in the temperature range 25–100°C. A study of the physical and chemical properties of RPOF reveals that the red pigment is also stable in the presence of Na_2SO_3, NaCl, amino acids, organic acids, sugars, starch, or metal ions (such as Ca^{2+}, Cu^{2+}, Fe^{3+}, Zn^{2+}, Al^{3+}, Mg^{2+}, and Na^+), but is bleached by strong oxidants ($KMnO_4$, $K_2Cr_2O_7$, and NaOCl).

Interestingly, RPOF shows excellent DPPH radical scavenging activity, and is superior to butylated hydroxytoluene (BHT). The scavenging effect is increased with increasing concentration and reaction times (Figure 59.1). RPOF also exhibits quite strong concentration-dependent inhibition of hydroxyl radicals at low concentrations, compared with ascorbic acid and quercetin. When the concentration of RPOF was 0.03 µg/ml, the scavenging percentage of hydroxyl radicals reached 92.3%.

Isolation, characterization, lipid peroxidation inhibition activities, total antioxidant activity, reducing power, metal chelating, and total phenolic compounds of natural melanin from OFS

Natural melanin from plants or animals possesses a broad spectrum of biological activities (Barr, 1983; Lukiewicz, 1972). Immunopharmacological properties of melanin are of great interest, because it may hold promise for AIDS treatment (Montefiori & Zhou, 1991). However, due to its rarity, natural melanin cannot be produced in sufficient amounts; thus, its application is restricted. In 2006, we reported that natural melanin from *O. fragrans* (NMOF) was isolated from OFS by alkaline extraction, acid hydrolysis, and repeated precipitation (Figure 59.2). The physical and chemical properties of NMOF revealed that the melanin obtained directly from OFS is similar to traditional melanin. NMOF is stable under ultraviolet light or room-light, stable in the range 25–100°C, and relatively stable in alkaline solution, reducer, and salt, but is bleached by strong oxidants ($KMnO_4$, $K_2Cr_2O_7$, and NaOCl). Metal ions, such as Ca^{2+}, Cu^{2+}, Fe^{3+}, and Zn^{2+}, will either increase the color or provide color preservation to NMOF solutions. Although Mg^{2+}, Al^{3+}, and Na^+ reduce pigment color, it is not obvious. Amino acids and organic acids do not affect NMOF, while sugars, starch, and glucose slightly affect it (Wang *et al.*, 2006).

The lipid peroxidation inhibition activities, total antioxidant activity, reducing power, metal chelating ability, and total phenolic compounds of NMOF have also been studied. The IC_{50} value for NMOF-inhibited lipid peroxidation is 0.135 mg/ml (Pan *et al.*, 2005). The total

TABLE 59.1 Effect of pH and Solvent on Color and λ_{max} of RPOF

Solvents or pH	Color	λ_{max} (nm)
Water, pH 1	Sanguine	539.5
Water, pH 2	Red	539.5
Water, pH 3	Red	538.5
Water, pH 5	Red	538.5
Water, pH 7	Henna	538.5
Water, pH 10	Light henna	538.5
Water, pH 12	Green	535.5
Methanol	Red	528.0
Ethanol	Red	537.0
Acetone	Henna	524.0

Red pigment of *Osmanthus fragrans* seed (RPOF) was dissolved in water with different pH, different colors were assumed, and slight changes in the maximum absorption wavelengths (λ_{max}).
Reprinted from Pan *et al.* (2009), *Food Chem.*, 112, 909–913, with permission.

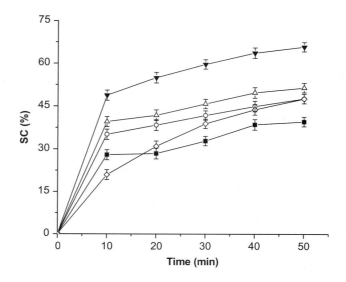

FIGURE 59.1

DPPH free radical scavenging activities of RPOF and BHT. Red pigment of *Osmanthus fragrans* seeds (RPOF) possessed significant scavenging activity on the DPPH radical, and the scavenging effect increased with increasing concentration and reaction time. ■, 0.2 mg/ml RPOF; ○, 0.5 mg/ml RPOF; △, 0.8 mg/ml RPOF; ▼, 1.2 mg/ml RPOF; ◇, 0.5 mg/ml BHT. SC % (percentage of scavenging activity of DPPH radical) = [1 − (absorbance of sample)/(absorbance of control)] × 100. Results are mean ± SD of three parallel measurements. *Reprinted from Pan* et al. *(2009),* Food Chem., 112, *909–913, with permission.*

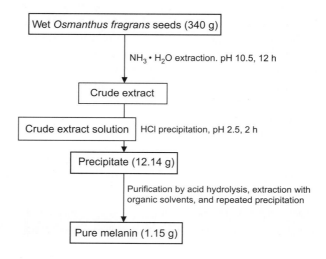

FIGURE 59.2

Procedure for extraction of melanic pigment from OFS. Melanic pigment from *Osmanthus fragrans* seeds (OFS) was isolated by alkaline extraction, and could be purified by acid hydrolysis, organic solvent (chloroform, ethyl acetate, and ethanol) treatment, and repeated precipitation. *Reprinted from Wang* et al. *(2006)* LWT-Food Sci. Technol., 39, *496–502, with permission.*

antioxidant capacity of NMOF is 0.25 g ascorbic acid/g (Wang *et al.*, 2004). NMOF shows higher antioxidant activity than BHT. Polyphenolic compounds have an important role in stabilizing lipid oxidation, and are associated with antioxidant activity (Yen *et al.*, 1993). The total phenolic content of OFSM (1 mg) is 12.5 mg pyrocatechol equivalent (Pan *et al.*, 2005). There is a correlation between antioxidant activity, reducing power, metal chelating ability, and total phenolic compounds. Similar to antioxidant activity, the reducing power and the metal chelating ability of NMOF depend on concentration, and increase with increasing concentration of NMOF.

In vitro sun protection factor of NMOF-bearing gel formulation

It is a well-known that over-exposure of human skin to ultraviolet light may lead to sunburn, premature skin aging, and an increased risk for skin cancers (Diffey *et al.*, 2000). There has been increasing interest in the use of antioxidants in sunscreens to provide supplemental photoprotective activity. Antioxidants from natural sources may provide new possibilities for the treatment and prevention of UV-mediated diseases (Bonina *et al.*, 1996; Saija *et al.*, 1998).

Recently, we have determined the *in vitro* sun protection factor (SPF) of an NMOF-bearing gel formulation using the method of Sayre and colleagues (1979). Results from the sun protection factor (SPF) *in vitro* determination of melanin-bearing gel formulations indicate that the SPF value of every formulation increased with the amount of NMOF, suggesting the presence of additional compounds with sunscreen activity in NMOF.

Chemical composition of OFS

Salidroside was isolated as an active component with radical scavenging activity from RPOF (Pan *et al.*, 2009).

Recently, we have isolated six compounds from OFS: oleanolic acid, β-sitosterol, stigmasterol, β-sitosterol-3-glucoside, (8E)-nüzhenide, and secoiridoid glucoside GL3.

ADVERSE EFFECTS AND REACTIONS (ALLERGIES AND TOXICITY)

No adverse effects have been reported in the literature.

SUMMARY POINTS

- *Osmanthus fragrans* is a traditional and famous ornamental plant in China, the seeds of which are used for folk medicine.
- RPOF is a potential natural, functional antioxidant food favor additive.
- NMOF is similar to traditional melanin.
- NMOF with photo-protective activity may be a good choice for the treatment and prevention of UV-mediated diseases.
- OFS could be used as a readily accessible source of a natural antioxidant for the food and pharmaceutical industries.
- Though many novel pharmacological properties of OFS have been identified, some bioactive substances in OFS responsible for specific bioactivities are still unknown. Further investigation should be carried out in order to optimize the utilization of OFS.

503

References

Barr, F. E. (1983). Melanin: The organizing molecule. *Medical Hypotheses, 11*, 1−139.

Bonina, F., Lanza, M., Montenegro, L., Puglisi, C., Tomaino, A., Trombetta, D., et al. (1996). Flavonoids as potential protective agents against photo-oxidative skin damage. *International Journal of Pharmaceutics, 145*, 87−94.

Diffey, B. L., Tanner, P. R., Matts, P. J., & Nash, J. F. (2000). *In vitro* assessment of the broad-spectrum ultraviolet protection of sunscreen products. *Journal of the American Academy of Dermatology, 43*, 1024−1035.

Duke, J. A., & Ayensu, E. S. (1984). *Medicinal plants of China.* Algonac, China: Reference Publications, Inc.

Es-Safi, N. E. (2004). Colour of a xanthylium pigment in aqueous solutions at different pH values. *Food Chemistry, 88*, 367−372.

Jin, H. X., Zheng, H., Jin, Y. J., Chen, J. Y., & Wang, Y. (2006). Research on major volatile components of 4 *Osmanthus fragrance* cultivars in Hangzhou Maolong Guiyu Park. *Forest Research, 19*, 612−615.

Lukiewicz, S. (1972). The biological role of melanin. I. New concepts and methodological approaches. *Folia Histochemica et Cytobiologica, 10*(1), 93−108.

Manandhar, N. P., & Manandhar, S. (2002). *Plants and people of Nepal*. Portland, Oregon: Timber Press.

Montefiori, D. C., & Zhou, J. Y. (1991). Selective antiviral activity of synthetic soluble L-tyrosine and L-dopa melanin against human immunodeficiency virus *in vitro*. *Antiviral Research, 15*(1), 11–25.

Pan, Y. M., Li, H. Y., Zhang, Y., Zhu, Z. R., & Wang, K. (2005). Study on total phenolic content and lipid peroxidation inhibition activities of melanin derived from *Osmanthus fragrans* seeds. *Food Research & Development, 26* (5), 145–148.

Pan, Y. M., Zhu, Z. R., Huang, Z. L., Wang, H. S., Liang, Y., Wang, K., et al. (2009). Characterisation and free radical scavenging activities of novel red pigment from *Osmanthus fragrans'* seeds. *Food Chemistry, 112*, 909–913.

Saija, A., Tomaino, A., Trombetta, D., Giacchi, M., Pasquale, A. D., & Bonina, F. (1998). Influence of different penetration enhancers on *in vitro* skin permeation and *in vivo* photoprotective effect of flavonoids. *International Journal of Pharmaceutics, 175*, 85–94.

Sayre, R. M., Agin, P. P., LeVee, G. J., & Marlowe, E. (1979). A comparison of *in vivo* and *in vitro* testing of sunscreening formulas. *Photochemistry and Photobiology, 29*, 559–566.

Wang, H. S., Pan, Y. M., Li, H. Y., Tang, L. D., Tang, X. J., & Huang, Z. Q. (2004). Study on extraction technology of melanin from *Osmanthus fragrans'* seeds and its antioxidant activity. *Journal of Yunnan University, 26*(6A), 55–57.

Wang, H. S., Pan, Y. M., Tang, X. J., & Huang, Z. Q. (2006). Isolation and characterization of melanin from *Osmanthus fragrans'* seeds. *LWT-Food Science and Technology, 39*, 496–502.

Wang, L. M., Li, M. T., Jin, W. W., Li, S., Zhang, S. Q., & Yu, L. J. (2009). Variations in the components of *Osmanthus fragrans* Lour. essential oil at different stages of flowering. *Food Chemistry, 114*, 233–236.

Yen, G. C., Duh, P. D., & Tsai, C. L. (1993). Relationship between antioxidant activity and maturity of peanut hulls. *Journal of Agricultural and Food Chemistry, 41*, 67–70.

Zang, D.-K., & Xiang, Q.-B. (2004). Studies on *Osmanthus fragrans* cultivars. *Journal Nanjing Forestry University (Natural Sciences Edition), 28*, 7–13.

Zang, D. K., Xiang, Q. B., Liu, Y. L., & Hao, R. M. (2003). The study of history and the application to international cultivar registration authority of sweet Osmanthus (*Osmanthus fragrans* Lour). *Journal of Plant Resources and Environment, 12*, 49–53.

Zhu, Z. R., Wang, K., Pan, Y. M., Wang, H. S., & Liang, M. (2007). The reducing power, metal chelating ability and metal content of melanin from *Osmanthus fragrans* seeds. *Shipin Gongye Keji, 28*(1), 179–180.

CHAPTER 60

Gallnuts (*Quercus infectoria* Oliv. and *Rhus chinensis* Mill.) and Their Usage in Health

Nalan Yilmaz Sariozlu, Merih Kivanc
Department of Biology, Faculty of Science, Anadolu University, Eskisehir, Turkey

505

LIST OF ABBREVIATIONS

EHEC, enterohemorrhagic *Escherichia coli*
MBC, minimal bactericidal concentration
MIC, minimal inhibitory concentration
MRSA, methicillin-resistant *Staphylococcus aureus*
MSSA, methicillin-susceptible *Staphylococcus aureus*
PGE_2, prostaglandin E_2

INTRODUCTION

Tannins are water-soluble polyphenolic secondary metabolites in plants. They occur naturally, and differ from most other natural phenolic compounds in their ability to precipitate proteins from solutions. Two main groups of tannins are distinguished according to their structures: hydrolyzable tannins, and condensed tannins (proanthocyanidins). Hydrolyzable tannins are composed of esters of gallic acid (gallotannins) or ellagic acid (ellagitannins), with a sugar core

Nuts & Seeds in Health and Disease Prevention. DOI: 10.1016/B978-0-12-375688-6.10060-X

which is usually glucose. Major commercial gallotannin (tannic acid) sources are Chinese gallnuts, Turkish gallnuts, sumac leaves, and tara pods. Both Turkish and Chinese gallnuts contain a large amount of tannin (gallotannin, 50—70%), and a small amount of free gallic acid and ellagic acid. The gallotannin is derived from 1,2,3,4,6-penta-O-galloyl-β-D-glucose. This main compound consists of a central polyol, such as glucose, esterified with several gallic acid molecules.

BOTANICAL DESCRIPTION

Gallnuts (or nutgalls) are pathological excrescences formed on young branches or twigs of plants, induced by insect attacks, as a result of deposition of the eggs (Figure 60.1). The larvae pass through the pupa stage in gallnuts. The maturating insect emerges, and leaves the gallnut by boring a hole. There are two significant gallnuts as a source of tannin (gallotannin): Turkish gallnuts (*Gallae Turcicae*) and Chinese gallnuts (*Gallae Chinensis*).

Turkish gallnuts are formed on *Quercus infectoria* Oliv. (Fagaceae) by the gall wasp, *Cynips gallae-tinctoriae* Oliv. These gallnuts are usually referred to as "gallnuts of *Q. infectoria*" in the literature. *Q. infectoria* is a small tree or shrub, native to and distributed throughout Turkey, Greece, Iran, and Syria. The gallnuts from *Q. infectoria* are about 1—2.5 cm in diameter, being almost spherical in shape, with quite a smooth surface that is tuberculated on the upper part. They can be dark bluish-green, olive green, or white-brown in color.

Chinese gallnuts are formed on several *Rhus* species, mainly *Rhus chinensis* Mill. (Anacardiaceae), by different species of gallnut aphids, particularly the Chinese sumac aphid, *Schlechtendalia chinensis* Bell. The aphid parasitizes the leaves or petioles of the plant. *R. chinensis* is native to and distributed in areas of East Asia, including China (mainly the southern provinces), Malaysia, Indochina, Japan, and Sumatra. Chinese gallnuts are approximately 4—5 × 1.5 cm in diameter, horned, and reddish-brown in color, covered with velvety down. These gallnuts are usually harvested in the autumn after removal of the larvae.

HISTORICAL USAGE

Gallnuts have historically been used mainly for tanning leather, and in the manufacture of inks and dyes. Medicinal uses of gallnuts were known by the ancient Greeks (500 BC); Hippocrates mentioned them as ingredients for the treatment of certain diseases. During an excavation in 1961, 3.5 kg of carbonized oak galls were recovered in a shop of ancient Herculaneum.

FIGURE 60.1
Gallnuts of *Quercus infectoria* Oliv.

These oak galls, found in a container called a *dolia* and preserved by the eruption of Mount Vesuvius in 79 AD, provide evidence of them as a trade article probably used in medicines (Larew, 1987).

Preparations such as decoctions, ointments, and tinctures of gallnuts used as an astringent were described in detail in the *The British Pharmaceutical Codex* (Galla, 1911). These preparations were used to arrest hemorrhage from the nose or gums, to lessen discharge from mucous membranes, as in leukorrhea, and as an astringent in painful hemorrhoids.

Gallnuts, in the form of a powder or decoction, were used for many centuries in traditional medicine for various other disorders, including as an astringent, as treatment for inflammatory conditions, and as a remedy for toothache and dental caries. They were also used for the treatment of intestinal discharges, such as diarrhea and dysentery.

PRESENT-DAY CULTIVATION AND USAGE

Gallnuts (Figure 60.2) are used for treatment mainly in China, India, and The Far East. They are applied in the form of decoctions, infusions, ointments, or powders for the treatment of inflammatory diseases. A decoction of gallnuts is generally employed as an astringent gargle, wash, or injection. Gargling a decoction or infusion of gallnuts is very effective for tonsillitis, stomatitis, and relaxing the throat, while local application of boiled and bruised gallnuts to the skin cures swelling or inflammation (Aroonrerk & Kamkaen, 2009). Powdered gallnuts mixed with vinegar are used for ringworm and alopecia (Khare, 2004). An ointment including powdered gallnuts is applied for hemorrhoids and diseases of the anus, and a powder or decoction of gallnuts is also used for diarrhea or dysentery. Injections of the decoction are applied to lessen mucous discharges of the vagina, such as in leukorrhea and gonorrhea. Gallnuts are used to remedy skin lesions and oral ulcerations. A dental powder including gallnuts is popularly used in Thailand for the treatment of aphthous ulcers (Aroonrerk & Kamkaen, 2009), and gallnuts are routinely prescribed as antidiarrheal drugs in Thai traditional medicine.

507

FIGURE 60.2
Gallnuts of *Q. infectoria* in herbal market (Eskisehir).

Gallnuts are used generally in Turkey as an astringent and antidiarrheal. Malay women still uses gallnuts to restore the elasticity of the uterine wall after childbirth. An extract of gallnuts is also used in some Asian countries to blacken gray hairs. In Korea, an aqueous extract of the Chinese gallnut, known as *Obaeja*, is used for the treatment of patients with so-called "So-Gal" symptom, typified by excessive urine excretion and water intake (Shim *et al.*, 2003).

APPLICATIONS TO HEALTH PROMOTION AND DISEASE PREVENTION

Gallnuts have great medicinal and pharmacological value. They demonstrate astringent, anti-inflammatory, local anesthetic, antipyretic, antiparkinsonian, antidiabetic, antitremorine, anti-aging, antioxidant, antimutagenic, anticaries, antibacterial, antiviral, antifungal, larvicidal, and antivenom effects.

Gallnuts are strongly astringent due to their high tannin content. Astringency provides protection of the sub-adjacent layers of mucosa from microorganisms and irritant chemicals, and also has an antisecretory effect (Mills & Bone, 2000). They are used for arrest haemorrhage from the nose or gums, and haemorrhoids. They are also used for treatment of diarrhoea, dysentery, and leucorrhea.

In the small intestine, gallic acid is released from hydrolyzed gallotannins. As gallic acid does not have an astringent effect, tannic acid is used therapeutically in its albumin tannate form. The tannin is released gradually from the albumin tannate in the alkaline medium of the intestine, and causes an astringent effect in the small intestine and colon (Schulz *et al.*, 2004).

Gallnuts have been used for centuries for treating inflammatory diseases. Gargling a hot water extract of galls is very effective against inflamed tonsils, while direct application of boiled and bruised gallnuts to the skin effectively cures any swelling or inflammation. The topical administration of ointment including powdered gallnuts also treats hemorrhoids, anal fissures, and chapped nipples (Khare, 2004).

Kaur and colleagues (2004) reported the anti-inflammatory effect of an alcoholic extract of *Q. infectoria* gallnuts. Their study showed that oral administration of the extract effectively inhibited carrageenan-, serotonin-, histamine-, and prostaglandin E_2 (PGE_2)-induced paw edemas, while topical administration of a gallnut extract inhibited phorbol-12-myristate-13-acetate (PMA)-induced ear inflammation. The extract also inhibited various functions of macrophages and neutrophils relevant to the inflammatory response, and significantly scavenged nitric oxide (NO) and superoxide ($O_2^{\bullet-}$) in the cell-free system. The gallnut extract significantly inhibited formyl-Met-Leu-Phe (fMLP)-stimulated neutrophil degranulation in neutrophils.

Powder of gallnuts *Q. infectoria* is a traditional herbal recipe that has been used in Thailand for aphthous ulcers for more than 70 years. It is known as "Ya-Kwad-Samarn-Lin Khaolaor Bhaesaj," and is still the manufacturer's best-seller (Aroonrerk & Kamkaen, 2009). This powder consists of gallnuts of *Q. infectoria*, rhizomes of *Kaempferia galanga*, roots of *Glycyrrhiza uralensis*, and roots of *Coptis chinensis*. Aroonrerk and Kamkaen (2009) investigated the role of these plants in the aphthous powder for anti-inflammatory activity. Of the four plant powders, *Q. infectoria* gallnut powder showed moderate anti-IL-6 activity with an IC_{50} value of 0.31 ± 0.02 µg/ml. The traditional herbal recipe had significantly higher anti-inflammatory activity than each of the single plants, by inhibiting IL-6 and PGE_2 production. The anti-inflammatory activity was significantly higher than that of prednisolone and the cyclo-oxgenase-2 (COX-2) inhibitor. The plant powders and the traditional herbal recipe had no growth-inhibitory effect on the human gingival fibroblast cells, even at the highest concentration. This study shows that the traditional Thai herbal recipe for aphthous ulcers can be used for local inflammation in humans without any skin irritation effects.

Gallnuts demonstrated significant antibacterial activity against a broad range of bacteria, including clinical isolates. Voravuthikunchai *et al.* (2008) reported that ethanol extracts of *Q. infectoria* gallnuts demonstrated significant antibacterial activity against many important pathogens, including enterohemorrhagic *Escherichia coli* (EHEC) and methicillin-resistant *Staphylococcus aureus* (MRSA). The extracts showed remarkable activity against MRSA, with minimal inhibitory concentrations (MICs) ranging from 0.02 to 0.4 mg/ml, and minimal bactericidal concentrations (MBCs) from 0.4 to 1.6 mg/ml. Gallnut extracts also showed significant activity against EHEC, with MICs of 0.05 to 0.1 mg/ml, and MBCs of 0.8 to 1.6 mg/ml. The results from this study indicate that gallnut extracts are a significant source of antibacterial substances with a broad spectrum of activity against antibiotic-resistant bacteria. Ethanolic extracts of gallnuts have significant antibacterial activity against multiple antibiotic-resistant *Helicobacter pylori*, which causes chronic gastritis, peptic ulcer, gastric cancer, and lymphoma (Voravuthikunchai & Mitchell, 2008).

Chusri and Voravuthikunchai (2008) reported that acetone, ethylacetate, ethanol, and aqueous extracts of *Q. infectoria* gallnuts have significant antibacterial activity against all MRSA and methicillin-susceptible *Staphylococcus aureus* (MSSA) strains. In this study, both MRSA and MSSA strains exhibited MIC and MBC values of 0.13 and 0.13–1.00 mg/ml, respectively.

The gallnuts of *Q. infectoria* are used as dental powder in traditional medicine for treatment of toothache and gingivitis. Vermani and Prabhat (2009) studied the antibacterial activity of four solvent extracts (petroleum ether, chloroform, methanol, and water) of *Q. infectoria* gallnuts against dental pathogens, including *Streptococcus mutans*, *Streptococcus salivarius*, *Staphylococcus aureus*, *Lactobacillus acidophilus*, and *Streptococcus sanguis*. The authors found that methanolic extract showed maximum antibacterial activity against all the bacteria. The MIC values of the methanolic and aqueous extracts against *S. sanguis* were 0.0781 mg/ml and 0.1563 mg/ml, respectively.

Hussein and colleagues (2000) demonstrated that methanol and water extracts of gallnuts have significant inhibitory activity ($> 90\%$ inhibition at 100 µg/ml) of Hepatisis C virus (HCV) protease, with IC_{50} values of 1 and 15 µg/ml, respectively.

Extracts of *Q. infectoria* gallnuts have larvicidal activity against various mosquitoes. Redwane and colleagues (2002) tested extracts (aqueous, methanol, ethylacetate, *n*-butanol, acetone, and gallotannins) and fractions of *Q. lusitania* var. *infectoria* gallnuts against *Culex pipiens* mosquito larvae. The aqueous extract was the most active on the second instar period of *C. pipiens* larvae (LC_{50} (24 h) $= 256$ ppm) while the gallotannins were most active on the fourth instar (LC_{50} (24 h) $= 373$ ppm).

Aivazi and Vijayan (2009) reported that ethylacetate extract was the most effective of all the five extracts tested for larvicidal activity against *Anopheles stephensi* mosquito for the fourth instar larvae, with an LC_{50} of 116.92 ppm, followed by gallotannin, *n*-butanol, acetone, and methanol, with LC_{50} values of 124.62, 174.76, 299.26, and 364.61 ppm, respectively.

Gallnuts are used in traditional treatment of wounds or burns associated with bacterial infections. The ethanolic extract of gallnuts from *Q. infectoria* has shown a wound healing effect, with a significant increase in the levels of the antioxidant enzymes superoxide dismutase and catalase in granuloma tissues (Umachigi *et al.*, 2008).

Kaur et al. (2008) demonstrated that ethanolic extract from gallnuts of *Q. infectoria* possess strongly potent free radical scavenging and antioxidant activities. The extract scavenged free radicals such as DPPH ($IC_{50} \sim 0.5$ µg/ml), ABTS ($IC_{50} \sim 1$µg/ml), hydrogen peroxide (H_2O_2) ($IC_{50} \sim 2.6$ µg/ml) and hydroxyl ($^{\bullet}OH$) radicals ($IC_{50} \sim 6$ µg/ml). It also chelated metal ions and inhibited Fe^{3+}-ascorbate induced oxidation of protein and peroxidation of lipids.

509

The methanolic extract of gallnuts *Q. infectoria* has an inhibitory activity on tyrosinase, which is a key enzyme in the synthesis of melanin, and intervenes in several intermediate stages of pigment formation (Khazaeli *et al.*, 2009). Tyrosinase inhibition is important for skin hypopigmentation. The major phenolic acids demonstrating potent inhibitory effects on tyrosinase activity in the gallnuts were found to be gallic acid and ellagic acid.

Ingested infusions of gallnuts or oak bark are very beneficial as an antidote to some poisonous substances with alkaloids, copper and lead salts.

Chinese gallnuts are used for treating many conditions, such as bleeding, inflammation, dysentery, toxicosis, bloody urine, painful joints, sores, etc., as a traditional folk medicine (Tian *et al.*, 2009). They have potent antioxidant, antibacterial, antiviral, anticaries, antimutagenic, anti-aging, and astringent activities. The ether, ethylacetate, ethanol, and water extracts of Chinese gallnuts presented remarkable antioxidant and antibacterial activities (Tian *et al.*, 2009).

Shim and colleagues (2003) reported that aqueous extracts from the Chinese gallnut exert an antidiabetic effect by inhibiting *Bacillus* alpha-glucosidase activity with an IC_{50} of 0.9 μg/ml.

Yang et al. (2009) showed that acqueous extracts of gallnuts (*Q. infectoria* and Chinese gallnuts) have anticyanobacterial activity against *Microcystis aeruginosa*, both with the same MIC value (0.39 mg/ml).

ADVERSE EFFECTS AND REACTIONS (ALLERGIES AND TOXICITY)

Gallnuts contain large amounts (50—70%) of gallotannin. Long-term medicinal intake of high doses of hydrolyzable tannins may cause negative health effects. Large amounts of tannins lead to an excessive astringent effect on mucous membranes, and oral administration may cause irritation of the gastric mucosa, nausea, and vomiting (Schulz *et al.*, 2004). High doses of tannic acid can be cytotoxic. Tannic acid can cause fatal liver damage if used on burns or as an ingredient of enemas. Tannins are metal ion chelators, and inhibit absorption of minerals such as iron; therefore, they can cause anemia in cases of long-term usage. Chronic intake of tannins inhibits digestive enzymes, particularly membrane-bound enzymes of the small intestinal mucosa (Mills & Bone, 2000). Hence, the use of gallnuts over a long period at high doses, either orally or topically, is not recommended.

SUMMARY POINTS

- Gallnuts contain 50—70% gallotannin, and small amounts (2—4%) of free gallic acid and ellagic acid.
- The gallnuts of *Q. infectoria* and Chinese gallnuts and their extracts have been widely used for centuries. They are used for tanning leather, dyeing textiles, as inks, food and feed additives, and for medicinal purposes.
- The gallnuts, as a raw material, are also used to obtain tannic acid (gallotannin) and gallic acid.
- Gallnuts are used in traditional medicine for the treatment of inflammatory diseases, dental caries, and wound healing. As they have strong astringent effects, they are also used for the treatment of hemorrhage, and intestinal disorders such as diarrhea, dysentery, and cholera.
- Astringent, anti-inflammatory, local anesthetic, antipyretic, antiparkinsonian, antidiabetic, antitremorine, anti-aging, antioxidant, antimutagenic, anticaries, antibacterial, antiviral, antifungal, larvicidal, and antivenom activities of the gallnuts have been reported in various studies.
- Further *in vivo* studies are needed to demonstrate the bioactivity properties and toxic effects of compounds separated and purified from gallnuts.

- Gallnuts, as a natural source with potential health-giving properties, hold promise for their use in human health in the future, as a food supplement, an antimicrobial agent, an antioxidant, for diabetic therapy, in the pharmaceutical and dermatology industries, as an antipigment agent in the cosmetic industry, or as a natural insecticide, among others.

References

Aivazi, A. A., & Vijayan, V. A. (2009). Larvicidal activity of oak *Quercus infectoria* Oliv. (Fagaceae) gall extracts against *Anopheles stephensi* Liston. *Parasitology Research, 104*, 1289—1293.

Aroonrerk, A., & Kamkaen, N. (2009). Anti-inflammatory activity of *Quercus infectoria, Glycyrrhiza uralensis, Kaempferia galanga* and *Coptis chinensis*, the main components of Thai herbal remedies for aphthous ulcer. *Journal of Health Research, 23*, 17—22.

Chusri, S., & Voravuthikunchai, S. P. (2008). *Quercus infectoria*: a candidate for the control of methicillin-resistant *Staphylococcus aureus* infections. *Phytotherapy Research, 22*, 560—562.

Galla, B. P. (1911). Galls. The British Pharmaceutical Codex. *Council of the Pharmaceutical Society of Great Britain*.

Hussein, G., Miyashiro, H., Nakamura, N., Hottori, M., Kakiuchi, N., & Shimotohno, K. (2000). Inhibitory effects of Sudanese medicinal plant extracts on hepatitis C virus (HCV) protease. *Phytotherapy Research, 14*, 510—516.

Kaur, G., Hamid, H., Ali, A., Alam, M. S., & Athar, M. (2004). Anti-inflammatory evaluation of alcoholic extract of galls of *Quercus infectoria*. *Journal of Ethnopharmacology, 90*, 285—292.

Kaur, G., Athar, M., & Alam, M. S. (2008). *Quercus infectoria* galls possess antioxidant activity and abrogate oxidative stress-induced functional alterations in murine macrophages. *Chemico Biological Interactions, 171*, 272—282.

Khare, C. P. (2004). *Indian herbal remedies: Rational Western therapy, Ayurvedic and other traditional usage, Botany* (pp. 395—396). Berlin, Germany: Springer-Verlag.

Khazaeli, P., Goldoozian, R., & Sharififar, F. (2009). An evaluation of extracts of five traditional medicinal plants from Iran on the inhibition of mushroom tyrosinase activity and scavenging of free radicals. *International Journal of Cosmetic Science, 31*, 375—381.

Larew, H. G. (1987). Oak galls preserved by the eruption of Mount Vesuvius in AD 79, and their probable use. *Economic Botany, 41*, 33—40.

Mills, S., & Bone, K. (2000). *Principles and practice of phytotherapy: Modern herbal medicine* (pp. 35—37). London, UK: Churchill Livingstone, Harcourt Publishers Ltd.

Redwane, A., Lazrek, H. B., Bouallam, S., Markouk, M., Amarouch, H., & Jana, M. (2002). Larvicidal activity of extracts from *Quercus lusitania* var. *infectoria* galls (Oliv.). *Journal of Ethnopharmacology, 79*, 261—263.

Shim, Y. J., Doo, H. K., Ahn, S. Y., Kim, Y. S., Seong, J. K., Park, I. S., et al. (2003). Inhibitory effect of aqueous extract from the gall of *Rhus chinensis* on alpha-glucosidase activity and postprandial blood glucose. *Journal of Ethnopharmacology, 85*, 283—287.

Schulz, V., Hänsel, R., Blumenthal, M., & Tyler, V. E. (2004). *Rational phytotherapy: A reference guide for physicians and pharmacists* (5th ed.). Berlin, Germany: Springer-Verlag. pp. 260—261.

Tian, F., Li, B., Ji, B., Yang, J., Zhang, G., Chen, Y., et al. (2009). Antioxidant and antimicrobial activities of consecutive extracts from *Galla chinensis*: the polarity affects the bioactivities. *Food Chemistry, 113*, 173—179.

Umachigi, S. P., Jayaveera, K. N., Ashok Kumar, C. K., Kumar, G. S., Vrushabendra Swamy, B. M., & Kishore Kumar, D. V. (2008). Studies on wound healing properties of *Quercus infectoria*. *Tropical Journal of Pharmaceutical Research, 7*, 913—919.

Vermani, A., & Prabhat, N. (2009). Screening of *Quercus infectoria* gall extracts as antibacterial agents against dental pathogens. *Indian Journal of Dental Research, 20*, 337—339.

Voravuthikunchai, S. P., Chusri, S., & Suwalak, S. (2008). *Quercus infectoria* Oliv *Pharmaceutical Biology, 46*, 367—372.

Voravuthikunchai, S. P., & Mitchell, H. (2008). Inhibitory and killing activities of medicinal plants against multiple antibiotic-resistant *Helicobacter pylori*. *Journal of Health Science, 54*, 81—88.

Yang, J. D., Hu, L. B., Zhou, W., Yin, Y. F., Chen, J., & Shi, Z. Q. (2009). Lysis of *Microcystis aeruginosa* with extracts from Chinese medicinal herbs. *International Journal of Molecular Sciences, 10*, 4157—4167.

511

Garden Cress (*Lepidium sativum*) Seeds in Fracture-induced Healing

Abdullah bin Habeeballah bin Abdullah Juma[1], Colin R. Martin[2]
[1] Department of Orthopaedics & Trauma, Faculty of Medicine, King Abdul Aziz University, Jeddah, Saudi Arabia
[2] Faculty of Education, Health and Social Sciences, University of the West of Scotland, Ayr, UK

513

LIST OF ABBREVIATIONS

CM, circumferential callus formation at the induced-fracture site
LL, longitudinal lateral callus formation at the induced-fracture site
LM, longitudinal medial callus formation at the induced-fracture site

INTRODUCTION

Lepidium sativum seeds have positive effects on accelerating fracture healing *in vivo* in rabbits, which by itself supports the observation noticed in traditional medicine (Juma, 2007). Fracture healing and its pathophysiological process have been the axis of a number of studies, and the factors accelerating or hindering healing are diverse and unpredictable (Ketchen *et al.*, 1978). However, the use of nutritional elements to treat some ailments and fractures is as ancient as the history of human beings. One of the plants used in traditional medicine is *Lepidium sativum* (Ageel *et al.*, 1987; Qudamah, 1995; Juma, 2007), which was given the common name of *Le Cresson* (the cress) and identified as a division of crucifers. The plant is well recognized in European communities as Herba Lepidii Sativi, and its consumption has increased in the former Soviet Union and Western European countries as a source of vitamins, for its diuretic effect, as a stimulant of bile function, and as a cough reliever (Czimber & Szabo,

Adapted from "The Effects of Lepidium Sativum Seeds on Fracture-Induced Healing in Rabbits" by Abdullah bin Habeeballah bin Abdullah Juma, FRCSEd, (Medscape General Medicine, 2007; 9(2):23).

1988). This plant is used in the community of Saudi Arabia as an important element in traditional medicine for multiple applications, but commonly in fracture healing (Ageel *et al.*, 1987; Ahsan *et al.*, 1989; Juma, 2007). Different Arabic names, such as *Rashad*, *Hurf*, and *Thuffa*, have been given to *Lepidium sativum* in Arabic countries, including Saudi Arabia, where the plant is grown in Al-Hijaz, Al-Qaseem, and the Eastern province (Ageel *et al.*, 1987). The roots of the plant, its leaves, and its seeds are used traditionally, and the effect of seeds on fracture healing, noticed clearly in Saudi folk medicine, has been reported in rats (Ahsan *et al.*, 1989) and rabbits (Juma, 2007).

BOTANICAL DESCRIPTION

A herb is a flowering, seed-bearing plant that has no permanent woody stem but dies back at the end of each growing season. Hence, garden cress (*Lepidium sativum*) is a fast-growing edible herb (Figure 61.1) that belongs to the family Cruciferae (Brassicaceae; mustard family) and is related to watercress and mustard, sharing their peppery, tangy flavor and aroma (Cassidy & Hall, 2002). The important components of *Lepidium sativum* come from the fresh green leaves, the shoots, and the seeds. The plants, however, belong to the kingdom Plantae (Haeckel, 1866), which includes trees, herbs, bushes, grasses, vines, ferns, mosses, and green algae. The scientific study of plants, known as botany, has identified about 350,000 species of plants, defined as seed plants, bryophytes, ferns, and fern allies. The naming of plants is governed by the International Code of Botanical Nomenclature (ICBN) (Greuter, 2003) and the International Code of Nomenclature for Cultivated Plants (ICNCP) (Brickell *et al.*, 2004).Therefore, *Lepidium sativum* has a tree number, B06.388.100.157.600.500, and different vernacular names according to the language and culture of the countries cultivating and consuming it.

The many synonyms of *Lepidium sativum* include the following (Ageel *et al.*, 1987; Seidemann, 2005):

- Ancient Greek Ἑλληνικόϲαρχαια: Kardamon, Lepidion, Κάρδαμον, Λεπίδιον
- Arabic (العربية): Abu khanjar, Barbeen, Nabatu al-kabbusin, Rashad, Thifa, أبو خنجر, ثفاء, ثِفَاء, نَبَاتُ الكَبُّوسِين, نبات الكبوسين, رَشَاد, رشاد
- Armenian (Հայերէն): Chri Godem, Godem, Jhri Kotem, Kotem, Կոտեմ, Ջրի Կոտեմ
- Assamese (অসমীয়া):Halim-shak
- Basque (Euskara): Beatze krechua, Beatze krexu, Berro, Bruminka
- Bengali (বাংলা): Bilrai, Halim-shak
- Bulgarian (Български): Gorukha posevna, Kreson, Latinka, Горуҳа посевна, Кресон, Латинка

FIGURE 61.1
Garden cress plant showing fresh green leaves, shoots, and sprouts.

- Catalan (Català): Bequera, Caputxina, Créixecs, Créixens, Llaguera, Morrissà, Morritort d'aigua, Morritort d'indies
- Croatian (Hrvatski): Dragoljub, Dragušac, Potočarka, Sjetvena grbica, bobovnjak vrtni, garbać vrtnji, grbak vrtni, grbica usjevna, kres pitomi, kreša, sjetvena grbica
- Czech (Česky): Lichořeřišnice větší, Potočnice, Řeřicha zahradní
- Danish (Dansk): Baerkarse, Blomsterkarse, Havekarse, Kapuciner karse, Landloeber, Nasturtie
- Dhivehi (ﺩﺮﻮﺭ): Asalhiyaa
- English: common nasturtium, creasy greens, cress, garden cress, garden nasturtium, garden pepper-grass, garden pepperwort, gardencress pepperweed, Indian cress, nasturtium, upland cress, water cress
- Esperanto (Esperanto): Akvokreso, Granda tropeolo, Tropeolo, Ĝardena kreso
- Estonian (Eesti keel): Salatkress, Suur mungalill, salatkress, Ürt-allikkerss
- Finnish (suomen kieli): Isovesikrassi, Koristekrassi, Köynnöskrassi, Ruokarassi, Vihanneskrassi, ryytikrassi, vihanneskrassi, vihantakrassi
- French (Français): Capucine, Cresson alénois, Cresson d'Inde, Cresson de fontaine, Passerage cultivée, cresson alénois
- Gallegan (Galego): Agrión
- Georgian (ქართული): Chichmati, Cicmati, წიწმატი
- German (Deutsch): Garten-Kresse, Gartenkresse, Indische Kresse
- Greek (Ελληνικά): Kardamo, Nerokardamo, Κάρδαμο Νεροκάρδαμο
- Gujarati (ગુજરાતી): Asaliya
- Hebrew (עברית): Gargar ha-nazir, Kova ha-nazir, Rashad, Shakhalayim Tarbutyim, שחליים תרבותיים , רשד , ראשד , כובע הנזיר , גרגר , הנחלים ,גרגר הנזיר
- Hindi (हिन्दी): Aselio, Halim
- Hungarian (Magyar): Borsfű, Borsika, Bécsi rosmaring, Csombor, Hurkafű, Kerti sarkantyúka, Kerti zsázsa, Pereszlén, Sarkantyúka, Sarkantyúvirág, Vízitorma
- Icelandic (Íslenska): Karsi, Skjaldflétta, Vætukarsi
- Indonesian (Bahasa Indonesia): Cencil, Selada air
- Irish (Gaeilge): Biolair
- Italian (Italiano): Agretto, Cappuccina, Crescione, Crescione d'acqua, Nasturzio del Perù, Nasturzio indiano, Nastuzio
- Japanese (日本語): Kinrenka, Koshyoso, Koshyōsō, Nasutachumu, Nozenharen, Nōzenharen, Oranda-garashi, Uotakuresu, ウォータークレス, キンレンカ, コショウソウ, ノウゼンハレン, 金蓮花
- Kannada (ಕನ್ನಡ): Allibija, Kurthike
- Korean (한국어): Kundadag-naengi, Kuraesson, Kuraessong, Mul-laengi, Nasuteochyum, Nasuteotium, Weota-kuresu, Wota-kuresu, 나스터츔, 크랫송, 콘다닥냉이
- Lao (ພາສາລາວ): Nha kat hon
- Latvian (latviešu valoda): Avotu krese, Krese, Kressalāti, dārza cietķērsa
- Lithuanian (lietuvių kalba): Mažoji nasturtė, Sėjamoji pipirnė, Vaistinio rėžiuko, sėjamoji pipirnė
- Marathi (मराठी): अळीव, अहळीव
- Persian (فارسی): Ladan, Shahi, Tazeh alaf cheshmeh, لادن ,تازه علف چشمه
- Spanish (Español): Berro di agua, Capuchina, Crenchas, Creson, Espuela de Galán, Lepido, Mastuerzo, Nasturcia

HISTORICAL CULTIVATION AND USAGE

Cultural links of human societies in the use of different plants in foods, medicine, divination, cosmetics, dyeing, and textiles, as construction tools, currency, and clothing, and in literature, rituals, and social life have been documented. The study of how plants have been perceived,

515

categorized and used by societies for food, medicine, and rituals is known as ethnobotany, and its investigation is the activity of the ethnobotanist.

The medicinal properties of garden cress were emphasized in the old Oriental and Mediterranean cultures. Its vermifugal powers were noted by Columela, and the antihistaminic properties were referred to Ibn al-Awwam. However, most information on its properties was collected by an Andalusian botanist called Ibn al-Baytar. It was mentioned that *Lepidium sativum* can be administered for leprosy, renal "cooling," and purification of hair. Seeds were used as an aphrodisiac in Iran and Morocco, and its application as an edible oil from the seeds was noted in Abyssinia. The reputation of garden cress among Muslims has been attributed to the direct recommendation by the Prophet. It was stated that Prophet Mohammed, "Peace be upon him," said:

الثفاء والصَبر (فى :

ن عبدالله بن عباس رضى الله عنه أن النبى صلى الله عليه وسلم قال: ماذا فى الأمرين من الشفاء
الجامع الاصول: أخرجه رزين, واثبته الحافظ الذهبي من اخراج الترمذي) [What's in the two of healing:
Al-Thuffa and Al-Sabir.]

(Abdullah bin Abbas (2000))

Al-Thuffa here is *Lepidium sativum*, or garden cress.

PRESENT-DAY CULTIVATION AND USAGE

Garden cress is grown commercially in Europe and other parts of the world. It needs moist conditions, and can reach a height of 50 cm, with many branches on the upper part. The white to pinkish flowers are clustered in branched racemes. The fresh green leaves, sprouts, and shoots are cultivated and consumed within a short time, the edible shoots being harvested in 1—2 weeks, when they are 4—10 cm tall. However, the mature seeds (Figure 61.2) can cope with dryness and be consumed over a longer period of time. Seeds have an embryo that is enclosed by one or two cell layers of endosperm (which serves as a food storage organ) and

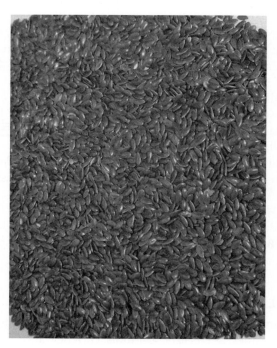

FIGURE 61.2
Lepidium sativum seeds, with brownish and glistening features.

the surrounding testa (seed coat). This makes *Lepidium sativum* a promising model system for endosperm weakening (Müller *et al.*, 2006). The mucilage is generated from the outer testa during imbibition. The edible whole seed is known to have health-promoting properties, and its chemical composition has been studied using it as nutraceutical food ingredient in a dietary fiber (DF) formulation (Gokavi *et al.*, 2004).

The usage of garden cress has been tested over time, and its benefits have been documented and observed throughout different communities and cultures. It is used in food and culinary practice for its spicy aroma, pungent taste, and tangy flavor. It is consumed in salads, sandwiches, and spreads (such as cottage cheese), and its raw leaves and shoots are used in making soap. The cut shoots are typically used in sandwiches with boiled eggs, salt, and mayonnaise, as seen in the United Kingdom, and the fresh or dried seed pods can be used as a peppery seasoning. Its green leaves are used as a garnish. The esthetic use of garden cress as an indoor plant in interior decoration is due to its easy cultivation and its limited requirement for water. The research effectiveness of this plant for the Natural Medicines Comprehensive Database was enabled by scientific work-ups and the development of new methods and applications. The components of the parts of garden cress, including its seeds, have been studied to reveal interesting results for general and clinical applications in both current therapeutic and alternative medicine fields.

Lepidium sativum has been used by people of different cultures and societies throughout the world. It was popular and widely used, giving rise to different beliefs regarding its properties, and diverse results in the hands of practitioners, patients, and consumers.

Medical applications of *Lepidium sativum* have been revealed from a review of its uses in different communities (Ravindra *et al.*, 2007). These medical usages of *Lepidium sativum* have been reported in folk medicine since time immemorial. Recently, medical applications and efficacy have observed in research, including effects on fracture healing (Ahsan *et al.*, 1989; Atasan, 1989; Juma, 2007), a role in diabetes as a blood glucose lowering agent (Eddouts *et al.*, 2005), a role in bronchial asthma (Paranjape & Mehta, 2006), and its activity as an antihypertensive (Maghrani *et al.*, 2005) and as an oral contraceptive (Sharief & Gani, 2004).

APPLICATIONS TO HEALTH PROMOTION AND DISEASE PREVENTION

Applications of *Lepidium sativum* seeds to health promotion and disease prevention were reviewed both clinically and scientifically in laboratories; in particular, the fracture healing phenomenon, and factors causing delay or non-union. Induction and acceleration of fracture healing was the scope of many works, as elicited by many methods and procedures. Induced fractures in the femur of rabbits by an open surgical technique, and their fixation with an intramedullary K-wire (Figure 61.3), was methodically demonstrated and relevant to the clinical setting (Juma, 2007). The operated rabbits were divided into control (fed with normal diet) and test groups (fed daily with a normal diet in addition to 6 g of *Lepidium sativum* seeds). These seeds were obtained from the local market, and were of the type grown in the Al-Qaseem area in Saudi Arabia. Radiographs were taken of the fracture sites 6 and 12 weeks' postoperatively (Figure 61.4). Callus formed at the fracture sites was the indicator of the healing of fractures radiologically. Measurements (in millimeters) of the callus were taken from the radiographs, from three aspects (LM, longitudinal medial; LL, longitudinal lateral; and CM, circumferential) were noted both at 6 and 12 weeks (Table 61.1).

A statistically significant difference between test and control rabbits was noted at 6 and 12 weeks' postoperatively (Table 61.2). Marked callus formation, indicating accelerated healing of induced fractures, was noted in the test group, which was fed with *Lepidium sativum* seeds. Circumferential (CM) callus at 6 and 12 weeks ($P < 0.001$ and $P < 0.004$, respectively) was predominant, as well as longitudinal medial callus (LM) at 12 weeks ($P < 0.043$).

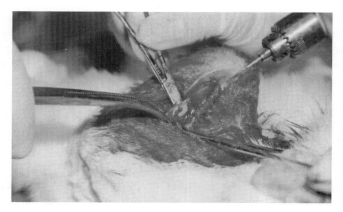

FIGURE 61.3
Open induced transverse fracture at midshaft left femur fixed with intramedullary K-wire in control and test rabbits.

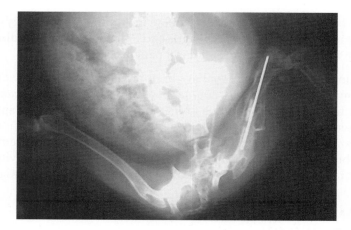

FIGURE 61.4
Radiograph of induced fracture of left femur with excess callus in test rabbits.

TABLE 61.1 Callus Measurements at Fracture-induced Sites (mm) in Control (C) and Test (T) Rabbits 6 and 12 Weeks Postoperatively

No. of Cases	Callus Formation at 6 Weeks' Postoperatively (mm)			Callus Formation at 12 Weeks' Postoperatively (mm)		
	Longitudinal Lateral (LL)	Longitudinal Medial (LM)	Circumferential (CM)	Longitudinal Lateral (LL)	Longitudinal Medial (LM)	Circumferential (CM)
C1	30	18	7	28	17	9
C2	23	30	2	32	15	2
C3	25	21	4	22	11	4
T1	23	26	22	25	23	17
T2	40	42	18	43	49	18
T3	33	33	21	34	39	19
t	−1.121	−1.817	−8.485	−1.12	−2.915	−6.018
$P <$	NS	NS	0.001	NS	0.043	0.004

TABLE 61.2 Comparison of Callus Formation at Fracture-induced Sites Between Control and Test Rabbits at 6 and 12 weeks Postoperatively

Groups	Longitudinal Lateral (LL)	Longitudinal Medial (LM)	Circumferential (CM)	Total (T)
Control	26.0 ± 2.1 (6w)	23.0 ± 3.6 (6w)	4.3 ± 1.5 (6w)	53.3 ± 1.7 (6w)
	27.3 ± 2.9 (12w)	14.3 ± 1.8 (12w)	5.0 ± 2.1 (12w)	46.7 ± 5.0 (12w)
Test	32.0 ± 4.9 (6w)	33.7 ± 4.6 (6w)	20.3 ± 1.2 (6w)	86.0 ± 8.4 (6w)
	34.0 ± 5.2 (12w)	37.0 ± 7.6 (12w)	18.0 ± 0.6 (12w)	89.0 ± 13.1 (12w)
t	-1.121 (6w)	-1.817 (6w)	-8.485 (6w)	-3.82 (6w)
	-1.12 (12w)	-2.915 (12w)	-6.018 (12w)	-3.02 (12w)
$P <$	NS (6w)	NS (6w)	0.001 (6w)	0.019 (6w)
	NS (12w)	0.043 (12w)	0.004 (12w)	0.039 (12w)

Values are presented as mean $\pm$ SEM; $n = 3$; NS, non-significant.

Fracture healing is a major issue that has been discussed widely in clinical, experimental, and traditional practices aiming at facilitating the healing phenomenon, and hindering factors that prevent or delay its occurrence. Consequently, natural elements have been observed to have effects on different diseases, and have been used extensively over time (Qudamah, 1995). Garden cress has been used for the treatment of many ailments in different societies (Ageel *et al.*, 1987; Czimber *et al.*, 1988). However, *Lepidium sativum*, and its seeds in particular, were publicly and commonly used in Saudi Arabia as a traditional alternative medicine practice, mostly for the treatment of recent traumatic fractures and, less commonly, in delayed or non-united fractures. Good results of healing of fractures were observed over decades in the hands of traditional practitioners in this community. Its application in fracture healing was noted in rats fed with *Lepidium sativum* seeds, which showed an increase of collagen deposition and tensile strength at the fracture sites (Ahsan *et al.*, 1989), and accelerated callus formation was noted radiologically in rabbits (Juma, 2007). The molecular activity of the fracture exudates was the most decisive factor for bone healing. It is worth noting that the emulsifying properties of the mucilage of *Lepidium sativum* seeds were possibly important factors having effects on one or more of these fracture exudate constituents. Consequently, fracture healing was accelerated. Further studies of *Lepidium sativum* seeds and their effects on fracture healing are required, and can be clinically applied, supporting the outcomes of traditional medicine and the *in vivo* study of rabbits (Juma, 2007).

ADVERSE EFFECTS AND REACTIONS (ALLERGIES AND TOXICITY)

Although garden cress seeds have many benefits and usages, it can be harmful by causing allergy or toxicity in moderate or high doses. It should be consumed with moderation, since digestive problems can arise due to its content of mustard oil.

SUMMARY POINTS

- *Lepidium sativum* is a herbal plant known by different names, such as Al-Thuffa and garden cress, in different communities, with a wide spectrum of usage both medically and non-medically.
- The effects of the plants are through their different anatomical parts, such as the green leaves, sprouts, shoots, and particularly seeds.
- Seeds are consumed fresh and dry with an outer layer of mucilage, which is possibly an important factor in accelerating fracture healing in rabbits.
- Fracture-healing efficacy has been demonstrated in rabbits. Further research study of the mechanism of *Lepidium sativum* seeds on callus formation is necessary.
- As their clinical application is so significant, it is vitally important that the seeds be approved pharmacologically and applied medically in the community.

References

Abdullah bin Abbas (579). (2000). Narrated by Razeen and proved by Al-Hafiz Al-Zahabi, directed by Al-Tirmidhi. In Jame Al-Osool., & A. Ibn Al-Atheer (Eds.), *Hadith No. 5663, Vol. 7* (pp. 535). Beirut, Lebanon: Dar Al-fikr.

Ageel, A. M., Tariq, M., Mossa, J. S., Al-Yahya, M. A., & Al-Said, M. S. (1987). *Plants used in Saudi folk medicine.* Riyadh, Saudi Arabia: King Saud University Press. 245—415.

Ahsan, S. K., Tariq, M., Ageel, M., Al-Yahya, M. A., & Shah, A. H. (1989). Studies on some herbal drugs used in fracture healing. *International Journal of Crude Drug Research, 27,* 235—239.

Atasan, S. K. (1989). Studies on some herbal drugs used in fracture healing. *International Journal of Crude Drug Research, 27,* 235—239.

Brickell, C. D., Baum, B. R., Hetterscheid, W. L. A., Leslie, A. C., McNeill, J., Trehane, P., et al. (2004). *International code of nomenclature for cultivated plants — Code International pour la Nomenclature des Plantes Cultivées* (7th ed.). Leuven, Belgium: ISHS Acta Horticulturae. 647.

Cassidy, F. G., & Hall, J. H. (2002). *Dictionary of American regional English.* Cambridge, MA: Harvard University Press. p. 97.

Czimber, G., & Szabo, L. G. (1988). Therapeutical effect and production of garden cress (*Lepidium Sativum* L.). *Gyogyszereszet, 32,* 79—81.

Eddouts, M., Maghrani, M., Zeggwagh, N. A., & Michel, J. B. (2005). Study of the hypoglycaemic activity of *Lepidium sativum* L. aqueous extract in normal and diabetic rats. *Journal of Ethnopharmacology, 97,* 391—395.

Gokavi, S. S., Malleshi, N. G., & Guo, M. (2004). Chemical composition of garden cress (*Lepidium sativum*) seeds and its fractions and use of bran as a functional ingredient. *Plant Foods for Human Nutrition, 59,* 105—111.

Greuter, W. (2003). *International Code of Botanical Nomenclature (St Louis Code).* International Association for Plant Taxonomy. Available online at: http://www.bgbm.org/iapt/nomenclature/code/saintlouis/0000st.luistitle.htm (accessed 21 January 2009).

Haeckel, G. (1866). *Generale Morphologie der Organismen.* Berlin, Germany: Verlag von Georg Reimer. Vol. 1, pp. i—xxxii, 1—574, pls I—II; Vol. 2, pp. i—clx, 1—462, pls I—VIII.

Juma, Abdullah bin Habeeballah bin Abdullah (2007). The effects of *Lepidium sativum* seeds on fracture-induced healing in rabbits. *Medscape General Medicine, 9,* 23.

Ketchen, E. E., Porter, W. E., & Bolton, N. E. (1978). The biological effects of magnetic fields on man. *American Industrial Hygiene Association Journal, 39,* 1.

Maghrani, M., Zeggwagh, N.-A., Michel, J.-B., & Eddonks, M. (2005). Antihypertensive effect of *Lepidium sativum* L. in spontaneously hypertensive rats. *Journal of Ethnopharmacology, 100,* 193—197.

Müller, K., Tintelnot, S., & Leubner-Metzger, G. (2006). Endosperm-limited Brassicaceae seed germination: abscissic acid inhibits embryo-induced endosperm weakening of *Lepidium sativum* (cress) and endosperm rupture of cress and *Arabidopsis thaliana. Plant Cell Physiology, 47,* 864—877.

Paranjape, A. N., & Mehta, A. A. (2006). A study on clinical efficacy of *Lepidium sativum* seeds in treatment of bronchial asthma. *Iranian Journal of Pharmacology & Therapeutics, 5,* 55—59.

Qudamah, A. (1995). *Dictionary of food and treatment by plants.* Beirut, Lebanon: Dar Alnafaes. 241—244.

Ravindra, G. M., Mahajan, S. G., & Mehta, A. A. (2007). *Lepidium sativum* (Garden cress): a review of contemporary literature and medicinal properties. *International Journal Oriental Pharmacy and Experimental Medicine, 7*(4), 331—335.

Seidemann, J. (2005). *World spice plants.* Berlin, Germany: Springer-Verlag.

Sharief, M., & Gani, Z. H. (2004). Garden cress *Lepidium sativum* seeds as oral contraceptive plant in mice. *Saudi Medical Journal, 25,* 965—966.

CHAPTER 62

Health Benefits of Garden Cress (*Lepidium sativum* Linn.) Seed Extracts

Mahavir H. Ghante[1], Sachin L. Badole[2], Subhash L. Bodhankar[2]
[1] Department of Chemistry, J. L Chaturvedi College of Pharmacy, MIDC, Nagpur, India
[2] Department of Pharmacology, Poona College of Pharmacy, Bharati Vidyapeeth University, Maharashtra, India

521

LIST OF ABBREVIATIONS

ACh, acetylcholine
HCTZ, hydrochlorothiazide
LS, *Lepidium sativum* Linn.
LSS, *Lepidium sativum* seeds
p.o., per oral

INTRODUCTION

Lepidium sativum Linn. (garden cress) is a shrub belonging to the family Cruciferae, and is cultivated as a salad plant and culinary herb throughout India (Wealth of India, 1998). *L. sativum* (LS) is a traditionally well-known and widely used herb in Africa, China, and other far Eastern countries (Al-Yahya *et al.*, 1994; Wealth of India, 1998). LS is documented as an official herb in *The Ayurvedic Pharmacopoeia of India* (Governmant of India, 2001). *L. sativum* seeds (LSS) are regularly consumed as food, and are considered to be effective in various pathological conditions (Wealth of India, 1998).

The current chapter encompasses certain important phytochemical and pharmacological aspects of LSS.

Nuts & Seeds in Health and Disease Prevention. DOI: 10.1016/B978-0-12-375688-6.10062-3

BOTANICAL DESCRIPTION

Lepidium sativum Linn. is a member of the Cruciferae family. It is an annual glabrous herb, which grows up to 15—45 cm in height. The stem is erect and smooth skinned. Leaves are entire, and the uppermost leaves are linear and sessile. The basal leaves have long petioles, pinnate to bipinnate, and cauline. Flowers are white and in elongated racemes. The fruit is siliquose and deeply notched at the top. The seeds are reddish-brown, with two per pod. The testa is smooth and shiny, with an incumbent cotyledon, fleshy, white, and with an abundance of mucilage, solitary in each cell (Figure 62.1; Wealth of India, 1998; Singh & Karthikeyan, 2000).

HISTORICAL CULTIVATION AND USAGE

Lepidium sativum thrives on any good light soil or moist loam, and grows at all elevations, throughout the year. Its seeds are sown thickly and covered until germination begins. Mature plants are pulled out, dried, and thrashed to obtain seeds for further germination (Wealth of India, 1998).

Ethnopharmacologically, the seeds have been reported to cure breast cancer, nasal polyps, uterine tumors, wound healing, rheumatism, sexual debility, and gastrointestinal and menstrual disorders. In African countries, LSS are chewed to cure throat diseases, asthma, and headaches (Al-Yahya *et al.*, 1994; Wealth of India, 1998; Khare, 2007).

PRESENT-DAY CULTIVATION AND USAGE

Lepidium sativum seed extracts are used as an antirheumatic, diuretic, and febrifuge, in abdominal discomfort, in fracture healing, and in the treatment of gout. These medicinal uses of LSS are practiced routinely in Ayurveda. *Kasturyadi (Vayu) gutika* is an important formulation containing LSS extract which is prescribed by Ayurvedic practitioners for

FIGURE 62.1
Seeds of Lepidium sativum Linn. (garden cress). Authenticated By: Post Graduate Teaching Department of Botany, Rashtra Sant, Tukdoji Maharaj Nagpur University, Nagpur-440001, Maharashtra, India; Authentication No. 9497.

dysentery *(Atisara)*, hiccoughs *(Hikka)*, and gout *(Vatrakatta)* (Government of India, 2001). Practitioners of the Unani system of medicine are using roasted LSS for its anti-inflammatory effect (Khare, 2007).

APPLICATIONS TO HEALTH PROMOTION AND DISEASE PREVENTION

Lepidium sativum seeds have been implicated in the treatment and management of a plethora of diseases, some of which are discussed below.

1. *Asthma.* The crude ethanolic extract (500 mg/kg) of LSS and its fractions, namely n-butanol (100 mg/kg) or methanol (100 mg/kg), have been reported to exhibit significant bronchoprotection, as compared to a control group, against acetylcholine (ACh) and histamine-induced bronchospasm in guinea pigs. In addition, the n-butanol fraction has shown significantly increased preconvulsive time induced by either ACh or histamine aerosol in guinea pigs. Moreover, bronchoprotection offered by n-butanol fractions was comparable to that of reference standard drugs such as ketotifen and atropine sulfate. This preclinical investigation indicates its anti-asthmatic potential, which is well corroborated by clinical investigation (Mali *et al.*, 2008).

2. *Pain, inflammation, and fever.* LSS extract exhibited analgesic activity in mice, and attenuated carrageenan-induced inflammation as well as yeast-induced hyperpyrexia. Thus, LSS extract has therapeutic value in treating nociception, inflammation, and hyperthermia (Al-Yahya *et al.*, 1994).

3. *Blood coagulation.* Hematological studies of methanolic extract of LSS (500 mg/kg) showed a significant increase in fibrinogen levels, while prothrombin time was unaffected, suggesting its effect on blood coagulation and related disorders (Al-Yahya *et al.*, 1994).

4. *Diuretic effect.* Diuretic effects of aqueous and methanolic extracts of LSS have been reported in rats at doses of 50 and 100 mg/kg (p.o.). Both extracts were reported to increase sodium excretion, while aqueous extract at 100 mg/kg was found to be responsible for increased potassium excretion. The effects of these extracts were comparable to that of the standard drug, hydrochlorothiazide (HCTZ). Moreover, methanolic extract of LSS showed potassium-conserving property (Tables 62.1, 62.2; Patel *et al.*, 2009). Therefore, LSS might be useful in the treatment of hypertension and related kidney disorders.

523

TABLE 62.1 Effect of Oral Administration of Aqueous and Methanol Extracts of *L. Sativum* and HCTZ on Sodium and Potassium Excretion in Rats

Treatment	Dose (mg/kg, p.o.)	Sodium (meq/100 g per 8 h) $\times 10^{-2}$	Potassium (meq/100 g per 8 h) $\times 10^{-2}$
Control	—	54.16 ± 1.72	17.00 ± 1.37
HCTZ	10	91.50 ± 1.12 [***]	29.66 ± 1.75[***]
L. sativum (Aq.)	50	62.50 ± 1.76[**]	16.83 ± 1.45
L. sativum (Aq.)	100	87.61 ± 1.25[***]	25.33 ± 1.16[**]
L. sativum (MeOH)	50	60.00 ± 1.37	17.83 ± 1.70
L. sativum (MeOH)	100	78.66 ± 1.76[***]	18.83 ± 1.07

50 mg/kg and 100 mg/kg of LSS aqueous (Aq) extract showed significant increase in Na^+ excretion whereas the 100 mg/kg of the methanol (MeOH) extract produced a significant increase in Na^+ excretion ($P < 0.001$) when compared to the control group. Only HCTZ and the 100-mg/kg dose of the aqueous extract produced significant increases in potassium excretion.

The results are expressed as mean values ± SEM (standard error of mean) of six pairs of rats. Statistical comparison was carried out by analysis of variance (ANOVA). The differences between the means of treated and non-treated control groups were evaluated by the Bonferroni Multiple Comparisons Test. The results were considered statistically significant when $P < 0.05$, **Source:** Patel *et al.* (2009), *Trop. J. Pharm. Res.*, *8*, 215–219.

[**]$P < 0.01$.
[***]$P < 0.001$.

TABLE 62.2 Effect of Oral Administration of Aqueous and Methanol Extracts of *L. sativum* and HCTZ on Urine Volume, Diuretic Index, Conductivity, and pH

Treatment	Dose (mg/kg)	Urine Volume (ml/100 g per h)	Diuretic Index	Conductivity	pH
Control	—	4.75 ± 0.13	—	12.80 ± 0.54	7.43 ± 0.18
HCTZ	10	7.48 ± 0.18[***]	1.5747	14.73 ± 0.73	7.40 ± 0.32
L. sativum (Aq.)	50	6.15 ± 0.18[***]	1.2947	13.77 ± 1.36	7.23 ± 0.24
L. sativum (Aq.)	100	7.12 ± 0.12[***]	1.4989	14.68 ± 1.14	7.22 ± 0.21
L. sativum (MeOH)	50	5.61 ± 0.13[**]	1.1810	13.89 ± 0.99	6.88 ± 0.22
L. sativum (MeOH)	100	6.70 ± 0.134[***]	1.4105	14.43 ± 0.77	6.50 ± 0.27

Oral administration of HCTZ showed 54% increase in urine volume. For aqueous (Aq) extract of LSS, the increase was 29% ($P < 0.001$) and 49% ($P < 0.001$) at doses of 50 and 100 mg/kg body weight, respectively. For methanolic (MeOH) extract, the corresponding changes were 18% ($P < 0.01$) and 41% ($P < 0.001$). Non-significant changes in other parameters (diuretic index, conductivity, and pH) were observed by all test groups when compared to control group. The results are expressed as mean values ± SEM (standard error of mean) of six pairs of rats. Statistical comparison was carried out by analysis of variance (ANOVA). The differences between the means of treated and non-treated control groups were evaluated by the Bonferroni Multiple Comparisons Test. The results were considered statistically significant when $P < 0.05$,

Source: Patel *et al.* (2009), *Trop. J. Pharm. Res.*, 8, 215–219.

[**]$P < 0.01$.

[***]$P < 0.001$. Diuretic index = volume treated group/volume control group.

5. *Estrogenic activity*. Volatile oil obtained by steam distillation of LSS is reported to exhibit an estrogenic effect. When added (3–4 drops) to the diet of immature rats, they showed development and increase in the weight of ovaries as compared to the control group (Wealth of India, 1998). This report indicates the efficacy of LSS as an estrogenic agent.

6. *Antioxidant*. The oil obtained from LSS is reported to possess antioxidant potential. This has been attributed to the presence of α-tocopherol and β-sitosterol in LSS (Wealth of India, 1998).

The reported and identified phytochemical constituents of LSS are alkaloids (lepidine along with the sinapic acid ethyl ester of N,N'-dibenzylthiourea), α-tocopherol, flavonoids, cardiotonic glycosides, coumarins, glucosinolates (glucotropaeolin-compound A and phenyl ethyl glucosinolate-compound B), proteins, saponins, sterols, sinapic acid (4-hydroxyl-3:5-dimethoxycinnamic acid), tannins, triterpene (sinapin), and uric acid (choline ester of sinapic acid). LSS is also reported to contain mucilaginous matter that contains a mixture of cellulose (18.3%) and uronic acid-containing polysaccharides, which on acid hydrolysis yield L-arabinose, D-galactose, L-rhamnose, D-galacturonic acid, and D-glucose. In addition, seedlings of LS contain benzyl thiocynate (Al-Yahya *et al.*, 1994; Wealth of India, 1998; Khare, 2007; Nayak *et al.*, 2009). These phytoconstituents of LSS might be responsible for its known as well as unknown pharmacological potential, which may provide novel phytoconstituents that can be evaluated for probable bioactivity.

ADVERSE EFFECTS AND REACTIONS (ALLERGIES AND TOXICITY)

Toxicity studies of methanolic extracts of LSS in mice and ethanolic extracts in rats have shown no symptoms of toxicity or mortality. However, significant gain in body weight of mice was reported after the administration of methanolic extract for a prolonged duration (90 days) at 100 mg/kg per day (Al-Yahya *et al.*, 1994). Similar effects have been reported after the administration of methanolic extract in rats. Therefore, LSS seems to be safe for use in various diseases.

SUMMARY POINTS

- *L. sativum* (garden cress) is an erect, herbaceous, glabrous annual, 15–45 cm tall, classified as shrub and belonging to the family Cruciferae.
- Traditionally, LS is a well-known and widely used herb in India, China, far Eastern countries, and Africa.

- Preclinical results suggest the effectiveness of LSS extracts in asthma, pain, inflammation, nociception, blood coagulation, oxidative stress, anuresis, and related disorders.
- The presence of various important phytochemical constituents in LSS (namely, alkaloids, flavonoids, cardiotonic glycosides, coumarins, glucosinolates, saponins, sterols, sinapic acid, tannins, triterpene, and uric acid) might be responsible for its ethnopharmacological, preclinical, and clinical recognition.
- LSS extracts does not produce any adverse effects or mortality in mice and rats, which suggests its wide margin of safety in acute as well as chronic toxicity studies.
- Extensive work has to be carried out for the isolation of novel bioactive compounds from LSS to ascertain their ethnopharmacological and new therapeutic potential.

References

Al-Yahya, M. A., Mossa, J. S., Ageel, A. M., & Rafatullah, S. (1994). Pharmacological and safety evaluation studies on *Lepidium sativum* L., seeds. *Phytomedicine, 1,* 155−159.

Government of India. (2001). *The Ayurvedic Pharmacopoeia of India.* In: *Part I, Vol. I.* New Delhi, India: Government of India, Ministry of Health and Family Welfare, Department of Indian System of Medicine and Homeopathy. p. 28.

Khare, C. P. (2007). *Indian medicinal plants, an illustrated dictionary.* New York, NY: Springer Science and Business Media. p. 370.

Mali, R. G., Mahajan, S. G., & Mehta, A. A. (2008). Studies on bronchodilatory effect of *Lepidium sativum* against allergen induced bronchospasm in guinea pigs. *Pharmacognosy Magazine, 4*(15), 189−192.

Nayak, P. S., Upadhyaya, S. D., & Upadhyaya, A. (2009). A HPTLC densitometer determination of sinapic acid in Chandrasur (*Lepidium sativum*). *Journal of Scientific & Industrial Research, 1,* 121−127.

Patel, U., Kulkarni, M., Undale, V., & Bhosale, A. (2009). Evaluation of diuretic activity of aqueous and methanol extracts of *Lepidium sativum* garden cress (Cruciferae) in rats. *Tropical Journal of Pharmaceutical Research, 8,* 215−219.

Singh, N. P., & Karthikeyan, S. (2000). *Flora of Maharashtra State, Dicotyledons, Vol. I, Ranunculaceae to Rhizophoraceae. Flora of India, Series 2.* Kolkata, India: Botanical Survey of India. pp. 196−197.

Wealth of India. (1998). *A dictionary of raw materials and industrial products. Raw Materials,Vol. VI.* New Delhi, India: National Institute of Science and Communication, CSIR. *L−M*pp. pp. 71−72.

CHAPTER 63

Antifungal and Lipid Transfer Proteins from Ginkgo (*Ginkgo biloba*) Seeds

Ken-ichi Hatano[1], Takuya Miyakawa[2], Yoriko Sawano[2], Masaru Tanokura[2]
[1] Department of Chemistry and Chemical Biology, Faculty of Engineering, Gunma University, Kiryu, Gunma, Japan
[2] Department of Applied Biological Chemistry, Graduate School of Agricultural and Life Sciences, The University of Tokyo, Bunkyo-ku, Tokyo, Japan

LIST OF ABBREVIATIONS

ASK1, apoptosis signal-regulating kinase 1
BAD, Bcl-2/Bcl-XL-agonist causing cell death
cDNA, complementary deoxyribonucleic acid
CRK, cysteine-rich receptor-like kinase
DUF26, domain 26 of unknown function
HIV-1, human immunodeficiency virus 1
nsLTP1, type-I non-specific lipid-transfer protein
PP2C, type-2C protein phosphatase
RT-PCR, reverse-transcription polymerase chain reaction
SDS-PAGE, sodium dodecyl sulfate-polyacrylamide gel electrophoresis

Nuts & Seeds in Health and Disease Prevention. DOI: 10.1016/B978-0-12-375688-6.10063-5

INTRODUCTION

Ginkgo biloba is the oldest living plant species. Dating back more than 250 million years, it is often referred to as "a living fossil." Ginkgo is a very hearty plant, and in fact was the only flora to resprout in the spring following the atomic bombing of Hiroshima. It is thought that ginkgo trees generated naturally in America, Asia and Europe, but then disappeared from America and Europe until being cultivated again in the 18th century. Over the past few years, ginkgo has ranked first among herbal medicines sold in health food stores in the United States (Blumenthal, 2000).

BOTANICAL DESCRIPTION

Ginkgo has no close living relatives in the plant kingdom and is classified in a separate division, the Ginkgophyta. Ginkgo is a tree with a long lifespan, and is highly resistant to insects, bacterial and viral infections, and air pollution (Singh *et al.*, 2008). It is a dioecious gymnosperm with male and female reproductive organs on separate trees, and its seeds are not protected by an ovary wall. The ginkgo tree has a central trunk with a maximum size of about 4 m in diameter and 30–40 m in height (Mahadevan & Park, 2008). The leaves are a unique fan-shape, 5–10 cm in length and grow on short shoots during the spring, then turn yellow and fall in autumn. The pollen and seeds are produced on short shoots in the male and the female organs, respectively. The seeds are coated with a light yellow-brown outer layer and contain nuts (Figure 63.1). Different from other plants, the fertilization process in this plant begins with the joining together of the egg and swimming sperm (Vaughn & Renzaglia, 2006).

HISTORICAL CULTIVATION AND USAGE

Ginkgo has long been cultivated in China. The medicinal uses of ginkgo can be traced back to the Chinese herbal database known as the *Materia Medica*, which was written about 2800 BC. The ginkgo leaves were used for brain disorders, circulatory disorders, and respiratory diseases as traditional Chinese medicine. Fallen leaves of ginkgo were used as insecticides and fertilizer. Ginkgo nuts were traditionally used for getting rid of coughs, sputum and fever, for stopping diarrhea and toothaches, for healing skin diseases and gonorrhea, and for reducing frequency of micturition (Hori & Hori, 1997). In Japan, ginkgo nuts have been eaten as a side dish since the Edo Period (1600–1867).

PRESENT-DAY CULTIVATION AND USAGE

Ginkgo trees are now cultivated extensively in Asia, Europe, North America, New Zealand and Argentina (Chan *et al.*, 2007) and produce 6000–7000 tonnes of dry nuts and over 8000 tonnes of dried leaves per year to meet the commercial demand (Singh *et al.*, 2008). The ideal plantation conditions are full sunlight and well drained soil at temperatures of 15–27°C. For commercial production of nuts, the high-yield cultivars are selected and propagated by a grafting technique using seedling rootstocks. For the purpose of leaf production, trees have been planted on a large scale in France and the United States, with the leaves being harvested in the summer while green (Singh *et al.*, 2008). Both the leaves and seeds (nuts) have been consumed as food and medicine throughout the world for the past several centuries. However,

FIGURE 63.1
Structure of ginkgo seeds. The structure consists of an outer seed coat (left), inner seed coat (center) and endosperm (right).

in the past 20–30 years, the leaf extract, EGb711, has become very popular for its purported medicinal effects and is used extensively as a supplement in Europe and the United States (Mahadevan & Park, 2008).

APPLICATIONS TO HEALTH PROMOTION AND DISEASE PREVENTION

Ginkgo is presently considered a medicinal plant in Europe and the United States and has been recognized as such since ancient times in East Asia. The tree produces a variety of secondary metabolites, among which the terpene trilactones, ginkgolides and bilobalide are the most characteristic. Ginkgolides are diterpenes that have been detected both in roots and leaves, and the sesquiterpene bilobalide is the major terpene trilactone in leaves and a minor component in roots. Ginkgolides are a structurally unique family of diterpenoids that are highly specific platelet-activating factor receptor antagonists (Guinot & Braquet, 1994), and bilobalide has neuroprotective effects and is effective in neurodegenerative diseases (Thompson, 1995). Therefore, extracts from the ginkgo leaf are the most widely sold as phytomedicines to treat early-stage Alzheimer's disease, vascular dementia, and so on. On the other hand, the nuts have been used as traditional Chinese herbal remedies for frequent urination, enuresis, asthmatic response and lung tuberculosis. However, mass intake of raw ginkgo nuts is bad for human health, since the nuts contain 4′-O-methylpyridoxine (gink-gotoxin; Figure 63.2). Ginkgotoxin is structurally related to vitamin B_6, and likely interferes with its biosynthesis, metabolism or function (see the following section). Accordingly, it is adequate and favorable for an adult to take 5–10 heat-treated nuts (6–15 g) per day. Several storage proteins have so far been identified in ginkgo seeds, including a 30-kDa glycoprotein, ginnacin, legumin and α-mannosidase. In addition, Wang and Ng found that a protein from the seeds exerts potent antifungal activity, moderate antibacterial activity, inhibitory activity toward HIV-1 reverse transcriptase and the proliferation of murine splenocytes (Wang & Ng, 2000). Similar findings have been reported for other antifungal proteins, such as those from the field bean *Dolichos lablab* (Ye *et al.*, 2000a) and the cowpea *Vigna unguiculata* (Ye *et al.*, 2000b). Recently, we characterized an antifungal protein and the other bioactive proteins in ginkgo nuts, in order to shed light on the relationship between these proteins and the pharmaceutical properties of the nuts.

As shown in Figure 63.3A, we fractionated extracts with proteinase-inhibitory activities from 775 g of shelled ginkgo seeds by using gel-filtration chromatography. The inhibitory fractions B (0.52 g) and C (1.06 g) were purified by ion exchange chromatography, and they appeared as a protein band with a molecular mass of approximately 10 kDa on SDS-PAGE (Figure 63.3B). Samples B and C were estimated to be more than 95% pure and were found to have molecular masses of 9319 and 11,654 Da, respectively, by matrix-assisted laser desorption/ionization time-of-flight mass spectrometry. The final yields from the starting material were 80 mg for sample B and 20 mg for sample C (Sawano *et al.*, 2005; Sawano *et al.*, 2008). Sample B exhibited partial non-competitive inhibition of the aspartic-acid proteinase pepsin and the cysteine proteinase papain ($K_i = 10^{-5}–10^{-4}$ M). The cDNA of the inhibitor contains a 357-bp open reading frame encoding a 119-amino acid protein with a potential signal peptide (27 residues), which indicates that this protein is synthesized as a preprotein and is secreted

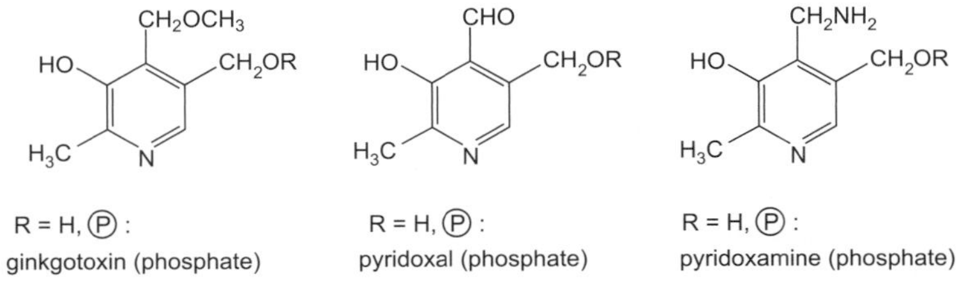

529

FIGURE 63.2

The B_6 antivitamin ginkgotoxin of (phosphate) and B_6 vitamers pyridoxal (phosphate) and pyridoxamine (phosphate).

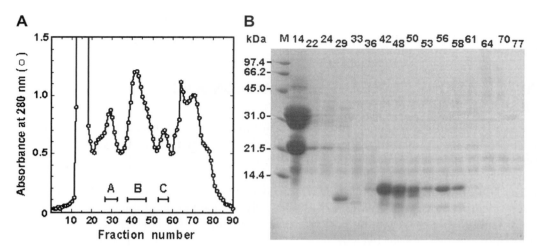

FIGURE 63.3

Fractionation and purification of proteinase-inhibitory activities from ginkgo seeds. (A) Second gel-filtration chromatography of the crude extracts (4.5 g) on a Sephadex G-50 column. The fraction size and flow rate were 5 ml/min and 0.33 ml/min, respectively. The fractions with a bar were pooled as samples A, B and C. (B) Glycine SDS-PAGE of the fractionation of the second gel-filtration chromatography. Molecular mass markers are shown in lane M.

outside the cells (Sawano *et al.*, 2008). Semi-quantitative RT-PCR revealed that this gene is expressed only in seeds, and not in stems, leaves and roots, suggesting that the protein is involved in seed development and/or germination (Sawano *et al.*, 2008). The inhibitor shows about 40% sequence homology with nsLTP1 from other plant species (Figure 63.4). This inhibitor (Gb-nsLTP1) does not show any antifungal or antibacterial activities, but is capable of exerting both lipid transfer activity and lipid binding activity (Sawano *et al.*, 2008). In addition, the site-directed mutagenesis study using the recombinant form revealed that the Pro79 and Phe80 residues are important for phospholipid transfer activity, and that the Pro79 and Ile82 residues are essential for binding activity toward *cis*-unsaturated fatty acids. On the other hand, the α-helical content of the P79A and F80A mutants was significantly lower than that of the wild-type protein (Sawano *et al.*, 2008). It is worth noting that the papain-inhibitory activity of the P79A and F80A mutants was twice that of the wild-type protein (Sawano *et al.*, 2008). In summary, we conclude that the Pro79 residue of Gb-nsLTP1 plays a critical role in both lipid transfer and binding activities.

```
gymnosperm            1        10        20        30        40        50        60        70        80        90
Ginkgo (100%)         APGCDTVDTDLAPCISYLQTGTGNPTVQCCSGVKTLAGTAQTTEDRKAICECIKTAAIRVKPV-ANAVKSLPGLCSVTLPFPISIA-TDQNKIV
Loblolly pine (36%)   AISCNQVVSAMTPCATYLIGNAATPAATCCPSIRGLDSQVKATPDRQAVCNCLKTQAKSYGVKLGKAAN-LPGLCKVTDLNVPISPNVDCSKVH
Monterey pine (40%)   ALDCNTIIQQITSCATVLTTGTPVPQEESSCCQGVQSLYGDATTTEEIQQICTCLKNEAINYNLN-DRALQSLPSNCGLQLSFTITRD-IDCSSIS
Red pine (36%)        GAISCNQVVSAMTPCATYLLGNAATPAAACCPSIRGLDSQVKATPDRQAVCNCFKTQARSYGVKLGKAAN-LPGLCKVTDLNVPISPNVDCSKVH

dicotyledonous(angiosperm)
Peach (40%)           ITCGQVSSALAPCIPYVRGGGAVP-PACCNGIRNVNNLARTTPDRQAACNCLKQLSASVPGVNPNNAAALPGKCGVHIPYKISAS-TNCATVK
Tomato (46%)          LTCGQVTAGLAPCLPYLQGRG--PLGGCCGGVKNLLGSAKTTADRKTACTCLKSAANAIKGIDLNKAAGIPSVCKVNIPYKISPS-TDCSTVQ
Tobacco (45%)         APPSCPTVTTQLAPCLSYIQG-GGDPSVPCCTGINNIYELAKTKEDRVAICNCLKTAFTHAGNVNPTLVAQLPKKCGISFNMPPIDKNYDCNTISMY

monocotyledonous(angiosperm)
Maize (42%)           ISCGQVASAIAPCISYARGQGSGPSAGCCSGVRSLNNAARTTADRRAACNCLKNAAAGVSGLNAGNAASIPSKCGVSIPYTISTS-TDCSRVN
Rice (39%)            ITCGQVNSAVGPCLTYARG-GAGPSAACCSGVRSLKAAASTTADRRTACNCLKNAARGIKGLNAGNAASIPSKCGVSVPYTISAS-IDCSRVS
Barley (38%)          LNCGQVDSKMKPCLTYVQG-GPGPSGECCNGVRDLHNQAQSSGDRQTVCNCLKGIARGIHNLNLNNAASIPSKCNVNVPYTISPD-IDCSRIY
                        *  * *   *     *                    *       *  *  **            *           ** *  * ** * *
```

FIGURE 63.4

Sequence alignment of mature Gb-nsLTP1 from ginkgo seeds with gymnosperm and angiosperm mature LTP1s. The numbering system corresponds to that of Gb-nsLTP1 (GenBank accession number, DQ836633). Dots and asterisks below the sequences indicate the well-conserved residues and the residues involved in ligand binding, respectively. Conserved cysteines are boxed. Sequence similarities (within parentheses) were analyzed by BLAST. The following plant LTP1 sequences were used for comparisons: loblolly pine (*Pinus taeda*), Q41073; monterey pine (*Pinus radiata*), AAB80805; red pine (*Pinus resinosa*), AAK00625; peach (*Prunus persica*), P81402; tomato (*Lycopersicon esculentum*), CAA39512; tobacco (*Nicotiana tabacum*), AAA21438; maize (*Zea mays*), ABA33852; rice (*Oryza sativa*), ABA96284; barley (*Hordeum vulgare*), AAA32970.

In addition to inhibitory activity against pepsin, sample C inhibits the growth of plant pathogenic fungi (*Fusarium oxysporum*, *Trichoderma reesei*, and *Rhizoctonia solani*) and human pathogenic fungi (*Aspergillus fumigatus*, *Candida albicans*, and *Mucor spinescens*), but the sample does not exhibit antibacterial action against *Bacillus subtillis* and *Escherichia coli* (Miyakawa, 2007). The cDNA contains a 402-bp open reading frame encoding a 134-amino acid protein with a potential signal peptide (26 residues), suggesting that this protein is synthesized as a preprotein and is secreted outside the cells (Sawano *et al.*, 2005). This antifungal protein (Gnk2) shows approximately 85% identity with embryo-abundant proteins from *Picea abies* and *Picea glauca* at the amino acid level (Figure 63.5A); however, there is no homology between Gnk2 and other plant antifungal proteins, such as the defensin, cyclophilin, miraculin and thaumatin proteins. Meanwhile, as shown in Figure 63.5A, Gnk2 shows a high sequence homology with DUF26 in the extracellular domain of the CRK protein. The CRK family from *Arabidopsis thaliana* is involved in the hypersensitive reaction by infection of phytopathogens, and most members of this family include two copies of DUF26 in the extracellular domain (Figure 63.5B). Recently, we have determined the crystal structure of Gnk2, including the full-length sequence of its DUF26 (Miyakawa *et al.*, 2009). The tertiary structure contains two α-helices and a five-stranded β-sheet in which the order of the secondary structure elements is βαββαββ, and Gnk2 shows a unique α + β fold unlike other antifungal proteins (Figure 63.6A). For instance, the antifungal protein most homologous to Gnk2, defensin, is composed of an α-helix and a two-stranded β-sheet, and the disulfide-bond arrangement is quite different from that of Gnk2. The positively-charged surface of defensin contributes to the interaction with negatively-charged phospholipids, while the first α-helix of Gnk2 also includes four arginine residues: Arg23, Arg26, Arg33 and Arg47 (Figure 63.5A). As shown in Figure 63.6B, these guanidino groups do not have any ionic interaction or salt bridge with the other groups but form a positively-charged surface around the helix structure. We speculate that the antifungal activity of Gnk2 may require an association between the positively-charged surface on the Gnk2 molecule and the negatively-charged phospholipids and/or phosphomannan on the fungal-cell surface. These findings would provide new insight into the molecular function of the CRK family.

Seed germination of plants begins in an environment rich in pathogens; therefore, many plant seeds contain antifungal and antibacterial proteins. We have here demonstrated that Gnk2 is able to inhibit the growth of several human pathogenic fungi, indicating that ginkgo nuts

531

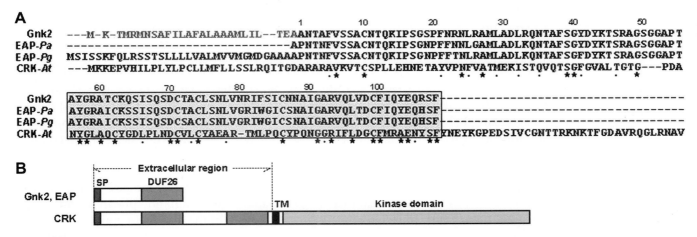

FIGURE 63.5

(A) Sequence alignment of Gnk2 and the homologous proteins with DUF26. The amino acid sequences were retrieved from the GenBank database: *Picea abies* embryo-abundant protein (EAP) (DQ120094), *Picea glauca* EAP (L47671) and *Arabidopsis thaliana* CRK (AC010796). The numbering system corresponds to that of Gnk2. Conserved and semi-conserved residues are indicated by asterisks and dots below the sequences, respectively. The signal peptide sequence of Gnk2 is written in gray, and the boxed sequence corresponds to DUF26. **(B) Schematic diagram of the domain structure of Gnk2, EAP and CRK proteins.** SP and TM represent the signal peptide and transmembrane domains, respectively.

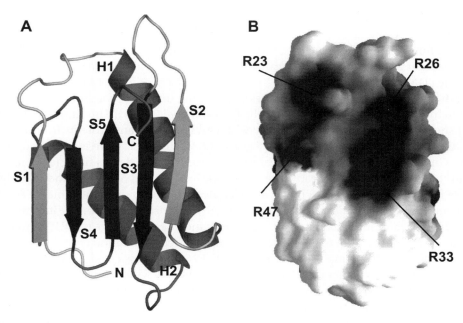

FIGURE 63.6

(A) Ribbon diagram of the Gnk2 structure with two α-helices (H1 and H2), and a five-stranded antiparallel β-sheet (S1—S5). The N- and C-termini are indicated by N and C, respectively. (B) Electrostatic surface potential map of Gnk2. The charged areas are colored from white (0 kT/e) to black (+10 kT/e), and arginine residues are labeled with their residue numbers.

could be a therapeutic agent for patients with bacterial pneumonia. On the other hand, Gb-nsLTP1 and Gnk2 exhibit a weak but clear inhibition of the aspartic proteinase pepsin. In general, it has been established that many pathogenic fungi and bacteria produce extracellular aspartic-acid proteinases and/or other proteinases; accordingly, these inhibitors would suppress the activity of the proteinases secreted by the fungi and bacteria and would reduce the resulting inflammation. At present, the characterization of other proteins (e.g., sample A) and site-specific mutagenesis experiments of Gnk2 are in progress, and we expect these studies to contribute to our understanding of the abovementioned therapeutic effect of ginkgo nuts.

ADVERSE EFFECTS AND REACTIONS (ALLERGIES AND TOXICITY)

Ginkgo nuts have been eaten as an ordinary food in China, Korea and Japan. However, it must be noted that ginkgo nuts have toxic effects (Wada *et al.*, 1985). It has been reported that ginkgotoxin (4-O-methoxypyridoxine) in ginkgo nuts causes symptoms such as seizures, vomiting and loss of consciousness, which are referred to as "the cytotoxicity of ginkgo nuts" (Wada *et al.*, 1985). Ginkgotoxin is structurally related to vitamin B_6, and physiologically acts as a B_6 "antivitamin" (Figure 63.2). The B_6 vitamers, pyridoxal phosphate and pyridoxamine phosphate, are essential for many enzymatic reactions, including amino acid metabolism and the synthesis of neurotransmitters (e.g., dopamine, serotonin, norephedrine and γ-aminobutyric acid). In fact, one plausible mechanism of the toxicity of ginkgotoxin is that it inhibits the pyridoxal kinase which plays a crucial role in the regulation of pyridoxal phosphate homoeostasis (Kästner *et al.*, 2007). Ginkgotoxin serves as an alternate substrate for this enzyme, with a lower K_m value than pyridoxal and pyridoxamine (Kästner *et al.*, 2007). The external seed coat of *Ginkgo biloba*, as well as its leaves, contains ginkgolic acids, which are a group of alkylphenols, and are suspected to have cytotoxic, allergenic, mutagenic and carcinogenic properties (Hecker *et al.*, 2002; Pan *et al.*, 2006). Ginkgolic acids also induce neuronal death through apoptosis (Ahlemeyer *et al.*, 2001). This neurotoxic effect is caused by activating PP2C, which acts as a regulator of apoptosis-related proteins such as ASK1, BAD and p53 (Ahlemeyer *et al.*, 2001).

SUMMARY POINTS

- *Ginkgo biloba* is the oldest living plant, and is highly resistant to insects, bacterial and viral infections, and air pollution.
- Ginkgo leaves and nuts have been traditionally used as food and phytomedicine in East Asia. Over the past few decades, the leaf extract has seen wide use as a supplement in Europe and the United States.
- Ginkgo leaves contain a variety of terpene trilactones, ginkgolides and bilobalide, which are effective in neuronal diseases such as vascular dementia and early-stage Alzheimer's disease.
- Ginkgo seeds contain several storage proteins, including proteins with antibiotic activities.
- Gnk2 inhibits the growth of human and plant pathogenic fungi. In a plausible antifungal mechanism, Gnk2 may require an association between the positively charged surface and the negatively charged phospholipids and/or phosphomannan on the fungal-cell surface.
- Gb-nsLTP1 exhibits binding activity toward *cis*-unsaturated fatty acids, and inhibits the aspartic protease pepsin. This inhibitor would suppress the proteinase activity secreted by pathogens and reduce the resulting inflammation.
- Ginkgo seeds and their external seed coat contain some toxic compounds, including ginkgotoxin and ginkgolic acids. The toxic mechanisms of these compounds have been elucidated in detail at the molecular level.

References

Ahlemeyer, B., Selke, D., Schaper, C., Klumpp, S., & Krieglstein, J. (2001). Ginkgolic acids induce neuronal death and activate protein phosphatase type-2C. *European Journal of Pharmacology, 430*, 1–7.

Blumenthal, M. (2000). Market report. *Herbalgram, 51*, 69.

Chan, P. C., Xia, Q., & Fu, P. P. (2007). *Ginkgo biloba* leaf extract: biological, medicinal, and toxicological effects. *Journal of Environmental Science and Health*, Part C, *25*, 211–244.

Guinot, P., & Braquet, P. (1994). Effects of the PAF antagonists, ginkgolides (bn-52063, bn-52021), in various clinical indications. *Journal of Lipid Mediators and Cell, 10*, 141–146.

Hecker, H., Johannisson, R., Koch, E., & Siegers, C. P. (2002). *In vitro* evaluation of the cytotoxic, potential of alkylphenols from *Ginkgo biloba* L. *Toxicology, 177*, 167–177.

Hori, S., & Hori, T. (1997). A cultural history of *Ginkgo biloba* in Japan and the generic name Ginkgo. In T. Hori, R. W. Ridge, W. Tulecke, P. Del Tredici, J. Trémouillaux-Giller, & H. Tobe (Eds.), Ginkgo Biloba — *A global treasure* (pp. 385–411). Tokyo, Japan: Springer-Verlag.

Kästner, U., Hallmen, C., Wiese, M., Leistner, E., & Drewke, C. (2007). The human pyridoxal kinase, a plausible target for ginkgotoxin from *Ginkgo biloba*. *FEBS Journal, 274*, 1036–1045.

Mahadevan, S., & Park, Y. (2008). Multifaceted therapeutic benefits of *Ginkgo biloba* L.: chemistry, efficacy, safety, and uses. *Journal of Food Science, 73*, R14–19.

Miyakawa, T. (2007). *Structure–function analysis of antifungal protein from ginkgo seeds. PhD thesis*. University of Tokyo. pp. 20–42.

Miyakawa, T., Miyazono, K., Sawano, Y., Hatano, K., & Tanokura, M. (2009). Crystal structure of ginkbilobin-2 with homology to the extracellular domain of plant cysteine-rich receptor-like kinases. *Proteins, 77*, 247–251.

Pan, W., Luo, P., Fu, R., Gao, P., Long, Z., Xu, F., et al. (2006). Acaricidal activity against *Panonychus citri* of a ginkgolic acid from the external seed coat of Ginkgo biloba. *Pest Management Science, 62*, 283–287.

Sawano, Y., Miyakawa, T., Yamazaki, H., Tanokura, M., & Hatano, K. (2005). Purification, characterization, and molecular cloning of an antifungal protein gene from *Ginkgo biloba* seeds. *Biological Chemistry, 388*, 273–280.

Sawano, Y., Hatano, K., Miyakawa, T., Komagata, H., Miyauchi, Y., Yamazaki, H., et al. (2008). Proteinase inhibitor from ginkgo seeds is a member of plant nonspecific lipid transfer protein gene family. *Plant Physiology, 146*, 1909–1919.

Singh, B., Kaur, P., Gopichand, Singh, R. D., & Ahuja, P. S. (2008). Biology and chemistry of *Ginkgo biloba*. *Fitoterapia, 79*, 401–418.

Thompson, C. B. (1995). Apoptosis in the pathogenesis and treatment of disease. *Science, 267*, 1456–1462.

Vaughn, K. C., & Renzaglia, K. S. (2006). Structural and immunocytochemical characterization of the *Ginkgo biloba* L. sperm motility apparatus. *Protoplasma, 227*, 165–173.

Wada, K., Ishigaki, S., Ueda, K., Sakata, M., & Haga, M. (1985). An antivitamin B6, 4′-O-methoxypyridoxine, from the seed of *Ginkgo biloba* L. *Chemical & Pharmaceutical Bulletin, 33*, 3555–3557.

Wang, H., & Ng, T. B. (2000). Ginkbilobin, a novel antifungal protein from *Ginkgo biloba* seeds with sequence similarity to embryo-abundant protein. *Biochemical and Biophysical Research Communications, 279*, 407–411.

Ye, X. Y., Wang, H. X., & Ng, T. B. (2000a). Dolichin, a new chitinase-like antifungal protein isolated from field beans (*Dolichos lablab*). *Biochemical and Biophysical Research Communications, 269*, 155–159.

Ye, X. Y., Wang, H. X., & Ng, T. B. (2000b). Structurally dissimilar proteins with antiviral and antifungal potency from cowpea (*Vigna unguiculata*) seeds. *Life Sciences, 67*, 3199–3207.

CHAPTER 64

Therapeutic Effects of Grains of Paradise (*Aframomum melegueta*) Seeds

Solomon Umukoro, Aderemi Caleb Aladeokin
Department of Pharmacology and Therapeutics, University of Ibadan, Ibadan, Nigeria

CHAPTER OUTLINE

LIST OF ABBREVIATIONS

AM, *Aframomum melegueta*
ASA, acetylsalicylic acid
CNS, central nervous system
COX-2, cyclo-oxgenase-2
FDA, Food and Drug Administration
GP, grain of paradise
MDA, malonyldialdehyde
NSAIDs, non-steriodal anti-inflammatory drugs
PHZ, phenylhydrazine
PW, paw withdrawal
RA, rheumatoid arthritis
WBC, white blood cell

INTRODUCTION

Aframomum melegueta roscoe K. Schum belongs to the ginger family, Zingiberaceae, commonly found in the coastal regions of Africa. AM is cultivated essentially for its valuable

Nuts & Seeds in Health and Disease Prevention. DOI: 10.1016/B978-0-12-375688-6.10064-7

seed, which earned it the name "grain of paradise," indicating its high value as a spice and medicine (Iwu *et al.*, 1999). The seed is extensively used in ethno-medicine for a variety of ailments. It possesses active constituents with several biological activities, especially against inflammation and infectious diseases (Iwu *et al.*, 1999). Studies have shown that the seeds of AM might turn out to be a source of medicines for health promotion and disease control (Dybas & Raskin, 2007; Umukoro & Ashorobi, 2008). In this chapter, the cultivation, applications to health and disease prevention, and adverse effects of *Aframomum melegueta* seeds are described.

BOTANICAL DESCRIPTION

Aframomum melegueta is a perennial herb with a palm-like appearance, growing to a height of 1.0–1.5 m (Figure 64.1). It has smooth, narrow leaves, about 23 cm long, and a scaly rhizome of a surface root system. The large purple flowers are trumpet-shaped, with one stamen. The mature fruits are red in color, ovoid, about 5–7 cm long, and contain numerous small, hard, shiny seeds of about 3 mm in diameter (Figures 64.2–64.4). The seeds possess a strong aromatic and pungent odor, and a peppery, hot, and slightly bitter taste (Iwu *et al.*, 1999). The seed resembles cardamom in appearance and pungency; however, AM seeds are reddish-brown in color, whereas cardamom is pale buff-colored and tends to be less pungent once cooked or heated. Grain of paradise (GP) seeds are also known as melegueta pepper, alligator pepper, and Guinea grains or Guinea pepper. Botanically, AM is classified as a member of the Zingiberaceae family.

536

FIGURE 64.1
Aframomum melegueta plant.

FIGURE 64.2
Fresh *Aframomum melegueta* pods.

FIGURE 64.3
The dried fruit of *Aframomum melegueta*.

HISTORICAL CULTIVATION AND USAGE

Historically, AM was grown in temporary clearings in the forest along with other crops and the fruits are usually harvested from the wild when red in color (Lock *et al.*, 1977). AM seeds have long been used in the cuisines of West and North Africa, where they were traditionally imported via caravan routes in a series of trans-shipments through the Sahara Desert to Europe. They were a very fashionable substitute for black pepper in 14th and 15th century Europe, especially in northern France, which was one of the most populous regions in Europe at that time. The seeds were used for flavoring sausages, and to give a fictitious strength to malt liquors, gin, and cordials. In the 18th century the importation of AM seeds to Great Britain collapsed after a Parliamentary Act forbade its use in malt liquor, aqua vitae, and cordials (Lock *et al.*, 1977). AM seeds have also been used by people on certain diets, such as a raw-food diet, because they are less irritating to digest than black pepper. In the southern regions of Nigeria

FIGURE 64.4
The seeds of *Aframomum melegueta*.

AM seeds were used for divination and in trials to determine guilt (Simmons, 1956). The seeds are served with kola nuts to entertain guests among the Igbo people of Nigeria.

PRESENT-DAY CULTIVATION AND USAGE

GP is usually propagated by seed, but planting of the rhizome parts appears to be easier and quicker. The seeds are sown in the rainy season, in the shade of other crops, and transplanted during the next rainy season. The herb starts to produce fruits 3 years after sowing, and may continue to do so for the next 4 years (Lock *et al.*, 1977). The fruits are harvested when they are red, and are carefully dried before removing the seeds for storage. The seeds are employed as a spice, and are also chewed in cold weather to warm the body. GP is currently use in ethno-medicine for the treatment of hemorrhoids, measles, leprosy, tumors, sleeping sickness, infections, abdominal pain, and inflammatory disorders (Rafatullah *et al.*, 1995; Iwu *et al.*, 1999). It is also used as a tonic for sexual stimulation, taken for excessive lactation, post-partum hemorrhage, and as a purgative, a galactogogue, and a hemostatic agent (Iwu *et al.*, 1999). It serves as medicine for intestinal discomfort, pain during pregnancy, and nausea and vomiting.

APPLICATIONS TO HEALTH PROMOTION AND DISEASE PREVENTION

Studies have revealed the potential of GP as a valuable medicinal agent for promoting good health, and prevention of a variety of ailments. In ethno-medicine, the macerated seeds are often applied to swollen parts of the body to relieve inflammation and pain (Akendengue & Louis, 1994). Experimental studies showed that the crude extract of AM seeds exhibited anti-inflammatory and analgesic activity in rats (Umukoro & Ashorobi, 2007, 2008).The extract inhibited both acute and chronic inflammatory responses in rats, which provides the basis for its use in traditional medicine for acute and chronic inflammatory disorders (Umukoro & Ashorobi, 2008). The anti-inflammatory effect of the seed might have resulted from inhibition of prostaglandin synthesis, as Bingol and Sener (1995) previously reported that extracts from the plants of the Zingiberaceae family, to which AM belongs, contain active constituents that are potent inhibitors of prostaglandin formation. Furthermore, the usefulness of GP in

TABLE 64.1 Effect of Aqueous Seed Extract of *Aframomum melegueta* on the Volume of Exudative Fluid and Number of Leukocytes Induced by Carrageenan in Rats

Treatment	Dose	Volume of Exudates (ml)	Decrease in WBC Counts (%)
Saline	10 ml/kg	5.54 ± 0.26[*]	—
A. melegueta	50 mg/kg	2.87 ± 0.20[*]	40.21 ± 5.2
A. melegueta	100 mg/kg	2.80 ± 0.19[*]	53.65 ± 3.40
A. melegueta	200 mg/kg	2.13 ± 0.25[*]	61.57 ± 6.20
Indomethacin	10 mg/kg	1.81 ± 0.25[*]	71.34 ± 2.60

Each value represents the mean ± SEM of 10 animals.
[*]$P < 0.05$ compared with saline-control group (Student's t-test) (Umukoro & Ashorobi, 2008).

chronic inflammatory disorders like rheumatoid arthritis (RA) may be due to its ability to inhibit the migration of white blood cells (WBC) to the site of injury. This has been corroborated by the findings of Umukoro and Ashorobi (2008), who reported that extract of the seeds of AM reduced the number of inflammatory cells (WBCs) in the granuloma air-pouch model of chronic inflammation in rats (Table 64.1). The granuloma air-pouch model provides a more suitable cavity for eliciting chronic inflammatory reactions, as found in the synovial cavity during RA (Zurier, 1988). It is worthy of note that the pathogenesis of RA has been linked to WBC-mediated release of cytotoxic products that have been shown to be responsible for the progression of the disease (Hascelic *et al.*, 1994; Patisson *et al.*, 2004).

In another study, an active compound obtained from the AM seed was described as the most potent likely anti-inflammatory agent ever discovered (Dybas & Raskin, 2007). The compound was shown to more selectively inhibit the COX-2 enzyme relevant in inflammation, indicating lack of the gastric ulceration that is often associated with the use of most NSAIDs. This compound has been licensed to biotechnology companies, and might turn out to be a source of new drugs for the treatment of arthritis, heart disease, and other conditions that have inflammation as their root cause (Dybas & Raskin, 2007). A clinical trial conducted in humans revealed that AM inhibited a component of the immune system known as cytokine modulators, such as C-reactive protein, which are involved in the regulation of inflammatory responses (Dybas & Raskin, 2007). AM at a very low concentration was found to significantly inhibit the production of C-reactive protein (Dybas & Raskin, 2007). These findings further support the notion that AM might be useful in the treatment of diseases, such as cardiovascular conditions, arthritis, migraines, and osteoporosis, which have inflammation as their hallmarks.

The analgesic effect observed in animal studies also appears to be related to the inhibition of the formation of prostaglandin, a chemical mediator known to cause pain through sensitization of nociceptors (Umukoro & Ashorobi, 2007). These studies provide evidence that AM seed might possess a phytoactive substance(s) with potent activity against inflammatory pains, justifying its use in relieving pain in ethno-medicine. The seeds are usually chewed to relieve body pains, especially stomach ache and abdominal discomfort, in traditional medical practice. Rafatullah and colleagues (1995) screened the extract of the seed of AM for anti-ulcer and cytoprotective properties. The study showed that the extract exerts anti-ulcerative effects and cytoprotective activity in rats. The extract also shows cytoprotection against the cellular-damaging effect of phenylhydrazine (PHZ), a potent oxidant compound. PHZ is known to trigger inflammatory reactions through direct membrane damage accompanied by the depletion of intracellular glutathione and release of free radicals, which further enhance tissue injury (Ferrali *et al.*, 1992). Elevated levels of MDA, which indicate increased free radical-mediated lipid peroxidative cellular damage induced by PHZ, were significantly reduced by AM in rats, suggesting antioxidant activity (Umukoro & Ashorobi, 2008; Table 64.2). Elevated

TABLE 64.2 Effect of *Aframomum melegueta* Seed Extract on Lipid Peroxidation Induced by Phenylhydrazine (1 mM) in Rat Erythrocytes		
Sample	Concentration	MDA Levels (μmol/ml RBC)
PHZ-control	1.0 mM	4.54 ± 1.85
A.melegueta	0.5 mg/ml	1.52 ± 0.06[*]
A.melegueta	1.0 mg/ml	0.69 ± 0.03[*]
A.melegueta	2.0 mg/ml	0.48 ± 0.02[*]
α-tocopherol	100 μg/ml	0.75 ± 0.02[*]

Each value represents the mean ± SEM of six experiments.
[*]$P < 0.05$ compared with PHZ (ANOVA) (Umukoro & Ashorobi, 2008).

levels of MDA have been found in a variety of human diseases, such as neurodegeneration, ischemia, and arthritis, that have inflammation as an underlying factor (Halliwell, 1995). Furthermore, increased levels of free radicals have been found in the synovial fluid of inflamed rheumatoid joints of patients, and the progression of bone destruction seen in these patients has been attributed to increased free radical activity (Patisson *et al.*, 2004). The finding that AM exerts cytoprotective and antioxidant activities might play a significant role in its anti-ulcer property, and in mitigating chronic inflammatory diseases.

Aframomum melegueta roscoe has also been reported to play a considerable role in the traditional healthcare delivery system of local inhabitants, where it is being used to treat infections due to microorganisms, as well as helminthic infections such as schistosomiasis (Iwu *et al.*, 1999). Experimental studies have confirmed the efficacy of AM seeds as an anti-infective agent against bacteria and fungi infections (Galal, 1995; Iwu *et al.*, 1999). Galal (1995) further reported that paradol and shogoal isolated from AM seeds exhibited potent antimycobacterial activity. It is pertinent to note that AM may serve as a source of new medicines for the treatment of tropical diseases.

In recent times, sexual disorders amongst males have increased, and plant-derived medicines have continued to provide remedies for males who desire to improve their sexual life (Neychev & Mitev, 2005). The AM seed is one such ingredient that is used for enhancing sexual performance. The sexual performance-enhancing effect of AM seeds has been confirmed by experimental findings, as it was shown to produce an increase in sexual arousal in rats. The extract was found to enhance penile erection, and to increase the frequency of intromission in rats. It also enhanced the orientation of male rats towards female counterparts — a behavior that was interpreted as an increase in sexual arousal (Kamtchouing *et al.*, 2002). Clinical trials revealed that the seeds of AM contain active ingredients that produce an improvement in sexual activity in both males and females (Soraya *et al.*, 1999). The study also revealed that it produced an improvement in penile rigidity, and may provide beneficial effects in the treatment of male erectile dysfunction and premature ejaculation. AM was shown to prevent premature ejaculation, and allowed men to maintain both libido and penile rigidity at such a level that no erection failure occurred during sexual performance (Soraya *et al.*, 1999). It did not cause priapism or any other sexual side effects during or after its use (Soraya *et al.*, 1999).

AM seed has been shown to contain several phytochemically active substances that might be responsible for its health-promoting effects, or its use in the treatment of a variety of ailments in ethno-medicine. Studies revealed that AM seed is rich in phytochemicals such as flavoniods, alkaloids, saponins, tannins, and phenolic compounds (Okwu, 2005). Flavoniods are known to exhibit protection against inflammation, allergies, microbes, tumors, ulcers, and viruses (Uruquiaga & Leighton, 2000). Flavoniods are the most widely distributed phytochemicals,

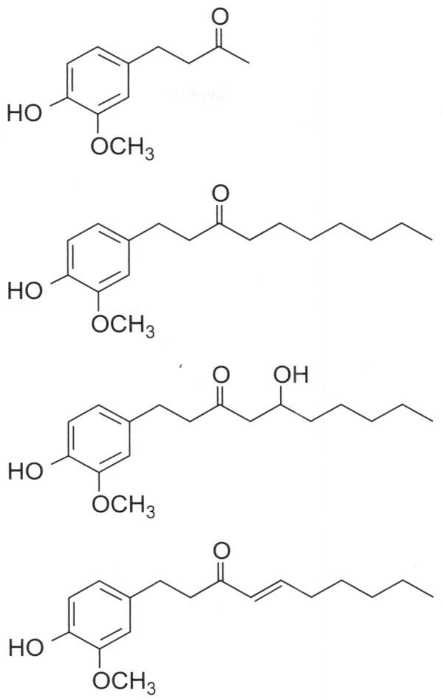

FIGURE 64.5
Structure of the major constituents of *A. melegueta* seed: zingerone (**1**), 6-paradol (**2**), 6-gingerol (**3**), and 6-shogaol (**4**).

and are potent free-radical scavengers, which prevent oxidative cell damage, with a strong antitumor property capable of inhibiting all stages of carcinogenesis (Uruquiaga & Leighton, 2000). In recent times, flavonoids have attracted much attention as dietary cancer-chemo-preventive agents. It is likely, as observed in wild gorillas, that regular consumption of the seeds of AM may lower the risk of cancers and heart disease in humans. Wild gorillas have been found to have a lower risk of developing heart disease compared to their counterparts in zoos (Dybas & Raskin, 2007). The presence of flavonoids in AM seed further confirmed its usefulness in the treatment of RA in ethno-medicine.

Phenols and phenolic compounds are widely used in disinfection, and remain the standard against which other antiseptics are compared. They are known to exhibit a potent antimicrobial property. The anti-infective activity shown by AM seed may be ascribed to the presence of phenols, and also provide a basis for its traditional uses as a remedy for wound healing and prevention of infections. Tannins are one of the major active metabolites found in AM, and these phytochemicals are known to have astringent properties, promoting the healing of wounds and inflamed tissues (Okwu, 2005). The efficacy of AM seed in the treatment of wounds, ulcers, and burns may be attributed to the presence of tannins. The discovery of alkaloids in GP further confirms its usefulness in the health delivery system, as these phytochemicals have provided clinically useful medicines such as analgesics, and are known to exhibit a wide range of CNS activities. Moreover, chemical analysis of AM seed revealed the presence of hydroxyalkanones such as zingerone, paradol, gingerol, shogoal (Figure 64.5), and essential oils, which occur only in traces (Galal, 1996). The antimicrobial activity of paradol and shogoal has been reported in the literature (Galal, 1996). Zingerone and gingerol are well known for their anti-inflammatory properties (Kiuchi *et al.*, 1992); indeed, gingerols are known to be chemically similar to other anti-inflammatory compounds, and have been found to inhibit prostaglandin and leukotriene biosynthesis (Kiuchi *et al.*, 1992).

ADVERSE EFFECTS AND REACTIONS (ALLERGIES AND TOXICITY)

Aframomum melegueta roscoe is included in the Food and Drug Administration's (FDA) list of botanicals that are generally recognized as safe. No side effects have been reported in the many years that people have been eating the seeds in Africa and Europe (Dybas & Raskin, 2007). Also, no interactions have been observed with standard prescription of AM seeds. There is as yet no recorded history of allergic reactions associated with the use or consumption of the seeds; rather, they are useful against allergic manifestations. Pharmaceutical preparations containing AM seeds were shown to exhibit anti-allergic effects, and were most potent in counteracting skin irritation by down-regulating the production of IgE and inhibiting the release of allergic mediators. There is strong evidence that AM seeds can serve as a skin-care product for the treatment of skin disorders associated with allergic reactions (Dybas & Raskin, 2007). However, people who are allergic to cardamom or ginger preparations are advised to use GP seeds with caution. The acute toxicity test showed that oral doses of the alcoholic extract of the seed also have a high safety profile, as it is well tolerated by laboratory animals and does not cause any toxic symptoms in mice (Rafatullah *et al.*, 1995).

SUMMARY POINTS

- There is increased awareness of the usefulness of GP seed as a spice for foods, and as a potential source of new medicines for a variety of diseases, ranging from inflammation and infections to CNS disorders.
- The discovery of an active compound that blocks COX-2 enzyme more selectively suggests that it might be more tolerable as an anti-inflammatory and analgesic than most conventional NSAIDs.
- As a potent anti-infective agent, GP may offer new medicines for the treatment of tropical diseases, which are one of the most common global health burdens.
- GP may serve as a remedy for the treatment of male erectile dysfunction and premature ejaculation.
- AM seed is generally recognized as safe, as no side effects have been reported from its consumption or usage over the years.

References

Akendengue, B., & Loius, A. (1994). Medicinal plants used by Masongo people of Gabon. *Journal of Ethnopharmacology, 41*, 193–200.

Bingol, F., & Sener, D. (1995). Review of terrestrial plants and marine organisms having anti-inflammatory activity. *International Journal of Pharmacognosy, 33*, 81–97.

Dybas, C. L., & Raskin, I. (2007). Out of Africa: a tale of gorillas, heart disease… and a swamp plant. *BioScience, 57*, 392–397.

Ferrali, M., Signorni, C., Ciccoli, L., & Comporti, M. (1992). Iron release and membrane damage in erythrocytes exposed to oxidizing agents, phenylhydrazine, divicine and isouramil. *Biochemical Journal, 285*, 295–301.

Galal, A. M. (1996). Anti-microbial activity of 6-paradol and related compounds. *International Journal of Pharmacognosy, 31*, 64–69.

Halliwell, B. (1995). Antioxidant characterization: methodology and mechanism. *Biochemical Pharmacology, 49*, 1341–1348.

Hascelic, G., Sener, B., & Hascelic, Z. (1994). Effect of some anti-inflammatory drugs on human neutrophil chemotaxisis. *Journal of International Medical Research, 22*, 100–106.

Iwu, W., Duncan, A. R., & Okuw, C. O. (1999). New antimicrobials of plant origin. In J. Janick (Ed.), *Perspectives on new crops and new uses* (pp. 457–462). Alexandria, VA: ASHS Press.

Kamtchouing, P., Mbonge, G. Y. F., Dimo, T., Watcho, P., Jatsa, H. B., & Sokeng, S. D. (2002). Effects of *Aframomum melegueta* and *Piper guineense* on sexual behaviour of male rats. *Behavioural Pharmacology, 13*, 243–247.

Kiuchi, F., Iwakami, S., Shibuya, M., Hanaoka, A., & Sankawa, U. (1992). Inhibition of prostaglandin and leukotriene biosynthesis by gingerols and diarylheptaniods. *Chemical & Pharmaceutical Bulletin, 40*, 387–391.

Lock, J. M., Hall, J. B., & Abbiw, D. K. (1977). The cultivation of melegueta pepper. *Economic Botany, 31*, 321–330.

Neychev, V. K., & Mitev, V. I. (2005). The aphrodisiac herb *Tribulus terrestris* does not influence the androgen production in young men. *Journal of Ethnopharmacology, 101*, 319–323.

Okwu, D. E. (2005). Phytochemicals, vitamins and mineral contents of Nigerian medicinal plants. *International Journal of Molecular Medicine and Advanced Sciences, 1*, 375–381.

Pattison, D. J., Silman, A. J., Goodson, N. J., Lunt, M., Bunn, D., Weich, A., et al. (2004). Vitamin C and the risk of developing inflammatory polyarthritis: prospective case–control study. *Annals of the Rheumatic Diseases, 63,* 843–847.

Rafatullah, S., Galal, A. M., Al-Yahya, M. A., & Al-Said, M. S. (1995). Gastric and duodenal anti-ulcer and Cyto-protective effects of *Aframomum melegueta* in rats. *International Journal of Pharmacognosy, 33,* 311–316.

Simmons, D. C. (1956). Efik divination, ordeals and omens. *Southwestern Journal of Anthropology, 12,* 223–228.

Soraya, A., Victor, N., Hartman, N. G., Owassa, S., & Ibea, M. (1999). *Aframomum* seeds for improving penile activity. US Patent: US 5879682. Patent. Patent Storm. September 28, 2009; available at http://www.patentstorm.us/patents/5879682/description.html.

Umukoro, S., & Ashorobi, R. B. (2007). Further studies on the antinociceptive action of aqueous seed extract of Aframomum melegueta. *Journal of Ethnopharmacology, 109,* 501–504.

Umukoro, S., & Ashorobi, R. B. (2008). Further pharmacological studies on aqueous seed extract of *Aframomum melegueta* in rats. *Journal of Ethnopharmacology, 115,* 489–493.

Uruquiaga, I., & Leighton, F. (2000). Plant polyphenol antioxidants and oxidative stress. *Biological Research, 33,* 159–165.

Zurier, R. B. (1988). Prostaglandins and inflammation. In P. B. Curtis-Prior (Ed.), *Prostaglandins: Biology and chemistry of prostaglandins and related eicosanoids* (pp. 595–615). Oxford, UK: Churchill Livingstone.

CHAPTER 65

Antibacterial Activity of Grape (*Vitis vinifera, Vitis rotundifolia*) Seeds

Cecile Badet
Laboratoire Odontologique de Recherche, UFR d'Odontologie, Université Victor Segalen, Bordeaux Cedex, France

545

LIST OF ABBREVIATIONS

GSE, grape seed extract
MIC, minimum inhibitory concentration

INTRODUCTION

Since the advent of antibiotics, only a few of the many pharmaceuticals of plant origin have been used as antimicrobials. However, to overcome the side effects of these molecules, such as the emergence of resistant bacteria, various researches have focused on the antimicrobial activity of plant extracts, including grape seed extracts (GSEs).

The antimicrobial properties of red and white wines have been known since antiquity, and, more recently, GSEs have shown a broad range of pharmacological activities, including antioxidant properties and the ability to serve as free radical scavengers (Pastrana-Bonilla *et al.*, 2003).

It is now well known that antimicrobial phytochemicals are mainly polyphenols, and that grape seeds are considered to be a rich source of these compounds (Mayer *et al.*, 2008, Nassiri-Asl & Hosseinzadeh, 2009).

Nuts & Seeds in Health and Disease Prevention. DOI: 10.1016/B978-0-12-375688-6.10065-9

BOTANICAL DESCRIPTION

All *Vitis* are lianas, or woody, climbing vines. *Vinifera* grapes have loose, flaky bark on older wood, but smooth bark on 1-year-old wood. Muscadine vines (*Vitis rotundifolia* Michx.) have smooth bark on wood of all ages. The leaves vary in shape and size, depending on the species and cultivar. Muscadine grapes have small, round, unlobed leaves with dentate margins. *Vinifera* has large cordate-to-orbicular leaves, which may be lobed. The flowers are small and green; muscadine grapes have only 10–30 flowers per cluster.

Flowering in grape occurs at the basal nodes of the current season's growth in all species. Grapes are true berries, small, round to oblong, with up to four seeds. The skin is generally thin, and is the source of the anthocyanin compounds, giving rise to red, blue, purple, and black (dark purple)-colored grapes.

HISTORICAL CULTIVATION AND USAGE

Vitis vinifera is thought to be native to the area near the Caspian Sea in southwestern Asia. The Phoenicians carried wine cultivars to Greece, Rome and southern France, and the Romans spread the grape throughout Europe. Spanish missionaries brought *Vinifera* grapes to California in the 1700s and found that they grew very well there. The medicinal and nutritional value of grapes has been known for thousands of years. Egyptians consumed grapes at least 6000 years ago, and several ancient Greek philosophers praised their healing power. European folk used an ointment from the sap of grapevines to cure skin and eye diseases. Grape leaves were used to stop bleeding, inflammation and pain. Unripe grapes were used to treat sore throats and dried grapes were used to heal consumption, constipation, and thirst. The round, ripe, sweet grapes were used to treat a range of health problems including cancer, cholera, smallpox, nausea, eye infections, and skin, kidney and liver diseases.

PRESENT-DAY CULTIVATION AND USAGE

There are at least 5000 cultivars of *Vinifera* grapes grown worldwide, and some estimates put the number of known cultivars as high as 14,000.

Today, healthcare professionals use standardized extracts of grape seed to treat a range of health problems related to free radical damage, including blood sugar regulation problems, heart disease, and cancer. The antioxidants contained in GSE have also been reported to be beneficial in cancer prevention.

Flavonoids found in red wine have been reported to protect the heart. They may inhibit the oxidation of LDL cholesterol. Studies have demonstrated a relationship between flavonoid intake and reduced risk of death from coronary heart disease.

APPLICATIONS TO HEALTH PROMOTION AND DISEASE PREVENTION

Various polyphenolic compounds have shown an antibacterial effect.

Composition of GSEs

The seeds contain 60–70% of the extractable polyphenols in grapes. The phenolic content of seeds may range from 5% to 8% by weight. The most abundant phenolic compounds are flavonoids, essentially catechins (catechin, epicatechin, and procyanidins) and their polymers.

The content of phenolic compounds in grape seeds is affected by various factors. There is a clear difference between red and white grapes. The phenolic molecules found in white grapes are esters of hydroxycinnamic acid, catechins, catechin-gallate, procyanidins, catechin-catechin gallate, and β-1,3,6-tri-O-galloyl-D-glucose, while those occurring in red grapes are hydroxycinnamic acid, esters of tartric acids, procyanidins, flavonol glucosides, and anthocyanins. The polyphenolic content of grapes also depends on the variety, climatic conditions, site of production and degree of maturity.

The procyanidin composition of grape seeds has already been determined (Lee & Jaworski, 1987). Escribano *et al.* (1992) have reported 17 chemical constituents in *Vitis vinifera* grape seeds. Gabetta *et al.* (2000) reported the presence of monomers, dimers, trimers, tetramers, pentamers, hexamers, heptamers and their gallates in grape seeds.

Pastrana-Bonilla *et al.* (2003) studied the phenolic content of seeds from *Vitis rotundifolia* grapes. They found that the seeds contained five times more polyphenols than the skin, and 80-fold more than grape pulp. The main phenolics in muscadines are ellagic acid, kaempfrol, myricetin, and quercetin. Muscadine grapeseeds contain high concentrations of catechin and epicatechin, as confirmed by Brown and colleagues, who also found resveratrol in muscadine grape seeds (Brown *et al.*, 2009).

Antibacterial activity

The antimicrobial activity of GSE has been checked against various bacterial species.

Jayaprakhapha and colleagues (2003) tested the activity of two kinds of grape (*Vitis vinifera*) seed extract on six bacterial species: acetone : water : acetic acid extract, and methanol : water : acetic acid extract (Figure 65.1). Monomeric procyanidins were found to be the major compound present in both extracts.

These two extracts showed higher antibacterial activity against gram-positive bacteria than gram-negative bacteria. The MIC with methanol : water : acetic acid extract showed higher activity for *Bacillus sp.*, and the same for other species (*Escherichia coli, Pseudomonas aeruginosa*), than with acetone : water : acetic acid extract.

Baydar and colleagues used acetone : water : acetic acid extract, and ethyl acetate : methanol : water extract for antimicrobial tests on 15 bacteria. They showed that the extract at 20% was more effective against 14 bacteria. The acetone : water : acetic acid extract was more active than the ethyl acetate : methanol : water extract. The most inhibited bacteria were *Listeria monocytogenes*. At the concentration of 4%, the effects of extracts were different. The acetone : water : acetic acid extract was effective against 11 bacteria. The highest antibacterial effect was also obtained with *Listeria monocytogenes*. The ethyl acetate : methanol : water showed no antibacterial activity at this concentration (Baydar *et al.*, 2004).

Listeria monocytogenes belongs to a group of foodborne pathogens. It causes listeriosis, which has a 20–30% mortality rate. Some authors have investigated the antimicrobial activity of GSE against this species. A significant reduction in bacteria numbers (2 log) was obtained very quickly after exposure to ethanol extract of Ribier grape seed (Rhodes *et al.*, 2006).These results confirm those obtained by Ahn *et al.*, and were confirmed by Over *et al.*, who reported a strong antilisterial activity of grape seed extract (Ahn *et al.*, 2004; Over *et al.*, 2009).

Recently, Anatasiadi and colleagues (2009) compared the antilisterial properties of GSE vs individual polyphenols. They showed that the antibacterial activity of the isolated polyphenolic molecules was significantly lower compared to the extracts.

When subjected to GSE, another foodborne pathogen, *Salmonella typhimurium*, which is responsible for systemic diseases, was susceptible to these extracts (Ahn *et al.*, 2007; Over *et al.*, 2009).

Results obtained on strains of *Escherichia coli* were very similar to those described above. GSE exhibited strong antibacterial activity against this bacterium (Jayaprakasha *et al.*, 2003, Ahn

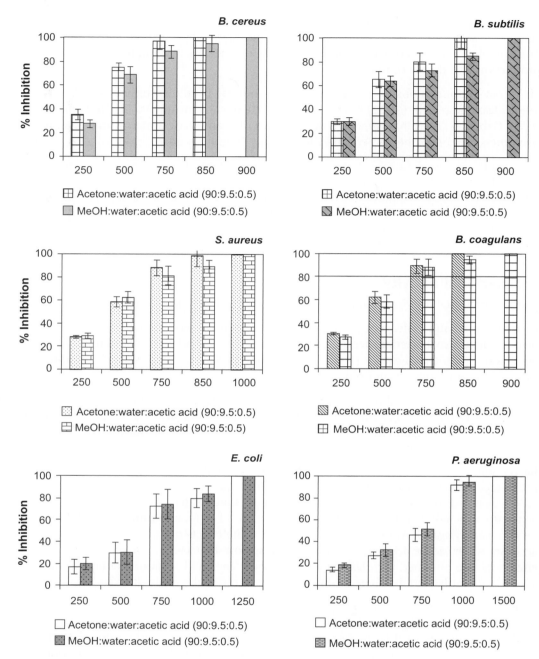

FIGURE 65.1

Effect of grape seed extracts on growth of different bacteria at different concentrations (ppm). The results show that the grape seed extracts exhibited antibacterial effect against all bacteria tested. Both extracts were found to be the most effective against gram-positive bacteria when compared to gram-negative bacteria. *Reprinted from Jayaprakasha et al. (2003), Food Res. Intl, 36, 117—122, with permission.*

et al., 2007, Over *et al.*, 2009, Yigit *et al.*, 2009). Baydar and colleagues (2006) showed that GSE at the concentrations of 0.5% and 1% had a bacteriostatic effect on *E. coli* O157-H7, while the extracts at the concentrations of 2.5% and 5% were bactericidal (Figure 65.2).

According to some authors, the compound involved in this antibacterial activity is gallic acid. Structure—activity correlation assays showed that three hydroxyl groups of the compound were effective for antibacterial action against *E. coli* (Tesaki *et al.*, 1999).

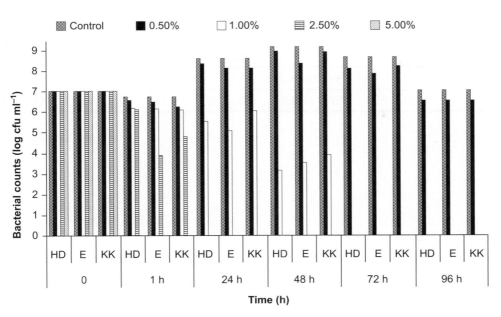

FIGURE 65.2
Inhibition curves of various concentrations of Hasandede (HD), Emir (E), and Kalecik Karasi (KK) seed extracts against
E. coli 0157:H7. The extracts at 0.5% and 1% concentrations had a bacteriostatic effect, while the extracts at 2.5% and 5%
concentrations had a bactericidal effect against *E. coli* 0157:H7. *Reprinted from Baydar* et al. *(2006),* Intl J. Food Sci.
Technol., 41, *799–804, with permission.*

Different results were obtained when comparing two fractions of white grape seeds. The
extract that contains aliphatic aldehydes, fatty acids and their derivatives, and sterols was
more effective than the fraction composed of catechin, epicatechin, and gallic acid (Palma
et al., 1999; Table 65.1). On the other hand, Rhodes and colleagues (2006) did not
observe any antibacterial activity against *E. coli* of GSE derived from the *Vitis vinifera*
variety Ribier.

Staphylococcus aureus is a commensal bacterium which could be implicated in opportunistic
infections. Many strains are resistant to various antibiotics, and research has been undertaken
in order to find molecules that could have an antibacterial activity against them.

GSEs seemed to have interesting properties. At the concentrations of 0.5% and 1%, they
inhibited the growth of *S. aureus* (bacteriostatical effect), while at the concentrations of
2.5% and 5%, they showed bactericidal activity (Baydar *et al.*, 2006; Figure 65.3). These
results were confirmed on strains of methicillin-resistant *Staphylococcus aureus*. High antibacterial activity was found with a fraction that contained oligomeric units of catechins and
epicatechins (Mayer *et al.*, 2008). More recently, it was found that water and methanol
extracts of *Vitis vinifera* seeds had the highest antibacterial activities against *Staphylococcus
aureus*, with 0.312–0.156 mg/ml MIC values (Yigit *et al.*, 2009).

Oral diseases are dental plaque-mediated pathologies. Microorganisms present in such
biofilm are more resistant to antimicrobial agents. However, GSE showed a bacteriostatic
effect on *Streptococcus mutans*, *Porphyromonas gingivalis*, and *Fusobacterium nucleatum*, not only
in their planktonic state but also when embedded in a biofilm (Smullen *et al.*, 2007, Furiga
et al., 2009).

Only a few studies have been carried out on *Vitis rotundifolia* grape seed extract.

Water-soluble extracts prepared from purple and bronze muscadine seeds showed strong
antibacterial activity against *Escherichia coli* and *Enterobacter sakazakii* strains. This effect is due
to polar substances such as tartaric, malic, and tannic acids (Kim *et al.*, 2008, 2009).

TABLE 65.1 Effects of Fractions A and B, Catechin, and Epicatechin against Human Pathogen Microorganisms

	μg/disk[a]											
	Fraction A			Fraction B			Catechin			Epicatechin		
Organism	1500	1000	500	1500	1000	500	1500	1000	500	1500	1000	500
Bacillus cereus (+)	−	−	−	−	−	−	−	−	−	−	−	−
Staphylococcus aureus (+)	+	−	−	+	+	−	−	−	−	+	+	−
Staphylococcus coagulans niger (+)	++	++	++	+	+	−	−	−	−	−	−	−
Citrobacter freundii (−)	++	+	−	+	+	−	−	−	−	+	−	−
Escherichia cloacae (−)	++	++	+	+	+	+	−	−	−	+	+	+
Escherichia coli (−)	+	+	−	+	+	−	+	+	+	+	+	−
Aspergillus flavus	−	−	−	−	−	−	−	−	−	−	−	−

The highest activities exhibited by fraction A were against *S. coagulans niger*, *E. cloacae*, *C. freundii*, and *E. coli*. These bacteria were susceptible at the concentration assayed, even at the lowest concentration. On *S. aureus*, there was only moderate activity at the highest concentration. *A. flavus* was resistant to fraction A at all concentrations assayed.
Reprinted from Palma et al. (1999), J. Agri. Food Chem., 47, 5044–5048.
[a]4-mm disk impregnated with selected concentration of fraction A or B. ++, high inhibition (diameter > 15 mm); +, inhibition (15 mm > diameter > 10 mm); −, no inhibition (diameter < 10 mm).

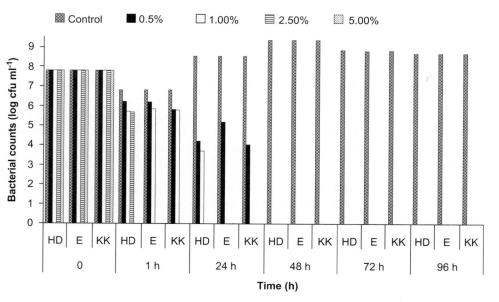

FIGURE 65.3
Inhibition curves of various concentrations of Hasandede (HD), Emir (E), and Kalecik Karasi (KK) seed extracts against *S. aureus*. All the extracts at all concentrations had bactericidal effects against *S. aureus* at the end of 48 hours. *Reprinted from Baydar et al. (2006), Intl J. Food Sci. Technol., 41, 799–804, with permission.*

Helicobacter pylori is considered the etiological agent of peptic ulcers and gastritis. Grape extracts and pure compounds have differing effects against this species, extracts being the most effective (Brown *et al.*, 2009).

ADVERSE EFFECTS AND REACTIONS (ALLERGIES AND TOXICITY)

Numerous studies have reported a lack of toxicity of GSE. However, grape seed proanthocyanidin extract at high doses has been shown to exhibit the cytotoxicity that is associated with increased apoptotic cell death (Shao *et al.*, 2006).

SUMMARY POINTS

- Grape seed extracts are rich in polyphenolic compounds.
- Various polyphenolic compounds have shown an antibacterial effect.
- Grape seed extracts exhibit strong antibacterial activity against many bacteria.
- Acetone : water : acetic acid and methanol : water : acetic acid extracts of grape seeds showed higher antibacterial activity against gram-positive bacteria than gram-negative bacteria.
- Numerous studies have reported grape seed extracts not to be toxic.

References

Ahn, J., Grün, I. U., & Mustapha, A. (2004). Antimicrobial and antioxidant activities of natural extracts *in vitro* and in ground beef. *Journal of Food Protection, 67,* 148–155.

Ahn, J., Grün, I. U., & Mustapha, A. (2007). Effects of plant extracts on microbial growth, color change, and lipid oxidation in cooked beef. *Food Microbiology, 24,* 7–14.

Anastasiadi, M., Chorianopoulos, N. G., Nychas, G.-J. E., & Karoutounian, S. A. (2009). Antilisterial activities of polyphenol-rich extracts of grapes and vinification by-products. *Journal of Agricultural and Food Chemistry, 57,* 457–463.

Baydar, N. G., Özkan, G., & Sağdiç, O. (2004). Total phenolic contents and antibacterial activities of grape (*Vitis vinifera* L.) extracts. *Food Control, 15,* 335–339.

Baydar, N. G., Sağdiç, O., Özkan, G., & Cetin, S. (2006). Determination of antibacterial effects and total phenolic contents of grape (*Vitis vinifera L.*) seed extracts. *International Journal of Food Science & Technology, 41,* 799–804.

Brown, J. C., Huang, G., Haley-Zitlin, V., & Jiang, X. (2009). Antibacterial effects of grape extracts on *Helicobacter pylori*. *Applied and Environmental Microbiology, 75,* 848–852.

Escribano-Baiton, T., Guttierez-Fernandez, Y., Rivas-Gonzalo, J., & Santo-Buelga, C. (1992). Characterization of procyanidins of *Vitis vinifera* variety tintal del pais grape seeds. *Journal of Agricultural and Food Chemistry, 40,* 1794–1799.

Furiga, A., Lonvaud-Funel, A., & Badet, C. (2009). *In vitro* study of antioxidant capacity and antibacterial activity on oral anaerobes of a grape seed extract. *Food Chemistry, 113,* 1037–1040.

Gabetta, B., Fuzzati, N., Griffin, A., Lolla, E., Pace, R., & Ruffilli, T. (2000). Characterization of proanthocyanidins from grape seeds. *Fitoterapia, 71,* 162–175.

Jayaprakasha, G., Selvi, T., & Sakariah, K. (2003). Antibacterial and antioxidant activities of grape (*Vitis vinifera*) seed extracts. *Food Research International, 36,* 117–122.

Kim, T. J., Weng, W. L., Stojanovic, J., Lu, Y., Jung, Y. S., & Silva, J. L. (2008). Antimicrobial effect of muscadine seed extracts on *E. coli* O157: H7. *Journal of Food Protection, 71,* 1465–1468.

Kim, T. J., Silva, J. L., Weng, W. L., Chen, W. W., Corbitt, M., Jung, Y. S., et al. (2009). Inactivation of *Enterobacter sakazakii* by water-soluble muscadine seed extracts. *International Journal of Food Microbiology, 129,* 295–299.

Lee, C., & Jaworski, A. (1987). Fractionation and HPLC determination of grape wine. *Journal of Agricultural and Food Chemistry, 35,* 257–259.

Mayer, R., Stecher, G., Wuerzner, R., Silva, R. C., Sultana, T., Trojer, L., et al. (2008). Proanthocyanidins: target compounds as antibacterial agents. *Journal of Agricultural and Food Chemistry, 56,* 6959–6966.

Nassiri-Asl, M., & Hosseinzadeh, H. (2009). Review of the pharmacological effects of *Vitis vinifera* (Grape) and its bioactive compounds. *Phytotherapy Research, 23,* 1197–1204.

Over, K. F., Hettiarachchy, N., Johnson, M. G., & Davis, B. (2009). Effect of organic acids and plant extracts on *Escherichia coli* O157:H7, *Listeria monocytogenes,* and *Salmonella typhimurium* in broth culture model and chicken meat systems. *Journal of Food Science, 74,* 515–521.

Palma, M., Taylor, L. T., Varela, R. M., Cutler, S. J., & Cutler, H. G. (1999). Fractional extraction of compounds from grape seeds by supercritical fluid extraction and analysis for antimicrobial and agrochemical activities. *Journal of Agricultural and Food Chemistry, 47,* 5044–5048.

Pastrana-Bonilla, E., Akoh, C. C., Sellappan, S., & Krewer, G. (2003). Phenolic content and antioxidant capacity of muscadine grapes. *Journal of Agricultural and Food Chemistry, 51,* 5497–5503.

Rhodes, P. L., Mitchell, J. W., Wilson, M. W., & Melton, L. D. (2006). Antilisterial activity of grape juice and grape extracts derived from *Vitis vinifera* variety Ribier. *International Journal of Food Microbiology, 107,* 281–286.

Shao, Z. H., Hsu, C. W., Chang, W. T., Waypa, G. B., Li, J., Li, D., et al. (2006). Cytotoxicity induced by grape seed proanthocyanidins: role of nitric oxide. *Cell Biology and Toxicology, 22,* 149–518.

Smullen, J., Koutsou, G. A., Foster, H. A., Zumbé, A., & Storey, D. M. (2007). The antibacterial activity of plant extracts containing polyphenols against *Streptococcus mutans*. *Caries Research, 41*, 342–349.

Tesaki, S., Tanabe, S., Moriyama, M., Fukushi, E., Kawabata, J., & Watanabe, M. (1999). Isolation and identification of an antibacterial compound from grape and its application to foods. *Nippon Nogeikagaku Kaishi, 73*, 125–128.

Yigit, D., Yigit, N., Mavi, A., Yildirim, A., & Güleryüz, M. (2009). Antioxidant and antimicrobial activities of methanol and water extracts of fruits, leaves and seeds of *Vitis vinifera L. cv.* Karaerik. *Asian Journal of Chemistry, 21*, 183–194.

CHAPTER 66

Gastroprotective Activity of Grapefruit (*Citrus paridisi*) Seed Extract Against Acute Gastric Lesions

Danuta Drozdowicz[1], Tomasz Brzozowski[1], Robert Pajdo[1], Slawomir Kwiecien[1], Michal Pawlik[2], Peter C. Konturek[1], Stanislaw J. Konturek[1], Wieslaw W. Pawlik[1]
[1] Department of Physiology, Jagiellonian University Medical College, Cracow, Poland
[2] Department of Internal Medicine, Thuringia-Clinic Saalfeld, Teaching Hospital of the University of Jeng Saalfeld, Germany

CHAPTER OUTLINE

LIST OF ABBREVIATIONS

COX-1, cyclooxygenase-1
COX-2, cyclooxygenase-2
GSE, grapefruit seed extract
GBF, gastric blood flow
HDC, histidine decarboxylase
I/R, ischemia—reperfusion
MDA, malonyldialdehyde
NO, nitric oxide
PG, prostaglandin
ROM, reactive oxygen metabolites
SOD, superoxide dismutase
WRS, water immersion restraint stress

Nuts & Seeds in Health and Disease Prevention. DOI: 10.1016/B978-0-12-375688-6.10066-0

INTRODUCTION

Previous studies documented that grapefruit seeds are the major depositories for limonoids (triterpenoid dilactones chemically related to limonin). Of these, 77% are neutral while 2% are acidic limonoids (Heggers *et al.*, 2002). Grapefruit also contains many flavonoid glycosides, naringenin, quercetin, kaempferol, hesperidin, and apigenin being the most abundant among their aglycones. Grapefruit seed extracts (GSEs), containing flavonoids, have been shown to possess antibacterial, antiviral, and antifungal properties (Reager *et al.*, 2002). The beneficial actions of GSE were attributed to the antioxidative activity of grapefruit, which contains citrus flavonoids (Lea & Reidenberg, 1998). For instance, these flavonoids exhibited *in vitro* the potent inhibitory influence against the activity of *Helicobacter pylori*, which is considered to be the major cause of ailments such as gastritis, ulcerations, MALT lymphoma, and other pathologies, particularly being linked to gastric cancer (Geil *et al.*, 1995; Bae *et al.*, 1999). Furthermore, grapefruit-containing flavonoids such as naringenin were also recently implicated in cytoprotection against injury induced by algal toxins in isolated hepatocytes (Blankson *et al.*, 2000). Interestingly, GSE in a formulation of Citricidal was demonstrated to be effective against more than 800 bacterial and viral strains, 100 strains of fungus, and a large number of single- or multi-celled parasites.

BOTANICAL DESCRIPTION

Grapefruit seed is prepared in extract form from the seeds, pulp, and white membranes of grapefruits from grapefruit trees. The grapefruit tree, first discovered on the Caribbean island of Barbados in the 17th century, was brought to Florida in 1823 for commercial cultivation. The origin of the name of the plant is not fully documented, but it is postulated that it was named "grapefruit" because its fruits grow in bunches or clusters. GSE has a broad spectrum of uses, and is considered to be a non-toxic substance with significant antimicrobial properties.

HISTORICAL CULTIVATION AND USAGE

The history of GSE started with its development by Dr Jacob Harich, a physicist originally born in Yugoslavia and who was educated in nuclear physics in Germany. After surviving World War II, Harich decided to devote the rest of his life to improving the human conditions. He expanded his educational pursuits to include medicine; specifically, gynecology and immunology. He migrated to the United States in 1957, and was interested in studying natural substances that might help to protect the body from pathological substances. In 1963 he moved to Florida, which is considered by many to be the heart of grapefruit country, and began research on the practical biocidal aspects of grapefruit seeds. Nowadays GSE, also known as citrus seed extract, is a liquid derived from the seeds, pulp, and white membrane of grapefruit. The extract is commercially distributed in two forms, as a liquid or a powder.

PRESENT-DAY CULTIVATION AND USAGE

GSE has been used as effective medication against many types of internal and external infections caused by parasites, viruses, bacteria, and fungi. Furthermore, GSE has been shown to correct yeast imbalances. Several researchers have observed that GSE helps to eliminate *Candida* yeast infections, thrush, gingivitis, oral infections, colds and flu, sore throat, strep throat, and sinusitis. The beneficial mechanism of the prophylactic activity of GSE depends on the disruption of the organism's cytoplasmic membrane, thereby not allowing the viral or bacterial organism to develop resistance. GSE has been shown also to stimulate the immune system, and was reported to be highly effective against food poisoning and against cholera or dysentery infections, particularly when traveling to Third World countries. It is believed to rid the body of worms and parasites without affecting the beneficial so-called "friendly" bacteria. Consumption of the bioactive compounds found in grapefruit seed and pulp is believed to suppress the development of colon cancer.

APPLICATIONS TO HEALTH PROMOTION AND DISEASE PREVENTION

The antimicrobiological properties of GSE against a wide range of gram-negative and gram-positive organisms were attributed to the disruption it caused to the bacterial membrane, and the subsequent liberation by this extract of the bacterial cytoplasmic contents within a relatively short time (e.g. 15–20 minutes). Moreover, the flavonoid naringenin, the bioactive component of GSE, has showed anticancer activity against various human breast cancers (So *et al.*, 1996). The underlying mechanism of the therapeutic efficacy of citrus seed extracts such as from grapefruits and red grapes seems to depend upon the presence of different classes of polyphenolic flavonoids, which were shown to inhibit platelet aggregation, thus decreasing the risk of coronary thrombosis and myocardial infarction. However, the involvement of GSE containing various flavonoids in the mechanism of gastric mucosal integrity and mucosal defense has not been extensively studied.

Previous studies demonstrated that flavonoids — namely, quercetin and meciadanol, which is a synthetic flavonoid that inhibits histidine decarboxylase (HDC) and decreases the histamine content in the stomach — attenuated gastric mucosal lesions produced by ethanol and aspirin via a mechanism unrelated to gastric acid secretion and endogenous prostaglandins (PG) (Konturek *et al.*, 1986). It remained unknown to what extent GSE influences the gastric mucosal injury induced by topical (ethanol) and non-topical ulcerogens (I/R and stress), and, if it does, what the mechanism of gastroprotection induced by GSE is. Therefore, using animal models of gastric lesions induced by I/R, water immersion and restraint stress (WRS) and 75% ethanol, we determined the influence of GSE on gastric lesions induced by these three ulcerogens, and the accompanying changes in the gastric blood flow (GBF) in the rat stomach. An attempt was also made to assess the contribution of the activity of superoxide dismutase (SOD) and lipid peroxidation as expressed by the malonyldialdehyde (MDA) concentration PG/COX system, nitric oxide (NO), and sensory nerves in the gastroprotective effect of GSE. Herein, we will provide evidence that the pretreatment with GSE applied orally attenuates the gastric lesions caused by I/R or cold stress, and those induced by the intragastric application to the stomach of noxious substance such as 75% ethanol. This protective effect of GSE is accompanied by the increase in the GBF and SOD activity and a reduction of MDA concentration, which is widely considered an index of lipid peroxidation (Figures 66.1, 66.2). Previous studies have demonstrated that the damaging action of ethanol, I/R, and WRS could be attributed to the enhancement in the reactive oxygen metabolites (ROM), and the ROM- and neutrophil-dependent increase in the lipid peroxidation and inhibition of anti-oxidizing enzyme activity (Kwiecien *et al.*, 2003). We found that GSE greatly attenuated the rise in MDA content in the gastric mucosa injured by I/R, WRS, and 75% ethanol (Figure 66.2), indicating that this extract can attenuate the process of neutrophil-dependent lipid peroxidation implicated in the pathogenesis of I/R, WRS, and ethanol-induced gastric damage. Ethanol decreased the gene expression of SOD in the gastric mucosa (Brzozowski & Konturek, 2005), suggesting that the suppression of key mucosal antioxidizing enzymes along with the elevation of lipid peroxidation play an important role in the pathogenesis of these lesions. These increases in mucosal lipid peroxidation, as well as the decrease in SOD expression and its activity, were attenuated by GSE. This would suggest that the reduction in lipid peroxidation by this seed extract might contribute to the attenuation of the harmful effects of noxious agents on the gastric mucosa. This is supported by the fact that the GBF was elevated and gastric mucosal generation of PGE$_2$ were enhanced in animals treated with GSE as compared to those treated with vehicle (Figure 66.3). This finding is in keeping with observations that some flavonoids stimulated PGE$_2$ production by isolated gastric mucosal cells while suppressing gastric acid secretion via a direct inhibitory effect on H$^+$/K$^+$-ATPase activity (Beil *et al.*, 1995). It is understood that the NO/NOS system is also involved in GSE-induced protection against I/R injury, due to the fact that both GSE-induced protection and hyperemia were counteracted by

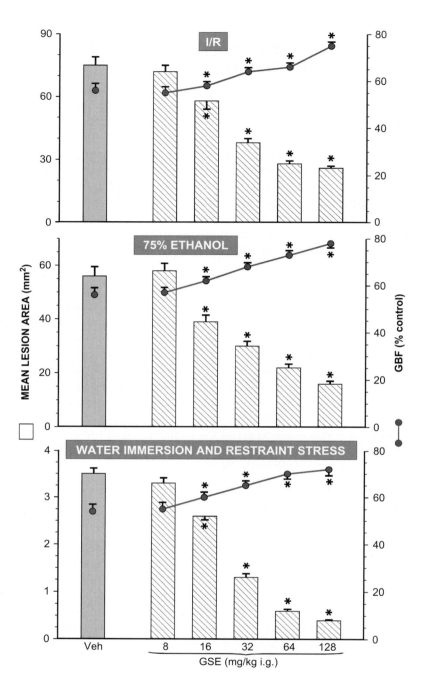

FIGURE 66.1

Effect of intragastric (i.g.) pretreatment with GSE applied in graded doses ranging from 8 mg/kg up to 128 mg/kg on the mean area of gastric lesions induced by ischemia—reperfusion (I/R), 75% ethanol, and 3.5 h of water immersion and restraint stress (WRS), and accompanying changes in the gastric blood flow (GBF). GSE significantly attenuates the area of I/R-, 75% ethanol-, and WRS-induced gastric lesions while raising GBF. Results are expressed as mean ± SEM, $n = 8$. Asterisk indicates a significant change vs vehicle (Veh, saline) control.

L-NNA, a non-specific inhibitor of NO-synthase. This effect was restored in these animals by the combined treatment with L-arginine and GSE, but not with the L-arginine stereoisomer, D-arginine (Figure 66.4). Thus, these sets of data suggest that some natural products of citrus fruits, such as GSE, afford protection against acute gastric lesions due to an increase in the gastric microcirculation involving endogenous NO release, and the preservation of the expression and activity of major antioxidizing enzymes such as SOD.

The mechanism of gastroprotective activity of GSE appears to be dependent on endogenous PG, and the functional activity of sensory nerves releasing CGRP. This notion is supported by our findings that indomethacin, a non-selective inhibitor of COX-1 and COX-2 activity, reversed this protection and the accompanying hyperemia (Figure 66.3). In addition, co-treatment of exogenous CGRP with GSE, administered to replace the deficit of this peptide

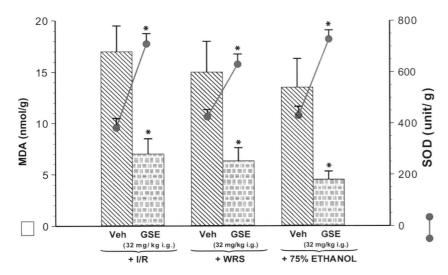

FIGURE 66.2

Effect of vehicle (saline) and GSE on the MDA concentration and SOD activity in the gastric mucosa in rats exposed to I/R or 75% ethanol, or subjected to 3.5 h of WRS. GSE significantly inhibits the gastric MDA content while producing a significant increase in SOD activity. Results are expressed as mean $\pm$ SEM, $n = 8$. Asterisk indicates a significant change ($P < 0.05$) vs vehicle (Veh, saline) control.

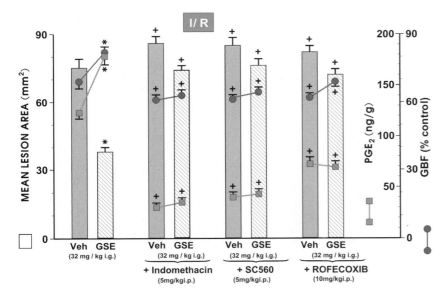

FIGURE 66.3

Effect of GSE on the mean area of I/R-induced gastric lesions and accompanying alterations in the gastric mucosal PGE$_2$ generation and GBF in rats with or without pretreatment with indomethacin, the non-selective inhibitor of COX activity; SC-560, the selective inhibitor of COX-1 activity; and rofecoxib, the selective inhibitor of COX-2 activity. The gastroprotective action and the accompanying increase in PGE$_2$ and GBF by GSE are abolished by indomethacin, SC-560, and rofecoxib. Results are expressed as mean $\pm$ SEM, $n = 8$. Asterisk indicates a significant change ($P < 0.05$) vs vehicle (Veh, saline) control; cross indicates a significant change ($P < 0.05$) vs Veh and GSE-alone groups.

557

in capsaicin-treated animals, restored the protective efficacy of GSE against WRS-induced gastric damage (Figure 66.5). Interestingly, both, SC-560 and rofecoxib, the highly selective COX-1 and COX-2 inhibitors (Brzozowski *et al.*, 1999), also inhibited the gastroprotective and hyperemic activities of GSE, which in turn suggests the involvement of COX-1 and COX-2 derived products in gastroprotection and the increase in the GBF induced by this seed extract (Figure 66.3). This gastroprotective activity of the GSE could be attributed to naringenin, a major GSE flavonoid. This flavonoid was reported to exhibit gastroprotection against the gastric injury induced by absolute ethanol, predominantly due to the increase in the secretion of mucus (Motilova *et al.*, 1994). It is of interest that this gastroprotective effect of naringenin and accompanying increase in the mucus secretion were, in part, attenuated by indomethacin, supporting the contribution of endogenous PG to the mechanism of gastroprotection by GSE. On the other hand, lack of the complete destruction of GSE protection by indomethacin suggests that PG might not be the primary mediator of this protection, and other

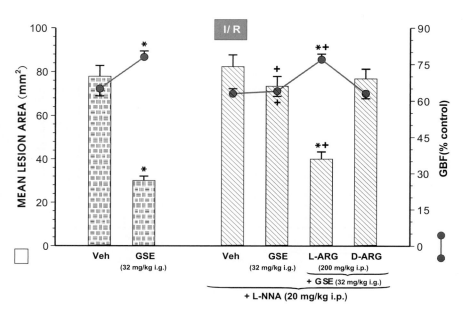

FIGURE 66.4
Effect of GSE on the mean area of I/R-induced gastric lesions and accompanying alterations in the GBF in rats with or without pretreatment with L-NNA, the NO-synthase inhibitor, applied alone or co-administered with L-arginine (L-ARG) or D-arginine (D-ARG). The gastroprotective and hyperemic activities of GSE are significantly attenuated by L-NNA and restored with L-ARG. Results are expressed as mean ± SEM, $n = 7$. Asterisk indicates a significant change ($P < 0.05$) vs vehicle (Veh, saline) control; cross indicates a significant change ($P < 0.05$) vs GSE-alone group; asterisk and cross indicate a significant change ($P < 0.05$) vs animals treated with the combination of GSE and L-NNA.

FIGURE 66.5
Effect of GSE on the WRS-induced gastric lesions and accompanying alterations in the GBF in rats with intact sensory nerves and those with capsaicin inactivation of sensory nerves with or without the administration of exogenous CGRP. Functional ablation of sensory nerves by capsaicin reduces the gastroprotective and hyperemic actions of GSE against WRS-induced gastric lesions. Results are expressed as mean ± SEM, $n = 8$. Asterisk indicates a significant change ($P < 0.05$ vs vehicle (Veh, saline) control; cross indicates a significant change ($P < 0.05$) vs GSE-alone group; asterisk and cross indicate a significant change ($P < 0.05$) vs capsaicin denervated animals treated with GSE.

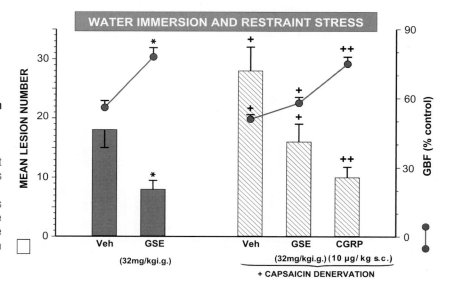

gastroprotective factors such as NO and/or neuropeptides released from sensory afferent nerves could be involved. This is the reason we carried out the study with L-NNA, a potent NO-synthase inhibitor, and with capsaicin, applied in a dose that causes functional ablation of sensory nerves, leading to the release of vasoactive neuropeptides such as CGRP. We found that L-NNA and capsaicin denervation inhibited the GSE-induced protection against I/R- and WRS-induced gastric lesions and accompanying gastric mucosal hyperemia (Figures 66.4, 66.5). Naringenin could be one of the important components of GSE exhibiting gastric protection and hyperemia, because pretreatment with naringenin applied in gradual concentrations counteracted the gastric mucosal lesions, and the accompanying fall in GBF induced by the mucosal corrosive agent (75% ethanol) (Figure 66.6).

Another candidate involved in GSE-induced protection could be gastrin, which is known to exhibit both gastroprotective and hyperemic activities. The contribution of gastrin was something we addressed in our previous study by directly determining the plasma gastrin

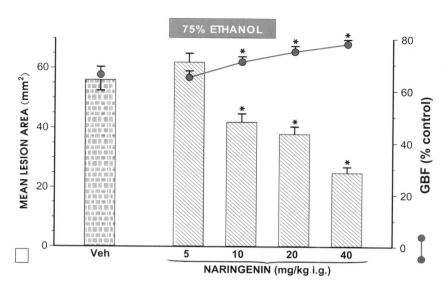

FIGURE 66.6
Effect of intragastric (i.g.) pretreatment with vehicle (saline) or naringenin applied in graded doses ranging from 5 mg/kg up to 40 mg/kg on the mean area of 75% ethanol-induced gastric lesions and the accompanying changes in the GBF. Naringenin dose-dependently reduced the area of ethanol-induced gastric lesions. Results are expressed as mean ± SEM, $n = 7$. Asterisk indicates a significant change ($P < 0.05$) vs vehicle (Veh, saline) control.

levels by specific RIA (Brzozowski & Konturek, 2005). Suppression of gastric secretion by GSE might contribute to the protective activity of this extract against WRS and J/R injury, because such injuries depend upon gastric acidity and ultimately become exaggerated by the acidic conditions in the stomach.

In summary, GSE exerts a potent gastroprotective activity against gastric lesions caused by acid-dependent (I/R, WRS) and acid-independent (ethanol) causes, due to a mechanism involving preservation of the activity of antioxidizing enzymes (SOD), reduction of free radical-dependent lipid peroxidation, enhancement in the GBF, and plasma gastrin levels exerting trophic influence on gastric mucosa. The GSE-induced protection and hyperemia were mimicked by intragastric application of naringenin, suggesting that this flavonoid could be involved in the gastroprotection and hyperemia afforded by GSE. This protective and hyperemic activity of GSE against I/R ulcerogenesis was abolished by COX-1 and COX-2 nonselective and selective inhibitors such as indomethacin, SC-560, and rofecoxib, and significantly ameliorated by L-NNA. Thus, this protective effect in the stomach may involve endogenous PG derived from COX-1 and COX-2 activity, suppression of lipid peroxidation, and gastric hyperemia mediated by NO, and neuropeptides released from afferent sensory nerves.

ADVERSE EFFECT AND REACTIONS (ALLERGIES AND TOXICITY)

GSE has been shown to be non-toxic at quantities many times greater than the recommended dosages. Even when taken daily, GSE seldom produces a significant allergic reaction. However, those individuals who suffer from allergy to citrus fruits should exercise caution in the use of GSE. Studies on commercial preparations have demonstrated that GSE may contain the compound benzalkonium chloride, which is a synthetic antimicrobial commonly used in disinfectants and cleaning products. From this, it was concluded that the universal antimicrobial activity of GSE is merely the result of its contamination with synthetic antimicrobials.

SUMMARY POINTS

- Grapefruit seed extract (GSE) is a product of the grapefruit seed and pulps, and has been found to exert antibacterial, antiviral, and antiparasitic activities.
- GSE has been reported to be safe and effective to use internally and externally in a variety of conditions, including acne, allergies, body odor, *Candida* infections, gastrointestinal infections, gingivitis, sinusitis, and sore throat.

559

- Some commercial GSEs might be contaminated by benzalkonium chloride, which causes toxicity and allergenicity.
- Recently, GSE was implicated in the mechanism of gastroprotection against the lesions induced by topical necrotizing substances such as ethanol, and non-topical ulcerogens such as ischemia—reperfusion and stress.
- The major mechanism of protection against gastric mucosal damage involves enhancement of the gastric microcirculation mediated by increased generation of endogenous prostaglandins (PGs) and nitric oxide (NO), and involving an activity of sensory neuropeptides such as CGRP.

References

Bae, E. A., Han, M. J., & Kim, D. H. (1999). *In vitro* anti-*Helicobacter pylori* activity of same flavonoids and their metabolites. *Planta Medica, 65,* 442—443.

Beil, W., Birkholz, C., & Sewing, K. F. (1995). Effects of flavonoids on parietal cell acid secretion, gastric mucosal prostaglandin production and *Helicobacter pylori* growth. *Arzneimittel Forschung, 45,* 697—700.

Blankson, H., Grotterod, E. M., & Seglen, P. O. (2000). Prevention of toxin-induced cytoskeletal disruption and apoptotic liver cell death by the grapefruit flavonoid naringenin. *Cell Death Different, 7,* 739—746.

Brzozowski, T., & Konturek, P. C. (2005). Grapefruit-seed extract attenuates ethanol- and stress-induced gastric lesions *via* activation of prostaglandin, nitric oxide and sensory nerve pathways. *World Journal of Gastroenterology, 11*(41), 6450—6458.

Brzozowski, T., Konturek, P. Ch., Konturek, S. J., Sliwowski, Z., Drozdowicz, D., Stachura, J., et al. (1999). Role of prostaglandins generated by cyclooxygenase-1 and cyclooxygenase-2 in healing of ischemia—reperfusion-induced gastric lesions. *European Journal of Clinical Pharmacology, 385,* 47—61.

Heggers, J. P., Cottingham, J., Gusman, J., Reagor, L., McCoy, L., Carino, E., et al. (2002). The effectiveness of processed grapefruit-seed extract as an antibacterial agent: II. Mechanism of action and *in vitro* toxicity. *Journal of Alternative and Complementary Medicine, 8,* 333—340.

Konturek, S. J., Kitler, M. E., Brzozowski, T., & Radecki, T. (1986). Gastric protection by meciadanol. A new synthetic flavonoid inhibiting histidine decarboxylase. *Digestive Diseases and Sciences, 3,* 847—852.

Kwiecien, S., Brzozowski, T., Konturek, P. C., Pawlik, M. W., Pawlik, W. W., Kwiecien, N., et al. (2003). The role of reactive oxygen species and capsaicin-sensitive sensory nerves in the pathomechanism of gastric ulcers induced by stress. *Journal of Physiology and Pharmacology, 54,* 423—437.

Lea, Y. S., & Reidenberg, M. M. (1998). A method for measuring naringenin in biological fluids and its disposition from grapefruit juice by man. *Pharmacology, 56,* 314—317.

Motilova, V., Alarcon dela Lastra, C., & Martin, M. J. (1994). Ulcer promoting effects of naringenin on gastric lesions induced by ethanol in rat; role of endogenous prostaglandins. *Journal of Physiology and Pharmacology, 46,* 91—94.

Reager, L., Gusman, J., McCoy, L., Carino, E., & Heggers, J. P. (2002). The effectiveness of processed grapefruit — seed extract as an antibacterial agent: I. An *in vitro* agar assay. *Journal of Alternative and Complementary Medicine, 8,* 325—332.

So, F. V., Guthrie, N., Cliombers, A. F., Moussa, M., & Carroll, K. K. (1996). Inhibition of human breast cancer cell proliferation and delay of mammary tumorigenesis by flavonoids and citrus juices. *Nutrition and Cancer, 26,* 167—181.

CHAPTER 67

Antimicrobial Activities of Gray Nickerbean (*Caesalpinia bonduc* Linn.)

Tusharkanti Mandal, Jitendra Bhosale, Gajendra Rao, Madan Mohan Padhi, Rajesh Dabur
National Research Institute of Basic Ayurvedic Sciences, CCRAS, Nehru Garden, Kothrud, Pune, India

561

LIST OF ABBREVIATIONS

CBSC, *Caesalpinia bonduc* seed coat
CBSK, *Caesalpinia bonduc* seed kernel
CFU, colony forming unit
i.p., intraperitoneal injection
LD_{50}, lethal dose 50%
WHO, World Health Organization

INTRODUCTION

Medicinal plants, which constitute a considerable segment of the flora, provide raw material for use in all the traditional systems of medicine. According to the World Health Organization (WHO), 80% of the population in developing countries relies on traditional medicine, mostly in the crude form of plant drugs, for their health care needs. Modern medicines also contain plant derivatives, to the extent of about 25%.

Nuts & Seeds in Health and Disease Prevention. DOI: 10.1016/B978-0-12-375688-6.10067-2

Caesalpinia bonduc (Linn.) Roxb. is a climber of the Caesalpiniaceae family. The Ayurvedic system of medicine prescribes *C. bonduc* seeds to treat tumors, cysts, and cystic fibrosis (Sharma, 1999). The seeds are traditionally used for the treatment of fever, inflammation, and liver disorders. Various parts of the plant have also been reported to possess multiple therapeutic properties, such as antipyretic, antidiuretic, anthelmintic, antibacterial, anticonvulsant, antiviral, anti-asthmatic, anti-amebic, and anti-estrogenic activities. Antibacterial and antifungal activities of diterpene bondenlide obtained from *C. bonduc* plant were reported by Simin and colleagues (Simin *et al.*, 2000). Recent studies have shown that subcutaneous administration of cortisone in the mouse model of cystic fibrosis has no effect on *Pseudomonas aeruginosa*-induced pneumonia. On the other hand, the seed kernel of *C. bonduc*-treated group showed significant improvement in lung pathology. It was observed that the use of hydroalcoholic extract of *C. bonduc* (Linn.) Roxb. showed promising activity against *Pseudomonas* chronic lung infections in cystic fibrosis (Arif *et al*, 2009).

The plant seeds contain bonducins, e.g., bonducellpins A, B, C (Figure 67.1) (Peter *et al.*, 1997), saponins, fatty oils, sucrose, phytosterols, phytosterenin, fixed oil, starch, triterpenoids, fatty acid, triglycerides, sterols, sucrose, and α, β, γ, δ, and ζ caesalpins. Several reports on the presence of flavonoids, triterpenoids, and steroids showed that these molecules have multiple biological effects due to their antioxidant and free radical scavenging abilities.

BOTANICAL DESCRIPTION

Caesalpinia bonduc (Linn.) Roxb. (synonyms *Caesalpinia crista*, *Caesalpinia bonducella*) is an extensive climber, with finely downy gray branches armed with both hooked and straight hard yellow prickles. The leaves (Figure 67.2) are bipinnate, 30—60 cm long with short prickly petioles; the stipules a pair of reduced pinnae at the base of the leaf, each furnished with a long mucronate point. There are six to eight pairs of pinnae, each 5.0—7.5 cm long, with a pair of hooked stipulary spines at the base. Leaflets are in six to nine pairs, 2.0—3.8 cm long and 1.3—2.2 cm wide, membranous, elliptic to oblong, obtuse, strongly mucronate, glabrous above and more or less puberous beneath. The calyx is 6—8 mm long, fulvous and hairy; the lobes are obovate—oblong and obtuse. The petals are yellow and oblanceolate; the filaments are declinate, flattened at the base, and clothed with long white silky hairs. Flowering and fruiting has been reported (Figure 67.3) in the months July—April, the authors have observed it during November to December in the geoclimatic conditions of Pune (Maharashtra, India). Flowers are produced in dense terminal racemes (usually spicate), with long peduncles and

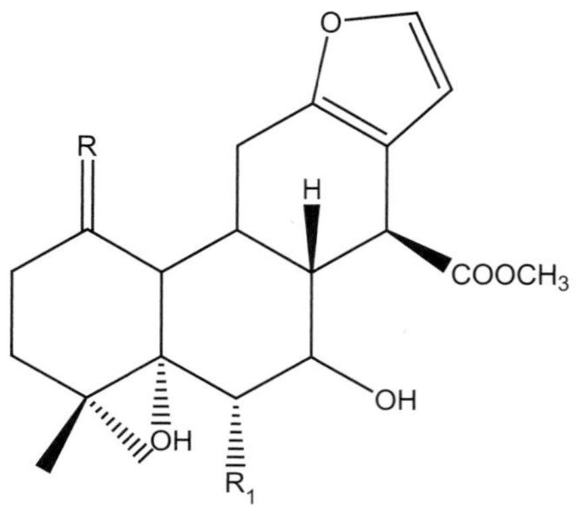

Bonducellpin A	R=H, a OAc	R₁=OAc
Bonducellpin B	R=O	R₁=OAc
Bonducellpin C	R=H, a OAc	R₁=H

FIGURE 67.1
Structure of bonducellpins A, B, C.

FIGURE 67.2
Fresh leaves of *Caesalpinia bonduc*.

supra-auxiliary racemes which are close at the top and looser lower down, 15—25 cm in length. The pedicles are very short in the buds, elongating to 5 mm in flowers and 8 mm in fruits, brown and downy; the bracts are squarrose, linear, acute, reaching up to 1 cm in length. Pods are shortly stalked, oblong, 5.0—7.5 cm by about 4.5 cm, and densely armed with wiry prickles. Seeds are oblong, dark gray, and up to 1.3 cm long (Figure 67.3) (Howard, 1988).

HISTORICAL BACKGROUND AND USAGE

According to descriptions in Ayurveda, *C. bonduc* is used to cure various diseases. In *Sushruta Samhita* it is described as kantakikaranja, a potherb for producing heat, and curing swellings and elephantiasis. According to Bhaavaprakaash, the fruits of the plant are considered to be hot, and can cure vomiting, biliousness, parasitic infections, piles, skin diseases, and urinary disorders. Commercial cultivation has not, so far, been reported in any of the archives, but seeds of *C. bonduc* were apparently carried to sea by floods and storm surges, where they floated until they were deposited on the shore. The scarifying action of sand-weathering, insects, or rodents eventually allowed water to enter the seeds, and thus their germination. They established themselves in new environments, producing roots whenever they came in contact with the ground, and sprouted. Plants thrived best, as they do even today, in coastal areas, inland in scrub jungle, in the hedges of crop fields, roadsides, and ditches, and sometimes as thickets on vacant plots as ruderal habitat.

563

FIGURE 67.3
Unripe and ripe fruits and seeds of *C. bonduc*.

The seeds of *C. bonduc* have been used for centuries and are still used as jewelry, prayer beads, good luck charms, and worry stones (Leona, 2002). They were used in ancient times as standards of weight in India (*Coins*, 2002). The ancient African and Indian games traditionally employed gray nickerbeans as game pieces (Driedges, 1972). The species is sometimes planted as a hedge to prevent undesired entry into property, and can be planted for dune stabilization (Nelson, 1996).

Siddha-bheshaja-manimaalaa, an ancient book, recommends the use of the seed kernel of *C. bonduc*, along with other herbs, in fever. For colic, two or three seed kernels should be fried and taken with salt. According to Vaidya Manoramaa (Physician), the seed kernel is recommended in digestive disorders, dysentery, vomiting, and internal abscesses. In the Unani system of medicine, karanjawaa (*C. bonduc*) seed kernel is burnt in sesamum oil and applied to infected wounds and cutaneous afflictions.

In other places, the plant is used in traditional medicine for the treatment of skeletal fractures, and to control blood sugar and asthma. The seed oil is an emollient, and is used for treating discharge from the ear, for removing facial freckles, and for treating skin diseases. The seed powder is given for menstrual disorders by the tribal communities of Madhya Pradesh (India). In homoeopathy, the plant is considered an excellent remedy for chronic fever with headache.

According to folklore, the seed powder is a household remedy, in Nicobar Island, for the treatment of diabetes. The plant juice is also given for 2 weeks after meals to cure intermittent fever. The seed kernel-based ointments of *C. bonduc* are applied to hydroceles; seed kernels in an infusion are prescribed in hemorrhages. Seeds are also administered as an anthelmintic, mixed with honey or castor oil.

PRESENT-DAY CULTIVATION AND USAGE

Cultivation of medicinal plants in fields is a recent phenomenon. Industries prefer to procure raw material from cultivated sources for reasons of authentication, reliability, and continuity. However, the lack of availability of quality planting material, coupled with poor development and extension of support for processing, has been a limiting factor in the cultivation of many herbs of medicinal value. Unorganized markets are another major constraint against the commercialization of cultivation. Therefore, concentrated efforts are required in both the collection and cultivation of medicinal plants, in order to ensure sustainability of the plant-material based industry.

C. bonduc is mostly propagated by seeds. The seeds are soaked overnight, and are sown in the field just after the first showers. Irrigation is carried out immediately after sowing. Dormancy of the seeds can be overcome by acid scarification, light and temperature treatment, or treatment with concentrated sulfuric acid for 30–90 minutes. The seeds treated with acid for 90 minutes, if exposed to blue-spectrum light for 72 hours at 30°C, exhibit 100% germination.

Preparations of the seeds and other plant parts are used to treat a large range of ailments (Parrotta, 2001). Extract from the seed is reported to lower blood sugar (Biswas *et al.*, 1997) and effectively suppress or cure infections due to several species of round worms (Amarsinghe *et al.*, 1993). The fat content of seeds has been measured at $34 \pm 0.83\%$. The semi-drying oil extracted would be useful for manufacture of high-quality alkalyd resins, polishes, and paint (Ajiwe *et al.*, 1996).

APPLICATIONS TO HEALTH PROMOTION AND DISEASE PREVENTION

In a recent study, the seed coat extracts of *C. bonduc* (CBSC) exhibited better *in vitro* antimicrobial activity than did those prepared from seed kernel (CBSK). Methanol and water extracts of CBSC showed activity against nine bacterial species used in the study, with the exception of

TABLE 67.1 Antimicrobial Activity of Seed Extracts by Microbroth Dilution Assay

Microorganisms	Extracts	Water Extract	Methanol Extract	Pet. Ether Extract	Amp	Met
P. aeruginosa MTCCB 741	CBSC	88	22	175	2.5	ND
	CBSK	350	175	—		
P. aeruginosa 135 (AR)	CBSC	175	22	350	—	ND
	CBSK	350	175	—		
P. aeruginosa 1124 (AR)	CBSC	88	22	175	—	ND
	CBSK	175	175	—		
P. aeruginosa 325 (AR)	CBSC	88	22	175	—	ND
	CBSK	175	175	—		
S. aureus MTCCB 737	CBSC	175	44	88	ND	2.5
	CBSK	88	—	175		
S. aureus 187 (MR)	CBSC	878	22	175	ND	—
	CBSK	44	—	350		
S. aureus 349 (MR)	CBSC	175	22	88	ND	—
	CBSK	88	—	88		
S. aureus 674 (MR)	CBSC	88	88	—	ND	—
	CBSK	350	175	350		
E. coli MTCCB 82	CBSC	—	88	88	2.5	ND
	CBSK	88	—	175		
P. mirabilis MTCCB 564	CBSC	22	88	175	5.0	ND
	CBSK	—	—	—		
K. pneumoniae MTCCB 109	CBSC	350	88	175	2.5	ND
	CBSK	—	350	88		
S. typhi MTCCB 733	CBSC	22	44	44	2.5	ND
	CBSK	88	—	175		
V. cholera MTCCB 458	CBSC	88	88	350	2.5	ND
	CBSK	175	88	22		
B. thuringiensis MTCCB 1824	CBSC	175	175	88	5.0	ND
	CBSK	175	175	175		
B. cereus MTCCB 1272	CBSC	—	22	22	2.5	ND
	CBSK	ND	ND	ND		

Amp, ampicillin; Met, methicillin; —, no activity; ND, not detected; CBSK, *C. bonduc* seed kernel; CBSC, *C. bonduc* seed coat; MR, methicillin resistant; AR, ampicillin resistant.

Escherichia coli, in the range of 22.0−350 μg/ml (Table 67.1). Petroleum ether extract also exhibited broad range activity at the higher concentrations. CBSK extracts exhibited variable activity at higher concentrations than did CBSC (Arif *et al.*, 2009). It was observed that the extracts of *C. bonduc* were effective against the methicillin- and ampicillin-resistant strains of *S. aureus* and *P. aeruginosa*. The results of *in vivo* experiments revealed no mortality in any of the infected groups. *P. aeruginosa* was cultured from the lungs of most of the surviving rats 2 weeks after challenge, and colony forming units (CFU) were counted (Table 67.2). The CBSK extract

TABLE 67.2 Macroscopic Pathology, Abscess Incidence and Median Numbers of CFU of *P. aeruginosa* in Rat Lungs after 14 Days' Intratracheal Challenge ($n = 9$ in Each Group)

Treatment Group	Median (Range) Bacterial Count (CFU)/Lung	No. (%) of Rats with < 100 CFU	No. of Rats with Lung Pathology Score			Lung Abscess Incidence (%)
			Grade 1	Grade 2	Grade 3	
Control	2.4×10^2 (1.3×10^1−5.6×10^4)	2 (22.2)	1	2	6	66.6
CBSK	1.9×10^1 (0−2.4×10^3)	6 (66.6)	5	3	1	11.1
Cortisone	5.6×10^3 (0.4×10^2−6.7×10^5)	2 (22.2)	0	2	7	77.7

Two-tailed *P* values showed significant difference ($P < 0.05$) in CBSK-treated group as compared to control.

reduced bacterial load significantly in the treated rats as compared to the controls ($P < 0.046$). Of nine CBSK-treated rats, six were found to have less than 100 CFU in their lungs. Abscesses, atelectases, hemorrhage, and fibrinous adhesion to the thoracic wall or diaphragm were found in all groups after 2 weeks from challenge. However, the lung pathology in the CBSK-treated group was found to be significantly milder as compared to the control and the cortisone-treated groups (Arif *et al.*, 2009).

Seed-coat as well as kernel extracts were reported to have anti-stress activity when administered orally at a dose of 300 mg/kg (see Kannur *et al.*, 2006, for further information). Ethanolic extract (70%) of *C. bonduc* seed kernel has been reported as having antipyretic and anti-nociceptive activities in doses of 30, 100, and 300 mg/kg (Archana *et al.*, 2005). Aqueous and ethanol extracts of *C. bonduc* seeds produced a hypoglycemic effect and anti-hyperglycemic activity as well as hypolipidemic activity (Sharma *et al.*, 1997).

C. bonduc seeds are one of the ingredients of the Ayurvedic drug *Ayush-64*, used for treating malaria and filaria. It has also been reported to effectively suppress or cure infections due to several species of *Ascaris* (Amarsinghe *et al.*, 1993; Rastogi *et al.*, 1996). Recently, hepatoprotective and antioxidant roles of *C. bonduc* were reported (Gupta *et al.*, 2003). Preliminary screening for antimalarial activity carried out on a rodent system infected with *Plasmodium berghei* gave promising results. It is used effectively as an antidote against opium, aconite, arsenic, and copper poisoning. Powdered seeds of *C. bonduc* were reported to have anti-estrogenic and contraceptive effects. In a clinical trial, seeds of *C. bonduc* were used for treating patients with griping pain and loose bowel motions mixed with mucus and blood. It was observed that the drug, at a dose of 2.0 g three times daily for 15 days, showed 100% improvement in the symptoms, and eradication of *Entamoeba histolytica* infection from the stool, indicating it to be an effective drug for dysentery. Seeds or nuts of *C. bonduc* are valuable in simple, continued, and intermittent fevers, asthma, colitis, etc., at a dose of 10–30 grains of the powdered seeds or kernel mixed with an equal quantity of powdered black pepper.

The seed extract of *C. bonduc* inhibited human trypsin and chymotrypsin in intestinal secretion. The seeds showed a 50% inhibitory activity against carrageenan-induced edema in rats' hind paw, at an oral dose of 1000 mg/kg, when given 24 hours and 1 hour prior to intra-peritoneal (IP) carrageenan injection. The activity (66.67% inhibition) was comparable with that of phenylbutazone at a dose of 100 mg/kg.

ADVERSE EFFECTS AND REACTIONS (ALLERGIES AND TOXICITY)

C. bonduc has been reported to be non-toxic up to an LD_{50} dose of 3000 mg/kg (Ali *et al.*, 2008). Assessment of *in vitro* cytotoxicity of aqueous plant extract on human leukocytes using Trypan blue test revealed that all aqueous plant extracts at concentrations less than or equal to 500 μg/ml were not toxic to cells. An oral dose 500 mg/kg for 30 days of *Ayush-64* (a drug having C. *bonduc* as a major component) did not produce any toxicity.

SUMMARY POINTS

- *Caesalpinia bonduc* (Linn.) Roxb. is an extensive woody, thorny climber that belongs to the Caesalpiniaceae family.
- The plant grows wild and mostly propagated by seeds.
- In Ayurvedic medicine, and also in traditional medicine, C. bonduc seeds are used for the treatment of different diseases, such as fever, diabetes, inflammation, liver disorders, cysts, tumors, etc.
- The reported chemical constituents of the plant include flavonoids, triterpenoids, diterpenoids, and steroids.
- The seed of the plants has been reported to possess *in vitro* as well as *in vivo* antimicrobial activity.

- The plant possesses antimalarial, antimicrobial, antitumor, hepatoprotective, hypoglycemic, and antioxidant activities.
- The seed extract of *C. bonduc* inhibited human trypsin and chymotrypsin in intestinal secretion, and also has promising activity for treating chronic lung infections in cystic fibrosis.
- Toxicity studies demonstrated the LD_{50} dose of *C. bonduc* to be 3000 mg/kg, indicating the use of this plant to be quite safe.

References

Ajiwe, V. I. E., Okeke, C. A., Agbo, H. U., Ogunleye, G. A., & Ekwuozor, S. C. (1996). Extraction, characterization and industrial uses of velvet tamarind, physic-nut and nicker-nut seed oils. *Bioresource Technology, 57,* 297–299.

Ali, A., Rao, N. V., Shalam, M., & Gouda, T. S. (2008). Anxiolytic activity of seed extract of *C. bonducella* (Roxb) in laboratory animals. *International Journal of Pharmacology, 5,* 1531–1537.

Amarsinghe, A. P. G., Sharma, R. D., Chaturvedi, C., & Agarwal, D. K. (1993). Anthelmintic effect of Ayurvedic recipe Kuberakshadi yoga in intestinal worms among children. *Journal of Research and Education in Indian Medicine, 12,* 27–31.

Archana, P., Tandan, S. K., Chandra, S., & Lal, J. (2005). Antipyretic and analgesic activities of *Caesalpinia bonducella* seed kernel extract. *Phytotherapy Research, 19,* 376–381.

Arif, T., Mandal, T. K., Kumar, N., Bhosale, J. D., Hole, A., Sharma, G. L., et al. (2009). *In vitro* and *in vivo* anti-microbial activities of seeds of Caesalpinia bonduc (Lin.) Roxb. *Journal of Ethnopharmacology, 123,* 177–180.

Biswas, T. K., Bandyopadhyay, S., Mukherjee, B., Sangupta, B. R., & Mukherjee, B. (1997). Oral hypoglycemic effect of *Caesalpinia bonducella*. *International Journal of Pharmacognosy, 35,* 261–264.

Coins (2002). Coinage. Available at http://www.jayanagaracoins.com/htm/coinage.htm.

Driedges, W. (1972). The game of Boa, or Mankala, in East Africa. *Mila, 3*(1), 7–19.

Gupta, M., Mazumder, U. K., & Kumar, R. S. (2003). Hepatoprotective and antioxidant role of *Caesalpinia bonducella* on paracetamol induced liver damage in rats. *Natural Product Science, 9,* 186–191.

Howard, R. A. (1988). *Flora of the Lesser Antilles, Leeward and Windward Islands. Dicotyledoneae, Part 1. Vol. 4. Arnold Arboretum.* Jamaica Plain, MA: Harvard University. p. 673.

Kannur, D. M., Hukkeri, V. I., & Akki, K. S. (2006). Adaptogenic activity of *Caesalpinia bonduc* seed extracts in rats. *Journal of Ethnopharmacology, 108,* 327–331.

Leona, R. (2002). Rancho Leona jewelry samples: rainforest seed chart. Available at http://rancholeona.com/seed.html.

Nelson, G. (1996). *The shrubs and woody vines of Florida.* Sarasota, FL: Pineapple Press. p. 391.

Parrotta, J. A. (2001). *Healing plants of Peninsular India.* Wallingford, UK: CABI. p. 917.

Peter, S. R., Tinto, W. F., McLean, S., Reynolds, W. F., & Yu, M. (1997). Bonducellpins A–D, new cassane furano-diterpenes of *Caesalpinia bonduc. Journal of Natural Products, 60*(12), 1219–1221.

Rastogi, S., Shaw, A. K., & Kulshreshtha, D. K. (1996). Characterization of fatty acids of antifilarial triglyceride fraction from *Caesalpinia bonduc. Fitoterapia, 67,* 63–64.

Sharma, P. V. (Ed.). (1999). *Sushruta Samhita, Chikitisa Sthana 5/37.* Varanasi, India: Chaukhamba Visvabharati.

Sharma, S. R., Dwivedi, S. K., & Swarup, D. (1997). Hypoglycemic, antihyperglycemic and hypolipidemic activities of *Caesalpinia bonducella* seeds in rats. *Journal of Ethnopharmacology, 58,* 39–44.

Simin, K., Khaliq-uz-Zaman, S. M., & Ahmad, V. U. (2000). Antimicrobial activity of seed extract and bondenolide from *Caesalpinia bonduc. Phytotherapy Research, 15,* 437–440.

CHAPTER 68

Anti-platelet Aggregation Effect of the Amazonian Herb Guarana (*Paullinia cupana*) Seeds

M.T. Ravi Subbiah
Department of Internal Medicine, University of Cincinnati Medical Center,
Cincinnati, Ohio, USA

CHAPTER OUTLINE

LIST OF ABBREVIATIONS

ADP, adenosine diphosphate
HPLC, high pressure liquid chromatography
MS, mass spectrometry

INTRODUCTION

The herbal supplement industry represents a multi-billion dollar business in the United States and around the world. The seeds of guarana, originating from the Amazon Basin (Erickson *et al.*, 1984) represents one of the most commonly used herbal products. Guarana is approved as a food additive in the US and is widely consumed as a high-caffeine stimulant mixed with soft drinks, energy drinks, and, more recently, beer (Houghton, 1995). Despite its wide use, relatively little is known with regard to other active components of guarana seeds and their specific use for a given metabolic effect (Subbiah, 2005). In this chapter we will briefly review the metabolic and cellular effects of guarana seed extract, with particular focus on its ability to inhibit platelet aggregation, which has considerable clinical potential for use as a nutraceutical.

Nuts & Seeds in Health and Disease Prevention. DOI: 10.1016/B978-0-12-375688-6.10068-4

BOTANICAL DESCRIPTION

Guarana, *Paullinia cupana* (Family: Saponidaceae; Genus: *Paullinia*), is a woody spiny or sprawling shrub native to the Amazon Basin. The guarana fruit, which approximates the size of a coffee berry, is brown to red in color, with a black seed with white arils. The caffeine content of guarana seeds is twice that found in coffee beans. Synonyms include guarana, guaranastraik, quarana, and Brazilian cocoa.

HISTORICAL CULTIVATION AND USAGE

According to early records, guarana was used extensively by Guarani and Tupi Indians in the Amazon region. The seeds of guarana fruits are crushed and commonly used as a stimulant and for other health benefits. Guarana seeds are one of the richest natural sources of caffeine, which amounts to 2–5% of the dry weight of its seeds (Belliardo *et al.*, 1985). For some time there was considerable debate regarding whether the stimulatory substance in guarana seeds was actually caffeine, or a similar compound called "guaranine" (Erickson *et al.*, 1984). This was due to the notion that the effects of guarana seeds reflected a steady and sustained release of "guaranine," unlike the immediate high noted with caffeine derived from coffee. Guarana has been extensively used as a tonic, an astringent, and against gastrointestinal ailments in the Amazon region.

PRESENT-DAY CULTIVATION AND USAGE

Further studies showed that "guaranine" was actually caffeine, and no significant difference in its absorption rate when compared to caffeine was evident (Bempong *et al.*, 1991; Bempong & Houghton, 1992). Apart from caffeine, recent analysis has identified smaller amounts of other methyl xanthines (theophylline and theobromine) in guarana seeds (Belliardo *et al.*, 1985). Guarana seeds also contain significant amounts of carbohydrates, resins, and tannins (Marx, 1990). Other compounds, such as methyl benzenes, cyclic sesqui-terpene hydrocarbons, fatty acids, and commonly encountered plant sterols, have been isolated from the oil of guarana seeds (Subbiah, 2005). Since 1980, guarana cultivation has continued to thrive in the Amazon region with the help of cooperative projects under the umbrella of the FUNAI (Indian National Foundation), and there is great demand for it as a high-stimulant food additive in Europe and the USA.

APPLICATIONS TO HEALTH PROMOTION AND DISEASE PREVENTION

Although guarana has been in the nutritional supplement market for 40 years, limited knowledge exists with regard to its pharmacology and metabolic effects. The indigenous people of the Amazon Basin believe that guarana is a source of energy and helps maintain good health. Further research has unveiled a number of metabolic effects of guarana seed extract. In human volunteers aged 18–24 years, an intake of 75 mg of guarana or a guarana/ginseng (75 mg/200 mg) combination significantly improved cognitive performance (Kennedy *et al.*, 2004), strongly suggesting potential psychoactive properties of guarana. Haskell and colleagues (2007) carried out a systematic study of dosage-related effect of guarana seeds (37–300 mg) on acute mood and cognitive parameters in humans. They noted that lower doses of guarana (up to 75 mg) seeds produced far greater improvement in cognitive parameters than did higher doses, suggesting that the beneficial effects cannot be attributed solely to higher caffeine content. Recently there has been a great deal of interest in using herbal mixtures containing guarana seeds for weight loss. Boozer *et al.* (2001) noted that taking a herbal supplement with Ma huang/guarana resulted in a significant short-term drop in body weight and fat content in both sexes. Berube-Parent *et al.* (2005) observed significant increases in energy expenditure in men after the administration of guarana seed and green tea mixtures (with variable amounts of epigallocatechin gallate). Opala *et al.* (2006) examined the effect of a botanical extract (containing guarana, asparagus, black tea, yerba mate, and kidney beans) in healthy overweight subjects. While no changes in body weight were evident after 12 weeks, positive changes on the body composition improvement index due to greater loss of body fat and increased metabolic activity was observed. Based on the current literature, the role of guarana seeds in promoting weight loss is still inconclusive and deserves further study. Animal and *in vitro* studies on guarana seed extracts have generally provided support for its role in a number of metabolic and cellular activities including: (1) antidepressive activity; (2) stimulation of fat oxidation; (3) exhibition of chemoprotective effects, and (4) the ability to protect against DNA damage and lipid peroxidation, as reviewed previously (Subbiah, 2005).

Anti-platelet aggregation effect of guarana seeds

Of all the claims regarding the metabolic and health effects of guarana seeds, only the ability to inhibit platelet aggregation has been investigated to a greater degree. Studies performed almost two decades ago in our laboratory (Bydlowski *et al.*, 1988) showed that extracts of guarana seeds possessed strong inhibitory properties against ADP and arachidonic-acid induced platelet aggregation. Further studies also showed that the anti-platelet aggregatory property of guarana seeds was due to its ability to inhibit thromboxane synthesis (Bydlowski *et al.*, 1991). Recent studies have focused on determining the nature of compounds in guarana seeds responsible for this inhibitory activity (Subbiah & Yunker, 2008). Experiments were carried out by differential extraction of guarana seeds with various solvent combinations (Figure 68.1). These studies showed that extraction of guarana seeds with butanol did not preserve its platelet aggregation inhibitory activity, but chloroform : methanol (2 : 1, v/v) extraction to remove lipids had no effect, indicating butanol solubility of active components. Previous studies (Bydlowski *et al.*, 1988) had

571

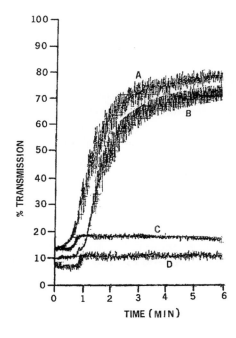

FIGURE 68.1

Effect of guarana seed fractions on response to human platelets to ADP. Response of human platelet-rich plasma to ADP (5 μg) and the effect of guarana fractions. A, control; B, guarana extract after butanol extraction; C, lipid-free guarana extract; D, guarana extract free of methyl xanthines. Experimental conditions as described. *Reproduced with permission from Subbiah & Yunker (2008),* Intl J. Vitam. Nutr Res., 78, 96—101.

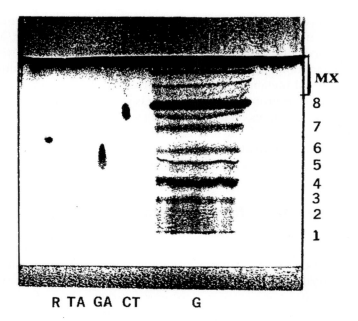

FIGURE 68.2

Thin-layer chromatographic separation of guarana seed extracts. Thin layer chromatographic separation of guarana extract on silica gel G. solvent system: chloroform : methanol : acetic acid (60 : 40 : 10, v/v/v). Standards: R, rutin, TA, tannic acid; GA, gallic acid; CT, catechins; G, guarana extract; MX, methyl xanthine area. The bands are numbered 1—8. *Reproduced from with permission from Subbiah & Yunker (2008),* Intl J. Vitam. Nutr Res., 78, 96—101.

572

shown that even after removal of methyl xanthine fractions (caffeine) by thin-layer chromatography of guarana seed extracts, platelet aggregation inhibitory activity still persisted. When total guarana seed extract was subjected to thin layer chromatography, about eight major bands and a large area containing methyl xanthine were noted (Figure 68.2). While some platelet aggregation inhibitory activity was evident in Bands 2 and 4, the bulk of the activity was associated with Band 8 (just below the caffeine-rich fraction, MX). HPLC/MS analysis of the purified Band 8 noted significant amounts of catechin, epicatechin along with caffeine, and other materials, eluting at 25—30 minutes' retention time, that ionized extensively (Figure 68.3). These compounds, at a retention time of 15 minutes, correspond to dimers of catechins (m/e 575). It is important to note that Band 8, even after repeated purification, still contained caffeine, along with catechin and epicatechin. Further subfractionation of Band 8 resulted in three bands (Bands A, B, and C) by thin layer chromatography. While Bands A and B contained catechin, epicatechin, and their dimers in different proportions, B and C predominantly contained caffeine. Bands A and B were strong inhibitors of platelet thromboxane synthesis, while Band C had a weak effect (Table 68.1).

ADVERSE EFFECTS AND REACTIONS (ALLERGIES AND TOXICITY)

In general, the available studies indicate no significant toxic effects of guarana seed extracts when administered alone to animals and humans (Oliveira *et al.*, 2005; Subbiah, 2005). However, strong adverse reactions have been reported when guarana seeds were co-administered with other selected herbs. In adult humans, taking a combination of ephedra-like alkaloids with guarana seeds prolonged the plasma half-life of ephedrine (Haller *et al.*, 2002). In some subjects, increases in systolic pressure and mean heart rate were noted 90 minutes after ingestion of a guarana seed/ephedrine combination (Baghkhani & Jafari, 2002). In epileptic patients, marked seizure exacerbation was noted upon consumption

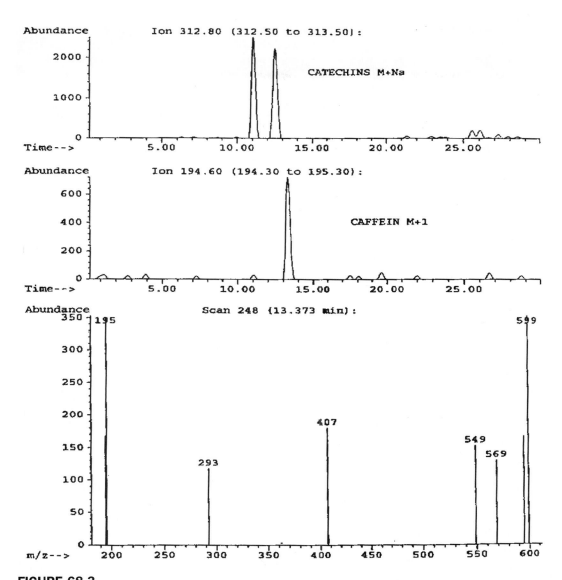

FIGURE 68.3

HPLC/MS analysis of Band B isolated from guarana seed extracts. HPLC/MS analysis of Band B using positive electrospray conditions. Note major ions representing molecular weights 312 (289 + 23 for Na) of catechin (first peak) and epicatechin (peak 2) in the top panel. In the middle panel, caffeine with a mass of 195 (mass + 1) is identified. The lower panel shows the presence of other ions notably at the 549—599 range, perhaps corresponding to dimers of catechins. *Reproduced from with permission from Subbiah & Yunker (2008), Intl J. Vitam. Nutr Res., 78, 96—101.*

of guarana with either ephedra or ma huang (Spinella, 2001). Other studies have reported episodes of premature ventricular contraction and cardiac arrhythmia, myoglobinuria, rhabdomyolysis, and increases in serum creative kinase in humans after consuming large amounts of guarana seed-containing supplements and health drinks, as reviewed recently (Subbiah, 2005). It is notable that alcoholic beverages (beer) containing guarana have been introduced in the USA. These products are aimed at counteracting the effect of alcohol with caffeine. Well-controlled studies in humans are needed to examine clearly any adverse side effects of the guarana/alcohol combination. This is especially important considering reports of cardiac arrhythmia and premature ventricular contraction following the consumption of health foods with high amounts of guarana seed extracts, as reviewed previously (Subbiah, 2005).

TABLE 68.1 Effect of Sub-fractions of Band 8 Isolated from Guarana Extracts on Prostaglandin E2 and Thromboxane B2 Synthesis in Washed Human Platelets[*]

Fraction	Thromboxane (dpm/mg protein)	B2 PGE2 (dpm/mg protein)
Control	13534 ± 474	1645 ± 498
Guarana fractions		
Band A	8201 ± 298[**]	1052 ± 331
Band B	2989 ± 534[**]	607 ± 82
Band C	9495 ± 738[**]	1894 ± 462

Reproduced with permission from Subbiah & Yunker (2008), *Intl J. Vitam. Nutr Res.*, 78, 96–101.
[*]Human platelets were pre-incubated with guarana fractions for 2 min at 37°C prior to the addition of 0.1 μCi of ^{14}C-arachidonic acid. After 10 min further incubation, the prostaglandins were extracted after the addition of authentic standards of PGE2 and thromboxane B2 as carriers, separated by TLC, and radioactivity was determined by scintillation counting.
[**]All data are presented as mean ± SEM, significantly different (P < 0.05) from control as estimated by one-way ANOVA with Benferroni's multiple comparison post test.

SUMMARY POINTS

- Guarana seed extracts possess a strong ability to inhibit human platelet aggregation in response to a variety of agents.
- Current research supports the concept that unique complexes of catechins, epicatechins, their dimers, and caffeine might be responsible for the anti-platelet aggregation properties of guarana seeds.
- The anti-platelet aggregation effect of guarana seeds is of interest because increased platelet aggregation and reactivity increases the risk of thrombosis, heart disease, and stroke, which is very important in young women taking oral contraceptives and older women on hormone replacement therapy (Subbiah, 2002).
- Active fractions isolated from guarana seeds might be of use as a nutraceutical in managing thrombosis, pending well-controlled *in vivo* studies in humans.
- Guarana seed extract also exhibits other properties related to health benefits, including improving cognition, stimulating fat oxidation and weight loss, and chemoprotective and antioxidant effects, that deserve further study.
- Guarana seed extract, when taken alone, has very few side effects, similar to those observed after consumption of products with a high caffeine content.
- It is strongly recommended that combining guarana seed with supplements containing Ma huang and ephedra-like products should be avoided.

References

Baghkhani, L., & Jafari, M. (2002). Cardiovascular adverse reactions associated with guarana: is this a causal effect? *Journal of Herbal Pharmacotherapy, 2,* 57–61.

Belliardo, F., Martelli, A., & Valle, M. G. (1985). HPLC determination of caffeine and theophyline in *Paullinia cupana* Kunth (guarana) and cola spp. samples. *Z Lebensm Unters Forrsch, 180,* 398–401.

Bempong, D., & Houghton, P. J. (1992). Dissolution and absorption of caffeine from guarana. *Journal of Pharmacy and Pharmacology, 44,* 769–771.

Bempong, D., Houghton, P. J., & Steadman, K. (1991). The caffeine content of guarana. *Journal of Pharmacy and Pharmacology, 43*(Suppl.), 125.

Berube-Parent, S., Pelletier, C., Dore, J., & Trembley, A. (2005). Effects of encapsulated green tea and guarana extracts containing a mixture of epigallocatechin-3-gallate and caffeine on 24 hr energy expenditure and fat oxidation in men. *British Journal of Nutrition, 94,* 432–436.

Boozer, C. N., Nasser, J. A., Heymsfield, S. B., Wang, V., Chen, G., & Soloman, J. L. (2001). A herbal supplement containing Ma Huang-Guarana for weight loss: a randomized double-blind trial. *International Journal of Obesity and Related Metabolic Disorders, 25,* 316–324.

Bydlowski, S. P., Yunker, R. L., & Subbiah, M. T. R. (1988). A novel property of aqueous guarana extract (*Paullinia cupana*): inhibition of platelet aggregation. *Brazilian Journal of Medical and Biological Research, 21*, 535–538.

Bydlowski, S. P., D'Amico, E. A., & Chamone, D. A. (1991). An aqueous extract of guarana (*Paullinia cupana*) decreases platelet thromboxane synthesis. *Brazilian Journal of Medical and Biological Research, 24*, 421–424.

Erickson, H. T., Correa, M. P. F., & Escobar, J. R. (1984). Guarana (*Paullinia cupana*) as a commercial crop in Brazil in the Amazon region. *Guarana Economic Botany, 38*, 273–286.

Haller, C. A., Jacob, P., & Benowitz, N. L. (2002). Pharmacology of ephedra alkaloids and caffeine after single dose dietary supplement use. *Clinical Pharmacology & Therapeutics, 71*, 421–432.

Haskell, C. F., Kennedy, D. O., Wesnes, K. A., Milne, A. L., & Scholey, A. B. (2007). A double-blind, placebo-controlled, multi-dose evaluation of the acute behavioral effects of guarana in humans. *Journal of Psychopharmacology, 21*, 65–70.

Houghton, P. (1995). Guarana. *Journal of Pharmaceutical Sciences, 254*, 435–436.

Kennedy, D. O., Haskell, C. F., Wesnes, K. A., & Scholey, A. B. (2004). Improved cognitive performance in human volunteers following administration of guarana (*Paullinia cupana*) extract: comparison and interaction with *Panex ginseng. Pharmacology Biochemistry and Behavior, 79*, 401–411.

Marx, F. (1990). Analysis of guarana seeds. II. Studies on the composition of tannin fraction. *Journal of Lebensm Unters Forsch, 190*, 429–431.

Oliveira, C. H., Moraes, M. E., Moraes, M. O., Bezerra, F. A., Abib, E., & De Nucci, G. (2005). Clinical toxicology study of an herbal medicinal extract of *Paullinia cupana, Trichilia catigua, Ptycopetalum olacoides* and *Zingiber officinale* in healthy volunteers. *Phytotherapy Research, 19*, 54–57.

Opala, T., Rzymski, P., Pischel, I., Wilczak, M., & Wozniak, J. (2006). Efficacy of 12 weeks supplementation of a botanical extract based weight loss formula on body weight, body composition and blood chemistry in healthy overweight subjects: a randomized double–blind placebo controlled clinical trial. *European Journal of Medical Research, 11*, 342–350.

Spinella, M. (2001). Herbal medicines and epilepsy: the potential for benefit and adverse effects. *Epilepsy & Behavior, 2*, 524–532.

Subbiah, M. T. R. (2002). Cardiovascular risk associated with estrogen replacement therapy: some critical questions about the type of estrogen used in most clinical trials. *Atherosclerosis, 173*, 373–374.

Subbiah, M. T. R. (2005). Guarana consumption. A review of health benefits and risks. *Alternative and Complementary Therapies, 4*, 212–213.

Subbiah, M. T. R., & Yunker, R. L. (2008). Studies on the nature of anti-platelet aggregatory factors in the seeds of the Amazonian herb guarana (*Paullinia cupana*). *International Journal of Vitamin and Nutrition Research, 78*, 96–101.

CHAPTER 69

Antibacterial Glycine-rich Peptide from Guava (*Psidium guajava*) Seeds

Patrícia Barbosa Pelegrini, Octavio Luiz Franco
Centro de Analises Proteomicas e Bioquímicas, Pós-Graduação em Ciências Genômicas e Biotecnologia, Universidade Católica de Brasília, Brasília-DF, Brazil

577

LIST OF ABBREVIATIONS

Ala, alanine amino acid residue
Arg, arginine amino acid residue
D1, distinctin protein from *Phyllomedusa distinct*
Glu, glutamic acid amino acid residue
GRP, glycine-rich protein
LTP, lipid transfer protein
Phe, phenylalanine amino acid residue
Pg-AMP1, *Psidium guajava* antimicrobial peptide 1
Pro, proline amino acid residue
PR-protein, pathogenesis-related protein
RIP, ribosome inactivating protein
SEP10, defensin peptide from *Pachyrrhizus erosus*
Ser, serine amino acid residue
TMV, tomato mosaic virus

INTRODUCTION

Plants are constantly exposed to a wide range of pathogenic organisms. Although they do not present an immune system, plants have, over the years, developed several defense mechanisms,

Nuts & Seeds in Health and Disease Prevention. DOI: 10.1016/B978-0-12-375688-6.10069-6

including the synthesis of proteins with antibacterial and antifungal properties (Bull *et al.*, 1992). Many proteins with antimicrobial activity have been described in recent years, and they are now classified according to their overall charge, molecular weight, and tertiary structure. Among the antimicrobial proteins already identified are the PR-1 proteins, (1,3) β-glucanases, chitinases, 2S albumins, chitin-binding proteins, defensins and thionins, thaumatin-like proteins, cyclophilin-like proteins, glycine-rich proteins, ribosome inactivating proteins (RIPs), lipid transfer proteins (LTPs), "killer" toxins, and enzyme inhibitors (Selitrennikoff, 2001). However, glycine-rich proteins (GRPs) are a new group of plant macromolecules with antimicrobial activity (Egorov *et al.*, 2005). They were first described as storage proteins, being used essentially as amino acid sources for the plant (Beyer *et al.*, 1977). Further studies revealed that genes encoding GRPs can be found in vascular tissues, especially the xylem, as well as in the cotyledon and other parts of the plant (Keller *et al.*, 1989; Ye & Ng, 1991; Ryser & Keller, 1992). The GRP family can be characterized by the high content of glycine residues in their amino acid sequences, but the concentration of this specific amino acid residue varies among different organisms (Fusaro & Sachetto-Martins, 2007).

Recently, GRPs have shown activity against plant pathogens. Previous studies revealed that eight proteins isolated from *Trititcum kiharae* seeds did not demonstrate activity against bacteria, but were able to inhibit the growth of filamentous fungi, such as *Helmintrhosporium sativum* and *Fusarium culmorum* (Egorov *et al.*, 2005). There are several reports describing the presence of GRPs related to plant defense. It has been observed that a binding-RNA GRP from *Nicotiana glutinosa* can also be related to plant–pathogen interaction, since the gene expression of this GRP increased after the plant was infected with TMV virus, indicating that the presence of the pathogen was important for the plant's production of this protein (Naqvi *et al.*, 2005). Another GRP from pumpkin, called cucurmoschin, has demonstrated antifungal activity against *Botrytis cinerea*, *F. oxysporum*, and *Mycosphaerella oxysporum* (Wang & Ng, 2003). Furthermore, although some GRPs are not considered antimicrobial molecules, they nevertheless present a chitin-binding domain, which enables them to inhibit the growth of filamentous fungi. This is the case of two proteins isolated from oat (*Avena sativa*) and *Ginkgo biloba* (Huang *et al.*, 2000).

BOTANICAL DESCRIPTION

Known as an important fruit tree, guava (*Psidium guajava* L.) belongs to the myrtle family (Myrtaceae). The guava tree is small (10 m) with spreading branches, and its young branches are quadrangular and downy. The leaves are extremely aromatic when compressed, evergreen, reverse, short-petioled, oval, and irregular in outline. The tree produces a flower with four or five white petals, which rapidly drop. The fruit, which has a soft green peel, exudes a strong sugary smell, and the central pulp, which can be red or yellowish-white, is juicy, and normally filled with hard, yellowish, unpalatable seeds.

HISTORICAL CULTIVATION AND USAGE

Guava has been cultivated for so many centuries that its place of origin is uncertain. However, it is known that the Egyptians grew it over a long period, and that it may have been transported from Egypt to Palestine and thence widely distributed throughout Asia and Africa. Years later, during the 15th century, this fruit arrived in the Bahamas and Bermuda, taken there by European colonizers, who had obtained it in the East Indies and Guam. Only during the 18th century was the guava reported in Florida, and today it can be found all over the tropical and subtropical Americas.

PRESENT-DAY CULTIVATION AND USAGE

Nowadays, several guava cultivars are planted around the world, including Red Land, Supreme, and Ruby. They can commonly be found in Latin America, India, and Africa. Guava trees are able to grow fast, and produce fruit between 2 and 4 years after seed germination. Trees aged 30–40 years have been found, but it is known that productivity declines around the fifteenth year. The guava tree is commonly considered a drought-tolerant plant; nevertheless, in regions with low humidity, lack of irrigation during the period of fruit development can reduce the size of the fruits. Ripe guavas bruise easily, and are very fragile. Fruits for processing may be harvested by automatic tree-shakers and plastic nets, but for fresh-fruit marketing and shipping, the fruits must be collected when full grown but under-ripe, and handled with immense care. After grading for size, fruits should be individually wrapped.

Guava fruits are commonly consumed *in natura*, but also in the form of jellies, juices, and ice-creams, and, in Brazil, a candy known as goiabada. Moreover, since guava is extremely rich in vitamins A, B, and C (180–300 mg of vitamin C per 100 g of fruit), showing higher concentrations than oranges and lemons, this fruit is highly appreciated by populations in developing countries both as a functional food and as a natural medicine.

APPLICATIONS FOR HEALTH PROMOTION AND DISEASE PREVENTION

Guava fruits are considered to be ethno-pharmacological medicines, and are commonly utilized to control diarrhea in children. Furthermore, tea made from the leaves of guava trees is used to reduce throat inflammation (Jaiarj *et al.*, 1999), and the aqueous extracts from guava buds are known for their intense activity toward *Salmonella*, *Serratia*, and *Staphylococcus*, which are responsible for diarrhea (Jaiarj *et al.*, 1999).

Recently, a study has identified a GRP with antimicrobial activity against gram-negative bacteria causing gastrointestinal infections in humans (Pelegrini *et al.*, 2008). This new peptide was isolated from guava seeds (*Psidium guajava*), and, unlike many members of the glycine-rich family, it is characterized by its low molecular weight and three-dimensional structure similar to antimicrobial peptides from several other families. Hence, Pg-AMP1, as it has been called, has about 6.0 kDa represented by a length of 55 amino acid residues. When Pg-AMP1 was assayed against gram-negative bacteria, it showed the ability to reduce the growth of *Proteus* and *Klebsiella* spp. by 30 and 90%, respectively, at a concentration of 6.5 μM. However, it did not show any antimicrobial activity against gram-positive bacteria, such as *Staphylococcus aureus*, revealing that Pg-AMP1 has specificity towards gram-negative microorganisms (Pelegrini *et al.*, 2008). Further studies with Pg-AMP1 were performed in order to analyze its structure display, as well as its mechanism of action, since it was the first GRP described to present activity toward pathogenic bacteria.

Therefore, a comparison of the primary sequence of Pg-AMP1 with another nine GRPs from several sources was carried out using BioEdit Software (Hall, 1999). In the alignment, it was observed that, although GRP amino acid sequences are highly diverse, they demonstrate conserved regions full of glycine residues (Figure 69.1). In addition, Pg-AMP1 seems to belong to the group of 3 GRPs, as it has a high content of glycine residues but no conserved domains. Proteins from *P. pattens*, *B. napus*, *S. tuberosum*, *M. sativa*, *R. obtusifolius*, and *E. esula* were also shown to be members of the group of 3 GRPs (Vermel *et al.*, 2002). Two peculiar conservations were also observed: two arginine residues at positions 50 and 89 were presented in most of the nine proteins analyzed, as well as a tyrosine residue at position 104. However, the presence of cysteine residues and disulfide bonds is not conserved between GRPs, which is an important feature of defense peptides from other families.

The tertiary structure of Pg-AMP1 is theoretically composed of two α-helices at the N-terminal and C-terminal regions, respectively, and a spiral grip between them (Figure 69.2A). A grip shape is found in most GRPs, which confers flexibility to the molecules. On the extreme outer edges of the α-helices there are arginine residues, providing a positive charge to these regions. Arg^7 and Arg^{45} form a water-mediated binding with Glu^{108} and Ser^{88}, respectively

```
Colicin E3   MSGGDGRGHNT------------------------------------GAHSTSGNING-GTGLGV----GGGASD-GS-GWSSENNPWGGSGSGI-HWGGSGHGN--------------G
Pg-AMP1      RESPSSRMECY-----------------------EQAERYGYG-GY----GGGR-----GGGYGG----YGGGYGSGR-------------------------------------G
BnGRP10      RETGRSRGFGFVTF--KDEKSMKDAIDEMNGKELDGRTITVNEAQSRGGG-GGGG-R---GGGYGG----RGGGGY-GGGGGGGYGSGR-R-------------------------
GRRBP2       RETGRSRGFGFVTF--MNSQELDGRNITVNEAQSRGSG-GGGG-----GGGYSR-GS-G----GGGGGGGYSRGGGGGGYGGGGR---------------------------
MsGRP        RETGRSRGFGFVTF--ANEKSMNDVIEAMNGQDLDGRNITVNQAQSRGSG-SGGG-----GGGYSR-GS-G----GGG-G-GRGGGG------------------------
PpGRP1       RETGRSRGFGFVTF---MNGKELDGRNITVNQAQSRGGG-SGGG-----GGGGYNN--------R----QGGGGG-------------------------------------
RoGRP        RETGRSRGFGFVTF--SSEQAMRDAIEGMNGQDLDGRNITVNEAQSRSG-GGGGGYRGGGGGGGYGG-RREG----GYNRGGGGY--GGGGGYGG-------------------
StGRP        RETGRSRGFGFVTF--KDEQAMRDAIEGMNGQDLDGRNITVNEAQSRGGG-GGGGGGR----REGGGGGY--GG-GGGYGGGREG--------------------------G
OsGRP        KPTGRSGVED-------QKWGGA------HGGGYGYG-GY----GGGYGYG-PG-Y-----GGGYGGGYGHPGYGGGYGGGYGQGYGGYGQGCGYGHPGHSGGYG
ZmGrp3       RELTEANGSGLKNNVKPAGEPGLKDE-KWFGGRYKHG---------GGYGNNQPGY-GGGGNSQP----GYGGGYKRHHPGGGYGSGQ
```

```
Colicin E3   GGNGNSGGGSGTGGNLSAVAAPVAFGFPALSTPGAGGLAVSISAGALSAAIADIMAALK
Pg-AMP1      QPVGQGVERSHDDN------------RNQRR
BnGRP10      GGGGYGSGGGGR-------------GGGG
GRRBP2       REGGYGGGGGGYNS-----------RSSGG
MsGRP        ----YGGG------------------GGG
PpGRP1       ----YG--------------------GGG
RoGRP        GGGGYGGGGGG-------------YGGG
StGRP        GGGGYSGGGGG-------------YGGG
OsGRP        GGGGYGGG-GGYGGG---------HGGGWP
ZmGrp3       GGPGCGC-GGGY-------------GGGN
```

FIGURE 69.1

Alignments of Pg-AMP1 with other GRPs from plant sources and Colicin E3 from *E. coli*. Bold residues correspond to the glycine-rich region; residues in black boxes correspond to the conserved RRM motif. Colicin E3, AAA88406; Pg-AMP1, P86030; BnGRP10, CAA78513; GRRBP2, AAC61786; MsGRP, CAA42554; PpGRP1, Nomata *et al.* (2004); RoGRP, Vermel *et al.* (2002); StGRP, CAA89058; OsGRP, CAA38315; ZmGrp3, ABB76252. Alignment was done using ClustalW Software (www. http://www.ebi.ac.uk/Tools/clustalw2/).

(A)

(B)

(C)

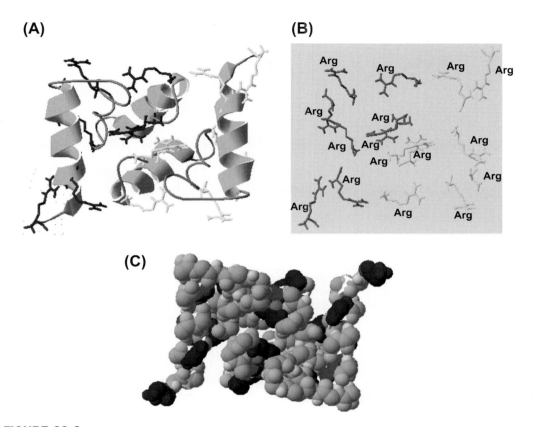

FIGURE 69.2
Structural analyses of Pg-AMP1. (A) Dimeric structure of Pg-AMP1 showing arginine residues (dark and light gray). (B) Diagram showing only arginine residues from Pg-AMP1 on its dimeric form. Residues in dark gray correspond to one of the monomers, while the residues in light gray correspond to the second monomer. (C) Dimeric structure of Pg-AMP1 showing positively-charged (dark gray) residues.

(Figure 69.2B). The other arginine residues seem to be important for interaction with the cell membrane, acting on antimicrobial activity. Some hydrophobic residues can also be observed along with the structure, which are important for dimer formation, as the loops of Pg-AMP1 are composed mainly of glycine residues (Pelegrini *et al.*, 2008).

Similar structures were found in other antimicrobial proteins, such as distinctin (D1) from *Phyllomedusa distincta*, which showed activity against *E. coli*, *P. aeruginosa*, and *Enterococcus faecalis* growth (Raimondo *et al.*, 2005). The action mechanism of this protein seems to involve the interaction of some amino acids with the lipidic layer of the cells, leading to an increase in permeation and cell death. The structure of D1 was obtained by nuclear magnetic resonance (NMR), showing two α-helices well stabilized when in the form of a homodimer (Raimondo *et al.*, 2005). Furthermore, a defensin isolated from *Pachyrrhizus erosus*, called SPE10, also showed similarities with Pg-AMP1. The antifungal activity of SPE10 against *A. flavus*, *A. niger*, and *Bipolaris maydis* seems to be mediated by the formation of a homodimer, which occurs at loop 2, between the amino acids Ala[7], Phe[10], Pro[13], Phe[15], and Phe[39] (Song *et al.*, 2005). It has been observed that the dimeric structure is also important for the antimicrobial activities of mammal defensins (Schibli *et al.*, 2002).

Therefore, the dimeric structure of Pg-AMP1 was analyzed by molecular dynamics in six different overlapped positions and compared between each of these, revealing that it remains in a stable conformation for long enough to interact with the membrane surface of the pathogenic cells (Pelegrini *et al.*, 2008). Hence, it was observed that, when exposed to an environment containing water molecules, Pg-AMP1 behaves in the same way for all six

(A)   **(B)**   **(C)**

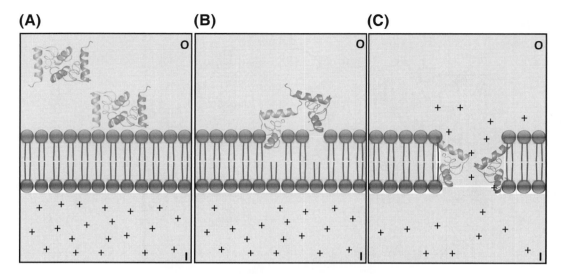

FIGURE 69.3
Probable Pg-AMP1 mechanism of action. (A) Interaction of dimeric Pg-AMP1 with the phospholipids from the membrane surface of the cell. (B) Pore formation. (C) Increase in membrane permeability, leading to cell death. O, outer environment; I, cytoplasmic environment.

different periods of dynamics. The three-dimensional structure of the Pg-AMP1 in dimer conformation is shown in Figures 69.2 and 69.3. Analyzing the electrostatic layer, two charged regions could be observed in the molecule: one positively-charged at the N-terminal, and one negatively-charged region at the C-terminal (Figure 69.2C). These regions are responsible for the dimer formation. In addition, there are non-polar residues along the structure, which seem to be important for the interaction between the two monomers of Pg-AMP1. The three-dimensional structure of Pg-AMP1 also demonstrated high similarity with Colicin E3, a protein isolated from *E. coli* and associated with antibiotic proteins that are toxic to bacteria (Soelaiman *et al.*, 2001). Domain A of Colicin E3 is composed of two α-helices that interact with the surface of cell membranes, playing a key role in its antimicrobial activity. Hence, the external α-helices from Pg-AMP1 also seem to interact in a similar way, being able to bind to the phospholipids in the cell membrane of the pathogens.

Thus, comparing Colicin E3 with Pg-AMP1 and observing the position of the arginine residues in the dimeric molecule, four mechanisms of action can be proposed for the guava antimicrobial peptide. The first hypothesis is based on the interaction of arginine residues presented at the external region of the dimer, with the negatively-charged side of the phospholipids at the membrane surface of the bacterial cell. This interaction would lead to the exposure of the hydrophobic region of the phospholipids, allowing the dimer to dissociate into two monomers and use its hydrophobic region to interact with the phospholipids. In this way, the peptides would be able to insert themselves in the cell cytoplasm, causing cell intoxication and, consequently, cell death (Jenssen *et al.*, 2006).

The second theory is based on the interaction of the dimer with sodium and potassium channels, blocking the passage of ions, which causes an osmotic misbalance. The positively charged external side of the dimeric structure of Pg-AMP1 would interact in the regions where it should be bound by Na^{2+} and K^{2+}, and lead to cell death (Jenssen *et al.*, 2006). The third hypothesis describes the interaction of arginine residues at the external part of the dimer with the negatively-charged region of the phospholipids of the cell membrane (Figure 69.3). Consequently, the solvent in the environment would facilitate dragging the phospholipids to a pore formation, leading to an increase in permeability and cell death (Jenssen *et al.*, 2006). The last theory is that, instead of intoxication or osmotic destabilization, the dimer would bind to RNA molecules, inactivating them and leading to cell death. In this case, Pg-AMP1 would work as an inhibitor of protein synthesis (Soelainman *et al.*, 2001).

Comparing the four mechanisms of action for Pg-AMP1 with Colicin E3 and other antimicrobial peptides that work in the form of a dimer, it is suggested that Pg-AMP1 would function as described by the third theory, above. Through comparative analyses of amino acid sequence and tertiary structure, as well as functional similarities, it seems that Pg-AMP1 might be able to bind with the surface of the cell membrane from the pathogen, leading to pore formation. Therefore, the formation of ion-permeable multimeric pores would be the main factor in cell death.

Hence, the discovery of novel GRPs with antimicrobial activity towards bacteria could provide, in the near future, new insights regarding their action mechanism. On the other hand, bactericidal proteins from guava seeds can be used as biotechnological molecules in the control of pathogens, by using genetic engineering tools. A new door has been opened for antimicrobial peptides; the use of storage proteins, whose antimicrobial activity was previously unknown, highlights the secondary function of these macromolecules in plant defense. It is probable that many other proteins already described in the literature can demonstrate this feature, but their antimicrobial activities have not yet been tested, so this function is still unknown. The hypothesis that storage proteins are static has, therefore, been overturned, since these molecules can present a variety of functions, some of which are still undiscovered.

ADVERSE EFFECTS AND REACTIONS (ALLERGIES AND TOXICITY)

To date, no adverse effects have been reported for *Psidium guajava* seeds. However, guava fruits have recently demonstrated a cardiac-depressant activity, and should be used with caution by those with heart diseases or using medication. Moreover, guava fruits have been seen to reduce blood sugar levels, and should be avoided by people with hypoglycemia.

SUMMARY POINTS

- Guava plants are found worldwide and are used medically to, for example, control diarrhea, reduce inflammation, and decrease antibacterial activity.
- A new antimicrobial peptide from guava seeds has been identified and characterized.
- Guava seed peptide belongs to the glycine-rich protein group, a family earlier described only as storage proteins.
- *In vitro* bioassays showed that *P. guajava* antimicrobial peptide was able to inhibit the growth of gram-negative bacteria causing gastrointestinal infections in human.
- The peptide herein described could be used, in the near future, as a novel tool in the treatment of infections caused by *Proteus* sp. and *Klebsiella* sp., also delaying the process of bacterial resistance toward currently used medicines.
- The action mechanism of *P. guajava* antimicrobial peptide seems to be by pore formation in the cell membrane when in a homodimer conformation.

References

Beyer, A. L., Christensen, M. E., Walker, B. W., & LeStourgeon, W. M. (1977). Identification and characterization of the packaging proteins of core 40S hnRNP particles. *Cell, 11*(1), 127–138.

Bull, J., Mauch, F., Hertig, C., Rebmann, G., & Dubler, R. (1992). Sequence and expression of a wheat gene that encodes a novel protein associated with pathogen defense. *Molecular Plant Microbe Interactions, 5*(6), 516–519.

Egorov, T. A., Odintsova, T. I., Pukhalsky, V. A., & Grishin, E. V. (2005). Diversity of wheat anti-microbial peptides. *Peptides, 26*, 2064–2073.

Fusaro, A. F., & Sachetto-Martins, G. (2007). Blooming time for plant glycine-rich proteins. *Plant Signalling & Behavior, 5*, 386–387.

Hall, T. A. (1999). BioEdit: a user-friendly biological sequence alignment editor and analysis program for Windows 95/98/NT. *Nucleic Acids Symposium Series, 41,* 95—98.

Huang, X., Xie, W., & Gong, Z. (2000). Characteristics and antifungal activity of a chitin binding protein from *Ginkgo biloba. FEBS Letters, 478,* 123—126.

Jaiarj, P., Khoohaswanb, P., Wongkrajang, Y., Peungvicha, P., Suriyawong, P., Saraya, M. L. S., et al. (1999). Anti-cough and antimicrobial activities of *Psidium guajava* Linn. leaf extract. *Journal of Ethnopharmacology, 67*(2), 203—212.

Jenssen, H., Hamill, P., & Hancock, R. E. W. (2006). Peptide antimicrobial agents. *Clinical Microbiology Reviews, 19* (23), 491—511.

Keller, B., Templeton, M. D., & Lamb, C. J. (1989). Specific localization of a plant cell wall glycine-rich protein in protoxylem cells of the vascular system. *Proceedings of the National Academy of Sciences, 86*(5), 1529—1533.

Naqvi, S. M. S., Harper, A., Carter, C., Ren, G., Guirfis, A., York, W. S., et al. (2005). Nectarin IV, a potent endo-glucanase inhibitor secreted into the nectar of ornamental tobacco plants. Isolation, cloning and character-ization. *Plant Physiology, 139,* 1389—1400.

Nomata, T., Kabeya, Y., & Sato, N. (2004). Cloning and characterization of glycine-rich RNA-binding protein cDNAs in the moss *Physcomitrella patens. Plant Cell Physiology, 45,* 48—56.

Pelegrini, P. B., Murad, A. M., Silva, L. P., dos Santos, R. C., Costa, F. T., Tagliari, P. D., et al. (2008). Identification of a novel storage glycine-rich peptide from guava (*Psidium guajava*) seeds with activity against Gram-negative bacteria. *Peptides, 29,* 1271—1279.

Raimondo, D., Andreotti, G., Saint, N., Amodeo, P., Renzone, G., Sanseverino, M., et al. (2005). A folding-dependent mechanism of antimicrobial peptide resistance to degradation unveiled by solution structure of distinctin. *Proceedings of the National Academy of Sciences, 102*(18), 6309—6314.

Ryser, U., & Keller, B. (1992). Ultra-structural localization of a bean glycine-rich protein in unlignified primary walls of protoxylem cells. *Plant Cell, 4*(7), 773—783.

Schibli, D. J., Hunter, H.-N., Aseyev, V., Dtarner, T. D., Wiencek, J. M., McCray, P. B., et al. (2002). The solution structures of the human beta-defensins lead to a better understanding of the portent bactericidal activity of HBD3 against *Staphylococcus aureus. Journal of Biological Chemistry, 277,* 8279—8289.

Selitrennikoff, C. P. (2001). Antifungal proteins. *Applied and Environmental Microbiology, 67,* 2883—2884.

Soelaiman, A., Jakes, K., Wu, N., Li, C., & Shonam, M. (2001). Crystal structure of colicin E3: implications for cell entry and ribosome inactivation. *Molecular Cell, 8,* 1053—1062.

Song, X., Wang, J., Wu, F., Li, X., Teng, M., & Gong, W. (2005). cDNA cloning, functional expression and antifungal activities of a dimeric plant defensin SPE10 from *Pachyrrhizus erosus* seeds. *Plant Molecular Biology, 57,* 13—20.

Vermel, M., Guermann, B., Delage, L., Grienenberger, J. M., Marechal-Drouard, L., & Gualberto, J. M. (2002). A family of RRM-type RNA-binding proteins specific to plant mitochondria. *Proceedings of the National Academy of Sciences, 99*(9), 5866—5871.

Wang, H. X., & Ng, T. B. (2003). Isolation of a novel antifungal peptide abundant in arginine, glutamate and glycine residues from black pumpkin seeds. *Peptides, 24,* 969—972.

Ye, X. Y., & Ng, T. B. (1991). Mungin, a novel cyclophilin-like antifungal protein from the mung bean. *Biochemical and Biophysical Research Communications, 273*(22), 1111—1115.

CHAPTER 70

Medicinal and Pharmacological Potential of Harmala (*Peganum harmala* L.) Seeds

Sarfaraz Khan Marwat[1], Fazal ur Rehman[2]
[1] University WENSAM College, Gomal University, Dera Ismail Khan, KPK, Pakistan
[2] Department of Pharmacognosy, Faculty of Pharmacy, Gomal University, Dera Ismail Khan, KPK, Pakistan

CHAPTER OUTLINE

LIST OF ABBREVIATIONS

CNS, central nervous system
DNA, deoxyribonucleic acid
GI, gastrointestinal
HIV, human immunodeficiency virus
MAO, monoamine oxidase
MAOI, monoamine oxidase inhibitor

INTRODUCTION

Harmala (*Peganum harmala* L.) is a medicinally important perennial herb of the family Zygophyllaceae (Goel *et al.*, 2009). It has various common names, including harmala, Syrian

Nuts & Seeds in Health and Disease Prevention. DOI: 10.1016/B978-0-12-375688-6.10070-2

rue, wild rue, harmal, and Africa rue. It is widely distributed throughout semi-arid areas of India, Pakistan, Iran, Africa, and Central Asian countries (Sultana, 1987), and has been introduced into America and Australia (Mahmoudian *et al.*, 2002). Nearly all parts of the plant are used in the traditional system of medicine, for the treatment of a number of diseases (Sultana, 1987). Medicinally, the fruits and seeds have digestive, diuretic, hallucinogenic, hypnotic, antipyretic, antispasmodic, nauseant, emetic, and uterine stimulant activities (Goel *et al.*, 2009). Various authors have described the possible therapeutic applications of the harmala seed alkaloids, such as protozoacidal agents and coronary dilators, embolic as well as for nervous diseases (Bukhari *et al.*, 2008).

BOTANICAL DESCRIPTION

Harmala (*Peganum harmala*) is a much branched, densely leafy, glabrous, perennial herb, with a spread of 120 cm and a height of up to 60 cm, and generally appears round and bushy in habit (Most, 1993) (Figure 70.1). Leaves are sessile, 4—8 cm long, irregularly and pinnatisectly dissected into 3—5 cm long, 2—3 mm broad, linear-lanceolate or sub-elliptic in shape, with acute segments. The flowers are white or yellowish white, 2—2.5 cm across; the filiform pedicle is 1.2 cm long; there are five linear sepals, often exceeding the petals, and five oblong or oblong—elliptic, subequal petals. There are generally 15 stamens, rarely less; the filament is 4—5 mm long, with the anther longer than the filament, and dorsifixed; the ovary is 8—10 mm long, the upper 6 mm being triangular or three-keeled. The fruit capsule is 6—10 mm across, trigonous, and depressed or retuse at the apex. Seeds are blackish-brown, triangular, and about 2 mm long. The harmala flowers between April and October (Nasir & Ali, 1974).

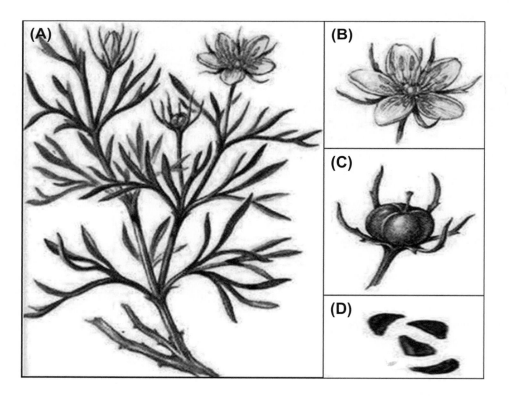

FIGURE 70.1

(A) Harmala (*Peganum harmala*) plant bearing flowers and fruit; (B) flower; (C) fruit; (D) seeds. *Source: Adapted from http://healthyhomegardening.com/Plant.php?pid=638.*

Chemical constituents

The pharmacologically active compounds of harmala are several alkaloids, which are found especially in the seeds and roots. The alkaloids present in the seeds include β-carbolines, such as harmine, harmaline, harmalol, harman (Mahmoudian *et al.*, 2002), harmalidine, *ruine*, and tetrahydroharmine (Khan, 1990), and the quinazoline derivatives vasicine (peganine) and vasicinone. The alkaloidal contents of unripe seeds are less than in ripe ones (Mahmoudian *et al.*, 2002). In addition to the above, some other important alkaloids, steroidal components, fatty acids, amino acids, and carbohydrates of harmala seeds have been reported (Table 70.1). The seeds also contain protein, lipids, and mineral elements (Khan, 1990).

TABLE 70.1 Chemical Constituents of Harmala Seeds

S. No.	Compound	Molecular Structure	Group	Reference(s)
1	Alanine $C_3H_7NO_2$	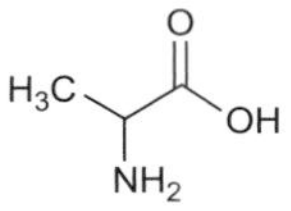	Amino acid	Khan, 1990
2	Arginine $C_6H_4N_4O_2$	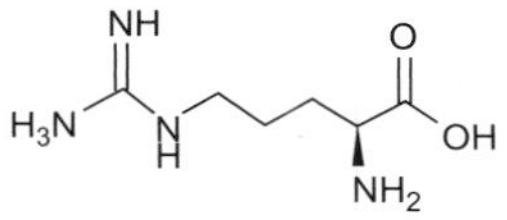	Amino acid	Khan, 1990
3	Aspartic acid $C_4H_7NO_4$	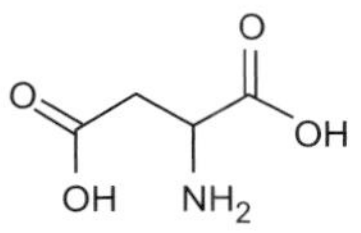	Amino acid	Khan, 1990
4	Behenic acid $C_{22}H_{44}O_2$	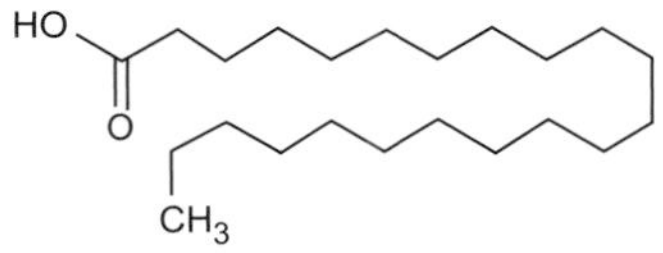	Fatty acid	Khan, 1990
5	Deoxypeganidine $C_{14}H_{16}N_2O$	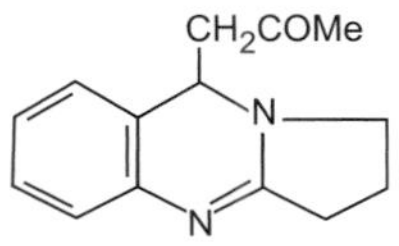	Alkaloid	Khan, 1990
6	Deoxipeganine $C_{11}H_{12}N_2$	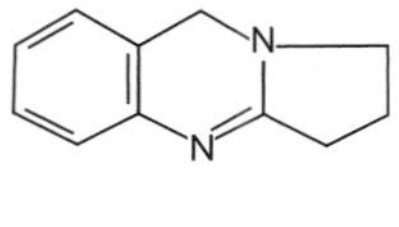	Alkaloid	Khan, 1990
7	Deoxivasicinone $C_{11}H_{10}N_2O$	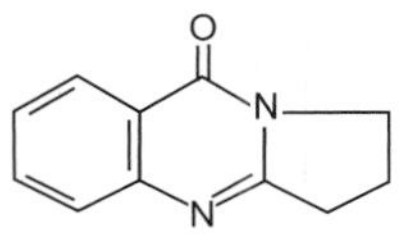	Alkaloid	Sultana, 1987; Khan, 1990

587

Continued

TABLE 70.1 Chemical Constituents of Harmala Seeds—continued

S. No.	Compound	Molecular Structure	Group	Reference(s)
8	Dihydroruine $C_{19}H_{24}N_2O_7$		Alkaloid	Sultana,1987; Khan, 1990
9	3,6-dihydroxy-8-methoxy-2-methyl anthraquinone		Anthraquinone	Pulpati *et al.*, 2008
10	Dipegine $C_{22}H_{20}N_4O$		Alkaloid	Sultana,1987; Khan, 1990
11	D-galactose $C_6H_{12}O_6$		Carbohydrate	Khan, 1990
12	D-glucose $C_6H_{12}O_6$		Carbohydrate	Khan, 1990
13	Glutamic acid $C_5H_9O_4$		Amino acid	Khan, 1990
14	Glycine $C_2H_5NO_2$		Amino acid	Khan, 1990
15	Harmacine $C_{12}H_{13}N_3O$		Alkaloid	Sultana, 1987
16	Harmalicine $C_{13}H_{14}N_2O_2$		Alkaloid	Sultana, 1987

TABLE 70.1   Chemical Constituents of Harmala Seeds—continued

S. No.	Compound	Molecular Structure	Group	Reference(s)
17	Harmalidine $C_{16}H_{18}N_2O$		Alkaloid	Sultana, 1987; Khan, 1992
18	Harmalacidine $C_{12}H_{12}N_2O_2$		Alkaloid	Sultana, 1987; Khan,1990
19	Harmalacinine $C_{16}H_{14}N_2O_2$		Alkaloid	Khan, 1990
20	Harmalanine $C_{16}H_{14}N_2O_2$		Alkaloid	Khan, 1990
21	Harmalicine $C_{13}H_{14}N_2O_2$		Alkaloid	Sultana, 1987; Pulpati et al., 2008
22	Harmaline $C_{13}H_{15}N_2O$		Alkaloid	Sultana, 1987; Khan, 1990;
23	Harmalol $C_{12}H_{12}N_2O$		Alkaloid	Khan, 1990
24	Harman $C_{12}H_{10}N_2$		Alkaloid	Khan, 1990

Continued

TABLE 70.1 Chemical Constituents of Harmala Seeds—continued

S. No.	Compound	Molecular Structure	Group	Reference(s)
25	Harmine $C_{13}H_{12}N_2O$		Alkaloid	Sultana, 1987; Khan, 1990
26	Harmol $C_{12}H_{10}N_2O$		Alkaloid	Sultana,1987; Khan, 1990
27	Histidine $C_9H_6N_2O_3$		Amino acid	Khan, 1990
28	5-hydroxytrypamine $C_{10}H_{12}N_2O$		Alkaloid	Sultana,1987; Khan, 1990
29	6-hydroxytrypamine $C_{10}H_{12}N_2O$		Alkaloid	Sultana,1987; Khan, 1990
30	Isoharmine $C_{13}H_{12}N_2O$		Alkaloid	Pulpati et al., 2008
31	Kaempferol $C_{15}H_{10}O_6$		Flavonoid	Pulpati et al., 2008
32	Kryptogenin $C_{27}H_{42}O_4$		Steroid	Khan, 1990

TABLE 70.1   Chemical Constituents of Harmala Seeds—continued

S. No.	Compound	Molecular Structure	Group	Reference(s)
33	Lanosterol $C_{30}H_{50}O$	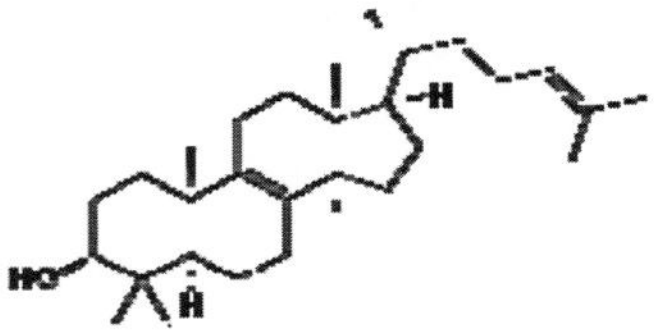	Steroid	Khan, 1990
34	Leucine $C_6H_{13}NO_2$	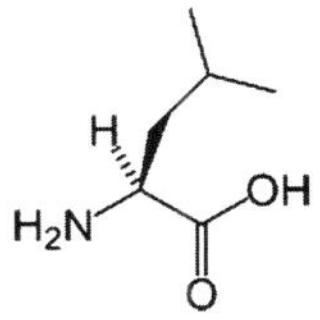	Amino acid	Khan, 1990
35	Linoleic acid $C_{18}H_{32}O_2$	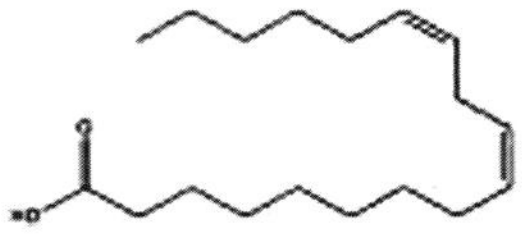	Fatty acid	Khan, 1990
36	Lysine $C_6H_{14}N_2O_2$	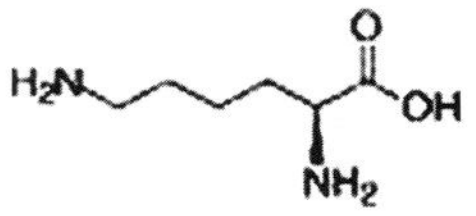	Amino acid	Khan, 1990
37	Norharmine $C_{12}H_{10}N_2O$	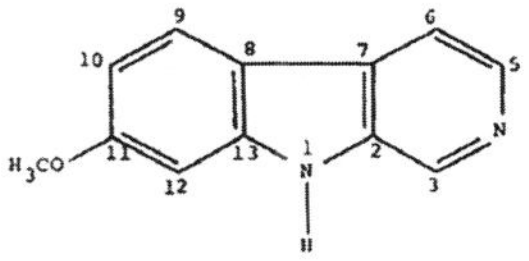	Alkaloid	Khan, 1990
38	Oleic acid $C_{18}H_{34}O_2$	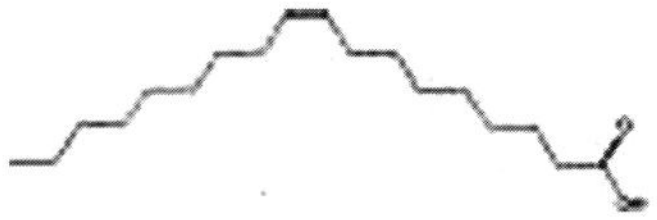	Fatty acid	Khan, 1990
39	8-Hydroxy-7-methoxy-2-methyl anthraquinone	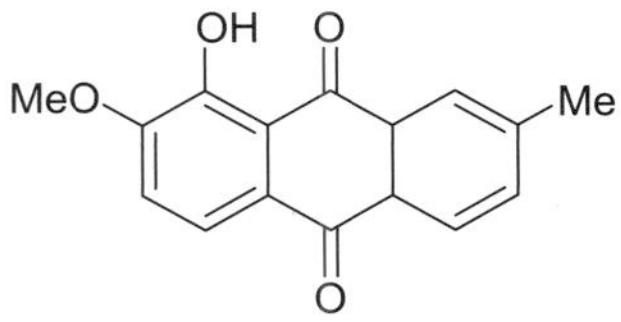	Anthraquinone	Pulpati *et al.*, 2008
40	Palmitic acid $C_{16}H_{32}O_2$		Fatty acid	Khan, 1990
41	Pegamine $C_{11}H_{12}N_2O$	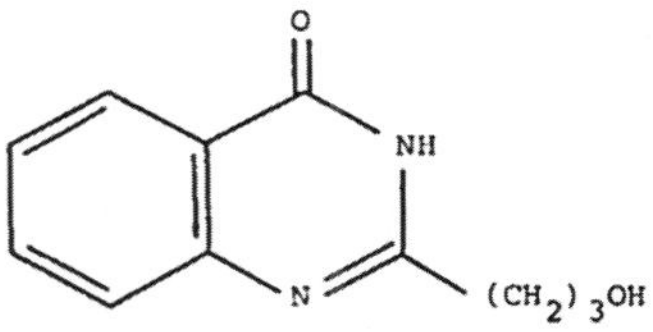	Alkaloid	Sultana, 1987; Khan, 1990

591

Continued

TABLE 70.1   Chemical Constituents of Harmala Seeds—continued

S. No.	Compound	Molecular Structure	Group	Reference(s)
42	Peganine (vasicine) $C_{11}H_{12}N_2O$		Alkaloid	Sultana, 1987; Khan, 1990
43	Peganidine $C_{14}H_{16}N_2O_2$		Alkaloid	Sultana, 1987; Khan, 1990
44	Peganole $C_{11}H_{12}N_2O$		Alkaloid	Sultana, 1987; Khan, 1990
45	Phenylalanine $C_9H_{11}NO_2$		Amino acid	Khan, 1990
46	Proline $C_5H_9NO_2$		Amino acid	Khan, 1990
47	Quercetin $C_{15}H_{10}O_7$		Flavonoid	Pulpati et al., 2008
48	Quinaldine $C_{10}H_9N$		Derivative of quinoline	Pulpati et al., 2008

TABLE 70.1   Chemical Constituents of Harmala Seeds—continued

S. No.	Compound	Molecular Structure	Group	Reference(s)
49	Quinoline C_9H_7N	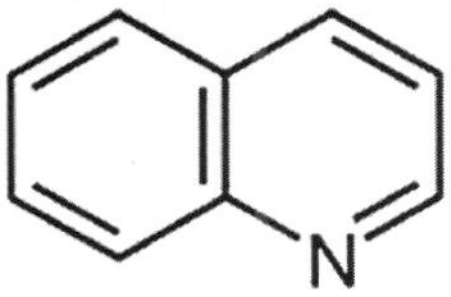	Aromatic nitrogen compound	Pulpati *et al.*, 2008
50	Ruine $C_{14}H_{22}N_2O_7$	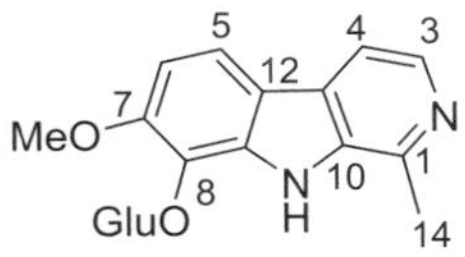	Alkaloid	Khan, 1990; Pulpati *et al.*, 2008
51	Serine $C_3H_7NO_3$	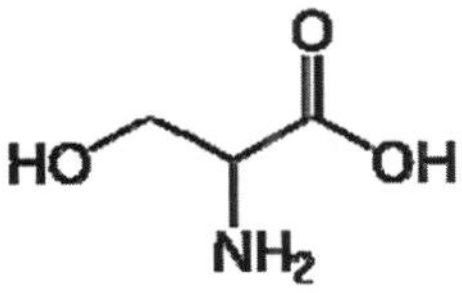	Amino acid	Khan, 1990
52	Stearic acid $C_{18}H_{36}O_2$		Fatty acid	Khan, 1990
53	β-sitosterol $C_{29}H_{50}O$	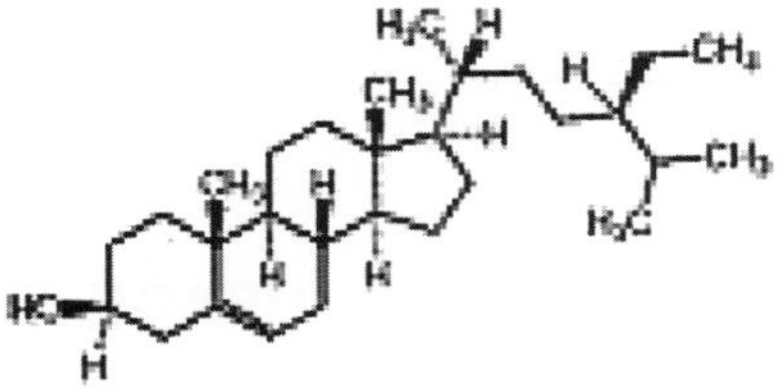	Steroid	Khan, 1990
54	Sucrose $C_{12}H_{22}O_{11}$	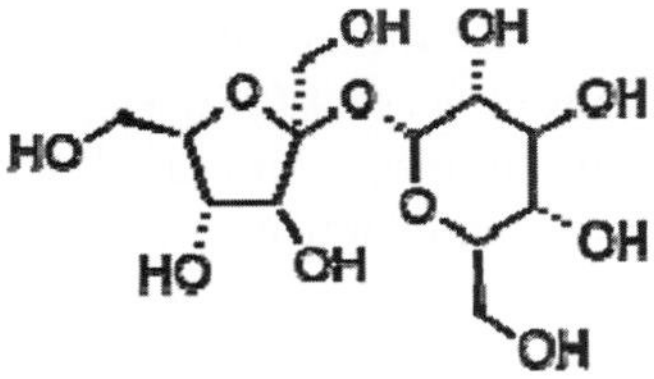	Carbohydrate	Khan, 1990
55	Tetrahydroharmine $C_{13}H_{16}N_2O$	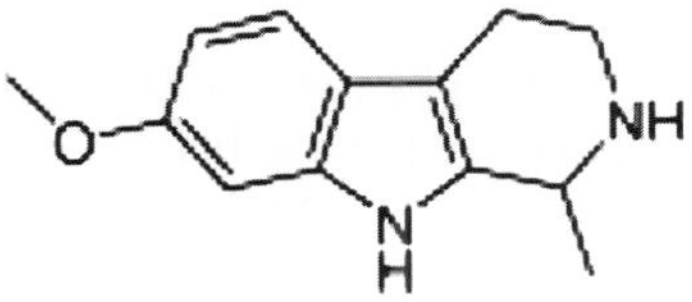	Alkaloid	Sultana, 1987; Khan, 1990
56	Threonine $C_4H_9NO_3$	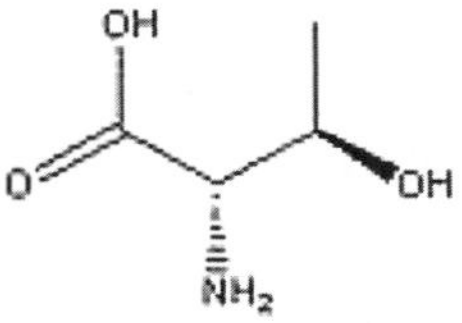	Amino acid	Khan, 1990

593

Continued

TABLE 70.1	Chemical Constituents of Harmala Seeds—continued			
S. No.	Compound	Molecular Structure	Group	Reference(s)
57	Tyrosine $C_9H_{11}NO_3$		Amino acid	Khan, 1990
58	Valine $C_5H_{11}NO_2$		Amino acid	Khan, 1990
59	Vasicinone $C_{11}H_{10}N_2O$		Alkaloid	Sultana, 1987; Khan, 1990; Pulpati et al., 2008

594

HISTORICAL CULTIVATION AND USAGE

Harmala (*Peganum harmala*) is a wild-growing herb. There are no historical references available regarding its cultivation. It has a long history of use as a psychoactive drug, and has been used as an enthogen in the Middle East for thousands of years. It was known to Dioscorides (40−90 AD), Galien (131−200 AD), and Avicenna (980−1037 AD). Harmala is thought to be the famous "Haoma," the sacred inebriating plant cited in an ancient Iranian religious text, the *Avesta*, in part attributed to Zoroaster (Zarathustra), and composed in the first millennium BC. The main economic value of the plant in historical times has been as the source of vegetable dyes extracted from the seeds, especially in Turkey, where the seeds are made into a red dye much employed for dying ornate rugs, although it has traditional ethno-medicinal uses as well. In Africa, harmala seeds have been used as incense since ancient times (Maya, 2009). Powdered seeds of harmala were used by the Ancient Greeks against tapeworms (Khan, 1990). From the very beginning, harmala seeds have been claimed to possess hypothermic and hallucinogenic properties (Monsef *et al.*, 2004).

PRESENT-DAY CULTIVATION AND USAGE

In 1935, harmala seeds were planted for the first time in the state of New Mexico, USA, by a farmer. Since then it has spread invasively to Arizona, California, Montana, Nevada, Oregon, Texas, and Washington (Most, 1993).

Harmala has been used as a traditional herbal remedy, mainly as an emmenagogue and an abortifacient agent. Its use in tribal rites has been reported. In Iran, Iraq, Tajikistan, Afghanistan, Pakistan, and parts of Turkey, dried seeds mixed with other ingredients are burnt to produce a scented smoke that is used as an air- as well as a mind-purifier. This

Persian practice seems to date to pre-Islamic Zoroastrian times (Frison *et al.*, 2008). The smoke of harmala seeds is traditionally used as a disinfectant (Arshad *et al.*, 2008). A red dye from seeds is widely used in Turkey and Iran for coloring carpets (Goel *et al.*, 2009). In modern Western culture, it is often used as an analog of *Banisteriopsis caapi* to create an *ad hoc* Ayahuasca (a psychoactive infusion or decoction prepared from *Banisteriopsis caapi*). Harmala, however, has distinct differences from *Banisteriopsis caapi*, and a unique enthogenic signature (Maya, 2009).

APPLICATIONS TO HEALTH PROMOTION AND DISEASE PREVENTION

Various parts of the harmala plant are highly renowned in the traditional systems of medicine for the treatment of a variety of human ailments, including lumbago, asthma, colic and jaundice, and for use as a stimulant emmenagogue (Sultana, 1987).

In the Indo-Pak subcontinent, harmala is used as a domestic medicine. Harmaline is reported to have narcotic, aphrodisiac, stimulant, sedative, vermifugal and soporific activities (Bukhari *et al.*, 2008).

The β-carboline alkaloids, harmine and harmaline, which show monoamine oxidase inhibition, are used as a psychoactive drug to treat Parkinson's disease (Astulla *et al.*, 2008). In Moroccan traditional medicine, harmala seeds have been used for the empiric treatment of cancers. Seeds are used in traditional medicine as an anthelmintic, lactogenic, antispasmodic, and emetic (Yousefi *et al.*, 2009). Powdered seeds of harmala are helpful in the treatment of remittent and intermittent fevers. They are also beneficial in chronic malaria, but are less effective in acute cases (Khan, 1990).

Harmala seeds are used as a galactogogue. They are also effective in painful and difficult menstruation, and for regulating menstruation. Here, a decoction of the seeds is given in 15—30 ml doses. The decoction is also useful as a gargle in laryngitis, and as a mouthwash (Panda, 1999).

The fruits and seeds of harmala are digestive, diuretic, antipyretic, antispasmodic, nauseant, and emetic (Goel *et al.*, 2009).

There are several reports in the literature indicating the great variety of pharmacological activities for harmala seeds (Table 70.2).

Abortifacient

It is believed that the quinazoline alkaloids (e.g., vasicine and vasicinone) of harmala seeds are responsible for their abortifacient activity. It has been reported that these chemicals have uterine stimulatory effect, apparently through the release of prostaglandins (Mahmoudian *et al.*, 2002).

Acetylcholinesterase inhibitor (AChEI) or Anti-cholinesterase

Cholinesterase inhibitors are used for the treatment of Alzheimer's disease. Deoxypeganine from harmala seeds has been reported to have activity similar to reversibly acting cholinesterase inhibitors. It inhibits acetylcholinesterase and monoamine oxidase, thereby preventing the degradation of acetylcholine and dopamine, and is therefore said to be helpful in treating Alzheimer's dementia (Singh, 2003).

Antileishmanial

Leishmaniasis has been identified as a major public health problem caused by *Leishmania* spp. Parasites of the genus *Leishmania* are transmitted by sandflies, which ingest the

TABLE 70.2 Biological and Pharmacological Activities of Active Compounds of Harmala Seeds

Trivial/formal Names of Compounds	Activities	References
Deoxypeganine: 1,2,3,9-Tetrahydropyrrolo(2,1-b) quinazoline	Toxic, anticholinesterase (acetylcholinesterase inhibitor), monoamine oxidase inhibitor	Singh, 2003
Harmaline: 4,9-Dihydro-7-methoxy-1-methyl-3H-pyrido [3,4-b]indole	Vasorelaxant, anticancer, hypothermic Antibacterial, antiprotozoal Antinociceptive Antiplasmodial Toxic, central nervous system (CNS) stimulator Human monoamine oxidase (MAOA) Inhibitor	Pulpati et al., 2008, Arshad et al., 2008 Monsef et al., 2004 Astulla et al., 2008 Frison et al., 2008 Herraiz et al., 2010
Harmalol: 1-Methyl-4,9-dihydro-3H-pyrido[3,4-b] indol-7-ol 3,4-dihydro, 7-hydroxy, 1-methyl-Bcs	Antibacterial, antiprotozoal	Arshad et al.,2008),
Harmine: 7-Methoxy-1-methyl-9H-pyrido[3,4-b] indole	Vasorelaxant, hypothermic, antinociceptive, anticancer, antileishmanial Antibacterial, antiprotozoal Antiplasmodial Toxic, CNS stimulator Human monoamine oxidase (MAO-A) Inhibitor	Pulpati et al., 2008), Arshad et al.,2008 Astulla et al.,2008 Frison et al., 2008 Herraiz et al., 2010
Tetrahydroharmine: 7-Methoxy-1,2,3,4-tetrahydro-harmine	CNS stimulator	Frison et al., 2008
Vasicine (peganine): (R)-1,2,3,9-Tetrahydropyrrolo(2,1-b) quinazolin-3-ol	Toxic	Frison et al., 2008
Vasicinone: (3S)-3-hydroxy-2,3-dihydro-1H-pyrrolo [2,1-b]quinazolin-9-one;	Anti-allergic, bronchodilator, thrombopoietic Anti-allergic Vasorelaxant Bronchodilator	Pulpati et al., 2008 Pulpati et al., 2008 Astulla et al., 2008 Mahmoudian et al., 2002

parasite in the amastigote stage residing in macrophages, then inoculate the promastigote stage into other hosts. It has been investigated that harmala seeds extract inhibits the growth of promastigote forms of *Leishmania major*, a parasite causing cutaneous Leishmaniasis (Yousefi *et al.*, 2009).

Antimicrobial

The antimicrobial potential of various extracts from harmala seeds has been investigated *in vitro* on multiple antibiotic-resistant pathogens (19 bacteria) and some selected protozoa (*Histomonas meleagridis, Tetratrichomonas gallinarum,* and *Blastocystis sp.*) isolated from poultry. Harmala seed extract was found to inhibit the growth of all bacteria and protozoa. The potential role of the known four β-carboline alkaloids in crude extracts of harmala was also investigated, for their antimicrobial activity. The activity of the pure alkaloids was in the order harmane > harmaline > harmalol > or = harmine for all bacteria, while for protozoa, it was different depending on the microorganism. It is concluded that harmala or its alkaloids could probably be used for the control of antibiotic-resistant isolates of bacteria as well as protozoa (Arshad *et al.*, 2008).

Antinociceptive

In one study, the effect of seed extract of harmala on formalin-induced pain response in mice was evaluated. Alkaloid extract was found to induce a significant reduction in the pain response. Harmaline — the last step in extraction — was the main effective antinociceptive agent of the harmala alkaloid extract (Monsef *et al.*, 2004).

Antiplasmodial and vasorelaxant activities

Malaria, caused by parasites of the genus *Plasmodium*, is one of the leading infectious diseases in many tropical and some temperate regions. Since harmine and harmaline have already been reported to have an inhibitory activity against some parasites, their inhibitory effect on the *Plasmodium* parasite was evaluated. Bioassay-guided purification from the seeds of harmala led to the isolation of harmine, harmaline, vasicinone, and deoxyvasicinone. Harmine and harmaline showed moderate *in vitro* antiplasmodial activity against *Plasmodium falciparum*. A quinazoline alkaloid, vasicinone, showed vasorelaxant activity against phenylephrine-induced contraction of isolated rat aorta (Astulla *et al.*, 2008).

Antitumor

The alkaloidic fraction of the methanol extract of harmala seeds was tested *in vitro* on three tumoral cell lines: UCP-Med and Med-mek carcinoma, and UCP-Med sarcoma. Proliferation was significantly reduced at all tested concentrations (20–120 µg/ml) during the first 24 hours of contact. A cell lysis effect occurred after 24 hours, and increased thereafter to complete cell death within 48–72 hours, depending on the tested concentration. Harmala alkaloids thus possess significant antitumor potential, which could prove useful as a novel anticancer therapy (Lamchouri *et al.*, 1999).

Antiviral

Harmala seed extract provides a new possibility for the treatment for viruses, and an entirely new method of blocking the viral life cycle. In assays of HIV inhibition, it induces a marked decrease in HIV activation, and shows promising therapeutic value. The seeds containing the active compounds could simply be ingested. The variety of compounds in these seeds inhibits HIV replication to varying degrees (Belzil, 2006).

Human monoamine oxidase inhibitors (MAOIs)

Harmala alkaloids are short-term monoamine oxidase inhibitors. An MAOI acts to inhibit a key enzyme, monoamine oxidase (MAO), in the human body, responsible for processes in the brain and throughout the body. The harmala alkaloids temporarily prevent biogenic amine from binding to the active site of the MAO molecule and undergoing deamination. The harmala alkaloids interfere with the protective enzyme MAO for 3–6 hours, before their action is reversed and MAO activity restricted (Most, 1993). Inhibition of MAOA by seed extracts has been quantitatively attributed to harmaline and harmine. The potent inhibition of MAOA by harmala seeds containing β-carbolines should contribute to the psychopharmacological and toxicological effects of this plant, and could be the basis for its purported antidepressant actions (Herraiz *et al.*, 2010).

ADVERSE EFFECTS AND REACTIONS (ALLERGIES AND TOXICITY)

All parts of the harmala plant are thought to be toxic, but toxicity is not severe. The β-carboline alkaloids harmine, harmaline, and tetrahydroharmine can stimulate the central nervous system by inhibiting the metabolism of amine neurotransmitters, or by direct interaction with specific receptors. The ingestion of harmala seed extract may result in toxic effects; namely, visual and auditory hallucinations, locomotor ataxia, nausea, vomiting, confusion, and agitation (Frison *et al.*, 2008).

Harmine induces a fall in blood pressure, chiefly due to weakening of cardiac muscle (Bukhari *et al.*, 2008). It also has depressant action on the CNS. In large toxic doses, it causes tremor and convulsions (Sultana, 1987).

A case report in humans has demonstrated that a 150-g oral dose of harmala seeds can result in severe gastrointestinal distress, including vomiting of blood, convulsions, and gastric ulcers (Mahmoudian *et al.*, 2002).

In another human case report, drinking 100 g of a harmala seed decoction resulted in unconsciousness, hypertension, tachycardia, and tachyhypnea, and elevation of hepatic and renal function markers. Poisoning with high doses of harmala can be life-threatening, although patients usually recover with supportive therapy alone (Yuruktumen *et al.*, 2008).

SUMMARY POINTS

- Harmala (*Peganum harmala*) is a medicinally important wild-growing flowering herb of the family Zygophyllaceae, which is widely distributed throughout semi-arid areas of India, Pakistan, Iran, Africa, Central Asian countries, America, and Australia.
- Harmala was known to Dioscorides (40–90 AD), Galien (131–200 AD), and Avicenna (980–1037 AD).
- The pharmacologically active compounds of harmala seeds are several alkaloids, including β-carbolines, such as harmine, harmaline, harmalol, harman and quinazoline derivatives, and vasicine (peganine) and vasicinone.
- Harmala seeds are used as an abortifacient, anthelmintic, antiseptic, aphrodisiac, and emmenagogue, as a galactogogue and diuretic, and as a remedy for painful and difficult menstruation, and for brain or nervous disorders.
- A red dye from seeds is widely used in Turkey and Iran for coloring carpets.
- All parts of the plant are thought to be toxic, but toxicity is not severe.
- Monoamine oxidase inhibitors act to inhibit a key enzyme in human body responsible for processes in the brain and throughout the body.

References

Arshad, N., Neubauer, C., Hasnain, S., & Hess, M. (2008). *Peganum harmala* can minimize *Escherichia coli* infection in poultry, but long-term feeding may induce side effects. *Poultry Sciences, 87*, 240–249.

Astulla, A., Zaima, K., Matsuno, Y., Hirasawa, Y., Ekasari, W., Widyawaruyanti, A., et al. (2008). Alkaloids from the seeds of *Peganum harmala* showing antiplasmodial and vasorelaxant activities. *Journal of Natural Medicines, 62*(4), 470–472.

Belzil, C. (2006). Therapeutic potential for inhibition of HIV activation. *Lethbridge Undergraduate Research Journal, 1*(2).

Bukhari, N., Choi, J. H., Jeon, C. W., Park, H. W., Kim, W. H., Khan, M. A., et al. (2008). Phytochemical studies of the alkaloids from *Peganum harmala*. *Applied Chemistry, 12*, 101–104.

Frison, G., Favretto, D., Zancanaro, F., Fazzin, G., & Ferrara, S. D. (2008). A case of beta-carboline alkaloid intoxication following ingestion of *Peganum harmala* seed extract. *Forensic Science International, 6*(179 2–3), 37–43.

Goel, N., Singh, N., & Saini, R. (2009). Efficient *in vitro* multiplication of Syrian rue (*Peganum harmala* L.) using 6-benzylaminopurine pre-conditioned seedling explants. *Nature Sciences, 7*(7), 129–134.

Herraiz, T., González, D., Ancín-Azpilicueta, C., Arán, V. J., & Guillén, H. (2010). β-Carboline alkaloids in *Peganum harmala* and inhibition of human monoamine oxidase (MAO). *Food and Chemical Toxicology, 48*(3), 839–845.

Khan, O. Y. (1990). Studies in the chemical constituents of *Peganum harmala* L. PhD Thesis, H.E.J. Research Institute of Chemistry University of Karachi, Pakistan, pp. 1–191.

Lamchouri, F., Settaf, A., Cherrah, Y., Hassar, M., Zemzami, M., Atif, N., et al. (2000). *In vitro* cell toxicity of *Peganum harmala* alkaloids on cancerous cell-lines. *Fitoterapia, 71*(1), 50–54.

Mahmoudian, M., Jalilpour, H., & Salehian, P. (2002). Toxicity of *Peganum harmala*. *Iranian Journal of Pharmacology & Therapeutics, 1*, 1–4.

Maya (2009). *Peganum harmala*. Maya ethnobotanicals.com. Available at: http://www.maya-ethnobotanicals.com/ buy Peganum_harmala_(accessed September 20, 2009).

Monsef, H. R., Ghobadi, A., Iranshahi., M., & Abdollahi, M. (2004). Antinociceptive effect of *Peganum harmala* L. alkaloid extract on mouse formalin test. *Journal of Pharmacy & Pharmceutical Sciences, 7*, 65–69.

Most, A. (1993). *Peganum harmala: The hallucinogenic herb of the American Southwest*. Plano, TX: Venom Press.

Nasir, E., & Ali, S. I. (1974). *Flora of Pakistan, Vol. 76*. Karachi, Pakistan: University of Karachi. 1–35.

Panda, H. (1999). *Herbs: cultivation and medicinal uses*. Delhi, India: National Institute of Industrial Research. 434–435.

Pulpati, H., Biradar, Y. S., & Rajani, M. (2008). High-performance thin-layer chromatography densitometric method for the quantification of harmine, harmaline, vasicine, and vasicinone in *Peganum harmala*. *Journal of AOAC International, 91*(5), 1179–1185.

Singh, A. P. (2003). *The role of natural products in pharmacotherapy of Alzheimer's disease.* Ethnobotanical Leaflets. Available at: http://www.ethnoleaflets.com/.

Sultana, N. (1987). Studies in the derivatives of harmine series of alkaloids and a reinvestigation of the chemical constituents of *Datura metel and Bryophyllum pinnatum*. PhD thesis. *H.E.J. Research Institute of Chemistry.* Karachi: University of Karachi. 1–170.

Yousefi, R., Ghaffarifar, F., & Asl, A. D. (2009). The effect of *Alkanna tincturia* and *Peganum harmala* extracts on *Leishmania major* (MRHO/IR/75/ER) *in vitro*. *Indian Journal of Parasitology, 4*, 40–47.

Yuruktumen, A., Karaduman, S., Bengi, F., & Fowler, J. (2008). Syrian rue tea: a recipe for disaster. *Clinical Toxicology (Philadelphia.), 46*(8), 749–752.

CHAPTER 71

Therapeutic Potential of Harmala (*Peganum harmala* L.) Seeds with an Array of Pharmacological Activities

Anuradha Dube[1], Pragya Misra[1], Tanvir Khaliq[2], Sriniwas Tiwari[2], Nikhil Kumar[3], Tadigoppula Narender[2]
[1] Division of Parasitology, Central Drug Research Institute, Lucknow, India
[2] Division of Medicinal & Process Chemistry, Central Drug Research Institute, Lucknow, India
[3] Betel Vine Biotechnology Laboratory, National Botanical research Institute, Lucknow, India

LIST OF ABBREVIATIONS

cAMP, cyclic adenosine monophosphate
cGMP, cyclic guanosine monophosphate

Nuts & Seeds in Health and Disease Prevention. DOI: 10.1016/B978-0-12-375688-6.10071-4

MAOI-A (RIMA), monoamine oxidase inhibitor-A (reversible inhibitor of monoamine oxidase)
NO, nitric oxide
Pgh, *Peganum harmala*
PLCγ2, phospholipase C-γ2

INTRODUCTION

Human awareness of plants and various plant parts (such as the root, stem, flowers, fruits, and seeds) as therapeutics dates back to prehistoric times. The use of ethnobotanical preparations for various ailments or diseases is still being practiced by various cultures worldwide. Considering the vast structural and biological diversity amongst terrestrial plants, they offer a unique renewable resource for the discovery of potential new drugs. Recent years have witnessed a shift towards a strategy for drug discovery, involving studies on plants and plant materials based on their ethnobotanical usage. This change, however, is still not adequately reflected in modern textbooks, which usually pay lesser attention to "traditional cultural drugs," and the average medical practitioner is always not aware of the usage and toxicity of these remedies, especially if such use is not part of their own cultural domain. In the era of globalization the disappearance of cultural boundaries is imminent, and it is very likely that a medical practitioner may visit a patient who has perhaps used a drug that is unknown to the attending practitioner. This can be illustrated by harmal (*Peganum harmala* L., family Zygo-phyllaceae), which is one such plant that is widely used as a drug in traditional systems in many countries, but has yet to find a place in textbooks.

This chapter provides an account of the pharmacological properties of *P. harmala* L. (*Pgh*) seeds, which are used in traditional medicine in regions where it occurs naturally.

BOTANICAL DESCRIPTION

Peganum harmala is a small, herbaceous plant (Figure 71.1A) with a bushy appearance. The plant is perennial, and can attain a height of 80 cm, though the common range is 30—50 cm. In mature plants, the stem is stiff, erect, highly branched, angled above, and glabrous. Leaves are alternate, fleshy, bright green, 2—5 cm long, and irregularly divided three times or more into linear segments. The plant flowers (Figure 71.1A) in late spring to early fall, and in India flowering has been reported from March to October. Flowers are bisexual, white or pale yellow in color, $\sim$2.5 cm in diameter, and solitary on stalks 2—5 cm long or more in the leaf axils. The fruits are spherical, leathery capsules (Figure 71.1B), 7—15 mm in diameter, green when unripe and orange-brown at maturity. Capsules have three chambers, which open by three valves at the apex to release numerous dark brown to black, angular, 3—4 mm long seeds (Figure 71.1C). The seeds are economically important, and have several uses (http://www.pfaf.org/database/plants.php/Peganum+harmala).

HISTORICAL CULTIVATION AND USAGE

Peganum harmala is native to the Middle East, Africa, and the Mediterranean, and it has been reported historically from Turkey, Iran, Iraq, Uzbekistan, Tajikistan, Russia, China, Mongolia, Afghanistan, India, and Pakistan. There are no reports of its cultivation in ancient times, and it was known to grow as invasive weed on poor soils. In Turkey, dried capsules from *Pgh* are used as a talisman to protect against the "evil eye." The use of fragrant smoke, produced by burning dried seeds mixed with other ingredients, in curing persons suffering from mental affliction is well known. In the Middle East it is used as an enthnogen, and in modern Western culture it is often used as an analog of *Banisteriopsis caapi*, to create an *ad hoc* Ayahuasca. Some scholars identify harmal with the enthnogenic haoma of pre-Zoroastrian Persian religions. There is speculation that this plant was considered to be sacred "Soma" plant, which was used by the ancients of Persia and India as a hallucinogenic aid

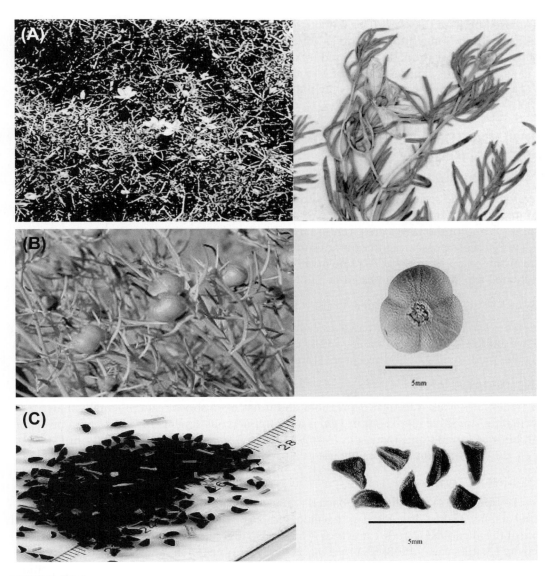

FIGURE 71.1
Plant of Peganum harmala. (A) Flower; (B) fruits: (C) seeds. The plant is a small herbaceous weed, with white or pale yellow flowers. Small, dark-colored triangular seeds are present in a three-chambered capsule. *(A) Reproduced courtesy of The Indian Institute of Integrative Medicine, Jammu, India. (B) and (C), reproduced from http://www.pfaf.org/database/plants.php/Peganum+harmala.*

to understanding the deeper meaning of life (http://www.pfaf.org/database/plants.php/Peganum+harmala).

PRESENT-DAY CULTIVATION AND USAGE

Although the plant is considered to be native to the Mediterranean and Central and Southwest Asia, it was also introduced into America and Australia in the beginning of the last century. Presently it is worldwide in distribution, except those regions where climatic conditions do not support its growth. It can tolerate temperatures as low as −20°C if the soil is dry, and grows very well on poor, dry soils. Due to its perennial nature and vigorous growth, it is invasive and has become an obnoxious weed. The bitter taste of the plant protects it against grazing, and thus contributes to its spread. The seeds are used as a spice and purifying agent. Some caution is advised because the seeds have narcotic properties, inducing a sense of euphoria and releasing inhibitions (http://www.pfaf.org/database/plants.php/Peganum+harmala). Ground *Pgh* seeds have occasionally been used in Morocco as a traditional treatment for skin cancer and subcutaneous cancers (Lamchouri *et al.*, 1999). The ancient Greeks used powdered *Pgh* seeds to treat recurring fevers (possibly malaria), and also to get rid of tapeworms (Panda, 2000).

APPLICATIONS TO HEALTH PROMOTION AND DISEASE PREVENTION

Traditional medicinal usages

Peganum harmala seeds are antispasmodic, hypnotic, antiperiodic, emetic, alterative, anthelmintic, and narcotic. Powdered seed is recommended as an anthelmintic, and decoctions of the seeds and of the leaves are given in laryngitis (Panda, 2000) and in rheumatism, respectively. *Pgh* seeds possess significant antitumor activity, having high cell toxicity and reducing the proliferation of tumoral cell line *in vitro*. Some other important uses that have been evaluated experimentally are listed in Tables 71.1 and 71.2.

ANALGESIC/ANTI-INFLAMMATORY EFFECT

Peganum harmala is used as an analgesic and anti-inflammatory agent, as its alkaloids exhibit active analgesic properties that act on the central and peripheral nervous systems and which may be mediated by opioid receptors (Farouk *et al.*, 2008).

ANTIDEPRESSANT

In Yemen, *Pgh* is used to treat depression, and it has been shown experimentally that harmaline, an active ingredient in *Pgh*, is a central nervous system stimulant and a reversible inhibitor of monoamine oxidase-A (MAO-A) (RIMA), a category of antidepressant (Monsef *et al.*, 2004), indicating its possible use as a psychoactive drug.

ANTI-INFECTIVE

Smoke from *Pgh* seeds kills algae, bacteria, intestinal parasites, and molds. *Pgh* has antibacterial activity against drug-resistant bacteria and protozoa (Arshad *et al.*, 2008). The potential role of the alkaloids in crude extracts of *Pgh* was also investigated regarding their antimicrobial activity (Astulla *et al.*, 2008; Khaliq *et al.*, 2009). Harmine (1) and harmaline (2) showed moderate *in vitro* antiplasmodial activity against *Plasmodium falciparum*. Vascinine (peganine) has been found to be safe and effective against *Leishmania donovani*, a protozoan parasite that can cause potentially fatal visceral leishmaniasis (Lala *et al.*, 2004; Khaliq *et al.*, 2009). Powdered seeds of *Pgh* have also been used as an anthelmintic agent (Panda, 2000).

ANTICANCER /ANTITUMOR ACTIVITY

Pgh has antioxidant and antimutagenic properties (Moura *et al.*, 2007). The β-carboline alkaloids present in *Pgh* have recently drawn attention due to their antitumor activities. They have shown effectiveness against various tumor cell lines both *in vitro* and *in vivo* (Jahaniani *et al.*, 2005).

EFFECT ON CARDIOVASCULAR ACTIVITY

Potent antiplatelet activity of *Pgh* was observed in harmane and harmine (Im *et al.*, 2009), which selectively affected collagen-induced platelet aggregation. This may be mediated by inhibiting PLCγ2 and protein tyrosine phosphorylation, with sequential suppression of cytosolic calcium mobilization and arachidonic acid liberation, indicating their potential as novel agents for atherothrombotic diseases (Berrougui *et al.*, 2006; Astulla *et al.*, 2008). Methanolic extract of *Pgh* seeds (MEP) also exerts a vasodilatory effect, and the main mechanism may be related to the inhibition of cAMP phosphodiesterase.

TABLE 71.1 Major Biological (Medicinal) Activities of Harmala Seeds, and Their Constituents

Biological Activities	Extract/Chemical Constituents	Remarks	Reference(s)
Analgesic/ antinociceptive	Seed extract Harmaline	Acts both centrally and peripherally; mediated by opioid receptors	Monsef *et al.*, 2004; Farouk *et al.*, 2008
Antidepressant	Seed extract Harmaline, harmine	Central nervous system stimulant and a "reversible inhibitor of MAO-A (RIMA)"	Monsef *et al.*, 2004
Anti-infective:			
Antibacterial and antifungal	Seed extract/ methanolic fraction Harmine, harmaline, harmalol	Against multiple antibiotic resistant pathogens — namely, *Staphylococcus aureus*, *Histomonas meleagridis, Tetratrichomonas gallinarum, Blastocystis* spp.; *Proteus vulgaris, Bacillus subtilis*, and *Candida albicans*	Prashanth & John, 1999; Arshad *et al.*, 2008; Moghadam *et al.*, 2010; Nenaah, 2010)
Antiprotozoal (antimalarial and antileishmanial)	Harmine Harmaline	*Plasmodium falciparum*: moderate *in vitro* antiplasmodial activity	Astulla *et al.*, 2008
	Peganine Harmine	*Leishmania donovani*: induce apoptosis in both the stages of *L. donovani* via loss of mitochondrial transmembrane potential	Lala *et al.*, 2004; Misra *et al.*, 2008; Khaliq *et al.*, 2009
Anthelmintic	Powdered seed extract	Cestodes	Panda, 2000
Anticancer/ antimutagenic	Methanolic seed extract and β-carboline alkaloids	High cell toxicity, and reduces the proliferation of tumoral cell lines *in vitro* and *in vivo* Antitumor activities by inhibiting DNA topoisomerases and interfering with DNA synthesis	Lamchouri *et al.*, 2000; Lamchouri *et al.*, 1999, 2000; Jahaniani *et al.*, 2005
Antioxidant	Harmine, harmaline	Antioxidant/scavenging or preventive capacity against free radicals as well as inhibiting the aggregation of the LDL protein moiety (apolipoprotein B) induced by oxidation	Berrougui *et al.*, 2006; Moura *et al.*, 2007
Effect on cardiovascular system:			
Antiplatelet	β-carboline alkaloids in particular, harmane, and harmine	Activities mediated by inhibiting PLCγ2 and protein tyrosine phosphorylation with sequential suppression of cytosolic calcium mobilization and arachidonic acid release	Im *et al.*, 2009
Vasorelaxant	Methanolic extract of seeds (MEP) Harmine, harmaline and harmalol	Vasodilatory activity related to the inhibition of cAMP phosphodiesterase; effects of harmine and harmaline are attributed to their actions on the endothelial cells to release NO and on the vascular smooth muscles to inhibit the contractions induced by the activation of receptor-linked and voltage-dependent Ca^{2+} channels	Berrougui *et al.*, 2006; Astulla *et al.*, 2008

605

TABLE 71.2 The Pharmacologically Active Alkaloids and Other Chemical Constituents of Harmala Seeds

Active Chemical Constituents		Effects	Test System	Reference
β-carboline alkaloids	Harmine or banisterine $C_{13}H_{12}N_2O$	Inhibition of actin dynamics	NIH 3T3 EF tumor cells extract	Patent: EP2050747 A1, 2009
	Harmane $C_{12}H_{10}N_2$	Effect on transmembrane potential	Adrenal pheochromocytoma PC12 cells of rat	Yang et al., 2008
	Harmalol $C_{12}H_{12}N_2O$	Effect on ventricular repolarization	Heart	Patent:WO2007/ 137380, 2007
	Harmaline (harmidine) $C_{13}H_{14}N_2O$	Inhibition of enzyme activity	Trypanothione reductase of Trypanosoma cruzi	Galarreta et al., 2008
	Harmalacidine $C_{12}H_{12}N_2O_2$	Cytotoxic	Solid tumor KB cell line (epithelioma of the nasopharynx) of human	Patent: US2008/ 69899, 2008
	Harmol $C_{12}H_{10}N_2O$	Modulation of actin dynamics	NIH 3T3 EF tumor cells extract	Patent: EP2050747 A1, 2009
	Isoharmine $C_{13}H_{12}N_2O$	Cytotoxic	Human embryonic kidney 293 cells	Wernicke et al., 2007
	Norharmin or norharmine $C_{12}H_{10}N_2O$	Inhibition of enzyme activity	IκBα-kinase complex from HeLa S3 cells	Patent: EP1209158 A1, 2002
Quinazoline alkaloids	Deoxyvasicinone or oxypeganine (quinazoline alkaloid) $C_{11}H_{10}N_2O$	Bronchodilatoric	Guinea pig trachea	Glasby, 1978
	Vasicinone $C_{11}H_{10}N_2O_2$	Cytotoxic	Human umbilical vein endothelial cells	Zhang et al., 2000
	Vasicine or Peganine $C_{11}H_{12}N_2O$	Antileishmanial	Promastigotes of Leishmania donovani genetically modified/ infected with: green fluorescent protein gene	Khaliq et al., 2009
		Cytotoxic	Human umbilical vein endothelial cells	Zhang et al., 2000
Sterols	Kryptogenin $C_{27}H_{42}O_4$	Antiproliferative	HeLa cells	Liu et al., 2007
	Lanosterol $C_{30}H_{50}O$	Cytotoxic	Human carcinoma lines	Chung et al., 2010
		Increase of utrophin protein level	Duchenne muscular dystrophy-patient derived muscle cells of human	Patent:WO2006/7910 A1, 2006
	β-sitosterol $C_{29}H_{50}O$	Anti-cancer	Breast, colon, and prostate cancer	Park et al., 2007
Flavonoids	Kaempferol $C_{15}H_{10}O_6$	Activation of peroxisome proliferator-activated receptor-γ (PPAR-γ)	Human embryonic kidney (HEK) 293 cells transfected with tK-PPREx3-Luc plasmid, pSG5-hPPAR-γ plasmid	Patent: WO2009/ 26657; 2009
	Quercetin $C_{15}H_{10}O_7$	Inhibition of enzyme activity	Erythrocyte calpain I of human	Kim et al., 2009
Amino acid	4-hydroxy-pipecolic acid $C_6H_{11}NO_3$	Inhibition of TNF-α converting enzyme (TACE)	TNF-α, pro-inflammatory cytokine from human macrophages	Letavic et al., 2002

Pharmacologically active constituents of harmala seeds

Analysis of chemical constituents of different species of *Peganum* revealed the presence of alkaloids, alkaloid glycosides, anthraquinones, anthraquinone glycosides, oxamides, flavonoids, flavonoid glycosides, steroids, triterpenoids, carbohydrates, amino acids, and the volatile constituents (essential oils). Major alkaloids found in seeds (2–7% of total dry weight) are localized in the seed coat, comprising large amounts of harmine. The alkaloidal content of unripe seeds is less than ripe seeds. The pharmacologically active compounds of *Pgh* are several alkaloids, which are found especially in the seeds. These include the β-carbolines harmine, harmaline (identical with harmidine), harmalol, and harman, and the quinazoline derivatives vasicine, vasicinone, peganol, peganidine, and deoxypeganine. The active alkaloids of harmal seeds are the MAOI-A (monoamine oxidase inhibitor A) compounds. Harmine and harmaline are reversible inhibitors of MAO-A (RIMA). Harmine inhibits phosphodiesterase and increases cAMP and cGMP levels, thus reducing the level of free radicals. It possesses cytotoxic and genotoxic activity, helpful in curing tumor cells. Harmaline is a potent antinociceptive agent. It shows a significant vasorelaxant effect, and inhibits phosphodiesterase. Experimental pharmacological evidences and characteristics of the active compounds are summarized in Table 72.2.

ADVERSE EFFECTS AND REACTIONS (ALLERGIES AND TOXICITY)

While *P. harmala* has been traditionally used in Bedouin medicine as an emmenagogue and an abortifacient agent (Saha & Kasinathan, 1961), there are few reports on its human toxic effects and syndromes. The ingestion of plant preparations containing the β-carboline alkaloids harmine, harmaline, and tetrahydroharmine may result in toxic effects, namely visual and auditory hallucinations, locomotor ataxia, nausea, vomiting, confusion, and agitation. β-Carboline alkaloids can stimulate the central nervous system by inhibiting the metabolism of amine neurotransmitters, or by direct interaction with specific receptor. Frison *et al.* (2008) reported the first case of *P. harmala* intoxication corroborated by toxicological findings following intentional ingestion of a *P. harmala* seed infusion. Symptoms of toxicity mainly consist of neurosensorial symptoms, hallucinations, a slight elevation in body temperature, and cardiovascular disorders such as bradycardia and low blood pressure; however, the signs of intoxication disappeared a few hours after ingestion.

607

SUMMARY POINTS

- Harmala (*Peganum harmala* L.; *Pgh*) seeds are associated with the culture of the Middle East and Mediterranean, and are also a part of widely used traditional and ethnobotanical system of plant medicine.
- Harmala grows in semi-arid and subtropical regions, on soils low in nutrients and water. Due to its perennial nature, *Pgh* is an invasive weed, currently worldwide in distribution.
- The seed is the main part used as a traditional medicine for mental afflictions, as a recreational drug, and as a cure for skin cancer, fever, tape worm, etc.
- The plant contains a number of bioactive compounds, mainly alkaloids. So far, about 34 compounds have been purified from harmala, of which 20 are alkaloids and others are sterols, anthraquinones, flavonoids, triterpenes, oxamide, amino acids, and fatty acids.
- In recent years, an array of pharmacological activities, such as analgesic, antidepressant, anti-infective, anticancer, antioxidant, antiplatelet, and vasorelaxant, have been demonstrated; these are mainly due to the different alkaloids present in harmala seeds.
- Although harmala has traditionally been used as an abortifacient agent, there are few reports regarding its human toxic effects and syndromes. Symptoms of *Pgh* toxicity mainly consist of neurosensorial symptoms, hallucinations, a slight elevation of body temperature, and cardiovascular disorders such as bradycardia and low blood pressure; however, signs of intoxication disappear a few hours after ingestion.

ACKNOWLEDGEMENTS

We are grateful to Dr R.A. Vishwakarma, Director, Indian Institute of Integrative Medicine, (CSIR), Jammu Tawi, India, for providing the photograph of a herbarium specimen of *P. harmala* with pale yellow flowers.

References

Arshad, N., Zitterl-Eglseer, K., Hasnain, S., & Hess, M. (2008). Effect of *Peganum harmala* or its beta-carboline alkaloids on certain antibiotic resistant strains of bacteria and protozoa from poultry. *Phytotherapy Research, 22,* 1533–1538.

Astulla, A., Zaima, K., Matsuno, Y., Hirasawa, Y., Ekasari, W., Widyawaruyanti, A., et al. (2008). Alkaloids from the seeds of *Peganum harmala* showing antiplasmodial and vasorelaxant activities. *Journal of Natural Medicines, 62,* 470–472.

Berrougui, H., Martin-Cordero, C., Khalil, A., Hmamouchi, M., Ettaib, A., Marhuenda, E., et al. (2006). Vasorelaxant effects of harmine and harmaline extracted from *Peganum harmala* L. seeds in isolated rat aorta. *Pharmacological Research, 54,* 150–157.

Choi, J. Y., Na, M., Hyun Hwang, I., Ho Lee, S., Young Bae, E., Yeon Kim, B., et al. (2009). Isolation of betulinic acid, its methyl ester and guaiane sesquiterpenoids with protein tyrosine phosphatase 1B inhibitory activity from the roots of *Saussurea lappa* C.B.Clarke. *Molecules, 14,* 266–272.

Chung, M. J., Chung, C. K., Jeong, Y., & Ham, S. S. (2010). Anticancer activity of subfractions containing pure compounds of Chaga mushroom (*Inonotus obliquus*) extract in human cancer cells and in Balbc/c mice bearing Sarcoma-180 cells. *Nutrition Research and Practice, 4,* 177–182.

Farouk, L., Laroubi, A., Aboufatima, R., Benharref, A., & Chait, A. (2008). Evaluation of the analgesic effect of alkaloid extract of *Peganum harmala* L.: possible mechanisms involved. *Journal of Ethnopharmacology, 115,* 449–454.

Frison, G., Favretto, D., Zancanaro, F., Fazzin, G., & Ferrara, S. D. (2008). A case of beta-carboline alkaloid intoxication following ingestion of *Peganum harmala* seed extract. *Forensic Science International, 179,* e37–43.

Galarreta, B. C., Sifuentes, R., Carrillo, A. K., Sanchez, L., Amado Mdel, R., & Maruenda, H. (2008). The use of natural product scaffolds as leads in the search for trypanothione reductase inhibitors. *Bioorganic & Medicinal Chemistry, 1–6,* 6689–6695.

Glasby, J. S. (1978). *Encyclopedia of the alkaloids.* London, UK: Plenum Press. 1367.

Im, J. H., Jin, Y. R., Lee, J. J., Yu, J. Y., Han, X. H., Im, S. H., et al. (2009). Antiplatelet activity of beta-carboline alkaloids from *Peganum harmala*: a possible mechanism through inhibiting PLCgamma2 phosphorylation. *Vascular Pharmacology, 50,* 147–152.

Jahaniani, F., Ebrahimi, S. A., Rahbar-Roshandel, N., & Mahmoudian, M. (2005). Xanthomicrol is the main cytotoxic component of *Dracocephalum kotschyii* and a potential anti-cancer agent. *Phytochemistry, 66,* 1581–1592.

Khaliq, T., Misra, P., Gupta, S., Reddy, K. P., Kant, R., Maulik, P. R., et al. (2009). Peganine hydrochloride dihydrate an orally active antileishmanial agent. *Bioorganic & Medicinal Chemistry Letters, 19,* 2585–2586.

Kim, H. J., Lee, J. Y., Kim, S. M., Park, D. A., Jin, C., Hong, S. P., & Lee, Y. S. (2009). A new epicatechin gallate and calpain inhibitory activity from *Orostachys japonicus*. *Fitoterapia, 80,* 73–76.

Lala, S., Pramanick, S., Mukhopadhyay, S., Bandyopadhyay, S., & Basu, M. K. (2004). Harmine: evaluation of its antileishmanial properties in various vesicular delivery systems. *Journal of Drug Target, 12,* 165–175.

Lamchouri, F., Settaf, A., Cherrah, Y., Zemzami, M., Lyoussi, B., Zaid, A., et al. (1999). Antitumour principles from *Peganum harmala* seeds. *Therapie, 54,* 753–758.

Lamchouri, F., Settaf, A., Cherrah, Y., Hassar, M., Zemzami, M., Atif, N., et al. (2000). *In vitro* cell-toxicity of *Peganum harmala* alkaloids on cancerous cell-lines. *Fitoterapia, 71,* 50–54.

Letavic, M. A., Axt, M. Z., Barberia, J. T., Carty, T. J., Danley, D. E., Geoghegan, K. F., et al. (2002). Synthesis and biological activity of selective pipecolic acid-based TNF-alpha converting enzyme (TACE) inhibitors. *Bioorganic & Medicinal Chemistry Letters, 12,* 1387–1390.

Liu, Y., Zhao, D. M., Lu, X. H., Wang, H., Chen, H., Ke, Y., et al. (2007). Synthesis of bisdesmosidic kryptogenyl saponins using the 'random glycosylation' strategy and evaluation of their antitumor activity. *Bioorganic & Medicinal Chemistry Letters, 17,* 156–160.

Misra, P., Khaliq, T., Dixit, A., SenGupta, S., Samant, M., Kumari, S., et al. (2008). Antileishmanial activity mediated by apoptosis and structure-based target study of peganine hydrochloride dihydrate: an approach for rational drug design. *Journal of Antimicrobial Chemotherapy, 62,* 998–1002.

Moghadam, M. S., Maleki, S., Darabpour, E., Motamedi, H., & Nejad, S. M. S. (2010). Antibacterial activity of eight Iranian plant extracts against methicillin and cefixime restistant *Staphylococcous aureus* strains. *Asian Pacific Journal of Tropical Medicine, 3,* 262–265.

Molnar, T., Visy, J., Simon, A., Moldvai, I., Temesvari-Major, E., Dornyei, G., et al. (2008). Validation of high-affinity binding sites for succinic acid through distinguishable binding of gamma-hydroxybutyric acid receptor-specific NCS 382 antipodes. *Bioorganic & Medicinal Chemistry Letters, 18,* 6290–6292.

Monsef, H. R., Ghobadi, A., Iranshahi, M., & Abdollahi, M. (2004). Antinociceptive effects of *Peganum harmala* L. alkaloid extract on mouse formalin test. *Journal of Pharmacy & Pharmaceutical Sciences, 7,* 65–69.

Moura, D. J., Richter, M. F., Boeira, J. M., Pegas Henriques, J. A., & Saffi, J. (2007). Antioxidant properties of beta-carboline alkaloids are related to their antimutagenic and antigenotoxic activities. *Mutagenesis, 22,* 293–302.

Nenaah, G. (2010). Antibacterial and antifungal activities of (beta)-carboline alkaloids of *Peganum harmala* (L) seeds and their combination effects. *Fitoterapia,* April 14 [Epub ahead of print].

Panda, H. (2000). *Herbs: Cultivation and medicinal uses.* Delhi, India: National Institute of Industrial Research. 435.

Park, C., Moon, D. O., Rhu, C. H., Choi, B. T., Lee, W. H., Kim, G. Y., et al. (2007). Beta-sitosterol induces anti-proliferation and apoptosis in human leukemic U937 cells through activation of caspase-3 and induction of Bax/Bcl-2 ratio. *Biological & Pharmaceutical Bulletin, 30,* 1317–1323.

Prashanth, D., & John, S. (1999). Antibacterial activity of *Peganum harmala. Fitoterapia, 70,* 438–439.

Saha, J. C., & Kasinathan, S. (1961). Ecbolic properties of Indian medicinal plants. II. *Indian Journal of Medical Research, 49,* 1094–1098.

Wernicke, C., Schott, Y., Enzensperger, C., Schulze, G., Lehmann, J., & Rommelspacher, H. (2007). Cytotoxicity of beta-carbolines in dopamine transporter expressing cells: structure-activity relationships. *Biochemical Pharmacology, 74,* 1065–1077.

Yang, Y. J., Lee, J. J., Jin, C. M., Lim, S. C., & Lee, M. K. (2008). Effects of harman and norharman on dopamine biosynthesis and L-DOPA-induced cytotoxicity in PC12 cells. *European Journal of Pharmacology, 587,* 57–64.

Zhang, Y. W., Morita, I., Zhang, L., Shao, G., Yao, X. S., & Murota, S. (2000). Screening of anti-hypoxia/reoxygenation agents by an *in vitro* method. Part 2: Inhibition of tyrosine kinase activation prevented hypoxia/reoxygenation-induced injury in endothelial gap junctional intercellular communication. *Planta Medica, 66,* 119–123.

CHAPTER 72

Antioxidants in Hazelnuts (*Corylus avellana* L.)

Marina Contini, Maria Teresa Frangipane, Riccardo Massantini
Dipartimento di Scienze e Tecnologie Agroalimentari, Tuscia University, Viterbo, Italy

CHAPTER OUTLINE

611

LIST OF ABBREVIATIONS

BHA, butylated hydroxyanisole
BHT, butylated hydroxytoluene
DPPH, 2,2-diphenyl-1-picrylhydrazyl radical
MUFA, monounsaturated fatty acid
ORAC, oxygen radical absorbance capacity
Trolox, 6-hydroxy-2,5,7,8-tetra-methylchroman-2-carboxylic acid

INTRODUCTION

Endogenous or exogenous free radical compounds can damage biological molecules such as proteins, lipids, and DNA. The human body can neutralize them, but its own defense systems are not fully efficient; hence, as time proceeds, free radicals accumulate. Free radical load causes cell damage and body ageing, and has been postulated to induce a variety of pathological events, such as atherogenesis and carcinogenesis.

Hazelnuts contain a series of antioxidants that may cooperate in concert, providing the body with potential help in hindering the free radical threat, thus improving human well-being by countering the initiation and progression of oxidative stress-mediated disorders and diseases.

Nuts & Seeds in Health and Disease Prevention. DOI: 10.1016/B978-0-12-375688-6.10072-6

BOTANICAL DESCRIPTION

Hazel is the common name for the flowering plant genus *Corylus*, usually placed in the Betulacee family, although some botanists consider it a separate family, Corylaceae. The genus *Corylus* comprises about 15 species, including the European *Corylus avellana* L., the common commercially grown hazelnut. It is a monoecious species, growing as large shrubs or small trees, usually 2–5 m high.

The edible portion of the hazelnut is the roughly spherical to oval kernel of the seed, which is 1.0–2.5 cm long and 1.2–2.0 mm broad. The kernel is covered by a dark brown perisperm (skin or pellicle), varying in thickness and appearance between varieties, and protected by a smooth, hard, woody shell. The seed grows in a bristly leafy outer husk that opens in autumn, when it ripens (about 7–8 months after pollination).

The most popular commercial hazelnut varieties are Tombul (Turkey), Tonda Gentile (Italy), Negret (Spain), Barcellona and Segorbe (Portugal), and Ennis, Daviana, and Butler (USA).

HISTORICAL CULTIVATION AND USAGE

The hazelnut is reputedly native to Asia Minor. It has been known since prehistory, and extensively eaten since pre-agricultural times.

The hazelnut is also known as a filbert, probably because its harvesting began on Saint Philbert's Day, August 22. "Filbert" may also derive from "full beard," for its long, leafy husk.

Throughout Europe, the hazelnut has long been revered not only as a source of food but also as a plant with mystical and magical powers. The ancients thought hazelnuts had medicinal properties, using them as remedies for various disorders and diseases, including sore throat, phlegm, chronic cough, impotency, and baldness.

PRESENT-DAY CULTIVATION AND USAGE

The world production of unshelled hazelnut amounts to nearly 1 million tonnes per year. Turkey is the leading producer and exporter, accounting for about 70% of world production, followed by Italy, with around 13%. The USA is third (about 4%), and Spain fourth (near 3%). Other minor producers are Azerbaijan, Georgia, Iran, China, Greece, and France.

About 90% of the world crop is absorbed by the food industry. Shelled hazelnuts are commercialized mainly after roasting, which provides a more intense, pleasant typical flavor and a crisper texture, in addition to allowing for removal of the slightly bitter and astringent skin. Hazelnuts are available in many forms (whole, chopped, crumbled, ground into a paste) and are extensively employed in confectionery. Hazelnut oil is used not only as food, but also in cosmetics, for its astringent and emollient properties, or as a carrier in aromatherapy.

APPLICATIONS TO HEALTH PROMOTION AND DISEASE PREVENTION
Nutrient and nutraceutical compounds in hazelnut and hazelnut oil

Hazelnuts are well known and appreciated for their organoleptic properties; in addition, they are very nutritious and healthful because of their favorable composition of nutrients and nutraceutical compounds.

Hazelnuts are a very rich source of fat (about 60%) and fiber (around 10%), as well as a good source of protein and carbohydrates; the major minerals are potassium, phosphorus, calcium, and magnesium, while significant amounts of copper, manganese, and selenium are also present (Table 72.1). The selenium content given in Table 72.1 (2.4–4.1 μg/100 g) is much lower than that reported by Dugo *et al.* (2003), which in Italian (Sicilian) hazelnuts was about 90 μg/100 g.

TABLE 72.1 Proximate Composition and Micronutrient Content in 100 g of Hazelnuts

Proximates	Unit	Dry Roasted	Unroasted	Blanched, Unroasted
Water	g	2.52	5.31	5.79
Protein (N × 5.3)	g	15.03	14.95	13.70
Total lipid (fat)	g	62.40	60.75	61.15
Ash	g	2.45	2.29	2.36
Carbohydrate, by difference	g	17.60	16.70	17.00
Fiber, total dietary	g	9.4	9.7	11.0
Sugars, total	g	4.89	4.34	3.49
Sucrose	g	4.75	4.20	3.35
Glucose (dextrose)	g	0.07	0.07	0.07
Fructose	g	0.07	0.07	0.07
Starch	g	1.10	0.48	0.93
Energy	kcal	646	628	629
Energy	kJ	2703	2629	2630
Minerals				
Potassium	mg	755	680	658
Phosphorous	mg	310	290	310
Magnesium	mg	173	163	160
Calcium	mg	123	114	149
Manganese	mg	5.550	6.175	12.650
Iron	mg	4.38	4.70	3.30
Zinc	mg	2.50	2.45	2.20
Copper	mg	1.750	1.725	1.600
Selenium	μg	4.1	2.4	4.1
Vitamins				
Vit E (α-tocopherol)	mg	15.28	15.03	17.50
β-Tocopherol	mg	0.33	0.33	0.35
γ-Tocopherol	mg	0	0	2.15
δ-Tocopherol	mg	0	0	0.14
Vit. C (total ascorbic acid)	mg	3.8	6.3	2.0
Niacin	mg	2.050	1.800	1.550
Pantothenic acid	mg	0.923	0.918	0.815
Vit B$_6$	mg	0.620	0.563	0.585
Thiamin	mg	0.338	0.643	0.475
Riboflavin	mg	0.123	0.113	0.110
Folate, total	μg	88	113	78
Vit A, RAE[a]	μg RAE	3	1	2
Vit A, IU[b]	IU	61	20	40
β-carotene	μg	36	11	23
α-carotene	μg	1	3	2
Choline, total	mg	n.r.	45.6	n.r.
Betaine	mg	n.r.	0.4	n.r.
Lutein + zeaxanthin	μg	n.r.	92	n.r.
Vitamin K (phylloquinone)	μg	n.r.	14.2	n.r.

Source: United States Department of Agriculture (USDA) National Nutrient Database for Standard Reference, Release 22 (2009), www.nal.usda.gov/fnic/foodcomp/search.
[a]RAE, Retinol Activity Equivalent;
[b]IU, Internationan Unit; n.r., not reported.

Selenium, though not an antioxidant itself, has an essential role in constructing the endogenous antioxidant defense system, protecting the human body against oxidative disorders and diseases, including cancer (Awad & Bradford, 2005).

Hazelnuts also contain appreciable amounts of vitamins (Table 72.1); in particular, they are an excellent source of α-tocopherol (over 15 mg/100 g), a fat-soluble phenolic compound that is very important for human health because of both its vitamin E function and its powerful biological antioxidant properties.

It is well recognized that a diet high in MUFAs (especially oleic acid) and phytosterols (especially β-sitosterol) tends to improve the cholesterol balance and triglyceride levels, reducing the risk of atherosclerosis and coronary heart disease. The hazelnut lipid fraction is close to that of olive oil; it has a high content of MUFAs (about 80%, essentially comprising oleic acid) and phytosterols, mainly β-sitosterol (almost 1 mg/g oil) (Table 72.2), the most active among phytosterols in reducing serum LDL cholesterol levels. Moreover, many epidemiologic/experimental studies have provided evidence that dietary phytosterols may offer protection against certain typologies of cancer, such as colon, breast, and prostate cancer (Awad & Bradford, 2005).

It has been demonstrated that hazelnut supplementation favorably changes plasma lipid profiles (VLDL cholesterol, triacylglycerol, and apolipoprotein B reduction; HDL cholesterol increase) in hypercholesterolemic adult men (Mercanligil et al., 2007). In young, healthy humans, a hazelnut-enriched diet was found to be beneficial in lowering total cholesterol and LDL cholesterol levels, elevating HDL cholesterol, and significantly increasing the HDL/LDL ratio; in addition, an enhancement in the plasma antioxidant potential and a reduction in plasma lipidic peroxidation levels were noted (Durak et al., 1999). In rabbits fed on a high cholesterol diet, Hatipoğlu et al. (2004) observed that hazelnut oil administration not only lowered aortic cholesterol accumulation, but also reduced plasma, liver, and aortic peroxide levels. These findings suggested that hazelnut bioactive substances (i.e., phytosterols, tocopherols, and/or other antioxidants) were able to ameliorate the cellular antioxidant defenses, thus improving the cardioprotective effects of the MUFA-rich hazelnut lipids.

Among five different nuts (hazelnut, macadamia, peanut, walnut, and almond), hazelnuts and macadamias were by far the best sources of squalene (186.4 and 185.0 μg/g oil, respectively); furthermore, hazelnut and almond oils were found to have very high vitamin E contents, mainly α-tocopherol (310.1 and 439.5 μg/g oil, respectively) (Table 72.2). Gordon and colleagues found 161 ± 6 mg/kg oil of phenolics (as caffeic acid equivalents) in the polar fraction of unrefined hazelnut commercial oils; myricetin and chrysin were tentatively identified as minor components, but the main phenolic compounds could not be recognized (Gordon et al., 2001).

Antioxidant characteristics of hazelnut and hazelnut oil

Recently, Arranz and colleagues found that hazelnut and pistachio oils exhibit a high antioxidant capacity, close to that of extra virgin olive oil, and higher than those of walnut, almond, and peanut oils (Arranz et al., 2008). They observed that after methanolic extraction of the polar fraction, a noticeable reduction of antioxidant capacity was recorded in the remaining nut oils. The authors found good correlation between the DPPH-scavenging activity of the stripped nut oils and tocopherol content, while proving that the antioxidant activity registered in the methanolic fraction was essentially due to phospholipids, the contribution of phenolics being negligible.

Currently hazelnut oil is mainly employed in the cosmetic and sometimes in the confectionery industry, but it is becoming increasingly popular as an edible oil, owing to its beneficial nutritional composition and health-promoting characteristics. It can be found on the market in both crude and refined forms. Cold-pressed hazelnut oil may be conveniently employed for cooking or frying, due to its high resistance to oxidative degradation, which is close to that of

TABLE 72.2 Total Oil of Five Edible Nuts, and Relative Content of Squalene, Tocopherols, and Phytosterols in Oils

| Nut | Total Oil (g/100 g) | Squalene (μg/g oil) | Tocopherols[a] | | | Phytosterols | | | |
			α-Tocopherol (μg/g oil)	γ-Tocopherol (μg/g oil)	Sum of Tocopherols (μg/g oil)	β-Sitosterol (μg/g oil)	Campesterol (μg/g oil)	Stigmasterol (μg/g oil)	Sum of Phytosterols (μg/g oil)
Hazelnut	49.2 ± 1.6	186.4 ± 11.6	310.1 ± 31.1	61.2 ± 29.8	371.3	991.2 ± 73.2	66.7 ± 6.7	38.1 ± 4.0	1096.0
Macadamia	59.2 ± 1.5	185.0 ± 27.2	122.3 ± 24.5	Trace	122.3	1506.7 ± 140.5	73.3 ± 8.9	38.3 ± 2.7	1618.3
Peanut	37.9 ± 1.8	98.3 ± 13.4	87.9 ± 6.7	60.3 ± 6.7	148.2	1363.3 ± 103.9	198.3 ± 21.4	163.3 ± 23.8	1724.9
Walnut	50.8 ± 1.4	9.4 ± 1.8	20.6 ± 8.2	300.5 ± 31.0	321.1	1129.5 ± 124.6	51.0 ± 2.9	55.5 ± 11.0	1236.0
Almond	40.8 ± 2.5	95.0 ± 8.5	439.5 ± 4.8	12.5 ± 2.1	452.0	2071.7 ± 25.9	55.0 ± 10.8	51.7 ± 3.6	2178.4

Hazelnut oil emerges as having high squalene and tocopherol (mainly α-tocopherol) contents; phytosterols (mainly β-sitosterol) are present in consistent amounts, although at lower concentrations than in all the other nuts.

Results are the mean ± SEM from three independent experiments.

Source: Maguire, O'Sullivan, Galvin, O'Connor, and O'Brien (2004), *Intl J. Food Sci. Nutr.* 55, 171–178.

[a]Traces of δ-tocopherol were detected in all nuts (data not shown).

extra virgin olive oil (Contini *et al.*, 1997). Of the five nut oils, hazelnut oil showed the best oxidative stability (Rancimat test) (Arranz *et al.*, 2008).

Plant phenolics are not nutrients for humans. Nevertheless, their inclusion in the diet is beneficial, because many could potentially play a major role in human health promotion and disease risk reduction. Many phenolic compounds exhibiting antioxidant properties have been studied and proposed for protection against numerous pathologies associated with oxidative damage. Plant phenols have been reported to show anticarcinogenic, anti-atherogenic, anti-ulcer, antithrombotic, anti-inflammatory, anti-allergic, immune modulating, antimicrobial, vasodilatatory, and analgesic effects (Wollgast & Anklam, 2000). Wu and colleagues (2004) found that one serving (28.4 g) of hazelnuts contained about 237 mg of total phenols (as gallic acid equivalent) and had a total antioxidant capacity (ORAC test) of 2739 μmol Trolox equivalent. Interestingly, the principal fraction responsible for antioxidant capacity was not the lipophilic fraction (comprising tocopherols), but the hydrophilic counterpart. Oliveira *et al.* (2008) demonstrated that boiling-aqueous extracts of hazelnut showed DPPH-radical scavenging ability and antioxidant activity in the β-carotene/linoleate model system, but their efficiency was much lower with respect to reference antioxidants; contrarily, the extracts showed a better reducing power than BHA and α-tocopherol. It was recently shown that consistent amounts of tree nut phenolics exist in a non-soluble (bound) form; in particular, the bound fraction accounted for close to 93% of the total phenols in hazelnuts (Yang *et al.*, 2009). Therefore, both the content and antioxidant activity of nut phenolics are probably underestimated in the literature.

Antioxidant characteristics of hazelnut skin

Most of the hazelnut phenolics are located in the seed skin (Shahidi *et al.*, 2007); this evidence suggested exploiting the hazelnut skin by-product (the industrial residue of pellicle removal) as a source of natural and efficient antioxidants. With respect to other hazelnut by-products, it was demonstrated that the skin waste provides a much higher yield of phenolic extract, which has a very high phenolic content and multiple antioxidant properties — i.e., DPPH-radical, hydrogen peroxide, and superoxide radical scavenging activities; reducing and chelating powers; the ability to inhibit oxidation of human LDL-cholesterol; and the capability to reduce DNA damage induced by hydroxyl radical (Shahidi *et al.*, 2007; Contini *et al.*, 2008; Contini *et al.*, 2009). In particular, hazelnut skin phenolic extract showed high Fe^{3+} reducing activity and excellent radical scavenging ability (DPPH test) (Figure 72.1A, 72.1B); moreover, an *in vivo* test demonstrated that it was biologically active in rats (Figure 72.1C) (Contini *et al.*, 2009).

Hazelnut skin extract was found to be very rich in tannins (over 60% of total phenols) (Contini *et al.*, 2008). Historically, tannins were considered to be antinutrients, but recently the recognition of their very effective antioxidative capacity and probable protective actions (such as cardioprotective, anticarcinogenic, gastroprotective, anti-inflammatory) has led to second thoughts towards their effect on human health (Santos-Buelga & Scalbert, 2000). Tannins resulted in much more powerful antioxidants than simple monomeric phenols, and may have unique roles in the human digestive metabolism, both as preservers of other biological antioxidants and as protectors of nutrients from oxidative damage. In particular, non-absorbed high polymerized tannins may exert important local activities in the gastrointestinal tract, especially in the colon (Halliwell, 2007), which, being particularly exposed to oxidizing agents, is prone to several pathologies, such as inflammation and cancer.

Because of their potential antioxidant and nutraceutical properties, phenolic extracts obtained from hazelnut skin might satisfy the demand for new natural phenolic antioxidants, useful in the food industry alongside or as a substitute for synthetic antioxidants, or as ingredients in the preparation of innovative foods with a high dietetic/functional value. One outcome of these findings is that it would be better to eat hazelnuts with intact skins, and the industry/consumer preference for peeled kernels should be revised.

(A)

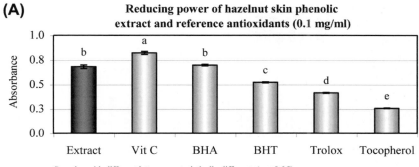

Reducing power of hazelnut skin phenolic extract and reference antioxidants (0.1 mg/ml)

Samples with different letters are statistically different (p < 0.05)

(B)

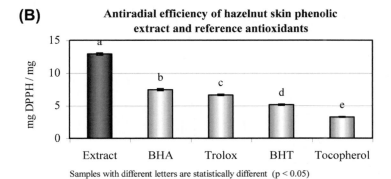

Antiradial efficiency of hazelnut skin phenolic extract and reference antioxidants

Samples with different letters are statistically different (p < 0.05)

(C)

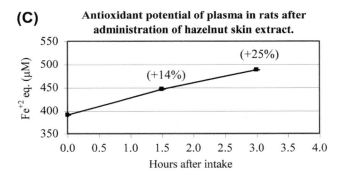

Antioxidant potential of plasma in rats after administration of hazelnut skin extract.

FIGURE 72.1

In vitro and *in vivo* antioxidant power of hazelnut skin crude phenolic extract. The extract showed ferric reducing activity similar to that of BHA (butylated hydroxyanisole), superior to that of BHT (butylated hydroxytoluene), Trolox (6-hydroxy-2,5,7,8-tetra-methylchroman-2-carboxylic acid) and α-tocopherol, and inferior only to ascorbic acid (A); its activity against the DPPH (2,2-diphenyl-1-picrylhydrazyl) radical was better than the reference antioxidants, including α-tocopherol (B). Plasma antioxidant potential of rats which consumed 10 mg of extract increased by 14% and 25% at 1.5 and 3 hours after administration, respectively (C). Reprinted from Contini *et al.* (2009), *Acta Hortic.*, *845*, 717—722, with permission.

Phenolic compounds of hazelnut kernel and skin

Some works have been published reporting the chemical composition of phenolics in hazelnut kernel and skin. A comprehensive bibliographic data compilation in this respect is summarized in Tables 72.3—72.5. The hazelnut has elevated contents of flavan-3-ols, especially highly polymerized ones (tannins over decamers). In addition to high amounts of flavan-3-ol monomers and dimers, tannin forms mainly consisting of B-type proanthocyanidins up to nonamers were detected (but not quantified) in roasted hazelnut skin. Alasalvar *et al.* (2009) were the first to show that the tannin fraction isolated from roasted hazelnut skin phenolic extract manifested the highest antioxidant activity (DPPH test) with respect to both the crude phenolic extract and the low molecular weight phenolic fraction. Several phenolic acids, such as benzoic acid and cinnamic acid derivatives, were also found in hazelnut and its skin. Regarding flavonols, myricetin, kaempferol, and quercetin were identified in the skin by-product, while the presence of quercetin in the kernel is controversial. A single anthocyanin (cyanidin) was found, but just in the kernel. Furthermore, some phytoestrogens (isoflavones, lignans and coumestrol) were found in the kernel. Phytoestrogens are phenolics that have important antioxidant properties which are thought able to protect against a wide range of diseases (Cornwell *et al.*, 2004).

TABLE 72.3 Phenolic Compounds Identified in Hazelnuts

Reference	Yurttas et al., 2000	Gu et al., 2003	Gu et al., 2004; USDA, 2004	Alasalvar et al., 2006	Harnly et al., 2006; USDA, 2007	Shahidi et al., 2007	Amarowicz et al., 2008	Prosperini et al., 2009
Extraction solvent	MeOH/H$_2$O (2:1)	Acetone/H$_2$O/acetic acid (70:29.5:0.5)	Acetone/H$_2$O/acetic acid (70:29.5:0.5)	EtOH/H$_2$O (80:20) or acetone/H$_2$O (80:20)	None	EtOH/H$_2$O (80:20)	Acetone/H$_2$O (80:20)	EtOH/H$_2$O (80:20) or acetone/H$_2$O (80:20), at different temperatures
Treatment[a]	Acid hydrolysis	Sephadex LH-20 column, then HPLC analysis before and after thiolytic degradation	Sephadex LH-20 column, then HPLC analysis before and after thiolytic degradation	Partition in diethyl ether Basic hydrolysis plus partition in diethyl ether	Acid hydrolysis	Partition in diethyl ether before and after basic hydrolysis	Oxidative depolymeri-zation (n-butanol/HCl)	Partition in diethyl ether Basic hydrolysis plus partition in diethyl ether
Analytical technique	HPLC-UV	HPLC-MS/MS	HPLC-MS/MS	HPLC/DAD	HPLC/DAD	HPLC/DAD	HPLC/DAD	HPLC/DAD HPLC/DAD
Phenolic class analyzed	Simple phenolics	Flavan-3-ols	Flavan-3-ols	Free phenolic acids Esterified phenolic acids	Flavan-3-ols; anthocyanins Phenolic acids	Phenolic acids	Flavan-3-ols	Free phenolic acids Esterified phenolic acids
Unit	Qualitative data[b,c]	Qualitative data[b]	mg/100 g of kernel	μg/g of extract[d]	mg/100 g of kernel	μg/g of extract	Qualitative data[b]	μg/g of extract[e]

Phenolic acids	Yurttas et al., 2000	Gu et al., 2003	Gu et al., 2004; USDA, 2004	Alasalvar et al., 2006	Harnly et al., 2006; USDA, 2007	Shahidi et al., 2007	Amarowicz et al., 2008	Prosperini et al., 2009
p-OH benzoic acid	✓							
Caffeic acid	✓[f]		n.d	n.d		81		n.d. 0–2.18
p-coumaric acid			5[g]	13 (21)		208		0–5.14 0–3.71
Ferulic acid			n.d.	n.d.		105		0–8.10 n.d.
Gallic acid	✓✓		n.d.	158 (204)		127		0–34.50 0–18.40
Sinapic acid	✓✓		n.d.	39 (52)		93		0–9.09 0–7.70
Total phenolic acids			**0–5**	**210 (277)**		**614**		**0–39.64 7.31–18.40**

Flavonols	Yurttas et al., 2000				Harnly et al., 2006; USDA, 2007			
Quercetin	✓				n.d.			

Anthocyanins					Harnly et al., 2006; USDA, 2007			
Cyanidin					6.7			

Flavan-3-ols monomers and polymers (condensed tannin)							
(+) catechin		✔[f]		1.2			
(−) epicatechin				0.2			
Epigallo-catechin				2.8			
Epigallo-catechin gallate				1.1			
Gallocatechin gallate				0.4			
Proantho-cyanidins (B-type)		✔					
Procyanidins[h]		✔[i]	✔[i]		✔	✔	
Prodel-phinidins[k]		✔[i]	✔[i]				
(epi)catechin glycoside		✔					
Monomers	9.8						
DP[j]							
2	12.5						
3	13.6						
4–6	67.7						
7–10	74.6						
>10	322.4						
Average DP[m]	14.0						
Total flavan-3-ols	**500.7**			**5.7**			
Total phenols quantified	500.7	0–5	210 (277)	12.4	614	0–39.64	7.31–18.40

Hazelnuts contain mainly flavan-3-ols, especially highly polymerized (tannins over decamers). Several phenolic acids and cyanidin were also found, while the presence of quercetin is controversial.

n.d., not detected.

[a] Treatment of sample/phenolic extract before analysis;
[b] ✔, reported presence;
[c] tentative identification;
[d] first value, ethanolic extract; second value (in brackets), acetonic extract;
[e] min and max amount dosed, depending on the solvent and temperature of extraction;
[f] unresolved peaks;
[g] present only in the acetonic extract;
[h] (epi)catechin polymers;
[i] presence of 3-O-gallate;
[j] types of flavan-3-ols found;
[k] (epi)gallocatechin polymers;
[l] DP, degree of polymerization;
[m] excluding monomers.

TABLE 72.4 Phenolic Compounds Identified in Hazelnut Skin

		Senter et al., 1983	Coisson et al., 2002	Travaglia et al., 2006	Shahidi et al., 2007	Monagas et al., 2009
Reference						
Sample		Skin	Skin by-product	Skin from roasted kernel	Skin by-product	Skin by-product
Extraction solvent		MeOH-HCl	MeOH	Several polar solvents	EtOH/H$_2$O (80:20)	Acetone/H$_2$O (80:20)
Treatment[a]		Acid hydrolysis	None	None	Partition in diethyl ether before and after basic hydrolysis	None / Sephadex LH-20 column
Analytical technique		GLC/MS	HPLC/UV	LC/MS	HPLC/DAD	LC-DAD-Fluorescence; LC-DAD/ESI-MS; MALDI-TOF MS
Phenolic class analyzed		Phenolic acids	Simple phenolics	Simple phenolics	Phenolic acids	Flavan-3-ols
Unit		µg/g of skin	µg/g of extract	Qualitative data[b]	µg/g of extract	µg catechin equivalent/g of extract / Qualitative data[b]
Phenolic acids	p-OH benzoic acid	tr	453.6	n.d.		
	Caffeic acid	tr		n.d.	tr	
	Chlorogenic acid		49.4		231	
	p-coumaric acid		6.9		124	
	Ferulic acid	n.d.	n.d.		387	
	Gallic acid	tr	472.4	✓		
	Protocatechuic acid	0.36	183.5	✓		
	Sinapic acid					
	Vanillic acid	tr	11.3		124	
	Total phenolic acids	**0.36**	**1177.1**		**866**	
Flavonols	Myricetin	91.2		✓		
	Kaempferol	66.3		✓		
	Quercetin	52.2		✓		
	Total flavonols	**209.7**				

Flavan-3-ols monomers and polymers (condensed tannins)			
(+) catechin	993.2	1168.00	✓
(−) epicatechin	342.2	n.d.	✓
Dimer B3		104.62	
Unknown dimer B		28.37	
Trimers[c]			
3(epi)catechin (2B)			✓
2(epi)catechin, 1(epi)gallocatechin (1A,1B)			✓
2(epi)catechin, 1(epi)gallocatechin (2B)			✓
1(epi)catechin, 2(epi)gallocatechin (1A,1B)			✓
1(epi)catechin, 2(epi)gallocatechin (2B)			✓
3(epi)catechin, 1gallate (2B)			✓
Tetramers[c]			
4(epi)catechin (3B)			✓
3(epi)catechin, 1(epi)gallocatechin (3B)			✓
2(epi)catechin, 2(epi)gallocatechin (3B)			✓
3(epi)catechin, 1(epi)gallocatechin, 1gallate (3B)			✓
Pentamers[c]			
5(epi)catechin (4B)			✓
4(epi)catechin, 1(epi)gallocatechin (4B)			✓
3(epi)catechin, 2(epi)gallocatechin (4B)			✓
4(epi)catechin, 1(epi)gallocatechin, 1gallate (4B)			✓
Hexamers[c]			
6(epi)catechin (5B)			✓
5(epi)catechin, 1(epi)gallocatechin (5B)			✓
4(epi)catechin, 2(epi)gallocatechin (5B)			✓

Continued

621

TABLE 72.4 Phenolic Compounds Identified in Hazelnut Skin—continued

Reference	Senter et al., 1983	Coisson et al., 2002	Travaglia et al., 2006	Shahidi et al., 2007	Monagas et al., 2009
Heptamers[c]					
7(epi)catechin (6B)					✓✓
6(epi)catechin, 1(epi) gallocatechin (6B)					✓✓
5(epi)catechin, 2(epi) gallocatechin (6B)					✓
Octamers[c]					
7(epi)catechin, 1(epi) gallocatechin (7B)					✓
6(epi)catechin, 2(epi) gallocatechin (7B)					✓
Nonamers[c]					
7(epi)catechin, 2(epi) gallocatechin (8B)					✓
Total flavan-3-ols		1335.4			1300.99
Other phenols					
Pyrocatechol		119.8			
Total phenols quantified	0.36	2842.0		866	1300.99

Hazelnut skin emerges as having very high contents of (+) catechin and (−) epicatechin. Flavan-3-ol polymers (tannins) were also found, but not quantified. Several phenolic acids and flavonols (myricetin, kaempferol, quercetin) are also present.

[a]Treatment of phenolic extract before analysis;
[b]✓, reported presence;
[c]in brackets, linkage; tr, traces; n.d., not detected.

TABLE 72.5 Phytoestrogens (µg/100 g) in Hazelnuts

	Reference	Liggins et al., 2000		Thompson et al., 2006
		Dry weight	Wet weight	Wet weight[a]
Isoflavones	Formononetin			1.2
	Daidzein	5.8	5.52	3.6
	Genistein	19.4	18.47	24.8
	Glycitein			0.5
	Total isoflavones	**25.2**	**23.99**	**30.2**
Lignans	Matairesinol			1.2
	Lariciresinol			14.3
	Pinoresinol			1.1
	Secoisolariciresinol			60.5
	Total lignans			**77.1**
Coumestan	Coumestrol			0.3
	Total phytoestrogens			**107.5**

The major hazelnut phytoestrogens are lignans (mainly secoisolariciresinol); genistein is the major isoflavone. Small amounts of coumestrol are also present.
[a]Hazelnut with skin (specified by the authors).

Studies on the composition and *in vitro* antioxidant activity of hazelnut phenolics are intensifying, symptomatic of the great interest of the scientific community regarding their potential beneficial effects. However, more research is needed to fully identify and characterize hazelnut phenolics, to evaluate their bioavailability, to establish the *in vivo* antioxidant action and metabolism of tannin and non-tannin fractions, as well as to elucidate the relationships between antioxidant properties and health benefits.

ADVERSE EFFECTS AND REACTIONS (ALLERGIES AND TOXICITY)

Like other tree nuts, hazelnuts can induce allergic reactions. The allergic responses depend strongly on individual sensitivity, ranging from mild symptoms to severe, life-threatening anaphylactic forms. Hazelnut Cor a 1 (a protein that shows high homology to the major birch pollen allergen Bet v 1) and Cor a 2 (porphyrin) are responsible for mild oral syndromes correlated to pollen sensitivity. More severe clinical symptoms (dermal, abdominal, respiratory, and cardiovascular) are manifestations of non-pollen related allergies involving other hazelnut protein allergens (Cor a 8, Cor a 9, Cor a 11, 2S albumin, and oleosin) (Flinterman *et al.*, 2008).

Nuts, including hazelnuts, are liable to infestation by aflatoxin-producing *Aspergillus*. Mould development may occur during crop growth, but mainly takes place during storage under unhygienic, unventilated, hot, humid conditions. The major aflatoxins produced by the mould are designated B_1, B_2, G_1, and G_2; among them, B_1 is by far the most toxic. The intense carcinogenicity of aflatoxin B1 is related to its oxidation to highly reactive products inducing DNA, RNA, and protein modifications. Thus, aflatoxin contamination is an important issue in the areas of food safety and international trade. To protect consumer health, the maximum aflatoxin content is rigorously controlled by national and international regulations, with limits varying in regulations for different countries.

SUMMARY POINTS

- Hazelnuts are very nutritious and healthful.
- The oil fraction is rich in oleic acid, phytosterols (mainly β-sitosterol), vitamin E, and squalene; this special composition may help diminish the risk of coronary and oxidative stress-induced diseases.

- The nut contains a precious mix of synergistically acting antioxidants, among which are tocopherols (mainly α-tocopherol) and phenolics (especially tannins, essentially located in the brown skin); phytoestrogens, selenium, and squalene contribute towards improving the antioxidant network efficiency.
- The hazelnut skin by-product, being remarkably rich in phenolics, is proposed as an excellent source of natural and powerful antioxidants.
- Dietary antioxidants counteract free radicals, promoters of oxidative deteriorations; the assumption is that hazelnut antioxidants might be a valuable aid against oxidative stress-mediated disorders and diseases, such as inflammatory, cardiovascular, and neurodegenerative diseases, and cancer.
- Because of their richness in nutrients and bioactive health-promoting compounds, there are good reasons for profitably including hazelnuts as part of a nutritious and functional diet.

References

Alasalvar, C., Karamać, M., Amarowicz, R., & Shahidi, F. (2006). Antioxidant and antiradical activity in extracts of hazelnut kernel (*Corylus avellana* L.) and in hazelnut green leafy cover. *Journal of Agricultural and Food Chemistry, 54*, 4826—4832.

Alasalvar, C., Karamać, M., Kosińska, A., Rybarczyr, A., Shahidi, F., & Amarowicz, R. (2009). Antioxidant activity of hazelnut skin phenolics. *Journal of Agricultural and Food Chemistry, 57*, 4645—4650.

Amarowicz, R., Troszyńska, A., Kosińska, A., Lamparski, G., & Shahidi, F. (2008). Relation between sensory astringency of extracts from selected tannin-rich foods and their antioxidant activity. *Journal of Food Lipids, 15*, 28—41.

Arranz, S., Cert, R., Pérez-Jiménez, J., Cert, A., & Saura-Calixto, F. (2008). Comparison between free radical scavenging capacity and oxidative stability of nut oils. *Food Chemistry, 110*, 985—990.

Awad, A. B., & Bradford, P. G. (2005). *Nutrition and Cancer Prevention*. Boca Raton, FL: CRC Press.

Coisson, J. D., Capasso, M., Travaglia, F., Piana, G., Arlorio, M., & Martelli, A. (2002). *Proprietà antiossidanti di estratti fenolici da sottoprodotti di lavorazione di cacao e nocciola. Proceedings of Vth Congresso Nazionale di Chimica degli Alimenti (V CISETA).* In *Ricerche ed innovazioni nell'industria alimentare.* Pinerolo, Italy: Chiriotti. 154—158.

Contini, M., Cardarelli, M. T., De Santis, D., Frangipane, M. T., & Anelli, G. (1997). Proposal for the edible use of cold pressed hazelnut oil. Note 2: evaluation of frying stability. *La Rivista Italiana delle Sostanze Grasse, 74*, 97—100.

Contini, M., Baccelloni, S., Massantini, R., & Anelli, A. (2008). Extraction of natural antioxidants from hazelnut (*Corylus avellana* L.) shell and skin by-products by long maceration at room temperature. *Food Chemistry, 110*, 659—669.

Contini, M., Baccelloni, S., Massantini, R., Anelli, G., Manzi, L., & Merendino, N. (2009). *In vitro* and *in vivo* antioxidant potential of phenolic extracts obtained from hazelnut skin by-products. *Acta Horticulturae, 845*, 717—722.

Cornwell, T., Cohick, W., & Raskin, I. (2004). Dietary phytoestrogens and health. *Phytochemistry, 65*, 995—1016.

Dugo, G., La Pera, L., Lo Turco, V., Mavrogeni, E., & Alfa, M. (2003). Determination of selenium in nuts by cathodic stripping potentiometry (CSP). *Journal of Agricultural and Food Chemistry, 51*, 3722—3725.

Durak, İ., Köksal, İ., Kaçmaz, M., Büyükkoçak, S., Çimen, B. M. Y., & Öztürk, H. S. (1999). Hazelnut supplementation enhances plasma antioxidant potential and lowers plasma cholesterol levels. *Clinica Chimica Acta, 284*, 113—115.

Flinterman, A. E., Akkerdaas, J. H., Knulst, A. C., van Ree, R., & Pasmans, S. G. (2008). Hazelnut allergy: from pollen-associated mild allergy to severe anaphylactic reactions. *Current Opinion in Allergy and Clinical Immunology, 8*, 261—265.

Gordon, M. H., Covell, C., & Kirsch, N. (2001). Detection of pressed hazelnut oil in admixtures with virgin olive oil by analysis of polar components. *Journal of the American Oil Chemists Society, 78*, 621—624.

Gu, L., Kelm, M. A., Hammerstone, J. F., Beecher, G., Holden, J., Haytowitz, D., et al. (2003). Screening of food containing proanthocyanidins and their structural characterization using LC-MS/MS and thiolytic degradation. *Journal of Agricultural and Food Chemistry, 51*, 7513—7521.

Gu, L., Kelm, M. A., Hammerstone, J. F., Beecher, G., Holden, J., Haytowitz, D., et al. (2004). Concentration of proanthocyanidins in common foods and estimations of normal consumption. *Journal of Nutrition, 134*, 613—617.

Halliwell, B. (2007). Dietary polyphenols: good, bad or indifferent for your health? *Cardiovascular Research, 73*, 341—347.

Harnly, J. M., Doherty, R. F., Beecher, G. R., Holden, J. M., Haytowitz, D. B., Bhagwat, S., et al. (2006). Flavonoid content of US fruits, vegetables, and nuts. *Journal of Agricultural and Food Chemistry, 54*, 9966—9977.

Hatipoğlu, A., Kanbağli, Ö., Balkan, J., Küçük, M., Çevikbaş, U., Aykaç-Toker, G., et al. (2004). Hazelnut oil administration reduces aortic cholesterol accumulation and lipid peroxides in the plasma, liver, and aorta of rabbits fed a high-cholesterol diet. *Bioscience Biotechnology and Biochemistry, 68*, 2050–2057.

Liggins, J., Bluck, L. J. C., Runswick, S., Atkinson, C., Coward, W. A., & Bingham, S. A. (2000). Daidzein and genistein content of fruits and nuts. *Journal of Nutritional Biochemistry, 11*, 326–331.

Maguire, L. S., O'Sullivan, S. M., Galvin, K., O'Connor, T. P., & O'Brien, N. M. (2004). Fatty acid profile, tocopherol, squalene and phytosterol content of walnuts, almonds, peanuts, hazelnut and macadamia nut. *International Journal of Food Sciences & Nutrition, 55*, 171–178.

Mercanligil, S. M., Arlsan, P., Alasalvar, C., Okut, E., Akgül, E., Pinar, A., et al. (2007). Effects of hazelnut-enriched diet on plasma cholesterol and lipoprotein profiles in hypercholesterolemic adult men. *European Journal of Clinical Nutrition, 61*, 212–220.

Monagas, M., Garrido, I., Lebrón-Aguilar, R., Gómez-Cordovés, M. C., Rybarczyk, A., Amarowicz, R., et al. (2009). Comparative flavan-3-ol profile and antioxidant capacity of roasted peanut, hazelnut, and almond skins. *Journal of Agricultural and Food Chemistry, 57*, 10590–10599.

Oliveira, I., Sousa, A., Morais, J. S., Ferreira, I. C. F. R., Bento, A., Estevinho, L., et al. (2008). Chemical composition, and antioxidant and antimicrobial activities of three hazelnut (*Corylus avellana* L.) cultivars. *Food and Chemical Toxicology, 46*, 1801–1807.

Prosperini, S., Ghirardello, D., Scursatone, B., Gerbi, V., & Zeppa, G. (2009). Identification of soluble phenolic acids in hazelnut (*Corylus avellana* L.) kernel. *Acta Horticulturae, 845*, 677–680.

Santos-Buelga, C., & Scalbert, A. (2000). Proanthocyanidins and tannin-like compounds — nature, occurrence, dietary intake and effects on nutrition and health. *Journal of the Science of Food and Agriculture, 80*, 1094–1117.

Senter, S. D., Horvat, R. J., & Forbus, W. R. (1983). Comparative GLC-MS analysis of phenolic acids of selected tree nuts. *Journal of Food Science, 48*, 798–799.

Shahidi, F., Alasalvar, C., & Liyana-Pathirana, C. M. (2007). Antioxidant phytochemicals in hazelnut kernel (*Corylus avellana* L.) and in hazelnut by-products. *Journal of Agricultural and Food Chemistry, 55*, 1212–1220.

Thompson, L. U., Boucher, B. A., Liu, Z., Cotterchio, M., & Kreiger, N. (2006). Phytoestrogen content of foods consumed in Canada, including isoflavones, lignans, and coumestan. *Nutrition and Cancer, 54*, 184–201.

Travaglia, F., Locatelli, M., Stévigny, C., Coïsson, J. D., Bennett, R., Arlorio, M., et al. (2006). Chemical characterization and antioxidant activity of polyphenolic fraction from *Corylus avellana* L. seed husk. Proceedings of VIIth Congresso Nazionale di Chimica degli Alimenti (VII CISETA). In *Ricerche e innovazioni nell'industria alimentare* (pp. 786–791). Pinerolo, Italy: Chiriotti.

United States Department of Agriculture (USDA) Database for the Proanthocyanidin content of selected foods (2004). Available at: www.nal.usda.gov/fnic/foodcomp.

United States Department of Agriculture USDA Database for the Flavonoid Content of Selected Foods, Release 2.1 (2007). Available at: www.ars.usda.gov/nutrientdata.

United States Department of Agriculture (USDA) National Nutrient Database for Standard Reference, Release 22 (2009). Available at: www.nal.usda.gov/fnic/foodcomp/search.

Wollgast, J., & Anklam, E. (2000). Review on polyphenols in *Theobroma cacao*: changes in composition during the manufacture of chocolate and methodology for identification and quantification. *Food Research International, 33*, 423–447.

Wu, X., Beecher, G. R., Holden, J. M., Haytowitz, D. B., Gebhardt, S. E., & Prior, R. L. (2004). Lipophilic and hydrophilic antioxidant capacities of common foods in the United States. *Journal of Agricultural and Food Chemistry, 52*, 4026–4037.

Yang, J., Liu, R. H., & Halim, L. (2009). Antioxidant and antiproliferative activities of common edible nut seeds. *LWT Food Science and Technology, 42*, 1–8.

Yurttas, H. C., Schafer, H. W., & Warthesen, J. J. (2000). Antioxidant activity of nontocopherol hazelnut (*Corylus* spp.) phenolics. *Journal of Food Science, 65*, 276–280.

625

CHAPTER 73

Hazelnut (*Corylus avellana* L.) Cultivars and Antimicrobial Activity

Elsa Ramalhosa, Teresa Delgado, Letícia Estevinho, José Alberto Pereira
CIMO/Escola Superior Agrária, Instituto Politécnico de Bragança, Bragança, Portugal

627

LIST OF ABBREVIATIONS

FAO, Food and Agriculture Organization of the United Nations
HDL, high density lipoprotein
LDL, low density lipoprotein
MIC, minimal inhibitory concentration
MUFA, monounsaturated fatty acid
PUFA, polyunsaturated fatty acid
SFA, saturated fatty acid
WHO, World Health Organization

INTRODUCTION

The hazelnut is one of the most cultivated and consumed nuts in the world, not only as a fruit but also incorporated as an ingredient into a diversity of manufactured food products.

Turkey is the world's largest hazelnut producer, contributing to approximately 65% of the total world global production (around 815,361 MT in 2007) (FAO, 2009), followed by Italy (16%), the United States (4.0%), and Azerbaijan (3.4%). Other countries, such as Georgia, Iran, and Spain, among others, contribute only 10% of the total production (FAO, 2009).

Nuts & Seeds in Health and Disease Prevention. DOI: 10.1016/B978-0-12-375688-6.10073-8

FIGURE 73.1
Green fruit with leaf cover and different varieties of hazelnuts. The photographs show the green fruit with leaf cover, and three varieties of hazelnuts, namely Daviana, Fertile de Coutard, and Merveille de Bollwiller.

BOTANICAL DESCRIPTION

Hazelnut (*Corylus avellana* L.) belongs to the *Betulaceae* family. The hazel is a tree or shrub that may grow to 6 m high, exhibiting deciduous leaves. These are rounded, 6—12 cm in length and width, softly hairy on both surfaces, and with a double-serrate margin. Hazel is mainly distributed on the coasts of the Black Sea region of Turkey, Southern Europe (Italy, Spain, Portugal, France, and Greece), and in some areas of the USA (Oregon and Washington). It can also be cultivated in other countries, such as New Zealand, China, Azerbaijan, Chile, Iran, and Georgia, among others. Green, ready-to-harvest hazelnut fruit are shown in Figure 73.1.

Several hazelnut varieties exist worldwide. The most common European varieties may be classified according their main use (Table 73.1). In Turkey, the main world producer, the most frequent are Acı, Cavcava, Çakıldak, Foşa, İncekara, Kalınkara, Kan, Kara, Karafındık, Kargalak, Kuş, Mincane, Palaz, Sivri, Tombul, Uzunmusa, Yassıbadem, Yuvarlak Badem, Yassi Badem, and Imperial de Trebizonde (Açkurt *et al.*, 1999; Özdemir *et al.*, 2001; Özdemir & Akinci, 2004; Köksal *et al.*, 2006). Daviana, Fertile de Coutard, and M. Bollwiller are hazelnut cultivars widely distributed.

HISTORICAL AND PRESENT-DAY CULTIVATION AND USAGE

Hazelnuts were and are still typically consumed as the whole nut (raw or roasted), or used as an ingredient in a variety of foods, especially in bakery products, snacks, chocolates, cereals, dairy products, salads, entrees, sauces, ice cream, and other dessert formulations. After cracking the hazelnut's hard shell, the hazelnut kernel may be consumed raw (with its skin)

TABLE 73.1 Classification of Main Hazelnut Varieties According to their Final Use

Table	Industry	Dual Use
Butler	Camponica	Da veiga
Cosford	Casina	I. Eugénie
Daviana	Comum	Ribet
Ennis	Couplat	San Giovani
Fertile de Coutard	Dawton	Sta. M. di Gesu
Grada de Viseu	Gironela	Segorbe
Griffol	Morell	Tonda di Giffoni
Grosse de Espanha	Mortarella	
Gunslebert	Negret	
Lansing	Pauetet	
L. d'Espagne	R. de Piémont	
M. Bollwiller	Tonda G. Romana	
Mollari		
Provence		

or roasted (without its skin). Hazelnut oil is also becoming increasingly popular in Turkey and elsewhere, being widely utilized for cooking, deep frying, salad dressings, and flavoring ingredients, among others uses (Alasalvar *et al.*, 2006a).

Several by-products, such as the leaves, the hazelnut green leafy cover, the hazelnut hard shell, and hazelnut skin, are obtained through the harvesting, shelling/hulling, cracking, and roasting processes, respectively. These by-products have a lower commercial value than hazelnut kernels. However, among these by-products, the hazelnut's hard shell is currently used for burning as a heat source, for mulching, and as a raw material for the production for furfural in the dye industry. Moreover, the hazelnut green leafy covers and tree leaves are sometimes used as organic fertilizers for hazelnut trees and other crops upon composting. Hazel leaves are also widely used in folk medicine, in the preparation of infusions for the treatment of hemorrhoids, varicose veins, phlebitis, and edema of the lower limbs, as consequence of its astringency, vasoprotective, and anti-edema properties (Valnet, 1992).

APPLICATIONS TO HEALTH PROMOTION AND DISEASE PREVENTION

Hazelnuts are highly nutritious, due to their high fat content. Table 73.2 shows their composition in terms of protein (10–20%) and oil (> 50%), which are the major constituents. The most important fatty acids in hazelnuts are the monounsaturated and polyunsaturated fatty acids (MUFAs and PUFAs, respectively) (Parcerisa *et al.*, 1998; Köksal *et al.*, 2006; Venkatachalam & Sathe, 2006; Oliveira *et al.*, 2008). The presence of MUFAs and PUFAs, notably the ω-3 and ω-6 fatty acids, is considered more desirable in terms of nutritional quality than the saturated fatty acids, and because of their possible health benefits (Venkatachalam & Sathe, 2006), as there is evidence that a MUFA-rich diet can lower the risk of coronary heart disease and may have a preventive effect against atherosclerosis. Oleic acid is the most important (> 70%), followed by linoleic acid. Taking into account the lipid classes that are present in hazelnuts — namely, non-polar lipids (triacylglicerols) and polar lipids (monogalactosyldiacylglycerols and phospholipids, the latter including phosphatidylinositol and phosphatidylcholine) — oleic acid is the main fatty acid in these classes, followed by linoleic acid (Parcerisa *et al.*, 1997). In terms of triacylglycerol composition, trioleoylglycerol is predominant, followed by dioleoyl-linoleoylglycerol, at 48.32–71.31% and 12.36–18.10%, respectively (Alasalvar *et al.*, 2006a).

629

TABLE 73.2 Energy, Oil, Protein and Fatty Acid Composition of Hazelnut (*Corylus avellana* L.) Kernels

Reference	Energy (kcal/g)	Oil (% Dry Weight)	Protein (% Dry Weight)	Fatty Acid Composition (% of Total Oil)								
				Oleic (18:1ω-9)	Linoleic (18:2 ω-6)	Palmitic (16:0)	Stearic (18:0)	Linolenic (18:3)	Myristic (14:0)	SFA (%)	MUFAs (%)	PUFAs (%)
Oliveira et al., 2008	6.49–6.77	58.0–64.2	15.6–16.2	80.67–82.63	9.84–11.26	4.95–5.76	1.74–1.87	0.15–0.19	0.03–0.04	6.97–7.77	81.12–83.05	9.99–11.43
Alasalvar et al., 2006a				82.78	8.85	4.81	2.69	0.12	0.04	7.79	83.24	8.97
Balta et al., 2006		57.5–74.1	10.7–19.2	73.48–81.57	10.46–15.61	4.39–8.85	1.67–3.18	0.02–0.34	0.01–0.16			
Köksal et al., 2006		56.07–68.52	12.2–21.9	74.2–82.8	9.82–18.7	4.72–5.87	0.86–2.49	0.029–0.076		7.1	92.8	13.1
Venkatachalam & Sathe, 2006				82.95	7.55	5.78	3.12	0.24	0.03	9.11	83.10	7.79
Özdemir & Akinci, 2004	6.49–6.80	57.39–62.90	18.25–22.06									
Özdemir et al., 2001		67.1–72.0		75.7–80.7	10.1–13.8	6.6–8.3	1.8–3.8					
Parcerisa et al., 1998				74.13–82.83	8.17–17.78	4.66–6.04	1.38–3.36	0.10–0.17		6.87–9.00	74.50–83.16	8.27–17.89

Hazelnut kernels also contain essential amino acids — for example, arginine, histidine, izoleucine, leucine, lysine, methionine, phenylalanine, threonine, and valine — with arginine and leucine being the most abundant (Köksal *et al.*, 2006; Venkatachalam & Sathe, 2006). Non-essential amino acids (alanine, aspartic acid/asparagine, glutamic acid/glutamine, glycine, proline, serine, and tyrosine) also exist in hazelnut kernels, with glutamic acid/glutamine found in the greatest quantity, followed by aspartic acid/asparagine, and alanine (Köksal *et al.*, 2006; Venkatachalam & Sathe, 2006). Venkatachalam and Sathe (2006) performed a study on the chemical composition of several edible nut seeds, including hazelnuts, and verified that, with the exception of almonds and peanuts (which were deficient in the sulfur amino acids, methionine and cysteine), all other tree nuts appear to contain adequate amounts of all of the essential amino acids when compared to the FAO/WHO recommended essential amino acid pattern for an adult.

Regarding hazelnut sugars, few studies have involved these compounds. In 2006, Venkatachalam and Sathe (2006), when analyzing the chemical composition of several edible nut seeds, determined a sugar percentage in hazelnuts of $1.41 \pm 0.05\%$ (wet weight). However, these values are indicative, as the sugar content of nut seeds varies considerably, depending on the growing conditions, seed maturity, cultivar, and location.

Hazelnuts also contain vitamins. The predominant soluble vitamins are vitamins B_1, B_2, B_6, niacin, ascorbic acid, and folic acid (only determined in some varieties) (Açkurt *et al.*, 1999; Köksal *et al.*, 2006). The major insoluble vitamins detected are α-tocopherol (Parcerisa *et al.*,1998; Açkurt *et al.*, 1999; Köksal *et al.*, 2006), followed by retinol and δ-tocopherol (Köksal *et al.*, 2006). Alasalvar *et al.* (2006a) and Amaral *et al.* (2006) analyzed hazelnuts of the Tombul variety and cultivars collected in several Portuguese localities, respectively, in relation to tocol (tocopherol and tocotrienol) composition, and detected and quantified seven tocols: α-tocopherol, β-tocopherol, γ-tocopherol, δ-tocopherol, α-tocotrienol, β-tocotrienol, and γ-tocotrienol. Once more, α-tocopherol was the most abundant.

There is great interest in determining the mineral composition of hazelnuts, due to some metals having pro-oxidant activity and health benefits. The mineral composition of hazelnut kernels depends not only on the variety but also on the growing conditions, such as the soil type and geographical factors. In the studies performed by Açkurt *et al.* (1999), Ozdemir *et al.* (2001), and Köksal *et al.* (2006), potassium was the predominant mineral, followed by calcium and magnesium. Moreover, the presence of iron, zinc, and copper, and a high potassium: sodium ratio, mean that hazelnuts are of interest for human diets, and especially for electrolyte balance. However, iron and copper are considered to be pro-oxidant minerals that might reduce both the shelf-life and the sensory characteristics of hazelnuts, because they are involved in rancidity (Açkurt *et al.*, 1999). Therefore, varieties with low unsaturated: saturated ratios, and that are low in pro-oxidant components, rich in antioxidant components, and low in enzymatic activities, are preferred, because they minimize post-harvest quality losses, and packaging and refrigeration costs (Açkurt *et al.*, 1999).

In recent years, some studies have been performed in order to evaluate the potential of hazelnut by-products as sources of natural antioxidants and as functional food ingredients (Alasalvar *et al.*, 2006b; Shahidi *et al.*, 2007; Contini *et al.*, 2008) — which may be of great importance to the hazelnut industry. Several phenolic compounds have been detected in hazelnut kernels and leaves. Gallic acid, caffeic acid, *p*-coumaric acid, ferulic acid, and sinapic acid (Figure 73.2) were observed in kernels (Alasalvar *et al.*, 2006b; Shahidi *et al.*, 2007). In addition to these, the leaves were found to contain 3-caffeoylquinic acid, 4-caffeoylquinic acid, 5-caffeoylquinic acid, caffeoyltartaric acid, *p*-coumaroyltartaric acid, myricetin 3-rhamnoside, quercetin 3-rhamnoside, and kaempferol 3-rhamnoside (Figure 73.2) (Oliveira *et al.*, 2007). The former group of compounds has also been detected in hazelnut skins, hard shells, and green leafy covers (Shahidi *et al.*, 2007).

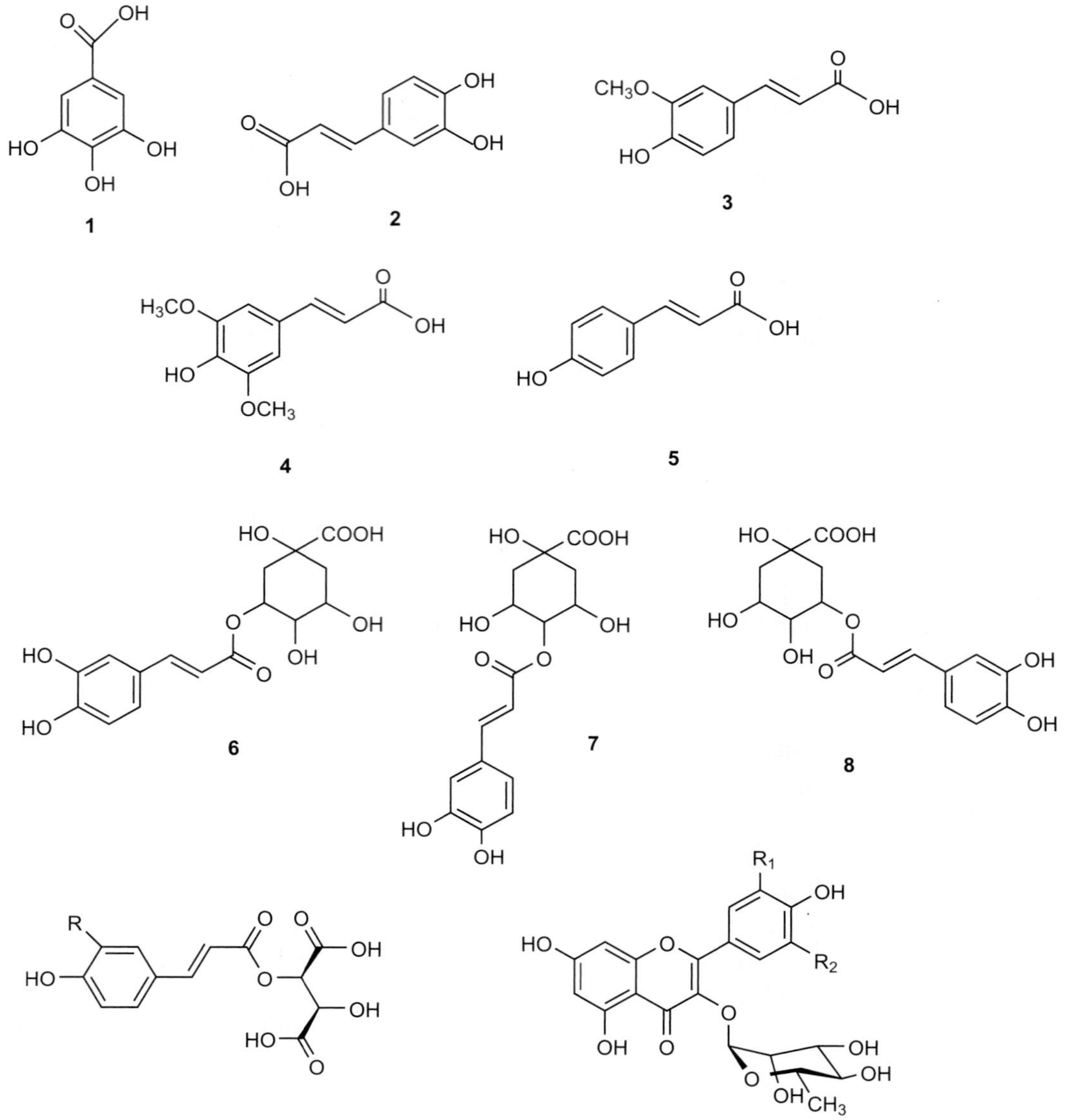

FIGURE 73.2

Chemical structures of identified phenolic compounds in hazelnut kernels and leaves. (1) Gallic acid, (2) caffeic acid, (3) ferulic acid, (4) sinapic acid, (5) *p*-coumaric acid, (6) 3-caffeoylquinic acid, (7) 4-caffeoylquinic acid, (8) 5-caffeoylquinic acid, (9) caffeoyltartaric acid, (10) *p*-coumaroyltartaric acid, (11) myricetin 3-rhamnoside, (12) quercetin 3-rhamnoside, (13) kaempferol 3-rhamnoside.

The composition of the hazelnut has led some recent research studies to indicate that enriching the diet with this nut may improve human health. The ingestion of hazelnuts might have benefits related to the blood serum lipid profile, as they protect low density lipoprotein (LDL) against oxidation, and decrease the plasma oxidized LDL level (Durak *et al.*, 1999; Orem *et al.*, 2008). Moreover, hazelnut supplementation may prevent peroxidation reactions, which are related to inflammatory and ischemic diseases, cancer, hemochromatosis, acquired immunodeficiency syndrome, emphysema, organ transplantation issues, gastric ulcers, hypertension and pre-eclampsia, neurologic diseases, alcoholism, smoking-related diseases, and others, due to the presence of antioxidant compounds.

Antimicrobial activity of hazelnuts

Nowadays, there is growing interest in discovering natural antimicrobial compounds that might be used as alternatives to the chemical preservatives used in food industry, and to antibiotics, thus decreasing their use, and lowering the probability of the occurrence of human resistance to these chemicals. Many studies have therefore focused on the antimicrobial agents and properties of plant-derived active principles, which have been used for some time in traditional medicine to overcome infections. As far as we know, only two studies regarding determination of the hazelnut's antimicrobial potential have been carried out; one with hazelnut kernels (Oliveira *et al.*, 2008), and one with leaves (Oliveira *et al.*, 2007), and involving three cultivars: M. Bollwiller, Fertile de Coutard, and Daviana. In both studies, the minimal inhibitory concentration (MIC) values were determined for gram-positive bacteria (*Bacillus cereus*, *Bacillus subtilis*, and *Staphylococcus aureus*), gram-negative bacteria (*Escherichia coli*, *Pseudomonas aeruginosa*, and *Klebsiella pneumoniae*), and fungi (*Candida albicans* and *Cryptococcus neoformans*), using the agar streak dilution method based on radial diffusion (Figure 73.3). MICs are normally used to evaluate antimicrobial activity, and are considered to be the lowest concentration of the tested sample needed to inhibit the growth of bacteria or fungi after 24 hours (Oliveira *et al.*, 2007, 2008). For each MIC, the diameter of the inhibition zones are generally measured, linking an inhibition zone of less than 1 mm to no antimicrobial activity (−), an inhibition zone of 2−3 mm to slight antimicrobial activity (+), an inhibition zone of 4−5 mm to moderate antimicrobial activity (++), an inhibition zone of 6−9 mm to high antimicrobial activity (+++), and an inhibition zone of higher than 9 mm to strong antimicrobial activity (++++).

When comparing both studies (Table 73.3), the hazelnut kernels and leaves showed different antimicrobial activities. In the case of leaves, the samples revealed antimicrobial activity against all microorganisms with the exception of *P. aeruginosa* and *C. albicans*, which were resistant to the extracts at a concentration of 100 mg/ml. In the case of aqueous extracts of hazelnut kernels, high antimicrobial activity was only found against gram-positive bacteria; namely, *B. cereus*, *B. subtilis*, and *S. aureus*. On the contrary, gram-negative bacteria and fungi were resistant to the tested extracts at all the assayed concentrations.

Regarding hazelnut cultivars, the most promising is M. Bollwiller for leaves, and Daviana for aqueous extracts of hazelnut kernels, since smaller MICs and larger zones of inhibition of growth were determined for these varieties.

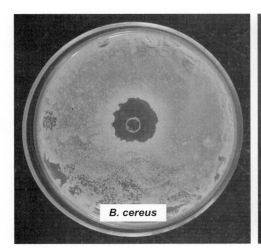
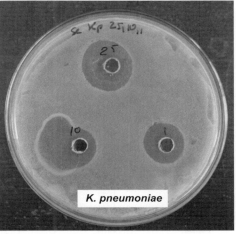

FIGURE 73.3

Antimicrobial activity of hazelnut extracts against *Bacillus cereus* and *Klebsiella pneumoniae*. The photographs show the inhibition zones obtained after application of hazelnut extracts to cultures of a gram-positive bacteria (*Bacillus cereus*) and a gram-negative bacteria (*Klebsiella pneumoniae*).

TABLE 73.3 Antimicrobial Activity (MICs and Inhibition Zones) of Hazelnut Kernels and Leaves*

Cultivar	Gram-Positive Bacteria						Gram-Negative Bacteria						Fungi			
	B. cereus		B. subtilis		S. aureus		P. aeruginosa		E. coli		K. pneumoniae		C. albicans		C. neoformans	
	Kernel	Leaf	Kernel	Leaf	Kernel	Leaf	Kernel	Leaf	Kernel	Leaf	Kernel	Leaf	Kernel	Leaf	Kernel	Leaf
M. Bollwiller	0.1 (++++)	0.1 (++++)	1 (++++)	1 (++++)	0.1 (++++)	0.1 (++)	100 (−)	100 (−)	100 (−)	50 (++++)	100 (−)	1 (++)	100 (−)	100 (−)	100 (−)	100 (+++)
F. Coutard	0.1 (++)	0.1 (++)	0.1 (++++)	1 (+++)	0.1 (+++)	0.1 (++++)	100 (−)	100 (−)	100 (−)	100 (−)	100 (−)	10 (+++)	100 (−)	100 (−)	100 (−)	100 (−)
Daviana	0.1 (++++)	0.1 (++)	0.1 (+++)	1 (++)	0.1 (++++)	0.1 (++++)	100 (−)	100 (−)	100 (−)	100 (++++)	100 (−)	50 (++++)	100 (−)	100 (−)	100 (−)	100 (−)

MIC, minimal inhibitory concentration, considered as the lowest concentration of the tested sample able to inhibit the growth of bacteria or fungi after 24 hours.
No antimicrobial activity (−), inhibition zone < 1 mm; slight antimicrobial activity (+), inhibition zone 2−3 mm; moderate antimicrobial activity (++), inhibition zone 4−5 mm; high antimicrobial activity (+++),
inhibition zone 6−9 mm; strong antimicrobial activity (++++), inhibition zone > 9 mm.

ADVERSE EFFECTS AND REACTIONS (ALLERGIES AND TOXICITY)

In areas where birch trees are endemic, hazelnut allergy might be observed. This is often the result of primary sensitization to the cross-reactive birch pollen allergen Bet v 1, usually resulting in a mild oral allergy syndrome in both adults and children (Flinterman *et al.*, 2006). On the other hand, sensitization to hazelnut in early childhood has been related to sensitization to other tree nuts and peanuts, which can all cause serious reactions. Moreover, hazelnuts, like other nuts, may, if not stored correctly, favor the development of aflatoxin fungi producers, such as *Aspergillus flavus*. This type of metabolite, due to its toxicity to humans, is of great concern regarding guaranteeing food safety and human health.

SUMMARY POINTS

- Hazelnuts are well adapted to different agro-climatic conditions, and a great number of varieties are known.
- The fruit is used in different forms, and in a large number of products.
- Hazelnuts possess a rich chemical composition, being beneficial to health mainly because of their high monounsaturated fatty acids content.
- Among the phytochemicals with biological properties, tocopherols and phenolic compounds are of great importance.
- Extracts of different *C. avellana* parts, such as leaves and fruits, and of different varieties, show the ability to inhibit the growth of different important pathogenic microorganisms.
- The consumption of these fruits, beyond other important biological properties, may promote human protection against infections, explaining their wide use in traditional medicine and showing their great potential as a source of bioactive substances.

635

References

Açkurt, F., Özdemir, M., Biringen, G., & Mahmut, L. (1999). Effect of geographical origin and variety on vitamin and mineral composition of hazelnut (*Corylus avellana* L.) varieties cultivated in Turkey. *Food Chemistry, 65,* 309–313.

Alasalvar, C., Amaral, J. S., & Shahidi, F. (2006a). Functional lipid characteristics of Turkish Tombul hazelnut (*Corylus avellana* L.). *Journal of Agricultural and Food Chemistry, 54,* 10177–10183.

Alasalvar, C., Karamac, M., Amarowicz, R., & Shahidi, F. (2006b). Antioxidant and antiradical activities in extracts of hazelnut kernel (*Corylus avellana* L.) and hazelnut green leafy cover. *Journal of Agricultural and Food Chemistry, 54,* 4826–4832.

Amaral, J. S., Casal, S., Alves, M. R., Seabra, R. M., & Oliveira, B. P. P. (2006). Tocopherol and tocotrienol content of hazelnut cultivars grown in Portugal. *Journal of Agricultural and Food Chemistry, 54,* 1329–1336.

Balta, M. F., Yarilgaç, T., Aşkin, M. A., Kuçuk, M., Balta, F., & Özrenk, K. (2006). Determination of fatty acid compositions, oil contents and some quality traits of hazelnut genetic resources grown in eastern Anatolia of Turkey. *Journal of Food Composition and Analysis, 19,* 681–686.

Contini, M., Baccelloni, S., Massantini, R., & Anelli, A. (2008). Extraction of natural antioxidants from hazelnut (*Corylus avellana* L.) shell and skin by-products by long maceration at room temperature. *Food Chemistry, 110,* 659–669.

Durak, I., Köksal, I., Kaçmaz, M., Büyükkoçak, S., Çimen, B. M. Y., & Öztürk, H. S. (1999). Hazelnut supplementation enhances plasma antioxidant potential and lowers plasma cholesterol levels. *Clinica Chimica Acta, 284,* 113–115.

FAO. (2009). http://faostat.fao.org/faostat (accessed October 2009).

Flinterman, A. E., Hoeskstra, M. O., Meijer, Y., Ree, R. V., Akkerdaas, J. H., Bruijnzeel-Koomen, C. A., et al. (2006). Clinical reactivity to hazelnut in children: association with sensitization to birch pollen or nuts. *Journal of Allergy and Clinical Immunology, 118,* 1186–1188.

Köksal, A. I., Artik, N., Şimşk, A., & Günes, N. (2006). Nutrient composition of hazelnut (*Corylus avellana* L.) varieties cultivated in Turkey. *Food Chemistry, 99,* 509–515.

Oliveira, I., Sousa, A., Valentão, P., Andrade, P. B., Ferreira, I. C. F. R., Ferreres, F., et al. (2007). Hazel (*Corylus avellana* L.) leaves as source of antimicrobial and antioxidative compounds. *Food Chemistry, 105,* 1018–1025.

Oliveira, I., Sousa, A., Morais, J. S., Ferreira, I. C. F. R., Bento, A., Estevinho, L., et al. (2008). Chemical composition and antioxidant and antimicrobial activities of three hazelnut (*Corylus avellana* L.) cultivars. *Food and Chemical Toxicology, 46,* 1801–1807.

Orem, A., Balaban, F., Kural, B. V., Orem, C., & Turhan, I. (2008). *Hazelnut consumption protect low density lipoprotein (LDL) against oxidation and decrease plasma oxidazed LDL level.* Istanbul, Turkey: 77th Congress of the European Atherosclerosis Society. April 26–29, p. 215.

Özdemir, F., & Akinci, I. (2004). Physical and nutritional properties of major commercial Turkish hazelnut varieties. *Journal of Food Engineering, 63,* 341–347.

Özdemir, M., Açkurt, F., Kaplan, M., Yildiz, M., Löker, M., Gürcan, T., et al. (2001). Evaluation of new Turkish hybrid hazelnut (*Corylus avellana* L.) varieties: fatty acid composition, α-tocopherol content, mineral composition and stability. *Food Chemistry, 73,* 411–415.

Parcerisa, J., Richardson, D. G., Rafecas, M., & Codony, R. (1998). Fatty acid, tocopherol and sterol content of some hazelnut varieties (*Corylus avellana* L.) harvested in Oregon (USA). *Journal of Chromatography A, 805,* 259–268.

Parcerisa, J., Richardson, D. G., Rafecas, M., Codony, R., & Boatella, J. (1997). Fatty acid distribution in polar and nonpolar lipid classes of hazelnut oil (*Corylus avellana* L.). *Journal of Agricultural and Food Chemistry, 45,* 3887–3890.

Shahidi, F., Alasalvar, C., & Liyana-Pathirana, C. M. (2007). Antioxidant phytochemicals in hazelnut kernel (*Corylus avellana* L.) and hazelnut by-products. *Journal of Agricultural and Food Chemistry, 55,* 1212–1220.

Valnet, J. (1992). *Phytothérapie: Traitment des maladies par les plantes, 6th edn).* Paris: Maloine. pp. 473–475.

Venkatachalam, M., & Sathe, S. K. (2006). Chemical composition of selected edible nut seeds. *Journal of Agricultural and Food Chemistry, 54,* 4705–4714.

CHAPTER 74

Medicinal Use of Hempseeds (*Cannabis sativa* L.): Effects on Platelet Aggregation

Delfín Rodríguez Leyva[1], Richelle S. McCullough[2], Grant N. Pierce[2]
[1] Cardiovascular Research Division, V.I. Lenin University Hospital, Holguin, Cuba
[2] Institute of Cardiovascular Sciences, St Boniface Hospital Research Centre, and Department of Physiology, University of Manitoba, Winnipeg, Manitoba, Canada

CHAPTER OUTLINE

LIST OF ABBREVIATIONS

ALA, omega-3 α-linolenic acid
CDB, cannabidiol
CE, cholesteryl esters
CRP, C reactive protein
DHA, docosahexaenoic acid
EFA, essential fatty acid
EPA, eicosapentaenoic acid
HDL, high density lipoprotein
LA, omega-6 linoleic acid
TC, total cholesterol

Nuts & Seeds in Health and Disease Prevention. DOI: 10.1016/B978-0-12-375688-6.10074-X

TG, triglycerides
PUFA, polyunsaturated fatty acid
THC, tetrahydrocannabinol

INTRODUCTION

Dietary interventions can exert important effects impacting health and disease. Hempseed (*Cannabis sativa* L.) is an excellent source of nutrition. The nutrient composition of hempseed is shown in Table 74.1. Principally, it is very rich in essential fatty acids (EFAs) and other polyunsaturated fatty acids (PUFAs), but it contains almost as much protein as soybean and is also rich in vitamin E and minerals such as phosphorus, potassium, sodium, magnesium, sulfur, calcium, iron, and zinc (Callaway, 2004).

TABLE 74.1 Nutrient Profile of Hempseed

Nutrient	Units	Value per 100 g
Proximates		
Energy	kcal	567
Energy	kJ	2200
Protein	g	24.8
Total lipid (fat)	g	35.5
Ash	g	5.6
Carbohydrates	g	27.6
Fiber, total dietary	g	27.6
Digestable fiber	g	5.4
Non-digestable fiber	g	22.2
Moisture	g	6.5
Glucose	g	0.30
Fructose	g	0.45
Lactose	g	< 0.1
Maltose	g	< 0.1
Minerals		
Calcium, Ca	mg	145
Iron, Fe	mg	14
Magnesium, Mg	mg	483
Phosphorus, P	mg	1160
Potassium, K	mg	859
Sodium, Na	mg	12
Zinc, Zn	mg	7
Copper, Cu	mg	2
Manganese, Mn	mg	7
Selenium, Se	mcg	<0.02
Vitamins		
Vitamin C	mg	1.0
Thiamin	mg	0.4
Riboflavin	mg	0.11
Niacin	mg	2.8
Vitamin B-6	mg	0.12
Vitamin A	IU	3800
Vitamin D	IU	2277.5
Vitamin E	mg	90.00

TABLE 74.1   Nutrient Profile of Hempseed—continued

Nutrient	Units	Value per 100 g
Lipids		
Saturated fat	g	3.3
16:0	g	3.44
18:0	g	1.46
20:0	g	0.28
Monounsaturated fat	g	5.8
18 : 1 (Omega-9)	g	9
Total polyunsaturated	g	36.2
18 : 2 (Omega-6)	g	56
18 : 3 (Omega-6)	g	4
18 : 3 (Omega-3)	g	22
Cholesterol	mg	0
Amino acids		
Tryptophan	g	0.20
Threonine	g	0.88
Isoleucine	g	0.98
Leucine	g	1.72
Lysine	g	1.03
Methionine	g	0.58
Cystine	g	0.41
Phenylalanine	g	1.17
Tyrosine	g	0.82
Valine	g	1.28
Arginine	g	3.10
Histidine	g	0.71
Alanine	g	1.28
Aspartic acid	g	2.78
Glutamic acid	g	4.57
Glycine	g	1.14
Proline	g	1.15
Serine	g	1.27

Data represent mean values of the nutritional content of the Finola variety of hempseed.
Adapted from West (1998) and Leizer *et al.* (2000).

BOTANICAL DESCRIPTION

The Swedish naturalist Carl Linnaeus (1754) was the first to describe the *Cannabis* genus using the "modern" system of taxonomic nomenclature. *Cannabis sativa* is a mono- and dioecious annual plant in the Cannabaceae family (Kingdom, Plantae; Division, Magnoliophyta; Class, Magnoliopsida; Order, Rosales) (Hillig, 2005). It is a flowering herb with palmately compound or digitate leaves and serrate leaflets. Generally, cannabis has imperfect flowers with staminate and pistillate flowers occurring on separate plants (Lebel-Hardenack & Grant, 1997). It is a wind-pollinated plant that produces seeds. Cannabis is a diploid organism, having a chromosome complement of 2n = 20.

Two main types of *Cannabis sativa* must be distinguished: the industrial type and the drug type. Industrial hemp can be found as fiber or seed oil. The main cannabinoids found in this type are cannabidiol (CDB) and tetrahydrocannabinol (THC). The THC concentrations are < 0.3%, so it has no psychoactivity. The second type of *Cannabis sativa* L. is also known as marijuana, hashish, or cannabis tincture, and contains THC as one of its main cannabinoids (Ross *et al.*, 2000; Holler *et al.*, 2008). THC is found in concentrations between 1 and 20% in this type of cannabis (Ross *et al.*, 2000; Holler *et al.*, 2008). Therefore, it possesses potent psychoactivity (Ross *et al.*, 2000; Holler *et al.*, 2008).

HISTORICAL CULTIVATION AND USAGE

Cannabis sativa is among the earliest plants thought to be cultivated by humans. Archeological findings indicate that it has been cultivated for fibers (strings, textiles, ropes, and paper) in China since 4000 BC (Li & Lin, 1974), and in India about 3000 years ago, where it was used as an analgesic, anticonvulsant, anti-inflammatory, antibiotic, antispasmodic, appetite stimulant, digestive, diuretic, and aphrodisiac, among other uses (Aldrich, 1997). Cannabis was considered sacred in Tibet, where it was used to facilitate meditation (Touwn, 1981). Assyrians also used it as incense, and the Persians were aware of many of the plant's effects (Touwn, 1981).

PRESENT-DAY CULTIVATION AND USAGE

It is not legal to cultivate hempseed, in the United States. This ban is mostly due to concerns that the legalization of hemp may make it easier to legalize marijuana (West, 1998). Other governments have accepted the distinction between the two types of cannabis and, while continuing to penalize the growing of marijuana, have legalized the growing of industrial hemp (West, 1998). Canada, France, Australia, China, Great Britain, Austria, Russia, and Spain have been among the most important producers of hempseed. Within the past 10 years, hempseed has been legally used as food for humans and animals in some countries, like Canada. It is also an important component in paint and varnish production (Callaway, 2004). Some countries have used the fiber from hempseed in the past to produce fabrics or specialty papers, such as canvas, linen, tea bags, paper money, and others (Callaway, 2004). Hempseed and hempseed meals are also edible products that are proposed to have health-related benefits. Hempseed contains proteins and polyunsaturated fatty acids (PUFAs), plus considerable quantities of vitamins and minerals (Table 74.1). It is the health benefits of hempseed upon which this chapter focuses.

APPLICATIONS TO HEALTH PROMOTION AND DISEASE PREVENTION

Dietary hempseed contains omega-6 fatty acid, linoleic acid (LA), and omega-3 fatty acid, α-linolenic acid (ALA), in a 4 : 1 ratio. This ratio has been found to be ideal for a healthy diet (Holub, 2002). Other rich sources of LA are shown in Table 74.2.

PUFAs originate from the diet. Longer chain PUFAs can be created from the elongation and desaturation of their dietary precursors, ALA and LA (Figure 74.1). Both families of fatty acids, n-3 and n-6, share and compete for the same enzymes (Δ^6-desaturase, Δ^5-desaturase, and elongases) in their biosynthesis, and Δ^6-desaturase is the rate-limiting step (Simopoulos, 2008). Following its metabolism, LA can be converted into arachidonic acid, while ALA will be converted into the long chain fatty acids, eicosapentaenoic acid (EPA) and docosahexaenoic acid (DHA) (Figure 74.2). Therefore, concurrent administration of LA in the diet will reduce ALA accumulation, and *vice versa*. Theoretically, a lower ratio of omega-6/omega-3 fatty acids is more advantageous in reducing the risk of many of the chronic diseases of high prevalence in Western societies, as well as in the developing countries (Simopoulos, 2008).

Based on the close relationship between the biochemical pathways of ALA and LA, and the capacity of both to be converted into long chain fatty acids, plant sources of ALA (like flax seed and canola) have attracted scientific attention for their health-related potential. Unfortunately, because of legal regulations, lack of knowledge, confusion, and controversies about the differences between fiber hemp and marijuana, hempseed research has been limited and slow to develop. Not only is the nutritional value of hempseed important, but the effects of LA as an

TABLE 74.2 Rich Sources of the Essential fatty Acid, Linoleic Acid

Source of LA	LA (g/100 g)[a]
Safflower oil	73
Corn oil	57
Hempseed oil	56
Cottonseed oil	50
Soybean oil	50
Sesame oil	40
Black walnuts	37
English walnuts	35
Sunflower seeds	30
Brazil nuts	25
Margarine	22
Pumpkin and squash seeds	20
Spanish peanuts	16
Peanut butter	15
Almonds	10

These foods are common dietary sources of linoleic acid.
Adapted from Ensminger & Konlande (1993).

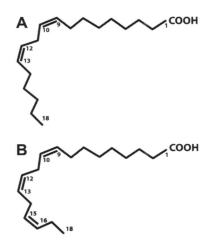

FIGURE 74.1

Linoleic and α-linoleic acid. Chemical structures of (A) linoleic acid, an 18-carbon omega-6 fatty acid, and (B) α-linolenic acid, its omega-6 counterpart. Carbons are labeled, indicating the locations of double bonds and their effects on the physical structure of the lipid. These essential fatty acids are presented in hempseed in a 4:1 ratio favoring linoleic acid.

essential fatty acid also demand a better understanding of the appropriate doses and presentation (oil, nuts, etc.), as well as the group of patients (age, health condition, comorbidities) that can obtain better benefits from a hempseed supplemented diet.

Medicinal use of hempseeds (*cannabis sativa* L.): effects on platelet aggregation

ANIMAL DATA

The biochemical metabolism of omega-6 fatty acids can produce eicosanoids (arachidonic acid) and stimulate thromboxane A2 synthesis (Figure 74.3). This can induce platelet aggregation (Simopoulos, 2008). Because of this, hempseed has been studied for its effects on platelet aggregation in normal and hypercholesterolemic animals. Hypercholesterolemia

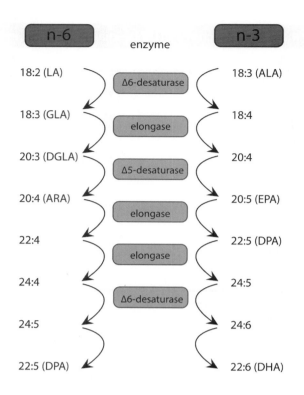

FIGURE 74.2

Competitive use of enzymes for the bioconversion of omega-3 and omega-6 fatty acids. This biochemical pathway is used for the metabolism of linoleic acid and α-linolenic acid into longer chain polyunsaturated fatty acids in animals and humans. LA, linoleic acid; GLA, γ-linolenic acid; DGLA, dihomo γ-linolenic acid; ARA, arachidonic acid; DPA, docosapentaenoic acid; ALA, α-linolenic acid; EPA, eicosapentaenoic acid; DHA, docosahexaenoic acid. Adapted from Chen & Nilsson (1993), *Biochim. Biophys. Acta, 1166,* 193—201.

FIGURE 74.3

The effects of omega-3 and omega-6 long chain fatty acid metabolism on inflammatory pathways and platelet aggregation, through the production of prostaglandins and eicosanoids. The production of thromboxane A2 through omega-6 fatty acids is believed to be the main mechanism for linoleic acid's effects on platelet aggregation. ARA, arachidonic acid; EPA, eicosapentaenoic acid; DHA, docosahexaenoic acid; COX, cyclooxygenase; LOX, lipoxygenase; LTB, leukotriene B; PGG_2, prostaglandin G_2; PGH_2, prostaglandin H_2. Adapted from Simopoulos (2008), *Exp. Biol. Med. (Maywood, NJ, US), 8,* 674—688.

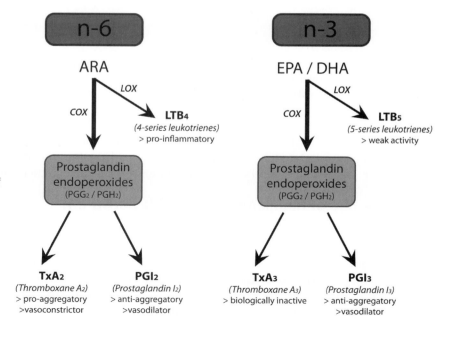

indirectly increases the risk of cardiovascular disease by enhancing the ability of platelets to aggregate (Prociuk et al., 2008). Prociuk and colleagues have shown that if male New Zealand White rabbits with an underlying hypercholesterolemic condition are supplemented with 10% of hempseed for 8 weeks, there is a return of the accelerated cholesterol-induced platelet aggregation to normal levels. The authors found that this normalization was not related to a reduction in plasma cholesterol levels, but was associated with an increased level of plasma γ-linolenic acid — a unique PUFA found in hempseed (Prociuk et al., 2008). The direct addition of γ-linolenic acid to platelet samples obtained from animals fed a cholesterol-supplemented diet blocked the cholesterol-induced stimulation of platelet aggregation (Prociuk et al., 2008). In this study, animals co-supplemented with hempseed and cholesterol significantly increased their plasma levels of cholesterol and triglycerides, and induced a non-selective, significant elevation of all fatty acids.

In another study examining the anti-aggregatory effects of hempseed, Richard and colleagues supplemented male Sprague-Dawley rats with diets that included 5 and 10% hempseed (Richard et al., 2008). Platelet aggregation and the rate of aggregation were significantly inhibited by both diets after 12 weeks of dietary intervention (Richard et al., 2008). Despite its similar total fat content, this effect was not achieved when palm oil was included in the diet. These data support the hypothesis that the specific fatty acid composition of hempseed was responsible for its beneficial effects on platelet aggregation. Total plasma PUFAs from rats fed with hempseed were significantly increased in comparison to control (Richard et al., 2008). ALA and LA levels also increased significantly in a concentration-dependent manner (Richard et al., 2008). These authors suggested that the elevation of ALA levels was the most likely factor involved in the inhibitory effect on platelet aggregation. ALA can be metabolized into the anti-aggregatory eicosanoids EPA (eicosapentaenoic acid) and/or DHA (docosahexaenoic acid). However, in this study EPA and DHA levels were unaltered after hempseed supplementation. Thus, the most probable explanation for the anti-aggregatory effects of hempseed was that ALA itself was responsible. This study also concluded that LA did not possess potent pro-thrombotic effects when delivered with ALA in the form of hempseed (Richard et al., 2008). Platelet aggregation plays a central role in thrombotic coronary syndromes as well as in thrombotic vascular disease. Because most patients at high risk of coronary heart disease have a hypercholesterolemic condition, these findings may have important clinical relevance. It also emphasizes the potential of dietary hempseed for treating or preventing cardiovascular diseases, like myocardial infarctions and strokes, that have thrombotic episodes as a central feature of the pathology.

HUMAN DATA

Hempseed actions in humans have only been studied to a limited extent. In a recent investigation aimed at comparing the effects of three different dietary oils (fish, flaxseed and hempseed), 86 healthy subjects completed a 12-week study where they were supplemented with 2 g/day of these oils. The oils were given in the form of two large capsules. This dosage was chosen as one that most of the general public would ingest under the belief that it would provide health-related benefits. Fish oil is enriched in the n-3 fatty acids, EPA and DHA. Flaxseed oil is enriched in ALA, and hempseed oil is enriched in ALA and the n-6 fatty acid LA (Kaul et al., 2008). In this investigation, intervention with 2 g/day of hempseed did not significantly increase the concentrations of any fatty acids, including ALA or LA. It also did not alter the level of plasma total cholesterol (TC), high density cholesterol, low density cholesterol, or triglycerides (TG). Hempseed ingestion at this dosage and in this form also did not induce any changes in collagen- or thrombin-stimulated platelet aggregation, nor the levels of inflammatory markers (Kaul et al., 2008). It is difficult to argue that 2 g/day of hempseed oil will ever have a physiological impact upon the body if the PUFA content in the blood is not increased in response to this dietary regimen. The lack of effects was likely attributable to the relatively low doses used, but this was chosen as one that most of the public would likely use.

643

We could conclude that such use of hempseed oil may, therefore, be a waste of money. However, this conclusion is tempered by the knowledge that the volunteers in this study were healthy subjects with no underlying pathology. Supplementation with this dose and form of hempseed oil may have more dramatic effects in patients who already exhibit clinical symptoms.

Higher doses of hempseed oil (30 ml/day) have also been tested (Schwab *et al.*, 2006). Using a randomized, double-blinded, crossover design, hempseed and flaxseed oils at the same doses were compared for their effects on serum lipids and fasting concentrations of total and lipo-protein lipids, plasma glucose, and insulin. The authors also investigated hemostatic factors in healthy humans. After 4 weeks of dietary intervention, hempseed supplementation induced higher levels of both LA and γ-linolenic acid in serum cholesterol esters (CE) and TG. The flaxseed intervention resulted in a higher proportion of ALA in both serum CE and TG compared to the hempseed oil. Importantly, the proportion of arachidonic acid in CE was lower after the flaxseed diet than after the hempseed period, but hempseed produced a lower total-to-HDL cholesterol ratio, a predictor of coronary heart disease. No significant differences were found between the interventions in fasting serum total or lipoprotein lipids, plasma glucose, insulin, or hemostatic factors (D-dimer, fibrinogen, FVIIa, and PAI-1 activity), or C reactive protein concentrations (Schwab *et al.*, 2006).

In summary, hempseed is a rich source of linoleic acid and other nutrients that may have beneficial health properties. However, the specific pathologies or conditions in which it can be used are in need of more research. Effects of LA on platelet aggregation are not completely clear. There remain a great number of unanswered questions that demand research before we can be convinced of the use of hempseed as a preventive or therapeutic dietary intervention.

644

ADVERSE EFFECTS AND REACTIONS (ALLERGIES AND TOXICITY)

In most countries where industrial hemp is licensed for cultivation, plants must not exceed a level of 0.3% THC, the principal psychoactive constituent of the species (Small & Marcus, 2003). Generally, the THC content of hempseed and hemp food is so low that psychotropic or even pharmacological effects can be excluded with certainty, even if larger quantities of food are consumed. THC limits for food have only been adopted in Switzerland since THC-containing food resulted in side effects for some consumers (Grotenhermen *et al.*, 2009). Canada legislated THC limits for raw and semi-finished hemp products in 1998. THC limits for food are listed in Table 74.3. CBD, the main compound in industrial hemp, has been shown to have sleep-inducing properties (Small & Marcus, 2003). Dietary hempseed has been associated with platelet inhibition (Prociuk *et al.*, 2008), so there is a possibility that bleeding times would be longer. This effect would be a significant concern in those patients under treatment with anticoagulant drugs or during surgical interventions. In general, drug interactions with hempseed ingestion are not well known. Diarrhea may also be an adverse effect of whole hempseed consumption; however, this remains to be tested. To date, there have been no reported cases of toxicity from the ingestion of hempseed oil or its other constituents (Leizer *et al.*, 2000).

SUMMARY POINTS

- *Cannabis sativa* L. (hemp) is a mono and dioecious annual plant in the Cannabaceae family, rich in linoleic acid (LA) and α-linolenic acid (ALA), proteins, vitamin E, and minerals.
- In industrial hemp (fiber type), the main cannabinoids are cannabidiol and tetrahydrocannabinol (THC); however, the concentrations are < 0.3%, so it has no appreciable psychoactivity.

TABLE 74.3 Legal Tetrahydrocannabinol (THC) Limits for Food in Switzerland

Food	Limit (mg/kg)
Hemp oil	50
Hempseeds	20[a]
Breads and pastries	5[a]
Vegetable food	2[a]
Spirits	5[b]
Non-alcoholic drinks	0.2[c]
Alcoholic drinks	0.2[d]
Herb and fruit teas	0.2[e]

Adapted from Grotenhermen et al. (1998).
[a]On a dry weight basis;
[b]mg per liter of pure alcohol;
[c]mg per liter of finished product;
[d]excluding spirits;
[e]mg per liter of finished product. Assumes 15 g of plant parts per kg of boiling water, steeped for 30 minutes at above 85°C.

- Archeological findings indicate that hemp has been cultivated in China since 4000 BC, and in India since about 1000 BC, and used as an analgesic, anticonvulsant, anti-inflammatory, antibiotic, antispasmodic, appetite stimulant, digestive, diuretic, and aphrodisiac.
- Animal data have shown that after 8 weeks of ingestion, a diet supplemented with 10% of hempseed returns cholesterol-induced platelet aggregation to normal levels.
- Another animal study concluded that LA has no pro-thrombotic effects when delivered with ALA in the form of hempseed.
- Following 12 weeks of dietary supplementation with 2 g/day of hempseed oil, no change was found in collagen or thrombin stimulated platelet aggregation in healthy humans.
- No significant differences in hemostatic factors (D-dimer, fibrinogen, FVIIa, and PAI-1 activity) were found after 4 weeks of dietary supplementation with 30 ml/day of hempseed oil as compared with flax-seed oil at the same dose.
- The THC content of hempseed and hemp food is so low that psychotropic or even pharmacological effects can be excluded with certainty, even if larger quantities of food are consumed.
- Drug interactions with dietary hempseed are not well known.

ACKNOWLEDGEMENTS

This work was supported through a grant from the Canadian Institutes for Health Research. The indirect costs of this research were supported by the St Boniface Hospital and Research Foundation. Dr Rodriguez Leyva was a Visiting Scientist of the Heart and Stroke Foundation of Canada.

References

Aldrich, M. (1997). History of therapeutic cannabis. In M. L. Mathre (Ed.), *Cannabis in medical practice Jefferson, NC* (pp. 35–55), McFarland.

Callaway, J. C. (2004). Hempseed as a nutritional resource: an overview. *Euphytica, 140*, 65–72.

Chen, Q., & Nilsson, A. (1993). Desaturation and chain elongation of n-3 and n-6 polyunsaturated fatty acids in the human CaCo-2 cell line. *Biochimica Biophysica Acta, 1166*, 193–201.

Ensminger, A. H., & Konlande, J. E. (1993). Fats and other lipids. *Foods & nutrition encyclopedia.* Boca Raton, FL: CRC Press. p. 691.

Grotenhermen, F., Karus, M., & Lohmeyer, D. (2009). THC limits for food: A scientific study. Hürth, Germany: Nova-Institute. Available at: http://www.naihc.org/hemp_information/content/nova_report/part1.html (accessed August 20, 2009).

Hillig, K. W. (2005). Genetic evidence for speciation in cannabis (Cannabaceae). *Genetic Resources and Crop Evolution, 52,* 161–180.

Holler, J. M., Bosy, T. Z., Dunkley, C. S., Levine, B., Past, M. R., & Jacobs, A. (2008). D9-tetrahydrocannabinol content of commercially available hemp products. *Journal of Analytical Toxicology, 32,* 428–432.

Holub, B. J. (2002). Clinical nutrition: 4. Omega-3 fatty acids in cardiovascular care. *Canadian Medical Association Journal, 166,* 608–615.

Kaul, N., Kreml, R., Austria, J. A., Landry, M. N., Edel, A. L., Dibrov, E., et al. (2008). A comparison of fish oil, flaxseed oil and hempseed oil supplementation on selected parameters of cardiovascular health in healthy volunteers. *Journal of the American College of Nutrition, 27,* 51–58.

Lebel-Hardenack, S., & Grant, S. R. (1997). Genetics of sex determination in flowering plants. *Trends in Plant Science, 2,* 130–136.

Leizer, C., Ribnicky, D., Poulev, A., Dushenkov, S., & Raskin, I. (2000). The composition of hempseed oil and its potential as an important source of nutrition. *Journal of Nutraceuticals Functional and Medical Foods, 2,* 35–53.

Li, H. L., & Lin, H. (1974). An archaeological and historical account of cannabis in China. *Economic Botany, 28,* 437–447.

Linnaeus, C. (1753). *Salvius* [*Facsimile edition.* In: *Species Plantarum, 2.* London, UK: Ray Society]. 1027Stockholm1957–1959.

Prociuk, M. A., Edel, A. L., Richard, M. N., Gavel, N. T., Ander, B. P., Dupasquier, C. M. C., et al. (2008). Cholesterol-induced stimulation of platelet aggregation is prevented by a hempseed-enriched diet. *Canadian Journal of Physiology and Pharmacology, 86,* 153–159.

Richard, M. N., Ganguly, R., Steigerwald, S. N., Al-Khalifa, A., & Pierce, G. N. (2007). Dietary hempseed reduces platelet aggregation. *Journal of Thrombosis and Haemostasis, 5,* 424–425.

Ross, S. A., Mehmedic, Z., Murphy, T. P., & Elsohly, M. A. (2000). GC-MS analysis of the total D9-THC content of both drug- and fiber-type cannabis seeds. *Journal of Analytical Toxicology, 24,* 715–717.

Schwab, U. S., Callaway, J. C., Erkkilä, A. T., Gynther, J., Uusitupa, M. I. J., & Jarvinen, T. (2006). Effects of hempseed and flaxseed oils on the profile of serum lipids, serum total and lipoprotein lipid concentrations and haemostatic factors. *European Journal of Nutrition, 45,* 470–477.

Simopoulos, A. P. (2008). The importance of the omega-6/omega-3 fatty acid ratio in cardiovascular disease and other chronic diseases. *Experimental Biology and Medicine (Maywood, NJ, US), 8,* 674–688.

Small, E., & Marcus, D. (2003). Tetrahydrocannabinol levels in hemp (*Cannabis sativa* L.) germplasm resources. *Economic Botany, 57,* 545–558.

Touwn, M. (1981). The religious and medicinal uses of cannabis in China, India and Tibet. *Journal of Psychoactive Drugs, 13,* 23–34.

West, D. P. (1998). Hemp and marijuana: Myths & realities. North American Industrial Hemp Council, Inc. Available at: http://www.naihc.org/hemp_information/content/hemp.mj.html (accessed July 12, 2009).

Hongay Oil Tree (*Pongamia pinnata* Linn.) Seeds in Health and Disease Benefits

Sachin L. Badole, Subhash L. Bodhankar
Department of Pharmacology, Poona College of Pharmacy, Bharati Vidyapeeth University, Maharashtra, India

LIST OF ABBREVIATIONS

HSV-1, Herpes simplex virus type-1
HSV-2, Herpes simplex virus type-2
PPSM, methanolic extract of *P. pinnata* Linn. seed
w/v, weight/volume

INTRODUCTION

The Indian beech, Pongam seed oil tree or Hongay seed oil tree is a fast-growing, medium-sized spreading tree that forms a broad canopy casting moderate shade. The plant grows well in tropical areas with a warm humid climate and well-distributed rainfall. Though it grows in almost all types of soils, silty soils on river banks are ideal. The tree is tolerant to drought and salinity. Originally an Indo-Malaysian species, it is now found in many countries. It is a shade bearer, and is considered to be a good tree for planting in pastures, as grass grows well in its shade. The tree is suitable for afforestation, especially in watershed areas and in drier parts of the country (Joy *et al.*, 1998).

FIGURE 75.1
Authenticated sample of seeds of *Pongamia pinnata* (L.).

BOTANICAL DESCRIPTION

Pongamia pinnata (Linn.) Pierre (Synonyms: *Pongamia glabra* Vent., *Derris indica* (Lam.) Bennet, *Cystisus pinnatus* Lam.) is a member of the Fabacae family (Papilionacae; Leguminasae). *Pongamia pinnata* is a medium-sized, glabrous, semi-evergreen tree, growing upto 18 m or more in height, with a short bole, spreading crown, and grayish green or brown bark. The leaves are imparipinnate, alternate, with five to seven leaflets, ovate and opposite; flowers are lilac or pinkish white and fragrant, in axillary racemes. The calyx is cup-shaped, four- to five-toothed, with a papilionaceous corolla. There are 10 monadelphous stamens, and the ovary is subsessile, two-ovuled, with an incurved, glabrous style ending in a capitate stigma. The seed pod is compressed, woody, indehiscent, and yellowish-gray when ripe; it varies in size and shape, being elliptic to obliquely oblong, 4.0—7.5 cm long and 1.7—3.2 cm broad, with a short curved beak. It usually contains one or, rarely, two elliptical or reniform seeds, 1.7—2.0 cm long and 1.2—1.8 cm broad, which are wrinkled with a reddish-brown leathery testa (Figure 75.1) (Joy *et al.*, 1998).

HISTORICAL CULTIVATION AND USAGE

The tree starts bearing at the age of 4—7 years. White and purplish flowers in auxiliary racemes appear from April to July. The fruits come to harvest at different periods of the years in various parts of the country, but the harvest period is from November—December to May—June. The pods are collected from April to June, and the shells removed by hand. The yield of seed is said to range from 9 to 90 kg per tree (Krishnamurthi, 1998). The pods are dried in the sun, and the seeds extracted by thrashing the fruits.

During the 16th century, fruits of Karanja were used for parasitic infections, obstinate urinary diseases, and diabetes. Sushruta (an ancient Indian surgeon) prescribed Karanja seeds with honey for hemorrhage; expressed oil from seeds, for use as a laxative, for intestinal parasites, and for skin afflictions; and used the fruits for urinary and vaginal discharges. The seed oil was an ingredient of hair oil, prescribed by Sushruta for baldness, and was applied externally for dermatitis, rheumatic diseases, and muscular atrophy (Khare, 2004).

PRESENT-DAY CULTIVATION AND USAGE

Propagation of *P. pinnata* is by direct seedlings, or by planting nursery raised seedlings. Propagation by branch cuttings and root suckers is also possible.

Seeds can be sown immediately after their removal from mature pods. Their natural longevity is 6 months, although they can be stored for up to a year if they are not removed from their pods; once removed from their pods, they can be stored for longer in an airtight box. Seeds are sown in raised seedbeds or in polybags, with a spacing of 7.5 cm × 1.5 cm, at the beginning of the hot season, with the micropyle facing downwards, and start to germinate from the seventh day of sowing. Soaking the seeds in hot water (50°C) for 15 minutes before sowing improves the germination percentage, and also vigor. The germination in this species is hypogeal — i.e., in normal seedlings the cotyledons remain beneath the germination medium, while the plumage pushes upwards and emerges above.

The seedlings are raised in seedbeds or polypots. Mulching of the beds is helpful. Seedlings attain a height of 25–30 cm in a year. When they have reached a height of 60 cm, they are planted out entire, with the ball of earth, or in the form of stumps. For a plantation, 30 cm^3 pits in a space 5 × 5 m are appropriate. Small agencies and forest contractors organize the collection and transport and seedlings.

Fruit-setting of *Pongamia* starts from the fifth year onwards in plantations. The tree flowers in April–May, and fruits mature in January–February. Each pod bears single seed, and the average fresh weight of a matured seed is 1.2 g. Commercial production of seeds from plantations starts 10 years after planting, and a full-grown tree may yield up to 100 kg or more of fresh seeds per annum for up to 60–70 years (Kureel *et al.*, 2008). The removal of seeds from pods is done manually (Joy *et al.*, 1998).

The seeds have many industrial and medicinal uses. Seed oil has antiseptic, stimulant, and healing properties, and is applied in skin diseases, scabies, sores, herpes, and eczema. The seeds are said to have antidyslipidemic, anti-inflammatory, antiviral, antifilarial, and anti-ulcerogenic activities.

649

APPLICATIONS TO HEALTH PROMOTION AND DISEASE PREVENTION

Seeds and seed oil both contain karanjin, pongamol, pongapin, and kanjone. The seeds also contain lanceolatin B, iso-pongaflavone, and pongol, while the seed oil contains iso-pongachromene and pongaglabrone. Glabrachalcone was isolated from the seed oil, which also gave the number of fatty acids, oleic and linolenic acid being major constituents (Chopade *et al.*, 2008).

The mature seed consists of about 5% of shell and 95% oleaginous kernel. The kernels are tough, white in color, and covered with thin, reddish, brittle skin. The main composition of a sample of air-dried kernels is as follows: moisture 19.0%, fatty oil 27.5%, protein 17.4%, starch 6.6%, crude fiber 7.3%, and ash 2.4%. The seeds contain mucilage (13.5%), traces of an essential oil, and a complex amino acid named glabrin. Six compounds (two sterols, three sterol derivatives, and one disaccharide) together with eight fatty acids (three saturated and five unsaturated) have been isolated from the seeds of *Pongamia pinnata*. Furanoflavones (i.e., karanjin, pongapin, kanjone, pongagalabrone, diketone pongamol, and pinnatin) have been isolated and characterized from the Indian seeds. Seeds from Australia contained only pongapin. Immature seeds contain a flavone derivative, "pongol." Another flavonoid isolated from the seeds was "glabrachalcone isopongachromene." Karanjin and pongamol, the major flavones of the seeds, are associated with the non-glyceride portion of the oil and have a bitter taste (Krishnamurthi, 1998; Chopade *et al.*, 2008).

Fresh extracted seed oil is yellowish-orange to brown, becoming darker during storage, and has a disagreeable odor and bitter taste. Solvent extraction yields a better quality of seed oil that bears specific characteristics (Table 75.1). The fatty acids present in the seed oil include

TABLE 75.1	Specific Characteristics of Hongay Seed Oil
Specific gravity at 30°C	0.925—0.940
Iodine value	80—96
Saponification value	185—195
Unsaponification value	2.6—3.0
Refractive index at 40°C	1.434—1.479
Acid value	6.3
Acetate value	20.9

Source: Krishnamurthi (1998).

TABLE 75.2	Fatty Acids Present in Hongay Seed Oil
Fatty Acids	**% Present in Oil**
Myristic acid	0.23
Palmitic	6.06
Stearic	2.19
Oleic	61.30
Linolenic	0.46
Linolic	9.72
Lignoceric	3.22
Dihydroxysteraric	4.36
Eicosenoic	9.5—12.4
Arachidic	4.30%
Behenic	4.2—5.3%

Source: Krishnamurthi (1998).

myristic, palmitic, stearic, oleic, linolenic, linolic, lignoceric, dihydroxysteraric, eicosenoic, arachidic, and behenic acids (Table 75.2). The seeds contain traces of an essential oil. The non-fatty components of the seed oil include karanjin (1.25%) and pongamol (0.85%) (Krishnamurthi, 1998).

Karanjin is the active principle responsible for the curative effect of the seed oil in skin diseases. Clinical experiments have indicated that it is free from the highly irritating and inflammatory effects of coumarin compounds. Its application in solution with vegetable oil or groundnut oil is better than when incorporated in a paraffin base (Khare, 2004). Seed oil shows antibacterial activity against both gram-positive and gram-negative organisms, and is also used in scabies, leprosy, hemorrhoids, ulcers, chronic fever, liver pain, and lumbago. Powdered seed is valued as a febrifuge and tonic, and is also used in bronchitis, chronic fever, whooping cough, and painful rheumatic joints (Kirtikar & Basu, 1987). The seeds, crushed to paste, are used for leprous sores, skin diseases, and painful rheumatic joints. Seed extracts of *Pongamia pinnata* have hypotensive effects, and produce uterine contractions. An anticon-vulsant effect, central nervous system depressant activity, and increased sensitivity to sound and touch have been reported (Basu *et al.*, 1994).

Seed cake is rich in protein (30—35%), but it is unpalatable and toxic to cattle even at a level of 4%, due to the presence of several toxic principles (karanjin, pongamol, etc.). These toxins are oil-soluble, and most of them are removed during solvent extraction of seed oil from cake, using hexane. Therefore, de-oiled cakes could be used in compound cattle feed. Essential oil of the seeds injected intravenously into experimental animals elevated their blood pressure and slightly relaxed the bronchioles. The pulp of the seeds is applied in leprosy. The powdered seeds, after decortication, are given as a specific remedy for whooping cough. The powdered seeds are also supposed to be value as a febrifuge and tonic in asthenic and debilitating conditions. Seed oil has antiseptic and stimulant healing properties, and is applied for skin diseases, scabies, sores, and herpes. For eczema, a mixture of the oil and zinc oxide is beneficial. An embrocation made of equal parts of the seed oil and lemon juice is applied in skin diseases due to fungus growth. Mixed with lime or lemon juice, it has been reported to be useful in the treatment of rheumatism. Internally, it has sometimes been used as stomachic and cholagogue in cases of dyspepsia and a sluggish liver (Kirtikar & Basu, 1987). The essential seed oil from *Pongamia pinnata* has shown mild antifungal activity against keratinophillic fungi, namely *Verticillium tenuipers*, *Malbranchea pulchella*, *Keratinophytontereum*, and *Chrysosporium tropicum*. Fruits are used to treat urinary disease and piles. The seeds are used in inflammations, otalgia, lumbago, pectoral diseases, chronic fevers, hydrocele, hemorrhoids, and anemia. The seed oil is recommended for ophthalmia, hemorrhoids, herpes, and lumbago. Seeds and seed oils are carminative, antiseptic, anthelmintic, and antirheumatic. Seeds are hematinic, bitter, and acrid. The seed oil is styptic and depurative. The seed contains pongam oil, which is used for lubrication, while its twigs are used as chewing sticks for cleaning the teeth, and its fruits and sprouts are used for the treatment of abdominal tumors in India (Buccolo & David, 1973).

Bhatia and colleagues (2008) reported antidyslipidemic activity of different solvent fractions of *Pongamia pinnata* fruits in a triton and high-fat diet fed hamster model. The chloroform and butanol extracts of *Pongamia pinnata* at concentrations of 500 µg significantly inhibited the generation of superoxide anions and hydroxyl radicals, as well as the reaction of lipid peroxidation, in rat liver microsomes induced by Fe^{+2}. The chloroform fraction caused significant reversal in the plasma level of triglycerides, total cholesterol, high density lipoprotein, cholesterol, glucose, and glycerol.

A methanolic extract of *P. pinnata* Linn. seed (PPSM) showed a dose-dependent (12.5−50 mg/kg, p.o., for 5 days) anti-ulcer effect against gastric ulcer induced by 2 hours of cold resistance stress. PPSM (25 mg/kg) showed anti-ulcerogenic activity against acute gastric ulcer induced by pylorus ligation and aspirin, and duodenal ulcer induced by cystamine, but not ethanol-induced gastric ulcer (Prabha *et al.*, 2009). Different extracts of seeds of *P. pinnata* have been reported to have anti-inflammatory activity (Singh & Pandey, 1996). Elanchezhiyan and colleagues (1993) reported the antiviral effect of an extract of *P. pinnata* seeds against herpes simplex virus type-1 (HSV-1) and type-2 (HSV-2) in Vero cells. A crude aqueous seed extract of *P. pinnata* completely inhibited the growth of HSV-1 and HSV-2 at concentrations of 1 and 20 mg/ml (w/v), respectively, as shown by complete absence of cytopathic effect. Fruit extracts of *P. pinnata* showed antifilarial potential on the cattle filarial parasite *Setaria cervi* (Uddin *et al.*, 2003).

ADVERSE EFFECTS AND REACTIONS (ALLERGIES AND TOXICITY)

The results of sub-acute toxicity studies have shown no abnormalities in body weight, hematological and biochemical parameters of blood, and histopathological studies. The biological effect of pongamol and its present toxicological studies suggest that it may be used safely in chronic toxicological studies and clinical trial (Baki *et al.*, 2007). Karanjin is highly toxic to fish; pongamol exhibits comparatively mild toxicity. The seeds are reported to be used as fish poison (Krishnamurthi, 1998).

SUMMARY POINTS

- Hongay oil tree (*Pongamia pinnata*) is a common plant found in many countries.
- The plant is cultivated from the seed present in the pods.
- Seeds yield oils, which have been traditionally used for gastrointestinal and skin disorders.
- Karanjin is the main active principle of the seed oil.

- Another ingredient, pongamol, has been reported to have anticonvulsant and central nervous system depressant activity.
- Different extracts of seeds of *Pongamia pinnata* are said to have antidyslipidemic, anti-inflammatory, antiviral, antifilarial, and anti-ulcerogenic activities.
- Pongamol is safe in human beings, whereas karanjin is highly toxic to fish.

References

Baki, M. A., Khan, A., Al-Bari, M. A. A., Mosaddik, A., Sadik, G., & Mondal, K. A. M. S. H. (2007). Sub-acute toxicological studies of pongamol isolated from *Pongamia pinnata*. *Journal of Medical Sciences Research, 2,* 53—57.

Basu, S. P., Mandal, J. K., & Mehdi, N. S. (1994). Anticonvulsant effect of pongamol. *Indian Journal of Pharmaceutical Sciences, 56,* 163—167.

Bhatia, G., Puri, A., Maurya, R., Yadav, P. P., Khan, M. M., Khanna, A. K., et al. (2008). Anti-dyslipidemic and antioxidant activities of different fractions of *Pongamia pinnata* (L.) fruit. *Medicinal Chemistry Research, 17,* 281—289.

Buccolo, G., & David, H. (1973). Quantitative determination of serum triglycerides by the use of enzymes. *Clinical Chemistry, 19,* 476—480.

Chopade, V. V., Tankar, A. N., Pande, V. V., Tekade, A. R., Gowekar, N. M., Bhandari, S. R., et al. (2008). *Pongamia: pinnata* phytochemical constituents, traditional uses and pharmacological properties: a review. *International Journal of Green Pharmacy, 2,* 72—75.

Elanchezhiyan, M., Rajarajan, S., Rajendran, P., Subramanian, S., & Thyagarajan, S. P. (1993). Antiviral properties of the seed extract of an Indian medicinal plant, *Pongamia pinnata* Linn., against herpes simplex viruses: *in vitro* studies on Vero cells. *Journal of Medical Microbiology, 38,* 262—264.

Joy, P. P., Thomos, J., Mathew, S., & Skaria, B. P. (1998). *Medicinal plants.* In: *Pongamia pinnata*. Kerala, India: Kerala Agriculture University, Aromatic and Medicinal Plant Research Station, 73—74.

Khare, C. P. (2004). *Encyclopedia of Indian medicinal plants.* New York, NY: Springer-Verlag. pp. 378—379.

Kirtikar, K. R., & Basu, B. D. (1987). In E. Blatter, J. F. Caius & K. S. Mhaskar (Eds.) (2nd ed.). *Indian medicinal plants, Vol. 1* (pp. 830—832) Allahabad, India: Basu, L.M.

Krishnamurthi, A. (1998). *The Wealth of India, Vol. 8,* Raw materials. New Delhi, India: Council of Scientific and Industrial Research. 206—211.

Kureel, R. S., Singh, C. B., Gupta, A. K., & Pandey, A. (2008). *Karanja: A potential source of biodiesel.* New Delhi, India: Government of India, Ministry of Agriculture National Oilseeds and Vegetable Oils Development Board. Available at: www.novodboard.com1—16.

Prabha, T., Dorababu, M., Goel, S., Agarwal, P. K., Singh, A., Joshi, V. K., et al. (2009). Effect of extract of *Pongamia pinnata* Linn on gastro-duodenal ulceration and mucosal offensive and defensive factor in rats. *Indian Journal of Experimental Biology, 47,* 649—659.

Singh, R. K., & Pandey, B. L. (1996). Anti-inflammatory activity of seed extracts of *Pongamia pinnata* in rats. *Indian Journal of Physiology and Pharmacology, 40,* 355—358.

Uddin, Q., Parveen, N., Khan, N. U., & Singhal, K. C. (2003). Antifilarial potential of the fruits and leaves extract of *Pongamia pinnata* on cattle filarial parasite *Setaria cervi*. *Phytotherapy Research, 17,* 1104—1107.

CHAPTER 76

Seeds of Horse Chestnut (*Aesculus hippocastanum* L.) and Their Possible Utilization for Human Consumption

Giorgia Foca[1], Alessandro Ulrici[1], Marina Cocchi[2], Caterina Durante[2], Mario Li Vigni[2], Andrea Marchetti[2], Simona Sighinolfi[2], Lorenzo Tassi[2]
[1] Department of Agricultural and Food Sciences, University of Modena and Reggio Emilia, Reggio Emilia, Italy
[2] Department of Chemistry, University of Modena and Reggio Emilia, Modena, Italy

CHAPTER OUTLINE

LIST OF ABBREVIATIONS

A, *Aesculus*
AH, *Aesculus hippocastanum*
CNS, central nervous system
SDS-PAGE, sodium dodecyl sulphate-polyacrylamide gel electrophoresis
SRXRF, synchrotron radiation X-ray fluorescence
ETAAS, electrothermal atomic absorption spectrometry
SEM-EDS, scansion electron microscopy—energy dispersive X-ray spectroscopy
TG, thermogravimetry
DSC, differential scanning calorimetry

Nuts & Seeds in Health and Disease Prevention. DOI: 10.1016/B978-0-12-375688-6.10076-3

INTRODUCTION

The ancestral practice of therapeutical medicine, based on empirical knowledge rather than rational investigation, is of great importance for less developed countries, but this approach also has a great impact on the applicative studies of medicine in more developed nations. In order to face the hazardous endemic pathologies present in sub-developed countries and to control chronic pathologies, which are more widespread in population groups with an high number of elderly people, new active principles are necessary, analogs to those potentially obtainable from the long experience of the traditional medicines.

BOTANICAL DESCRIPTIONS

The sub-family of *Hippocastanaceae* (genus, *Aesculus*; family, Sapindaceae) comprises 20 species, whose habitat is mainly represented by the temperate zones of the northern hemisphere. Among these species, *A. chinensis* and *A. turbinata* Blume (the Japanese horse chestnut) seem to be the most widespread and well-known species in the Eurasian continent, while *A. pavia* is probably the most common in North America. From the physiological point of view, *A. hippocastanum* tolerates low temperatures, prefers fertile soil, and has a low resistance towards atmospheric contaminants. Unfortunately, a new parasite called *Cameraria ohridella* (a Lepidopter, an exclusive leaf miner for AH), first observed in Macedonia about 30 years ago, is responsible for the serious attacks on AH, in particular towards the white flower species, causing dry leaves and defoliation at the beginning of summer.

HISTORICAL CULTIVATION AND USAGE

A. hippocastanum originated from the Balkan and Caucasian regions, before spreading to the European countries (*c.* 1600). Today it is widespread, at up to 1200 m in altitude.

This plant has been used in common human and veterinary medicines — the decoctions from leaves and seeds as cardiotonics and anti-inflammatories, those from bark and twigs as febrifuges, and both to treat dermatitis. The cotyledons release saponins in hot water, acting as tensioactive agents with some antimicrobial and bacteriostatic properties. The seed meal is inedible for humans because of its strong bitterness and the relevant toxicity of escin fractions. However, when roasted, the seeds can be used as a coffee substitute.

PRESENT-DAY CULTIVATION AND USAGE

In Europe, AH is so widespread that intensive cultivation is unnecessary. The fruits harvested from wild plants are sufficient for industrial processes. The most important usages are in the field of health products (in herboristic remedies and pharmaceutical formulations). The seed extracts are also used as formulating agents in various cosmetic applications, such as shampoos, shower foams, creams, lotions, and toothpastes (Thornfeldt, 2005).

It should be noted that some other varieties of *Aesculaceae* growing in Asian countries (A. *indica* and A. *turbinata* Blume) are more suitable for human consumption, following some preliminary preparation. For example, the seeds from A. *indica* represent a foodstuff still largely consumed by local populations (Parmar & Kaushal, 1982). The starch contained within the seeds ($\sim$40% in fresh fruits, ratio amylopectin : amylose $\cong$ 3:1) is blended with other flours, after removal of its bitter taste due to saponins by extended soaking in water.

Horse chestnuts: chemical composition and characterization

The chemical characterization of the seeds of genus *Aesculus* is rather incomplete. The literature provides partial information, with many papers being standpoints of excellence in this field.

The composition of AH seeds can be summarized as follows. The main classes of components are:

- starch and non-starch saccharides
- proteins (Azarkovich & Gumilevskaya, 2006)
- lipids (essential oils)
- minerals (ashes).

Among minor compounds, the following classes are the most important bioactively:

- escin (saponin and sapogenin fractions), which are the most abundant
- coumarin-derived compounds (aesculin, fraxin, and scopolin, among others)
- tannins (leucocyanidine, proanthocyanidin A_2)
- non-protein nitrogen compounds (i.e., adenine, adenosine, guanine, uric acid)
- vitamin-B complex, methionine, and holine
- sterols.

Escin is a generic mixture of saponins (triterpenoid glycosides) and sapogenins. The latter family is more complex, having two molecular groups, triterpenic and steroydic (Figure 76.1), which show a different basic structure and are variously substituted.

Typically, different classes of the main bioactive principles are present in different organs and parts of *Aesculaceae*: escin fractions predominate in the seeds, essential oils in the leaves and flowers, tannins and coumarin-derived compounds in the bark. In addition, the crude extracts of the seeds contain significant amounts of some other families of different compounds, such as condensed tannins (mainly flavonoids), and sterols (Stankovic *et al.*, 1984).

The oils extracted from seeds generally contain both saponifiable and unsaponifiable matter. The unsaponifiable constituents of horse chestnut still remain largely unidentified. Although they represent a very small fraction of the essential oil, approximately 2–3% (equivalent to 0.08–0.15% on the dry-meal basis), they include sterols, triterpenes, aliphatic alcohols, vitamins, chlorophylls, and pigments, among others.

Hricoviniova and Babor (1991) analyzed the saccharide constituents of different parts of AH seeds. They reported that starch, arabinans, and glucoarabinans are the main constituents of the cotyledon. A series of monosaccharides (D-glucose, L-arabinose, D-glucuronic acid, D-xylose, D-galactose, and fucose) were found in the hydrolyzed extract of the seeds. Episperm, the external protective cellulosic layer, mainly contains xylans, associated to glucoxylans.

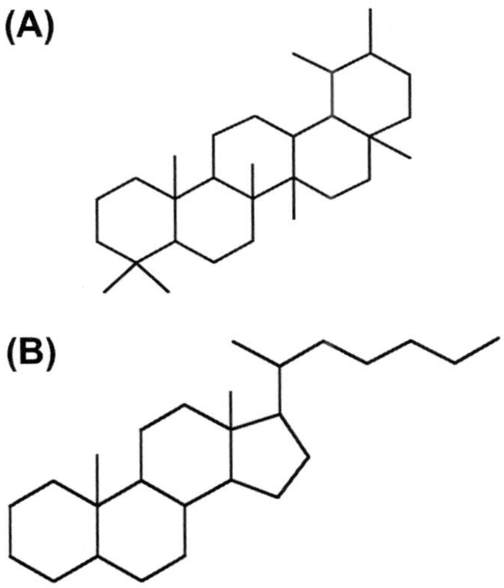

655

FIGURE 76.1
Base structure of (A) triterpenic and (B) steroydic groups, the most representative molecular families of sapogenins.
Triterpenic and steroydic molecular structures, variously substituted and derivatized, are the main components of the sapogenins family present in *Aesculaceae* seeds.

Because of the scarcity of experimental data regarding the chemical composition of AH seeds, in a recent paper we partially characterized these natural products by applying some analytical techniques (Baraldi *et al.*, 2007). We compared the seeds of the two most common Mediterranean varieties: the *pure* species (AHP, with white flowers), and a *hybrid* species (AHH, with soft pink flowers). Some specific information on the morphological structure of the seeds was also obtained. Surface analysis by SEM-EDS (Figure 76.2) showed no significant differences in the meal samples (wild-type) from the two different botanical origins. Both of them appear to be composed of a complex matrix in which platelet-form starch particles are dispersed. However, thermal analysis (TG, DSC) has outlined some significant differences between them. Study of the analytical composition reveals some differences in residual moisture, protein, lipid, glucide, and ash contents. Probably these fractions modulate other undifferentiated chemical parameters, such as Cold Water Solubility (CWS), and Total Inorganic Soluble Salts (TISS). The quantitative results are summarized in Table 76.1.

For comparison purposes, the last column of Table 76.1 summarizes some analytical data reported by Parmar and Kaushal (1982) regarding the composition of *A. indica* seeds. Unfortunately, only limited data of this type are available in the literature. As far as the total glucidic content in *A. indica* is concerned, 11.0% (9.1% reducing and 1.9% non-reducing

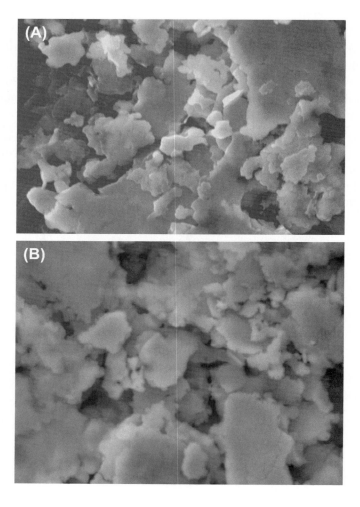

FIGURE 76.2
SEM images (×5000) of flour samples (wild type) of seeds of different botanical origin: (A) AHP, (B) AHH. SEM images of floured seeds of different botanical origin (AHP and AHH) from Modena showed no significant differences in the meal samples (wild type), with irregular platelet forms for the starch-based particles.

TABLE 76.1 Chemical Composition Values of Horse Chestnut Samples of Different Botanical Origins from Modena, and Comparison with Literature Data for Another Species

	AHP (Pure; White Flowers)	AHH (Hybrid; Pink Flowers)	A. *indica* (Parmar & Kaushal, 1982)
Moisture % (fresh seeds)	50.8 ± 0.6	50.1 ± 0.6	50.5
Humidity % (residual)	6.97 ± 0.35	6.59 ± 0.24	
Proteins % (d.s.)	2.64 ± 0.42	1.82 ± 0.34	0.77
Lipids % (d.s.)	4.13 ± 0.36	5.10 ± 0.50	
Glucids % (d.s.)	15.2 ± 0.66	14.3 ± 0.80	11.0
Ashes % (d.s.)	2.51 ± 0.12	2.19 ± 0.09	3.83
CWS % (d.s.)	53.9 ± 0.79	48.6 ± 0.65	
TISS % (d.s.)	2.18 ± 0.10	1.92 ± 0.11	

Uncertainties are expressed as standard deviation of five samples (s_5). d.s., dry sample.

sugars) is reported — a result 35% lower than that determined for the AHP and AHH seeds. Also, the protein content reported for A. *indica* is smaller, at −240% (AHP) and −140% (AHH). On the contrary, the mineral content is higher, the differences being +53% (AHP) and +75% (AHH).

The lipid fraction of AH seeds is 4−5% of the dry mass. Some fatty acids — the most important of which are oleic, linolenic, palmitic, and stearic acids — have been identified by Leung and Foster (1996), and confirmed by us (Baraldi *et al.*, 2007). Surprisingly, we observed that the main components are differently distributed in the two varieties (AHP and AHH): oleic acid prevails in AHH (49.7%, compared to 43.2% in AHP), while linoleic acid strongly prevails in AHP (35.2%, compared to 23.0% in AHH).

The oleic acid content in AH and A. *indica* seeds oils seems to be very different (Table 76.2), as claimed by Kapoor and colleagues (2009). Unfortunately, no further information on this topic can be obtained from the literature at this time.

TABLE 76.2 Fatty Acid Content of Horse Chestnut Samples of Different Botanical Origin from Modena, Compared with Literature Data for Anther Species

Acid	AHP (Pure, White Flowers)	AHH (Hybrid, Pink Flowers)	A. *indica* (Kapoor et al., 2009)
Myristic (C14 : 0)	0.6 ± 0.3	0.6 ± 0.2	
Myristoleic (C14 : 1)	0.2 ± 0.1	0.4 ± 0.1	
Pentadecanoic (C15 : 0)	0.1 ± 0.1	0.2 ± 0.1	
Pentadecenoic (C15 : 1)	0.1 ± 0.1	0.1 ± 0.1	
Palmitic (C16 : 0)	7.1 ± 0.5	6.8 ± 0.6	
Palmitoleic (C16 : 1)	0.7 ± 0.3	0.9 ± 0.3	
Heptadecanoic (C17 : 0)	0.1 ± 0.1	0.4 ± 0.2	
Heptadecenoic (C17 : 1)	0.1 ± 0.1	0.3 ± 0.1	
Stearic (C18 : 0)	0.8 ± 0.2	2.9 ± 0.3	
Oleic (C18 : 1)	43.2 ± 1.1	49.7 ± 1.3	65 ÷ 70
Linoleic (C18 : 2)	35.2 ± 1.2	23.0 ± 0.8	
Linolelaidic (C18 : 2)	2.2 ± 0.4	2.1 ± 0.2	
Linolenic (C18 : 3)	5.9 ± 0.5	7.5 ± 0.5	
Arachidic (C20 : 0)	0.2 ± 0.1	0.5 ± 0.2	
Behenic (C22 : 0)	0.1 ± 0.1	0.3 ± 0.1	
Erucic (C22 : 1)	2.9 ± 0.3	4.1 ± 0.2	

Uncertainties are expressed as standard deviation of five samples (s_5).

In a recent paper, de Magalhaes and Arruda (2007) provided an interesting overview regarding some analytical procedures focusing on experimental investigation of the metallo-protein content in AH seeds, by applying techniques such as SDS-PAGE, SRXRF, and ETAAS.

A significant contribution to the literature on the chemical composition of AH was made by Kapusta *et al.* (2007), who traced the flavonoid profile of the seeds and the wastewater obtained as a by-product during industrial processing of AH seeds. It was concluded that flavonoids present in these fractions can be safely used to obtain quercetine and kaempferol glycoside for the cosmetic, nutraceutical, and supplement industries.

APPLICATIONS TO HEALTH PROMOTION AND DISEASE PREVENTION

Escin, the major bioactive principle in AH seeds, has shown satisfactory evidence of clinically significant activity in the treatment of chronic venous insufficiency, hemorrhoids, post-operative edema, and mammary induration, and in cancer therapy. In some controlled trials, escin was demonstrated to be as effective as traditional medical treatments for the same pathologies, in addition to having excellent tolerability (Sirtori, 2001).

Horse chestnut extract has a higher antioxidant activity than vitamin E, showing one of the highest "active-oxygen" scavenging abilities compared to other natural products; moreover, it manifests a potent cell-protective effect towards radicals of any type. These properties are strictly connected to the presence of antioxidant molecules, which develop an anti-aging effect (Kapoor *et al.*, 2009). Among these molecules, the flavonoids are particularly active, exerting protective and beneficial effects in several ways.

These observations support the results of Fujimura *et al.* (2006), who reported the activity of horse chestnut extracts on cutaneous tissues, stimulating the contraction of non-muscle cells such as fibroblasts. These contractions play an important role in determining cell morphology, vasoconstriction, and wound healing. The results are very encouraging, suggesting that *Aesculaceae* extracts are potent anti-aging ingredients.

There is some evidence that various escin molecules (saponins and sapogenins) show beneficial effects when administered at the right concentration, exhibiting an ethanol absorption inhibitory effect and hypoglycemic activity in the oral glucose tolerance test *in vivo* (Sirtori, 2001). These observations are strictly related to the anti-obesity effects of horse chestnut extracts. Hu and colleagues (2008), working with the seeds of *A. turbinata* Blume, gave a very good rationale for these evidences, completing the description of the mechanisms of the involved species on the metabolic processes observed both *in vitro* and *in vivo*.

In some recent papers (Niu *et al.*, 2008, and references therein), the apoptotic and anti-proliferative activity of β-escin towards some human acute myeloid leukemia cell lines (i.e., HL-60, K562) were examined and evaluated by applying different assays and experimental methodologies. The results confirm that β-escin is a potent natural inhibitor of leukemic cell proliferation and an apoptosis inducer, depending on the dose and time of administration. These desirable effects are generally associated with good pharmacological tolerability, and indicate that β-escin may be a useful candidate for exploring new potential antileukemic drugs.

β-Escin from AH extracts was also tested to evaluate the chemopreventive efficacy of its dietary intake on azoxymethane-induced colonic aberrant crypt foci. The cell-growth inhibitory effects and the induction of apoptosis in the HT-29 human colon carcinoma cell line has also been investigated, giving positive results with either wild-type or mutant p53. These novel features make β-escin a candidate as an effective agent for both colon cancer chemoprevention and treatment (Patlolla *et al.*, 2006).

Zhang and Li (2007) revealed the presence and confirmed the structures of a number of triterpenoid saponins from the seeds of *A. pavia*, a small tree species naturally growing throughout the southeastern United States. Some differences were observed when comparing these structures with those of saponins isolated from Eurasian AH and *A. chinensis* in their oligosaccharide moieties, suggesting a different chemotaxonomic feature among the species.

A set of triterpenoid saponins was systematically tested *in vitro* for their activity against 59 cell lines from nine different human cancers, and some of the molecules inhibited all cell proliferation activities. In view of the possible relevance of β-escin in anticancer therapy, some formulations were tested in order to establish their effectiveness on the basis of bioavailability and bioequivalence (Bassler *et al.*, 2003). The positive results strongly encourage pursuing new cure modalities, with treatments based on well-tolerated natural products.

Fant and colleagues report that antimicrobial protein-1 (Ah-AMP1), isolated from horse chestnuts, is a very effective plant defense, since it inhibits the growth of a broad range of fungine species (Fant *et al.*, 1999). This knowledge may imply a possible technological transferability, by using natural products in the integrated biological struggle to preserve vegetation and foodstuffs as well. Moreover, it is well known that natural antibacterials, antimicrobials, antivirals, and antifungins, and any other inhibitors of damaging species, are generally active at moderate dosages and concentrations, associated with an excellent tolerability for upper organisms. The substitution of synthetic phytotherapeutics by natural products with therapeutic activity in many industrial formulations can provide significant advantages and benefits, especially in terms of biocompatibility and environmental preservation.

ADVERSE EFFECTS AND REACTIONS (ALLERGIES AND TOXICITY)

Adjuvant and hemolytic activities of many saponins purified from medicinal and food plants were examined by Oda *et al.* (2000). In particular, escins showed a weak adjuvant but strong hemolytic activity. Oral administration is generally well tolerated because some digestive processes partially destroy the original molecules, forming less toxic species. However, the abuse or misuse of saponins is strongly discouraged.

On the contrary, parentheral administration manifests the most negative effects due to hemolytic activity, and can lead to irreversible destruction of red corpuscles of the hematic tissue.

At the moment, the hemolytic activity seems to be the most important adverse effect associated with therapeutic formulations based on AH extracts.

SUMMARY POINTS

- *Aesculus hippocastanum* seeds and their derived products can be considered as an opencast mine of natural different compounds that playing a number of roles related to several biological activities in several ways.
- Escins, the most important class of bioactive molecules in *Aesculaceae* seeds, show wide-ranging mechanisms of therapeutic activity both in peripheral treatments for clinical disorders and in targeted treatments towards cancerous cells, by inhibiting proliferation or inducing apoptosis. The excellent tolerability of escins indicates that these treatments offer benefits for patients with a case history.
- The main adverse effects of escins in humans are due to their hemolytic activity. Research efforts in this field are devoted to improving the selectivity for aberrant red corpuscles, promoting the β-escin fraction as a useful candidate agent for exploring new potential antileukemic drugs.

- Concerning nutritional use, fresh or naturally desiccated seeds are usually treated by long leaching with water or wooden ashes to remove harshness and bitterness. These treatments cause a variation in the molecular structures of escin fractions, reducing the toxicity but maintaining their nutraceutical potential and anti-obesity effects. Alternatively, the slow-roasting of nuts makes the escins harmless and the seeds edible.
- Seed extracts show more potent antioxidant activity than vitamin E. Considering the potent cell-protective effects due to this activity, they can be safely used as radical scavengers, with some anti-aging properties associated with the presence of antioxidant molecules such as flavonoids.
- The claimed toxicity of these extracts makes them natural antibacterials, antimicrobials, antivirals, and antifungals, to some extent, that also act as environmentally biocompatible phytotherapeutics.

ACKNOWLEDGEMENTS

The financial support by Italian Ministero dell'Istruzione, Università e Ricerca (MIUR) is gratefully acknowledged.

References

Azarkovich, M. I., & Gumilevskaya, N. A. (2006). Proteins of cotyledons of mature horse chestnut seeds. *Russian Journal of Plant Physiology, 53*, 629−637.

Baraldi, C., Bodecchi, L. M., Cocchi, M., Durante, C., Ferrari, G., Foca, G., et al. (2007). Chemical composition and characterisation of seeds from two varieties (pure and hybrid) of *Aesculus hippocastanum. Food Chemistry, 104*, 229−236.

Bassler, D., Okpanyi, S., Schrodter, A., Loew, D., Schurer, M., & Schulz, H. U. (2003). Bioavailability of beta-aescin from horse chestnut seed extract: comparative clinical studies of two galenic formulations. *Advances in Therapy, 20*, 295−304.

de Magalhaes, C. S., & Arruda, M. A. Z. (2007). Sample preparation for metalloprotein analysis: A case study using horse chestnuts. *Talanta, 71*(5), 1958−1963.

Fant., F., Vranken, W. F., & Borremans, F. A. M. (1999). The three-dimensional solution structure of *Aesculus hippocastanum* antimicrobial protein 1 determined by H-1 nuclear magnetic resonance. *Protein Structure Function and Genetics, 37*, 388−403.

Fujimura, T., Tsukahara, K., Moriwaki, S., Hotta, M., Kitahara, T., & Takema, Y. (2006). A horse chestnut extract, which induces contraction forces in fibroblasts, is a potent anti-aging ingredient. *Journal of Cosmetic Science, 57*, 369−376.

Hricoviniova, Z., & Babor, K. (1991). Saccharide constituents of horse chestnut (*Aesculus-hippocastanum* L.) seeds. 1. Monosaccharides and their isolation. *Chemical Papers − Chemicke Zvesti, 45*, 553−558.

Hu, J. N., Zhu, X. M., Han, L. K., Saito, M., Sun, Y. S., Yoshikawa, M., et al. (2008). Anti-obesity effects of escins extracted from the seeds of *Aesculus turbinata* BLUME (Hippocastanaceae). *Chemical & Pharmaceutical Bulletin, 56*, 12−16.

Kapoor, V. K., Dureja, J., & Chadha, R. (2009). Herbals in the control of ageing. *Drug Discovery Today, 14*, 992−998.

Kapusta, I., Janda, B., Szajwaj, B., Stochmal, A., Piacente, S., Pizza, C., et al. (2007). Flavonoids in horse chestnut (*Aesculus hippocastanum*) seeds and powdered waste water byproducts. *Journal of Agricultural and Food Chemistry, 55*, 8485−8490.

Leung, A. Y., & Foster, S. (1996). *Encyclopedia of common natural ingredients ised in food, drugs, and cosmetics* (2nd ed.). New York, NY: John Wiley & Sons.

Niu, Y. P., Wu, L. M., Jiang, Y. L., Wang, W. X., & Li, L. D. (2008). β-escin, a natural triterpenoid saponin from Chinese horse chestnut seeds, depresses HL-60 human leukaemia cell proliferation and induces apoptosis. *Journal of Pharmacy and Pharmacology, 60*, 1213−1220.

Oda, K., Matsuda, H., Murakami, T., Katayama, S., Ohgitani, T., & Yoshikawa, M. (2000). Adjuvant and haemolytic activities of 47 saponins derived from medicinal and food plants. *Biological Chemistry, 381*, 67−74.

Parmar, C., & Kaushal, M. K. (1982). Aesculus indica. In *Wild fruits*. New Delhi, India: Kalyani Publishers.

Patlolla, J. M. R., Raju, J., Swamy, M. V., & Rao, C. V. (2006). β-Escin inhibits colonic aberrant crypt foci formation in rats and regulates the cell cycle growth by inducing p21(waf1/cip1) in colon cancer cells. *Molecular Cancer Therapeutics, 5*, 1459−1466.

Sirtori, C. M. (2001). Aescin: pharmacology, pharmacokinetics and therapeutic profile. *Pharmaceutical Research, 44,* 183–193.

Stankovic, S. K., Bastic, M. B., & Jovanovic, J. A. (1984). Composition of the sterol fraction in horse chestnut. *Phytochemistry, 23,* 2677–2679.

Thornfeldt, C. (2005). Cosmaceuticals containing herbs: fact, fiction, and future. *Dermatologic Surgery, 31,* 873–880.

Zhang, Z., & Li, S. (2007). Cytotoxic triterpenoid saponins from the fruits of *Aesculus pavia* L. *Phytochemistry, 68,* 2075–2086.

CHAPTER 77

Usage of Indian Gooseberry (*Emblica officinalis*) Seeds in Health and Disease

Y. Vimala[1], K. Vijaya Rachel[2], Y. Pramodini[3], A. Umasankar[4]
[1] Department of Microbiology, GITAM University, Visakhapatnam, Andhra Pradesh, India
[2] Department of Biochemistry, GITAM University, Visakhapatnam, Andhra Pradesh, India
[3] Government Dental hospital, Hyderabad, Andhra Pradesh, India
[4] Department of Biochemistry, Faculty of Medicine, Al Tahadi University, Sirt, Libya

663

LIST OF ABBREVIATIONS

MTCC, microbial type culture collection
ATCC, American type culture collection
MIC, minimum inhibitory concentration
USDA, United States Department of Agriculture

INTRODUCTION

The increasing failure of chemotherapeutics and the antibiotic resistance exhibited by pathogenic microbial infectious agents has led to the screening of several medicinal plants for their potential antimicrobial activity (Colombo & Bosisio, 1996; Ritch-kro *et al.*, 1996; Chung *et al.*, 1998; Rastogi & Mehrotra, 1999; Khilare & Saindanshiv, 2004; Srinivasu *et al.*, 2004).

Nuts & Seeds in Health and Disease Prevention. DOI: 10.1016/B978-0-12-375688-6.10077-5

FIGURE 77.1
Seeds of *Emblica officinalis*. Reprinted from Vimala *et al.* (2009), *J. Pure Appl. Microbiol.*, *3*(1), 219—222, with permission.

Increased and indiscriminate use of drugs has led to the development of drug resistance in human pathogens against commonly used antibiotics. This has necessitated the search for new antimicrobial substances from other sources, including plants. Screening of medicinal plants for antimicrobial and elementological activities is important for finding potentially new compounds for therapeutic use (Figure 77.1).

BOTANICAL DESCRIPTION

Emblica officinalis is commonly called the 'Indian gooseberry'. It belongs to the family Euphorbiaceae, and is known as Amla in Hindi, and Amalaki in Sanskrit. It is a small to medium-sized tree with a crooked trunk and spreading branches, and grayish-green bark that peels off in flakes. The branchlets are glabrous or finely pubescent, 10—20 cm long, usually deciduous, with the leaves simple, subsessile, and closely set along the branchlets. The leaves are light green, resembling pinnate leaves. The flowers are greenish-yellow, borne in axillary fascicles, and give way to globose fruit. The fruits are depressed globose in shape, 1—2.5 cm in diameter, fleshy, and obscurely six-lobed, containing six trigonous seeds. They are green when unripe, and turn light yellow or brick red when mature.

The seeds are acrid and sweet, with aphrodisiac and antipyretic properties, and are useful in treating biliousness, leukorrhoea, vomiting, and vata (in Indian medicine, vata is associated with cold; increased vata is obtained by cooling). They yield about 16% of brownish yellow oil containing 44% linoleic acid, 28.4% oleic acid, 4.8% linolenic acid, 2.2% stearic acid, 3.0% palmitic acid, and 1.0% myristic acid. The seed weight is about 570 g/1000 seeds. The seed oil resembles linseed oil.

HISTORICAL CULTIVATION AND USAGE

Amla, or Indian gooseberry (*Emblica officinalis*), is a native of tropical South-eastern Asia, particularly Central and Southern India, Pakistan, Bangladesh, Sri Lanka, Malaya, China, and the Mascerene Islands. It is a common species in dry deciduous forests. In 1901 the USDA received seeds from Reasonia Brothers in Florida, and distributed them to Florida, Bermuda, Cuba, Puerto Rico, Trinidad, Panama, Hawaii, and the Philippines. The high level of ascorbic acid in amla aroused some interest in 1940.

According to Ayurveda Rasayana, herbs slow aging, are revitalizing, restorative, and prevent disease. They can also be taken over long periods of time without causing side effects. There are only 13 herbs in this category, and amla is one of these. Adaptogens are remarkable, natural substances that help the body restore balance. They increase the body's resistance to physical, biological, emotional, and environmental stressors, and provide a defense response to acute or chronic stress (Winston, 2004). There are only 21 known adaptogens, including amla.

PRESENT-DAY CULTIVATION AND USAGE

Amla grows wild as well as being cultivated. Propagation of *Emblica officinalis* is by its seed. Its area of cultivation extends from the Himalayas in the North to Sri Lanka, south of India, and from Malacca to South China. It is widely cultivated in Uttar Pradesh in Northern India. The fruit yield has been reported as 19.6–20.2 kg for fruit harvested from the wild in India. Cultivated trees can yield 187–299 kg/year, and 200 kg can be obtained from grafted trees.

Amla (or aonla) is an indigenous plant used for its medicinal properties, in a decoction, as a powder in milk, and as chyawanprash in India. The fruit is either pickled or eaten raw. Literature on the edible portion of the fruit is available (Gopalan *et al.*, 1971; Shankuntala Manay & Shadakshar Swamy, 1987; Barthakur & Arnold, 1991). An ointment made from the burnt seeds and oil is applied to skin afflictions. The seeds are used in treating asthma, bronchitis, diabetes, and fevers. They contain proteolytic and lipolytic enzymes, phosphatides, and a small amount of essential oil. Approximately 16% of the seeds is a brownish-yellow fixed oil. Limited literature is available on the medicinal properties of seeds, and this chapter attempts to throw light on these.

The fruits are sour and astringent, and are occasionally eaten raw. They are greatly valued for making pickles, preserves, and jellies. The fruit is a rich source of pectin. Amla fruit is probably the richest known natural source of vitamin C; the fruit juice contains nearly 20 times as much vitamin C as orange juice. Some antioxidants present in the fruit prevent or retard the oxidation of the vitamin, rendering the fruit a valuable antiscorbitic both when fresh and dried. The fruit (pulp) is known to have antipyretic, spasmolytic, antifungal, antibacterial (Elizabeth & Karthik, 2001), antiviral, anticytotoxic, immuno-modulatory, and immunostimulatory activities (Cruckshank *et al.*, 1975). In this chapter, the aqueous, cold methanol, distillate, and residual extracts of the seeds are evaluated against various human pathogens.

Seed collection and storage

The seeds are extracted by drying the ripe fruits, collected during January, until they burst with a cracking sound. When the seeds emerge, they have a very short viability. It takes 80–85 kg of fruit to give 1 kg of seeds; there are 65,000–90,000 seeds per kilogram, and 24–27 seedlings result from each kilogram of seed. The proportion of seeds that develop into seedlings is approximately 40%.

The time taken for germination is 40–50 days; no pretreatment of seeds is necessary, but placing them in cow dung slurry for 48 hours improves the outcome. Treatment of seeds with *Azospirillum* biofertilizer slurry for 24 hours increases the germination rate of *Emblica officinalis* to 79% as compared to 50% in controls. The treatment also increases the shoot height of a 4-week seedling to 4.5 cm and its root length to 5.3 cm, compared to 2.4 cm for both height and length in controls. The total length of seedling almost doubled as compared to control.

Pretreated or pre-germinated seeds are put in bags in March, under cover. Germination begins 24 days after sowing, with regular watering. The plants are ready for planting out in July and August.

A high percentage of rooting (84%) has been reported from semi-hard wood cuttings from the middle portions of vigorous shoots of young trees planted in beds at 33°C. Budding and soft wood grafting may also give good results.

Seedlings are susceptible to a root rot disease caused by *Rhizoctonia solani*. Trees are also affected by rusts such as leaf rust, caused by *Phakopspora phyllanthi*, and ring rust, caused by *Ravenilia emblicae*. The fruits are susceptible to rot diseases as a result of infection by *Penicillium* spp.,*Glomerella cingulata*, *Phoma putamium*, and *Aspergillus niger*.

Nutritional aspects of seeds

Amla fruit has a number of medicinal uses in indigenous medicine. The edible fruit tissue of *Emblica officinalis* was found to contain about 160 times as much ascorbic acid as there in apples (*Malus pumila*). The fruits also contained higher concentrations of most minerals and of amino acids than in apples. Glutamic acid, proline, aspartic acid, alanine, and lysine constituted 29.6, 14.6, 8.1, 5.4, and 5.3% of the total amino acids, respectively. The concentration of each amino acid, except cystein, was much higher than in apples (Barthakur & Arnold, 1991). *Emblica* is therefore highly nutritious, and could be an important source of vitamin C, minerals, and amino acids (Tables 77.1-77.3).

TABLE 77.1 Mineral and Trace Elements (per 100 g of Edible Portion)

Moisture (g)	81.8
Protein (g)	0.5
Fat (g)	0.1
Minerals (g)	0.5
Crude fiber (g)	3.4
Carbohydrates (g)	13.7
Energy (kcal)	58
Calcium (mg)	50
Phosphorus (mg)	20
Iron (mg)	1.2
Magnesium	—
Sodium (mg/g)	5.0
Potassium (mg/g)	225
Copper	—
Manganese	—
Molybdenum	—
Zinc	—
Chromium	—
Sulfur	—
Chlorine	—

TABLE 77.2 Oxalic Acid and Phytin Phosphorus in *Emblica officinalis*

Oxalic acid (mg/100 g)	296 mg/100 g
Phytin P	—
Phytin P as percentage of total phosphorus	—

TABLE 77.3 Vitamin Content in *Emblica officinalis*

Carotene (μg)	09
Thiamine (mg)	0.03
Riboflavin (mg)	0.01
Niacin (mg)	0.2
Total B6 (mg)	—
Free folic acid (μg)	—
Total folic acid (μg)	—
Vitamin C (mg)	21
Choline (mg)	—

APPLICATIONS TO HEALTH PROMOTION AND DISEASE PREVENTION

Emblica fruit is one of the three myrobalans, a term derived from the Greek for acorn. *Emblica officinalis*, *Terminalia chebula*, and *Terminalia bellerica* are three fruits that are combined to make a popular remedy; triphala (meaning three fruits) is a rejuvenating formula that is applied in treating intestinal disorders (inflammation, infection, diarrhea, and constipation).

Heavy metal toxicity studies

Several animal studies have shown that amla can help prevent a toxic build-up of heavy metals caused by frequent exposure to metals like aluminum, lead, and nickel. When vitamin C alone was used, equivalent to that found in amla fruit, only partial protection from heavy metals was provided; however, when the whole amla fruit was used, almost complete protection was achieved. This indicates that it is the combined action of various ingredients found in the fruit and seed that effectively helps to shield DNA from heavy metal poisoning (Giri, 1986; Dhir *et al.*, 1991).

Vitamin C and oral health

Vitamin C increases salivation, which decreases the activity of microorganisms that cause dental caries. Saliva plays an important role in fighting and preventing caries. Salivation is increased by the use of artificial salivary substitutes such as salivart, which usually consist of salt ions, flavoring agents, and preservatives like paraben, cellulose derivatives, animal mucin, and fluoride. In place of chemical substances as salivary substitutes, nature has provided foods that contain vitamin C. Intake of natural foods rich in vitamin C, like *Emblica*, shows a significant reduction in the occurrence of scurvy, xerostomia, Sjögren's syndrome, and dental caries.

Diabetes

A fruit infusion of the seeds (Nadkarni, 1982) and decoctions of leaves and seeds are used in the treatment of diabetes mellitus. The seeds are given internally as a cooling remedy in bilious infections and nausea, and as an infusion they make a good drink for fevers (Nadkarni, 1982).

Other medical uses

For nosebleeds, seeds are fried in ghee and ground in conjee (the liquid from boiled rice), then applied to the forehead. To calm pruritis, seeds are burnt, powdered, and mixed in oil, as a useful application for scabies or itching.

A preparation of 11.7 g of seeds, soaked in a tin vessel overnight and ground with cow's milk the following morning, is taken as a remedy for nausea and biliousness.

The fruit pulp of amla has been reported to increase the survival rate and inhibit radiation-induced weight loss in mice (Singh *et al.*, 2005).

Herbal eye-drop preparations made from *Emblica officinalis*, *Carum copticum*, *Terminalia bellerica*, *Curcumalonga*, *Ocimum sanctum*, *Cinnamomum camphora*, *Rosa damascene*, and *meldespumapum* have been used in ophthalmic disorders, namely conjunctivitis, conjunctivitis xerosis, degenerative conditions, and following cataract surgery. It was found to have a useful role in infective, inflammatory, and degenerative ophthalmic disorders (Mitra *et al.*, 2000).

Antimicrobial activity

Studies have been conducted to evaluate the antimicrobial activity of *Emblica* on human pathogens. The seeds were collected from the Rytu bazaar, Gopalapatnam, Visakhapatnam, Andra Pradesh, in India. The seeds were dried and ground to a fine powder, and mixed with

667

sterile distilled water to give a concentration of 100 mg/ml stock solution. The extracts were stored in the refrigerator untill further use. The residual methanol extract was prepared by mixing 100 g of dried and ground powder in 1 liter of methanol in an aspirator bottle for 48 hours. The solution was subsequently collected and subjected to six cycles of distillation, until a thick brown-colored paste was obtained; 500 mg of the residual methanol extract was then mixed in 5 ml of methanol to give a concentration of 100 mg/ml *Emblica officinalis* seeds.

The microbial strains used in this study were obtained from MTCC Chandigarh, India. These were *Salmonella typhi* (ATCC 10749), *Pseudomonas aeruginosa* (ATCC 25619), *Salmonella typhimurium* (ATCC 23564), *Yersinia enterocolitica* (ATCC 9610), *E. coli* (uropathogen), *Candida albicans* (ATCC 2091), and *Rhizopus*, *Geotrichum*, *Fusarium*, and *Aspergillus flavus*.

The bacterial cultures were maintained on nutrient agar slants or plates (peptone 0.5%, beef extract 0.3%, NaCl 0.5%, agar 2.0%), and fungal cultures on sabourauds agar slants or plates (mycological peptone 1%, dextrose 2–4%, agar 2%). Overnight cultures were used in all experiments; a single colony of each type of culture was inoculated in each respective 5-ml broth, and incubated at 37°C for 18–24 hours (bacteria), or at room temperature for 48 hours (fungi). Antimicrobial reference substances used were ampicillin (10 μg/disc) and nystatin (100 units/disc), procured from Himedia, Mumbai, India. Nutrient agar plates and sabouraud agar plates were inoculated with 0.1 ml of overnight liquid culture of each type, containing 1.0×10^7 cells, and spread uniformly using a spreader. Subsequently, five sterile filter-paper discs (5 mm) and reference antibiotic discs such as ampicillin (10 μg/disc) and nystatin (100 units/disc) were placed on corresponding plates. Four sterile paper discs were prepared, which were placed on nutrient agar and sabouraud agar plates on each filter paper disc; the aqueous extracts and residual extracts were added at 2 mg/disc, and the cold methanol, distillate, methanol, and control were added at 20 μl/disc.

The bacterial cultures were incubated at 37°C for 18–24 hours, and fungal cultures at 28°C for 48 hours. Zones of inhibition were measured. The microbes were plated in duplicates, and the average zone diameter was noted. The minimal inhibitory concentrations (MICs) were determined by tube dilution techniques. Different concentrations of residual extracts of *Emblica officinalis* seeds were serially diluted in duplicates. The control test tube did not receive any extract. Later, 1000 cells of microorganisms were added to each test tube and incubated at 37°C or at room temperature for 18–24 hours. The lowest concentrations that inhibited the growth were considered to be the MIC.

The results of the study are shown in Tables 77.4–77.6. The aqueous extract showed a 9-mm zone of inhibition for *E. coli* and an 8-mm zone of inhibition for *Candida albicans*. All the microorganisms tested showed resistance with this extract. The cold methanol extract showed

TABLE 77.4 Antimicrobial Activity of *Emblica officinalis* Seed Extract: Zones of Inhibition (in mm) on Certain Human Pathogens (Bacteria)

S. No	Microorganisms	Aqueous Extract (2 mg/disc)	Cold Methanol Extract (20 μl/disc)	Distillate (20 μl/disc)	Residual (2 mg/disc)	Methanol Control (20 μl/disc)	Ampicillin (10 μg/disc)
1	ST ATCC 10749	R	R	R	12	11	6
2	PA ATCC 25619	R	9	14	16	13	R
3	STM ATCC 23564	R	6	6	14	6	13
4	YE ATCC 9610	R	9	15	13	17	16
5	ECU	9	10	R	R	R	11

R, resistant; ST, *Salmonella typhi*; PA, *Pseudomonas aeruginosa*; STM, *Salmonella typhimurium*; YE, *Yersinia enterocolitica*; ECU, *E. coli* uropathogen.
Uniyal, 2004.

TABLE 77.5 Antimicrobial Activity of *Emblica officinalis* Seed Extract: Zones of Inhibition (in mm) on Fungal Pathogens

S. No	Microorganisms	Aqueous Extract (2 mg/disc)	Cold Methanol Extract (20 µl/disc)	Distillate (20 µl/disc)	Residual (2 mg/disc)	Methanol control (20 µl/disc)	Nystatin (100 units/disc)
1	*Rhizopus*	R	8	12	13	16	29
2	*C. albicans*	8	10	9	24	12	27
3	*Geotrichum*	R	12	17	16	12	37
4	*Fusarium*	R	12	11	19	19	27
5	*A. flavus*	R	8	8	15	24	20

R, resistant.

TABLE 77.6 Minimum Inhibitory Concentrations of Residual Extracts of *Emblica officinalis* Seeds on Bacterial and Fungal Pathogens

S. No.	Microorganism Tested	MIC (µg/ml)
1	*Salmonella typhi* ATCC 10749	100
2	*Pseudomonas aeruginosa* ATCC 25619	100
3	*S. typhimurium* ATCC 23564	800
4	*Yersinia enterocolitica* ATCC 9610	800
5	*E. coli* (uropathogen)	200
6	*Rhizopus* (isolate)	800
7	*Candida albicans* ATCC 2091	200
8	*Geotrichum* (isolate)	400
9	*Fusarium* (isolate)	200
10	*A. flavus* (isolate)	50

zones of inhibition ranging from 6 mm to 12 mm; *Geotrichum* and *Fusarium* were sensitive, showing 12-mm zones of inhibition. The distillate showed zones of inhibition ranging from 6 to 17 mm; *Geotrichum* was sensitive and *S. typhimurium* resistant. The residual extract showed zones of inhibition from 12 to 24 mm; *Candida albicans* showed the maximum zone of inhibition, and *S. typhi* showed the smallest zone of inhibition. The results clearly indicate that the seeds possess potent antimicrobial compounds that can be used to kill pathogens, especially *Candida albicans* and *Pseudomonas aeruginosa* (which is commonly a hospital acquired organism). The active compounds need to be isolated and their efficacy tested against other microbial pathogens.

Conclusion

Only the pulp, leaves, and stem of *Emblica officinalis* have been tested for their antimicrobial properties, as cited by the literature. From the present investigation, it is clear that the seeds of this fruit are also active against a different pathogenic microorganism, which shows that the whole fruit can be used in isolating bioactive compounds. Seeds also have high nutritive value and are effective against oral diseases, pruritis, etc. The efficacy is profound with long-term usage of seeds and fruits.

ADVERSE AFFECTS AND REACTIONS (ALLERGIES AND TOXICITY)

No toxicity or allergies are associated with the usage of seeds as food and medicine; rather, seeds are found to protect against heavy metal toxicity, and shield DNA from heavy metal poisoning. The fruit is not at all toxic, and is widely used in India as a health-promoting food. Fruit alone is contraindicated in diarrhea and dysentery.

SUMMARY POINTS

- Seeds of *Emblica officinalis* are of high nutritive value, showing remarkable amounts of ascorbic acid, potassium, calcium, and amino acids.
- The seeds are health promoting and disease preventive.
- *E. officinalis* seeds play an important role in heavy metal toxicity.
- The presence of satisfactory amounts of vitamin C means the seeds safeguard oral health.
- *E. officinalis* seeds are also used against diabetes mellitus, pruritis, nose bleeds, nausea, and biliousness.
- The seeds have been tested for antimicrobial properties on human pathogens. Results clearly indicate that they possess potent antimicrobial compounds that can be used to kill pathogens, especially *Candida albicans* and *Pseudomonas aeruginosa* (which is commonly a hospital acquired organism).
- Seed extracts have also been found to have both antibacterial and antifungal properties, although further study is needed regarding the antiviral properties of seed extracts.
- Bioactive compounds can be isolated from the seeds.
- No toxicity or allergies are associated with usage of seeds as food or medicine.

References

Barthakur, N. N., & Arnold, N. P. (1991). Chemical analysis of *Phyllanthus emblica* L. and its potential as a food source. *Scientia Horticulturae, 47*(1–2), 99–105.

Chung, K. T., Wong, T. Y., Wei, C. I., Huang, Y. W., & Lin, Y. (1998). Tannins and human health: a review. *Critical Reviews in Food Science and Nutrition, 38*, 421–464.

Colombo, M. L., & Bosisio, E. (1996). Pharmacological activities of *Chelidonium magus* L. (Papaveraceae). *Pharmacological Research, 33*, 127–134.

Cruckshank, R., Dugrid, I. P., Marmoin, B. P., & Swain, R. H. A. (1975). *Microbiologia Medica*, Vol. II, (4th ed). Lisboa, Portugal: Fundacao Calouste Gulbenkian Lisboa, pp. 393–403.

Dhir, H., Agarwal, A., Sharma, A., & Talukder, G. (1991). Amla research and clinical studies. *Cancer Letters, 59*, 9–18.

Elizabeth, K. M., & Karthik, K. (2001). The antimicrobial activity of Emblica officinalis *on certain human pathogenic microorganisms. Paper presented at The National Symposium on Environmental Strategies and Action Plan.* Visakhapatnam: Andhra University. October 12, 2001.

Giri, A. K. (1986). Amla research and clinical studies. *Cytologia, 51*, 375–380.

Gopalan, C., Ramasastry, B. V., & Balasubramanian, S. C. (1971). Nutritive value of foods. Revised and updated by B.S. Narasinga Rao, Y.G. Deosthale, & K.C. Pant (1987). Hyderabad, India: National Institute of Nutrition ICMR.

Khilare, C. J., & Saindanshiv, S. E. (2004). Survey of medicinal plants and their conservation for sustainable health, hygiene and environment in India. *Journal of Environment and Ecoplan, 8*(3), 737–740.

Mitra, S. K., Sundaram, R., Venkataranganna, M. V., Gopu Madhavan, S., Prakash, N. S., Jayaram, H. D., et al. (2000). Anti-inflammatory, antioxidant, antimicrobial activity of Opthacare brand, an herbal eye drops. *Phytomedicine, 7*, 123–127.

Nadkarni, K. M. (1982). Emblica officinalis. In Indian Material Medica, Vol. 1, Ayurvedic, Unani, and Home remedies (p. 480). Mumbai, India: Popular Prakashan Pvt.

Rastogi, P., & Mehrotra, B. N. (1999). Compendium of Indian medicinal plants. *Drug Research Perspectives, 2*, 1–859.

Ritch-kro, E. M., Turner, N. J., & Towers, G. H. (1996). Carrier herbal medicine: an evaluation of the antimicrobial and anticancer activity in some frequently used remedies. *Journal of Ethnopharmacology, 5*, 151–156.

Shankuntala Manay, N., & Shadakshar Swamy, M. (1987). *Food facts and techniques*. New Delhi, India: New Age International Ltd.

Singh, I., Sharma, A., Nunia, U., & Goyal, P. K. (2005). Radioprotection of swiss albino mice by *Emblica officinalis*. *Phytotherapy Research, 19*, 444–446.

Srinivasu, T., Pathan, S. N., & Pardish, S. N. (2004). Biodiversity of medicinal plants – Mumbai railway track-sides. *Asian Journal of Microbiology Biotechnology and Environmental Sciences, 6*(4), 625–633.

Uniyal, M. R. (2004). Noida effective Ayurvedic Medicinal plants used in rasayan therapy. *Maharishi Ayurvedic Products*.

Vimala, Y., Vijaya Rachel, K., & Uma Sankar, A. (2009). Antimicrobial activity of *Emblica officinalis* seeds on human pathogens. *Journal of Pure and Applied Microbiology, 3(1)* 219–222.

Winston, D. (2004). Native American, Chinese and Ayurvedic Materia Medica. *HTSBM*.

CHAPTER 78

Indian Mustard (*Brassica juncea* L.) Seeds in Health

Reka Szollosi
Department of Plant Biology, University of Szeged, Hungary

671

LIST OF ABBREVIATIONS

GST, glutathione-S-transferase
LDL, low density lipoprotein
NADPH, nicotinamide adenine dinucleotide phosphate
VLDL, very low density lipoprotein

INTRODUCTION

Indian or brown mustard, belonging to the mustard family, has been consumed for centuries all over the world. Its leaves, roots, and shoot are known as vegetables, while its seed can be used as a condiment or a source of oil. In Ancient Greece mustard was used in cases of scorpion or snake bites, while in India and China mustard seeds have been used in traditional medicine since time immemorial for numerous diseases and disorders (Peter, 2004). In the past four decades, the beneficial effects of mustard seed consumption have been rediscovered due to its special components, the glucosinolates (Cartea & Velasco, 2008).

Moreover, since the 1990s interest has focused on Indian mustard because of its ability to accumulate toxic and non-toxic heavy metals. It has become one of the most important test plants for phytoremediation — the science that deals with recovery of contaminated areas by plants (Szőllősi et al., 2009).

Nuts & Seeds in Health and Disease Prevention. DOI: 10.1016/B978-0-12-375688-6.10078-7

BOTANICAL DESCRIPTION

Brassica juncea L. belongs to the mustard family (Brassicaceae or Cruciferae) and has numerous common names, such as brown mustard, Chinese mustard, and oriental mustard. The leaves are ovate or obovate, simple and petioled; the flowers of the raceme inflorescences are bisexual, with four free sepals and four yellow petals, along with two longer and two shorter stamens. In China, "brown mustard" includes the hybrids of *Brassica nigra* and other *Brassica* species that have brown seeds, while "oriental mustard" has yellow seeds.

This annual herb originates from the natural hybridization between black mustard (*Brassica nigra* L. Koch) and turnip mustard (*Brassica rapa* L.), and retains the whole genome of both parents; therefore, it is amphidiploid. Although it is widespread in Europe, Africa, North America, and Asia, several authors believe that Eastern India, the Caucasus, and China are the main centers for *Brassica juncea* (Fu *et al.*, 2006; Small, 2006; Dixon, 2007).

HISTORICAL CULTIVATION AND USAGE

Ancient texts from India, mainly the Ayurveda, contain prescriptions for applications of brown mustard (Sarson, in Hindi; *Raajikaa*, in Sanskrit) seeds for internal and external diseases. The recipes, written 5000 years ago, recommended Raajikaa internally as an appetizer, or for enlargement of the liver and spleen. The seeds were applied externally mainly for inflammations, skin diseases, catarrh, or rheumatic problems (Khare, 2004). Other famous authors, including Pythagoras (6 BC) and Plinius Secundus (1 AD), also emphasized the medicinal values of mustards, including Indian mustard (Peter, 2004).

The leaves, seeds, and seed oils of this plant have all been used in traditional Chinese medicine for centuries (Small, 2006).

In Western Europe, mostly in France, where mustard producers have had secret recipes and special rights since the Middle Ages, Dijon mustard — one of the most famous mustards containing brown mustard — has been prepared since the 19th century and protected by law since 1937 (Small, 2006).

PRESENT-DAY CULTIVATION AND USAGE

In the Far East, especially in China, where *Brassica juncea* is traditionally used as a vegetable, spice, or herb, several types and varieties are known (Hu, 2005) For example, var. *rugosa* is grown for its large leaves and is very popular with the Chinese population of America, while var. *strumata*, *linearifolia*, *tumida*, and *tsatsai* have swollen stems, var. *multisecta* (so called "thousand leaves") has many sessile shoots, and var. *megarrhiza* has remarkable fleshy roots. The types that are grown for their vegetative organs (root, stem, and leaves) are usually harvested before development of the flowering shoot (Hu, 2005; Larkcom, 2007).

The seeds are also applied in both intact and powdered forms (mustard flour). In Ayurvedic cooking, as described as early as 3000 BC, whole seeds are important ingredients in salads, pickles, curries, or other vegetable dishes, while the powder, because of its strong flavor, is used as a component of sauces, curries, and salad dressings (Small, 2006).

APPLICATIONS TO HEALTH PROMOTION AND DISEASE PREVENTION

Though brown mustard seed has been a well-known and generally used spice in India for centuries (Kapoor, 1990), it has been employed in recent medication not only in India and China, but also in Europe and North America. Traditional and present-day medicinal applications of seeds are given in Table 78.1.

TABLE 78.1 Traditional and Present-Day Medicinal Applications of *Brassica juncea* Seeds

Symptoms/ Diseases	Effects	Form of Application	Geographic Area of Application	Reference(s)
External				
Abscesses	Anti-inflammatory	Mustard paste (poultice, plaster)	China, India, North America	Duke, 2002; Small, 2006; Khare, 2004
Backache	Analgesic	Mustard paste (poultice, plaster)	India	Duke, 2002; Khare, 2004
Foot ache	Analgesic	Mustard paste (poultice, plaster)	India	Duke, 2002; Khare, 2004
Lumbago	Analgesic	Mustard paste (poultice, plaster)	India	Duke, 2002; Khare, 2004
Rheumatism	Analgesic	Mustard paste (poultice, plaster)	China, India	Duke, 2002; Khare, 2004
Snake bite	Anti-inflammatory, antiseptic	No data	No data	Duke, 2002
Skin cancer/diseases	Antitumor	No data	China, India	Duke, 2002; Khare, 2004
Swelling	Anti-inflammatory	Mustard paste (poultice, plaster)	No data	Duke, 2002
Ulcer	Anti-inflammatory	No data	China	Small, 2006
Internal				
Anorexia	Appetizer, digestive, gastrostimulant	Whole and ground seeds	India	Duke, 2002; Khare, 2004
Chest infections	Stimulant, expectorant	Seed oil (embrocation)	India	Kapoor, 1990
Cold	Anti-inflammatory, antibacterial	Mustard paste (poultice, plaster)	China, India	Khare, 2004
Dysentery	Antidysenteric	Whole and ground seeds	India	Duke, 2002; Khare, 2007
Dyspepsia	Appetizer, digestive, gastrostimulant	Whole and ground seeds	India	Duke, 2002
Eruptions	Anti-inflammatory	Mustard paste (poultice, plaster)	China	Small, 2006
Enlargement of liver/ spleen	Anti-inflammatory	No data	India	Khare, 2004
Fever	Febrifuge	No data	No data	Duke, 2002
Parasites (fungi, vermin)	Parasiticide, vermifuge, fungicide	Whole and ground seeds	India	Duke, 2002; Khare, 2007
Pneumonia	Stimulant, expectorant	Mustard paste (poultice, plaster)	No data	Duke, 2002
Stomach disorders	Digestive, gastrostimulant, laxative, emetic	Whole and ground seeds	China, India	Kapoor, 1990; Duke, 2002; Small, 2006; Khare, 2007
Stomachache	Digestive, gastrostimulant, laxative	Whole and ground seeds	India	Duke, 2002; Khare, 2007
Tumor of throat	Antitumor	No data	China	Duke, 2002
Tumor of uterus	Antitumor	No data	China	Duke, 2002
Tumor of wrist	Antitumor	No data	China	Duke, 2002

To understand the effects of Indian mustard seeds (whether in whole or powdered form), it is necessary to know the chemical composition. Plants of the *Brassica* genus contain special sulfur compounds called glucosinolates (e.g., sinigrin and sinalbin), which are very important secondary metabolites. These glucosinolates are inactive until they react with water or any moisture. At this point, they are hydrolyzed to isothiocyanates (resulting in a pungent, irritating odor and characteristic flavor), glucose, and potassium bisulfate (Peter, 2004; Small, 2006; Cartea & Velasco, 2008). Mustard seed oil comprises mainly erucic, arachidic, α-linolenic, oleic, and palmitic acids (Kapoor, 1990; Peter, 2004; Table 78.2).

The above-mentioned glucosinolates, which are generally present in the highest concentrations in seeds (Cartea & Velasco, 2008), may have a key role in cancer prevention, mainly in the initiation or promotion phases and in cell apoptosis. Results have shown that products of the enzymatic breakdown of glucosinolates can regulate cancer development through the induction of detoxification enzymes and/or inhibition of activation enzymes. Detoxification enzymes such as glutathione-S-transferase (GST) or NADPH reductase can conjugate with carcinogens, transforming them into a water-soluble inactive form that can later be excreted through urine. At the same time, enzymes like cytochrome P450s that activate carcinogens in animal and human cells can become inhibited due to glucosinolates and their degradation products (Fahey *et al.*, 2001; Cartea & Velasco, 2008). The literature also describes antibacterial, insecticidal, nematocidal, and antifungal activities of glucosinolates and isothiocyanates (Fahey *et al.*, 2001).

TABLE 78.2 Main Chemical Components of *Brassica juncea* Seeds

Main Chemical Components	Bioactivities for Human Health Known	Reference(s)
Glucosinolates		
Sinigrin	After enzymatic hydrolysis of sinigrin volatile components (allyl isothiocyanates) cause external or internal irritation and local vasodilation; regulating cancer cell development	Kapoor, 1990; Duke, 1983, 2002; Fahey *et al.*, 2001; Peter, 2004; Cartea & Velasco, 2008
Progoitrin	Precursor of goitrin which inhibits thyroperoxidase	Cartea and Velasco, 2008
Sterols (phytosterols)		
Brassicasterol	All have cholesterol-decreasing properties	Duke, 1983
Campesterol		Duke, 1983
Sitosterol		Duke, 1983
Avenasterol		Duke, 1983
Stigmasterol		Duke, 1983
Glycerid ester of		
Arachidic acid	No data	Peter, 2004
Erucic acid	No data	Peter, 2004
α-Linolenic acid	Reduces the risk of ischemic heart disease	Rastogi *et al.*, 2004
Oleic acid	No data	Peter, 2004
Palmitic acid	No data	Peter, 2004
Other components		
Sinapic acid	No data	Peter, 2004
Sinapine	No data	Peter, 2004
Proteins	No data	Peter, 2004

It is not only the glucosinolates and their breakdown products that have chemoprotective effects in several types of cancers; sterols of mustard seeds do, too. Sterols (or phytosterols, Table 78.2), which are cholesterol-like molecules, are used as dietary supplements due to their cholesterol-reducing ability (Yadav, Vats, Ammini, & Gover, 2004). Some researchers have focused on the influences of plant sterols on stomach cancer, and a strongly negative relationship was found between total phytosterol intake and gastric cancer risk (De Stefani *et al.*, 2000).

When the beneficial effects of mustard oil, which comprises a remarkable amount of α-linolenic acid, was examined in India, a two-fold lower risk of ischemic heart disease was observed (Rastogi *et al.*, 2004).

To date, several studies have already confirmed its beneficial effects on serum glucose levels and kidney function in diabetic rats. Although a diet containing mustard powder did not reduce the blood glucose level significantly, serum creatinin values, a marker of diabetic nephropathy, decreased. Results show that the seeds of *Brassica juncea* should potentially be applied as a food adjuvant for non-genetic diabetes (Grover *et al.*, 2002, 2003). In rats fed a special diet containing curry leaf and Indian mustard seeds, a decrease in total serum cholesterol and low or very low density lipoprotein (LDL, VLDL) levels was observed (Khan *et al.*, 1996). Besides the antihyperglycemic and antihypercholesterolemic capacities, there are some data regarding the antihyperlipidemic activity of *Brassica juncea* seeds. At the same time, it is still not obvious whether this effect is a consequence of direct or indirect mechanisms (Yadav *et al.*, 2004). Moreover, seeds have antioxidant power, too. When lipid peroxidation as a marker of oxidative stress induced by a high-fat diet in rats was examined, amelioration appeared with curry leaf and brown mustard seed application (Khan *et al.*, 1997).

ADVERSE EFFECTS AND REACTIONS (ALLERGIES AND TOXICITY)

Despite the numerous advantages of consuming Indian mustard seeds, side effects and prejudicial interactions may occur. When a poultice made of mustard flour is used externally on adults, it must be removed after 10—15 minutes because allyl isothiocyanates (volatile components) can cause burns on the skin, as well as local vasodilation and blushing. The components mentioned above may also cause thyroid disorders. During internal application, overdose must be avoided, especially in cases of ulcer or nephrosis, since this can cause or aggravate inflammation (Duke, 2002).

Although glucosinolates are considered to have chemoprotective influences in several types of cancer, some of them, and their breakdown products, may cause damage or physiological disorders. One of the most notable disadvantageous glucosinolates is progoitrin, which is toxic and may result in goiter or growth retardation, but these effects have only been observed in animals and not in humans. However, progoitrin is the precursor of goitrin, which is an inhibitor of thyroperoxidase (a key enzyme of thyroid hormone synthesis), and it has been discovered that goitrin can get through nitrosation in the human stomach if the nitrate level of water consumed is high, and form a mutagen molecule, N-nitroso-oxazolidone (Cartea & Velasco, 2008).

SUMMARY POINTS

- *Brassica juncea* is a natural hybrid of black mustard (*Brassica nigra* L. Koch) and turnip mustard (*Brassica rapa* L.), and has the whole genome of both parents.
- Traditional Chinese and Indian medicine has used Indian mustard seeds for centuries to cure internal and external diseases, but it is employed in recent medications in Europe and North America, too.

- The main chemical constituents of *B. juncea* seeds are glucosinolates, sterols, and glyceride esters of fatty acids.
- Antihyperglycemic, antihyperlipidemic, and antihyperchlolesterolemic activities of the seeds have already been shown.
- Glucosinolates have antibacterial, insecticidal, and antifungal effects in animals and humans, but show chemoprotective capacities in many cancers.

References

Cartea, M. E., & Velasco, P. (2008). Glucosinolates in *Brassica* foods: bioavailability in food and significance for human health. *Phytochemistry Reviews, 77*, 213–229.

De Stefani, E., Boffetta, P., Ronco, A. L., Brennan, P., Deneo-Pellegrini, H., Carzoglio, J. C., et al. (2000). Plant sterols and risk of stomach cancer: a case-control study in Uruguay. *Nutrition and Cancer, 37*, 140–144.

Dixon, G. R. (2007). Vegetable brassicas and related crucifers. *Crop Production Science in Horticulture, 14*, 30–33.

Duke, J. A. (1983). *Handbook of Energy Crops.* Unpublished. Available at http://www.hort.purdue.edu/newcrop/duke_energy/Brassica_juncea.html.

Duke, J. A. (2002). *Handbook of medicinal herbs* (2nd ed.). Boca Raton, FL: CRC Press. 516–517.

Fahey, J. W., Zalcmann, A. T., & Talalay, P. (2001). The chemical diversity and distribution of glucosinolates and isothiocyanates among plants. *Phytochemistry, 56*, 5–51.

Fu, J., Zhang, M.-F., & Qi, X.-H. (2006). Genetic diversity of traditional Chinese mustard crops *Brassica juncea* as revealed by phenotypic differences and RAPD markers. *Genetic Resources and Crop Evolution, 53*, 1513–1519.

Grover, J. K., Yadav, S., & Vats, V. (2002). Hypoglycemic and antihyperglycemic effect of *Brassica juncea* diet and their effect on hepatic glycogen content and the key enzymes of carbohydrate metabolism. *Molecular and Cellular Biochemistry, 241*, 95–101.

Grover, J. K., Yadav, S., & Vats, V. (2003). Effect of feeding *Murraya koenigii* and *Brassica juncea* diet kidney functions and glucose levels in streptozotocin diabetic mice. *Journal of Ethnopharmacology, 85*, 1–5.

Hu, S.-Y. (2005). *Food plants of China.* Hong Kong, China: The Chinese University Press. 407–409.

Kapoor, L. D. (1990). *CRC Handbook of Ayurvedic Medicinal Plants.* Boca Raton, FL: CRC Press, 84.

Khan, B. A., Abraham, A., & Leelamma, S. (1996). Biochemical response in rats to the addition of curry leaf (*Murraya koenigii*) and mustard seeds (*Brassica juncea*) to the diet. *Plant Food for Human Nutrition, 49*, 295–299.

Khan, B. A., Abraham, A., & Leelamma, S. (1997). Antioxidant effects of curry leaf, *Murraya koenigii* and mustard seeds, *Brassica juncea* in rats fed with high fat diet. *Indian Journal of Experimental Biology, 35*, 148–150.

Khare, C. P. (2004). *Indian herbal remedies: rational Western therapy, Ayurvedic, and other traditional usage, botany.* Berlin, Germany: Springer-Verlag. 109–111.

Larkcom, J. (2007). *Oriental vegetables.* London, UK: Francis Lincoln Ltd. 43–49.

Peter, K. V. (2004). *Handbook of herbs and spices, Vol. 2.* Boca Raton, FL: CRC Press. 196–204.

Rastogi, T., Reddy, K. S., Vaz, M., Spiegelman, D., Prabhakaran, D., Willett, W. C., et al. (2004). Diet and risk of ischemic heart disease in India. *American Journal of Clinical Nutrition, 79*, 582–592.

Small, E. (2006). *Culinary herbs.* Ottawa, Canada: National Research Council Canada, 221–228.

Szőllősi, R., Varga, Sz., I., Erdei., L., & Mihalik, E. (2009). Cadmium-induced oxidative stress and antioxidative mechanisms in germinating Indian mustard (*Brassica juncea* L.) seeds. *Ecotoxicology and Environmental Safety, 72*, 1337–1342.

Yadav, S. P., Vats, V., Ammini, C., & Gover, J. K. (2004). *Brassica juncea* (Rai) significantly prevented the development of insulin resistance in rats fed fructose-enriched diet. *Journal of Ethnopharmacology, 93*, 113–116.

CHAPTER 79

Use of Jackfruit (*Artocarpus heterophyllus*) Seeds in Health

Ibironke Adetolu Ajayi
Industrial Unit, Chemistry Department, Faculty of Science, University of Ibadan, Ibadan, Oyo State, Nigeria

677

LIST OF ABBREVIATIONS

α2-HSG, α2-HS glycoprotein
AIDS, acquired immunodeficiency syndrome
CD, cluster of differentiation
Gal, galactose
GalNAc, N-acetyl-galactosamine
HIV, human immunodeficiency virus
HRP, horse radish peroxidase
IgA, immunoglobin A
IgAN, immunoglobin A nephropathy
TF, Thomsen-Freiedenreich

INTRODUCTION

The jackfruit (*Artocarpus heterophyllus* Lam.) is a member of the mulberry family (Moraceae). It is a monoecious evergreen tree grown in several tropical countries. It is also called *jak-fruit, jak,* and *jaca* in Malaysia, *nangka* in the Philippines, *khanum* in Thailand, *khnor* in Cambodia, *mak mi* or *may m*i in Laos, and *mit* in Vietnam (Babitha *et al.*, 2006). The jackfruit tree is handsome

Nuts & Seeds in Health and Disease Prevention. DOI: 10.1016/B978-0-12-375688-6.10079-9

and stately. In the tropics it grows to an enormous size, like a large eastern oak, and it bears the largest edible fruit in the world (Molla *et al.*, 2008). Jackfruit trees generally attain a height of 8—25 m, and a trunk diameter of 30—80 cm.

The tree produces a large pear- or barrel- shaped fruit which is considered the largest tree-borne fruit in the world, reaching up to 90 cm long and 50 cm in diameter, and weighing 20—49 kg (Kabir, 1998).

BOTANICAL DESCRIPTION

Based on its botanical structure, jackfruit is a multiple fruit of which about 8—15% of the fruit weight is seed (Mukprasirt & Sajjaanantakul, 2004). There may be 100—500 seeds in a single fruit, which are viable for 3—4 days or more (John & Narasimham, 1993). A single seed is enclosed in a white aril encircling a thin brown spermoderm, which covers the fleshy white cotyledon. The heavy fruit is borne primarily on the trunk and interior part of main branches; the exterior of the compound fruit is green or yellow when ripe. The interior consists of large edible bulbs of yellow, banana-flavored flesh enclosing smooth, oval, light-brown seeds. The fruits take 90—180 days to reach maturity, and all parts of the tree except the outer bark are edible. Jackfruits are broadly classified into two groups: one has small, fibrous, mushy but very sweet carpels with a texture somewhat akin to raw oyster, while the other variety is crisp and almost crunchy, though not quite as sweet. The latter form is the more important commercially, and is more palatable to Western tastes. There are also intermediate types (Rahman *et al.*, 1999).

HISTORICAL CULTIVATION AND USAGE

No one knows the jackfruit's place of origin, but it is believed to be indigenous to the rain-forests of the Western Ghats. It is common in South-east Asia, and is occasionally found in Pacific island home gardens. The plants of *Artocarpus* species have since been distributed throughout tropical and subtropical regions (Wei *et al.*, 2005). It is often planted in central and eastern Africa, and is fairly popular in Brazil and Surinam.

The jackfruit has played a significant role in Indian agriculture for centuries. Archeological findings in India have revealed that it was cultivated in India 3000—6000 years ago. The plant is mainly known in the jackfruit-growing area as a source of food; poorer people in these areas used to eat its fruit for one of their daily meals instead of rice and other starchy foods. The seeds of the fruit in particular contain high levels of carbohydrate (~50%), protein (~20%), fat (~11%), and fiber (~7%) (Ajayi, 2008). People consume it mostly as a fruit when ripe, but also as a vegetable in its unripe stages. Jackfruit is a major part of the diet of the people of South-east Asia (where it is known as "poor man's" food), both as a vegetable and as a nutritious food during the season (Chowdhury *et al.*, 1997).

PRESENT-DAY CULTIVATION AND USAGE

Jackfruit, which probably originated in India, is now widely cultivated in South and South-east Asia, including Bangladesh, Malaysia, Myammar, Indonesia, the Philippines, Sri Lanka, South China, Thailand, Vietnam, in the Caribbean and Latin America, particularly Brazil, and parts of Africa (Rahman *et al.*, 1999). It is cultivated in some countries in the evergreen forest zone of West Africa. In Nigeria its cultivation has not been encouraged, though it is found in the south coastal parts of the country, where it grows wild or semi-conserved (Odoemelam, 2005). The jackfruit tree is a multipurpose species, providing food, fuel, timber and medicinal extracts. The oil from the seeds can serve nutritional purposes (Ajayi, 2008). Jackfruit is used as a shade tree for coffee, pepper, betel nut, and cardamom, and, due to its beautiful foliage, many products, and bountiful production, it has become an excellent tree for home gardens. The wood of the tree is used for building furniture and for house construction in India, and it has also been used for the production of musical instruments.

APPLICATIONS TO HEALTH PROMOTION AND DISEASE PREVENTION

In Indonesia, the plants of *Artocarpus* species have been used as traditional folk medicine against inflammation and malaria fever. The pulp and seeds have been used as a cooling tonic; the roots in diarrhea and fever; the wood as a sedative in convulsions; the leaves for the activation of milk supply in women and animals, and also as an antisyphilic and a vermifuge; while leaf ash has been applied to treat ulcers and wounds (Khan *et al.*, 2003).

As food, both the tender and ripe fruits, as well as the seeds, are rich in minerals and vitamins (Molla *et al.*, 2008). The pulp of the young fruit is cooked as a starchy food, while the ripe fruit is eaten fresh or processed into numerous delicacies, including jam, jelly, and chutney. The seeds and young leaves are also used as food. The seeds serve as an excellent addition to curries, or can be eaten freshly cooked or dried with salt as snacks. The cooked and dried seeds are milled to a flour-like consistency and added to bread dough. The tender young leaves are cooked as a vegetable, while green leaves are used as fodder for cattle, goats, and other small ruminants.

Health-wise, all the parts of the tree are said to have medicinal properties. The pulp and seeds are used as a tonic, the warmed leaves have healing properties if applied to wounds, and the latex, mixed with vinegar, promotes healing of abscesses, snake bites, and glandular swellings. The wood has a sedative effect, and its pith is said to cause abortion. The root is used as a remedy against skin diseases and asthma, and its extract is taken in cases of fever and diarrhea (Fang *et al.*, 2008). Jackfruit cotyledons are fairly rich in starch and protein (Singh *et al.*, 1991).

Antimicrobial activity

The antimicrobial activity of extracts from *Artocarpus heterophyllus* seed has been investigated. Reports showed that *A. heterophyllus* seed extracts did not exhibit good inhibitory activity, despite the traditional use of the plant against antibacterial activity. This indicates that the active compounds are mainly distributed in aerial parts, roots, and rhizomes, but not in the seeds. Karthy *et al.* (2009) showed that the crude ethanol and methanol extracts of the seed exhibited minimal (< 12 mm) antibacterial activity against methicillin-resistant *Staphylococcus aureus* (MRSA); however, Khan and colleagues (2003) reported that the crude methanolic extract of the stem and root barks, stem and root heartwood, leaves, fruits, and seeds of *A. heterophyllus*, and their subsequent partitioning with petrol, dichloromethane, ethyl acetate, and butanol, gave fractions that exhibited a broad spectrum of antibacterial activity, with the butanol fractions of the root bark and fruits being found to be most active. Extracts from *A. heterophyllus* plant have been reported to show intensive activity cariogenic bacteria. Extracts of *A. heterophyllus* leaves have been reported to significantly improve glucose tolerance in normal subjects and diabetic patients when investigated at oral doses equivalent to 20 g/kg of starting material (Fernando *et al.*, 1991).

According to reports, the only medicinal compound isolated from the *A. heterophyllus* seed so far is jacalin, the major protein from jackfruit seeds (Kabir, 1998). Jacalin is a tetrameric two-chain lectin which combines a heavy α-chain of 133 amino acid residues with a light β-chain of 20−21 amino acid residues. Jacalin inhibits the proliferation of HT29 colon cancer cells, as assessed by both thymidine incorporation and cell counting (Zhou *et al.*, 1993). The inhibition is non-cytotoxic, partial, and probably mediated through the TF antigen, Gal β1-3GalNAcα, expressed on HT29 cells. It is reversed by carbohydrate moieties such as MeαGal, GAlNAc, melibiose, and Gal β1-3GalNAc. These findings suggest that jacalin may have potential as a therapeutic agent for cancer. Jacalin has been used as a histochemical reagent to study tissue-binding properties in benign and malignant lesions of the breast and thyroid (Vijayakumar *et al.*, 1992). The binding of horse radish peroxidase (HRP) conjugated jacalin was found to be focally strong in neoplastic cells compared to normal and hyperplastic cells. It

TABLE 79.1 Proximate Composition of Jackfruit (*Artocarpus heterophyllus*) Seeds in Comparison with Some Other Seeds

Parameter	Species[a]			
	A. heterophyllus	T. africana	Groundnut	Palm kernel
Moisture	2.78	3.78	4.45	14.26
Ash	6.72	5.56	2.77	1.50
Crude protein (N × 6.25)	20.19	27.44	26.50	6.94
Crude fat	11.39	18.54	40.83	54.18
Crude fiber	7.10	8.20	—	—
Carbohydrate	51.82	36.48	25.40	23.10

Source: Ajayi (2008).
[a]*Mean of triplicate determination*

TABLE 79.2 Physicochemical Properties[a] of Jackfruit (*Artocarpus heterophyllus*) Seed Oil in Comparison with Some Other Seed Oils

Properties	A. heterophyllus	T. africana	Groundnut	Palm kernel
State at RT	Liquid	Liquid	Liquid	Liquid
Color	Yellow	Yellow	Yellow	Yellow
Refractive index	1.452	1.463	—	—
Acid value (mg NaOH/g oil)	24.96	10.04	2.77	16.60
Saponification number (mg KOH/g oil)	89.76	85.71	362.00	732.00
Iodine value (mg/100 g oil)	104.06	—	11.20	33.30
%FFA as oleic acid	12.48	5.06	0.44	0.57
Peroxide value (mg/g oil)	15.00	6.67	20.00s	20.00

Source: Ajayi (2008).
[a]*Mean of triplicate determination; RT, room temperature; FFA, free fatty acid.*

TABLE 79.3 Antibacterial Activity of *Artocarpus heterophyllus* Seed Extracts Against 12 MDR MRSA in Comparison With activity of other seed extracts

S No.	MRSA Isolates	Mean Zone of Inhibition (mm) 30 mg/ml									Methicillin	DMSO
		Artocarpus heterophyllus			Eletattaria cardamomum			Moringa oleifera				
		E	M	A	E	M	A	E	M	A	5 µg	5%
1	S_aW_1	9	10	8	17	16	16	20	19	20	—	—
2	S_aW_2	8	10	—	17	16	16	17	15	17	—	—
3	S_aW_3	10	10	—	13	14	12	20	21	22	—	—
4	S_aW_4	7	11	—	11	12	11	20	20	18	11	—
5	S_aW_5	11	11	8	17	16	16	21	20	18	10	—
6	S_aW_6	8	12	—	11	12	13	26	20	20	—	—
7	S_aW_7	11	11	—	11	12	13	21	21	20	—	—
8	S_aW_8	7	11	—	15	14	10	20	20	20	13	—
9	S_aW_9	12	10	—	12	14	14	19	19	19	14	—
10	S_aW_{10}	12	10	—	15	14	16	19	20	18	—	—
11	S_aW_{11}	12	10	—	16	16	16	20	19	18	—	—
12	S_aW_{12}	12	11	—	15	15	15	19	21	20	—	—

MRSA, methicillin-resistant *Staphylococcus aureus*; S_aW, *Staphylococcus aureus* wound; E, ethanol; M, methanol; A, acetone; —, no inhibition of the concentrated tested; DMSO, dimethyl sulfoxide.
Source: Karthy *et al.* (2009).

TABLE 79.4 Binding of Various Glycoproteins to Jacalin

Glycoproteins	Reference(s)*
Human plasma glycoproteins	
IgAl	Hagiwara *et al.* (1998), Kondoh *et al.* (1986)
IgD	Aucouturier *et al.* (1987), Zehr & Litwin (1987)
Cl-inhibitors	Hiemstra *et al.* (1987), Pilate *et al.* (1989)
C4b-binding gp120	Pilatte *et al.* (1995)
Hemopexin	Pilatte *et al.* (1995)
Plasminogen	Hortin & Trimpe (1990)
α1- antitrypsin	Pekelharing & Animashaun (1989)
α2-macroglobulin	Pekelharing & Animashaun (1989)
8Sα3 glycoprotein	Pekelharing & Animashaun (1989)
αHSG	To *et al.* (1995)
Other human proteins	
Leukosialin	Maemura & Fukuda (1992)
CD4	Favero *et al.* (1993), Lafont *et al.* (1996)
Urinary IL-1 inhibitors (30 kDa)	Kabir & Wigzell (1989)
Pro-insulin like growth factor	Hudgin *et al.* (1992)
Bone alkaline phosphatase	Miura *et al.* (1994)
Milk bile salt activated lipase	Wang *et al.* (1994)
Miscellaneous	
Rabbit IgA	Kabir (1993)
Rabbit IgG	Kabir (1993), Kabir *et al.* (1995), Kabir & Gerwig (1997)
Fetuin	Hortin & Trimpe (1990)
Bovine protein z	Hortin & Trimpe (1990)
Bovine coagulation factor x	Hortin & Trimpe (1990)

Source: Kabir (1998), and * references therein.

has been used to monitor the progression of cervical intraepithelial neoplasia (Remani *et al.*, 1994). Jacalin has been used to distinguish malignant cells from benign reactive cells in serious effusions (Sujathan *et al.*, 1996). It is suitable for studying various O-linked glycoproteins, particularly human 1gA1, and is a useful tool for the evaluation of the immune status of patients infected with the human immunodeficiency virus (HIV)-1. It has found applications in diverse areas, including the isolation of IgA1 from human serum, the binding of a number of human plasma glycoproteins (1gA1, C1-inhibitor, hemopexin, α2-HSG), the investigation of 1gA nephropathy (IgAN), the identification of O-linked glycoproteins, and the detection of tumors (Kabir, 1998).

There are two areas in AIDS research where jacalin can make important contributions. First, jacalin is unique, as it is mitogenic only for $CD4^+$ cells. Therefore, jacalin-induced T cell proliferation can provide information on the quantitative as well as qualitative deficiency of $CD4^+$ T cells in HIV-1 infection. Secondly, native jacalin as well as jacalin-derived peptide have been found to prevent HIV-1 infection *in vitro*. However, the mechanism of action of jacalin in preventing HIV-1, as well as jacalin's mode of interaction with CD4, is not well understood. Further work is needed to explore whether jacalin can be used as a therapeutic agent for the prevention of HIV-1 infection (Kabir, 1998). Using jacalin histochemistry, the sTF has been detected in various normal human tissues, such as stomach, liver, trachea, lung, uterus, brain, erythrocytes, macrophages, and lymphocytes (Cao *et al.*, 1996). Moreover, examination of several samples of paraffin-embedded tissues suggests that jacalin is a good marker for the identification of free macrophages (Urdiales-Viedma *et al.*, 1995).

ADVERSE EFFECTS AND REACTIONS (ALLERGIES AND TOXICITY)

No adverse effects have been reported in the literature consulted.

SUMMARY POINTS

- *A. heterophyllus*, popularly known as jackfruit, is a tree that attains a height of 8–25 m and a stem diameter 30–80 cm.
- The fruits can weigh of up to 20–49 kg.
- The tree seems to be indigenous to the rainforests of the Western Ghats, but has since been distributed throughout tropical and subtropical regions.
- The jackfruit tree is a multipurpose species; it provides food, fuel, timber, and medicinal extracts. All the parts of the tree are said to have medicinal properties.
- The compound isolated from the seed of *A. heterophyllus*, jacalin, has potential as a therapeutic agent for cancer.
- Jacalin could make important contributions in two areas in AIDS research.

References

Ajayi, I. A. (2008). Comparative study of the chemical composition and mineral element content of *Artocarpus heterophyllus* and *Treculia africana* seeds and seed oils. *Bioresource Technology, 99*, 5125–5129.

Babitha, S., Soccol, C. R., & Pandey, A. (2006). Jackfruit seed for production of Monascus pigments. *Food Technology and Biotechnology, 44*, 465–471.

Cao, Y., Stosiek, P., Springer, G. F., & Karsten, U. (1996). Thomsen-Friedenreich-related carbohydrate antigens in normal adult human tissues: a systematic and comparative study. *Histochemistry and Cell Biology, 106*, 197.

Chowdhury, F. A., Raman, M. A., & Mian, A. J. (1997). Distribution of free sugars and fatty acids in jackfruit (*Artocarpus heterophyllus*). *Food Chemistry, 60*, 25–28.

Fang, S., Hsu, C., & Yen, G. (2008). Anti-inflammatory effects of phenolic compounds isolated from the fruits of *Artocarpus heterophyllus*. *Journal of Agricultural and Food Chemistry, 56*, 4463–4468.

Fernando, M. R., Naline Wickramasinghe, S. M. D., Thabrew, M. I., Ariyananda, P. I., & Karunanayake, E. H. (1991). Effect of *Artocarpus heterophyllus* and *Asteracanthus longifolia* on glucose tolerance in human subjects and in maturity-onset diabetic patients. *Journal of Ethnopharmacology, 31*, 277–282.

John, P. J., & Narasimham, P. (1993). Processing and evaluation of carbonated beverage from jackfruit waste (*Artocarpus heterophyllus*). *Journal of Food Processing and Preservation, 16*, 373–380.

Kabir, S. (1998). Jacalin: a Jackfruit (*Artocarpus heterophyllus*) seed-derived lectin of versatile applications in immunobiological research. *Journal of Immunological Methods, 212*, 193–211.

Karthy, E. S., Ranjitha, P., & Mohankumar, A. (2009). Antimicrobial potential of plant seed extracts against multidrug resistant methicillin resistant Staphlococcus aureus (MDR-MRSA). *International Journal of Biology, 1*, 34–40.

Khan, M. R., Omoloso, A. D., & Kihara, M. (2003). Antimicrobial activity of *Artocarpus heterophyllus*. *Fitoterapia, 74*, 501–505.

Molla, M. M., Nasrin, T. A. A., Islam, M. N., & Bhuyan, M. A. J. (2008). Preparation and packaging of jackfruit chips. *International Journal of Sustainable Crop Production, 3*, 41–47.

Mukprasirt, A., & Sajjaanantakul, K. (2004). Physico-chemical properties of flour and starch from jackfruit seeds (*Artocarpus heterophyllus Lam.*) compared with modified starches. *International Journal of Food Science and Technology, 39*, 271–276.

Odoemelam, S. A. (2005). Functional properties of raw and heat processed jackfruit flour. *Pakistan Journal of Nutrition, 4*, 366–370.

Rahman, M. A., Nahar, N., Mian, A. J., & Mosihuzzaman, M. (1999). Variation of carbohydrates composition of two forms of fruit from jack tree (*Artocarpus heterophyllus* L) with maturity and climatic conditions. *Food Chemistry, 65*, 91–97.

Remani, P., Augustine, J., Vijayan, K. K., Ankathil, R., Vasudevan, D. M., Nair, K. M., et al. (1989). Jackfruit lectin binding pattern in benign and malignant lesions of the breast. *In Vivo (Attiki), 3*, 275.

Singh, A., Kumar, S., & Singh, I. S. (1991). Functional properties of jackfruit seed flour. *Lebensmittel Wissenschaft und Technologie, 24*, 373–374.

Sujathan, K., Kannan, S., Remani, P., Pillai, K. R., Chandralekha, B., Amma, N. S., et al. (1996). Differential expression of jackfruit-lectin-specific glycoconjugates in metastaic adeno-carcinoma and reactive mesothelial cells – a diagonistic aid in effusion cytology. *Journal of Cancer Research & Clinical Oncology, 122*, 433.

Urdiales-Viedma, M., Haro-Monoz, T. D., Martos-Padilla, S., Ahad-Ortega, J., Varela-Duran, J., & Granda-Paez, R. (1995). Jacalin, another marker for histiocytes in paraffin-embedded tissues. *Histopathology, 10,* 597.

Vijayakumar, T., Augustine, J., Matthew, L., Aleykutty, M. A., Nair, B. M., Remani, P., et al. (1992). Tissue binding pattern of plant lectins of thyroid. *Journal of Experimental Pathology, 6,* 11.

Wei, B., Weng, J., Chiu., Hung, C., Wang, J., & Lin, C. (2005). Anti-inflammatory flavonoids from *Artocarpus heterophyllus* and *Artocarpus communis. Journal of Agricultural and Food Chemistry, 53,* 3867−3871.

Zhou, Z. Q., Yu, L. G., Milton, J. D., Fernig, D. G., & Rhodes, J. M. (1993). Jacalin causes non-cytotoxic inbihition of proliferation of HT29 colon cancer cells. *Clinical Science, 85,* 11P.

683

CHAPTER 80

Biological Activities of *Eugenia jambolana* (Family Myrtaceae) Seeds

Siham Mostafa Ali El-Shenawy
Department of Pharmacology, National Research Centre, Dokki, Cairo, Egypt

685

LIST OF ABBREVIATIONS

CNRS, National Centre of Scientific Research

HIV, human immunodeficiency virus

IMRA, Malagasy Institute of Applied Research

NCCAM, National Centre for Complementary and Alternative Medicine

SGOT, serum glutamic oxaloacetic transaminase

SGPT, serum glutamic pyruvic transaminase

WHO, World Health Organization

INTRODUCTION

The world is facing an explosive increase in the incidence of many systemic diseases, and cost-effective complementary therapies are needed (Sridhar *et al.*, 2005). Medicinal plants have been used for centuries as remedies for human diseases (Jasmine *et al.*, 2007). India, with its rich culture, traditions, and natural biodiversity, offers a unique opportunity for drug discovery researchers. *Eugenia jambolana* Lam., of the family

Nuts & Seeds in Health and Disease Prevention. DOI: 10.1016/B978-0-12-375688-6.10080-5

Myrtaceae (myrtle family), is an edible plant long used as a remedy in folk medicine. It is extensively cultivated in most regions of the world, in tropical and subtropical zones. The family Myrtaceae comprises about 75 genera and 3000 species. It is valued in the Ayurveda and Unani systems of medicine for its many therapeutic properties (Sagrawat *et al.*, 2006).

BOTANICAL DESCRIPTION

Eugenia jambolana Lam. (syn: *Syzygium cumini*, Java plum) is a member of the Mytreae tribe of the Myrtoideae subfamily of the family Myrtaceae (Willis, 1973). It is an evergreen tree of up to 25 m tall, with young grayish-white stems, and coarse and discolored lower bark (Figure 80.1). The leaves are opposite, simple, entire, elliptic to broadly oblong, smooth, glossy, somewhat leathery, 5–10 cm in length, and shortly pointed at the tips. Petioles are up to 3 cm long, the leaf midrib is prominent and yellowish, and the blades have many closely parallel lateral veins. The flowers are white to pinkish, about 1 cm across, and found in branched clusters at the stem tips. The calyx is cuplike, and there are four petals fused into a cap, and many stamens. The fruit is an ovoid, dark purplish-red, shiny, one-seeded berry up to 2 cm long, with white to lavender flesh (Lawrence, 1958) (Figure 80.2).

HISTORICAL CULTIVATION AND USAGE

The jambolan (*E. jambolana* Lam.) is native to India, Myanmar, Sri Lanka, and the Andaman Islands, and is naturalized in Malaya. It is thought to be of prehistoric introduction to the Philippines, where it is widely planted and naturalized, as it is in Java and elsewhere in the East. The tree flourishes in areas of the USA, such as California, especially in the vicinity of Santa Barbara. In southern Florida, there are numerous fruiting specimens (Morton, 1963).

E. jambolana trees require a humid climate, and the fruits mature during the rainy season. The plant does not grow well in arid, dry areas such as southern Madagascar. The fruit season runs from February to April, and, depending on its age and growth stage, each tree can produce 15–180 kg of fruit, representing 5–60 kg of seeds.

FIGURE 80.1
E. jambolana fam (evergreen tree).

FIGURE 80.2
Eugenia jambolana plant (leaves, fruits, seeds). Reproduced from indusextracts.com/eugenia-jambolana.htm.

The jambolan has received far more recognition in folk medicine and in the pharmaceutical trade than in any other field. Medicinally, the extracts of the seed have antidiabetic activity (Steinmetz, 1960).

PRESENT-DAY CULTIVATION AND USAGE

E. jambolana grows naturally in tropical as well as subtropical zones. It is native to Bangladesh, India, Nepal, Pakistan, and Indonesia, and is also grown in other areas of southern and southeastern Asia, including the Philippines, Myanmar, and Afghanistan. In Brazil, where it was introduced from India during Portuguese colonization, it has dispersed spontaneously in the wild in some places. Most of the plant's parts are used in the traditional medicine of India. *Syzygium cumini* is a medicinal plant whose seeds have been pharmacologically proved to possess hypoglycemic, antibacterial, anti-inflammatory, anti-human immunodeficiency virus (HIV), and antidiarrheal effects (Sharma *et al.*, 2008). *E. jambolana* seeds also have gastro-protective and hepatoprotective properties (Chaturvedi *et al.*, 2007).

APPLICATIONS TO HEALTH PROMOTION AND DISEASE PREVENTION

There are numerous medicinal plants in current use all over the world in both Ayurvedic and traditional medicine. These plants have many uses, and deserve to be exhibited to enthusiasts who wish to know more about their stem, flower, fruit, and seed characteristics, in addition to the research that has been undertaken. However, it is impossible to exhibit all, or even the majority, of them. For this reason, one of the common medicinal plants, *E. jambolana*, that has received research attention both locally and internationally is discussed here. In addition, some medicinal plants that have so far attracted little attention from researchers have also been included owing to their recognition in Ayurvedic and traditional medicine in some Asian countries (Chopra & Doiphode, 2002).

Ayurveda is a natural healthcare system that originated in India more than 5000 years ago. Its main objective is to achieve optimal health and well-being through a comprehensive approach that addresses mind, body, behavior, and environment. Ayurveda emphasizes prevention and health promotion, and provides treatment for disease. It considers the development of consciousness to be essential for optimal health, and meditation to be the main technique for achieving this. Treatment of disease is highly individualized, and depends on the psycho-physiological constitution of the patient. There are different dietary and lifestyle recommendations for each season of the year. It may be projected from Ayurveda's comprehensive approach, emphasis on prevention, and ability to manage chronic disorders that its widespread use would improve the health status of the world's population (Sharma *et al.*, 2007). Not surprisingly, the World Health Organization's (WHO) concept of health propounded in the modern era is in close approximation with the concept of health defined in Ayurveda (Kurup, 2004).

Antidiabetic activity

With regard to the medicinal uses of *E. jambolana* in general, and the seeds in particular, its main and most effective pharmacological action relates to its hypoglycemic effects (Mukherjee *et al.*, 2006), since diabetes, one of the most common chronic diseases, is associated with high economic costs to both patients and their governments. From 1967 to 1985, *E. jambolana* seeds were studied by researchers at both the Malagasy Institute of Applied Research (IMRA) and the National Centre of Scientific Research (CNRS) of the Faculty of Medicine at the University of Paris. The IMRA is a private foundation, recognized by the government of Madagascar and WHO as a collaborative center. It is committed to drug regulation, quality control of imported and local drugs, and legislation to protect and authorize the use of new drugs. It guarantees free access to and exchange of information and products with national and international partners.

The formulation studied here, called Madeglucyl® ("the natural way to maintain healthy blood sugar levels"), is prepared from the seeds of *E. jambolana*, and contains 1% active ingredient. Tests have established that the drug is stable and consistently effective as a treatment for diabetes. Trials in laboratory animals were carried out over several generations of rats (particularly the sand rat), which were found to be the most appropriate laboratory animal for this stage of research. The trials showed that Madeglucyl did indeed alleviate the symptoms of diabetes. In humans, clinical trials have been held to study the effects of Madeglucyl in thousands of diabetes patients in Madagascar, and, more recently, in Germany and the United States. On December 18, 1997, Madeglucyl was registered as a licensed medicine in Madagascar (Denis *et al.*, 2008).

The following are some of the main antidiabetic features of Madeglucyl:

- When used in patients with type 2 diabetes, glycemia rates return to normal levels within 3—6 months in 75% of cases. It has proven particularly effective for obese patients, whose glycemia rates start to decline after 15 days and return to normal within 3 months (again in 75% of cases).
- It reduces, by nearly 40%, the daily insulin required by type 1 insulin-dependent patients.
- In some cases, it has improved the functioning of the kidneys, and relieved some of the eye complications related to diabetes.
- It acts by improving the ability of the body's tissues to absorb glucose, thus enhancing the effectiveness of insulin.

The compounds that make Madeglucyl effective against diabetes were extracted, isolated and developed in project partnership with the French-based Rhône Poulenc Rorer pharmaceutical company. In addition to WHO, partners and collaborators in the project include the following: private companies, including SOAMADINA Limited Company (Madagascar), CFAO-EURA-PHARMA (France), Afrique Initiatives (Belgium), and the Carlesimo Foundation (France); the Madagascar public health sector, including the Ministry of Health, National Drug Agency, Faculty of Medicine, Faculty of Science, Polytechnic School, and Medical Practitioner Association; and universities and international institutes, including the Catholic University of Louvain (Belgium), Faculties of Pharmacy at Rheims, Chateney and Lille (France), the Natural History Museum (France), and the Institute of Health (Italy).

The findings of the *E. jambolana* project could easily be applied to other countries, especially those bordering the Indian Ocean, where the plant grows naturally and extensively. The process for producing Madeglucyl from the seeds of the plant is protected by international licence; however, commercial agreements could be made with the public and private sectors of other developing and developed countries to allow this safe and affordable diabetes treatment to be made available to patients worldwide.

Hepatoprotective effect

Another significant medicinal application of the *E. jambolana* seed is its use as a hepatoprotective agent, which has been reported by Sisodia and Bhatnagar (2009). In that study, the ethanol extracts of *E. jambolana* seeds showed hepatoprotective effects in carbon tetrachloride-treated rats. In addition, another study has reported on the hepatoprotective and antioxidant activity of *E. jambolana* seeds (El-Shenawy, 2008). Hepatoprotective effects are attributed to its antioxidant activity, which restores the activity of superoxide dismutase, catalase, and glutathione peroxidase to normal levels, and increases glutathione content and levels of lipid peroxidation and hydroperoxides in the liver (Ravi *et al.*, 2004). Seed content includes glycosides, traces of pale yellow essential oil, fat, resin, albumin, chlorophyll, the alkaloid jambosine, gallic acid, ellagic acid, corilagin and related tannins, 3,6-hexahydroxydiphenoyl glucose and its isomer 4,6-hexahydroxydiphenoyl glucose, 1-galloyl glucose, 3-galloyl glucose, quercetin, and elements such as zinc, chromium, vanadium, potassium, and sodium. Unsaponifiable matter of the seed fat contains β-sitoterol. Dried seeds of *E. jambolana* have been reported, with 11.67% alcohol-soluble extractive fiber, 3.397% inorganic fiber, 40% water-soluble gummy fiber, and 15% water-insoluble neutral detergent fiber (Bhatia & Bajaj, 1975). Kumar and colleagues (2009) state that the ethyl acetate and methanol extracts of the seeds of *S. cumini* show the presence of alkaloids, amino acids, flavonoids, glycosides, phytosterols, saponins, steroids, tannins, and triterpenoids. Since the current worldwide morbidity and mortality due to liver disease is increasing every year, with corresponding increases in expenditure for drug treatment, alternative plant therapies may be beneficial.

Anticancer activity

A recent study reports that berry extract of *E. jambolana* inhibits growth and induces apoptosis of human breast cancer cells (Li *et al.*, 2009). Future research on this topic would help to identify safe and effective anticancer drugs, and further the exploration of their mechanisms of action. Ayurvedic practitioners and researchers in the medical sciences can help to improve the clinical practice of medicine by increasing their involvement and contribution. As a research design, the case study can serve as the basis for future research directions, and can provide valuable contributions to the medical field at minimal cost. Case studies have also been suggested by the National Center for Complementary and Alternative Medicine (NCCAM, Bethesda, Maryland, USA) as a means to determine whether a complementary anticancer therapy demonstrates potential efficacy against a particular cancer, and whether clinical development of the therapy should continue (Bhushan & Kadian, 2009). It is no longer an option to ignore Ayurvedic drugs or to treat them as something unconventional with respect to regular medical practice.

Additionally, new uses of *E. jambolana* seeds are potentially sustainable, as long as appropriate laws, regulations, standard operating procedures and good manufacturing practices are put in place and supported by ministries of health, drug regulatory agencies, and environmental protection institutions at the national and international levels.

ADVERSE EFFECTS AND REACTIONS (ALLERGIES AND TOXICITY)

An acute toxicity study in mice after oral administration of various doses (100, 200, 500, 1000, and 2000 mg/kg) of *E. jambolana* seed extract showed no behavioral changes or mortality up to 72 hours after treatment. Subacute toxicity studies with extract (1 g/kg) in rats (4 weeks of treatment) showed no significant change in general physiological parameters (body weight, food and water intake), organ weight (liver, kidney, adrenals, and testis), hematological parameters (hemoglobin and white blood cell count), liver function (total bilirubin, serum glutamic oxaloacetic transaminase (SGOT), serum glutamic pyruvic transaminase (SGPT), alkaline phosphatase, total protein and albumin), and renal function (blood urea and creatinine). Histological studies of the liver, kidney, testis, and ovaries did not indicate any gross or microscopic changes (Chaturvedi *et al.*, 2007).

Additional tests showed that the extract is not toxic or carcinogenic, does not lead to the malformation of embryos, and has no other detrimental side effects. In rats, graded oral doses of up to 5 g/kg (single dose) of ethanol extracts of the seeds and pericarp of *E. jambolana* induced no mortality 24 hours after administration (El-Shenawy, 2008). As a requirement for Ethics Committee clearance, subacute toxicity studies were performed using a single administration of 2.5 and 5.0 g/kg *E. jambolana* seed powder; animals were observed for 14 days, with no evidence of mortality or abnormalities (Sridhar *et al.*, 2005).

SUMMARY POINTS

- *E. jambolana*, of the family Myrtaceae, is an edible plant long used as a remedy in folk medicine.
- It is extensively cultivated in most regions of the world, and can be cultivated at relatively low cost.
- The use of *E. jambolana* seed extract in traditional medicine suggests its action as possible chemopreventive agents or phytoceuticals.
- Evidence shows that the oral administration of *E. jambolana* seed extract or powder is useful for the treatment of many diseases, without any reported adverse effects or systemic toxicity in experimental animals and in humans, based on the results of experimental studies.
- *E. jambolana* seeds can be used for its antibacterial, antifungal, aphrodisiac, anti-inflammatory, antioxidant, antipediculosis, and antiulcerative properties.
- Its most important uses may be in chronic disorders, because of its antimutagenic and hepatoprotective effects.
- This chapter provides further support for the traditional use of *E. jambolana seeds* as a safe medicinal plant in some systemic diseases.
- Additional research should be carried out to identify the active principle(s) which can be used for treatment of each illness.

References

Bhatia, I. S., & Bajaj, K. L. (1975). Chemical constituents of the seeds and bark of Syzygium cumini. *Planta Medica, 28*, 346−352.

Bhushan, S., & Kadian, S. S. (2009). Cancer therapy in Ayurveda — a systemic review. *Pharmaceutical Reviews, 7*(4).

Chaturvedi, A., Kumar, M. M., Bhawani, G., Chaturvedi, H., Kumar, M., & Goel, R. K. (2007). Effect of ethanolic extract of *Eugenia jambolana* seeds on gastric ulceration and secretion in rats. *Indian Journal of Physiology and Pharmacology, 51*, 131−140.

Chopra, A., & Doiphode, V. V. (2002). Ayurvedic medicine. Core concept, therapeutic principles, and current relevance. *Medical Clinics of North America, 86*, 75−89.

Denis, R., Mamy, R., Christian, R., Francisco, R., Kiban, C., Anne-Marie, C., et al. (2008). *Ex situ* conservation and clonal propagation of the Malagasy *Syzygium cuminii*, an antidiabetic plant. *Belgian Journal of Botany, 141*, 14−20.

El-Shenawy, S. M. A. (2008). Evaluation of some pharmacological activities of ethanol extracts of seeds, pericarp and leaves of *Eugenia Jamolana* in rats. *Inflammopharmacology, 16*, 1−8.

Jasmine, R., Daisy, P., & Selvakumar, B. N. (2007). *In vitro* efficacy of flavonoids from *Eugenia jambolana* seeds against ESβL-producing multidrug resistant enteric bacteria. *Research Journal of Microbiology, 4*, 369−374.

Kumar, A., Ilavarasan, R., Jayachandran, T., Decaraman, M., Aravindhan, P., Padmanabhan, N., et al. (2009). Phytochemicals investigation on a tropical plant, *Syzygium cumini* from Kattuppalayam, Erode District, Tamil Nadu, South India. *Pakistan Journal of Nutrition, 18*, 83−85.

Kurup, P. N. V. (2004). Ayurveda — a potential global medical system. In L. C. Mishra (Ed.), *Scientific basis for Ayurvedic therapies* (pp. 1−15). New York, NY: CRC Press.

Lawrence, G. H. M. (1958). *Taxonomy of vascular plants*. New York, NY: Macmillan.

Li, L., Adams, L. S., Chen, S., Killian, C., Ahmed, A., & Seeram, N. P. (2009). *Eugenia jambolana* Lam. berry extract inhibits growth and induces apoptosis of human breast cancer but not non-tumorigenic breast cells. *Journal of Agricultural and Food Chemistry, 57*, 826−831.

Morton, J. F. (1963). The jambolan (*Syzygium cumini* Skeels) − its food, medicinal, ornamental and other uses. Proceedings of the Florida State Horticultural Society, pp. 328−338.

Mukherjee, P. K., Maiti, K., Mukherjee, K., & Houghton, P. J. (2006). Leads from Indian medicinal plants with hypoglycemic potentials. *Journal of Ethnopharmacology, 106*, 1−28.

Ravi, K., Ramchandran, B., & Subramanian, S. (2004). Protective effect of *Eugenia jambolana* seed kernel on tissue antioxidants in streptozotocin induced diabetic rats. *Biological Pharmaceutical Bulletin, 78*, 1212−1217.

Sagrawat, H., Mann, A. S., & Kharya, M. D. (2006). Pharmacological potential of *Eugenia jambolana*: a review. *Pharmacognosy, 6*, 96−105.

Sharma, H., Chandola, H. M., Singh, G., & Basisht, G. (2007). Utilization of Ayurveda in health care: an approach for prevention, health promotion, and treatment of disease. Part 1 − Ayurveda, the science of life. *Journal of Alternative and Complementary Medicine, 13*, 1011−1020.

Sharma, B., Viswanath, G., Salunke, R., & Roy, P. (2008). Effects of flavonoid-rich extract from seeds of *Eugenia jambolana* (L.) on carbohydrate and lipid metabolism in diabetic mice. *Food Chemistry, 110*, 697−705.

Sisodia, S. S., & Bhatnagar, M. (2009). Hepatoprotective activity of *Eugenia jambolana* Lam. in carbon tetrachloride treated rats. *Indian Journal of Pharmacology, 41*, 23−27.

Sridhar, S. B., Sheetal, U. D., Pai, M. R. S. M., & Shastri, M. S. (2005). Preclinical evaluation of the antidiabetic effect of *Eugenia jambolana* seed powder in streptozotocin-diabetic rats. *Brazil Journal of Medical and Biological Research, 38*, 463−468.

Steinmetz, E. F. (1960). A botanical drug from the tropics used in the treatment of diabetes mellitus. *Acta Phytotherapeutica, 7*, 23−25.

Willis, J. C. (1973). *Dictionary of the flowering plants and ferns* (2nd ed.). Cambridge, UK: Cambridge University Press.

CHAPTER 81

Kersting's Nut (*Kerstingiella Geocarpa*): A Source of Food and Medicine

Folake Lucy Oyetayo, Olubunmi Bolanle Ajayi
Department of Biochemistry, University of Ado-Ekiti, Nigeria

693

INTRODUCTION

Kersting's nut (*Kerstingiella geocarpa*) is an underutilized legume indigenous to West Africa. It is an annual non-climbing herb, used as food and drink. Though it is found in other parts of Africa, Asia, and India it has not attained national or worldwide importance (Chickwendu, 2007). Many cultivars of *K. geocarpa* are tolerant of poor soil conditions. It grows on sandy loam soils in the savannah areas of West Africa. It is an important crop in the fixation of atmospheric nitrogen to enrich the soil for the growth of other crops, through symbiosis with *Rhizobium* bacteria. Hence, apart from its importance as a food legume *Kerstingiella geocarpa* seed will be beneficial in crop rotations and intercropping. However, the seed is mostly produced and consumed in few rural communities. This chapter provides useful information that could further popularize the production and utilization of the seed.

BOTANICAL DESCRIPTION

Kerstingiella geocarpa belongs to the family Fabaceace, subfamily Leguminosae, genus Macrotyloma, and tribe Phaseoleae. A herbaceous annual non-climbing herb, it is self-pollinating, producing fruit in the form of a pod containing one to three seeds. The morphological diversity is low, with three seed colors (Pasquet *et al.*, 2002) varying from white, to brown-speckled, to black (Amuti, 1980), although Chickwendu (2007) identified four colors (reddish-brown, brown, light brown, and deep cream). The pods develop on or near the soil surface, and immature seeds are green in color. Seed germination takes 24 hours and seeds exhibit epigeal germination. The plant has either a bunched or an open growth habit, and there is rooting at every node. Two days after the flower has opened, a peg carrying the ovary elongates and grows down toward the soil surface. The pod develops either on the soil surface or about 1–2 cm below it; seed development accompanies pod development. The fruit matures about 6–9 weeks after the flower opens.

HISTORICAL CULTIVATION AND USAGE

Kersting's nut is cultivated on a small scale in West Africa (Pasquet *et al.*, 2002), especially in the indigenous population of the northern savannah zones. It is grown for home consumption only, by the elderly (Amuti, 1980) in most populations, and hence is going out of production. *Kerstingiella geocarpa*, as described by Harms (1908), was collected by a German, Kersting, in Togo between 1905 and 1907; a few years later Chevalier (1910), from Benin, also recognized the plant. The plant was recorded in Nigeria by Stapf (*1931*), while in Northern Cameroon a wild relative of *Kerstingiella geocarpa*, *K. geocarpa* var. *tisserantii* (Pellegrin) Hepper, was identified (Hepper, 1963).

The seeds can be preserved for about 2 years by mixing them with wood ash.

PRESENT-DAY CULTIVATION AND USAGE

At present, *Kerstingiella geocarpa* seed has not attained national or worldwide importance compared to staple legumes such as cowpea, groundnut, and soybean, because of its low production and scanty information about its nutritive value. It is cultivated in the rainy season, either in pure stands or as mixed crop with yams. After maturation and harvest, seeds are obtained from pods by gently pounding the dried fruit in a mortar and pestle, followed by winnowing to separate seeds from husks, and whole from broken seeds (Amuti, 1980). The total production is low, since it is cultivated over a small acreage.

The seed is a tasty food, and contains appreciable content of nutrients such as vegetable protein (19.91–22.36%) (Chikwendu, 2007; Oyetayo & Ajayi, 2005; Table 81.1). This compares favorably with other legumes, such as beans *Vigna unguiculata* (20.5%), and groundnut (*A. hypogea*) (23.2%) (Ononogbu, 1988). The seed serves as a source of protein in the diet of populations of the dry savannah zone. The water in which the seed is boiled is traditionally used as a panacea for diarrhea. Powdered dry seed flour is used in cake-making, and in cases of poisoning it can be mixed with water and used as an emetic (Amuti, 1980). It is useful in the formulation of weaning diets.

TABLE 81.1 Proximate Composition (%) of *Kerstingiella geocarpa* Seed	
Nutrient	**Concentration**
Moisture	6.17 ± 0.20
Crude protein	21.33 ± 0.25
Fat	0.98 ± 0.00
Ash	3.79 ± 0.10
Carbohydrate (%)	61.53 ± 0.01

Values means of triplicate determinations.
Reproduced from Oyetayo and Ajayi (2005), *Biosci. Biotech. Res. Asia*, 3, 47–50, with permission.

APPLICATIONS TO HEALTH PROMOTION AND DISEASE PREVENTION

The earliest report on the nutritional value of *Kerstingiella geocarpa* shows that it has high nutritional value and a high nitrogen content (Irvine, 1969). It contains appreciable quantities of nutrients such as protein (21.5%) and carbohydrate (73.9%). Its protein is composed of lysine 6.2% and methionine 1.4% (NAS, 1979). Although total protein content is important in the biological utilization of protein, the level and balance of the essential amino acids primarily determines the nutritional value. The protein of *Kerstingiella geocarpa* seed is a rich source of essential amino acids, which have beneficial health implications, such as their role in growth and blood formation. Reports by Ajayi and Oyetayo (2009) showed that rats fed *Kerstingiella geocarpa* seed diets had a high plasma protein concentration, and their white blood count and hemoglobin concentrations were significantly higher ($P \leq 0.05$), than in the casein-fed controls. Hence, the seed could be found useful in diet formulations for children and anemic patients.

Arginine, an essential amino acid for pediatric growth, is the most concentrated amino acid in the *Kerstingiella geocarpa* seed, followed by leucine and phenylalanine (Ajayi & Oyetayo, 2009). Some of the essential amino acids in the seed flour — arginine (9.5%) and histidine (2.1%) — were found to occur in higher concentrations than the FAO/WHO amino acid requirement pattern for the preschool age group (2- to 5-year-olds) (Table 81.2). Both amino acids are particularly essential for children (FAO/WHO/UNU, 1985). If one or more essential amino acids is in inadequate supply, the rate at which protein synthesis occurs will be reduced by the same ratio (Crisan & Sands, 1978). Lysine, an essential amino acid limited in most vegetable proteins such as staple cereals, was also found in a fair concentration (2.9 g/100 g) in the seed; thus, the seed is a rich source of essential amino acids which are important for protein synthesis. The practice of consuming the legume alongside cereals which have a low concentration of essential amino acids is encouraged.

The amino acid ratio plays a significant role in the atherogenicity of a protein. A high lysine/arginine ratio increases the atherogenic potential of a diet, while a lower one exerts a hypocholesterolemic effect (Ajayi & Oyetayo, 2009). The ratio has been suggested to influence cholesterol metabolism by affecting the synthesis of the apoprotein, which is the constituent of the lipoprotein that causes atherogenicity in animals (Sen & Bhattacharya, 2001). We reported a lower Lys/Arg ratio for *Kerstingiella geocarpa* seed-fed rats (Ajayi & Oyetayo, 2009) than for the casein control. Hence, the seed would be beneficial for the management and prevention of atherogenic-related disorders.

TABLE 81.2 FAO/WHO Recommendations for Preschool Age (2–5 Years) Children	
Amino Acid	**FAO/WHO Recommendation**
Isoleucine	2.8
Leucine	6.6
Lysine	5.8
Phenyalanine	2.8
Phenylalanine + tyrosine	6.3
Threonine	3.4
Valine	3.5
Arginine	2.0
Histeine	1.9
Methionine + cysteine	2.5

TABLE 81.3 Major Mineral Composition (mg/g) of *Kerstingiella geocarpa* Seed

Element	Concentration
Potassium (mg/g)	7.67 ± 0.01
Calcium (mg/g)	8.84 ± 0.10
Phosphorus (mg/g)	9.30 ± 0.02
Sodium (mg/g)	5.67 ± 0.02

Values means of triplicate determinations.
Reproduced from Oyetayo and Ajayi (2005), *Biosci. Biotech. Res. Asia*, 3, 47–50, with permission.

While the *Kerstingiella geocarpa* seed is also an important source of nutritive elements such as iron, zinc, magnesium, and calcium (Table 81.3), it has a low sodium concentration (Oyetayo & Ajayi, 2005). Some of these elements are hematinics, essential in blood-building, and a low concentration of sodium is important for the formulation of diets for hypertensives. Also, the seed has a low fat concentration, which is an important factor in weight-reducing diets. Apart from these, the seed contains tannin (Table 81.4), a plant polyphenol that has the ability to inhibit the oxidation of unsaturated lipids and prevent the formation of oxidized LDL, which induces cardiovascular disease (Amic *et al.*, 2003). Hence, the *K. geocarpa* seed is important in the health and management of disease.

Conclusion

Although the *Kerstingiella geocarpa* seed had not attained international significance, it is a fair source of inexpensive protein and nutritive elements that can enrich the human diet. The seed has potential in the prevention and management of protein energy malnutrition, atherosclerosis, and other related disorders.

ADVERSE EFFECTS AND REACTIONS (ALLERGIES AND TOXICITY)

The acceptability and utilization of unconventional and lesser known legumes in the developing world have been limited due to the presence of certain antimetabolic factors, some of which are antinutrients. They elicit their toxic properties by inhibiting or reducing the utilization of important nutrients in the body, or by inhibiting some metabolic pathways, leading to adverse effects and toxicity. Tannin inhibits the activity of digestive enzymes, and interacts with proteins to make them insoluble (Singh, 1984), while cyanide acts as a potent inhibitor of the electron transport chain by binding to the Fe^{3+} of heme a_3 in cytochrome c oxidase, which catalyzes the terminal step of the electron transport chain.

The total cyanide concentration of the seed (0.26 ± 0.02) (Table 81.4) is lower than that of kidney beans (Liener, 1983), and the phytate concentration is considered moderate compared with that obtained for some bean cultivars, which ranged between 240 and

TABLE 81.4 Antinutrient Composition of *Kerstingiella geocarpa* Seed

Antinutrient	Concentration
Phytate (mg/100 g)	225.64 ± 0.02
Tannin (%)	0.98 ± 0.00
Cyanide (%)	0.28 ± 0.02

Values means of triplicate determinations.
Reproduced from Oyetayo and Ajayi (2005), *Biosci. Biotech. Res. Asia*, 3, 47–50, with permission.

TABLE 81.5 Calculated Phytate : Zn, Ca : Phytate and [Ca][Phytate] : [Zn] Molar Ratios of *Kerstingiella geocarpa* Seed[*]

Molar Ratio	Value
Phytate : Zn	0.08
Ca : phytate	63.10
[Ca][phytate] : [Zn]	0.02

Reproduced from Oyetayo and Ajayi (2005), *Biosci. Biotech. Res. Asia, 3*, 47–50, with permission.

1390 mg/100 g (Onwuliri & Obu, 2002). Moreover, the calculated [Ca][phytate] : [Zn] molar ratio obtained for *Kerstingiella geocarpa* seed (0.02; Table 81.5) was far below the critical value of 0.5 mol/kg. Hence, the phytate concentration in the seed will not adversely affect zinc bioavailability. We found that the concentrations of these antinutrients were not as high as those found in some other legumes, and that they are unlikely to produce any undesirable effect on the consumer.

SUMMARY POINTS

- The *Kerstingiella geocarpa* seed is an underutilized legume grown mostly in northern Nigeria.
- The seed has potential nutritional and medicinal properties.
- It is an important source of inexpensive vegetable protein, with a good distribution of essential amino acids, and has a low atherogenic index and hypocholesterolemic potential.
- Alongside its nutrients *Kerstingiella geocarpa* seed contains antinutrients, which occur in concentration below critical levels.
- The *Kerstingiella geocarpa* seed will find use as a protein supplement, especially in the formulation of weaning diets.

References

Ajayi, O. B., & Oyetayo, F. L. (2009). Potential of *Kerstingiella geocarpa* as a health food. *Journal of Medicinal Food, 12,* 184–187.

Amic, D., David-Amic, D., Belso, D., & Trinajstic, N. (2003). Structure–radical scavenging activity relationship of flavonoids. *Croatia Chemical Acta, 76,* 55–61.

Amuti, K. (1980). *Kerstingiella geocarpa* in Ghana. *Economic Botany, 34,* 358–361.

Chevalier, A. (1910). *Kerstingiella (Kerstingiella geocarpa* Harms). *Veg. Utility of African. Trypanosomiasis, 8,* 358–361.

Chikwendu, N. J. (2007). Chemical composition of four varieties of ground beans of *Kerstingiella geocarpa. Journal of Agricultural Food Environment Extension, 6,* 73–87.

Crisan, E. V., & Sands, A. (1978). Nutritional value of edible mushrooms. In S. T. Chang, & W. A. Hayes (Eds.), *The biology and cultivation of edible mushrooms* (pp. 137–168). New York, NY: Academic Press.

FAO/WHO/UNU. (1985). *World Health Organization (WHO) Technical Report Series 224.* Geneva: WHO.

Harms, H. (1908). Uber Geokarpie bei einer afrikanischen Leguminosae. Bericht Deutsch. Bot. *Gesellschaft, 26,* 225–231.

Hepper, F. N. (1963). The bambara groundnut (*Voandzeia subterranea*) and Kersting's groundnut (*Kerstingiella geocarpa*) wild in West Africa. *Kew Bulletin, 16,* 395–407.

Irvine, P. R. (1969) *West African crops,* Vol. 2. Oxford, UK: Oxford University Press. 204–210.

Liener, I. E. (1981). Toxic constituent in legumes. In S. K. Arora (Ed.), *Chemistry and biochemistry of legumes* (pp. 217–257). London, UK: Edwards & Arnold.

NAS. (1979). *Tropical legumes: Resources for the future.* Washington, DC: National Academy of Science.

Ononogbu, J. C. (1988). *Lipids lipoprotein: chemistry methodology, metabolism, biochemical and physiochemical importance.* Owerri, Nigeria: New Africa Publishing.

Onwuliri, V. A., & Obu, J. A. (2002). Lipids and other constituents of *Vigna unguliculata* and *Phaseolus vulgaris* grown in northern Nigeria. *Food Chemistry, 78,* 1–7.

Oyetayo, F. L., & Ajayi, O. B. (2005). Chemical profile and zinc bioavailability studies on Hausa groundnut (*Kerstingiella geocarpa*). *Biosciences Biotechnology Research Asia, 3*, 47–50.

Pasquet, R. S., Mergeai, G., & Baudon, J. P. (2002). Genetic diversity of the African geocarpic legume Kersting's groundnut, *Macrotyloma geocarpum* (Tribe Phaseoleae; Fabaceae). *Biochemical Systematics and Ecology, 30*, 943–952.

Sen, M., & Battacharya, D. K. (2001). Nutritional quality of seasame seed protein fraction extracted with isopropanol. *Journal of Agricultural and Food Chemistry, 49*, 2641–2646.

Singh, U. (1984). The inhibition of digestive enzymes by polyphenols of chicken pea and pigeon pea. *Nutrition Reports International, 21*, 745–753.

Stapf, O. (1913). Kerstingiella geocarpa. *Bulletin of Miscellaneous Information, Kew* 93–94.

CHAPTER 82

Use of Litchi (*Litchi sinensis* Sonn.) Seeds in Health

Bao Yang[1], Hengshan Wang[2], Nagendra Prasad[1], Yingming Pan[2], Yueming Jiang[1]
[1] South China Botanical Garden, Chinese Academy of Sciences, Guangzhou, China
[2] School of Chemistry and Chemical Engineering, Guangxi Normal University, Guilin, China

699

LIST OF ABBREVIATIONS

SGC-7901, human gastric carcinoma
A-549, human lung adenocarcinoma
HepG-2, human hepatocellular liver carcinoma cell line

INTRODUCTION

Litchi (*Litchi chinensis* Sonn.) is a tropical to subtropical crop that originates in South-east Asia. Litchi fruits are prized on the world market for their flavor, semi-translucent white aril, and attractive red skin. Litchi is now grown commercially in many countries, and production in Australia, China, Israel, South Africa, and Thailand has expanded markedly in recent years. Increased production has made significant contributions to the economic development in these countries, especially those in South-east Asia. The litchi seed accounts for 10–20% of the whole fruit in weight. In combination with the pericarp and pulp tissues, litchi fruit has been used for health care and treatment of disease since ancient times (Zhao *et al.*, 2007). In recent years, some effort has been made to explore the pharmacological potential and novel

Nuts & Seeds in Health and Disease Prevention. DOI: 10.1016/B978-0-12-375688-6.10082-9

health-beneficial effects of litchi seed. This chapter describes the cultivation of litchi fruit, and reviews the pharmacological properties of the litchi seed.

BOTANICAL DESCRIPTION

Litchi (*Litchi chinensis*) belongs to the family of *Sapindaceae*. The plant is a dense, symmetrical, evergreen tree with a dark brown trunk. The tree has erect or drooping branches, depending on the cultivar. The plant can live for many years, and grows to 6–12 m in height. A well-managed litchi orchard usually has about 70 trees per hectare (http://gears.tucson.ars.ag.gov/book/chap5/litchi.html). There are more than 100 commercial cultivars throughout the world (Li, 2008). Litchi fruits are oval to round, *ca.* 25 mm in diameter, and range in color from green to red (Zhuang *et al.*, 1998). The outer pericarp encloses a white translucent pulp and a dark brown seed.

HISTORICAL CULTIVATION AND USAGE

Litchi has been grown for more than 2000 years, and wild litchi has been found in Guangxi, Yunnan, and Guangdong provinces. The fruit tree spread to Burma in the 17th century, and then to India, the UK, France, South Africa, the USA, and other countries over the next several centuries. Commercial orchards have been established in the past 20 years. Litchi is largely eaten fresh, and is also dried for preservation. Table 82.1 gives the major nutritional composition of litchi pulp. The fruit also exhibits good pharmacological properties, such as beneficial effects against gastralgia, tumor, and cough. In traditional Chinese medicine, litchi seed is used to relieve neuralgic pain (Li, 2008).

PRESENT-DAY CULTIVATION AND USAGE

Despite increasing cultivation of litchi in other countries, China still has the highest production output of this fruit. In China, the area of cultivation of litchi was 580,800 ha and the annual production 1.44 million tonnes in 2005, accounting for about 70% of the world production. With the improvement of postharvest technology, fresh litchi fruit can be stored for about 30 days at a low temperature (Jiang *et al.*, 2006), which makes it possible for consumers throughout the world have a chance to taste this special fruit. In addition, more

TABLE 82.1 Anticancer Activities of 50% (v/v) Ethanol Seed Extract Obtained from Litchi Seeds

Sample	Concentration (µg/ml)	Cancer Cell Lines					
		SGC−7901		A−549		HEPG-2	
		Inhibition (%)	IC_{50} (µg/ml)	Inhibition (%)	IC_{50} (µg/ml)	Inhibition (%)	IC_{50} (µg/ml)
50% Ethanol seed extract	100	68.2 ± 1.9	77.92 ± 1.4	45.9 ± 3.2	116.40 ± 1.8	39.57 ± 0.95	137.87 ± 3.2
	50	25.5 ± 6.6		29.5 ± 0.3		25.2 ± 1.8	
	25	10.1 ± 0.1		25.3 ± 6.3		18.8 ± 3.3	
	12.5	NA		20.5 ± 1.8		NA	
Cisplatin*	100	63.9 ± 4	58.03 ± 2.2	49.57 ± 1.3	101.3±3.2	55.37 ± 2.1	87.9 ± 2.3
	50	34.26 ± 0.3		29.78 ± 1		33.27 ± 2.1	
	25	19.44 ± 1.8		19.89 ± 1.8		22.22 ± 2.5	
	12.5	12.03 ± 2.1		14.94 ± 2		18.0 ± 2.7	

The 50% ethanol seed extract showed good cytotoxity against three cancer cell lines. Data were expressed as means ± standard deviations of three replicate determinations. IC_{50} value represents the concentration of the tested sample to kill 50% of the cancer cells. NA, no activity;
*Positive control; SGC-7901, human gastric carcinoma; A-549, human lung adenocarcinoma; and HepG-2, human hepatocellular liver carcinoma cell line.

litchi-derived food is being produced in the market, including litchi wine, juice, ice-cream, and yoghurt (Li, 2008). Litchi seed is also processed into traditional Chinese medicines, for medical treatment.

APPLICATIONS TO HEALTH PROMOTION AND DISEASE PREVENTION

Litchi seed is an important part of litchi fruit, accounting for 10—20% of the total fruit weight. Due to its pharmacological activities, this seed extract has been used in traditional Chinese and Indian medicines since ancient times (Li, 2008). Litchi seed was traditionally used to relieve neural pain, swelling, and other hernia-like conditions. In recent years, many other pharmacological activities of the fruit, such as hypoglycemic, anticancer, antibacterial, and antiviral, have been discovered. Phenolics, fatty acids, triterpenes, and proteins in litchi seed could be responsible for these bioactivities.

Phenolic compounds have attracted much attention due to their beneficial effects on human health. Previous work has identified five phenolic compounds (Prasad *et al.*, 2009) — gallic acid, (−)-gallocatechin, procyanidin B2, (−)-epicatechin, and (−)-epicatechin-3-gallate — from the ethanolic extract of litchi seed (Figure 82.1). In addition, 5-O-coumaroyl methyl quinate, protocatechuic aldehyde, protocatechuic acid, and daucosterol have been identified from litchi seed (Liu *et al.*, 2003; Liang & Liu, 2009). Phenolic compounds have been confirmed to possess strong antioxidant activity due to the occurrence of the phenolic hydroxyl moiety. Assays have indicated that the ethanolic extract from litchi seeds has strong DPPH radical scavenging and inhibitory activity against lipid peroxidation, and good antityrosinase activity (Prasad *et al.*, 2009). Through removing reactive oxygen species from cells, phenolics can maintain the metabolic function, and show anti-inflammatory activity. Gong and colleagues (2008) reported that the phenolic compounds of the litchi seed can significantly inhibit the special protease, with an IC_{50} value of 40 µg/ml. Moreover, polyphenols from litchi seeds can inhibit the proliferation of respiratory syncytial virus, with an IC_{50} value of 59 µg/ml (Liang *et al.*, 2006). The results indicate its potential for treating influenza and other respiratory diseases.

701

Triterpenes with well-characterized biological activities are a large group of natural products derived from C_{30} precursors. Significant amounts of triterpenes have been detected in litchi seeds. Grondin and colleagues reported six triterpene alcohols, four 4-methylsterols, and six 4-desmethylsterols in litchi seed oils (Grondin *et al.*, 1995). Hemolytic activity is one of the most important bioactivities of triterpenes. The triterpene fraction from the litchi seed has been proved to show good regulation capabilities regarding blood glucose and serum lipid levels, to significantly inhibit gluconeogenesis and decrease triacylglycerol and cholesterol

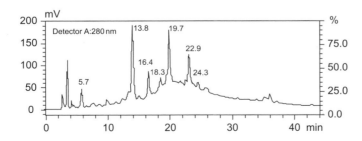

FIGURE 82.1
HPLC profile of ethanolic extract of litchi seed. Five phenolic compounds were identified from this chromatogram. Retention times: 5.7 min, gallic acid; 16.4 min, (−)-gallocatechin; 18.3 min, procyanidin B2; 19.7 min, (−)-epicatechin and 24.3 min, (−)-epicatechin-3-gallate.

levels, and to increase liver glycogen levels in rats (Zhang *et al.*, 2005). The hypoglycemic effect of saponins from litchi seeds is apparently higher than that of berberine (Lou *et al.*, 2007).

Litchi seed oil is an important source of cyclopropanoic fatty acids. The oil contains dihydrosterculic acid, cis-7,8-methylenehexadecanoic acid, cis-5,6-methylenetetradecanoic acid, and cis-3,4-methylenedodecanoic acid (Gaydou *et al.*, 1993; Grondin *et al.*, 1997). These cyclopropanoic fatty acids generally occur in bacteria. Moreover, large amounts of vitamin E (9.4%, w/w) and unstaturated fatty acids are detected in litchi seed oil (Matthaus *et al.*, 2003). Thus, the diversity of litchi seed oil exhibits a range of pharmacological properties. In Ning and colleagues' study, litchi seed oil was fed to hypercholesterolemic rats and the effect on serum lipids was then determined (Ning *et al.*, 1996). The authors found that the levels of total serum cholesterol and low density lipoprotein cholesterol levels decreased, while the high density lipoprotein cholesterol and total cholesterol levels increased significantly. This investigation shows the beneficial effect of litchi seed oil against cardiovascular diseases. Though there is no direct report on the healthcare effects of litchi seed oil, much of the literature regarding the bioactivities of vitamin E and unsaturated fatty acids suggests that litchi seed oil should have great potential as an anti-aging, antioxidation, and anti-blood clotting agent.

Recently, using HepG-2 (human hepatocellular liver carcinoma), A-549 (human lung adenocarcinoma), and SGC-7901 (human gastric carcinoma) cancer cell lines, it has been found that 50% ethanol extract of litchi seeds exhibits anticancer activity comparable to that of the conventional anticancer drug cisplatin (Table 82.1). Ferreira *et al.* (2007) reported the existence of proteins in litchi seed that have antibacterial and antiviral activities. Nitrosamine is an effective carcinogen which may be implicated in gastric and hepatic cancers (Lin, 2004). Using the *in vitro* human gastric juice model, the alcoholic extract of litchi seed has an inhibition effect of 84.5% against the synthesis of nitrosamine (Liu *et al.*, 2003). By feeding rats with aqueous extract of litchi seed, fibrosarcoma S180 tumor was inhibited by 30–36%, while hepatic cancer was inhibited by 30–39% (Xiao *et al.*, 2004). Furthermore, the cytotoxicity test confirmed that a purified fraction of litchi seed by ion exchange chromatography exhibited antiproliferation activity on HepG-2 cells by accelerating the apoptosis rate (Tao *et al.*, 2008). It can change the cell cycle and hinder the cells' evolution from the synthesis phase into the G2/M phase. Thus, litchi seed can potentially be used as a readily accessible source of a natural anticancer agent.

ADVERSE EFFECTS AND REACTIONS (ALLERGIES AND TOXICITY)

Though the litchi seed has been reported to exhibit many significant pharmacological properties, its acute toxicity is very low. In rats, 3 days after intragastric administration of a dose of 20 g/kg there were no deaths. Other adverse effects are not reported in the literature. It is suggested that litchi seed as a herbal medicine exhibits a high safety.

SUMMARY POINTS

- Litchi seeds have been used for traditional Chinese medicine formulation, which mainly helps to relieve neural pain and swelling, since time immemorial.
- Litchi seed exhibits great antityrosinase, antioxidant, and anticancer activities.
- The combination of litchi seed extract with some commercial drugs can improve the treatment effect and facilitate the recovery of patient.
- The main phenolics in litchi seed have been identified.
- The acute toxicity of litchi seed is very low.

References

Ferreira, C. H., Alves, S. B. V., Ramos, L. A., Oliveira, L. D. R., Ribeiro, C. A., Coutinho, M. S., et al. (2007). Antibacterial and antiviral proteins from *Litchi chinensis* seeds. *Plant Medicine, 73*, 879.

Gaydou, E. M., Ralaimanarivo, A., & Bianchini, J. P. (1993). Cyclopropanoic fatty acids of litchi (*Litchi chinensis*) seed oil. A reinvestigation. *Journal of Agricultural and Food Chemistry, 41,* 886—890.

Gong, S., Su, X., Yu, H., Li, J., Qin, Y., Xu, Q., et al. (2008). A study on anti-SARS-CoV 3CL protein of flavonoids from *Litchi chinensis* core [in Chinese with English abstract]. *Chinese Pharmacology Bulletin, 24,* 699—700.

Grondin, I., Smadja, J., Farines, M., & Soulier, J. (1995). Study of the triterpene fraction of *Litchi sinensis* Sonn and *Euphorbia longana* Lam seed oils. *Ocl—Ol. Corps Gras Lipides, 2,* 229—235.

Grondin, I., Smadja, J., Farines, M., & Soulier, J. (1997). The tryacylglycerols of two Sapindaceae seed oils: study of *Litchi sinensis* Sonn. and *Euphoria longana* Lam. lipids. *Ocl-ol. Corps Gras Lipides, 4,* 295—300.

Jiang, Y., Wang, Y., Song, L., Liu, H., Lichter, A., Kerdchoechuen, O., et al. (2006). Production and postharvest characteristics and technology of litchi fruit: an overview. *Australian Journal of Experimental Agriculture, 46*(12), 1541—1556.

Li, J. G. (2008). The Litchi *[in Chinese].* Beijing: China: China Agricultural Press.

Liang, R., Liu, W., Tang, Z., & Xu, Q. (2006). Inhibition on respiratory syncytial virus *in vitro* by flavonoids extracted from the core of *Litchi chinensis* [in Chinese with English abstract]. *Journal of Fourth Military Medical University, 27,* 1881—1883.

Liang, Y. R., & Liu, Z. G. (2009). Isolation and characterization of polyphenols in seed of *Litchi chinensis* [in Chinese with English abstract]. *Zhong Yao Cai, 32,* 522—529.

Lin, J. K. (1990). Nitrosamines as potential enviromental carcinogens in man. *Clinical Biochemistry, 23,* 67—71.

Liu, A. W., Chen, Q., & Zheng, J. Y. (2003). Inhibition effect of litchi seed extract on nitrosamine synthesis [in Chinese with English abstract]. *Shi Ping Gong Ye Ke Ji, 24,* 27—29.

Lou, Z. M., Tian, J. X., Wang, W. X., Yuan, H., Luo, D. J., Yang, Y. H., et al. (2007). Effect of total saponin extract from litchi core on the blood glucose levels of diabetic mice [in Chinese with English abstract]. *Zhe Jiang Yi Xue, 29,* 548—550.

Matthaus, B., Vosmann, K., Pham, L. Q., & Aitzetmüller, K. (2003). FA and tocopherol composition of Vietnamese oilseeds. *Journal of the American Oil Chemists Society, 80,* 1013—1020.

Ning, Z., Peng, K., Yan, Q., Wang, J., & Tan, X. (1996). Effects of the kernel oil from *Litchi chinensis* Sonn. seed on the level of serum lipids in rats. *Acta Nutrimenta Sinica, 18,* 159—162.

Prasad, K. N., Yang, B., Yang, S. Y., Chen, Y. L., Zhao, M. M., Ashraf, M., et al. (2009). Identification of phenolic compounds and appraisal of antioxidant and antityrosinase activities from litchi (*Litchi sinensis* Sonn.) seeds. *Food Chemistry, 116,* 1—7.

Tao, X. H., Wang, H., Wang, Y., Huang, X. S., Guo, G. Q., & Shen, W. Z. (2008). Effect of litchi seed extract on proliferation and apoptosis of HepG-2 *in vitro* [in Chinese with English abstract[. *Natural Product of Research and Development, 20,* 988—992.

Xiao, L. Y., Zhang, D., Feng, Z. M., Chen, Y. W., Zhang, H., & Ling, P. Y. (2004). Antitumor effect of litchi seed *in vivo* [in Chinese with English abstract]. *Zhong Yao Cai, 27,* 517—518.

Zhang, Y. M., Hong, Y., Tian, J. X., Shen, L., Yu, H. Y., Yi, H. P., et al. (2005). Effects of saponin of litchi seed on gluconeogenesis and metabolism of blood lipid in mice [in Chinese with English abstract]. *Hang Zhou Shi Fan Xue Yuan Xue Bao (Natural Science), 4,* 435—436.

Zhao, M. M., Yang, B., Wang, J. S., Liu, Y., Yu, L. M., & Jiang, Y. M. (2007). Immunomodulatory and anticancer activities of flavonoids extracted from litchi (*Litchi chinensis* Sonn.) pericarp. *International Immunopharmacology, 7,* 162—166.

Zhuang, Y. M., Ke, K. W., Zeng, W. X., & Pan, X. W. (1998). Litchi. In Tropical and subtropical fruits in China *[in Chinese]* (pp. 103—107). Beijing, China: China Agricultural Press.

CHAPTER 83

Antioxidant Activity of Longan (*Dimocarpus Longan L.*) Seed Extract

Kai Wang[1,3], **Ying Liang**[2], **Ye Zhang**[3], **Xianghui Yi**[3], **Hengshan Wang**[1], **Yingming Pan**[1]
[1] School of Chemistry and Chemical Engineering, Guangxi Normal University, Guilin, China
[2] Department of Environmental Engineering, Guilin University of Electronic Technology, Guilin, China
[3] Department of Chemistry and Engineering Technology, Guilin Normal College, Guilin, China

LIST OF ABBREVIATIONS

BHT, butylated hydroxytoluene
DPPH, 2,2-diphenyl-1-picrylhydrazyl
SC_{50}, the concentration that scavenges 50% of the free radical

INTRODUCTION

The longan (*Dimocarpus Longan* Lour.) is an evergreen tree of the Sapindaceae family that originated in China. In China, longan is cultivated in the Guangdong, Guangxi, Fujian, Hainan, Taiwan, Sichuan, and Yunnan provinces. It is also grown commercially in many other countries throughout the tropics, including Thailand, India, and Vietnam. In recent years, a number of studies have been conducted to explore the pharmacological potential and novel health benefits of longan seed, particularly concerning its antioxidant activity. This chapter describes the cultivation of longan, and reviews the antioxidant properties of its seed.

Nuts & Seeds in Health and Disease Prevention. DOI: 10.1016/B978-0-12-375688-6.10083-0

BOTANICAL DESCRIPTION

Longan is the fruit produced on evergreen trees that are members of the Sapindaceae family, which normally grow to about 20 m in height. Wild longan can be discriminated from the cultivated tree by its morphological traits, including very small conical, heart-shaped, or spherical fruit with a thin, leathery and indehiscent pericarp, rather thin pulp, and very big seeds with a smooth testa (Wu *et al.*, 2007). As yellow pigment synthesis is initiated, the pericarp can vary in color from yellowish to light brown until the fruit matures, producing a smooth skin. The edible portion of the longan fruit is a fleshy, translucent-white aril, which is an extension of the funiculus or seed stalk that arises from the placenta and surrounds the seed. The outermost epicarp has a continuous cuticle, a uniseriate epidermis, and subepidermal sclerenchyma, while the middle mesocarp is parenchymatous tissue, and the inner endocarp is made up of small, thin-walled, un-suberized epidermal cells (Jiang *et al.*, 2002).

HISTORICAL CULTIVATION AND USAGE

Longan was cultivated as a native fruit and Chinese traditional medicine for more than 2000 years before spreading to India and other South Asian countries from the 18th century onwards. Currently, over 300 cultivars are known. The longan cultivated in Guangxi, Yunnan, Guangdong, Fujian, and Sichuan provinces is the same as the wild litchi found throughout China. Its fruit can be eaten fresh, or dried for preservation to be used in a tonic called "Guiyuan." In traditional Chinese medicine, this fruit has been used for the treatment of anemia, heart palpitations, insomnia, forgetfulness, neurasthenia, and postpartum frail embolism, as well as for promoting beauty and longevity.

PRESENT-DAY CULTIVATION AND USAGE

Due to its sweet juicy taste and health benefits, longan has become a desirable fruit gradually accepted by consumers all over the world. Dried longan is the most popular form, and can be eaten directly as a dessert or snack food. It can also be used to prepare refreshing drinks and longan tea. Its cultivation is now increasing in both tropical and subtropical countries of the world where climates have distinct wet/dry periods with a cool, non-freezing fall/winter period. China is the largest producer of longan, with over 400,000 ha under cultivation and an annual production of around 1 million tonnes (Yang *et al.*, 2009a). In recent years, production of longan fruits in China has dramatically increased due to the continuous development of plantations and improvement of agricultural management techniques.

At the beginning of the century, potassium chlorate ($KClO_3$) was found to solve the problem of alternate bearing, and enabled the grower to produce off-season longan by flower induction (Yen, 2000). Currently, $KClO_3$ is still used to induce off-season flowers and fruits in longan trees worldwide (Manochai *et al.*, 2005; Potchanasin *et al.*, 2009a, 2009b). Other cultivation methods, such as ozone in combination with some organic acids to control postharvest decay and pericarp browning of longan fruit, and improved bagging, have also been developed to improve quality and production in the cross-winter off season (Whangchai *et al.*, 2006; Yang *et al.*, 2009b).

APPLICATIONS TO HEALTH PROMOTION AND DISEASE PREVENTION

Studies have been completed on the biochemical and physiological activities of longan seed. The longan seed, a traditional folklore medicine, is administered to counteract heavy sweating, and the pulverized kernel serves as a styptic. It has also been reported to have antibacterial, antiviral, anti-inflammatory, anticarcinogenic (Bravo, 1998), antiglycated (Yang *et al.*, 2009c), and anticancer activities (Prasad *et al.*, 2009a). High amounts of bioactive compounds, such as phenolic acids, flavonoids, hydrolyzable tannins, and polysaccharides, have been found in longan seed.

It is well known that plant phenolic compounds are highly effective free radical scavengers and antioxidants, so there should be a close correlation between the content of phenolic compounds and a plant's antioxidant activity. Significant amounts of phenolic compounds have also been detected in the longan seed, which may explain its excellent antioxidant activity. The five previously uncharacterized polyphenols — ethyl gallate (**2**), 1-β-O-galloyl-D-glucopyranose (**3**), methyl brevifolin carboxylate (**4**), brevifolin (**5**), and 4-O-α-L-rhamno-pyranosyl-ellagic acid (**8**) — and the previously identified polyphenols — gallic acid (**1**), corilagin (**6**), and ellagic acid (**7**) — were also isolated from longan seed by Zheng *et al.* (2009) (Figure 83.1). These polyphenols exhibited a scavenging activity towards DPPH radicals with SC$_{50}$ values of 0.80–5.91 µg/ml, and towards superoxide radicals with SC$_{50}$ values of 1.04–7.03 µg/ml (Table 83.1). In addition, Sun and colleagues identified two compounds from longan pericarp as 4-O-methylgallic acid and (−)-epicatechin (Sun *et al.*, 2007). The results of antioxidant activity showed that 4-O-methylgallic acid had higher 2,2-diphenyl-1-picrylhydrazyl (DPPH), hydroxyl radical, and superoxide radical scavenging activities than (−)-epicatechin.

As compounds with antioxidant activity in other plants, gallic acid and ellagic acid have also been identified in the longan seed (Figure 83.2) by other researchers (Soong & Barlow, 2006; Zheng *et al.*, 2009). However, free gallic acid and ellagic acid were not the major contributors to the antioxidant activity of longan pericarp. Other uncharacterized polyphenolic/flavonoid glycosides and ellagitannins may form a synergistic multilevel defense system against oxidation which results in the antioxidant activity of longan.

Aside from the polyphenol compounds, polysaccharides (e.g., hemicellulose and cellulose) in longan fruit pericarp tissues might be responsible for the antioxidant effects of longan. Yang and colleagues studied the effect of ultrasonic treatment (Yang *et al.*, 2009a) and methylation

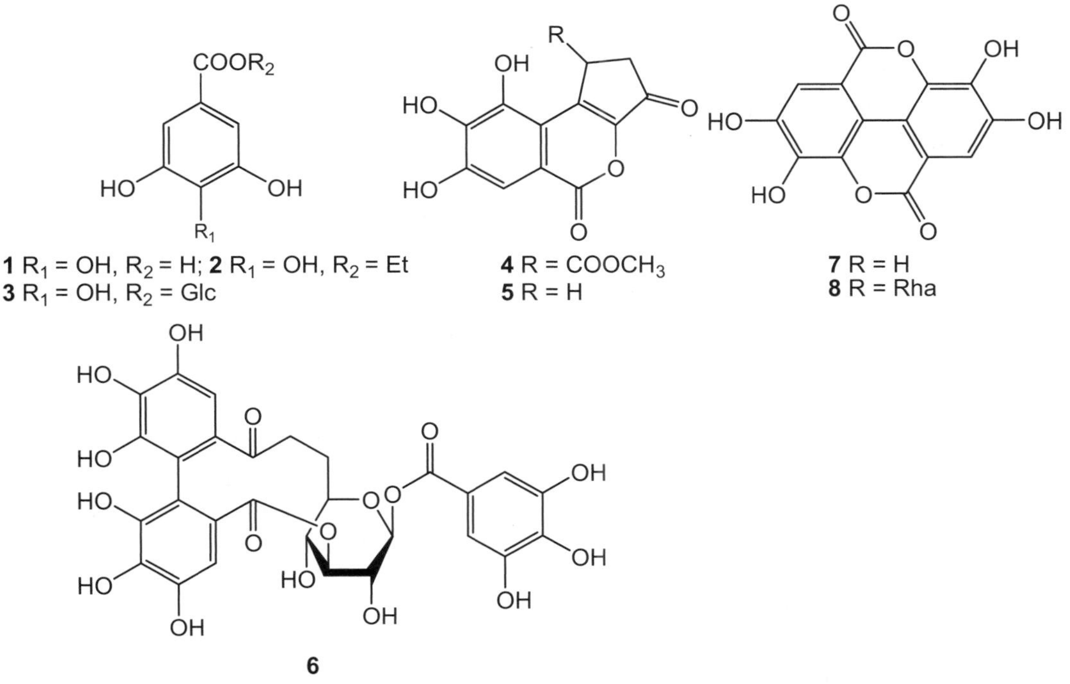

707

FIGURE 83.1
Structures of polyphenols from longan seed. Gallic acid (**1**), ethyl gallate (**2**), 1-β-O-galloyl-D-glucopyranose (**3**), methyl brevifolin carboxylate (**4**), brevifolin (**5**), corilagin (**6**), ellagic acid (**7**), and 4-O-α-L-rhamnopyranosyl-ellagic acid (**8**) were isolated from longan seed. Reprinted from Zheng *et al.* (2009), *Food Chem.*, 116, 433–436, with permission.

TABLE 83.1 SC$_{50}$ Values of Compounds 1–8 Towards DPPH and Superoxide Radicals

Compound	DPPH Radical µg/ml (µM)	Superoxide Radical µg/ml (µM)
1	0.80 ± 0.02 (4.71 ± 0.10)	1.04 ± 0.11 (6.12 ± 0.67)
2	1.09 ± 0.03 (5.53 ± 0.13)	1.42 ± 0.16 (7.17 ± 0.80)
3	3.84 ± 0.10 (11.56 ± 0.31)	3.24 ± 0.43 (9.76 ± 1.13)
4	1.79 ± 0.01 (5.84 ± 0.03)	2.62 ± 0.26 (8.56 ± 0.84)
5	2.84 ± 0.02 (11.45 ± 0.06)	3.40 ± 0.29 (13.71 ± 1.17)
6	2.01 ± 0.03 (3.17 ± 0.05)	2.37 ± 0.27 (3.74 ± 0.42)
7	2.20 ± 0.10 (7.29 ± 0.32)	2.56 ± 0.35 (8.48 ± 1.16)
8	5.91 ± 0.20 (13.19 ± 0.44)	7.03 ± 0.79 (15.69 ± 1.76)
Vitamin C	2.13 ± 0.02 (12.10 ± 0.09)	Not determined

In the DPPH radical-scavenging assay, SC$_{50}$ values of the compounds **1–8** ranged from 0.80 to 5.91 µg/ml in a decreasing antioxidant order of **1 > 2 > 4 > 6 > 7 > 5 > 3 > 8**, of which the first four compounds (**1, 2, 4,** and **6**) showed better scavenging activity than vitamin C. In the superoxide anion radical-scavenging assay, the SC$_{50}$ values of the eight compounds were between 1.04 and 7.03 µg/ml in a similar order (**1 > 2 > 6 > 7 > 4 > 3 > 5 > 8**) to that in the DPPH radical-scavenging assay. Reprinted from Zheng *et al.* (2009), *Food Chem.*, 116, 433–436, with permission.

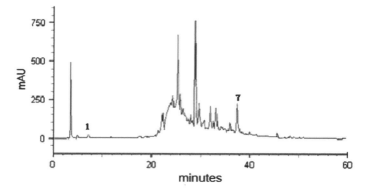

FIGURE 83.2
HPLC chromatogram (monitored at 280 nm) of gallic acid (1) and ellagic acid (7) extracted from freeze-dried longan seed with ethanol at 70°C for 1 hour. As compounds with antioxidant activity in other plants, gallic acid and ellagic acid were also identified in longan seed. Reprinted from Soong and Barlow (2006), *Food Chem.*, *97*, 524–530, with permission.

(Yang *et al.*, 2010) on the recovery and free radical scavenging activity of polysaccharides from longan fruit pericarp. Methylation resulted in a decrease in the DPPH radical scavenging activity of the polysaccharides from longan fruit pericarp, while the superoxide anion radical scavenging activity of polysaccharides from longan fruit pericarp decreased with an increasing degree of methylation.

The synergistic effects of different active compounds on the antioxidant activity of longan extracts have also been studied. In our previous work (Pan *et al.*, 2008), longan seed peel was extracted using various methods, then the total phenolic content of the extracts was determined, as well as their antioxidant properties. Results indicated that the antioxidant activity of microwave-assisted extract of longan seed peel was superior to the synthetic antioxidant, butylated hydroxytoluene (BHT). This provides support for microwave-assisted extraction as an effective method for investigating the possible antioxidant compounds from longan seed peel. Using a 50% ethanol high-pressure (500 MPa) extraction (Jiang *et al.*, 2009) along with ultrasonic assisted extraction (Prasad *et al.*, 2009b), the resulting longan fruit pericarp exhibited strong antioxidant activity in a dose-dependent manner. Therefore, this could be a readily accessible source of a natural antioxidant for the food and pharmaceutical industries.

ADVERSE EFFECTS AND REACTIONS (ALLERGIES AND TOXICITY)

No adverse effects have been reported in the literature, suggesting that longan seed, when used as a herbal medicine, is safe.

SUMMARY POINTS

- Longan is the fruit produced on evergreen trees from the Sapindaceae family.
- The longan seed has great potential as a medicine for its antibacterial, antiviral, anti-inflammatory, and anticarcinogenic properties, and antiglycated and anticancer activities.
- The longan seed shows excellent antioxidant activity, with high amounts of bioactive compounds, including phenolic acids and polysaccharides.
- The longan seed could be used as a readily accessible source of natural antioxidants for the food and pharmaceutical industries.
- The mechanism involved in the synergistic multilevel defence system in longan extract that was proven to play an important role in the longan seed's antioxidant activity is still unclear. Further investigation should be carried out to in order to optimize utilization of the longan seed.

References

Bravo, L. (1998). Polyphenols: chemistry, dietary sources, metabolism, and nutritional significance. *Nutrition Reviews, 56*(11), 317–333.

Jiang, G. X., Jiang, Y. M., Yang, B., Yu, C. Y., Rong, T., Zhang, H., et al. (2009). Structural characteristics and antioxidant activity of oligosaccharides from longan fruit pericarp. *Journal of Agricultural and Food Chemistry, 57*(19), 9293–9298.

Jiang, Y. M., Zhang, Z. Q., Joyce, D. C., & Ketsa, S. (2002). Postharvest biology and handling of longan fruit (*Dimocarpus longan* Lour.). *Postharvest Biology and Technology, 26*, 241–252.

Manochai, P., Sruamsiri, P., Wiriya-alongkorn, W., Naphrom, D., Hegele, M., & Bangerth, K. F. (2005). Year round off season flower induction in longan (*Dimocarpus Longan*, Lour.) trees by KClO$_3$ applications: potentials and problems. *Scientia Horticulturae, 104*, 379–390.

Pan, Y. M., Wang, K., Huang, S. Q., Wang, H. S., Mu, X. M., He, C. H., et al. (2008). Antioxidant activity of microwave-assisted extract of longan (*Dimocarpus Longan* Lour.) peel. *Food Chemistry, 106*, 1264–1270.

Potchanasin, P., Sringarm, K., Naphrom, D., & Bangerth, K. F. (2009a). Floral induction in longan (*Dimocarpus Longan*, Lour.) trees. Part IV: The essentiality of mature leaves for potassium chlorate induced floral induction and associated hormonal changes. *Scientia Horticulturae, 122*, 312–317.

Potchanasin, P., Sringarm, K., Sruamsiri, P., & Bangerth, K. F. (2009b). Floral induction (FI) in longan (*Dimocarpus Longan*, Lour.) trees. Part I: Low temperature and potassium chlorate effects on FI and hormonal changes exerted in terminal buds and sub-apical tissue. *Scientia Horticulturae, 122*, 288–294.

Prasad, K. N., Hao, J., Shi, J., Liu, T., Li, J., Wei, X. Y., et al. (2009a). Antioxidant and anticancer activities of high pressure-assisted extract of longan (*Dimocarpus Longan* Lour.) fruit pericarp. *Innovative Food Science Emerging Technologies, 10*, 413–419.

Prasad, K. N., Yang, E., Yi, C., Zhao, M. M., & Jiang, Y. M. (2009b). Effects of high pressure extraction on the extraction yield, total phenolic content and antioxidant activity of longan fruit pericarp. *Innovative Food Science Emerging Technologies, 10*, 155–159.

Soong, Y. Y., & Barlow., P. J. (2006). Quantification of gallic acid and ellagic acid from longan (*Dimocarpus Longan* Lour.) seed and mango (*Mangifera Indica* L.) kernel and their effects on antioxidant activity. *Food Chemistry, 97*, 524–530.

Sun, J., Shi, J., Jiang, Y. M., Xue, S. J., & Wei, X. Y. (2007). Identification of two polyphenolic compounds with antioxidant activity in longan pericarp tissues. *Journal of Agricultural and Food Chemistry, 55*(14), 5864–5868.

Whangchai, K., Saengnil, K., & Uthaibutra, J. (2006). Effect of ozone in combination with some organic acids on the control of postharvest decay and pericarp browning of longan fruit. *Crop Protection, 25*, 821–825.

Wu, Y. L., Yi, G. J., Zhou, B. R., Zeng, J. W., & Huang, Y. H. (2007). The advancement of research on litchi and longan germplasm resources in China. *Scientia Horticulturae, 114*, 143–150.

Yang, B., Jiang, Y. M., Wang, R., Zhao, M. M., & Sun, J. (2009a). Ultra-high pressure treatment effects on polysaccharides and lignins of longan fruit pericarp. *Food Chemistry, 112*, 428–431.

Yang, W. H., Zhu, X. C., Bu, J. H., Hu, G. B., Wang, H. C., & Huang, X. M. (2009b). Effects of bagging on fruit development and quality in cross-winter off-season longan. *Scientia Horticulturae, 120*, 194–200.

Yang, B., Zhao, M. M., & Jiang., Y. M. (2009c). Anti-glycated activity of polysaccharides of longan (*Dimocarpus longan* Lour.) fruit pericarp treated by ultrasonic wave. *Food Chemistry, 114,* 629−633.

Yang, B., Zhao, M. M. K., Prasad, N., Jiang, G. X., & Jiang, Y. M. (2010). Effect of methylation on the structure and radical scavenging activity of polysaccharides from longan (*Dimocarpus Longan* Lour.) fruit pericarp. *Food Chemistry, 118,* 364−368.

Yen, C. R. (2000). Firecracker ingredient found to produce flowering: From boom to bloom. *Agr. Hawaii, 1*(3), 26−27.

Zheng, G.., Xu, L. X., Wu, P., Xie, H. H., Jiang, Y. M., Chen, F., et al. (2009). Polyphenols from longan seeds and their radical-scavenging activity. *Food Chemistry, 116,* 433−436.

CHAPTER 84

Lupine (*Lupinus caudatus* L., *Lupinus albus* L.) Seeds: History of Use, Use as an Antihyperglycemic Medicinal, and Use as a Food

David H. Kinder[1], **Kathryn T. Knecht**[2]
[1] College of Pharmacy, Ohio Northern University, Ada, Ohio, USA
[2] School of Pharmacy, Loma Linda University, Loma Linda, California, USA

711

INTRODUCTION

Lupine species, in particular *Lupinus albus*, produce edible seeds which, in addition to their nutritional value, are also hypoglycemic. This member of the pea family has an advantage over other food and medicinal crops in that it restores soil as it fixes nitrogen, and is tolerant of drought conditions. We previously determined that the *L. caudatus* has similar properties to *L. albus*, which suggests that more lupine species plants might also have hypoglycemic properties.

Being from a Third World country does not preclude someone from having diseases associated with the more developed world, such as type 2 diabetes. What distinguishes these groups from developed societies is their ability (or lack thereof) to treat the disease. Because natural

Nuts & Seeds in Health and Disease Prevention. DOI: 10.1016/B978-0-12-375688-6.10084-2

products can often be grown and prepared in somewhat agrarian areas, the ability to produce plant materials that can be used crudely for control of disease is attractive. Disease treatment is therefore accessible to the poorer regions of the world, and, in the case of the lupine, the plant also provides a source of nourishment that might not otherwise be available. Type 2 diabetes lends itself to this practice, as one of the several plant materials which have been shown to control diabetes is the lupine.

BOTANICAL DESCRIPTION

Lupines are legumes (pea family) that are distributed worldwide. Despite the presence of alkaloids, which can be problematic both from the toxicity viewpoint as well as having an unpleasant taste, the lupine seeds have been used nutritionally, and, more recently, have been studied for their potential for replacement of conventional food crops such as soybeans. The historical use of lupine for food is well documented for the Mediterranean region, as well as in South America in the Andes highlands (Ballester *et al.*, 1980). In the Mediterranean region, the lupine that has been used is *Lupinus albus*, which produces a large white seed that is about the size of a lima bean (1 cm in diameter, 0.4 mm). In Egypt, the seeds are soaked in water to extract the alkaloids, then toasted prior to sale in the marketplace. During the soaking process the seed coat can be removed, but, as the seed coat of the white lupine is thin, this need not be done. The seeds are typically ground to make flour for bread. The proteins of legumes are generally considered good sources of lysine, but are low in the sulfur-containing amino acids, making the protein availability low unless supplemented by sulfur-containing amino acids such as methionine (Ballester *et al.*, 1980). By combining lupine flour with corn or other grains that are high in methionine, the nutritional value of lupine is greatly enhanced.

HISTORICAL CULTIVATION AND USAGE

L. albus is known to have hypoglycemic properties, and is used in Egypt for treating type 2 diabetes. While the first use of lupine for this purpose is obscure, it is generally thought that it occurred in Egypt. This activity is not associated with the quinolizidine or pyrolizidine alkaloids found in the lupine, since alkaloid-extracted seeds, or, more recently, the cultivars known as sweet lupines, are active as hypoglycemic agents. Additionally, there are other species of lupine which possess antiglycemic activity. The authors showed that *L. caudatus*, which was collected in Mesa Verde National Park (Cortez, Colorado) and is found in the western United States, possesses hypoglycemic properties, and has a similar chemical fingerprint to *L. albus* (Knecht *et al.*, 2006). The inference from this is that ancestral Puebloans may have had knowledge of the antihyperglycemic effects of lupines, which were abundantly available for them to eat.

In addition to its value as a food crop, lupine decoctions or extracts have been used in other medicinal indications. The earliest reports of domesticated lupine seeds are from Egyptian tombs dating from the 22 BC, with the use of seeds spreading to other areas of the Mediterranean. Evidence of lupines in the New World dates to the 6−7 BC. Though references mention use of the plant as food and green manure, Kurlovitch cites sources referring to the ancient use of lupine in treating conditions as diverse as abscesses, parasites, cardiac conditions, and rheumatism (Kurlovich, 2002). A role in epilepsy is also proposed.

One advantage of the lupine, as both a medicinal plant and food crop, is that it can be grown on poorer soils and under drier conditions than the soybean, and does not require the same fertilization to produce a significant crop. Indeed, one of the authors (DHK) has observed lupine growing and producing seed in sandy soil in northwestern New Mexico, where rainfall in the area amounted to only 5 cm over a 4-month period.

Lupines are found worldwide, and more than 240 species have been identified. There are 12 species common to Europe and the Middle East, with the bulk of the different species being spread across North America. The lupines are showy plants, producing a spectacular raceme on which blue, yellow, or white blossoms occur. Several cultivars have been developed solely for their colorful blooms. However, no medicinal or nutritional values have been studied in these decorative lupines.

As with most pea-related plants, the seeds are found in pods. Some lupine species will spread their seed as the seed pod pops when mature and warmed in the sunshine. This is especially true of the mountain or tailcup lupine (*L. caudatus*). The seeds range in size from a few millimeters in diameter, to 1 cm in diameter for the white lupine, which has seeds that are the size of a lima bean. Seeds have been used for both culinary and medicinal purposes.

The name lupine (lupin) derives from Anglo-French, in about the 14th century (Dendle, 2001). The root of the name comes from the Latin for "wolf," and two reasons for the name can be found: the *Encyclopædia Britannica* (2010) suggests the name was given because the plant is known for "wolfing" minerals from the ground, while a more plausible reason for naming the lupine as such is the use of the plant parts for treating epilepsy (Dendle, 2001). In the Middle Ages, a decoction of the plant was given to those who were possessed by the "devil" — e.g., were epileptic — and were often brought to a normal state. This is presumed to be in part because the lupine (in particular, the white lupine) is known to concentrate various metals from the soil, including manganese. Manganese depletion has been implicated in triggering epileptic fits, and replacing the manganese would tend to decrease or ameliorate the seizures, thus removing the "devil" possession (wolves being associated with the "devil") (Dendle, 2001).

PRESENT-DAY CULTIVATION AND USAGE

There has been interest in recent times in developing the lupine as a food crop for both human and farm animals, to replace other beans — notably, the soybean. Nutritionally, lupine seeds are quite similar to soybeans, containing similar quantities of high-quality protein, oils, and metals. Unlike the soybean, the lupine will grow on poor soils and is drought tolerant; like the soybean, it is a legume that fixes nitrogen. Bitter lupines contain quinolizidine alkaloids that can be removed by aqueous extraction of the seed, but "sweet" lupines have been found to occur and can be cultivated, making this extraction step no longer necessary (Goggin *et al.*, 2008). Several countries have attempted to develop the sweet lupine as a food crop, but with limited success. The cost of producing the lupine, the lower yield of seeds per hectare of land compared to soybeans, and the generally unknown nature of this crop has prevented significant progress in producing lupine as a food crop. However, if lupine were to be considered as a medicinal crop for controlling diabetes, there would be a greater interest in this crop without tying it to food as a replacement for soy.

APPLICATIONS TO HEALTH PROMOTION AND DISEASE PREVENTION

Diabetes as a condition was known by its symptoms in the ancient world. The frequent urination characteristic of diabetes was mentioned in 3 AD in the Ebers Papyrus (Carpenter *et al.*, 1998), along with various herbal treatments; Greek sources of a similar time period also mention both symptoms and treatments (Gemmill, 1972). Lupine is one of several traditional remedies used by Jordanian and Greek diabetes patients, reflecting the folk history of this plant product in diabetes therapy (Kubo *et al.*, 2006; Otoom *et al.*, 2006), while an Italian pharmacopeia of the early 20th century also includes lupine (Ferranini & Pirolli, 1937).

ADVERSE EFFECTS AND REACTIONS (ALLERGIES AND TOXICITY)

Lupinus species are broadly divided into bitter lupins, which contain high levels of alkaloids in their seeds, and sweet lupines, which are a naturally occurring variant of lupine that do not produce significant quantities of bitter alkaloids. The primary alkaloids found in most lupine

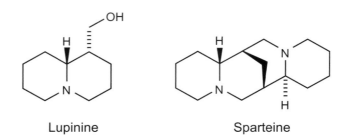

FIGURE 84.1
Structures of toxic quinolizidine alkaloids found in bitter lupine seeds. The quinolizidine alkaloids lupinine and sparteine. These alkaloids are also found in the "sweet" lupines, albeit in greatly reduced quantities, and are thought to be responsible for hepatotoxicity of the white lupine.

species are lupinine and sparteine (Figure 84.1), which are part of the quinolizidine alkaloid class. This class of alkaloid can be hepatotoxic in large quantities. Lupine seeds are usually soaked in water or salt water to remove excess quantities of the alkaloids, for both improved taste and decreased toxicity. The sweet lupines contain lower levels of alkaloids, sufficiently low to consume without pre-soaking in water, although they are still soaked to remove residual quantities of the alkaloids in order to eliminate bitter flavors. They were found growing with the bitter lupines, and have since been cultivated for human consumption (Kurlovich, 2002; Kubo et al., 2006).

Most work related to the hypoglycemic effects of lupine has been conducted in animals. Recent work on streptozoticin-induced diabetic rats has shown that extracts of the seed can lower blood glucose levels in these rats to normal levels, and can blunt the glucose spike following a meal. The mechanism by which white lupine has this effect is unknown, but the effect is seen both in the wild-type white lupines (with alkaloids present) and in the "sweet" lupines that are being cultivated in Egypt and Australia. The alkaloids are not the component responsible for the antihyperglycemic effect.

Quinolizidine alkaloids are restricted in lupin-based food sources, and can result in toxicity. Symptoms include trembling, convulsions, and, potentially, respiratory arrest, as well as a reduction in liver weight. The fact that sweet lupines used as food can have antihyperglycemic effects suggests that quinolizidine alkaloids, minimal in sweet lupines, are not essential for this therapeutic role. Nonetheless, antihyperglycemic effects have been reported for several alkaloids. Kubo reported that the phenylquinolizidine alkaloid (7R*,9aS*)-7-phenyl-octahydroquinolizin-2-one (Figure 84.2) and its 4-pyridone-type derivatives have been shown to lower glucose levels in streptozotocin-treated mice (Kubo et al., 2006).

The alkaloid 2-thionosparteine, but not lupanine, has also been reported as lowering blood glucose in diabetic animals in a manner comparable to that of the conventional antidiabetic

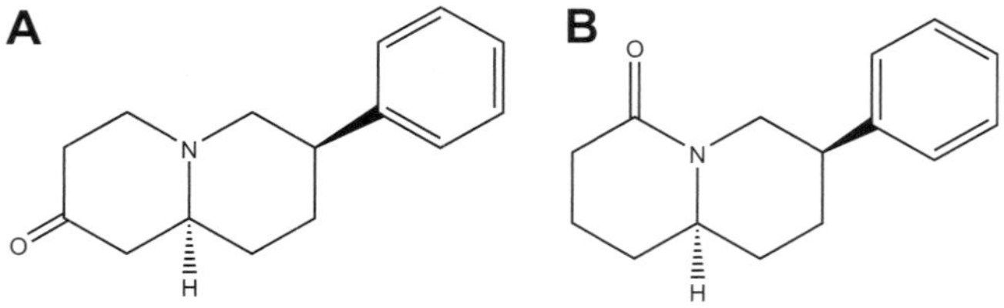

FIGURE 84.2
Structure of phenyl-quinolizidine alkaloids showing structural similarities to those in Figure 84.1. Structure of the phenylquinolizidine alkaloid (7R*,9aS*)-7-phenyl-octahydroquinolizin-2-one (A) and its congener (7R*,9aS*)-7-phenyl-octahydroquinolizin-4-one. While other quinolizidine alkaloids do not have antihyperglycemic activity, these two have been reported to be active in a mouse model.

sulfonylurea glyburide (Glibenclamide®) (Bobkiewicz-Kozlowska *et al.*, 2007). Lupanine may be effective at high glucose concentrations (García López *et al.*, 2004). Although increasing insulin secretion appeared to play a role in decreasing blood glucose in these animals (Bobkiewicz-Kozlowska *et al.*, 2007), other as yet unspecified mechanisms may also be involved. In our studies with mice, we did not find an increase in insulin secretion, though considerable variation in insulin measurements was seen.

We have demonstrated that another species of lupine, *L. caudatus* (tailcup lupine), which is native to the US southwest and Mexico, also demonstrates antihyperglycemic activity when given to experimentally-induced pre-diabetic mice. The activity was similar on a weight-to-weight basis of the extract. High pressure liquid chromatography (HPLC) fingerprints were developed which showed that extracts of the tailcup lupine and the white lupine were quite similar, without significant variation in any of the major components of the extract.

A major protein component of the seed is conglutins, which, in addition to their food value, have potentially therapeutic effects on cholesterol and blood glucose. Conglutin-γ has been shown to bind endogenous insulin and, more importantly, to demonstrate insulin-mimetic properties in murine myoblasts, despite a lack of clear structural similarity (Terruzzi *et al.*, 2009). The protein leginsulin, which bears some structural similarity to human insulin, has been found in at least one lupine species (Ilgoutz *et al.*, 1997; Dun *et al.*, 2008). The similar soy protein β-conglycinin lowers both blood glucose and insulin in mice, as well as lipids (Moriyama *et al.*, 2004). However, 35 g of lupin protein per day for 6 weeks did not lower fasting blood glucose in hypercholesterolemic subjects (Weiße *et al.*, 2010).

Lupine, like other legumes, contains fiber as well as protein. High-fiber lupine-enriched bread promoted satiety relative to white bread, possibly by modulating ghrelin levels (Lee *et al.*, 2006), although trials with lupine fiber did not show effects on blood glucose (Feldman *et al.*, 1995). The fact that, in our studies, lupine effected lowered tolerance of orally-administered glucose but not glucose administered i.p. suggests that lupine's effects may be mediated in the digestive tract. We also found that lupine added bulk to gastrointestinal contents, possibly indicating a slowing of glucose absorption. *In vitro*, lupine fibers have demonstrated greater water binding and viscosity than soy, pea hull, cellulose, or wheat fiber. (Turnbull *et al.*, 2005). It is also possible that gastrointestinal effects of lupine could secondarily enhance phase 1 insulin release, as in the case of gut-derived incretins.

SUMMARY POINTS

- Lupine seeds are useful as food sources for developing countries.
- Lupine seeds offer an alternative for diabetes therapy for undeveloped countries, or where managing cost of therapy is advantageous.
- Lupine seed flour should be combined with high-methionine containing grains for the best nutritional effects.
- Lupine seeds contain materials that will lower blood glucose in patients and experimental animals.
- The mechanism by which blood glucose is lowered is not known.
- Quinolizidine and pyrolizidine alkaloids are present in the lupine seeds, and must be removed before consumption to avoid toxicity.
- Sweet lupines have been cultivated for their use as a feed in countries where arid conditions preclude or hamper the production of soybean.

References

Ballester, D., Yanez, E., Garcia, R., Erazo, S., Lopez, F., Haardt, E., et al. (1980). Chemical composition, nutritive value, and toxicological evaluation of two species of sweet lupine (*Lupinus albus* and *Lupinus luteus*). *Journal of Agricultural and Food Chemistry, 28*, 402−405.

Bobkiewicz-Kozlowska, T., Dworacka, M., Kuczynski, S., Abramczyk, M., Kolanos, R., Wysocka, W., et al. (2007). Hypoglycaemic effect of quinolizidine alkaloids lupanine and 2-thionosparteine on non-diabetic and strep-tozotocin-induced diabetic rats. *European Journal of Pharmacology, 565,* 240–244.

Carpenter, S., Rigaud, M., Barile, M. P., Priest, T. J., Perez, L. F., & Ferguson, J. B. (1998). An interlinear transliteration and English translation of portions of the Ebers Papyrus possibly having to do with diabetes mellitus. Annandale-on-Hudson, NY: Bard College. Available at http://biology.bard.edu/ferguson/course/bio407/Carpenter_et_al_(1998).pdf (retrieved September 4, 2008).

Dendle, P. (2001). Lupines, manganese, and devil-sickness: an Anglo-Saxon medical response to epilepsy. *Bulletin of the History of Medicine, 75,* 91–101.

Dun, X.-P., Li, F.-F., Wang, J.-H., & Chen, Z.-W. (2008). The effect of pea albumin 1F on glucose metabolism in mice. *Peptides, 29,* 891–897.

Encyclopædia Britannica (2010). Lupine. *Encyclopædia Britannica,* Encyclopædia Britannica Online.

Feldman, N., Norenberg, C., Voet, H., Manor, E., Berner, Y., & Madar, Z. (1995). Enrichment of an Israeli ethnic food with fibres and their effects on the glycaemic and insulinaemic responses in subjects with non-insulin-dependent diabetes mellitus. *British Journal of Nutrition, 74,* 681–688.

Ferranini, A., & Pirolli, M. (1937). L'azione del decotto di semi di Lupinus albus sulla curva glicemica da carico di glucosio nei soggetti normali e diabetici. *Folia Medica, 23,* 729–748.

García López, P., de la Mora, P., Wysocka, W., Maiztegui, B., Alzugaray, M., Del Zotto, H., et al. (2004). Quinolizidine alkaloids isolated from *Lupinus* species enhance insulin secretion. *European Journal of Pharmacology, 504,* 139–142.

Gemmill, C. L. (1972). The Greek concept of diabetes. *Bulletin New York Academy of Medicine, 48,* 1033–1036.

Goggin, D. E., Mir, G., Smith, W. B., Stuckey, M., & Smith, P. M. C. (2008). Proteomic analysis of lupin seed proteins to identify Conglutin $\hat{I}^2$ as an allergen. *Journal of Agricultural and Food Chemistry, 56,* 6370–6377.

Ilgoutz, S. C., Knittel, N., Lin, J. M., Sterle, S., & Gayler, K. R. (1997). Transcription of genes for conglutin γ and a leginsulin-like protein in narrow-leafed lupin. *Plant Molecular Biology, 34,* 613–627.

Knecht, K. T., Nguyen, H., Auker, A. D., & Kinder, D. H. (2006). Effects of extracts of lupine seed on blood glucose levels in glucose resistant mice. Antihyperglycemic effects of *Lupinus albus* (white lupine, Egypt) and *Lupinus caudatus* (Tailcup lupine, Mesa Verde National Park). *Journal of Herbal Pharmacotherapy, 6*(3), 107–121.

Kubo, H., Inoue, M., Kamei, J., & Higashiyama, K. (2006). Hypoglycemic effects of multiflorine derivatives in normal mice. *Biological & Pharmaceutical Bulletin, 29,* 2046–2050.

Kurlovich, B. S. (2002). The history of lupin domestication. In: Lupins (geography, classification, genetic resources, and breeding). In B. S. Kurlovich (Ed.). St Petersburg, Russia: OY International North Express.

Lee, Y. P., Mori, T. A., Sipsas, S., Barden, A., Puddey, I. B., & Burke, V. (2006). Lupin-enriched bread increases satiety and reduces energy intake acutely. *American Journal of Clinical Nutrition, 84,* 975–980.

Moriyama, T., Kishimoto, K., Nagai, K., Urade, R., Ogawa, T., & Utsumi, S. (2004). Soybean beta-Conglycinin diet suppresses serum triglyceride levels in normal and genetically obese mice by induction of beta-oxidation, down-regulation of fatty acid synthase, and inhibition of triglyceride absorption. *Bioscience Biotechnology and Biochemistry, 68,* 352–359.

Otoom, S. A., Al-Safi, S. A., Kerem, Z. K., & Alkofahi, A. (2006). The use of medicinal herbs by diabetic Jordanian patients. *Journal of Herbal Pharmacotherapy, 6,* 31–41.

Terruzzi, I., Senesi, P., Magni, C., Montesano, A., Scarafoni, A., & Luzi, L. (2009). Insulin-mimetic action of conglutin-γ, a lupin seed protein, in mouse myoblasts. *Nutrition Metabolism and Cardiovascular Diseases.* in press.

Turnbull, C. M., Baxter, A. L., & Johnson, S. K. (2005). Water-binding capacity and viscosity of Australian sweet lupin kernel fibre under *in vitro* conditions simulating the human upper gastrointestinal tract. *International Journal of Food Sciences & Nutrition, 56,* 87–94.

Weiße, K., Brandsch, C., Zernsdorf, B., Nkengfack Nembongwe, G., Hofmann, K., & Eder, K. (2010). Lupin protein compared to casein lowers the LDL cholesterol:HDL cholesterol–ratio of hypercholesterolemic adults. *European Journal of Nutrition, 49,* 65–71.

CHAPTER 85

Macadamia Nuts (*Macadamia integrifolia* and *tetraphylla*) and their Use in Hypercholesterolemic Subjects

Lisa G Wood[1], Manohar L. Garg[2]
[1] Centre for Asthma and Respiratory Diseases, Level 3, Hunter Medical Research Institute, John Hunter Hospital, Newcastle, NSW, Australia
[2] Nutraceuticals Research Group, School of Biomedical Sciences, The University of Newcastle, Callaghan, NSW, Australia

717

LIST OF ABBREVIATIONS

AHA, American Heart Association
CVD, cardiovascular disease
GAE, gallic acid equivalents
HDLC, high density lipoprotein cholesterol
LDLC, low density lipoprotein cholesterol
MUFA, monounsaturated fatty acid

Nuts & Seeds in Health and Disease Prevention. DOI: 10.1016/B978-0-12-375688-6.10085-4

PUFA, polyunsaturated fatty acid

SFA, saturated fatty acid

TAG, triacylglycerols

TC, total cholesterol

TG, triglyceride

USDA, United States Department of Agriculture

INTRODUCTION

Macadamia nuts are being increasingly recognized as a valuable source of nutrients, which confer a range of health benefits. One of the most significant features of macadamia nuts is their high fat content, which varies by region of origin, but is typically reported to be approximately 72%, or 718 kCal/100 g of edible nuts (USDA, 2009; Table 85.1). Importantly, over 77% of the fat is unsaturated, predominantly in the form of monounsaturated fatty acids (MUFAs), while only around 16% is present as saturated fatty acids (SFAs) (USDA, 2009; Table 85.2). As a result, macadamia nuts can be useful in improving the quality of fats in the diet, which is important for a variety of health outcomes.

Macadamia nuts are a good source of plant protein, containing around 8% protein (USDA, 2009; Table 85.1), including all of the essential amino acids except tryptophan (Venkatachalam & Sathe, 2006). Other beneficial components of macadamia nuts include various essential micronutrients, such as potassium, magnesium, calcium, and phosphorus, as well as small amounts of iron, zinc, selenium, manganese, and copper (Table 85.3). Macadamia nuts also contain B complex vitamins — niacin, thiamin, riboflavin, vitamin C, pantothenic acid, folate and vitamin E (Maguire *et al.*, 2004; Table 85.4).

Other bioactive constituents of macadamia nuts include phytosterols, which are present in macadamia nuts at around 120 mg/100 g of edible nuts, predominantly as β-sitosterol, with small amounts of campesterol (Table 85.5). Levels in macadamia nut oil are approximately 10-fold higher, and in addition to campesterol and β-sitosterol, stigmasterol and squalene can be detected (Maguire *et al.*, 2004). The total phenolic content of macadamia nuts is approximately 46 mg gallic acid equivalents (GAE)/100 g (fresh weight) (Kornsteiner *et al.*,

TABLE 85.1 Proximate Composition of Macadamia Nuts

Nutrient	Units	Value Per 100 g Nuts	Std Error
Water	g	1.36	0.068
Energy	kJ	3004	
	kcal	718	
Protein	g	7.91	0.351
Fat	g	75.77	1.147
Ash	g	1.14	0.034
Carbohydrate, by difference	g	13.82	
Dietary Fibre	g	8.6	0.911
Sugars, total	*g*	*4.57*	*0.180*
Sucrose	g	4.43	0.180
Glucose (dextrose)	g	0.07	0.000
Fructose	g	0.07	0.000
Lactose	g	0.00	0.000
Maltose	g	0.00	0.000
Starch	g	1.05	0.024

Source: USDA National Nutrient Database for Standard Reference, Release 22 (2009), published online at: http://www.nal.usda.gov/fnic/foodcomp/search/ (accessed January 2010).

TABLE 85.2 Fatty Acid Composition of Macadamia Nuts

Fatty Acid	Units	Value Per 100 g	Std Error
Total SFAs	g	12.061	
4:0	g	0.000	
6:0	g	0.000	
8:0	g	0.000	0.000
10:0	g	0.000	0.000
12:0	g	0.076	0.013
14:0	g	0.659	0.100
16:0	g	6.036	0.035
18:0	g	2.329	0.238
20:0	g	1.940	0.094
22:0	g	0.616	0.002
24:0	g	0.281	0.013
Total MUFAs	g	58.877	
14:1	g	0.000	0.000
16:1	g	12.981	0.682
18:1	g	43.755	1.251
20:1	g	1.890	0.047
22:1	g	0.233	0.020
24:1	g	0.018	0.012
Total PUFAs	g	1.502	
18:2	g	1.296	0.090
18:3	g	0.206	0.018
18:4	g	0.000	
20:2 n-6	g	0.000	0.000
20:3	g	0.000	0.000
20:4	g	0.000	0.000
20:5 n-3	g	0.000	0.000
22:5 n-3	g	0.000	0.000
22:6 n-3	g	0.000	0.000

The weights of individual fatty acids, saturated, monounsaturated, polyunsaturated, and total fatty acids, are listed.
Source: USDA National Nutrient Database for Standard Reference, Release 22 (2009), published online at: http://www.nal.usda.gov/fnic/foodcomp/search/ (accessed January 2010).

2006). Phenolic compounds include 2,6-dihydroxybenzoic acid, 2′-hydroxy-4′-methoxyacetophenone, 3′5′-dimethoxy-4′-hydroxyacetophenone and 3,5-dimethoxy-4-hydroxycinnamic acid. Thus, these highly palatable and accessible nuts provide a nutrient-rich food source that may benefit a wide range of health outcomes.

TABLE 85.3 Mineral Content of Macadamia Nuts

Mineral	Units	Value Per 100 g	Std Error
Calcium, Ca	mg	85	11.267
Iron, Fe	mg	3.69	0.728
Magnesium, Mg	mg	130	50.738
Phosphorus, P	mg	188	11.570
Potassium, K	mg	368	10.067
Sodium, Na	mg	5	1.066
Zinc, Zn	mg	1.30	0.093
Copper, Cu	mg	0.756	0.228
Manganese, Mn	mg	4.131	4.216
Selenium, Se	µg	3.6	

Source: USDA National Nutrient Database for Standard Reference, Release 22 (2009), published online at: http://www.nal.usda.gov/fnic/foodcomp/search/ (accessed January 2010).

TABLE 85.4 Vitamin Content of Macadamia Nuts

Vitamin	Units	Value Per 100 g	Std error
Vitamin C	mg	1.2	0.255
Thiamin	mg	1.195	0.255
Riboflavin	mg	0.162	0.322
Niacin	mg	2.473	0.337
Pantothenic acid	mg	0.758	0.078
Vitamin B_6	mg	0.275	0.032
Folate, total	µg	11	1.433
Vitamin B_{12}	µg	0.00	
Vitamin A	µg_RAE	0	
Retinol	µg	0	
Vitamin A	IU	0	
Vitamin E			
α-tocopherol	mg	0.54	0.090
β-tocopherol	mg	0.00	0.000
γ-tocopherol	mg	0.00	0.000
δ-tocopherol	mg	0.00	0.000
Vitamin D ($D_2 + D_3$)	µg	0.0	

RAE, retinoic acid equivalents; IU, international units.
Source: USDA National Nutrient Database for Standard Reference, Release 22 (2009), published online at: http://www.nal.usda.gov/fnic/foodcomp/search/ (accessed January 2010).

TABLE 85.5 Phytosterol Composition of Macadamia Nuts

Nutrient	Units	Value Per 100 g
Total phytosterols	*mg*	*116*
Stigmasterol	mg	0
Campesterol	mg	8
Beta-sitosterol	mg	108

Source: USDA National Nutrient Database for Standard Reference, Release 22 (2009), published online at: http://www.nal.usda.gov/fnic/foodcomp/search/ (accessed January 2010).

BOTANICAL DESCRIPTION

The macadamia nut tree comes from the Proteaceae family, and originates from subtropical Eastern Australia. While there are at least five species of macadamia trees, there are only two that produce edible nuts: the "smooth-shelled macadamia" (*Macadamia integrifolia* Maiden & Betche) and the "rough-shelled macadamia" (*M. tetraphylla* L. Johnson). In addition, hybrid forms exist between the two species. The *Macadamia integrifolia* is the species most commonly grown for its crops of nuts. Mature nuts are encased in a very hard, woody shell surrounded by a green-brown fibrous husk.

HISTORICAL CULTIVATION AND USAGE

Commercial cultivation of the macadamia nut began in Australia in the 1870s, through the establishment of plantations. Prior to their commercial cultivation, macadamias were eaten by Australian Aborigines, who harvested nuts that had fallen to the ground. In the 1890s, the macadamia nut was introduced to some regions of the USA, such as Hawaii. Commercial processing of macadamia nuts was limited, however, until 1954, when the first mechanized processing plant was established, which was to break the shells open safely without damaging the nuts. Today, macadamias are commercially important in the USA, Australia, New Zealand, South Africa, and parts of South and Central America.

PRESENT-DAY CULTIVATION AND USAGE
Cultivation

The macadamia is an evergreen tree that grows in rainforests and in moist places with an annual rainfall of 1000—2000 mm. Macadamias prefer deep, fertile, well-drained soils with a pH of 5.0—6.5. As they are shallow-rooted, they do not suit areas of high winds. They are slow-growing, medium-sized trees, reaching a mature height of up to 12—15 m. With high rainfall, plentiful sunshine, and frost-free conditions, they will produce nuts after 4 or 5 years. Full production is reached in 12—15 years, and can continue for over 100 years when grown in suitable conditions.

Macadamia nuts are usually harvested manually after they have fallen, although some varieties need to be picked from the tree when ripe. Fallen nuts need to be harvested at least every 2—4 weeks, to prevent losses due to mold, germination, and pig or rat damage. The husks need to be removed within 24 hours of harvesting in order to reduce heat respiration, facilitate drying, and reduce the risk of developing mold. The nuts are dried until the moisture content is below 15%, and the shell is then removed.

Uses

Macadamia nuts are primarily cultivated for the kernel, which is usually oil-roasted or dry-roasted. Kernels are sold as snack nuts, and as an ingredient for ice cream and baked products. Macadamia nut oil has also become very popular in cooking. Its high smoke-point makes it very suitable as a frying oil, and it is a very stable oil, with an unrefrigerated shelf life of 1—2 years. Macadamia oil's smooth texture and high oxidative stability make it especially suitable for use in soaps, sunscreens, skin-care formulations, and shampoos.

The kernel only accounts for approximately one-third of the macadamia nut, with the rest of the nut being the inedible shell and husk. These contain high proportions of lignin and cellulose, and can be utilized in products such as mulch, activated carbon used for water purification, and fuel for processing macadamia nuts; in plastic manufacturing; and as a substitute for sand in the sand-blasting process.

APPLICATIONS TO HEALTH PROMOTION AND DISEASE PREVENTION

As macadamia nuts are a rich source of nutrients, they can form an important part of a healthy diet, useful for promoting good health and preventing disease. While macadamia nuts are very high in fat and energy content, the fat is primarily in the monounsaturated form, which has been shown to have a variety of health benefits. In addition, macadamia nuts contain other important dietary constituents, including proteins, dietary fiber, vitamins, minerals, and phytochemicals. Research regarding the health benefits of macadamia nuts is limited. However, there is a growing body of evidence suggesting that macadamia nuts are effective in improving various risk factors for CVD. They have been shown to improve circulating lipid profiles, and are thus particularly useful in hypercholesterolemic subjects. Other beneficial effects of macadamia nuts are their ability to reduce inflammation and oxidative stress, as well as possibly contributing to weight loss (Figure 85.1).

Circulating Lipid Profiles

Intervention studies have been used to demonstrate the beneficial effects of macadamia nuts on circulating lipid levels. These include two animal feeding studies (Yan *et al.*, 2003; Matthan *et al.*, 2009) and six human clinical trials using macadamia nuts or oil (Nestel *et al.*, 1994; Colquhoun *et al.*, 1996; Curb *et al.*, 2000; Garg *et al.*, 2003; Hiraoka-Yamamoto *et al.*, 2004; Griel *et al.*, 2008). Matthan and colleagues (2009) undertook a study in hamsters, to investigate the effect of diets enriched with macadamia, palm (SFA, 16:0), canola (MUFA, 18:1), or

721

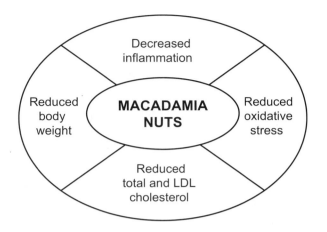

FIGURE 85.1

Potential health benefits of macadamia nuts. This figure summarizes the key physiological changes that can be induced with the consumption of macadamia nuts. Each of these changes, including reduced inflammation and oxidative stress, reduced total and LDL cholesterol, and reduced body weight, lead to a reduced risk of cardiovascular disease.

safflower (PUFA, 18:2) oils on lipoprotein profiles. After 12 weeks, macadamia oil-fed hamsters had lower non-high density lipoprotein cholesterol (HDLC) and triglyceride (TG) concentrations compared with the palm and coconut oil-fed hamsters. Furthermore, HDLC levels were higher in the macadamia-oil fed hamsters compared with the coconut, canola, and safflower oil-fed hamsters (Matthan *et al.*, 2009). In another animal study, hyperlipidemic rats were fed macadamia nuts at doses of between 12.5 and 25.0% of total energy for 6 weeks (Yan *et al.*, 2003). The macadamia nuts led to significantly lower levels of serum total cholesterol (TC) and TG, and significantly higher levels of serum HDLC levels, compared to the control group.

The human clinical intervention studies report a consistent reduction in serum/plasma TC and LDLC levels, while the effects on HDLC are variable (Table 85.6). While the details of the supplementation trials are not consistently reported, Table 85.6 summarizes the data that are available. The duration of the studies ranged from 3 to 5 weeks, and the dose of nuts varied from 20 to 100 g/day, accounting for an estimated 10—20% of total daily energy intake. Following a 3-week macadamia nut intervention, Colquhoun and colleagues (1996) reported an 8% reduction in serum TC, an 11% reduction in LDLC, and a 21% reduction in plasma TAG. HDLC was unchanged in this study. Curb and colleagues (2000) conducted a randomized crossover trial of three 30-day diets, in 30 volunteers aged 18—53 years. Each was fed a "typical American" diet high in saturated fat (37% energy from fat), an American Heart Association (AHA) Step 1 diet (30% energy from fat), and a macadamia nut-based monounsaturated fat diet (37% energy from fat), in random order. Mean total cholesterol, LDL cholesterol, and HDL cholesterol levels were significantly lower following the macadamia nut-based diet and the AHA Step 1 diet when compared with a typical American diet (Curb *et al.*, 2000). In the study by Garg and colleagues (2003), hyper-cholesterolemic men were given macadamia nuts (40—90 g/d), equivalent to 15% energy intake, for 4 weeks. Plasma MUFAs 16:1(n-7), 18:1(n-7), and 20:1(n-9) were elevated after the intervention. Plasma TC and LDLC concentrations decreased by 3 and 5%, respectively, and HDLC levels increased by 8%, after macadamia nut consumption. Hiraoka-Yamamoto *et al.* (2004) studied the effect of a 3-week intervention of macadamia nuts, in young, healthy Japanese female students. Serum concentrations of TC and LDLC were significantly decreased (6 and 7%, respectively), and body weight and body mass index were also decreased in the group fed macadamia nuts, compared to baseline. In this study, plasma HDL cholesterol was unchanged. More recently a study by Griel and colleagues (2008), in

TABLE 85.6 Summary of Macadamia Nut Intervention Studies Researching the Beneficial Effects of Macadamia Nuts on Circulating Lipid Levels

Study	n	Male/ Female	Age (years)	Duration (days)	Quantity of Nuts (g)	% Energy from Nuts	Effect on TC	Effect on LDLC	Effect on HDLC
1. Colquhoun et al., 1996)	14	7/7	25—59	21	50—100	20	↓	↓	NC
2. (Curb et al., 2000)	30	15/15	18—53	30	NR	NR	↓	↓	↓
3. (Garg et al., 2003)	17	17/0	Mean 54	28	40—90	15	↓	↓	↑
4. (Hiraoka-Yamamoto et al., 2004)	24	0/24	19—23	21	20	NR	↓	↓	NC
5. (Griel et al., 2008)	25	10/15	25—65	35	42.5	NR	↓	↓	↓

Intervention studies consistently show that macadamia nuts consumption causes a decrease in TC and LDLC. The effect on HDLC varies, with the intervention studies showing an increase, a decrease, or no change in HDLC levels.

↓, Decrease; ↑, increase; NC, no change; NR, not reported

mildly hypercholesterolemic subjects, compared a macadamia nut-rich diet (33% total fat, 18% MUFAs) to an average American diet (33% total fat, 11% MUFAs). Serum concentrations of TC, LDLC, and non-HDLC were lower following the macadamia nut-rich diet than the average American diet. In summary, all of these intervention studies consistently show that macadamia nut consumption causes a decrease in TC and LDLC. The effect on HDLC varies, with the intervention studies showing an increase, a decrease, or no change in HDLC levels. It is uncertain why this variability in the data occurs. However, it has been suggested that a decrease in HDLC levels may occur when the proportion of SFA in the diet is simultaneously decreased, as SFA has a cholesterol-raising effect (Griel *et al.*, 2008).

While the cholesterol-lowering properties of macadamia nuts are attributed to the presence of a high proportion of MUFAs, it is interesting to consider the role of individual fatty acids. Macadamia nuts are a rich source of the fatty acids 18:1n-9 (43.8% by weight) and 16:1n-7 (13% by weight), and contain small amounts of 20:1n-9. Interestingly, it has previously been suggested that 16:1n-7 may have a detrimental effect on blood lipids (Smith *et al.*, 1996), having a similar effect to SFA (Nestel *et al.*, 1994). However, there is a now a substantial body of evidence demonstrating that macadamia nuts are beneficial to blood lipid levels. Thus, it must be concluded that the combination of nutrients found in whole macadamia nuts has an overall beneficial effect, irrespective of the effect of 16:1n-7. While it is not possible to elucidate the role of individual nutrients from whole food interventions, it is clear that there is a complementary or synergistic interaction between the nutrients found in macadamia nuts. Furthermore, it is evident that other components, such as dietary fiber, plant sterols, and phytochemicals, are contributing to the modification of lipid profiles in these studies, as the extent of the changes observed cannot be explained by the fat content of the nuts alone (Kris-Etherton *et al.*, 1999).

Body Weight

Interestingly, despite their very high fat content (approximately 72% by weight), regular consumption of macadamia nuts has been shown to have no effect on body weight (Colquhoun *et al.*, 1996; Curb *et al.*, 2000; Griel *et al.*, 2008), or may even lead to a small reduction in body weight (Garg *et al.*, 2003; Hiraoka-Yamamoto *et al.*, 2004). This is in agreement with other investigations which demonstrate that nut consumption among free-living individuals is not associated with higher BMI compared with non-nut consumers, despite the fact that nuts are fat- and energy-dense foods (Sabate, 2003). Suggested reasons why nut consumption is not associated with increased BMI in free-living individuals include higher energy expenditure through physical activity or increased resting metabolic rate (due to high protein and unsaturated fat content), enhanced satiety and corresponding decreased intake of other foods, and incomplete absorption of energy from nuts (Sabate, 2003). While one or more of these factors may have contributed to the maintenance or loss of body weight in the macadamia nut interventions, there is evidence from one of these studies that subjects self-adjusted for the increased energy consumed as a result of the nuts that were added to their diets, resulting in no net increase in energy intake (Garg *et al.*, 2003).

Inflammation and Oxidative Stress

Short-term macadamia nut consumption has been shown to favorably modify oxidative stress and inflammation, despite an increase in dietary fat intake. A recent study in hypercholesterolemic males used macadamia nuts (40–90 g/d), equivalent to 15% energy intake, for a period of 4 weeks (Garg *et al.*, 2007). Plasma levels of monounsaturated fatty acids (16:1n-7, 18:1n-9 and 20:1n-9) were elevated following the intervention, and this was associated with significantly lower levels of inflammation (leukotriene, B4) and oxidative stress (8-isoprostane) (Garg *et al.*, 2007). These data further suggest that regular consumption of macadamia nuts may play a role in the prevention of coronary artery disease. This whole food intervention cannot determine which nutrient, or combination of nutrients, was responsible for the anti-inflammatory effects. However, it has previously been demonstrated that various components of macadamia nuts, such as MUFAs (Baer *et al.*, 2004), vitamin E (van Tits *et al.*, 2000), and phenolic compounds (Visioli *et al.*, 2000), can reduce levels of inflammation and oxidative stress.

ADVERSE EFFECTS AND REACTIONS (ALLERGIES AND TOXICITY)

Tree nuts are one of the most common foods causing allergic reactions in children and adults. Macadamia nuts may occasionally cause serious IgE-mediated allergic reactions in sensitized individuals. Symptoms may include angioedema, dysphagia, chest tightness and pain, palpitations, dyspnea, dizziness, and severe itching. Very severe reactions may result in anaphylaxis. Cases are becoming more frequent, probably due to the increasing use of macadamia nuts. In an American study of 13,493 individuals, tree nut allergy was self-reported in 166 (1.2%) subjects. Of the subjects with tree nut allergy, for which information regarding specific nuts was available, 24% had an allergy to macadamia nuts (Sicherer *et al.*, 2003).

Several non-commercial species of macadamias produce poisonous and/or inedible nuts, such as *M. whelanii* and *M. ternifolia*. The toxicity of these nuts is due to the presence of cyanogenic glycosides, which can be removed by prolonged leaching. While this makes these species of nuts commerically unviable, the practice has been used by Australian Aborigines.

SUMMARY POINTS

- Macadamia nuts are a valuable source of nutrients, in particular monounsaturated fatty acids.
- Modern cultivation practices have resulted in a reliable and plentiful supply of macadamia nuts, from countries such as the United States of America and Australia.
- Macadamia nuts have been shown to improve various risk factors for cardiovascular disease, including hypercholesterolemia, body weight, oxidative stress, and inflammation.
- Macadamia nuts may be useful when consumed alone, or in combination with traditional pharmacotherapy to reduce risk of cardiovascular disease.
- Long-term health benefits of macadamia nut consumption warrant further investigation.

References

Baer, D. J., Judd, J. T., Clevidence, B. A., & Tracy, R. P. (2004). Dietary fatty acids affect plasma markers of inflammation in healthy men fed controlled diets: a randomized crossover study. *American Journal of Clinical Nutrition, 79*, 969—73.

Colquhoun, D. M., Humphries, J. A., Moores, D., & Somerset, S. M. (1996). Effects of a macadamia nut enriched diet on serum lipids and lipoproteins compared to a low fat diet. *Food Australia, 48*, 216—221.

Curb, J. D., Wergowske, G., Dobbs, J. C., Abbott, R. D., & Huang, B. (2000). Serum lipid effects of a high-mono-unsaturated fat diet based on macadamia nuts. *Archives of Internal Medicine, 160*, 1154—1158.

Garg, M. L., Blake, R. J., & Wills, R. B. (2003). Macadamia nut consumption lowers plasma total and LDL cholesterol levels in hypercholesterolemic men. *Journal of Nutrition, 133*, 1060—1063.

Garg, M. L., Blake, R. J., Wills, R. B., & Clayton, E. H. (2007). Macadamia nut consumption modulates favourably risk factors for coronary artery disease in hypercholesterolemic subjects. *Lipids, 42*, 583—587.

Griel, A. E., Cao, Y., Bagshaw, D. D., Cifelli, A.., Holub, B., & Kris-Etherton, P. M. (2008). A macadamia nut-rich diet reduce total and LDL-cholesterol in mildly hypercholesterolemic men and women. *Journal of Nutrition, 138*, 761—767.

Hiraoka-Yamamoto, J., Ikeda, K., Negishi, H., Mori, M., Hirose, A., Sawada, S., et al. (2004). Serum lipid effects of a monounsaturated (palmitoleic) fatty acid-rich diet based on macadamia nuts in healthy, young Japanese women. *Clinical and Experimental Pharmacology and Physiology, 31*(Suppl. 2), S37—38.

Kornsteiner, M., Wagner, K., & Elmadfa, I. (2006). Tocopherols and total phenolics in 10 different nut types. *Food Chemistry, 98*, 381—387.

Kris-Etherton, P. M., Yu-Poth, S., Sabate, J., Ratcliffe, H. E., Zhao, G., & Etherton, T. D. (1999). Nuts and their bioactive constituents: effects on serum lipids and other factors that affect disease risk. *American Journal of Clinical Nutrition, 70*, 504S—511S.

Maguire, L. S., O'Sullivan, S. M., Galvin, K., O'Connor, T. P., & O'Brien, N. M. (2004). Fatty acid profile, tocopherol, squalene and phytosterol content of walnuts, almonds, peanuts, hazelnuts and the macadamia nut. *International Journal of Food Sciences & Nutrition, 55*, 171—178.

Matthan, N. R., Dillard, A., Lecker, J. L., Ip, B., & Lichtenstein, A. H. (2009). Effects of dietary palmitoleic acid on plasma lipoprotein profile and aortic cholesterol accumulation are similar to those of other unsaturated fatty acids in the F1B golden Syrian hamster. *Journal of Nutrition, 139*, 215—221.

Nestel, P., Clifton, P., & Noakes, M. (1994). Effects of increasing dietary palmitoleic acid compared with palmitic and oleic acids on plasma lipids of hypercholesterolemic men. *Journal of Lipid Research, 35*, 656—662.

Sabate, J. (2003). Nut consumption and body weight. *American Journal of Clinical Nutrition, 78*, 647S—650S.

Sicherer, S. H., Munoz-Furlong, A., & Sampson, H. A. (2003). Prevalence of peanut and tree nut allergy in the United States determined by means of a random digit dial telephone survey: a 5-year follow-up study. *Journal of Allergy and Clinical Immunology, 112*, 1203—1207.

Smith, D. R., Knabe, D. A., Cross, H. R., & Smith, S. B. (1996). A diet containing myristoleic plus palmitoleic acids elevates plasma cholesterol in young growing swine. *Lipids, 31*, 849—858.

USDA. (2009). National Nutrient Database for Standard Reference, Release 22. Available at: www.nal.gov/fnic/foodcomp/search (accessed January 2010).

van Tits, L. J., Demacker, P. N., de Graaf, J., Hak-Lemmers, H. L., & Stalenhoef, A. F. (2000). Alpha-tocopherol supplementation decreases production of superoxide and cytokines by leukocytes *ex vivo* in both normolipidemic and hypertriglyceridemic individuals. *American Journal of Clinical Nutrition, 71*, 458—464.

Venkatachalam, M., & Sathe, S. K. (2006). Chemical composition of selected edible nut seeds. *Journal of Agricultural and Food Chemistry, 54*, 4705—4714.

Visioli, F., Caruso, D., Galli, C., Viappiani, S., Galli, G., & Sala, A. (2000). Olive oils rich in natural catecholic phenols decrease isoprostane excretion in humans. *Biochemical and Biophysical Research Communications, 278*, 797—799.

Yan, S., Xiao, Y., Wang, J., & Liang, X. (2003). Effect of nuts enrich in monounsaturated acid on serum lipid of hyperlipidemia rats. *Wei Sheng Yan Jiu, 32*, 120—122.

CHAPTER 86

Use of Magnolia (*Magnolia grandiflora*) Seeds in Medicine, and Possible Mechanisms of Action

Yang Deok Lee
Department of Internal Medicine, Eulji University School of Medicine, Daejeon, Korea

727

LIST OF ABBREVIATIONS

COX-2, cyclooxygenase 2
mTOR, mammalian target of rapamycin
PI3K, phosphoinositide 3-kinase
PTEN, phosphatase and tensin homolog

INTRODUCTION

Magnolia is a large genus of about 210 flowering species in the family Magnoliaceae (Table 86.1). The American evergreen species *M. grandiflora*, commonly known as the southern magnolia or bull bay, is of horticultural importance in the United States, and over 100 cultivars have been developed (Table 86.2). This species is native to the southeastern United States and Mexico, and occurs along the coastal plains from North Carolina to eastern Texas. *M. grandiflora* grows to about 5—20 m in height (Vazquez, 1990).

This species has been widely used in traditional medicine, and as a decorative tree because of its large beautiful flowers. This plant has been reported to have beneficial effects on several

Nuts & Seeds in Health and Disease Prevention. DOI: 10.1016/B978-0-12-375688-6.10086-6

TABLE 86.1 Scientific Classification of *Magnolia grandiflora*

Kingdom	Order	Family	Genus	Subgenus	Section	Species
Plantae	Magnoliales	Magnoliaceae	*Magnolia*	*Magnolia*	*Magnolia*	*Magnolia grandiflora*

Magnolia grandiflora is one of the many species that belong to the genus *Magnolia*. *Magnolia* is a large genus of about 210 flowering species in the family Magnoliaceae.

TABLE 86.2 Characteristics of Representative Cultivars of *Magnolia grandiflora*

Cultivar	Characteristics
Little Gem	Reblooming compact form
Overton	Smaller, growing to only 7.2 m tall
Vitoria	Hardy green-back forms, without rusty-colored tomentose leaves
Angustifolia	Narrow spear-shaped leaves
Exmouth	Huge flowers with up to 20 petals, and vigorous growth
Ferruginea	Dark green leaves with rust-brown undersides

Over 100 cultivars of *Magnolia grandiflora* have been developed. The newly developed cultivars have a diversity of characteristics, including cold-hardiness, and the cultivated range has hence been extended to northern areas.

ailments, including high blood pressure, heart disturbances, dyspnea, abdominal discomfort, muscle spasm, infertility, and epilepsy (Mellado *et al.*, 1980).

BOTANICAL DESCRIPTION

M. grandiflora is a medium-to-large evergreen tree which may grow to 27.5 m in height (Gardiner, 2000). The tree typically has a single stem and a pyramidal shape. The leaves are thick, shiny, and broadly ovate, 12—25 cm long and 6—12 cm broad, with smooth margins (Zion, 1995). The leaves are dark green, stiff, and leathery, with tomentose undersides that vary from pale green to rusty red in color. The flowers are large, showy, powerfully citronella-scented, white, up to 30 cm wide, have 6—12 petals with a waxy texture, and emerge from the ends of nearly every twig on mature trees during late spring. Flowering is followed by production of a rose-colored cone-like fruit. The individual seed compartments split open to reveal bright flame-red seeds along each seam (Sternberg *et al.*, 2004).

HISTORICAL CULTIVATION AND USAGE

The plant collector Mark Catesby first brought *M. grandiflora* from North America to Britain in 1726, where it became a popular cultivated plant. Specimens collected in the vicinity of the Mississippi River of Louisiana were also introduced to France at about the same time (Aitken, 2008). This species was glowingly described by Philip Miller in his 1731 work *The Gardeners' Dictionary* (Gardiner, 2000). Sir John Colliton was one of the earliest people to cultivate this plant in Europe. He arranged scaffolding and tubs around his tree, so that gardeners could propagate branches by layering (Gardiner, 2000).

In North American traditional medicine, species of Magnoliaceae have been used for treating many conditions (Song & Fischer, 1999; Schuhly *et al.*, 2001; Maruyama *et al.*, 2002). In particular, natives have used species of *Magnolia* for the treatment of several illnesses associated with inflammation. Fever and rheumatism have been considered indications for the use of *Magnolia* (Schuhly *et al.*, 2001). In Asia, practitioners of traditional medicine still use several *Magnolia* species to treat nasal congestion, sinusitis, and asthma (Baek *et al.*, 2009).

PRESENT-DAY CULTIVATION AND USAGE

M. grandiflora is a very popular ornamental tree throughout the southeastern United States, grown for its attractive foliage and flowers. It is the state tree of Mississippi, and the state flower of Mississippi and Louisiana. On the east coast of the United States, cold-hardy cultivars can grow north of the Ohio River, although large specimens in this region are rare and they are usually found only as shrubs. This "subtropical indicator" tree is grown in some gardens as far north as southern Connecticut. When grown near the northern limit of cultivation, the tree may suffer from dieback during very hard freezes, but typically survives normal freezes very well. Specimens grown under severe winter conditions, especially in full sun, develop sun scald and foliage damage that can reduce their ornamental value or even result in tree death. On the west coast, this species is common as far north as Vancouver, British Columbia (Gardiner, 2000).

In the United States, M. grandiflora, sweetbay (Magnolia virginiana), and cucumbertree (Magnolia acuminata) are commercially harvested for their wood. Wood from all three species is termed "magnolia," and is used in the construction of furniture, boxes, pallets, venetian blinds, sashes, doors, and veneers. Southern magnolia has yellowish-white sapwood and light-to-dark brown heartwood that is tinted yellow or green. The usually straight-grained wood has a uniform texture with closely spaced rings. The wood is considered moderate in heaviness, hardness, and stiffness; moderately low in shrinkage, bending, and compression strength; and moderately high in shock resistance (Miller, 2007). Its use in the southeastern United States has been supplanted by the availability of harder woods (Callaway, 1994).

APPLICATIONS TO HEALTH PROMOTION AND DISEASE PREVENTION

Asians and Native Americans have used the leaves, bark, buds, and seeds of Magnolia species for food and medicine. The seeds of some species, such as M. grandiflora and M. officinalis, have known effects on human diseases (Table 86.3).

Bioactive extracts of M. grandiflora seeds contain honokiol and magnolol (Figure 86.1, Table 86.4). Honokiol, magnolol, and 4'-O-methylhonokiol strongly inhibit cyclooxygenase-2 (COX-2), and 3-O-methylmagnolol moderately inhibits COX-2 (Schuhly et al., 2009). The anti-inflammatory activity of Magnolia species is well known, and has been attributed to the presence of sesquiterpene lactones. The lignans honokiol and magnolol have been reported to have a variety of pharmacological effects related to their anti-inflammatory, antibacterial, and antioxidant properties (Park et al., 2004).

Reactive oxygen species produced by neutrophils contribute to the pathogenesis of focal cerebral ischemia/reperfusion injury and trigger the inflammatory response. Thus, the reported protective effects of honokiol against focal cerebral ischemia/reperfusion injury in rats are presumably related to its antioxidant effects (Liou et al., 2003). Another study reported that the hepatoprotective effects of honokiol and magnolol were attributable to their antioxidant activities (Park et al., 2003). The anti-inflammatory activity of magnolol, honokiol, and methylhonokiol can be explained by their inhibition of prostaglandin biosynthesis (Table 86.5).

Honokiol was first characterized with respect to antimicrobial activity, but has more recently been shown to have anti-angiogenic properties (Crane et al., 2009). For example, honokiol isolated from an extract of M. grandiflora seed cones has been reported to decrease phosphoinositide 3-kinase (PI3K)/mammalian target of rapamycin (mTOR)-mediated immunoresistance in glioma, breast, and prostate cancer cells, but with no significant affect on pro-inflammatory T cell function (Table 86.5). In addition, honokiol has been reported to induce caspase-dependent apoptosis in B cell chronic lymphocytic leukemia cells, and to inhibit bone metastatic growth of human prostate cancer cells (Crane et al., 2009). The honokiol-mediated inhibition of PI3K/mTOR can enhance T cell-specific tumor responses by

TABLE 86.3 Biological Effects of *Magnolia grandiflora* Seed Extracts

Area of Activity	Effects
Central nervous system	Sedative and hypnotic
	Control of sleep and body temperature
	Anticonvulsant
Cancer	Overcome immunoresistance in glioma, breast and prostate carcinoma
Immune system	Antimicrobial, and antiinflammatory
Gastrointestinal system	Antispasmodic

Magnolia grandiflora has been used in traditional medicine, where its seed extracts are proven to have beneficial effects on several systems.

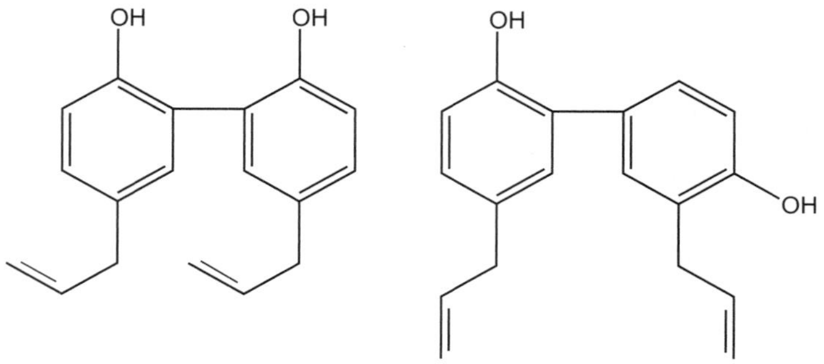

Magnolol   Honokiol

FIGURE 86.1

Compounds from *Magnolia grandiflora* seed extracts. Seed extracts of *Magnolia grandiflora* contain two bioactive compounds. The magnolol and honokiol are structural isomer and biphenolic compounds.

TABLE 86.4 Active Compounds from *Magnolia grandiflora* Seed Extracts

Compound	IUPAC Name	Molecular Formula	Molar Mass
Magnolol	4-Allyl-2-(5-allyl-2-hydroxy-phenyl)phenol	$C_{18}H_{18}O_2$	266.334 g/mol
Honokiol	2-(4-Hydroxy-3-prop-2-enyl-phenyl)-4-prop-2-enyl-phenol	$C_{18}H_{18}O_2$	266.334 g/mol

Magnolol and honokiol are two major bioactive compounds of *Magnolia grandiflora* seed extracts; the two compounds are isomers.
IUPAC, International Union of Pure and Applied Chemistry.

increasing T cell survival and by maintenance of pro-inflammatory T cell functions. Thus, the available evidence suggests that honokiol may be useful for augmentation of T cell-mediated immunotherapy, and that the material is unlikely to negatively impact the immune system.

Honokiol may also reduce the expression of other immunoresistant proteins resulting from the loss of the phosphatase and tensin homolog (PTEN) protein in cancer cells. In contrast to other conventional inhibitors, honokiol has a minimal effect on the proliferation of T cells, expression of activation markers, or pro-inflammatory cytokine production. Thus, honokiol treatment may decrease tumor cell survival, and the PI3K/mTOR-mediated immunoresistance to adaptive immune responses. It has been suggested that honokiol should be considered as an adjunct to active immunotherapy (Crane *et al.*, 2009).

Traditional Mexican medicine has used *M. grandiflora* for antispasmodic and other effects. *M. grandiflora* seeds have substances that apparently act on neurons of the hypothalamus that

TABLE 86.5 Possible Mechanisms of Action of *Magnolia grandiflora* Seed Extracts

Compound	Mechanism	Potential Usage in Human Disease
Honokiol	Decreases PI3K/mTOR-mediated immunoresistance in cancer without significantly affecting pro-inflammatory T cell function	T cell-mediated cancer immunotherapy
	Inhibits COX-2	Inflammatory diseases such as arthritis, Alzheimer's disease, gastrointestinal tumors
	Reduces reactive oxygen species production by neutrophils	Focal cerebral ischemia/reperfusion injury
Magnolol	Inhibits COX-2	Arthritis
	Interferes with the action of glucosyltransferase	Prevention of the formation of bacterial plaque
Whole extracts	Acts on the neurons of the hypothalamus	Prevention/reduction of convulsive seizures
	Reduces reactive oxygen species and apoptosis in cardiomyocytes	Amelioration of cardiotoxicity of anticancer drugs

Seed extracts of *Magnolia grandiflora* have been known to have beneficial effects in human disease. The beneficial effects can be explained by diverse mechanisms, but more research is needed before patients can be treated in practice.

are involved in the control of sleep and body temperature, and extracts seem to reduce the level of convulsive seizures in epileptic patients (Table 86.5). *M. grandiflora* seed extracts may also contain components with sedative or hypnotic properties (Bastidas Ramirez *et al.*, 1998). Thus, extracts may be potentially valuable, because alterations in sleep caused by modern lifestyle changes can adversely affect physiological and behavioral processes and lead to health problems. It is known that a modest amount of sleep loss can alter molecular processes that drive cellular immune activation and induce a pro-inflammatory cytokine response (Lee *et al.*, 2009).

ADVERSE EFFECTS AND REACTIONS (ALLERGIES AND TOXICITY)

Allergy to sesquiterpene lactones has been reported, but *Magnolia*-specific allergy is rare. Seeds of *M. grandiflora* have no known adverse effects. In particular, compounds derived from *Magnolia* species showed no cytotoxicity to macrophages or kidney fibroblasts (Schuhly *et al.*, 2009). *Magnolia* seed extract has also been reported to substantially attenuate doxorubicin-induced cardiac damage (Park *et al.*, 2008).

SUMMARY POINTS

- *M. grandiflora* is an evergreen tree commonly known as the southern magnolia or bull bay.
- *M. grandiflora* has been widely used in traditional medicine and as a decorative tree because of its beautiful appearance.
- The bioactive extracts from seeds of *M. grandiflora* are honokiol and magnolol.
- Seed extracts of *M. grandiflora* have anticonvulsant, antimicrobial, and anti-inflammatory effects.
- Seed extracts of *M. grandiflora* have no known adverse effects.

References

Aitken, R. (2008). *Botanical riches: Stories of botanical exploration*. Melbourne, Australia: Miegunyah Press. p. 112.

Baek, J. A., Lee, Y. D., Lee, C. B., Go, H. K., Kim, J. P., Seo, J. J., et al. (2009). Extracts of *Magnoliae flos* inhibit inducible nitric oxide synthase via ERK in human respiratory epithelial cells. *Nitric Oxide, 20*(2), 122–128.

Bastidas Ramirez, B. E., Navarro Ruiz, N., Quezada Arellano, J. D., Ruiz Madrigal, B., Villanueva Michel, M. T., & Garzon, P. (1998). Anticonvulsant effects of *Magnolia grandiflora* L. in the rat. *Journal of Ethnopharmacology, 61*(2), 143–152.

Callaway, D. J. (1994). *The world of magnolias.* Portland, OR: Timber Press. 11–234.

Crane, C., Panner, A., Pieper, R. O., Arbiser, J., & Parsa, A. T. (2009). Honokiol-mediated inhibition of PI3K/mTOR pathway: a potential strategy to overcome immunoresistance in glioma, breast, and prostate carcinoma without impacting T cell function. *Journal of Immunotherapy, 32*(6), 585–592.

Gardiner, J. (2000). *Magnolias: A gardener's guide.* Portland, OR: Timber Press. 1–330.

Lee, Y. D., Kim, J. Y., Lee, K. H., Kwak, Y. J., Lee, S. K., Kim, O. S., et al. (2009). Melatonin attenuates lipopoly-saccharide-induced acute lung inflammation in sleep-deprived mice. *Journal of Pineal Research, 46*(1), 53–57.

Liou, K. T., Shen, Y. C., Chen, C. F., Tsao, C. M., & Tsai, S. K. (2003). The anti-inflammatory effect of honokiol on neutrophils: mechanisms in the inhibition of reactive oxygen species production. *European Journal of Pharmacology, 475*, 19–27.

Maruyama, Y., Ikarashi, Y., & Kurihara, H. (2002). Bioactivity and pharmacological aspects of *Magnolia*. In S. D. Sarker & Y. Maruyama (Eds.), Magnolia. *Medicinal and aromatic plants – industrial profiles, Vol. 28* (pp. 75–88). New York, NY: Taylor & Francis.

Mellado, V., Chavez, S. M. A., & Lozoya, X. (1980). Pharmacological screening of the aqueous extracts of *Magnolia grandiflora* L. *Archives of Medical Research, 11*(3), 335–346.

Miller, R. B. (2007). Characteristics and availability of commercially important woods. In *The Encyclopedia of Wood* (pp. 1–7). New York, NY: Skyhorse Publishing Inc.

Park, E. J., Zhao, Y. Z., Na, M., Bae, K., Kim, Y. H., Lee, B. H., et al. (2003). Protective effects of honokiol and magnolol on tertiary butyl hydroperoxide- or D-galactosamine-induced toxicity in rat primary hepatocytes. *Planta Medica, 69*, 33–37.

Park, J., Lee, J., Jung, E., Park, Y., Kim, K., Park, B., et al. (2004). *In vitro* antibacterial and anti-inflammatory effects of honokiol and magnolol against *Propionibacterium* sp. *European Journal of Pharmacology, 496*, 189–195.

Park, K. H., Kim, S. Y., Gul, R., Kim, B. J., Jang, K. Y., Chung, H. T., et al. (2008). Fatty acids ameliorate doxorubicin-induced intracellular ca^{2+} increase and apoptosis in rat cardiomyocytes. *Biological & Pharmaceutical Bulletin, 31*(5), 809–815.

Schuhly, W., Khan, S. I., & Fischer, N. H. (2001). The ethnomedicinal uses of Magnoliaceae from the southeastern United States as leads in drug discovery. *Pharmaceutical Biology, 39*, 63–69.

Schuhly, W., Khan, S. I., & Fischer, N. H. (2009). Neolignans from North American Magnolia species with cyclo-oxygenase 2 inhibitory activity. *Inflammopharmacology, 17*(2), 106–110.

Song, Q., & Fischer, N. H. (1999). Biologically active lignans and neolignans from *Magnolia* species. *Revista de la Sociedad Quimica de Mexico, 43*, 211–218.

Sternberg, G., Wilson, J., & Wilson, J. (2004). Magnolia grandiflora. In G. Sternberg (Ed.), *Native trees for North American landscapes: from the Atlantic to the Rockies* (pp. 266–270). Portland, OR: Timber Press.

Vazquez, G. J. A. (1990). *Taxonomy of the genus* Magnolia *(Magnoliaceae) in Mexico and Central America*, pp. 1–60. MSc thesis. University of Wisconsin-Madison.

Zion, R. L. (1995). Evergreen trees. In *Trees for architecture and landscape* p. 224. New York, NY: John Wiley & Sons Inc.

732

Use of Seeds of Malay Apple (*Ziziphus mauritiana*) and Related Species in Health and Disease

Tulika Mishra, Alistair G. Paice, Aruna Bhatia
Immunology and Immunotechnology Laboratory, Department of Biotechnology,
Punjabi University, Punjab, India
Department of Nutrition and Dietetics, King's College London, London, UK

733

LIST OF ABBREVIATIONS

AIDS, acquired immunodeficiency syndrome
TH, T-helper cells
HL-60, human promyelocytic leukemia cell line

INTRODUCTION

Ziziphus belongs to the *Rhamnaceae* family, named after the genus *Rhamnus*. The fruits are drupes, which are dry. The name *Ziziphus* is related to the North African Coastal Arabic word *zizoufo*, the ancient Persian words *zizfum* or *zizafun*, and the ancient Greek word *ziziphon*, all of

Nuts & Seeds in Health and Disease Prevention. DOI: 10.1016/B978-0-12-375688-6.10087-8

which were which used to describe jujube. There are two major domesticated jujubes; *Z. mauritiana* Lam. (the Indian jujube, or ber), and *Z. jujuba* Mill. (the Chinese or common jujube). These two species have been cultivated over vast areas of the world. The fruits of many *Ziziphus* species are edible, and are prepared for consumption in various ways.

BOTANICAL DESCRIPTION

The plants of genus *Ziziphus* are erect trees, or small to large shrubs, usually spiny, glabrous, or relatively hairy. Leaves are alternate or, rarely, sub-opposite, simple and coriaceous, or membranous, acuminated and toothed, three- to five-nerved from the base, and petiolated with stipules. The leaf base is either asymmetrical or symmetrical. Flowers are actinomorphic and hermaphroditic, borne sometimes solitarily or grouped two or three together in axillary cymes or in umbels arranged in terminal panicles or thyrses. Inflorescences may be pedunculated or sessile. The calyx is usually arranged with triangular acute lobes or dentates. The petals are long, unguiculated at the base, and clasp the stamens or the filaments. The stamens are partly adnate to the petal bases, and insert under the edge of the disk. The ovary is superior or sub-inferior, and sunk into the disk. It is two- to four-celled, with two to four styles. When a two-celled ovary is present it produces one seed. The fruits are subglobose, ovoid, or oblong (Figure 87.1). The flesh of the drupe is usually a juicy pulp, but may rarely be relatively dry. The seeds contain large embryos with a sparse or absent endosperm. They are enclosed within a hard, woody endocarp known as the stone, which is sometimes erroneously referred to as the "seed" (Figure 87.2). At the middle of each fruit, one stone is embedded in the pulp. Stones can be round, sub-ovate, or ovate in shape, with more or less pronounced ridges on the outer surface. Cleaned stones are about 1–2 cm in size, containing one or two seeds (Pareek, 2001). The complete botanical description of the genus *Ziziphus* is provided in Table 87.1.

HISTORICAL CULTIVATION AND USAGE

Ziziphus mauritiana ("Ber") has been a useful fruit in India since antiquity — it was even mentioned in *Yajurveda* (Macdonell & Keith, 1958). All the scripted sources from the period 1000 BC–400 AD have been mentioned by Pareek (2001). The main ancient

FIGURE 87.1
Fruit of *Ziziphus mauritiana,* Complete mature fruit set, *Ziziphus mauritiana* var. Umran. *Source: Botanical Gardens, Punjabi University, Patiala, Punjab, India; reproduced with permission.*

(A)   **(B)**   **(C)**

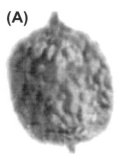

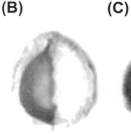

FIGURE 87.2
Endocarp and seed of *Ziziphus mauritiana* var. Umran. Single seed subovate endocarp in an intact form showing ridges on the outer surface, and in a broken form showing the shallow groove where the actual seed (one or two) is situated. (A) Subovate complete endocarp (stone); (B) broken endocarp; (C) seed.

TABLE 87.1 Botanical Description of *Ziziphuz mauritania* Lam.

Kingdom	Plantae — Plants
Subkingdom	Tracheobionta — Vascular plants
Superdivision	Spermatophyta — Seed plants
Division	Magnoliophyta — Flowering plants
Class	Magnoliopsida — Dicotyledons
Subclass	Rosidae
Order	Rhamnales
Family	Rhamnaceae — Buckthorn family
Genus	*Ziziphus* Mill. Jujube
Species	*Ziziphus mauritiana* Lam.— Indian jujube

cultivation areas included the Deccan Plateau, which predates Gangetic civilization (Sankalia, 1958).

Ziziphus jujuba Mill. is thought to be as old as the Bronze Age Shang Dynasty, and has been cultivated since 2000 BC (Qu, 1983). Excavations have indicated that the earliest farmers of the Indus Valley collected the fruits from native trees and propagated them to other areas (Zohary & Hopf, 1988). Table 87.2 highlights their native places of cultivation and subsequent migration.

Traditionally, the plant has been of high medicinal value. The roots have been used to treat coughs and headaches, whilst the bark has been used on boils, and for dysentery. The leaves are antipyretic, whilst the fruit has been used to assist digestion and to treat tuberculosis. The seeds help to cure eye diseases, and are helpful in leukorrhea, and as an astringent tonic to the heart and brain. The seeds also help to relieve thirst, and have a sedative and hypnotic effect, which is helpful in insomnia, pain, physical weakness, and rheumatic symptomology (P. Oudhia, personal communication).

PRESENT-DAY CULTIVATION AND USAGE

The historical value of *Ziziphus* led to its cultivation in countries such as India. However, worldwide, jujubes remain a relatively minor crop. Table 87.2 demonstrates the major cultivation areas for this plant.

TABLE 87.2 Historical and Present Day Cultivation Areas (Pareek, 2001)

Species	Major Cultivation Areas
Z. mauritiana	Distributed throughout the warm subtropics and tropics of South Asia (India). Cultivation spread south-eastwards through Malaysia, and eastwards through IndoChina and southern China, Africa, and Southern Arabia.
Z. jujube	Native spread to temperate Asia — China, Mongolia, and the Central Asian Republics. Cultivation spread to the Mediterranean, throughout the near East and South-West Asia, and eastwards to Korea and Japan.
Z. spina Christi	Native and cultivated spread to the Middle East through Arabia, from South Africa to North Africa, Ethiopia, Eastern Africa, and Turkey.
Z. lotus	Native to Asia minor, and cultivation spread to Arabia, Egypt, along the North African coast, to Cyprus and Greece in Europe. Also cultivated in southern Portugal and Spain, parts of Italy, France, and Sicily.

The major cultivars were selected from heterogeneous populations, with the best genotypes being protected and subsequently utilized by local people. The most widespread cultivars included "Umran" in *Ziziphus mauritiana*, and "Sui Menor Liin" in *Ziziphus jujuba*. Nowadays, many optimization techniques are employed, including hybridization, mutation, polyploidy (traditional methods), tissue culture, genetic engineering, molecular methods, and embryo rescue (biotechnological methods). Most of the breeding work was carried out in India and China. The major cultivars of *Ziziphus mauritiana* and *Ziziphus jujube* are described in Table 87.3.

In addition to its traditional roles, *Ziziphus* now acts as a source of income not only from its fruit, but also in honey production and bee-keeping in its trees, lac insect rearing, silkworm feeding, wood production for boat-making and agricultural implement manufacturing, environmental protection, etc. In addition, the plant contains many important ethano-pharmaceutical compounds utilized for medicinal purposes.

APPLICATIONS TO HEALTH PROMOTION AND DISEASE PREVENTION

As a rich source of many compounds like ascorbic acid, riboflavin, alkaloids, glycoside, saponins, and triterpenoic acid, *Ziziphus* is used in the prevention and cure of many diseases. The seeds of *Ziziphus* species possess anxiolytic and sedative activities. They are known to suppress the central nervous system, resulting in reduced anxiety and the induction of sleep. The flavonoids, saponins, and alkaloids present in seeds can also be used for sedation (Zhang *et al.*, 2003).

Patients suffering from Alzheimer's disease have depleted acetylcholine levels. The choline acetyltransferase enzymes are crucial in the production of this hormone. *In vitro* studies have demonstrated that the oleamide component of the *Ziziphus jujube* seed extract enhances the activity of this enzyme (Heo *et al.*, 2003).

Although the leaf extract of *Ziziphus jujuba* has been shown to enhance the phagocytic and intracellular killing potency of neutrophils (Ganachari *et al.*, 2004), little literature exists describing the effect of seed extract on the overall immune response. Recently, it has been shown that the aqueous ethanolic seed extract of *Ziziphus mauritiana* enhanced both the humoral and the cell-mediated immune response in Swiss albino mice (Mishra & Bhatia, 2010). Similarly, cytokine studies have shown enhanced IFN-γ activity, reflecting augmentation of TH-1 mediated immunity (Mishra & Bhatia, 2010). Seed extract may therefore have applications in the treatment of tuberculosis, cancer, and AIDS, as alterations to TH-1 mediated immunity are implicated in these diseases. Preliminary work by the current authors has

TABLE 87.3 Major Cultivars of *Ziziphus* (FAO, 1988; Pareek, 2001)

India	Chinese	Russia
Umran	Sovetskii	Yuzhanin
Banarsi Karaka	Tavrika	Khurman
Kakrola	Tayantszao	Burnim
Gola	Suontszao	Kitaiskii-93
Mundia Murhara	Druzhba	Finik
Sanori	Kitaiskii	
Illaichi		
Safeda Selected		
Kaithali		
Reshmi		
Chhuhara		
Seb		

shown that seed extract collected from different geographical locations shows variable bioactivity.

The methanolic root extracts of *Ziziphus spina-christi* and *Ziziphus jujube* have been shown to possess antidiabetic activity (Hussein, 2006). The present authors found that the aqueous ethanolic seed extract of *Ziziphus mauritiana* showed antidiabetic potential against alloxan-induced diabetes in Swiss albino mice. This study demonstrated the beneficial effect of this extract in acute treatment, as well as after 28 days. There was appreciable synergy with the effects of standard medications targeting diabetes (Bhatia & Mishra, 2010).

Ziziphus is a good source of betulinic acid, which, when extracted, shows selective toxicity against human melanoma (Kim *et al.*, 1998). In our own study, betulinic acid was found in the seed extract of *Ziziphus mauritiana* (Mishra & Bhatia, 2010). This is a very valuable phytocompound, which can be used against melanoma (Patočka, 2003). One of the most astonishing features of this compound is its selective cytotoxicity against cancer cells, unlike most chemotherapy agents. The present authors studied the effect of seed extract on HL-60 cells *in vitro*, and against Ehrlich ascites carcinoma in animal studies. The seed extract induced apoptosis in HL-60 cells, and also helped the animals to recover from Ehrlich ascites carcinoma (Mishra & Bhatia, unpublished data).

Ziziphus seeds are rich in saponins, known to bind cholesterol. These compounds are widely used for cholesterol control. Kim (2002) reported that the seed extract possessed anti-hyperlipidemic activity in rats. In a preliminary study, the present authors have found that the aqueous ethanolic seed extract helps to lower the cholesterol level in Swiss albino mice after 15 days of treatment (unpublished data). The aqueous ethanolic seed extract of *Ziziphus mauritiana* also has potent antioxidant and hepatoprotective effects that help combat alcohol-induced oxidative stress (Bhatia & Mishra, 2010).

Peng and colleagues investigated the anxiolytic effect of *Ziziphus jujuba* seed in a mouse model of anxiety (Peng *et al.*, 2000). Their studies revealed that seed extract might help combat insomnia or other anxiety-related disorders.

The pericarp and the seeds of *Z. jujuba* contain two main classes of phospholipids: phosphatidylcholines and phosphatidylglycerols. Fractionation of these seeds demonstrated the presence of the triglyceride 1,3-di-O-[9(Z)-octadecenoyl]-2-O-[9(Z),12(Z)-octadecadienoyl] glycerol, and a fatty acid mixture of linoleic, oleic, and stearic acids (Su *et al.*, 2002). These are all pharmacologically active compounds.

Certain classes of drugs, such as peptides, are poorly transported across cell membranes. This diminishes their subsequent bioavailability as a drug. To overcome this barrier, permeability enhancers are used to ease passage of these drugs across the cell membrane. Enhancers belonging to the alkylglycoside family are commonly used. The aqueous extract of *Z. jujuba* seeds was compared with alkylglycosides in a cell culture system. Measurements of transepithelial electrical resistance showed that the seed extract of *Z. jujuba* lowered cell resistance more rapidly over a given time than alkylglycosides, and allowed full recovery of cells in a shorter time period. It appears that the seed extract is a more efficient permeability enhancer than the two alkylglycosides used (Eley & Hossein, 2002).

Research into the potential medicinal applications of compounds within jujubes is at an early stage. However, there appears to be a broad range of benefits from a seed extract that is usually thrown away. The authors are keen to combat this waste of a potentially valuable resource. The seeds of the *Ziziphus* offer many possible medicinal opportunities (Table 87.4). The hardy nature of jujubes, and their wide geographical distribution, means that they could provide a potentially cheap and accessible source of such compounds for medicine.

737

TABLE 87.4 Pharmacological Action of Seeds of the Genus *Ziziphus*

Bioactivity	Model	Pharmacological Action	Reference
Anxiolytic and sedative action	Mice	Reduction of anxiety, and induces sleep	Zhang *et al.* (2003)
Modulates choline acetyltransferase enzymes	Mice	Helpful in Alzheimer's disease, to increase acetylcholine levels	Heo *et al.* (2003)
Immunomodulation	Mice	Enhances Th-1 mediated immune response, and can be helpful in immune-related disorders and diseases like cancer, AIDS, etc.	Mishra & Bhatia, (2010)
Antimalarial activity	Mice	Effective against *P. berghei* infection	Mishra & Bhatia (unpublished)
Antidiabetic effect	Mice	Shows synergistic behavior with Glyburide®	Bhatia & Mishra (2010)
Anticancer potential	Cell culture	Induces apoptosis in HL-60 cell line with no such effect on normal cells	Mishra and Bhatia (unpublished)
	Mice	Reduces tumor volume and increases life span in Ehrlich ascite carcinoma bearing animals	
Antihyperlipidemic effect	Rats	Helpful in improvement of blood glucose and lipid composition in serum of dietary hyperlipidemic animals	Kim (2002)
Hypocholesteremic effect	Mice	Reduces total cholesterol level within 15 days	Mishra & Bhatia (unpublished)
Antioxidant and hepatoprotective effects	Mice	Helpful against alcohol induced oxidative stress	Bhatia & Mishra (2009)
Permeability enhancer	Cell culture	Helpful for easy passage of drugs across the cell membrane	Eley & Hossein, 2002

738

ADVERSE EFFECTS AND REACTIONS (ALLERGIES AND TOXICITY)

The seed extract did not show any cytotoxicity. Our own studies have demonstrated that seed extract administration does not cause any toxicity symptoms. However, at the higher doses animals showed more lethargic behavior (Mishra & Bhatia, 2010).

SUMMARY POINTS

- *Ziziphus* is a fruit tree whose cultivation dates back to ancient Indian and Chinese civilizations.
- Historically *Ziziphus* has had many pharmacological applications, which may have relevance in the present day.
- Cultivation of the plant started in central Asia, and moved to a wide array of geographical locations. This demonstrates its adaptability to different soils and climatic conditions.
- The plant's adaptable and resistant nature makes it an economical source for medication.
- The woody endocarp protects the *Ziziphus* seed, helping it to sustain its biological and cultivation properties under adverse conditions.
- Being a rich source of different phytocompounds, the seed of the plant may be a potential source of treatments for diseases such as AIDS or diabetes.
- It contains the valuable phytocompound betulinic acid, which may be useful in new drug designs.
- This adaptable plant has a wide range of potential economic roles, such as bee-keeping, fodder, or fuel.

References

Bhatia, A., & Mishra, T. (2009). Free radical scavenging and antioxidant potential of *Ziziphus mauritiana* (Lamk.) seed extract. *Journal of Complementary Integrative Medicine, 8,* 42–46.

Bhatia, A., & Mishra, T. (2010). Hypoglycemic activity of *Ziziphus mauritiana* aqueous ethanol seed extract in alloxan-induced diabetic mice. *Pharmaceutical Biology, 48*(6), 604–610.

Eley, J. G., & Hossein, D. (2002). Permeability enhancement activity from *Ziziphus jujuba*. *Pharmaceutical Biology, 40,* 149–153.

FAO (1988). *Traditional fruit plants*. Food and Nutrition Paper 42. Rome, Italy: FAO.

Ganachari, M. S., Shiv, K., & Bhat, K. G. (2004). Effect of *Ziziphus jujube* leaves extract on phagocytosis by human neutrophils. *Journal of Natural Remedies, 41,* 47–51.

Heo, H. J., Park, Y. J., Suh, Y. M., Choi, S. J., Kim, M. J., Cho, H. Y., et al. (2003). Effects of oleamide on choline acetyltransferase and cognitive activities. *Bioscience Biotechnology and Biochemistry, 67,* 23–29.

Hussein, H. M., El-Sayed, E. M., & Said, A. A. (2006). Antihyperglycemic, antihyperlipidemic and antioxidant effects of *Zizyphus spina christi* and *Zizyphus jujube* in alloxan diabetic rats. *International Journal of Pharmacology, 2,* 563–570.

Kim, D. S. H. L., Pezzuto, J. M., & Pisha, E. (1998). Synthesis of betulinic acid derivatives with activity against human melanoma. *Bioorganic & Medicinal Chemistry Letters, 8,* 1707–1712.

Kim, H. S. (2002). Effects of the *Zizyphus jujuba* seed extract on the lipid components in hyperlipidemic Rats. *Journal of Food Sciences and Nutrition, 7,* 72–77.

Macdonell, A. A., & Keith, A. B. (1958). *A Vedic index of names and subjects, Part II*. Varanasi, India: Motilal Banarasidas & Co.

Mishra, T., & Bhatia, A. (2010). Augmentation of expression of immunocytes functions by seed extract of *Ziziphus mauritiana* (Lamk.). *Journal of Ethanopharmacology, 127,* 341–345.

Pareek, O. P. (2001). *Fruits for the future 2: Ber*. Southampton, UK: International Centre for Underutilised Crops.

Patočka, J. (2003). Biologically active pentacyclic triterpenes and their current medicine signification. *Journal of Applied Biomedicine, 1,* 7–12.

Peng, W. H., Hsieh, M. T., Lee, Y. S., Lin, Y. C., & Liao, J. (2000). Anxiolytic effect of seed of *Ziziphus jujuba* in mouse models of anxiety. *Journal of Ethanopharmacology, 72,* 435–441.

Qu, Z. Z. (1983). Jujubes in ancient China *[in Chinese]*. Zhanhua, China: Jujube Institute, Heberi Agricultural University.

Sankalia, H. D. (1958). The excavations at Navdaloli – Maheshwar. *Technical Reports on Archaeological Objects*. Baroda, India: Deccan College Research Institute, Maharaja Sayajirao University, Publication No. 1.

Su, B. N. C. M., Farnsworth, N. R., Fong, H. H., Pezzuto, J. M., & Kinghorn, A. D. (2002). Activity-guided fractionation of the seeds of *Ziziphus jujuba* using a cyclooxygenase-2 inhibitory assay. *Planta Medica, 68,* 1125–1128.

Zhang, M., Ming, G., Shou, C., Lu, Y., Hang, D., & Zhang, X. (2003). Inhibitory effect of Jujuboside A on glutamate mediated exictatory signal pathway in hippocampus. *Planta Medica, 69,* 692–695.

Zohary, D., & Hopf, M. (1988). *Domestication of plants in the Old World. The origin and spread of cultivated plants in West Asia, Europe and the Nile Valley*. Oxford, UK: Clarendon Press, ix. 249.

CHAPTER 88

Mango (*Mangifera indica* L.) Seed and Its Fats

Julio A. Solís-Fuentes[1], **María del Carmen Durán-de-Bazúa**[2]
[1] Food Science Area, Instituto de Ciencias Básicas, Universidad Veracruzana, Xalapa, Ver., México
[2] Chemical Engineering Department, Facultad de Química, Universidad Nacional Autónoma de México, México

CHAPTER OUTLINE

LIST OF ABBREVIATIONS

FA, fatty acid
FFA, free fatty acid
MKF, mango kernel fat
MSK, mango seed kernel
POP, 2-oleodipalmitin
POS, 2-oleopalmitostearine
SFC, solid fat content
SMKO, sapote mamey kernel oil
SOS, 2-oleodistearine
TG, triacylglycerols

INTRODUCTION

Lipids, and particularly fats and oils, are a large group of important compounds in the structure and functioning of cells, essential in the diet for their nutritional value, and highly desirable for their effect on the functional properties of food. Unlike oil, natural vegetable fats

Nuts & Seeds in Health and Disease Prevention. DOI: 10.1016/B978-0-12-375688-6.10088-X

are scarce, and have multiple applications. Mango kernel fat (MKF) has a composition and physical characteristics that make it a consumer alternative to processed semi-solid fats high in *trans* FA content, with serious adverse effects on human health maintenance.

BOTANICAL DESCRIPTION

Economically speaking, the mango is the most important fruit crop of the family *Anacardiaceae* (Cashew or poison ivy family) in the order of *Sapindales*. The family contains 73 genera and between 600 and 700 species, well-known for the presence of caustic resins in the leaves, bark, and fruits. Several of these, including mango, may cause some type of dermatitis in humans. The genus *Mangifera* contains about 60 species, of which about 15 produce edible fruits, among them *M. sylvatica*, a possible ancestor of *M. indica*. Currently there are over 1000 known varieties of mango, whose nomenclature is sometimes complicated because of certain regionalisms. In the world, only a few varieties are grown on a commercial scale and traded. The fruit has a large, central stone, flattened, with a woody cover containing a nucleus or kernel with a single embryo, or two to five embryos (Hindu and Indo-Chinese varieties, respectively) (Morton, 1987; Vasanthaiah *et al.*, 2007).

HISTORICAL CULTIVATION AND USAGE

The mango has been cultivated since prehistoric times. Apparently, it is endemic to North-Eastern India and Myanmar (Burma); possibly also to Sri Lanka (Ceylon). It was distributed throughout South-east Asia and the Malay Archipelago, from where it spread to Africa and to the New World through the first Portuguese and Spanish maritime routes and colonization. The first permanent planting in Florida, however, dates from the 1860s (Vasanthaiah *et al.*, 2007).

Historically, its flesh has been used almost exclusively as fresh and processed fruit. Various plant parts have been used in traditional medicine as a cure for a number of diseases. The kernel seed has been consumed by humans and animals in some Asian groups, and, in different preparations, has been used as a vermifuge and as an astringent in diarrhea, hemorrhages, and bleeding hemorrhoids. The kernel fat has been administered in cases of gastritis (Morton, 1987).

PRESENT-DAY CULTIVATION AND USAGE

At present, the mango is cultivated on a commercial scale throughout tropical and subtropical regions, in over 3.7 million ha in the world. Mango fruit is one of the most important crops. It is grown in over 90 countries, representing about 50% of the tropical fruits produced worldwide. World production in 2005 was 28.5 million tonnes (Evans, 2008). India produces nearly half of the world output, followed by China, Thailand, and Mexico; all in all, the 10 countries with the highest production of this fruit contain about 80% of the world production. Currently, the mango is still mainly used as food. The lipids' therapeutic properties, and particularly the fat of the seed kernel, have been subjects of extensive research for the past decade (Puravankara *et al.*, 2000; Yella Reddy & Jeyarani, 2001; Abdalla *et al.*, 2007), with a promising outlook.

APPLICATIONS TO HEALTH PROMOTION AND DISEASE PREVENTION

Mango seed kernel is being re-evaluated not only for its qualities as a natural food (edible, non-toxic, with a high quality protein and amino acid score) and for the biological activity of some of its constituents (lipid fractions and phenolic compounds identified as active anti-oxidants and modulators of lipid metabolism) which have shown positive impacts, directly or indirectly, on health and nutrition, but also because its fat in the natural state has shown significant functional and physicochemical characteristics that could lead to it replacing fats like cocoa butter or others that are widely used in industry and in food processing, but are now facing serious objections because of their adverse effects on nutrition and human health.

Chemical composition and lipids of the mango seed kernel

By its biological nature, MSK has a composition that responds to varietal and phenotypic variations. The available data are for a substantial but still minority number of commercially important varieties from major producing regions. Table 88.1 gives the range of values most often reported for the relevant chemical constituents of the kernel. The seed, depending on the variety, can constitute 3—25% of the total mass of the fruit, and the kernel occupies 54—85% of the seed; it has a moisture content of 33—86%, and the solid dried matter has proteins (4.0—8.1%), crude fiber (1.7—7.6%), ash (1.0—3.7%), total carbohydrates (70—76%), around 0.1—6.4% of phenolic compounds, and 3.7—12.6% of crude fat. MSK proteins have high scores of essential amino acids (78) and protein quality (177—189 adults), and an *in vitro* digestibility of 26.7—29.8% (Dhingra & Kapoor, 1985, Abdalla *et al.*, 2007).

TABLE 88.1 Mango Seed Kernel and Mango Kernel Fat Composition Ranges of Mango (*Mangifera indica*) Seed

Kernel and Fat Composition	Values	References
Seed in fruit	3—25[a]	Lakshminarayan *et al.*, 1983
Kernel in seed	54—85	
Moisture	33—86[b]	
Protein	4.0—8.1[c]	
Crude fiber	1.7—7.6	Dhingra & Kapoor, 1985; Solís-Fuentes and Duran de Bazúa, 2004
Ash	1.0—3.7	Lakshminarayan *et al.*, 1983
Total carbohydrates	70—76	Solís-Fuentes & Duran de Bazúa, 2004
Phenolic compounds	0.1—6.4	Abdalla *et al.* 2007; Parmar & Sharma, 1990
Crude fat	3.7—12.6	Lakshminarayan *et al.*, 1983
Simple lipids	94.8—97.5[d]	Van Pee *et al.*, 1981
Triacylglycerols	55.6—91.5	Ali *et al.*, 1985
Partial glycerides	2.3—4.0	
myristic (C14:0)	0.7—8[e]	Baliga & Shitole, 1981
palmitic (C16:0)	3—18	Lakshminarayan *et al.*, 1983
stearic (C18: 0)	24—57	
oleic (C18:1)	34—56	
linoleic (C18:2)	0—13	
linolenic (C18:3)	0.2—5.3	Van Pee *et al.*, 1981; Dhingra & Kapoor, 1985
arachidic (C20:0)	< 4.0	Lakshminarayan *et al.*, 1983
oleic *sn*-2 position	85.2—89.9	Van Pee *et al.*, 1981
linoleic *sn*-2 position	8.3—13.0	
Free fatty acids	0.8—1.4	Ali *et al.*, 1985; Abdalla *et al.*, 2007
Unsaponifiables	0.9—5.3[d]	Lakshminarayan *et al.*, 1983; Gaydou & Bouchet, 1984
Squalene	1.0—1.1	Abdalla *et al.*, 2007
Sterol fractions	0.5—1.2	Ali *et al.* 1985; Abdalla *et al.* 2007
Tocopherol fractions	0.3—0.4	Abdalla *et al.*, 2007
Complex lipids	2.5—5.2	Van Pee *et al.*, 1981
Phospholipids	0.1—2.8	Van Pee *et al.*, 1981; Ali *et al.*, 1985
Glycolipids	0.6—2.6	

[a]%, as-is basis;
[b]% of kernel, as-is basis;
[c]% of kernel, dry basis;
[d]% of total lipids;
[e]% of total FA.

Although phenolic compounds act as anti-nutritive factors, they have recently become the subject of intense research because of their high antioxidant activity. Tannins, gallic acid, coumarins, caffeic acid, vanillin, mangiferin, ferulic acid, and cinnamic acid have been identified in the MSK and analyzed for their antioxidant activity (Puravankara *et al.*, 2000; Abdalla *et al.*, 2007; Maisuthisakul & Gordon, 2009).

Lipids are important components of food, and also basic structural and functional constituents of cells; therefore, they are decisive in states of health and illness of individuals. MKF has been studied regarding their yield during extraction, its toxicological safety, fatty acid and glyceride composition, and chemical, physical, thermal and phase properties; these are important aspects for its applicability as a substitute or a supplement to fats from other sources. Some studies have been guided by its similarity to cocoa butter and its potential use, mainly in the food industry or in pharmaceuticals and cosmetics.

Fat yields from the MSK fluctuate widely among varieties. Van Pee and colleagues (1981) reported values between 6.8 and 12.6% (db) for African varieties; others, such as Lakshminarayanan *et al.* (1983), found lower yields, with levels ranging from 3.7% for some varieties from India, and Gaydou and Bouchet (1984) reported atypical ranges of 27–38% of fat on a dry basis for Malagasy varieties. The extracted fat was solid at room temperature, and cream-colored.

Simple lipids in the MSK make up 94.8–97.5% of the total lipids; the major constituents of these are triacylglycerols (55.6–91.5%), followed by partial glycerides (2.3–4%) and FFAs (0.8–1.42%). Ali and colleagues (1985) found varieties with a FFA content of up to 37% in India. Unsaponifiable compounds, one minor component group in vegetable fats and oils known to have significant effects on fat metabolism (Lau *et al.*, 2005), range from 0.9 to 5.3%; whereas the complex lipids, phospholipids and glycolipids, whose important effect has been studied extensively in humans, are found in the order of 2.5–5.2% of the total in the MSK. Table 88.1 also shows the amounts of unsaponifiable fractions, phospholipids, and glycolipids in MSK, and Table 88.2 contains ranges of the reported values for the main physical and chemical characteristics of MKF, with a saponification number between 185.6 and 198, and an iodine value between 34.0 and 57.7.

Mango kernel fat composition

Oleic acid (18:1) is the most abundant in MKF, fluctuating between 34 and 56%; it is followed by stearic (18:0) (24–57%) and palmitic (16:0) (3–18%) acids. Other fatty acids in smaller quantities are linoleic (18:2) (up to 13%), and linolenic (20:0) and arachidonic acids with smaller amounts. A few reports indicate the presence of myristic (0.7–8%) and lauric (12:0) acids, and tridecanoic (13:0), pentadecane (15:0), palmitoleic (16:1), margaric (17:0),

TABLE 88.2 Main Physical and Chemical Properties of Mango Kernel Fat

Physical	Range	References
Specific gravity	0.87–0.93[a]	Abdalla *et al.*, 2007
Refractive index at 40°C	1.359–1.559[a]	Ali *et al.*, 1985; Abdalla *et al.*, 2007
Melting point,	25.0–47.0[b]	Ali *et al.*, 1985
Chemical		
Peroxide value,	0.1–1.21[c]	Ali *et al.*, 1985; Abdalla *et al.*, 2007
Iodine number	34.0–57.7	Dhingra & Kapoor, 1985; Van Pee *et al.*, 1981
Saponification number	185.6–198.6	Abdalla *et al.*, 2007
Acid value	3.8–4.9[d]	Van Pee *et al.*, 1981

[a]At 40°C ;
[b]°C;
[c]meq. O_2/kg fat;
[d]mg KOH/g fat.

nonadecane (19:0), and gondoic acids (20:1) in small and trace amounts in some varieties (Gaydou & Bouchet, 1984).

The high predominance of oleic, stearic, and palmitic acids (over 85% of the total) has shown that MKF is similar to other major natural fats, like cocoa butter (*Theobroma cacao*), Sal (*Shorea robusta*), illipe butter (*Shores etenoptera*), Shea butter (*Butyrospermum parkii*), and Mowrah (*Bassia latifolia*), all widely used in food processing. In the case of cocoa butter, these three fatty acids constitute about 95% of the total, and provide a relatively simple glyceride composition; in cocoa butters from different regions of the world, three triacylglycerols (POP, POS, and SOS) make up more than 80% of the total, giving them unique properties.

To diversify their use, natural fats and oils are processed industrially through inter-, trans- and direct esterification, selective or homogeneous hydrogenation, wet or dry fractionation (with solvents or surfactants in an aqueous medium), and mixed treatments (Baliga & Shitole, 1981). These methods yield fats with the desired physical characteristics in terms of their composition, melting points, and consistency, as well as stability against oxidation. Some, like cocoa butter equivalents, and others used as margarine, shortenings, frying fats, etc., have, however, presented significant drawbacks, because if they are made with severe heat treatment the positional isomerization of the *cis* double bond of unsaturated FA may take place, leading to the occurrence of trans-isomers as by-products.

Van Pee *et al.* (1981) found that the triglyceride composition of the MKS shows a FA distribution in the glycerol molecule characterized by the location of the saturated FA in the *sn*-1 and *sn*-3 positions, the remaining sites being proportionately occupied by unsaturated oleic, linoleic, and linolenic acids. In 10 varieties grown in Zaire, MKF oleic and linoleic acids occupied between 85.2 and 89.9%, and 8.3 to 13.0%, respectively, of the *sn*-2 position in TG molecules.

Mango kernel fat thermal behavior and polymorphism

745

It is known that natural fats have, in general, a complex thermal and structural behavior derived from their TG composition. TGs and fats are characterized by multiple melting behaviors due to their polymorphism; basically, these exhibit three different polymorphic forms (α, β′, and β). The amounts and structure of the TGs determine their nutritional (TG postprandial response), physical (spreadability, resistance to water/oil loss, etc.), and sensory (melting, graininess, etc.) properties.

The thermal behavior and polymorphism of MKF, alone and in mixtures with other natural fats, have been studied by Baliga and Shitole (1981), Yella Reddy and Jeyarani (2001), Solis-Fuentes and Durán-de-Bazúa (2003, 2004), and Solis-Fuentes *et al.* (2005), among others. In general, such behavior has been compared with that of cocoa butter, showing that, among the variety of obtained fats and behaviors studied, some of them closely resemble cocoa butter or other fats widely used in the food manufacturing processes that involve the partial hydrogenation of vegetable oils.

Solis-Fuentes and Durán-de-Bazúa (2004) showed that the melting curves of the MKF var. Manila, grown widely in Mexico and other countries, are relatively simple, and show a great resemblance to those of cocoa butter; however, the MKF fusion curve is usually wider than the cocoa butter curve. Two stable polymorphic forms of MKF with the possibility of assuming two other unstable forms have been identified through DSC and X-ray diffraction: the α form, with a low melting point, and the β form, with a high melting point. Figure 88.1 shows the SFC profiles of the MKF and cocoa butter polymorphs. The compatibility and product characteristics of MKF blends with other fats and oils and their fractions have been analyzed.

In MKF and cocoa butter blends, it has been shown that MKF is slightly softer than cocoa butter and has a more evident softening effect in mixtures when it comprises between 60% and 80% of the weight. If its participation in the mixtures is small, the softening effect is negligible; if it is

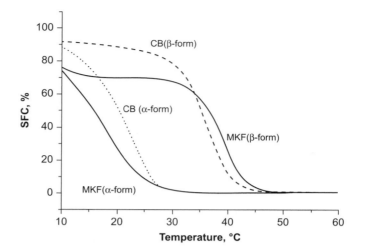

FIGURE 88.1

SFC profiles for mango kernel fat (var. Manila) and cocoa butter in their non-stabilized and stabilized polymorphs. The figure shows the similarity of the most stable polymorphs of mango kernel fat in their solid/liquid relationships with those appreciated in cocoa butter for its wide use in confectionary products.

higher, the effect is compensated when the solids content is small. Additionally, MKF requires higher temperatures than cocoa butter in order to melt.

When MKF and cocoa butter blends are taken to achieve β forms, mixture compatibility is improved, non-softening effects appear in compositions with less than 20% of mango fat, and a harder mixture is obtained for compositions with more than 80% of MKF. However, the wider MKF fusion profile requires lower temperatures than cocoa butter to remain 80% solid, and higher temperatures to melt the last 5% of solids of the fat in its pure state. Isosolid diagrams have shown a lower compatibility between MKF and cocoa butter than between cocoa butter and other fats such as *Coberine*, with isolines more parallel and horizontal during its mixing with cocoa butter (Talbot, 1995). MKF is, however, much more compatible with cocoa butter than milk fat, lauric fats, and hydrogenated cottonseed oil are (Solís-Fuentes & Durán-de-Bazúa, 2004).

The analysis of the phase behavior of ternary blends of MKF, cocoa butter, and sapote mamey (*Pouteria sapota*) kernel oil (SMKO) has shown that these fats can support the preparation of mixtures with different compositions that can become like cocoa butter equivalents or other useful mixtures for food, pharmaceutical, and cosmetic uses (Solís-Fuentes & Durán-de-Bazúa, 2003). In other studies, Yella-Reddy and Jeyarani (2001) showed that it is possible to prepare bakery shortenings with no trans fatty acids by using MKF and mahua (*Madhuca latifolia*) fats and their fractions.

ADVERSE EFFECTS AND REACTIONS (ALLERGIES AND TOXICITY)

MSK and MKF are edible, and thus far there have been no studies showing that they contain any toxic or allergenic compounds. In times of food shortages and famine, poor people from some producing regions have consumed boiled kernels. Rukmini and Vijayaraghavan (1984) studied the nutritional value and toxicological safety of kernel fat by feeding rats with MKF and groundnut oil in balanced diets with fat contents of 10%, and making multi-generation breeding evaluations. The food efficiency ratio and growth rate of the rats fed with the MKF diet were comparable with those of the control group. The retention of nutrients (calcium, phosphorus, and nitrogen) was not affected adversely by MKF intake. The serum and liver total cholesterol, total lipids levels, and liver TG type were alike in MKF and control-fed animals. The histo-pathological evaluations of the rats' organs showed no abnormalities.

An important advantage of MKF, along with other natural fats, is that it contains no trans fatty acids, which have been proven to contribute to development of various serious diseases, and to have adverse effects on human maintenance.

SUMMARY POINTS

- Mango fruit is one of the most important crops worldwide, with more than 28.5 million tonnes produced and traded in the world annually. Its seed is an easily available waste product of the mango-processing industry.
- Mango seed kernel is being re-evaluated as a potential food source of functional ingredients due to its composition of proteins, antioxidant compounds, and lipids.
- The consumption of trans-fatty acids in dietary hydrogenated fats is a worldwide public health problem because of their implications in the development of some major diseases.
- The physical and chemical characteristics of natural mango kernel fat make it a viable consumer alternative to high trans-fatty acid dietary fats.
- Dietary fat composition is an issue of great importance in the health and disease development of individuals.

References

Abdalla, A. E. M., Darwish, S. M., Ayad, E. H. E., & El-Hamahmy, R. M. (2007). Egyptian mango by-product 1. Compositional quality of mango seed kernel. *Food Chemistry, 103*, 1134–1140.

Ali, M. A., Gafur, M. A., Rahman, M. S., & Ahmed, G. M. (1985). Variations in fat content and lipid class composition in ten different mango varieties. *Journal of the American Oil Chemists Society, 62*, 520–523.

Baliga, B. P., & Shitole, A. D. (1981). Cocoa butter substitutes from mango fat. *Journal of the American Oil Chemists Society, 58*, 110–114.

Dhingra, S., & Kapoor, A. (1985). Nutritive value of mango seed kernel. *Journal of the Science of Food and Agriculture, 36*, 752–756.

Evans, E. A. (2008). *Recent trends in world and US mango production, trade, and consumption.* Gainesville, FL: EDIS document FE718, Food and Resource Economics Department, University of Florida.

Gaydou, E. M., & Bouchet, P. (1984). Sterols, methyl sterols, triterpene alcohols and fatty acids of the kernel fat of different Malagasy mango (*Mangifera indica*) varieties. *Journal of the American Oil Chemists Society, 61*, 1589–1592.

Lakshminarayana, G., Chandrasekhara-Rao, T., & Ramalingaswamy, P. A. (1983). Varietal variations in content characteristics and composition of mango seed and fat. *Journal of the American Oil Chemists Society, 60*, 88–89.

Lau, V. W. Y., Journoud, M., & Jones, P. J. H. (2005). Plant sterols are efficacious in lowering plasma LDL and non-HDL cholesterol in hypercholesterolemic type 2 diabetic and nondiabetic persons. *American Journal of Clinical Nutrition, 81*, 1351–1358.

Maisuthisakul, P., & Gordon, M. H. (2009). Antioxidant and tyrosinase inhibitory activity of mango seed kernel by product. *Food Chemistry, 117*, 332–341.

Morton, J. (1987). Mango. In J. F. Morton (Ed.), *Fruits of Warm Climates* (pp. 221–239). Miami, FL: Purdue University.

Parmar, S. S., & Sharma, R. S. (1990). Effect of mango (*Mangifera indica* L.) seed kernels pre-extract on the oxidative stability of Ghee. *Food Chemistry, 35*, 99–107.

Puravankara, D., Bohgra, V., & Sharma, R. S. (2000). Effect of antioxidant principles isolated from mango (*Mangifera indica* L.) seed kernels on oxidative stability of buffalo ghee (butter-fat). *Journal of the Science of Food and Agriculture, 80*, 522–526.

Rukmini, C., & Vijayaraghavan, M. (1984). Nutritional and toxicological evaluation of mango kernel oil. *Journal of the American Oil Chemists Society, 61*, 789–792.

Solís-Fuentes, J. A., & Durán-de-Bazúa, M. C. (2003). Characterization of eutectic mixtures of different natural fat blends by thermal analysis. *European Journal of Lipid Science and Technology, 105*, 742–748.

Solís-Fuentes, J. A., & Durán-de-Bazúa, M. C. (2004). Mango seed uses: thermal behaviour of mango seed almond fat and its mixtures with cocoa butter. *Bioresource Technology, 92*, 71–78.

Solís-Fuentes, J. A., Hernández-Medel, M. R., & Durán-de-Bazúa, M. C. (2005). Determination of predominant polymorphic form of mango (*Mangifera indica*) almond fat by differential scanning calorimetry and X-ray diffraction. *European Journal of Lipid Science and Technology, 107*, 395–401.

747

Talbot, G. (1995). Fat eutectics and crystallisation. In S. T. Beckett (Ed.), *Physico-chemical aspects of food processing* (pp. 142–166). London, UK: Blackie Academic and Professional.

Van Pee, W. M., Boni, L. A., Foma, M. N., & Hendrikx, A. (1981). Fatty acid composition and tropical fruit of the kernel fat of different mango (*Mangifera indica*) varieties. *Journal of the Science of Food and Agriculture, 32,* 485–488.

Vasanthaiah, H. K. N., Ravishankar, K. V., & Mukunda, G. K. (2007). Mango. In C. Kole (Ed.), *Genome mapping and molecular breeding in plants, Vol. 4* (pp. 303–323). Berlin, Germany: Springer-Verlag.

Yella Reddy, S., & Jeyarani, T. (2001). *Trans*-free bakery shortenings from Mango kernel and Mahua fats by fractionation and blending. *Journal of the American Oil Chemists Society, 78,* 635–640.

CHAPTER 89

Antioxidant Potentials and Pharmacological Activities of Marking Nut (*Semecarpus anacardium* L.f.)

Dilipkumar Pal
College of Pharmacy, Institute of Foreign Trade & Management, Lodhipur Rajput, Moradabad, India

749

LIST OF ABBREVIATIONS

ABP, androgen binding protein
Ach, acetylcholine
Ach E, acetylcholine esterase
AD, Alzheimer's disease
CAM, complementary and alternative medicine
GOT, glutamate oxaloacetate transaminase
GPT, glutamate pyruvate transaminase
LDH, lactate dehydrogenase
LH, luteinizing hormone
NSAIDS, non-steroidal anti-inflammatory drugs
PFC, plaque-forming cells
ROS, reactive oxygen species
SA, *Semecarpus anacardium*
SAE, *S. anacardium* extract
SDH, succinate dehydrogenase

Nuts & Seeds in Health and Disease Prevention. DOI: 10.1016/B978-0-12-375688-6.10089-1

INTRODUCTION

Free radicals were of major interest to early physicists, radiologists, and scientists, and much later were found to be a product of normal metabolism. Although oxygen is essential for aerobic forms of life, oxygen metabolites are highly toxic in nature. As a consequence, reactive oxygen species (ROS) are known to be implicated in many cellular disorders and abnormalities and found in the development of many diseases, including cardiovascular diseases, atherosclerosis, cataracts, chronic inflammation, carcinoma, neurodegenerative diseases, and immature aging. Antioxidants, which can inhibit, prevent, or delay the oxidation of an oxidizable substrate in a chain reaction, therefore appear to be very important in the prevention and control of many diseases. However, because of the toxic and carcinogenic effects of synthetic antioxidants, their use is limited and restricted. Therefore, interest in finding natural antioxidants, without undesirable side effects or with lesser adverse reactions, has increased greatly (Pal & Dutta, 2006; Pal and Nimse, 2006; Pal *et al.*, 2008a, 2008b; Zhang *et al.*, 2009).

The number of antioxidant compounds synthesized by plants as secondary products, mainly phenolics, flavonoids, ascorbic acids, vitamins, etc., serve in plant defense mechanisms to counteract ROS (Havsteen, 2002). To find new natural sources of antioxidants, we have reviewed the antioxidant potential of the marking nut, *S. anacardium* L.f. (SA), and its pharmacological activities.

BOTANICAL DESCRIPTION

Semecarpus anacardium Linn.f. has a number of synonyms, including Bhela, Bhelatuki (Hindi and Beng); Balia (Oriya), and marking nut tree or Oriental cashew (English). It is a moderate-sized deciduous tree that grows to a maximum height of 12–15 m and a girth of 1.25 m. The bark is rough, and dark brown in color. Leaves are 18–60 × 10–30 cm, obovate–oblong, rounded at the apex, coreaceous glabrous on the upper surface, ashy gray or buff and more or less pubescent beneath, and with cartilaginous margins. The base of the leaf is rounded, cordate or cuneate, sometimes shortly auricled; there are 12–25 pairs of main nerves making a wide angle with the coasts; the petioles are 1.2–3.8 cm long. The flowers are greenish white, subsessile, fascicled in pubescent panicles, with short pedicles and lanceolate bracts, pilose, with calyse segments about 1 mm long. Petals are 4–5 mm long by 2 mm broad, ovate, and acute. The ovary in male flowers is rudimentary and hairy; in female flowers it is subglobose, densely pilose, and crowned with three styles. Drupes are 2.5 cm long, obliquely ovoid or oblong, smooth and shining, and black when ripe; they are seated on a fleshy receptacle about 2 cm long, which is smooth and yellow when ripe (Figures 89.1–89.3). The fruit has an acrid scent, and is hot and sweetish. Its active principles include anacardic acid, cardol, catechol, anacardoside, fixed oil, semecarpol, bhilawanol, biflavonoids, biflavones, etc. (Figure 89.4; Chopra *et al.*, 1992; Kirtikar and Basu 2000; Majumder *et al.*, 2008).

HISTORICAL CULTIVATION AND USAGE

The tree is not cultivated, but is common in forests, often found existing with Sal. It is distributed in the sub-Himalayan tract from the Bias eastwards, ascending in the outer hills to a height of 1100 m; in Assam, the Khasia Hills, Chittagang, North Orissa, Central India and the western Peninsula, the Eastern Archipelago, and N. Australia.

The tree seeds appear at an early stage. They have poor viability, and should be sown soon after collection. Seedlings are frost-sensitive, but have good powers of recovery (*Wealth of India*, 1999).

Telinga physicians use Bhallataka (SA) as a specific agent in different kinds of venereal afflictions (Roxburgh). A brown gum that exudes from the bark is utilized by Hindus in

FIGURE 89.1

Marking nut (SA) tree — fruiting branch and leaf. *Source: Majumdar* et al. *(2008). Medicinal potentials of* Semecarpus anacardium *nut — a review. J. Herbs Med. Toxicol., 2,* 9—13. Reproduced with permission.

scrofulous and leprous afflictions. Oil from the nuts acts as a vesicant in rheumatism and sprains. The bruised nut is applied to the os uteri by native women to produce abortion. The ashes of the plant are prescribed in combination with other drugs in snake bite; the nut is used similarly for scorpion sting (Kitrikar & Basu 2000).

The smoke obtained from the burning pericarp is good for tumors. The nut oil is used as an aphrodisiac and to blacken the hair; it is also good for leukoderma, and epilepsy and other

FIGURE 89.2

S. anacardium with flowers and fruits.

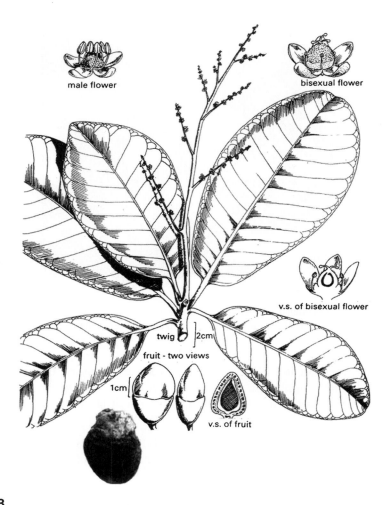

FIGURE 89.3

Vertical section of different parts of *S. anacardium*. *Source: Arya Vaidya Sala, Kottakal, Indian Medicinal Plants, Vol. 5, p. 99. Chennai, India: Orient Longman.* Reproduced with permission.

nervous diseases. It lessens inflammation, and is useful in paralysis, for superficial pain, and for ulcers. Detoxified nuts of SA have been incorporated in prescriptions for toxic conditions, skin diseases, malignant growth, fever, menorrhagia, hemoptysis, and intestinal parasites (Kirtikar & Basu 2000).

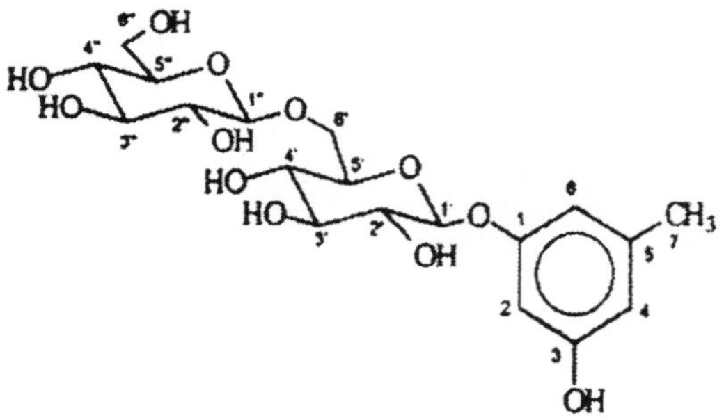

FIGURE 89.4

Anacardoside — active principle of *S. anacardium*. *Source: Majumdar et al. (2008). Medicinal potentials of Semecarpus anacardium nut — a review. J. Herbs Med. Toxicol., 2, 9—13.* Reproduced with permission.

PRESENT-DAY CULTIVATION AND USAGE

SA is used nowadays as a constituent in drugs in many alternative medicines. Ayurvedic preparations that have Bhallataka (SA) as an ingredient include Narasimha Ghrita (Ashtaanga Hridaya), Bhallataka Vati (Bhaishaya Ratnaavali), which is used as a blood purifier, and hematinic tonics. Angaruya-e-kabir is prescribed for neurological affections (Majumder *et al.*, 2008).

Nowadays, the resinous juice is applied to cracked heels. It is highly valued for the treatment of malignant tumors, adjuvant arthritis, leukoderma, hemorrhoids, and ulcers. Bruised nuts are given as a vermifuge, and applied to the os uteri, to cause abortion. Nut oil is used externally in rheumatism and leprous nodules. Gum from the bark is used in scrofulous, venereal, and leprous afflictions, and in nervous debility. Ashes of the plant are used, mixed with other drugs, in snake bite and scorpion stings. The juice of the pericarp, mixed with lime, forms an indelible ink; the juice is a component of some marking inks sold and used in Europe. It is also used for dyeing cloth (Chopra *et al.*, 1992; *Wealth of India*, 1999).

APPLICATIONS TO HEALTH PROMOTION AND DISEASE PREVENTION

Different SA nut extract preparations are effective against many diseases, such as arthritis, tumors, infections, etc., and they are non-toxic even at a high dose of 2000 mg/kg. Although many studies have been carried out, the detailed pharmacology of SA is not yet clearly understood. A few of the reported and established activities are discussed below.

1. *Antioxidant activity*. It is of timely interest to search for new antioxidants from plant sources, and attention has been directed towards the bioactivity of flavonoids, ascorbic acid, and polyphenols as dietary sources of antioxidants (Pal & Dutta, 2006; Pal & Nimse, 2006). The *in vitro* antioxidant activity of SA was investigated using different extracts of SA nuts. It was found that petroleum ether and ethanol extracts have more potent activity than other extracts. The antioxidant activity was concentration-dependent (Table 89.1; Figure 89.5). The detailed chemical nature of the active principles responsible for such activity is not known; however, preliminary phytochemical screening has confirmed the presence of flavonoids, ascorbic acid, and tannins, which might be responsible for antioxidant activity (Mohanta *et al.*, 2007; Pal *et al.*, 2008b).

2. *Anticancer activity*. In traditional medicine, SA nut is highly valued in the treatment of tumors and malignant growth. Studies have been performed to prove the anticancer

753

TABLE 89.1 Antioxidant Activity of Different Extracts of *S. anacardium* (SA) Nuts Using Non-enzymatic Hemoglycosylation Method		
Samples	Final Concentration of the Tested Compound (mg/ml)	
	0.5	1.0
PE	29.20 ± 0.87	56.25 ± 1.04
CE	22.75 ± 0.74	49.50 ± 1.13
EA	25.60 ± 0.90	57.29 ± 0.95
EE	29.10 ± 0.94	34.38 ± 0.79
AE	18.0 ± 0.68	17.2 ± 0.66
D-α-tocopherol	11.0 ± 0.52	43.75 ± 1.19
Ascorbic acid	4.8 ± 0.19	7.3 ± 0.31

Percent inhibition of hemoglobin glycosylation was measured at two concentrations of petroleum ether extract (PE), chloroform extract (CE), ethyl acetate extract (EA), ethanol extract (EE), and aqueous extract (AE). The activities were compared with those of D-α-tocopherol and ascorbic acid. Values are mean ± SD (n = 3).
Source data: Pal *et al.* (2008b), J. Nat. Remed., 8, 162.

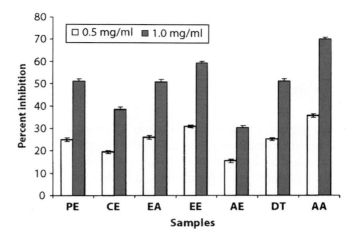

FIGURE 89.5
DPPH free radical scavenging effects of different extracts of *S. anacardium* nuts. Percent inhibition of DPPH free radicals was measured at two concentrations of petroleum ether extract (PE), chloroform extract (CE), ethyl acetate extract (EA), ethanol extract (EE), and aqueous extract (AE). The activities were compared with those of D-α-tocopherol and ascorbic acid. Values are mean ± SD (n = 3). *Source: Pal et al. (2008b). A study on the antioxidant activity of* Semecarpus anacardium. *L.f. nuts. J. Nat. Rem., 8, 160–163.* Reproduced with permission.

and hepatoprotective activities of SA milk extract against alfatoxin B1 (AFB1)-induced hepatocarcinoma in rats, and established its protective role on deranged cell membranes. A detailed analysis regarding its effects on biochemical abnormalities during cancer shows that SA modulates the abnormalities of all biochemical pathways, including carbohydrate, lipid, cytochrome P-450 mediated microsomal drug metabolism, cancer markers, and membrane proteins, during cancer progression. Recently, promising results have also been found in the treatment of cancers of esophagus, urinary bladder and liver, and of leukemia, with an Ayurveda marking nut preparation. The dramatic reduction in α-feroprotein level, a specific marker of hepatocellular carcinoma, and histopathological studies have also assured its anticancer efficacy (Majumder *et al.*, 2008).

3. *Anti-inflammatory and immunomodulatory activity.* SA has significant anti-inflammatory and immunomodulatory activities. Phenolic compounds such as semicarpol and bhilawanol present in nuts inhibit the acute tuberculin reaction in sensitized rats, and also the primary phase of adjuvant arthritis. Again, in rheumatoid arthritis some flavonoids present in the drug are related to their anti-inflammatory activities. The probable mechanism of action is due to the inhibition of release of early mediators (histamine and serotonin) and in phase by inhibition of cyclo-oxygenase. Furthermore, SA has the capacity to inhibit monocyte infiltration and fibroblast proliferation. It has an immunomodulatory effect in inflammatory conditions. Its effect is observed on plaque-forming cells (PFC) and antibody titer in arthritis. These two parameters, which were increased in arthritic animals, reverted significantly ($P < 0.005$) on administration of SA (Ramprasath *et al.*, 2006). NO radicals, which are highly fat-soluble, are greatly amplified in inflammation. SA causes a significant reduction in the nitrate/nitrite level, which may be attributed to its antioxidant potential (Majumder *et al.*, 2008).

4. *Neuroprotective activity:* It is found that SA has neuroprotective activity especially to the hippocampus region in stress induced neurodegeneration like Alzheimer's disease (AD). Saline cognitive decline in AD is generally found in dysfunction of cholinergic neurotransmission in the brain. Loss of cholinergic cells, particularly in the basal forebrain is associated with the loss of neurotransmitter Ach. One of the most accepted

strategies in AD treatment is the therapy of cholinesterase (AchE) inhibitors. SA is effective to prolong the half life of acetylcholine through inhibition of AchE. Hence, SA is known to be useful for treatment of cognitive decline, improvement of memory or related CNS activities (Vinutha *et al.*, 2007).

5. *Antispermatogenic effect.* Oral administration of SA extract caused antispermatogenic activity, evidenced by the reduction in number of spermatogenic cells and spermatozoa. Reduction in sperm density in the caudal epididymis may be due to a disorder in androgen metabolism. Impaired sperm maturation is found by the alteration in the secretion and function of proteins, which are synthesized by principal cells of the epididymis. Consumption of SA fruit extract caused alteration in Leydig cell function, evidenced by reduced Leydig cell area and nuclear dimensions, and fewer mature Leydig cells. Atrophic Leydig cells have been found in the testes of SA-treated animals; this may be due to a decline in LH secretion. The reduction in the number of secondary spermatocytes and spermatids suggested the non-availability of androgen binding protein (ABP) from Sertoli cells, which control gonadotropin and testosterone. The blood parameters remained within the normal range after SA treatment (Sharma *et al.*, 2003).

6. *Anti-atherogenic effect.* The main cause of development of atherosclerosis is an imbalance between the peroxidants and antioxidants. To counteract such a condition, antioxidant therapy is beneficial. SA, having good antioxidant properties, has the capacity to scavenge the superoxide and hydroxyl radicals at low concentrations. The process of atherogenesis is initiated by peroxidation of lipids in low density lipoproteins, which was also prevented by SA. It is possible that SA's beneficial anti-atherogenic effect may be connected with its antioxidant, anticoagulant, hypolipidemic, platelet anti-aggregation, and lipoprotein lipase releasing activities. It has also been shown that the hypotriglyceridemic effect of SA may be partly due to stimulation of lipoprotein lipase activity (Majumder *et al.*, 2008).

7. *Hypoglycemic and antiglycemic effect.* The effects of ethanol extract of dried SA nuts (EESA) on blood sugar level was investigated in both normal (hypoglycemic) and steptozotacin- and alloxan-induced diabetic (antihyperglycemic) rats. The blood sugar level was estimated at 0, 1, 2, and 3 hours after treatment. It was observed that 3 hours after oral administration of the extract at a dose of 100 mg/kg, the blood glucose of normal animals was reduced from 84 to 67 mg/dl ($P < 0.05$). Similarly, in alloxan-induced diabetic rats, it significantly lowered the blood sugar level from 325 to 144 mg/dl ($P < 0.05$). The exact mechanism of such activity is not clear, but the flavonoid-containing constituents may be responsible for lowering blood sugar levels (Kothai *et al.*, 2005).

8. *Antibacterial and fungistatic activity.* Chloroform extract of SA at a dose of 100 mg/kg showed considerable antibacterial activity against *V. cholerae*, *P. aeruginosa*, and *Streptococcus aureus*. The alcoholic extract of SA showed *in vitro* antifungal activity against *A. fumigatus* and *C. albicans*, in a dose-dependent manner. At an extract concentration of 400 mg/ml, growth of both fungi was totally inhibited, and considerable reduction in size of cells and hyphae was noted. The flavonoid present in SA produces antifungal activity (Sharma *et al.*, 2002; Mohanta *et al.*, 2007).

9. *Anthelmintic activity.* Petroleum and chloroform extracts of SA at a concentration of 10 mg/ml showed significant anthelmintic activity. The activity was concentration dependent (Pal *et al.*, 2008c). Anacardic acid is lethal to earthworms.

10. *Hypocholesterolaemic activity.* Administration of SA nut shell extract to cholesterol-fed rabbits caused a significant reduction in serum cholesterol levels and serum LDL cholesterol. It also prevented the deposition of cholesterol/ triglycerides in the liver, heart muscle, and aorta (Sharma *et al.*, 1995).

11. *Analgesic and antipyretic activity.* The analgesic effect of SA at dose of 150 mg/kg was observed in rats, possibly acting through peripheral and central pathways. It also exerted strong antipyretic effects, and supports the view that nut extract has some influence on

prostaglandin biosynthesis, because prostaglandin is believed to be a regulator of body temperature (Ramprasath *et al.*, 2006). Although gastric discomfort and ulcers are common major side effects of currently used NSAIDS, no significant ulceration was found in the SA-treated animals.

In alternative medicine, medicinal plant preparations have found widespread use, particularly in the case of diseases not amenable to treatment by modern methods. Chemical and phytochemical analyses of SA nut reveal the presence of biflavonoids, phenolic compounds, bhilawanols, anacardoside, minerals, vitamins and amino acids. A variety of SA nut extract preparations are effective against variety of diseases, and are relatively non-toxic in nature. However, understanding of the mechanism of action of SA nut would be greatly aided by the isolation of its active constituents and determination of structure–function relationships. Moreover, the potent curative effects of SA nut extract against human ailments needs to be verified by controlled clinical studies.

ADVERSE EFFECTS AND REACTIONS (ALLERGIES AND TOXICITIES)

A toxicity study of SA was conducted in albino rats at acute and subchronic levels. SAE from ripe nuts was administered orally at sub-lethal doses (250 mg, 500 mg, and 750 mg/kg body weight) for 7 days, and two sub-lethal doses (83.3 mg and 166.6 mg/kg body weight) were administered for 21 days. Acute and subchronic studies showed adverse effects on the glutamate oxaloacetate transaminase, glutamate pyruvate transaminase, lactate dehydrogenase, and succinate dehydrogenase activities of the liver, indicating liver disorders. Histopathological studies confirmed these findings (Choudhary & Deshmukh, 2004).

Toxic symptoms of overdose with SA include high-colored, scanty urine, sometimes tinged with blood; irritable and loose bowels with griping pain; and an erythomatous skin eruption with itching and burning. Anacardic acids are allergic in nature. The maximum tolerated dose of a 50% alcoholic extract of SA in mice was found to be 250 mg/kg body weight, i.p. (Nadkarni 1993; Kirtikar & Basu 2000).

Restrictions when taking preparations of SA include avoiding walking in the sun, excessive sexual intercourse, and indulgence in nitrogenous foods, salt, and water. Plenty of ghee, milk, starchy and saccharine foods should be taken. Winter is the best season for the use of SA (Nadkarni 1993).

SUMMARY POINTS

- Since time immemorial, nuts and seeds have been playing an important role in health care and the prevention of disease.
- Marking nut (*Semecarpus anacardium*, SA) preparations have found widespread use in alternative medicine, particularly in the case of diseases not amenable to treatment by modern methods. Different parts of SA have been used classically in the Ayurveda, Siddha, and Unani systems of medicine.
- *S. anacardium* nut has good antioxidant properties, and a variety of nut-extract preparations have many significant therapeutic activities, including anticancer, anti-inflammatory, neuroprotective, anti-atherogenic, hypoglycemic and antiglycemic, antibacterial and fungistatic, and anthelmintic effects. Nuts are bruised and applied to the os uteri to procure abortion, and also given as a vermifuge.
- The detailed chemical nature of the active principle(s) responsible for such activities of SA is not known. Chemical and phytochemical analyses of SA reveal the presence of biflavonoids, phenolic compounds, bhilawanols, anacardoside, minerals, vitamins, and amino acids.
- The medicinal potentiality of SA needs to be established and verified by the controlled clinical studies.

References

Chopra, R. N., Nayer, S. L., & Chopra, I. C. (1992). *Glossary of Indian medicinal plants*. New Delhi, India: Publication and Information Directorate, CSIR. p. 225.

Choudhary, C. V., & Deshmuku, P. B. (2004). Toxicology study of *Semecarpus anacardium* on histology of liver and some of its enzymes in albino rat. *Abstracts of Articles Presented at the 23rd Annual Conference of STOX, 12*, 29–45.

Havsteen, B. H. (2002). The biochemistry and medical significance of the flavonoids. *Pharmacology & Therapeutics, 96*, 67–202.

Kirtikar, K. R., & Basu, B. D. (2000). *Semecarpus anacardium*. In K. S. Mhaskar, & J. F. Caius (Eds.), *Indian medicinal plants, Vol. 8* (pp. 939–941). Delhi, India: Sri Satguru Publications.

Kothai, R., Arul, B., Kumar, K. S., & Christina, A. J. (2005). Medicinal potentials of *Semecarpus anacardium* nut – a review. *Journal of Herbal Pharmacotherapy, 5*, 49–56.

Majumdar, S. H., Chakrabarty, G. S., & Kulkarni, K. S. (2008). Medicinal potentials of *Semecarpus anacardium* nut – a review. *Journal of Herbal Medicine Toxicology, 2*, 9–13.

Mohanta, T. K., Patra, J. K., Rath, S. K., Pal, D. K., & Thatoi, H. N. (2007). Evaluation of antimicrobial activity and phytochemical screening of oils and nuts of *Semecarpus anacardium* L.f. *Scientific Research and Essays, 2*, 486–490.

Nadkarni, A. K. (1993). *Indian Materia Medica, Vol. 1*. Mumbai, India: Popular Prakashan. pp. 1119–1125.

Pal, D. K., & Dutta, S. (2006). Evaluation of the antioxidant activity of the roots and rhizomes of *Cyperus rotundus* L. *Indian Journal of Pharmaceutical Sciences, 68*, 256–258.

Pal, D. K., & Nimse, S. B. (2006). Screening of the antioxidant activity of *Hydrilla verticillata* plant. *Asian Journal of Chemistry, 18*, 3004–3008.

Pal, D. K., Kumar, M., Chakrabarty, P., & Kumar, S. (2008a). Evaluation of the antioxidant activity of aerial parts of *Cynodon dactylon*. *Asian Journal of Chemistry, 20*, 2479–2481.

Pal, D. K., Kumar, S., Chakrabarty, P., & Kumar, M. (2008b). A study on the antioxidant activity of *Semecarpus anacardium*. L.f. nuts. *Journal of Natural Remedies, 8*, 160–163.

Pal, D. K., Mahapatra, T. K., & Das, A. (2008c). Evaluation of anthelmintic activity of nuts of *Semecarpus anacardium*. *Ancient Science of Life, 17*, 41–44.

Ramprasath, V. R., Shanthil, P., & Sachdanandam, P. (2006). Immunomodulatory and anti-inflammatory effects of *Semecarpus anacardium* Linn. milk extract in experimental inflammatory conditions. *Biological & Pharmaceutical Bulletin, 29*, 693–700.

Sharma, K., Mathur, R., & Dixit, V. P. (1995). Hypocholesterolemic activity of nut shell extract of *Semecarpus anacardium* (Bhilawa) in cholesterol feed rabbits. *Indian Journal of Experimental Biology, 33*, 444–448.

Sharma, K., Shukla, S. D., Mehta, P., & Bhatnagar, M. (2002). Fungistatic activity of *Semecarpus anacardium* Linn. *Indian Journal of Experimental Biology, 40*, 314–318.

Sharma, A., Verma, P. K., & Dixit, V. P. (2003). Effect of *Semecarpus anacardium* fruits on reproductive functions of male albino rats. *Asian Journal of Andrology, 5*, 121–124.

Vinutha, B., Prashanth, D., Salma, K., & Deepak, M. (2007). Medicinal potentials of *Semecarpus anacardium* nut – a review. *Journal of Ethnopharmacology, 109*, 359–363.

Wealth of India. (1999). In Y. R. Chadha (Ed.), *Wealth of India: A Dictionary of Indian Raw Material and Industrial Products, Vol. 2B* (pp. 271–273). New Delhi, India: Publication and Information Directorate, CSIR.

Zhang, W. M., Li, B., Han, L., & Zhang, H. D. (2009). Antioxidant activities of extracts from areca (*Areca catechu* L.) flower, husk, and seed. *African Journal of Biotechnology, 8*, 3887–3892.

CHAPTER 90

Milk Thistle (*Silybum marianum* L. Gaert.) Seeds in Health

Sanjib Bhattacharya
Bengal School of Technology (A College of Pharmacy), Sugandha, Hooghly, West Bengal, India

759

LIST OF ABBREVIATIONS

DNA, deoxyribonucleic acid
Hep3B, hepatitis B virus positive p53 matured
HepG2, Hepatitis B virus negative p53 intact

INTRODUCTION

Milk thistle (*Silybum marianum* L. Gaert., Asteraceae) seeds have been used for centuries as a herbal medicine, mainly for the treatment of liver diseases. The common name, milk thistle, is derived from the milky-white veins on the leaves, which, when broken open, yield a milky sap. The active constituents of milk thistle seeds are three isomeric flavonolignans, namely

Nuts & Seeds in Health and Disease Prevention. DOI: 10.1016/B978-0-12-375688-6.10090-8

silibinin (silybin), silychristin, and silidianin, collectively known as silymarin, which is extracted from the dried milk thistle seeds. Silibinin is the most biologically active. The seeds also contain other flavonolignans, betaine, apigenin, silybonol, proteins, fixed oil, and free fatty acids, which may contribute to the health-giving effects of milk thistle seeds (DerMarderosian, 2001; Evans, 2002).

BOTANICAL DESCRIPTION

Milk thistle is an erect, stout, annual or biennial plant, that grows to 1.5—3 m in height. It has large prickly leaves, large purple flowering heads, and strongly spinescent stems (Figure 90.1). When broken, the leaves and stems exude a milky sap. The glabrous leaves are dark green, oblong, sinuate-lobed or pinnatified, with spiny margins. The leaves have milk-white veins that give the leaves, which initially form a flat rosette, a diffusely mottled appearance. During the flowering season, from June to September, each stem bears a terminal head containing a single large, purple, slightly fragrant flower ridged with sharp spines. The achenes, 6—7 mm in length and transversely wrinkled, are dark in color, gray-flecked with a yellow ring at the apex. Attached to the achene is a long white pappus. The fruits are glossy brown or gray, with spots (Evans, 2002).

HISTORICAL CULTIVATION AND USAGE

Milk thistle was once cultivated in Europe as a vegetable. The de-spined leaves were used in salads, and similarly to spinach. The stalk, root, and flowers were also consumed, and the roasted seeds were used as a coffee substitute. Preparations of milk thistle seeds have been used medicinally since as early as 4 BC, and were first reported by Theophrastus. Traditionally, the seeds have been used in Europe as a galactogogue in nursing mothers, as a bitter tonic, and as antidepressant; and in liver complications (including gallstones), dyspepsia, splenic

FIGURE 90.1
Milk thistle (*Silybum marianum*) flowering head

congestion, varicose veins, diabetes, amenorrhea, uterine hemorrhage, and menstrual problems. Its use as a liver protectant can be traced back to Greek and Roman references in 1 AD (DerMarderosian, 2001; Barceloux, 2008).

PRESENT-DAY CULTIVATION AND USAGE

Milk thistle is indigenous to Kashmir, Southern Europe, Southern Russia, North Africa, and Asia Minor. It was introduced to most areas of Europe, North and South America, and Southern Australia, and is cultivated mainly in the dry rocky soils of European countries, Australia, Canada, China, North and South America as a medicinal plant. It is also grown as ornamental plant for its attractive foliage. The seeds are collected when ripe, during late summer. Presently, milk thistle seed, its purified extracts and its active constituents are mainly used in liver diseases. It is the most widely used hepatoprotective agent in chronic inflammatory hepatic disorders, including hepatitis, jaundice, alcohol abuse, fibrosis, cirrhosis, and fatty infiltration; and in hepatotoxicity by mushroom poisoning and by industrial pollutants. It is also widely used as nutraceutical agent. In homoeopathy, the seed tincture is used in liver disorders, jaundice, gall stones, peritonitis, hemorrhage, bronchitis, and varicose veins. Extracts, tablets, or capsules containing standardized extract of milk thistle seeds are available commercially (DerMarderosian, 2001; Barceloux, 2008).

APPLICATIONS TO HEALTH PROMOTION AND DISEASE PREVENTION

The seeds of milk thistle can be consumed raw (usually freshly milled), made into a tea, or used as a hydro-alcoholic extract for medicinal use. Silymarin is included in the pharmacopoeia of many countries. The average adult dose of powdered seed is 12–15 g/day; of dry standardized seed extract, (silymarin), 200–400 mg/day; and of liquid seed extract, 4–9 ml/day. Silymarin is very poorly soluble in water, so milk thistle seed is not very effective in the form of tea. Extracts from the seed are generally marketed as tablet and encapsulated forms for oral use, usually containing a concentrated seed extract standardized to 70–80% of silymarin. Silymarin is also administered by the parenteral route (DerMarderosian, 2001). The effects of silymarin (the standardized extract from milk thistle seed) are discussed below.

Milk thistle seeds' therapeutic and health-promoting efficacy involves a variety of molecular mechanisms. Its primary activities are of use as an antioxidant and a hepatoprotective.

Antioxidant activity

Silymarin has been reported to act as an excellent antioxidant, scavenging free radicals (reactive oxygen species) and inhibiting lipid peroxidation, thereby protecting cells against oxidative stress. It augments the non-enzymatic and enzymatic antioxidant defense systems of cells involving reduced glutathione, superoxide dismutase, and catalase. It can protect the liver, brain, heart, and other vital organs from oxidative damage by its ability to prevent lipid peroxidation, and by replenishing the reduced glutathione levels. Silibinin exhibits membrane-protective properties, and it may protect blood constituents from oxidative damage (Das *et al.*, 2008; Kshirsagar *et al.*, 2009).

Hepatoprotective activity

Use of milk thistle seeds as a liver protectant dates back to 1 AD. Antioxidant activity is one of the important factors in hepatoprotection.

1. *Antihepatotoxic potential.* Silymarin protects liver cells against many hepatotoxins in humans and animals. Some mushrooms (e.g., *Amanita phalloides*, the death cup fungus, and *A. virosa*) contain two toxins, phalloidine and α-amanatine, which destroy the hepatocyte cell membrane and block hepatic protein synthesis, leading to severe liver damage and death. Silymarin effectively prevents both of these effects by blocking the toxin's binding sites, and increasing the regenerative capacity of liver cells. Silibinin was found to be an effective measure against liver damage if it is administered intravenously within 24 hours after mushroom ingestion. In one study, 60 patients with severe *Amanitia* poisoning were treated with infusions of 20 mg/kg of silibinin with excellent results, showing no deaths among the patients treated. Silymarin is often used as supportive therapy in food poisoning due to fungi (Desplaces *et al.*, 1975; Vogel *et al.*, 1984).

Silymarin also offers liver protection against tetracycline-, D-galactosamine-, and thallium-induced liver damage, and erythromycin estolate, amitryptiline, nortryptiline and tert-butyl hydroperoxide exposure of neonatal hepatocytes. It reduces liver damage due to long-term treatment with phenothiazine or bytyrophenone. Silibinin significantly inhibits concanavalin A-induced liver disease. It also provides heparoprotection against poisoning by phalloidin, halothane, thioacetamide, acetaminophen, and carbon tetrachloride, and protects the liver from ischemic injury, iron overload, and radiation (DerMarderosian, 2001; Kshirsagar *et al.*, 2009).

Silymarin is used for the treatment of several liver diseases characterized by degenerative necrosis and functional impairment, including chronic liver disorders. The German Commission E endorses use of silymarin for the treatment of liver diseases, including hepatitis A, alcoholic cirrhosis, and chemically-induced hepatitis (Barceloux, 2008).

2. *Alcoholic liver disease/cirrhosis.* Ethanol metabolism involves the formation of free radicals, leading to oxidative stress in the liver. Silymarin successfully opposes alcoholic cirrhosis, with its antioxidant and hepatoprotective mechanisms restoring the normal liver biochemical parameters. Silymarin also ameliorates cytolysis in active cirrhosis patients. However, use of silymarin is inadvisable in decompensated cirrhosis (Ferenci *et al.*, 1989).

3. *Hepatitis.* In patients with acute viral hepatitis, silymarin shortens the treatment time and shows improvement in serum bilirubin and serum liver enzymatic levels. Biochemical values are restored to normal sooner in silymarin-treated patients. In chronic active hepatitis, silymarin treatment improves liver function tests. Histological improvement is observed in silymarin-treated patients with chronic hepatitis. Silymarin causes stable remission of alcoholic hepatitis, normalizing the liver biochemical parameters (Magliulo *et al.*, 1978).

4. *Liver fibrosis.* Liver fibrosis can result in remodeling of the liver architecture, leading to hepatic insufficiency, portal hypotension, and hepatic encephalopathy. The conversion of hepatic stellate cells into myofibroblasts is considered a central event in fibrogenesis. Silymarin treatment markedly inhibits this process in liver fibrosis patients, showing antifibrotic potential (Kshirsagar *et al.*, 2009).

5. *Liver tissue regeneration.* Silymarin stimulates liver tissue regeneration by increasing protein synthesis in the injured liver. In *in vivo* and *in vitro* experiments performed in rats, from which part of the organ (liver) was removed, silibinin produces a significant increase in the formation of ribosomes and in DNA synthesis, as well as an increase in protein synthesis. Interestingly, the increase in protein synthesis is induced by silibinin only in injured livers and not in healthy ones (Kshirsagar *et al.*, 2009).

There are several systematic reviews regarding the applications of milk thistle seeds in liver diseases, but most of the human studies done to date are of such variable design, quality, and results — that no definitive conclusions about degrees of effectiveness in the treatment or prevention of alcoholic cirrhosis and hepatitis can yet be made. Better-quality clinical trials are necessary (DerMarderosian, 2001; Jacobs *et al.*, 2002).

Anti-inflammatory activity

Milk thistle seed and its active extract silymarin have anti-inflammatory and anti-arthritic effects due to excellent antioxidant properties, scavenging free radicals that act as pro-inflammatory agents. Silymarin was found to be more effective in cases of developing arthritis compared to developed arthritis. Silymarin and silibinin hinder the inflammatory process by inhibiting neutrophil migration and Kupffer cell inhibition. They also inhibit the formation of inflammatory mediators, especially prostaglandins and leukotrienes (by inhibiting the 5-lipoxigenase pathway), and release of histamine from basophils. Therefore, milk thistle seed may possess anti-allergic and anti-asthmatic activities (Fiebrich & Koch, 1979; Dixit *et al.*, 2009).

Immunomodulatory activity

Silymarin's immunomodulatory activity in liver disease patients may also be involved in its hepatoprotective action. Silymarin has been found to protect experimental rodents from ultraviolet radiation-induced immunosuppression (Meeran *et al.*, 2006). Silibinin inhibits the activation of human T-lymphocytes and human polymorphonuclear leukocytes. Silymarin significantly suppresses the inflammatory mediators, the expression of histocompatibility complex molecules, and nerve cell damage. Long-term administration improves immunity by increasing T-lymphocytes and interleukins, and reducing all types of immunoglobulins. Silymarin could be useful in the development of a therapeutic adjuvant in which immunosuppression is required, including autoimmune and infectious diseases (DerMarderosian, 2001; Das *et al.*, 2008).

Liver lipidemic control

It was found that silymarin and silibinin reduce the synthesis and turnover of phospholipids in the liver. Silibinin neutralizes ethanol-induced inhibition of phospholipid synthesis, and reduction in glycerol incorporation into lipids of isolated hepatocytes. Furthermore, silibinin stimulates phosphatidylcholine synthesis and increases the activity of cholinephosphate cytidyltransferase in rat liver, both in normal conditions and after galactosamine intoxication. Silymarin significantly inhibits hepatic lipid peroxidation, and may diminish triglyceride synthesis in the liver. Impairments in the liver lipid profile caused as a result of prolonged effects of ethanol, antitubercular drugs (isoniazid, rifampacin), and liver toxicants (acetaminophen, halothane, microcystin) are effectively improved by silymarin (Kshirsagar *et al.*, 2009).

Blood (Plasma) lipidemic control

Administration of silymarin to type II hyperlipidemic patients resulted in slightly decreased total cholesterol and high-density lipoprotein levels in blood plasma. Silymarin reduces plasma levels of cholesterol and low-density lipoprotein levels in hyperlipidemic rats, whereas silibinin does not reduce plasma levels of cholesterol in normal rats, although it does reduce total phospholipid levels. Biliary cholesterol and phospholipid concentrations in rats are also slightly reduced. Silymarin-induced reduction of biliary cholesterol and phospholipids in both rats and humans may be in part due to decreased liver cholesterol synthesis. Silymarin could represent a novel agent in the prevention and therapy of hypercholesterolemia and atherosclerosis (Kreeman, 1998; Kshirsagar *et al.*, 2009).

Biliary effect

Silymarin undergoes excessive enterohepatic circulation, which allows a continuous loop between intestine and liver. It prevents the disturbance of bile secretion, thereby increasing bile secretion, and cholate and bilirubin excretion (DerMarderosian, 2001).

Antiviral effect

Although silymarin does not affect viral replication, it has a beneficial role in viral hepatitis through its inhibitory action on inflammatory and cytotoxic processes induced by viral infection. Silibinin strongly inhibits the growth of both HepG2 (hepatitis B virus negative; p53 intact) and Hep3B (hepatitis B virus positive; p53 matured) cells, with relatively more cytotoxicity in Hep3B cells, which is associated with apoptosis induction. Silymarin also showed inhibitory activity against other viruses in different cell lines (Das *et al.*, 2008).

Antitumor and anticarcinogenic effects

Silymarin significantly inhibits tumor growth, and also cause regression of established tumors. It is associated with *in vitro* antiproliferative, pro-apoptopic, and anti-angiogenic efficacy in prostate tumors. Silymarin feeding during the promotion phase of 4-nitroquinoline-1-oxide-induced rat tumorigenesis exerts chemopreventive activity against tongue squamous cell carcinoma. The cancer chemopreventive and anticarcinogenic effects of silymarin in long-term animal tumorigenesis models and in human prostate, breast, and cervical carcinoma cells are also reported. Treatment with silibinin results in a highly significant inhibition of both cell growth and DNA synthesis, with loss of cell viability in the case of cervical carcinoma cells (Das *et al.*, 2008; Dixit *et al.*, 2009).

It is well demonstrated that ultraviolet light-induced immunosuppression and oxidative stress play an important role in the induction of skin cancers. Topical or dietary administration of silymarin to mouse skin prevents photocarcinogenesis by significantly inducing apoptosis, and increase in catalase activity, and induction of cyclo-oxygenase and ornithine decarboxylase activity. Similar results are also obtained in other chemical-induced skin carcinogenesis models. Prevention of ultraviolet light-induced immunosuppression and oxidative stress by silymarin may be associated with the prevention of photocarcinogenesis (Katiyar *et al.*, 1997).

Silibinin significantly induces growth inhibition, a moderate cell cycle arrest, and a strong apoptotic cell death in small-cell and non-small-cell human lung carcinoma cells. Silibinin inhibits the growth of human prostate cancer cells both *in vitro* and *in vivo*. Silymarin and silibinin have strong anti-angiogenesis effects on the colon cancer cell line, and are effective against chemical-induced bladder carcinogenesis in mice, and hepatocellular carcinoma in rats (Das *et al.*, 2008).

Neuroprotective effect

Silymarin was found to be useful in the prevention and treatment of neurodegenerative and neurotoxic processes due to its antioxidant effects. Silymarin can effectively protect dopaminergic neurons against lipopolysaccharide-induced neurotoxicity in brain (Wang *et al.*, 2002).

Cardioprotective effect

During cancer therapy, the use of certain chemotherapeutic agents like doxorubicin is limited by cardiotoxicity, which is known to be mediated by oxidative stress and apoptosis induction. Silibinin has cardioprotective properties due to its antioxidant and membrane-protective actions (Chlopeikova *et al.*, 2004).

Miscellaneous

Silymarin helps to maintain normal renal function, and silibinin reduces oxidative damage to kidney cells *in vitro*. In rats, silibinin prevents cisplatin-induced nephrotoxicity, but does not prevent cyclosporine-induced glomerular damage. As an antioxidant, silymarin can protect the pancreas against certain forms of damage. In a controlled trial of human diabetics treated with silymarin, patients experienced decreases in blood glucose and insulin requirements. It exhibits anti-ulcer activity in rats (DerMarderosian, 2001; Das *et al.*, 2008; Dixit *et al.*, 2009). In one study of post-parturient cattle given milk thistle seed meal, milk production was increased and ketonuria reduced, as compared to controls (Vojtisek *et al.*, 1991).

The value of silymarin in the treatment of psoriasis may be due to its ability to improve endotoxin removal by the liver, inhibition of cyclic adenosine monophosphate phospho-diesterase, and leukotriene synthesis. Abnormally high levels of cyclic adenosine monophosphate and leukotrienes are observed in patients with psoriasis, and normalization of these levels may improve the condition (Fiebrich & Koch, 1979; Kock *et al.*, 1985).

Conclusion

Milk thistle seed shows great promise as a superior herbal drug. Its good safety profile, better standardization and quality control, easy availability, and low cost are added advantages. Further research on milk thistle seed may make a breakthrough as a new approach in disease prevention, in addition to liver complications.

ADVERSE EFFECTS AND REACTIONS (ALLERGIES AND TOXICITY)

Human studies performed with milk thistle seeds indicated little need for concern regarding adverse effects, demonstrating that milk thistle seed extract (silymarin) is safe and well tolerated. It is generally non-toxic, and causes no side effects when administered to adults in a dose range of 200–900 mg/day in two or three divided doses. Higher doses (> 1500 mg/day) could produce minor gastrointestinal disturbances involving a slight laxative effect, which may be due to increased bile secretion and flow. Mild allergic reactions (pruritus, urticaria, arthralgia) are observed, but are rarely severe enough to cause discontinuation. Commonly noted adverse effects, such as bloating, dyspepsia, epigastric pain, flatulence, nausea, irregular stools, and laxation, were observed in 2–10% of patients in a clinical trial. Headaches and dermatological symptoms were also noted (Anonymous, 1999; Barceloux, 2008).

Silymarin was found to be non-toxic in rats and mice after oral doses of 2500 or 5000 mg/kg body weight, producing no unwanted symptoms. Similar reports were also obtained for rabbits and dogs. No evidence of ante- or postnatal toxicity in animals was reported. These data reveal that the acute toxicity of silymarin is very low (Desplaces *et al.*, 1975; Barceloux, 2008).

It was found that silymarin at higher concentrations has an inhibitory effect on both phase I and phase II hepatic drug metabolizing cytochrome enzyme systems. However, the plasma concentrations at therapeutic doses are very much less than that needed for inhibition, so it exhibits no beneficial or harmful drug interactions at normal doses (Barceloux, 2008).

Safety of milk thistle seed in pregnancy and lactation has not been studied in humans. Traditionally considered safe in lactation, no clinical studies have been performed in this respect. Safety in children has also not yet been studied. No known contraindications have been reported (Barceloux, 2008).

765

SUMMARY POINTS

- Milk thistle (*Silybum marianum*) seeds have been used for over 2000 years as a natural remedy for the treatment of several diseases, especially those of the liver, and are still widely used for the same.
- The active constituents of milk thistle seed are three isomeric flavonolignans (silibinin or silybin, silychristin, and silidianin), collectively known as silymarin, which is extracted from the milk thistle seeds and available commercially as a standardized extract.
- Milk thistle seed extract (silymarin) and its constituents (mainly silibinin) are antioxidant and hepatoprotective, and effective in treating toxin poisoning, hepatitis, cirrhosis, and fibrosis of the liver; they also stimulate liver regeneration. However, human studies regarding the management of alcoholic cirrhosis and hepatitis are equivocal.
- Milk thistle seed demonstrates anti-inflammatory, immunomodulatory, lipid, and biliary effects. It also has antiviral, antitumor, and other health-beneficiary properties.
- Milk thistle seed preparations are safe, well tolerated, and cause no serious side effects in humans, although there may be mild gastrointestinal and allergic reactions.
- Milk thistle seed is a very promising natural drug. More research is warranted to validate its wide range of health-promoting effects.

References

Anonymous. (1999). *Silybum marianum* (Milk thistle). *Alternative Medicine Review, 4,* 272–274.

Barceloux, D. G. (2008). *Medical toxicology of natural substances.* Hoboken, NJ: John Wiley & Sons.

Chlopeikova, A., Psotova, J., Miketova, P., & Simanek, V. (2004). Chemopreventive effect of plant phenolics against anthracycline-induced toxicity on rat cardiomyocytes. Part I. Silymarin and its flavonolignans. *Phytotherapy Research, 18,* 107–110.

Das, S. K., Mukherjee, S., & Vasudevan, D. M. (2008). Medicinal properties of milk thistle with special reference to silymarin: an overview. *Natural Product Radiance, 7,* 182–192.

DerMarderosian, A. (2001). *The review of natural products* (1st ed.). St Louis, MO: Facts and Comparisons.

Desplaces, J., Choppin, G., Vogel, G., & Trost, W. (1975). The effects of silymarin on experimental phalloidine poisoning. *Arzneimittelforschung, 25,* 89–96.

Dixit, N., Baboota, S., Kohli, K., Ahmad, S., & Ali, J. (2009). Silymarin: a review of pharmacological aspects and bioavailability enhancement approaches. *Indian Journal of Pharmacology, 39,* 172–179.

Evans, W. C. (2002). *Trease and Evans pharmacognosy* (15th ed.). New Delhi, India: Reed Elsevier India Pvt. Ltd.

Ferenci, P., Dragosics, B., & Dittrich, H. (1989). Randomised controlled trial of silymarin treatment in patients with cirrhosis of the liver. *Journal of Hepatology, 9,* 105–113.

Fiebrich, F., & Koch, H. (1979). Silymarin, an inhibitor of lipoxygenase. *Experentia, 35,* 150–152.

Jacobs, B., Dennehy, C., Ramirez, G., Sapp, J., & Lawrence, V. (2002). Milk thistle for the treatment of liver disease; a systematic review and meta-analysis. *American Journal of Medicine, 113,* 506–515.

Katiyar, S. K., Korman, N. J., Mukhtar, H., & Agarwal, R. (1997). Protective effects of silymarin against photo-carcinogenesis in a mouse skin model. *Journal of National Cancer Institute, 89,* 556–566.

Kock, H. P., Bachner, J., & Loffler, E. (1985). Silymarin: potent inhibitor of cyclic AMP phosphodiesterase. *Methods & Findings in Experimental & Clinical Pharmacology, 7,* 409–413.

Kreeman, V., Skottova, N., & Walterova, D. (1998). Silymarin inhibits the development of diet-induced hyper-cholesterolemia in rats. *Planta Medica, 64,* 138–142.

Kshirsagar, A., Ingawale, D., Ashok, P., & Vyawahare, N. (2009). Silymarin: a comprehensive review. *Pharmacognosy Reviews, 3,* 116–124.

Magliulo, E., Gagliardi, B., & Fiori, G. P. (1978). Results of a double blind study on the effect of silymarin in the treatment of acute viral hepatitis, carried out at two medical centres. *Medizinische Klinik, 73,* 1060–1065.

Meeran, S. M., Katiyar, S., Elmets, C. A., & Katiyar, S. K. (2006). Silymarin inhibits UV radiation-induced immunosuppression through augmentation of interleukin-12 in mice. *Molecular Cancer Therapeutics, 7,* 1660–1668.

Vogel, G., Tuchweber, B., Trost, W., & Mengs, U. (1984). Protection by silibinin against *Amanita phalloides* intoxication in beagles. *Toxicology and Applied Pharmacology, 73,* 355–362.

Vojtisek, B., Hronova, B., Hamrik, J., & Jankova, B. (1991). Milk thistle (*Silybum marianum,* L., Gaertn.) in the feed of ketotic cows. *Veterinární Medicína (Praha), 36,* 321–330.

Wang, M. J., Lin, W. W., Chen, Y. H., Ou, H. C., & Kuo, J. S. (2002). Silymarin protects dopaminergic neurons against lipopolysaccharide-induced neurotoxicity by inhibiting microglia activation. *European Journal of Neuroscience, 16,* 2103–2112.

CHAPTER 91

Extracts of Mobola Plum (*Parinari curatellifolia* Planch ex Benth, Chrysobalanaceae) Seeds and Multiple Therapeutic Activities

Steve Ogbonnia
Department of Pharmacognosy, University of Lagos, Lagos, Nigeria

LIST OF ABBREVIATIONS

ALT, aminoalanine transferase
AST, aspartate amino transferase
HDL, low density lipoprotein
LD_{50}, median lethal dose
LDL, low density lipoprotein
TC, total cholesterol
TG, total triglyceride

INTRODUCTION

For centuries, seeds and nuts have served as invaluable sources of food and nutrients for both humans and wildlife. The discovery of their health potential, in addition to their food

Nuts & Seeds in Health and Disease Prevention. DOI: 10.1016/B978-0-12-375688-6.10091-X

value, as important therapeutic agents against diseases made them *a sine qua non* to life. A seed may be defined as a plant member derived from a matured and fertilized ovule, consisting of a kernel with stored food surrounded by one, two, or three protective seed coats derived from integuments of the ovule (Wallis, 2005). "Nut" is a general term for a large, dry, oily seed or fruit of some plants. Nuts also contain a significant proportion of vitamins, formerly known as "accessory food factors" (Evans, 1989), and secondary metabolites (especially polyphenols) that account for their therapeutic activities. Polyphenols are the most commonly found secondary products in seeds and nuts that form an important part of both human and animal diets. They are also associated with useful therapeutic activities, as well as protective biological functions.

Some seeds and nuts, including *P. curatellifolia* seeds, are used mostly for their medicinal values and may contain other secondary metabolites, such as alkaloids, terpenoids, and glycosides. *P. curatellifolia* seed contains alkaloids and anthraquinone glycosides in addition to polyphenols, and are used either alone or in combination with other herbs in the treatment of diabetes and several other diseases. (Ogbonnia *et al.*, 2009).

In this chapter, the health benefits, adverse effects, and reactions of *P. curatellifolia* seed are outlined. The botanical description and cultivation of the plant, and the chemical constituents of the seeds, are also discussed.

BOTANICAL DESCRIPTION

Parinari curatellifolia (Family Chrysobalancaceae, Order Malphighiales, Class Magnoliopsida; Figures 91.1, 91.2) is very variable in size and shape, ranging from a small shrub of 3 m tall to a large tree of up to 15−20 m tall. It is an evergreen, with pale-green spreading foliage forming a dense roundish to mushroom-shaped crown. The tree has a single bare stem, with dark gray, deeply fissured bark. The wood is light brown to yellowish-red, with tensile strength varying from location to location (Burkill, 1985; Sanogo *et al.*, 2006). The leaves have an entire margin, are simple with a spiral but sometimes alternate arrangement, oval to oblong, 3−17 × 2−6 cm in size, leathery, and dark green on top; they are finely velvety when young, losing the hair later. The underside is gray to yellow, and densely hairy. The leaves often have small galls, and up to 20 pairs of fusing lateral veins. The white flowers are sweetly scented and tinged with pink, 4−6 mm in diameter, with five petals and sepals, and seven or more stamens joined at the base in a short ring inserted in the mouth of the receptacle, with two chambered ovaries. The stalk and calyces are densely covered with yellowish woolly hair.

FIGURE 91.1
Mature *P. curatellifolia* plants. *Reproduced with permission from: www.zimbabweflora.co.zw*

FIGURE 91.2
A young *P. curatellifolia* plant.

The fruit of *P. curatellifolia* (Figure 91.3) is an ovoid or sub-globose drupe, measuring up to 5.0 × 3.5 cm, russet-yellow to grayish, and becoming orange-yellow when ripe. It has a pale yellow to reddish mealy pulp, which is edible and is said to be very palatable, with an edible kernel (Maraji & Glen, 2008). It has rough scaly skin, pitted with golden-colored warts on the surface, and is usually harvested when it becomes yellow-orange. The volatile flavor components have been isolated from the fruits, and a total of 88 components identified, including 2-aminobezaldehyde and phenylacetaldoxime, which were detected for the first time in edible fruits (Joulain *et al.*, 2004). Two unusual nitrate compounds were also identified, including optically active (2-nitrobutyl)-benzene, which is a new natural product. Ent-kaurene diterpenoids have been isolated, and their rearrangement has been found to mediate cell-cycle specific cytotoxicity (Lee *et al.*, 1996).

The seed (Figure 91.4) has a hard and woody endocarp about 1−2 cm in diameter, and contains one or two embryos (kernels). They are obtained by removing the fruit skin and pulp

FIGURE 91.3
Ripe *P. curatellifolia* fruits.

FIGURE 91.4
P. curatellifolia seeds, (A) with coat; (B) without coat.

with a knife, and washing with water. Extraction of large quantities of seeds can be achieved by soaking the fruits for 1–2 days and pounding them with a pestle and mortar, sometimes with coarse sand (Maraji & Glen, 2008). After mixing well with large quantities of water, the fruit skin and pulp are separated, leaving clean seeds. The seeds are difficult to decorticate, and may be used fresh or dried. The results of recent phytochemical screening revealed the presence of alkaloids (Ogbonnia *et al.*, 2009), though alkaloids were initially reported absent in Nigerian species (Burkill, 1985). The seeds also contained free and bound anthraquinones and polyphenols as constituents. Saponins were conspicuously absent.

HISTORICAL CULTIVATION AND USES

The timber characteristics of Nigerian species of *P. curatellifolia* have made the plant a cherished and valuable economic plant, which deserves further examination. The timber is used in the northern and western regions of the country for making mortars and building houses, for firewood, and for conversion into charcoal. Some communities cherish the tree for its economic value, and cultivate it for this reason. A special nursery to raise young seedlings may not be required; house backyards are used initially, and the seedlings may then be transferred to farms or designated planting sites.

PRESENT-DAY CULTIVATION AND USAGE

P. curatellifolia propagation is carried out using both asexual and sexual methods. All the basic farming tools, including a scythe, hoe, digger, shovel, and so on, are needed for both pre-planting and post-planting maintenance (Kadiri, 2008). The species regenerates naturally from seed, coppicing, and suckers. Most of the trees are propagated asexually from the root suckers. The results of asexual reproduction are always like the parents, although there may rarely be mutations in the offspring (Kadiri, 2008). The sexual method of reproduction involves allowing the flower to grow and pollinate, and the ovules to be fertilized to give rise to new seed. *P. curatellifolia* plant parts are used in traditional medicines. The wood is very hard, and especially suited for objects requiring great durability. It is used as pegs, rafters, and beams, in building, mortars and canoes, and as railway sleepers.

P. curatellifolia seeds show prolonged dormancy, and germination may take months. The seeds have hard seed coats and rarely germinate artificially, even after pretreatment. They may be pretreated by boiling or immersion in boiling water or dilute sulfuric acid, to speed up germination (Maraji & Glen, 2008). When boiled for 15 minutes, allowed to cool, and soaked for 24 hours, the seed may still take up to 6 months to germinate. The germination rate is poor, with the best being about 34%, reported when the seed coats are completely removed and the seeds are stored for 60 days before sowing.

Seeds lying on the ground may be infected by parasites; therefore, fresh seeds collected from the plant should be sown in river sand in flat seedling trays, pressed down till they are level with soil surface, and covered with a thin layer of sand. The use of seed, which is the product of sexual reproduction, may cause variations in the quality of the drugs produced from such plants, because of the high incidence of genetic exchange (Kadiri, 2008; Maraji & Glen, 2008). The seedlings are transplanted at the three-leaf stage, and care is taken to avoid damaging the taproot, which is very susceptible.

To ensure that the propagation process is successful, healthy seed or plant parts must be used for propagation, and the plant parts divided or cut with a clean, sharp knife.

APPLICATIONS TO HEALTH PROMOTION AND DISEASE PREVENTION

P. curatellifolia seed is widely used in traditional medicine due to its multiple therapeutic activities. The health and nutritive benefits of the seeds may be attributed to their secondary products and vitamin content. Vitamins constitute a range of different types of organic molecules that are essential for the proper functioning of humans (Evans, 1989). Polyphenols including flavonoids are aromatic hydroxylated compounds, and are among the most potent and therapeutically useful bioactive substances (Siess *et al.*, 1996; Apak *et al.*, 2007). *P. curatellifolia* seed extract is used as bitters and a stomachic for the promotion of good health, by virtue of its antioxidant properties (Fathiazad *et al.*, 2006; Shafaghat & Salimi, 2008). The antioxidant activity of seed extracts has been utilized by traditional herbalists in treating diverse degenerative diseases of the central nervous system (Zafra-Stone *et al.*, 2007; L. Adebisi, personal communication, October 17, 2009). The seed extract also exhibits a wide spectrum of therapeutic activities in many other disease conditions, including inflammation, and is used in the treatment of inflammation of the prostate gland (Demiray *et al.*, 2009; Agu and Ndimkora, personal communication, September 20, 2009). As an antimicrobial agent, the seed extract is used in traditional medicine for the treatment of dysentery, and has remarkable reducing effects on the plasma glucose, total cholesterol (TC), and triglyceride (TG) levels, especially in high doses, giving credence to the use of the extract as a hypoglycemic agent in diabetes management. The increase in HDL cholesterol levels and a reduction in LDL cholesterol levels observed in all the treated animals are indicators that the seed extract could reduce the cardiovascular risk factors that contribute to the death of diabetic subjects, supporting its traditional use as an antidiabetic agent (Ogbonnia *et al.*, 2009). The extract acts as a cardioprotective agent, and is used for the treatment of periodontal and liver diseases; it is also speculated to have anticancer, anti-asthma, and anticataract activities (Demiray, 2009). The seed extract increases hemoglobin levels due to the increased absorption of iron, and this, coupled with an increase in the white blood corpuscle (WBC) count, emphasizes its beneficial effect on the general well-being of experimental animals. *P. curatellifolia* seed is widely used traditionally because of its multiple therapeutic activities. Some of the ethnobotanical uses of the seed extracts in disease management are summarized in Table 91.1.

ADVERSE EFFECTS AND REACTIONS (ALLERGIES AND TOXICITY)

Acute toxicity investigation of the seed extract in albino mice showed that it is slightly toxic (Table 91.2). The median acute toxicity (LD_{50}) value of the extract was determined to be between 5 and 15 g/kg body weight (Ogbonnia *et al.*, 2009). The viscera of the dead animals did not show any macroscopic changes, compared with the control, for possible clues to the cause of death. However, the animals did not convulse before dying. It could be postulated that mortality did not result from some action of the extract on the nervous system.

The effects of the extract on biochemical parameters in subchronic investigations are summarized in Table 91.3. The extract in high doses decreased the plasma protein level, which

TABLE 91.1 Summary of some Ethnobotanical Uses of *Parinari curatellifolia* Seed Extracts in Disease Management

	Used Singly or in Combination with Other Herbs	Uses	Reference
1	Singly, or with *Alstonia boonei* bark or other herbs	In diabetes	Ogbonnia *et al.*, 2009
2	With *Gongronema latifolia*, climbing stem, and *Cocos nucifera*	As bitters and stomachic for promotion of good health; as an antioxidant	
3	Used singly	As an antiprotozoan in the treatment of dysentery; as an anti-inflammatory agent in the treatment of inflammation of the prostate gland	L. Adebisi (personal communication, October 17, 2009)
4	With *Xylopia ethiopica*	As an antioxidant, antibacterial, and anticoagulant agent following childbirth	L. Adebisi (personal communication, October 17, 2009); Agu and Ndimkora (personal communication, September 20, 2009)
5	With *Rauwolfia vomitoria* root	To increase libido and as treatment for men with erectile problems	
6	Used singly	In the control of body weight and fluid (seed extracts) In worm infections, especially in women experiencing miscarriage	
7	In combination with other herbs	In liver diseases, and in asthmatic conditions	
8	With *Piper guinese*	In periodontal disease To reduce advancing age-induced oxidative stress	

772

could be a sign of impaired renal function (Ogbonnia *et al.*, 2009). There was an elevation in plasma creatinine concentration, which indirectly suggests kidney damage, specifically of the renal filtration mechanism (Wasan *et al.*, 2001). The seed extract in high doses may cause kidney damage. Also, there was an increase in the AST level only in the animals treated with high doses of the extract, while a decrease in ALT level was observed in all treated animals. This implies that the extract might not have caused any toxic effect on the liver and heart tissues at low and moderate doses, but could have some deleterious effects on the heart tissue in high doses (Ogbonnia *et al.*, 2009). Calcium levels were not affected in any of the treated animals, while a significant increase in the level of phosphorus was observed only in the animals treated with the highest dose of the extract, which could be associated with renal failure. Since there was an increase in creatinine and phosphorus levels and a decrease in

TABLE 91.2 Acute Toxicity of the Aqueous Ethanol Extract of *P. curatellifolia* in mice

Group	No. of Mice	Doses of Extract (g/kg body weight)	No. of Dead Mice	%Cumulative of Dead Mice
1	5	Control	0	0.0
2	5	1.0	0	0.0
3	5	2.5	0	0.0
4	5	5.0	1	9.1
5	5	10.0	2	27.3
6	5	15.0	3	54.5
7	5	20.0	5	100.0

Animals in the control group each received a daily oral dose of 0.3 ml 2% Tween 80 solution.
The LD_{50} value of the seed extract was determined by plotting the percentage cumulative death against the dose, and was the dose that killed 50% of the population calculated graphically ($n = 5$ animals in each group).

TABLE 91.3 Effects of the Seed Extract on Biochemical Parameters in Subchronic Studies in Rats

Parameter	Control	Dose mg/kg Body Weight		
		100	250	500
Glucose (mg/dl	90.6 ± 0.12	87.6 ± 0.62	$85.2 \pm 0.1^{**}$	$81.1 \pm 0.3^{*}$
Cholesterol (mg/dl)	37.6 ± 0.6	$23.8 \pm 0.03^{*}$	$30.6 \pm 0.42^{**}$	$26.0 \pm 1.4^{*}$
Triglyceride (mg/dl)	24.6 ± 0.4	$12.7 \pm 1.5^{*}$	$18.3 \pm 0.02^{*}$	24.4 ± 0.02
HDL (mg/dl)	140.3 ± 0.02	$178.5 \pm 0.01^{*}$	$223.7 \pm 0.07^{*}$	$155.4 \pm 0.5^{*}$
LDL (mg/dl)	81.4 ± 2.5	30.8 ± 2.6	Very low	Very low
Protein (g/dl)	3.6 ± 0.1	3.6 ± 0.3	3.3 ± 0.6	$1.5 \pm 0.20^{*}$
Creatinine (mg/dl)	2.5 ± 0.03	2.2 ± 0.01	$1.3 \pm 0.05^{**}$	$4.7 \pm 0.01^{*}$
AST (IU/l)	8.2 ± 1.2	$7.69 \pm 2.5^{**}$	$7.1 \pm 0.03^{**}$	$6.5 \pm 2.3^{*}$
ALT (IU/l)	52.5 ± 1.80	$10.60 \pm 2.5^{**}$	36.20 ± 2.6	12.00 ± 3.6

Values are mean $\pm$ SEM of three determinations ($n = 5$),
Three groups of rats were fed with 100, 250, and 500 mg/kg body weight of the seed extract orally, respectively, and the control group 0.5 ml 2% Tween 80 solution orally, for minimum of 30 days. Thereafter, the animals were sacrificed and the plasma tested for various biochemical parameters.
*Significant difference at $P < 0.05$;
**Significant difference at $P < 0.01$ vs control group. Control group received 0.5 ml 2% Tween 80 solution.

the protein level observed in the animals that received the high doses of the extract, it implies that the extract in high doses could cause kidney damage and subsequent renal failure (Wasan *et al.*, 2001).

SUMMARY POINTS

- *Parinari curatellifolia* is very variable in size and shape, ranging from a small shrub to a large tree. It is widely spread in varying climatic conditions, and can be cultivated by both asexual and sexual means.
- *Parinari curatellifolia* seed extracts have curative potential, with multiple therapeutic activities attributed to the presence of vitamins and natural products such as polyphenols.
- Polyphenols are antioxidants, and are implicated in the management of diverse disease conditions.
- Seed extracts also exhibit a broad spectrum of therapeutic activities, including anti-inflammatory effects, antidiabetic effects, reduction of advancing age-induced oxidative stress, and antimicrobial activity.
- The seed extract in high doses could cause kidney damage, and has some deleterious effects in the heart tissue but not in the liver.
- The extract contains hypolipidemic agents, and reduces the cardiovascular risk factors that contribute to the death of diabetic subjects.

References

Apak, R., Güçlü, K., Demirata, B., Özyürek, M., Çelik, E. S., Bektaşoğlu, B., et al. (2007). A review: comparative evaluation of various total antioxidant capacity assays applied to phenolic compounds with the CUPRAC assay. *Molecules, 12,* 1496–1547.

Burkill, M.H. (1985). *The useful plants of West Tropical Africa,* (2nd ed.). Vol. 1: *Families A–D.* London, UK: Royal Botanic Garden, Kew, pp. 382–385.

Demiray, S., Pintado, M. E., & Castro, P. L. M. (2009). Evaluation of phenolic and antioxidant activities of Turkish medicinal plants: *Tilia argentea, Crataegi folium* leaves and *Polygonum bistorta* roots. *World Academy of Sciences, Engineering and Technology, 54,* 312–317.

Evans, C. (1989). Vitamin, hormones and antibiotics. In *Trease and Evan's pharmacognosy* (13th ed.). (pp. 657–668) London, UK: Baillière Tindall.

Fathiazad, F., Delazar, A., Amiri, R., & Sarker, D. S. (2006). Extraction of flavonoids and quantification of rutin from waste tobacco leaves. *Iranian Journal of Pharmaceutical Research, 3,* 222–227.

Joulain, D., Casazza, A., Laurent, R., Potier, D., Guillamon, N., Pandya, R., et al. (2004). Volatile flavour constituents of fruits from Southern Africa: Mobola plum (*Parinari curatellifolia*). *Journal of Agricultural and Food Chemistry, 52*, 2322–2325.

Kadiri, B. A. (2008). Cultivation and propagation of medicinal plants in Nigeria. In T. Odugbemi (Ed.), *A textbook of medicinal plants from Nigeria* (pp. 151–157). Lagos, Nigeria: University of Lagos Press.

Lee, K.-S., Chai, H.-B., Chagwedera, E. T., Besterman, M. J., Farnsworth, R. N., Cordell, A. G., et al. (1996). Cell-cycle specific cyctotoxicity mediated by rearranged ent-kaurene diterpenoids isolated from *Parinari curatellifolia*. *Chemico-Biological Interactions, 99*, 193–204.

Maraji, V., & Glen, H. F. (2008). *Parinari curatellifolia* Planch ex Bench. Available at: http://www.plantzafrica.com/plantnop/parinaricurat.htm.

Ogbonnia, S. O., Olayemi, S. O., Anyika, E. N., Enwuru, V. N., & Poluyi, O. O. (2009). Evaluation of acute toxicity in mice and subchronic toxicity of hydroethanolic extract of *Parinari curatellifolia* Planch (Chrysobalanaceae) seeds in rats. *African Journal of Biotechnology, 8*, 1800–1806.

Sanogo, S., Gondwe, D., Ronne, C., & Sacande, M. (2006). *Parinari curatellifolia* Planch ex Blanch. London, UK: Seed leaflet, Royal Botanic Garden, Kew.

Shafaghat, A., & Salimi, F. (2008). Extraction and determining of chemical structure of flavonoids in *Tanacetum parthenium* (L.) Schultz. Bip. *Iranian Journal of Science, 18*, 39–42.

Siess, M.-H., Le Bon, A.-M., Canivenc-Lavier, M.-C., Amiot, M.-J., Sabatier, S., Aubert, Y. S., et al. (1996). Flavonoids of honey and propolis: characterization and effects on hepatic drug-metabolizing enzymes and benzo[*a*] pyrene–DNA binding in rats. *Journal of Agricultural and Food Chemistry, 44*, 2297–2301.

Wallis, E. T. (2005). Seeds. In *Textbook of pharmacognosy* (5th ed.). (pp. 188–233) New Delhi, India: CBS Publishers & Distributors.

Wasan, K. M., Najafi, S., Wong, J., & Kwong, M. (2001). Assessing plasma lipid levels, body weight, and hepatic and renal toxicity following chronic oral administration of a water soluble phytostanol compound FM1806 VP4, to gerbils. *Journal of Pharmaceutical Sciences, 4*, 228–234.

Zafra-Stone, S., Yasmin, T., Bagchi, M., Chatterjee, A., Vinson, A. J., & Bagchi, D. (2007). Berry anthocyanins as novel antioxidants in human health and disease control. *Molecular Nutrition & Food Research, 51*, 675–683.

CHAPTER 92

Moringa (*Moringa oleifera*) Seed Extract and the Prevention of Oxidative Stress

Swaran J.S. Flora, Vidhu Pachauri
Division of Pharmacology and Toxicology, Defence Research and Development Establishment, Gwalior, India

CHAPTER OUTLINE

LIST OF ABBREVIATIONS

AMS, ancient medicinal system
As, arsenic
CCl_4, carbon tetrachloride
GSH, glutathione
H_2O_2, hydrogen peroxide
LE, leaf extract
MOS, *Moringa oleifera* seeds
MOSE, *Moringa oleifera* seed extract
MOSEE, *Moringa oleifera* seed ethanolic extract

Nuts & Seeds in Health and Disease Prevention. DOI: 10.1016/B978-0-12-375688-6.10092-1

ROS, reactive oxygen species
SE, seed extract
SOD, superoxide dismutase

INTRODUCTION

Moringa oleifera Lam. (Moringaceae) is a well-known tree in folk medicine in its native land, with nutritional and traditional medicinal claims, but scientific awareness has only recently been acknowledged. The tree has an impressive range of medicinal claims, with practically every part used in the Indian traditional system of medicine, Ayurveda. Different parts of this plant contain important minerals, and are a good source of proteins, vitamins, β-carotene, amino acids, and various phenolics. It provides a rich and rare combination of zeatin, β-sitosterol, caffeoylquinic acid, and kaempferol. The plant is reported to have various biological activities. It is used traditionally to purify water in Asia and Africa, due to its effective coagulation properties.

BOTANICAL DESCRIPTION

Moringa oleifera is a small deciduous tree with sparse foliage. The tree grows to 8 m in height, and has a smooth, dark gray bark with thin yellowish slashes; its crown is wide, open, typically umbrella-shaped, and usually has a single stem. The leaves are alternate, large, with opposite pinnae, spaced about 5 cm apart up the central stalk, usually with a second lot of opposite pinnae, and bears leaflets in opposite pairs. Leaflets are dark green above and pale on the undersurface, and variable in size and shape, but often rounded elliptic, and up to 2.5 cm long. Flowers are produced throughout the year, in loose axillary panicles up to 15 cm long, and possess a very sweet smell. The fruit is large and distinctive, light brown, up to 90 cm long and 12 mm broad, slightly constricted at intervals, gradually tapering to a point, three- or four-angled, with two grooves on each face. It splits along each angle to expose the rows of rounded blackish oily seeds, each with three papery wings.

HISTORICAL CULTIVATION AND USAGE

Moringa oleifera Lam. is most widely cultivated species of the monogeneric family in many locations in the tropics. It is an important crop in West, East, and South Africa, tropical Asia, Latin America, the Caribbean, Florida USA, and the Pacific Islands. All parts of the Moringa tree have long been consumed by humans. Young unripe pods are cooked with curries as a preventive against intestinal worms. Unani practitioners prescribe soup made from the pods in sub-acute cases of enlarged liver and spleen, articular pain, debility, neurological disorders, and skin disease. Seeds from the pods, ground with water and instilled into the nostrils, cure headaches due to colds and rhinitis.

PRESENT-DAY CULTIVATION AND USAGE

The plant is cultivated in various regions of world, including Afghanistan, Bangladesh, Ethiopia, Gambia, Ghana, India, Indonesia, Iran, Malaysia, Myanmar, Nepal, Oman, Pakistan, the Philippines, Qatar , Saudi Arabia, Sudan, Tanzania, Thailand, Togo, Uganda, the United Arab Emirates, and Vietnam. The plant readily colonizes on stream banks and in savannah areas, where the soils are well drained and the water table remains fairly high all the year round. Fairly drought-tolerant, it yields much less foliage where continuously under water stress. It is not harmed by frost, but can be killed back to ground level by a freeze. It quickly sends out new growth from the trunk when cut, or from the ground when frozen. The young leaves, flowers, and pods are common vegetables in the Asian diet, considered to be of high nutritional value. Seeds are used for water purification. Coagulant protein and various phytochemicals identified for medicinal use have been isolated and investigated (Table 92.1).

TABLE 92.1 List and Structural Moiety of Major Phytochemicals from the Seeds of *Moringa oleifera*

Principal Phytochemicals (Isolated/Reported from Seed Extract)	Reported Activity	Note	Structure
Alkaloid: Moringine	Relaxes bronchioles	1	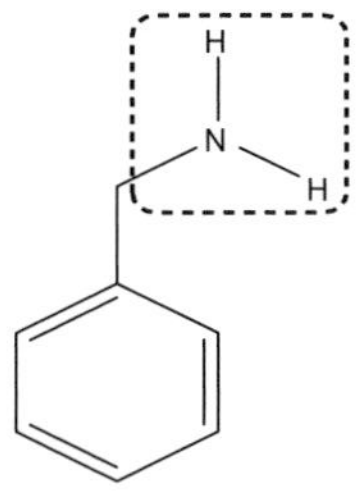
Flavonoids *Glycosides:* Benzylisothiocyanate and its derivative: 4-(4-acetyl-α-L-rhamnosyloxy) benzoisothiocyanate	Antioxidant Anti-inflammatory, antimicrobial, anticancer, antihypertensive	2	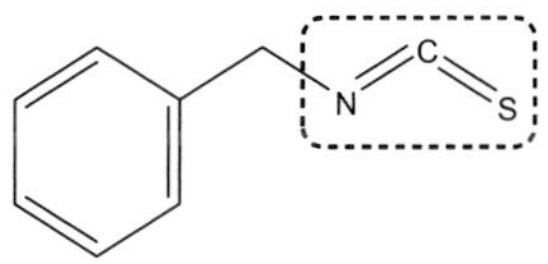
4 [α-L-rhamnosyloxy], phenylacetonitrile, 4-hydroxyl phenyl-acetonitrile, and 4-hydroxyphenyl-acetamide	Mutagens	3	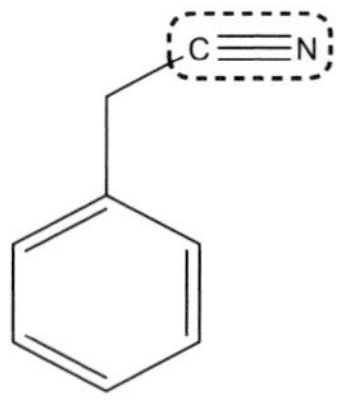
4-[α-L-rhamnosyloxy], 4-[α-L-rhamnosyloxy] phenylacetonotrile β-carotene], sterols. and lecithin	Antibiotic, antimicrobial	4	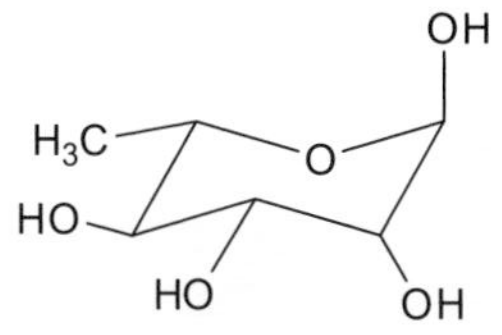

Continued

777

TABLE 92.1 List and Structural Moiety of Major Phytochemicals from the Seeds of *Moringa oleifera*—continued

Principal Phytochemicals (Isolated/Reported from Seed Extract)	Reported Activity	Note	Structure
O-ethyl-4-(α-L-rhamnosyloxy) benzylcarbamate with seven derivatives: niazimicin, niazirin, β-sitostrol, glycerol-1-1-(9-octadecanoate), 3-O-(6-α-O-oleoyl-β-D-glucopyranosyl)-β-sitosterol, and β-sitosterol-3-O-β-D-glucopyranoside	Antitumor activity, antihypertensive	5	
		6	
Nutrients: Vitamins B1, B6, riboflavin, nicotinic acid, folic acid, Vitamins C and E	As nutritives, antioxidants		
Fats: In seed oil	Undefined properties		

1, Moringine; **2,** benzylisothiocyanate; **3,** phenylacetonitrile; **4,** 4-(α-L-rhamnosyloxy); **5,** X = S and Y = OH (Niazimicin), X = S and Y = OAc (Niaziminin), X = O and Y = OH (O-ethyl-4-(α-L-rhamnosyloxy) benzyl carbamate); **6,** R = H (β-sitosterol), R = 6′-O-oleoyl-β-D-glucopyranosyl (3-O-(R)-β-sitosterol), R = β-D-glucopyranosyl (β-sitosterol-3-O-β-D-glucopyranoside).

APPLICATIONS TO HEALTH PROMOTION AND DISEASE PREVENTION

Almost all diseases are directly or indirectly linked to oxidative stress, either as an underlying cause or as its manifestation. The cell utilizes reactive oxygen species (ROS), which play a crucial role in the defense mechanism. Oxidative stress adversely affects every cell, thus lowering the body's defense and contributing to the pathological conditions, justifying the usefulness of antioxidants in treating chronic diseases.

Ayurveda and the Chinese ancient medicinal system, etc., are known for their exhaustive experience in herbal remediation. *Moringa oleifera* is one plant used in these ancient systems. Although all parts of this tree possess medicinal value, this chapter is restricted to the therapeutic potential of seeds, and especially their antioxidant value.

Protection against metal-induced oxidative stress

Moringa oleifera seed extracts are rich sources of a number of vitamins and minerals (Table 92.1), which may contribute significantly to its antioxidant, therapeutic, and nutritional values. Extending its utility as a water purifier, we reported therapeutic potential of MOS in arsenic poisoning *in vivo* (Mishra *et al.*, 2009). MOS contain natural coagulants useful for chelation and for their antioxidant property. These beneficial effects are both direct and indirect, and may be attributed to high concentrations of methionine, cysteine, and antioxidants such as vitamins C, E, β-carotene, etc. Methionine and cysteine are rich in the —SH group, which provides a binding site for arsenic. In the oxidative methylation reaction, where trivalent forms of arsenic are sequentially methylated to form mono- and di-methylated products, S-adenosyl methionine (SAM) functions as a methyl donor and glutathione (GSH) as an essential co-factor. SAM is synthesized from condensation of ATP with L-methionine, and thus later plays an important role in removal of arsenic from the site. Two potential cellular defense mechanisms against oxidative stress are the cysteinyl-containing polypeptide and tripeptide known as metallothionein (MT) and glutathione (GSH). Arsenic is known to increase hepatic MT, which is lowered by MOS, and is thus correlated with arsenic elimination from the liver. It is not clear which constituent of MOS plays the role of a chelator. We proposed a chemical hypothesis for the interaction between arsenic and MOS. Arsenic exists in anionic and neutral forms, while oxyanions of arsenic exist in three different arsenite species; namely, H_3AsO_3, $H_2AsO_3^-$, and $HAsO_3^{2-}$, in the pH ranges of 7—8, 10—11, and 12—13, respectively. An aqueous extract of MOS is a heterogeneous complex mixture, having cationic polypeptides with various functional groups, particularly of low molecular weight amino acids. Thus, there is a possibility that proteinacious amino acids with a variety of structurally related pH-dependent properties of generating either positively- or negatively-charged sites may attract the anionic and cationic species of arsenic. With an increase in pH range, the carboxylic group of the amino acids would progressively be deprotoneated as carboxylate ligands, thus simultaneously protonating the amino group. The possibility of protein/amino acid—arsenic interaction was further supported by IR spectrometry of MOS powder alone and in combination with arsenic, which confirmed the cationic nature of polypeptides present in MOS powder. Such positively charged (NH_3^+) ions facilitate the MO—arsenic complexation, and might be responsible for excretion of arsenic from *in vivo* sites. MOS are rich in magnesium, sodium, phosphorus, and calcium, which might compete with metals/metalloids. Moreover, glycoprotein, a main native acidic protein that has hemagglutinating activity, is capable of forming a complex with arsenic that can be excreted from the body (Gupta *et al.*, 2007). A significant recovery in blood and liver enzymatic and non-enzymatic oxidative stress markers suggests a dual mechanism whereby MOS, by lowering the arsenic burden, indirectly reduces arsenic-induced oxidative stress while executing direct free radical scavenging and antioxidant-enhancing mechanisms (Gupta *et al.*, 2005) (Figures 92.1, 92.2, 92.3).

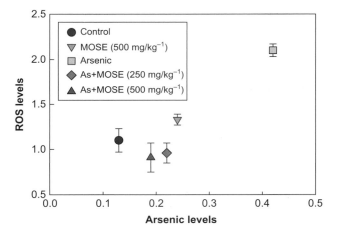

FIGURE 92.1
Correlation between blood arsenic and ROS levels after treatment with *Moringa oleifera* seed extract in arsenic-exposed animals. Graph depicts that levels of arsenic in blood are in direct correlation with ROS levels. Units: blood arsenic, μg/ml; ROS, ηmoles/min per ml blood.

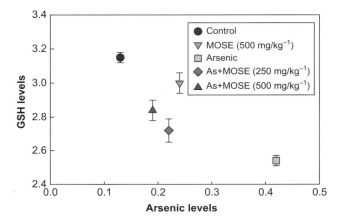

FIGURE 92.2
Inverse correlation between blood arsenic and GSH levels after treatment with *Moringa oleifera* seed extract in arsenic-exposed animals. Graph depicts that levels of arsenic in blood are in inverse correlation with GSH levels. Units: blood arsenic, μg/ml; GSH, ηmoles/min per ml blood.

Except in the case of arsenic, there is not much evidence in the literature to suggest the protective efficacy of MOS against metal toxicity. MOSE proved effective in mitigating fluoride toxicity (Stanley *et al.*, 2000; Rajan *et al.*, 2009). Urinary fluoride excretion was significantly enhanced, as well as protection against nephro- and hepatotoxicity in experimental animals. The authors concluded that lowered fluoride absorption at the gastrointestinal level due to flocculant action and reduced oxidative stress owing to the antioxidant potential of MOSE may be the key factors in its efficacy against fluoride toxicity.

Anti-inflammatory activity

Inflammation is another underlying cause for various diseases, or their manifestation. There is a close association between inflammatory processes and the generation of ROS as part of the body's defense pathway. MOSE has shown protection against several inflammatory pathological conditions, whether chemically induced or immune-mediated. Researchers suggest that MOSE provides significant alleviation in bronchoalveolar inflammation by

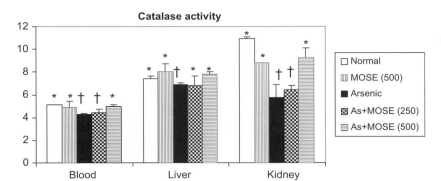

FIGURE 92.3

Effect of *Moringa oleifera* seed extract on catalase activity in blood and tissue after arsenic toxicity. Graph shows that arsenic toxicity reduced catalase activity that is recovered after *Moringa oleifera* seed extract therapy dose-dependently. Unit: catalase, μmoles of H_2O_2 consumed/min per mg protein.

inducing decreased infiltration of inflammatory cells, and in the secretion of inflammatory mediators into the airways. Interleukin-4 (IL-4) and IL-6 were found to be ameliorated significantly by treatment with MOS ethanolic extract (MOSEE). Inflammatory asthma is reported to increase tissue malondialdehyde levels, and decrease activity of antioxidant enzymes like GSH, SOD, and catalase. MOSEE (200 mg/kg) could reverse cellular biochemical symptoms, and overcome oxidative stress in chemical-induced asthma. Similarly, traditional use of MOSE in allergic and immune-mediated asthma was supported by the fact that it suppresses most inflammatory cells and cytokines predominant in ovalbumin-induced airway inflammation, and inhibits histamine release by down-regulation of histamine-releasing cells (Mahajan & Mehta, 2008).

Fractionation and isolation of active phytochemicals from MOS is still awaited, with conclusive observations regarding their mechanism of action. Studies confirm that certain rare classes of phytochemicals in seed extracts of this plant include various alkaloids, glucosinolates, and isothiocyanates. Reports suggest that one of the Moringa alkaloids has activity resembling that of ephedrine, and that it may be useful in asthma therapy, where *Moringine* has shown bronchiole relaxation activity (Kirtikar & Basu, 1975).

Inflammation and tissue injury-related oxidative stress have been recognized as having implications for the pathogenesis of rheumatoid arthritis. MOSEE (200 mg/kg p.o.) is reported to be effective in suppressing various inflammatory mediators along with IL-6, showing an immunosuppressant-like effect in chronic arthritis (Mahajan & Mehta, 2008). The authors reported elevated SOD and catalase activity in diseased animals, which was reduced by MOSEE, indicating lowered generation of ROS. Further, flavonoids in Moringa have been credited with having osteoporosis-preventive activity, by causing an increase in bone density (Nijveldt *et al.*, 2001).

Protection against oxidative DNA damage

Regular and extensive exposure to environmental toxins and xenobiotics raises serious concerns regarding the oxidative insult to DNA, more for its mutagenic and carcinogenic consequences than apoptotic or necrotic cell death. *In vitro* studies report that MO leaves, fruit, and seed extracts protect DNA from oxidative stress by virtue of their antioxidant properties. MOSE at the concentration of 10−25 μg/ml could provide significant partial and complete protection of supercoiled DNA, and mitigate the oxidative stress produced by Fenton's reaction mediated by OH. The polyphenolic antioxidant compounds of MOSE are suggested to diminish the reduction potential of Fe ions by virtue of their ability to bind with redox active metals in solution, to form complexes. This prevents reduction of Fe ions by H_2O_2, thus inhibiting the Fenton-like reaction. Moreover, possible direct free radical scavenging ability of the extract may also contribute in providing protection to supercoiled DNA against OH-mediated strand breaks (Figure 92.4). MO extracts also show distinctly significant synergistic antioxidant efficacy with trolox and protection against OH mediated DNA damage (Singh, 2009).

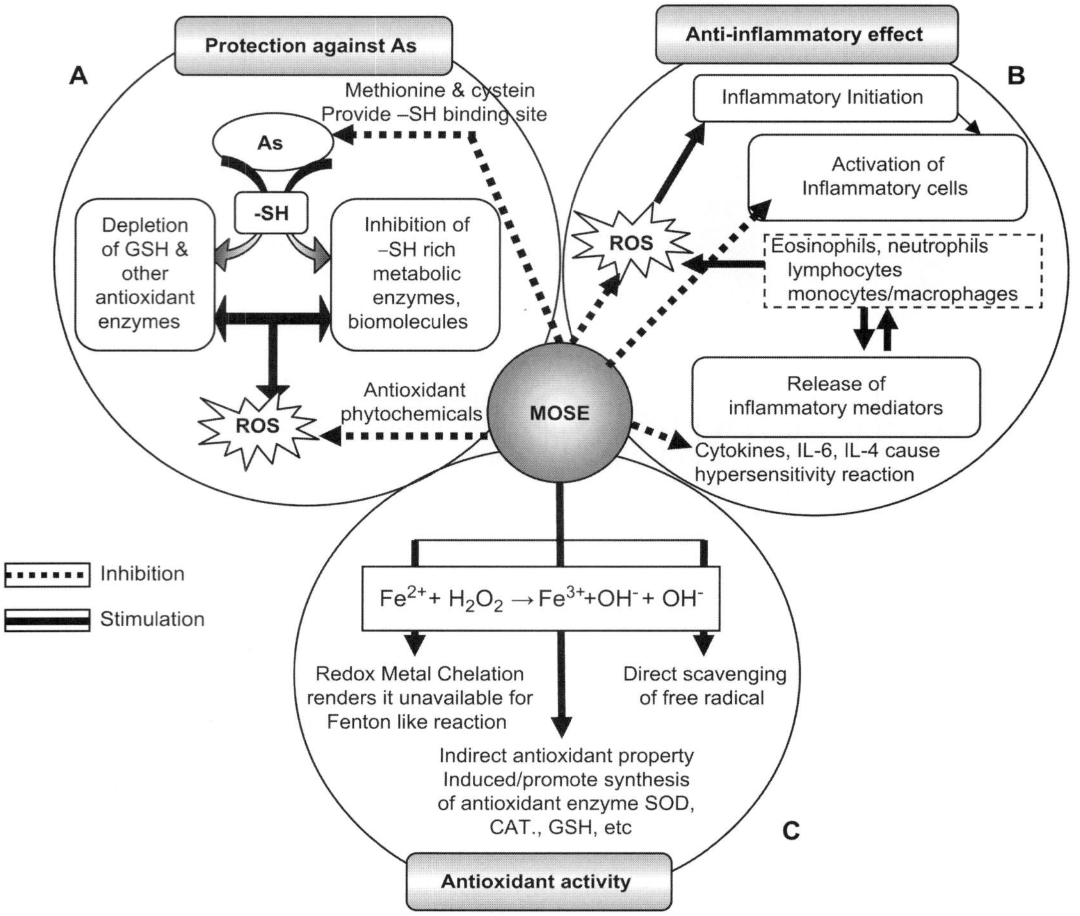

FIGURE 92.4

Major mechanisms involved in pharmacological effects of *Moringa oleifera* seed extract with respect to its antioxidant potential. The beneficial effect of MOSE is attributed mainly to its antioxidant and anti-inflammatory potential. Circle **A** depicts possible mechanisms involved in protection against arsenic-induced toxicity by MOSE by providing an —SH binding site for chelation or via antioxidant potential; Circle **B** depicts the ability of MOSE to directly inhibit inflammatory cell activation and mediator release, or indirect inhibit inflammatory processes by scavenging ROS; Circle **C** suggests antioxidant mechanisms exhibited by MOSE that may include chelating redox metal (Fe) to inhibit Fenton's reaction, and/or scavenging ROS, and/or promoting antioxidant enzyme generation.

Cancer prevention

Chemoprevention of cancer involves the use of natural or synthetic agents to inhibit, delay, or reverse the development of cancer in normal or pre-neoplastic conditions. Although oxidative stress may be one of the mechanisms for cancer, its prevention requires a balance between Phase I and II biotransformation enzymes, and good antioxidant defense status. Bharali *et al.* (2003) reported cancer-chemopreventive potential of MOSE, by virtue of its positive effects on hepatic cytochrome (b5 and P450) and glutathione-S-transferase, and Phase I and II enzyme systems. MOSE is a promising candidate for a possible cancer chemopreventive agent, based on its ability to significantly reduce DMBA-induced skin carcinogenesis in mice. Furthermore, 4-(4′-O-acetyl-α-L-rhamnopyranosyloxy) benzyl isothiocyanate and niazimicin (Table 92.1) have been identified as potent inhibitors of phorbol ester (TPA)-induced Epstein-Barr virus early antigen activation in lymphoblastoid cells. Niazimicin also suppressed tumor promotion in the mouse two-stage DMPA-TPA tumor model (Guevara *et al.*, 1999).

Antimicrobial activity

Bacteriological studies reveal that MOSE shows active antimicrobial activity. Interestingly, the recombinant protein in seed has been reported to flocculate gram-positive and gram-negative bacteria in a manner similar to removal of colloids in water. Seeds may act directly as a bacteriostatic by disruption of the cell membrane or by inhibiting essential enzymes (Suarez *et al.*, 2003). Inhibition of bacteriophage replication has also been reported (Sutherland *et al.*, 1990). The principal ingredient contributing to the activity was identified as 4-(4′-O-acetyl-α-L-rhamnopyranosyloxy) benzyl isothiocyanate.

Antihypertensive activity

Aqueous and ethanol extracts of the whole pod and its parts (seeds) revealed a pronounced blood-pressure lowering effect. Isolation of thiocarbamate, isothiocyanate glycosides, β-sitosterol, and methyl p-hydroxybenzoate from MOSE identified them as the principal hypotensive constituents (Faizi *et al.*, 1998). It may be noted that, owing to its anti-hypercholesterolemic activity (leaves) and pronounced antioxidant function, Moringa qualifies as an important source of cardiovascular drugs.

Antioxidant potential and phytochemicals

Study of the specific phenolic composition of MOS by HPLC reveals the presence of phenolic compounds, including gallic acid, ellagic acid, and kaempferol. The identified phytomolecules have well-documented antioxidant potential, free radical scavenging activity, and metal ion-chelating capacity (Singh *et al.*, 2009). The total phenolic content of MOSE is 45.81 mg gallic acid equivalents/g dry extract, and 9.93 mg quercetin equivalents/g for total flavonoid content, with the ascorbic acid content being 62.11 mg/100 g fresh tissue. The effects of oxidative stress on other biomolecules, like lipid peroxidation and protein carbonyl formation, respectively, were protected by MOSE *in vitro* (Singh, 2009) and *in vivo*, accompanied by elevated liver SOD level against CCl_4-induced liver toxicity and fibrosis. The authors concluded that phenolic compounds quench ROS, chelate metal, and regenerate membrane-bound antioxidants, and that these might be the major mechanisms responsible for the antioxidant potential exhibited by MOS. The reversal in liver fibrosis associated with neutrophil infiltration in CCl_4-treated animals was attributed to the anti-inflammatory activity of MOS. Hepatic stellate cells are also activated by both chronic inflammation and/or ROS, suggesting that MOS possibly exerts a dual effect, with its efficacy as anti-fibrotic agent (Hamza, 2010).

ADVERSE EFFECTS AND REACTIONS (ALLERGIES AND TOXICITY)

Santos and colleagues (2005) reported *M. oleifera* seeds to be non-toxic, and recommended their use as an effective coagulant. However, they also concluded that heat treatment of MOS-purified water may be necessary for denaturation of lectin protein, which is known to be an antinutritional factor. Oral acute and chronic toxicity tests in rats with both Moringa steno-petala and MOS (dosages 50 and 500 mg/kg body weight, respectively) have been reported to produce no toxic effects, but rather increased the weights of the rats (Jahn, 1988). On the contrary, the toxicity and mutagenic effects of MOSE at 200 mg/l in guppies, protozoa, and bacteria were reported, which accounts for its therapeutic utility more than environmental disadvantage (Ndabigengesere, 1995). Oluduro and Aderiye (2009) reported the high activities of aspartate amino transferase, alkaline transferase, and alkaline phosphatase in kidney and heart tissues of rats, and their corresponding decreased activities in the serum, indicative of non-toxicity of MOSE in these organs. Therefore, most toxicity reports confirm high safety margins for plant (seed as well as other parts) extracts, yet more studies may be required. Moreover, it is important to consider the wide clinical exposure of all parts of the plant as part of the regular diet and in medicinal utility for generations, without any reported adverse effects.

SUMMARY POINTS

- The class of phytochemicals identified in *Moringa oleifera* seed extracts (aqueous and ethanolic) is well documented as antioxidants.
- *Moringa oleifera* seed has proved an excellent natural coagulant, used traditionally for water purification, and studies support its benefits over synthetic coagulants.
- *Moringa oleifera* seed extract provides protection against toxic metal induced oxidative stress by virtue of its chelation and antioxidant properties.
- *Moringa oleifera* seed extract has a high ferric reducing antioxidant power value, which is known to exhibit high scavenging activity.
- The high phenolic content, ascorbic acid, and isothiocynate derivatives in *Moringa oleifera* seed extract contribute to protection against inflammatory diseases, directly or indirectly.
- The cancer-preventive effects of *Moringa oleifera* seed extract relate to its activity in modulating biotransformation enzymes to facilitate detoxification.
- *Moringa oleifera* seed extract shows pronounced antimicrobial and hepatoprotective potential.
- Phytochemicals in *Moringa oleifera* may provide treatment for cardiovascular disorders, owing to their antihypertensive and hypercholesterolemic activities.

ACKNOWLEDGEMENTS

The authors thank Dr R. Vijayaraghavan, Director of the establishment, for his support and encouragement

References

Bharali, R., Tabassum, J., & Azad, M. R. H. (2003). Chemomodulatory effect of *Moringa oleifera*, Lam, on hepatic carcinogen metabolizing enzymes, antioxidant parameters and skin papillomagenesis in mice. *Asian Pacific Journal of Cancer Prevention, 4*, 131–139.

Faizi, S., Siddiqui, B. S., Saleem, R., Aftab, K., Shaheen, F., & Gilani, A. H. (1998). Hypotensive constituents from pods of *Moringa oleifera. Planta Medica, 3*, 957–963.

Guevara, A. P., Vargas, C., & Sakurai, H. (1999). An antitumor promoter from *Moringa oleifera* Lam. *Mutat. Res., 440*, 181–188.

Gupta, R., Kannana, G. M., Sharma, M., & Flora, S. J. S. (2005). Therapeutic effects of *Moringa oleifera* on arsenic-induced toxicity in rats. *Environmental Toxicology and Pharmacology, 20*, 456–464.

Gupta, R., Dubey, D. K., Kannan, G. M., & Flora, S. J. S. (2007). Concomitant administration of *Moringa oleifera* seed powder in the remediation of arsenic-induced oxidative stress in mouse. *Cell Biology International, 31*, 44–56.

Hamza, A. A. (2010). Ameliorative effects of Moringa oleifera Lam seed extract on liver fibrosis in rats. *Food and Chemical Toxicology, 48*, 345–355.

Jahn, S. A. A. (1988). Using Moringa seeds as coagulants in developing countries. *Journal of the American Water Works Association, 80*, 45–50.

Kirtikar, K. R., & Basu, B. D. (1975). *Moringa oleifera*. In D. Dun, B. Singh & M. P. Singh (Eds.), *Indian medicinal plants, Vol. 1* (pp. 676–683). Dehra Dun, India: Bishen Singh Mahendrapal Singh Publishers.

Mahajan, S. G., & Mehta, A. A. (2008). Effect of *Moringa oleifera* Lam. seed extract on ovalbumin-induced airway inflammation in guinea pigs. *Inhalation Toxicology, 20*, 897–909.

Mishra, D., Gupta, R., Pant, S. C., Kushwah, P., Satish, H. T., & Flora, S. J. (2009). Co-administration of mono-isoamyl dimercaptosuccinic acid and *Moringa oleifera* seed powder protects arsenic-induced oxidative stress and metal distribution in mice. *Toxicology Mechanisms and Methods, 19*, 169–182.

Ndabigengesere, A., Narasiah, K. S., & Talbot, B. E. (1995). Active agents and mechanism of coagulation of turbid waters using *Moringa oleifera. Water Research, 29*, 703–710.

Nijveldt, R. J., Nood, E. V., Hoorn, D. E., Boelens, P. G., Norren, K. V., Paul, A. M., et al. (2001). Flavonoids: a review of probable mechanism of action and potential applications. *American Journal of Clinical Nutrition, 74*, 418–425.

Oluduro, A. O., & Aderiye, B. I. (2009). Effect of *Moringa oleifera* seed extract on vital organs and tissue enzymes activities of male albino rats. *African Journal of Microbiology Research, 3*, 537–540.

Rajan, R., Swarup, D., Patra, R. C., & Chandra, V. (2009). *Tamarindus indica* L. and *Moringa oleifera* M. extract administration ameliorates fluoride toxicity in rabbits. *Indian Journal of Experimental Biology, 47*, 900–905.

Santos, A. F. S., Argolo, A. C. C., Coelho, L. C. B. B., & Paiva, P. M. G. (2005). Detection of water soluble lectin and antioxidant component from *Moringa oleifera* seeds. *Water Research, 39,* 975–980.

Singh, B. N., Singh, B. R., Singh, R. L., Prakash, D., Dhakarey, R., Upadhyay, G., et al. (2009). Oxidative DNA damage protective activity, antioxidant and anti-quorum sensing potentials of Moringa oleifera. *Food and Chemical Toxicology, 47,* 1109–1116.

Stanley, V. A., Kumar, T., Lal, A. A. S., Pillai, K. S., & Murthy, P. B. K. (2000). *Moringa oleifera* (Radish tree) seed extract as an antidote for fluoride toxicity. *Fluoride, 35,* 251.

Suarez, M., Entenza, J. M., & Doerries, C. (2003). Expression of a plant-derived peptide harboring water-cleaning and antimicrobial activities. *Biotechnology and Bioengineering, 81,* 13–20.

Sutherland, J. P., Folkard, G., & Grant, W. D. (1990). Natural coagulants appropriate water treatment: a novel approach. *Water-lines, 8,* 30–32.

CHAPTER 93

Moringa (*Moringa oleifera*) Seed Oil: Composition, Nutritional Aspects, and Health Attributes

Hasanah Mohd Ghazali, Abdulkarim Sabo Mohammed
Department of Food Science, Faculty of Food Science and Technology, Universiti Putra Malaysia, Serdang, Selangor D.E., Malaysia

INTRODUCTION

The *Moringa oleifera* seed contains about 22–40% by weight of pale yellow, non-drying oil which is used for lighting, as it burns without smoke. It is also utilized in hairdressing and for soap-making. It is valued in perfume manufacture because of its capacity to absorb and retain floral fragrances (Chopra *et al.*, 1965), and formerly was highly prized for lubricating fine watches and machinery in Europe (Lowell, 1999). The oil from the seeds, known commercially as "Ben" or "Behen" oil, is edible and pleasant-tasting; it resembles olive oil and is widely used as a salad dressing in Haiti and the Caribbean. The seed oil contains all the fatty acids found in olive oil, except for linolenic acid.

BOTANICAL DESCRIPTION

M. oleifera Lam belongs to the genus *Moringaceae*, which consists of 14 known species. Of these, *M. oleifera* is the most widely known and utilized. The plant is a native of the

Nuts & Seeds in Health and Disease Prevention. DOI: 10.1016/B978-0-12-375688-6.10093-3

sub-Himalayan regions of North-West India, and is now indigenous to many countries in Africa, Arabia, South East Asia, the Pacific, the Caribbean islands, and South America. The tree is widely known as the drumstick or horseradish tree. In Malaysia, the plant is known as "kelor". The tree ranges in height from 5 to 12 m, with an open, umbrella-shaped crown, and a straight short trunk with corky, whitish bark and soft, spongy wood. It has slender, wide-spreading, drooping, fragile branches. The foliage can be evergreen or deciduous, depending on climatic conditions. It is attractive and gracefully lacy, the alternate twice- or thrice-pinnate leaves being spirally arranged, mostly at the branch tips. The leaves are about 20—50 cm long, with a long petiole and four to six pairs of pinnae bearing two pairs of opposite leaflets that are elliptical or obovate (Figure 93.1A). The fruits (or pods) are initially light green (Figure 93.1B), becoming dark green, and grow up to 60 cm long. When mature, they become dry and brown in color (Figure 93.1C). Mature seeds are round or triangular-shaped, and the kernel is surrounded by a light wooded shell with three papery wings (Figure 93.1D).

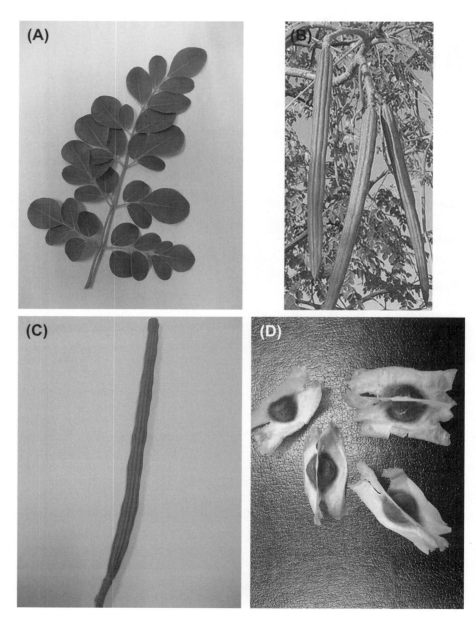

FIGURE 93.1

Parts of the *M. oleifera* plant: (A) leaves; (B) mature green pods; (C) mature dry pod; (D) mature seeds surrounded by papery wings.

HISTORICAL CULTIVATION AND USAGE

Ancient usage of *M. oleifera* has been widely reported. The use of oil extracted from mature seeds by Ancient Egyptians has been well documented. The oil was treasured by the Egyptians for skin protection against both infections and damage due to extremes of desert conditions. The benefits of the healthful attributes of the oil were later relayed to the Ancient Greeks and Romans, who also used the oil in skin protection. The light and non-drying nature of the oil makes it a good massage oil, and also for aromatherapy applications. Due to its tremendous cosmetic value, it was used extensively by the Ancient Egyptians in body and hair care, as a moisturizer and conditioner. It was also used by Ancient Egyptians, Greeks, and Romans in extracting floral fragrances used in perfumes. The oil has an excellent ability to retain fragrances extracted from flowers.

PRESENT-DAY CULTIVATION AND USAGE

Over the centuries, more uses of the oil were discovered in all parts of the world where the plant is grown. Almost every part of *M. oleifera* is valued as food. In India, Malaysia, the Philippines, and tropical Africa, the tree is prized mainly for its edible fruits, leaves, flowers, roots, and seed oil. Tender young plants less than 50 cm high, young leaves, and even mature leaflets and the flower, stripped from their stems, are cooked and consumed in salads, soups, and sauces, or simply as greens. When mature, the pods yield seeds that can be extracted and treated like green peas, and fried or roasted and eaten like peanuts. Powdered seeds have been used in parts of Africa for water clarification, and disinfection during water treatment. In Malaysia, immature pods are cut into small pieces and added to curries. In Haiti and elsewhere, the seeds are browned in a skillet, then crushed and boiled in water. The oil, which floats, is skimmed off for use as a general culinary and salad oil, and for the treatment of skin diseases (Foidl *et al.*, 2001).

APPLICATIONS IN HEALTH PROMOTION AND DISEASE PREVENTION

Proximate composition of *M. oleifera* seed

The proximate composition of *M. oleifera* seed oil has been reported by Abdulkarim *et al.* (2005) as follows: moisture, $7.9 \pm 1.00\%$; crude protein, $38.3 \pm 1.03\%$; crude oil, $30.8 \pm 2.19\%$, crude fiber, $4.5 \pm 0.38\%$; and ash, $6.5 \pm 0.15\%$. The oil content is comparable to levels reported previously (Makkar & Becker, 1997). Anwar and Bhanger (2003) reported higher oil contents (38–42%) in seeds produced by *M. oleifera* grown in temperate regions of Pakistan.

Properties of *M. oleifera* seed oil

The high degree of unsaturation (75.2%) of the oil is due to the high percentage of oleic acid ($\sim70\%$) (Abdulkarim *et al.*, 2005). Apart from oleic acid, other prominent fatty acids include palmitic (7.8%), stearic (7.6%), and behenic (6.2%) acids (Table 93.1). Anwar and Bhanger (2003), in their study on *M. oleifera* grown in temperate regions, reported that the oleic acid content tended to be higher (up to 78.5%) compared to plants grown in the tropics (Abdulkarim *et al.*, 2005).

The oil is liquid at room temperature, and pale yellow in color. Electronic nose analysis showed that the unrefined oil has a flavor similar to that of peanut oil. The melting point, estimated by differential scanning calorimetry, was found to be 19.0°C (Abdulkarim *et al.*, 2005). The oil contains 36.7% triolein as the main triacylglycerol, followed by other oleic acid-containing triacylglycerols, such as palmito-diolein and stearo-diolein. Table 93.2 shows some other properties of the oil.

TABLE 93.1 Fatty Acid Composition, and Degrees of Saturation and Unsaturation, of *M. oleifera* Seed Oil	
Type of Fatty Acid	**Percent**
Myristic/tetradecanoic acid (C14:0)	0.2
Palmitic/hexadecanoic acid (C16:0)	6.8
Palmitoleic/hexadecenoic acid (C16:1)	2.9
Stearic/octadecanoic acid (C18:0)	6.5
Oleic/octadecenoic acid (C18:1)	70.0
Linoleic/octadecadienoic acid (C18:2)	0.9
Linolenic/octadecatrienoic acid (C18:3)	—
Arachidic/eicosanoic acid (C20:0)	4.2
Gadoleic/eicosaenoic acid (C20:1)	1.4
Behenic/docosanoic acid (C22:0)	5.8
Arachidic/eicosanoic acid (C24:0)	1.3
Unsaturated fatty acid	50.9
Saturated fatty acid	49.1

Source: Abdulkarim *et al.* (2005).

It has also been reported that unrefined extracts of the oil contain sterols such as campesterol (16.0%), stigmasterol (19.0%), β-sitosterol (46.65%), Δ^5-avenasterol (10.70%), and clerosterol (1.95%) (Anwar & Bhanger, 2003). There were also minute amounts of 24-methylenecholesterol, Δ^7-campestanol, stigmastanol, and 28-isoavenasterol. Some sterols, specifically cholesterol, were not detected. The oil was also reported to contain important minor components such as tocopherols α, γ, and δ (up to 123.50−161.30, 84.07−104.00, and 41.00−56.00 mg/kg, respectively).

The application of enzymes for the extraction of the oil has been studied and reported by Abdulkarim *et al.* (2006). In this study, four commercial enzymes, namely neutral protease, α-amylase, cellulose, and pectinases, were used to extract the oil from ground seeds. The conditions of extraction reflected the optimum temperature and pH of the enzyme in use. It was found that the neutral protease was able to produce the highest percentage oil recovery (72%) of the total yield obtained from solvent-assisted extraction. Greater effectiveness of the protease compared to the other enzymes examined can be ascribed to its ability to hydrolyze proteins in the seeds, disrupting any possible association between protein and oil in the seed matrix and cell membrane. Most of the properties shown in Table 93.2 were not affected by the use of enzymes during extraction, except that the oil that was slightly more intense in color when solvent-extracted. Combining all four enzymes at 2% (w/v) each improved recovery to 74%. The oleic acid content of the oil extracted may be elevated to

TABLE 93.2 Physical and Chemical Properties of *M. oleifera* Seed Oil	
Properties	**Values**
Melting point (°C)	19.0
Solid fat content (%) (at room temperature)	11.1%
Color:	
Red index	0.7
Yellow index	5.9
Iodine value (g I$_2$/100 g)	65.4
Saponification value (mg KOH/g oil)	164
Unsaponifiable matter content (%)	0.74
Viscosity (cP)	17.7

Sources: Abdulkarim *et al.* (2005, 2006).

75.2% oleic through judicious use of temperature-assisted fractionation (Abdulkarim *et al.* 2007a).

The high content of oleic acid (67—74.5%, or up to 75.2% after fractionation of oil initially containing 70% oleic acid) allows the oil to be classified along with other high-oleic acid oils, such as olive oil, and genetically modified high-oleic sunflower (> 80%), high-oleic safflower (> 77%), and high-oleic canola (> 75%) oils. Olive oil, which has been termed the most ancient functional food in history (Ruiz-Gutiérrez & Perona, 2007), paves the way for the natural *Moringa* oil to be similarly considered. Research and development are underway worldwide to increase the number of plants capable of producing high oleic-acid contents through various means, including genetic modification, such as in cotton, oil palm, and soybean (Watkins, 2009a, 2009b). The high percentage of oleic acid in the oil makes it desirable in terms of nutrition and high-stability cooking and frying oil.

Much research has focused attention on high oleic-acid vegetable oils. It has been demonstrated that a higher dietary intake of "bad" fats (saturated and *trans* fatty acids) is associated with an increased risk of coronary heart disease caused by high cholesterol levels in the blood (Mensink & Katan, 1990), whereas a higher intake of "good" fats (monounsaturated acids/oleic acid) is associated with a decreased risk (Corbett, 2003; Lopez-Huertas, 2010). Mensink and Katan (1990) reported that monounsaturated fatty acids (MUFAs) such as oleic acid were capable of reducing blood cholesterol levels in non-hypertriglyceridemic individuals. Further, Allman-Farinelli *et al.* (2005) reported that when foods rich in oleic acid, including margarine, were used as substitutes for foods rich in saturated fatty acids, human subjects under test were found to have lower levels of blood serum low density lipoprotein cholesterol, triglycerides, and factor VII coagulant activity — a component of the hemostatic system that is a demonstrated risk factor for either primary or secondary atherothrombotic events. It was also reported that diets rich in monounsaturated fatty acids may lower fibrinogen, and decrease both the activity and antigenic PAI-1 in healthy persons.

High oleic-acid vegetable oils have been found to have enough oxidative stability to be used in demanding applications such as deep-fat frying (Corbett, 2003). In addition, high-oleic oils have low saturated fatty acid levels. Therefore, high-oleic oils can be viewed as a healthy alternative to partially hydrogenated vegetable oils. Studies regarding the effect of frying foods on the properties of *M. oleifera* oil have shown the oil to be stable to oxidation. Tsaknis and Lalas (2002) reported that the cold-pressed oil had better frying stability compared to the hexane-extract oil, and that virgin olive oil had the highest resistance to thermal degradation. Abdulkarim *et al.* (2007b) reported that the frying performance of the oil was better than that of regular canola and soybean oils, and comparable to that of palm olein. The data obtained showed lower conversion of the oil to the rather toxic polymers that are often formed when oils high in polyunsaturated fatty acids are used for deep-fat frying.

The use of *M. oleifera* for the prevention and treatment of various ailments has been reported in many folk medicines throughout the world. In India Ayurvedic medicine, *Moringa* is used as a preventative agent against about 300 diseases. Most of the health claims are attributed to the leaves and roots, and their decoctions. In the past, the medicinal value of the oil extracted from the seeds was restricted to the treatment of few conditions, such as skin disorders. More recently, however, the benefits of using the oil for the prevention and cure for other diseases have been studied and reported. Some of the therapeutic and nutritional uses of the oil are listed in Table 93.3.

ADVERSE EFFECTS AND REACTIONS (ALLERGIES AND TOXICITY)

There have not been any reported cases of adverse effects, allergies, or toxicity due to *M. Oleifera* seed oil.

TABLE 93.3 Therapeutic and Nutritional Uses of *M. oleifera* Seed Oil

Conditions	Reference
Medical:	
Fungal infections	Chuang *et al.* (2007)
Skin infections (pyodermia)	Makkar & Becker (1997)
Constipation (as purgative)	Fuglie (2001)
General disorders, prostate function, bladder function, gout, scurvy	Fuglie (2001)
Nutritional	
Antioxidant	Siddhuraju & Becker (2003)

SUMMARY POINTS

- *Moringa oleifera* Lam seeds contain a high proportion of oil that can launch the plant as a major source of plant oil for edible and non-edible purposes. Application of enzymes for oil extraction, though giving lower yields compared to solvent-assisted extraction, may allow labeling of the oil as organic oil.
- The fatty acid composition of *M. oleifera* seed oil is similar to that of olive oil.
- Like olive oil and other genetically modified high-oleic acid oils, such as sunflower, safflower, and canola oils, *M. oleifera* oil is rich in monounsaturated fatty acids, particularly oleic acid.
- The high oleic acid content of the oil confers on it a functional food property, as oils containing high oleic content have been shown to reduce the risk of coronary heart disease.
- The oil is also an excellent frying medium, with a low degradation rate when applied at high cooking temperatures.
- The unrefined oil contains minor components, including tocopherols and other metabolites, that has led to its use in treating medical disorders and the like.

References

Abdulkarim, S. M., Lai, O. M., Muhammad, S. K. S., Long, K., & Ghazali, H. M. (2005). Some physico-chemical properties of *Moringa oleifera* seed oil extracted using solvent and aqueous enzymatic methods. *Food Chemistry, 93*, 253–263.

Abdulkarim, S. M., Lai, O. M., Muhammad, S. K. S., Long, K., & Ghazali, H. M. (2006). Use of enzymes to enhance oil recovery during aqueous extraction of M *oringa oleifera* seed oil. *Journal of Food Lipids, 13*, 113–130.

Abdulkarim, S. M., Lai, O. M., Muhammad, S. K. S., Long, K., & Ghazali, H. M. (2007a). Oleic acid enhancement of *Moringa oleifera* seed oil by enzymatic transesterification and fractionation. *ASEAN Food Journal, 14*, 89–100.

Abdulkarim, S. M., Lai, O. M., Muhammad, S. K. S., Long, K., & Ghazali, H. M. (2007b). Frying quality and stability of high-oleic *Moringa oleifera* seed oil in comparison with other vegetable oils. *Food Chemistry, 105*, 1382–1389.

Allman-Farinelli, M. A., Gomes, K., Favaloro, E. J., & Petocz, P. (2005). Diet rich in high-oleic-acid sunflower oil favorably alters low-density lipoprotein cholesterol, triglycerides, and Factor VII coagulant activity. *Journal of the American Dietetic Association, 105*, 1071–1079.

Anwar, F., & Bhanger, M. I. (2003). Analytical characterization of *Moringa oleifera* seed oil grown in temperate regions of Pakistan. *Journal of Agricultural and Food Chemistry, 51*, 6558–6563.

Chopra, R. N., Badhwar, R. L., & Ghosh, S. (1965). *Poisonous Plants of India, Vol. 1.* New Delhi, India: Council of Agricultural Research. pp. 412–427.

Chuang, P.-H., Lee, C.-W., Chou, J.-C., Murugan, M., Shieh, B.-J., & Chen, H.-M. (2007). Anti-fungal activity of crude extracts and essential oil of *Moringa oleifera* Lam. *Bioresource Technology, 98*, 232–236.

Corbett, P. (2003). It is time for an oil change? Opportunities for high-oleic vegetables oils. *Inform, 14*, 480–481.

Foidl, N., Makkar, H. P. S., & Becker, K. (2001). The potential of Moringa oleifera for agricultural and industrial uses. In J. F. Lowell (Ed.), The miracle tree: The multiple uses of *Moringa* (pp. 45–76). Wageningen, The Netherlands: CTA.

Fuglie, L. J. (2001). *Moringa oleifera:* Natural nutrition for the tropics. Dakar, Senegal: Church World Service. p. 68. Revised and published as The miracle tree: The multiple uses of Moringa. Wageningen, The Netherlands: CTA, p. 172.

Lopez-Huertas, E. (2010). Health effects of oleic acid and long chain omega-3 fatty acids (EPA and DHA) enriched milks. A review of intervention studies. *Pharmacological Research, 61,* 200–207.

Lowell, J. F. (1999). *Moringa oleifera: Natural Nutrition for the tropics.* Dakar, Senegal: Church World Service.

Makkar, H. P. S., & Becker, K. (1997). Nutrients and anti-quality factors in different morphological parts of the *Moringa oleifera* tree. *Journal of Agricultural Science Cambridge, 128,* 311–322.

Mensink, R. P., & Katan, M. B. (1990). Effect of dietary trans-fatty acids on high-density and low-density lipoprotein cholesterol levels in healthy subjects. *Journal of Clinical Nutrition, 323,* 439–445.

Ruiz-Gutiérrez, V., & Perona, J. V. (2007). Olive oil. The most ancient functional food in history. *Inform, 18,* 151–153.

Siddhuraju, P., & Becker, K. (2003). Antioxidant properties of various solvent extracts of total phenolic constituents from three different agro-climatic origins of drumstick tree (*M. Oleifera Lam.*). *Journal of Agricultural and Food Chemistry, 51,* 2144–2155.

Tsaknis, J., & Lalas, S. (2002). Stability during frying of *Moringa oleifera* seed oil variety Periyakulam 1. *Journal of Food Composition and Analysis, 15,* 79–101.

Watkins, C. (2009a). Oilseeds of the future: Part 2. *Inform, 20,* 342–344.

Watkins, C. (2009b). Oilseeds of the future: Part 3. *Inform, 20,* 408–410.

CHAPTER 94

Mucanã (*Canavalia grandiflora*) Seeds and Their Anti-inflammatory and Analgesic Effects

Marcelo H. Napimoga[1,2], Edson H. Teixeira[3], Afrânio G. Fernandes[4], Benildo S. Cavada[5]

[1] Laboratory of Biopathology and Molecular Biology, University of Uberaba, Uberaba/MG, Brazil

[2] Laboratory of Immunology and Molecular Biology, São Leopoldo Mandic Institute and Research Center, Campînas, SP, Brazil

[3] Faculty of Medical School of Sobral, Federal university of Ceara, Sobral/Ceará, Brazil

[4] Center of Sciences and Technology, State University of Ceará, Fortaleza/Ceará, Brazil

[5] BioMol-Lab, Federal University of Ceará, Fortaleza/Ceará, Brazil

795

LIST OF ABBREVIATIONS

ConGF, *Canavalia grandiflora* lectin

CINC-1, cytokine-induced neutrophil chemoattractant-1

IL-1β, interleukin-1β

IL-6, interleukin 6

ICAM, intercellular adhesion molecules

LTB4, leukotriene B4

PECAM-1, platelet/endothelial cell adhesion molecule-1

TNF-α, tumor necrosis factor-α

INTRODUCTION

The *Canavalia* genus of Papilionoideae vines includes about 51 species that exhibit wonderful variation and distribution patterns (see Sauer, 1964, for review), and is popularly known as

Nuts & Seeds in Health and Disease Prevention. DOI: 10.1016/B978-0-12-375688-6.10094-5

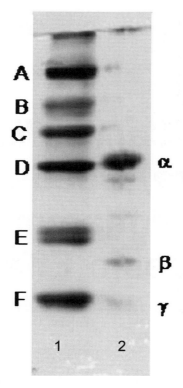

FIGURE 94.1
SDS-PAGE of ConGF. Lane 1: Molecular mass standards: A, 66 kDa; B, 45 kDa; C, 36 kDa; D, 29 kDa; E, 20 kDa; F, 14.2 kDa. Lane 2: purified lectin from *C. grandiflora* (ConGF). α, β, γ = fragments.

Mucunã or "Feijão Bravo" (wild bean). *Canavalia* is palatable to livestock and high in protein; however, it contains toxic compounds and must be used with caution. Usually *Canavalia* is used as a green adubation to improve soil productivity. The seeds of various *Canavalia* species are used as a source of important lectins that have already been purified and characterized, such as concanavalin A from *Canavalia ensiformis* seeds, and ConGF from *Canavalia grandiflora* seeds.

ConGF is a D-glucose/D-mannose-specific lectin with physicochemical and structural properties similar to those of lectins from *Canavalia* and related genera from the subtribe Diocleinae. ConGF was purified by affinity chromatography on Sephadex G-50 (Ceccatto *et al.*, 2002). By SDS polyacrylamide gel electrophoresis, ConGF yielded three protein bands with apparent molecular masses of 29–30 kDa (alpha chain), 16–18 kDa (beta fragment), and 12–13 kDa (gamma fragment) (Figure 94.1). The amino acid composition analysis of ConGF showed that the protein is rich in aspartic acid and serine (Table 94.1). Methionine and cysteine residues were not detected. This profile is very similar to the amino acid composition shown by other *Diocleinae* lectins, and it is characteristic of most plant stored seed proteins. Furthermore, ConGF demonstrates important effects as a modulator of key inflammatory events accruing at the interface of leukocytes and the vascular endothelium during inflammatory responses.

BOTANICAL DESCRIPTION

Canavalia grandiflora is a weakly perennial, prostrate, herbaceous leguminous plant. Its leaves are trifoliolate; leaflets are ovate with an acute apex, 12–15 cm long and 6–8 cm wide (i.e., about twice as long as wide), and glabrous. Inflorescences are axillary racemes, with up to two purple flowers. The pods are glabrous, 6–10 cm long and approximately 1 cm wide, brown to dark brown in color, dehiscent, and carry an average of eight seeds. The petals are about 3.5 cm long,

TABLE 94.1 Amino Acid Composition of *Canavalia grandiflora* Lectin (ConGF)

Amino Acid	Nmol%	Residues/Mol
Lys	7.12	17.9
His	0.87	2.1
Arg	1.23	2.9
Asx	16.20	38.4
Thr	6.85	16.2
Ser	10.83	24.6
Glx	7.36	17.4
Pro	4.52	10.7
Gly	8.51	20.2
Ala	7.62	18.1
Val	6.21	14.7
Met	-	-
Ile	4.98	11.8
Leu	8.83	20.9
Tyr	2.01	4.8
Phe	6.85	16.2
Trp	-	-
Cys	-	-

Amino acid and amino sugar composition of ConGF were carried out with an AlphaPlus (Pharmacia) after sample hydrolysis in sealed, evacuated ampoules at 106°C with 4N HCl for 4 h (determination of amino sugars), and with 6N HCl for 18 h for amino acid composition.
Adapted from Ceccato *et al.* (2002).

with a coiled nail on top, spreading and reflexed, and wide ovate in shape. The wings are about 2.5 cm long, narrow and tortuous, dilated above the nail, with a rounded auricle. The seeds are light brown to brown, approximately 8 mm long and 5 mm wide, with a 4-mm long black hilum; the 1000-seed weight is 140–160 mg. There is a high level of hard seededness.

HISTORICAL CULTIVATION AND USAGE

The genus *Canavalia* (Leguminosae; Papilionoideae; Phaseoleae; Diocleinae) comprises 51 species of leguminous vines, which are widespread in Brazil, Guiana, and the Atlantic coastal regions of tropical Central America and Mexico. The *Canavalia grandiflora* is a wild leguminous vine found in the Northeast of Brazil. There is no commercial or domestic cultivation, and the wild plant is only used for animal fodder.

PRESENT-DAY CULTIVATION AND USAGE

Although seeds of other *Canavalia* species, like jackbeans (*Canavalia ensiformis*), are commercially grown, the species *Canavalia grandiflora* Benth. is not grown commercially and occurs only in the wild in the state of Ceará, Brazil. *Canavalia* plants are used for animal fodder, although they contain some antinutritional factors, such as the non-proteic amino acid canavanin, trypsin inhibitors, canatoxin, and lectins. The seeds are used as potential sources of the urease enzyme, and also of concanavalin A, a lectin with many biotechnological applications, such as chromatography for glicoconjugate purification. The whole jackbean plant is largely used to improve soil fertility.

APPLICATIONS TO HEALTH PROMOTION AND DISEASE PREVENTION

In the past 30 years, evidence from several studies has dramatically increased the knowledge regarding several human diseases caused by an excessive inflammatory response. Examples include multiple sclerosis, rheumatoid arthritis, asthma, arteriosclerosis, etc. Those pathologies indicate a critical role for tissue infiltration by leukocytes in inflammatory disease pathogenesis. For immigration of circulating leukocytes into tissues, transmigration through the vascular endothelial layer involves two independently regulated events: binding to vessel endothelium, followed by diapedesis. For this purpose, cell arrest is mediated by activation of adhesion receptors on the moving cell, followed by attachment to counter-receptors on other cells or endothelial cells, leading to an immobilized cell. Furthermore, the inflammatory response is mediated by a broad spectrum of mediators able to promote vascular events, edema, and, finally, the recruitment of inflammatory cells (Friedl & Weigelin 2008).

Many new anti-inflammatory drugs aim at interfering with transendothelial migration by antagonizing surface integrins, selectins, or chemokines involved in leukocyte homing. However, despite the billions of dollars that have been spent on research by the pharmaceutical industry, few effective anti-inflammatory drugs have been developed. Thus, constituents isolated from plants, seeds, and legumes have evoked great interest worldwide regarding their ability to combat many different inflammatory disorders.

Nevertheless, in inflammation the resident cells, such as dendritic cells, macrophages, mast cells, and lymphocytes, are tissue cells that, after recognizing the inflammatory stimuli, release cytokines. These cytokines play an essential role in the development of inflammatory pain, as well as other inflammatory events such as leukocyte migration. Certainly pain is an important symptom of inflammatory diseases, and is the principal reason why patients with inflammatory diseases look for specialized treatment. Sensory systems have the role of informing the brain about the state of the external environment and the internal milieu of the organism. Pain is a perception, and as such is one of the outputs of a system in more highly evolved animals — the nociceptive system — which itself is a component of the overall set of controls responsible for homeostasis. In this way, pain constitutes an alert that helps to protect the organism by triggering reactions and inducing learned avoidance behaviors, which may decrease whatever is causing the pain and, as a result, limit the (potentially) damaging consequences.

Following tissue injury and inflammation, nociceptors are sensitized in such a way that previously slight or ineffective stimulation becomes painful. The sensitization of primary afferent nociceptors is a common denominator of all kinds of inflammatory pain that leads to a state of hyperalgesia and/or allodynia in humans, better described as hypernociception in animal models. Hypernociception is induced by the direct action of inflammatory mediators, such as prostaglandins and sympathetic amines, on the peripheral nociceptors. These direct acting hyperalgesic mediators are ultimately released in the inflamed tissue following a cascade of cytokines by the resident and migratory cells: TNF-α is released early and stimulates the release of IL-6 and IL-1β, which triggers eicosanoid release. TNF-α also induces CINC-1 (rat IL-8 related chemokine) release, which stimulates sympathomimetic amine production (Cunha et al., 2005). Besides the nociceptive direct-acting mediators described above, there is evidence that endothelins, substance P, and LTB4 also sensitize peripheral nociception. Furthermore, it has been demonstrated that leukocytes — in particular neutrophils — are strongly implicated in inflammatory nociception. Considering that activated neutrophils produce and release pro-inflammatory cytokines, including TNF-α, IL-1β, and CINC-1/CXCL1, and mediators such as prostaglandin E$_2$, which are essential to the development of inflammatory hypernociception, it is conceivable that neutrophils could be a relevant source of hypernociceptive cytokines or, alternatively, of the direct-acting hypernociceptive mediators, such as prostaglandin (Cunha et al., 2008).

In this line, lectins constitute a group of widely distributed and structurally heterogeneous carbohydrate-binding proteins that comprises distinct but evolutionary related families. Several biological activities of lectins have been described, such as its effect in cancer therapy, insecticidal effects, bacterial inhibition, etc. (Pusztai et al., 2008). Regarding to the inflammatory process, some previous works have reported the useful application of lectins as an anti-inflammatory approach by inhibiting leukocyte migration (Napimoga *et al.*, 2007), since lectins are recognized among adhesion molecules that actively participate in those responses, such as selectins (L-, P-, and E-selectin) and integrins CD11b/CD18 (Mac-1), CD31 (PECAM-1), and CD44. Specifically, it has been demonstrated that *C. grandiflora* lectin, a leguminosae lectin, was able to inhibit carrageenan-induced peritonitis by decreasing leukocyte rolling and adhesion to the endothelium (Nunes *et al.*, 2009). This result is partially explained by the inhibition of pro-inflammatory cytokines IL-1β and TNF-α, since they are necessary to activate the endothelial cells to present β2-integrin in an active conformation, which in turn allows intercellular adhesion molecule (ICAM) expression on cytokine-activated endothelium, as well as on some circulating blood cells. Thus, these data suggest that the *in vivo* anti-inflammatory effect of this lectin works through the inhibition of pro-inflammatory cytokine production at inflammatory sites, which in turn inhibits leukocyte interaction with endothelial cells (Figure 94.2). However, the reduction in rolling and adhesion of leukocytes is also a consequence of the interference of ConGF in the interaction of leukocytes with the endothelium so cannot be excluded. In fact, several lectins reduce these interactions of leukocytes in *in vitro* and *in vivo* experiments.

ConGF can also be a useful pharmacological tool as an analgesic agent, since it has been demonstrated that ConGF is able to inhibit mechanical hypernociception, and this effect correlated with lower neutrophil influx to mice hind paws (Nunes *et al.*, 2009). Furthermore, it was demonstrated that ConGF administration was able to inhibit mouse paw edema after carrageenan administration. The edema induced by carrageenan is a temporal and multi-mediated phenomenon, involving the participation of a diversity of mediators. In the initial phase (1/2–1 hour), bradykinin, histamine, and serotonin are released by local cells. In a later phase (3–4 hours), the release of prostaglandins takes place. Since ConGF potently inhibited the edema evoked by carrageenan, until the fourth hour after challenge – the time coincident with prostaglandin release – it is conceivable that this lectin can inhibit prostaglandin E_2 synthesis. Pretreatment of mice with ConGF also inhibited the second phase of hypernociception induced by formalin in mice paws. Importantly, intraplantar injections of formalin produce a biphasic behavioral reaction in which the first phase results essentially

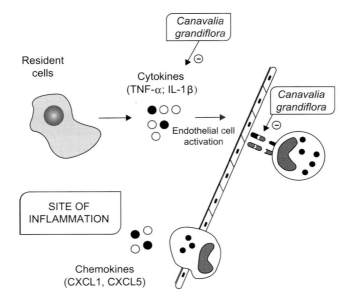

FIGURE 94.2

Schematic representation of the mechanisms of C. grandiflora lectin (ConGF) in inflammatory migration. ConGF is able to inhibit the leukocyte adhesion on venular endothelial cells due to competitive inhibition of leukocyte–endothelium interaction. Furthermore, neutrophils migrate in response to cytokines, and this effect is also inhibited by treatment with ConGF.

from the direct stimulation of nociceptors, whereas the second involves a period of sensitization during which inflammatory phenomena occur. These results reinforce the proposal that ConGF's antihyperalgesic effect is mostly due to a reduction in neutrophils which mediate the release of direct-acting hypernociceptive mediators (e.g., prostaglandins and sympathetic amines).

Thus, the development of therapies using ConGF as prototypes to inhibit neutrophil migration could be an important aim in the management of inflammatory diseases, as well as pain-associated processes. Nevertheless, the challenge for the future is to identify lectins or natural compounds that block specific trafficking molecule(s) which will most specifically inhibit that subset of cells while leaving most leukocytes unaffected, so the host is not left more susceptible to certain infections. In this regard, we are still working to evaluate which adhesion molecule the lectin ConGF is interacting with in order to inhibit specific leukocyte migration.

ADVERSE EFFECTS AND REACTIONS (ALLERGIES AND TOXICITY)

Ricin is a potent toxin extracted from castor beans (*Ricinus communis*). This protein is the better understood lectin with RIP (ribosome inactivating protein) activity (Spivak *et al.*, 2005). Several lectins have been described as being potential inflammatory proteins, because these molecules stimulate and bind to cells of the immune system (Assreuy *et al.*, 2003; Alencar *et al.*, 2007; Lee *et al.*, 2007), specifically mast cells (Alencar *et al.*, 2005; Lopes *et al.*, 2005) including histamine induction release (Gomes *et al.*, 1994; Ferreira *et al.*, 1996). Recently, our group has published results about the use of plant lectins as potential immunomodulatory agents (Reis *et al.*, 2008; Nunes *et al.*, 2009). Despite these findings, the utilization of plant lectins as anti-inflammatory agents is an important research field, and a number of publications about this theme can easily be found. Some lectins, like *Vatairea macrocarpa* lectin, when injected peritoneally (Alencar *et al.*, 2003) or intraplantarly (Alencar *et al.*, 2004) show inflammatory effects. However, when the same lectin is administered endovenously, nothing is observed (Alencar *et al.*, 1999). The utilization of ConGF as a biological tool to control inflammatory mechanisms has already been observed (Nunes *et al.*, 2009); however, its therapeutic use requires more toxicity studies *in vitro* and *in vivo*, as *C. grandiflora* possesses toxic compounds such as non-proteic amino acid (canavanine), toxic enzymes (urease), and neurotoxic protein (canatoxin). Besides, it is important to note that when we administered a high dose of *C. grandiflora* (10 mg/kg) in a single-dose scheme over 7 consecutive days, there were no signs of toxicity, such as a change in the animal's corporal mass, and altered urea, aspartate transaminase, and alanine transaminase values (Table 94.2). To reduce the immunogenicity of proteins like ConGF, the drug design field may be an interesting way ahead.

TABLE 94.2 Sub-chronic Treatment of Animals with *C. grandiflora* Lectin (ConGF)

Parameters	Treatment (0.1 mL; i.v.)	
	Vehicle	**ConGF (10 mg/kg)**
Corporal mass (g) after	35.5 ± 2.9	30.4 ± 1.8
Corporal mass (g) before	31.7 ± 1.4	28.4 ± 3.5
Liver	1.83 ± 0.2	1.61 ± 0.1
Kidney	0.23 ± 0.0	0.20 ± 0.0
Heart	0.21 ± 0.0	0.21 ± 0.0
Blood Neutrophils $\times 10^6$/mL	1.36 ± 0.0	1.57 ± 0.2
Urea (mg/dl)	35.1 ± 3.3	32.5 ± 6.4
ALT (UI/L)	58.6 ± 5.0	42.4 ± 2.5
AST (UI/L)	203.6 ± 26.0	192.2 ± 33.9

Animals were injected daily with single doses for 7 consecutive days. After 7 days of treatment, animals were weighed, blood samples were collected for leukogram count and biochemical dosage, animals were sacrificed, and the organs' wet weight recorded. Values are reported as mean ± standard deviation.

SUMMARY POINTS

- *Canavalia grandiflora* has the common names of Mucunã or Feijão Bravo (wild bean) in Brazil.
- *C. grandiflora* is a weakly perennial, prostrate, herbaceous leguminous plant.
- The plants are used for animal fodder, although they contain some antinutritional factors. The seeds are used as potential sources of urease enzyme, and also as a source of concanavalin A, a lectin with many biotechnological applications.
- ConGF is a D-glucose/D-mannose-specific lectin with three protein bands with molecular masses of 29–30 kDa (alpha chain), 16–18 kDa (beta fragment), and 12–13 kDa (gamma fragment), and is rich in aspartic acid and serine.
- ConGF possesses an important anti-inflammatory activity by decreasing leukocyte rolling and adhesion to the endothelium due to the inhibition of pro-inflammatory cytokines, which are important in activating the adhesion molecules in the endothelium.
- ConGF also demonstrates an antihypernociceptive action by inhibiting neutrophil migration and edema formation.

References

Alencar, N. M., Teixeira, E. H., Assreuy, A. M., Cavada, B. S., Flores, C. A., & Ribeiro, R. A. (1999). Leguminous lectins as tools for studying the role of sugar residues in leukocyte recruitment. *Mediators of Inflammation, 8*, 107–113.

Alencar, N. M., Assreuy, A. M., Alencar, V. B., Melo, S. C., Ramos, M. V., Cavada, B. S., et al. (2003). The galactose-binding lectin from *Vatairea macrocarpa* seeds induces *in vivo* neutrophil migration by indirect mechanism. *International Journal of Biochemistry & Cell Biology, 35*, 1674–1681.

Alencar, N. M., Assreuy, A. M., Criddle, D. N., Souza, E. P., Soares, P. M., Havt, A., et al. (2004). *Vatairea macrocarpa* lectin induces paw edema with leukocyte infiltration. *Protein & Peptide Letters, 11*, 195–200.

Alencar, V. B., Brito, G. A., Alencar, N. M., Assreuy, A. M., Pinto, V. P., Teixeira, E. H., et al. (2005). *Helianthus tuberosus* agglutinin directly induces neutrophil migration, which can be modulated/inhibited by resident mast cells. *Biochemistry and Cell Biology, 83*, 659–666.

Alencar, N. M., Assreuy, A. M., Havt, A., Benevides, R. G., de Moura, T. R., de Sousa, R. B., et al. (2007). *Vatairea macrocarpa* (Leguminosae) lectin activates cultured macrophages to release chemotactic mediators. *Naunyn Schmiedebergs Archives of Pharmacology, 374*, 275–282.

Assreuy, A. M., Alencar, N. M., Cavada, B. S., Rocha-Filho, D. R., Feitosa, R. F., Cunha, F. Q., et al. (2003). Porcine spermadhesin PSP-I/PSP-II stimulates macrophages to release a neutrophil chemotactic substance: modulation by mast cells. *Biology of Reproduction, 68*, 1836–1841.

Ceccatto, V. M., Cavada, B. S., Nunes, E. P., Nogueira, N. A., Grangeiro, M. B., Moreno, F. B., et al. (2002). Purification and partial characterization of a lectin from *Canavalia grandiflora* benth. seeds. *Protein & Peptide Letters, 9*, 67–73.

Cunha, T. M., Verri, W. A., Jr., Silva, J. S., Poole, S., Cunha, F. Q., & Ferreira, S. H. (2005). A cascade of cytokines mediates mechanical inflammatory hypernociception in mice. *Proceedings of the National Academy of Sciences of the United States of America, 102*, 1755–1760.

Cunha, T. M., Verri, W. A., Jr., Schivo, I. R., Napimoga, M. H., Parada, C. A., Poole, S., et al. (2008). Crucial role of neutrophils in the development of mechanical inflammatory hypernociception. *Journal of Leukocyte Biology, 83*, 824–832.

Ferreira, R. R., Cavada, B. S., Moreira, R. A., Oliveira, J. T., & Gomes, J. C. (1996). Characteristics of the histamine release from hamster cheek pouch mast cells stimulated by lectins from Brazilian beans and concanavalin A. *Inflammation Research, 45*, 442–447.

Friedl, P., & Weigelin, B. (2008). Interstitial leukocyte migration and immune function. *Nature Immunology, 9*, 960–969.

Gomes, J. C., Ferreira, R. R., Cavada, B. S., Moreira, R. A., & Oliveira, J. T. (1994). Histamine release induced by glucose (mannose)-specific lectins isolated from Brazilian beans. Comparison with concanavalin A. *Agents Actions, 41*, 132–135.

Lee, J. Y., Kim, J. Y., Lee, Y. G., Byeon, S. E., Kim, B. H., Rhee, M. H., et al. (2007). *In vitro* immunoregulatory effects of Korean mistletoe lectin on functional activation of monocytic and macrophage-like cells. *Biological & Pharmaceutical Bulletin, 30*, 2043–2051.

Lopes, F. C., Cavada, B. S., Pinto, V. P., Sampaio, A. H., & Gomes, J. C. (2005). Differential effect of plant lectins on mast cells of different origins. *Brazilian Journal of Medical and Biological Research, 38*, 935–941.

Napimoga, M. H., Cavada, B. S., Alencar, N. M., Mota, M. L., Bittencourt, F. S., Alves-Filho, J. C., et al. (2007). *Lonchocarpus sericeus* lectin decreases leukocyte migration and mechanical hypernociception by inhibiting cytokine and chemokines production. *International Immunopharmacology, 7,* 824–835.

Nunes, B. S., Rensonnet, N. S., Dal-Secco, D., Vieira, S. M., Cavada, B. S., Teixeira, E. H., et al. (2009). Lectin extracted from *Canavalia grandiflora* seeds presents potential anti-inflammatory and analgesic effects. *Naunyn Schmiedeberg's Archives of Pharmacology, 379,* 609–616.

Pusztai, A., Bardocz, S., & Ewen, S. W. (2008). Uses of plant lectins in bioscience and biomedicine. *Frontiers in Bioscience, 13,* 1130–1140.

Reis, E. A., Athanazio, D. A., Cavada, B. S., Teixeira, E. H., de Paulo Teixeira Pinto, V., Carmo, T. M., et al. (2008). Potential immunomodulatory effects of plant lectins in *Schistosoma mansoni* infection. *Acta Tropica, 108,* 160–165.

Sauer, J. (1964). Revision of Canavalia. *Brittonia, 16,* 106–181.

Spivak, L., & Hendrickson, R. G. (2005). Ricin. *Critical Care Clinics, 21,* 815–824.

CHAPTER 95

Antibacterial Potential of Neem Tree (*Azadirachta indica* A. Juss) Seeds

Shyamapada Mandal[1], Manisha Deb Mandal[2]
[1] Department of Microbiology, Bacteriology and Serology Unit, Calcutta School of Tropical Medicine, Kolkata, India
[2] Department of Physiology and Biophysics, KPC Medical College and Hospital, Jadavpur, Kolkata, India

LIST OF ABBREVIATIONS

CFU, colony forming unit
MIC, minimum inhibitory concentration
MDR, multidrug resistant
NSEE, neem seed ethanol extract
RSTI, respiratory tract infection
RTI, reproductive tract infection
SIC, sub-inhibitory concentration
UTI, urinary tract infection
ZDI, zone diameter of inhibition

INTRODUCTION

Neem has two closely related species, *Azadirachta indica* A. Juss and *Melia azadirachta*; the former is popularly known as the Indian neem (margosa tree) or Indian lilac, and the latter as

Nuts & Seeds in Health and Disease Prevention. DOI: 10.1016/B978-0-12-375688-6.10095-7

the Persian lilac. *A. indica* has been well known in India and its neighboring countries for more than 2000 years as a versatile medicinal plant with a broad spectrum of biological activity, including antibacterial properties. Currently, it is regarded as the most useful traditional medicinal plant in India, and known as the "village pharmacy." The Sanskrit name for the neem tree is *Arishtha*, meaning "reliever of sickness," and all parts of the tree have been in use in Ayurvedic, Unani, and homeopathic medicine; hence, it is considered to be *Sarbaroganibarini* ("one that can cure all bodily ailments").

In recent years, interest in neem has focused on its repellent, anti-feedant, and growth-disrupting effects on insects, as well as its ability to provide a cheap natural pesticide. Neem oil, produced from the seed kernels of neem fruit, is the most commonly used neem product. It is generally regarded as possessing the highest concentration of active components with regard to the manufacture of pesticides. It is also used in soap manufacture, and as a treatment for various skin disorders. Some recent studies have also demonstrated activity of the plant against bacteria, viruses, and fungal infections. Originating in India, neem is now widespread throughout arid, tropical, and subtropical areas of the world. In Nepal, it mostly grows in the southern Terai region.

BOTANICAL DESCRIPTION

Neem is a fast-growing evergreen tree of 15—20 m in height. It has a straight trunk and its bark is hard, rough, and scaly, fissured even in small trees. The leaves, 20—40 cm long, are alternate with several leaflets, and serrated margins. The tree blossoms in the spring, with small white bisexual axillary flowers; the inflorescences branch upwards and bear 150—250 flowers. The fruit is a smooth olive-like drupe, which varies in shape from elongate oval to round, and when ripe is 1.4—2.8 × 1.0—1.5 cm in size (Figure 95.1). The fruit skin (exocarp) is thin, and the bitter-sweet pulp (mesocarp) is yellowish-white and very fibrous. The white, hard inner shell (endocarp) of the fruit encloses a single elongated seed (kernel) with a brown seed coat. The neem tree was described as *Azadirachta indica* as early as 1830, by De Jussieu. Taxonomically, it is a member of the Melieae tribe of the subfamily Meliodeae (Family: Meliaceae; Suborder: Rutenae; Order: Rutales).

FIGURE 95.1
Neem fruits and leaves. *Photograph by Shyamapada Mandal.*

HISTORICAL CULTIVATION AND USAGE

All parts of neem have been in use since ancient times, to treat several human ailments, and also as a household pesticide. The commercial use of neem was known to exist in the Vedic period in India (over 4000 years BC), and domestic uses were mentioned by Kautilya in his *Arthasastra* (4 BC). Under natural conditions, neem seeds ordinarily fall onto the ground and the viable seeds germinate within a week or two, in the presence of rain water. The seedlings that grow in the shade of the tree are transplanted and propagated in empty fields, since neem is renowned for good growth on dry, infertile sites. Neem seed oil, bark, and leaf extracts have been therapeutically used as folk medicine to control diseases such as leprosy, intestinal helminthiasis, respiratory disorders, constipation, and skin infections (Biswas *et al.*, 2002). Neem's excellent antibacterial and anti-allergic properties make it effective in fighting most epidermal dysfunctions, such as acne, psoriasis, and eczema. Ancient ayurvedic practitioners believed high blood sugar levels caused skin diseases, and the bitter principles of neem were said to counteract the excess sugar. This tree, originally cultivated in India, occupies a prominent place in Asian popular medicine, as well as in bodily hygiene.

PRESENT-DAY CULTIVATION AND USAGE

Neem is widely distributed in India, and is now becoming cultivated throughout the world. It can be propagated through the seeds and shoot cuttings. Tissue culture techniques have also been established for plant propagation; the plants are regenerated both from directly formed somatic embryos, and from somatic embryos derived from cell suspensions.

Modern research confirms the curative powers of neem for many diseases. Multiple active substances from different parts of the neem tree have been identified, and the components are reputed to have medicinal, spermicidal, antiviral, antibacterial, antiprotozoal, insecticidal, insect-repellent, antifungal, and antinematodal properties. They are also believed to remove toxins from the body, neutralize free radicals, and purify the blood. Recently, neem has been used in anticancer treatment, and it has hepato-renal protective activity and hypolipidemic effects.

Current scientific study results support the use of neem, as it is employed in India, for the treatment of diarrhea and cholera. Neem is now being used by leading pharmaceutical companies throughout the world for producing new medicines in order to combat many life-threatening diseases.

APPLICATIONS TO HEALTH PROMOTION AND DISEASE PREVENTION

Medicinal plants, from the very dawn of civilization, have been part and parcel of human society in combating diseases. Many of the drugs that are in use in modern medicine were initially used in a crude form in traditional or folk healing practices, or for other purposes that suggest potentially useful biological activity. Neem is a very important traditional medicinal plant in India that has been used extensively in Ayurveda, Unani, and homoeopathic medicine, and has become a focus of modern medicine. Every part of it has been used, since antiquity, in traditional household remedies for a number of human ailments (Biswas *et al.*, 2002). There are many reports on the biological activities and pharmacological actions of neem based on modern scientific investigations, but since the details of these studies are beyond the scope of this review, some of the important medicinal attributes of various parts of neem, as cited in Ayurveda, are summarized in Table 95.1.

Neem comprises a vast array of chemically diverse and structurally complex biologically active compounds. Neem chemistry dates back to 1880–1890, when, influenced by its folklore medicinal values, chemists attempted the isolation of the active principle from its seed and other parts; however, Siddiqui (1942) was the first to report the isolation of three products — nimbin,

TABLE 95.1 Some Important Medicinal Attributes of Various Parts of Neem

Plant Part(s)	Medicinal Uses
Mixture of root, bark, leaf, flower and fruit	Leprosy, blood morbidity, biliary afflictions, skin ulcers, burning sensation, and itching
Fruit	Leprosy, diabetes, hemorrhoids, intestinal worms, urinary disorders, skin diseases, eye problems, and wounds
Seed pulp and oil	Leprosy, and intestinal worms
Flower	Bile suppression, and intestinal worms
Leaf	Leprosy, intestinal worms, eye problems, anorexia, biliousness, skin ulcers
Twig	Cough, asthma, hemorrhoids, intestinal worms, diabetes
Gum	Skin diseases, such as ringworm, scabies, wounds, and ulcers
Bark	Used as an analgesic, and to cure fever

nimbidin, and nimbinin — from its oil. Since then a large number of compounds have been isolated from the different parts of neem, and many reviews have appeared in the literature regarding the chemistry and structural diversity of these compounds (Biswas *et al.*, 2002). Among those compounds, the bitterness of neem is due to limonoids, which are tetranor-triterpenoids, and the antibacterial properties are attributed to the presence of such bitter substances as azadirachtin, salannin, meliantriol, mahmoodin, and nimbin.

The consensus is that the general antimicrobial activity of neem extracts is mainly due to azadirachtin. Nimbidin, a crude bitter principle extracted from neem seed oil, has demonstrated several biological activities; *in vitro*, it can completely inhibit the growth of *Mycobacterium tuberculosis*, and has also been found to be bactericidal; nimbolide also shows antibacterial activity against *Staphylococcus aureus* and *S. coagulase*, and mahmoodin, a deoxy-gedunin isolated from seed oil, has been reported to possess antibacterial action against human pathogenic bacteria (Biswas *et al.*, 2002). Neem seed extracts have shown excellent antibacterial activity against *Salmonella enterica* serovar Typhi associated with enteric fever in and around Kolkata, India (Mandal *et al.*, 2007). Pritima and Pandian (2008) tested the antibacterial potential of neem extract against reproductive tract infection (RTI)-causing bacteria like *Escherichia coli*, *Enterococcus faecalis*, *Proteus mirabilis*, *Bacillus cereus*, *Klebsiella pneumoniae*, *S. aureus*, and *Neisseria gonohorreae* seen among women, and found all to be susceptible to the extracts, apart from *N. gonohorreae*, which was resistant. SaiRam *et al.* (2000) reported antimicrobial activity of a new vaginal contraceptive from neem oil. Okemo *et al.* (2001) reported the antibacterial activity of *A. indica* against bacterial isolates such as *S. aureus*, *E. coli*, and *Pseudomonas aeruginosa*. Thus, neem seed extract is effective in controlling the bacteria that cause watery diarrhea and cholera, typhoid, RSTI, UTI, RTI, secondary infections in wounds, and many other bacterial life-threatening diseases.

Neem seed ethanol extract (NSEE) (500 μg/ml) showed excellent antibacterial activity against different pathogenic bacteria; the results recorded in our laboratory (Mandal *et al.*, unpublished data) in terms of the ZDI and MIC of the extract are represented in Figures 95.2 and 95.3. The kill kinetics (based upon the author's own study results) of NSEE against gram-positive (*S. aureus*) and gram-negative (*E. coli*) bacterial pathogens are represented in Figures 95.4 and 95.5. At the SIC ($1/2 \times$ MIC, 32 μg/ml) the NSEE started to show killing activity, but at this concentration the extract did not show bactericidal action against *S. aureus* even after 24 hours; the higher extract concentrations ($1 \times$ MIC, 64 μg/ml and $2 \times$ MIC, 128 μg/ml) completely wiped out all viable bacteria at 6 and 24 h, respectively (Figure 95.4).

NSEE at the MIC (256 μg/ml) level and at a concentration of 512 μg/ml ($2 \times$ MIC) did not completely kill the *E. coli* cells, even at 24 h; the concentrations were found to be bactericidal at 24 and 6 h, respectively, since the initial inocula of 5×10^5 CFU/ml (5.698 $\log_{10}$ CFU/ml) was

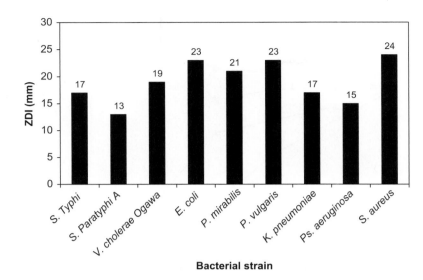

FIGURE 95.2
Zone diameter of inhibition (ZDI), obtained following the agar diffusion technique, of neem (*A. indica*) seed ethanol extract for clinical bacterial strains.

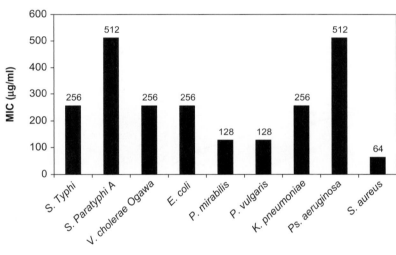

FIGURE 95.3
Minimum inhibitory concentration (MIC) values of neem (*A. indica*) seed ethanol extract for clinical bacterial strains.

807

reduced to 0.00131×10^5 CFU/ml ($2.12 \log_{10}$ CFU/ml) and 0.0003×10^5 CFU/ml ($1.48 \log_{10}$ CFU/ml), respectively; the SIC ($1/2 \times$ MIC, 128 µg/ml) of the extract caused initial reduction of the bacterial population, while at 24 h the population exceeded the initial inocula (Figure 95.5). Thus, for gram-positive (*S. aureus*) and gram-negative (*E. coli*) bacteria, the killing was both concentration- and time-dependent, and this is a more rational basis for determining optimal dosage for antimicrobial treatment regimens (Chalkley & Koornholf, 1985).

Banso and Adeymo (2003) recorded higher antibacterial activity of plant extracts with increased concentration against gram-negative bacteria, namely *Ps. aeruginosa*, *K. pneumonia*, and *E. coli*, and stated that the microorganisms need higher extract concentrations to inhibit growth or kill them depending on their cell wall components; thus, the antibacterial substances in the extracts might affect synthesis of the peptidoglycan layer of the cell wall. Okemo *et al.* (2001) showed concentration-dependent killing of gram-negative (*Ps. aeruginosa* and *E. coli*) and gram-positive (*S. aureus*) bacteria with neem extracts, and mentioned that the

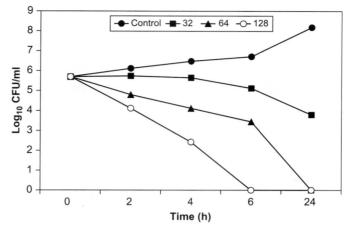

FIGURE 95.4

Killing activity of neem (*A. indica*) seed ethanol extract against *S. aureus* SA3 strain isolated from urinary tract infection case at the Calcutta School of Tropical Medicine, India. The bacterial strain, with initial inocula of 5.698 $\log_{10}$ CFU/ml in Mueller Hinton broth, was exposed to the extract at different concentrations: ½ × MIC, 32 µg/ml; 1 × IC, 64 µg/ml; and 2 × MIC, 128 µg/ml, as listed in the key to the figure.

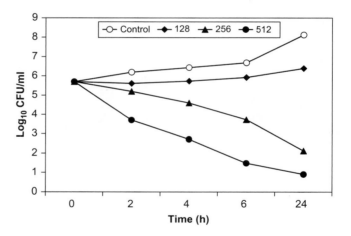

FIGURE 95.5

Killing activity of neem (*A. indica*) seed ethanol extract against *E. coli* EC2 strain isolated from urinary tract infection case at the Calcutta School of Tropical Medicine, India. The bacterial strain, with initial inocula of 5.698 $\log_{10}$ CFU/ml in Mueller Hinton broth, was exposed to the extract at different concentrations: ½ × MIC, 128 µg/ml; 1 × MIC, 256 µg/ml; and 2 × MIC, 512 µg/ml, as listed in the key to the figure.

mode of action of the extracts is strongly cell-wall related. Mandal *et al.* (2007) reported an interesting finding of concentration- as well as time-dependent killing of the gram-negative bacterial pathogen *S. enterica* serovar Typhi. Thus, data presented in this chapter support the view put forward by Banso and Adeymo (2003), Mandal *et al.* (2007), and Okemo *et al.* (2001) regarding the mode of action of neem extracts. Further, the antibacterial activity of neem seed oil *in vitro* against pathogenic bacteria has been assessed, and the activity was found to be bactericidal, which was probably due to the inhibition of cell-membrane synthesis in the bacteria, as has been reported by Baswa *et al.* (2001). Neem oil has been reported to suppress various species of pathogenic bacteria, such as *S. aureus* and *S. typhosa*, and strains of *M. tuberculosis* (Chaurasia & Jain, 1978; Rao *et al.*, 1986); the growth of *Salmonella enterica* serovar Paratyphi and *V. cholerae* was also inhibited (Rao, 2005).

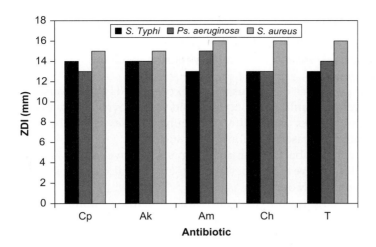

FIGURE 95.6
Zone diameter of inhibition of antibiotics for the clinical isolates of multidrug resistant *S. enterica* serovar Typhi, *Ps. aeruginosa*, and *S. aureus* isolates following disc diffusion antibiotic susceptibility test.

The findings of the current *in vitro* study have revealed that NSEE is synergistic with some of the antibiotics — namely, ciprofloxacin, amikacin, ampicillin, chloramphenicol, and tetracycline — against *Ps. aeruginosa*, *S. enterica* serovar Typhi, and *S. aureus* (Figure 95.6; Mandal *et al.*, unpublished data). In this study, all clinical bacterial isolates were MDR, having a ZDI ranging from 8 mm to none at all (6-mm antibiotic discs from Hi-Media, India); the ZDIs for the bacterial isolates increased in the presence of NSEE (250 µg/ml), ranging from 13 mm to 16 mm.

ADVERSE EFFECTS AND REACTIONS (ALLERGIES AND TOXICITY)

Toxicity studies performed by Arnason *et al.* (1989) showed that neem plant concoctions can be used in fair quantities without apparent hazardous consequences. However, neem seed oil and/or extracts demonstrated lethal toxicity to the fish *Oreochromis niloticus* and *Cyprinus carpio*, acute toxicity to rats and rabbits, a severe hypoglycemic effect on rats, and spermicidal activity in rhesus monkeys and humans; it also produces vomiting, diarrhea, drowsiness, acidosis, and encephalopathy in humans (Jacobson, 1995). However, neem has long been consumed on a daily basis by humans, and the seed kernel cake has been successfully used as a protein supplement for livestock. Thus, neem has extremely low mammalian toxicity, and is relatively safe to non-target organisms. No hazard has been documented under conditions of normal usage, although Dhongade *et al.* (2008) reported a case of neem-oil poisoning in which a 5-year-old child presented with refractory seizures and metabolic acidosis; a late neurological disorder in the form of auditory and visual disturbances, and ataxia, was found. Due to its spermicidal property, neem oil may be an effective contraceptive; however, possible impairment of male fertility, and unintentional contraceptive and abortive effects, must be considered in long-term use. Presence of aflatoxin in crude neem oil preparations may cause acute toxic effects with several clinical symptoms, the liver being the main target organ, and a high dose causes hepatic failure.

Azadirachtin ($C_{35}H_{44}O_{16}$) is the most active principle present in neem extract, and structural analysis suggests its genotoxicity and carcinogenicity, as well predicting that it may cause cellular toxicity (Rosenkranz & Klopman, 1995a, 1995b). Alterations in testicular DNA (Evenson *et al.*, 1986) or the predicted cellular toxicity might be the cause of germ cell genotoxicity of azadirachtin, and hence the potential danger of the neem extract component cannot be ignored as a possible long-term genetic hazard in humans.

SUMMARY POINTS

- Neem is a versatile medicinal plant; almost every part of the plant has been in use since ancient times in folklore and traditional systems of medicine, for the treatment of a variety of human ailments.
- Neem is a unique source of various types of compounds with diverse chemical structure, many of which have antibacterial activity, and thus provide a very rich source of non-antibiotic drugs in order to treat life-threatening diseases caused by bacterial infection, as well as to combat increased multidrug resistance.
- Neem seed extracts are bactericidal against gram-negative as well as gram-positive pathogens, and thus have broad spectrum activity; they also have a synergistic interaction in combination with antibiotics.
- As the global scenario is now changing towards the use of non-toxic plant products that have traditional medicinal use, the development of modern drugs from neem should be emphasized for the control of various diseases.
- Since plant seeds are easily available and highly effective, unguided usage should be discouraged, and, apart from investigating their antimicrobial properties and mechanism of action, clinical trials must be carried out with a view to determining their safety for human consumption, and dose determination.
- Although crude extracts from various parts of neem have been used for medicinal applications since time immemorial, modern drugs can be developed after extensive investigation of its pharmacotherapeutics and toxicity, as well as proper standardization.
- The cultivation and conservation of such an important medicinal plant are also recommended.

References

Arnason, J. T., Morand, P., & Pinogene, R. T. (1989). *Insecticides of plant origin. American Chemical Society Symposium, Script 387*. Washington, DC: American Chemical Society. p. 110.

Banso, A., & Adeymo, S. (2003). Phytochemical screening and antimicrobial assessment of *Abutilon mauritianum, Bacopa monnifera* and *Datura stramonium. Biokemistri, 18*, 39–44.

Baswa, M., Rath, C. C., Dash, S. K., & Mishra, R. K. (2001). Antibacterial activity of Karanj (*Pongamia pinnata*) and Neem (*Azadirachta indica*) seed oil: a preliminary report. *Microbios, 105*, 183–189.

Biswas, K., Chattopadhyay, I., Banerjee, R. K., & Bandyopadhyay, U. (2002). Biological activities and medicinal properties of neem (*Azadirachta indica*). *Current Science, 82*, 1336–1345.

Chalkley, L. J., & Koornholf, H. J. (1985). Antimicrobial activity of ciprofloxacin against *Pseudomonas aeruginosa, Escherichia coli* and *Staphylococcus aureus* determined by the kill curve method: antibiotics comparison and synergistic interactions. *Antimicrobial Agents and Chemotherapy, 28*, 331–342.

Chaurasia, S. C., & Jain, P. C. (1978). Antibacterial activity of essential oils of four medicinal plants. *Indian Journal of Hospital Pharmacy, 15*, 166–168.

Dhongade, R. K., Kavade, S. G., & Damle, R. S. (2008). Neem oil poisoning. *Indian Pediatrics, 45*, 56–57.

Evenson, D. P., Bear, R. K., Jost, L. K., & Gesch, R. W. (1986). Toxicity of thiotepa on mouse spermatogenesis as determined by dual-parameter flow cytometry. *Toxicology and Applied Pharmacology, 82*, 151–163.

Jacobson, M. (1995). Toxicity to vertebrates. In H. Schmutterer (Ed.), *The neem tree: Source of unique natural products for integrated pest management, medicine, industry and other purposes* (pp. 484–495). Weinheim, Germany: VCH.

Mandal, S., Mandal, M., & Pal, N. K. (2007). Antibacterial potential of *Azadirachta indica* seed and *Bacopa monniera* leaf extracts against multidrug resistant *Salmonella enterica* serovar Typhi isolates. *Archives of Medical Science, 3*, 14–18.

Okemo, P. O., Mwatha, W. E., Chhabra, S. C., & Fabry, W. (2001). The kill kinetics of *Azadirachta indica* a.juss. (meliaceae) extracts on *Staphylococcus aureus, Escherichia coli, Pseudomonas aeruginosa* and *Candida albicans. African Journal of Science and Technology, 2*, 113–118.

Pritima, R., & Pandian, R. (2008). Antibacterial potency of crude extracts of *Azadirachta indica* A. juss (Leaf) against microbes causing reproductive tract infections among women. *Current Biotica, 2*, 193–205.

Rao, K. (2005). Neem tree — a profile. Available at: http://max pages.com/neem /articles/neem tree profile.

Rao, D. V. K., Singh, I., Chopra, P., Chhabra, P. C., & Ramanujalu, G. (1986). *In vitro* antibacterial activity of neem oil. *Indian Journal of Medical Research, 84*, 314–316.

Rosenkranz, H. S., & Klopman, G. (1995a). An examination of the potential genotoxic carcinogenicity of a biopesticide derived from neem tree. *Environmental and Molecular Mutagenesis, 26,* 255–260.

Rosenkranz, H. S., & Klopman, G. (1995b). Relationship between electronegativity and genotoxicity. *Mutation Research, 328,* 215–227.

SaiRam, M., Ilavazhagan, G., Sharna, S. K., Dhanraj, S. A., Suresh, B., Parida, M. M., et al. (2000). Anti-microbial activity of a new vaginal contraceptive from neem oil (*Azadirachta indica*). *Journal of Ethnopharmacology, 71,* 377–382.

Siddiqui, S. (1942). A note on the isolation of the three new bitter principles from nim oil. *Current Science, 11,* 278–279.

CHAPTER 96

Biological Properties of a Methanolic Extract of Neem Oil, A Natural Oil from the Seeds of the Neem Tree (*Azadirachta indica* var. A. Juss)

Cecilia Aiello, Valerio Berardi, Francesca Ricci, Gianfranco Risuleo
Dipartimento di Genetica e Biologia Molecolare, Sapienza Università di Roma, Roma Italy

813

LIST OF ABBREVIATIONS

cccDNA, circular covalently closed DNA
COXV, coxsackieviruses B
DNV, dengue virus
ER, electrorotation

Nuts & Seeds in Health and Disease Prevention. DOI: 10.1016/B978-0-12-375688-6.10096-9

MDA, malonaldialdehyde
MEX, methanol extract
MTT, cell survival test
PCR, polymerase chain reaction
Py, murine polyomavirus
ssRNA, single stranded RNA

INTRODUCTION

Neem oil is a natural mixture obtained from the seeds of *Azadirachta indica* (A. Juss), also known as the neem tree. The oil composition is heterogeneous, depending upon the growth habitat and environmental situation (Schmutterer, 2002). It contains high levels of many different organic compounds. Azadirachtin, a much studied component, is a tetra-nor-triterpenoid, active as pest and insect control. We prepared a methanolic extract of the whole oil, deprived of the terpenoid/limonoid moiety, and tested its bioactivity. Our conclusion is that the biological effects exhibited by MEX are not attributable to azadirachtin, but rather to other bioactive molecules (Di Ilio *et al.*, 2006).

BOTANICAL DESCRIPTION

The neem tree, *Azadirachta indica*, belongs to the mahogany family Meliaceae, and thrives best in subtropical arid areas. It is native to Burma (now Myanmar), and has spread to India, Africa, Central America, the Caribbean Islands, and the Philippines. The tree is of medium size, with a straight trunk reaching 15—20 m; the bark is ribbed and silver-gray to reddish-brown in color. The neem is evergreen and its crown has a round shape; the leaves are rather large and typically pinnate, with a size of about 30 cm. The efflorescence, comprising small, scented white flowers, is abundant in spring time. At 5 years of age the tree starts producing glabrous yellow olive-like fruits of about 2 cm (Figure 96.1), which contains a hard white capsule that encloses one or two seeds resembling an almond. The seeds contain high concentrations of azadirachtin, possibly the most thoroughly investigated and used active principle present in the neem tree (Schmutterer, 2002).

FIGURE 96.1
Details of leaves, flowers and fruit of *Azadiractha indica*.

HISTORICAL CULTIVATION AND USAGE

The neem tree produces a solid hard wood, to which in India, magic properties are traditionally attributed. Ritual reasons prevented its usage before 100 years of age; the tree may live up to three centuries. The termite-resistant wood is used for housing, furniture, and handicraft art items. The bark is rich in tannins, and the amber-colored rubbery latex is used to dye clothes and in popular medicinal ointments. The oil prepared from seeds has been extensively used in Ayurveda, Unani, and homoeopathic medicines for centuries (Schmutterer, 2002; Brahmachari, 2004). The neem tree is cited in religious Hindu works; the Sanskrit name *sarva roga nivarini* means "universal healer of all illnesses" (http://www.natureneem.com/index. htm). Mahatma Gandhi supported neem's beneficial effects: he prayed under the tree, and consumed leaves in his daily diet. Its Indian nickname is "the village pharmacy," since fruits, seeds, leaves, and roots have different bioactivities (Subapriya & Nagini, 2005).

PRESENT-DAY CULTIVATION AND USAGE

Siddiqui, an Indian biochemist, was the first scientist to discover the phytopharmacological potentials of the neem tree. He isolated nimbin, nimbinin, and nimbidin from neem oil, which contains a *plethora* of organic compounds such as terpenoids, limonoids, oily sulfur compounds, unsaturated fatty acids, and a complex secondary metabolite, azadirachtin (Figure 96.2), which is important in agriculture as a biological pest control. Neem derivatives are used worldwide to control about 500 different pests, including insects and nematodes. These preparations do not kill, but rather repel parasites, inhibit their growth, and alter their behavior and physiology; thus, pests become unable to feed, breed or metamorphose. Neem products are non-toxic for animals and beneficial insects; being cheap, they are ideal tools for pest control in disadvantaged countries. In conclusion, neem products find usages as diverse as agriculture, the cosmetics industry, and nutrition, since the shoots, flowers and leaves are eaten in India and South East Asia (Puri, 1999; Koul & Wahab, 2004).

815

APPLICATIONS TO HEALTH PROMOTION AND DISEASE PREVENTION

General bioactivity of neem tree components

The fruit, leaves, bark, and roots of *A. indica* contain compounds with many different biological properties. Recent scientific literature reports on the ability of neem extracts to combat fungal infections and inflammation, as well as viral and bacterial infections (Badam *et al.*, 1999; Parida *et al.*, 2002). Antitumor and antiproliferative activities have also been ascribed to extracts of *A. indica* (Bose *et al.*, 2007; Harish Kumar *et al.* 2009). Neem tree extracts seem also to exert a negative control on type 2 diabetes (Saxena & Vikram, 2004). In this chapter we shall review recent data from our laboratory, where we demonstrate that a derivative of the

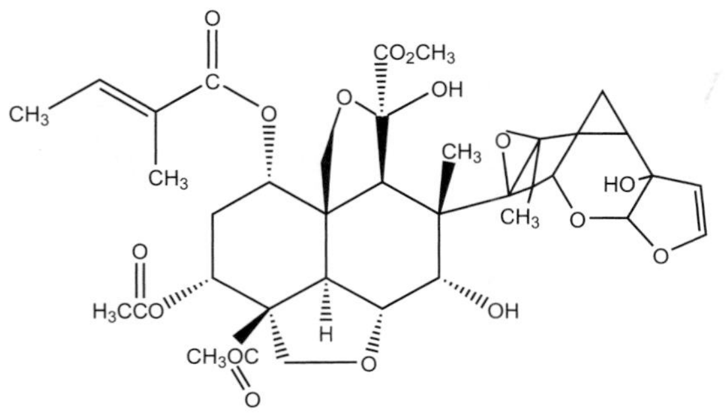

FIGURE 96.2
Structure of azadirachtin.

whole-seed neem oil, MEX, obtained by methanol extraction, exhibits differential cytotoxicity on tumor cells as compared to normal fibroblasts in culture (Ricci *et al.*, 2008). We shall also focus on the antiviral action of MEX that strongly inhibits the proliferation of murine polyomavirus (Py). This experimental approach was conducted by the spectroscopic technique known as electrorotation (Bonincontro *et al.*, 2007; Berardi *et al.*, 2009; see also below).

Cytoxicity of MEX obtained from whole neem oil

We evaluated the cytotoxicity of MEX both on 3T6 and HeLa cells by the MTT assay (Mosmann, 1983). Cells were treated with four different concentrations of extract (see Figure 96.3 and legend, for details). Exposure to relatively low concentrations of MEX (0.1 and 0.3 mg/mL) does not show dramatically cytotoxic effects, while viability is strongly reduced at higher concentrations (1.0 mg/mL) with almost the whole cell population failing to survive. Data shown in Figure 96.3 clearly show that MEX, at the same concentration, is more toxic for HeLa than for 3T6 cells; these results were further confirmed by vital dye exclusion with trypan blue (not shown; Ricci *et al.*, 2008).

Molecular assessment of the selective sensitivity to MEX

The higher sensitivity of the tumor cells was further explored by RT-PCR. In this experiment, the two cells lines were co-cultured. Due to the faster duplication rate of 3T6 cells, different amounts of the two cell types were plated, assuring that at the end of the treatment an identical number of cells from the same mixed culture was obtained (parallel cultures of both cell types were grown separately and, at the end of the experiment, the cell count never differed by more than 5%). Analysis by RT-PCR demonstrated that, as expected, HLM, a tumor-specific protein, is expressed only in HeLa cells cultures (Figure 96.4A, lane 2). The result in Figure 96.4B shows a fainter band for the tumor-specific antigen after treatment with MEX, while the actin band shows the same intensity. This supports the idea that MEX is selectively toxic for tumor cells.

Cellular and molecular targets of MEX

We also addressed the mechanism of cell death and the cellular/molecular target(s) of this neem oil extract. As already demonstrated in our laboratory treatment with MEX, it alters the structure of the plasma membrane (Bonincontro *et al.*, 2007). Therefore, the level of membrane lipoperoxidation was evaluated, since this phenomenon is indicative of membrane damage deriving from oxidative stress (Chancerelle & Kergonou, 1995). The assay is based on the quantitative measurement of malonaldialdehyde, which is produced by the decomposition of polyunsaturated fatty acids induced during oxidative stress. The result in Figure 96.5A

816

FIGURE 96.3
Viability of cells evaluated by the Mossman assay (treatment time 24 hours). Exposure to different concentrations of MEX (0.1 and 0.3 mg/ml). No dramatic cytotoxic effects were monitored. Viability is strongly reduced at higher concentrations (1.0 mg/ml). MEX, at the same concentration, is more toxic for HeLa than for 3T6 cells.

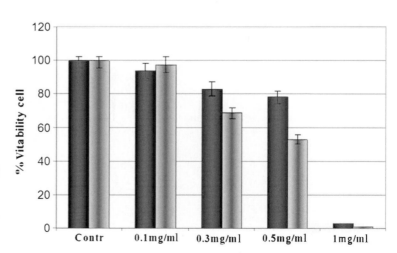

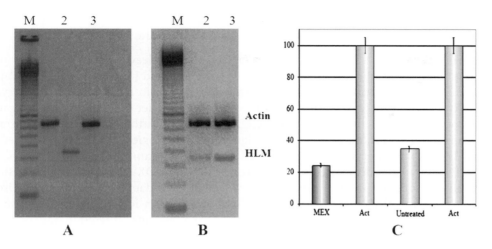

FIGURE 96.4

Higher sensitivity of the tumor cells explored by RT-PCR. The two cells lines were co-cultured. RT-PCR demonstrates that HLM, a tumor-specific protein, is expressed only in HeLa cells cultures (Panel A, lane 2). The band for the tumor-specific antigen after treatment with MEX is fainter, while the actin band shows the same intensity. Panel C shows a quantitative evaluation of data shown in panels A and B.

shows that in untreated control cells, very limited (if any) production of MDA occurs. On the contrary, the intracellular concentration of this compound (which is normally not present in the cell) increases significantly after treatment with MEX: this is direct evidence of oxidative stress. This conclusion is further corroborated by the results presented in Figure 96.5B, where treatment with MEX in the presence of three known peroxide scavengers and/or antioxidants (Calandrella *et al.*, 2007; de la Lastra and Villegas, 2007; Bakhshi *et al.*, 2008) strongly reduces the intracellular formation of MDA. In conclusion, this result, combined with the lipoperoxidation assays, strongly suggests that apoptosis caused by MEX is attributable to oxidative stress, is reversed by contemporary treatment with antioxidants, and derives essentially from structural and functional alterations of the plasma membrane.

MEX alters the cell membrane fluidity and has an antiviral effect

The effect of treatment with MEX on the fluidity of the plasma membrane was well established in our laboratory by electrorotation. This powerful biophysical technique allows single cell examination, leading to the evaluation of the dielectric parameters of the plasma membrane: specific capacitance and specific conductance (for a more detailed illustration of the physical basis of electrorotation and its biological significance, see Bonincontro *et al.*, 2007, and Berardi *et al.*, 2009). The alteration of these parameters is suggestive of a change in the structure/function of the plasma membrane. Therefore, we also adopted this strategy to correlate the variation of the dielectric parameters in response to the viral proliferation in the presence of

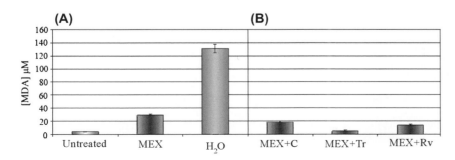

FIGURE 96.5

Treatment with MEX in the presence of three known peroxide scavengers and/or antioxidants. (A) Induction of oxidative stress in the presence of MEX or hydrogen peroxide; (B) same experiment performed in the presence of C, curcumin; Tr, trolox; Rv, resveratrol. See text for further details.

MEX. The rationale of this experiment was to explore possible membrane damage deriving from virion entry into the cell, and at the same time to assess the antiviral effect on Py.

The inhibitory effect on virus proliferation by extracts of neem leaves against coxsackievirus B (COXV) and type II dengue (DNV) viruses was already reported (Badam *et al.*, 1999; Parida *et al.*, 2002). However, these viruses are drastically different from Py as far as their life cycle, host range, and biomolecular features are concerned. As a matter of fact, Py belongs to the group of polyomaviruses that have a small cccDNA genome; furthermore, these viruses are "silent passengers" — i.e., they do not cause significant pathologies in immunocompetent individuals, although they assume an increased risk factor in immunosuppressed individuals such as HIV or transplanted patients. On the contrary, both COXV and DNV have an ssRNA genome, and belong, respectively, to picornaviridae and flaviridae; finally, both these viruses are causative agents of serious diseases. In the light of these considerations, a comparison between the antiviral effects of neem leaf extracts and MEX may be of interest in understanding the mechanisms of viral inhibition.

In previous electrorotation studies performed in our laboratory, MEX was shown to alter the membrane fluidity after administration to cultured cells (Bonincontro *et al.*, 2007); in another study, it was clearly elucidated that the cell membrane is one of the main targets of this natural mixture (Ricci *et al.*, 2008). The results of these new dielectric measurements are reported in Table 96.1. However, cells infected by Py, in the presence of MEX, showed an unaltered parameter C. This implies that the membrane essentially maintains its structural features: in other words, the data resemble those obtained in control cultures where cells were not exposed to MEX (Berardi *et al.*, 2009). A logical interpretation is that in the absence of the extract, a significant impairment of viral DNA production occurs. In fact, Figure 96.6 does show that after 48 hours of infection in the presence of MEX, the

TABLE 96.1 Dielectric Parameters of the Plasma Membrane at Different Infection Times with Murine Polyomavirus

Infection Time	Specific Capacitance C (μF/cm^2)	Specific Conductance G (S/cm^2)
Mock infection	0.8 ± 0.1	0.26 ± 0.06
24 hours	0.9 ± 0.1	0.20 ± 0.04
48 hours	1.2 ± 0.2	0.25 ± 0.04
72 hours	1.3 ± 0.2	0.45 ± 0.03

FIGURE 96.6

Effect of MEX on polyomavirus DNA *de novo* replication. (A) After 48 hours of infection in the presence of MEX, the *de novo* viral DNA synthesis decreases dramatically, as indicated by the arrow. (B) Relaxation frequency f* as a function of the solvent conductivity. The straight line results from the best fit according to the equation published in Berardi *et al.* (2009).

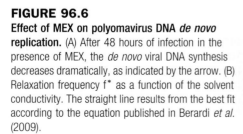

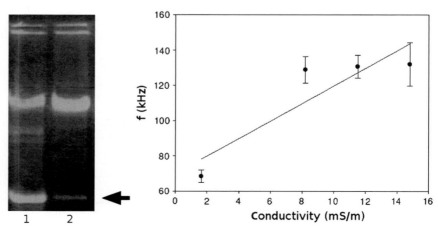

viral DNA synthesis decreases dramatically. The conclusion is that MEX actually has an antiviral action towards Py proliferation. This notion broadens the antiviral range of neem extracts, and possibly extends it to all viruses having a DNA genome, and similar infectious behavior and proliferation features.

Conclusions

Cell death follows two different pathways, necrosis and apoptosis, but the two mechanisms are not mutually exclusive. The data discussed in this work demonstrate that MEX induces cell death by activation of the apoptotic pathway. In addition, data from our laboratory, obtained by a biophysical approach, strongly suggest that MEX alters the dielectric properties of the membrane, thus inducing extensive damage. In conclusion, the cytotoxicity of MEX is attributable to the drastic structural alterations of the plasma membrane that involve destruction of the membrane lipids and consequent loss of function. This may also explain the reason for the differential toxicity exerted by MEX on tumor cells: it is known that tumor transformation induces modifications of the plasma membrane lipid bi-layer, and an increased level of lipid phosphorylation. Furthermore, the variation of specific dielectric parameters of the cell membrane, measured by electrorotation, validates the idea that the plasma membrane is actually one of the cellular/molecular targets of MEX, which causes a higher membrane polarizability with significant enhancement of ion permeability. Therefore, the overall conclusion drawn from the dielectric data is that treatment with this derivative of neem oil increases the fluidity of the membrane. Interestingly, the same dielectric parameters show no variation after Py infection performed in presence of this extract, and this is consistent with severe inhibition of the *de novo* synthesis of viral DNA. These results corroborate the idea that MEX may potentially find at least two clinical applications; first as a specific antiproliferative agent for the control of tumor cells, and second, for the treatment of infections by viruses with a DNA or an RNA genome and with phylogenetically distant origins. However we are aware that these data were essentially obtained on cultured cells, and therefore they must be validated by studies *in vivo*. In any case these results are stimulating, and urge further investigations into neem products and, more broadly, the field of natural bioactive agents.

819

ADVERSE EFFECTS AND REACTIONS (ALLERGIES AND TOXICITY)

Neem products are becoming increasingly popular, as they have positive biological responses with apparently relatively negligible side effects. The leaves and bark of the tree do not seem to have a significant toxicity; however, the seeds may be poisonous in large doses and oral administration of 5–30 ml of oil was reported to be toxic in 13 infants. Coma, hepato- and encephalopathy, metabolic acidosis, and were also episodically reported (Dhongade *et al.*, 2008; Brahamachari, 2004). The effects are symptomatic, and there is no specific antidote. In adults neem oil is relatively safe, but the oral LD_{50} is 14 ml/kg in rats and 24 ml/kg in rabbits. Higher dosages in rats cause also diarrhea, convulsions, motor and respiratory problems. In humans, possible allergic reactions should be tested prior to use of neem products. Neem's use as a pesticide (limited to non-food crops) has been approved by the Environmental Protection Agency. Azadirachtin, one of the most used neem products, is biodegradable, non-mutagenic, and non-toxic to mammals, fish, and birds (Gandhi *et al.*, 1988). In conclusion, neem products can be considered safe if used at a low dosage and for a short time.

SUMMARY POINTS

- Neem oil is a natural mixture of biological interest obtained from the seeds of *Azadirachta indica* (A. Juss), known as the neem tree. The products of this plant are used in popular medicine.

- Neem oil contains a large number of organic compounds; terpenoids, limonoids, oily sulfur compounds, unsaturated fatty acids, and azadirachtin, which is possibly the best known neem product.
- We prepared an azadichatin-free methanolic extract (MEX) of the whole neem oil, deprived of the terpenoid/limonoid moiety, and tested its biological activities on normal and tumor cells.
- MEX induces cell death by apoptosis, as shown by the assessment of several apoptotic markers subsequent to its administration.
- The cellular target of the extract is the plasma membrane.
- Tumor cells show a higher sensitivity to MEX, compared to normal ones.
- A potential usage of this neem product for the control of proliferative diseases is suggested.

ACKNOWLEDGEMENT

We are indebted to Professor Bonicontro, Department of Physics at the University of Sapienza, who introduced us to electrorotation and co-authored several works where this technique was adopted. Thanks are also due Riccardo Risuleo, who made the iconographic elaborations.

References

Badam, L., Joshi, S. P., & Bedekar, S. S. (1999). *In vitro* antiviral activity of neem (*Azadirachta indica. A. Juss*) leaf extract against group B coxsackieviruses. *Journal of Communication Disorders, 31*, 79–90.

Bakhshi, J., Weinstein, L., Poksay, K. S., Nishinaga, B., Bredesen, D. E., & Rao, R. V. (2008). Coupling endoplasmic reticulum stress to the cell death program in mouse melanoma cells: effect of curcumin. *Apoptosis, 13*, 904.

Berardi, V., Aiello, C., Bonincontro, A., & Risuleo, G. (2009). Alterations of the plasma membrane caused by murine polyomavirus proliferation: an electrorotation study. *Journal of Membrane Biology, 229*, 19–25.

Bonincontro, A., Di Ilio, V., Pedata, O., & Risuleo, G. (2007). Dielectric properties of the plasma membrane of cultured murine fibroblasts treated with a nonterpenoid extract of *Azadirachta indica* seeds. *Journal of Membrane Biology, 215*, 75–79.

Bose, A., Haque, E., & Baral, R. (2007). Neem leaf preparation induces apoptosis of tumor cells by releasing cytotoxic cytokines from human peripheral blood mononuclear cells. *Phytotherapy Research, 21*, 914–920.

Brahmachari, G. (2004). Neem tree – an omnipotent plant: a retrospection. *Chemistry and Biochemistry, 5*, 408–421.

Calandrella, N., Scarsella, G., Pescosolido, N., & Risuleo, G. (2007). Degenerative and apoptotic events at retinal and optic nerve level after experimental induction of ocular hypertension. *Molecular and Cellular Biochemistry, 301*, 155–163.

Chancerelle, Y., & Kergonou, J. F. (1995). Immunologic relevance of malonic dialdehyde. *Annales Pharmaceutiques Françaises, 53*, 241–250.

de la Lastra, C. A., & Villegas, I. (2007). Resveratrol as an antioxidant and pro-oxidant agent: mechanisms and clinical implications. *Biochemistry Society Transactions, 35*, 1156–1160, 914.

Dhongade, R. K., Kavade, S. G., & Damle, R. S. Neem oil poisoning. *Indian Pediatrics, 45*, 56–57.

Di Ilio, V., Pasquariello, N., van der Esch, S. A., Cristofaro, M., Scarsella, G., & Risuleo, G. (2006). Cytotoxic and antiproliferative effects induced by a non terpenoid polar extract of *A. indica* seeds on 3T6 murine fibroblasts in culture. *Molecular and Cellular Biochemistry, 287*, 69–77.

Gandhi, M., Lal, R., Sankaranarayanan, A., Banerjee, C. K., & Sharma, P. L. (1988). Acute toxicity study of the oil from *Azadirachta indica* seed (neem oil). *Journal of Ethnopharmacology, 23*, 39–51.

Harish Kumar, G., Chandra Mohan, K. V., Jagannadha Rao, A., & Nagini, S. (2009). Nimbolide, a limonoid from *Azadirachta indica*, inhibits proliferation and induces apoptosis of human choriocarcinoma (BeWo) cells. *Investigational New Drugs, 27*, 246–252.

Koul, O., & Wahab, S. (Eds.). (2004). *Neem: today and in new millenium*. Dordrecht, The Netherlands: Kluwer Academic Publishers.

Mosmann, T. (1983). Rapid colorimetric assay for cellular grow and survival: application to proliferation and cytotoxicity assay. *Journal of Immunological Methods, 65*, 55–63.

Parida, M. M., Upadhyay, C., Pandya, G., & Jana, A. M. (2002). Inhibitory potential of neem (*Azadirachta indica* Juss) leaves on dengue virus type-2 replication. *Journal of Ethnopharmacology, 79*, 273–278.

Puri, H. S. (1999). *The divine tree Azadirachta indica*. Amsterdam, The Netherlands: OPA Overseas Publishers Association, Harwood Academic Publishers.

Ricci, F., Berardi, V., & Risuleo, G. (2008). Differential cytotoxicity of MEX: a component of Neem oil whose action is exerted at the cell membrane level. *Molecules, 14*, 122–132.

Saxena, A., & Vikram, N. K. (2004). Role of selected Indian plants in management of type 2 diabetes: a review. *Journal of Alternative and Complementary Medicine, 10*, 369–378.

Schmutterer, H. (2002). *The neem tree and other meliaceous plants.* Mumbai, India: Neem Foundation.

Subapriya, R., & Nagini, S. (2005). Medicinal properties of neem leaves: a review. *Current Medicinal Chemistry Anticancer Agents, 5*, 149–156.

CHAPTER 97

Effects of Nigella (*Nigella sativa* L.) Seed Extract on Human Neutrophil Elastase Activity

Rachid Kacem
Department of Biology, Faculty of Sciences, Ferhat Abbas University, Setif, Algeria

823

LIST OF ABBREVIATIONS

α_1-AT, α_1-antitrypsin
ARDS, acute respiratory distress syndrome
CF, cystic fibrosis
COPD, chronic obstructive pulmonary disease
DMSO, dimethyl sulfoxide
DTQ, dithymoquinone
HIC, highest inhibitory concentration
HNE, human neutrophil elastase
IC_{50}, inhibitory concentration, 50%
LD_{50}, lethal dose, 50%
MMPs, matrix metalloproteinases
MPO, meloxy pero-oxydase
NSP, neutrophil serine protease

Nuts & Seeds in Health and Disease Prevention. DOI: 10.1016/B978-0-12-375688-6.10097-0

PMN, polymorphonuclear leukocyte
THQ, thymohydroquinone
THY, thymol
TQ, thymoquinone

INTRODUCTION

The *Nigella* genus comprises more than 20 species; all have been reported as medicinal plants in traditional medicine. Several research reports have confirmed that extracts from *N. sativa* (L.) seeds have anti-inflammatory activity, mainly due to the enrichment of this extract with many bioactive molecules. Essential oil produced from *N. sativa* (L.) seeds as a secondary metabolite has been intensively investigated because of this enrichment, and the metabolites include many types of monoterpenes (Ghosheh *et al.*, 1999; Kacem & Meraihi, 2006).

Neutrophil serine proteases (NSPs) are involved in many inflammatory diseases, such as COPD, acute respiratory distress syndrome (ARDS), and cystic fibrosis (CF). The production of emphysema by instilling HNE into the lungs of animal models was reported. HNE and MMPs were also shown to contribute to the matrix degradation that occurs in emphysema and other tissue inflammatory diseases caused by the recruitment of blood polymorphonuclear leukocytes (PMNs) to inflammatory sites. Thus, it became evident that HNE is involved in the pathogenesis of different inflammatory diseases, such as emphysema, CF, and COPD (Kanter *et al.*, 2004).

The anti-inflammatory mechanisms of action of the essential oil constituents at sites of inflammation are still not clear. Moreover, few research tests have been carried out to investigate the effects of bioactive molecules present in the extract of this medicinal plant on HNE activity. The aim of this chapter is to focus on the effects of *N. sativa* (L.) seed extracts, the essential oil, and its main components on HNE activity, and evaluate the bioactive molecules present in the extracts of this medicinal plant that play an inhibitory effect against HNE activity. The possibility of using this product as a natural anti-elastase agent for the treatment of injuries in emphysema and COPD is also examined (Kacem & Meraihi, 2006).

BOTANICAL DESCRIPTION

The word *Nigella* comes from *nigellus*, meaning black. The plant is known by many other names, such as nigelle, coriauder of trome, sweet flavor, fenuel flower, and black seed. *N. sativa* is native to Asia, specifically Syria and Iraq, but is now widely cultivated in many Mediterranean countries. It is a bushy, self-branching plant with white or pale to dark blue flowers. *N. sativa* (L.) is self-reproducing, and forms a fruit capsule that contains many white, trigonal seeds. Once the fruit capsule has matured, it opens up and the seeds contained within are exposed to the air, becoming black in color (El-Dakhakhny, 1965).

HISTORICAL CULTIVATION AND USAGE

According to Zohary and Hopf (2000), archeological evidence regarding the earliest cultivation of *N. sativa* is still scanty, but they report that *N. sativa* seeds have been found in several sites from ancient Egypt, including Tutankhamun's tomb. Although *N. sativa*'s exact role in Egyptian culture has not been established, it is known that items entombed with a pharaoh were carefully selected to assist him in the afterlife. The earliest written reference to *N. sativa* is thought to be in the book of Isaiah in the Old Testament, where the reaping of Nigella and wheat is contrasted (Isaiah 28: 25, 27). It was a traditional condiment of the Old World during classical times, and its black seeds were extensively used to flavor food (Zohary & Hopf, 2000). *N. sativa* has been used for medicinal purposes for centuries in Asia, the Middle East, and Africa. It has been traditionally used for a variety of conditions and treatments, as an analgesic, anti-inflammatory, anti-allergenic, antioxidant, anticancer, and antiviral, and for general well-being. In Islam, it is regarded as one of the greatest forms of healing medicine available. Muhammad (S.A.W.) once stated that the black seed can heal every disease, as recounted in the hadith. The seeds have traditionally been used in the Middle East and South-east Asian countries to treat ailments including asthma, bronchitis, rheumatism, and related inflammatory diseases; to increase milk production in nursing mothers; to promote digestion; and to fight parasitic infections. Its oil has been used to treat skin conditions such as eczema and boils, and to treat cold symptoms. Its many uses have earned nigella the Arabic approbation *Habbatul barakah*, meaning the seed of blessing (El-Dakhakhny, 1965).

PRESENT-DAY CULTIVATION AND USAGE

N. sativa (L.) is easily grown in any good garden soil, preferring a light soil in a warm, sunny position. This species is often cultivated, especially in western Asia and India, for its edible seed, which is aromatic, with a nutmeg scent. It is a greedy plant, inhibiting the growth of other plants nearby, especially legumes, and can be transplanted if necessary (Huxley, 1992).

Like many aromatic culinary herbs, the seeds of *N. sativa* are beneficial for the digestive system, soothing stomach pains and spasms, and easing wind, bloating and colic; it is a carminative, diuretic, emmenagogue, galactagogue, laxative, and stimulant. An infusion is used in the treatment of digestive and menstrual disorders, insufficient lactation, and bronchial complaints. The seeds are much used in India to increase the flow of milk in nursing mothers, and can be used to treat intestinal worms, especially in children. The seed is also ground into a powder, mixed with sesame oil, and used externally to treat abscesses, hemorrhoids, and orchitis (Huxley, 1992).

APPLICATIONS TO HEALTH PROMOTION AND DISEASE PREVENTION

Several research reports have confirmed that extracts from *N. sativa* (L.) seeds have anti-inflammatory activity; this is mainly due to the enrichment of this extract with many bioactive molecules. Essential oil produced from *N. sativa* (L.) seeds as a secondary metabolite has been intensively investigated because of this enrichment, and the metabolites include many types of monoterpenes (Kacem & Meraihi, 2006; Ghosheh *et al.*, 1999).

The anti-inflammatory mechanisms of action of the essential oil constituents at sites of inflammation are still not clear. Moreover, few research tests have been carried out to investigate the effects of bioactive molecules present in the extract of this medicinal plant on HNE activity. The aim of this section is to focus on the effects of *N. sativa* (L.) seed extracts, the essential oil, and its main components on HNE activity, and evaluate the bioactive molecules present in the extract of this medicinal plant that play an inhibitory effect against HNE activity. The possibility of using this product as a natural anti-elastase agent for the treatment of injuries in emphysema and COPD is also examined (Kacem & Meraihi, 2006).

Effects of *N. sativa* (L.) seed extracts on HNE activity
STRUCTURE AND INHIBITION OF HNE

Elastase is an enzyme belonging to the serine proteases family, which comprises hydrolases that break down peptide bonds. It is a glycoprotein, and the polypeptide chain constitutes 218 amino acids. Elastase is biosynthesized as a proelastase containing a terminal peptide of 29 radicals and 27 initiation signals; these are eliminated during biosynthesis, leading to activation; once activated, elastase is secreted as a glycoprotein that contains a serine residue in position 57 in its binding site. Although HNE is likely to be the major factor mediating elastolysis in patients with α_1-AT deficiency, few HNE inhibitors have been tested. The most studied is α_1-AT; it is a protein molecule, molecular weight 54 kDa, and possesses high inhibitory capacity against many serine proteases, including NE. Its binding site contains an active methionine residue, and although the exact mechanism by which elastase is inhibited by α_1-AT is still not clear, some scientists suggest that the reaction occurs between the active methionine of α_1-AT and the serine residue in the binding site of elastase. The evidence that methionine residue in α_1-AT is involved in the inhibition of elastase arose from the finding that oxidation of the methionine residue in position 358 near the COOH terminal by MPO led to inactivation of α_1-AT, and the inhibitory activity decreased by about 20-fold. As a result of α_1-AT–elastase binding, a peptide bond, Met–Ser (358–359), is hydrolyzed; this leads to inactivation of both elastase and α_1-AT. The evidence that α_1-AT is hydrolyzed by elastase emerged after many scientists detected a stable compound immediately following the reaction, with molecular weight equal to 80 kDa (Rust *et al.*, 1994).

HNE ASSAY

Briefly, pure HNE was dissolved in Tris-buffer solution, and the pH adjusted to 7.5. Volumes of HNE solution were added to each microassay tube in 96-well microplates at 37°C, and incubated with a range of concentrations of test compounds or controls. Test compounds were first diluted in DMSO, and a range of concentrations were prepared in buffer solution. After incubation, substrate solution was added. All tubes were reincubated with simple agitation (450 rpm, 37°C, 20 min). After incubation, absorbance was read at 405 nm after stopping the reaction, and the mean ± SEM of three repeated samples was calculated (Kacem & Meraihi, 2006).

CALCULATION OF IC$_{50}$

Inhibitory concentration (IC$_{50}$) values corresponding to the concentration of test compounds showing 50% inhibition of HNE activity were estimated based on the least squares regression line of three plots of the logarithmic concentrations versus HNE activity (Kacem & Meraihi, 2006).

EFFECTS OF ESSENTIAL OIL ON HNE ACTIVITY

Several reports have demonstrated that the main components of essential oil extracted from *N. sativa* (L.) seeds showing inhibitory effects on HNE activity are monoterpenes, including p-cymene (37.3 %), TQ (13.7 %), carvone (0.9 %), THY (0.33 %), and carvacrol (11.77 %) . The chemical structures of these test compounds are shown in Figure 97.1. These compounds comprise 65% of the total essential oil. We have recently reported that the active constituents isolated from *N. sativa* essential oil inhibit HNE with different IC$_{50}$ values (Table 97.1). TQ inhibits HNE activity with an IC$_{50}$ value of 30 μM, and the IC$_{50}$ value of carvacrol was the lowest (12 μM). Inhibition of HNE activity by total essential oil was found to be dose-dependent (Figure 97.2). The HIC of essential oil (5.8 mg/ml) caused total inhibition of HNE activity. The lowest inhibitory concentration was 0.4 mg/ml. When the HIC of different test compounds was added to HNE simultaneously with the specific substrate, a slow decrease in the rate of substrate hydrolysis was observed (Figure 97.3). From these results, it is clear that the test compounds inhibited HNE activity with different IC$_{50}$ values, but carvacrol inhibited HNE with a very low IC$_{50}$ value (12 μM). The inhibitory concentration of THY was the highest, at 104 μM. Although some studies have indicated that TQ is the most active molecule that could be responsible for the effects of essential oil, the results of our studies clearly indicate that this compound inhibits HNE activity with an IC$_{50}$ value (12 μM) about three-fold that of carvacrol. It is clear that carvacrol, which has a hydroxyl group in position 3 on the benzene ring, is the most bioactive molecule in inhibiting HNE activity. Because a specific substrate was used in this study, the inactivation of HNE could be explained by the competition of this compound and substrate for the same specific binding sites. We observed about two-fold progressive inhibition of HNE by carvacrol compared with THY. Inhibition of HNE activity by carvacrol was explained by its direct binding with the enzyme, forming an enzyme–inhibitor complex.

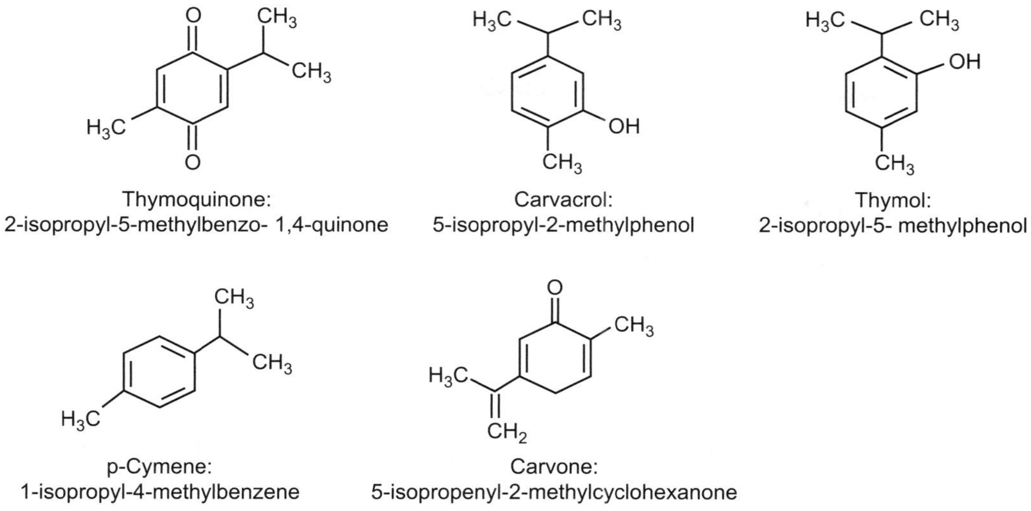

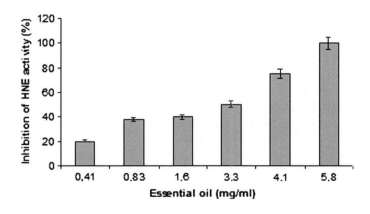

FIGURE 97.1

Chemical structure of test compounds showing inhibitory activity against HNE. The main components of essential oil extracted from *Nigella sativa* (L.) seeds showing inhibitory activity against human neutrophil elastase (HNE) are monoterpenes p-cymene (37.3%), thymoquinone (13.7%), carvone (0.9%), thymole (0.33%), and carvacrol (11.77%).

TABLE 97.1 Inhibition of Human Neutrophil Elastase (HNE) Activity By Essential Oil Components

Compound	Concentration (μM)	HNE inhibition (% of control)	IC$_{50}$ (μM)
Thymoquinone	11	18	30
Thymol	15	7	104
Carvacrol	11	45	12
Carvone	7	25	14
p-Cymene	13	27	25

827

FIGURE 97.2

Effects of different concentrations of essential oil extracted from N. sativa (L.) seeds on HNE activity. The inhibition of human neutrophil elastase (HNE) activity by essential oil extracted from *Nigella sativa* (L.) seeds is dose dependent; the highest inhibitory concentration (5.8 mg/ml) caused total inhibition of HNE activity. Results are expressed as mean $\pm$ SEM; $P < 0.01$

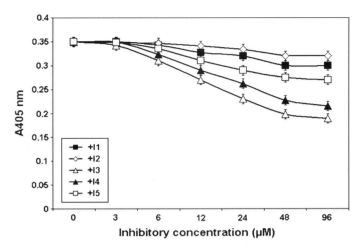

FIGURE 97.3

Inhibition of HNE activity by test compounds. After the addition of the highest inhibitory concentration (HIC) of different test compounds (I1, thymoquinone; I2, thymol; I3, carvacrol; I4, carvone; I5, p-cymene) to human neutrophil elastase (HNE) simultaneously with the specific substrate, a slow decrease in the rate of substrate hydrolysis was observed. Results are expressed as mean ± SEM; $P < 0.01$.

The inhibitory effects of total essential oil on HNE activity can be explained by the presence of bioactive molecules, mainly carvacrol. Taking into consideration the IC_{50} values of each compound, carvacrol is a potent inhibitor of HNE (Kacem & Meraihi, 2006).

ADVERSE EFFECTS AND REACTIONS (ALLERGIES AND TOXICITY)

The aqueous and oil extracts of the seeds have been shown to possess antioxidant, anti-inflammatory, anticancer, analgesic, and antimicrobial activities. TQ has been shown to be the active principle responsible for many of the seed's beneficial effects. The toxicity of the fixed oil of *N. sativa* (L.) seeds in mice has been examined; changes in key hepatic enzymes levels were not observed in rats, suggesting the safety of therapeutic doses of *N. sativa* (L.) fixed oil (Zaoui *et al.*, 2002). Several findings have provided clear evidence that TQ possesses reproducible antioxidant effects through enhancing the oxidant scavenger system. The oil and TQ have shown potent anti-inflammatory effects on several inflammation-based models, including experimental encephalomyelitis, colitis, peritonitis, edema, and arthritis, through suppression of the inflammatory mediator; prostaglandins and leukotrienes. The oil and certain active ingredients showed beneficial immunomodulatory properties, augmenting the T cell- and natural killer cell-mediated immune responses. One of the potential properties of *N. sativa* (L.) seeds is the ability of one or more of its constituents to reduce toxicity due to its antioxidant activities and antihistaminic properties (Bonnier, 1990). There was an indication from the traditional use of *N. sativa* seeds that its active ingredients have a substantial impact on the inflammatory diseases mediated by histamine. Components of *N. sativa* (L.) seeds have also been shown to have appreciable anti-inflammatory effects in several inflammatory diseases; the oil and active ingredients of *N. sativa* (L.) seeds have been reported to exert antimicrobial activities and antiviral effects. Several studies have indicated that both the oil and the active ingredients of *N. sativa* (L.) seeds possess antitumor effects. By investigating the effects of the volatile oil of *N. sativa* (L.) seeds on different human cancer cell lines, the oil expressed marked cytotoxic effects against a panel of human cancer cell lines. The reported *in vitro* antitumor effects of *N. sativa* (L.) oil and its active ingredients have also been confirmed *in vivo* in different tumor models (Bonnier, 1990; Haq *et al.*, 1999; Al-Rowais, 2002; Zaoui *et al.*, 2002).

Although several reports have indicated that *N. sativa* (L.) seeds are not toxic, as evidenced by high LD_{50} values, hepatic enzyme stability, and organ integrity, few studies have addressed the possible toxicity of *N. sativa* (L.) seeds and their components. Moreover, the route of administration of *N. sativa* (L.) seems to be crucial for its toxicity, since the LD_{50} values were higher with oral administration than via the intraperitoneal route, indicating that oral intake is safer than systemic intake. Furthermore, several toxicological studies have demonstrated that crude extracts of the seeds and some of its active constituents might have a protective affect against nephrotoxicity and hepatotoxicity induced by chemicals (Ali & Blunden, 2003). The administration of *N. sativa* (L.) seed extracts decreased the harmful effects of gentamicin sulfate-induced nephrotoxicity related to oxidative damage, both by inhibiting free radical formation and by restoration of the antioxidant systems (Raja, 2006).

SUMMARY POINTS

- Pure essential oil extracted from *Nigella sativa* (L.) seeds inhibits the activity of human neutrophil elastase in a dose-dependent manner.
- The results of several experiments have demonstrated that essential oil constituents showing inhibitory activity against human neutrophil elastase are monoterpenes.
- The inhibitory effects of total essential oil on human neutrophil elastase activity can be explained by the presence of bioactive molecules, mainly 5-isopropyl-2-methyl phenol.
- Taking into consideration the inhibitory concentration (IC_{50}) value of each compound, 5-isopropyl-2-methyl phenol is a potent inhibitor of human neutrophil elastase, and could be considered as a natural anti-elastase agent in the treatment of injuries that appear in chronic obstructive pulmonary disease and emphysema. Further *in vivo* studies are recommended.

References

Ali, B. H., & Blunden, G. (2003). Pharmacological and toxicological properties of *Nigella sativa*. *Phytotherapy Research, 17*, 299–305.

Al-Rowais, N. A. (2002). Herbal medicine in the treatment of diabetes mellitus. *Saudi Medical Journal, 23*, 1327–1331.

Bonnier, G. (1990). *La grande flore en couleurs* (Vol. 1). Paris, France: La librairie des Éditions Belin. pp. 17–77.

El-Dakhakhny, M. (1965). Studies on the Egyptian *Nigella sativa* (L.). Some pharmacological properties of the seeds' active principle in comparison to its dihydro compound and its polymer. *Arzneim.-Forsch., Beih, 15*, 1227–1229.

Ghosheh, P., Crooks, A. J., & Houdi, O. A. (1999). High performance liquid chromatographic analysis of the pharmacologically active quinones and related compounds in the oil of the black seed. *Journal of Pharmaceutical and Biomedical Analysis, 19*, 757–762.

Haq, A., Lobo, P. I., Al-Tufail, M., Rama, N. R., & Al-Sedairy, S. T. (1999). Immuno-modulatory effect of *Nigella sativa* proteins fractionated by ion exchange chromatography. *International Journal of Immunopharmacology, 21*(4), 283–295.

Huxley, A. (1992). *The new RHS dictionary of gardening*. London: MacMillan Press.

Kacem, R., & Meraihi, Z. (2006). Effects of essential oil extracted from *Nigella sativa* (L.) seeds and its main components on human neutrophil elastase activity. *Yakugaku Zasshi, 126*(4), 301–305.

Kanter, M., Coskun, O., Korkmaz, A., & Oter, S. (2004). Effects of *Nigella sativa* on oxidative stress and beta-cell damage in streptozotocin-induced diabetic rats. *The Anatomical Record Part A, Discoveries in Molecular, Cellular, and Evolutionary Biology, 279*, 685–691.

Raja, T. A. (2006). Alpha1-antitrypsin deficiency. *Respiratory Medicine, 1*, 80–87.

Rust, B. L., Messing, C. R., & Iglewski, B. H. (1994). *Methods in Enzymology, 235*, 554–563.

Zaoui, A., Cherrah, Y., Alaoui, K., Mahassine, N., Amarouch, H., & Hassar, M. (2002). Effects of *Nigella sativa* fixed oil on blood homeostasis in rat. *Journal of Ethnopharmacology, 79*(1), 23–26.

Zohary, D., & Hopf, M. (2000). *Domestication of plants in the Old World*. Oxford, UK: Oxford University Press, p. 206.

CHAPTER 98

Antioxidant and Antimicrobial Activity of Nutmeg (*Myristica fragrans*)

Ashish Deep Gupta[1], Deepak Rajpurohit[2]
[1] Mangalayatan University, Institute of Biomedical Education & Research, Department of Biotechnology, Uttar Pradesh, India
[2] College of Horticulture and Forestry, Department of Biotechnology, Rajasthan, India

831

LIST OF ABBREVIATIONS

BHA, butylated hydroxyanisole
BHT, butylated hydroxytoluene
MMDA, 3-methoxy-4,5-methylene-dioxyamphetamine
TMA, 3,4,5 trimethoxyamphetamine
ROS, reactive oxygen species

INTRODUCTION

Due to potential liver damage and carcinogenic effects, the most widely used synthetic antioxidant compounds, butylated hydroxytoluene (BHT) and butylated hydroxyanisole (BHA), have restricted usage (Senevirathne *et al.*, 2006). Therefore, much emphasis is currently being given to searching for new and natural antioxidants and antimicrobials from dietary plants, because they can safeguard the human body against the oxidative damage of biological

Nuts & Seeds in Health and Disease Prevention. DOI: 10.1016/B978-0-12-375688-6.10098-2

FIGURE 98.1
Nutmeg seeds. Nutmeg (*Myristica fragrans*) is an important spice. Nutmeg is the actual seed, and has been reported to have strong antioxidant and antimicrobial potential.

macromolecules. Many spices, such as nutmeg (*Myristica fragrans*), have been reported to have antioxidant and antimicrobial properties, apart from their traditional use in numerous medical conditions, and can thus offer a potential solution in the search for new and natural sources of antioxidants and antimicrobials.

BOTANICAL DESCRIPTION

The nutmeg is a member of the magnoliales order and Myristicaceae family (Figure 98.1). Nutmeg and mace are two important spices derived from the fruit. Nutmeg is the seed of the tree; it is dark brown, ovoid, 2–3 cm long, and weighs between 5 and 10 g. Nutmeg seeds possess ruminate endosperm, and are considered to be the most primitive among the flowering plants. Mace is the dried lacy reddish covering or aril of the seed. A nutmeg tree takes around 20 years to reach its full potential, but the first harvest can be carried out 7–9 years after planting. World production of nutmeg is estimated to be 10,000–12,000 tonnes per year, with annual world demand estimated at 9000 tonnes, whereas production of mace is estimated at 1500–2000 tonnes. The main producers and exporters are Indonesia and Grenada; India, Malaysia, Papua New Guinea, Sri Lanka, and the Caribbean Islands are other important producers.

HISTORICAL CULTIVATION AND USAGE

Nutmeg originated in the Banda Islands of Indonesia, and was discovered by the Portuguese in 1512. The importance of the nutmeg seed was propagated by the Dutch. The name nutmeg is derived from the Latin *nux muscatus*, meaning "musky nut." In India, nutmeg is known as *Jaiphal*. According to the ethno-medical literature, nutmeg seed oil was used for intestinal disorders by Indians, in embalming by Egyptians, and to cure plague by Italians. In ancient times, nutmeg seeds were used in medicines as an aphrodisiac, abortifacient, and anti-flatulent, a narcotic, and as a means to induce menses. The effect of the nutmeg seeds on the central nervous system was first observed in the early 19th century. Traditional uses of nutmeg seeds include treatment of hemorrhoids, chronic vomiting, rheumatism, cholera, psychosis, stomach cramps, nausea, and anxiety. Nutmeg seed oil also has antiseptic, analgesic, and antirheumatic properties.

PRESENT-DAY CULTIVATION AND USAGE

Both *in vitro* and *in vivo* studies have resulted in a wide array of pharmacological actions attributed to nutmeg seeds, including antioxidant, antifungal, antibacterial (Takikawa *et al.*, 2002), aphrodisiac (Tajuddin *et al.*, 2003), anti-inflammatory (Olajide *et al.*, 1999), and hepatoprotective activities. Nutmeg seeds are used to control diarrhea and Crohn's disease, as they inhibit prostaglandin activity in the intestine. Nutmeg seeds also improve Alzheimer's disease, as they directly affect acetylcholinesterase activity in the brain. Nutmeg seed essential oil is used in aromatherapy, because the three main constituents of nutmeg (myristicin, elemicin, and isoelemicin) act as stress relievers.

APPLICATION TO HEALTH PROMOTION AND DISEASE PREVENTION

Volatile essential oil and various extracts of nutmeg seeds have been reported to have many pharmacological properties, including antioxidant, antimicrobial, insecticidal, anti-amebic, and anticarcinogenic activity. Nutmeg seed oil is a colorless or pale yellow liquid with the characteristic odor and taste of nutmeg. The composition of the essential oil of nutmeg seeds and that of mace differ significantly. The essential oil of nutmeg seeds mainly contains sabinene (15–50%), α-pinene (10–22%), β-pinene (7–18%), myrcene (0.7–3%), 1,8-cineole (1.5–3.5%), α-phellandrene (0.3–6.2%), myristicin (0.5–13.5%), limonene (2.7–4.1%), euginol (0.1–1%), safrole (0.1–3.2%), and terpinen-4-ol (0–11%). It has been noted that the composition of nutmeg seed oil depends upon its source.

Antioxidant activity

In the human body there are various exogenous and endogenous sources for the spontaneous generation of free radicals and other reactive oxygen species (ROS), such as hydroxyl radicals ($\cdot OH$), superoxide anions ($\cdot O_2^-$), and hydrogen peroxide (H_2O_2), which can affect lipid, protein, and nucleic acid in various ways and thus play an important role in the initiation and/or progression of various diseases (Figure 98.2). Under pathological conditions and in immune-compromised hosts, there is an imbalance between the generation of ROS and their quenching by the host antioxidant system, which leads to oxidative stress. The exogenous supply of antioxidants could be very helpful in conquering oxidative stress. Many spices, including nutmeg, have been reported to possess good antioxidant capacity.

The antioxidant capacity of nutmeg seeds can be measured by various chemical assays, such as estimation of total phenolic concentration, capacity to scavenge the stable free radical DPPH (2,2-diphenyl-1-picrylhydrazyl), ferric reducing/antioxidant power assay (FRAP), inhibition of lipid peroxidation, inhibition of bleaching of β-carotene, etc. The antioxidant capacity of the essential oil of nutmeg seeds and various extracts has been established by much research. Jukic *et al.* (2006) found that the aglycone fraction, enzymatically isolated from glycosidically bound volatiles of nutmeg, possesses a higher antioxidant capacity compared with free volatiles from its essential oil. The variation was due to differences in the amounts of eugenol and isoeugenol. Tomaino *et al.* (2005) studied the effect of heating on the antioxidant effectiveness and chemical composition of nutmeg seed essential oil. They reported significantly higher free radical scavenger activity with heating, which could be due to the volatilization of the hydrocarbons of the oil at higher temperatures, resulting in the accumulation of phenolic constituents in the remaining oil.

Antioxidant properties are contributed by the variety of active phytochemicals, including vitamins, carotenoids, terpenoids, alkaloids, flavonoids, lignans, simple phenols, phenolic acids, etc. It has been reported that total phenolic content and antioxidant activity have a significant and positive correlation. In plants, phenolic antioxidants are mainly produced by secondary metabolism, and their antioxidant property largely depends on their redox properties and chemical structure (i.e., the number and position of the hydroxyl group). Shan *et al.* (2005) reported caffic acid and catechin as the major phenolic acids present in nutmeg seeds.

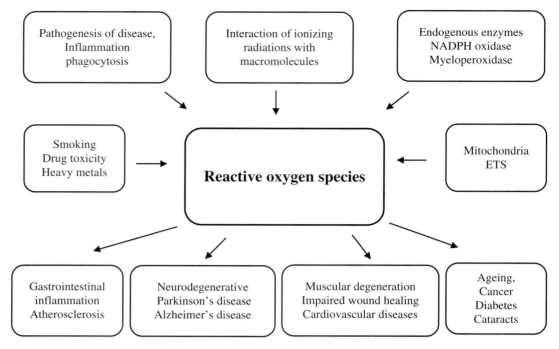

FIGURE 98.2
Synthesis of reactive oxygen species in the human body, and their consequences. In the biological system, there are many possible ways which can generate ROS. Synthesis of ROS can lead to various life threatening diseases.

Compounds like caffic acid, having a catechol structure, are considered to be good antioxidants because the catechol structure can easily donate phenolic hydrogen or electrons to the acceptors, such as reactive oxygen species or lipid peroxyl groups. Chatterjee and colleagues (2007) reported an antioxidant activity of acetone extract from mace aril, and found that acetone extract is mainly constituted of lignans. Calliste *et al.* (2010) reported lignan derivatives as a class of compounds that contribute to the antioxidant potential of nutmeg seeds. These lignan derivatives essentially belong to the dibenzyl butane group, with either a guaiacyl or piperonyl moieties on the aromatic rings. The principal compounds of this category are argenteane (bis-erythro-5,5′-bis [1-(4-hydroxy-3-methoxyphenyl)-4-(3,4-methylenedioxyphenyl)-2,3 dimethylbutane]), meso-dihydroguaiaretic acid, and erythro-austrobailignan-6 (Figure 98.3). Central moieties of these compounds are able to release one or two H atoms to the free radicals, which can be explained by density functional theory calculations of the O–H bond dissociation enthalpies. After absorption into the body, nutmeg seed lignans and their glycosides are metabolized to produce biologically active compounds containing the catechol structure, which could account for the high antioxidant potential of the nutmeg seeds (Nakai *et al.*, 2003). Besides their antioxidant activity, lignans are known to possess diverse pharmacological potentials, including antitumor, antiviral, and anti-atherosclerotic activities.

Antioxidant activity could be attributed to the occurrence and concentration of various chemical substances present in the plant. Many compounds that possess good antioxidant activity have been isolated from nutmeg seeds (Figure 98.4). Compounds such as eugenol and β-caryophyllene, which contain hydrogen atoms in the benzylic and/or allylic positions, could be good contributors for antioxidant activity. These compounds have high antioxidant activity because of the relatively easy abstraction of atomic hydrogen from these functional groups by peroxy radicals formed under oxidative stress. Another view that favors the antioxidant role of eugenol in nutmeg could be that it promotes the activities of catalase, superoxide dismutase, glutamine transferase, glutathione peroxidase, and glucose-6-phosphate dehydrogenase enzymes (Kumaravelu *et al.*, 1996).

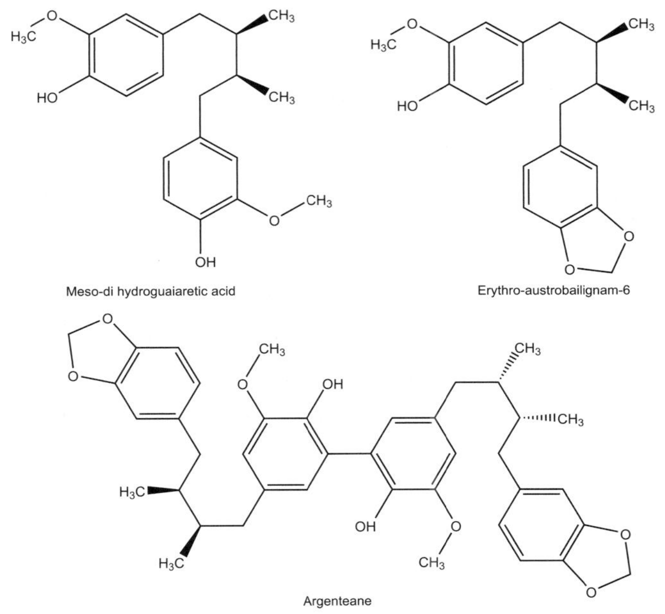

FIGURE 98.3
Important Lignans of the nutmeg seed. Meso-di hydroguaiaretic acid, erythro-austrobailignan-6 and argenteane (bis-erythro-5,5′-bis [1-(4-hydroxy-3-methoxyphenyl)-4-(3,4-methylene dioxyphenyl) 2,3 dimethylbutane]) are major lignans in nutmeg. Lignans are important antioxidants, as they can release H atoms to the free radicals.

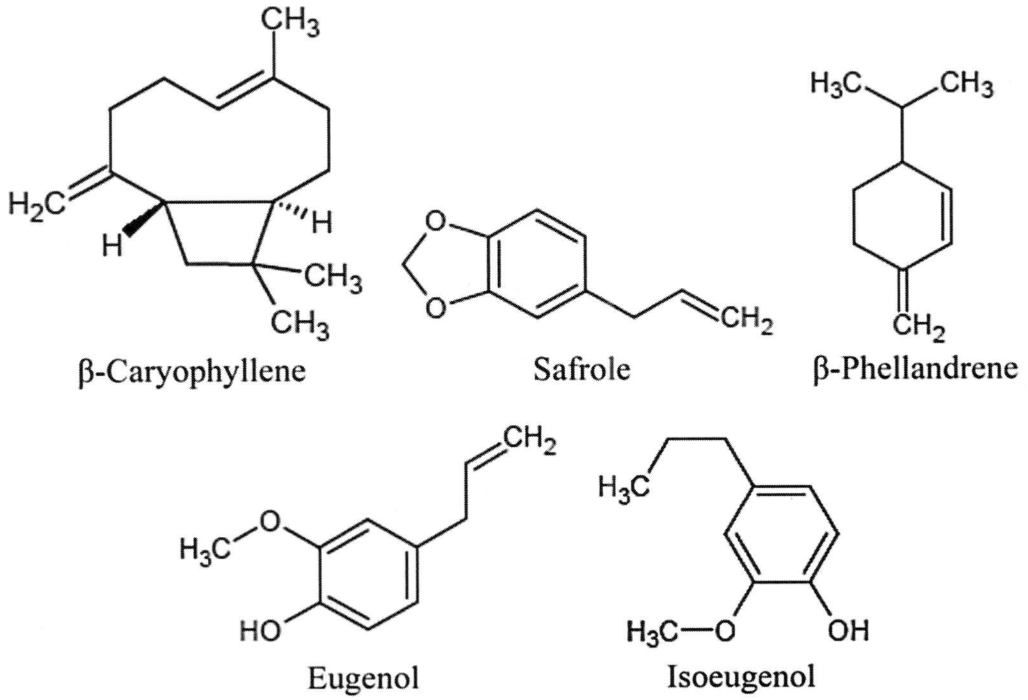

FIGURE 98.4
Important constituents of nutmeg seed oil. β-Caryophyllene [trans-(1R,9S)-8-Methylene-4,11,11-trimethyl bicycle [7.2.0] undec-4-ene], safrole [3,4-methylenedioxyphenyl-2-propene], eugenol [2-Methoxy-4-(2-propenyl) phenol], and isoeugenol [2-Methoxy-4-(1-propenyl) phenol] exert antioxidant activity via various mechanisms. β-Phellandrene [3-methylene-6-(1-methyl ethyl) cyclohexene] is the main aphrodisiac compound of nutmeg.

Antimicrobial activity

Various extracts and the essential oil of nutmeg seeds have presented strong antimicrobial activity against gram-positive and gram-negative bacteria, as well as a variety of fungi. Takikawa *et al.* (2002) reported antimicrobial activity of ethanolic extract of nutmeg seeds against entero-hemorrhagic *E. coli* O157, which was found to be highly sensitive to β-pinene. Narasimhan and Dhake (2006) reported potent antibacterial activity of chloroform extract of nutmeg seeds against both gram-positive and gram-negative bacteria. They found trimyristin and myristic acid to be the chief antibacterial principles isolated from nutmeg seeds. Cho *et al.* (2007) isolated three lignans (erythro-austrobailignan-6, meso-dihydroguaiaretic acid, and nectan-drin-B) from the methanolic extract of nutmeg seeds, which were reported to have antifungal activity. These three lignans were found to suppress the development of rice blast and wheat leaf rust. Some important antimicrobial compounds reported in nutmeg seeds are α-pinene, β-pinene, p-cymene, carvacrol, and β-caryophyllene, (Dorman & Deans, 2004) (Figure 98.5). Many plant phenolics have been reported to possess antimicrobial activity. β-Caryophyllene has been reported to have anti-inflammatory and antifungal activities (Sabulal *et al.*, 2006). α-Pinene and β-pinene (pinene-type monoterpene hydrocarbons) have been reported to have antimicrobial activity (Dorman & Deans, 2000), and are supposed to be involved in membrane disruption by the lipophilic compounds. Another important component for antimicrobial activity could be carvacrol. The mode of action of carvacrol on bacteria is similar to that of other phenolic compounds, and occurs via membrane damage, resulting in an increase in membrane permeability to protons and potassium ions, depletion of the intracellular ATP pool, and disruption of the proton-motive force. p-Cymene could also be an important component, because it is a precursor of carvacrol. It has been reported that p-cymene shows weak anti-bacterial activity when used alone, but works synergistically with carvacrol in expanding the membrane, which in turn causes destabilization of the membrane (Ultee *et al.*, 2002). It has been suggested that antimicrobial activity could be attributed to both major and minor components; it is possible that the antimicrobial activity of major components is regulated by

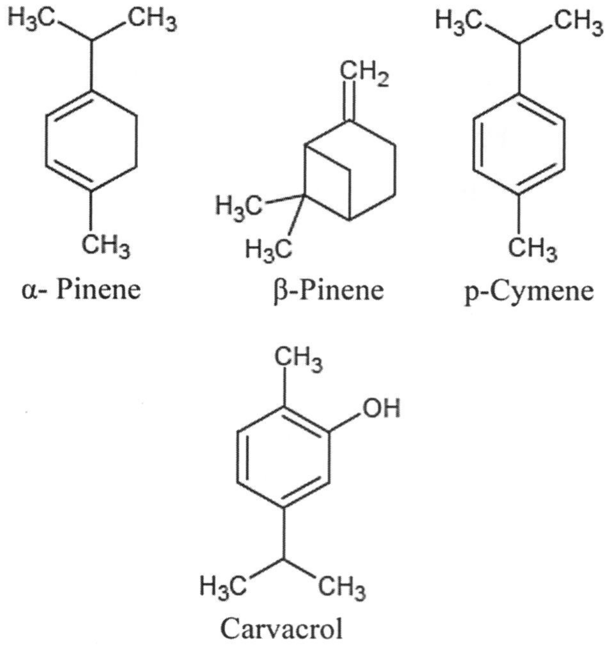

FIGURE 98.5

Important antimicrobial compounds of nutmeg seed. α-Pinene, β-pinene, p-cymene [1-methyl-4-(1-methylethyl)-benzene], and carvacrol [2-methyl-5-(1-methylethyl)-phenol] are the chief antimicrobial compounds of nutmeg. The main mechanisms for antimicrobial activity are membrane disruption, or depletion of the intracellular ATP pool.

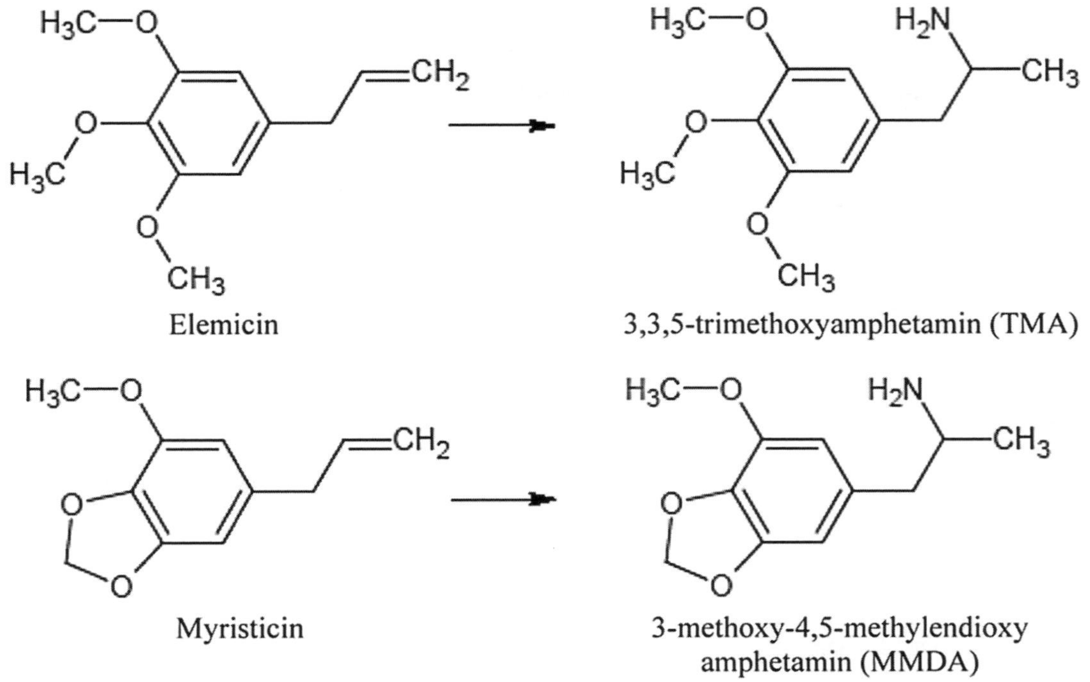

FIGURE 98.6

Metabolism of elemicin and myristicin in 3,3,5-trimethoxyamphetamin (TMA) and 3-methoxy-4,5-methylendioxy amphetamine (MMDA). 3,3,5-Trimethoxyamphetamin (TMA) and 3-methoxy-4,5-methylendioxy amphetamine (MMDA) are major compounds that cause hallucinogenic effects.

some other minor components, as well as these minor components being able to interact with other components to exert antimicrobial activity (Bounatirou *et al.*, 2007).

ADVERSE EFFECTS AND REACTIONS (ALLERGIES AND TOXICITY)

Consumption of nutmeg seeds in high dosage has been reported to lead to facial flushing, tachycardia, hypertension, dry mouth, blurred vision, psychoactive hallucinations, feelings of euphoria and unreality, and delirium. Recently, several cases of nutmeg seed ingestion have been reported in adolescents in particular, all of whom were attempting to achieve a euphoric state at low cost (Demetriades *et al.*, 2005). Symptoms usually begin about 3—6 hours after ingestion, and resolve by 24—36 hours. The medical literature does not cite any fatalities solely related to nutmeg intoxication. The possible cause for the psychoactivity of nutmeg seeds could be metabolic conversion of elemicin and myristicin into amphetamine-like compounds (Figure 98.6). Elemicin is observed to metabolize to 3,4,5 trimethoxyamphetamine (TMA), and myristicin to 3-methoxy-4,5-methylene-dioxy amphetamine (MMDA), which are amphetamine derivatives (Stein *et al.*, 2001). Moreover, myristicin is a weak inhibitor of monamine oxidase, which could be responsible for some cardiovascular symptoms.

SUMMARY POINTS

- Due to potential liver damage and carcinogenic effects, most widely used synthetic antioxidant compounds such as butylated hydroxytoluene and butylated hydroxyanisole have restricted usage.
- Nutmeg is the dried kernel of a broadly ovoid seed. Nutmeg seeds have been reported to have antioxidant and antimicrobial properties, besides their use as an important folk medicine.

- Nutmeg seeds possess a high antioxidant potential, which is attributed to caffic acid, catechin, eugenol, β-caryophyllene, argenteane, meso-dihydroguaiaretic acid, and erythro-austrobailignan-6.
- Nutmeg seeds show strong antimicrobial activity against gram-positive and gram-negative bacteria, as well as against various pathogenic fungi. Antimicrobial activity is contributed by β-caryophyllene, α-pinene, β-pinene, p-cymene, and carvacrol.
- Due to its high antioxidant and antimicrobial activities, nutmeg could be considered as a significant natural source of antioxidants and antimicrobials. Nutmeg, being a natural product, can offer more safety to people and the environment, and is considered to be less of a risk for resistance development by pathogenic microorganisms.

References

Bounatirou, S., Smiti, S., Miguel, M. G., Faleiro, L., Rejeb, M. N., Neffati, M., et al. (2007). Chemical composition, antioxidant and antibacterial activities of the essential oils isolated from Tunisian *Thymus capitatus* Hoff. et Link. *Food Chemistry, 105*, 146–155.

Calliste, C. A., Kozlowski, D., Duroux, J. L., Champavier, Y., Chulia, A. J., & Trouillas, P. (2010). A new antioxidant from wild nutmeg. *Food Chemistry, 118*, 489–496.

Chatterjee, S., Niaz, Z., Gautam, S., Adhikari, S., Variyar, P. S., & Sharma, A. (2007). Antioxidant activity of some phenolic constituents from green pepper (*Piper nigrum* L.) and fresh nutmeg mace (*Myristica fragrans*). *Food Chemistry, 101*, 515–523.

Cho, J. Y., Choi, G. J., Son, S. W., Jang, K. S., Lim, H. K., Lee, S. O., et al. (2007). Isolation and antifungal activity of lignans from *Myristica fragrans* against various plant pathogenic fungi. *Pest Management Science, 63*, 935–940.

Demetriades, A. K., Wallman, P. D., McGuiness, A., & Gavalas, M. C. (2005). Low cost, high risk: accidental nutmeg intoxication. *Journal of Emergency Medicine, 22*, 223–225.

Dorman, H. J. D., & Deans, S. G. (2000). Antimicrobial agents from plants: antibacterial activity of plant volatile oils. *Journal of Applied Microbiology, 88*, 308–316.

Dorman, H. J. D., & Deans, S. G. (2004). Chemical composition, antimicrobial and *in vitro* antioxidant properties of *Monarda citriodora* var. *citriodora, Myristica fragrans, Origanum vulgare* ssp. *hirsutum, Pelargonium* species, and *Thymus zygis* oils. *Journal of Essential Oil Research, 16*, 145–150.

Jukic, M., Politeo, O., & Milos, M. (2006). Chemical composition and antioxidant effect of free volatile aglycones from nutmeg (*Myristica fragrans* Houtt.) compared to its essential oil. *Croatica Chemica Acta, 79*, 209–214.

Kumaravelu, P., Subramanyam, S., Dakshinmurthy, D. P., & Devraj, N. S. (1996). The antioxidant effect of eugenol on carbon-tetrachloride induced erythrocyte damage in rats. *Journal of Nutritional Biochemistry, 7*, 23–28.

Nakai, M., Harada, M., Akimoto, K., Shibata, H., Miki, W., & Kiso, Y. (2003). Novel antioxidative metabolites in rat liver with ingested sesamin. *Journal of Agricultural and Food Chemistry, 51*, 1666–1670.

Narasimhan, B., & Dhake, A. S. (2006). Antibacterial principles from *Myristica fragrans* seeds. *Journal of Medicinal Food, 9*, 395–399.

Olajide, O. A., Ajayi, F. F., Ekhelar, A. L., Awe, S. O., Makinde, J. M., & Alada, A. R. A. (1999). Biological effect of *Myristica fragrans* (nutmeg) extract. *Phytotherapy Research, 13*, 344–345.

Sabulal, B., Dan, M., John, A., Kurup, R., Pradeep, N. S., Valsamma, R. K., et al. (2006). Caryophyllene-rich rhizome oil of *Zingiber nimmonii* from South India: chemical characterization and antimicrobial activity. *Phytochemistry, 67*, 2469–2473.

Senevirathne, M., Kim, S., Siriwardhana, N., Ha, J. H., Lee, K. W., & Jeon, Y. J. (2006). Antioxidant potential of *Ecklonia cava* on reactive oxygen species scavenging metal chelating, reducing power and lipid peroxidation inhibition. *Food Science and Technology International, 12*, 27–38.

Shan, B., Cai, Y. Z., Sun, M., & Corke, H. (2005). Antioxidant capacity of 26 spice extracts and characterization of their phenolic constituents. *Journal of Agricultural and Food Chemistry, 53*, 7749–7759.

Stein, U., Greyer, H., & Hentschel, H. (2001). Nutmeg (myristicin) poisoning – report on a fatal case and a series of cases recorded by a poison information centre. *Forensic Science International, 118*, 87–90.

Tajuddin, S., Ahmad, S., Latif, A., & Qasmi, I. A. (2003). Aphrodisiac activity of 50% ethanolic extracts of *Myristica fragrans* Houtt. (nutmeg) and *Syzygium aromaticum* (L.) Merr. & Perry. (clove) in male mice: a comparative study. *BMC Complementary and Alternative Medicine, 3*, 6–10.

Takikawa, A., Abe, K., Yamamoto, M., Ishimaru, S., Yasui, M., Okubo, Y., et al. (2002). Antimicrobial activity of nutmeg against *Escherichia coli* O157. *Journal of Bioscience and Bioengineering, 94*, 315–320.

Tomaino, A., Cimino, F., Zimbalatti, V., Venuti, V., Sulfaro, V., De Pasquale, A., et al. (2005). Influence of heating on antioxidant activity and the chemical composition of some spice essential oils. *Food Chemistry, 89*, 549—554.

Ultee, A., Bennink, M. H. J., & Moezelaar, R. (2002). The phenolic hydroxyl group of carvacrol is essential for action against the food borne pathogens *Bacillus cereus*. *Applied and Environmental Microbiology, 68*, 1561—1568.

Chemical and Antioxidant Properties of Okra (*Abelmoschus esculentus* Moench) Seed

Oluyemisi Elizabeth Adelakun, Olusegun James Oyelade
Department of Food Science and Engineering, Ladoke Akintola University of Technology, Ogbomoso, Oyo State, Nigeria

841

INTRODUCTION

In traditional societies nutrition and health care are interwoven, as many food plants are consumed for their health benefits. Food plants offer a variety of health benefits, either in the natural form or after processing. To identify the specific health value of food plants, studies on the nutritional and phytotherapeutic properties are essential. In this regard, the consumption of vegetables has been linked to the prevention of chronic diseases and cancers. The antioxidant activities in vegetables are essential in this field, because they are active in inhibiting free radical reactions, and consequently protect the human body against damage by reactive oxygen species (Yang *et al.*, 2006).

Okra (*Abelmoschus esculentus* (L.) Moench) is a vegetable widely cultivated for its tender fruits in many tropical countries, including Nigeria. The dried seed of okra has been found to be a good source of protein, fat, and, recently, of antioxidant activity. Thus, okra seed is an increasingly highly valued crop in the human diet.

Nuts & Seeds in Health and Disease Prevention. DOI: 10.1016/B978-0-12-375688-6.10099-4

BOTANICAL DESCRIPTION

Okra, also called "gumbo" in Africa, is a well-known horticultural crop. It is well distributed throughout the tropics, where its fruits are extensively used as a vegetable and for pickling. The harvested fruits are pale green, green, or purplish pods, which are ridged in many cultivars (Figure 99.1). When fully matured they are dark brown dehiscent or indehiscent capsules (Camcium *et al.*, 1998), and the seeds are embedded within these capsules. Okra seeds, according to Oyelade *et al.* (2003) and Calisir *et al.* (2005), vary in length between 4.5 and 5.5 mm, whilst the width, thickness, and weight are 4.2–5.5 mm, 3.5–4.5 mm, and 0.05–0.07 g, respectively. The volume ranges from 0.31 to 0.50 cm^3, with a surface area of 28.27–88.92 cm^2 at a moisture content range of 6.35–15.22%, dry basis.

HISTORICAL CULTIVATION AND USAGE

Wild varieties of okra exist in Ethiopia and in the Upper Nile in Sudan, and perennial varieties in West Africa. The presence of okra in the New World has been attributed to the slaves from Africa (Yamaguchi, 1983). Okra belongs to the Mallow family, Malvaceae, and is now grown in all parts of the tropics, and, during the summer, in the warmer parts of the temperate region. It is a herbaceous, shrub-like dicotyledonous annual plant, with woody stems, growing to a height of 1–2 m. It has alternate palmate broad leaves, and the flowers have five large yellow petals with a large purple area covering the base. Okra (*Abelmoschus esculentus* Moench) is one of the most important vegetables in Nigeria, where it is widely grown for its tender fruits and young leaves. It is easy to cultivate, and grows well in both tropical and temperate zones ranging from Africa to Asia, Southern European, and America (Calisir *et al.*, 2005).

PRESENT-DAY CULTIVATION AND USAGE

Okra is presently grown in large quantities in different parts of Nigeria and other West African countries because of the favorable climatic conditions. Okra, which is specially valued for its tender and delicious fruits, has been reported to have an average nutritive value of 3.21, which is higher than that of tomato, eggplant, and most cucurbits except bitter gourd (Sahoo & Srivastava, 2002). Okra fruit is commonly processed into soups and stews. The process involves slicing, drying, and grinding, and the resulting material can then be mixed with other ingredients to make soup. In addition to its use in stews, okra can be used in the processing of other food items, such as candies, salad dressings, and cheese spreads. Apart from its food usage, the tree and pods have a number of additional economic benefits. The tree can be used to make rope and paper, while the pod (fruits) can be used, in extract form, as a fat substitute in brownies.

FIGURE 99.1
Fresh okra fruits and dried seeds from Nigeria.

Nutritionally, the richest part of the okra plant is the seed. Although the oil of the okra seed after processing is edible, and the residual meal following oil extraction is significantly rich in protein, the seed is not traditionally used for either oil or protein, but rather for seedling and regeneration purposes. This probably explains why there are only fragmentary accounts of the use of okra seed as a food (Karakoltsidis & Constantinides, 1975). However, large quantities of seeds are discarded as unfit for seedling purposes. The seeds of mature okra pods, sometimes used as poultry feed, are also consumed after roasting, and as a coffee substitute. The volatile compounds in the coat of okra seeds, which can be released by simply rubbing the seeds, have been reported to have biological activity, and some of these compounds, including farnesol derivatives and aliphatic esters, seem to support the hypothesis of their role in defense against seed-eating insects (Camcium *et al.*, 1998). By-products of okra seed have also been found to have industrial usage. Biodiesel was recently derived from okra (*Hibiscus esculentus*) seed oil, by methanol-induced trans-esterification using an alkali catalyst. It was concluded that okra seed oil is an acceptable feedstock for biodiesel production (Anwar *et al.*, 2010).

APPLICATIONS TO HEALTH PROMOTION AND DISEASE PREVENTION

Nutrition and health care are interwoven, which means that many plant foods are consumed for their health benefits. Okra is generally used as a nutritional supplement, containing vitamins C and A, B complex vitamins, and iron and calcium. It is good for people suffering from renal colic, leukorrhea, and general weakness. Due to its high iodine content, the fruit is considered useful for the control of goiter. Okra leaves are used in Turkey for the preparation of medicine to reduce inflammation. A bland mucilage is also used for the control of dysentery, and as a clarifying agent in the preparation of gur (Sahoo & Srivastava, 2002). The dried seed of okra is essentially beneficial to health because of its chemical and antioxidant composition. These are discussed below.

843

Chemical composition of okra seed

Studies have confirmed the potential of okra seed (dried seed) as a good source of oil and protein for both the temperate regions and the tropics (Adelakun *et al.*, 2009a). The seed coat fraction, as reported by Oyelade *et al.* (2003), was found to have a protein content of about 20%, while the defatted fraction was 55%. The protein content in the whole seed flour fractions was as high as 45% for all the five cultivars studied (UI4-30, VI-104, LD-88, 47-4, and V-35). Also, when the endosperm was ground and defatted, the protein content was 55%. The fat content in the seed coat was 11%, in the endosperm was 34%, and in whole seed was 22%. Andras *et al.*, (2005) reported on okra seed from Greece as a potential source of oil, with concentrations varying from 15.9 to 20.7%. The seed oil mainly consisted of unsaturated fatty acids, especially linoleic acid (up to 47.4%). This seed oil also represents a potential source of palmitic acid, a chemical imported into the European Community, and an important raw material for soaps, esters, and plasticizers. It could also improve the quality of soybean oil, which has limitations as a shortening because it only contains around 11% palmitic acid (Karakoltsidis & Constantinides, 1975). Moreover, the linoleic acid, once separated, could be utilized for producing dyes, plastics, and resins. Good quality protein and fats are important for growth and repair of damage in the body, and they also serve as a source of energy. Okra seeds can supply these.

Despite the high protein content and fat of okra seed, the potential of the seed as a new source of vegetable protein and fat has yet to be fully exploited, probably because of the problem of removal of the okra seed hull from the kernel (Oyelade *et al.*, 2003), and the dearth of knowledge regarding okra seed protein characteristics. However, the removal of this seed hull from the kernel can be enhanced by pretreatments such as soaking, blanching, and roasting (Adelakun *et al.*, 2009a, 2009b). Table 99.1 lists the proximate composition of okra seeds from various studies.

TABLE 99.1 Proximate Composition of Okra Seeds (Dry Basis)							
	Whole Seed[a]	Whole Seed[b]	Whole Seed[c]	Seed Coat[c]	Whole Seed[d]	Whole Soaked (36 h) Seed[d]	Whole Roasted (40 min) Seed[d]
Total dietary fiber (%)	46.3	26.34	—	—	3.45	3.75	3.60
Protein (%)	18.7	19.10	45	20	41.11	44.03	38.10
Fat (%)	0.8	8.21	34	11	31.04	28.09	17.22
Ash (%)	9.0	4.63	—	—		3.75	4.06
Carbohydrates (%)	71.5		—	—	3.42	9.33	26.33
Moisture (%)	—	6.35	8.0	—	10.60	10.88	10.69
Crude energy (kcal/g)	—	25.4	—	—	—	—	—
Water-soluble extract (%)	—	2.6	—	—	—	—	—
Ether-soluble extract (%)	—	8.7	—				
Non-soluble HCl acid ash (%)		0.41×10^{-2}	—				

Sources:
[a]Kahlon et al. (2007);
[b]Calisir et al. (2005);
[c]Oyelade et al. (2003);
[d]Adelakun et al. (2009a, 2009b).

Antioxidant composition of okra seed

Antioxidants are an important class of compounds that quench reactive free radical intermediates formed during oxidative reactions. A major feature of free radicals is that they have an extremely high chemical reactivity, because of the presence of unpaired electrons, which explains how they inflict damage in cells (Mohajeri & Asemani, 2009).

The scavenging activity of an antioxidant is documented to be the first line of defense against free radicals, particularly reactive oxygen species (ROS), which are involved in the pathogenesis of several chronic and degenerative diseases, such as inflammation, cardiovascular disease, neurodegenerative diseases, cancer, and age-related disorders (Locatelli *et al.*, 2009). Reports also show that oxidation in food can induce rancidity and/or deterioration of nutritional quality and organoleptic properties (color, flavor, texture), and consequently becomes a safety concern (Antolovich, 2002). In order to prevent oxidative reactions in foods and to protect biological tissues against damage to molecular targets such as proteins, lipids, carbohydrates, and DNA, various synthetic or natural antioxidants can be used. Recently, the use of natural antioxidants to protect against oxidative stress due to their scavenging activity against ROS has been promoted because of concerns regarding the safety of synthetic antioxidants (Locatelli *et al.*, 2009). Plant foods serve as sources of a wide variety of dietary antioxidants, such as vitamins C and E, carotenoids, flavonoids, and other phenolic compounds. The increased consumption of vegetables is considered to be particularly important because of their beneficial effect in maintaining health, and their potential in the prevention of chronic diseases and cancers. Soong and Barlow (2004) have reported that the total phenolic content of seeds of several fruits, such as mango, avocado, and jackfruit, were higher than in their edible flesh. Therefore, these seeds could be considered as valuable source of phenolics.

Data regarding okra seeds' antioxidant activity is scanty. Recently, four quercetin derivatives were identified (Huang *et al.*, 2007). This prompted Arapitsas (2008) to focus on elucidation of the okra polyphenolic profile — an issue of which there is limited knowledge. Qualitative and quantitative analyses demonstrated that this vegetable is rich in polyphenolic compounds, with the seeds mainly composed of quercetin derivatives and catechins, and the skins of quercetin and hydroxycinnamic acid derivatives. Adelakun *et al.* (2009a, 2009b), however, worked on the effect of pretreatments, such as soaking, blanching, and roasting at 160°C for

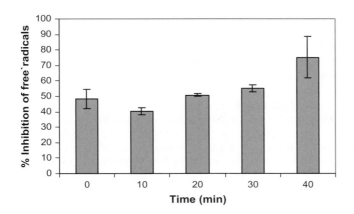

FIGURE 99.2
Effect of roasting on the antioxidant activity of okra seed flour. Values are the means ± SD of triplicate measurements. Adapted from Adelakun *et al.* (2009a), *Food Chem. Toxicol.*, *47*, 657—661, with permission.

varying times (10—40 minutes; Figure 99.2), on the antioxidant activity of okra seed, and reported that roasting significantly increased this activity, in a time-dependent manner. Thus, incorporation of the okra seed into the human diet would be of positive benefit in preventing chronic diseases such as cancer and age-related diseases.

Conclusion

Based on the available information on the proximate content of okra seed, and the effect of processing on this, consumption of okra seed can be promoted because of its potential health benefits. The recent reports on its antioxidant content validate this assertion.

ADVERSE EFFECTS AND REACTIONS (ALLERGIES AND TOXICITY)

Obstacles to the use of okra seeds as a food crop are the content of gossypol and cyclopropenoid fatty acids. However, the gossypol concentration is well below FAO limits, and the cyclopropenoid fatty acids can be eliminated by hydrogenation (Martin & Rhodes, 1983).

SUMMARY POINTS

- Okra is generally consumed in the form of immature pods.
- Dried seeds can serve as a good source of protein and fat.
- Depending on the variety, the protein content can be as high as 50%.
- Like other fruit-producing seeds, its health benefits have been established.
- Pre-treatment, such as roasting, can increase the antioxidant properties of okra seed.

References

Adelakun, O. E., Oyelade, O. J., Ade-Omowaye, B. I. O., Adeyemi, I. A., Van de Venter, M., & Koekemoer, T. C. (2009a). Influence of pretreatment on yield, chemical and antioxidant properties of a Nigerian okra seed (*Abelmoschus esculentus* Moench) flour. *Food and Chemical Toxicology, 47*, 657—661.

Adelakun, O. E., Oyelade, O. J., Ade-Omowaye, B. I. O., Adeyemi, I. A., & Van de Venter, M. (2009b). Chemical composition and the antioxidative properties of Nigerian okra seed (*Abelmoschus esculentus* Moench) flour. *Food and Chemical Toxicology, 47*, 1123—1126.

Andras, C. D., Simandi, B., Orsi, F., Lambrou, C., Tatla, D. M., & Panayiotou, C. (2005). Supercritical carbon dioxide extraction of okra (*Hibiscus esculentus* L.) seeds. *Journal of the Science of Food and Agriculture, 85*, 1415—1419.

Antolovich, M., Prenzeler, P. D., Pastilides, E., McDonald, S., & Robards, E. (2002). Methods for testing antioxidant activity. *Analyst, 127*, 183—198.

Anwar, F., Rashid, U., Ashraf, M., & Nadeem, M. (2010). Okra (*Hibiscus esculentus*) seed oil for biodiesel production. *Applied Energy, 87*, 779—785.

845

Arapitsas, P. (2008). Identification and quantification of polyphenolic compounds from okra seeds and skins. *Journal of Food Chemistry, 110*, 1041–1045.

Calisir, S., Ozcan, M., Haciseferogullari, H., & Yildiz, M. U. (2005). A study on some physicochemical properties of Turkey okra (*Hibiscus esculenta* L) seeds. *Journal of Food Engineering, 68*, 73–78.

Camcium, M., Bessiere, J. M., Vilarem, G., & Gaset, A. (1998). Volatile components in okra seed coat. *Phytochemistry, 48*, 311–315.

Huang, Z., Wang, B., Eaves, D. H., Shikany, J. M., & Pace, R. D. (2007). Phenolic compound profile of selected vegetables frequently consumed by African Americans in the southeast United States. *Food Chemistry, 103*, 1395–1402.

Kahlon, T. S., Chapman, M. H., & Smith, G. E. (2007). *In vitro* binding of bile acids by okra, beets, asparagus, eggplant, turnips, green beans, carrots, and cauliflower. *Food Chemistry, 103*, 676–680.

Karakoltsidis, P. A., & Constantinides, S. M. (1975). Okra seed: a new protein source. *Journal of Agricultural and Food Chemistry, 23*, 1204–1207.

Locatelli, M., Gindro, R., Travaglia, F., Coïsson, J., Rinaldi, M., & Arlorio, M. (2009). Study of the DPPH scavenging activity: development of a free software for the correct interpretation of data. *Food Chemistry, 114*, 889–897.

Martin, F. W., & Rhodes, A. M. (1983). Seed characteristics of okra and related Abelmoscus species. *Plant Foods for Human Nutrition, 33*, 41.

Mohajeri, A., & Asemani, S. S. (2009). Theoretical investigation on antioxidant activity of vitamins and phenolic acids for designing a novel antioxidant. *Journal of Molecular Structure, 930*, 15–20.

Oyelade, O. J., Ade-Omowaye, B. I. O., & Adeomi, V. F. (2003). Influence of variety on protein, fat contents and some physical characteristics of okra seeds. *Journal of Food Engineering, 57*, 111–114.

Sahoo, P. K., & Srivastava, A. P. (2002). Physical properties of okra seed. *Biosystems Engineering, 83*, 441–448.

Soong, Y., & Barlow, P. J. (2004). Antioxidant activity and phenolic content of selected fruit seeds. *Food Chemistry, 88*, 411–417.

Yamaguchi, M. (1983). *World vegetables*. Westport, CT: Ellis Hardwood Ltd.

Yang, R., Tsou, S., Lee, T., Wu, W., Hanson, P. M., Kuo, G., et al. (2006). Distribution of 127 edible plant species for antioxidant activities by two assays. *Journal of the Science of Food and Agriculture, 86*, 2395–2403.

CHAPTER 100

Olive (*Olea europaea* L.) Seeds, From Chemistry to Health Benefits

Alam Zeb[1], Michael Murkovic[2]
[1] Department of Biotechnology, University of Malakand, Chakdara, Pakistan
[2] Institute of Biochemistry, Graz University of Technology, Graz, Austria

847

CHAPTER OUTLINE

LIST OF ABBREVIATIONS

FFA, free fatty acid
ESI-MS, electrospray ionization mass spectrometry
APPI-MS, atmospheric pressure photo-ionization mass spectrometry
NMR, nuclear magnetic resonance spectroscopy
LDL, low density lipoprotein
HDL, high density lipoprotein
CCK, cholecystokinin
TAG, triacylglycerol

Nuts & Seeds in Health and Disease Prevention. DOI: 10.1016/B978-0-12-375688-6.10100-8

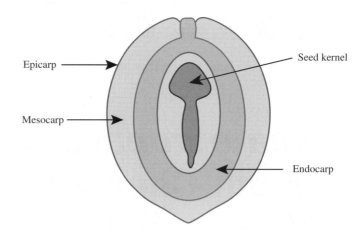

FIGURE 100.1
Typical representation of a dissected olive fruit showing an oval-shaped seed in the center.

INTRODUCTION

The olive fruit containing the seed generally consists of three parts: the epicarp, mesocarp, and endocarp. The epicarp, which is the skin, peel, or epidermis, is covered with wax, and usually remains green throughout the growth phase. During ripening, or when ripe, it may turn purple, brown, or black, depending on the variety. The mesocarp, which is the pulp or flesh, usually has a sugar content of 3—7.5% and a high oil content in the range of 15—30%. The composition of the fruit is mainly dependent on the variety, environment, and degree of ripeness. The endocarp is the hard part of the fruit, and is made of fibrous lignin. Its ovoid shape and the extent to which it is furrowed are varietal characteristics, as shown in Figure 100.1. The endocarp encloses the olive seed, which accounts for approximately 3% of the fruit weight and contains 2—4% of the total fruit oil. The olive seed weighs from 1 to 5 g, and has an average chemical composition of 30% water, 27% oil, 27% sugars, 2% cellulose, 10% proteins, 1.5% minerals, and 2% ash (Niaounakis & Halvadakis, 2006). A huge amount of data is available regarding olive oil, but less information is available on the olive seeds. This chapter provides an overview of the chemical composition and health applications of the olive oil obtained from olive seeds, regardless of any differences in variety, origin, and methods of extraction and analysis.

BOTANICAL DESCRIPTION

The olive tree (*Olea europaea* L.) belongs to the *Oleaceae* family, and is a polymorphous, medium-sized tree with a furrowed trunk. Generally, the grayish-green leaves are about 5—6 cm long and about 1—1.5 cm wide at the middle, with smooth edges and a short peduncle. The plant is widely cultivated for oil and food uses, and also grows wild. The fruit of wild olives generally matures during the summer, varying in color from light green to violet-red to black. Green maturation occurs in the autumn, and complete ripening in the winter. The fruit reaches its optimal size in warm, bright sites with aerated soil. It has a natural alternation in fruit production, which can be regulated with good cultivation practices. Harvesting for oil begins with the autumn rains (Ramírez-Tortosa *et al.*, 2006).

HISTORICAL CULTIVATION AND USAGE

The olive tree is one of the oldest known cultivated trees in the world. It is considered to be a cultural marker and a compass to explore the development of civilizations. From the very beginning to the present, olive oil has been a staple food in the Mediterranean region. The olive tree possesses an astonishing ability to survive, having strong resistance to unfavorable conditions, including extreme hot weather. On the other hand, if it is to be well cared for it is a demanding crop for cultivation. A suitable environment and proper cultural care are necessary for the full development of agronomic characteristics, and steady production conditions.

PRESENT-DAY CULTIVATION AND USAGE

Nowadays the olive tree is mainly cultivated in Mediterranean countries, including Spain, Italy, Greece, Tunisia, Turkey, Morocco, Syria, Algeria, Egypt, Israel, Libya, Jordan, Lebanon, Cyprus, Croatia, and Slovenia, as well as in Argentina, Chile, Mexico, Peru, the United States, and Australia (Boskou, 2009). Olive seed is used for feed and food uses. Olive oil is obtained from the seed or the whole berries, and seed oil is obtained from the whole seed.

APPLICATIONS TO HEALTH PROMOTION AND DISEASE PREVENTION

Health promoting chemical components

Triacylglycerols (TAG) are the major component of olive oil. The fatty acid composition of triacylglycerols includes oleic acid, which is present in much higher concentrations than the other well-known fatty acids, such as linoleic, palmitic, and stearic acid. The oleic acid (O) accounts for 68–83%, palmitic acid (P) for 8–16%, stearic acid (S) for 1.7–3.0%, and linoleic acid (L) for 5–13%. Low levels of myristic acid (M) are also present. The common triacylglycerols are MOP, MOS, POP, SOS, POP, SOP, SSO, LOO, OLO, POO, SOO, PLP, SLP, OOO, POL, SOL, MLO, PLO, and SLO (Ollivier *et al.*, 2003). Reversed phase liquid chromatography coupled to mass spectrometry is the most valuable tool to differentiate among the regio-isomers in olive oil. Triacylglycerols have been used for authentication of olive oil using mass spectrometry. It was found that mixtures of olive oils with other vegetable oils can be easily discriminated using both ESI-MS and APPI-MS spectra. Recent results obtained from NMR analysis and isotopic ratios indicate the possibility of determining the geographical origin of the olive oil.

Other minor components are important in terms of health applications. These include esters of fatty acids, aliphatic alcohols, terpenes, hydrocarbons, sterols, and phenolic compounds (Table 100.1). In olive oil, fatty acids are also esterified with different alcoholic compounds, forming non-glyceride esters. The ester-forming alcohols include methanol, ethanol, β-sitosterol, campesterol, stigmasterol, tricycloartenol, and 24-methyl-cycloartenol

TABLE 100.1 Typical Chemical Composition of Olive Seed Oil

S. No.	Chemical Component	Amount (mg/kg)
1	Aliphatic alcohols	60–200
2	Triterpenes	500–3000
3	Phytosterols	1800–5000
4	Total hydrocarbons	1500–8000
5	β-Carotene	0.33–4.0
6	Tyrosol	1.4–29.0
7	Vanillic acid	6.7–4.0
8	Luteolin	0.20–7.0
9	Apigenin	0.70–2.0
10	Tocopherols/tocotrienols	12.0–400
11	Hydrophilic phenols	40–1000
12	Lignans	40–60

Range values represent variations due to variety, origin, and methods of extraction and analysis.

(Kiritsakis & Markakis, 1987). The total amounts of non-glyceride esters are in the range of 100–250 mg/kg. Waxes are esters of long-chain aliphatic alcohols (C27–C32) made of more or less 58 carbon atoms. Generally waxes are present in the skin of the olive fruits and seed, which prevent water loss.

Different types of alcohols are present in olive oil, including aliphatic and terpene alcohols. Aliphatic alcohols are a fraction mainly composed of long-chain saturated alcohols (C18–C30). The total amount of these compounds is in the range of 60–200 mg/kg (Ranalli *et al.*, 1999). Triterpene alcohols are present in a range of 500–3000 mg/kg. Two of the most important are two dihydroxy triterpenes named erythrodiol and uvaol. Erythrodiol occurs in the free and esterified form. The levels of these compounds in virgin olive oil are strongly affected by cultivar and origin. These two components are very important, as they provide the basis for recognizing the purity of olive oils (Kiritsakis & Markakis, 1987; Ramírez-Tortosa *et al.*, 2006).

Plant sterols (phytosterols) are very important for human health. They are tetracyclic compounds biosynthesized from squalene. The squalene contents ranges from 3900–9600 mg/kg. In olive oil, sterols are present in the range of 1800–5000 mg/kg (Mulinacci *et al.*, 2005). The amount of phytosterols can be used as an indicator for identification oil origin as well as its purity. The major sterols are β-sitosterol, Δ-5-avenasterol, and campesterol. Other sterols present in smaller quantities are stigmasterol, cholesterol, brassicasterol, chlerosterol, ergosterol, sitostanol, campestanol, Δ-7-avenasterol, Δ-7-cholesterol, Δ-7-campesterol, Δ-7-stigmastenol, Δ-5,23-stigmastadienol, Δ-5,24-stigmastadienol, Δ-7,22-ergastadienol, Δ-7,24-ergostadienol, 24-methylene-cholesterol, and 22,23-dihydrobrassicasterol (Boskou, 2009), present in olive oils of different origins and varieties.

In olive oil, total hydrocarbons are in the range of 1500–8000 mg/kg. These include carotenoids and squalenes. Carotenoid hydrocarbons are present in minor quantities. The most important carotenoids present in olive oil are β-carotene and lutein, which contribute to the color. Virgin olive oils produced from mature olives contain β-carotene in the range of 0.33–3.69 mg/kg (Rahmani & Csallany, 1991). During refining, most of the carotenoids are degraded and reduced to a minimal level. Several other components, like chlorophyll and its oxidized products, are also responsible for olive oil coloration.

Olive oil can be rich in phenolic compounds. Phenolic acids were the first group of phenolic compounds identified in virgin olive oil. The aromatic acids comprise benzoic acid, p-hydroxybenzoic acid, protocatechuic acid, gallic acid, vanillic acid, and syringic acid. The cinnamic acid derivatives include cinnamic acid, *o*-coumaric acid, *o*-coumaric acid, caffeic acid, ferulic acid, and sinapic acid. Tyrosol, vanillic acid, luteolin, and apigenin were identified and quantified by LC-MS (Murkovic *et al.*, 2004). Tyrosol was found to be the major phenol, in the range of 1.4–29 mg/kg, followed by vanillic acid (0.67–4.0 mg/kg), luteolin (0.22–7.0 mg/kg), and apigenin (0.68–1.6 mg/kg), of seven different olive oils.

Other important components are tocopherols and tocotrienols, in a range of 12–400 mg/kg. Hydrophilic phenols have been identified as being responsible for most of the antioxidant properties of virgin olive oil. Hydrophilic phenols (40–1000 mg/kg) may be divided into different classes, such as phenolic acids, phenolic alcohols, secoiridoids, lignans, and flavones (Servili *et al.*, 2004). Lignans were found in virgin olive oil at a mean level of 42 mg/kg. Two compounds, namely 1-pinoresinol and 1-acetoxypinoresinol, were the major identified lignans.

Health benefits and disease prevention
ANTIOXIDANT ACTIVITY

In biological systems, free radical reactions are associated with aging, cancer, cardiovascular disease, optical disease, and neurodegenerative disease. Antioxidants are used to prevent such reactions, and are therefore supporting health. Among the chemical components of olive oils, most of the minor components are responsible for the antioxidant potential. Phenolic components of olive oil are widely studied for their antioxidant activity. It was found that the radical scavenging actions of some olive oil phenols, such as hydroxytryrosol and oleuropein, result in reduced formation of superoxide anions, reduced neutrophil respiratory burst, and hypochlorous acid. This could explain the observed lower incidence of coronary heart disease and cancer associated with the Mediterranean diet (Visioli *et al.*, 1998). In addition to its antioxidant potential in biological systems, olive oil is more resistant to thermal oxidation during frying than other edible oils (Zeb *et al.*, 2008). During thermal oxidation, tocopherols and phenols are the major compounds contributing to the stability. The high concentration of oleic acid in the oil also contributes to the stability.

CARDIOVASCULAR DISEASE

Cardiovascular disease is the main cause of death in the modern world. Diet and life style play an important role in prognosis and prevention. Due to the presence of high levels of unsaturated fatty acids, olive oil improves the lipid profile by reducing the LDL/HDL ratio in humans. The phenolic compounds provide resistance against the oxidation of LDL, and thus play an important role in preventing ischemic heart disease. Olive oil also improves endothelial function, thus improving the hemostatic system in hypercholesterolemic individuals (Perez-Jimenez *et al.*, 2002). Furthermore, olive oil markedly lowers the blood pressure and reduces the daily antihypertensive dosage requirement among hypertensive subjects. The low level of cardiovascular mortality in the Mediterranean region has been attributed to olive oil being the principal fat source. Generally, a Mediterranean diet is characterized by high consumption of olive oil, legumes, unrefined cereals, fruits, and vegetables, moderate consumption of dairy products (mostly as cheese and yogurt), moderate to high consumption of fish, low consumption of meat and meat products, and moderate wine consumption.

IMMUNE FUNCTION

The protective functions of the immune system against infections involve the killing or destroying of viruses and bacteria appearing in the body. Monounsaturated fatty acids like oleic acid were found to have a positive effect on natural killing cell activity. Yaqoob *et al.*, (1998) showed that the monounsaturated fatty acids in olive oil play an important role in the immune function. A study demonstrated that diets containing 50 or 100 g/kg of olive oil completely suppressed increases in tissue zinc content, liver protein synthesis, and serum ceruloplasmin levels in response to subcutaneous *Escherichia coli* endotoxin, when compared with a maize oil diet or standard laboratory chow (Besler & Grimble, 1995). From these results, they concluded that olive oil can protect against the toxicity of endotoxins.

GASTROINTESTINAL DISEASE

Studies on animals showed that olive oil increases the concentration of some important hormones in blood (Yago *et al.*, 1997). Results showed that adaptation of the pancreas to the type of dietary fat is associated with differences in the circulating levels of several gastrointestinal hormones, either at rest or in response to food ingestion, which in turn act on the synthesis or secretion of enzymes and other constituents of the digestive juices. In humans, an olive oil meal stimulates CCK, and the high level of oleic acid in olive oil was found to be responsible for this stimulation (Manas *et al.*, 2006).

CANCER

Cancer is one of the leading causes of death in developing countries. Risk factors include diet, life style, toxic chemicals, radiation, and some occupations. Recent epidemiological studies have revealed the beneficial role of olive oil against cancer (Gallus *et al.*, 2004). Studies conducted in Mediterranean regions — especially Greece, Italy, and Spain — showed a clear association between olive oil consumption and a reduced cancer risk. The protective role of olive oil against different malignancies, including breast, ovarian, endometrial, colorectal, laryngeal, esophageal, lung, and pancreatic cancers, has been reported (Boskou, 2009).

FATTY LIVER DISEASE

Fatty liver disease is another major mortal disease in developing countries. Among such liver diseases, non-alcoholic fatty liver disease and non-alcoholic steatohepatitis are very important, and occur mostly in the general population of countries where alcohol is banned or not allowed for religious reasons. The disease is clinically important, as it progresses to fibrosis, cirrhosis, and, consequently, to hepatocellular carcinoma. An olive oil-rich diet was found to decrease the accumulation of TAGs in the liver, to improve postprandial TAGs, glucose, and glucagon (like peptide-1) responses in insulin-resistant subjects, and to upregulate glucose transporter-2 expression in the liver (Assy *et al.*, 2009). The beneficial effect of the Mediterranean diet in curing fatty liver diseases is a consequence of the uptake of oleic acid from olive oil.

OTHER HEALTH APPLICATIONS

The presence of important phenols and other components in olive oil is a hot research topic in the field of health and diseases, as shown in Figure 100.2. Much work is still necessary to expand the health applications. Religious beliefs may also contribute to increased dietary use of olive oil in some parts of the world where it is currently rarely grown or used. Nowadays, other beneficial health effects of olive oil include the treatment of diabetes, inflammatory bowel disease, thrombosis, and atherosclerosis.

ADVERSE EFFECTS AND REACTIONS (ALLERGIES AND TOXICITY)

Because of the presence of high levels of phenolic compounds, ingestion of excessive amounts of olive oil can cause temporary mild diarrhea. However, no other adverse effects or allergic reactions from external use of the olive oil have been reported.

852

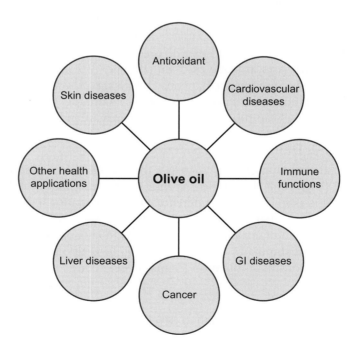

FIGURE 100.2
Graphical representation of applications of olive seed oil in different health areas.

SUMMARY POINTS

- Olive oil obtained from olive seeds or fruits contains large amounts of monounsaturated and polyunsaturated fatty acids.
- Olive oil is a good source of esters of fatty acids, hydrocarbons (including carotenoids and squalene), and phytosterols.
- Tocopherols and tocotrienols are also present in olive oil.
- The high levels of phenolic compounds result in high antioxidant activity.
- Olive oil influences positively cardiovascular disease, immune function, and gastrointestinal (GI) disease.
- Olive oil improves the curing of mortal diseases like cancer and liver disease.
- Dietary use of olive oil is therefore helpful in maintaining good health.

References

Assy, N., Nassar, F., Nasser, G., & Grosovski, M. (2009). Olive oil consumption and non-alcoholic fatty liver disease. *World Journal of Gastroenterology, 15*, 1809–1815.

Besler, H. T., & Grimble, R. F. (1995). Comparison of the modulatory influence of maize and olive oils and butter on metabolic responses to endotoxin in rats. *Clinical Science, 88*, 59–66.

Boskou, D. (2009). *Olive oil chemical constituents and health.* New York, NY: Taylor & Francis Group, LLC.

Gallus, S., Bosetti, C., & Vecchia, C. L. (2004). Mediterranean diet and cancer risk. *European Journal of Cancer Prevention, 13*, 447–452.

Kiritsakis, A., & Markakis, P. (1987). Olive oil: a review. *Advances in Food Research, 31*, 453–483.

Manas, M., Yago, M. D., & Victoria, E. M. (2006). Olive oil and regulation of gastrointestinal function. In J. L. Quiles, M. C. Ramírez-Tortose, & P. Yaqoob (Eds.), *Olive oil and health* (pp. 284–309). Cambridge, MA: CABI International.

Mulinacci, N., Giaccherini, C., Innocenti, M., Romani, A., Vincieri, F. F., Marotta, F., et al. (2005). Analysis of extra virgin olive oils from stoned olives. *Journal of the Science of Food and Agriculture, 85*, 662–670.

Murkovic, M., Lechner, S., Pietzka, A., Bratakos, M., & Katzogiannos, M. (2004). Analysis of minor constituents in olive oil. *Journal of Biochemical and Biophysical Methods, 61*, 155–160.

Niaounakis, M., & Halvadakis, C. P. (2006). *Olive processing waste management literature: Review and patent survey* (2nd ed.). Oxford: Elsevier Ltd.

Ollivier, D., Artaud, J., Pinatel, C., Durbec, J. P., & Re, M. G. (2003). Triacylglycerol and fatty acid compositions of French virgin olive oils: characterization by chemometrics. *Journal of Agricultural and Food Chemistry, 51*, 5723–5731.

Perez-Jimenez, F., Lopez-Miranda, J., & Mata, P. (2002). Protective effect of dietary monounsaturated fat on arteriosclerosis: beyond cholesterol. *Atherosclerosis, 163*, 385–398.

Rahmani, M., & Csallany, A. S. (1991). Chlorophyll and β-carotene pigments in Moroccan virgin olive oils measured by high-performance liquid chromatography. *Journal of the American Oil Chemists Society, 68*, 672–674.

Ramírez-Tortosa, M., Granados, S., & Quiles, J. L. (2006). Chemical composition, types and characteristics of olive oil. In J. L. Quiles, M. C. Ramírez-Tortose, & P. Yaqoob (Eds.), *Olive oil and health* (pp. 45–62). Cambridge, MA: CABI International.

Ranalli, A., Ferrante, M. L., Mattia, D. G., & Costantini, N. (1999). Analytical evaluation of virgin olive oil of first and second extraction. *Journal of Agricultural and Food Chemistry, 47*, 417–424.

Servili, M., Selvaggini, R., Esposto, S., Taticchi, A., Montedoro, G., & Morozzi, G. (2004). Health and sensory properties of virgin olive oil hydrophilic phenols: agronomic and technological aspects of production that affect their occurrence in the oil. *Journal of Chromatography A, 1054*, 113–127.

Visioli, F., Bellomo, G., & Galli, C. (1998). Free radical–scavenging properties of olive oil polyphenols. *Biochemical and Biophysical Research Communications, 247*, 60–64.

Yago, M. D., Martinez-Victoria, E., Manas, M., Martinez, M. A., & Mataix, J. (1997). Plasma peptide YY and pancreatic polypeptide in dogs after long-term adaptation to dietary fats of different degrees of saturation: olive and sunflower oil. *Journal of Nutritional Biochemistry, 8*, 502–507.

Yaqoob, P., Knapper, J. A., Webb, D. H., Williams, C. M., Newsholme, E. A., & Calder, P. C. (1998). The effect of olive oil consumption on immune functions in middle-aged men. *American Journal of Clinical Nutrition, 67*, 129–135.

Zeb, A., Khan, S., Khan, I., & Imran, M. (2008). Effect of temperature, UV, sun and white lights on the stability of olive oil. *Journal of the Chemical Society of Pakistan, 30*, 790–794.

CHAPTER 101

Use of Fermented Papaya (*Carica papaya*) Seeds as a Food Condiment, and Effects on Pre- and Post-implantation Embryo Development

Mansurah A. Abdulazeez, Ibrahim Sani
Department of Biochemistry, Faculty of Science, Ahmadu Bello University Zaria, Kaduna State, Nigeria

855

LIST OF ABBREVIATIONS

AGD, anogenital distance
CRL, crown–rump length

Nuts & Seeds in Health and Disease Prevention. DOI: 10.1016/B978-0-12-375688-6.10101-X

INTRODUCTION

According to Hayes (1945), *Carica papaya* is included in a list made in 1831 of plants introduced by Don Martin. Its name is supposed to be a corruption of the "Carib" name for the fruit "ababai." Its common names include pawpaw, *kapaya, kepaya,* lapaya,tapaya, etc. (Morton, 1987). The fruits, leaves, and seeds of *Carica papaya* are used as sources of food for humans and animals. They have also economic and health benefits (Hayes, 1945), just like many other fruits. The advantage of fermenting papaya seeds as a food condiment is immense, but research on its safety cannot be overlooked, as the seeds have also been demonstrated to have adverse effects.

BOTANICAL DESCRIPTION

Carica papaya belongs to the family Caricaceae. It originated from Central America, and is now grown in all tropical countries and many subtropical regions of the world (Morton, 1987). It lives for about 5–10 years, and normally grows with a single unbranched trunk (Morton, 1987).

The leaves are palmately-lobed, up to 75 cm across, on long, hollow petioles. The blades are divided into five to nine main segments, bearing prominent yellowish ribs and veins (Morton, 1987) (Figure 101.1). The flowers are born on inflorescences which appear in the axils of the leaves (Hayes, 1945). Generally, the fruit is melon-like, round or long, and may contain more than 1000 seeds (Figure 101.2). The skin is smooth and green, but turns yellow or orange when ripe (Hayes, 1945).

HISTORICAL CULTIVATION AND USAGE

Propagation by seeds used to be the most common and preferred method of reproduction, while cutting and grafting were condemned because they yielded unsatisfactory plants. Also, cuttings developed more slowly than seedlings, and both cutting and grafted plants seemed to lack vigor and bear poorer fruits than their parents; moreover, it was noted that the size and quality of the fruits deteriorated as the plants become old (Hayes, 1945). Harvesting was done manually, and using pickers.

Originally, papaya was grown for consumption of its fruits. Its young leaves, shoots, and fruits were also cooked as a vegetable, while the seeds were used as a substitute for black pepper. The fruits were also used in making jams and candies (Morton, 1987). Papain, an enzyme

FIGURE 101.1

A pawpaw tree with fruits. A large single-stemmed pawpaw tree, with large palmately lobed, dark green leaves and thick middle veins. The fruit is oval-shaped and unripe. *Photographed at the Faculty of Education, Ahmadu Bello University, Zaria, Kaduna State, Nigeria.*

FIGURE 101.2
Seeds (A) and fruits (B) of *Carica papaya.* The seeds are black, round, and enclosed inside a transparent gelatinous aryl. The fruit is ripe and ready for consumption.

present in the latex of unripe pawpaw, was used to tenderize meat by rubbing or even cooking with it. It was also used in the treatment of skin blemishes (Hayes, 1945).

PRESENT-DAY CULTIVATION AND USAGE

The ecology, propagation, and harvesting methods known historically are still applicable today; however, studies have demonstrated an improvement in the use of cuttings and grafted plants (Hayes, 1945). The potential of rapid propagation of papaya selections by tissue culture is being explored, and promises to be feasible even for the establishment of commercial plantations of superior strains. Harvesting is aided by hydraulic lifts in developed countries (Morton, 1987).

Papain is now produced in commercial quantities for use in the pharmaceutical, cosmetics, and leather industries (Hayes, 1945; Morton, 1987). In Nigeria, the seeds are used as a food condiment (Dakare, 2005), and the ripe fruit is cooked as a soup with melon seeds and other spices (Adeneye & Olagunju 2009).

APPLICATIONS TO HEALTH PROMOTION AND DISEASE PREVENTION

Carica papaya seeds as a food condiment

In a study by Dakare (2005), papaya seeds were used to produce an indigenous Nigerian food condiment called "daddawa;" this is the most important condiment used in soups and stews in the entire savannah region of west and central Africa. It has a strong, ammonia-like smell, characteristic of the latter stages of its fermentation. However, although daddawa is used as a flavoring agent, it also contributes to the calorie and protein component of food (Dakare, 2005), serving as a nutritious non-meat protein substitute in many parts of West Africa. Most diets in this part of the continent are rich in carbohydrates, such as cassava, yam, maize, and cocoyam, for example. Hence, seeds of legumes used in making such food condiments become the only source of protein for such groups (Achi, 2005). Countries where daddawa is important and commonly used include Nigeria, Ghana, the Ivory Coast, Chad, Sierra Leone, Gambia, Guinea, Togo, Mali, Senegal, and the Republic of Niger (Dakare, 2005).

The safety of daddawa produced from papaya seeds has been a subject of contention, but recently studies carried out by Abdulazeez *et al.* (2009a) have confirmed its safety in females, using female Wistar rats (*Rattus norvegicus*) as models.

Carica papaya is a source of calcium, and an excellent source of vitamins A and C. The plant contains many biologically active compounds, whose concentrations differ in the root, fruit, latex, and leaves. The complex chemical composition of its latex, together with results from various research studies, explains the use of its leaves, fruits, and root extracts in traditional medicines, and suggests potential for its uncharacterized effects on the health of humans and other organisms (Adeneye & Olagunju, 2009).

In Asian folk medicine, the latex is used as an abortificant and antiseptic, and as a cure for dyspepsia; in Africa, it is used for treating venereal diseases and hemorrhoids. In Cuba, the latex is used in the treatment of psoriasis, ringworm, and cancerous growths (Adeneye & Olagunju 2009). It has also been reported to aid protein digestibility and break down clots after surgery (Oduola *et al.*, 2007).

C. papaya's fruit and seed extracts have been shown to possess bactericidal activity against *Staphylococcus aureus*, *Bacillus cereus*, *Escherischia coli*, *Pseudomonas aeruginosa*, and *Shigella flexneri* (Adeneye & Olagunju, 2009). The seed extracts are used in parts of Asia and South America as a vermifugal agent, killing worms like *Toxascaris transfuga*, *Ascaris lumbricoides*, *Pheretima* sp., and *Caenorhabditis elegans*, *in vitro* (Adebiyi *et al.*, 2003). The seeds have also been used in folk medicine to facilitate good menstrual flow (Adebiyi *et al.*, 2003), and were recently reported to have hypoglycemic and hypolidemic effects (Adeneye & Olagunju, 2009).

Fermentation of *Carica papaya* seeds

Fermentation is one of the oldest methods of food processing. It is widely utilized as a means of food preservation in developing countries, particularly in areas where refrigeration, canning, and freezing facilities are either inaccessible or unavailable. Fermentation enhances nutritional quality, through the biosynthesis of vitamins, essential amino acids, and proteins. It improves protein digestibility, enhances micronutrient bioavailability, and degrades anti-nutritional factors (Achi, 2005).

The composition of papaya seeds differ from region to region, depending on the geography, climate, and even the analytical techniques used. However, the fermented seeds have been found to have higher levels of crude lipid and protein content (Dakare, 2005).

Methods of processing

- *Acquisition and pre-treatment.* Here, seeds or beans (depending on the raw material to be used) are cooked for about 30 minutes to 12 hours, depending on the strength of the testa, and dehulling efficiency. *Carica papaya* seeds are dehulled to obtain the kernels before boiling for 2–3 hours (Dakare, 2005).
- *Incubation.* This follows dehulling, and allows the characteristic aroma of the condiment to develop. The fermentation is achieved by indigenous micro-flora, or the addition of fermented material from a previous production through back-slopping (Achi, 2005). As for fermented *Carica papaya* seeds, the kernels are wrapped in *Carica papaya* leaves and incubated for 72 hours (Dakare, 2005).
- *Post-processing.* Condiments have various post-processing methods. Daddawa produced from *Carica papaya* seeds is sun dried. It is important to note that fermented condiments do not keep well if over-fermented; this is because unacceptable levels of volatile acids are generated, and the environment also becomes suitable for yeast growth (Achi, 2005).

Effect of fermented and unfermented papaya seeds on pre-implantation embryo development

Implantation is a unique part of reproduction that begins when the blastocyst assumes a fixed position in the uterus and establishes a more intimate relationship with the endometrium. For successful implantation to occur the uterus must be receptive, due to the blastocyst's sensitivity. This occurs during the window of implantation, which is determined by the state of activity of the blastocyst, and lasts for a limited time (Paria *et al.*, 1993).

In a recent study by Abdulazeez *et al.* (2009a, 2009b), the use of fermented papaya seeds as a condiment was found to be safe in female mammals after carrying out fertility experiments. This confirmed the work of Oderinde *et al.* (2002), Adebiyi *et al.* (2003), and Chinoy *et al.* (2006), who found that the unfermented extract of *Carica papaya* seeds affects reproductive function in female animals. The study demonstrated a progressive increase in weight gain of rats in all groups tested, apart from those administered the unfermented extract at a higher dose (Figure 101.3). The increase in weight was attributed to pregnancy (USEPA, 1996), as well as palatability of the fermented seeds. There was a decrease in implantation sites in animals administered the unfermented extract at 1500 mg/kg compared to the control group, the unfermented extract at 500 mg/kg group, or those given the fermented extract at 1500 mg/kg, even though there was no significant difference in the number of corpora lutea in all groups. As shown in Table 101.1, there was a dose–response effect in the percentage pre-implantation loss between animals administered the unfermented extract at 1500 mg/kg and 500 mg/kg. There was also an increase in the percentage pre-implantation loss in rats treated with 500 mg/kg and 1500 mg/kg unfermented extract, compared to rats treated at 500 mg/kg and 1500 mg/kg of fermented extract.

Effect of fermented and unfermented papaya seeds on post-implantation embryo development

In most of the dams given the unfermented extract at a higher dose there was vaginal bleeding in the week preceding parturition, and at birth there were two dead litters. This showed that the unfermented extract affected fetal well-being and integrity. This finding agrees with those of Oderinde *et al.* (2002) and Chinoy *et al.* (2006), who demonstrated the abortifacient properties of the unfermented extract of papaya seeds.

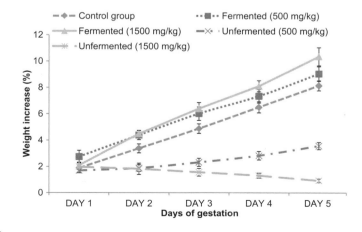

FIGURE 101.3

Effect of fermented and unfermented extracts of *Carica papaya* seeds on body weight before implantation (pre-implantation) in female Wistar rats. The body weights of animals gradually increased before implantation in all groups, except in those administered the unfermented extract at a dose of 1500 mg/kg, where the body weight decreased. Results are expressed as mean ± SEM, N = 40. *Reprinted from Abdulazeez* et al. *(2009b),* Sci. Res. Essays, 4, *1080–1084, with permission.*

TABLE 101.1 Effects of Fermented and Unfermented Seeds of *Carica papaya* on Pre-implantation Embryo Development

	Control Group	Unfermented Extract (500 mg/kg b.w.)	Unfermented Extract (1500 mg/kg b.w.)	Fermented Extract (500 mg/kg b.w.)	Fermented Extract (1500 mg/kg b.w.)
Implantation sites	12.17 ± 1.01	11.57 ± 0.81^b	7.13 ± 1.06^a	10.38 ± 0.56	11.86 ± 1.08^b
Corpora lutea	15.00 ± 1.21	16.43 ± 1.08	16.63 ± 0.91	15.75 ± 2.40	14.71 ± 1.15
Preimplantation loss (%)	19.00 ± 1.65	29.29 ± 1.85^b	57.75 ± 5.00^a	20.87 ± 3.00^b	20.00 ± 1.23^b

The number of implantation sites and percentage pre-implantation loss was significantly higher in the rats dosed at 1500 mg/kg with the unfermented extract when compared with the control group. There was no difference in the number of corpora lutea in all groups. Results are expressed as mean $\pm$ SEM; $n = 40$. Reprinted from Abdulazeez et al. (2009b), *Sci. Res. Essays*, 4, 1080–1084, with permission.
[a]Significant at $P < 0.05$ vs control group;
[b]significant at $P < 0.05$ vs Unfermented 1500 mg/kg b.w. group.

The increase in weight of rats in all groups (Figure 101.4) is evidence of gestation. Also, the higher percentage post-implantation loss (Table 101.2) in animals treated with the unfermented extract at a higher dose indicates that papaya seeds either altered the intra-uterine hormonal level, or compromised the uterine integrity. This may have affected the well-being of the developing fetus, leading to fetal death.

Rats given the fermented extract of papaya seeds had offspring with higher birth weights than did those given the unfermented extract at a higher dose. Also, the AGD/CRL ratio was significantly higher in the group treated with the unfermented extract compared to the control. This shows that it is likely that the unfermented extract at a very high dose alters the hormonal level in the neonates, since AGD is a sensitive index used in estimating intrauterine hormonal levels (Ronis *et al.*, 1998). It is well established that the effect of AGD can be masked by body size (USEPA, 1996; Ronis *et al.*, 1998), hence the need to evaluate the AGD/CRL ratio to determine possible hormonal alterations in animals. In normal individuals, AGD is shorter in females but increases during maternal exposure to androgens or androgenic substances; in males, maternal exposure to anti-androgens decreases the AGD (Ronis *et al.*, 1998). Therefore, this study demonstrated that the unfermented extract of *Carica papaya* seeds has androgenic effects, and so may have adverse effects on sexual development and reproductive function.

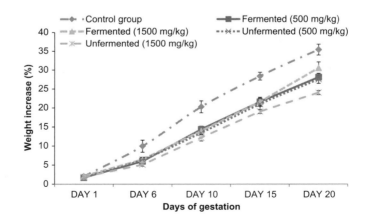

FIGURE 101.4
Effect of fermented and unfermented extracts of *Carica papaya* seeds on body weight during gestation in female Wistar rats. The body weights of dams increased as gestation days progressed in animals of all groups. The control group had the highest increase in weight, while rats dosed at 1500 mg/kg with the unfermented extract had the least increase in weight during gestation. Results are expressed as mean $\pm$ SEM, $n = 40$. *Reprinted from Abdulazeez* et al. *(2009a)*, Afr. J. Biotechnol., 8, 854–857, with permission.

TABLE 101.2 Effect of Fermented and Unfermented Extracts of *Carica papaya* Seeds on Some Gestational Parameters

	Control Group	Unfermented Extract (500 mg/kg)	Unfermented Extract (1500 mg/kg)	Fermented Extract (500 mg/kg)	Fermented Extract (1500 mg/kg)
Litter size	6.16 ± 0.54	$4.67 \pm 0.33^{a,b}$	3.33 ± 0.21^a	5.16 ± 0.31^b	5.83 ± 0.31^b
AGD/CRL (mm)	1.22 ± 0.02	1.35 ± 0.06	1.52 ± 0.04^a	$1.13 \pm 0.04^{b,c}$	$1.17 \pm 0.03^{b,c}$
Litter bodyweight (g)	5.57 ± 0.05	4.85 ± 0.10^a	4.72 ± 0.17^a	$5.41 \pm 0.12^{b,c}$	$5.56 \pm 0.08^{b,c}$
Dead Litter	–	–	2	–	–
Post-implantation loss (%)	39.33 ± 3.85	51.67 ± 2.11^b	66.50 ± 2.01	44.00 ± 4.30^b	46.00 ± 1.53^b
Implantation sites	10.17 ± 0.48	9.67 ± 0.42	9.67 ± 0.56	9.33 ± 0.42	10.83 ± 0.43

Results are expressed as mean $\pm$ SEM, $n = 40$.
Reprinted from Abdulazeez *et al.* (2009a), *Afr. J. Biotechnol.*, 8, 854–857, with permission; and Abdulazeez (2006), Thesis submitted to the Ahmadu Bello University, Zaria, Kaduna State, Nigeria.
[a]Significant at P < 0.05 vs control group;
[b]Significant at P < 0.05 vs 1500 mg/kg b.w. unfermented extract group;
[c]Significant at P < 0.05 vs 500 mg/kg b.w. unfermented extract group;
[d]Significant at P < 0.05 vs 500 mg/kg b.w. fermented extract group.

The authors concluded that the high number of pre-implantation sites and low post-implantation losses recorded for rats given the fermented extract of the seeds of *C. papaya*, when compared to those given the unfermented extract, was a result of the fermentation process, where the husk was removed and only the kernel of the seed used; hence it is possible that any compound with antifertility effects may have been reduced, and the residual concentration in the kernel after fermentation may not have had the potency to produce any measurable effect.

ADVERSE EFFECTS AND REACTIONS (ALLERGIES AND TOXICITY)

The latex found in *Carica papaya* plays a significant role in defending the plant against pathogenic microorganisms and herbivores, but it causes skin irritations and extreme allergic reactions if meats tenderized using papain are undercooked (Morton, 1987). Several studies have demonstrated that the unripe fruit and seeds of *Carica papaya* affect the reproductive system, resulting in reproductive toxicity.

Reproductive toxicity

According to the USEPA (1996), reproductive toxicity refers to the occurrence of adverse effects on the reproductive system of females or males that may result from exposure to environmental agents, affecting the female or male reproductive organs, related endocrine system, or pregnancy outcome. The manifestation of such toxicity may include (but is not limited to) adverse effects on the onset of puberty, gamete production and transport, reproductive cycle normality, sexual behavior, fertility, gestation, parturition, and lactation; developmental toxicity; premature reproductive senescence; or modifications in other functions that are dependent on the integrity of the reproductive systems.

Substances with antifertility effects in female animals are known to exert their actions on one or more of the following structures: hypothalamus, anterior pituitary, ovary, oviduct, uterus, and vagina (Farnsworth *et al.*, 1975).

Effect of *Carica papaya* on reproduction

A complete loss of fertility has been reported in studies using animal models (Lohiya *et al.*, 2005), suggesting that ingestion of papaya may adversely affect the fertility of male and female mammals, including humans. In India, parts of South-east Asia, and Indonesia, the fruit is considered harmful in pregnancy (Adebiyi *et al.*, 2002). Most studies carried out with various formulations of the papaya plant, such as extracts of seeds or pulp of the unripe fruits, and crude papain, demonstrated the following effects:

- Antifertility (Adebiyi *et al.*, 2002; Oderinde *et al.*, 2002; Chinoy *et al.*, 2006)
- Anti-implantation and abortifacient (Adebiyi *et al.*, 2003; Chinoy *et al.*, 2006)
- Increases in post-implantation loss (Oderinde *et al.*, 2002, Chinoy *et al.*, 2006).

In males, the seeds of papaya have been emerging as a potential antifertility drug. Various studies have shown that extracts from the seed induce variable responses, depending on the dose, duration, and route of administration, in laboratory animals (Lohiya *et.al.*, 2005).

Early studies showed that the compound 5-hydroxytryptamine, present in the seeds of papaya, was responsible for its antifertility effects (Farnsworth *et al.*, 1975). However, more recent studies have demonstrated that benzylisothiocyanate (BITC) is the chief bioactive ingredient in the seeds (Kermanshai *et al.*, 2001) whose damaging effect can lead to infertility (Adebiyi *et al.*, 2003).

SUMMARY POINTS

- *Carica papaya* can be found in all tropical and subtropical regions of the world.
- It is capable of growing on most soil types, provided there is adequate drainage.
- Its fruits, leaves and seeds can be consumed, and used in industries and as source of medicine.
- Although the seed is believed to have antifertility and contraceptive effects, it is used in the production of a fermented condiment called daddawa in northern Nigeria.
- Studies using female rats have demonstrated that, unlike the unfermented seed, the fermented seed has no contraceptive effects and does not affect litter size, litter body weight, and AGD/CRL ratio; hence, it does not affect the developing fetus or alter the intra-uterine environment.
- Thus, the fermented seeds of *Carica papaya* may be safe for consumption in female mammals. However, more studies need to be carried out to determine its effect in male animals.

References

Abdulazeez, M. A., Ameh, D. A., Sani, I., Ayo, J. O., & Ambali, S. F. (2009a). Effect of fermented seed extract of *Carica papaya* on litters of female wistar rats (*Rattus norvegicus*). *African Journal of Biotechnology, 8*, 854—857.

Abdulazeez, M. A., Ameh, D. A., Sani, I., Ayo, J. O., & Ambali, S. F. (2009b). Effect of fermented and unfermented seed extracts of *Carica papaya* on pre-implantation embryo development in female Wistar rats (*Rattus norvegicus*). *Scientific Research and Essays, 4*, 1080—1084.

Achi, O. K. (2005). Traditional fermented protein condiments in Nigeria. *African Journal of Biotechnology, 4*, 1612—1621.

Adebiyi, A., Adaikan, G. P., & Prasad, R. N. (2002). Papaya (*Carica Papaya*) consumption is unsafe in pregnancy: fact or fable? Scientific evaluation of a common belief in some part of Asia using a rat model. *British Journal of Nutrition, 88*, 199—203.

Adebiyi, A., Adaikan, P. G., & Prasad, R. N. V. (2003). Tocolytic and toxic activity of papaya seed extract on isolated rat uterus. *Life Sciences, 74*, 581—592.

Adeneye, A. A., & Olagunju, J. A. (2009). Preliminary hypoglycemic and hypolipidemic activities of the aqueous seed extract of *Carica papaya Linn.* in Wistar rats. *Biology and Medicine, 1*, 1—10.

Chinoy, N. J., Dilip, T., & Harsha, J. (2006). Effect of *Carica papaya* seed extract on female rat ovaries and uteri. *Phytotherapy Research, 9*, 165—169.

Dakare. (2005). *Biochemical assessment of "daddawa" (food seasoning) produced by fermentation of pawpaw (Carica papaya) seeds.* Zaria: MSc thesis, Department of Biochemistry, Ahmadu Bello Univeristy.

Farnsworth, N. R., Bingel, A. S., Cordell, A. G., Crane, A. F., & Fong, H. S. (1975). Potential value of plants as source of raw antifertility agents. *Journal of Pharmaceutical Sciences, 64*, 535—592.

Hayes, W. B. (1945). The papaya. In *Fruit growing in India* (pp. 170—179). Lahore, India: Kitabistan Publishing Company.

Kermanshai, R., McCarry, B. E., Rosenfeld, J., Summers, P. S., Weretilnyk, E. A., & Sorger, G. J. (2001). Benzyliso-thiocyanate is the chief or sole anthelminthic in papaya seed extracts. *Phytochemistry, 57,* 427–435.

Lohiya, N. K., Mishra, P. K., Pathak, N., Manivannan, B., Bhande, S. S., Panneerdoss, S., et al. (2005). Efficacy trial on the purified compounds of the seeds of *Carica papaya* for male contraception in albino rat. *Reproductive Toxicology, 20*(10), 135–148.

Morton, J. F. (1987). Papaya (*Carica papaya* L.). In *Fruits of warm climates* (pp. 336–346). Winterville, NC: Creative Resource Systems.

Oderinde, O., Noronha, C., Oremosu, A., Kusemiju, T., & Okanlawon, O. A. (2002). Abortifacient properties of aqueous extract of *Carica papaya* (Linn) seeds on female Sprague-Dawley rats. *Nigerian Postgraduate Medical Journal, 9,* 95–98.

Oduola, T., Adeniyi, F. A. A., Ogunyemi, E. O., Bello, I. S., Idowu, T. O., & Subair, H. G. (2007). Toxicity studies on an unripe *Carica papaya* aqueous extract: biochemical and haematological effects in wistar albino rats. *Journal of Medicinal Plants Research, 1,* 1–4.

Paria, B. C., Huet-Hudson, Y. M., & Dey, S. K. (1993). Blastocyst's state of activity determines the "window" of implantation in the receptive mouse uterus. *Proceedings of the National Academy of Sciences of the United States of America, 90,* 10159–10162.

Ronis, M. J., Gandy, J., & Badger, T. (1998). Endocrine mechanisms underlying reproductive toxicity in the developing rat chronically exposed to dietary lead. *Journal of Toxicology and Environmental Health, 54,* 77–99.

United States Environmental Protection Agency (USEPA). (1996). Guidelines for reproductive toxicity risk assessment. *Federal Register, 61,* 56274–56322.

CHAPTER 102

Antifungal Protein from Passion Fruit (*Passiflora edulis*) Seeds

Tzi-Bun Ng[1], Sze-Kwan Lam[1], Randy C.F. Cheung[1], Jack H. Wong[1], He-Xiang Wang[2], Patrick H.K. Ngai[3], Xiujuan Ye[1], Evandro F. Fang[1], Yau-Sang Chan[1]
[1] School of Biomedical Sciences, Faculty of Medicine, The Chinese University of Hong Kong, Hong Kong, China
[2] State Key Laboratory of Agrobiotechnology, Department of Microbiology, China Agricultural University, Beijing, China
[3] School of Life Sciences, Faculty of Science, The Chinese University of Hong Kong, Hong Kong, China

865

INTRODUCTION

Passion fruits of the genus *Passiflora* are used in wine, desserts, ice-cream toppings, cakes, tarts, flavoring for cheesecakes, jams to flavor yogurt, fruit juice, and fruit salads. Because of its unique, intense, aromatic flavor characteristics, passion fruit is often an ingredient for juice mixes. In Germany, one of the largest juice-consuming countries in the world, passion fruit concentrate and banana puree often constitute the base of the highly popular "multivitamin" juices produced.

BOTANICAL DESCRIPTION

Passion fruit has a strong aroma. It is spherical or ovate in shape, has waxy purple skin with white spots, and is filled with membranous sacs. The juice is orange, and the fruit contains numerous hard, dark-colored seeds. The two types of passion fruit are morphologically distinct. The golden passion fruit is bright yellow in color and as large as a grapefruit, whereas purple passion fruit is much smaller in size. The skin is moderately toxic, due to a small amount of cyanogenic glycosides. The two types grow in different climatic regions, and flower

Nuts & Seeds in Health and Disease Prevention. DOI: 10.1016/B978-0-12-375688-6.10102-1

in different seasons. The purple-fruit type is self-fertile, while the yellow-fruit type requires two clones for pollination.

HISTORICAL CULTIVATION AND USAGE

It is believed that purple passion fruit originated in South America, whereas golden passion fruit had its origin in the Amazon Basin or in Australia. Passion fruit first appeared in Florida and Hawaii in the 1880s. By 1958, the plantations and associated industry in Hawaii were flourishing. Nevertheless, viral infections, high labor costs, and escalating land values subsequently suffocated the industry. Nowadays, commercial passion-fruit plantations have vanished from Hawaii.

Passion fruit was flourishing in Australia before 1900, but in 1943, a devastating viral infection wiped out the vines. Despite reconstruction of some plantations, the harvest cannot meet demand, and imports are necessary.

PRESENT-DAY CULTIVATION AND USAGE

Passion fruit is cultivated extensively in California, and in the tropical belt of South America to Australia, Asia, and Africa. South America is the biggest producer of passion fruit in the world, followed by Ecuador, Australia, New Zealand, Kenya, South Africa, and India.

In New Zealand, Australia, and South America, fresh passion fruit is used to make juice and jams, or added to fruit salads, and fresh fruit pulp or sauce is used in desserts as cake and ice-cream toppings, and as cheesecake flavoring. In Vietnam, passion fruit is mixed with honey and ice to make refreshing smoothies. In Indonesia, passion fruit juice is cooked with sugar to generate a thick syrup.

APPLICATIONS TO HEALTH PROMOTION AND DISEASE PREVENTION

Passion fruit has an abundance of vitamins A and C, folic acid, niacin, calcium, iron, potassium, and other nutrients. The leaves have been used for centuries by Latin American tribes as a sedative or calming tonic. The fruit has been utilized in Brazil as a cardiac tonic and medicine, and, as the beverage maracuja grande, is often employed in the treatment of asthma, bronchitis, whooping cough, and other recalcitrant coughs. Passion fruit plays an important role in South American traditional medicine. In Peru, the juice is used to treat urinary infections, and also as a mild diuretic. In Madeira, the juice is used to stimulate digestion, and for treatment of gastric cancer.

Passicol is a member of the polyacetylenic group of compounds, which have antibacterial and antifungal activities, found in passion-fruit rinds (Birner & Nicolls, 1973).

Two steroidal saponins with three and four glucose moieties, respectively, have been found to exert fungicidal effects on *Aspergillus* and *Fusarium* spp. (De Lucca *et al.*, 2006).

An antifungal protein has been isolated from seeds of the passion fruit (*Passiflora edulis*) (Lam & Ng, 2009). The isolation procedure involved ion-exchange chromatography on Q-Sepharose, hydrophobic interaction chromatography on Phenyl-Sepharose, ion-exchange chromatography on an DEAE-cellulose, and FPLC-gel filtration on Superdex 75. Ion-exchange chromatography of *P. edulis* seed extract on Q-Sepharose produced a very large unadsorbed fraction (Q1), and two adsorbed fractions (Q2 eluted with 0.1 M NaCl, and Q3, Q4, and Q5, eluted with 0.5 M NaCl). Antifungal activity resided only in fraction Q4. This fraction was separated on Phenyl-sepharose into an unadsorbed fraction (PS1) devoid of antifungal activity, and an adsorbed fraction (PS2) with antifungal activity. Fraction PS2 was subsequently resolved on DEAE-cellulose into a large unadsorbed fraction (D1) and two smaller adsorbed fractions (D2 and D3). Antifungal activity was confined to the adsorbed fraction D2 eluted within the 0–0.6 M NaCl gradient. This active fraction D2 was subjected to final purification on Superdex 75. Four fractions, S1 to S4, were obtained. Antifungal activity resided in the first fraction (S1). The first fraction demonstrated a single 34-kDa band in SDS-PAGE, and a single 67-kDa peak upon rechromatography on Superdex 75. The 67-kDa protein, named as passiflin, demonstrated an N-terminal amino acid sequence highly homologous to that of bovine β-lactoglobulin. It exhibited a β-lactoglobulin-like N-terminal sequence. Its dimeric nature is rarely found in antifungal proteins. It inhibited mycelial growth in *R. solani* with an IC_{50} value of $16 \pm 0.9\,\mu M$ ($n = 3$), but not in *M. arachidicola* and *F. oxysporum* when tested up to 100 μM. The species-specific antifungal activity of passiflin is reminiscent of findings on antifungal proteins from asparagus seeds and shallot bulbs, where the latter inhibits only one out of the several fungi examined. The antifungal protein inhibited proliferation of MCF-7 tumor cells with an IC_{50} near $15 \pm 1.2\,\mu M$ ($n = 3$), in line with the antiproliferative action of some antifungal proteins such as ribosome-inactivating proteins and defensins. However, there was no inhibitory action on hepatoma HepG2 cells, signifying a specificity of action. Similarly, the ribosome-inactivating proteins trichosanthin and momorcharin exert highly potent inhibitory activity against choriocarcinoma cells but are much less active toward hepatoma cells.

Western blotting of bovine β-lactoglobulin using a rabbit-antibovine-β-lactoglobulin antiserum yielded positive results. In contrast, there was no cross-reactivity of passiflin with the same antiserum. Passiflin is unique in its N-terminal amino acid sequence with pronounced resemblance to bovine β-lactoglobulin. This is noteworthy because passiflin is of botanical origin, while β-lactoglobulin is a mammalian whey protein. In spite of this structural homology, there are many differences between the two proteins, indicating that they are distinct proteins. Passiflin (67 kDa) has a higher molecular mass than does β-lactoglobulin (36.6 kDa). Furthermore, intact β-lactoglobulin does not manifest antifungal or antiproliferative activity, but passiflin does demonstrate these activities. However, β-lactoglobulin hydrolysate exhibits antifungal activity (Hernandez-Ledesma *et al.*, 2008). In addition, passiflin does not cross-react with a rabbit-antibovine-β-lactoglobulin antiserum, revealing that they are immunologically distinct.

Passiflin lacks ribonuclease and hemagglutinating (lectin) activities when tested up to 100 μM, in contrast to some ribonucleases (like ginseng ribonucleases) and some lectins that display antifungal activity. It lacks HIV-1 reverse transcriptase inhibitory activity, unlike some protease inhibitors, lectins, and antifungal proteins that manifest this antiretroviral activity. β-Lactoglobulin is also devoid of ribonuclease and HIV-1 reverse transcriptase inhibitory activities. Another distinctive feature of passiflin is that it is adsorbed on Q-Sepharose and DEAE-cellulose, whereas most of the antifungal proteins are unadsorbed on these anion exchangers.

Passiflin differs from the 2S albumin-like antifungal protein and peptide from seeds of passion fruit (Agizzio *et al.*, 2003; Pelegrini *et al.*, 2006), in molecular mass, N-terminal amino acid sequence, and species-specificity of antifungal activity. The 2S albumin-like antifungal protein and peptide, but not passiflin, exhibit antifungal activity toward *F. oxysporum* (Tables 102.1–102.3).

Passiflin is a distinctive antifungal protein for the following reasons. To date, only a few antifungal proteins, such as those from sanchi ginseng, Chinese ginseng, and American ginseng, have been shown to be dimeric. Passiflin exhibits a β-lactoglobulin-like N-terminal sequence. However, there is no cross-reactivity with an anti-β-lactoglobulin antiserum. It manifests antiproliferative and antifungal activities which are lacking in β-lactoglobulin. Thus, passiflin is biologically and immunologically unrelated to β-lactoglobulin. In this context, it deserves mention that thaumatin-like proteins inhibit fungal growth activity but have no sweet taste, while the opposite is true of thaumatin, although they closely resemble each other in structure.

TABLE 102.1 Characteristics of Antifungal Protein (Passiflin) in Seeds of Passion Fruits (Lam & Ng, 2009)

Molecular mass	67 kDa (dimeric)
N-terminal sequence	AFLDIQKVAGTWYSLA
Chromatographic behavior on:	
DEAE-cellulose	Unadsorbed
Affi-gel blue gel	Adsorbed
SP-Sepharose	Adsorbed
FPLC Superdex-75	
Antifungal activity against:	
Fusarium oxysporum	$IC_{50} = 1.8\,\mu M$
Mycosphaerella arachidicola	IC_{50} not determined
Antiproliferative activity against:	
L1210 cells	$IC_{50} = 4\,\mu M$
MBL2 cells	$IC_{50} = 9\,\mu M$
HIV-1 reverse transcriptase inhibitory activity	Absent
RNase	Absent
DNase	Absent
Lectin	Absent
Protease inhibitory activity	Absent

TABLE 102.2 Characteristics of 2S-albumin-like Antifungal Protein in Seeds of Passion Fruits (Pelegrini *et al.*, 2006)

Molecular mass	5 kDa
N-terminal sequence	QSERFEQQMQGQDFSHDERFLSQAA
Chromatographic behavior on:	
Red-Sepharose	Adsorbed
Reverse phase HPLC on Vydac C18-TP	
Antifungal activity against:	
Trichoderma harzianum	$IC_{50} = 6.4\,\mu M$
Fusarium oxysporum	$IC_{50} = 6.8\,\mu M$
Aspergillus fumigatus	$IC_{50} = 8\,\mu M$
Other biological activities not tested	

TABLE 102.3 Characteristics of Another 2S-albumin-like Protein in Seeds of Passion Fruits (Agizzio *et al.*, 2003)

Molecular mass	One 3.5-kDa subunit and one 8-kDa subunit
N-terminal sequence	PSERCRRQMQGDFS
Chromatographic behavior on:	
Gel filtration on Sephadex G50	Adsorbed
CM-Sepharose	
HPLC Vydac 14 column	
Antifungal activity against:	
Fusarium oxysporum	24% inhibition
Colletotrichum lindemuthianum	78% inhibition
Colletotrichum musae	32% inhibition
Saccharomyces cerevisiae	32% inhibition
Other biological activities not tested	

In addition to the aforementioned antifungal proteins and peptides, passion fruit has some other therapeutic effects, which are described below. The total dietary fiber present in alcohol-insoluble material from yellow passion fruit rind exceeds 70% dry matter, of which over 60% is insoluble dietary fiber. Non-starch polysaccharides, especially cellulose, are the major components. Hence, dietary fiber from passion-fruit rind may be promising for treatment of ailments such as diabetes, colon cancer, and other diverticular diseases (Yapo & Koffi, 2008).

The incidence of wheezing, coughing, and shortness of breath decreased significantly, and the forced vital capacity increased, in asthmatic patients receiving passion-fruit peel extract (150 mg/d), but no similar changes were seen in subjects treated with the placebo. No adverse consequences were observed. Thus, peel extract may be used to alleviate symptoms in asthmatic subjects (Watson *et al.*, 2008).

Passion-fruit juice (PFJ) was capable of diminishing the number, size, and invasiveness of transformed foci in a BALB/c 3T3 neoplastic transformation model. PFJ failed to alter cell-cycle kinetics, but was capable of inducing caspase-3 in the mammalian cell line MOLT-4, suggesting that it brought about the changes in transformed foci by inducing apoptosis rather than by an antiproliferative action. These changes were observed at concentrations that would be achieved in the blood after ingestion of the juice (Rowe *et al.*, 2004).

Both passion flower (*Passiflora incarnata*) extract (45 drops/day) in a double-blind randomized trial plus placebo drops for a 4-week trial, and oxazepam (30 mg/day) were efficacious in treating generalized anxiety disorder. However, in subjects on oxazepam therapy, there were more problems related to deterioration of job performance (Akhondzadeh *et al.*, 2001)

Oral pre-treatment of carbon tetrachloride-treated rats with the leaf extract of *P. alata* (0.18 mg/kg per day) exerted a significant hepatoprotective effect, as evidenced by reduction of hepatic necrosis, hepatic and cardiac lipid peroxidation, and elevated activities of the hepatic antioxidant enzymes catalase and superoxide dismutase (Rudnicki *et al.*, 2007)

A tri-substituted benzoflavone from *Passiflora incarnata* enhanced libido, sperm count, and sexual fertility in 2-year-old male rats, and expedited the recovery of sexuality in rats after discontinuation of administration of alcohol and nicotine, which exert an adverse action on male sexuality and fertility (Dhawan & Sharma, 2002).

ADVERSE EFFECTS AND REACTIONS (ALLERGIES AND TOXICITY)

Passiflora alata (passion flower) caused IgE-mediated asthma and rhinitis in pharmacy workers involved in manual preparation of products (Giavin a-Bianchi *et al.*, 1997).

Five patients were hospitalized following consumption of the herbal product Relaxir, produced mainly from the passion fruit (*Passiflora incarnata*), for insomnia and restlessness (Solbakken *et al.*, 1997).

A 34-year-old female was admitted to hospital after development of severe nausea, vomiting, drowsiness, and non-sustained ventricular tachycardia following intake of a herbal remedy composed of *Passiflora incarnate* (Fisher *et al.*, 2000).

Passiflora alata extract (containing 2.6% flavonoids) exerted a genotoxic effect, as evidenced by DNA damage in peripheral blood and bone marrow cells, in comet and in micronucleus tests (Boeira *et al.*, 2010).

A phytotherapeutic product, CPV (made from dried extracts of *Crataegus oxyacantha*, *Passiflora incarnata*, and *Valeriana officinalis*), was devoid of both acute and chronic toxicity when administered to rats, mice, and dogs, as judged by LD_{50}, motor coordination and motor activity, weight increase/reduction, behavioral parameters, estrus cycle, fertility effects, and teratogenicity studies (Tabach *et al.*, 2009).

The hydroethanol extract, methanol extract, aqueous crude fractions, and chromatographic fractions obtained from *Passiflora actinia* induced catalepsy in mice (Santos *et al.*, 2005)

SUMMARY POINTS

- The seeds of passion fruit produce distinct proteins with antifungal activity.
- One of them, designated passiflin, specifically inhibits the fungus *Rhizoctonia solani*, with an IC_{50} of 16 μM, but no effect is seen on two other fungi, *Fusarium oxysporum* and *Mycosphaerella arachidicola*.
- Passiflin suppresses proliferation of breast cancer cells with an IC_{50} of 15 μM, but there is no effect on hepatoma HepG2 cells. It is devoid of ribonuclease, hemagglutinating, and HIV-1 reverse transcriptase inhibitory activities which may be present in some antifungal proteins.
- Unlike the majority of other antifungal proteins, passiflin is dimeric, and adsorbed on DEAE-cellulose and Q-Sepharose.
- Despite a substantial N-terminal sequence resemblance to β-lactoglobulin, passiflin is immunologically unrelated to β-lactoglobulin, which lacks antifungal and antiproliferative activities.
- 2S albumin-like antifungal proteins and peptides distinct from passiflin in species-specificity of antifungal activity have also been isolated from passion fruit.

References

Agizzio, A. P., Carvalho, A. O., Ribeiro, Sde F., Machado, O. L., Alves, E. W., Okorokov, L. A., et al. (2003). A 2S albumin-homologous protein from passion fruit seeds inhibits the fungal growth and acidification of the medium by *Fusarium oxysporum*. *Archives of Biochemistry and Biophysics, 416*, 188−195.

Akhondzadeh, S., Naghavi, H. R., Vazirian, M., Shayeganpour, A., Rashidi, H., & Khani, M. (2001). Passionflower in the treatment of generalized anxiety: a pilot double-blind randomized controlled trial with oxazepam. *Journal of Clinical Pharmacy and Therapeutics, 26*, 363−367.

Birner, J., & Nicolls, J. M. (1973). Passicol, an antibacterial and antifungal agent produced by Passiflora plant species: preparation and physicochemical characteristics. *Antimicrobial Agents and Chemotherapy, 3*, 105−109.

Boeira, J. M., Fenner, R., Betti, A. H., Provensi, G., Lacerda, L. D., Barbosa, P. R., et al. (2010). Toxicity and geno-toxicity evaluation of *Passiflora alata Curtis* (Passifloraceae). *Journal of Ethnopharmacology, 128*, 526−532.

De Lucca, A. J., Boue, S., Palmgren, M. S., Maskos, K., & Cleveland, T. E. (2006). Fungicidal properties of two saponins from *Capsicum frutescens* and the relationship of structure and fungicidal activity. *Canadian Journal of Microbiology, 52*, 336−342.

Dhawan, K., & Sharma, A. (2002). Prevention of chronic alcohol and nicotine-induced azospermia, sterility and decreased libido, by a novel tri-substituted benzoflavone moiety from *Passiflora incarnata* Linneaus in healthy male rats. *Life Sciences, 71*, 3059−3069.

Fisher, A. A., Purcell, P., & Le Couteur, D. G. (2000). Toxicity of *Passiflora incarnata* L. *Journal of Toxicology-Clinical Toxicology, 38*, 56−63.

Giavin a-Bianchi, P. F., Jr, Castro, F. F., Machado, M. L., & Duarte, A. J. (1997). Occupational respiratory allergic disease induced by *Passiflora alata* and *Rhamnus purshiana*. *Annals of Allergy Asthma & Immunology, 79*, 449−454.

Hernández-Ledesma, B., Recio, I., & Amigo, L. (2008). Beta-lactoglobulin as source of bioactive peptides. *Amino Acids, 35*, 257−265.

Lam, S. K., & Ng, T. B. (2009). Passiflin, a novel dimeric antifungal protein from seeds of the passion fruit. *Phytomedicine, 16*, 172−180.

Pelegrini, P. B., Noronha, E. F., Muniz, M. A., Vasconcelos, I. M., Chiarello, M. D., Oliveira, J. T., et al. (2006). An antifungal peptide from passion fruit (*Passiflora edulis*) seeds with similarities to 2S albumin proteins. *Biochimica Biophysica Acta, 1764*, 1141114−1141116.

Rowe, C. A., Nantz, M. P., Deniera, C., Green, K., Talcott, S. T., & Percival, S. S. (2004). Inhibition of neoplastic transformation of benzo[alpha]pyrene-treated BALB/c 3T3 murine cells by a phytochemical extract of passionfruit juice. *Journal of Medicinal Food, 4*, 402−407.

Rudnicki, M., Silveira, M. M., Pereira, T. V., Oliveira, M. R., Reginatto, F. H., Dal-Pizzol, F., et al. (2007). Protective effects of *Passiflora alata* extract pretreatment on carbon tetrachloride induced oxidative damage in rats. *Food and Chemical Toxicology, 45*, 656−661.

Santos, K. C., Santos, C. A., & de Oliveira, R. M. (2005). *Passiflora actinia* Hooker extracts and fractions induce catalepsy in mice. *Journal of Ethnopharmacology, 100,* 306–309.

Solbakken, A. M., Rørbakken, G., & Gundersen, T. (1997). Nature medicine as intoxicant. *Tidsskrift for den Norske laegeforening, 117,* 1140–1141.

Tabach, R., Rodrigues, E., & Carlini, E. A. (2009). Preclinical toxicological assessment of a phytotherapeutic product — CPV (based on dry extracts of *Crataegus oxyacantha* L., *Passiflora incarnata* L. *Valeriana officinalis* L.). *Phytotherapy Research, 23,* 33–40.

Watson, R. R., Zibadi, S., Rafatpanah, H., Jabbari, F., Ghasemi, R., Ghafari, J., et al. (2008). Oral administration of the purple passion fruit peel extract reduces wheeze and cough and improves shortness of breath in adults with asthma. *Nutrition Research, 28,* 166–171.

Yapo, B. M., & Koffi, K. L. (2008). Dietary fiber components in yellow passion fruit rind — a potential fiber source. *Journal of Agricultural and Food Chemistry, 56,* 5880–5883.

CHAPTER 103

Health Benefits of Peanut (*Arachis hypogaea* L.) Seeds and Peanut Oil Consumption

Katarzyna Suchoszek-Łukaniuk[1], Anna Jaromin[2], Mariola Korycińska[2], Arkadiusz Kozubek[2]
[1] Oleofarm, Producer of Pharmaceutical & Cosmetic Raw Materials, Dietary Supplements and Healthy Food, Pietrzykowice, Poland
[2] Faculty of Biotechnology, University of Wroclaw, Department of Lipids and Liposomes, Wrocław, Poland

873

LIST OF ABBREVIATIONS

AAD, average American diet
AIP, atherogenic index of plasma
BMI, body mass index
CHD, coronary heart disease
CVD, cardiovascular disease
FDA, Federal Drug Administration
FFPF, fat-free peanut flour
HDL, high density lipoprotein
INS-1, the rat pancreatic beta cell line
LDL, low density lipoprotein
MUFA, monounsaturated fatty acid
OO, olive oil
PO, peanut oil
PPB, peanuts and peanut butter
PUFA, polyunsaturated fatty acids
TAG, triacylglycerol

Nuts & Seeds in Health and Disease Prevention. DOI: 10.1016/B978-0-12-375688-6.10103-3

TC, total cholesterol
TNF-α, tumor necrosis factor α
USDA, United States Department of Agriculture

INTRODUCTION

The peanut is the edible seed of the plant *Arachis hypogaea* L. Originating in South America, peanuts are now widely cultivated in warm countries throughout the world, and are known by many local names. Although the fruit of this plant is considered a "nut" in the culinary sense, in the botanical sense the fruit is a legume or pod. Moreover, the seed-containing pods mature underground instead of aerially, as in most legumes. Peanuts vary in size, from the small, rounder nuts, which are used mainly for peanut butter and oil, to the larger oval nuts that are usually roasted.

BOTANICAL DESCRIPTION

Arachis hypogaea L., a member of the legume family (*Fabaceae*), is an annual that measures about 30—50 cm in height (Figure 103.1). The leaves are alternate and pinnate with four leaflets (two opposite pairs; no terminal leaflet); each leaflet is 1—7 cm long and 1—3 cm broad. Peanut flowers are borne in axillary clusters above the ground. After the flowers have been pollinated, a short, thick stem at the flower base, termed the gynophore, grows downward and penetrates into the soil, so the fruiting body develops entirely underground. The pods, usually containing from one to three seeds, develop only underground. Each seed is covered with a thin papery seed coat. The peanut has a well-developed taproot, with numerous lateral roots that extend several inches into the ground. Most roots have nodules. The plant prefers light, well-drained, sandy loams, but will grow in heavier soils.

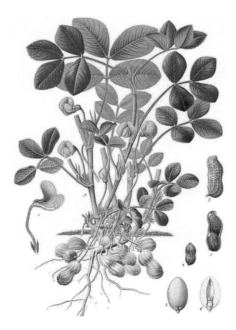

FIGURE 103.1
Peanut (*Arachis hypogaea*) (Köhler, 1887).

HISTORICAL CULTIVATION AND USAGE

Wild peanuts were first grown by the Incas of ancient Peru, who offered them to the sun god as part of their religious ceremonials. Peanut cultivation was also active in Ecuador, Bolivia, Brazil, and the Caribbean. The Spanish explorers of the New World took peanuts back to Spain, and traders and explorers distributed them from Spain to Asia and Africa. Peanuts came to North America as a consequence of the slave trade, and were planted throughout the southern United States. Until 1900, peanuts were not extensively grown because growing and harvesting techniques were slow and difficult. The development of equipment for production, harvesting, and shelling, and of processing techniques, resulted in expansion of the peanut industry.

PRESENT-DAY CULTIVATION AND USAGE

Peanuts are among the most important food crop of the world, being an important source of edible oil and vegetable protein. The pods ripen 120 to 150 days after the seeds are planted. In harvesting, the entire plant, with adhering seed pods, is lifted from the soil, mainly mechanically, before being dried in windrows or stacks and then threshed to remove the seeds. China is the major producer, followed by India and the United States of America (Table 103.1).

Peanuts are mainly used directly for food (oil, flour, roasted peanuts, peanut butter, etc.), with about half the crop turned into peanut butter, as in the United States. The shells, skins, and kernels of peanuts may be also used to make a vast variety of non-food products, being used as an ingredient in such things as cosmetics, paints, and nitroglycerin.

APPLICATIONS TO HEALTH PROMOTION AND DISEASE PREVENTION

The consumption of either peanuts or processed peanuts has been shown to be beneficial to health. This is primarily due to their desirable lipid profile, which is higher in unsaturated fatty acids than in saturated fatty acids. Peanuts have a high lipid content, ranging from 47% to 50%, and for that reason they are a good source of oil. The oil extracted from peanuts by cold-pressing has the most desirable nutritional values, a very light nutty aroma, and a light to dark yellow color. Usually it is used as salad dressing oil, but when refined it's suitable for frying because of its high smoke point of 232.22°C (450°F). The chemical and physical characteristics of peanut oil, according to the International Codex Alimentarius Commission Standard, are given in Table 103.2.

Peanut oil is naturally trans fat-free, cholesterol-free, and low in saturated fats. It consists mostly of oleic acid (n-9), a monounsaturated fatty acid (MUFA) (52%), and linoleic acid

875

TABLE 103.1 Top Ten Producers of Peanuts, 2008/2009	
Country	**Production (million tonnes)**
People's Republic of China	14.30
India	6.25
United States	2.34
Nigeria	1.55
Indonesia	1.25
Myanmar	1.00
Sudan	0.85
Senegal	0.71
Argentina	0.58
Vietnam	0.50
World	*34.43*

Source: USDA Foreign Agricultural Service: Table 13, Peanut Area, Yield and Production.

TABLE 103.2 Chemical and Physical Characteristics of Peanut Oil	
Characteristic Index	**Range of Values**
Relative density ($\times$ °C/water at 20°C)	0.912—0.920
Refractive index (N_D40°C)	1.460—1.465
Iodine value (I) (g/100 g)	86—107
Saponification value (mg KOH/g oil)	187—196
Unsaponifiable matter (g/kg)	10
Peroxide value (mmol/kg)	6
Acid value (KOH) (mg/g)	1
Insoluble impurities (%)	0.05
Moisture and volatile substance (%)	0.1

Source: International Codex Alimentarius Commission Standard CODEX-STAN 210-1999, Codex Standards for Named Vegetable Oils.

(n-6), a polyunsaturated fatty acid (PUFA) (32%) (Beare-Rogers et al., 2001; Özcan *et al.*, 2010). The fatty acid composition of peanut oil is shown in Table 103.3, and may vary due to the geographic area of cultivation, genetics, and environmental factors — especially the temperature during seed formation (Vassillou *et al.*, 2009; Özcan *et al.*, 2010). The oil is also a source of natural occurring compounds such as antioxidants, vitamin E, phytosterols, squalene, and p-coumaric acid, which are all beneficial in maintaining health.

Peanut oil shows many positive biological effects, which are mostly connected with its high oleic acid content. A number of studies have shown the unique properties of this fatty acid, and the importance of maintaining its intake at as high a level as possible. Oleic acid has been shown to have a positive influence on cardiovascular risk factors, such as lipid profiles, blood pressure, and glucose metabolism. These beneficial effects were first observed for olive oil, which is also rich in oleic acid, but they are now also being reported for peanut oil, as well as peanut seeds alone, or even products made of peanuts.

Cardiovascular disease (CVD) is the leading cause of death in the USA. Many studies have revealed that consumption of peanuts or peanut oil is associated with reduced CVD risk, and may improve serum lipid profiles, decrease LDL oxidation, and exert a cardioprotective effect. A pooled analysis of four large prospective epidemiological studies showed that subjects consuming nuts at least four times a week showed a 37% reduced risk of coronary heart disease compared to those who never or seldom ate nuts. Each additional serving of nuts per week was associated with an average 8.3% reduced risk of coronary heart disease (Kelly & Sabate, 2006).

In a controlled, randomized, double-blind human study conducted by Kris-Etherton *et al.* (1999), the effect of the average American diet (AAD) on the CVD risk profile was compared

TABLE 103.3 Fatty Acid Composition of Peanut Oil			
Fatty Acid (Common Name)	**Systematic Name**	**Shorthand Formula**	**% (By Weight of Total Fatty Acids)**
Lauric acid	Dodecanoic acid	C12:0	0.0—0.1
Myristic acid	Detradecanoic acid	C14:0	0.0—0.1
Palmitic acid	Hexadecanoic acid	C16:0	8.3—14.0
Stearic acid	Octadecanoic acid	C18:0	1.9—4.4
Oleic acid	9-Octadecaenoic acid	C18:1 (n-9)	36.4—67.1
Linoleic acid (ω6)	9,12-Octadecadienoic acid	C18:2 (n-6)	14.0—43.0
Alpha linolenic acid (ω3)	9,12,15-octadecatrienoic acid	C18:3 (n-3)	0.0—0.1
Arachidic acid	Icosanoic acid	C20:0	1.1—1.7
Behenic acid	Docosanoic acid	C22:0	2.1—4.4
Lignoceric acid	Tetracosanoic acid	C24:0	1.1—2.2

Source: International Codex Alimentarius Commission Standard CODEX-STAN 210-1999, Codex Standards for Named Vegetable Oils.

TABLE 103.4 Lipid and Lipoprotein Endpoint Results for the Experimental Diets[1] (Kris-Etherton *et al.*, 1999).

	AAD	Step II	OO	PO	PPB
Total cholesterol (mmol/l)	5.41 ± 0.23^a	4.92 ± 0.23^a	4.79 ± 0.23^b	4.93 ± 0.23^b	4.82 ± 0.23^b
LDL cholesterol (mmol/l)	3.52 ± 0.20^a	3.01 ± 0.20^b	2.98 ± 0.20^b	3.13 ± 0.20^b	3.03 ± 0.20^b
HDL cholesterol (mmol/l)	1.29 ± 0.11^a	1.24 ± 0.11^b	$1.28 \pm 0.11^{a,b}$	$1.26 \pm 0.11^{a,b}$	$1.26 \pm 0.11^{a,b}$
Triacylglycerol (mmol/l)	1.33 ± 0.13^a	$1.48 \pm 0.13^{a,2}$	1.15 ± 0.13^b	1.18 ± 0.13^b	1.16 ± 0.13^b
Apolipoprotein A (g/l)	1.54 ± 0.10	1.50 ± 0.10	1.52 ± 0.10	1.49 ± 0.10	1.48 ± 0.10
Apolipoprotein B (g/l)	1.01 ± 0.05^a	0.95 ± 0.05^b	0.92 ± 0.05^b	0.95 ± 0.05^b	0.92 ± 0.05^b
Lipoprotein(a) (g/l)[3]	$0.12 \pm 0.06^{a,b}$	0.14 ± 0.06^a	$0.12 \pm 0.06^{a,b}$	0.12 ± 0.06^b	$0.13 \pm 0.06^{a,b}$
Total : HDL cholesterol	4.5 ± 0.3^a	$4.3 \pm 0.3^{a,b}$	4.1 ± 0.3^b	4.2 ± 0.3^b	4.1 ± 0.3^b
LDL : HDL cholesterol	3.0 ± 0.3^a	2.7 ± 0.3^b	2.6 ± 0.3^b	2.7 ± 0.3^b	2.7 ± 0.3^b

AAD, average American diet; Step II, American Heart Association/National Cholesterol Education Program Step II diet; OO, olive oil diet; PO, peanut oil diet; PPB, peanut and peanut butter diet.

Reprinted from Kris-Etherton *et al.* (1999), *Am. J. Clin. Nutr.*, 70(6), 1009–1015, American Society for Nutrition, with permission.

[1]*Least-squares mean $\pm$ SE; n = 22. Within a row, values with different superscript letters are significantly different, P < (with Tukey-Kramer adjustment).*
[2]*Different from AAD, P = 0.06.*
[3]*n = 20; subjects whose values were < 0.05 g/l for all diet treatments were dropped from the analysis (J. Judd, personal communication, 1998).*

with the effect of four cholesterol-lowering diets: an American Heart Association/National Cholesterol Education Program Step II diet (a low-fat diet), and three high-MUFA diets — olive oil (OO), peanut oil (PO), and peanuts and peanut butter (PPB). Serum lipids and lipoproteins were examined, and the results are shown in Table 103.4. All three high-MUFA diets lowered total cholesterol by 10% and LDL cholesterol by 14%, and this result was comparable with that observed for the Step II diet. However, unlike the Step II diet, the PO and PPB diets decreased the triacylglycerol (TAG) level and did not decrease HDL cholesterol. The OO, PO, and PPB diets decreased CVD risk by an estimated 25%, 16%, and 21%, respectively, whereas the Step II diet lowered CVD risk by 12% (Figure 103.2) (Kris-Etherton *et al.*, 1999). Similar effects were also observed in postmenopausal hypercholesterolemic women placed on a low fat–MUFA rich diet. In this group, serum cholesterol decreased by 10% and LDL cholesterol level decreased by 12% compared to women on a low fat diet (O'Byrne *et al.*, 1997). These results clearly indicate that a high-MUFA, cholesterol-lowering diet may be preferable to a low-fat diet regarding CVD protection, because of more favorable effects on the CVD risk profile. Moreover, peanut products (i.e., peanuts, peanut butter, and peanut oil) can be used in designing high-MUFA diets.

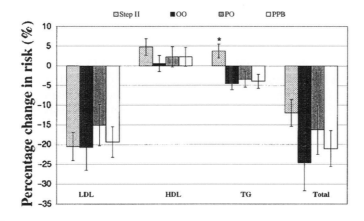

FIGURE 103.2

Effects of LDL-cholesterol, HDL-cholesterol, and triacylglycerol (TAG) concentrations on cardiovascular disease (CVD) risk reduction in response to Step II, olive oil (OO), peanut oil (PO), and peanut and peanut butter (PPB) diets (n = 22) (Kris-Etherton et al., 1999). ∗Significantly different from the other three diets, P = 0.005. Values are least-squares means ± SEs. *Reprinted from Kris-Etherton et al. (1999), Am. J. Clin. Nutr., 70(6), 1009–1015, American Society for Nutrition, with permission.*

The form in which peanuts are consumed and the method of processing (such as roasting or adding flavor) seem not to affect their CVD protective properties, as shown in a study conducted by McKiernan *et al.* (2010). Results of this randomized study, with 118 subjects from different countries, showed that consumption of 56 g of whole raw, roasted unsalted, roasted salted or honey-roasted peanuts, or ground peanut butter daily for 4 weeks resulted in a significant increase in the HDL cholesterol level and decrease in the total cholesterol, LDL cholesterol, and TAG levels in individuals classified as having elevated fasting plasma lipids, compared with those with normal fasting plasma lipids. These results suggest that the processing attributes do not compromise the lipid-lowering effects of peanuts (McKiernan *et al.*, 2010).

A study conducted by Ghadimi and colleagues (2010) revealed that the addition of peanuts to the diet, without any other dietary modification, can also favorably modify lipid profiles, the atherogenic index of plasma (AIP), and the estimated risk of coronary heart disease (CHD) in hypercholesterolemic men. Consumption of 77 g of peanuts for 4 weeks significantly reduced the TC/HDL cholesterol ratio and the LDL/HDL ratio, and at the same time increased the HDL and total antioxidant capacity. Similar protective effects were observed in a cohort study in which frequent nut and peanut butter consumption was associated with a significantly lower CVD risk in women with type 2 diabetes. Consumption of at least five servings per week of nuts or peanut butter (serving size 28 g (1 ounce) for nuts and 16 g (1 tablespoon) for peanut butter) was significantly associated with lower LDL cholesterol, non-HDL cholesterol, total cholesterol, and apolipoprotein-B-100 concentrations (Li *et al.* 2009).

A study on male Syrian golden hamsters showed the protective effects against atherosclerosis of fat-free peanut flour (FFPF), peanuts, and peanut oil. In this experiment, each diet group of animals was fed a high fat, high cholesterol diet with various peanut components (FFPF, peanut oil, or peanuts). All of three groups had significantly lower total plasma cholesterol and non-HDL than the control group. These results showed that the non-lipid portion of peanuts (FFPF), as well as peanut oil, retarded the development of atherosclerosis in animals consuming an atherosclerosis-inducing diet (Stephens *et al.*, 2010).

Peanuts were also proved to be beneficial in lowering the risk of type 2 diabetes. People with this type of diabetes do not produce adequate amounts of insulin for the needs of the body, and/or cannot use insulin effectively. A study conducted on INS-1 (a rat pancreatic beta cell line) showed that oleic acid and peanut oil high in oleic acid were able to enhance insulin production. Pre-treatment with oleic acid reversed the inhibitory effect of TNF-α on insulin. Peanut oil ultimately reversed the negative effects of inflammatory cytokines observed in obesity and non-insulin dependent diabetes mellitus. Type 2 diabetic mice that were administered a high oleic acid diet derived from peanut oil had decreased glucose levels compared to animals given a high fat diet with no oleic acid (Vassiliou *et al.*, 2009). Researchers from the Harvard School of Public Health, in a prospective cohort study of more than 83,000 women, found that women who consumed nuts or peanut butter had a significantly lower risk for type 2 diabetes. Women who reported eating nuts at least five times per week reduced their risk of type 2 diabetes by almost 30%, and those who ate peanut butter reduced their risk for type 2 diabetes by almost 20%, compared to women who rarely or never ate nuts or peanut butter. The reduced risk was independent of known risk factors for type 2 diabetes, such as body mass index (BMI), family history of diabetes, physical activity, smoking, alcohol use, and dietary factors (Jiang *et al.*, 2002).

Processed peanuts and by-products in the production of peanut products also exhibit health-promoting and preventive effects. A study conducted by Hwang *et al.* (2008) showed that roasted and defatted peanut dregs (a by-product in peanut oil production) exhibit anti-mutagenic and antiproliferative effects at 100 µg/ml concentration. At this concentration, they inhibited the proliferation of leukemia U937 and HL-60 cells by 56% and 52%, respectively, showing anticancer activity (Hwang *et al.*, 2008). Consumption of peanuts was also shown to have cancer-protective effects. A prospective cohort with a 10-year follow-up study, conducted by Yeh *et al.* (2006), showed that peanut consumption may help to reduce colorectal cancer risk in women. This anticancer effect is suggested to be a result of the action of nutrients found in peanuts, such as folic acid, phytosterols, phytic acid, and resveratrol, which have been reported to have anticancer effects (Yeh *et al.*, 2006). These results suggest potential applications for peanuts and their by-products as natural chemotherapeutic or chemopreventive agents.

ADVERSE EFFECTS AND REACTIONS (ALLERGIES AND TOXICITY)

It is estimated that peanut allergy, a type of food allergy, affects 0.4—0.6% of the population. For these people, eating a single peanut or just breathing the dust from peanuts can cause a fatal reaction. Moreover, an allergic reaction can also be triggered by eating foods that have been processed with machines that have previously processed peanuts. A strict exclusion diet and avoidance of foods that may be contaminated with peanuts is the only way to avert an allergic reaction. For this reason, the FDA Food Allergen Labeling and Consumer Protection Act requires that all food containing peanuts be clearly labeled.

Peanuts can be infected by the mold *Aspergillus flavus*, releasing the carcinogenic substance aflatoxin. Lower-quality specimens and damaged or spoiled seeds, particularly where mold is evident, are more likely to be contaminated. Raw peanuts are tested for aflatoxin levels by the USDA. Peanuts are also processed at a high temperature to ensure any microorganisms are killed.

SUMMARY POINTS

- Peanuts are one of the most important food crops of the world.
- Originating in South America, they are widely cultivated in warm countries throughout the world.
- The seed-containing pods only develop underground.
- Peanuts have a high lipid content, ranging from 47% to 50%.
- Peanut oil is naturally trans fat-free and cholesterol-free, and is rich in monounsaturated fatty acid (MUFA), especially oleic acid (n-9).
- Peanuts and peanut oil are beneficial to cardiovascular health, helping to reduce total and LDL cholesterol levels, and decrease CVD risk
- Peanut oil retards the development of atherosclerosis in animals consuming an atherosclerosis-inducing diet.
- Peanut and peanut products (peanuts, peanut butter, and peanut oil) can be used in designing a high-MUFA, cholesterol-lowering diet that is preferable to a low-fat diet regarding CVD protection.
- Peanuts are beneficial in lowering the risk of type 2 diabetes.
- Frequent intake of peanut and its products may reduce the risk of colorectal cancer.
- Some people have allergic reactions to peanuts; exposure can cause severe physical symptoms.
- Peanuts may be contaminated with the mold *Aspergillus flavus*, which produces carcinogenic aflatoxin.

References

Beare-Rogers, J., Dieffenbacher, A., & Holm, J. V. (2001). Lexicon of lipid nutrition (IUPAC Technical Report). *Pure and Applied Chemistry, 73*(4), 685—744.

Hwang, J. Y., Wang, Y. T., Shyu, Y., & Wu, J. S. (2008). Antimutagenic and antiproliferative effects of roasted and defatted peanut dregs on human leukemic U937 and HL-60 cells. *Phytotherapy Research, 22*(3), 286—290.

Ghadimi, N. M., Kimiagar, M., Abadi, A., Mirzazadeh, M., & Harrison, G. (2010). Peanut consumption and cardiovascular risk. *Public Health Nutrition, 13*(10), 1581—1586.

Hwang, J. Y., Wang, Y. T., Shyu, Y., & Wu, J. S (2008). Antimutagenic and antiproliferative effects of roasted and defatted peanut dregs on human leukemic U937 and HL-60 cells. *Phytotherapy Research, 22*(3), 286–290.

Jiang, R., Manson, J. E, Stampfer, M. J., Liu, S., Willett, W. C., & Hu, F. B. (2002). Nut and peanut butter consumption and risk of type 2 diabetes in women. *Journal of the American Medical Association, 288*(20), 2554–2560.

Kelly, J. H., & Sabate, J. (2006). Nuts and coronary heart disease: an epidemiological perspective. *British Journal of Nutrition, 96*(Suppl. 2), 61–67.

Köhler, F. E. (1887). *Köhler's Medizinal-Pflanzen.* Germany: Gera-Untermhaus.

Kris-Etherton, P. M., Pearson, T. A., Wan, Y., Hargrove, R. L., Moriarty, K., Fishell, V., et al. (1999). High-monounsaturated fatty acid diets lower both plasma cholesterol and triacylglycerol concentrations. *American Journal of Clinical Nutrition, 70*(6), 1009–1015.

Li, T. Y., Brennan, A. M., Wedick, N. M., Mantzoros, C., Rifai, N., & Hu, F. B. (2009). Regular consumption of nuts is associated with a lower risk of cardiovascular disease in women with type 2 diabetes. *Journal of Nutrition, 139*(7), 1333–1338.

McKiernan, F., Lokko, P., Kuevi, A., Sales, R. L., Costa, N. M., Bressan, J., et al. (2010). Effects of peanut processing on body weight and fasting plasma lipids. *British Journal of Nutrition, 11*, 1–9.

O'Byrne, D. J., Knauft, D. A., & Shireman, R. B. (1997). Low fat-monounsaturated rich diets containing high-oleic peanuts improve serum lipoprotein profiles. *Lipids, 32*(7), 687–695.

Özcan, M. M. (2010). Some nutritional characteristics of kernel and oil of peanut (*Arachis hypogaea* L.). *Journal of Oleo Science, 59*(1), 1–5.

Stephens, A. M., Dean, L. L., Davis, J. P., Osborne, J. A., & Sanders, T. H. (2010). Peanuts, peanut oil, and fat free peanut flour reduced cardiovascular disease risk factors and the development of atherosclerosis in Syrian golden hamsters. *Journal of Food Science, 75*(4), 116–122.

Vassiliou, E. K., Gonzalez, A., Garcia, C., Tadros, J. H., Chakraborty, G., & Toney, J. H. (2009). Oleic acid and peanut oil high in oleic acid reverse the inhibitory effect of insulin production of the inflammatory cytokine TNF-α in both *in vitro* and *in vivo* systems. *Lipids Health Disease, 26*(8), 25.

Yeh, C. C., You, S. L., Chen, C. J., & Sung, F. C. (2006). Peanut consumption and reduced risk of colorectal cancer in women: a prospective study in Taiwan. *World Journal of Gastroenterology, 12*(2), 222–227, 14.

CHAPTER 104

Antioxidants in Pecan Nut Cultivars [*Carya illinoinensis* (Wangenh.) K. Koch]

Ana G. Ortiz-Quezada, Leonardo Lombardini, Luis Cisneros-Zevallos
Department of Horticultural Sciences, Fruit and Vegetable Improvement Center, Texas A&M University, College Station, Texas, USA

LIST OF ABBREVIATIONS

AC, antioxidant capacity
CAE, chlorogenic acid equivalents
CE, catechin equivalents
CT, condensed tannins
DPPH, 2, 2-diphenyl-1-picrylhydrazyl
GAE, gallic acid equivalents
HPLC-MS, high performance liquid chromatography-mass spectrometry
ORAC, oxygen radical absorbance capacity assay
PDA, photodiode array
TE, Trolox equivalents
TP, total phenolics

INTRODUCTION

Pecan [*Carya illinoinensis* (Wangenh.) K. Koch] is the most valuable nut tree native to North America. Over 1000 different pecan varieties have been described, although 90% of cultivated acreage is represented by only a few dozen varieties. Pecan kernels contain about 70% lipids, namely oleic (over 60%), linoleic, palmitic, stearic, and linolenic acids. Pecans' antioxidant capacity, considered one of the highest among nut crops, comes from the non-lipid portion, and is cultivar-dependent. Defatted pecan kernels contain mainly condensed and hydrolyzable tannins. After a consecutive base/acid hydrolysis, phenolics released are mainly gallic acid, catechin, epicatechin, ellagic acid, and ellagic acid derivatives.

BOTANICAL DESCRIPTION

Pecan has been known for centuries for its edible nuts. The species is distributed over a broad geographic area, encompassing tremendous climatic variation. The native range of pecan extends for about 26° in latitude, from northern Iowa (lat. 42°20′N) to Oaxaca in Mexico (lat. 16°30′N) (Figure 104.1) (Thompson & Grauke, 1991). Pecan trees grow abundantly along the Mississippi River, the rivers of central and eastern Oklahoma, and on the Edwards plateau of Texas. The long tap roots and the frequent presence of shallow water tables allow native trees to survive the severe hot and dry summers that characterize the area of native distribution. However, the deeper the water source, the greater is the energy expended to obtain it, which leads to diversion of valuable energy from the developing leaves and nuts.

HISTORICAL CULTIVATION AND USAGE

The pecan is the only nut crop native to the North American continent that has significant commercial importance. All other major nut crops have been imported from other areas of the world. The term "pecan" derives from the word *pacane*, which is the Algonquin word for "nut that must be cracked with a stone" (Brison, 1974). Spanish explorers in the 16th century came to refer to them as "pacanos," or sometime simply *nueces* (nuts) or *nogales*, due to their resemblance to the fruit of the Persian walnut, *Juglans regia* L. The botanical name of the species was chosen because *Carya* is the ancient Greek name for walnut, and because one of the earliest indications of pecan use by Native Americans goes back to archaeological evidence found in present-day Illinois, dating from 9000 years ago. The archeological findings, combined with the descriptions made by the first Europeans during the 16th century, show that pecans were an essential ingredient for most Native American tribes of the southern part of what is now the United States. There is also evidence that some migration patterns traced the

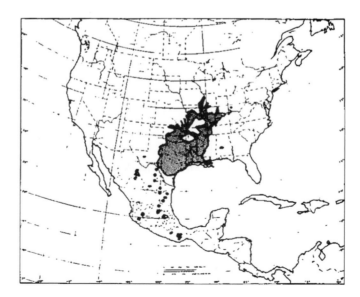

FIGURE 104.1
Native distribution of pecan. Native distribution of pecans includes south central states of the United States and different regions of Mexico (from Thompson & Grauke, 1991).

pecan season along alluvial plains and other fertile areas of modern Mexico and the United States. Pecan trees were revered by Native Americans not only as a food source. Members of the Ojibwa tribe utilized the wood of pecan and other hickories to make bows and finish off baskets. Additionally, pecan paste and oil, as well as leaf or bark infusions, were used by other tribes as abatement for several illnesses, such as intestinal worms, constipation, skin eruptions, rheumatism, gastrointestinal problems, and colds, or as facilitator in abortions. Despite the importance of pecans for Native Americans, it is almost certain that the Natives only relied on large groves of uncultivated or wild pecan trees, without making use of any widespread horticulture techniques to propagate and cultivate these trees. It was not until the mid-1800s that farmers began to realize the commercial value, plus the culinary importance, of pecans, and started developing horticultural techniques to maximize production. In 1847, it took a slave gardener (remembered only by his first name of Antoine) at the Oak Alley Plantation, Louisiana, to develop a technique to graft an improved variety ("scion") onto a different tree ("rootstock"). This episode marked the beginning of the modern pecan industry, because it allowed mass propagation of those trees which showed desirable characteristics, such as high kernel percentage (high meat content), resistance to diseases and insects, and reduced alternate bearing characteristics (Worley, 1994).

PRESENT-DAY CULTIVATION AND USAGE

A little bit more than a century and a half after Antoine's breakthrough discovery, pecan is today an economically important tree-nut crop for the United States and Mexico, with an annual economic value of about $300 million for the United States alone. There are over 1000 different pecan varieties that have been described, classified as either native or improved varieties.

Native trees or seedlings are those that have not been grafted with improved varieties. The latter are those that have been genetically altered through selection, and controlled, and are usually associated with more intensive horticultural practices, but sell at a premium price compared with native varieties because they usually produce larger kernels and are perceived as higher quality products. The majority of the improved acreage in the United States comprises only four varieties (Stuart, Western Schley, Desirable, and Wichita), and about 90% of the acreage comprises 33 varieties. In recent years, other varieties, such as Pawnee, have been extensively planted in newly established orchards; however, official data are not available (T.E. Thompson, personal communication).

A survey conducted recently in Texas reported that most pecan consumers prefer to purchase shelled pecans to use them as ingredient in food dishes (Figure 104.2) (Lombardini *et al.*, 2008). The United States and Mexico are the greatest producers of this nut. In the United States, Georgia, New Mexico, and Texas are the major producers for improved varieties, whereas fruits from native and seedlings trees are mainly from Oklahoma and Texas (Table 104.1).

APPLICATIONS TO HEALTH PROMOTION AND DISEASE PREVENTION

Pecan kernels are sources of protein, dietary fiber, vitamins, minerals, and many other bioactive substances, also called phytochemicals, which are known to provide health benefits. According to the US Department of Agriculture Nutrient Database (USDA, 2009), pecan kernels contain 72% lipids, 14% carbohydrates, 9% protein, 3.5% water, and 1.5% ash. Regarding the vitamins and minerals, pecan kernels are a good source of vitamins A and E, the B vitamins, folic acid, calcium, magnesium, potassium, and zinc.

Clinical studies with human subjects have shown that consuming pecans and other tree nuts may play an important role in reducing the risk of heart disease by improving the serum lipid

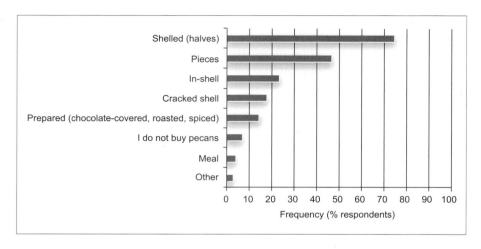

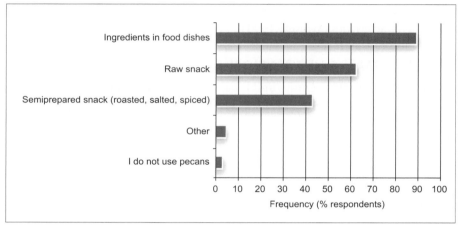

FIGURE 104.2
Purchasing behavior (top) and consuming preference (bottom) in regards to pecan, as emerged from a survey conducted in Texas. Pecan halves and pieces are preferentially purchased by consumers and used mainly as ingredients in food dishes and snacks (from Lombardini *et al.*, 2008).

profile. These benefits are mainly due to their high unsaturated fatty acid content (Rajaram *et al.*, 2001). Little information is available about the phytochemicals contained in the non-fatty portion of pecan kernels, which is believed to protect the oil. Around 97% of the total antioxidant capacity of pecans, measured by oxygen radical absorbance capacity assay (ORAC), comes from the hydrophilic portion. Wu *et al.* (2004) screened pecan kernels from unidentified cultivar(s), and nine other tree nuts, and found that pecans had the highest antioxidant capacity by ORAC (179.40 µmol TE/g) and total content of phenolics (20.16 mg of GAE/g). In another study, 98 common foods were screened for their proanthocyanidin content, or condensed tannin. It was found that within the nut group, pecans from unidentified cultivar(s) had the second highest content (494 mg/100 g fresh weight), after hazelnuts (500 mg/100 g fresh weight) (Gu *et al.*, 2004).

Recently, the phytochemical constituents in defatted pecan kernels were investigated (Villarreal-Lozoya *et al.*, 2007) in six cultivars (Desirable, Kanza, Kiowa, Nacono, Pawnee, and Shawnee) chosen for their commercial relevance. Results showed significant differences due to cultivar and, to a lesser extent, to orchard location. The study also revealed that a large portion of the antioxidant capacity was attributable to the total phenolics (TP), and especially to the condensed tannins (CT), by a ratio ranging from 0.31 to 0.56 CT/TP (Figure 104.3). Total phenolics measured by Folin-Ciocalteau spectrophotometric assay as chlorogenic acid equivalents ranged from 62 to 102 mg CAE/g defatted pecan. Condensed tannins were measured with the vanillin assay as catechin equivalents per gram defatted pecan, and their values were between 25 and 47 mg CE/g defatted pecan. After base and acid hydrolysis of an aqueous acetone extract of the defatted matrix, catechin, epicatechin, gallic acid, ellagic acid,

TABLE 104.1 Utilized Pecan Production (× 1000 lb) in the United States by Variety and State, 2004–2008

Improved Varieties	Utilized Production (In-Shell Basis)				
	2004	2005	2006	2007	2008
AL	1,000	3,200	5,400	10,000	7,400
AZ	14,000	22,000	14,000	23,000	17,500
AR	1,000	1,100	1,150	1,500	1,000
CA	3,500	3,900	3,400	4,400	3,750
FL	400	300	200	1,700	1,400
GA	42,000	72,000	36,000	135,000	66,000
LA	2,500	1,000	3,500	3,000	1,000
MS	700	800	2,000	2,200	900
MO		200	160	2	110
NM	39,000	65,000	47,000	74,000	43,000
NC	70	1,650	420	160	600
OK	6,000	6,000	5,000	3,000	1,000
SC	800	1,500	900	1,500	3,000
TX	28,000	50,000	33,000	44,000	20,000
US	*138,970*	*228,650*	*152,130*	*303,462*	*166,660*

Native and Seedlings	2004	2005	2006	2007	2008
AL	100	800	600	2,000	600
AR	700	1,200	1,050	800	500
FL	100	700	300	200	300
GA	3,000	8,000	6,000	15,000	4,000
KS	1,800	3,200	2,000	500	1,900
LA	6,500	4,000	17,500	11,000	4,000
MS	300	200	500	800	600
MO		2,400	940	3	830
NC	30	350	80	40	100
OK	22,000	15,000	12,000	27,000	4,000
SC	300	700	200	500	400
TX	12,000	15,000	14,000	26,000	10,000
US	*46,830*	*51,550*	*55,170*	*83,843*	*27,230*

Georgia, New Mexico, and Texas are the leading states in pecan production for improved varieties, whereas Texas, Oklahoma, Louisiana, and Georgia lead native and seedling production.
Source: USDA Economics, Statistics and Market Information System (http://usda.mannlib.cornell.edu).

and an ellagic acid derivative were identified (Villarreal-Lozoya *et al.*, 2007). The compounds found in greater concentrations were gallic acid (651–1300 µg/g defatted pecan) and ellagic acid (2505–4732 µg/g defatted pecan), but no significant differences were detected among cultivars. ORAC values ranged from 373 to 817 µmol TE/g defatted pecan, and strong correlations were found between antioxidant capacity and TP, as well as with condensed tannins (Villarreal-Lozoya *et al.*, 2007). The range values obtained for CT/TP and AC_{ORAC}/TP imply that proportions of condensed and hydrolyzable tannins differ for each cultivar, and this proportion determines the specific antioxidant activity of the phenolics present in each cultivar (Villarreal-Lozoya *et al.*, 2007).

The main fatty acids found in the lipid fraction of pecan kernels were oleic (over 60%), linoleic, palmitic, stearic, and linolenic. Pecans are considered to have high amounts of γ-tocopherol, along with walnuts, while almonds and hazelnuts are rich in α-tocopherol. Tocopherol content in pecans varies by cultivar, and other factors such as genetics, environment, maturity, and storage conditions. Tocopherols (vitamin E) are fat-soluble antioxidants naturally present in vegetable oils, such as those extracted from olive, almond, and hazelnut. Gamma-tocopherol values for the six pecan cultivars ranged from 72 to 135 µg γ-tocopherol/g oil.

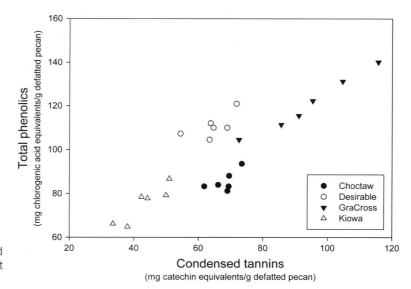

FIGURE 104.3

Positive correlation between total phenolic content and condensed tannins from defatted pecan kernels. Condensed tannins are the major components of the total phenolics present in defatted pecan kernels.

Nut shells were also analyzed for total phenolic content and condensed tannins for the six cultivars. Interestingly, their TP and CT values were 6 and 18 times higher than the ones found in the defatted kernels. It was concluded that phenolics from shells are mainly formed by condensed tannins (TP ≈ CT), and their presence may affect the content of kernel phenolics. It has been suggested that tannins leach from shells to kernels during soaking and preconditioning during commercial processing. Another possible source of leaching could be cold-room storage prior to cracking and shelling, due to water condensation inside the nuts as a result of small temperature fluctuations. Shells represent a large by-product of the pecan industry, with the shell percentage in pecan nuts varying from 40% to 50% (Worley, 1994). Processing plants have found only a limited market for pecan shells, with minimal profit. Thus, the high antioxidant capacity observed shows a potential alternative use of pecan shells as a novel source of antioxidants.

The presence of high contents of phenolic compounds, tocopherol, and mononounsaturated fatty acid suggest several health benefits. Phenolic compounds have been reported to protect against atherosclerosis, hypertension, cardiovascular diseases, cancer, and viral infections, and to act as general antioxidants. Tannins are water-soluble phenolic compounds of high molecular weight, and are classified as condensed (proanthocyanidins) or hydrolyzable (gallotannins or ellagitannins). The prevention of several chronic diseases, including cancer, cardiovascular and neurological diseases, and inflammation, have been associated with the intake of tannins. In general, tannins are known to have certain health benefits, such as antioxidant, anti-allergy, antihypertensive, and antitumor, as well as antimicrobial activities (Okuda, 2005). Condensed tannins are associated with foods of high antioxidant capacity, such as wine, cocoa, and grape seed (Gu *et al.*, 2004). Hydrolyzable tannins have been identified in several fruits and nuts, such as pomegranate juice (Seeram *et al.*, 2007) and walnuts (Fukuda *et al.*, 2003). Ellagitannins have recently attracted attention because of their antioxidant capacity and their antiproliferative activity, which inhibit prostate cancer growth (Seeram *et al.*, 2007). Upon hydrolysis, ellagitannins release ellagic acid, which is also of particular interest, as it reportedly has antiviral properties (Corthourt *et al.*, 1991) and provides protection against cancers of the colon, lung, and esophagus (Rao et al., 1991; Stoner & Morse, 1997).

A study was conducted to evaluate the differences in phenolic compounds between organically and conventionally grown pecan cultivars, and it concluded that the effects on cultivation method vary by cultivar (Malik *et al.*, 2009). Three pecan cultivars were analyzed: Cheyenne, Desirable, and Wichita. Nine free phenolic compounds were identified: gallic acid, catechol, *m*-coumaric acid, catechin, caffeic acid, epicatechin, chlorogenic acid, ellagic

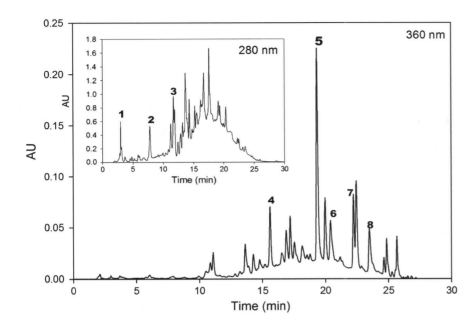

FIGURE 104.4
HPLC chromatogram of the main hydrolyzable tannins from defatted Choctaw pecan kernels at 280 and 360 nm. Peaks were defined by mass spectrometry, including (1) 2, 3 HHDP-glucose (RT, 2.96 min; λ_{max} 247; 481 m/z); (2) Pedunculagin isomer (RT, 7.81 min; λ_{max} 247; 783 m/z); (3) Galloyl pedunculagin (RT, 11.68 min; λ_{max} 248; 951 m/z); (4) Glausrin C (RT, 15.54 min; λ_{max} 270; 933 m/z); (5) Ellagic acid pentose conjugate (RT, 19.35 min; λ_{max} 249, 360; 433 m/z); (6) Ellagic acid (RT, 20.42 min; λ_{max} 246, 271, 365; 301 m/z); (7) Ellagic acid galloyl pentose conjugate (RT, 22.24 min; λ_{max} 247, 359; 585 m/z); (8) Ellagic acid galloyl pentose conjugate (RT, 23.5 min; λ_{max} 246, 360; 585 m/z).

acid, and an ellagic acid derivative. Only catechin, gallic acid, and ellagic acid were present in sufficient amounts to be quantified. Results showed that organically grown Desirable had a higher concentration of catechin and ellagic acid (86% and 311%, respectively) than the conventionally grown; however, there were no significant differences in the levels of gallic acid. In the other two organically grown varieties, Cheyenne and Wichita, the differences were smaller, but followed the same trend. It was concluded that growing organic pecans could increase the phenolic content in the kernel, but that results may depend on the cultivar investigated.

In another study, the condensed and hydrolyzable tannins present in kernels from four cultivars (Choctaw, Desirable, GraCross, and Kiowa) were identified (Ortiz-Quezada *et al.*, unpublished data) (Figure 104.4). Extracts of defatted pecan powder were obtained to measure total phenolics (TP) and antioxidant activities, using 2, 2-diphenyl-1-picrylhydrazyl (DPPH) and the ORAC assay. GraCross and Desirable had a significantly higher TP content, but only GraCross showed a high CT content and antioxidant capacity by DPPH and ORAC (Figure 104.5). Nonetheless, the ellagitannin content, as measured after a 2-M HCl hydrolysis by HPLC, showed inverse results, as Desirable and Kiowa had the highest ET value, followed by Choctaw, and lastly by GraCross. Extracts were analyzed by HPLC-MS, and tannins were identified in the negative ion mode according to their retention times, PDA and mass spectra, daughter ions, and fragmentation patterns. Thirteen ellagitannins were identified, but three other ellagitannins have not yet been characterized (599, 951 and 1085 m/z); free ellagic acid was also found. The masses of these ellagitannins ranged from 433 to 1207 m/z. Now that the chemistry of defatted pecan has been characterized, health benefit assays can be performed in order to find alternative uses for pecan and understand their mechanisms of action.

A study in progress will explore the effects of pecan metabolites on adipogenesis, as an anti-obesity assay at the molecular level.

ADVERSE EFFECTS AND REACTIONS (ALLERGIES AND TOXICITY)

Pecan kernels can be allergenic for sensitive population. A 2S albumin, Car i 1, has been characterized as the allergen contained in pecan protein. Another group of allergens

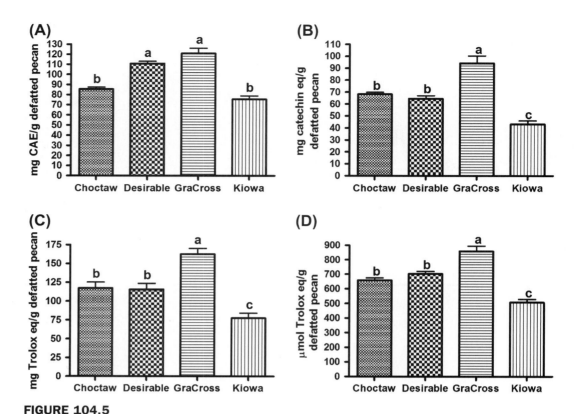

FIGURE 104.5
(A) Total phenolic content; (B) condensed tannin content; (C) antioxidant capacity by DPPH; (D) antioxidant capacity by ORAC of four pecan cultivars. The effects of pecan variety on phenolic content and the corresponding antioxidant activity. GraCross variety showed the overall highest phenolic and antioxidant activity.

(neoallergens), characterized by the same molecular weight as the original allergen, can develop during storage or after a thermal treatment, such as baking or roasting (Malanin et al., 1995). These neoallergens can derive from protein degradation due to autolysis that involves the interaction with hydrolyzed sugars (Maillard or browning reaction) (Berrens, 1996). Consequently, several individuals that are not allergic to raw or fresh pecans may develop symptoms in response to heated or stored pecan kernels. Another study found no toxicity for pecan color after feeding at a concentration of 5% to female rats for 90 days (Sekita et al., 1998).

Indirect toxicity caused by ingestion of pecan kernels may be also caused by mold developing during storage at relatively high humidity. Even without apparent shell damage, spores of *Aspergillus* spp. and *Penicillium* spp. may be present in the kernel (Doupnik & Bell, 1971). *Aspergillus* is of special concern, since it produces aflatoxins, which are toxic and carcinogenic, especially to the liver. If pecans are moldy, aflatoxins may be present above safe levels of < 20 ppb.

SUMMARY POINTS

- Pecan is the most valuable nut tree native to North America.
- Pecan use by Native Americans goes back to archaeological evidence found in present-day Illinois, dating from 9000 years ago.
- Over 1000 different pecan varieties have been described.
- The United States and Mexico are the greatest producers of this nut.
- Pecan's high antioxidant capacity comes from the non-lipid portion.

- Defatted pecan kernels contain phenolic compounds; the main ones are gallic acid, catechin, epicatechin, ellagic acid, and ellagic acid derivatives.
- These individual phenolics are arranged in polymers as condensed and hydrolyzable tannins.
- Pecan nut shells contain mainly condensed tannins.
- These phenolic compounds have been proven to be beneficial against several cancers and inflammation, and to have antiviral and antihypertensive activity, among others.
- The identified allergen is a 2S albumin called Car i 1.
- Pecans can become toxic when they get moldy.

References

Berrens, L. (1996). Neoallergens in heated pecan nut: products of Maillard-type degradation? *Allergy, 51*, 277–278.

Brison, F. R. (1974). *Pecan culture.* Austin, TX: Capital Printing.

Corthourt, J., Pieters, L. A., Claeys, M., Vanden Berghe, D. A., & Vlietinck, A. J. (1991). Antiviral ellagitannins from *Spondias mombin. Phytochemistry, 30*, 1129–1130.

Doupnik, B., & Bell, D. K. (1971). Toxicity to chicks of *Aspergillus* and *Penicillium* species isolated from moldy pecans. *Applied Microbiology, 21*, 1104–1106.

Fukuda, T., Ito, H., & Yoshida, T. (2003). Antioxidative polyphenols from walnuts (*Juglans regia* L.). *Phytochemistry, 63*, 795–801.

Gu, L., Kelm, M. A., Hammerstone, J. F., Beecher, G., Holden, J., Haytowitz, D., et al. (2004). Concentrations of proanthocyanidins in common foods and estimations of normal consumption. *Journal of Nutrition, 134*, 613–617.

Lombardini, L., Waliczek, T. M., & Zajicek, J. M. (2008). Consumer knowledge of nutritional attributes of pecans and factors affecting purchasing behavior. *HortTechnology, 18*, 481–488.

Malanin, K., Lundberg, L., & Johansson, S. G. O. (1995). Anaphylactic reaction caused by neoallergens in heated pecan nut. *Allergy, 50*, 988–991.

Malik, N. S. A., Perez, J. L., Lombardini, L., Cornacchia, R., Cisneros-Zevallos, L., & Braford, J. (2009). Phenolic compounds and fatty acid composition of organic and conventional grown pecan kernels. *Journal of the Science of Food and Agriculture, 89*, 2207–2213.

Okuda, T. (2005). Systematics and health effects of chemically distinct tannins in medicinal plants. *Phytochemistry, 66*, 2012–2031.

Rajaram, S., Burke, K., Connel, B., Myint, T., & Sabaté, J. (2001). A monounsaturated fatty acid-rich pecan-enriched diet favorably alters the serum lipid profile of healthy men and women. *Journal of Nutrition, 131*, 2275–2279.

Rao, C. V., Tokumo, K., Rigotty, J., Zang, E., Kelloff, G., & Reddy, B. S. (1991). Chemoprevention of colon carcinogenesis by dietary administration of piroxicam, α-difluoromethylornithine, 16α-fluoro-5-androsten-17-one, and ellagic acid individually and in combination. *Cancer Research, 51*, 4528–4534.

Seeram, N. P., Aronson, W. J., Zhang, Y., Henning, S. M., Moro, A., Lee, R., et al. (2007). Pomegranate ellagitannin-derived metabolites inhibit prostate cancer growth and localize to the mouse prostate gland. *Journal of Agricultural and Food Chemistry, 55*, 7732–7737.

Sekita, K., Saito, M., Uchida, O., Ono, A., Ogawa, Y., Kaneko, T., et al. (1998). Pecan nut color: 90-days dietary toxicity study in F344 rats. *Journal of Food Hygienic Society of Japan, 39*, 375–382.

Stoner, G. D., & Morse, M. A. (1997). Isothiocyanates and plant polyphenols as inhibitors of lung and esophageal cancer. *Cancer Letters, 114*, 113–119.

Thompson, T. E., & Grauke, L. J. (1991). Pecans and other hickories (*Carya*). *Acta Horticulturae, 290*, 839–904.

US Department of Agriculture (USDA). (2009). *USDA National Nutrient Database for Standard Reference, Release 22.* Washington, DC: USDA Agricultural Research Service, USDA Nutrient Data Laboratory.

Villarreal-Lozoya, J. E., Lombardini, L., & Cisneros-Zevallos, L. (2007). Phytochemical constituents and antioxidant capacity of different pecan [*Carya illinoinensis* (Wangenh.) K. Koch] cultivars. *Food Chemistry, 102*, 1241–1249.

Worley, R. (1994). Pecan production. In C. R. Santerre (Ed.), *Pecan technology* (pp. 12–38). New York, NY: Chapman & Hall.

Wu, X., Beecher, G. R., Holden, J. M., Haytowitz, D. B., Gebhardt, S. E., & Prior, R. L. (2004). Lipophilic and hydrophilic antioxidant capacities of common foods in the United States. *Journal of Agricultural and Food Chemistry, 52*, 4026–4037.

CHAPTER 105

Health Effects of a Pecan [*Carya illinoinensis* (Wangenh.) K. Koch] Nut-rich Diet

Ella H. Haddad
Department of Nutrition, School of Public Health, Loma Linda University, Loma Linda, California, USA

891

LIST OF ABBREVIATIONS

AHS, Adventist Health Study
CAE, chlorogenic acid equivalents
CE, catechin equivalents
DV, daily value
FRAP, ferric reducing activity of plasma
GAE, gallic acid equivalents
ORAC, oxygen radical absorbance capacity
TE, Trolox equivalents
TEAC, Trolox equivalent antioxidant capacity

INTRODUCTION

Pecan nuts (*Carya illinoinensis*) are fruits of a species of hickory tree indigenous to North America, and were used as a food by Native Americans. Early settlers appreciated the nut, and

Nuts & Seeds in Health and Disease Prevention. DOI: 10.1016/B978-0-12-375688-6.10105-7

commercial cultivation of pecans began in the late 19th century. Currently, the US produces most of the world's pecans. Although pecans have a high fat density (94% of energy), their fat is mainly monounsaturated (53% of energy) and polyunsaturated (28% of energy), with a relatively small amount of saturated fat (8% of energy). Pecans do not, however, just provide fat to the diet. As shown in Tables 105.1 and 105.2, pecans are a complex food, and a source of multiple nutrients and bioactive components. For example, a typical 28-g (1-oz) portion of pecans provides 2.6 g of protein, which represents 5% of the daily value (DV) of 50 g protein needed by a healthy adult. Other prominent nutrients in pecans are the vitamin thiamin, and the minerals magnesium, phosphorus, zinc, copper, and manganese. Pecans are a particularly rich source of redox-active metabolites, such as tocopherols, flavonoids, and phytosterols. The aim of this chapter is to present the contribution of pecans to human nutrition, and review epidemiological and clinical research on pecan consumption and health.

APPLICATIONS TO HEALTH PROMOTION AND DISEASE PREVENTION

Epidemiology of nuts and heart disease

As with other nuts, pecans were considered a peripheral food — a snack food, or an ingredient in cookies and desserts. This all changed based on findings from the Adventist Health Study (AHS), the first epidemiological cohort to show that eating nuts was

TABLE 105.1 Nutrient Composition of Pecans per Typical 28-g (1-oz) Serving Compared to the Daily Value and Percentage of Daily Value (% DV)

Nutrient	Amount	Daily Value (% DV)
Energy, kJ	819	
Energy, kcal	196	2000 (10)
Protein, g	2.6	50 (5)
Carbohydrate, g	3.9	300 (1)
Fat, g	20.4	65 (31)
Saturated, g	1.75	—
Monounsaturated, g	11.57	—
Polyunsaturated, g	6.13	—
18:2, g	5.85	—
18:3, g	0.28	—
Dietary fiber, g	2.7	25 (11)
Thiamin, mg	0.19	1.5 (12)
Riboflavin, mg	0.04	1.7 (2)
Niacin, mg	0.33	20 (1.6)
Vitamin B6, mg	0.06	2 (3)
Folate, µg	6	400 (2)
Alpha-tocopherol, mg	1.40	—
Gamma-tocopherol, mg	24.44	—
Calcium, mg	20	1000 (2)
Iron, mg	0.72	18 (4)
Magnesium, mg	34	400 (9)
Phosphorus, mg	79	1000 (8)
Zinc, mg	1.28	15 (9)
Copper, mg	0.34	2 (17)
Manganese, mg	1.28	2 (64)

The composition of pecans is from the USDA Nutrient Database at http://www.nal.usda.gov/fnic/cgi-bin/nut_search.pl. Daily Values are set by the US Food and Drug Administration (FDA) for food labeling purposes. The daily values listed are for adults and children aged 4 or more years, and are based on a caloric intake of 2000 kilocalories. Daily values of ≥ 10% and ≥ 20% designate a good source and a high source of a nutrient, respectively. Pecans are a good source of dietary fiber, thiamin, and copper, and a high source of manganese.
See FDA Food Labeling Guide at: http://www.fda.gov/Food/GuidanceComplianceRegulatoryInformation/GuidanceDocuments/FoodLabelingNutrition/FoodLabelingGuide/ucm064928.htm

TABLE 105.2 Flavonoid and Phytosterol Composition of 100 g of Pecans

Component	Units	Amount
Total flavonoids	**mg**	**526**
Anthocyanidins:		
Cyanidin	mg	10.74
Delphinidin	mg	7.28
Total proanthocyanidins:	mg	494
Monomers	mg	17.22
Dimers	mg	42.13
Trimers	mg	26.03
4–6mers	mg	101.43
7–10mers	mg	84.23
Polymers	mg	223.01
Flavan-3-ols:		
Epicatechin	mg	0.82
Epicatechin 3-gallate	mg	0
Epigallocatechin	mg	5.63
Epigallocatechin 3-gallate	mg	2.30
Catechin	mg	7.24
Total Phytosterols	**mg**	**102**
β-sitosterol	mg	89

Flavonoids have antioxidant properties, and phytosterols interfere with intestinal absorption of cholesterol and lower blood cholesterol levels. Flavonoid data are from the USDA Database for the Flavonoid Contect of Selected Foods, Release 2.1, 2007 (available at: http://www.ars.usda.gov/Services/docs.htm?docid=6231). Phytosterol data are from USDA Nutrient Database Standard Reference, Release 22, 2009 (available at http://www.nal.usda.gov/fnic/cgi-bin/nut_search.pl).

associated with reduced risk of cardiovascular mortality. The AHS is a prospective cohort investigation of approximately 31,000 California Adventists on whom extensive dietary and lifestyle data were recorded. One characteristic of this population was the wide range in frequency of nut consumption. Participants who ate nuts five or more times per week showed a 48% reduction in risk of a myocardial infarction (P for trend < 0.01), and 38% reduction in risk of death from cardiovascular disease (P for trend < 0.05) (Fraser et al., 1992).

The findings of the AHS were substantiated by subsequent observational cohorts. Briefly, in the Iowa Women's Health Study, postmenopausal women in the highest quartile of nut consumption had a 40% reduction in risk of fatal coronary heart disease compared to the lowest quartile that seldom or never ate nuts (RR 0.60, 95% CI, 0.36–1.01, P for trend 0.016) (Ellsworth et al., 2001). Data from the Nurses Health Study showed a 39% reduction in fatal coronary heart disease in those who ate nuts five or more times per week, compared to those eating nuts less than once per month (RR 0.61, 95% CI 0.35–1.05, P for trend 0.007) (Hu et al., 1998). Furthermore, nut consumption was inversely associated with risk of type 2 diabetes in this cohort of nurses (Jiang et al., 2002). In the Physicians Health Study, those consuming nuts more than once a week had a 23% reduction in fatal coronary heart disease (P for trend 0.04) (Albert et al., 2002). A recent pooled analysis of epidemiologic studies demonstrated that subjects in the highest intake group for nut consumption have an approximately 35% reduced risk of heart disease incidence (Kris-Etherton et al., 2008).

The epidemiological evidence is strong for the cardioprotective effect of nut consumption. Although none of the studies separated out the specific nuts that were eaten, the studies were conducted in the US, and as pecans are the second most commonly consumed tree nut in the US they were included in the mix consumed by the studied cohorts.

Pecans and serum lipids

Two feeding trials tested the lipid-lowering effects of pecans. In the first study, Morgan and Clayshulte (2000) used a randomized parallel study design to compare serum lipid concentrations of free-living subjects who consumed self-selected diets plus a 68-g per day supplement of pecans for 8 weeks, with subjects who consumed self-selected diets but no nuts. They observed a 10% reduction in LDL cholesterol at week 4 and a 6% reduction at week 8 ($P<0.05$) in the treatment group.

Rajaram *et al.* (2001) applied a crossover design to investigate the effect of a pecan-enriched diet on blood lipid. Using a controlled metabolic protocol, subjects were randomized to either the Step I diet as control, or a pecan-enriched Step I with pecans contributing 20% of energy as the intervention diet. Subjects were men and women with normal to moderately high serum cholesterol, and the amount of pecans consumed was approximately 60 g per 2000 kcal of food.

Although both diets lowered plasma lipids, the pecan-enriched diet altered the lipid profile more favorably than the Step I diet. Compared to the Step I diet, total cholesterol decreased by 6.7%, LDL cholesterol by 10.4%, triglycerides by 11.1%, apolipoprotein B by 11.6%, and lipoprotein(a) by 11.1%. Protective markers such as HDL increased by 5.6%, and apolipoprotein A1 by 2.2%.

It is important to note that observed alterations in blood lipids were greater than those expected when calculated by predictive equations based on changes in dietary fatty acids and cholesterol. These effects may be due to components in pecans such as dietary fiber, plant protein, and phytosterols, all of which have cholesterol-lowering properties.

Pecans and antioxidant status

There is great interest in the role of bioactive constituents of foods in modulating oxidative stress. Oxidative stress is implicated in atherosclerosis, coronary heart disease, and other degenerative conditions. The great diversity and complexity of these constituents makes the process of characterizing them challenging. Pecan cultivars exhibit a high degree of variability in chemical composition and polyphenolic content due to genetic species and growth conditions (Venkatachalam *et al.*, 2007; Villarreal-Lozoya *et al.*, 2007). Crude estimates are obtained using the Folin-Ciocalteau reagent to measure total phenolic content, and the vanillin assay to quantify tannins (Table 105.3).

TABLE 105.3 Total Phenols, Total Tannins, and Antioxidant Capacity of Pecans Assayed As Ferric Reducing Ability of Plasma (FRAP), Lipophilic Oxygen Radical Absorbance Capacity (ORAC), and Hydrophilic ORAC

Test	Unit	Reference	Value
Total phenols	mg GAE/g	Wu *et al.*, 2004	20.16 ± 0.98
Total phenols	mg GAE/g	Kornsteiner *et al.*, 2006	12.84
Total phenols	mg CAE/g defatted nut	Villarreal-Lozoya *et al.*, 2007	76 ± 1.9
Total tannins	mg CE/g	Venkatachalam *et al.*, 2007	8.8
Ferric reducing ability of plasma (FRAP)	mmol/100 g	Halvorsen *et al.*, 2006	9.67 ± 2.67
Lipophilic ORAC	micromol TE/g	Wu *et al.*, 2004	4.16 ± 0.98
Hydrophilic ORAC	micromol TE/g	Wu *et al.*, 2004	175.24 ± 10.36

Phenols and tannins are estimated by the Folin-Ciocalteau and vanillin reagent, respectively. Phenols are expressed as gallic acid equivalents (GAE) per gram, and chlorogenic acid equivalents (CAE) per gram of defatted nut; tannins are expressed as catechin equivalents (CE) per gram, and ORAC as Trolox equivalents (TE) per gram. The antioxidant capacity of pecans was the tenth highest among the more than 1000 foods tested by the FRAP assay, and the sum of hydrophilic and lipophilic ORAC of pecans was the highest among the commonly consumed tree nuts.

The development of the analytical technique of normal phase HPLC coupled with tandem mass spectrometry (LC-MS/MS) enabled the quantification of complex molecules such as proanthocyanidins (Gu *et al.*, 2004, 2006). Proanthocyanidins are polymers of flavan-3-ols, and are present in pecans as mixtures of monomers, oligomers, and polymers. When commonly-eaten portion sizes are compared, the amount of these compounds found in pecans is comparable to that in berries, dried beans, and black chocolate. Anthocyanidins (cyanidin, delphinidin) and flavan-3-ol monomers (epicatechin, epicatechin gallate, epigallocatehin, epigallocatechin gallate, and catechin) found in pecans have also been quantified (Table 105.2).

It is important not only to identify and quantify various bioactive metabolites in a food, but also to evaluate the effect of consuming the food on *in vivo* antioxidant status. Halvorsen *et al.* (2006) applied the ferric reducing ability of plasma (FRAP) assay of Benzie and Strain (1996) to evaluate the antioxidant potential of 1120 foods, and pecans ranked tenth in antioxidant activity among the foods that were tested. The lipophilic and hydrophilic antioxidant capacities of pecans were determined using the oxygen radical absorbance capacity (ORAC) assay by Wu *et al.* (2004). Total antioxidant capacity (sum of hydrophilic and lipophilic ORAC of pecans at 179 micromols of Trolox equivalents (TE) per gram) was highest among the nuts. These data demonstrate that pecans are a rich source of redox-active compounds, and may contribute to antioxidant protection provided in the diet by plant foods.

Effect of pecans on *in vivo* oxidative stress

That bioactive constituents in pecans possess antioxidant activity has been clearly demonstrated. What has not been evaluated is their action in humans following pecan consumption. Little is known about their absorption, bioavailability, or transport to tissues. Despite the fact that extracts of pecans exhibit high antioxidant activity when tested *in vitro*, few studies have examined the impact of pecan consumption on physiological oxidative protection.

Haddad and colleagues (2006) reported that a diet in which pecans contributed 20% of energy reduced fasting plasma concentrations of the malondialdehyde-thiobarbituric acid adduct (MDA-TBA) by 7% ($P < 0.05$) compared to the Step I diet in normal and mildly lipidemic individuals. The MDA-TBA adduct measured by HPLC is a biomarker of lipid peroxidation. However, fasting plasma antioxidant capacity concentrations, measured by the FRAP assay of Benzie and Strain (1996) and by the Trolox equivalent antioxidant capacity (TEAC) assay of Miller and Rice-Evans (Miller *et al.*, 1993), of participants on the pecan-enriched diet did not differ from those of the Step I diet. This may be explained by the fact that plasma antioxidant capacity responds within a few hours following the consumption of foods rich in polyphenols, such as tea, wine, and chocolate (López-Uriarte *et al.*, 2009). Subsequent metabolism and elimination of these compounds returns plasma antioxidant capacity to baseline levels.

Recently, Hudthagosol *et al.* (2011) investigated the effect of a pecan test meal on postprandial antioxidant capacity. Sixteen healthy volunteers consumed test meals of either 90 g of pecans, or an isocaloric meal of equivalent macronutrient composition but formulated of refined ingredients, in a crossover design. Blood samples were drawn at baseline, and at 1, 2, 3, 5, and 8 hours following the test meals. Plasma antioxidant capacity was estimated by the FRAP method, uric acid by colorimetric assay, and catechins by HPLC with coulometric array electrochemical detection.

As shown in Figure 105.1, plasma FRAP concentrations increased following the pecan meal, and were significantly higher ($P < 0.05$) at 2 and at 3 hours post-pecan ingestion. This effect cannot be attributed to increases in urate, since fluctuations in plasma uric acid were not different from baseline. The observed effect on the FRAP parameter is due to the bioactive components found in pecans.

895

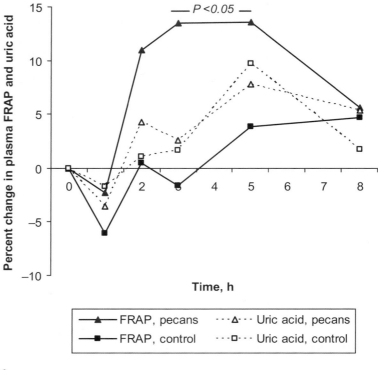

FIGURE 105.1

Percent change in postprandial ferric acid reducing ability of plasma (FRAP) and uric acid concentration following consumption of test meals containing 90 g of pecans compared to a control test meal of similar composition formulated of refined ingredients in 16 healthy subjects. Compared to baseline, the pecan test meals resulted in an increase in FRAP at 3 and 5 hours postprandially ($P < 0.05$). The data suggest that pecan consumption may enhance postprandial antioxidant capacity. Plasma uric acid concentrations did not differ from baseline, and may not have contributed to this effect.

Finally, although pecans contain relatively small amounts of catechin monomers (catechin, epicatechin, epigallocatechin gallate), preliminary data from the same study show that these compounds are bioavailable and appear in plasma following pecan consumption. Figure 105.2 shows that the concentration of total catechins post-ingestion of

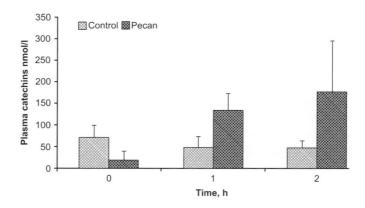

FIGURE 105.2

Plasma catechin concentrations (sum of catechin, epicatechin, epigallocatechin, epigallocatechin gallate and epicatechin gallate) at baseline (0 hour) and at 1 hour and 2 hours following consumption of whole pecans (90 g) and control test meals. Mean $\pm$ SE, $n = 12$. Concentrations of catechins was significantly higher following the pecan test meal at 2 hours compared to baseline, and compared to control meal at baseline and at 2 hours postprandially ($P < 0.05$). The data suggest that catechin monomers in pecans are bioavailable, and appear in plasma following ingestion of pecans.

a test meal containing 90 g of pecans was higher at 2 hours than at baseline ($P < 0.05$), and higher than that observed at 2 hours following the catechin-free control test meal ($P < 0.05$).

In conclusion, despite clear evidence of a link between nut consumption and heart disease, little is known about the mechanisms that mediate this effect. Preliminary research suggests that pecans may lower lipid oxidation and enhance antioxidant capacity. The field is ripe for intensive study on the extent to which bioactive components in pecans contribute to health benefits.

SUMMARY POINTS

- Prospective epidemiological studies consistently show a relationship between frequent nut consumption and reduced risk of cardiovascular disease.
- When incorporated into meals, pecans reduce cardiovascular disease through a reduction in lipid and lipoprotein risk factors.
- Pecan extracts show high antioxidant capacity when tested by various assays *in vitro*.
- There is a paucity of clinical intervention trials examining the effect of pecan consumption on *in vivo* oxidative stress biomarkers.
- Preliminary data indicate that some pecan phenolics are bioavailable and may influence postprandial antioxidant capacity.

References

Albert, C. M., Gaziano, J. M., Willett, W. C., & Manson, J. E. (2002). Nut consumption and decreased risk of sudden cardiac death in the Physicians' Health Study. *Archives of Internal Medicine, 162*, 1382–1387.

Benzie, I. F. F., & Strain, J. J. (1996). The ferric reducing ability of plasma (FRAP) as a measure of "antioxidant power": the FRAP assay. *Analytical Biochemistry, 239*, 70–76.

Ellsworth, J. L., Kushi, L. H., & Folsom, A. R. (2001). Frequent nut intake and risk of death from coronary heart disease and all causes in postmenopausal women: the Iowa Women's Study. *Nutrition Metabolism & Cardiovascular Diseases, 11*, 372–377.

Fraser, G. E., Sabato, J., Beeson, W. L., & Strahan, T. M. (1992). A possible protective effect of nut consumption on risk of coronary heart disease. The Adventist Health Study. *Archives of Internal Medicine, 152*, 1416–1424.

Gu, L., Kelm, M. A., Hammerstone, J. F., Beecher, G., Holden, J., Haytowitz, D., et al. (2004). Concentrations of proanthocyanidins in common foods and estimations of normal consumption. *Journal of Nutrition, 134*, 613–617.

Gu, L., House, S. E., Wu, X., Ou, B., & Prior, R. L. (2006). Procyanidin and catechin contents and antioxidant capacity of cocoa and chocolate products. *Journal of Agricultural and Food Chemistry, 54*, 4057–4061.

Haddad, E., Jambazian, P., Karunia, M., Tanzman, J., & Sabato, J. (2006). A pecan-enriched diet increases γ-tocopherol/cholesterol and decreases thiobarbituric acid reactive substances in plasma of adults. *Nutrition Research, 26*, 397–402.

Halvorsen, B. L., Carlsen, M. H., Phillips, K. M., Bohn, S. K., Holte, K., Jacobs, D. R., Jr., et al. (2006). Content of redox-active compounds (i.e., antioxidants) in foods consumed in the United States. *American Journal of Clinical Nutrition, 84*, 95–135.

Hu, F. B., Stampfer, M. J., Manson, J. E., Rimm, E. B., Colditz, G. A., Rosner, B. A., et al. (1998). Frequent nut consumption and risk of coronary heart disease in women: prospective cohort study. *British Medical Journal, 317*, 1341–1345.

Hudthagosol, C., Haddad, E. H., McCarthy, K., Wang, P., Oda, K., & Sabaté, J. (2011). Pecans acutely increase plasma postprandial antioxidant capacity and catechins and decrease LDL oxidation in humans. *Journal of Nutrition, 141*, 56–62.

Jiang, R., Manson, J. E., Stampfer, M. J., Liu, S., Willett, W. C., & Hu, F. B. (2002). Nut and peanut butter consumption and risk of type 2 diabetes in women. *Journal of the American Medical Association, 288*, 2554–2560.

Kornsteiner, M., Wagner, K. H., & Elmadfa, E. (2006). Tocopherols and total phenolics in 10 different nut types. *Food Chemistry, 98*, 381–387.

Kris-Etherton, P. M., Hu, F. B., Ros, E., & Sabato, J. (2008). The role of tree nuts and peanuts in the prevention of coronary heart disease: multiple potential mechanisms. *Journal of Nutrition, 138*, 1746S–1751S.

López-Uriarte, P., Bulló, M., Casas-Augustench, P., Babio, N., & Salas-Salvadó, J. (2009). Nuts and oxidation: a systematic review. *Nutrition Reviews, 67*, 497–508.

Miller, N. J., Rice-Evans, C. A., Davies, M. J., Gopinathan, V., & Milner, A. (1993). A novel method for measuring antioxidant capacity and its application to monitoring the antioxidant status in premature neonates. *Clinical Science, 84*, 407–412.

Morgan, W. A., & Clayshulte, B. J. (2000). Pecans lower low-density lipoprotein cholesterol in people with normal lipid levels. *Journal of the American Dietetic Association, 100,* 312–318.

Rajaram, S., Burke, K., Connell, B., Myint, T., & Sabato, J. (2001). A monounsaturated fatty acid-rich pecan-enriched diet favorably alters the serum lipid profile of healthy men and women. *Journal of Nutrition, 131,* 2275–2279.

Venkatachalam, M., Kshirsagar, H. H., Seeram, N. P., Heber, D., Thompson, T. E., Roux, K. H., et al. (2007). Biochemical composition and immunological comparison of select pecan [*Carya illinoinensis* (Wangenh.) K. Koch] cultivars. *Journal of Agricultural and Food Chemistry, 55,* 9899–9907.

Villarreal-Lozoya, J. E., Lombardini, L., & Cisneros-Zevallos, L. (2007). Phytochemical constituents and antioxidant capacity of 6 different pecan [*Carya illinoinensis* (Wangenh.) K. Koch] cultivars. *Food Chemistry, 102,* 1241–1249.

Wu, X., Beecher, G. R., Holden, J. M., Haytowitz, D. B., Gebhardt, S. E., & Prior, R. L. (2004). Lipophilic and hydrophilic antioxidant capacities of common foods in the United States. *Journal of Agricultural and Food Chemistry, 52,* 4026–4037.

898

Perennial Horse Gram (*Macrotyloma axillare*) Seeds: Biotechnology Applications of its Peptide and Protein Content — Bowman-Birk Inhibitors and Lectin

899

Marcos Aurélio de Santana[1], William de Castro Borges[2], Larissa Lovatto Amorin[2], Alexandre Gonçalves Santos[2], Sonaly Cristine Leal[2], Milton Hércules Guerra de Andrade[2]
[1] Universidade Federal dos Vales do Jequitinhonha e Mucuri, Departamento de Farmácia, Diamantina, Minas Gerais, Brasil
[2] Universidade Federal de Ouro Preto, Departamento de Ciências Biológicas, Ouro Preto, Minas Gerais, Brasil

LIST OF ABBREVIATIONS

BBI, Bowman-Birk inhibitors
DBL, Dolichos biflorus seed lectin
GalNAc, N-acetyl-α-D-galactosamine
MaL, Macrotyloma axillare seed lectin

Nuts & Seeds in Health and Disease Prevention. DOI: 10.1016/B978-0-12-375688-6.10106-9

INTRODUCTION

Macrotyloma axillare is a leguminous plant widely found in tropical areas of the world. Its main use is restricted to food for cattle but it is also popularly utilized by some communities to treat diverse ailments of humans and animals. Although it has been poorly explored by the scientific community, *M. axillare* seeds constitute a potential source of important biomolecules of therapeutic and research value. This chapter describes the isolation, characterization, and functional studies related to two important bio-products extracted from *M. axillare* seeds. First, we report on the isolation and function of a 28-kDa lectin (monomer) isolated from the seed plant, demonstrating its potential for blood-typing and glycoprotein research. Subsequently, we describe the biochemical features of the small 8-kDa peptides, called Bowman-Birk inhibitors (BBIs), isolated from *M. axillare* and other leguminous plants, highlighting their potential as cancer preventive agents.

BOTANICAL DESCRIPTION

Due to its main use as forage, *M. axillare* data can be found in specialized webpages of recognized scientific accuracy. The following botanical and morphological description can be found at the *Tropical Forages* website (Cook *et al.*, 2005).

Macrotyloma axillare is a member of the Fabaceae (pea) family, of the order Fabales. It is a trailing and twining perennial plant (Figure 106.1) with an erect basal stem usually about 1 cm in diameter, developing to 3 cm in unrestricted stands, and with the ability to climb to > 10 m up an appropriate framework such as trees. It has a strong, woody taproot and rootstock. Stems are cylindrical, glabrescent to pubescent with appressed hairs, and no tendency to develop adventitious roots. Leaves are trifoliolate, with leaflets ovate-lanceolate, up to 7.5 cm long and 4 cm across; they are glabrous to pubescent, slightly glossy on the upper surface, paler and matt below, with stipules to 5 mm long. The inflorescence is an axillary raceme, comprising 2—4 (sometimes up to 10) whitish to greenish-yellow papilionate flowers, standard oblong-elliptical, 1—2.4 cm long and 0.6—1.5 cm across. The pod is linear oblong, shortly stipulate, laterally flattened, 3—8 cm long and 5—8 mm broad, glabrous to pubescent, with a terminal point up to 7 mm long. It contains between five and nine seeds, but usually seven or eight. The seeds are subovoid, 3—4 mm long, 2.5—3 mm broad, hard and smooth, buff to reddish brown, with sparse to dense black mottling; there are 50,000—200,000 seeds/ kg (Cook *et al.*, 2005).

FIGURE 106.1
M. axillare **plant and seeds.** Note the trailing pattern of growth. *Photograph from Tropical Forages website (http://www.tropicalforages.info/key/Forages/Media/html/Macrotyloma_axillare.htm).*

HISTORICAL CULTIVATION AND USAGE

As a native plant in Africa, *M. axillare* has been employed in a variety of veterinary applications. Apart from its main use as forage, *M. axillare*, among other leguminous plants, is considered to display antimicrobial, antifungal, and antihelminthic properties (McGaw & Eloff, 2008). In addition, *M. axillare* has been included in a list of 33 medicinal plants that are utilized by men for managing sexual impotence and erectile dysfunction in western Uganda (Kamatenesi-Mugisha & Oryem-Origa, 2005). More recently, Morris (2008) analyzed the anthocyanin content of *M. axillare* and *M. uniflorum* and detected the presence of D-pinitol, a carbohydrate that was shown to reduce postprandial glycemic levels in type 2 diabetes patients. This finding places *M. axillare* in a category of leguminous plants from which novel and potential biomolecules with pharmacological interest can be isolated. However, despite its potential use, *M. axillare* remains poorly investigated by the scientific community. In this chapter we report on the isolation, structural characterization, and biological activities of two constituents of *M. axillare* seeds that have constituted the subject of our investigation over the past few years: the lectin, and the Bowman-Birk type inhibitors (Figure 106.2).

PRESENT-DAY CULTIVATION AND USAGE

Notwithstanding the biotechnological potential of its seeds, the main use of *Macrotyloma axillare* remains as forage for cattle. In addition, its use in the recovery of eroded soils has been appreciated.

APPLICATIONS TO HEALTH PROMOTION AND DISEASE PREVENTION

Biotechnology applications of the protein seed content of *Macrotyloma axillare* seed lectin (MaL)

901

The availability of a great number of lectins displaying distinct glycidic specificities has resulted in the utilization of such molecules as useful tools in medical and biological research. Lectins

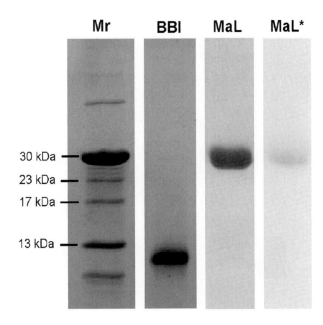

FIGURE 106.2
1D gel separation of two bio-products isolated from *M. axillare*. BBI, Bowman-Birk inhibitors at approximately 8 kDa after reduction and alkylation of cysteine residues; MaL, *M. axillare* lectin at approximately 28 kDa (monomer form). BBI and MaL are stained in Coomassie. MaL*, *M. axillare* lectin stained using the PAS method to reveal its carbohydrate content.

can be used to explore the cell surface through its ability to recognize and bind the glycidic moieties of glycoproteins and glycolipids of the cell glycocalix. Alternatively, immobilization of lectins onto inert matrices offers great potential for affinity chromatography and isolation of glycoproteins, an important primary step towards the characterization of the glycoproteome. Due to their inherent functions, these versatile molecules are also used for blood-typing, are employed as mitotic agents (e.g, lectins such as concanavalin A stimulate mitosis in certain types of lymphocytes), and can act as probes to detect the alterations associated with cell development and transformation. Nowadays, lectin derivatives are also commercially available. These include the covalent attachment of fluorescein isothyocianide, biotin, or radioactive isotopes to lectins to facilitate their detection whilst preserving their biological activity (Kennedy *et al.*, 1995).

Lectins from leguminous plants are used as model systems to study the molecular events associated with the interaction between a protein and a specific carbohydrate in biological research. This choice is based on the relative ease in purifying and isolating significant amounts of lectins from seed extracts. Once isolated, the lectin specificity can be evaluated by biophysical techniques such as X-ray crystallography, nuclear magnetic resonance, and microcalorimetry. Although a great variety of glycidic specificities have been observed for the isolates, a remarkable conservation of amino acid sequence is reported among leguminous lectins (Loris *et al.*, 1998).

The lectin isolated from *M. axillare* (MaL) seeds, and its counterpart, purified from *Dolichos biflorus* (DBL) seeds, are specific to the carbohydrate N-acetyl-α-D-galactosamine (GalNAc). This finding offers biotechnological relevance, given that this sugar is an antigenic determinant of blood type A from the ABO system. Routinely, DBL is used to differentiate subgroups A_1 from A_2, the difference residing in the number of antigens on the erythrocyte surface (greater in A_1 compared to A_2) (Santana, 2008). Subgroup A_1 is then identified by positive hemagglutination, and the test is negative for subgroup A_2 under the use of a standard specific activity (Figure 106.3). Considering that DBL and MaL are closely related to each other and share the same sugar specificity (Haylett & Swart, 1982), methodologies to obtain the lectin isolated from *M. axillare* can be useful, as this leguminous plant can be cultivated and obtained in vast amounts in tropical countries.

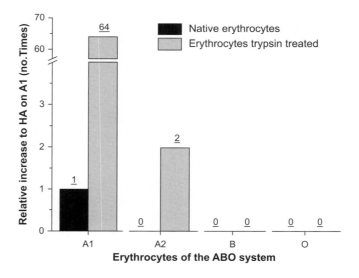

FIGURE 106.3
Macrotyloma axillare seed Lectin (MaL) specificity on the ABO system. MaL specificity study on native and trypsin-treated erythrocytes for the human ABO antigen system. *Reproduced from Santana* et al. *(2008), Intl J. Biol. Macromol., 43, 352–358, with permission.*

MaL was first isolated by Haylett and Swart (1982) by affinity chromatography, essentially as proposed for the isolation of DBL. The protocol revealed a good protein yield; however, the method used for elution involved the utilization of a low pH buffer as an alternative to the use of GalNAc, as initially described by Etzler and Kabat (1970) during isolation of DBL. Further studies from our research group, at the Federal University of Ouro Preto, have revealed that a low pH can interfere with MaL activity as a tretamer form, which led us to propose a new protocol for the isolation of MaL. The procedure involved the utilization of fewer and simple steps resulting in high lectin yields and maximum retention of biological activity, and therefore suitable for scaling up for industrial production (Santana *et al.*, 2008). The detailed technique was patented, and is now deposited under the accession BR200401517-A (Guerra de Andrade *et al.*, 2004).

The Bowman-Birk inhibitors from *M. axillare*

Bowman-Birk type inhibitors, commonly called BBI, are cytoplasmic protein inhibitors of serine proteases. BBI were first isolated from soybean seeds by Bowman in 1946, and then characterized by Birk in 1961. Since then, BBI from other leguminous plants have been isolated and characterized (Norioka & Ikenaka, 1983). In general, BBI are low molecular mass proteins, of approximately 8 kDa, consisting of a single amino acid chain exhibiting a significant content of cysteine residues which confer a rigid tridimensional folding highly conserved among the different isolates of the Plantae Kingdom, to which they are exclusive. The main enzymes inhibited by BBI are trypsin and chymotrypsin. BBI are classified as bivalent inhibitors, as their tridimensional folding defines two inhibition loops or "heads," one being specific for the binding of trypsin and the other for the binding of chymotrypsin (Prakash *et al.*, 1996). Other enzymes reported to be inhibited by BBI are cathepsin G, elastase, and chymase. Crystallographic analyses have shown that the tridimensional structure of this protein is mainly stabilized by seven disulfide bridges, followed by a number of intramolecular hydrogen bondings and a contribution of the hydrophobic core of the molecule (Chen *et al.*, 1992). As a result, BBI are highly resistant molecules supporting extremes of temperature, ionic strength, and pH, being soluble in the pH range 1.5–12.

Serine protease inhibitors react with the protease active site at the serine residue, forming derivatives of inactivated enzymes. This class of proteases is inhibited by *Kunitz* type inhibitors, BBI, the potato inhibitors type I and II, and the *kazal* type inhibitors (Losso, 2008). In this context, BBI isolated from *M. axillare* merit attention due to their highly specific inhibitory activity, especially considering BBI isolated from germinated seeds (Cesar, 2009). The interaction of BBI with either trypsin or chymotrypsin is able to strongly inhibit the activity of these proteases; in addition, simultaneous binding of these two proteases with a single BBI monomer has been reported. The conformational loop of the BBI reactive site is fully complementary to the active site of the enzyme to be inhibited, allowing for a strong protein–protein interaction to take place (Chen *et al.*, 1992). Although this interaction is reversible, it occurs with an affinity compared to that observed for the formation of the complex resulting from the interaction between the protease and its protein substrate. The perfect docking of the inhibitor at the enzyme active site prevents necessary conformational changes to occur, resulting in an unfavorable energetic barrier to hydrolysis.

BBI have been the subject of scientific research involving protease inhibition, mainly due to its antichymotrypsin activity. The rationale relies on the observation that the inhibition of a cell's chymotrypsin activity is linked to an anticarcinogenic effect. BBI have also been shown to possess anti-inflammatory and radioprotective properties, also being able to inhibit the production of free radicals and some carcinogen-induced transformations. Although a wide spectrum of anticancer activities has been reported, the exact mechanisms these inhibitors utilize to promote anticarcinogenic effects remain to be elucidated (Billings *et al.*, 1990; Kennedy, 1994; Losso, 2008).

More recently, the activity of BBI has been shown to extend beyond their classical chymotrypsin/trypsin targets. In this regard, Chen *et al.* (2005), using BBI isolated from soybeans, demonstrated that these molecules also inhibited the chymotrypsin-like acitivity of the 26S proteasome, a major nanomachine involved in intracellular proteolysis. *In vitro* and *in vivo* experiments confirmed the specific inhibition of the proteasome's chymotrypsin-like activity, using MCF7 cells isolated from breast cancer tissue. It was also verified that such inhibition is linked to an accumulation of ubiquitylated proteins, the natural proteasome substrates, and a reduction in the levels of regulatory cyclins involved in cell division. Altogether, the authors suggested that inhibition of the 26S proteasome activity by BBI could greatly contribute to their preventive effect on cancer.

Further evidence of the interaction of BBI and the proteasome has been reported by Saito *et al.* (2007), who also showed inhibition of the chymotryspin-like activity in osteosarcoma cells. More specifically, BBI inhibited the degradation of connexin 43 by the ubiquitin-proteasome system in these tumor cells. It was suggested that the antiproliferative effect observed was due to maintenance of connexin function through homeostatic balance promoted by GAP junctions.

M. axillare seed germination and BBI

The practice of seed germination and consumption as a health-promoting activity has been observed in different cultures. In this context, it was verified that germinated *M. axillare* seeds exhibit a significant increase in the activity of BBI, reaching up to a four-fold increase when compared to the activity of the inhibitors isolated from non-germinated seeds (Cesar, 2009). The activation process requires hydrolysis and reduction of the molecular mass of BBI, with the monomer found on 5-day germinated seeds being approximately 6 kDa, as opposed to the 8-kDa monomers found on dormant seeds (Figure 106.4). This reduction in size produces a notable effect on the pharmacokinetic properties of these inhibitors when they are administered to mice. The most evident effect observed was a significant increase in their distribution volume, allowing for a better diffusion of BBI throughout the body (Table 106.1).

The higher distribution volume and the increased activity of BBI present in the cotyledon strengthen the possibility of a better efficacy of these inhibitors on cancer prevention. The pharmacokinetic data of tissue distribution revealed a marked accumulation of BBI isolated

904

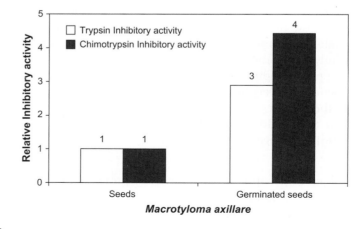

FIGURE 106.4

Inhibitory activity of BBI from *M. axillare*. Relative inhibitory activity of BBI isolated from *M. axillare* dormant seeds compared to those isolated from 5-day germinated seeds. Up to a four-fold increase is observed for antichymotrypsin activity.

TABLE 106.1 Pharmacokinetic Parameters of BBI Isolated from *M. axillare* Seeds and Labeled with Radioactive Isotope (I^{125}) Determined after Oral Administration to *Swiss* Mice[3].

Pharmacokinetic parameter	Seeds	Cotyledons
α (distribution constant)	0.0405/min	0.0938/min
β (elimination constant)	0.0008/min	0.0016/min
VD (distribution volume)	4.76 ml	5.41 ml
$t_{1/2}$ (plasma half life)	14.4 hours	7.2 hours

Andrade, M.H.G. and Santos, A.G. (2007), Patent number: BRPI0601377 (A), available at http://v3.espacenet.com/publicationDetails/originalDocument?CC=BR&NR=PI0601377A&KC=A&FT=D&date=20071127&DB=EPODOC&locale=en_EP.

from *M. axillare* in the gastric tissue when compared to other organs when BBI are administered orally (Figure 106.5). This constitutes a striking feature of *M. axillare* BBI when compared with those isolated from soybeans (*Glycine max*).

ADVERSE EFFECTS AND REACTIONS (ALLERGIES AND TOXICITY)

To our knowledge, no specific study on the toxicity and adverse effects promoted by *M. axillare* seeds has been reported to date. However, the two aforementioned constituents isolated from this plant, when administered to animals, particularly at high doses, are expected to provoke adverse effects due to their intrinsic biochemical function. First, the lectin, by binding to specific cell receptors, might interfere with normal cell physiology, and in this context there are a number of reports demonstrating the harmful effects of lectins, especially in the gastrointestinal tract of recipients (Vasconcelos *et al.*, 2004; Losso, 2008). In parallel, purified BBI are expected to produce some degree of antinutritional effect by inhibiting the enzymes trypsin and chymotrypsin, as they play a key role in protein digestion and therefore amino acid utilization. Nevertheless, it is worth emphasizing that BBI and lectin molecules are consumed daily as constituents of a variety of food and grains belonging to the animal and human diet. In this scenario, new studies are needed to fully evaluate the therapeutic potential of isolated BBI and lectin from *M. axillare*, aiming at a better understanding of their benefits and risks to human and animal health.

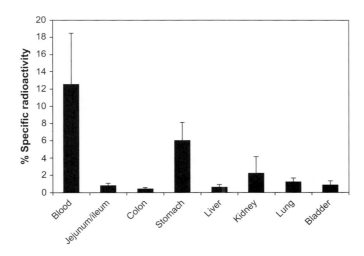

FIGURE 106.5

Biodistribution of I^{125}-labeled BBI in mice. Specific radioactivity of I^{125}-labeled BBI after oral administration to mice. The data are represented by average $\pm$ one standard deviation ($n = 6$). Note that apart from the blood compartment, the stomach is the preferential site of accumulation of BBI isolated from *M. axillare* seeds. *Andrade, M.H.G. and Santos, A.G. (2007), Patent number: BRPI0601377 (A), available at http://v3.espacenet.com/publicationDetails/originalDocument?CC=BR&NR=PI0601377A&KC=A&FT=D&date=20071127&DB=EPODOC&locale=en_EP.*

SUMMARY POINTS

- *M. axillare* is a leguminous plant widely found in tropical areas of the world.
- Presently, the main uses of *M. axillare* include forage for cattle, recovery of eroded soil, and as a primary source of biomolecules from its seeds, such as D-pinitol, the Anti-A_1 lectin (MaL), and Bowman-Birk inhibitors.
- The 28-kDa lectin monomer isolated from *M. axillare* (MaL) is specific to the carbohydrate N-acetyl-α-D-galactosamine (GalNAc, Anti-A_1 antigen).
- MaL constitutes a relevant clinical tool, as it allows discrimination between blood groups A_1 and A_2 of the ABO system.
- *M. axillare* seeds are a source of the classic known trypsin and chymotrypsin plant inhibitors called Bowman-Birk inhibitors.
- Pharmacokinetic data obtained after oral administration of Bowman-Birk inhibitors isolated from *M. axillare* revealed considerable accumulation of the inhibitors in the stomach.
- Bowman-Birk inhibitors isolated from 5-day germinated seeds of *M. axillare* display increased inhibitory activity over trypsin and chymotrypsin.
- A number of reports have demonstrated the potential of Bowman-Birk inhibitors as cancer-preventive agents.

References

Billings, P. C., Newberne, P. M., & Kennedy, A. R. (1990). Protease inhibitor suppression of colon and anal gland carcinogenesis induced by dimethyl-hidrazine. *Carcinogenesis, 11*, 1083−1086.

Cesar, J. J., Santana, M. A., Oliveira, M. E., Santos, A. G., Miranda, A. A. C., Santos, A. M. C., et al. (2009). A new extraction and purification methodology of Bowman-Birk inhibitors from seeds and germinated seeds of *Macrotyloma axillare*. *Chromatographia, 69*, 357−360.

Chen, P., Rose, J., Love, R., Wei. C. H., & Wang, B. C. (1992). Reactive sites of an anticarcenogenic Bowman-Birk Protease inhibitor are similar to other trypsin inhibitions. *Journal of Molecular Biology, 267*, 1990−1994.

Chen, Y. W., Huang, S. C., Lind-Shiau, S. Y., & Lin, J. K. (2005). Bowman-Birk inhibitors abates proteasome function and suppresses the proliferation of MCF7 breast cancer cells through accumulation of MAP kinase phosphate-1. *Carcinogenisis, 26*, 1296−1306.

Cook, B. G., Pengelly, B. C., Brown, S. D., Donnelly, J. L., Eagles, D. A., Franco, M. A., et al. (2005). Tropical forages: an interactive selection tool *[CD-ROM]*. Brisbane, Australia: CSIRO, DPI&F(Qld), CIAT and ILRI.

Etzler, M. E., & Kabat, E. A. (1970). Purification and characterization of a lectin (plant hemagglutinin) with blood group A specificity from *Dolichos biflorus*. *Biochemistry, 9*, 869−877.

Guerra de Andrade, M. H., Baba, E. H., & Santana, M. A. (2004). Purification and isolation of lectin of *Macrotyloma axillare* by obtaining lectin of *Macrotyloma axillare* through isolation of lectin by thermally treating brute extract of leguminous plant seeds. Universidade Federal de Ouro Preto, Brazil, Patent number BR200401517-A.

Haylett, T., & Swart, L. S. (1982). Isolation and characterization of an Anti-A1 lectin from *Macrotyloma axillare*. *South African Journal of Chemistry, 35*, 33−36.

Kamatenesi-Mugisha, M., & Oryem-Origa, H. (2005). Traditional herbal remedies used in the management of sexual impotence and erectile dysfunction in western Uganda. *African Health Sciences, 5*(1), 40−49.

Kennedy, A. R. (1994). Prevention of carcinogenesis by protease inhibitors. *Cancer Research, 54*, 1999−2005.

Kennedy, J. F., Paiva, P. M. G., Correia, M. T. S., Cavalcanti, M. S. M., & Coelho, C. B. B. (1995). Lectins, versatile proteins of recognition − a review. *Carbohydrate Polymers, 26*, 219−230.

Loris, R., Hamelryck, T., Bouckaert, J., & Wyns, L. (1998). Legume lectin structure. *Biochimica Biophysica Acta, 1383*, 9−36.

Losso, J. N. (2008). The biochemical and functional food properties of the Bowman-Birk Inhibitor. *Critical Reviews in Food Science and Nutrition, 48*, 94−118.

McGaw, L. J., & Eloff, J. N. (2008). Ethnoveterinary use of southern African plants and scientific evaluation of their medicinal properties. *Journal of Ethnopharmacology, 119*, 559−574.

Morris, J. B. (2008). *Macrotyloma axillare* and *M. uniflorum*: descriptor analysis, anthocyanin indexes, and potential uses. *Genetic Resources and Crop Evolution, 55*, 5−8.

Norioka, S., & Ikenaka, T. J. (1983). Aminoacid sequences of trypsin-chymotrypsin inhibitors (A-I, A-II, B-I and B-II) from peanut (*Arachis hypogaea*): a discussion on the molecular evolution of legume Bowman-Birk type inhibitors. *Journal of Biochemistry, 94*, 589−599.

Prakash, B., Selvaraj, S., Murthy, M. R. N., Sreerama, Y. N., Rao, D. R., & Gowda, L. R. (1996). Analysis of the aminoacid sequences of plant Bowman-Birk inhibitors. *Journal of Molecular Evolution, 42*, 560−569.

Saito, T., Sato, H., Virgona, N., Hagiwara, H., Kashiwagi, K., Suzuki., K., et al. (2007). Negative growth control of osteosarcoma cell by Bowman-Birk protease inhibitor from soybean; involvement of connexin 43. *Cancer Letters, 253*, 249–257.

Santana, M. A., Santos, A. M. C., Oliveira, M. E., Oliveira, J. S., Baba, E. H., Santoro, M. M., et al. (2008). A novel and efficient and low-cost methodology for purification of *Macrotyloma axillare* (Leguminosae) seed lectin. *International Journal of Biological Macromolecules, 43*, 352–358.

Vasconcelos, I. M., & Oliveira, J. T. (2004). Antinutritional properties of plant lectins. *Toxicon, 44*, 385–403.

CHAPTER 107

Compounds with Antioxidant Properties in Pistachio (*Pistacia vera* L.) Seeds

Marcello Saitta, Daniele Giuffrida, Giuseppa Di Bella, Giovanna Loredana La Torre, Giacomo Dugo
Dipartimento di Scienze degli Alimenti e dell'Ambiente "G. Stagno d'Alcontres", Università di Messina, Messina, Italy

CHAPTER OUTLINE

LIST OF ABBREVIATIONS

AOP, antioxidant potential
CoA , coenzime A
DNA, deoxyribonucleic acid
GSH, glutathione
HDL, high density lipoprotein
LDL, low density lipoprotein
MDA, malondialdehyde
RNS, reactive nitrogen species
ROS, reactive oxygen species

Nuts & Seeds in Health and Disease Prevention. DOI: 10.1016/B978-0-12-375688-6.10107-0

SOD, superoxide dismutase
TAA, total antioxidant activity
TBARS, thiobarbituric acid-reactive substances
TEAC, Trolox equivalent antioxidative capacity

INTRODUCTION

Oxidative stress plays a role in the pathology of cancer, arteriosclerosis, neurodegenerative diseases, and aging processes. Fruits and vegetables provide protection against several diseases, such protection being attributed to antioxidants. In living systems, dietary antioxidants and endogenous enzymes protect against oxidative damage. Phenols are able to reduce *in vitro* oxidation of LDL; scientific studies have proved that antioxidants protect cells from free radical damage. The use of antioxidants as chemopreventive agents by inhibiting radical generation has been suggested, since free radicals are responsible for DNA damage and scavengers are probably important in cancer prevention (Moure *et al.*, 2001). We herein report on the antioxidants found in pistachio seeds and their benefits to human health.

BOTANICAL DESCRIPTION

Pistacia vera L. is a diploid member of the order Sapindales, family Anacardiaceae. This plant is a tree that grows up to 10 m tall, with a single or several trunks. Leaves are deciduous, compound-pinnate, 10—20 cm long, with three to five oval leaflets. Pistachio is dioecious, with staminate and pistillate inflorescences; both types of flowers are apetalous, and wind is the pollinating agent. Mature pistillate flowers consist of two to five tepals and a pistil with three stigmas. Although commercial pistachios are known as nuts, the pistachio fruit is actually a drupe with an exocarp, a mesocarp (hull), and a hard, dehiscent endocarp (shell) that splits longitudinally when the fruit has ripened. Commercial pistachios comprise the shell and the edible kernel, which has a papery seed coat (skin), the color of which ranges from yellow to green (Hormaza & Wunsch, 2007).

HISTORICAL CULTIVATION AND USAGE

The pistachio probably originated in central and southwestern Asia. Local human populations used the wild trees as a source of fuel, and for heavy pasturing of cattle. The presence of pistachio nuts in archeological excavations provides evidence that the pistachio has long been associated with human activities. Pistachio cultivation is very ancient; in fact, remnants of pistachio nuts dating from 6 BC have been found in Afghanistan and Iran. From its presumed center of origin, pistachio cultivation was extended within the ancient Persian Empire, from where it gradually expanded westward. The name "pistachio" probably derives from the word *pistak* in the ancient Persian language, Avestan. Pistachio nuts are mentioned in the Bible as precious gifts carried from Canaan to Egypt by the sons of Jacob. Pistachio cultivation spread into Mediterranean countries between 1 BC and 1 AD, with the advent of the Roman empire (Hormaza & Wunsch, 2007).

PRESENT-DAY CULTIVATION AND USAGE

Trees are planted in orchards, and take approximately 7—10 years to reach significant production. Production is alternate-bearing or biennial-bearing, meaning that the harvest is heavier in alternate years. Peak production is reached at approximately 20 years. Trees are usually pruned to a size to make harvesting easier. One male tree produces enough pollen for 8—12 nut-bearing females. The kernels are often consumed as a snack food and eaten whole, either fresh or roasted and salted, and are also used in the food industry as an ingredient in ice cream, pastries, fermented meats, puddings, and in confections such as baklava and cold cuts such as mortadella (Hormaza & Wunsch, 2007).

APPLICATIONS TO HEALTH PROMOTION AND DISEASE PREVENTION

Antioxidants: classification and activity evaluation

During lipid oxidation, antioxidants with various functions act in different ways, binding metal ions, scavenging and quenching radicals, and decomposing peroxides. Often many mechanisms are involved, and this can cause synergistic effects. Antioxidants either prevent reactive oxygen species (ROS) and reactive nitrogen species (RNS) from being formed, or remove them before they can damage vital components of the cell. ROS include singlet oxygen, superoxide anions, hydroxyl radicals, and hydrogen peroxide; RNS include nitric oxide and peroxynitrite. Antioxidant classification depends on whether they are soluble in water (hydrophilic) or in lipids (hydrophobic, lipophilic). Another classification considers primary (chain-breaking) and secondary (reducing the rate of chain initiation) antioxidants; some compounds possess both activities. Antioxidant efficiency is determined by several factors (intrinsic chemical reactivity toward the radicals, site of radical generation, fate of antioxidant derived radicals, concentration and mobility of the antioxidant, biological absorption, and interactions with other antioxidants). The chemical reactivity towards the radicals is very important, because the antioxidants must scavenge the radicals before they can attack the target molecule. The site of radical generation may be the aqueous phase or the lipophilic domain of membranes and lipoproteins; in the first case hydrophilic antioxidants can act more efficiently, while in the second the lipophilic ones are preferred. The antioxidant activity must be evaluated with different tests for different mechanisms. Frequently used methods for measuring the oxidative damage evaluate total oxidative DNA damage, levels of antioxidant enzymes, levels of low molecular weight antioxidants (catalase, superoxide dismutase, glutathione peroxidases, uric acid, glutathione, flavonoids, catechins, anthocyanins, vitamin C, β-carotene, and vitamin E), oxidative damage to lipids (isoprostanes, thiobarbituric acid-reactive substances (TBARS)), and protein damage (numbers of protein carbonyl and modified tyrosine residues). Most of the chemical methods are based on the ability to scavenge different free radicals, but UV absorption and chelation ability are also responsible for the antioxidant activity in oily systems. The redox potential, reducing power, and degradation rate of the antioxidant substance have also been positively correlated with antioxidant activity (Moure *et al.*, 2001).

Antioxidants in pistachio seeds

There has been limited research on the composition of pistachio seeds over the past 30 years; Table 107.1 summarizes the literature data regarding the antioxidants found.

An anthocyanin, cyanidin-3-galactoside, was found in pistachio kernels by Miniati (1981). Cyanidin-3-galactoside and cyanidin-3-glucoside (Figure 107.1) were found by other authors in kernels and skins (Wu & Prior, 2005; Seeram *et al.*, 2006; Bellomo & Fallico, 2007). In particular, Seeram *et al.* (2006) showed that these anthocyanins are exclusively present in the skins; the authors found values of 696 mg/kg for cyanidin-3-galactoside, 209 mg/kg for cyanidin-3-glucoside in raw nuts, and 462 and 87 mg/kg, respectively, in roasted nuts. Anthocyanin levels decreased when the nuts were subjected to bleaching using hydrogen peroxide. The antioxidant activity was evaluated with the TEAC method, and the results showed that the oxygen-radical absorbing capacity of the nuts was correlated to the anthocyanin concentration. Bellomo and Fallico (2007) found a correlation between cyanidin-3-galactoside concentration in skins and ripeness: values were very low or under the detection limit in unripe pistachios, reached 107–145 mg/kg in intermediate samples, and finally were 287–426 mg/kg in ripe pistachios. Furthermore, Wu and Prior (2005) affirmed that pistachios are the only tree nut that contains anthocyanins.

Yalpani and Tyman (1983) investigated the presence of phenolic acids in the outer green shell of the pistachio nut. They found about 1.5% of anacardic acids (6-alkylsalicylic acids with 13, 15, and 17 carbon atoms in the alkyl chains, saturated and monounsaturated) in dried shells

TABLE 107.1 Antioxidant Compounds Found in Pistachio Seeds (Literature Data 1981–2009)

Compounds	Seed Part	References
Cyanidin-3-galactoside	Kernel	Miniati (1981)
Various anacardic acids (6-alkylsalicylic acids)	Hull	Yalpani & Tyman (1983)
Resveratrol (3,5,4′-trihydroxystilbene)	Kernel	Tokusoglu et al. (2005)
Anthocyanins (cyanidin-3-galactoside, cyanidin-3-glucoside)	Kernel	Wu & Prior (2005)
Lutein, violaxanthin, neoxanthin, luteoxanthin, β-carotene, chlorophylls, pheophytins	Kernel	Giuffrida et al. (2006)
Lutein, β-carotene, β-tocopherol, γ-tocopherol, δ-tocopherol	Kernel	Kornsteiner et al. (2006)
Rutin, eriodictyol, quercetin, luteolin, naringenin, apigenin, cianidyn-3-galactoside, cyanidin-3-glucoside	Skin and kernel	Seeram et al. (2006)
Cyanidin-3-galactoside, cyanidin-3-glucoside, chlorophylls, lutein, β-carotene	Skin and kernel	Bellomo & Fallico (2007)
Vitamin C, α-tocopherol, γ-tocopherol, trans-resveratrol, proanthocyanidins, daidzein, genistein	Kernel	Gentile et al. (2007)
Trans-resveratrol, trans-resveratrol-3-O-β-glucoside	Kernel	Grippi et al. (2008)
Various cardanols (3-alkylphenols)	Kernel	Saitta et al. (2009)

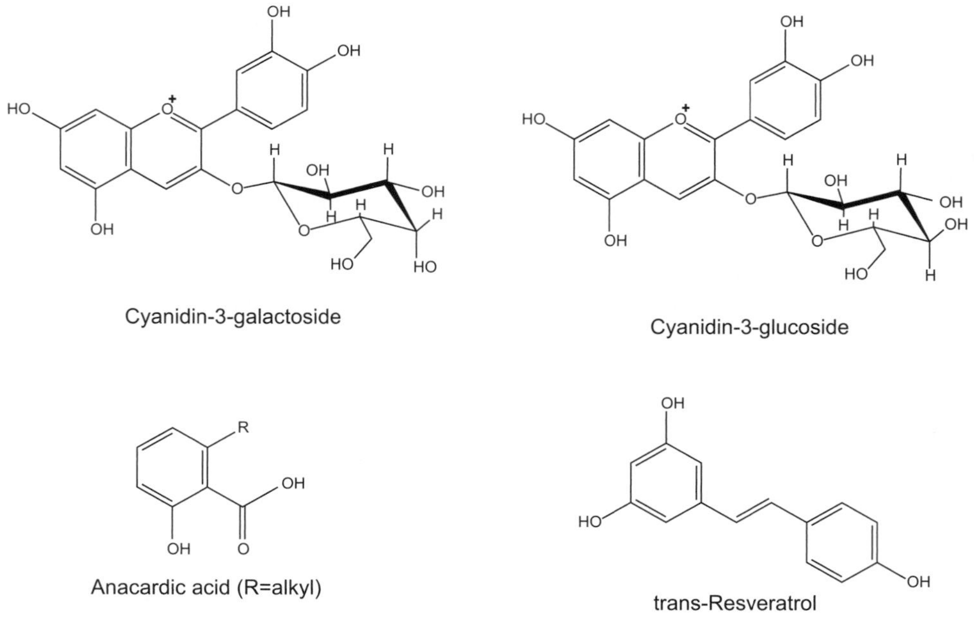

912

Cyanidin-3-galactoside

Cyanidin-3-glucoside

Anacardic acid (R=alkyl)

trans-Resveratrol

FIGURE 107.1

Structures of antioxidants found in pistachio seeds: anthocyanins, anacardic acids, resveratrol.

(Figure 107.1). Goli *et al.* (2005) evaluated the total phenolic content of the pistachio hull in 32—34 mg/g (dry weight) of the samples, and suggested the use of the extracts as alternative natural antioxidants. The detection of the products of ROS was used in the hypoxanthine/xanthine oxidase assay to evaluate the antioxidant properties of various anacardic acids (Trevisan *et al.*, 2006); authors found that these compounds are more potent antioxidants than hydroxytyrosol or caffeic acid.

Resveratrol, *trans*-3,5,4′-trihydroxystilbene (Figure 107.1), was found by some authors in pistachio kernels (Tokusoglu *et al.*, 2005; Gentile *et al.*, 2007; Grippi *et al.*, 2008). This compound has been associated with a reduction in cardiovascular disease by inhibiting or altering platelet aggregation and coagulation, or modulating lipoprotein metabolism (Grippi *et al.*, 2008). The resveratrol contents found by Tokusoglu *et al.* (2005) were higher (0.09—1.67 mg/kg, mean value 1.15 mg/kg) than those reported by Gentile *et al.* (2007) and Grippi *et al.* (2008); both found a mean value of 0.12 mg/kg, but the presence of the glycosidic derivative *trans*-piceid (*trans*-resveratrol-3-O-β-glucoside) was reported in higher concentrations with respect to free resveratrol, with a mean value of 6.97 mg/kg (Grippi *et al.*, 2008).

Lutein, violaxanthin, neoxanthin, luteoxanthin, and β-carotene (Figure 107.2) were found in pistachio kernels by Giuffrida *et al.* (2006); other authors reported only the presence of lutein and β-carotene (Kornsteiner *et al.*, 2006; Bellomo & Fallico, 2007). It is known that the potential health benefits of carotenoid-rich diets are due to their role (antioxidants and agents preventing cardiovascular diseases and degenerative eye pathologies) (Giuffrida *et al.*, 2006). Carotenoids are correlated to pistachio ripeness, and the highest lutein concentrations were found in unripe samples, 41.3—52.1 mg/kg, while intermediate samples showed concentrations of 17.9—34.7 mg/kg, and ripe sample concentrations of 18.1—37.7 mg/kg (Bellomo & Fallico, 2007); the same authors found β-carotene contents below 1.8 mg/kg. Kornsteiner *et al.* (2006) found similar concentrations (lutein mean value 44 mg/kg, β-carotene mean value 4 mg/kg). Giuffrida *et al.* (2006) found higher concentrations of β-carotene (mean value 7.1 mg/kg) and intermediate levels of lutein (mean value 29.14 mg/kg); violaxanthin, neoxanthin, and luteoxanthin had mean values of 2.81, 3.04, and 2.75 mg/kg, respectively.

Chlorophylls and chlorophyll-derived compounds were identified in Sicilian pistachio kernels by Giuffrida *et al.* (2006). In this work, chlorophyll *a*, chlorophyll *b*, and pheophytin *a* were reported to be present at 54.14, 30.2, and 25.68 mg/kg, respectively. Different contents of chlorophyll *a* and *b* in pistachio kernels from various countries were reported by Bellomo and Fallico (2007); the chlorophyll *a* content range was 18.3—150.6 mg/kg and the chlorophyll *b* content range was 7.1—49.7 mg/kg . Chlorophylls and related molecules' chemical structures present a conjugated tetrapyrrole ring which allows them to absorb light, and this is directly involved in the color and oxidative stability of the food products that contain these pigments (Giuffrida *et al.*, 2006).

Three tocopherols (β, γ, and δ) were found in pistachio kernels by Kornsteiner *et al.* (2006), while α-tocopherol and γ-tocopherol (Figure 107.3) were detected by Gentile *et al.* (2007). Tocopherols are usually called vitamin E, although this name is more correctly assigned to α-tocopherol only (the best antioxidant in this class). Tocopherols react to break the lipid peroxidation that destroys both lipids and neighboring molecules such as proteins and nucleic acids. In pistachio, the sum of β- and γ-tocopherols was 293 mg/kg (mean value), and the δ-tocopherol mean level was 5 mg/kg (Kornsteiner *et al.*, 2006); however, different values were found in the samples analyzed by Gentile *et al.* (2007) — means of 0.51 mg/kg for α-tocopherol and 105.4 mg/kg for γ-tocopherol.

Some flavonoids (Figure 107.4) were found in pistachio kernels and skins by Seeram *et al.* (2006): two flavones (apigenin and luteolin), two flavanones (naringenin and eriodictyol), and one flavonol and its glycoside (quercetin and rutin). Authors showed that the flavonoids are mainly present in the skins; mean levels were 0.2 and 0.03 mg/kg (apigenin, skins, and

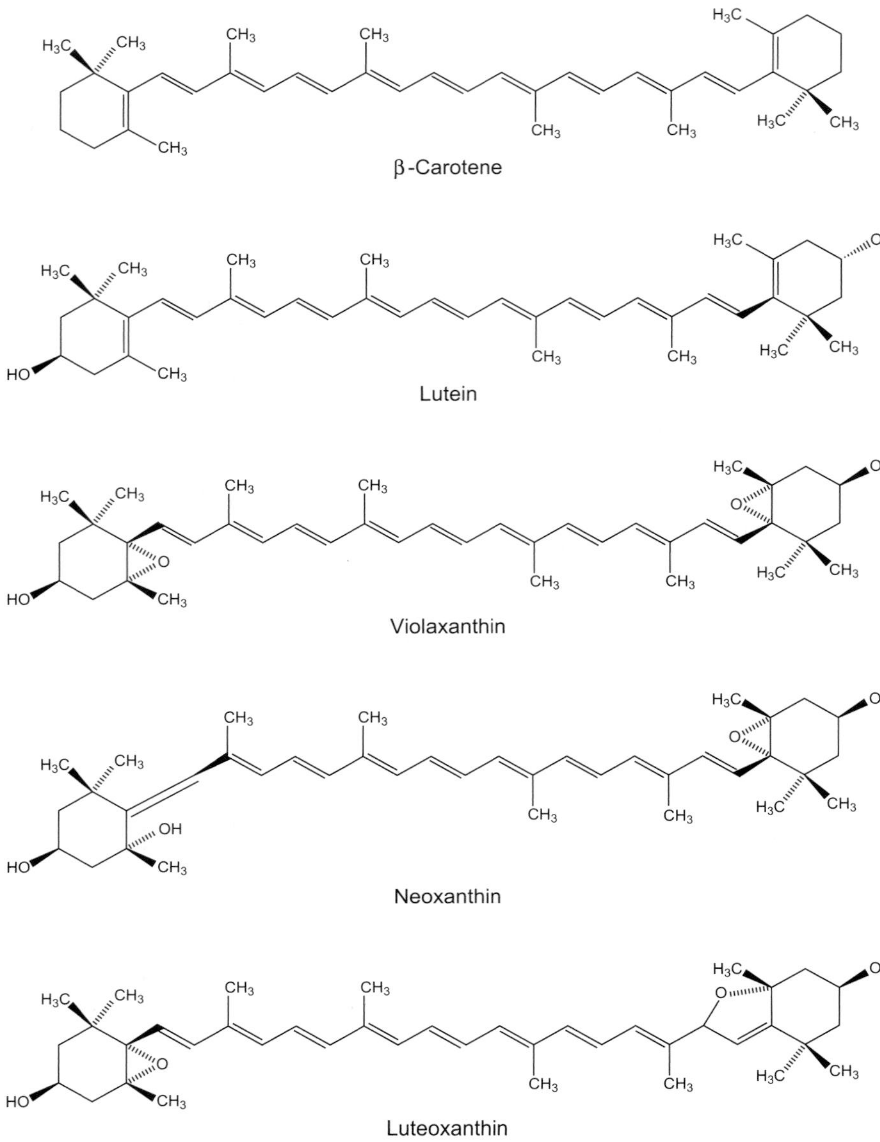

FIGURE 107.2
Structures of antioxidants found in pistachio seeds: carotenoids.

whole kernels), 10.0 and 1.04 mg/kg (luteolin, skins, and whole kernels), 1.2 and 0.12 mg/kg (naringenin, skins, and whole kernels), 10.2 and 1.1 mg/kg (eriodictyol, skins, and whole kernels), 14.9 and 0.29 mg/kg (quercetin, skins, and whole kernels), and 1.6 and 0.55 mg/kg (rutin, skins, and whole kernels). Flavonoids like quercetin, rutin, luteolin, and naringenin act as antioxidants and free radical scavengers, and naringenin can reduce oxidative damage to DNA *in vitro* (Moure *et al.*, 2001).

The isoflavones daidzein and genistein (Figure 107.5) were found in pistachio kernels by Gentile *et al.* (2007). Their amounts were 36.8 and 34.0 mg/kg, respectively; isoflavones are capable of interacting with estrogenic receptors. The same authors reported the presence of vitamin C (Figure 107.5, ascorbic acid; mean value 34.8 mg/kg) and proanthocyanidins (flavan-3-ol polymers) in pistachio kernels. Ascorbate can neutralize several radicals (such as hydroxyl, alcoxyl, peroxyl, glutathione) by hydrogen donation; proanthocyanidins can bind

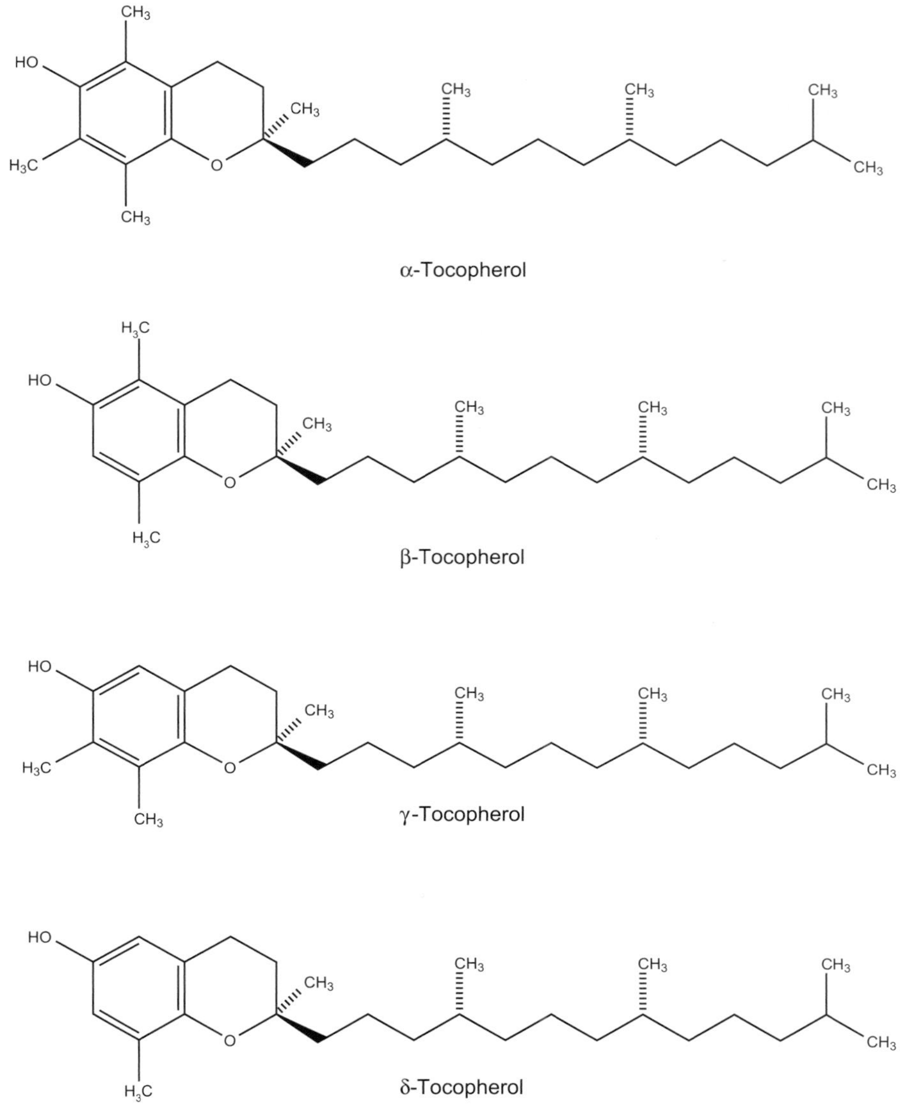

FIGURE 107.3
Structures of antioxidants found in pistachio seeds: tocopherols.

metals through complexation involving their *o*-diphenol groups and then inhibit the metal-induced lipid oxidation (Moure *et al.*, 2001).

Sixteen different 3-alkylphenols (cardanols, Figure 107.5), mainly monounsaturated, were found in pistachio kernels (Saitta *et al.*, 2009); the total amount of these compounds was evaluated at 213.4 mg/kg (mean value of the kernels). Cardanols are not such strong antioxidants as anacardic acids (Trevisan *et al.*, 2006).

Effects of the use of pistachio nuts on health

A study on the antioxidant potential of pistachio nuts showed that hydrophilic extracts inhibited both the metal-dependent and the metal-independent lipid oxidation of bovine liver microsomes, and also the Cu^{2+}-induced oxidation of human low density lipoproteins, suggesting that these inhibitions are due to a peroxyl radical-scavenging and chelating activity of nut components (Gentile *et al.*, 2007).

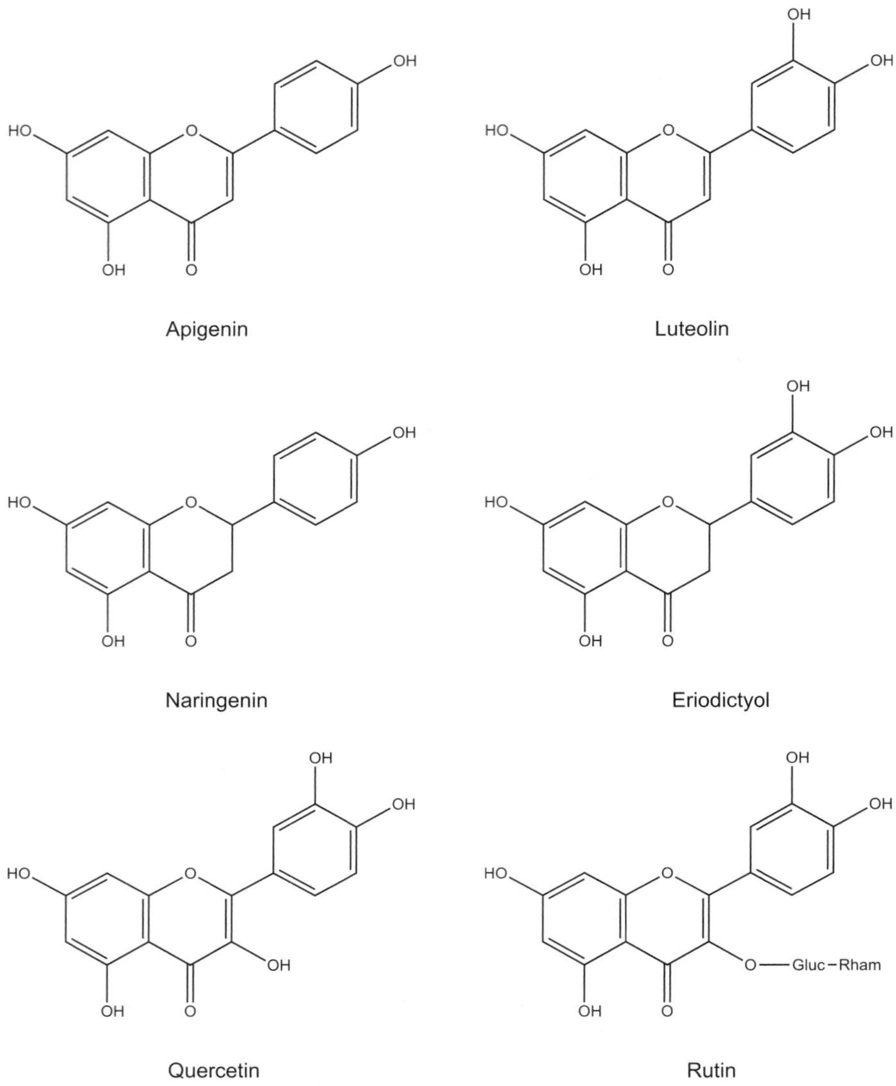

FIGURE 107.4
Structures of antioxidants found in pistachio seeds: flavonoids.

Effective studies of the benefits due to pistachio consumption have been conducted on humans and animals (Edwards *et al.*, 1999; Kocyigit *et al.*, 2006; Alturfan *et al.*, 2009). Edwards *et al.* (1999) evaluated a diet where 20% of the daily caloric intake was pistachio nuts in human patients with moderate hypercholesterolemia; after 3 weeks of this diet, the results showed a moderate decrease in the total cholesterol content, the total cholesterol/HDL ratio, and the LDL/HDL ratio, and an increase in HDL. Kocyigit *et al.* (2006) conducted a similar study on humans who consumed a regular diet for 1 week before being randomly divided into two groups; one group then used 20% of the daily caloric intake as pistachio nuts in their diet for 3 weeks. In these patients, the mean plasma total cholesterol, malondialdehyde levels, total cholesterol/HDL ratio, and LDL/HDL ratio significantly decreased, while HDL, AOP, and the AOP/MDA ratio significantly increased. Alturfan *et al.* (2009) evaluated the effects of pistachio consumption in rats fed a high-fat diet for 8 weeks. The rats were divided in two groups: the first group showed increases in total cholesterol, triglycerides, sialic acid, and TBARS, and a decrease in TAA, glutathione, and total thiol levels; the second group, subjected to pistachio consumption, showed significantly decreased levels of triglycerides and TBARS,

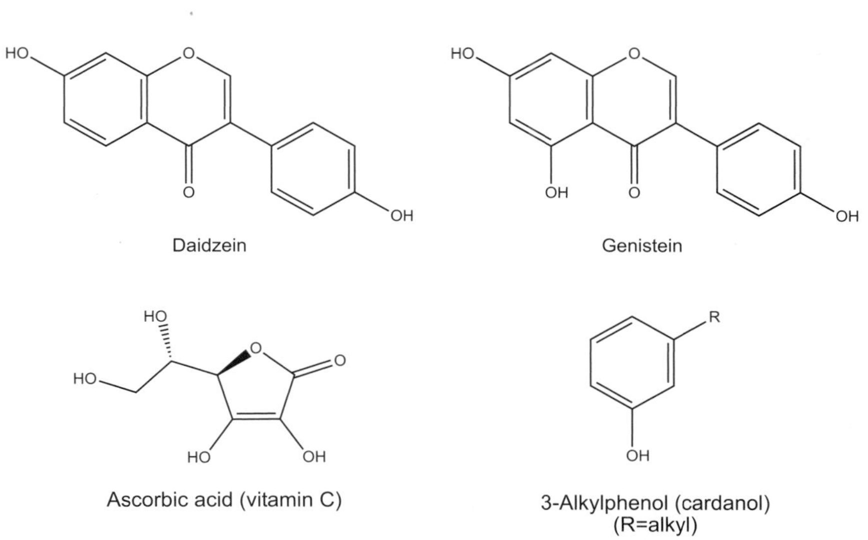

FIGURE 107.5
Structures of antioxidants found in pistachio seeds: isoflavones, vitamin C, cardanols.

and increased TAA, demonstrating an improvement in oxidative stress in experimental hyperlipidemia.

ADVERSE EFFECTS AND REACTIONS (ALLERGIES AND TOXICITY)

Various nuts and seeds have been reported to contain allergens and produce clinical symptoms of allergy. Pistachio nuts have been cited as causes of urticaria, angioedema, and even anaphylaxis (Perkin, 1990). Members of the *Anacardiaceae* family characteristically produce a series of compounds based on unsaturated fatty acid precursors. Biosynthetically, usually the CoA ester of palmitoleic acid serves as a precursor of this group of compounds, by chain elongation with three acetate groups; the side chain structure of active compounds can be mono-, di-, or triunsaturated, suggesting that a number of unsaturated acyl-CoA starter units can also be employed. Some pure compounds are very active human allergens, such as the urushiols (3-alkyl-pyrocatechols, biosynthetically derived by anacardic acids after a decarboxylation), which are present in other *Anacardiaceae* (Saitta *et al.*, 2009).

SUMMARY POINTS

- Dietary antioxidants act against oxidative damage, binding metal ions, scavenging radicals, and decomposing peroxides, so these compounds are able to protect cells from free radical damage.
- Different antioxidant classes are present in pistachio seeds: anthocyanins, tocopherols, carotenoids, chlorophylls, flavonoids, isoflavones, proanthocyanidins, anacardic acids, and cardanols, as well as resveratrol and vitamin C.
- These antioxidants can interact in very different ways: the synergistic effects can improve the protection against oxidative damage, and play an important role in the biological antioxidant network.
- Pistachio nut extracts inhibit both the metal-dependent and the metal-independent lipid oxidation of bovine liver microsomes, and also the Cu^{2+}-induced oxidation of human low density lipoproteins, suggesting these inhibitions are due to a peroxyl radical-scavenging and chelating activity of nut components.
- Studies of the benefits of pistachio consumption in humans show that the mean plasma total cholesterol and malondialdehyde levels, total cholesterol/high density lipoprotein ratio, and low density lipoprotein/high density lipoprotein ratio significantly decrease,

while high density lipoproteins, antioxidant potential, and the antioxidant potential/malondialdehyde ratio significantly increase.

References

Alturfan, A. A., Emekli-Alturfan, E., & Uslu, E. (2009). Consumption of pistachio nuts beneficially affected blood lipids and total antioxidant activity in rats fed a high-cholesterol diet. *Folia Biologica, 55,* 132–136.

Bellomo, M. G., & Fallico, B. (2007). Anthocyanins, chlorophylls and xanthophylls in pistachio nuts (*Pistacia vera*) of different geographic origin. *Journal of Food Composition and Analysis, 20,* 352–359.

Edwards, K., Kwaw, I., Matud, J., & Kurtz, I. (1999). Effect of pistachio nuts on serum lipid levels in patients with moderate hypercholesterolemia. *Journal of the American College of Nutrition, 18,* 229–232.

Gentile, C., Tesoriere, L., Bufera, D., Fazzari, M., Monastero, M., Allegra, M., et al. (2007). Antioxidant activity of Sicilian pistachio (*Pistacia vera* L. var. Bronte) nut extract and its bioactive components. *Journal of Agricultural and Food Chemistry, 55,* 643–648.

Giuffrida, D., Saitta, M., La Torre, G. L., Bombaci, L., & Dugo, G. (2006). Carotenoid, chlorophyll and chlorophyll-derived compounds in pistachio kernels (*Pistacia vera* L.) from Sicily. *Italian Journal of Food Science, 18,* 309–316.

Goli, A. H., Barzegar, M., & Sahari, M. A. (2005). Antioxidant activity and total phenolic compounds of pistachio (*Pistacia vera*) hull extracts. *Food Chemistry, 92,* 521–525.

Grippi, F., Crosta, L., Aiello, G., Tolomeo, M., Oliveti, F., Gebbia, N., et al. (2008). Determination of stilbenes in Sicilian pistachio by high-performance liquid chromatographic diode array (HPLC-DAD/FLD) and evaluation of eventually mycotoxin contamination. *Food Chemistry, 107,* 483–488.

Hormaza, J. I., & Wunsch, A. (2007). Pistachio. In C. Kole (Ed.), Genome mapping and molecular breeding in plants. *Vol. 4,* Fruits and nuts (pp. 243–252). Berlin, Germany: Springer–Verlag.

Kocyigit, A., Koylu, A. A., & Keles, H. (2006). Effects of pistachio nuts consumption on plasma lipid profile and oxidative status in healthy volunteers. *Nutrition Metabolism & Cardiovascular Diseases, 16,* 202–209.

Kornsteiner, M., Wagner, K. H., & Elmadfa, I. (2006). Tocopherols and total phenolics in 10 different nut types. *Food Chemistry, 98,* 381–387.

Miniati, E. (1981). Anthocyanin pigment in the pistachio nut. *Fitoterapia, 52,* 267–271.

Moure, A., Cruz, J. M., Franco, D., Dominguez, J. M., Sineiro, J., Dominguez, H., et al. (2001). Natural antioxidants from residual sources. *Food Chemistry, 72,* 145–171.

Perkin, J. E. (1990). Major food allergens and principle of dietary management. In J. E. Perkin (Ed.), *Food allergies and adverse reactions* (pp. 51–68). Gaithersburg, MD: Aspen Publishers.

Seeram, N. P., Zhang, Y., Henning, S. M., Lee, R., Niu, Y., Lin, G., et al. (2006). Pistachio skin phenolics are destroyed by bleaching resulting in reduced antioxidative capacities. *Journal of Agricultural and Food Chemistry, 54,* 7036–7040.

Saitta, M., Giuffrida, D., La Torre, G. L., Potortì, A. G., & Dugo, G. (2009). Characterisation of alkylphenols in pistachio (*Pistacia vera* L.) kernels. *Food Chemistry, 117,* 451–455.

Tokusoglu, O., Unal, M. U., & Yemis, F. (2005). Determination of the phytoalexin resveratrol (3,5,4′-trihydroxy-stilbene) in peanuts and pistachios by high-performance liquid chromatographic diode array (HPLC-DAD) and gas chromatography-mass spectrometry (GC-MS). *Journal of Agricultural and Food Chemistry, 53,* 5003–5009.

Trevisan, M. T. S., Pfundstein, B., Haubner, R., Wurtele, G., Spiegelhalder, B., Bartsch, H., et al. (2006). Characterization of alkyl phenols in cashew (*Anacardium occidentale*) products and assay of their antioxidant capacity. *Food Chemistry and Toxicology, 44,* 188–197.

Wu, X., & Prior, R. L. (2005). Identification and characterization of anthocyanins by high-performance liquid chromatography-electrospray ionization-tandem mass spectrometry in common foods in the United States: vegetables, nuts, and grains. *Journal of Agricultural and Food Chemistry, 53,* 3101–3113.

Yalpani, M., & Tyman, J. H. P. (1983). The phenolic acids of *Pistacia vera*. *Phytochemistry, 22,* 2263–2266.

Potential Health Benefits of Pomegranate (*Punica granatum*) Seed Oil Containing Conjugated Linolenic Acid

Kazunori Koba[1], Teruyoshi Yanagita[2]
[1] Department of Nutritional Science, University of Nagasaki, Siebold, Nagasaki, Japan
[2] Department of Applied Biochemistry and Food Science, Saga University, Saga, Japan

919

LIST OF ABBREVIATIONS

CLA, conjugated linoleic acid
CLN, conjugated linolenic acid
PPAR, peroxisome proliferator-activated receptor

INTRODUCTION

The oil in pomegranate seeds comprises about 20% of the total seed weight; more than 95% of the oil is triglyceride (Lansky & Newman, 2007). Pomegranate seed oil is characterized by its fatty acid composition, which consists of approximately 80% punicic acid, one of the conjugated fatty acids.

Nuts & Seeds in Health and Disease Prevention. DOI: 10.1016/B978-0-12-375688-6.10108-2

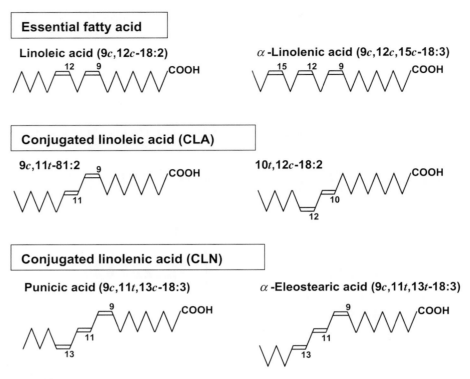

FIGURE 108.1

Chemical structure of essential fatty acids and conjugated fatty acids. Essential fatty acids, such as linoleic acid and α-linolenic acid, have a methylene interrupted *cis* double bond structure in the carbon chains (−CH=CH−CH₂−CH=CH−). Conjugated fatty acids, such as conjugated linoleic acid and conjugated linolenic acid, have conjugated double bonds (−CH=CH−CH=CH−) with multiple positional and geometric isomers.

Figure 108.1 shows the structure of punicic acid and other conjugated fatty acids. The structure is characterized by conjugated double bonds (−CH=CH−CH=CH−), having either a *cis* or *trans* configuration. The most representative conjugated fatty acid is conjugated linoleic acid (CLA; 9*c*,11*t*-18:2 and 10*t*,12*c*-18:2), which has been shown to exert various physiological functions (Pariza *et al.*, 2001). Punicic acid (9*c*,11*t*,13*c*-18:3) is an isomer of α-linolenic acid (a so-called conjugated linolenic acid, CLN), and has two pairs of conjugated double bonds (triene-type). In nature, at least four other triene-type CLNs have been found in specific plant seeds (Suzuki *et al.*, 2001). Pomegranate seeds as well as other plant seeds have their own conjugase enzyme, which converts linoleic acid into a CLN (Figure 108.2). Therefore, the CLN in these plant seeds consists of a single isomer. Unlike CLA, the number of reports on physiological properties of punicic acid and other CLNs is still low. The limited evidence, however, indicates potential benefits of punicic acid. In this chapter, we describe the potential health benefits of pomegranate seed oil, focusing on punicic acid.

BOTANICAL DESCRIPTION

The pomegranate plant, *Punica granatum*, belongs to the Punicaceae family, and is native to a region extending from eastern Iran to the Himalayas in northern India. *Punica granatum* shares its botanical family only with *Punica protopunica*, which is found solely on Socotra, an island off the Yemenite coast (Lansky & Newman, 2007).

The pomegranate tree grows 3−5 m high, has many spiny branches, and can be extremely long-lived − over 200 years (Jurenka, 2008). Pomegranate fruit is grenade-shaped, and is up to 12−13 cm wide, with a deep-red, leathery skin. The fruit is often considered to be a large berry.

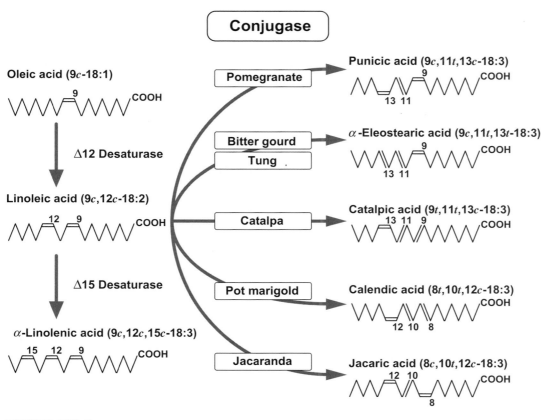

FIGURE 108.2
Biosynthesis of conjugated linolenic acid in plant seeds. In nature, there are specific conjugated linolenic acid (CLN) isomers in specific plant seeds as one of the major fatty acids. Each of these plant seeds has a specific conjugase enzyme, which converts linoleic acid into CLN. Therefore, each plant seed accumulates only a single CLN isomer. For example, pomegranate (*Punica granatum*) seeds accumulate punicic acid ($9c,11t,13c$-18:3) as a major fatty acid component in seed lipids (Suzuki *et al.*, 2001).

Pomegranate seeds are rich in oil; the seeds are located in the fruit, separated by a white, membranous pericarp, and each seed is covered with small amounts of tart, red juice.

HISTORICAL CULTIVATION AND USAGE

The pomegranate plant originated in the Middle East, and has been cultivated for its fruit throughout the whole of the Mediterranean region and Asia since ancient times. This plant was one of the first known sources of medicines, and was used by Ancient Egyptians and Greeks to treat various ailments (Baumann, 2007). Folk-medicine use of the pomegranate in the Middle East, India and Iran included treatment for skin inflammation, rheumatism, and sore throats. In Ayurvedic medicine, pomegranates reputedly nourish and restore balance to the skin. Pomegranate features prominently in Judaism, Christianity, Islam, Buddhism, and Zoroastrianism (Lansky & Newman, 2007). In Japan, the pomegranate has long been a popular ornamental garden or bonsai tree.

PRESENT-DAY CULTIVATION AND USAGE

Today, the pomegranate is cultivated for its fruit worldwide: in the Middle East and extending throughout the Mediterranean regions, eastward to India, China, and Japan, and to the southwest regions of North America and Mexico. The pomegranate is widely consumed as fresh fruit and juice. The seeds have been recognized as inedible, but pomegranate fruit extracts containing seeds are available.

Recently, there has been interest in the pomegranate as a medical and nutritional product. Over the past decade, significant progress has been made in establishing the physiological and pharmacological functions of pomegranate fruit extracts containing seeds. The evidence is still at the experimental basis, but it suggests that the extracts have therapeutic properties, such as antioxidant, anticarcinogenic, and anti-inflammatory properties, which could be due to a series of polyphenols found mostly in the fruit (Jurenka, 2008). Oil extracts from pomegranate seeds, on the other hand, have been suggested to exert anticarcinogenic and anti-obesity properties.

APPLICATIONS TO HEALTH PROMOTION AND DISEASE PREVENTION

Pomegranate seeds are rich in oil (up to 20% of seed weight), and more than 95% of the oil is triglyceride (Lansky & Newman, 2007). As mentioned previously, pomegranate seed oil is characterized by its fatty acid composition, of which approximately 80% is punicic acid ($9c,11t,13c$-18:3), one of the CLNs. Punicic acid is believed to exert several physiological functions as an active biological component, as described below.

Anticarcinogenicity

Cytotoxicity of pomegranate seed oil and other plant seed oils containing CLNs was examined in mouse tumors and human monocytic leukemia cells (Suzuki *et al.*, 2001). In this report, the cytotoxic effect of pomegranate seed oil (containing $9c,11t,13c$-18:3), as well as tung seed oil (containing $9c,11t,13t$-18:3) and catalpa seed oil (containing $9t,11t,13c$-18:3), was much stronger than that of pot marigold seed oil (containing $8t,10t,12c$-18:3), suggesting that the position of the double bond could be an important determinant for the cytotoxicity of CLNs. The anticarcinogenic activity of pomegranate seed oil was also observed in rats (Kohno *et al.*, 2004). Because both $9c,11t,13c$-18:3 and $9c,11t,13t$-18:3 are considered to be metabolized to $9c,11t$-18:2 (Tsuzuki *et al.*, 2006), the effects of these CLNs could be attributed, in part, to the effect of the metabolite CLA ($9c,11t$-18:2) (Pariza *et al.*, 2001). However, not all effects of CLNs can be explained by the metabolite CLA. Indeed, α-eleostearic acid but not a CLA was demonstrated to exert anticarcinogenicity in mice by an oxidation-dependent mechanism (Tsuzuki *et al.*, 2004). Although the mechanism underlying the effect is not clear, induction of apoptosis, activation of peroxisome proliferator-activated receptor-γ (PPARγ), modulation of cytokines, and inhibition of cancer cell proliferation could be involved (Yasui *et al.*, 2005; Moon *et al.*, 2010); these effects are known to exert efficient anticarcinogenic activity.

Anti-obesity and modulation of lipid metabolism

Pomegranate seed oil has been suggested to lower visceral adipose tissue weight in rodents (Koba *et al.*, 2006, 2007). The effect was associated with an increase in hepatic fatty acid β-oxidation. Interestingly, the effect of lowering body fat was not observed with feeding of other CLNs: α-eleostearic acid ($9c,11t,13t$-18:3) in bitter gourd seed oil, catalpic acid ($9t,11t,13c$-18:3) in catalpa seed oil, and calendic acid ($8t,10t,12c$-18:3) in pot marigold seed oil (Koba *et al.*, 2006). In this case, the punicic acid in pomegranate seed oil, rather than its metabolite CLA, is likely responsible for this effect, because both $9c,11t,13c$-18:3 and $9c,11t,13t$-18:3 can be converted to $9c,11t$-18:2, but only the former (punicic acid) decreased adipose tissue weight. In separate studies, dietary pomegranate seed oil was also suggested to decrease dose-dependently the white adipose tissue weight of rats and mice (Koba *et al.*, 2007). Therefore, punicic acid ($9c,11t,13c$-18:3) in pomegranate seed oil could be at least one of the factors responsible for decreasing the visceral adipose tissue weight in rodents.

The effects of dietary pomegranate seed oil on serum and liver lipid concentrations are not always the same. Feeding of pomegranate seed oil decreased liver triglyceride concentration (Koba *et al.*, 2007), which could be partly due to a reduction of apoB 100 secretion and triglyceride synthesis (Arao *et al.*, 2004). In another study, dietary pomegranate seed oil increased liver triglyceride concentration without influencing serum triglyceride concentration (Koba *et al.*, 2006). The difference could be due to the experimental conditions, such as diet formulation, animal species, and so on.

Others

Dietary pomegranate seed oil may affect immune function (Yamasaki *et al.*, 2006). Recently, this oil was also demonstrated to ameliorate obesity-related inflammation and insulin resistance by activating PPARγ (Hontecillas *et al.*, 2009).

Conclusion

To date, pomegranate seed oil containing punicic acid has mainly been indicated to exert anticarcinogenic and anti-obesity activities. Although evidence of these effects in humans is lacking, the results observed *in vitro* and *in vivo* suggest potential health benefits. More studies to evaluate the physiological functions of punicic acid, as well as safety concerns, are required. In contrast to CLAs, oils rich in punicic acid are naturally available in pomegranate seeds. Pomegranate seed oil may therefore be useful, and its effects as one of the new types of functional oils will be considerable.

ADVERSE EFFECTS AND REACTIONS (ALLERGIES AND TOXICITY)

According to a report by Jurenka (2008), pomegranate and its constituents have safely been consumed for centuries without adverse effects. As for pomegranate seed oil, adverse effects are not known as yet. However, an extremely high dose of pomegranate seed oil may not be recommended, because punicic acid in pomegranate seed oil and conjugated fatty acids are known to exhibit lower oxidation stability than non-conjugated fatty acids. More data are needed to clarify this issue.

SUMMARY POINTS

- Pomegranate seeds contain punicic acid, a conjugated linolenic acid, as a major fatty acid component.
- Punicic acid in pomegranate seeds has exerted an anticarcinogenic effect in both *in vitro* and *in vivo* studies.
- The anticarcinogenic effect of punicic acid could be attributed to punicic acid and/or the metabolite conjugated linoleic acid.
- Punicic acid in pomegranate seeds could be a factor in the observed decrease of white adipose tissue weight in rodents.

References

Arao, K., Yotsumoto, H., Han, S.-Y., Nagao, K., & Yanagita, T. (2004). The 9cis, 11trans, 13cis isomer of conjugated linolenic acid reduces apolipoprotein B100 secretion and triacylglycerol synthesis in HepG2 cells. *Bioscience Biotechnology and Biochemistry, 68*, 2643–2645.

Baumann, L. S. (2007). Less-known botanical cosmeceuticals. *Dermatologic Therapy, 20*, 330–342.

Hontecillas, R., O'Shea, M., Einerhand, A., Diguardo, M., & Bassaganya-Riera, J. (2009). Activation of PPAR γ and α by punicic acid ameliorates glucose tolerance and suppresses obesity-related inflammation. *Journal of the American College of Nutrition, 2*, 184–195.

Jurenka, J. M. T. (2008). Therapeutic applications of pomegranate (*Punica granatum* L.): a review. *Alternative Medicine Review, 13*, 128–144.

Koba, K., Akahoshi, A., Tanaka, K., Miyashita, K., Iwata, T., Kamegai, T., et al. (2006). Dietary conjugated linolenic acid modifies body fat mass, and serum and liver lipid levels in rats. In Y.-S. Huang, T. Yanagita, & H. R. Knapp (Eds.), *Dietary fats and risk of chronic disease* (pp. 92–105). Champaign, IL: AOCS Press.

Koba, K., Imamura, J., Akahoshi, A., Kohno-Murase, J., Nishizono, S., Iwabuchi, M., et al. (2007). Genetically modified rapeseed oil containing cis-9, trans-11, cis-13 octadecatrienoic acid affects body fat mass and lipid metabolism in mice. *Journal of Agricultural and Food Chemistry, 55,* 3741−3748.

Kohno, H., Suzuki, R., Yasui, Y., Hosokawa, M., Miyashita, K., & Tanaka, T. (2004). Pomegranate seed oil rich in conjugated linolenic acid suppresses chemically induced colon carcinogenesis in rats. *Cancer Science, 95,* 481−486.

Lansky, E. P., & Newman, R. A. (2007). *Punica granatum* (pomegranate) and its potential for prevention and treatment of inflammation and cancer. *Journal of Ethnopharmacology, 109,* 177−206.

Moon, H. S., Guo, D. D., Lee, H. G., Choi, Y. J., Kang, J. S., Jo, K., et al. (2010). Alpha-eleostearic acid suppresses proliferation of MCF-7 breast cancer cells via activation of PPARgamma and inhibition of ERK1/2. *Cancer Sciences, 101,* 396−402.

Pariza, M. W., Park, Y., & Cook, M. E. (2001). The biologically active isomers of conjugated linoleic acid. *Progress in Lipid Research, 40,* 283−298.

Suzuki, R., Noguchi, R., Ota, T., Abe, M., Miyashita, K., & Kawada, T. (2001). Cytotoxic effect of conjugated trienoic fatty acids on mouse tumor and human monocytic leukemia cells. *Lipids, 36,* 477−482.

Tsuzuki, T., Tokuyama, Y., Igarashi, M., & Miyazawa, T. (2004). Tumor growth suppression by α-eleostearic acid, a linolenic acid isomer with a conjugated triene system, via lipid peroxidation. *Carcinogenesis, 25,* 1417−1425.

Tsuzuki, T., Kawakami, Y., Abe, R., Nakagawa, K., Koba, K., Imamura, J., et al. (2006). Conjugated linolenic acid is slowly absorbed in rat intestine, but quickly converted to conjugated linoleic acid. *Journal of Nutrition, 136,* 2153−2159.

Yamasaki, M., Kitagawa, T., Chujo, H., Koyanagi, N., Maeda, H., Kohno-Murase, J., et al. (2006). Dietary effect of pomegranate seed oil on immune function and lipid metabolism in mice. *Nutrition, 22,* 54−59.

Yasui, Y., Hosokawa, M., Sahara, T., Suzuki, R., Ohgiya, S., Kohno, H., et al. (2005). Bitter gourd seed fatty acid rich in 9c,11t,13t-conjugated linolenic acid induces apoptosis and up-regulates the GADD45, p53 and PPARγ in human colon cancer Caco-2 cells. *Prostaglandins Leukotrienes and Essential Fatty Acids, 73,* 113−119.

CHAPTER 109

Phenolic Acids in Pumpkin (*Cucurbita pepo* L.) Seeds

Vera Krimer-Malešević[1], Senka Mađarev-Popović[2], Žužana Vaštag[2], Ljiljana Radulović[2], Draginja Peričin[2]
[1] Public Health Institute, Sanitary Chemistry, Subotica, Serbia
[2] University of Novi Sad, Faculty of Technology, Department of Applied and Engineering Chemistry, Novi Sad, Serbia

925

LIST OF ABBREVIATIONS

BHP, benign prostatic hyperplasia
DHT, dihydrotestosterone

INTRODUCTION

Plants have been used as medicine since the beginning of human civilization. The World Health Organization reports that 80% of people worldwide use herbal medicine. Pumpkin is one of the plants that have been frequently used as a functional food or herbal drug. The seeds of *C. pepo* are excellent against intestinal parasites. Males affected by benign prostatic hyperplasia are treated by using the seeds of the pumpkin. Modern science gives explanation for such effects; *C. pepo* is abundant with phytochemicals, yet it is not easy to identify all the compounds that are responsible for its beneficial effects. In recent years there has been a growing interest in studying polyphenols, because they represent potentially health-promoting substances. *C. pepo*, like other plants, has ubiquitous phenolic acids, which may contribute to its beneficial effects.

BOTANICAL DESCRIPTION

Cucurbita pepo (see Table 109.1 for botanical classification) has creeping plants that are compact or semi-shrubby, annual, with broadly ovate-cordate to triangular-cordate leaves

TABLE 109.1 Botanical Classification of *Cucurbita pepo* L.		
Kingdom	Plantae	Plants
Subkingdom	Tracheobionta	Vascular plants
Superdivision	Spermatophyta	Seed plants
Division	Magnoliophyta	Flowering plants
Class	Magnoliopsida	Dicotyledons
Subclass	Dilleniidae	
Order	Violales	
Family	Cucurbitaceae	Cucumber family
Genus	*Cucurbita* L.	Gourd
Species	*Cucurbita pepo* L.	Field pumpkin

FIGURE 109.1

Images of *Cucurbita pepo L.* plant: (A) flower and the leaves; (B) fruit. The fruit can be very variable in size, shape and color.

(Figure 109.1A), with or without white spots, often with three to five deep lobules. Tendrils have two to six branchlets, or are simple and little developed. The flowers are pentamerous. The fruit is very variable in size and shape, with a rigid skin varying in color from light to dark green, plain to speckled, with cream or green contrasting with yellow or orange, or two-colored (Figure 109.1B). The flesh is cream to yellowish or pale orange, ranges from being soft and not bitter to fibrous and bitter. The pumpkin has numerous seeds, which are narrowly or broadly elliptical or, rarely, orbicular, and slightly flattened. Pumpkin seeds are dark green (because of the presence of protochlorophylls). They may be encased in a yellow-white husk (hulled or husked seeds), although some pumpkins produce seeds without shells (hull-less or naked seeds) which have only very thin dark-green skin. These two varieties of *C. pepo* seeds are shown in Figure 109.2(A, B).

HISTORICAL CULTIVATION AND USAGE

Cucurbita pepo is native to the Americas (originating from northeastern Mexico and Texas), where it has been cultivated for several thousand years (Paris, 1989). Pumpkins were dispersed to other countries by transoceanic voyagers at the turn of the 16th century. The image *Quegourdes de turquie* (completed no later than 1508) represents the earliest-known representation of *C. pepo* in Europe (Paris *et al.*, 2006). Native Americans dried strips of pumpkin and wove them into mats. They also roasted long strips of pumpkin and ate them. Galen, Hippocrates, Plinius and Dioscorides used pumpkin seeds, in the form of compresses, against swelling. Later, pumpkin was used for the management of nephritis, tuberculosis, and internal

FIGURE 109.2
Two varieties of *C. pepo* seeds: (A) hulled, and (B) naked (hull-less).

worms and parasites (Strobl, 2004). A famous painting entitled *Fruittivendola (The Fruit Seller)*, located in Milan and painted in 1580, depicts the flower buds of *C. pepo*, which were used for culinary purposes (Paris & Janick, 2005).

PRESENT-DAY CULTIVATION AND USAGE

Pumpkin (*Cucurbita pepo*) is a widely cultivated vegetable in America, Europe, Africa, and Asia, used for human consumption and in traditional medicine. In Austria and adjacent countries, pumpkins are grown for the production of highly esteemed salad oil. The seeds can be ground into a powder and mixed with cereals for making bread, or, after roasting, are consumed as a snack. In the therapy of small disorders of the prostate gland and the urinary bladder, seeds and oil have shown good results (Bombardelli & Morazoni, 1997). Pumpkin extracts, from different parts of the plant, have shown antidiabetic, antibacterial, hypo-cholesterolemic, antioxidant, anticancer, antimutagenic, immunomodulatory, antihelmintic, and anti-bladderstone activity, and other miscellaneous effects (Caili *et al.*, 2006).

APPLICATIONS TO HEALTH PROMOTION AND DISEASE PREVENTION

Mature, dried seeds of *C. pepo* are described by a positive monograph of the German Commission E as a drug. Naked pumpkin seeds are, compared with hulled seeds, richer in $\triangle^7$-sterols. Pumpkin seeds and seed extracts represent a complex mixture of substances, and their effects may not be attributed to just few ingredients. $\triangle^7$-Sterols are significant for the treatment of BHP, a condition that commonly affects men 50 years and older and involves enlargement of the prostate gland. One of the factors that contributes to BPH is over-stimulation of the prostate cells by testosterone and its conversion product, DHT. As aging occurs, the amount of DHT in the prostate gland remains high, even though the circulating testosterone level drops. Components in pumpkin seed and seed oil appear to interrupt this triggering of prostate cell multiplication by testosterone and DHT, although the exact mechanism for this effect is still a matter of discussion. The reason for $\triangle^7$-sterols efficiency may be in their structural similarity to DHT (Strobl, 2004). Pumpkin seeds also contain citrulline, which is attributed an anti-edematous effect. Furthermore, some as yet not fully described ingredients have antimicrobial and anti-inflammatory activity, as well as a regulatory influence on the bladder musculature. The carotenoids found in pumpkin seeds have been studied for their potential prostate benefits, and it was concluded that men with higher amounts of carotenoids in their diet have less risk for BPH (Brawley & Barnes, 2001). The same is valid for selenium and vitamin E. Other effective ingredients are zinc, linoleic, oleic, palmitic and stearic acids, and magnesium salts. They are ubiquitous in pumpkin seeds, but the question of whether they are significant in the relief from BHP-associated discomfort remains unanswered. Pumpkins also contain biologically active components that include polysaccharides, proteins, and peptides. However, the presence of antinutritients in pumpkin seeds limits their nutritional value. It was found that technologies such as germination and fermentation could reduce antinutritional materials and affect the pharmacological activities of pumpkin. Pumpkin and other species of the Cucurbita family possess an unusual amino acid, known as cucurbitin, which is the main and most active chemical principle which is essentially the reason for the antihelmintic activity displayed by a pumpkin seed remedy (Rybaltovskiĭ, 1966) .

In recent years, there has been a growing interest in studying phenolic compounds of oilseeds, their skins, hulls, and oil cake meals, because they represent potentially health-promoting substances and have industrial applications. Due to their antioxidant, antimicrobial, anti-proliferative, and preservative properties, phenolic acids, which account for 30% of total dietary plant polyphenols, have an important role in plant resistance towards diseases, in human health maintenance (inhibition of oxidative damage diseases such as coronary heart disease, stroke, cancers, diabetes, neurodegenerative diseases, cataracts, and age-related functional decline), and in prevention of food deterioration. In many cases, aldehyde analogs are also grouped with, and referred to, as phenolic acids. Phenolic compounds occur in free and bound forms. Bound phenolics may be linked to various plant components through ester, ether, or acetal bonds, and are typically involved in cell-wall structure. For this reason, hydrolytic procedures have been employed to quantify total phenolics. Phenolics behave as antioxidants, due to the reactivity of the phenol moiety (hydroxyl substituent on the aromatic ring). Although there are several mechanisms, the predominant mode of antioxidant activity is believed to be radical scavenging via hydrogen atom donation.

Other established antioxidant, radical-quenching mechanisms are through electron donation and singlet oxygen quenching. Substituents on the aromatic ring affect the stabilization and therefore the radical-quenching ability of these phenolic acids. Different acids therefore have different antioxidant activity. The antioxidant behavior of free, esterified, glycosylated, and non-glycosylated phenolics has been reported (Robbins, 2003). Phenolic acids are divided into two main groups: hydroxycinnamic (caffeic, *p*-coumaric, ferulic, sinapic) and hydroxybenzoic (protocatechuic, *p*-hydroxybenzoic, vanillic, syringic, gentisic, gallic) acids. Greater antioxidant capacity of hydroxycinnamic acid derivatives (compared to hydroxybenzoic acid derivatives) is linked to the presence of the propenoic side chain, instead of the carboxylic group. The conjugated double bond in the side chain could have a stabilizing effect by resonance on the phenoxyl radical, thus enhancing the antioxidant activity of the aromatic ring. The resonance stabilization of the phenoxyl radical is shown in Figure 109.3.

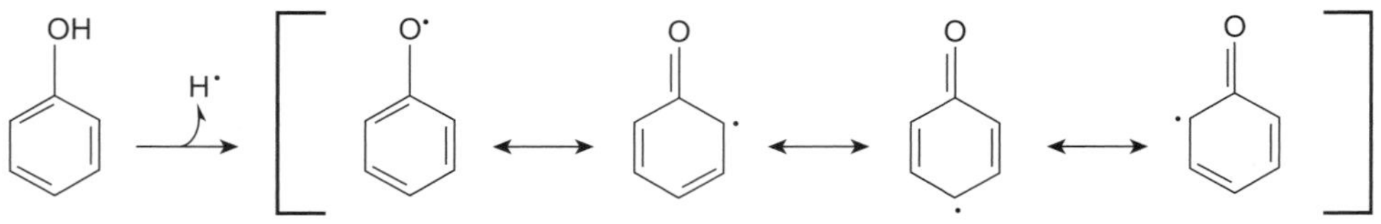

FIGURE 109.3
Phenoxyl radical resonance. The high reactivity of phenolic compounds to scavenge harmful free radicals may be explained by their ability to donate a hydrogen atom from their hydroxyl group. During this reaction the phenoxyl radical is formed, which undergoes a change to a resonance structure by redistributing (stabilizing) the unpaired electron on the aromatic core. Thus, the phenoxyl radical exhibits a much lower reactivity compared to free radicals.

However, despite the ubiquitous presence of phenolic acids in plant-based food, and their role as dietary antioxidants, until recently the literature about the presence of phenolic compounds in pumpkin *C. pepo* has been limited. In pumpkin peel, for example, small amounts of vanillic, p-coumaric and sinapic acids were reported; in puree, chlorogenic, syringic, and caffeic acid were found; while in seed oil, p-coumaric and *trans*-cinnamic acid were quantified. Peričin *et al.* (2009) investigated two varieties of *C. pepo* seeds (Figures 109.2A, B), and reported that p-hydroxybenzoic acid was the dominant phenolic acid, with 34.72, 67.38, and 51.80% of the total phenolic acid content in whole hull-less seed, kernels (from hulled seed), and hulls, respectively. Besides p-hydroxybenzoic acid, the most abundant phenolic compounds, in decreasing order of quantity, were caffeic, ferulic, and vanillic acids in whole hull-less seeds. In the hulled pumpkin variety, kernels were rich in trans-sinapic and protocatechuic acids, and p-hydroxybenzaldehyde; in the hulls, p-hydroxybenzaldehyde, vanillic, and protocatechuic acids were present in valuable quantities. The same authors also determined that the sum of phenolic acids present in three forms (free, esterified, and insoluble-bound) was much higher in the outer seed layers (skin and hull) than in the whole naked seed and in kernels from the hulled variety of *C. pepo*. That result was expected, because skins and hulls, as a rule, contain much higher concentrations of phenolic compounds than inner layers, as they represent the first line of the plant's defense against the environment. It is not surprising, then, that the total sum of phenolic acids in whole hull-less seed (kernel + thin green skin) was higher than in kernels from the hulled variety, at 77.02 versus 51.53 mg/kg dry weight, respectively.

The structures of phenolic acids identified in pumpkin seed are shown in Figure 109.4, while the content and distribution of phenolic acids in whole naked seed, and kernel from the hulled variety of *C. pepo*, are presented in Table 109.2. In all investigated samples the bound phenolic acids content (esterified and insoluble) was higher than the free phenolic acids content, indicating that the major phenolic acids were released upon alkaline hydrolysis — that is, they were present in the cell-wall structure. This was most noticeable in defatted flour from whole hull-less pumpkin seeds, where the esterified form comprised 63.06% and the insoluble-bound 21.07% (in sum 84.13%) of the total phenolic acids. Thus, without including the bound phenolic acids, the total phenolic content of pumpkin *Cucurbita pepo* seeds would clearly be underestimated.

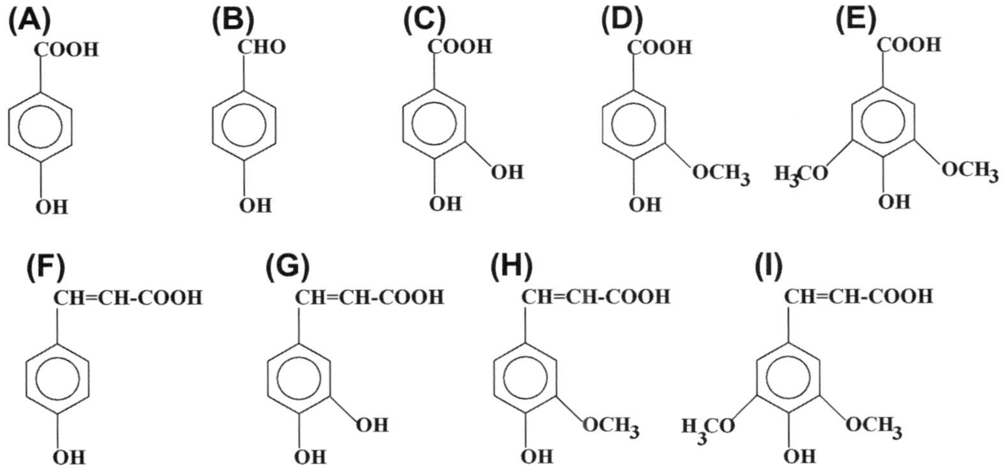

FIGURE 109.4
Hydroxybenzoic (A—E) and hydroxycinnamic (F—I) acid derivatives present in pumpkin seed. (A) p-Hydroxybenzoic acid, (B) p-hydroxybenzaldehide, (C) protocatechuic acid, (D) vanillic acid, (E) syringic acid; (F) p-coumaric acid, (G) caffeic acid, (H) ferulic acid, and (I) sinapic acid. Hydroxybenzoic acids possess a carboxylic group, while hydroxycinnamic acids have a propenoic side chain.

TABLE 109.2 Phenolic Acids in Defatted Whole Hull-less Pumpkin Seed, Kernels, and Hulls from Hulled Seed

Phenolic Acid	Whole Hull-less Pumpkin Seed			Kernels from Hulled Seed			Hulls		
	Free	Esterified	Insoluble-bound	Free	Esterified	Insoluble-bound	Free	Esterified	Insoluble-bound
Protocatechuic	ND	3.66 ± 0.35	0.72 ± 0.06	0.40 ± 0.03	1.49 ± 0.13	0.60 ± 0.05	6.64 ± 0.60	0.59 ± 0.03	1.34 ± .09
p-Hydroxybenzoic	3.64 ± 0.38	15.96 ± 1.20	7.14 ± 0.11	18.33 ± 1.73	5.81 ± 0.42	10.58 ± 1.02	52.28 ± 2.54	12.70 ± 1.12	17.24 ± 0.64
p-Hydroxybenzaldehyde	0.94 ± 0.09	0.35 ± 0.02	1.34 ± 0.12	1.15 ± 0.10	ND	1.23 ± 0.07	8.79 ± 0.82	1.04 ± 0.04	23.04 ± 1.71
Vanillic	ND	6.66 ± 0.57	1.37 ± 0.12	ND	0.72 ± 0.06	0.84 ± 0.06	5.72 ± 0.34	2.84 ± 0.26	14.61 ± 0.68
Caffeic	2.80 ± 0.25	12.20 ± 0.70	2.08 ± 0.20	ND	0.90 ± 0.08	0.86 ± 0.07	ND	ND	ND
Syringic	ND	0.36 ± 0.01	0.28 ± 0.05	ND	0.80 ± 0.06	NQ	ND	2.19 ± 0.13	2.13 ± 0.12
trans p-Coumaric	1.79 ± 0.16	1.82 ± 0.13	0.69 ± 0.01	1.00 ± 0.09	ND	0.87 ± 0.07	2.93 ± 0.20	0.49 ± 0.04	ND
Ferulic	1.01 ± 0.10	7.05 ± 0.16	1.78 ± 0.16	ND	ND	0.94 ± 0.08	ND	1.07 ± 0.09	ND
trans Sinapic	2.04 ± 0.18	0.51 ± 0.04	0.83 ± 0.08	1.98 ± 0.14	1.62 ± 0.12	1.38 ± 0.12	ND	3.09 ± 0.04	ND
Total phenolic acid content	12.22 ± 0.22	48.57 ± 2.08	16.23 ± 0.11	22.87 ± 1.80	11.34 ± 1.01	17.32 ± 1.52	76.36 ± 5.75	24.01 ± 2.33	58.36 ± 3.21

The table shows the mean ± SD (in mg/kg dry weight) of three sets of analysis of defatted whole hull-less pumpkin seed, kernels, and hulls from hulled seed. SD, standard deviation; ND, not detected; NQ, not quantified.

Cell wall materials associated with bound phenolic compounds may survive upper gastrointestinal digestion conditions, and finally reach the colon. Colonic digestion of such materials by intestinal microflora may release the bulk of the bound phytochemicals, which exert their health benefits locally and beyond after absorption.

In the future, it would be interesting to investigate to what extent the phenolic acids from *C. pepo* seeds contribute to curing different disorders in humans.

ADVERSE EFFECTS AND REACTIONS (ALLERGIES AND TOXICITY)

Allergy is a hypersensitive reaction by the body to foreign substances (antigens), which in similar amounts and circumstances are harmless within the bodies of other people. The allergic response develops when the natural immune defense mechanism, responsible for the correct reaction to environmental agents, is disturbed. Allergy to zucchini (*C. pepo*) is rare, although some reported cases do exist. After the intake of *C. pepo*, some individuals complained of allergic symptoms such as oral allergy syndrome, nausea, diarrhea, or pruritus. Zucchini allergens can cause systematic reactions, and are at least partially heat stable. Allergy to zucchini can occur as a result of primary sensitization, as well as cross-reactions to the panallergen profilin (an actin-binding protein) and cross-reacting carbohydrate determinants (Reindl *et al.*, 2000). When hydroalcoholic extracts of the 10 most frequently used taenidicidal herbs used in Ethiopia were examined for toxicity, *C. pepo* was the least toxic (Desta, 1995). *C. pepo* is generally considered as safe and non-toxic plant, and reported investigations regarding its toxicity are rare.

SUMMARY POINTS

- Pumpkin seeds and seed extracts represent a complex mixture of substances, and their effects on health may perhaps not be attributed to only a few ingredients.
- Among recognized ingredients of *Cucurbita pepo* L. that improve, maintain, and safeguard the health of humans are △7-sterols, carotenoids, selenium, and vitamin E, although the impact of other substances present is still insufficiently investigated.
- Phenolic acids are distributed in nature in their free and bound forms, and are ubiquitous in plants.
- They represent 30% of polyphenols taken by diet, and have an antioxidant capacity that is related to their health-promoting features.
- A certain profile of phenolic acids is also present in pumpkin seed, and may contribute to the prevention or cure of different disorders in humans.

931

References

Bombardelli, E., & Morazoni, P. (1997). *Cucurbita pepo* L. Fitoterapia, 68, 291–302.

Brawley, O. W., & Barnes, S. T. (2001). Potential agents for prostate cancer chemoprevention. *Epidemiologic Reviews*, 23, 168–172.

Caili, F., Huan, S., & Quanhong, L. (2006). A review on pharmacological activities and utilization technologies of pumpkin. *Plant Foods for Human Nutrition*, 61, 73–80.

Desta, B. (1995). Ethiopian traditional herbal drugs. Part I: Studies on the toxicity and therapeutic activity of local taenicidal medications. *Journal of Ethnopharmacology*, 45, 27–33.

Paris, H. S. (1989). Historical records, origins, and development of the edible cultivar groups of *Cucurbita pepo* (Cucurbitaceae). *Economic Botany*, 43, 423–443.

Paris, H. S., & Janick, J. (2005). Early evidence for the culinary use of squash flowers in Italy. *Chronica Horticulturae*, 45, 20–21.

Paris, H. S., Daunay, M.-C., Pitrat, M., & Janick, J. (2006). First known image of *Cucurbita* in Europe, 1503–1508. *Annals of Botany*, 98, 41–47.

Peričin, D., Krimer, V., Trivić, S., & Radulović, Lj (2009). The distribution of phenolic acids in pumpkin's hull-less seed, oil cake meal, dehulled kernel and hull. *Food Chemistry*, 113, 450–456.

Reindl, J., Anliker, M. D., Karamloo, F., Vieths, S., & Wütrich, B. (2000). Allergy caused by ingestion of zucchini *(Cucurbita pepo)*: characterization of allergens and cross-reactivity to pollen and other foods. *Journal of Allergy and Clinical Immunology, 106,* 379–385.

Robbins, R. J. (2003). Phenolic acids in foods: an overview of analytical methodology. *Journal of Agricultural and Food Chemistry, 51,* 2866–2887.

Rybaltovskiĭ, O. V. (1966). On the discovery of cucurbitin — a component of pumpkin seed with anthelmintic action. *Meditsinskaia parazitologiia, 35,* 487–488.

Strobl, M. (2004). Δ^7-Sterole und Δ^7-Sterolglykoside aus Samen von *Cucurbita pepo L.*: Isolierung und Strukturaufklärung. Doctoral Dissertation, Fakultät für Chemie und Pharmazie der Ludwig-Maximilians-Universität München.

Pumpkin (*Cucurbita moschata* Duchesne ex Poir.) Seeds as an Anthelmintic Agent?

Carine Marie-Magdeleine, Maurice Mahieu, Harry Archimède
INRA, UR143 Unité de Recherches Zootechniques, Centre INRA-Antilles-Guyane, Domaine de Duclos, Petit-Bourg, Guadeloupe (French West Indies)

933

INTRODUCTION

The tropical America native pumpkin *Cucurbita moschata* (Figure 110.1) belongs to the *Cucurbitaceae* family (Fournet, 2002). The fruit is used as a vegetable, and the seed is a high energy source (40–50% lipids and 30–37% proteins in dry embryo material) which is consumed throughout the world with increasing popularity (Table 110.1; Leung *et al.*, 1968; Caili *et al.*, 2006). It is a valuable source of potassium, phosphorus, iron, and β-carotene (TRAMIL, 1999; Caili *et al.*, 2006). The pumpkin seed is used as a vermifuge, galactogogue, and anti-emetic, and to treat various other medical issues, including prostate and bladder problems, in several countries (Caili *et al.*, 2006). It contains a wide range of bioactive compounds, some of which could possess anthelmintic properties, prompting further experimental studies (Figure 110.2).

BOTANICAL DESCRIPTION

Cucurbita moschata is an annual dicotyledonous vegetable, with creeping or climbing stems (growing up to 5 m) bearing tendrils. The stems are strong, cylindrical or pentangular, with petioles measuring 12–30 cm. The stems and leaves are mildly hairy. The leaves are circular, kidney-shaped, heart-shaped, or triangular, often deeply indented at the base, weakly lobed,

Mahieu, INRA-URZ, Guadeloupe

FIGURE 110.1
Longitudinal section of pumpkin *Cucurbita moschata* Duchesne ex Poir. The flesh is orange, and the hollow center contains pulpy loose fibers and flat, oval white seeds.

wavy and toothed, more or less white spotted, up to 20 cm long and 30 cm wide. The flowers are large, yellow, bell-shaped, five-lobed, and up to 12 cm long. The peduncle is strong, with a rounded pentangular base and large apex. Fruits are round, oblate, oval, oblong, or pear-shaped, variously ribbed, 15—60 cm in diameter, and weigh up to 45 kg. Their flesh is deep yellow, orange, pale green, or white, and the hollow center contains pulpy loose fibers and numerous seeds. The seeds (Figure 110.1) are oval, flat, white to brown, thin-shelled, irregularly margined, with a meaty kernel (Morton, 1981; Fournet, 2002).

HISTORICAL CULTIVATION AND USAGE

C. moschata is believed to be native to tropical America (Morton, 1981), where it has been used as a popular vegetable in cooking for several thousands of years. The seeds are edible, and have medicinal applications. In Austria and adjacent countries, pumpkins have been grown for oil production for about three centuries (Caili *et al.*, 2006). *C. moschata* has been used traditionally as a medicine in many countries, including China, Yugoslavia, Argentina, India, Mexico, Brazil, and America (Caili *et al.*, 2006). The anthelmintic activity of the seeds of the

TABLE 110.1 *Cucurbita moschata* Seed: Chemical Composition for 100 g of Husked Seeds	
Constituent	**Amount**
Water	5.5 g
Energy	2331 kJ (555 kcal)
Proteins	23.4 g
Lipids	46.2 g
Carbohydrates	21.5 g
Fibre	2.2 g
Ca	57 mg
P	900 mg
Fe	2.8 mg
Thiamine	0.15 mg
Niacine	1.4 mg

Data from Leung *et al.* (1968).

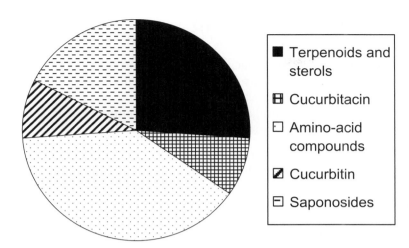

FIGURE 110.2
Qualitative phytochemical composition of *Cucurbita moschata* seed, suspected to have anthelmintic activity. The phytochemical determination was performed by thin layer chromatography methods. Qualitative composition and quantitative evaluation were assessed according to the intensity of the obtained spots. *Data from Marie-Magdeleine (2009), PhD thesis.*

Cucurbita species has long been known (Fang *et al.*, 1961). The Surinam and Cherokee Amerindians used pumpkin seed as an anthelmintic, and also as a pediatric urinary aid to treat bed-wetting (Vogel, 1990).

PRESENT-DAY CULTIVATION AND USAGE

Nowadays, *C. moschata* represents an economically important vegetable species, commonly cultivated worldwide (Taylor & Brant, 2002). The cultivation characteristics are summarized in Table 110.2.

C. moschata seeds are eaten in several countries as snacks, following salting and roasting (Koike *et al.*, 2005). They are commonly used for treating diabetes, prostate gland disorders, and parasites. Data from the ethno-pharmacological literature indicate that *C. moschata* seeds are a potential vermifuge. Eaten fresh or roasted, they help relieve abdominal cramps and distension due to intestinal worms (Caili *et al.*, 2006). Pumpkin seeds are also an important traditional Chinese medicine in the treatment of cestodiasis, ascariasis, and schistosomiasis (Koike *et al.*, 2005). Seeds are used in the Middle East as an effective taenifuge (Karamanukian, 1961). In Middle America, a vermifugal preparation containing shelled and powdered pumpkin seeds is used to expel intestinal worms, including tapeworms (Morton, 1981).

935

TABLE 110.2 *Cucurbita moschata* **Cultivation Characteristics**

Plant Needs	*C. moschata* Constraints
Location	Tropical countries, up to 1800 m altitude
Plant density	2–3 kg of seed/ha, 2–4 seeds per hill; planting distance 2 m × 2 m
Temperature	Temperature above 20°C (day) and 14°C (night).
Period of growth	Photoperiod-insensitive, year-round crop, but better growth during rainy season, if without irrigation.
Soil	Not very demanding with respect to soil conditions; can be cultivated on fertile and well-drained soil, pH 5.5–6.8.
Sensitivity	Drought-tolerant, but sensitive to frost and water-logging; excessive humidity stimulates fungal and bacterial development

Data from Grubben & Denton (2004).

APPLICATIONS TO HEALTH PROMOTION AND DISEASE PREVENTION

Over the years, scientists have studied the many pharmacological actions and potential uses of pumpkin and its extracts. Clearly, there is a lot to learn about the health effects of this plant. Popular medicinal uses of pumpkin seeds have motivated experimental studies on their anthelmintic properties. Such studies have been conducted on different helminth models, and have produced conclusions that call for further research.

Veen and Collier (1949) showed the efficacy of an aqueous extract of the seeds of *C. moschata* as an anthelmintic in humans. Methylene chloride and methanolic extracts of *C. moschata* seed were evaluated on *Caenorhabditis elegans*, and showed no efficacy at the tested doses (Atjanasuppat *et al.*, 2009). Beloin *et al.* (2005) reported anthelmintic activity against *Caenorhabditis elegans* at 500 μg/ml. Marie-Magdeleine *et al.* (2009) showed that aqueous, methanolic, and dichloromethane extracts of *C. moschata* seed greatly inhibited (> 90%) larval development of *Haemonchus contortus in vitro*. The dichloromethane and methanolic extracts had a marked effect on adult worm motility *in vitro* (inhibition of motility > 59% after 24 h of incubation). Atjanasuppat *et al.* (2009) evaluated the activity of methylene chloride and methanolic extracts of *C. moschata* seeds against *Paramphistomum epiclitum*, and found no efficacy at the tested doses. Seeds administered *per os* at a dose of 80 g/person showed high antischistosomial activity (TRAMIL, 1999). Clinical trials in Thailand confirmed that seed extracts were effective against schistosomes and tapeworms (Grubben & Denton, 2004). The use of aqueous extracts of pumpkin seeds in the treatment of puppies experimentally infected with heterophyasis gave promising results, with even better results when combining extracts of areca nut and pumpkin seeds than when giving either extract alone. An effect was reported at the minimum inhibitory concentration of 23 g of pumpkin seeds in 100 ml of distilled water in preclinical studies (Caili *et al.*, 2006).

Some of the bioactive compounds present in the pumpkin seeds thus appear to possess anthelmintic properties, prompting further studies. Pumpkin seed oil contains 9.5–13% palmitic, 6–7.93% stearic, 0.04% arachidic, 37–39% oleic, and 44% linoleic acid. The seed also contains a wide range of bioactive compounds.

The vermifugal active principle of pumpkin seeds has for many years eluded a large number of investigators (Mihranian & Abou-Chaar, 1968). In 1931, it was shown that this bioactive compound is soluble in 75% ethanol but not in petroleum ether, that it is dialyzable, and that its activity is reduced by boiling in dilute sulfuric acid. In 1937, it was also established that the compound is soluble in water and heat resistant. A purified deproteinized aqueous extract of the seeds of *C. moschata* was used as an anthelmintic in humans by Veen and Collier (1949). In 1961, after a comparative morphological, historical, and clinical study (Mihranian & Abou-Chaar, 1968), it was found that pharmacologically active strains of pumpkin seeds should be administered in doses of about 500 g.

The secondary metabolites suspected to be responsible for anthelmintic activity in *C. moschata* seed (Marie-Magdeleine, 2009; Marie-Magdeleine *et al.*, 2009) are a triterpenic compound named cucurbitacin B, a non-proteic amino acid named cucurbitin (3-amino-pyrrolidine-3-carboxylic acid), saponins, and sterols (Mihranian & Abou-Chaar, 1968; TRAMIL, 1999), but other compounds might also be involved, such as cucurmosin, a ribosome-inactivating protein present in the sarcocarp of the pumpkin and also in the seed (Morton, 1981). Researchers are currently working on the structural properties and function of bioactive compounds of pumpkin (Caili *et al.*, 2006) in order to elucidate their modes of action.

The role of non-proteic amino acids in plants is to protect seeds by intoxicating predators via an antimetabolite action; interference between non-proteic and normal amino acids during the biosynthesis of proteins by the predator causes toxicity because the enzymatic system of the predators cannot distinguish these non-proteic amino acids from the normal amino acids, due to their isostery. Consequently, the proteins biosynthesized by the predator may not be functional.

The non-proteic amino acid cucurbitin, which is only present in the seeds (0.4–0.84% of the whole seed; Mihranian & Abou-Chaar, 1968), has been focused on as the active principle responsible for anthelmintic, notably taenicidal and schistosomicidal, activity (Fang et al., 1961). Cucurbitin is also used as an anti-allergen for the preparation of cosmetics and pharmaceutical, particularly dermatological, products. The chemical structure of this amino acid is shown in Figure 110.3A.

In 1965, a study of the amino acids of the *Cucurbitaceae* in several species of *Cucurbita* (in Mihranian & Abou-Chaar, 1968) revealed that cucurbitin was present in all species examined. A further study by Mihranian and Abou-Chaar (1968) confirmed that it appears to be confined to the genius *Cucurbita*, a matter of chemotaxonomic importance. Cucurbitin was associated with a large number of free amino acids in the seed. The free water-soluble amino acid cucurbitin was isolated from *C. moschata* seed, and showed inhibition of the growth of immature *Schistosoma japonicum in vivo* (Fang et al., 1961). Preliminary human research conducted in China and Russia has shown that it may also help in the treatment of tapeworm infestations (Plotnikov et al., 1972). Nevertheless, a high dose of cucurbitin appears to be necessary for efficient anthelmintic action (Mihranian & Abou-Chaar, 1968).

The chemical structure of the cucurbitin compound is similar to that of the nematicidal compound named kainik acid (Figure 110.3B). Kainik acid has a neurodegenerative action on nematodes by substituting for glutamate. The analogy in chemical structure may underlie the similar mode of action of cucurbitin as an anthelmintic (Marie-Magdeleine et al., 2009). Several methods of detection, quantitative determination, and extraction of this amino acid from pumpkin seeds have been used: a two-dimensional paper chromatographic method, a separation method on ion-exchange cellulose phosphate paper, and HPLC and gas chromatography-mass spectrometry (Fang et al., 1961; Mihranian & Abou-Chaar, 1968; Schenkel et al., 1992).

Despite the focus on cucurbitin as the active principle, other secondary metabolites present in the seeds of *C. moschata* might be considered as anthelmintics.

The ribosome-inactivating protein (RIP) cucurmosin is a RNA N-glycosidase that inactivates ribosomes via site-specific deadenylation of the large ribosomal RNA (Marie-Magdeleine et al., 2009). Ribosome-inactivating proteins are also capable of inactivating many non-ribosomal nucleic acid substrates. Some RIPs are able to promote antitumor or antiviral activities. The cucurmosin RIP, which may be present in the seed, could also have an

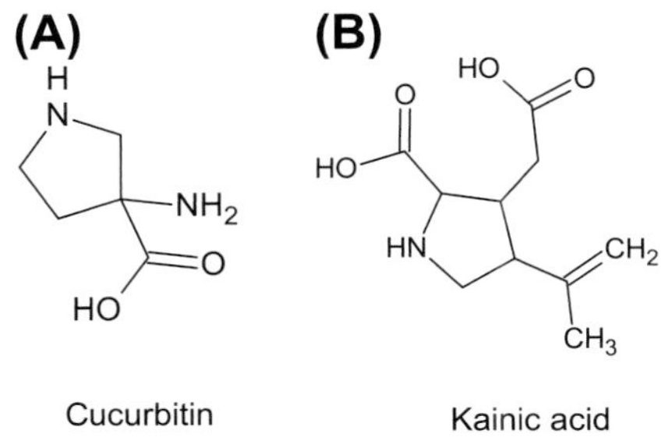

FIGURE 110.3
Comparison of chemical structures of (A) the amino acid cucurbitin (3-amino-pyrrolidine-3-carboxylic acid) and (B) the nematicidal compound kainik acid. The similarity in chemical structure (the pyrrolidine ring) may underlie the similar mode of action of cucurbitin as an anthelmintic.

anthelmintic action by inhibiting the synthesis of proteins and stopping development in helminths.

Terpenoid compounds (e.g., essential oils, saponins) are known to be active against a large range of organisms. Some of them are bioactive, whereas others can affect physical variables. Because they are a complex mixture of compounds, terpenoids can be effective against several targets. The triterpenic cucurbitacins can be toxic, purgative compounds, and are involved in insect resistance (Marie-Magdeleine *et al.*, 2009). Terpenoids reduced the mobility and the consequent migration ability of ovine nematode larvae (Marie-Magdeleine *et al.*, 2009). Furthermore, triterpenoids, saponins, and sterols are all antibacterial, antimicrobial, anticarcinogenic, and antifungal (Marie-Magdeleine *et al.*, 2009). The complexity of the nature of these compounds and their chemical structures could enable interaction with multiple molecular targets at the various developmental stages of the parasites, and be responsible for the anthelmintic activity of the *C. moschata* seeds. The mode of action of these groups of compounds against nematodes and helminths may involve synergy (Marie-Magdeleine *et al.*, 2009).

ADVERSE EFFECTS AND REACTIONS (ALLERGIES AND TOXICITY)

Ingestion of the internal part of seed given *per os* to 105 children infected with *Enterobius vermicularis* was effective, and had no side effects (TRAMIL, 1999). No side effects were observed when testing cucurbitin compared to a placebo on 53 persons for prostate hyperplasia over 3 months (TRAMIL, 1999). Low-level toxicity in dogs and humans was described for cucurbitin (TRAMIL, 1999).

SUMMARY POINTS

- The pumpkin *Cucurbita moschata* is an annual dicotyledonous vegetable, belonging to the *Cucurbitaceae* family. It is commonly cultivated worldwide.
- The seeds and fruits of *C. moschata* are edible; they are used as a medicinal plant for prostate and bladder problems, and as an anthelmintic, a galactogogue, and an anti-emetic.
- Pumpkin seed is used as a vermifuge in several countries, and *C. moschata* seed contains a wide range of bioactive compounds, some of which could possess anthelmintic properties, prompting experimental studies.
- Biological assays showed that the *C. moschata* seed has nematicidal, trematodicidal, taenicidal, and schistosomicidal effects.
- The non-proteic aminoacid cucurbitin (3-amino-pyrrolidine-3-carboxylic acid) is suspected to be the active principle. Cucurbitin was mostly reported to have no side effects, and only a weak level of toxicity was described in dogs and humans.

References

Atjanasuppat, K., Wongkham, W., Meepowpan, P., Kittakoop, P., Sobhon, P., Bartlett, A., et al. (2009). *In vitro* screening for anthelmintic and antitumour activity of ethnomedicinal plants from Thailand. *Journal of Ethnopharmacology, 123*, 475–482.

Beloin, N., Gbeassor, M., Akpagana, K., Hudson, J., De Soussa, K., Koumaglo, K., et al. (2005). Ethnomedicinal uses of *Momordica charantia* (*Cucurbitaceae*) in Togo and relation to its phytochemistry and biological activity. *Journal of Ethnopharmacology, 96*, 49–55.

Caili, F., Huan, S., & Quanhong, L. (2006). A review on pharmacological activities and utilization technologies of pumpkin. *Plant Foods for Human Nutrition, 61*, 73–80.

Fang, S. D., Li, L. C., Niu, C. I., & Tseng, K. F. (1961). Chemical studies on *Cucurbita moschata* Duch. I. The isolation and structural studies of cucurbitine, a new amino acid. *Scientia Sinica, X* 845–851.

Fournet, J. (2002). *Flore illustrée des phanérogames de Guadeloupe et de Martinique*. Trinité, Martinique: CIRAD-GONDWANA Editions.

Grubben, G. J. H., & Denton, O. A. (Eds.). (2004). *Plant resources of tropical Africa, Vol. 2: Vegetables*. Wageningen, The Netherlands: PROTA Edition.

Karamanukian. (1961). *Comparative study of the commercial varieties of* Cucurbita *and their possible relation to the taenifugal activity of their seeds. MS thesis.* Beirut, Lebanon: American University of Beirut.

Koike, K., Li, W., Liu, L., Hata, E., & Nikaido, T. (2005). New phenolic glycosides from the seeds of *Cucurbita moschata. Chemical & Pharmaceutical Bulletin (Tokyo), 53,* 225–228.

Leung, W.-T. W., Busson, F., & Jardin, C. (1968). *Food composition table for use in Africa.* Rome, Italy: FAO.

Marie-Magdeleine, C. (2009). Etude des propriétés anthelminthiques de quelques ressources végétales tropicales pour un usage vétérinaire. PhD thesis, Université Antilles Guyane (in press).

Marie-Magdeleine, C., Hoste, H., Mahieu, M., Varo, H., & Archimède, H. (2009). *In vitro* effects of *Cucurbita moschata* seed extracts on *Haemonchus contortus. Veterinary Parasitology, 161,* 99–105.

Mihranian, V. H., & Abou-Chaar, C. I. (1968). Extraction, detection and estimation of cucurbitin in *Cucurbita* seeds. *Lloydia, 31,* 23–29.

Morton, J. (1981). *Atlas of medicinal plants of Middle America.* Springfield, IL: Charles C. Thomas.

Plotnikov, A. A., Karnaukhov, V. K., Ozeretskovskaia, N. N., Stromskaia, T. F., & Firsova, R. A. (1972). Clinical trial of cucurbin (a preparation from pumpkin seeds) in cestodiasis. Klinicheskaia ispytaniia kukurbina (preparata iz semian tykvy) pri tsestodozakh. *Medical Parazitol Parazite Bolezni, 41,* 407–411.

Schenkel, E., Duez, P., & Hanocq, M. (1992). Stereoselective determination of cucurbitine in *Cucurbita spp.* seeds by gas chromatography and gas chromatography-mass spectrometry. *Journal of Chromatography, 625,* 289–298.

Taylor, M., & Brant, J. (2002). Trends in world cucurbit production, 1991 to 2001. In D. N. Maynard (Ed.), *Cucurbitaceae* (pp. 373–379). Alexandria, VA: ASHS Press.

TRAMIL (1999). *Pharmacopée Caribéenne.* Fort-de France, French West Indies: Editions Désormeaux, Edition L. Germosen-Robineau.

Veen, A. G., & Collier, W. A. (1949). Stable palatable preparation of Laboe Marah (*Cucurbita moschata*). Dutch Patent 63,333, June 15. *Chemical Abstracts, 43,* 7198.

Vogel, V. J. (1990). *American Indian medicine.* Oklahoma, OH: University of Oklahoma Press.

Role of Purple Camel's Foot (*Bauhinia purpurea* L.) Seeds in Nutrition and Medicine

V. Vadivel, H.K. Biesalski
Institute for Biological Chemistry and Nutrition, University of Hohenheim, Stuttgart, Germany

CHAPTER OUTLINE

941

LIST OF ABBREVIATIONS

BHT, butylated hydroxytoluene
FAO, Food and Agricultural Organization
L-dopa, L-3,4-dihydroxyphenylalanine
PCF, purple camel's foot
RDA, recommended dietary allowance
WHO, World Health Organization

INTRODUCTION

Legume grains have long played a key role in the traditional diets of people throughout the world, and are an excellent source of protein, dietary fiber, starch, micronutrients, and phytochemicals with a low fat content (Vadivel & Pugalenthi, 2008). The total per capita consumption of legume grains has markedly increased over the past two decades in the US, due to increased attention to beans being classified as functional foods

Nuts & Seeds in Health and Disease Prevention. DOI: 10.1016/B978-0-12-375688-6.10111-2

(Luthria & Pastar-Corrales, 2006). Accumulation of chemical, biochemical, clinical, and epidemiological evidences has suggested a positive correlation between consumption of food legumes and decreasing incidence of several chronic diseases, such as cancer, cardiovascular diseases, obesity, and diabetes (Siddhuraju & Becker, 2003).

Beside commonly consumed legume grains, earlier research works have indicated the promising nutritive quality of certain under-utilized/wild legume seeds, including the pulses of tribal utility (Janardhanan *et al.*, 2003). The seed material of *Bauhinia purpurea* L., commonly known as purple camel's foot (PCF), is one such promising under-utilized food that merits wider use in many tropical countries. Apart from having excellent nutritive value, PCF seed materials also possess various medicinal properties.

BOTANICAL DESCRIPTION

Bauhinia is a large and diverse tropical and subtropical genus comprising approximately 300 species, belonging to the family Leguminosae (Caesalpinioideae). Most of them possess typical bi-lobed leaves. The genus *Bauhinia* has recently been divided into four sub-genera: *Barklya* (1 species), *Bauhinia* (140 species), *Elayuna* (6 species), and *Phanera* (150 species). The latter are tendril-bearing species, while the three former taxa comprise tree or shrubby species (Duarte-Almeida *et al.*, 2004). The PCF is a small, evergreen avenue tree, found throughout peninsular India. It exhibits many favorable agro-botanical traits, such as fast vegetative growth and early flowering, along with a high fertility index with a high pod weight (8.3 ± 0.93 g), weight of seeds per pod (2.4 ± 0.14 g), number of seeds per pod (eight or nine), and seed recovery percentage (28.92%). The seeds are light in weight and the pods are woody in nature (Rajaram & Janardhanan, 1991).

HISTORICAL CULTIVATION AND USAGE

Purple camel's foot trees are widely distributed in most tropical countries, including Africa, Asia, and South America. In India, they are widespread in the forests of the Deccan plateau, Western Himalayas, and Khasi hills (Vijayakumari *et al.*, 1997; Janardhanan *et al.*, 2003). The cooked young pods and seeds of PCF are eaten by certain ethnic groups in India, particularly the Garo, Naga, Khatkharis, Lambady, and Gonds (Vijayakumari *et al.*, 2007). PCF seeds are used traditionally in indigenous medicine for curing pain, fever, stomach cancer, dropsy, rheumatism, convulsions, diarrhea, delirium, septicemia, ulcers, and indigestion in human beings (Janardhanan *et al.*, 2003).

PRESENT-DAY CULTIVATION AND USAGE

PCF trees have been cultivated for food use by the Lambady tribes in the Tamil Nadu, Kerala, Karnataka, and Andhra Pradesh states of India. Several therapeutic properties are assigned to *Bauhinia* species, including anti-amebic, antidiabetic, antidysenteric, anti-inflammatory, anti-analgesic/antinociceptive, antirheumatic, and hypocholesterolemic activities (Rajaram & Janardhanan, 1991). The seeds of PCF have been used frequently in folk medicine as a remedy for different kind of pathologies, including infections and inflammatory processes. Recently, it has been demonstrated that the PCF seeds have an anti-analgesic effect when evaluated against writhing, hot plate, and formalin tests, and the anti-inflammatory effect of PCF seeds in the carrageenan-induced paw edema test has also been scientifically proven (Pettit *et al.*, 2006). Folklore history has emphasized the use of PCF seeds in the treatment of cancer. Pettit *et al.* (2006) have isolated four new and very remarkable cancer cell growth inhibitors from PCF pods, designating them as Bauhiniastatins, which exhibited significant growth inhibition of P-388 cancer cell line.

APPLICATIONS TO HEALTH PROMOTION AND DISEASE PREVENTION

Purple camel's foot seeds are a rich source of crude protein (25.6–27.2%), crude lipid (12.3–4.3%), fiber (4.6– 5.8%), carbohydrates (51%), minerals, and essential amino acids (Rajaram & Janardhanan, 1991; Vijayakumari *et al.*, 1997). The crude protein content of raw PCF seeds (25.63%) (Table 111.1) was found to be higher than that of certain common legume seeds, such as chickpea (24%), cowpea (24.7%), and green pea (24.9%) (Iqbal *et al.*, 2006). The high protein content of PCF seeds has nutritional significance, since moderate intake of these grains will greatly increase the total dietary protein intake. Further, the PCF seeds

TABLE 111.1 Nutritional Profiles of PCF Seeds	
S. No. **Chemical Composition**	**PCF Seeds**
01 **Proximate Composition (g/kg DM)**	
Moisture content	84.00 ± 0.6
Crude protein	256.3 ± 2.1
Crude lipid	143.0 ± 0.8
Crude fiber	46.60 ± 0.7
Ash	36.90 ± 0.4
Nitrogen free extractives	517.2
Energy (kcal/kg DM)	4381
02 **Mineral Composition (mg/kg DM)**	
Sodium	148.2 ± 0.8
Potassium	24906.2 ± 57.3
Calcium	3420 ± 5.6
Magnesium	767.0 ± 0.9
Phosphorus	1725.0 ± 5.8
Iron	26.40 ± 0.7
Copper	4.800 ± 0.6
Zinc	19.40 ± 1.4
Manganese	2.100 ± 0.7
03 **Amino Acid Composition (g/kg protein)**	
Aspartic acid	110.0
Glutamic acid	123.5
Alanine	62.0
Valine	67.0
Glycine	42.8
Arginine	71.9
Serine	32.6
Cystine	Trace
Methionine	Trace
Threonine	38.8
Phenylalanine	67.9
Tyrosine	52.8
Isoleucine	13.3
Leucine	122.4
Histidine	28.4
Lysine	55.5
Tryptophan	ND
Proline	87.8

Mean $\pm$ standard deviation ($n = 5$) of analysis of nutrient profiles of PCF seeds.
Reproduced from Rajaram & Janardhanan (1991), *J. Sci. Food Agric.*, 55, 423–431, with permission.

registered an appreciable level of crude lipid (14.3%) when compared to certain conventional pulses such as *Cicer arietinum* (4.16%), *Vigna aconitifolia* (0.69%), *V. mungo* (0.45%), *V. radiata* (0.71%), and *Phaseolus vulgaris* (0.9%) (Bravo *et al.*, 1999).

The fiber content of PCF seeds (4.6%) could qualify them as a healthy food because fiber has been postulated to have many important physiological effects, such as reducing the transit time in the mammalian gut, and reducing the incidence of diabetes and obesity. Several studies have shown the hypoglycemic effect of fiber, and also indicated a correlation between low incidence of colon cancer and high fiber diets (Pugalenthi *et al.*, 2005). The ash content (3.69%) of PCF seed materials indicates the presence of marked level of minerals. Due to the presence of high levels of carbohydrates, proteins, and lipids, PCF samples exhibited remarkable calorific value (4381 kcal/kg DM) (Rajaram & Janardhanan, 1991).

When compared to the RDA, the PCF seeds were found to be a rich source of potassium, calcium, and iron (Table 111.1). Generally, a diet that meets two-thirds of the RDA is considered to be adequate for human consumption. Thus, wild PCF seeds might fulfil some of the dietary mineral needs of humans. The data on the amino acid profile of PCF seeds shown in Table 111.1 reveal the presence of adequate levels of all the essential amino acids, except for cystein, methionine, and tryphtophan, when compared with the WHO/FAO amino acid reference pattern. Further, the levels of essential amino acids in PCF seeds seem to be comparable with the amino acid composition of soy bean seeds. Further, lysine is also present in sufficient quantity in PCF seeds when compared to the WHO/FAO pattern (Vijayakumari *et al.*, 1997).

Purple camel's foot seeds yield a considerable amount of oil, which seems to be a good source of essential fatty acids and lipid-soluble bioactive compounds. The high linoleic acid content makes the oil nutritionally valuable. The PCF seed oil is found to contain 66% unsaturated fatty acids, which is also a desirable feature from a nutritional point of view for human food. The levels of tocopherols and sterols in PCF seed oil would be of nutritional importance. Thus, PCF seeds could nutritionally be considered as a new non-conventional supply for the pharmaceutical industries, and also for edible purposes (Ramadan *et al.*, 2006).

Recently, many researchers have attempted to analyze the bioactive principles from different parts of the PCF plant that are responsible for disease prevention/curative effects. The bark extract of PCF showed an increase in hepatic glucose-6-phosphatase activity and anti-peroxidative effect, as indicated by a decrease in hepatic lipid peroxidation and/or increase in the activity of antioxidant enzyme(s). It appears that the seed extracts are capable of stimulating thyroid function in female mice (Vadivel, 2009). The ethanolic extract of PCF seed has been demonstrated to possess antidiabetic activity and adrenergic properties (Vadivel, 2009). Recently, eleven new secondary metabolites, together with two known flavanones and five known bibenzyls, have been isolated from PCF root and found to have antimycobacterial, antimalarial, antifungal, and anti-inflammatory activities (Surat *et al.*, 2007).

The raw seed materials of PCF are found to possess appreciable levels of various bioactive compounds, such as phenolics, tannins, flavonols, anthocyanins, tartaric esters, flavonoids, L-dopa, and phytic acid (Vadivel, 2009) (Table 111.2). The level of bioactive compounds in PCF seeds is comparable with those in common legume grains, such as *Phaseolus vulgaris*, soybean, pea, mung bean, horse gram, cowpea, and moth bean. During cooking, a significant ($P < 0.05$) improvement in levels of phenolics, tannins, anthocyanins, and tartaric esters was noticed, which might be due to their release from the complex bound form with other major nutrients like proteins (Pugalenthi & Vadivel, 2007).

Various bioactive compounds extracted from both raw and processed PCF seeds have been found to exhibit more effective free radical inhibition activity against DPPH free radicals (Table 111.3). The observed results suggest that all the bioactive compounds will exert protective effects under *in vivo*, and also against oxidative and free radical injuries that appear

TABLE 111.2 Levels of Various Bioactive Compounds (mg/100 g seed flour) in Raw and Processed PCF Seeds

S. No.	Bioactive Compounds	Raw Seeds	Cooked Seeds
1	Total phenolics	$3900^b \pm 0.21$	$4260^a \pm 0.06$
2	Tannins	$66^b \pm 0.41$	$98^a \pm 0.04$
3	Flavonols	$386^a \pm 0.32$	$316^b \pm 0.11$
4	Anthocyanins	$65^b \pm 0.23$	$78^a \pm 0.14$
5	Tartaric esters	$81^b \pm 0.15$	$106^a \pm 0.23$
6	Flavonoids	$172^a \pm 0.24$	$144^b \pm 0.16$
7	L-dopa	$1330^a \pm 0.07$	$920^b \pm 0.42$
8	Phytic acid	$934^a \pm 0.22$	$837^b \pm 0.31$

Values are mean $\pm$ standard deviation of three separate determinations. Values in the same row with different superscripts are significantly different ($P < 0.05$).
Data from Vadivel (2009), PhD Thesis.

TABLE 111.3 Free Radical Inhibition Activity and Antioxidant Activity of Various Bioactive Compounds Extracted from PCF Seeds

S. No.	Bioactive Compounds	DPPH Free Radical Inhibition Activity (%)		Antioxidant Activity (%) by β-Carotene/Linoleic Acid Method	
		Raw Seeds	Cooked Seeds	Raw seeds	Cooked seeds
1	Total phenolics	$58.17^b \pm 0.24$	$72.12^a \pm 0.12$	$61.13^b \pm 0.12$	$66.22^a \pm 0.15$
2	Tannins	$54.46^b \pm 0.18$	$59.32^a \pm 0.15$	$48.25^a \pm 0.13$	$50.65^a \pm 0.04$
3	Flavonols	$71.36^b \pm 0.16$	$80.19^a \pm 0.21$	$53.68^a \pm 0.20$	$55.82^a \pm 0.10$
4	Anthocyanins	$67.47^a \pm 0.14$	$69.35^a \pm 0.06$	$15.06^b \pm 0.15$	$17.40^a \pm 0.06$
5	Tartaric esters	$36.25^b \pm 0.23$	$41.66^a \pm 0.22$	$11.38^b \pm 0.17$	$14.53^a \pm 0.05$
6	Flavonoids	$75.14^a \pm 0.17$	$69.38^b \pm 0.08$	$56.83^a \pm 0.22$	$53.98^b \pm 0.12$
7	L-dopa	$86.32^b \pm 0.13$	$93.56^a \pm 0.15$	$80.16^b \pm 0.14$	$88.75^a \pm 0.15$
8	Phytic acid	$58.42^b \pm 0.11$	$65.40^a \pm 0.20$	$45.75^b \pm 0.02$	$51.57^a \pm 0.14$
9	Standard (BHT)	98.4 ± 0.12	98.4 ± 0.12	92.8 ± 0.12	92.8 ± 0.12

Values are mean $\pm$ standard deviation of three separate determinations. Values in the same row with different superscripts are significantly different ($P < 0.05$).
Data from Vadivel (2009), PhD Thesis.

during different pathological conditions. The cooking treatment significantly ($P < 0.05$) improves the free radical scavenging ability of all the bioactive compounds except for anthocyanins, in which only a very small level of increase was noticed (Vadivel, 2009).

The antioxidant activity of bioactive compounds extracted from PCF seeds analyzed through the β-carotene/linoleic acid system showed that the L-dopa (80.16%) and total phenolics (61.13%) exhibited higher levels of antioxidant activity compared to the BHT standard (92.8%) (Table 111.3). Such higher levels of antioxidant activity registered by L-dopa may be due to its characteristic aromatic ring structure with two hydroxyl groups (Pugalenthi & Vadivel, 2007).

Regarding its antidiabetic effect, in Brazil, plants belonging to the genus *Bauhinia* are reported to be used by the rural population as an important antidiabetic agent; the seeds, leaves, and stem-bark of these plants are used in different phyto-preparations to lower the blood glucose level (Filho, 2009). Moreover, the flavonoid-containing fraction with hypoglycemic activity has been isolated from PCF leaves in Egypt (Arora, 2006). In India, PCF seeds have been consumed by certain rural populations, especially the Lambady tribes living in the Tamil Nadu, Kerala, Karnataka, and Andhra Pradesh states, as an important antidiabetic agent (Vadivel, 2009). The Siddha medicine system also prescribes PCF seeds for the management of type 2 diabetic patients (Pettit *et al.*, 2006). Different parts of the PCF plant have demonstrated remarkable hypoglycemic activity in a laboratory animal model (Muralikrishna *et al.*, 2008).

TABLE 111.4 *In Vitro* Hypoglycemic Effect of Various Bioactive Compounds Extracted from Raw and Processed PCF Seeds

S. No.	Bioactive Compounds	α-Amylase Inhibition Activity (%)		α-Glucosidase Inhibition Activity (%)	
		Raw Seeds	Cooked Seeds	Raw Seeds	Cooked Seeds
1	Total phenolics	$17.45^b \pm 0.14$	$20.51^a \pm 0.15$	$52.32^{ab} \pm 0.16$	$55.12^a \pm 0.18$
2	Tannins	$28.48^a \pm 0.08$	$32.82^a \pm 0.08$	$65.76^a \pm 0.21$	$68.66^a \pm 0.12$
3	Flavonols	$09.21^b \pm 0.14$	$10.14^a \pm 0.15$	$45.32^b \pm 0.06$	$48.56^a \pm 0.11$
4	Anthocyanins	$10.28^b \pm 0.05$	$11.46^a \pm 0.24$	$48.33^b \pm 0.17$	$52.62^a \pm 0.12$
5	Tartaric esters	$08.14^b \pm 0.24$	$09.15^a \pm 0.15$	$41.24^a \pm 0.20$	$38.49^b \pm 0.15$
6	Flavonoids	$19.16^b \pm 0.15$	$21.33^a \pm 0.21$	$55.19^a \pm 0.15$	$53.38^a \pm 0.18$
7	L-dopa	$12.73^b \pm 0.16$	$15.24^a \pm 0.06$	$46.41^a \pm 0.25$	$48.33^a \pm 0.14$
8	Phytic acid	ND	ND	ND	ND

Values are mean ± standard deviation of three separate determinations. Values in the same row with different superscripts are significantly different ($P < 0.05$). ND, not detected.
Data from Vadivel (2009), PhD Thesis.

To experimentally assess the *in vitro* hypoglycemic activity of PCF seeds, the α-amylase and α-glucosidase enzyme inhibition activities of different bioactive compounds of the seeds were analyzed, since these enzymes play a key role in the management of hyperglycemia (Pinto *et al.*, 2008). Among the various bioactive compounds, the tannins exhibited the highest levels of α-amylase and α-glucosidase inhibition activity (28.48 and 65.76%, respectively) in raw PCF seeds (Table 111.4) (Vadivel, 2009).

Purple camel's foot seeds were found to possess a lower level of α-amylase inhibition activity than that of α-glucosidase inhibition activity. It is a well-known fact that the dietary management of hyperglycemia-linked type 2 diabetes can be targeted through whole foods that have high α-glucosidase inhibition and moderate α-amylase inhibition activity. This is because excessive α-amylase inhibition leads to the accumulation of undigested starch in the intestine, and consequent stomach distension and discomfort (Pinto *et al.*, 2008). Hence, cooked PCF seeds with high α-glucosidase inhibition and moderate α-amylase inhibition activity could be considered as a potential candidate for further *in vivo* studies as a part of more comprehensive dietary designs to manage the early stages of hyperglycemia-linked type 2 diabetes, or in the prevention of diabetes.

Even though earlier clinical studies on the genus *Bauhinia* began in 1929, there are no concrete reports available with human subjects. In one toxicological study, where *B. forficata* seeds were evaluated for their hypoglycemic effect in randomized cross-over double-blind studies, no acute or chronic effect on plasma glucose levels or glycated hemoglobin was found when using a group of 10 normal human subjects and another group of 16 type 2 diabetic patients, indicating that the infusion had no hypoglycemic effect on either normal subjects or type 2 diabetic patients. Therefore, additional investigations are necessary to confirm the antidiabetic potential of this plant in human beings. Nonetheless, it is recognized that *Bauhinia* seed materials are not toxic, which has been confirmed by several experimental investigations using animal models (Filho, 2009).

ADVERSE EFFECTS AND REACTIONS (ALLERGIES AND TOXICITY)

Although legumes constitute one of the richest and least expensive source of protein in human and animal diets, their utilization has been limited largely because of the presence of certain antinutritional/antiphysiological compounds. Raw PCF seeds contain certain anti-nutritional compounds, such as hemagglutinins (lectins), trypsin inhibitors (41.08 TIU/mg protein), hydrogen cyanide, and saponins, and oligosaccharides such as raffinose (0.54 g/100 g), stachyose (1.17 g/100 g), and verbascose (0.95 g/100 g) (Table 111.5) (Vijayakumari *et al.*, 1997, 2007; Rajaram & Janardhanan, 1991).

TABLE 111.5 Major Antinutritional Compounds in PCF Seeds

S. No.	Antinutritional Compounds	PCF Seeds		
01	Trypsin Inhibitor Activity[1]	41.08		
02	Oligosaccharides[2]			
	Raffinose	0.54		
	Stachyose	1.17		
	Verbascose	0.95		
03	Haemagglutination Activity[3]	Human Blood Erythrocytes		
	Protein fractions	A	B	O
	Albumins	+	−	+
	Albumins	+	−	+
	Albumins	+	−	+
	Globulins	+	+	+
	Globulins	+	+	++
	Globulins	+	+	+

Values are mean ± standard deviation of three separate determinations.
Reproduced from Rajaram & Janardhanan (1991), *J. Sci. Food Agric.*, 55, 423–431, and Vijayakumari *et al.* (2007), *Food Chem.*
103, 968–975, with permission.
[1]*Half inhibition of trypsin/mg protein;*
[2]*g/100 g seed flour;*
[3]*−, no agglutination; +, partial agglutination; ++, strong agglutination.*

However, the hemagglutinating activity exhibited by PCF seeds is very weak, and the trypsin inhibitor activity in PCF seeds also falls within physiological limits, not exceeding the levels present in some commonly consumed legume grains such as chickpea, black gram, green gram, and pigeonpea (Rajaram & Janardhanan, 1991). Furthermore, the reduction of significant levels of various antinutritional compounds including oligosaccharides (raffinose by 63%, stachyose by 42%, and verbascose by 79%) was observed during processing of PCF seeds (Vijayakumari *et al.*, 2007). A significant level of improvement in the protein efficiency ratio, true digestibility, biological value, and net protein utilization of PCF seeds was also observed after pressure-cooking treatment (Figure 111.1), which might be due to inactivation of various antinutritional compounds (Vijayakumari *et al.*, 1997).

947

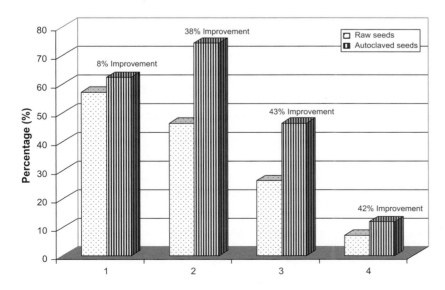

FIGURE 111.1
Protein quality of raw and autoclaved PCF seeds.
1, Biological value; 2, true digestibility; 3, net protein utilization; 4, utilizable proteins. Results are expressed as mean of triplicate determinations. *Reproduced from Vijayakumari et al., 1997, J. Sci. Food Agric., 73, 297–286, with permission.*

SUMMARY POINTS

- The seed materials of *Bauhinia purpurea* appear to be a rich source of major nutrients such as protein, starch, and fiber, with a desirable amino acid, fatty acid, and mineral composition.
- Apart from their good nutrient profile, purple camel's foot seeds also possess appreciable levels of certain bioactive compounds with potential antioxidant and free radical scavenging activities.
- Most of the investigated phytochemical compounds showed a high level of α-glucosidase inhibition activity and a moderate level of α-amylase inhibition activity, which might be a reason for the antidiabetic potential of purple camel's foot seeds.
- *In vitro* experiments indicated that cooking/pressure-cooking appears to improve the nutritional, antioxidant, and hypoglycemic properties of purple camel's foot seeds.
- Such processed purple camel's foot seed materials could be recommended for their versatile utilization as a nutraceutical/functional food with potential health benefits, after conducting extensive trials with human subjects.

References

Arora, R. (2006). Research highlights — Bauhinistatins: new and unusual cancer cell growth inhibitors from *Bauhinia purpurea. African Journal of Traditional Complementary and Alternative Medicines, 3,* 115—120.

Bravo, L., Siddhuraju, P., & Saura-Calixto, F. (1999). Composition of underexploited Indian pulses. Comparison with common legumes. *Food Chemistry, 64,* 185—192.

Duarte-Almeida, J. M., Negri, G., & Salatino, A. (2004). Volatile oils in leaves of *Bauhinia* (Fabaceae Caesalpinioideae). *Biochemical Systematics and Ecology, 32,* 747—753.

Filho, V. C. (2009). Chemical composition and biological potential of plants from the genus *Bauhinia. Phytotherapy Research, 23,* 1347—1354.

Iqbal, A., Khalil, I. A., Ateeq, N., & Khan, M. S. (2006). Nutritional quality of important food legumes. *Food Chemistry, 97,* 331—335.

Janardhanan, K., Vadivel, V., & Pugalenthi, M. (2003). Biodiversity in Indian under-exploited/tribal pulses. In P. K. Jaiwal, & R. P. Singh (Eds.), *Improvement strategies for Leguminosae biotechnology* (pp. 353—405). Dordrecht, The Netherlands: Kluwer Academic Publishers.

Luthria, D. L., & Pastar-Corrales, M. A. (2006). Phenolic acids content of fifteen dry edible bean (*Phaseolus vulgaris* L.) varieties. *Journal of Food Composition and Analysis, 19,* 205—211.

Muralikrishna, K. S., Latha, K. P., Shreedhara, C. S., Vaidya, V. P., & Krupanidhi, A. M. (2008). Effect of Bauhinia purpurea Linn. on Alloxan-induced diabetic rats and isolated frog's heart. *International Journal of Green Pharmacy, 2,* 83—86.

Pettit, G. R., Numata, A., Iwamoto, C., Usami, Y., Yamada, T., Ohishi, H., et al. (2006). Antineoplastic agents: isolation and structures of Bauhiniastatins 1—4 from *Bauhinia purpurea. Journal of Natural Products, 69,* 323—327.

Pinto, M. S., Kwon, Y., Apostolidis, E., Lajolo, F. M., Genovese, M., & Shetty, K. (2008). Functionality of bioactive compounds in Brazilian strawberry (*Fragaria* × *ananassa* Duch.) cultivars: evaluation of hyperglycemia and hypertension potential using *in vitro* models. *Journal of Agricultural and Food Chemistry, 56,* 4386—4392.

Pugalenthi, M., & Vadivel, V. (2007). L-Dopa (L-3,4-dihydroxyphenylalanine): a non-protein toxic amino acid in *Mucuna pruriens* seeds. *Food, 1,* 322—343.

Pugalenthi, M., Vadivel, V., & Siddhuraju, P. (2005). Alternative food/feed perspectives of an under-utilized legume *Mucuna pruriens* var. *utilis* — a review. *Plant Foods for Human Nutrition, 60,* 201—218.

Rajaram, N., & Janardhanan, K. (1991). Chemical composition and nutritional potential of the tribal pulses *Bauhinia purpurea, B. racemosa* and *B. vahlii. Journal of the Science of Food and Agriculture, 55,* 423—431.

Ramadan, M. F., Sharanabasappa, G., Seetharam, Y. N., Seshagiri, M., & Moersel, J. T. (2006). Characterisation of fatty acids and bioactive compounds of kachnar (*Bauhinia purpurea* L.) seed oil. *Food Chemistry, 98,* 359—365.

Siddhuraju, P., & Becker, K. (2003). Studies on antioxidant activities of *Mucuna* seed (*Mucuna pruriens* var. *utilis*) extract and various non-protein amino/imino acids through *in vitro* models. *Journal of the Science of Food and Agriculture, 83,* 1517—1524.

Surat, B., Pakawan, P., Apiwat, B., Chulabhorn, M., Somsak, R., & Prasat, K. (2007). Bioactive compounds from *Bauhinia purpurea* possessing antimalarial, antimycobacterial, antifungal, anti-inflammatory and cytotoxic activities. *Journal of Natural Products, 70,* 795—801.

Vadivel, V. (2009). Nutraceutical properties of certain under-utilized legume grains collected from South India PhD Thesis. Coimbatore, TN, India: Bharathiar University.

Vadivel, V., & Pugalenthi, M. (2008). Effect of various processing methods on the antinutritional constituents and protein digestibility of velvet bean seeds. *Journal of Food Biochemistry, 32,* 795—812.

Vijayakumari, K., Siddhuraju, P., & Janardhanan, K. (1997). Chemical composition, amino acid content and protein quality of the little-known legume *Bauhinia purpurea* L. *Journal of Science of Food and Agriculture, 73,* 279—286.

Vijayakumari, K., Pugalenthi, M., & Vadivel, V. (2007). Effect of soaking and hydrothermal processing methods on the levels of antinutrients and *in vitro* protein digestibility of *Bauhinia purpurea* L. seeds. *Food Chemistry, 103,* 968—975.

Purple Viper's Bugloss (*Echium plantagineum*) Seed Oil in Human Health

Soressa M. Kitessa[1], Peter D Nichols[2], Mahinda Abeywardena[1]
[1] CSIRO Food and Nutritional Sciences, Adelaide, South Australia, Australia
[2] CSIRO Marine and Atmospheric Research, Hobart, Tasmania, Australia

LIST OF ABBREVIATIONS

ALA, α-linolenic acid (18:3n-3)

CVD, dardiovascular disease

DGLA, dihomo γ-linolenic acid (20:3n-6)

DHA, docosahexaenoic acid (22:6n-3)

ESO, echium seed oil

EPA, eicosapentaenoic acid (20:5n-3)

ETA, eicosatetraenoic acid (20:4n-3)

GLA, γ-linolenic acid (18:3n-3)

LA, linoleic acid (18:2n-6)

n-3 LC PUFA, omega-3 long-chain ($\geq C_{20}$) polyunsaturated fatty acid

n-6 LC PUFA, omega-6 long-chain polyunsaturated fatty acid

RDI, recommended daily intake

PA, pyrrolizidine alkaloid

SDA, stearidonic acid (18:4n-3)

TAG, triacylglycerol

Nuts & Seeds in Health and Disease Prevention. DOI: 10.1016/B978-0-12-375688-6.10112-4

INTRODUCTION

The word *Echium* originated from the ancient Greek word *echis* (εχισ), which means "viper," due to its claimed use to cure viper's bite, or the resemblance of its nutlets to a viper's head, or both (Klemow *et al.*, 2002). The genus *Echium* encompasses over 50 species that belong to the *Boraginaceae* (borage family). Echium originated in the Mediterranean region, but many species are now found throughout Europe, North America, and Australia. This chapter is limited to the species known as purple viper bugloss, or Paterson's curse (*E. plantagineum*). Echium is a very interesting herb, because it is either admired for its spectacular floral colors and its medicinal value, or condemned as a noxious weed of incredible persistence. We summarize the agronomic description, health benefits, and toxicity associated with the use of echium seeds for human health.

BOTANICAL DESCRIPTION

The biology of *E. plantagineum* is very well described by Piggin (1982), and the following is a summary taken from that source. Echium is a bristly annual or biennial herb or shrub, varying in stature and floral colors. It stands 20–60 cm tall, and has a floral color dominated by purple. It exhibits thickened, short stems, with hairy rosette leaves that are 5–20 cm long and 1.5–10 cm wide (Figure 112.1). Mature plants can produce 1–20 branching flowering stems with a sparsely branched taproot system, which can extend up to a meter below ground. The fruits (nutlets) are gray and pyramidal, with a pointed tip and flat base which bears an attachment scar (Figure 112.1). The seeds inside have an embryo, a cotyledon, a conical radicle, and a seed coat (no endosperm). In laboratory conditions, seeds have maintained viability for up to 6 years. Figure 112.1 shows *E. Plantagineum* in its reproductive stage, both as an opportunistic plant occupying roadside verges and as a sown crop.

HISTORICAL CULTIVATION AND USAGE

Early medicinal uses included as a remedy for bites of serpents and stings of scorpions. Other herbal medicine applications include treatments for colds, coughs, fever, headache, water retention, kidney stones, inflammation, skin boils, and melancholia, and applications for pain relief and the promotion of wound healing (Klemow *et al.*, 2002). The cultivation history is largely for use as a garden and/or border plant, and its opportunistic invasion of marginal lands and sand dunes.

PRESENT-DAY CULTIVATION AND USAGE

Experimental plot studies have shown that seedling emergence is severely limited if seeds are buried 7.6 cm or deeper (Piggin, 1982). The seed yield per plant can be 60–260 (Piggin, 1982). Seed dispersal is by herbivores and other opportunistic means (e.g., runoff). Berti *et al.* (2007) reported on the agronomic performance of echium as a sown crop. Seed oil content was 272–298 g per kg, biomass yields ranged from 3 to 11 t/ha, and seed yields ranged from 63 to 425 kg/ha. They estimated oil yield per ha at 116 kg. For comparison, typical oil yields expressed per 100 kg seed and per ha for some common vegetable oils are given in Table 112.1.

Echium seed oil (ESO) yield per unit area is much inferior to that of other vegetable oils (Table 112.1). We could not find any historical or current data on the global area of land devoted to commercial cultivation of Echium.

APPLICATIONS TO HEALTH PROMOTION AND DISEASE PREVENTION

The modern use of ESO relates to its essential n-3 and n-6 fatty acids. These polyunsaturated fatty acids (PUFAs) are termed essential fatty acids because the human body is unable to synthesize adequate amounts endogenously, and therefore they must be obtained by dietary

FIGURE 112.1

(A) *Echium plantagineum* in rural western Australia, and a close-up on a pot plant; (B) sown *Echium plantagineum* stands in North Dakota at the onset of flowering and at full bloom; and (C) Echium seeds compared to barley (*Hordeum vulgare* L.) (left) and canola (*Brassica napus* L.) (right). (A) Reprinted courtesy of the Department of Agriculture and Food, Western Australia, Australia. *(B) & (C) Reprinted with permission from Berti et al. (2007), in J. Janick & A. Whipkey (eds), Proceedings of the 6th National New Crops Symposium — Issues in New Crops and New Uses, San Diego, October 14—18, 2007. Alexandria, VA: ASHS Press.*

TABLE 112.1 Typical Oil Yields per Hectare and per Unit Weight of Seeds for Common Oilseeds		
Oilseed	**kg per 100 kg Seeds**	**Litres per ha**
Rapeseed[a]	37	1190
Linseed[a]	35	478
Soybean[a]	14	446
Echium[b]	27	116

[a]*Extracted from Journey to Forever (www.journeytoforever.org/biodiesel_yield.html), accessed December 30, 2009.*
[b]*Approximation from Berti et al. (2007).*

means. In n-3 PUFAs, the first unsaturation occurs at the third carbon, counting from the methyl (omega) end of the fatty acid (Figure 112.2A), whilst in n-6 PUFAs, the first double bond appears in the sixth position. The first fatty acid in the n-3 PUFA series is α-linolenic acid (ALA, 18:3n-3). For n-6 PUFAs, the essential precursor is linoleic acid (LA, 18:2n-6). The first step in the n-3 biosynthetic pathway is stearidonic acid (SDA, 18:4n-3) through the enzyme Δ-6 desaturase (Figure 112.2B). This is thought to be the rate-limiting step. In lower marine organisms, the pathway directly culminates in the production of DHA. In mammals, the Δ-4 desaturase has not been isolated. Sprecher and colleagues (1995) proposed the side reaction which involves elongation to a C_{24} LC-PUFA followed by peroxisomal β-oxidation to create DHA.

Echium seed oil has two main distinguishing features from other vegetable oils (Table 112.2): (1) it is naturally rich in SDA (Figure 112.2B), and (2) its n-6 fatty acid content includes GLA (18:3n-6). The latter is thought to lead to production of anti-inflammatory eicosanoids similar to that of eicosapentaenoic acid (see below). The typical fatty acid profile of ESO derived from *E. plantagineum* and marketed by Croda (Croda Australia, Villawood, NSW 2163, Australia) together with profiles of other common vegetable oils are compared in Table 112.2. ESO is the only oil that is naturally rich in SDA.

A joint FAO/WHO expert consultation on fats and fatty acids in human nutrition has collated a number of elegant reviews on the health benefits of n-3 PUFAs in lowering the risk of CVD, modulating inflammation and immune responses, development of brain and visual acuity in infants, ameliorating age-related degenerative diseases, and modulating depression and mood disorders into a special issue of *Annals of Nutrition and Metabolism* (FAO/WHO, 2009). It is generally agreed that EPA and DHA lower the risk of CVD. It is also generally accepted that the role of vegetable oil-based n-3 PUFAs such as ALA is as a precursor for EPA and DHA (Figure 112.2B for pathway). Recent reviews on bioconversion of ALA to EPA and DHA in humans have shown that the very low activity of this conversion does not enable humans to get their recommended daily intake (RDI) of EPA and DHA from this precursor fatty acid (Brenna *et al.*, 2009). Due to low seafood intake, the per capita consumption of EPA and DHA in many developed nations is considerably below the RDI, which is generally set at around 500 mg EPA plus DHA per day. As shown in Figure 112.2B, the SDA-containing oils, such as

(A)

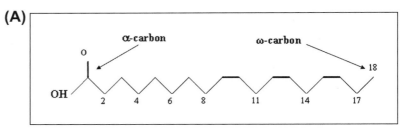

(B)

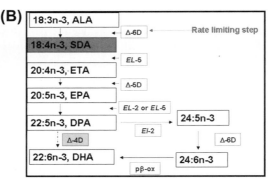

FIGURE 112.2
(A) Linear structure of α-linolenic acid (ALA, 18:3 n-3); (b) Biosynthesis of n-3 long-chain polyunsaturated fatty acids from α-linolenic acid (ALA). ALA, α-linolenic acid; DHA, docosahexaenoic acid; DPA, docosapentaenoic acid; ETA, eicosatetraenoic acid; SDA, stearidonic acid. Elongases extend chain lenght, EL-2 and EL-5; elongases extend chain length; desaturases add double bonds (Δ-4, Δ-5, and Δ-6); peroxisomal β-oxidation removes carbons (pβ-ox). DHA synthesis through Δ-4 desaturase is only applicable in lower marine organisms, not mammals.

TABLE 112.2 Fatty Acid Composition (g/100 g Total Fatty Acids) of Vegetable Oils

Fatty Acid	Canola Oil[a]	Soybean Oil[a]	Linseed Oil[a]	Blackcurrant Oil[b]	Echium Oil[c]
Palmitic (16:0)	4.80	11.0	6.40	7.4	6.60
Stearic (18:0)	1.90	3.80	3.10	0.8	3.50
Oleic (18:1 *cis*-9)	58.5	23.3	20.1	10.4	17.1
Linoleic (18:2 n-6)	23.0	54.5	18.2	48.1	19.4
γ-linolenic acid (GLA, 18:3 n-6)	—	—	—	17.1	9.94
α-linolenic acid (ALA, 18:3 n-3)	7.70	5.90	51.4	12.7	29.8
Stearidonic acid (SDA, 18:4 n-3)	—	—	—	2.6	13.2

[a]*Chouinard et al. (2001);*
[b]*Crozier et al. (1989);*
[c]*Kitessa & Young (2009).*

ESO, bypass the first rate-limiting step. Hence, a number of human and animal studies have considered the potential of SDA-containing oils as precursors of EPA and DHA (Whelan, 2009). In the following section we will consider the benefits of ESO for human health from two angles: (1) the indirect use of ESO in enriching animal-derived foods from livestock and aquaculture species with EPA and DHA to increase population access to n-3 PUFAs; and (2) the direct use of ESO as a nutritional supplement, and its health benefits in humans.

ESO in aquaculture and livestock diets

The aquaculture industry relies on fish oil as an ingredient in aquafeeds for growth and to maintain the health benefits of fish. The use of vegetable oils in aquafeeds is hampered by the very limited conversion of ALA to EPA and DHA. Miller *et al.* (2008) showed limited conversion of SDA to EPA, and in particular DHA, in Atlantic salmon smolt (seawater phase); the authors noted that the ability of salmon and other species to digest, accumulate, and biosynthesize SDA into EPA and DHA needs to be further assessed before it can be considered as an ingredient in aquafeeds. We also propose that, for the benefits of SDA-containing oils in aquafeeds to be maximized, the n-3:n-6 ratio may need to be considerably higher than occurs in ESO.

Enrichment of meat and milk from livestock with EPA and DHA has been pursued by various groups, although the use of vegetable oils in livestock feed to increase EPA and DHA in meat and milk has also been limited by the inefficient conversion of ALA into EPA and DHA. In poultry, Kitessa and Young (2009) showed that the amounts of EPA + DHA per 100 g thigh muscle were 32 and 49 mg for rapeseed oil- and ESO-supplemented broilers, respectively. Hence, there is some evidence that ESO has the potential to improve the n-3 PUFA content of animal-derived foods for better health outcomes for the consumer.

Despite this, the current supply of ESO is so limited and expensive that it is presently neither practical nor commercially viable to use it in livestock and aquafeeds. Biotechnology companies have already launched high-SDA oils through the insertion of Δ-6 desaturase into traditional oilseeds like soybean (Bernal-Santos *et al.*, 2010). Such approaches have the potential to overcome the yield limitations as well as the toxins and allergens associated with echium, and possibly the capacity to improve the n-3 : n-6 ratio mentioned earlier.

ESO in human nutrition

Anti-inflammatory relief can be considered the basis of both the herbal medicine and modern use of ESO as an essential oil and nutritional supplement. Regarding the latter, there is now a consensus on the mechanisms by which n-3 and n-6 PUFAs are involved in inflammatory processes in the body. The major diseases and conditions with an inflammatory component include acute cardiovascular events, acute respiratory distress syndrome, allergic diseases, asthma (childhood and adult), atherosclerosis, cancer cachexia, chronic obstructive pulmonary disease, cystic fibrosis, inflammatory bowel disease (Crohn's disease, ulcerative colitis), lupus, multiple sclerosis, neurodegenerative disease of aging, obesity, and psoriasis, rheumatoid arthritis; systemic inflammatory response to surgery, trauma, and critical illness; and type 1 and type 2 diabetes (Calder, 2006). In addition to the well-recognized benefits against CVD, conditions and diseases where the evidences of n-3 PUFAs are considered to be greater are asthma, inflammatory bowel disease, and rheumatoid arthritis (Calder, 2006). Figure 112.3 presents the production of pro- and anti-inflammatory compounds from n-3 and n-6 PUFAs. The presence of significant quantities of ALA, SDA, and GLA in ESO (Table 112.2) enables it to play a potentially valuable anti-inflammatory role in the body through the supply of anti-inflammatory mediators arising from both its n-6 and, to a larger extent, n-3 PUFAs (Figure 112.3). Consequently, ESO has an added advantage over other vegetable oils, because it has an n-3 PUFA with an advanced step in the biosynthetic pathway as well as an n-6 PUFA which is atypically a precursor of anti-inflammatory eicosanoids.

Studies in mice have shown decreases in plasma TAG and very low density lipoprotein concentrations, and decreases in hepatic liver TAG content in mice supplemented with ESO (Zhang *et al.*, 2008). In humans, Miles *et al.* (2004) showed enhanced EPA levels in blood lipids of healthy young male volunteers when supplemented with SDA from ESO. Similarly, Surette *et al.* (2004) reported increases in plasma n-3 PUFAs and a 21% reduction in serum TAG in hypertriglyceridemic subjects supplemented with 15 g of ESO per day for 4 weeks. These studies have shown that ESO, as a source of n-3 PUFAs, can play a pivotal role in the prevention of chronic diseases. Recently, Harris *et al.* (2008) reported that SDA-enriched GM soybean oil increased the omega-3 index (an emerging CVD risk marker) in a study with human volunteers.

The existing evidence from animal and human studies, although based on short-term observations, points to comparative improvements from using ESO in tissue deposition of n-3 PUFAs and some biomarkers of CVD over other vegetable oils. However, there are indications that ESO will be superseded by the development of oils with greater SDA content and higher n-3:n-6 ratios through plant biotechnology. For instance, the SDA-enhanced oil from

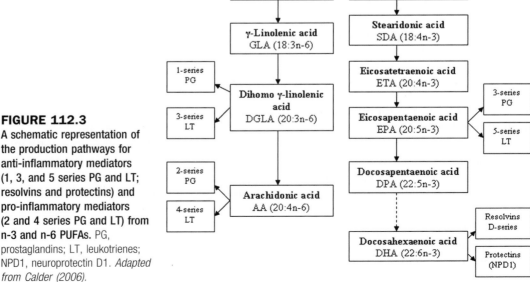

FIGURE 112.3

A schematic representation of the production pathways for anti-inflammatory mediators (1, 3, and 5 series PG and LT; resolvins and protectins) and pro-inflammatory mediators (2 and 4 series PG and LT) from n-3 and n-6 PUFAs. PG, prostaglandins; LT, leukotrienes; NPD1, neuroprotectin D1. *Adapted from Calder (2006).*

genetically modified soybeans used in the Bernal-Santos *et al.* (2010) dairy study had twice the SDA content — about 27% SDA in total fatty acids. Selection and breeding programs are needed to improve the oil yield per ha and SDA concentrations, and to decrease the anti-nutritional factors mentioned below.

ADVERSE EFFECTS AND REACTIONS (ALLERGIES AND TOXICITY)

Echium plants produce alkaloids as a chemical defense mechanism. The specific echium alkaloids are called pyrrolizidine alkaloids (PAs). PAs are hepatotoxic, and cause liver damage (Cheeke, 1988). In Australia, echium poisoning has been reported in sheep, cattle, horses, swine, and poultry (Cheeke, 1988). For humans, it is suggested that PA levels in herbal products of proven health benefit should be 1 µg per day for oral and 100 µg per day for external use, for a period of no more than 6 weeks in a year (Edgar *et al.*, 2002). The use of PA-containing herbal products in pregnant and lactating women is prohibited. Culvenor *et al.* (1981) reported 0.950 µg of PA per g of honey from *E. plantagineum*; hence, caution should be exercised in using honey from an area where bees largely rely on echium pollens and nectars to produce honey. With respect to ESO itself, the alkaloids are removed during extraction, and the PA content of echium seeds does not limit their use for the supply of n-3 oil.

SUMMARY POINTS

- *Echium* has a long history of use as a garden plant with some herbal medicine applications.
- ESO has a combination of specific n-3 and n-6 PUFAs that makes it unique among other vegetable oils.
- ESO is naturally rich (12—15%) in SDA, which is a superior precursor of EPA than ALA.
- Animal experiments and direct human supplementation studies show that ESO can bring about beneficial changes in biomarkers of chronic diseases, such as lowering TAG in plasma and tissue.
- ESO production is currently a niche activity, and widespread use is therefore limited by supply and cost.
- Selection and breeding programs are needed to move the role of echium in human health beyond niche-product status.

References

Bernal-Santos, G., O'Donnell, A. M., Vicini, J. L., Hartnell, G. F., & Bauman, D. E. (2010). Hot topic: enhancing n-3 fatty acids in milk fat of dairy cows by using SDA-enriched soybean oil from genetically modified soybeans. *Journal of Dairy Science, 93*, 32—37.

Berti, M., Johnson, B. L., Dash, S., Fischer, S., Wilckens, R., & Hevia, F. (2007). Echium: a source of SDA adapted to the northern great plains in the US. In J. Janick, & A. Whipkey (Eds.), *Proceedings of the 6th National New Crops Symposium — Issues in New Crops and New Uses, San Diego, October 14—18, 2007* (pp. 120—125). Alexandria, VA: ASHS Press.

Brenna, J., Salem, N., Jr., Sinclair, A. J., & Cunane, S. C. (2009). α-Linolenic acid supplementation and conversion to n-3 long-chain polyunsaturated fatty acids in humans. *Prostaglandins Leukotrienes and Essential Fatty Acids, 80*, 85—91.

Calder, P. C. (2006). n-3 polyunsaturated fatty acids, inflammation and inflammatory diseases. *American Journal of Clinical Nutrition, 83*(Suppl.), 1505S—1519S.

Cheeke, P. R. (1988). Toxicity and metabolism of pyrrolizidine alkaloids. *Journal of Animal Science, 66*, 2343—2350.

Chouinard, P. Y., Corneau, L., Butler, W. R., Chilliard, Y., Drackley, J. K., & Bauman, D. E. (2001). Effect of dietary lipid source on CLA concentrations in milk fat. *Journal of Dairy Science, 84*(3), 680—690.

Crozier, G. L., Fleith, M., Traitier, H., & Finot, P. A. (1989). Black currant seed oil feeding and fatty acids in liver lipid classes of guinea pigs. *Lipids, 24*(5), 460—466.

Culvenor, C. C. J., Edgar, J. A., & Smith, L. W. (1981). Pyrrolizidine alkaloids in honey from *Echium plantagineum* L. *Journal of Agricultural and Food Chemistry, 29*, 958—960.

Edgar, J. A., Roeder, E., & Molyneus, R. J. (2002). Honey from plants containing pyrrolizidine alkaloids: a potential threat to health. *Journal of Agricultural and Food Chemistry, 50*, 2719—2730.

FAO/WHO. (2009). Fats and fatty acids in human nutrition. *Annals of Nutrition and Metabolism (Special Issue)*, *55*, 1–3.

Harris, W. S., Lemke, S. L., Hansen, S. N., Goldstein, D. A., DiRienzo, M. A., Su, H., et al. (2008). SDA-enriched soybean oil increased the omega-3 index, an emerging cardiovascular risk marker. *Lipids, 43*, 805–811.

Kitessa, S. M., & Young, P. (2009). Echium oil is better than rapeseed oil in enriching poultry meat with n-3 polyunsaturated fatty acids, including EPA and DPA. *British Journal of Nutrition, 101*, 709–715.

Klemow, K. M., Clemens, D. R., Threadgill, P. F., & Cavers, P. B. (2002). The biology of Canadian weeds. 116. *Echium vulgare* L. *Canadian Journal of Plant Science, 82*, 235–248.

Miles, E. A., Banerjee, T., & Calder, P. C. (2004). The influence of different combinations of GLA, SDA and EPA on the fatty acid composition of blood lipids and mononuclear cells in human volunteers. *Prostaglandins Leukotrienes and Essential Fatty Acids, 70*, 529–538.

Miller, R. M., Bridle, A. R., Nichols, P. D., & Carter, C. G. (2008). Increased elongase and desaturase gene expression with SDA enriched diet does not enhance long-chain (n-3) content of seawater Atlantic salmon (*Salmo salar* L.). *Journal of Nutrition, 138*, 2179–2185.

Piggin, C. M. (1982). The biology of Australian weeds. 8. *Echium plantagineum* L. *Journal of the Australian Institute of Agricultural Science, 48*, 3–16.

Sprecher, H., Luthria, D. L., Mohammed, B. S., & Baykousheva, S. P. (1995). Re-evaluation of the pathways for the biosynthesis of polyunsaturated fatty acids. *Journal of Lipid Research, 36*, 2471–2477.

Surette, M. E., Edens, M., Chilton, F. H., & Tramposch, K. M. (2004). Dietary echium oil increases plasma and neutrophil long chain (n-3) fatty acids and lowers serum triacylglycerols in hypertriglyceridemic humans. *Journal of Nutrition, 34*, 1406–1411.

Whelan, J. (2009). Dietary SDA is a long chain (n-3) polyunsaturated fatty acid with potential health benefits. *Journal of Nutrition, 139*, 5–10.

Zhang, P., Boudyguina, E., Wilson, M. D., Gebre, A. K., & Parks, J. S. (2008). Echium oil reduces plasma lipids and hepatic lipogenic gene expression in ApoB100-only LDL receptor knockout mice. *Journal of Nutritional Biochemistry, 19*, 655–663.

CHAPTER 113

Extracts from Purple Wheat (*Triticum* spp.) and Their Antioxidant Effects

Trust Beta[1], Yang Qiu[1], Qin Liu[2], Anders Borgen[3]
[1] Department of Food Science, University of Manitoba, Winnipeg, Manitoba, Canada
[2] School of Food Science and Engineering, Nanjing University of Finance and Economics, Nanjing, Jiangsu, China
[3] Agrologica, Mariager, Denmark

959

LIST OF ABBREVIATIONS

ABTS, 2,2′-azinobis-(3-ethylbenzothiazoline-6-sulfonate)
CE, catechin equivalent
DiFA, diferulic acid
DPPH, 2,2′-diphenyl-1 picryl hydrazyl
FAE, ferulic acid equivalent
Fe(III)-TPTZ, ferric 2,4,6-tripyridyl-s-triazine
FRAP, ferric reducing antioxidant power
HPLC, high performance liquid chromatography
MS/MS, tandem mass spectrometry

Nuts & Seeds in Health and Disease Prevention. DOI: 10.1016/B978-0-12-375688-6.10113-6

ORAC, oxygen radical absorbance capacity

RSA, radical scavenging activity

TAC, total anthocyanin content

TE, Trolox equivalent

TEAC, Trolox equivalent antioxidant capacity

TFC, Total flavonoid content

TPC, Total phenolic content

INTRODUCTION

Wheat is an important cereal grain in the human diet, providing protein, carbohydrates, and dietary fiber. Commercial wheats are mostly white or red-colored grains. Although anthocyanin-pigmented wheat grains (i.e., purple, blue, and black wheats) are currently produced in small amounts, they have attracted more attention in recent years since they can be used to make special and colorful foods due to their phytochemical composition and distinctive colors. Purple wheat is one of the pigmented wheats derived from tetraploid Ethiopian wheat, having anthocyanins in the pericarp layer (Zeven, 1991).

The phytochemical profiles and antioxidant properties of different wheat varieties and their milled fractions have been investigated by several research groups (Adom *et al.*, 2003; Beta *et al.*, 2005), but most of them were focused on the common wheat, namely white or red wheat varieties. Purple wheat, due to its rare occurrence as a commercial crop in Western countries, has been less studied. The anthocyanin composition and other phytochemicals in purple wheat have been reported based on one or two samples (Abdel-Aal and Hucl, 2003; Siebenhandl *et al.*, 2007; Hosseinian *et al.*, 2008). Data obtained recently on three purple wheat genotypes have been included in this report.

BOTANICAL DESCRIPTION

Cultivated wheats (*Triticum* spp.) are characterized by purple-, red/brown- and white-colored pericarp in their kernels (Belay *et al.*, 1995). The taxonomic status of purple-grained tetraploid wheat remains unclear (Zeven, 1991). Only one accession of common wheat (*T. aestivum* L.), native to China, is known to have purple pericarp. The rest of the reported purple-grained tetraploid wheats are found in Ethiopian collections of the tetraploid (2n = 4x = 28) wheat taxa (included by some taxonomists within the broad species *T. turgidum* L. sensu lato) (Belay *et al.*, 1995). References have also been made to *T. dicoccum*, *T. durum*, *T. polonicum*, and "*T. aethiopicum* Jakubz" (syn. "*T. abyssinicum*"), which is *T. durum* from Ethiopia (Zeven, 1991). All tetraploid wheat taxa that are found in Ethiopia, with the possible exception of *T. dicoccon* Shrank (locally known as Adja), may possess the purple pericarp color, although in varying frequencies — very low in *T. polonicum* L., and high in *T. carthlicum* Nevski and *T. durum* Desf (Belay *et al.*, 1995). Sometimes the color might be deep purple or brownish-purple, and the latter may be almost black or violet, as in Chinese black-grained wheat (Li *et al.*, 2005). When the purple-grain character is transferred to hexaploid wheat, derivatives may show colors different from the tetraploid parent. Altitude, light intensity, and temperature may also influence the production of anthocyanins (Belay *et al.*, 1995).

HISTORICAL CULTIVATION AND USAGE

Purple-grained wheats originated from regions 2400 m above sea level; however, they are found mostly in the 2400—2800 m zone (Tesemma & Belay, 1991). With increased altitude, there is an associated waterlogging stress at earlier developmental stages of the wheat plant. Purple wheats are better equipped with favorable adaptive traits to highland growth conditions (Belay *et al.*, 1995). As reviewed by Zeven (1991), Charcoal, a purple wheat, was bred to incorporate genetic markers for feed or food. In 1938, BI6 pourpre d'Ngan-King (purple wheat from Ngan-King, the capital of the Ngan-Hwei province) was among the Chinese bread wheat accessions that were sent to France. Another accession, BI6 rouge de Han-Keou (that is, red wheat from Han-Keou, a part of the Hou-Pei province) comprised 20% purple grains.

PRESENT-DAY CULTIVATION AND USAGE

Purple-grain tetraploid wheats (*Triticum turgidum* L.) are widely cultivated in the Ethiopian highlands, despite the claim that they have lower industrial quality and market prices than the white or red grain types (Belay *et al.*, 1995). Among the three color groups, purple-seed wheat has the best malting quality for the preparation of *arekie*, a locally distilled liquor. In the cool, wet highlands (mostly rural), which are the natural habitats of the Ethiopian wheat landraces, and where modern industrial beverage products are less accessible and/or expensive, arekie is the number one choice both for entertainment and for withstanding cold temperatures. Arekie is also used for medicinal purposes, mainly as an aid to digestion and to relieve stomach discomfort (Belay *et al.*, 1995). Several special foods have recently been developed from these wheats by InfraReady Products (SK, Canada), including purple wheat bran muffins and antho-beers (Li *et al.*, 2007a). In follow-up studies, bread made from purple wheat flour showed the highest antioxidant capacity when compared with bread made from whole meal and refined flour prepared from common wheats.

APPLICATIONS TO HEALTH PROMOTION AND DISEASE PREVENTION

When their extracts are assessed for antioxidant effects, purple wheats have been found to have potential applications to health promotion and disease prevention.

Extraction of antioxidant components

The extraction is mainly focused on phenolic compounds (or phenolics), the largest group of hydrophilic antioxidants in cereal grains. The procedure was reported by Adom *et al.* (2003), with minor modifications. Purple wheat grain was ground and extracted with acidic methanol (methanol:1-M HCl = 85 : 15, v/v) and aqueous acetone (acetone : water = 70 : 30, v/v) to recover free and conjugated phenolic compounds. The residue was then subjected to alkaline hydrolysis and re-extracted with ethyl acetate to obtain bound phenolic acids.

Antioxidant properties of purple wheat extracts

The oxygen radical absorbance capacity (ORAC) assay measures the antioxidant capacity against peroxyl radicals. The ORAC values of purple wheat, determined according to the method reported by Li *et al.* (2007b), are shown in Figure 113.1 and expressed as Trolox equivalent (TE). The acidic methanol extracts exhibited approximately 3.5 times higher values than aqueous acetone extracts, indicating that the former could recover more extractable phenolics from purple wheat. The ORAC values of methanol extracts ranged from 3864 to 6898 μmol TE/100 g. The highest and lowest values were observed in the Charcoal and Indigo genotypes, respectively. The acetone extracts from purple wheat exhibited ORAC values ranging from 1126 to 1920 μmol TE/100 g. Significant differences in ORAC values can be attributed to several complex factors, including the wheat genotype and growing environment. The extraction solvents also significantly influence the ORAC values of purple wheat.

Free radical scavenging activity (RSA) against 2,2-diphenyl-1-picrylhydrazyl (DPPH) radical assay is a method used to evaluate hydrogen-donating antioxidants in plant extracts (Brand-Williams *et al.*, 1995). It evaluates the inhibitory activity of nitrogen radical induced oxidation (Karadag *et al.*, 2009). Figure 113.2 shows the DPPH RSA of different purple wheat extracts, and the results are expressed as μmol of Trolox equivalent (TE) per 100 g of grain on dry weight basis. Consistent with ORAC results, the acidic methanol extracts (715−856 μmol TE/100 g) exhibited higher DPPH RSA than the acetone extracts (492−616 μmol TE/100 g), but with

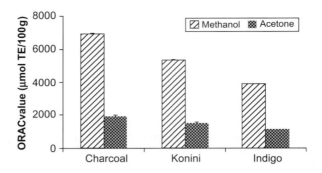

FIGURE 113.1
Oxygen radical absorbance capability (ORAC) of acidic methanol and aqueous acetone extracts from three purple wheat genotypes. Acidic methanol extracts from purple wheat exhibited higher ORAC values than acetone extracts. Charcoal genotype showed the highest ORAC values. Results are expressed as mean ± SD ($n = 3$).

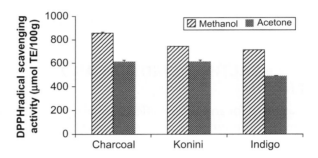

FIGURE 113.2
DPPH radical scavenging activity (RSA) of acidic methanol and aqueous acetone extracts from three purple wheat genotypes. Acidic methanol extracts exhibited higher DPPH RSA than acetone extracts. Charcoal purple wheat exhibited the highest DPPH RSA. Results are expressed as mean ± SD ($n = 3$).

smaller differences (1.2- to 1.5-fold). In both extracts, the RSA decreased in the order Charcoal > Konini > Indigo. Significant RSA was detected in all the purple wheat samples tested. Charcoal had the highest antioxidant activity using both ORAC and DPPH assays, indicating that it may provide more health benefits if grown in the right environment. However, the extracts above were composed of mainly free and soluble-conjugated phenolic compounds. In order to take into account unextractable (insoluble bound) phenolics, Siebenhandl *et al.* (2007) used sodium hydroxide to digest Ethiopian purple wheat samples, and then determined the antioxidant activity of wholemeal and milled fractions using the assays described below.

Trolox equivalent antioxidant capacity (TEAC) assay measures the scavenging ability of antioxidants to the long-lived 2,2′-azinobis-(3-ethylbenzothiazoline-6-sulfonate) (ABTS) radical cation. The total antioxidant activity of Ethiopian purple wheat wholemeal and its milled fractions were measured by this assay (Siebenhandl *et al.*, 2007) (Table 113.1). The TEAC of Ethiopian purple wheat wholemeal was 0.63 mmol/100 g. Among the milled fractions, the bran and shorts showed the highest TEAC (1.91 mmol/100 g), followed by the middling fraction (0.72 mmol/100 g). Purple wheat flour had the lowest TEAC, approximately four-fold less than the bran and shorts fraction. When considering the contribution to the total antioxidant activity of purple wheat, the relative amounts of each fraction produced from whole grain were taken into account. According to Siebenhandl *et al.*, (2007), Ethiopian purple wheat produced 0.9, 11.8, 44.6, and 42.7% of bran, shorts, middlings, and flour. Therefore, the middling fraction made the most contribution (42%) to the total antioxidant activity of purple wheat whole grain.

TABLE 113.1 Trolox Equivalent Antioxidant Capacity (TEAC) and Ferric Reducing Antioxidant Power (FRAP) of One Ethiopian Purple Wheat Wholemeal and Its Milling Fractions

Fraction	TEAC (mmol Trolox equiv/100 g)	FRAP (mmol Fe^{2+} equiv/100 g)
Bran + shorts	$1.91^a \pm 0.19$	$9.40^a \pm 1.05$
Middlings	$0.72^b \pm 0.11$	$2.74^b \pm 0.63$
Flour	$0.47^b \pm 0.02$	$2.75^b \pm 0.31$
Wholemeal	$0.63^b \pm 0.03$	$1.82^b \pm 0.15$

Significant differences in TEAC and FRAP were detected among purple wheat milling fractions, with the highest value observed for the bran and shorts fraction. Data are adapted from Siebenhandl *et al.* (2007) and expressed as mean $\pm$ SD. Means with the same superscripts are not significantly different.

Ferric reducing antioxidant power (FRAP) assay measures the ability of phenolics to reduce the yellow ferric 2,4,6-tripyridyl-s-triazine (Fe(III)-TPTZ) to a blue ferrous complex (Fe(II)-TPTZ). Similar to the TEAC results, the bran and shorts fraction exhibited the greatest FRAP (Table 113.1), approximately 3.5-fold higher than middling and flour fractions. The significant differences in FRAP among the milled fractions indicate that the antioxidants are not evenly distributed in purple wheat grain, but instead are predominantly located in the pericarp and aleurone layers (Beta *et al.*, 2005).

Phenolic compounds responsible for antioxidant effects of purple wheat extracts

TOTAL PHENOLIC CONTENT AND PHENOLIC ACID COMPOSITION

The extractable (free and soluble conjugated) phenolics in purple wheat were recovered by acidic methanol and aqueous acetone. Their total phenolic content (TPC) was evaluated by the method previously described by Singleton *et al.* (1965), and modified in our laboratory (Li *et al.*, 2007b). As shown in Table 113.2, methanol extracts contained more phenolics than the acetone ones. TPC ranged from 155 to 226 and 78 to 118 mg/100 g in methanol and acetone extracts, respectively. Charcoal had significantly higher TPC than Konini and Indigo purple wheat.

The unextractable (insoluble bound) phenolics were analyzed using high performance liquid chromatography (HPLC) coupled with tandem mass spectrometry (MS/MS) in extracts obtained by alkaline hydrolysis of purple wheat. Purple wheat contained a wild range of phenolic acids (Table 113.3). The total phenolic acid content in purple wheat genotypes varied

963

TABLE 113.2 Total Phenolic Content (TPC), Total Anthocyanin Content (TAC), and Total Flavonoid Content (TFC) of Different Purple Wheat Extracts

Cultivar*	ID** and Color	Acidified Methanol Extract		Aqueous Acetone Extract	
		TPC^a	TAC^b	TPC^a	TFC^c
*Charcoal	GRIN: Cltr 17422; dark purple	226.40 ± 5.95	23.45 ± 0.06	117.89 ± 2.29	102.95 ± 1.15
Konini	RICP 01C0203758; purple	181.27 ± 8.27	2.54 ± 0.04	108.66 ± 5.48	21.59 ± 0.58
Indigo	Commercial seed; purple	155.63 ± 2.16	7.24 ± 0.04	77.76 ± 2.03	35.79 ± 0.25

The extractable TPC, TAC, and TFC in purple wheat were influenced by wheat genotypes and extraction solvents. TPC, TAC, and TFC were expressed as mg of ferulic aid equivalents (FAE), anthocyanidin-3-glucoside equivalent, and catechin equivalent (CE) per 100 g of wheat grain (dry weight), respectively. Results are expressed as mean $\pm$ SD ($n = 3$).
*Triticumaestivum.
**Identification provided by Agrologica, Denmark.

TABLE 113.3 Phenolic Acids (mg/100 g on Dry Basis) in Purple Wheat Alkaline Hydrolyzates

Cultivar	Monomeric Phenolic Acids					Diferulic Acids			Total Phenolic Acids ($\sum$PA)
	Vanillic	Caffeic	*p*-Coumaric	Ferulic	Sinapic	5-5'	8-O-4'	8-5'	
Charcoal	3.35 ± 0.74	0.95 ± 0.07	2.87 ± 0.12	87.37 ± 1.42	3.04 ± 0.08	1.13 ± 0.15	6.62 ± 0.61	9.09 ± 0.44	114.43 ± 1.09
Konini	2.58 ± 0.64	1.4 ± 0.14	2.38 ± 0.26	81.38 ± 1.70	1.93 ± 0.21	1.62 ± 0.26	6.57 ± 0.81	11.7 ± 0.31	109.58 ± 1.29
Indigo	3.21 ± 0.65	0.84 ± 0.15	2.12 ± 0.20	86.59 ± 0.42	2.36 ± 0.11	0.89 ± 0.18	5.31 ± 0.99	8.26 ± 0.25	109.68 ± 1.31

Ferulic acid and 8-O-4' diferulic acid were the most dominant monomeric and dimeric phenolic acids in purple wheat. Results are expressed as mean ± SD ($n = 3$).

from 109 to 114 mg/100 g. Ferulic acid was the predominant acid (81–87 mg/100 g), constituting up to 79% of the total acids. Other phenolic acids identified included vanillic, *p*-coumaric, and sinapic, with caffeic acid present in trace amounts (< 1 mg/100 g). A high concentration of ferulic acid increases dimerization. Several ferulic acid dehydrodimers (also referred to as diferulic acids or diferulates) were identified in purple wheat, including 8-5', 8-O-4', and 5-5' diferulic acids (DiFA). They made up 13.2 to 17.9% of the total phenolic acids. The most abundant ferulic acid dehydrodimer was identified as 8-5' DiFA (benzofuran form) ranging from 8.3 to 11.7 mg/100 g. The concentration of 8-O-4' DiFA and 5-5' DiFA varied from 5.3 to 6.6 and 0.9 to 1.1 mg/100 g, respectively.

TOTAL ANTHOCYANIN CONTENT AND ANTHOCYANIN COMPOSITION

Anthocyanins are the primary pigments in purple wheat, and possess strong antioxidant activity. The total anthocyanin content (TAC) in purple wheat was extracted by acidic methanol, and determined as reported by Abdel-Aal *et al.* (1999). As seen in Table 113.2, the extracts from Charcoal, Konini, and Indigo exhibiting purple-red, pink, and red colors were found to contain significantly different levels, ranging from 2.5 to 23.5 mg/100 g. The highest TAC was found in Charcoal, followed by Indigo. Compared with previously published data for Konini, wholemeal TAC = 3.8 mg/100 g (Abdel-Aal *et al.*, 2006) and TAC = 6.1 mg/100 g (Abdel-Aal & Hucl, 2003), the values reported here were significantly lower, probably due to the variation in environmental factors (Siebenhandl *et al.*, 2007). According to Abdel-Aal *et al.* (2006), the environmental effect plays the most significant role in TAC of purple wheat due to the location of anthocyanin pigments in the outer pericarp layers. To date, 10 types of anthocyanins have been identified in purple wheat. Cyanidin 3-glucoside is the most abundant anthocyanin, followed by peonidin 3-glucoside. They represent approximately 31% and 16%, respectively of the total anthocyanins (Abdel-Aal *et al.*, 2006). Other anthocyanins include cyanidin-malonyl glucoside, peonidin-malonyl glucoside, peonidin-malonyl succinyl glucoside, cyanidin-succinyl glucoside, and peonidin-succinyl glucoside. The latter two were found as isomeric forms in purple wheat.

TOTAL FLAVONOID CONTENT AND PROANTHOCYANIDIN COMPOSITION

Flavonoids are an important class of phenolic compounds, accounting for approximately two-thirds of the dietary phenols. The total flavonoid content (TFC) of purple wheat was measured for 70% aqueous acetone extracts based on the method previously described by Adom *et al.* (2003). As shown in Table 113.2, purple wheat contained a wide range of TFC, varying from 21.6 (Charcoal) to 103.0 (Konini) mg/100 g. Indigo had a moderate TFC (35.8 mg/100 g). The aqueous acetone extracts of purple wheat were also used for detecting proanthocyanidins, which are known to exhibit antioxidant effects. Proanthocyanidins are polymerized phenolics comprising dimeric to polymeric flavan-3-ol units. Wheat proanthocyanidins (procyanidin B3, prodelphinidin B3, and monomeric unit (+)-catechin) were first reported in red-grained wheat bran (Rongotea and Oroua) by McCallum and Walker (1990). Recently, Dinelli *et al.* (2009) detected these compounds in several wheat varieties; procyanidin B3 in Iride and Kamut, and prodelphinidin B3 in Solex, Levante, Urris and Kamut wheat. These findings suggest that purple wheat may also comprise proanthocyanidins. By applying HPLC-MS/MS, no proanthocyanidins were detected in the purple genotypes Charcoal, Konini, and Indigo.

ADVERSE EFFECTS AND REACTIONS (ALLERGIES AND TOXICITY)

No reports are available on allergies and toxicity resulting from purple wheat products. However, it can be speculated that adverse effects and reactions associated with common wheat may also be applicable to purple wheat. Indeed, the peptides derived from gluten proteins present in wheat are known to be responsible for celiac disease, an intestinal disorder caused by T-cell responses to these peptides (Spaenij-Dekking *et al.*, 2005). Secondary intolerances including viral hepatitis and intestinal infections may also occur in predisposed individuals. However, there is potential for selection of non-toxic varieties for celiac-disease patients. High levels of wheat-specific immunoglobulin E (IgE) have been reported in patients with anaphylaxis (Pourpak *et al.*, 2004).

SUMMARY POINTS

- Extracts from purple wheat differ in their antioxidant effects, depending on the extraction methods employed as well as genotypic and environmental effects.
- Acidic methanol appears to recover more extractable antioxidants from purple wheat than does aqueous acetone.
- Among the purple wheats (Charcoal, Konini, and Indigo), Charcoal had the highest antioxidant activity, likely resulting from its high content of total phenolics, anthocyanins, and flavonoids.
- Knowledge of the antioxidant effects of several purple wheats would be useful for screening and selecting genotypes with higher antioxidant activity and potentially health-promoting properties.
- The adverse effects and reactions associated with common wheat may also be applicable to purple wheat.

965

References

Abdel-Aal, E.-S. M., & Hucl, P. (1999). A rapid method for quantifying total anthocyanins in blue aleurone and purple pericarp wheats. *Cereal Chemistry, 76,* 350—354.

Abdel-Aal, E.-S. M., & Hucl, P. (2003). Composition and stability of anthocyanins in blue-grained wheat. *Journal of Agricultural and Food Chemistry, 51,* 2174—2180.

Abdel-Aal, E.-S. M., Young, J. C., & Rabalski, I. (2006). Anthocyanin composition in black, blue, pink, purple, and red cereal grains. *Journal of Agricultural and Food Chemistry, 54,* 4696—4704.

Adom, K. K., Sorrells, M. E., & Liu, R. H. (2003). Phytochemical profiles and antioxidant activity of wheat varieties. *Journal of Agricultural and Food Chemistry, 51,* 7825—7834.

Belay, G., Tesemma, E. B., & Mitiku, D. (1995). Natural and human selection for purple-grain tetraploid wheats in the Ethiopian highlands. *Genetic Resources and Crop Evolution, 42,* 387—391.

Beta, T., Nam, S., Dexter, J. E., & Sapirstein, H. D. (2005). Phenolic content and antioxidant activity of pearled wheat and roller-milled fraction. *Cereal Chemistry, 82,* 390—393.

Brand-Williams, W., Cuvelier, M. E., & Berset, C. (1995). Use of a free radical method to evaluate antioxidant activity. *LWT-Food Science and Technology, 28,* 25—30.

Dinelli, G., Carretero, A. S., Di Silvestro, R., Marotti, I., Fu, S., Benedettelli, S., et al. (2009). Determination of phenolic compounds in modern and old varieties of durum wheat using liquid chromatography coupled with time-of-flight mass spectrometry. *Journal of Chromatography A, 1216,* 7229—7240.

Hosseinian, F. S., Li, W., & Beta, T. (2008). Measurement of anthocyanins and other phytochemicals in purple wheat. *Food Chemistry, 109,* 916—924.

Karadag, A., Ozcelik, B., & Saner, S. (2009). Review of methods to determine antioxidant capacities. *Food Analytical Methods, 2,* 41—60.

Li, W., Shan, F., Sun, S., Corke, H., & Beta, T. (2005). Free radical scavenging properties and phenolic content of Chinese black-grained wheat. *Journal of Agricultural and Food Chemistry, 53*, 8533–8536.

Li, W., Pickard, M. D., & Beta, T. (2007a). Evaluation of antioxidant activity and electronic taste and aroma properties of antho-beers from purple wheat grain. *Journal of Agricultural and Food Chemistry, 55*, 8958–8966.

Li, W., Pickard, M. D., & Beta, T. (2007b). Effect of thermal processing on antioxidant properties of purple wheat bran. *Food Chemistry, 104*, 1080–1086.

McCallum, J. A., & Walker, J. R. L. (1990). Proanthocyanidins in wheat bran. *Cereal Chemistry, 67*, 282–285.

Pourpak, Z., Mansouri, M., Mesdaghi, M., Kazemneja, A., & Farhoudi, A. (2004). Wheat allergy: clinical and laboratory findings. *International Archives of Allergy and Immunology, 133*, 168–173.

Siebenhandl, S., Grausgruber, H., Pellegrini, N., Rio, D. D., Fogliano, V., Pernice, R., et al. (2007). Phytochemical profile of main antioxidants in different fractions of purple and blue wheat, and black barley. *Journal of Agricultural and Food Chemistry, 55*, 8541–8547.

Singleton, V. L., & Rossi, J. A. (1965). Colorimetry of total phenolics with phosphomolybdic-phosphotungstic acid reagents. *American Journal of Enology and Viticulture, 16*, 144–158.

Spaenij-Dekking, L., Kooy-Winkelaar, Y., van Veelen, P., Wouter Drijfhout, J., Jonker, H., van Soest, L., et al. (2005). Natural variation in toxicity of wheat: potential for selection of nontoxic varieties for celiac disease patients. *Gastroenterology, 129*, 797–806.

Tesemma, T., & Belay, G. (1991). Aspects of Ethiopian tetraploid wheats with emphasis on durum wheat genetics and breeding research. In H. Gebre Mariam, D. G. Tanner, & M. Hulluka (Eds.), *Wheat research in Ethiopia: A historical perspective* (pp. 47–71). Addis Ababa, Ethiopia: IAR/CIMMYT.

Zeven, A. C. (1991). Wheat with purple and blue grains: a review. *Euphytica, 56*, 243–258.

CHAPTER 114

Rapeseed (*Brassica napus*) Oil and its Benefits for Human Health

Gerhard Jahreis, Ulrich Schäfer
Friedrich Schiller University, Institute of Nutrition, Jena, Germany

967

LIST OF ABBREVIATIONS

ALA, α-linolenic acid
ARA, arachidonic acid
COX, cyclooxygenases
CVD, cardiovascular disease
DGA, dihomo-γ-linolenic acid
DHA, docosahexaenoic acid
DPA, docosapentaenoic acid
EPA, eicosapentaenoic acid
FA, fatty acid
FAME, fatty acid methylester
LA, linoleic acid
LC-PUFA, long-chain polyunsaturated fatty acid
LOX, lipoxygenases
LT, leukotrienes
MUFA, monounsaturated fatty acid
PG, prostaglandin
PUFA, polyunsaturated fatty acid

Nuts & Seeds in Health and Disease Prevention. DOI: 10.1016/B978-0-12-375688-6.10114-8

RBC, red blood cell
SFA, saturated fatty acid
TAG, triacylglyceride
TXA, thromboxane

INTRODUCTION

To establish healthy and tasty oils for human consumption, a cooperation of plant breeding, food processing, and nutritional physiology is necessary. Such oils are expected to have a low percentage of SFAs, a high percentage of MUFAs and n-3 PUFAs, and a low n-6 : n-3 ratio. Furthermore, they should have low levels of adverse fatty acids (erucic acid) and other undesirable substances (e.g., pesticides), as well as high levels of nutritionally important compounds (e.g., tocopherols). The history of modern rapeseed oil reflects successful development in this direction.

BOTANICAL DESCRIPTION

Brassica napus Linnaeus is a yellow flowering member of the Brassicacae family (Figure 114.1), also known as the mustard or cabbage family. *B. napus* is a hybrid evolved from *B. rapa* (syn. *B. campestris*) and *B. oleracea* (wild mustard) via crossbreeding.

B. napus or rapeseed is cultivated because its seeds yield about 40% oil. All parts of the rapeseed plant are utilized:

- Seeds are used for the production of oil for human nutrition, or for biodiesel
- Leaves are consumed as vegetables (mainly in Asia) and as animal fodder
- Oilseed cake and rapeseed meal (after hexane extraction), being by-products of oil processing, are used as protein-rich animal feed
- Dried stalks are used as domestic fuel.

HISTORICAL CULTIVATION AND USAGE

The original wild-type oil of *B. napus* was formerly used as lamp oil and as a lubricant for steam engines. It had an unpleasant bitter taste, and was rich in erucic acid (about 50%; assumed to be toxic) and in glucosinolates.

The success story of rapeseed oil came in the 1960s and 1970s. Breeding efforts resulted in the development of *B. napus* strains low in erucic acid. In Canada in particular, breeders succeeded

FIGURE 114.1
Blooming rapeseed field in the northern part of Germany. The worldwide production of rapeseed amounts to about 50 million tonnes, of which about one-fifth is produced in China. ©*UFOP e.V.*

in reducing the content of erucic acid and glucosinolates. This improved rapeseed was provisionally labelled "Can. O., L.-A". (Canadian oilseed, low acid). From this abbreviation the artificial name "Canola" was created.

PRESENT-DAY CULTIVATION AND USAGE

Today, Canola-quality means "double low" or "double zero" or "00-quality" (low in erucic acid and in glucosinolates). This is the only type of rapeseed oil allowed for human nutrition in North America and Europe, as well as in most Asian countries. In just a few decades, the rapeseed oil has transformed from being lamp oil to being an edible oil with a physiologically excellent fatty acid distribution. It contains the two essential fatty acids: LA and ALA.

ALA is present only in small quantities, if at all, in most other edible oils, with the exception of flaxseed oil, which is rich in ALA (Table 114.1). Since the early 1990s, rapeseed production in the European Union has shifted to rapeseed 00-quality, which is genetically still unchanged. At present, three types of rapeseed oil for human usage are on the market:

1. *Refined edible oil.* This is an oil of 00-quality obtained after physical pressing and subsequent hexane extraction. The oil is in general degummed, dried, bleached, and deodorized. The crude oil has to be refined in order to remove water, phospholipids, free fatty acids, naturally colored substances (e.g., chlorophylls), flavorings, and natural breakdown and oxidation products. Rapeseed oil treated in this way is tasteless, consists of only TAGs (with small quantities of phytosterols and tocopherols), has a high smoke point (the temperature to which oil can be heated before it smokes and discolors) of > 200°C, and a long shelf life.

2. *Cold-pressed edible oil.* Small and medium-sized plants produce oil of 00-quality by carefully physical pressing at a low temperature. Because of the gentle pressing process, rapeseed cake with a high oil content is a by-product; this is used as a high-value animal feed. The

TABLE 114.1 Fatty Acid Content (% FAME) of Vegetable Edible Fats and Oils

Oil	SFA	MUFA	PUFA		
			Linoleic Acid (n-6)	α-Linolenic Acid (n-3)	n-6 : n-3 Ratio
Common edible oils					
Flaxseed	9	21	15	53	1 : 4
Rapeseed	6	59	22	10	2 : 1
Soybean	15	23	54	8	7 : 1
Olive	15	75	10	< 1	15 : 1
Corn	13	29	57	1	57 : 1
Sunflower	12	16	72	< 1	140 : 1
Safflower	10	14	76	0.1	760 : 1
Sesame	13	40	44	0	Not available
Palm	48	44	8	0	Not available
Coconut	89	8	2	0	Not available
Peanut	13	48	32	0	Not available
Special oils					
Perilla	9	20	14	55	1 : 4
Chia seed	10	8	19	61	1 : 3
Camelina	8	17	19	36	1 : 2
Hemp	8	13	57	19	3 : 1
Walnut	9	20	54	13	4 : 1

Edible rapeseed oil meets all requirements of a food lipid for the prevention of CVD. It has the lowest content of SFA among the commonly used edible oils and fats, and contains the second highest percentage of n-3 FA after flaxseed oil.

majority of phospholipids remain in this cake. After hydration, the rest of the phospholipids are removed by a sedimentation process. Commercially cold-pressed edible oil has a typical seed-like taste, and generally a higher content of tocopherols but a shorter shelf life compared to refined oil. Depending on producers' experience, there are sometimes problems with the sensory quality (adverse taste), oxidation stability (Matthäus & Brühl, 2003), and relatively low smoke point (130°C−190°C).

3. *HOLLi edible oil.* Oil from a new rapeseed cultivar with high oleic acid and low ALA contents (HOLLi) has been available since 2008. This variety contains only about 3% ALA, whereas conventional Canola oil contains 8−12% ALA. The lower ALA content guarantees a higher oxidative stability. This oil is recommended for frying and other types of food processing involving high-temperature treatment. The smoke point of refined HOLLi oil is comparable to that of refined edible rapeseed oil, and it has the highest oxidative stability among the described three types of edible oil. However, the extremely low ALA content disqualifies the HOLLi oil in comparison with the two other types of ALA-rich rapeseed oil. Thus, considering health aspects, it is strongly recommended that the declaration "edible rapeseed oil" has to be reserved exclusively for 00-quality with an ALA content of about 10% (Barth, 2009).

APPLICATIONS TO HEALTH PROMOTION AND DISEASE PREVENTION
Metabolism of n-3 fatty acids

DHA is the most abundant n-3 LC-PUFA in the membranes of human cells (Arterburn *et al.*, 2006). The retina contains more than 20 g/100 g, the cerebral cortex and sperm contain about 14 g/100 g, and all other tissues contain significantly less than 5 g/100 g of total fatty acids. ALA and EPA are generally present in marginal percentages in human tissues. ALA was only found in muscle tissues, in cheek, and in adipose tissues, in concentrations < 1 g/100 g (Arterburn *et al.*, 2006).

Among PUFAs, LA is the fatty acid with substantial consumption via seed oils in both Western countries and developing countries. The intake of LA has increased dramatically due to the use of soybean oil and sunflower oil, and, in some countries, safflower oil (Table 114.1). In contrast, the intake of n-3 fatty acids has remained relatively constant during recent years, though it has risen in countries where Canola oil has been introduced to the diet (Russo, 2009). The predominant source of n-3 LC-PUFAs is fish (EPA and DHA). Vegetable oils are the main source of ALA.

It is assumed that the intake of ALA is effective in increasing plasma EPA and DPA (C22:5n-3) concentrations, but has only little effect on DHA concentration. The extent of ALA conversion to DHA in humans seems to be minimal − utilizing stable isotopes, the conversion rate of ALA to DHA in omnivores has been estimated to be < 1%.

In humans, a desaturation-chain elongation pathway (Figure 114.2), predominantly in the liver but also in the brain, takes place, converting LA or ALA to the corresponding LC-PUFA. Thus, EPA and DHA should not strictly be considered as essential fatty acids. In agreement, from tracer studies and dietary supplementation studies it can be concluded that the conversion of ALA to EPA is low (< 5%), and to DHA even lower − not greater than 0.5% (Arterburn et al., 2006). This means that up to 85% of dietary ALA appears to be oxidized for energy generation (Barcelo-Coblijn & Murphy 2009).

During the past few years, the $\triangle$-6 desaturation product of ALA, namely stearidonic acid (C18:4n-3), has been used in studies as a more effective precursor for eicosanoids compared with ALA. Because the human DHA pool is greater than the EPA pool, supplementation periods of more than 6 months are necessary to ensure that the conversion rate contributes to an increase in the DHA pool.

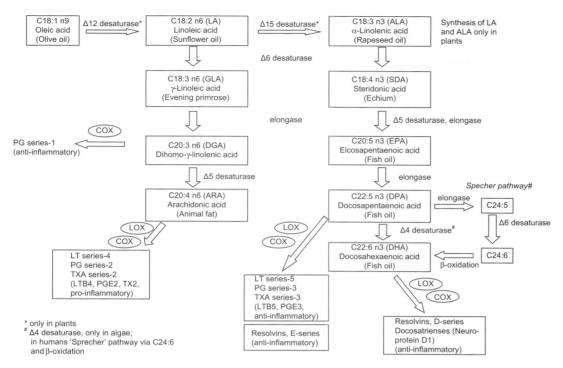

FIGURE 114.2

Conversion of long-chain n-6 and n-3 PUFAs to their corresponding very long-chain metabolites, eicosanoids and resolvins. The figure includes anti-inflammatory or pro-inflammatory properties of the important metabolites. The term eicosanoids summarizes biologically active signaling molecules which are oxygenated derivatives from three different kinds of PUFA, namely EPA, ARA, and DGA (C20:3n-6), all being 20 carbon atoms in length. The eicosanoid group includes leukotrienes, prostanoides, and thromboxanes. There is metabolic competition for elongation and desaturation of enzymes between *n*-6 and *n*-3 fatty acids. A higher intake of n-6 fatty acids inhibits the elongation of ALA and the formation of anti-inflammatory metabolites.

ALA supplementation does not result in appreciable changes of n-3 LC-PUFAs after short-term supplementation for 28 days (Table 114.2). Only EPA and DHA supplementation significantly reduces the ARA concentration of plasma phospholipids. Thus, the need to establish dietary reference intakes for individual LC-PUFAs was suggested (Kris-Etherton *et al.*, 2009), rather than recommendations on the total intake of, for example, EPA + DHA, as issued by numerous health authorities.

Whereas short-term supplementation with EPA and DHA is very effective in increasing n-3 LC-PUFAs in serum or RBCs in humans (Tables 114.2, 114.3), only long-term supplementation (1 year!) with ALA leads to similar results (Table 114.3). After long-term supplementation,

TABLE 114.2 Effects of Short-term Dietary Supplementation with ALA (3.7 g/d), EPA-ethyl Ester (4 g/d), or DHA-ethyl Ester (4 g/d) on Fatty Acids in Plasma Phospholipids (Arterburn *et al.*, 2006)

Dietary Supplementation	Effects on Fatty Acids in Plasma Phospholipids			
	ALA	EPA	DHA	ARA
ALA	—	↑	—	↑
EPA	—	↑ ↑	—	↓ ↓
DHA	—	↑	↑ ↑	↓ ↓

Dietary supplementation with ALA only marginally increases EPA in the plasma phospholipids. ARA is highly significantly (↓ ↓) reduced after EPA- and DHA-ethyl ester supplementation.

TABLE 114.3 Selected Fatty Acids and Fatty Acid Ratios of RBC After Supplementation with EPA + DHA (3 g/d) from Fish Oil for 15 Weeks, or One Spoonful of Flaxseed Oil per Day over 1 Year (5 g ALA/d) (unpublished own results)

	Short-Time Study (3 g EPA + DHA/d)		Long-Time Study (5 g ALA/d)
	Start	After 15 Weeks	After 1 Year
Fatty acids in RCB (% FAME)			
ALA	0.06	0.08	0.47
ARA	14.7	14.0	12.7
EPA	0.71	2.06	1.13
DPA	1.60	2.18	2.73
DHA	2.71	3.80	2.60
Ratio			
ARA/EPA	23.8	9.67	11.2
ARA/EPA + DPA + DHA	3.07	1.88	1.97
n-6 : n-3	4.64	2.80	3.26

Long-term supplementation with ALA (flaxseed oil) increases EPA and DPA in the RBC compared to start. The ratios of ARA/EPA + DPA + DHA and n-6 : n-3 are similarly decreased by short-term supplementation with fish oil fatty acids, and also by long-term supplementation with flaxseed oil.

a six-fold higher percentage of ALA in RBCs, an increase of EPA compared to start, and, most interestingly, a high increase in DPA (C22:5 n-3) compared to start were found (Figure 114.2). Also in our study, there was no change of the DHA percentage in RBCs (Table 114.3). As shown, there is a bottleneck from DPA to DHA. However, the ratio of ARA : EPA + DPA + DHA is improved after long-term ALA supplementation, as in the case of EPA + DHA supplementation.

The eicosanoids are crucial metabolites of PUFA. They are involved in the modulation of the intensity and duration of inflammatory responses. In recent years, the relevance of resolvins formed from both EPA and DHA has been intensively investigated. As the name indicates, resolvins appear to exert anti-inflammatory actions. As shown in Figure 114.2, the eicosanoids as well as the resolvins of n-3 LC-PUFAs exert mostly anti-inflammatory effects and help to prevent diseases in which the genesis of inflammatory processes plays a role, such as rheumatism, CVD, type 2 diabetes, and obesity (metabolic syndrome).

Health effects of ALA

A meta-analysis by Brouwer *et al.* (2004) underlines ALA's association with a reduced risk of fatal coronary heart diseases. An inverse relationship between ALA intake (estimated by adipose tissue analysis and a food frequency questionnaire) and reinfarction risk has recently been found (Campos *et al.*, 2008). On the other hand, Wendland *et al.* (2006) suggested that the impact of ALA in preventing CVD is unlikely to be mediated through metabolic products like EPA or DHA. Only a decrease in fibrinogen concentration and fasting plasma glucose was found due to ALA supplementation (review of 14 studies). Moderate amounts of dietary ALA (6.0 g/d), EPA (2.8 g/d), or DHA (2.9 g/d) administered to healthy volunteers did not significantly affect blood concentrations of glucose, insulin, fructosamine, and HbA1c in healthy men and women over a very short trial of 3 weeks (Egert *et al.*, 2009). The intake of these three n-3 LC-PUFAs only affects fasting serum TAG (EPA, −0.14 ml/l; DHA, −0.30 ml/l; ALA, −0.17 ml/l). The authors concluded that moderate amounts of ALA, EPA, and DHA are effective in improving lipid profiles of healthy humans (Egert *et al.*, 2009).

In a recent review of prospective cohort studies and randomized clinical trials, Mente *et al.* (2009) found causal links between dietary factors and CVD. The selected studies with high methodological quality give strong evidence for SFA intake and CVD risk. This association supports the beneficial health effect of the low-SFA rapeseed oil. The authors found insufficient evidence of an association of CVD with intake of ALA.

However, there is strong evidence, shown by meta-analysis, of a significant dose—response relationship between the risk of CVD death and n-3 LC-PUFA intake, with a relative risk reduction of 37% at an average EPA + DHA intake of 566 mg/d (Harris et al., 2008). In the USA and Germany, the daily intake of EPA and DHA is around 0.1—0.3 g. Recommendations regarding adequate ingestion of n-3 LC-PUFAs vary from 0.2 g/d to 0.65 g/d. Moreover, the American Heart Association and the European Society of Cardiology recommend a daily intake of 1 g n-3 LC-PUFA to prevent CVD.

In our study (Dawczynski et al., 2010), intervention with about 3 g of n-3 LC-PUFA resulted in a significant improvement of cardiovascular risk factors — for example, n-3 fatty acid index, AA/EPA ratio, total cholesterol, and TAG. The TAG concentration and LDL/HDL ratio were lower at the end of the intervention period in comparison with the control period, whereas HDL cholesterol was higher at the end of the intervention period. Most importantly, n-3 LC-PUFA intervention did not cause additional oxidative DNA damage, as shown by 7,8-dihydro-8-oxo-2′-deoxyguanosine excretion.

General scientific opinion is that the use of rapeseed oil is limited by the low conversion rate of ALA to n-3 LC-PUFAs. In addition, the high dietary intake of n-6 PUFAs impairs this conversion in populations of industrial countries. Nevertheless, the conversion cannot be neglected; otherwise, populations without meat and fish consumption (e. g., vegans) could not exist. The fatty acid elongation and desaturation in such people is not inhibited by dietary EPA and/or DHA. Thus, their conversion rate of ALA to n-3 LC-PUFAs is more effective (Figure 114.2). In Maasai, without fish and with a low meat intake, we found an ideal fatty acid distribution (ratio of AA : EPA) in the erythrocytes, regardless their high intake of saturated fatty acids (Knoll et al., 2010). Also, pregnant and breast-feeding women are high converters through estrogen-related enzyme stimulation.

Investigations show that long-term application of ALA-containing oils contributes to the enrichment of ALA and EPA in human cell membranes. This enrichment is the pre-condition for the formation of anti-inflammatory eicosanoids. Furthermore, and most importantly, the enrichment of ALA and EPA in the cell membranes leads to displacement of n-6 fatty acids, and consequently to a smaller formation of pro-inflammatory eicosanoids. While in one epidemiological study no association between ALA and CVD was found (Mente et al., 2009), both metabolic and clinical studies underline the preventive potential of ALA. Moreover, Nguemeni et al. (2010) conclude from their studies that rapeseed oil can be recommended as a functional food/nutraceutical aiding in stroke prevention.

ADVERSE EFFECTS AND REACTIONS (ALLERGIES AND TOXICITY)

The German Reference Values for Nutrient Intake (2008) recommend a supply of n-3 PUFAs at the level of 0.5% of the daily energy intake (ca. 1 g/d). The International Society for the Study of Fatty Acids and Lipids recommends an intake of at least 200 mg of EPA and/or DHA per day; the total intake of DHA + EPA per day should be about 650 mg, or 0.2% of the daily energy intake, for adults.

Studies investigating risk assessment established that the enrichment of food and feed with DHA- and/or EPA-rich sources may lead to an intake of n-3 fatty acids in humans fourfold greater than the recommended level. The following risks are recognized regarding excessive consumption of n-3 LC-PUFAs:

- A negative influence on the immune defense system, especially in the elderly
- A rise in serum LDL cholesterol
- An extended bleeding time, as in Inuits.

The authorities recommend that, on average, not more than 1.5 g n-3 LC-PUFAs should be consumed from all sources per day. Foods that usually do not contain any fat (e.g., water-based drinks) should not be enriched with n-3 fatty acids.

SUMMARY POINTS

- Rapeseed oils for human nutrition are offered as a refined edible oil with 00-quality, as a cold-pressed edible oil, and as HOLLi quality oil (very low α-linolenic acid of 3%).
- Refined and cold-pressed rapeseed oils are valuable because of their high α-linolenic acid content (about 10%), their low saturated fatty acid content (6%), and their optimal n-6 : n-3 ratio (2 : 1).
- In humans, α-linolenic acid is converted to eicosapentaenoic acid and docosahexaenoic acid; these two fatty acids are therefore not essential.
- The conversion efficiency of α-linolenic acid to long-chain n-3 PUFAs, important precursors of anti-inflammatory eicosanoids, is limited. High dietary intake of n-6 PUFA impairs the conversion rate.
- According to several studies, consumption of n-3 PUFA is positively correlated with cardiovascular and neuropsychiatric health effects, and results in a lower ratio of arachidonic acid/eicosapentaenoic acid in human cell membranes. These data support the health-beneficial effects of rapeseed oil.

References

Arterburn, L. M., Hall, E. B., & Oken, H. (2006). Distribution, interconversion, and dose response of n-3 fatty acids in humans. *American Journal of Clinical Nutrition, 83,* 1467—1476.

Barcelo-Coblijn, G., & Murphy, E. J. (2009). Alpha-linolenic acid and its conversion to longer chain n-3 fatty acids: benefits for human health and a role in maintaining tissue n-3 fatty acid levels. *Progress in Lipid Research, 48,* 355—374.

Barth, C. A. (2009). Nutritional value of rapeseed oil and its high oleic/low linolenic variety — a call for differentiation. *European Journal of Lipid Science and Technology, 111,* 856—953.

Brouwer, I. A., Katan, M. B., & Zock, P. L. (2004). Dietary alpha-linolenic acid is associated with reduced risk of fatal coronary heart disease, but increased prostate cancer risk: a meta-analysis. *Journal of Nutrition, 134,* 919—922.

Campos, H., Baylin, A., & Willett, W. C. (2008). Alpha-linolenic acid and risk of nonfatal acute myocardial infarction. *Circulation, 118,* 339—345.

Dawczynski, C., Martin, L., Wagner, A., & Jahreis, G. (2010). n-3 LC-PUFA-enriched dairy products are able to reduce cardiovascular risk factors: a double-blind, cross-over study. *Clinical Nutrition.* in press.

Egert, S., Kannenberg, F., Somoza, V., Erbersdobler, H. F., & Wahrburg, U. (2009). Dietary alpha-linolenic acid, EPA, and DHA have differential effects on LDL fatty acid composition but similar effects on serum lipid profiles in normolipidemic humans. *Journal of Nutrition, 139,* 861—868.

Harris, W. S., Kris-Etherton, P. M., & Harris, K. A. (2008). Intakes of long-chain omega-3 fatty acid associated with reduced risk for death from coronary heart disease in healthy adults. *Current Atherosclerosis Reports, 10,* 503—509.

Knoll, N., Kuhnt, K., & Jahreis, G. (2010). *Favourable ratio of arachidonic acid to eicosapentaenoic acid in the Maasai's erythrocytes despite the high intake of saturated fatty acids.* In preparation.

Kris-Etherton, P. M., Grieger, J. A., & Etherton, T. D. (2009). Dietary reference intakes for DHA and EPA. *Prostaglandins Leukotrienes and Essential Fatty Acids, 81,* 99—104.

Matthäus, B., & Brühl, L. (2003). Quality of cold-pressed edible rapeseed oil in Germany. *Nahrung, 47,* 413—419.

Mente, A., de Koning, L., Shannon, H. S., & Anand, S. S. (2009). A systematic review of the evidence supporting a causal link between dietary factors and coronary heart disease. *Archives of Internal Medicine, 169,* 659—669.

Nguemeni, C., Delplanque, B., Rovère, C., Simon-Rousseau, N., Gandin, C., Agnani, et al. (2010). Dietary supplementation of alpha-linolenic acid in an enriched rapeseed oil diet protects from stroke. *Pharmacological Research, 61,* 226—233.

Russo, G. L. (2009). Dietary n-6 and n-3 polyunsaturated fatty acids: from biochemistry to clinical implications in cardiovascular prevention. *Biochemical Pharmacology, 77,* 937—946.

Wendland, E., Farmer, A., Glasziou, P., & Neil, A. (2006). Effect of alpha-linolenic acid on cardiovascular risk markers: a systematic review. *Heart, 92,* 166—169.

CHAPTER 115

Use of Red Clover (*Trifolium pratense* L.) Seeds in Human Therapeutics

Hatice Çölgeçen[1], U. Koca[2], H.N. Büyükkartal[3]
[1] Zonguldak Karaelmas University, Faculty of Arts and Science, Department of Biology, İncivez, Zonguldak, Turkey
[2] Gazi University, Faculty of Pharmacy, Department of Pharmacognosy, Etiler, Ankara, Turkey
[3] Ankara University, Faculty of Science, Department of Biology, Tandoğan, Ankara, Turkey

LIST OF ABBREVIATIONS

HPLC-DAD, high performance liquid chromatography-diode array detection
SHR, spontaneously hypertensive rats

INTRODUCTION

Trifolium, a legume and a valuable feed plant, has long been providing significant contributions to agricultural and animal production in Europe and America. Red clover (*T. pratense* L.) grows naturally in southeast Europe and Anatolia; in fact, Anatolia has been accepted as the motherland of *T. pratense* (Çölgeçen & Toker, 2008). Seven diploid ($2n = 2x = 14$) varieties have been found in all over the world. *T. pratense* is a natural source of valuable isoflavonoids,

Nuts & Seeds in Health and Disease Prevention. DOI: 10.1016/B978-0-12-375688-6.10115-X

which are used in commercial (e.g., Menoflavon®, Rimostil®, Promensil®) products in Europe and America.

BOTANICAL DESCRIPTION

Trifolium, belonging to Fabaceae, has about 300 species worldwide, 25 of which have been cultured for a variety of purposes. *T. pratense* L. is a perennial herbaceous plant. The main root in red clover is not as thick as those of clover and melilot; the roots are thin and close to the soil surface. The plant has a semi-erect growth habit, reaching a length of about 60–70 cm. The stalks are fine and thin, and do not coarsen even in later stages of maturity. While the stalks and leaves are hairy in the American varieties, those of European varieties are hairless in general. *Trifolium pratense* L. has three leaflets (trifoliate). The flowers are generally in the form of capitulum. The fruits may have single or multiple seeds. Red clover is considered a diploid species, with a gametophytic self-incompatibility system (Ulloa *et al.*, 2003; Elçi, 2005).

HISTORICAL CULTIVATION AND USAGE

Red clover is one of the main forage species from temperate regions. Its life span is generally about 3–4 years, and it may be seeded alone for hay production, or, mixed with cereals, in crop rotation. It can be grown across a wide range of soil types, pH levels, and environmental conditions, and gives good yields (Ulloa *et al.*, 2003). The tetraploid *T. pratense* L., which is obtained by classical hybridization methods, is an important feed plant in Asia, Europe, America, New Zealand, and Austrialia. Records show that in 2000, clover seed was produced on 25,422 hectares in the member states of the European Union (red clover (*Trifolium pratense* L.), 11,031 hectares; white clover (*Trifolium repens* L.), 4346 hectares; and other clover species, 10,045 hectares). France has the largest red clover production area (5732 hectares), followed by Sweden (Boelt, 2002).

The plant is used not just for animal feeding, but also for medicinal purposes. Tea made from flowers, or tinctures (ethanolic extract), have been used for upper respiratory conditions such as coughs, asthma, and bronchial conditions, and as a mild sedative, antispasmodic, and remedy for rheumatism. Decoctions and ointments have also been used to treat burns and wounds (Sezik *et al,.* 1997).

PRESENT-DAY CULTIVATION AND USAGE

Red clover brings considerable benefits in terms of animal production and meat and milk quality, although monocultures are commonly grown as a companion grass for pest and disease resistance (Abberton & Marshall, 2005). Another tetraploid ($2n = 4x = 28$) variety, called "Hungarypoli" in Hungary and Elçi red clover in Turkey (Figures 115.1, 115.2), and superior to diploid varieties, is being cultivated at present (Elçi, 2005). The focus of our ongoing studies is on the cultivation and production of the plant as an industrial and a pharmaceutical crop. Understanding the genetic diversity and its partition within and among populations can speed up the development of improved red clover cultivars (Ulloa *et al.*, 2003).

APPLICATIONS TO HEALTH PROMOTION AND DISEASE PREVENTION

Legumes have been cultivated for thousands of years, and play an important role in the traditional diets of many regions throughout the world. In particular, beans have been recognized for their high protein content, and more recently for their soluble fiber. In recent years, functional compounds from legume crops, such as flavonoids, have raised a lot of interest (Prati *et al.* 2007).

FIGURE 115.1

Natural tetraploid of *Trifolium pratense* L. (Elçi red clover) in the experimentation gardens of Ankara University's Department of Biology in the Faculty of Science (Prof. Sahabettin Elçi (left) and Dr Hatice Çölgeçen (right)).

FIGURE 115.2

Natural tetraploid of *Trifolium pratense* L. (Elçi red clover): dried flowers (left); yellow and dark brown seeds of natural tetraploid of *T. pratense* L. (right).

Trifolium pratense contains isoflavonoids, flavonoids, coumarin derivatives, cyanogenic glycosides, volatile oil, saponins, and trace vitamins and minerals. Major isoflavonoids are biochanin A, daidzein, formononetin, and geinstein, which have estrogenic activities. It has therefore been used to ease menopausal symptoms, for the maintenance of the cardiovascular system and improvement of bone health, and for its beneficial effects in breast, ovarian, and prostate cancers. *In vitro* and *in vivo* studies, as well as clinical trials, have been conducted in an attempt to find a scientific basis for the usage of red clover extract and isoflavonoids isolated from it. However, studies with methodological faults in clinical trials, small sample sizes, lack of control groups, and vague statistical analyses of the results obtained from the extracts and placebos have caused conflicting outcomes. There are limited numbers of well-designed experiments proposing a decline in menopausal complaints such as hot flashes, and an improvement in bone density and arterial compliance, as well as a decrease in serum lipid levels.

In this chapter, we will first discuss clinical studies involving the application of red clover dietary supplements; this will be followed by discussion of selected *in vitro* studies (Krenn *et al.* 2002). Although there are numerous studies on the phytochemical composition and health benefits of *Trifolium* species, very few studies have dealt with the seed of red clover (Prati *et al.*, 2007).

Limited studies of the seed extract of *T. pratense* have indicated the presence of the flavonoid compounds quercetin, taxifolin (trans-dihydroquercetin), taxifolin-O-hexos derivatives, hyperoside (quercetin-3-O-galactoside), and isoquercitrin (quercetin-3-O-glucoside), and small amounts of other phenolic compounds, some of which had previously been revealed in aerial parts of the plant. The total flavonoid content in three cultivars of *T. pratense* was examined and compared with the soybean reference cultivar (37G). The highest amount of total flavonoid was determined in the 47F coded *T. pratense* cultivar as 1.5 mg diosmin/g fresh seed weight. High performance liquid chromatography-diode array detection (HPLC-DAD) analysis of the extract from *T. pratense* (21F, red clover) seeds indicated the presence of four main flavonoid compounds; however, small amounts of other phenolic compounds were also present (Prati *et al.* 2007).

Another study revealed that quercetin and its glycosides are the major compounds; additionally, soyasaponin I and its 22-O-glycoside, 22-O-diglycosides, and astrogaloside VIII were also determined (Oleszek & Stochmal, 2002). Although the seeds of red clover are not the main source of quercetin and taxifolin derivatives, those flavonoids are the major components of the seeds, along with soyasaponin and its glycosides. Bioactivity of the compound quercetin was studied extensively in *in vivo* and *in vitro* studies. In particular, antioxidant, anticarcinogenic, anti-inflammatory, and cardioprotective properties are several biological activities attributed to quercetin. Research on quercetin showed that it has a significant antioxidant effect, which increases the value of the compound, as a function of the chemical structure of quercetin. A number of authors have associated the gastroprotective, anti-inflammatory, anticarcinogenic, and cardiovascular effects to the antihistaminic and antioxidant and free radical scavenging properties of quercetin. Quercetin has been shown to modify eicosanoid biosynthesis and protect low density lipoprotein from oxidation, thus preventing atherosclerotic plaque formation and platelet aggregation, and promoting relaxation of cardiovascular smooth muscle by antihypertensive, anti-arrhythmic effects. Therefore, quercetin might be a significant aid in the prevention of certain diseases, such as cancer, atherosclerosis, and chronic inflammation (Murota & Terao 2003). Anxiolytic and cognitive-enhancing effects of intranasal quercetin liposomes on rats were also reported by Priprem *et al.* (2008), correlated with the radical scavenging activity of the compound. Quercetin reduces the severity of hypertension in spontaneously hypertensive rats (SHR). Taxifolin, 3,3′,4′,5,7-pentahydroxi-flavanon has been shown to exhibit anti-inflammatory effects in protection against oxidative cellular injury in rat peritoneal macrophage and human endothelial cells (Sendra *et al.*, 2007). The concentration of these glycosides in seeds of *Trifolium* species is similar to the concentration in other leguminous plants. Studies revealed multiple health-promoting properties, including plasma cholesterol lowering, anticarcinogenic, hepatoprotective, and antiviral activities (Hayashi *et al.*, 1997). Studies investigating the biological activities of soyasaponin derivatives have been limited to *in vitro* experiments and a few animal studies.

The high concentration of quercetin and the presence of soyasapogenol B glycosides make the seeds of some *Trifolium* species a promising plant material to be used in human nutrition as nutraceuticals or food additives.

ADVERSE EFFECTS AND REACTIONS, ALLERGIES AND TOXICITY

Application of red clover should be avoided in patients taking hormonal medications, since the chemical constituents of red clover and hormones will compete for the same hormonal receptor sites. It may interfere with the medication, leading to unpredictable consequences, such as inflammation of the eyes, mouth, and penis. Feeding red clover to animals results in decreased cervical mucus viscoelasticity, which leads to a cervix that is less accessible to sperm, thus causing reduced fertility.

Drug interactions with red clover, and formulas containing red clover, have been recorded. Users of thrombolytic agents and low molecular weight heparin should be aware of the possibility of increased bleeding, since the plant may contain coumarins that have anticoagulant effects. Simultaneous usage of red clover and contraceptives may alter the effectiveness or augment the side effects of contraceptives. Loss of effectiveness of progesterone may also occur when it is used with red clover. Animal experiments have demonstrated that the concomitant use of red clover and the anticancer drug Tamoxifen leads to decreased effectiveness of the drug. The plant, the plant extract, or semipurified isoflavonoids from the extract should not be used during pregnancy until further studies have confirmed that they are safe to be used (Adams, 1995).

SUMMARY POINTS

- *Trifolium* (Fabaceae) has about 300 species worldwide, 25 of which have been cultured for a variety of purposes. *Trifolium*, a legume, provides significant contributions to agricultural and animal production in Europe and America.
- Anatolia has been accepted as the motherland of *T. pratense*. Red clover (*Trifolium pratense* L.) is one of the main forage species from temperate regions.
- The lifespan of red clover is generally about 3−4 years; it may be seeded alone for hay production, or mixed with cereals, or used in crop rotation.
- *T. pratense* is a natural source of valuable isoflavones, which have indisputable health benefits.
- Limited studies of the seed extract of *T. pratense* have indicated the presence of the flavonoid compounds quercetin, taxifolin (trans-dihydroquercetin), taxifolin-O-hexos derivatives, hyperoside (quercetin-3-O-galactoside), and isoquercitrin (quercetin-3-O-glucoside), and small amounts of other phenolic compounds, some of which were previously revealed in aerial parts of the plant.
- Bioactivity of the compound quercetin has been studied extensively in *in vivo* and *in vitro* studies. Antioxidant, anticarcinogenic, anti-inflammatory, and cardioprotective properties in particular are some of the biological activities attributed to quercetin.
- Taxifolin has been shown to exhibit anti-inflammatory effects, plasma cholesterol lowering effects, and anticarcinogenic, hepatoprotective, and antiviral activities.

979

References

Abberton, M. T., & Marshall, A. H. (2005). Progress in breeding perennial clovers for temperate agriculture. *Journal of Agricultural Science, 143*, 117−135.

Adams, N. R. (1995). Organizational and activational effects of phytoestrogens on the reproductive tract of the ewe. *Proceedings of the Society for Experimental Biology and Medicine, 208*, 81−87.

Boelt, B. (2002). Legume seed production and research in Europe. *Forage Seed, 9*, 33−34.

Çölgeçen, H., & Toker, M. C. (2008). Plant regeneration of natural tetraploid *T. pratense* L. *Biological Research, 41*, 25−31.

Elçi, Ş. (2005). *Baklagil ve Buğdaygil yem bitkileri, TC tarım ve Köy İşleri Bakanlığı*, p. 486. Turkey.

Hayashi, K., Hayashi, H., Hiraoka, N., & Ikeshiro, T. (1997). Inhibitory activity of soyasaponin II on virus replication *in vitro*. *Planta Medica, 63*, 102−105.

Krenn, L., Unterrieder, I., & Ruprechter, R. (2002). Quantification of isoflavones in red clover by high-performance liquid chromatography. *Journal of Chromatography B, 777*, 123−128.

Murota, K., & Terao, J. (2003). Antioxidative flavonoid quercetin: implications of its intenstinal absorption and metabolism. *Archives of Biochemistry and Biophysics, 417*, 12−17.

Oleszek, W., & Stochmal, A. (2002). Triterpene saponins and flavonoids in the seeds of *Trifolium* species. *Phytochemistry, 61*, 165−170.

Prati, S., Baravelli, V., Fabbri, D., Schwarzinger, C., Brandolini, V., Maietti, A., et al. (2007). Composition and content of seed flavonoids in forage and grain legume crops. *Journal of Separation Science, 30*, 491–501.

Priprem, A., Watanatorn, J., Sutthiparinyanont, S., Phachonpai, W., & Muchimapura, S. (2008). Anxiety and cognitive effects of quercetin liposomes in rats. *Nanomed. Nanotechnology Biology and Medicine, 4*, 70–78.

Sezik, E., Yeşilada, E., Tabata, M., Honda, G., Takaishi, Y., Fujita, T., et al. (1997). Traditional medicine in Turkey VIII. Folk medicine in East Anatolia; Erzurum, Erzincan, Ağrı, Kars, Iğdır Provinces. *Journal of Economic Botany, 51*, 3195–3211.

Sendra, J. M., Sentandreu, E., & Navarro, J. L. (2007). Kinetic model for the antiradical activity of the isolated *p*-catechol group in flavanone type structures using the free stable radical 2,2-diphenyl-1-picrylhydrazyl as the antiradical probe. *Journal of Agricultural and Food Chemistry, 5*, 5512–5522.

Ulloa, O., Ortega, F., & Campos, H. (2003). Analysis of genetic diversity in red clover (*T. pratense* L.) breeding populations as revealed by RAPD genetic markers. *Genome, 46*, 529–535.

CHAPTER 116

Red Onion (*Allium caepa* L. var. *tropeana*) Seeds: Nutritional and Functional Properties

Irene Dini
Dipartimento di Chimica delle Sostanze Naturali, Università di Napoli "Federico II", Naples, Italy

981

INTRODUCTION

Onions are one of the most important vegetable crops, with world production of about 4,197,549 tonnes in 2007 (FAO, 2007). Their consumption is attributed to several factors, mainly heavy promotion linking flavor and health, and the popularity of onion-rich ethnic foods. Onion bulbs are the main edible part of the plant, with a distinctive strong flavor and pungent odor. Onion seeds are also eaten, especially in some Indian dishes, and although they do not affect the breath as strongly as bulbs, their commercial availability is currently limited. Perhaps if consumers were more acquainted with onion seeds' nutritional and functional properties, there would be an increase in trade of this product. The "Cipolla Rossa of Tropea Calabria" is known for its qualitative and organoleptic characteristics, including its mildness, sweetness, high digestibility, and very low level of sulfates in the bulbs, which make them less sharp and sour and more easily digestible than other onions. These characteristics enable Cipolla Rossa of Tropea Calabria to be even consumed raw, in greater quantities.

BOTANICAL DESCRIPTION

Cipolla Rossa of Tropea Calabria, sometimes called purple onions, are cultivars of the genus *Allium* of the Alliaceae family. In February 2008, the Cipolla Rossa of Tropea Calabria obtained Geographical Protected Indication (GPI) status, so only Tropea onions grown as described in the Official Journal of the European Union (2007/C 160/15) can be named as such. It denotes bulbs of the species *Allium caepa* exclusively from the local ecotypes "Tondo Piatta" (an early crop), "Mezza campana" (an early–mid-season crop), and "Allungata" (a later crop), which have a characteristic shape and are produced early, owing to the effect of the photoperiod. There are three types of products: *cipollotto* (color, white to pink or purple; taste, sweet and mild); *cipolla* for fresh consumption (color, white to pink or purple; taste, sweet and mild); and *cipolla* for storage (color, white to purple; taste, sweet and crunchy).

HISTORICAL CULTIVATION AND USAGE

Since ancient times, people from widespread areas of the world have been using red onions as a food, spice, and herbal remedy, especially in the northern hemisphere. Pliny the Elder, in his *Naturalis Historia*, spoke about its many therapeutic qualities. Historical and bibliographical references (Pagano, 1901; Griffiths *et al.*, 2002) credit the arrival of onions in the Mediterranean Basin and Calabria first to the Greeks and then to the Phoenicians. Highly appreciated in the Middle Ages and during the Renaissance, onions were considered a staple food, and key to the local economy, where they were traded, and exported by sea to Tunisia, Algeria, and Greece. The many travellers to Calabria between 1700 and 1800 who visited the Tyrrhenian coast between Pizzo and Tropea frequently refer to the red onions commonly found there. Onions have always been a feature of the rural diet and of local production. Dr Albert, who travelled to Calabria in 1905 and visited Tropea, was struck by the poverty of the local people, who ate only onions. In the early 1900s, onion-growing in Tropea moved from the family garden and vegetable patch and became a major crop. In 1929 the Valle Ruffa aqueduct paved the way for irrigation, and, as a result, better fields and quality. The product became more widespread during the Bourbon period, reaching northern European markets and swiftly becoming sought after and prized, as described in *Studi sulla Calabria* (Pagano, 1901), which refers to the shape of the bulb — the red oblong onions from Calabria. The first organized statistical information on onion-growing in Calabria is contained in the *Reda Agricultural Encyclopaedia* (1936–1939).

PRESENT-DAY CULTIVATION AND USAGE

Cipolla Rossa of Tropea Calabria is grown in medium, sand-rich loam or in heavy loam rich in clay or lime, in the coastal area or around alluvial rivers and streams, which, despite the gravel, does not restrict the growth and development of the bulbs. Coastal land is ideal for growing early onions for fresh consumption. Inland areas with heavier, clay-rich soil are ideal for growing late onions for storage. Today, as in the past, red onions are grown in family vegetable patches and in large crops, and form part of the rural landscape, local food and dishes, and traditional recipes. The soil and climate conditions in the defined area contribute to the high quality and uniqueness of the product, which is widely acclaimed the world over. The production area covers suitable land in all or part of the municipalities of the Provinces of Cosenza, Catanzaro, and Vibo Valentia in the region of Calabria.

Geographical Protected Indication status for Cipolla Rossa of Tropea Calabria was requested and obtained from the European Community in 2008, both to prevent imitation of the product and ensure the traceability of all stages in the production process because of its unique characteristics and international reputation; and because of its historical and cultural significance in the area concerned — a significance that is still reflected today in farming practices, in cooking, in everyday language, and in folkloric events. Producers and processors of this product and the land parcels on which it is grown are entered in registers managed by the inspection body. The production process is described in the Official Journal of the European Union (2007/C 160/15), and can be summarized as follows:

- The seeds of Cipolla Rossa of Tropea Calabria are sown from August onwards; planting distances vary according to the land and the method used, ranging from 4–20 cm in rows, with 10–22 cm between rows, and a density of between 250,000 and 900,00 plants/

hectare, the latter where there are four bulbs per hole for final radication. One of the regular operations is to provide irrigation, varying according to precipitation levels.

- After the *cipollotto* bulbs have been harvested, the earth-covered outer layer is removed, the foliage is cut to 40 cm, and the bulbs are then placed in trays in bundles.
- In the case of *cipolla da consumo fresco*, the outer layer of the bulbs is also removed and foliage is cut if it is longer than 60 cm. The onions are then tied in bundles of 5—8 kg, and placed in trays or crates.
- For *cipolla da serbo*, the bulbs are placed on the ground in windrows covered with their own foliage, and left for 8—15 days to dry, become more compact and resistant, and develop a bright red color. Once dried, the bulbs may be topped, or the foliage left for plaiting.

APPLICATIONS TO HEALTH PROMOTION AND DISEASE PREVENTION

Spices, in view of their many promising health beneficial physiological effects, have assumed the status of "nutraceuticals," and are considered natural and necessary components of our daily nutrition. Red onion seeds (Figure 116.1) are receiving attention for their nutritional and medicinal properties. In fact, they contain higher amounts of oil (20.4%), crude protein (24.8%), copper (0.92 mg/100 g of edible portion), and zinc (7.25 mg/100 g of edible portion), and lower amounts of sodium (11.2 mg/100 g of edible portion) than other typical spices of Italian cuisine, such as dill (*Anethum graveolens*), caraway (*Carum carvi*), and fennel (*Faeniculum vulgare*) seeds (Tables 116.1, 116.2). The high fat content suggests the possibility of using these seeds to obtain an aromatic oil useful to make dishes more flavorful. The seeds can be used in diets for people who take diuretics to control hypertension, and who suffer from excessive excretion of potassium, because of the high potassium/sodium ratio (90.2) contained in them. Unfortunately, the high fiber (22.4 g/100 g) and oxalate (0.209 g/100 g) (Tables 116.1, 116.3) contents can render some mineral nutrients unavailable by binding them to form insoluble salts that are not absorbed by the intestine (Dini *et al.*, 2008a). The chloride content in Tropea red onion seeds (0.88 g/100 g of edible portion) exceeds the daily intake recommended for American 25-year-old males (2300 mg/day). This is important because chloride determines the inhibitory action of glycine and some of the action of GABA in the central nervous system (Price *et al.*, 2009).

The proteins present in red onion seeds, as in other typical spices of Italian cuisine (*Anethum graveolens*, *Carum carvi*, and *Faeniculum vulgare* seeds, and *Sinapis alba*), are incomplete proteins (Table 116.4); all the essential amino acids are at lower concentrations when compared with the reference pattern. Nevertheless, these seeds reveal large amounts of free amino acids, which are probably responsible for their organoleptic and biological

FIGURE 116.1
Seeds of Cipolla Rossa of Tropea Calabria.

TABLE 116.1 Chemical Composition of Commercial *Allium caepa* L. var. *tropeana* Seeds and Some Typical Spices of Italian Cuisine

	Water (%)	Crude protein (%)	Fat (%)	Carbohydrate (%)	Fiber (%)
Allium caepa var. *tropeana*[1] (Tropea red onion seeds)	10.5 ± 0.1	24.8 ± 0.1	20.4 ± 0.7	21.9 ± 0.9	22.4 ± 0.4
Allium caepa bulb raw[2]	89.11	1.1	0.1	9.34	1.7
Allium caepa seeds[3]	6–8	27.9	23.6	20.4	16.1
Anethum graveolens (dill seeds)[2]	7.7	16.0	14.5	55.2	21.1
Carum carvi (caraway seeds)[2]	9.9	19.8	14.6	49.9	38.0
Faeniculum vulgare (fennel seeds)[2]	8.8	15.8	14.9	52.3	39.8
Sinapis alba (mustard)[2]	6.9	24.9	28.8	34.9	14.7

Reprinted from Dini *et al.* (2008a), *Food Chem.*, *107*, 613–621, with permission.
[1]Values are the mean ± SD of three different determinations (P < 0.0001).
[2]USDA National Nutrient Database for Standard Reference, Release 16.1 (2004);
[3]Udayasekhara, P.R. (1994).

TABLE 116.2 Microelements in *Allium caepa* L. var. *tropeana* Seeds (mg/100 g of Edible Portion) Compared to Some Typical Spices of Italian Cuisine

	Calcium	Magnesium	Potassium	Sodium	Iron	Copper	Zinc
Allium caepa L. var. *tropeana* seeds[1]	175.0 ± 1.2	325.0 ± 1.1	1010.0 ± 1.1	11.20 ± 0.3	6.60 ± 0.4	0.920 ± 0.04	7.25 ± 0.02
Allium caepa L. bulb raw[2]	23	10	146	4	0.21	0.039	0.17
Allium caepa seeds[3]	279	414			43	0.5	5.6
Anethum graveolens (dill seeds)[2]	1516	256	1186	20	16.33	0.780	5.20
Carum carvi (caraway seeds)[2]	689	258	1351	17	16.23	0.910	5.50
Faeniculum vulgare (fennel seeds)[2]	1196	385	1694	88	18.54	1.067	3.70
Sinapis alba (mustard)[2]	521	298	682	5	9.98	0.410	5.70

Reprinted from Dini *et al.* (2008a), *Food Chem.*, *107*, 613–621, with permission.
[1]Values are the mean ± SD of three different determinations (P < 0.0001).
[2]USDA National Nutrient Database for Standard Reference, Release 16.1 (2004);
[3]Udayasekhara, P.R. (1994).

TABLE 116.3 Anions in *Allium caepa* L. var. *tropeana* Seeds

Anions	mg/100 g
Cl^-	879.70 ± 0.7
F^-	203.33 ± 0.5
NO_3^-	35.66 ± 0.5
PO_4^{3-}	1712.39 ± 1.1
SO_4^{2-}	605.51 ± 0.5
$C_2O_4^{2-}$	209.40 ± 0.8

Values are the mean ± SD of three different determinations (P < 0.0001).
Reprinted from Dini *et al.* (2008a), *Food Chem.*, *107*, 613–621, with permission.

TABLE 116.4 Essential Amino Acid Pattern of *Allium caepa* L. var. *tropeana* Seeds Compared to Some Typical Spices of Italian Cuisine and the FAO/WHO Reference Pattern (1990) for Evaluating Proteins (mg/100 g)

Amino Acid	Allium Caepa var. tropeana (Tropea Red Onion Seeds)[1]	FAO[4]	Essential Amino Acid Score	Allium Caepa Bulbs Raw[2]	Allium Caepa Seeds[3]	Anethum Graveolens (Dill Seeds)[2]	Carum Carvi (Caraway Seeds)[2]	Faeniculum Vulgare (Fennel Seeds)[2]	Sinapis Alba (Mustard)[2]
Isoleucine	0.7 ± 0.4	2.8	0.25		4.4	0.8	0.8	0.7	1.1
Leucine	1.2 ± 0.1	6.6	0.18		6.1	0.9	1.2	1.0	1.8
Lysine	1.3 ± 0.2	5.8	0.22		5.0	1.0	1.0	0.8	1.5
Phenylalanine	0.8 ± 0.6				3.9	0.7	0.9	0.6	1.1
Tyrosine	0.5 ± 0.5	6.3	0.07	0.014	2.3	Missing	0.6	0.4	0.7
Cystine	0.9 ± 0.5	2.5	0.36			Missing	0.3	0.2	0.6
Threonine	0.8 ± 0.7	3.4	0.23	0.021	2.5	0.6	0.7	0.6	1.1
Valine	1.0 ± 0.6	3.5	0.28	0.021	4.4	1.1	1.0	0.9	1.3
Arginine	2.3 ± 0.4			0.104	9.4	1.3	1.2	0.7	1.7
Alanine	0.9 ± 0.4			0.021	4.1	Missing	0.9	0.8	1.2
Serine	1.5 ± 0.3			0.021	5.4	Missing	0.9	0.9	1.1
Proline	0.7 ± 0.5			0.012	1.9	Missing	0.9	0.9	1.9
Aspartic acid	1.7 ± 0.3			0.091	6.5	Missing	2.1	1.8	2.0
Glycine	1.0 ± 0.4			0.025	3.8	Missing	1.3	1.1	1.3
Histidine	0.5 ± 0.4	1.9	0.26	0.014	1.9	0.3	0.5	0.3	0.8
Glutamic acid	4.7 ± 0.6			0.258	21.3	Missing	3.2	2.9	5.0

Sources: Reprinted from Dini et al. (2008a), *Food Chem.,* 107, 613–621, with permission.

[1] Values are the mean ± SD of three different determinations (P < 0.0001).
[2] USDA National Nutrient Database for Standard Reference, Release 16.1 (2004);
[3] Udayasekhara, P.R. (1994);
[4] FAO/WHO (1990).

properties. In particular, four amino acids constitute 68.8% of the total free amino acids in Tropea red onion seeds: glutamic acid (97.3 mg/100 g) and arginine (88.9 mg/100 g), along with lesser amounts of tyrosine (69.3 mg/100 g) and asparagine (52.3 mg/100 g) (Dini *et al.*, 2008a) (Table 116.5). Furthermore, the content of red onion seeds includes sulfur-containing flavor compounds, whose precursors are odorless, non-protein sulfur amino acids, namely S-alk(en)yl cysteine sulfoxides. These compounds have been reported to improve glycemic control, delay oxidation damage, down-regulate inflammatory cytokines, and enhance anticoagulant activity in diabetic mice via their antioxidant activities, and induce adipose tissue cell death; therefore, they should be used for the treatment of obesity. During tissue damage, S-alk(en)yl cysteine sulfoxides are converted to their respective thiosulfinates or propanethial-S-oxide by the action of the enzyme alliinase. To date, four major and three minor cysteine sulfoxide derivatives have been identified in the genus *Allium*: (+)-S-methyl-l-cysteine sulfoxide (methiin), (+)-S-ethyl-l-cysteine sulfoxide (ethiin), (+)-S-propyl-l-cysteine sulfoxide (propiin), (+)-S-allyl-l-cysteine sulfoxide (alliin), (+)-S-(*trans*-propenyl)-l-cysteine sulfoxide (isoalliin), (+)-S-butyl-l-cysteine sulfoxide (butiin), and (S_SR_C)-S-(3-pentenyl)-L-cysteine sulfoxide (Figure 116.2) (Dini *et al.*, 2008b). These compounds are found together only in red onion seeds var. *tropeana*. Moreover, red onion seed contain S-methylcysteine, S-ethylcysteine, S-propylcysteine, S-allyl-cysteine, S-2-hydroxyethyl-cysteine, and S-propylmercapto-cysteine (Figure 116.2). Sulfur-containing flavor compounds can be used as an adjunct to current cancer therapies; as cardioprotective compounds (some of these partially protecting LDL or plasma against oxidative damage and glycative deterioration, and inhibiting fatty acid, triglyceride and cholesterol synthesis); and to alleviate the symptoms of dermatologic disorders related to the impairment of lipid metabolism (thus being suitable for the treatment of edematous-fibrosclerotic panniculopathy, ichthyosis, hyperkeratosis, Darier disease, lichen simplex chronicus, keloid, scar, acne, rosacea, and couparose) (Dini *et al.*, 2008b). Noteworthy is the presence of propylmercapto-cysteine, whose occurrence has never been reported in onions; this has the ability to decrease food intake (Hatono *et al.*, 1997) and reduce cancer risk (Pinto *et al.*, 2006). Mercapto derivatives yield persulfide products (RSSH) that are potential sources of sulfane sulfur, which may regulate cell proliferation and/or apoptotic responses, modifying protein function by reacting at important cysteinyl domains. Cysteine derivatives present

TABLE 116.5 Free Amino Acid Pattern of *Allium caepa* L. var. *tropeana* Seeds

Free Amino Acid	(mg/100 g of Red Onion Seeds)
Threonine	22.8 ± 0.3
Aspartic acid	25.1 ± 0.6
Valine	8.9 ± 0.7
Asparagine	52.3 ± 0.3
Glutamic acid	97.3 ± 0.2
Serine	6.7 ± 0.4
Glycine	11.0 ± 0.4
Glutamine	19.6 ± 0.6
Hystidine	3.0 ± 0.4
Arginine	88.9 ± 0.6
Alanine	30.5 ± 0.6
Proline	5.9 ± 0.4
Tyrosine	69.3 ± 0.3
Gaba	5.8 ± 0.5

Values are the mean $\pm$ SD of three different determinations ($P < 0.0001$).
Reprinted from Dini *et al.* (2008a), *Food Chem.*, *107*, 613–621, with permission.

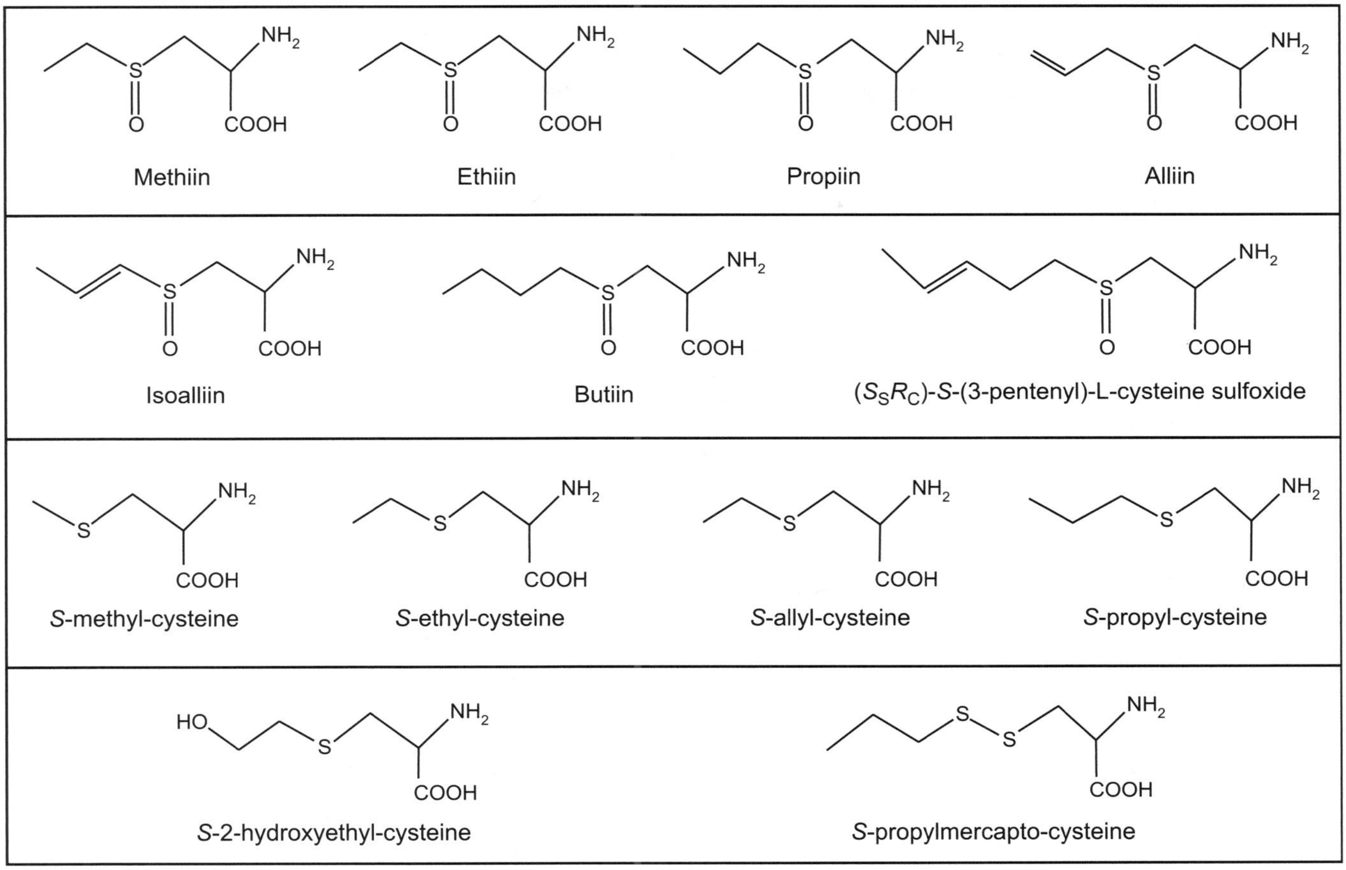

FIGURE 116.2

Cysteine derivatives in *Allium caepa* L. (var. *tropeana*) seeds. Two works of Dini and colleagues have described the presence of *S*-alk(en)yl cysteine derivatives in *Allium caepa* L. (var. *tropeana*) seeds (Dini *et al.*, 2008a, 2008b). Sulfur-containing flavor compounds can be used as an adjunct to current cancer therapies, as cardioprotective compounds, to decrease food intake, and to alleviate the symptoms of dermatologic disorders related to the impairment of lipid metabolism.

in red onion seeds reveal a discrete antioxidant effect that increases after boiling the seeds, making Tropea red onion seeds a good antioxidant additive for food, and increasing their potential usability as a functional food and in ethnomedicine (Dini *et al.*, 2008a) (Figure 116.3). Furthermore, the antioxidant effect of seed extracts containing cysteine derivatives increases after boiling the seeds, although cooking methods cause significant losses of antioxidant compounds in water. Dietary antioxidants protect the human body against free radicals and reactive oxygen species, which are known to be the major contributors to degenerative diseases of ageing, and are recognized as major factors causing cancer, cardiovascular disorders, and diabetes. Finally, Tropea red onion seeds contain significant concentrations of steroidal saponins, bioactive glycosides which typically occur in small quantities in the *Allium* genus (Dini *et al.*, 2005). These constituents possess some biological properties, such as hypocholesterolemic, antidiabetic (Wang & Ng 1999), antitumor, and antitussive activities (Hiromichi, 2001) (Figure 116.4).

ADVERSE EFFECTS AND REACTION (ALLERGIES AND TOXICITY)

There are no reports of adverse effects caused by *Allium caepa* L. var. *tropeana* seeds, but Navarro and colleagues (1995) defined *Allium caepa* var. Brunswick as an "occupational allergen"; therefore, the potential allergenicity of Tropea red onion seeds should be assessed.

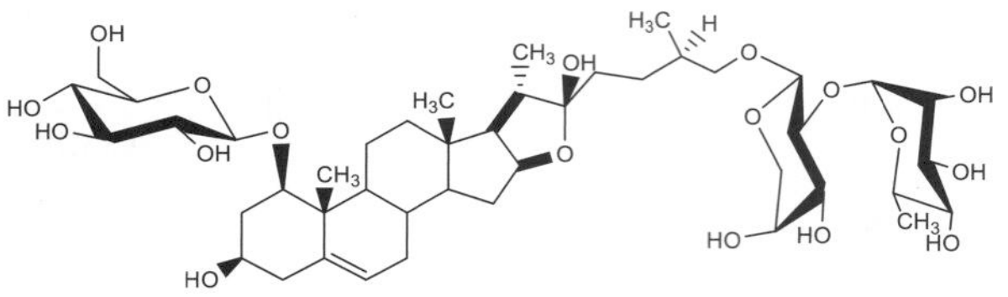

1-*O*-β-D-glucopyranosyl-(25R)-furost-5(6)-en-1β,3β,22α,26-tetraol-26-*O*-α-L-rhamnopyranosyl-(1‴→2″)-*O*-α-L-arabinopyranoside

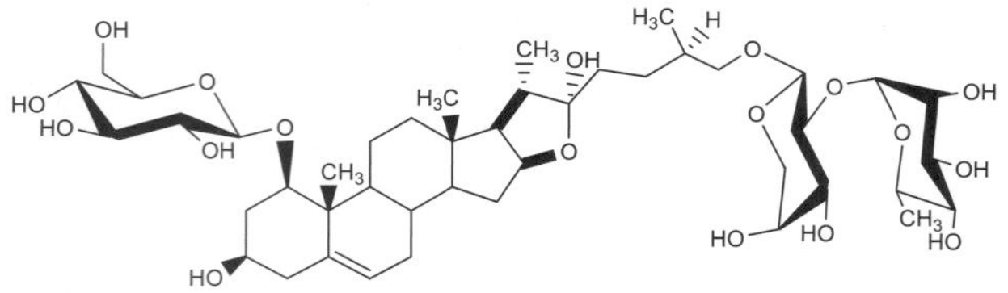

1-*O*-β-D-glucopyranosyl-(25R)-furost-5(6)-en-1β,3β,22β,26-tetraol-26-*O*-α-L-rhamnopyranosyl-(1‴→2″)-*O*-α-L-arabinopyranoside

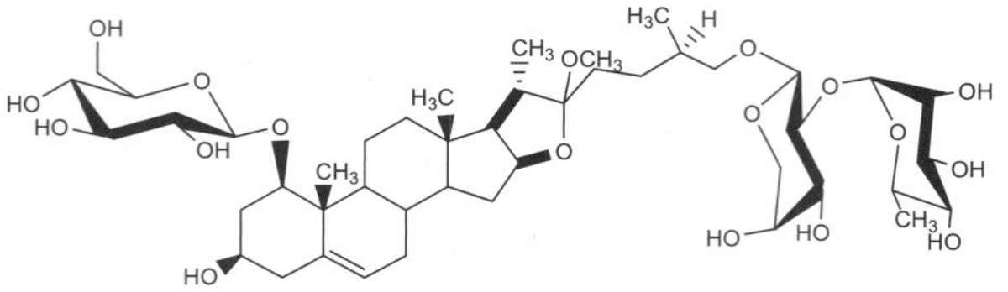

1-*O*-β-D-glucopyranosyl-22-*O*-methyl-(25R)-furost-5(6)-en-1β,3β,22ξ,26-tetraol-26-*O*-α-L-rhamnopyranosyl-(1‴→2″)-*O*-α-L-arabinopyranoside

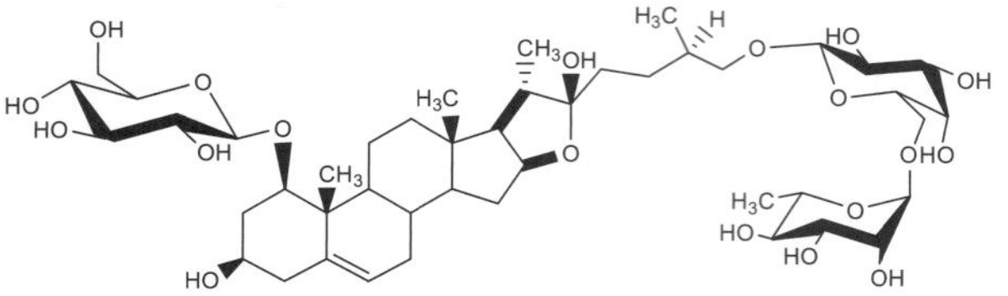

1-*O*-β-D-glucopyranosyl-(25R)-furost-5(6)-en-1β,3β,22β,26-tetraol-26-*O*-α-L-rhamnopyranosyl-(1‴→6″)-*O*-β-D-galactopyranoside

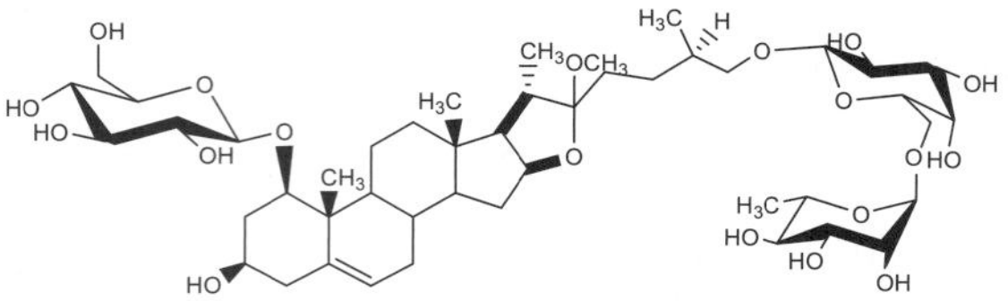

1-*O*-β-D-glucopyranosyl-22-*O*-methyl-(25R)-furost-5(6)-en-1β,3β,22ξ,26-tetraol-26-*O*-α-L-rhamnopyranosyl-(1‴→6″)-*O*-β-D-galactopyranoside

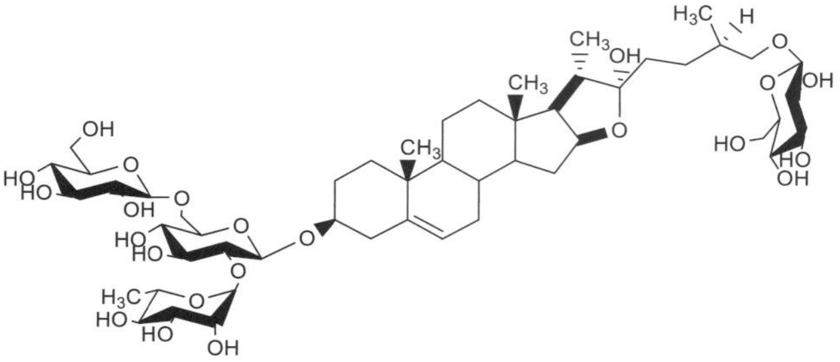

26-*O*-β-D-glucopyranosyl-(25R)-furost-5(6)-en-3β,22α,26-triol-3-*O*-α-L-rhamnopyranosyl-(1″→2′)-*O*-[β-D-glucopyranosyl-(1‴→6′)-*O*]-β-D-glucopyranoside

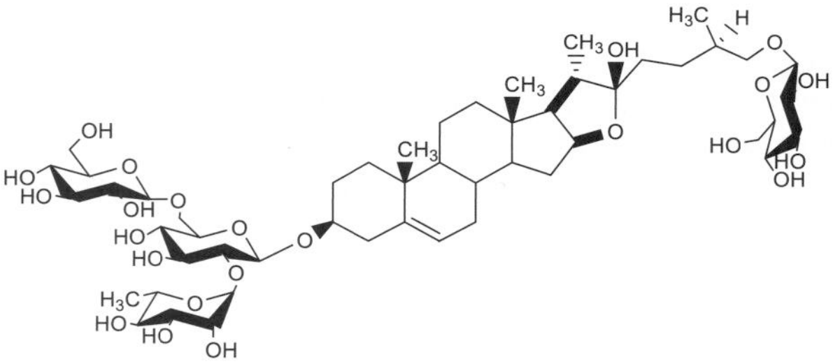

26-*O*-β-D-glucopyranosyl-(25R)-furost-5(6)-en-3β,22β,26-triol-3-*O*-α-L-rhamnopyranosyl-(1″→2′)-*O*-[β-D-glucopyranosyl-(1‴→6′)-*O*]-β-D-glucopyranoside

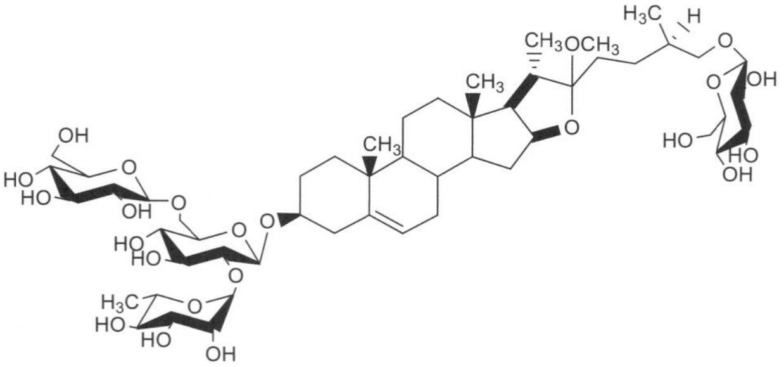

26-*O*-β-D-glucopyranosyl-22-*O*-methyl-(25R)-furost-5(6)-en-3β,22ξ,26-triol-3-*O*-α-L-rhamnopyranosyl-(1″→2′)-*O*-[β-D-glucopyranosyl-(1″→6′)-*O*]-β-D-glucopyranoside

FIGURE 116.3

Saponin quality evaluation of *Allium caepa* L. (var. *tropeana*) seeds (Dini *et al.*, 2005).

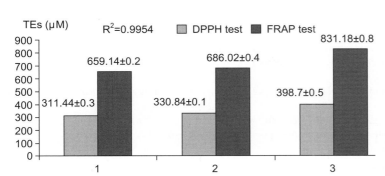

Values are the mean ± SD of three different determinations ($P < 0.0001$).

FIGURE 116.4

Antioxidant tests of seed extracts containing cysteine derivatives before and after boiling *Allium caepa* L. var. *tropeana* seeds, and of cooking water extracts containing cysteine derivatives. *Reprinted from Dini* et al. *(2008a), Food. Chem., 107, 613–621, with permission.*

Moreover, it is unlikely that toxic effects related to nitrates have been assessed, because the content of nitrates in these seeds is low (36 mg/kg vs 300–940 mg/kg in other vegetables). Nitrates and their metabolite nitrites can be found in a variety of plant-derived foods as naturally occurring compounds. Dietary exposure to both nitrates and nitrites is of interest from a human health perspective, in terms of both direct toxic effects (e.g., cyanosis and methemoglobin) and possible indirect effects as precursors of carcinogenic N-nitrosamines (Reinik *et al.*, 2008). Therefore the Joint FAO.WHO Committee on Food Additives and the European Commission's Scientific Committee on Food have set an acceptable daily intake for nitrates of 0–3.7 mg/kg body weight.

SUMMARY POINTS

- Cipolla Rossa of Tropea Calabria, sometimes called purple onions, are cultivars of the genus *Allium*, Alliaceae family.
- Cipolla Rossa of Tropea Calabria are cultivated in the Cosenza, Catanzaro, and Vibo Valentia provinces of the Calabria region
- Red onion seeds are receiving attention for their nutritional and medicinal properties.
- Red onion seeds contain higher amounts of oil, crude protein, copper, and zinc, and lower amounts of sodium, than other typical spices of Italian cuisine, along with phytochemicals such as sulfur-containing flavor compounds able to improve glycemic control, delay oxidation damage, down-regulate inflammatory cytokines, enhance anticoagulant activity, induce adipose tissue cell death; and steroidal saponins with hypocholesterolemic, antidiabetic, antitumor, and antitussive activities.
- There are no reports of adverse effects caused by *Allium caepa* L. var. *tropeana* seeds, but the potential allergenicity of Tropea red onion seeds should be assessed.

References

Dini, I., Trimarco, E., Tenore, G. C., & Dini, A. (2005). Furostanol saponins in *Allium caepa* L. var. *tropeana* seeds. *Food Chemistry, 93,* 205–214.

Dini, I., Tenore, G. C., & Dini, A. (2008a). Chemical composition, nutritional value and antioxidant properties of *Allium caepa* L. var. *tropeana* (red onion) seeds. *Food Chemistry, 107,* 613–621.

Dini, I., Tenore, G. C., & Dini, A. (2008b). S-Alkenyl cysteine sulfoxide and its antioxidant properties from *Allium caepa* var. *tropeana* (red onion) seeds. *Journal of Natural Products, 71,* 2036–2037.

FAO (2007). Data drawn from FAOSTAT (available at http://faostat.fao.org/).

FAO/WHO. (1990). *Protein quality evaluation. In: Report of Joint FAO/WHO expert consultation.* Rome, Italy: Food and Agricultural Organization of the United Nations. p. 23.

Griffiths, G., Trueman, L., Crowther, T., Thomas, B., & Smith, B. (2002). Onions — a global benefit to health. *Phytotherapy Research, 16*, 603—615.

Hatono, S., & Wargovich, M. J. (1997). Role of garlic in disease prevention — preclinical models. *Nutraceuticals: Designer foods III: Garlic, soy and licorice. Course on Designer Foods, Proceedings (3rd)*. Trumbull, CT: Food and Nutrition Inc. 139—151.

Hiromichi, M. (2001). Saponins in garlic as modifiers of the risk of cardiovascular disease. *Journal of Nutrition, 131*, 1000S—1005S.

Navarro, J. A., Del Pozo, M. D., Gastaminza, G., Moneo, I., Audicana, M. T., & De Corres, L. F. (1995). *Allium caepa* seeds: a new occupational allergen. *Journal of Allergy and Clinical Immunology, 96*, 690—693.

Pagano, N. (1901). Studi sulla Calabria. Calabria, Italy.

Pinto, J. T., Krasnikov, B. F., & Cooper, A. J. L. (2006). Redox-sensitive proteins are potential targets of garlic-derived mercaptocysteine derivatives. *Journal of Nutrition, 136*, 835S—841S.

Price, T. J., Cervero, F., Gold, M. S., Hammond, D. L., & Prescott, S. A. (2009). Chloride regulation in the pain pathway. *Brain Research Reviews, 60*, 149—170.

Reinik, M., Tamme, T., & Roasto, M. (2008). Naturally occurring nitrates and nitrites in foods. *Acs Symposium Series* 227—253.

USDA. (2004). *National Nutrient Database for Standard Reference, Release 16.1(2004)*. Washington, DC: USDA.

Udayasekhara, P. R. (1994). Nutrient composition of some less-familiar oil seeds. *Food Chemistry, 50*, 379—380.

Wang, H. X., & Ng, T. B. (1999). Natural products with hypoglycemic, hypotensive, hypocholesterolemic, anti-atherosclerotic and antithrombotic activities. *Life Science, 65*, 2663—2677.

CHAPTER 117

Sacha Inchi (*Plukenetia volubilis* L.) Nut Oil and Its Therapeutic and Nutritional Uses

Hans-Peter Hanssen[1], Markus Schmitz-Hübsch[2]
[1] University of Hamburg, Institute of Pharmaceutical Biology and Microbiology, Hamburg, Germany
[2] Rosen-Apotheke, Hamburg, Germany

991

LIST OF ABBREVIATIONS

ADHD, Attention Deficit Hyperactivity Disorder
FAO, Food and Agriculture Organization
WHO, World Health Organization

INTRODUCTION

Sacha Inchi has been — and still is — used as a traditional nutrient and remedy in the Peruvian Andes. The oil from the nuts shows an interesting fatty acid profile, predominated by unsaturated components. It has been successfully introduced to the cosmetics and dietary supplementation markets (Hanssen, 2008). In 2007, some 60,000 liters were sold; by 2008 this had already risen to 120,000 liters. A number of possible medicinal applications are discussed here, but significant clinical studies have yet to be completed.

BOTANICAL DESCRIPTION

Plukenetia volubilis L. (family Euphorbiaceae), commonly known as Sacha Inchi, Inca peanut or mountain peanut, is a perennial plant with somewhat hairy leaves. The plant reaches

Nuts & Seeds in Health and Disease Prevention. DOI: 10.1016/B978-0-12-375688-6.10117-3

FIGURE 117.1
Green and mature fruit of Sacha Inchi (*Plukenetia volubilis* L.). *Photograph courtesy of Roda, Lima (www.rodaperu.com).*

a height of 2 m, and has alternate, heart-shaped, serrated leaves, 10–12 cm long and 8–10 cm wide, with 2- to 6-cm petioles. The male flowers, arranged in clusters, are small and white, and two female flowers are located at the base of each inflorescence.

The fruit capsules (3–5 cm in diameter with four to seven points) are green, and ripen to a blackish-brown (Figure 117.1). They generally consist of four, but in some cases up to seven, lobes. The seeds are oval, dark brown, and 1.5–2 cm in diameter (Figure 117.2).

HISTORICAL CULTIVATION AND USAGE

Sacha Inchi is native to the Amazon rainforest, and grows in warm climates at altitudes of up to 1700 m. The plant prefers an environment in which water is continuously available, and well-drained acidic soil.

Sacha Inchi has been cultivated by indigenous people for centuries. Traditionally, the oil (mixed with flour) is used for cosmetic purposes by women of the Mayoruna, Campas, Huitotas, Shipibas, Yaguas, and Bora tribes. It has also been used medically in treating rheumatic problems and aching muscles.

PRESENT-DAY CULTIVATION AND USAGE

During recent years, considerable efforts have been made to introduce Sacha Inchi into the agricultural systems of the region as an alternative crop, in order to reduce local farmers' dependence on the cultivation of coca. In this respect, the successful introduction of Sacha Inchi oil to the cosmetics market has already shown some positive effects. Due to its composition (high in unsaturated fatty acids), the first dietary supplementation products are now available in the market.

FIGURE 117.2
The oil of Sacha Inchi (*Plukenetia volubilis* L.) derives from the seeds. *Photograph courtesy of Roda, Lima (www.rodaperu.com).*

APPLICATIONS TO HEALTH PROMOTION AND DISEASE PREVENTION

At the beginning of the 1990s, oil derived from the Sacha Inchi nut was investigated more intensely (Hamaker *et al.*, 1992). It was found that the oil contains remarkably high contents of α-linoleic acid (*ca.* 49%) and linolic acid (*ca.* 36%). These results were later confirmed by a further study (Guillén *et al.*, 2003). Sacha Inchi oil is therefore classified as an edible oil with the highest proportion of unsaturated fatty acids. A comparison of Sacha Inchi oil with other plant oils is given in Table 117.1.

As the oil also has a high content of γ- and δ-tocopherols, it is, despite its high proportion of unsaturated fatty acids, comparably stable against oxidation. Furthermore, vitamins A and E, and phytosterols, mainly stigmasterol, campesterol, and Δ5-avenasterol (*ca.* 250 mg/100 ml oil), have been identified.

The protein content of the seeds is again relatively high (*ca.* 33%). The most important component is a 3S storage protein — a water-soluble albumin that makes up about one-third of the total protein content. It consists of two glycosylated polypeptides with molecular weights of 32.8 and 34.8 kDa, respectively. Regarding its composition, the amino acid profile conforms to the recommendations of the FAO/WHO as a nutrient for adults. The compound contains unusually high contents of tryptophan (44 mg/g protein) and a comparably low content of phenylalanine (9 mg/g protein). It exhibits an exquisite digestibility *in vitro* (Sathe *et al.*, 2002).

Medicinal applications of the seeds and the oil, respectively, have been discussed for a variety of diseases. Besides their cholesterol and blood pressure lowering properties, it is also presumed that health improvements can be obtained in diabetes, arthritis, and even in the case of certain psychological disorders and cancers (breast cancer, prostate carcinoma).

As no data from clinical studies are available so far, specific health claims should be considered with caution. On the other hand, Sacha Inchi oil is, because of its high content of unsaturated fatty acids, predestined for medical applications. The intake of omega-3 fatty acids from other sources is widely used to lower high blood triglyceride levels. Moreover, omega-3 fatty acids influence blood flow and blood clotting positively. Epidemiological studies within Inuit, Japanese, and Dutch populations have shown a correlation between the intake of fish oil and a decreased incidence of cardiovascular disease. In this context, Sacha Inchi oil has been assumed to be an alternative to fish oil.

Further possible applications of oils rich in unsaturated fatty acids, like Sacha Inchi oil, include:

1. Treatment of patients with ADHD (Attention Deficit Hyperactivity Disorder). This is because low levels of unsaturated fatty acids in the plasma and membranes of red blood cells may increase symptoms of ADHD. Licensed preparations are used in Germany to treat ADHD symptoms in children with additional unsaturated fatty acids.

993

TABLE 117.1 Protein Content, Oil Content, and Fatty Acid Profiles of Different Commercial Oil Plants (%)

	Olive	Soy	Maize	Peanut	Sunflower	Palm	Sacha Inchi
Protein	*ca.* 2	28		23	24		33
Total oil	22	19		45	48		54
Palmitic acid	13	10.7	11	12	7.5	45	3.9
Stearic acid	3	3.3	2	2	5.5	4	2.5
Oleic acid	71	22.3	28	43.3	29.3	40	8.8
Linolic acid[*]	10	54.5	58	36.8	57.9	10	36.8
Linoleic acid[*]	1	8.3	1	—	—	—	48.6

Table modified www.newcenturyproducts.net.
[*]*Unsaturated fatty acid; —, no specification.*

2. Treatment of arthritis. Due to their positive influence on the metabolism of prostaglandins, unsaturated fatty acid components can be used to treat arthritis because less inflammatory prostaglandins are produced.

Possible future applications for medical purposes, however, remain speculative until reliable data from serious studies have been presented.

ADVERSE EFFECTS AND REACTIONS

So far, no data concerning adverse effects and reactions of Sacha Inchi oil or single components of the oil have been described.

SUMMARY POINTS

- *Plukenetia volubilis* L. (Euphorbiaceae) is one of the oldest cultivated plants of Peru.
- Sacha Inchi (Inca nut) oil is classified as an edible oil with the highest proportion of unsaturated fatty acids.
- Due to its high contents of omega-3 fatty acids (especially α-linoleic acid), omega-6 fatty acids, and omega-9 fatty acids, the oil is used in medical and cosmetic applications.
- Despite relatively high tocopherol concentrations, the oil is comparatively stable against oxidation.
- Prospective fields of application are coronary heart disease, arthritis, diabetes, ADHD, and inflammatory skin diseases.

References

Guillén, M. D., Ruiz, A., Cabo, N., Chirinos, R., & Pascual, G. (2003). Characterization of Sacha Inchi (*Plukenetia volubulis* L.) oil by FTIR spectroscopy and 1H NMR comparison with linseed oil. *Journal of the American Oil Chemist's Society, 80*, 755–762.

Hamaker, B. R., Valles, C., Gilman, C. R., Hardmeier, R. M., Clark, D., Garcia, H. H., et al. (1992). Amino acid and fatty acid profiles of the Inca peanut (*Plukenetia volubulis* L.). *Cereal Chemistry, 69*, 461–463.

Hanssen, H.-P. (2008). Sacha Inchi — die Inka-Nuss auf dem Weg zum Weltmarkt. *Deutsche Apotheker Zeitung, 148*, 60–61.

Sathe, S. K., Hamaker, B. R., Sze-Tao, K. W. C., & Venkatachalam, M. (2002). Isolation, purification, and biochemical characterization of a novel water soluble protein from Inca peanut (*Plukenetia volubulis* L.). *Journal of Agricultural and Food Chemistry, 50*, 4906–4908.

CHAPTER 118

Analgesic and Other Medicinal Properties of Safflower (*Carthamus tinctorius* L.) Seeds

Alexander M. Popov[1], Daein Kang[2]
[1] Pacific Institute of Bioorganic Chemistry, Far Eastern Branch of Russian Academy of Science, Vladivostok, Russia
[2] Korean Pharmacopuncture Institute (KPI), Yeoksam-Dong, Gangnam-Gu, Seoul, Republic of Korea

995

LIST OF ABBREVIATIONS

CS, N-(*p*-coumaroyl)serotonin
FS, N-feruloylserotonin
5-HT, serotonin (5-hydroxytryptamine)
LA, linoleic acid
LDL, low density lipoprotein
OSS, oil of safflower seed
SC, serotonin conjugates
SS, safflower seed
SSE, ethanol-ethyl acetate extract of safflower seeds
ST, serotonin transporter
TCA, tricyclic antidepressant
TCS, N^1,N^5-(Z)-N^{10}-(E)-tri-*p*-coumaroylspermidine

Nuts & Seeds in Health and Disease Prevention. DOI: 10.1016/B978-0-12-375688-6.10118-5

INTRODUCTION

Safflower (*Carthamus tinctorius* L.) is cultivated for the edible oil that can be obtained from its seeds. The oil of the safflower seed (OSS) contains polyunsaturated fatty acids, including a mixture of linoleic (LA) (more than 70%), oleic, and other acids, as well as serotonin and its conjugates, polyphenols, lignans, and other compounds.

The main new development for therapeutic use of OSS is its characteristic analgesic activity. OSS, prepared by pressing *Carthamus tinctorius* seeds, has moderate analgesic activity, and has been used for many years in Korean herbal acupuncture (Popov *et al.*, 2009). In this chapter, new data are presented from studies done on serotonin content in OSS, the analgesic properties of OSS in a model of functional blockade of the sciatic nerve in mice, and other pharmacological activities. The putative mechanisms of the actions of OSS components are discussed.

BOTANICAL DESCRIPTION

Safflower (*Carthamus tinctorius* L.) is a member of the family *Compositae* or *Asteraceae*. Plants are 30—150 cm tall, with globular flower heads (capitula) and flowers that are usually a brilliant yellow, orange, or red, blooming in July. Its glabrous, branching stem grows from 30—100 cm high, and bears alternate, sessile, oblong or ovate-lanceolate leaves armed with small spiny teeth. Each branch will usually have from one to five flower heads containing 15—20 seeds per head. Safflower seed (SS) is used mainly for oil and meal production, but recently it has been increasingly used in the birdseed trade. The seed oil content ranges from 30 to 45%. The meal usually contains about 24% protein, and a lot of fiber. The meal that remains after oil extraction is used as a protein supplement for livestock.

HISTORICAL CULTIVATION AND USAGE

Safflower is one of humanity's oldest domesticated crops, but generally remains a minor commercial crop, with world seed production at around 800,000 tonnes per year. It is believed to have originated in southern Asia, and has been cultivated in China, India, Persia, and Egypt since prehistoric times. During the Middle Ages it was cultivated in Italy, France, and Spain, and soon after the discovery of America the Spanish took it to Mexico, and then to Venezuela and Colombia. Over 60 countries grow safflower, but over half is produced in India (mainly for the domestic vegetable oil market). OSS has been produced commercially and for export for about 50 years as an oil source for the paint industry.

PRESENT-DAY CULTIVATION AND USAGE

Safflower seed is a minor commercial crop today, with about 800,000 tonnes being produced each year. OSS is used mainly as cooking oil, in salad dressings, and for the production of margarine. OSS is flavorless and colorless, and nutritionally similar to sunflower oil. It is also used in paint products in place of linseed oil, and in particular with white paint, as it does not have the yellow tint of linseed oil.

OSS has a very high percentage of linoleic acid (LA), and therefore it helps to moisturize, nourish, and restructure the skin when used in balms, creams, and lip care products. As an industrial oil, it is considered to be a drying or semidrying oil for use in manufacturing paints and other surface coatings. OSS can also be used as a diesel fuel substitute, but, like most vegetable oils, is currently too expensive for this use.

The safflower seed (SS) is well suited for organic skin care products, and has been clinically proven to be highly beneficial in lowering serum cholesterol levels. It is also used quite commonly as an alternative to sunflower seeds in birdfeeders. Today, SS — the source of OSS — is used for meal, birdseed, in the food and industrial products markets, and foots (the residue from oil processing) to manufacture soap, but it is primarily grown for its oil.

APPLICATIONS TO HEALTH PROMOTION AND DISEASE PREVENTION

Safflower has long been applied for empirically treating cerebral ischemia and depression in traditional Chinese medicine (Zhao *et al.*, 2009). OSS has been mainly used in Korea herbal acupuncture for many years as an analgesic remedy (Popov *et al.*, 2009). However, until recently we did not know which active components are in the oily preparation.

The production techniques of the sterile form of OSS called "CF," were designed in the Korean Pharmacupuncture Institute under the direction of Dr D. Kang. Analgesic actions of the product CF (i.e., a light yellow oil with no odor, obtained by pressing safflower seeds), were tested on mouse sciatic nerve. The tests performed showed that CF (OSS) is an optimal local natural anesthetic with moderate analgesic effects, as seen by a reduction of pain reactions (Figure 118.1), controlled by serotoninergic methods (Popov *et al.*, 2009). As a result of the research, Patent No. 2308283 from The Committee of the Russian Federation for Patents and Trademarks was received. The date of the invention is September 14, 2005, and the patent covers the remedy, possessing analgesic activity.

The OSS activities are probably based on the presence of serotonin (5HT). A typical chromatogram of an extract of OSS, without the addition of external serotonin, is shown in Figure 118.2B. Calculations based on analysis of samples showed that the mean serotonin concentration in OSS was 0.109 ± 0.005 nmol/ml.

It was suggested (Popov *et al.*, 2009) that administration of OSS to the region of the sciatic nerve led to a local increase in the quantity of serotonin in the synaptic cleft of the sciatic nerve, which also led to partial blockage of its functional activity. However, unlike the tricyclic

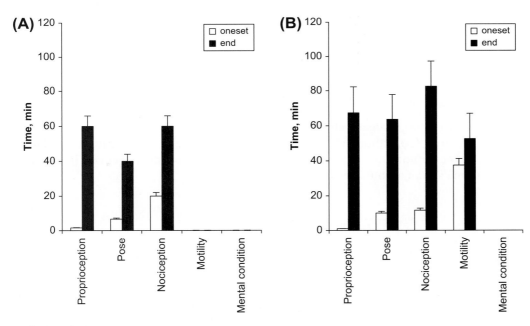

FIGURE 118.1

Time of onset of partial blockade and the complete recovery of the function of the sciatic nerve using different concentrations of CF ($n = 20$). CF doses: (A) 0.1 ml; (B) 0.2 ml. Within 10—15 minutes after injection of OSS at a dose of 0.1 ml, animals showed sharp reductions in nociceptive and proprioceptive functions. The state of the animals completely normalized over the next 30—40 minutes. At a dose of 0.2 ml, OSS strengthened partial blockade of pain, motor, and proprioceptive functions, with return to the normal state starting 0.8—1 hour after injection. Mental state (the mobility of the animals and their responses to mild external stimuli) remained normal (i.e., as in intact animals). Results are expressed as mean $\pm$ SEM, $n = 5$.

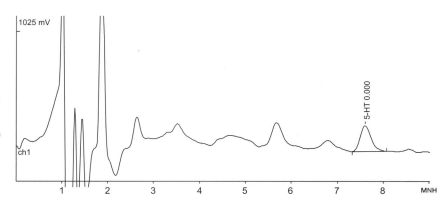

FIGURE 118.2
Determination of the contents of serotonin in OSS.
Chromatogram of a sample of OSS without the
addition of serotonin; detector sensitivity: 5 nA.
Serotonin was extracted from OSS by adding 200 μl of
propanol to 0.5 ml of study sample. The serotonin
concentration in the propanol fraction was measured
by high performance liquid chromatography using an
electrochemical detector.

antidepressant (TCA) amitriptyline (Figure 118.3), OSS at the doses studied lacked neurotoxic
activity.

There are some common features in the nature of the nerve-blocking actions of the TCA
amitriptyline and OSS. This similarity may result from the fact that OSS contains 5HT, which is
known to accumulate in the synaptic cleft when TCA acts on serotoninergic pathways.
Monoamine transmitters 5HT and dopamine, extracellular levels of which are limited by
Na^+/Cl^--dependent monoamine transporters in presynaptic neurons, play key roles in
regulating pain reactions (Sindrup *et al.*, 1999). TCA inhibits the serotonin transporter (ST) in
the synaptic cleft, thus blocking re-uptake of serotonin into the neuron, and subsequent
inactivation. Excessive serotonin accumulation leads to increases in the functional activity of
the serotoninergic antinociceptive system, and partial or complete blockade of neuromuscular
transmission (Sudoh *et al.*, 2003).

5HT and serotonin conjugates (SC) belong to a class of phenylpropanoid amides found at low
levels in SS (Figure 118.4A–C). Representative serotonin derivatives include feruloylserotonin
(FS) and 4-coumaroylserotonin (CS). Due to the strong antioxidant activity and other
therapeutic properties of serotonin derivatives, these compounds may have great potential in
treatment and prophylaxis, as cosmetic ingredients, and as major components of functional
foods or feeds that have health-improving effects (Kang *et al.*, 2009).

SC can possess a direct analgesic effect. N-feruloylserotonin (FS) did not influence anxiety in
low pain-threshold rats, although it reduced anxiety in the high pain-threshold rats. From
these results, the authors concluded that FS has selective stress-reducing effects in stress-
sensitive animals (Yamamotová *et al.*, 2007).

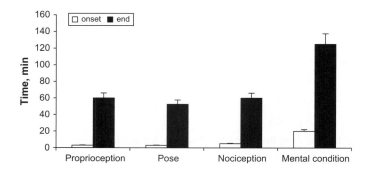

FIGURE 118.3
Time of onset of partial blockade and complete recovery of function of sciatic nerve using 0.1 ml of 1-mM amitriptyline
solution. Mental state (the mobility of the animals and their responses to mild external stimuli) is abnormal between
20 and 125 minutes, as compared with intact animals. Results are expressed as mean ± SEM, $n = 4$.

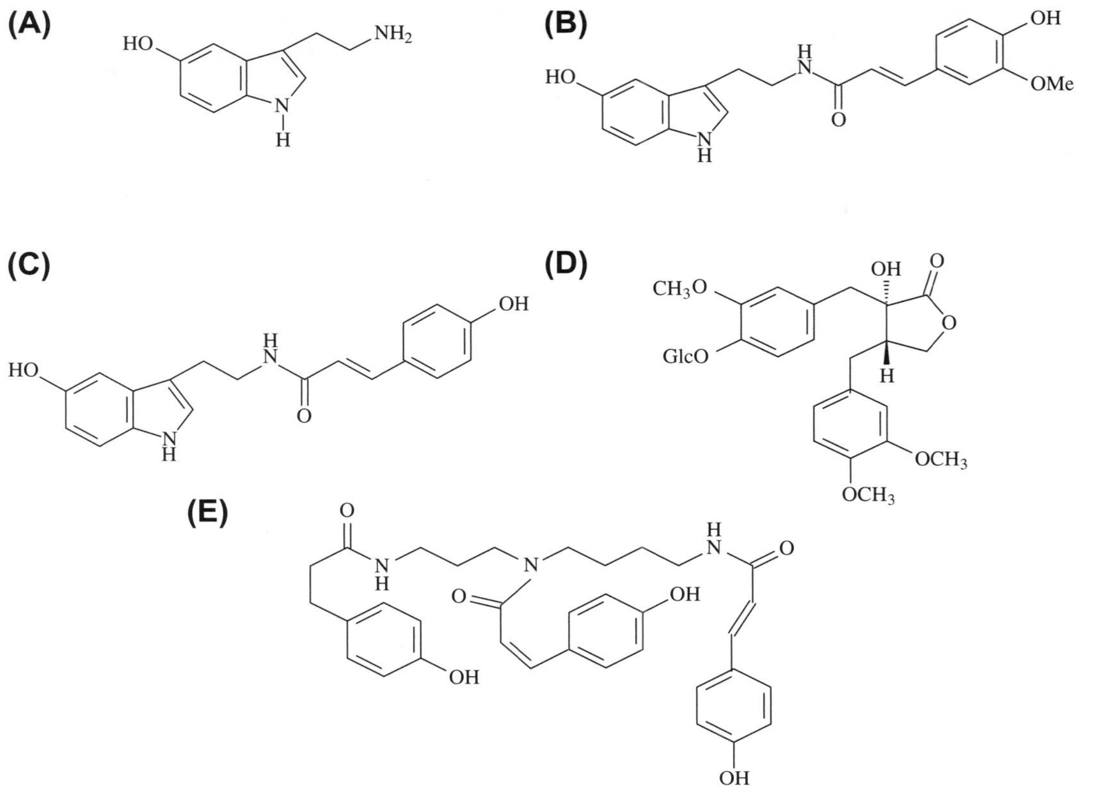

FIGURE 118.4

Chemical structures of biological activity substances from SS: (A) serotonin, (B) N-feruloylserotonin, (C) N-(*p*-coumaroyl) serotonin, (D) tracheloside, (E) N^1,N^5-(Z)-N^{10}-(E)-tri-*p*-coumaroylspermidine.

The analgesic effect of OSS may also be associated with the high antioxidant activity and strong inhibitory influence on tyrosinase by SC (i.e., FS and CS) that are present in OSS. Mono-aminergic neurotransmitters are substrates of the enzyme tyrosinase, which is present in all animals, is an oxidase, and can take part in the transformation and inactivation of serotonin. Thus, inhibition of tyrosinase by SC present in OSS may have indirect effects on the increased serotonin content in synaptic clefts, thus influencing the functional activity of the sciatic nerve (Roh *et al.*, 2004).

Furthermore, serotonin derivatives of FS and CS exerted a neuroprotective effect on high glucose-induced nerve cell death, and inhibited the activation of caspase-3, which represents the last and crucial step within the cascade of events leading to apoptosis. It inhibited over-production of the mitochondrial superoxide, which represents the most dangerous radical produced by hyperglycemia, by acting as a scavenger of the superoxide radical (Piga *et al.*, 2009).

The oil base of OSS consists of unsaturated (C18:2) linoleic acid (LA) (more than 70%) (Table 118.1). Attractively, the LA accelerates the proteolytic degradation of tyrosinase, and may thus indirectly inhibit the synthesis of the monoaminergic neurotransmitters that transmit the pain signal (Ando *et al.*, 1999).

OSS can be used as a local anesthetic in pharmacopuncture for the treatment of pain syndromes of various etiologies. Pharmacopuncture, the administration of drugs into acupuncture points, is widely used in eastern medicine. Injections are performed using very low doses, as the drugs are addressed directly into acupuncture points, which increases their therapeutic efficacy. In comparison with intramuscular injections, pharmacopuncture is not only more effective, but also allows a significant reduction in drug load on the patient's body,

TABLE 118.1 Comparative Compositions (per 100 g) of the Fatty Acids of Safflower and Linseed Oils

Fatty Acid	Linseed Oil	Safflower Oil
Palmitic acid (g)	5.9	5.7
Stearic acid (g)	3.6	2.4
Arachidonic acid (g)	—	0.5
Oleic acid (g)	18.2	11.5
Ecosapentaenic acid (g)	—	0.5
Linoleic acid (g)	13.9	74.4
α-Linolenic acid (g)	54.2	0.5

The fatty acid methyl esters were analyzed by us on a gas chromatograph using a procedure described by Christie (1982).

and has few or no side effects. One of the main advantages of pharmacopuncture is the rapid relief of pain. OSS is inexpensive, available, and has great potential for use in pharmaco- and acupuncture, and also in therapeutic massage (Popov et al., 2009).

Recently, Zhao and colleagues (2009) assessed whether safflower or its isolate would be effective in functionally regulating monoamine transporters using in vitro screening cell lines. They discovered that a novel coumaroylspermidine analog -N^1,N^5-(Z)-N^{10}-(E)-tri-p-coumaroylspermidine (TCS) from safflower (Figure 118.4—E), significantly inhibited serotonin uptake in Chinese hamster ovary cells stably expressing ST. Pharmacologically, this compound potently and selectively inhibited 5HT uptake in S6 cells or in synaptosomes, with an IC_{50} of 0.74 ± 0.15 µM for S6 cells or 1.07 ± 0.23 µM for synaptosomes, and with a reversible competitive property for the 5HT-uptake inhibition. Animals treated with this testing compound showed a significant decrease in synaptosomal 5HT uptake capacity. Thus, in the opinion of the authors, hydroxycinnamic acid amide TCS is a novel ST inhibitor, which could improve neuropsychological disorders through regulating serotoninergic transmission (Zhao et al., 2009).

Traditionally, OSS has been used in a medicinal oil to promote sweating and cure fevers in the Middle East, India, and Africa. In China, SS extracts are widely used in the treatment of many disorders and diseases, including menstrual problems, cardiovascular disease, pain and swelling associated with trauma, chronic and atrophic gastritis, rheumatism, and chronic nephritis. SS extracts have especially been used for the treatment of cardiovascular disease, because they invigorate circulation and reduce blood cholesterol levels. In Korea, SS extracts have traditionally been used for the treatment of blood stasis, promotion of bone formation, and prevention of osteoporosis. SS extracts might also have the potential for being used as a drug for bone regeneration and regeneration of periodontal defects. In addition, safflower seeds have a protective effect on bone loss caused by estrogen deficiency, without a substantial effect on the uterus (Kim et al., 2008).

Serotonin derivatives from SS inhibited the progress of atherosclerosis and ameliorated wall distensibility, which contributed, in part, to the lowering of local pulse wave velocity, used clinically as a direct measure of arterial stiffness. SC may be beneficial in improving vascular distensibility and in reducing cardiovascular risk (Katsuda et al., 2009).

Ethanol-ethyl acetate extract of SS (SSE) and the major antioxidant components of SC exhibit powerful immune-modulating properties, free-radical scavenging, and antioxidant effects, especially in the case of atherosclerotic dysfunctions. SSE and SC inhibited low density lipoprotein (LDL) oxidation induced in vitro by an azo-containing free-radical initiator V70 or copper ions. These findings demonstrate that SC of SSE are absorbed into the circulation and attenuate atherosclerotic lesion development possibly because of the inhibition of oxidized LDL formation through their strong antioxidative activity (Koyama et al., 2006).

Safflower seeds also contain numerous polyphenolic compounds, such as lignans, flavonoids, and glucosides. Some of these compounds, termed phytoestrogens, are known to have weak estrogenic or anti-estrogenic activity towards mammals (Dixon, 2004). The lignan glycoside, tracheloside (Figuer 118.4D) was isolated from seeds of *Carthamus tinctorius* as an anti-estrogenic principle. Tracheloside significantly decreased the activity of alkaline phosphatase, an estrogen-inducible marker enzyme, with an IC_{50} of 0.31 μg/ml — a level of inhibition comparable to that of tamoxifen ($IC_{50} = 0.43$ μg/ml) (Yoo *et al.*, 2006).

OSS contains nearly 75% LA. Comparative compositions of the fatty acids of OSS and linseed oil are shown in Table 118.1. OSS contains considerably more LA than linseed oil, and also corn, soybean, cotton seed, peanut, and olive oils. Researchers disagree on whether oils high in polyunsaturated acids like LA help decrease blood cholesterol levels and related heart and circulatory problems (Rodriguez-Cruz *et al.*, 2005).

LA is a polyunsaturated fatty acid used in the biosynthesis of arachidonic acid, and thus some prostaglandins. It is found in the lipids of cell membranes. LA is a member of the group of essential fatty acids called omega-6 fatty acids, so-called because they are an essential dietary requirement for all mammals. Epidemiological and clinical studies have established that the n-6 fatty acid LA, and the n-3 fatty acids linolenic acid, eicosapentaenoic acid, and docosa-hexaenoic acid, collectively protect against coronary heart disease. LA is the major dietary fatty acid regulating low density lipoprotein (LDL) metabolism by down-regulating LDL production and enhancing its clearance. The distinct functions of these two families make the balance between dietary n-6 and n-3 fatty acids an important consideration influencing cardiovascular health. This balance corresponds to an n-6:n-3 ratio of 6:1 (Wijendran & Hayes, 2004). However, the absolute mass of essential fatty acids consumed, rather than their n-6:n-3 ratio, should be the first consideration when contemplating lifelong dietary habits affecting cardiovascular benefit from their intake.

ADVERSE EFFECTS AND REACTIONS (ALLERGIES AND TOXICITY)

No adverse effects and reactions of biologically active substances in SS have been registered. Many years of using OSS in Korean herbal acupuncture has revealed no serious side effects. Administration of OSS at doses of 0.1–0.2 ml along the route of the sciatic nerve in mice produced no negative effects on the general state of the experimental animals, or skin irritation at the injection site (Popov *et al.*, 2009). In addition, there were no toxic reactions or side effects when SSE was given orally. TCS did not affect cell viability, thus suggesting that it would be safe and possess potential for development (Zhao *et al.*, 2009).

SUMMARY POINTS

- The bioactive components from SS possess analgesic properties.
- The actions of OSS, which contain 5HT and its derivatives, are associated with significant decreases in the nociceptive and proprioceptive functions.
- TCS, as a novel potent ST inhibitor, could possess antidepressant action, and would be helpful in improving neuropsychological disorders through regulating serotoninergic transmission.
- The lignan glycoside, tracheloside, has high anti-estrogenic activity comparable to that of tamoxifen, antagonist of the estrogen receptor in breast tissue.
- The analgesic, antioxidant, anti-atherogenic, immunomodulatory, free radical-scavenging, anti-inflammatory, and other medicinal properties of SC, bioflavonoids, and lignans can be used against various diseases and complications.
- OSS contains nearly 75% LA. LA is the major dietary fatty acid regulating LDL metabolism by down-regulating LDL production and enhancing its clearance.

References

Ando, H., Watabe, H., Valencia, J. C., Yasumoto, K., Furumura, M., Funasaka, Y., et al. (2004). Fatty acids regulate pigmentation via proteasomal degradation of tyrosinase: a new aspect of ubiquitin-proteasome function. *Journal of Biological Chemistry, 279*, 15427–15433.

Christie, W. W. (1982). A simple procedure for transmethylation of glycerolipids and cholesterol esters. *Journal of Lipid Research, 23*, 1073–1075.

Dixon, R. A. (2004). Phytoestrogens. *Annual Review of Plant Biology, 55*, 225–261.

Kang, K., Park, S., Kim, Y. S., Lee, S., & Back, K. (2009). Biosynthesis and biotechnological production of serotonin derivatives. *Applied Microbiology and Biotechnology, 83*, 27–34.

Katsuda, S. I., Suzuki, K., Koyama, N., Takahashi, M., Miyake, M., Hazama, A., et al. (2009). Safflower seed polyphenols (N-(p-coumaroyl)serotonin and N-feruloylserotonin) ameliorate atherosclerosis and distensibility of the aortic wall in Kurosawa and Kusanagi-hypercholesterolemic (KHC) rabbits. *Hypertension Research, 32*, 944–949.

Kim, K.-W., Suh, S.-J., Lee, T.-K., Ha, K.-T., Kim, J,-K, Kim, K.-H., et al. (2008). Effect of safflower seeds supplementation on stimulation of the proliferation, differentiation and mineralization of osteoblastic MC3T3-E1 cells. *Journal of Ethnopharmacology, 115*, 42–49.

Koyama, N., Kuribayashi, K., Seki, T., Kobayashi, K., Furuhata, Y., Suzuki, K., et al. (2006). Serotonin derivatives, major safflower (*Carthamus tinctorius* L.) seed antioxidants, inhibit low-density lipoprotein (LDL) oxidation and atherosclerosis in apolipoprotein E-deficient mice. *Journal of Agricultural and Food Chemistry, 54*, 4970–4976.

Piga, R., Naito, Y., Kokura, S., Handa, O., & Yoshikawa, T. (2009). Protective effect of serotonin derivatives on glucose-induced damage in PC12 rat pheochromocytoma cells. *British Journal of Nutrition, 14*, 1–7.

Popov, A. M., Lee, I. A., & Kang, D.-I. (2009). Analgesic properties of "CF" extracted from safflower (*Carthamus tinctorius* L.) seeds and potential for its use in medicine. *Pharmaceutical Chemistry Journal, 43*, 41–44.

Rodrigues-Cruz, M., Tovar, A. R., Del Prado, M., & Toress, N. (2005). Molecular mechanisms of action and health benefits of polyunsaturated fatty acids. *Revista de Investigación Clínica, 57*, 457–472.

Roh, J. S., Han, J. Y., Kim, J. H., & Hwang, J. K. (2004). Inhibitory effects of active compounds isolated from safflower (*Carthamus tinctorius* L.) seeds for melanogenesis. *Biological Pharmaceutical Bulletin, 27*, 1976–1978.

Sindrup, S. H., & Jensen, T. S. (1999). Efficacy of pharmacological treatments of neuropathic pain: an update and effect related to mechanism of drug action. *Pain, 83*, 389–400.

Sudoh, Y., Cahoon, E. E., Gerner, P., & Wang, G. K. (2003). Tricyclic antidepressants as long-acting local anesthetics. *Pain, 103*, 49–55.

Wijendran, V., & Hayes, K. C. (2004). Dietary n-6 and n-3 fatty acid balance and cardiovascular health. *Annual Review of Nutrition, 24*, 597–615.

Yamamotová, A., Pometlová, M., Harmatha, J., Rasková, H., & Rokyta, R. (2007). The selective effect of N-feruloylserotonins isolated from *Leuzea carthamoides* on nociception and anxiety in rats. *Journal of Ethnopharmacology, 112*, 368–374.

Yoo, H. H., Park, J. H., & Kwon, S. W. (2006). An anti-estrogenic lignan glycoside, tracheloside, from seeds of *Carthamus tinctorius*. *Bioscience, Biotechnology, and Biochemistry, 70*, 2783–2785.

Zhao, G., Gai, Y., Chu, W. J., Qin, G. W., & Guo, L. H. (2009). A novel compound N(1),N(5)-(Z)-N(10)-(E)-tri-p-coumaroylspermidine isolated from *Carthamus tinctorius* L. and acting by serotonin transporter inhibition. *European Neuropsychopharmacology, 19*, 749–758.

CHAPTER 119

Sea Buckthorn (*Hippophae rhamnoides* L. ssp. *turkestanica*) Seeds: Chemical and Physicochemical Properties

Alam Zeb
Department of Biotechnology, University of Malakand, Chakdara, Pakistan

CHAPTER OUTLINE

LIST OF ABBREVIATIONS

CA, cinnamic acid
Caf A, caffeic acid
FA, ferulic acid
GA, gallic acid
p-CA, *p*-coumaric acid
p-HBA, *p*-hydroxybenzoic acid
ProCA, protocatechuic acid
SDS-PAGE, sodium dodecyl sulfate-polyacrylamide gel electrophoresis
SEM, scanning electron micrograph
VA, vanillic acid

Nuts & Seeds in Health and Disease Prevention. DOI: 10.1016/B978-0-12-375688-6.10119-7

INTRODUCTION

Sea buckthorn is a medicinal and aromatic plant that is both cultivated and naturally grown in various parts of the world. The distribution of sea buckthorn ranges from the Himalayan regions, including India, Nepal, Bhutan, Pakistan and Afghanistan, to China, Mongolia, Russia, Kazakhstan, Hungary, Romania, Switzerland, Germany, France, and Britain, and northwards to Finland, Sweden, and Norway (Rousi, 1971; Zeb, 2004a). The wide distribution of sea buckthorn reflects its habitat-related variation in berry characteristics, such as fresh weight, and chemical and sensory attributes.

BOTANICAL DESCRIPTION

Sea buckthorn (*Hippophae rhamnoides* ssp. *turkestanica*) is a unique herb and aromatic plant that belongs to the Elaeagnaceae family. The plant is a small or medium deciduous tree or large shrub, growing to a height of 2–6 m. Thick, rough bark covers the main trunk, while young branches are smooth, gray to light-ash colored, with needle-shaped thorns (Zeb & Khan, 2008a). Almost all the *Hippophae* species start to develop thorns on 2- or 3-year-old plants, ranging in length from 2.5–5.0 cm. Staminate and pollinate flowers appear before the leaves. The leaves are alternate, narrow, and 4–6 cm long, and silver-gray on the upper side. Flower buds are formed mostly on 2-year-old wood, differentiated during the previous growing season.

The weight of a sea buckthorn berry ranges from 200–500 mg, depending on many factors, including the variety and environmental conditions. The pulpy fruit part is used for its juice. The berry or fruit also contains a single dicotyledonous brownish-black seed. A scanning electon micrograph (SEM) of dissected sea buckthorn ssp. *turkestanica* seed is shown in Figure 119.1 (Zeb & Malook, 2009). The seed is approximately 4.3 mm long and 2.4 mm wide, and slightly pointed at both ends, with a furrow all around the seed body. A cut section shows two distinct layers of the testa, derived from the outer and inner integument of the ovule. The outer layer of the seed is thick, hard, and brownish-black in color, while the inner layer is thin and fibrous in appearance. The total thickness of the seed coat is about 0.1–0.2 nm. The seed coat encloses the massive embryo, the axis of which is visible in the SEM. It appears that the seed morphology of the sea buckthorn ssp. *turkestanica* is in close agreement with the Canadian sea buckthorn var. Indian summer, as reported by Harrison and

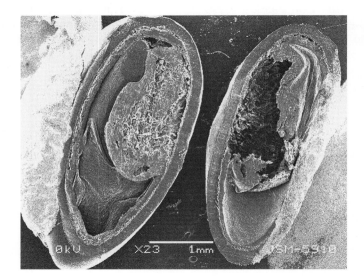

FIGURE 119.1
Scanning electron micrograph (SEM) of cross-section of sea buckthorn ssp. *turkestanica* seed. The seed was examined with scanning electron microscopy (JEOL JSM–5910, Japan) at a magnification of 23×. *Reproduced with permission from Zeb & Malook (2009).*

TABLE 119.1 Physicochemical Characteristics of Sea Buckthorn ssp. *turkestanica* Seed

Chemical Property	Value/Range
Seed	
Ash content (%)	1.76–2.05
1000-seed mass (g)	10.2
Moisture (%)	3.0–5.5
Seed Oil	
Specific gravity	0.91
Refractive index	1.47
Acid value (mg/g)	6.30
Anisidine value	43.34
Iodine value (g/100 b)	125.0
Saponification value (mg KOH/g)	181.0
Unsaponifiable matter (%)	0.90

Variation in these values may arise depending on the method of extraction, method of determination, variety, and origin.
Adapted from Zeb & Malook (2009), with permission.

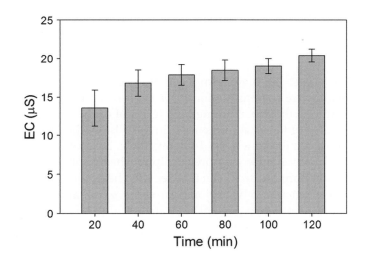

FIGURE 119.2

Changes in the electrical conductivity of sea buckthorn ssp. *turkestanica* seed of Pakistani origin. Changes in electrical conductivity were measured by imbibed seed in de-ionized water. *Reproduced with permission from Zeb and Malook (2009).*

Beveridge (2002). The chemical and physicochemical properties of the seeds, however, vary with climate, origin, and variety (Table 119.1). The small change in the electrical conductivity (Figure 119.2) during imbibition experiments shows the hardness of seed.

HISTORICAL CULTIVATION AND USAGE

Medicinal applications of sea buckthorn have been well-documented since ancient times, and are still in use in Europe and Asia (Zeb, 2004b). The presence of valuable components in sea buckthorn seeds or berries increases its use across the globe. Traditionally, the Chinese have used sea buckthorn as medicine for thousands of years. Some clinical studies on the medicinal uses of sea buckthorn were performed in Russia during the 1950s. It was officially listed in the *Chinese Pharmacopoeia* by the Ministry of Public Health in 1977, and its reputation as a medicinal plant has been established (Li and Beveridge, 2003).

PRESENT-DAY CULTIVATION AND USAGE

Hippophae rhamnoides (ssp. *turkestanica*) is distributed in the northern areas of Pakistan, including Kurram Agency, Chitral, Swat, Utror-Gabral, Gilgit, Astore, Skardu, Ganche, Baltistan, and Ladakh. The environment in northern areas of Pakistan provides tremendous potential for the growth and development of this wild type of sea buckthorn. Despite the wide occurrence, types, and composition of the sea buckthorn, it is used for medicinal purposes only. So far, more than 20 different drugs have been developed from sea buckthorn, and are available in different forms, including liquids, powders, pastes, aerosols, etc.

APPLICATIONS TO HEALTH PROMOTION AND DISEASE PREVENTION

Health-promoting chemical properties

The chemical and phytochemical composition of sea buckthorn has been reviewed by various researchers (Tiitinen *et al.*, 2005; Zeb & Khan, 2008b), and a recent book by Singh (2008) provides a good source of information on the chemical composition of this plant, which has been found to vary depending on the origin of the seed, the climate, and the method of extraction. The chemistry of sea buckthorn ssp. *turkestanica* has not been studied in detail yet (Shah *et al.*, 2007); however, a brief account of each component from the available literature is given below.

1. *Carotenoids and vitamin C.* Sea buckthorn ssp. *turkestanica* seed oil is similar in color to red palm oil. Carotenoids are responsible for the red color, and it has been found that sea buckthorn oil is among those containing high levels of carotenoids (Zeb, 2004a). The average reported value of the total carotenoids in the oils of other varieties has been found to be 1167 mg/100 g; we found it to be 1400 mg/100 g. β-Carotene was the major carotenoid; this is in accordance with previous published results, which showed that β-carotene is the dominant carotenoid in the seed oil. The detailed carotenoid composition of ssp. *turkestanica* still needs to be explored; however, α-, β-, and γ-carotenes, lycopene, pheopytin, zeaxanthin, and β-cryptoxanthin were found to be the major carotenoids in seed oils of other varieties. The vitamin C content of the seeds, measured using the standard AOAC method, was 35.4 mg/100 g (Table 119.2), and was higher in fresh seeds/berries (250−333 mg/100 g).

2. *Fatty acid composition.* The fatty acid composition of sea buckthorn ssp. *turkestanica* oil has been revealed by Abid *et al.*, (2007), using a standard gas chromatographic method with a Celite packed column. It was found that the seed oil is more unsaturated than pulp oil: saturated fatty acids account for 14.1%, and unsaturated fatty acids for 85.0%. Fatty acids include palmitic acid (11.1%), palmitoleic acid (9.8%), stearic acid (2.6%), oleic acid/vaccenic acid (22.1%), linoleic acid (29.6%), and α- and γ-linolenic acid (23.4%) (Table 119.2). The high level of unsaturation displayed by the fatty acid composition and high conjugated dienes, *p*-anisidine, and peroxide values suggest low stability towards oxidation.

3. *Phytosterols.* High blood cholesterol is one of the well-established risk factors for coronary heart disease across the globe. Interest in lowering blood cholesterol is a major research area, which can probably impact the incidence of heart disease. Phytosterols or plant sterols are capable of lowering plasma cholesterol upon consumption by humans. Phytosterols are the major constituents of the unsaponifiable fraction of sea buckthorn seed. Sabir *et al.* (2005b) showed that sea buckthorn ssp. *turkestanica* contains 3.3−5.5% of total plant sterols. The major sterols in sea buckthorn seed oil are β-sitosterol and 5-avenasterol. Other phytosterols are present in relatively minor quantities.

TABLE 119.2 Chemical Composition of Sea Buckthorn ssp. *turkestanica* Seed of Different Origins

Chemical Component	Amount	Reference
Carotenoids & Vitamins	**(mg/100g)**	**Zeb & Malook, 2009**
β-Carotene	1400	Zeb & Malook, 2009
Vitamin C	35.4	Zeb & Malook, 2009
Fatty Acids & Sterols	**(%)**	**Abid *et al.*, 2007**
Palmitic acid	11.1	Abid *et al.*, 2007
Palmitoleic acid	9.80	Abid *et al.*, 2007
Stearic acid	2.60	Abid *et al.*, 2007
Oleic/vaccenic acids	22.1	Abid *et al.*, 2007
Linoleic acid	29.1	Abid *et al.*, 2007
Linolenic acids	23.4	Abid *et al.*, 2007
Phytosterols (total)	3.3—5.5	Sabir *et al.*, 2005b
Phenolic Acids (seed kernel)	**(mg/kg)**	**Arimboor *et al.*, 2008**
Total	5741	Arimboor *et al.*, 2008
Gallic acid	3441	Arimboor *et al.*, 2008
p-Hydroxybenzoic acid	265	Arimboor *et al.*, 2008
Vanillic acid	368	Arimboor *et al.*, 2008
Cinnamic acid	82	Arimboor *et al.*, 2008
p-Coumaric acid	74	Arimboor *et al.*, 2008
Ferulic acid	156	Arimboor *et al.*, 2008
Caffeic acid	25	Arimboor *et al.*, 2008

Variation in these values may arise depending on the method of extraction, method of determination, variety, and origin.
Adapted from Zeb & Malook (2009), with permission.

4. *Phenolic acids.* A recent study by Arimboor *et al.* (2008) revealed free and bound phenolic acid (PA) composition in different parts of the sea buckthorn ssp. *turkestanica*, including the seed, using HPLC-DAD. The total phenolic acid content in the seed kernel (5741 mg/kg) was higher than in the seed coat (Table 119.2). The total phenolic acid content in the seed kernel was two-fold higher than that reported for sesame and cotton seed, and slightly lower than that reported for rapeseed and canola meals and flax seed. The major phenolic acids present in seed kernel were GA (3441 mg/kg), ProCA (1330 mg/kg), *p*-HBA (265 mg/kg), VA (368 mg/kg), FA (156 mg/kg), CA (82 mg/kg), *p*-CA (74 mg/kg), and Caf A (25 mg/kg). In the seed coat, GA (230 mg/kg), ProCA (82 mg/kg), CA (44 mg/kg), VA (39 mg/kg), *p*-HBA (26 mg/kg), FA (17 mg/kg), and Caf A (10 mg/kg) were identified. ProCA (43 mg/kg) was present as the major phenolic acid of the free phenolic acid fraction, whereas GA (200 mg/kg) was the major constituent in bound phenolic acid fractions from the seed coat.

5. *Elemental composition.* Elemental components like sodium, potassium, phosphorus, magnesium, calcium, zinc, and iron are important for nutritional purposes as well as the cultivation and growth of the plant. Sea buckthorn ssp. *turkestanica* seeds are a good source of these important minerals (Zeb & Malook, 2009). The mineral composition revealed high contents of calcium (912 mg/kg), magnesium (758 mg/kg), iron (290 mg/kg), zinc (96.6 mg/kg), and potassium (88.0 mg/kg), while the sodium content is lower, at 47.7 mg/kg, and phosphorus content is the lowest among the elements, at 0.43 mg/kg (Table 119.3). The investigated value of Na^+ was close to that given by Sabir *et al.* (2005a), while K^+ values were in close agreement with the value of 100 mg/kg given by Tong *et al.*, (1989) for other varieties. The phosphorus, magnesium, calcium, zinc and iron levels show a high degree of variation from those in the reported literature (Zeb, 2004a) for other

TABLE 119.3 Elemental Composition of Sea Buckthorn ssp. *turkestanica* Seed of Pakistani Origin	
Element	**Quantity (mg/kg)**
Calcium	912[*]
Magnesium	758[*]
Iron	290[*]
Zinc	96.6[*]
Potassium	88.0[**]
Sodium	47.7[**]
Phosphorus	0.43[***]

Reproduced with permission from Zeb & Malook (2009).
[*]*Determined with the help of flame atomic absorption spectrometer (Perkin Elmer, model Analyst 700) with air/acetylene flame at 2000–2400 K (photomultiplier tube detector).*
[**]*Flame atomic absorption spectrometer (Perkin Elmer, model Analyst 700) with air/acetylene flame at 2200–2400 K (photomultiplier tube detector).*
[***]*Flame photometer (Jenway PFP7); UV-visible spectrophotometer (Shimadzu UV-1700) at 660 nm.*

varieties. The variation may be due to the natural content of elements in the soil, geographical location, contamination of soil and air, and different methods and equipment used for analysis.

6. *Seed storage proteins*. The seed storage proteins represent an important source of food and energy, and can be analyzed by SDS-PAGE (Shuaib *et al.*, 2007) or other chromatographic methods. The SDS-PAGE profile is also helpful in measuring the genetic diversity in sea buckthorn genotypes. The molecular weights of protein subunits are measured by comparing seed sample bands to the standard protein molecular weight marker bands in the electropherogram. Ahmed & Kamal (2003) showed that there is a high variation in protein mobility in the gel, and the inter-relationship of sharing of at least one common band among all was mentioned; however, the exact molecular weight of each band was shown. In our study, we found a total of six bands in the 15- to 50-kDa range (Zeb & Malook, 2009). The molecular weight range of the protein bands indicate that there are four protein bands in the 15- to 25-kDa range, one band in the 30- to 40-kDa range, and one band in the 40- to 50-kDa range. Further analysis is, however, necessary to explore all the proteins that are important for their nutritional and medicinal values.

Health promotion and disease prevention

Sea buckthorn seed oil is widely used to promote recovery from skin injuries and skin diseases, including eczema, burns, and wounds, and the skin-damaging effects of sun, therapeutic radiation treatments, and cosmetic laser surgery (for details, see Chapter 120). It is also used in treating cancer, digestive ulcers, and cardiovascular disease (Yang & Kallio, 2002; Zeb, 2004b). A recent study suggests that oral supplementation with fruit extract from sea buckthorn ssp. *turkestanica* is effective in modifying haloperidol-induced behavioral deficits in rats (Batool *et al.*, 2009). It was suggested that increased precursor availability due to the extract in rats led to increased brain 5-hydroxytryptamine synthesis. In addition to the role of serotonin in schizophrenia, sea buckthorn may be used as a nutritive therapy in psychotic patients.

ADVERSE EFFECTS AND REACTIONS (ALLERGIES AND TOXICITY)

Most of the research work available regarding sea buckthorn seed concentrates on the medicinal values; there is a lack of literature regarding any adverse effects, allergies, and toxicity.

SUMMARY POINTS

- Sea buckthorn ssp. *turkestanica* is native to Asia, and bears yellow-reddish berries that contain an important black-brown hard seed.
- The seed is a source of highly unsaturated oil, containing saponifiable and unsaponifiable matter.
- The seed oil is rich in various carotenoids and vitamins, and contains high levels of oleic, linoleic, and linolenic acids.
- Sea buckthorn seeds are also a good source of important minerals, such as sodium, potassium, phosphorus, magnesium, calcium, zinc, and iron.
- The seed is rich in important phenolic acids, such as gallic acid.
- Seed storage proteins contribute to nutritional and food uses.
- β-Sitosterol is one of the major phytosterols present in the seed and seed oil.
- The presence of potentially health-promoting components in sea buckthorn seed makes it an important plant for use in the food and pharmaceutical industries.

References

Abid, H., Hussain, A., & Ali, S. (2007). Physicochemical characteristics and fatty acid composition of sea buckthorn (*Hippophae rhamnoides* L) oil. *Journal of the Chemical Society of Pakistan, 29*, 256–259.

Ahmed, S. D., & Kamal, A. (2003). Morpho-molecular characterization of local genotypes of *Hippophae rhamnoides* L. spp. *turkestanica*, a multipurpose plant from northern areas of Pakistan. *Journal of Biological Sciences, 2*, 351–354.

Arimboor, R., Kumar, K. S., & Arumughan, C. (2008). Simultaneous estimation of phenolic acids in sea buckthorn (*Hippophae rhamnoides*) using RP-HPLC with DAD. *Journal of Pharmaceutical and Biomedical Analysis, 47*, 31–38.

Batool, F., Shah, A. H., Ahmed, S. D., & Haleem, D. J. (2009). Oral supplementation of sea buckthorn (*Hippophae rhamnoides* L. spp. *turkestanica*) fruit extract modifies haloperidol induced behavioral deficits and increases brain serotonin metabolism. *Journal of Food and Drug Analysis, 17*, 257–263.

Harrison, J. E., & Beveridge, T. (2002). Fruit structure of *Hippophae rhamnoides* cv. Indian Summer (sea buckthorn). *Canadian Journal of Botany, 80*, 399–409.

Li, T. S. C., & Beveridge, T. H. J. (2003). (with contributions by B. D. Oomah. In W. R. Schroeder, & E. Small (Eds.), *Sea Buckthorn (Hippophae rhamnoides L.): production and utilization* (pp. 1–5). Ottawa, ON, Canada: NRC Research Press.

Rousi, A. (1971). The genus *Hippophae*; a taxonomic study. *Annales Botanici Fennici, 8*, 177–277.

Sabir, S., Maqsood, H., Hayat, I., Khan, M. Q., & Khaliq, A. (2005a). Elemental and nutritional analysis of sea buckthorn (*Hippophae rhamnoides* ssp. *turkestanica*) berries of Pakistani origin. *Journal of Medicinal Food, 8*, 518–522.

Sabir, S. M., Maqsood, H., Ahmed, S. D., Shah, A. H., & Khan, M. Q. (2005b). Chemical and nutritional constituents of sea buckthorn (*Hippophae rhamnoides* ssp. *turkestanica*) berries from Pakistan. *Italian Journal of Food Science, 17*, 455–462.

Shah, A. H., Ahmad, D., Sabir, M., Arif, S., Khaliq, I., & Batool, F. (2007). Biochemical and nutritional evaluations of sea buckthorn (*Hippophae rhamnoides* L. spp. *turkestanica*) from different locations of Pakistan. *Pakistan Journal of Botany, 39*, 2059–2065.

Shuaib, M., Zeb, A., Ali, Z., Ali, W., Ahmad, T., & Khan, I. (2007). Characterization of wheat varieties by seed storage protein electrophoresis. *African Journal of Biotechnology, 6*, 497–500.

Singh, V. P. (2008). *Sea buckthorn, a multipurpose wonder plant – advances in research & development*. New Delhi, India: Daya Publishing House.

Tiitinen, K. M., Hakala, M. A., & Kallio, H. P. (2005). Quality components of sea buckthorn (*Hippophae rhamnoides*) varieties. *Journal of Agricultural and Food Chemistry, 53*, 1692–1699.

Tong, T., Zhang, C., Zhang, Z., Yang, E., & Tian, K. (1989). Proceedings of the international Symposium on Sea Buckthorn (*H.rhamnoides* L.), Xian, China, pp. 132–137.

Yang, B., & Kallio, H. (2002). Composition and physiological effects of sea buckthorn (*Hippophae*) lipids. *Trends in Food Science & Technology, 13*, 160–167.

Zeb, A. (2004a). Chemical and nutritional constituents of sea buckthorn juice. *Pakistan Journal of Nutrition, 3*, 99–106.

Zeb, A. (2004b). Important therapeutic uses of sea buckthorn (*Hippophae*); a review. *Journal of Biological Sciences, 4*, 687–693.

Zeb, A., & Khan, I. (2008a). Pharmacological applications of sea buckthorn (Hippophae). In M. Eddouks (Ed.), *Advances in Phytotherapy Research* (pp. 1—16). New Delhi, India: Research Signpost.

Zeb, A., & Khan, I. (2008b). Composition and medicinal properties of sea buckthorn juice. In V. Singh (Ed.), *Sea buckthorn, a multipurpose wonder plant — advances in research & development* (pp. 205—213). New Delhi, India: Daya Publishing House.

Zeb, A., & Malook, I. (2009). Biochemical characterization of sea buckthorn (*Hippophae rhamnoides* L. spp. turkestanica) seed. *African Journal of Biotechnology, 8,* 1625—1629.

CHAPTER 120

Sea Buckthorn (*Hippophae rhamnoides* L.) Seed Oil: Usage in Burns, Ulcers, and Mucosal Injuries

Asheesh Gupta, Nitin K. Upadhyay
Department of Biochemical Pharmacology, Defence Institute of Physiology
and Allied Sciences, Timarpur, Delhi, India

1011

LIST OF ABBREVIATIONS

ALA, α-Linolenic acid
LA, linoleic acid
p.o., *per os*
PUFA, polyunsaturated fatty acid
SFE, supercritical fluid extraction
UV, ultraviolet

Nuts & Seeds in Health and Disease Prevention. DOI: 10.1016/B978-0-12-375688-6.10120-3

INTRODUCTION

Sea buckthorn (*Hippophae rhamnoides* L.), a multipurpose plant with exceptionally high contents of nutrients and phytochemicals, grows naturally in various regions of Asia and Europe. Sea buckthorn seed oil contains a number of bioactive substances, namely vitamins, carotenoids, sterols, and polyunsaturated fatty acids (PUFAs). A wide spectrum of pharmacological effects of sea buckthorn seed oil have been reported, including anti-oxidative, anti-microbial, anti-atherogenic and cardioprotective, hepatoprotective, radioprotective, and tissue regenerative. Recent studies have reconfirmed the traditional therapeutic use of sea buckthorn seed oil for burns, ulcers, mucosal injuries, and other skin diseases. No toxicity or adverse reactions have been reported after oral or topical application of sea buckthorn seed oil.

BOTANICAL DESCRIPTION

Sea buckthorn (genus *Hippophae*) belongs to the family Elaeagnaceae. Three species have been described in India, of which *Hippophae rhamnoides* L. is the major one. It is naturally distributed in dry temperate and cold desert areas up to an altitude of 5500 m in regions of the North-West Himalayas. Sea buckthorn is a deciduous, dioecious, branched, spiny plant. Its strong and complex root system with nitrogen-fixing nodules, having Frankia-actinorhizal symbiotic association, makes sea buckthorn an optimal pioneer plant for water and soil conservation in eroded areas. Leaves are alternate, narrow and lanceolate, with a silver-gray color. The male bud consists of four to six apetalous flowers, whereas the female bud usually consists of one single apetalous flower with one ovary and one ovule. The ripe barriers are drupe-like and orange/red in color, consisting of a single seed surrounded by a soft, fleshy outer tissue. Seeds are dark brown, glossy, ovoid to elliptical in shape, 2.8–4.2 mm in size, and contain 8–18% oil (Bartish *et al.*, 2002).

HISTORICAL CULTIVATION AND USAGE

Sea buckthorn is one of the most valuable bioresources, being locally used for centuries as fuel, fodder, small timber, food, and medicine. It was cultivated by some ancient plant-breeders. It is believed that the ancient Greeks used it in a diet for race horses; hence its botanical name *Hippophae*, meaning "shiny horse." References to medicinal use of sea buckthorn were found in ancient Tibetan medicinal texts, including the *RGyud Bzi* (The Four Books of Pharmacopoeia) dating to the times of Tang Dynasty (618–907 AD). Traditionally, it has been used to aid digestion and treat colds, coughs, and lung, circulatory, and skin disorders (Li and Schroeder, 1996; Zeb, 2004).

PRESENT-DAY CULTIVATION AND USAGE

Sea buckthorn has attracted the attention of researchers, industrialists, and ecologists for its multipurpose value. Many Asian and European countries have accorded ample importance to the potential of this plant by massive organized cultivation to improve afforestation, enhances soil fertility, and also help to uplift local rural economies. Sea buckthorn grows wild under natural conditions; however, it can be propagated by seeds, suckers, cuttings, layering, and micropropagation. Fruit harvesting is the most time-consuming operation in cultivating sea buckthorn. Efforts have been made recently to advance post-harvest technologies and develop thornless sea buckthorn cultivars with a higher yield, larger fruits, and a high vitamin content, using different breeding approaches — namely, selection, polyploidy, and irradiation. Sea buckthorn seed oil is widely used to promote the recovery of various skin conditions, including eczema, burns, and impaired wounds, sun damage, and radiation damage, and also in gastric ulcer protection (Upadhyay *et al.*, 2009; Zeb, 2004).

APPLICATION TO HEALTH PROMOTION AND DISEASE PREVENTION

In alternative and complementary systems of medicine, such as Ayurveda, Siddha, Amchi, Chinese, and aromatherapy, plants are used to combat several diseases and pathological conditions. Herbal products seem to possess moderate efficacy, with less (or no) toxicity and less expensive compared to synthetic drugs. Sea buckthorn seed oil combines high levels of beneficial unsaturated fatty acids (ALA, LA, oleic, palmitoleic, and vaccenic), natural antioxidants and vitamins (E, K), carotenoids, and phytosterols (campesterol, stigmasterol, and β-sitosterol) (Li *et al.*, 1996; Basu *et al.*, 2007). These phytochemicals make it ideal for medicinal and cosmetic industries, giving synergistic power to protect the cell membrane and enhance cell regeneration. Sea buckthorn seed oil can be extracted by different methods — cold-pressing, solvent extraction, and supercritical fluid extraction (SFE). Carbon dioxide is most commonly used in SFE of sea buckthorn seed oil at low temperature, in a microbe- and oxygen-free system (Basu *et al.*, 2007). Protective and curative effects of sea buckthorn seed oil against burns, scalds, ulcers, and mucosal injuries have been extensively investigated using different animal models, and by clinical trials (Zhao *et al.*, 1994; Xing *et al.*, 2002; Upadhyay *et al.*, 2009). Furthermore, sea buckthorn seed oil possesses anti-inflammatory, antiradiation, anti-atherogenic, and cardioprotective activities (Zhang *et al.*, 1988; Zeb, 2004; Basu *et al.*, 2007). In addition, sea buckthorn seed oil curtails hypoxia-induced cerebral vascular leakage, as demonstrated in laboratory studies (Purushothaman *et al.*, 2008). Sea buckthorn seed oil absorbs strongly in the UV-B range (290–320 nm), and may therefore be used as a natural sunscreen.

Healing efficacy of sea buckthorn seed oil on burns

Skin, the largest organ in the human body, plays a crucial role in the sustenance of life through the regulation of water and electrolyte balance, thermoregulation, and acting as a barrier to external noxious agents, including micro-organisms. When this barrier is disrupted for any reason — ulcers, burns, neoplasm, or trauma — these functions are no longer adequately performed. It is therefore vital to restore its integrity as soon as possible. Wound healing is a highly complex but well-orchestrated cascade of events, comprising three overlapping phases: inflammation, proliferation, and tissue remodeling (Figure 120.1). The normal healing process can be impeded at any step along its path by a variety of factors that can contribute to impaired healing. Impaired wound healing may be a consequence of pathologic states associated with diabetes, immune disorders, ischemia, and venous stasis, and injuries such as burns, frostbite, and gunshot wounds (Singer & Clark, 1999).

Burns are one of the most common and devastating forms of trauma. A burn is characterized by a hypermetabolic state which compromises the immune system, leading to chronic wound

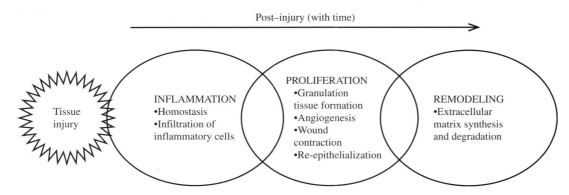

FIGURE 120.1
Schematic diagram showing the different overlapping phases of the wound healing process after injury.

healing. Thermal exposure to the body surface causes damage to the skin by membrane destabilization, protein coagulation, associated energy depletion, and hypoxia at the cellular level, which leads to extensive tissue necrosis and invasion of infectious agents. Healing impairment in burn injury is characterized by increased free-radical mediated damage, delayed granulation tissue formation, reduced angiogenesis, and decreased collagen reorganization. The use of synthetic drugs (for example, antibacterial agents and disinfectants) poses problems such as allergy, drug resistance, etc., forcing scientists to seek alternatives (Church *et al.*, 2006).

Recent studies have clearly shown that sea buckthorn seed oil is effective in treatment of burns of different etiology. Nearly 90% of sea buckthorn seed oil is unsaturated fatty acids, including ALA (ω-3) (18:3, n-3), LA (ω-6) (18:2, n-6), oleic acid (18:1, n-9), and palmitoleic acid (16:1, n-7), etc. The wound-healing activity of sea buckthorn seed oil is thought to be attributable to its high content of vitamin E, carotenoids, phytosterols, and PUFAs (ω-3 and -6) (Table 120.1). These phytoconstituents may be acting in an individual or a synergistic manner to provide protection and nutrition to the wound tissue. One ingredient, palmitoleic acid, is a component of skin that is considered a valuable topical agent in treating burns and healing wounds. This fatty acid can also nourish the skin when taken orally. A pre-clinical study performed on excision-type wounds in male calves concluded that wound healing progresses faster under sea buckthorn seed oil than under liquid paraffin and 5% povidone-iodine ointment (Rana, 2006). Sea buckthorn seed extracts possess growth-inhibiting effects on *Bacillus cereus, B. coagulans, B. subtilis, Listeria monocytogenes*, and *Yersinia enterocolitica*. The antibacterial effects of sea buckthorn seed implicate its potential for combating burn wound infection (Chauhan *et al.*, 2007).

A systematic pre-clinical study conducted by Upadhyay *et al.* (2009) demonstrated the safety and efficacy of sea buckthorn CO_2-SFE seed oil on experimental burn wounds in rats. It was reported that sea buckthorn seed oil co-administered by two routes at a dose of 2.5 ml/kg body weight (p.o.) and 200 µl (topical) for 7 days promotes significant re-epithelialization and wound closure (Figure 120.2). Treatment also enhanced formation of granulation tissue and collagen biosynthesis in burn wounds, in comparison with control and reference control treated with 1% silver-sulfadiazine ointment. Sea buckthorn seed oil has been observed to possess mitogenic potential, and is supposed to be involved in fibroblast and keratinocyte proliferation at the wound site. It has the capability to increase new blood capillary formation, thus contributing to structural repair through the formation of granulation tissue. Sea buckthorn seed oil treatment up-regulated the expression of matrix metalloproteinases-2 and 9, collagen type-III, and vascular endothelial growth factor at the wound site, suggesting that it plays an important role in the tissue-remodeling phase of wound repair. Seed oil treatment showed an increase in endogenous antioxidant and decrease in free radical

TABLE 120.1 Possible Role of Important Phytochemicals of Sea Buckthorn Seed Oil in Wound Healing and Skin Care

Sea Buckthorn Seed Oil (Phytoconstituents)	Medicinal Properties
Vitamin E	Acts as antioxidant; minimizes lipid oxidation; maintains tissue integrity; reduces skin toughening and wrinkling; helps to relieve pain
Vitamin K	Prevents bleeding; promotes wound healing; anti-ulcer effect
Carotenoids	Acts as antioxidant; heals ulcerative tissue; protects stomach lining from irritation; helps in epithelialization
Phytosterols	Improve microcirculation in the skin; reduce inflammation; antimicrobial and anti-ulcer effect
Polyunsaturated fatty acids	Antibacterial effect against wound pathogens; regulate the inflammatory process; enhance collagen synthesis; keep moist wound environment; enhance epithelialization; skin nourishing/conditioning effects

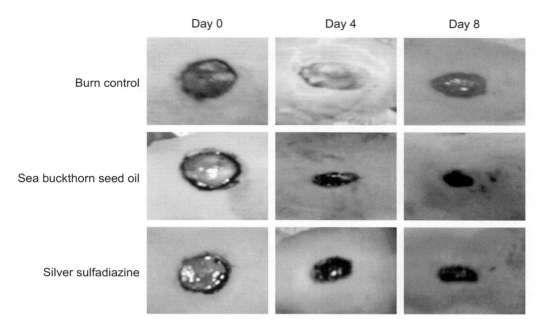

Day 0   Day 4   Day 8

Burn control

Sea buckthorn seed oil

Silver sulfadiazine

FIGURE 120.2
Photomicrograph showing faster wound contraction after sea buckthorn seed oil application on burn wounds in rats.
Sea buckthorn seed oil application resulted in faster wound closure in comparision to control and silver-sulfadiazine
treated burn wounds. *Reprinted from Upadhyay* et al. *(2009), Food Chem. Toxicol., 47, 1146–1153, with permission.*

production in burn wounds. The possible mechanism of sea buckthorn seed oil in promoting
the wound-healing process has been depicted in Figure 120.3.

In a clinical study, Zhao (1994) gave topical treatment of sea buckthorn seed oil to 32
patients with burned or scalded skin (12 cases of first-degree burns, 18 cases of superficial
second-degree burns, and 2 cases of deep second-degree burns). All the patients were cured
after 7 days of treatment, with no visible signs of scarring. It was suggested that sea buckthorn
seed oil has functions that improve immunity, remove blood stasis, and promote blood
circulation.

Gastric ulceration protection by sea buckthorn seed oil

The preventive and curative effects of sea buckthorn seed oil have been reported in
different types of gastric ulcers. Xing *et al* (2002) reported that oral administration of sea
buckthorn CO_2-SFE seed oil (7.0 ml/kg per day) significantly reduced ulcer formation in
water-immersion and reserpine-induced models in rats. In addition, seed oil administration

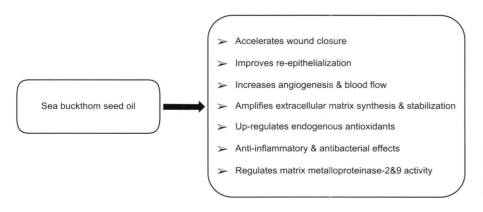

Sea buckthorn seed oil →

➤ Accelerates wound closure

➤ Improves re-epithelialization

➤ Increases angiogenesis & blood flow

➤ Amplifies extracellular matrix synthesis & stabilization

➤ Up-regulates endogenous antioxidants

➤ Anti-inflammatory & antibacterial effects

➤ Regulates matrix metalloproteinase-2&9 activity

FIGURE 120.3
Schematic diagram showing the possible effect
of sea buckthorn seed oil in promoting the
wound-healing process.

TABLE 120.2 Efficacy of Supercritical CO_2-extracted Sea Buckthorn Seed Oil on Different Gastric Ulcer-induced Models in Rats

Group (Dose)	Percentage Inhibition of Gastric Ulcer			
	Water-Immersion	Reserpine	Pylorus Ligation	Acetic Acid
Sea buckthorn seed oil (3.5 ml/kg per day)	21.8	11.5	22.2	44.0
Sea buckthorn seed oil (7.0 ml/kg per day)	38.9	69.5	20.3	45.1
Cimetidine (80 mg/kg per day)	40.9	44.6	30.5	51.8

Source: Xing *et al.* (2002), *Fitoterapia*, *73*, 644–650, with permission.

(3.5 ml/kg per day) reduced the index of pylorus-ligation induced gastric ulcer and, enhanced healing process of acetic acid-induced gastric ulcer (Table 120.2). The anti-ulcerative mechanism of sea buckthorn seed oil may be related to an increase in the surface hydrophobicity of the gastric mucosal membrane, inhibition of both gastric secretion and proteolytic activity of gastric juice, and promotion of the wound-regeneration process of mucosa. α-Tocopherol, β-sitosterol, β-sitosterol-β-D-glucoside, and PUFAs were suggested to be anti-ulcer components of sea buckthorn seed oil. Zhao *et al.* (1994) reported that sea buckthorn seed oil (200 mg/kg per day, intragastric injection) had anti-ulcer activity on pylorus-ligation, water-immersion, and acetic-acid induced ulcers.

Efficacy of sea buckthorn seed oil on mucosal injuries

Mucous membrane covers the digestive, respiratory, and urogenital tracts. Mucous membranes are constantly under challenge from bacterial invasion, external toxins and allergens, stress, aging, and side-effects of medical treatments. Sea buckthorn seed oil has been used clinically to treat chronic cervicitis. Wang (1995) treated 30 patients suffering from partial erosion of the cervix with topically sprayed sea buckthorn seed oil. All the patients were cured after 90 days of treatment. Sea buckthorn seed oil was used topically (three to four times a day) to treat 60 children (aged 4 months–12 years) with ulcerative stomatitis; 55 cases were cured after 3–5 days of treatment, and two severe cases were cured after 8 days of treatment (Wang, 1992). The positive effect of sea buckthorn seed oil on mucosal injuries may be related to its high content of natural carotenoids and tocopherols.

Effect of sea buckthorn seed oil on atopic dermatitis

Atopic dermatitis is an inflammatory skin disease characterized by dry, itchy, and lichenous skin. Treatment with sea buckthorn seed oil led to increased levels of ALA, LA, and eicosapentaenoic acid in plasma phospholipids, resulting in improved atopic dermatitis. Seed oil treatment slightly increased the level of docosapentaenoic acid (22:5, n-3) and decreased the level of palmitic acid (16:0) in skin glycerophospholipids. These results indicate a higher efficiency of incorporation and metabolism of ALA than LA in the form of sea buckthorn seed oil, and a relatively stable fatty acid composition of skin glycerophospholipids of patients with atopic dermatitis (Yang *et al.*, 2000).

ADVERSE EFFECTS AND REACTIONS (ALLERGIES AND TOXICITY)

No adverse effects are associated with topical or oral supplementation of sea buckthorn seed oil in experimental animals. There are no known drug interactions, contradictions, common allergic reactions, or toxicity to sea buckthorn seed oil. In an acute toxicity study, a single oral dose of sea buckthorn CO_2-SFE seed oil up to 10.0 ml/kg body weight did not show any adverse effects within 24 hours of treatment, and even after 14 days of treatment. In repeated-dose oral toxicity studies, there was no significant change in the organ/body weight ratio, clinical biochemistry, and hematological parameters of seed-oil treated rats, in comparison to controls. Furthermore, kidney, liver, spleen, adrenal, testis, heart, and lung of seed-oil treated

TABLE 120.3 Toxicity Studies of Supercritical CO_2-extracted Sea Buckthorn Seed Oil in Rats

Toxicity Test	Route of Administration	Animal	Dose Exposure	Mortality/Adverse Reactions
Acute	Oral	Rats	2.5, 5.0, 7.5, 10 ml/kg body weight (single administration)	None/none
Sub-acute	Oral	Rats	2.5 ml/kg, 5.0 ml/kg body weight, once a day for 14 days 2.5 ml/kg body weight, once a day for 28 days	None/none
Irritation	Dermal	Rabbits	0.5 ml for 4 hour (Single administration)	No erythema or edema observed

Source: Upadhyay *et al.* (2009), *Food Chem. Toxicol.*, 47, 1146–1153, with permission.

rats showed a normal histological appearance in both acute and repeated-dose oral toxicity studies (Upadhyay *et al.*, 2009) (Table 120.3).

SUMMARY POINTS

- Sea buckthorn seed oil has immense nutraceutical, pharmaceutical, and cosmaceutical uses.
- It is a rich source of bioactive molecules such as vitamins, carotenoids, phytosterols, and PUFAs.
- It enhances microcirculation and epidermal regeneration, and acts as a potent antioxidant.
- It has therapeutic potential for treating burns of different etiology, and inflammatory skin disorders.
- It has antibacterial and cholesterol lowering effects.
- It helps in the recuperation of mucous membranes of the stomach and other organs.
- It has light absorption, UV skin protection, and emollient properties.
- No toxicity or adverse reactions have been reported.

ACKNOWLEDGEMENTS

The authors are grateful to the Director, DIPAS, for providing useful suggestions; NKU is thankful to CSIR, India, for the Senior Research Fellowship.

1017

References

Bartish, J. V., Jeppson, N., Nybom, H., & Swenson, U. (2002). Phylogeny of *Hippophae* (Elaeagnaceae) inferred from parsimony analysis of chloroplast DNA and morphology. *System Botany*, 27, 41–54.

Basu, M., Prasad, R., Jayamurthy, P., Pal, K., Arumughan, C., & Sawhney, R. C. (2007). Anti-atherogenic effects of sea buckthorn (*Hippophaea rhamnoides*) seed oil. *Phytomedicine*, 14, 770–777.

Chauhan, A. S., Negi, P. S., & Ramteke, R. S. (2007). Antioxidant and antibacterial activities of aqueous extract of sea buckthorn (*Hippophae rhamnoides*) seeds. *Fitoterapia*, 78, 590–592.

Church, D., Elsayed, S., Reid, O., Winston, B., & Lindsay, R. (2006). Burn wound infections. *Clinical Microbiology Reviews*, 19, 403–434.

Li, T. S. C., & Schroeder, W. R. (1996). Sea buckthorn (*Hippophae rhamnoides* L.): a multipurpose plant. *Horticulture Technologies*, 6, 370–380.

Purushothaman, J., Suryakumar, G., Shukla, D., Malhotra, A. S., Kasiganesan, H., Kumar, R., Chand, S. R., et al. (2008). Modulatory effects of sea buckthorn (*Hippophae rhamnoides* L.) in hypobaric hypoxia induced cerebral vascular injury. *Brain Research Bulletin*, 77, 246–252.

Rana, R. K. (2006). Studies on the therapeutic potential of sea buckthorn (*Hippophae sp.*) seed and pulp oil in the healing of aseptic excisional cutaneous wounds in calves. *Indian Journal of Veterinary Surgery*, 27, 1–7.

Singer, A. J., & Clark, R. A. F. (1999). Cutaneous wound healing. *New England Journal of Medicine*, 341, 738–746.

Upadhyay, N. K., Kumar, R., Mandotra, S. K., Meena, R. N., Siddiqui, M. S., Sawhney, R. C., et al. (2009). Safety and wound healing efficacy of sea buckthorn (*Hippophae rhamnoides* L.) seed oil in experimental rats. *Food and Chemical Toxicology, 47,* 1146—1153.

Wang, L. J. (1992). Sea buckthorn oil and chymotrypsin are effective in treating ulcerative stomatitis of children. *Hippophae, 5,* 32—34.

Wang, J. (1995). A preliminary report on the clinical effects of sea buckthorn seed oil on partial erosion of cervix. *Hippophae, 8,* 37—38.

Xing, J., Yang, B., Dong, Y., Wang, B., Wang, J., & Kallio, H. P. (2002). Effects of sea buckthorn (*Hippophae rhamnoides* L.) seed and pulp oils on experimental models of gastric ulcer in rats. *Fitoterapia, 73,* 644—650.

Yang, B., Kalimo, K. O., Tahvonen, R. L., Mattila, L. M., Katajisto, J. K., & Kallio, H. P. (2000). Effect of dietary supplementation with sea buckthorn (*Hippophae rhamnoides*) seed and pulp oils on the fatty acid composition of skin glycerophospholipids of patients with atopic dermatitis. *Journal of Nutritional Biochemistry, 11,* 338—340.

Zeb, A. (2004). Important therapeutic uses of sea buckthorn (*Hippophae*): a review. *Journal of Biological Sciences, 4,* 687—693.

Zhang, W. L., Zhang, Z. F., Fan, J. J., Yang, S. Y., Li, Z. M., Deng, Z. C., et al. (1988). Experimental observation and clinical investigation effect of sea buckthorn oil on acute radiodermatitis. *Hippophae, 1,* 27—30.

Zhao, Y. (1994). Clinical effects of *Hippophae* seed oil in the treatment of 32 burn cases. *Hippophae, 7,* 36—37.

Zhao, Y., Jiang, J., Song, Y., & Sun, S. (1994). Research on the antigastric ulcer effect of sea buckthorn seed oil. *Hippophae, 7,* 33—36.

CHAPTER 121

Extracts of Sesame (*Sesamum indicum* L.) Seeds and Gastric Mucosal Cytoprotection

Dur-Zong Hsu, Pei-Yi Chu, Ming-Yie Liu
Department of Environmental and Occupational Health, National Cheng Kung University Medical College, Tainan, Taiwan

CHAPTER OUTLINE

LIST OF ABBREVIATIONS

COX, cyclooxygenase
GSH, glutathione
IL, interleukin
LPO, lipid peroxidation
NO, nitric oxide
NSAID, non-steroidal anti-inflammatory drug
PGE2, prostaglandin E2
ROS, reactive oxygen species
TNF, tumor necrosis factor

Nuts & Seeds in Health and Disease Prevention. DOI: 10.1016/B978-0-12-375688-6.10121-5

INTRODUCTION

Sesame (*Sesamum indicum* L.) is one of the oldest cultivated plants in the world. Sesame seed oil, made from the seeds of sesame, and its lignan, sesamol, have potent antioxidative and anti-inflammatory effects. Gastric mucosal damage is commonly caused by long-term ingestion of alcohol and non-steroidal anti-inflammatory drugs (NSAIDs) such as aspirin and diclofenac. Anti-acid agents (for neutralizing gastric acidity), histamine H_2 blockers (for decreasing acid secretion), and antibiotics (for killing *Helicobacter pylori* in gastric mucosa) used to be the only choices for alleviating gastric ulceration. Recently, however, sesame seed oil and sesamol have been shown to be mucosa-protective agents.

BOTANICAL DESCRIPTION

Sesame, an annual herb in the *Pedaliaceae* family (Sugano & Akimoto, 1993) (Figure 121.1), prefers a hot climate. It is a drought-resistant plant that gives its maximum yield when grown at 25−27°C. It takes about 125−135 days from planting to maturity. The large, bell-shaped flowers of the sesame are yellow, although they vary in color: some are blue and some are purple. Pollen grains of this family are 5-13-colpate and oblate-subprolate (van Haaster, 1990). Sesame can grow to over 2 m tall, although most varieties are 60−90 cm, depending upon growing conditions. Seed capsules are 2.5−3.5 cm long, with 8 rows of seeds in each capsule. Each kilogram of sesame contains approximately 33,000 seeds. Depending on the variety, sesame seeds vary in color (Figure 121.2). Sesame seeds contain a high percentage of oil — around 55% of their total mass.

HISTORICAL CULTIVATION AND USAGE

Sesame was the earliest plant to be used as a source of edible oil, and was a highly prized crop in Babylon and Assyria. The origin of the sesame plant is controversial, but some historians believe that it originated on the Indian subcontinent. The whole sesame seed is used in the cuisines of the Middle East and Asia. In India, sesame seed oil has long been used to combat skin fungi. In Europe and North America, the seeds are used to flavor and garnish various foods, particularly breads and other baked goods.

FIGURE 121.1

Sesame plants, flowers, and seed capsules. Sesame (*Sesamum indicum* L.) is an annual herb that belongs to the *Pedaliaceae* family. The large, bell-shaped flowers of the sesame vary in color. There are 8 rows of seeds in each capsule.

FIGURE 121.2
Sesame seeds vary in color, depending on the variety, and yield a high level of oil — around 55% of its total mass.

PRESENT-DAY CULTIVATION AND USAGE

China and India are currently the world's largest producers of sesame, followed by Burma, Sudan, Mexico, Nigeria, Venezuela, Turkey, Uganda, and Ethiopia. The total global production of sesame seeds is around 3 million tonnes annually, although the export market is dominated by China and India. Sesame seeds are used in baking, candy-making, and other food industries. Sesame seed oil is used for a number of purposes, including cooking and salad oils, and margarine, and for making soaps, pharmaceuticals, paints, exotic perfumes, cosmetics, and lubricants. It is also used as a massage oil in India's traditional Ayurveda herbal medicine system.

APPLICATIONS TO HEALTH PROMOTION AND DISEASE PREVENTION

Sesame seed extracts

The main constituents of sesame seed oil (Figure 121.3) include fatty acids, lignans, and antioxidants, such as γ-tocopherol (Simon *et al.*, 1984). The fatty acids in sesame seed oil include palmitic (16:0; 7.0–12%), palmitoleic (16:1; < 0.5%), stearic (18:0; 3.5–6.0%), oleic (18:1; 35.0–50.0%), linoleic (18:2; 35–50%), linolenic (18:3; < 1.0%), and eicosanoic (20:1; < 1.0%) acids. Sesame seed oil has been regarded as a daily nutritional supplement to increase cell resistance to lipid peroxidation (LPO) (Kaur & Saini, 2000). The antioxidants in sesame seed oil include sesamin, sesamolin, sesamol, and tocopherol. Sesamol (3,4-methylenedioxyphenol), a lignan in sesame seed oil, is formed from sesamin and sesamolin via biotransformation, and is thought to be sesame seed oil's most potent antioxidant (Figure 121.4).

Oxidative stress in gastric mucosal damage

Oxidative stress is imposed on cells as a result of one of two factors: (1) an increase in oxidant generation, or (2) a decrease in antioxidant protection. Under normal physiological conditions, a homeostatic balance exists between the formation of reactive oxygen species (ROS), such as hydroxyl radicals and superoxide anions, and their removal by endogenous antioxidant scavenging compounds. Oxidative stress occurs when this balance is disrupted by excessive ROS production, by inadequate antioxidative defense (Gutteridge & Mitchell, 1999) — for example, glutathione (GSH) depletion — or by both. ROS changes the redox balance through the oxidation of metabolically active compounds, which leads to LPO and degradation. However, ROS-initiated oxidative stress and LPO can be regulated by cell defense

FIGURE 121.3
Sesame seeds are used in a variety of food products: sesame seed oil, sesame cookies, sesame cake, sesame paste, etc.

mechanisms. GSH and nitric oxide (NO) are the main substances that protect against gastric mucosal oxidative stress and damage. GSH, a circulating antioxidant, mediates gastric cyto-protection by scavenging free radicals during gastric oxidative stress (Szabo *et al.*, 1981). GSH prevents interactions of reactive intermediates with critical cellular constituents, such as the phospholipids of biomembranes, nucleic acids, and proteins. Studies have indicated that GSH protects gastric mucosa from ethanol- or aspirin-induced damage (Shaw *et al.*, 1990; van Lieshout *et al.*, 1997). On the other hand, mucosal NO is cytoprotective and maintains gastric mucosal integrity. It also reduces the generation of oxyradicals, which potently attack gastric mucosa (Kanner *et al.*, 1991).

Protective effect of sesame seed oil on alcohol-associated gastric ulceration

It has been known since the 19th century that exposing gastric mucosa to concentrated ethanol causes acute gastritis (Beaumont, 1833). Increased gastric mucosal LPO is responsible for ethanol-induced gastric mucosal damage. Sesame seed oil significantly decreases gastric mucosal LPO and ethanol-induced gastric mucosal injury. Sesame seed oil also increases GSH and NO levels in gastric mucosa. We hypothesize that reducing gastric mucosal oxidative stress is the mechanism by which sesame seed oil protects gastric mucosa in ethanol-challenged stomachs.

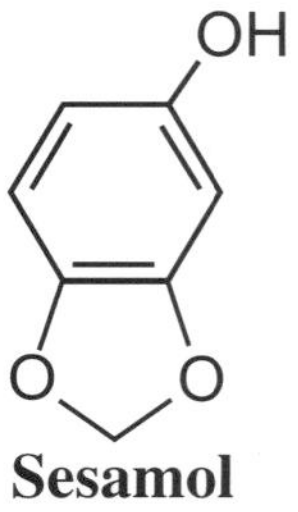

Sesamol

FIGURE 121.4
The antioxidative component sesamol in sesame seed oil. The main constituents of sesame seed oil include fatty acids and lignans. Antioxidants in sesame seed oil include sesamol, sesamin, sesamolin, and γ-tocopherol. Sesamol, a lignan in sesame seed oil, is formed from sesamin and sesamolin via biotransformation, and is suggested to be the most potent antioxidant in sesame seed oil.

Maintaining mucosal GSH and NO levels may be involved in the sesame seed oil-associated protection against acidified ethanol-challenged rats. NO modulation of GSH homeostasis may increase GSH stores in gastric mucosa. Although its mechanism for scavenging radicals has not been clearly delineated, the ability of NO to scavenge superoxide anions has been documented both *in vitro* (Rubanyi *et al.*, 1991) and *in vivo* (Gaboury *et al.*, 1993). NO also protects against H_2O_2-induced gastric damage by chelating stored Fe^{3+} and reduced Fe^{2+} to form iron-nitrosyl compounds. Additionally, NO is the termination of radical chain propagation by direct reaction with lipid radicals. Therefore, the NO pathway is involved in sesame seed oil's attenuation of acidified ethanol-induced gastric mucosal damage in rats (Hsu *et al.*, 2009a).

Sesame seed oil's gastric protection is not a result of its interfering with ethanol absorption in the stomach. Although it is possible that oils form a physical barrier between gastric acid and mucosa, thereby reducing gastric injury, sesame seed oil, but not mineral oil, is potently protective in the ethanol-treated stomach. In addition, sesame seed oil does not affect serum ethanol levels, which indicates that sesame seed oil does not interfere with the gastric absorption of ethanol. Therefore, sesame seed oil may attenuate ethanol-induced gastric mucosal damage by reducing mucosal oxidative stress, but not exclusively by forming a physical barrier.

NSAID-associated gastric mucosal injury

NSAIDs are commonly used to manage pain, fever, and various inflammatory diseases in clinical patients. The world market for NSAIDs exceeds US$7.5 billion per year (Wallace, 2007). However, aspirin and other NSAIDs almost invariably cause acute gastroduodenal injury, and probably account for approximately 12,000 episodes of bleeding ulcers and 1200 deaths per year in the United Kingdom. Gastric ulceration and bleeding occur not only in patients with long-term NSAID treatment, but also in patients who intentionally or accidently overdose on NSAIDs. Acute gastrointestinal symptoms can even be found within 2 hours after an NSAID overdose (MacDougall *et al.*, 1984).

PATHOGENESIS OF NSAID-INDUCED GASTRIC MUCOSAL INJURY

It is widely accepted that both the beneficial and detrimental effects of NSAIDs are attributable to their ability to inhibit prostaglandin synthesis by directly blockading cyclooxygenase (COX). Prostaglandin E2 (PGE2) is a gastroprotective substance in the gastric mucosa. Suppressing PGE2 leads to a decrease in mucus synthesis and secretion, both of which contribute to the pathogenesis of NSAID-induced gastric ulceration. Recently, however, PG-independent mechanisms have been associated with the pathogenesis of NSAID-induced gastrointestinal damage. Gastric mucosal LPO is involved in the development of NSAID-associated gastric mucosal injury.

In addition, inflammation is associated with the pathogenesis of NSAID-associated gastric mucosal damage (Jainu & Shyamala Devi, 2005). Even though NO protects gastric mucosal integrity (Wallace & Miller, 2000), it is also a pro-inflammatory mediator and the source of oxygen free radicals, such as peroxynitrite and hydroxyl radical, both of which are regulated by cytokines such as tumor necrosis factor (TNF)-α and interleukin (IL)-1β. On the other hand, the activation of neutrophils and their infiltration into the gastric mucosa is crucial for initiating aspirin-induced mucosal oxidative stress and inflammation, and contributes to gastric mucosal lesions induced by NSAIDs.

PROTECTIVE EFFECTS OF SESAMOL ON NSAID-ASSOCIATED ACUTE MUCOSAL INJURY

Antioxidants such as tocopherol and ascorbic acid reduce NSAID-induced gastric ulceration, which indicates that mucosal oxidative stress is crucial for the development of NSAID-induced gastric mucosal injury. Sesamol decreases LPO, which is consistent with two- to three-fold increases of LPO reported in various models of NSAID-induced gastric mucosal oxidative stress. In addition, sesamol decreases hydroxyl radical, but not superoxide anion, generation. Therefore, sesamol attenuates diclofenac-induced gastric mucosal damage by reducing hydroxyl radical-initiated mucosal LPO.

On the other hand, the COX pathway is not involved in sesamol's protection against diclofenac-induced gastric injury. It is well known that diclofenac interrupts the COX pathway and therefore inhibits the generation of PGE2, which is important for the production and release of gastric mucus, a mucosal barrier that protects against offensive factors such as gastric acid and ulcerogens (Johansson & Bergstrom, 1982). However, sesamol potently reduces diclofenac-induced mucosal damage, but neither increases nor maintains diclofenac-induced decreases of PGE2, mucus, or COX activity. Moreover, ulceration does not develop spontaneously in mice with a disrupted COX-1 gene. A mechanism other than the COX pathway may be important in NSAID-induced gastric mucosal injury (Figure 121.5). Therefore, sesamol may protect against diclofenac-induced gastric mucosal damage using a mechanism independent of the COX pathway (Hsu et al., 2008a).

PROTECTIVE EFFECTS OF SESAMOL ON LONG-TERM NSAID-INDUCED GASTRIC MUCOSAL INJURY

Inhibition of gastric mucosal oxidative stress and inflammation may be involved in sesamol's protection against long-term aspirin-induced gastric mucosal injury (Figure 121.6). Sesamol significantly inhibits aspirin-induced gastric hemorrhage and ulceration, as well as gastric mucosal LPO and pro-inflammatory cytokine levels. Therefore, the anti-inflammatory and antioxidative properties of sesamol are crucial in gastric protection against long-term aspirin-induced gastrotoxicity. Inhibiting the activation of neutrophils and their infiltration into gastric mucosa may be sesamol's anti-inflammatory and antioxidative mechanism. Neutrophil activation and infiltration are crucial in the pathogenesis of NSAID-induced gastric inflammation and oxidative stress (Souza et al., 2008). Activating neutrophils causes the expression of pro-inflammatory genes and the overproduction of pro-inflammatory mediators, including TNF-α and IL-1β, and initiates an inflammatory response. Sesamol inhibits neutrophil infiltration in the gastric mucosa of long-term aspirin-treated rats. Therefore, sesamol may decrease the risk of long-term aspirin-induced gastric injury by inhibiting neutrophil infiltration in gastric mucosa.

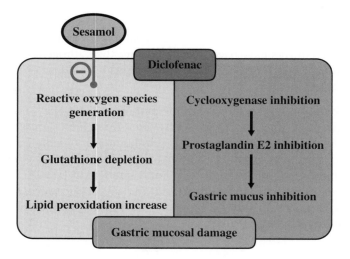

FIGURE 121.5

The mechanism underlying sesamol's gastroprotective activity. In acute diclofenac-induced gastric mucosal damage, both oxidative stress and cyclooxygenase pathways are involved. Sesamol decreases gastric mucosal damage by inhibiting the generation of reactive oxygen species; however, it does not affect the cyclooxygenase pathway.

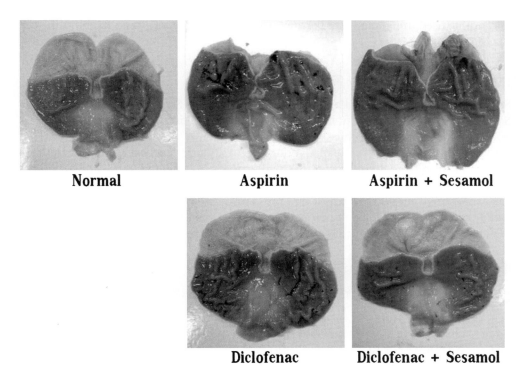

FIGURE 121.6
Morphological changes in gastric mucosa. Aspirin and diclofenac cause gastric mucosal ulceration. Sesamol is significantly gastroprotective against aspirin- and diclofenac-induced gastric mucosal damage.

The combination of aspirin and sesamol may be a feasible approach for decreasing the risk of gastric mucosal damage in patients who receive long-term aspirin therapy. Current strategies for decreasing the aspirin-associated risk of gastric damage are focused on COX and nitric oxide. NSAIDs non-selectively inhibit both COX-1 and COX-2. Inhibiting COX-1 decreases prostaglandin E2, a gastric mucosal protective substance, and is one of the major causes of NSAID-induced gastric mucosal damage. Recently, selectively inhibiting COX-2 decreased the risk of NSAID-induced gastric injury; however, the adverse effects of NSAIDs on the cardiovascular and renal systems limit their clinical use. On the other hand, although nitric-oxide-releasing aspirin has been suggested as a solution for decreasing the adverse effect of aspirin, its effect on NSAID-induced gastrotoxicity is paradoxical. In contrast, sesamol potently decreased aspirin-induced gastric mucosal injury without affecting normal physiology, for example, body weight (Hsu *et al.*, 2009b). In summary, sesame seed oil and sesamol protect against alcohol- and NSAID-associated acute and chronic gastric mucosal injury in animal studies. Furthermore, daily consumption of a recommended dose of sesame seed oil or sesamol may be beneficial in protecting against gastric mucosal damage induced by alcohol or NSAIDs.

ADVERSE EFFECTS AND REACTIONS (ALLERGIES AND TOXICITY)

Sesame seed oil is not known to be harmful when taken in recommended dosages; however, long-term ingestion of sesame seed oil has caused a lower body weight increase compared to normal in animal studies (Hsu *et al.*, 2008b). Moreover, long-term ingestion of a large dose of sesame seed oil not only fails to produce a cumulative antioxidative effect, but also significantly decreases the effect (Hsu *et al.*, 2008b). Therefore, we suggest that a small amount of sesame seed oil is beneficial for long-term use. Because the long-term effects of sesame seed oil have not been investigated in humans, the oil should be used with caution in children, and pregnant or breast-feeding women. In addition, because of its laxative effects, sesame seed oil should not be used in people with diarrhea. There is very little information about the adverse effects of sesamol.

SUMMARY POINTS

- Sesame seed oil and sesamol are extracted from sesame (*Sesamum indicum* L.).
- Sesame seed oil reduces alcohol-induced gastric mucosal injury by increasing mucosal GSH and NO, and therefore inhibits mucosal LPO.
- Sesamol protects against diclofenac-induced acute gastric mucosal damage using a mechanism independent of the COX pathway.
- Combining sesamol with aspirin is potentially a new approach for decreasing gastrointestinal injury caused by long-term aspirin use.
- Daily consumption of a recommended dose of sesame seed oil or sesamol may be beneficial for protecting against gastric mucosal damage induced by alcohol or NSAIDs.

References

Beaumont, W. (1833). *Experiment and observation on the gastric juice and the physiology of digestion*. New York, NY: Plattsburgh.

Gaboury, J., Woodman, R., Granger, D. N., Reinhardt, P., & Kubes, P. (1933). Nitric oxide prevents leukocyte adherence: role of superoxide. *American Journal of Physiology, 265*, H862—H867.

Gutteridge, J. M. C., & Mitchell, J. (1999). Redox imbalance in the critically ill. *British Medical Bulletin, 55*, 49—75.

Hsu, D. Z., Chu, P. Y., Li, Y. H., & Liu, M. Y. (2008a). Sesamol attenuates diclofenac-induced acute gastric mucosal injury via its cyclooxygenase-independent anti-oxidative effect in rats. *Shock, 30*, 456—462.

Hsu, D. Z., Chien, S. P., Li, Y. H., & Liu, M. Y. (2008b). Sesame oil does not show accumulatively enhanced protection against oxidative-stress-associated hepatic injury in septic rats. *Journal of Parenteral and Enteral Nutrition, 32*, 276—280.

Hsu, D. Z., Chu, P. Y., & Liu, M. Y. (2009a). Effect of sesame oil on acidified ethanol-induced gastric mucosal injury in rats. *Journal of Parenteral and Enteral Nutrition, 33*, 423—427.

Hsu, D. Z., Chu, P. Y., Chandrasekaran, V. R., & Liu, M. Y. (2009b). Sesamol protects against aspirin-induced gastric mucosal damage in rats. *Journal of Functional Foods, 1*, 349—355.

Jainu, M., & Shyamala Devi, C. S. (2005). Attenuation of neutrophil infiltration and pro-inflammatory cytokines by *Cissus quadrangularis*: a possible prevention against gastric ulcerogenesis. *Journal of Herbal Pharmacotherapy, 5*, 33—42.

Johansson, C., & Bergstrom, S. (1982). Prostaglandin and protection of the gastroduodenal mucosa. *Scandinavian Journal of Gastroenterology, 77*(Suppl), 21—46.

Kanner, J., Harel, S., & Granit, R. (1991). Nitric oxide as an antioxidant. *Archives of Biochemistry and Biophysics, 289*, 130—136.

Kaur, I. P., & Saini, A. (2000). Sesamol exhibits antimutagenic activity against oxygen species mediated mutagenicity. *Mutation Research, 470*, 71—76.

MacDougall, L. G., Taylor-Smith, A., Rothberg, A. D., & Thomson, P. D. (1984). Piroxicam poisoning in a 2-year-old child. A case report. *South African Medical Journal, 66*, 31—33.

Rubanyi, G. M., Ho, E. H., Cantor, E. H., Lumma, W. C., & Botelho, L. H. (1991). Cytoprotective function of nitric oxide: inactivation of superoxide radicals produced by human leukocytes. *Biochemical and Biophysical Research Communication, 181*, 1392—1397.

Shaw, S., Herbert, V., Colman, N., & Jayatilleke, E. (1990). Effect of ethanol-generated free radicals on gastric intrinsic factor and glutathione. *Alcohol, 7*, 153—157.

Simon, J. E., Chadwick, A. F., & Craker, L. E. (1984). Herbs: An indexed bibliography. 1971—1980. *The Scientific Literature on Selected Herbs, and Aromatic and Medicinal Plants of the Temperate Zone*. Hamden, CT: Archon Books.

Souza, M. H., Mota, J. M., Oliveira, R. B., & Cunha, F. Q. (2008). Gastric damage induced by different doses of indomethacin in rats is variably affected by inhibiting iNOS or leukocyte infiltration. *Inflammation Research, 57*, 28—33.

Sugano, M., & Akimoto, K. A. (1993). Multifunctional gift from nature. *Journal of Chinese Nutrition Society, 18*, 1—11.

Szabo, S., Trier, J. S., & Frankel, P. W. (1981). Sulfhydryl compounds may mediate gastric cytoprotection. *Science, 214*, 200—202.

van Haaster, H. (1990). Sesame (*Sesamum indicum* L.) pollen in 14th century cesspits. *Hertogenbosch. Circaea, 6*, 105—106.

van Lieshout, E. M., Tiemessen, D. M., Peters, W. H., & Jansen, J. B. (1997). Effects of nonsteroidal anti-inflammatory drugs on glutathione S-transferases of rat digestive tract. *Carcinogenesis, 18,* 485−490.

Wallace, J. L. (2007). Building a better aspirin: gaseous solutions to a century-old problem. *British Journal of Pharmacology, 152,* 421−428.

Wallace, J. L., & Miller, M. J. (2000). Nitric oxide in mucosal defense: a little goes a long way. *Gastroenterology, 119,* 512−520.

CHAPTER 122

Sesame (*Sesamum indicum* L.) Seeds in Food, Nutrition, and Health

Mohamed Elleuch[1], Dorothea Bedigian[2], Adel Zitoun[1]
[1] Unité de Valorisation des Résultats Scientifiques, Centre de Biotechnologie de Sfax, Sfax, Tunisia
[2] Missouri Botanical Garden, St Louis, Missouri, USA

LIST OF ABBREVIATIONS

MUFA, monounsaturated fatty acid
PUFA, polyunsaturated fatty acid
UFA, unsaturated fatty acid
UV, ultra-violet

INTRODUCTION

Sesame (*Sesamum indicum* L.) is a herbaceous annual plant belonging to the Pedaliaceae family. It is cultivated for its protein-rich seed and its edible oil, which is a rich source of UFAs (Elleuch *et al.*, 2007). A substantial amount of the world's sesame production is consumed as oil. Sesame is a major food source in many parts of the world. It has improved the nutritional status and prevented various diseases in China and Japan for several thousand years (Chen *et al.*, 2005). Sesame seeds are also an important source of dietary fiber and micronutrients such as minerals, lignans, tocopherols, and phytosterols.

Nuts & Seeds in Health and Disease Prevention. DOI: 10.1016/B978-0-12-375688-6.10122-7

BOTANICAL DESCRIPTION

Sesame belongs to the genus *Sesamum*, which comprises about 20 species (Bedigian, 2010); a few other species in the genus are occasionally cultivated for their edible seeds and leaves. Sesame grows primarily in tropical and subtropical areas of the world, and is well adapted to semi-arid regions. It is an annual self-pollinated plant with a strong woody, hairy and often many-branching stem that reaches 0.5–2 m in height. The flowers are tubular in shape, 2–2.5 cm long, and white, some with pink or purple markings. The fruit are capsules, which dehisce with a pop when the seeds mature. The seeds come in a variety of colors, including shades of brown, brick red, black, yellow, beige, gray, and white, depending on the cultivar.

HISTORICAL CULTIVATION AND USAGE

Sesame, domesticated on the Indian subcontinent during the Harappan era (Bedigian, 2000), was taken to Mesopotamia by the Early Bronze Age, emerged in the Armenian Highlands by the early Iron Age, where its cultivation became widespread among the Urartu in the Kingdom of Ararat, and expanded from there throughout the Greco-Roman world. Used for both its edible seed and its oil, Greek sources describe sesame cakes at wedding feasts. Sesame oil served as a common lamp oil in earlier days. The Chinese invented lamp-black by collecting its soot, and it offers the finest texture black pigment for ink blocks.

PRESENT-DAY CULTIVATION AND USAGE

During the past five decades, world production of sesame increased steadily, accompanied by a correspondingly increasing world population and worldwide demand for sesame oil. According to the latest available FAO statistics, the world's sesame production is 3,387,598 tonnes; overall, 55% is crushed for oil. Sesame is primarily cultivated in Africa and Asia, although interest is rising in Latin America. China, Ethiopia, India, Myanmar, Nigeria, Sudan, and Tanzania are currently the foremost producers.

Sesame has numerous domestic and industrial applications. Sesame seeds serve to decorate bread, biscuits, and hamburger buns; sesame paste flavors salads and sauces, and constitutes the basis of halva (a dense block of sweetened sesame paste). The oil extracted from sesame seeds is used as a salad and cooking oil, and in the production of margarine; in formulations of body massage creams, soap, and lubricants; as a solvent for intramuscular injections; and as an insecticide. Sesame oil meal, obtained after oil extraction, is a nutritious livestock feed, and may be composted.

Sesame seeds are usually dehulled and roasted before use; this improves their functional properties, releases more flavor and color, and improves their sensorial quality. Dehulling removes relatively high amounts of antinutritional oxalic acid and fiber contained in the testa (seed coat), resulting in lighter-colored, less bitter-tasting seeds. Roasted sesame seeds are used to garnish bread loaves, breadsticks, bagels, cookies, stuffed date rolls, and chocolate. It is also used in the preparation of condiments such as zahtar or dukka (roasted sesame seeds, sumac gunpowder, and milled thyme), gomasio (sesame salt), and shichimi togarashi.

Sesame seed butter or paste, known as tahin (tahina, tehineh, tahini), is a traditional food in the Middle East produced by grinding hulled or unhulled roasted sesame seeds. Tahin is widely used, incorporated into dishes such as fava beans, hummus bi tahin (with cooked chickpeas), baba gannouj (with mashed eggplant), salads, and desserts.

A dense, block-shaped confection made from sweetened tahin, known as halva (halaweh, halvah), is a common food in Greece and across South-west Asia. Halva is prepared from dehulled, roasted, and milled white sesame seeds (tahin), sugars (sucrose and glucose syrup), citric acid, and *Saponaria officinalis* L. root (Figure 122.1). Some manufacturers add natural

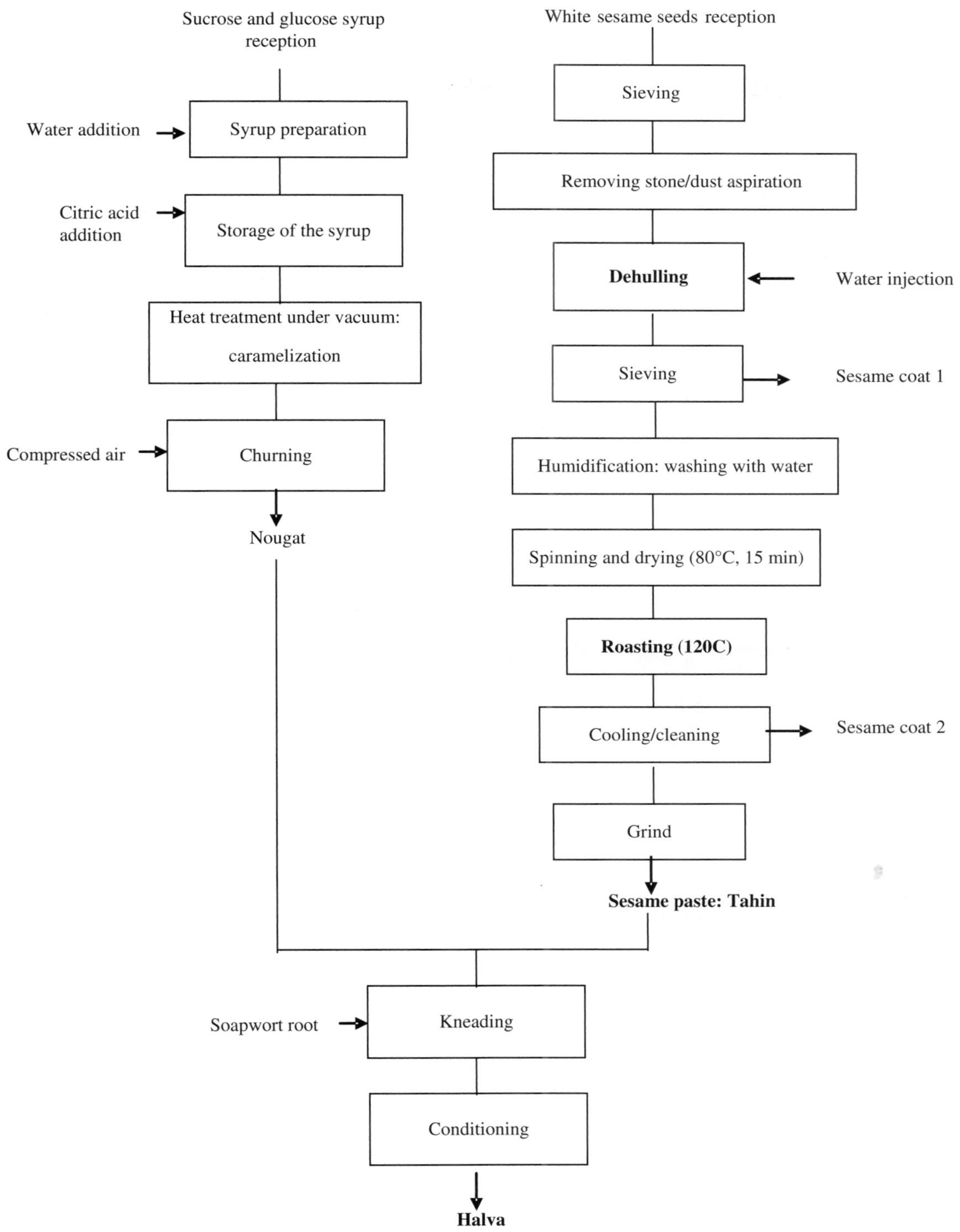

FIGURE 122.1

Flow diagram showing the different stages in the industrial production of sesame paste (tahin) and sweetened sesame paste (halva).

ingredients to boost flavor (e.g., vanilla, chocolate, almonds, pistachio nuts, and, in Greece, palm oil at the kneading stage). Alternative sweeteners such as honey, dates, and molasses obtained from grape or mulberry pekmez may be substituted. In Turkey, tahin and pekmez are available for sale separately, and the proportion of tahin in blending the mixture is in accordance with consumer preference.

Sesame oil has been valued since ancient times for its nutritional quality, resistance to oxidative deterioration, and good flavor (Elleuch *et al.*, 2007). The process for sesame oil preparation includes cleaning, optional roasting and dehulling, grinding, and oil extraction by press or with an organic solvent. Roasting conditions (temperature and duration) have an effect upon the color, composition, and quality of the oil. Sesame oil is a pale yellow liquid, but when the seed is roasted the oil becomes darker in color and has a more pronounced flavor. The oil is added to salads, and used for cooking and to produce margarine.

Table 122.1 indicates that sesame seeds are a very good energy source, providing 2340 kJ/100 g, mainly because of their high oil content. This oil is rich in MUFAs and PUFAs, and contains a significant amount of linoleic acid (an essential fatty acid) (Table 122.2). From linoleic and linolenic acids, the body can make all the other fatty acids it needs. Sesame seeds are also a good source of protein (high content of methionine, tryptophan, and valine), dietary fiber, and micronutrients such as lignans, tocopherol, minerals (calcium, phosphorus, potassium, magnesium, and zinc), and phytosterol (Tables 122.1–122.3).

TABLE 122.1 Nutrient Content of Sesame Seeds (per 100 g)

Sesame Seed Oil	Content
Energy (kcal/kJ)	559/2340
Oil (g)	49.77
Protein (g)	24.55
Soluble simple sugars (g)	2.36
Starch (g)	0.84
Total fiber (g)	18.41
Insoluble fiber (g)	13.30
Soluble fiber (g)	5.11
Ash (g)	4.46
Calcium (mg)	980
Potassium (mg)	501.13
Magnesium (mg)	333.42
Phosphorus (mg)	491.69
Sodium (mg)	14.56
Iron (mg)	10.85
Copper (mg)	2.05
Zinc (mg)	8.45
Manganese (mg)	3.29
Polyphenols (mg)	83.63
Total lignan (mg)[*]	405–1178
Lignans[*]	
Sesamin (mg)	167–804
Sesamolin (mg)	48–279
Lignan glycosides[*]	
Sesaminol (mg)	32–298
Sesamolinol (mg)	tr-58

Sesame seeds are a very good energy source, due to their high oil and protein contents. Sesame seed is also rich in dietary fiber, and micronutrients such as minerals and lignans.
Sources: Elleuch *et al.* (2007), *Food Chem.*, *103*, 641–650,
[*]*Moazzami* et al. *(2007), Eur. J. Lipid Sci. Tech., 109, 1022 –1027 with permission.*

TABLE 122.2 Fatty Acid Composition of Sesame Seed Oil (g/100 g of Total Fatty Acids)	
Fatty Acid	**Content (%)**
Saturated	
Palmitic C16:0	11.18
Stearic C18:0	6.40
Arachidic C20:0	0.70
Lignoceric C24:0	0.20
Monounsaturated	
Palmitoleic C16:1	0.21
Oleic C18:1(n-9)	44.06
Cis-vaccenic C18:1 (n-11)	0.97
Eicosenoic C20:1	0.18
Polyunsaturated	
Linoleic C18:2	35.56
Linolenic C18:3	0.50

Sesame oil is rich in monounsaturated and polyunsaturated fatty acids. The most abundant fatty acids in sesame oil were oleic, linoleic, palmitic, and stearic acids, which together comprised about 96% of the total fatty acids.
Source: Elleuch *et al.* (2007), *Food Chem.*, *103*, 641—650, with permission.

TABLE 122.3 Minor Constituents of Sesame Oil		
Compound	**Content (mg/kg oil)**	**Reference(s)**
Lignans		
Sesamin	6490	Shahidi *et al.*, 1997
Sesamolin	1830	Shahidi *et al.*, 1997
Polyphenol	23.06	Elleuch *et al.*, 2007
Sesamol	8.11	Elleuch *et al.*, 2007
Tocopherol		
γ-Tocopherol	358.0—663.5	Kamel-Eldin & Appelqvist, 1994; Shahidi *et al.*, 1997
α-Tocopherol	3.10—6.86	Kamel-Eldin & Appelqvist, 1994
δ-Tocopherol	7.84—13.02	Kamel-Eldin & Appelqvist, 1994
Phytosterol	5100—7600	Kamel-Eldin & Appelqvist, 1994
Sitosterol	2687—4132	Kamel-Eldin & Appelqvist, 1994
Campesterol	706.8—1001.0	Kamel-Eldin & Appelqvist, 1994
Δ5-Avenasterol	351—777.8	Kamel-Eldin & Appelqvist, 1994
Stigmasterol	338.0—405.8	Kamel-Eldin & Appelqvist, 1994

Sesame oil is an important source of unsaponifiable matter such as lignan, tocopherol, and phytosterol. The major lignan, tocopherol, and phytosterol found in sesame oils were sesamin and sesamolin, γ-tocopherol, and sitosterol, respectively.
Sources: Kamel-Eldin & Appelqvist (1994), Shahidi *et al.* (1997), Elleuch *et al.* (2007).

APPLICATIONS TO HEALTH PROMOTION AND DISEASE PREVENTION

Sesame seed

Sesame seeds have been used as a health food for disease prevention in Asian countries for several thousand years. The literature reports indicate many health benefits associated with the consumption of sesame seeds; for example, they significantly increase plasma γ-tocopherol and enhance vitamin E activity, which are believed to prevent human aging-related diseases such as cancer and heart disease (Cooney *et al.* 2001). Studies have also shown that including sesame in the diet can exert hypocholesterolemic effects, and improve antioxidant capacity (Chen *et al.*, 2005).

Sesame seed oil

Studies have shown that sesame oil can inhibit human colon cancer growth *in vitro* (Salerno & Smith 1991), lower blood pressure, decrease lipid peroxidation, and increase antioxidant status in hypertensive patients (Sankar *et al.*, 2006). Animal studies demonstrate that the oral administration of sesame oil can inhibit pro-inflammatory cytokine and nitric oxide production; this inhibition might be involved in sesame oil-associated protection against lead-plus-lipopolysaccharide induced acute hepatic injury (Hsu *et al.*, 2007). Sesame oil has been found to ameliorate multiple organ failure, attenuate oxidative stress, increase survival rates, and relieve hepatic disorders during endotoxemia in rats (Hsu & Liu, 2002), and to inhibit atherosclerosis lesion formation (Bhaskaran *et al.*, 2006). Elleuch *et al.* (2007) showed that sesame seed and sesame seed-coat oils (see Figure 122.1) shield against UV-B (290−320 nm) and UV-A (320−400 nm) radiation, which are responsible of most of the cellular damage caused by UV exposure (Figure 122.2).

These beneficial effects of sesame oil are due to its fatty acid and non-saponifiable components (lignans, tocopherol, etc.).

Sesame lignans

Several sesame lignans have been isolated; Figure 122.3 shows their chemical structures. *In vitro* and animal studies have shown that sesame seed is a rich source of mammalian lignan precursors (sesamin, etc.) (Liu *et al.*, 2006). These mammalian lignans may have protective effects against hormone-related diseases such as breast cancer.

High amounts of sesamin and sesamolin are present both in sesame seed and its oil (Tables 122.1, 122.3). Sesamol occurs in small quantities, and its content increases significantly (generated from sesamolin) during roasting or during the process of refining sesame oil (Elleuch *et al.*, 2007).

Sesamin, a major lignan of sesame seeds, exerts multiple functions, such as an antihypertensive effect, and cholesterol, lipid-lowering, and anticancer activities (Yokota *et al.*, 2007); these authors showed that sesamin induces growth inhibition in human cancer cells by regulating

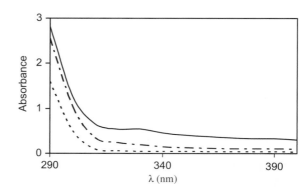

FIGURE 122.2
Ultra-violet spectra of sesame seeds and sesame coats oils. Figure derived from scan (λ = 290−400 nm) of oil diluted 1 : 100 in hexane. —, sesame seed; − - − , sesame coat 1; ——, sesame coat 2. Sesame coats 1 and 2 were by-products of tahin production (see Figure 122.1). Sesame seed and sesame coat oils showed some absorbance in the UV-B (290−320 nm) and UV-A (320−400 nm) ranges. *Reprinted from Elleuch* et al. *(2007), Food Chem., 103, 641−650, with permission.*

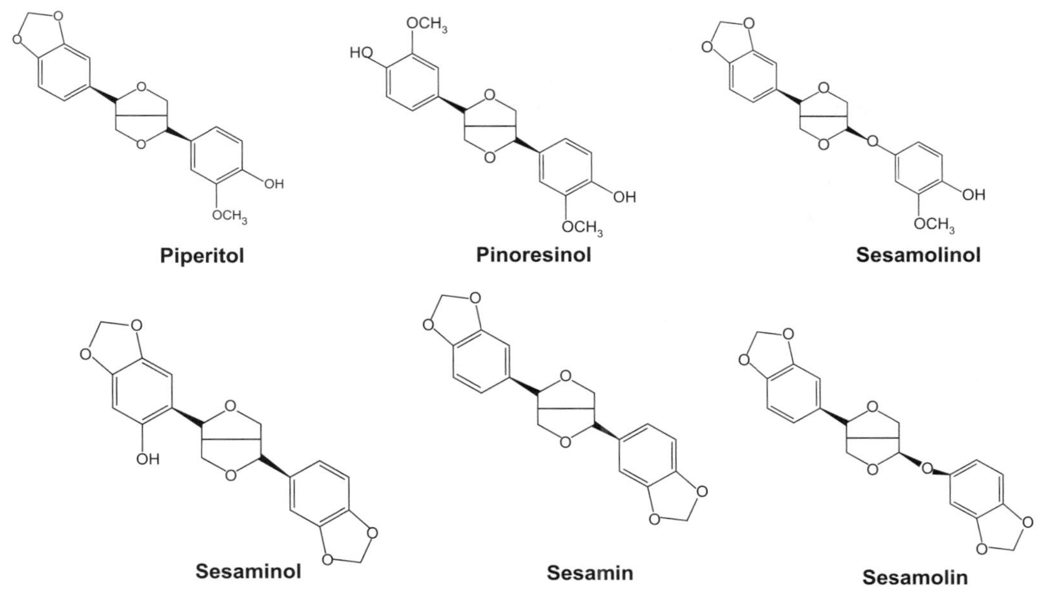

FIGURE 122.3
Chemical structure of sesame lignans. Sesamin and sesamolin are oil-soluble lignans, whereas sesaminol and sesamolinol are the major oil-insoluble lignans.

cyclin D1 protein expression in various kinds of human tumor cells. Sesamin can also induce neuronal differentiation, and regulates the metabolism of lipids, xenobiotics, and alcohol (Tsuruoka *et al.*, 2005; Hamada *et al.*, 2009).

Sesamol is reported to be a potent antioxidant with potential therapeutic benefit due to its capacity to scavenge the entire range of reactive oxygen species (ROS), and to exhibit powerful inhibitory effects on lipid peroxidation (Geetha *et al.*, 2009). An *in vivo* study indicated that sesamol showed an anticancer effect, and as an anti-aging agent prevented photodamage due to chronic UV exposure (Sharma & Kaur, 2006).

ADVERSE EFFECTS AND REACTIONS (ALLERGIES AND TOXICITY)

Allergy to sesame was first described in 1950. Wolffa *et al.* (2003) reported that the 14-kDa protein is the major sesame allergen, later identified as 2S albumin precursor. The symptoms of sesame allergy can include urticaria/angioedema, allergic rhinitis, asthma, atopic dermatitis, oral allergy syndrome, and even anaphylaxis. Some people have experienced contact dermatitis, a result of direct exposure to cosmetics or pharmaceutical products containing sesame allergens.

SUMMARY POINTS

- Sesame (*Sesamum indicum* L.) is cultivated for both its edible seed and its oil.
- Sesame was domesticated on the Indian subcontinent.
- Culinary use of sesame seed includes the decoration of bread and cookies, to produce paste (tahin) added to certain dishes (e.g., with chickpeas as hummus bi tahini, with eggplant as baba gannouj, with lemon juice as salad dressing, etc., and in desserts such as sweetened tahin (halva). Sesame oil is a cooking and salad oil.
- Nutritionally, sesame seeds are rich in oil (high levels of unsaturated fatty acids, mainly oleic and linoleic), protein (high levels of methionine), and micronutrients such as minerals, lignans, tocopherol, and phytosterol.

- Sesame seed exerts many health benefits, such as hypocholesterolemic effects, anticancer activity, oxidative stress attenuation, and blood pressure reduction.
- Sesame seed may induce allergenic symptoms such as urticaria/angioedema, allergic rhinitis, and asthma, and even anaphylaxis.

References

Bedigian, D. (2000). Sesame. In K. F. Kiple, & C. K. Ornelas-Kiple (Eds.), *The Cambridge World History of Food, Vol. 1* (pp. 411–421). New York, NY: Cambridge University Press.

Bedigian, D. (2010). Cultivated sesame, and wild relatives in the genus *Sesamum* L. In D. Bedigian (Ed.), Sesame: the genus *Sesamum*. *Medicinal and Aromatic Plants — Industrial Profiles series* (pp. 33–77). Boca Raton, FL: CRC Press.

Bhaskaran, S., Santanam, N., Penumetcha, M., & Parthasarathy, S. (2006). Inhibition of atherosclerosis in low-density lipoprotein receptor-negative mice by sesame oil. *Journal of Medicinal Food, 9,* 487–490.

Chen, P. R., Chien, K. L., Su, T. C., Chang, C. J., Liu, T.-L., et al. (2005). Dietary sesame reduces serum cholesterol and enhances antioxidant capacity in hypercholesterolemia. *Nutrition Research, 25,* 559–567.

Cooney, R. V., Custer, L. J., Okinaka, L., & Franke, A. A. (2001). Effects of dietary sesame seeds on plasma tocopherol levels. *Nutrition and Cancer, 39,* 66–71.

Elleuch, M., Besbes, S., Roiseux, O., Blecker, C., & Attia, H. (2007). Quality characteristics of sesame seeds and by-products. *Food Chemistry, 103,* 641–650.

Geetha, T., Rohit, B., & Pal, K. I. (2009). Sesamol: an efficient antioxidant with potential therapeutic benefits. *Journal of Medicinal Chemistry, 5,* 367–371.

Hamada, N., Fujita, Y., Tanaka, A., Naoi, M., Nozawa, Y., Ono, Y., et al. (2009). Metabolites of sesamin, a major lignan in sesame seeds, induce neuronal differentiation in PC12 cells through activation of ERK1/2 signaling pathway. *Journal of Neural Transmission, 116,* 841–852.

Hsu, D. Z., & Liu, M. Y. (2002). Sesame oil attenuates multiple organ failure and increases survival rate during endotoxemia in rats. *Critical Care Medicine, 30,* 1859–1862.

Hsu, D. Z., Chen, K. T., Chu, P. Y., Li, Y. H., & Liu, M. Y. (2007). Sesame oil protects against lead-plus-lipopolysaccharide-induced acute hepatic injury. *Shock, 27,* 334–337.

Kamel-Eldin, A., & Appelqvist, L. A. (1994). Variation in the composition of sterols, tocopherols and lignins in seed oils from four *Sesamum* species. *Journal of the American Oil Chemist's Society, 71,* 135–139.

Liu, Z., Saarinen, N. M., & Thompson, L. U. (2006). Sesamin is one of the major precursors of mammalian lignans in sesame seed (*Sesamum indicum*) as observed *in vitro* and in rats. *Journal of Nutrition, 136,* 906–912.

Moazzami, A. A., Haese, S. L., & Kamal-Eldin, A. (2007). Lignan contents in sesame seeds and products. *European Journal of Lipid Science and Technology, 109,* 1022–1027.

Salerno, J. W., & Smith, D. E. (1991). The use of sesame oil and other vegetable oils in the inhibition of human colon cancer growth *in vitro*. *Anticancer Research, 11,* 209–215.

Sankar, D., Rao, M. R., Sambandam, G., & Pugalendi, K. V. (2006). Effect of sesame oil on diuretics or beta-blockers in the modulation of blood pressure, anthropometry, lipid profile, and redox status. *Yale Journal of Biology and Medicine, 79,* 19–26.

Shahidi, F., Amarowicz, R., Abou-Garbia, H. A., & Shehata, Y. (1997). Endogenous antioxidants and stability of sesame oil as affected by processing and storage. *Journal of the American Oil Chemist's Society, 74,* 143–148.

Sharma, S., & Kaur, I. P. (2006). Development and evaluation of sesamol as an antiaging agent. *International Journal of Dermatology, 45,* 200–208.

Tsuruoka, N., Kidokoro, A., Matsumoto, I., Abe, K., & Kiso, Y. (2005). Modulating effect of sesamin, a functional lignan in sesame seeds, on the transcription levels of lipid- and alcohol-metabolizing enzymes in rat liver: a DNA microarray study. *Bioscience Biotechnology and Biochemistry, 69,* 179–188.

Wolffa, N., Cogana, U., Admonb, A., Dalalc, I., Katze, D. Y., Hodosf, N., et al. (2003). Allergy to sesame in humans is associated primarily with IgE antibody to a 14 kDa 2S albumin precursor. *Food and Chemical Toxicology, 41,* 1165–1174.

Yokota, T., Matsuzaki, Y., Koyama, M., Hitomi, T., Kawanaka, M., Enoki-Konishi, M., et al. (2007). Sesamin, a lignan of sesame, down-regulates cyclin D1 protein expression in human tumor cells. *Cancer Science, 98,* 1447–1453.

CHAPTER 123

Effect of Sour Date (*Semen ziziphi spinosae*) Seed Extract on Treating Insomnia and Anxiety

Cheng-Jie Chen[1,2], Min Li[1], Xiao-li Wang[1], Fan-Fu Fang[2], Chang-Quan Ling[2]
[1] Department of Naval Medicine, Second Military Medical University, Shanghai, China
[2] Department of Traditional Chinese Medicine, Changhai Hospital, Shanghai, China

INTRODUCTION

Insomnia, defined as persistent difficulty in falling or staying asleep, thus affecting daytime function, can induce significant psychological and physical disorders. Most patients engage in long-term use of benzodiazepine analogs to treat insomnia; however, these drugs have limited benefits with obvious side effects (Thomas & Christopher, 2004). Herbal drugs have been gradually accepted throughout the world because of their safety and efficiency. *Semen ziziphi spinosae* (referred to as *suan zao ren* in China), the seed of ziziphus jujube, is one of the most popular traditional Chinese medicines. Modern pharmacological studies show that *Semen ziziphi spinosae* possesses activities for treating insomnia, palpitations, anxiety, sweating problems, and poor memory problems (Xie, 1996).

BOTANICAL DESCRIPTION

Plants are usually low shrubs, 1—3 m in height. Branches are erect, not tortuous, and have spines (Figure 123.1). Leaves are alternate and small, 2—3.5 cm in length and 0.6—1.2 cm in width. The drupe is subglobose or broadly oblong, small, and 0.7—1.2 cm in diameter. The mesocarp is sour-tasting and thin, and the stone is obtuse at both ends. Flowers bloom in June—July, and the fruiting period is August—September (Qian & Chen, 2002).

Nuts & Seeds in Health and Disease Prevention. DOI: 10.1016/B978-0-12-375688-6.10123-9

FIGURE 123.1
Zizyphus spinosa bearing whole fruit (1) and seeds (2).

Semen ziziphi spinosae is the ripe kernel of *Ziziphus spinosa* Hu (family Rhamnaceae), which is produced mainly in Hebei, Henan, Shandong, Shanxi and Shaanxi provinces. It is collected in autumn, dried in sunlight after getting rid of the fruit pulp and atone shell, used unprepared or stir-baked, and crushed instantly before using (see Figure 123.1).

HISTORICAL CULTIVATION AND USAGE

Semen ziziphi spinosae first appeared in the Chinese pharmacopeia via *The Divine Farmer's Materia Medica Classic* (Shen Nong Ben Cao Jing) around 300 BC. The Shen Nong Ben Cao Jing includes a description of zizyphus (Yang, 1998):

> *Semen ziziphi spinosae* is sour and balanced (in nature; being neither too warming nor cooling, but combining both warming and cooling effects). It mainly treats heart and abdominal cold and heat, and evil binding qi, aching, pain in the limbs, and damp impediment. Protracted taking may quiet the five viscera, make the body light, and prolong life.

It grows in rivers and swamps. During the middle ages, this herb was believed to be so powerfully strengthening and health-giving, it was fed to infants and used as a nutritional aid. It is documented to help relieve frustration, irritability, anxiety, and even excessive night sweats.

PRESENT-DAY CULTIVATION AND USAGE

Originally native to southern Asia in the regions of Syria, northern India, and central and southern China, this small shrub has been widely cultivated for over 4000 years and has over 400 known cultivators. Most jujube cultivators produce fruit without cross-pollination. After the fruit ripens in late autumn or early winter, it is picked and dried in the sunlight, then pitted and shelled to obtain the ripe kernel of *Semen ziziphi spinosae*. Much of this process is done in the Hebei, Henan, Shandong, Shanxi and Shaanxi provinces of China. This herb is used in formulas to treat irritability, insomnia, palpitations, cold sweats, night sweats, spontaneous sweats, forgetfulness, amnesia, fright, and general weakness — especially due to long-term stress. It nourishes heart and liver and calms the spirit. *Semen ziziphi spinosae* is a nourishing sedative that supports the nervous system and reduces "leakages" that might weaken it (Ou, 1989).

APPLICATIONS TO HEALTH PROMOTION AND DISEASE PREVENTION

Zizyphus seeds are usually stir-fried prior to use; the seeds are turned rapidly in a hot wok (Figure 123.2), and then allowed to cool in straw baskets (Figure 123.3). The fried herb is said to be especially useful for nourishing the liver blood, calming the spirit, and stopping sweating; the raw herb may be used to drain the liver and gallbladder, and also calms the spirit, but is less nourishing. Pharmacology evaluations indicate that both raw and fried seeds have a similar sedative action (Lou, 1987). The reference in the ancient literature to the sour taste of the herb applies to the fruit pulp; the seed itself is deemed sweet — in fact, the taste is relatively bland.

Semen ziziphi spinosae is probably best known in the system of ancient Chinese medicine for its key role in the formula Suanzaoren Tang (Zizyphus Combination) of the *Jingui Yaolue* (220 AD). In that text, the formula is described simply as follows (Hsu & Wang, 1983): "*Semen ziziphi spinosae* Combination treats weakness fatigue, and distress due to weakness, which causes insomnia." *Semen ziziphi spinosae* is the main ingredient of the formula in terms of both the quantity used and its central action for the treatment of deficiency and insomnia, which are the formula's main indications. Even today, if one discusses sedative formulas with prominent physicians in China, this formula is mentioned as being particularly effective; in books about using traditional formulas, the *Semen ziziphi spinosae* Combination is among the first to be listed. A modern presentation of the formula is shown in Table 123.1 (Huang & Wang 1993):

The actions of the formula, as described in the modern text, are to nourish the blood, clear heat, relieve restlessness, and treat insomnia. Anemarrhena is a cold-natured herb that contributes to the action of clearing heat; as indicated in the Shen Nong Ben Cao Jing passage,

1039

FIGURE 123.2
Zizyphus seeds cooking in a wok.

FIGURE 123.3
Fried zizyphus seeds cooling in a basket.

Semen ziziphi spinosae also contributes to clearing heat, even though it doesn't have a cold nature. *Semen ziziphi spinosae* provides all the other actions attributed to the formula. Its blood-nourishing effect is complemented by that of cnidium, while both licorice and hoelen provide additional sedative effects and contribute to the overall tonification therapy by improving the spleen functions.

In the context of analyzing *Semen ziziphi spinosae* Combination, one can mention the use of the related herb *Semen ziziphi spinosae* in the well-known sedative formula Ganmai Dazao Tang (Wheat and *Semen ziziphi spinosae* Combination, shown in Table 123.2).

The dosage of jujube is usually measured by the number of fruits, in this case seven; the seeds are removed and *Semen ziziphi spinosae*, the remaining fruit, weighs 2 g. The actions of this small formula are to nourish the heart, normalize the function of the stomach and spleen, calm the mind, and treat insomnia and spasms. The role of wheat, which is normally consumed by many people in large quantities (mainly in the form of noodles or bread), may be to treat B vitamin deficiencies that can arise in people who do not normally eat wheat or other sources rich in these vitamins (B vitamins play an important role in nervous system functions). This type of deficiency has been a common problem in the Orient. For example, in Japan, until quite recently, beri beri (a vitamin B1 deficiency that causes leg edema) was a common disorder because the diet lacked good sources of the vitamin, which is found in whole grains. The combination of *Semen ziziphi spinosae* and licorice in this formula serves as a spleen tonic and sedative, which is a role similar to that of zizyphus and licorice in the *Semen ziziphi spinosae* Combination.

According to ancient prescriptions and theories, both raw and prepared *Semen ziziphi spinosae* can work as a natural herb for insomnia. Clinical practice has shown that the efficacy of these raw or prepared natural herbs is not opposing. According to traditional Chinese medicine

TABLE 123.1 A Modern Presentation of the Suanzaoren Tang Formula
Herb Weight
Semen ziziphi spinosae (*suan zao ren*) 18 g
Anemarrhena (*zhimu*) 10 g
Hoelen (*fuling*) 10 g
Cnidium (*chuanxiong*) 5 g
Licorice (*gancao*) 3 g

TABLE 123.2 A Well-known Sedative
Formula Ganmai Dazao Tang

Herb Weight

Wheat (*fuxiaomai*) 15 g
Semen ziziphi spinosae (*suanzaoren*) 14 g
Licorice (*gancao*) 3 g

theory, it makes sense that prepared *Semen ziziphi spinosae* should be used in the warming method and raw *Semen ziziphi spinosae* should be used in the clearing method; however, experimental results on animals and humans show that both raw and prepared *Semen ziziphi spinosae* have the same role and efficacy. There is no difference and no opposing effect of waking up between them. The sedative and hypnotic effects may be related to its soluble components, such as sugar and jujube acid, etc. There are no sedative and hypnotic effects found in its extract of ether or chloroform. Hence, we can conclude that there are different functions between *Semen ziziphi spinosae*'s kernel and pulp. However, this does not necessarily mean that the opposite effect would occur after *Semen ziziphi spinosae* is prepared. To save the procedure of preparation, raw *Semen ziziphi spinosae* is recommended. Both raw and prepared *Semen ziziphi spinosae* are very good natural herbs for insomnia; it is easier to extract the useful ingredients from *Semen ziziphi spinosae* which is fried slightly. Clinically, raw *Semen ziziphi spinosae* should be used in the syndrome of palpitation and insomnia that is caused by deficient heat of the liver and gall, while prepared *Semen ziziphi spinosae* should be used in patients with the syndrome of palpitation and insomnia caused by deficiency of both heart and spleen, especially with the symptoms of abnormal sweating due to general debility, and indigestion because of a weak spleen and stomach.

Laboratory animal studies of *Semen ziziphi spinosae* extract confirm a sedative effect, though the constituents that contribute this effect have not all been specifically identified. Some sedative effects were shown to be produced by several different components (Zhu, 1998). The only components of *Semen ziziphi spinosae* that are present in quantities likely to be responsible for the observed clinical effects are triterpenes. The unique triterpenes for this herb are known as jujubosides (see Figure 123.4). Additionally, there are related triterpene compounds (such as betulic acid and oleanolic acid) that are found in several other herbs. Structurally, the jujubosides are nearly identical to the active constituents found in ginseng, an herb that is traditionally used as a sedative and a tonic. According to one analysis, the triterpene level in zizyphus is 2.5% (Yen, 1992), so that an 18-g dose, as used in the *Semen ziziphi spinosae* Combination, would yield about 360 mg of triterpenes — a level typically relied upon for most triterpenes. The simple powder of *Semen ziziphi spinosae*, consumed orally, appears to be effective in relatively low dosages. One report indicates that good effects were obtained using 0.8–1.2 g of the powder before bed.

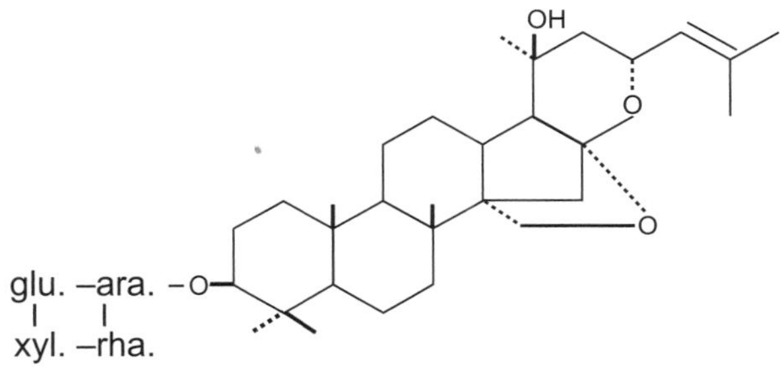

FIGURE 123.4
The jujuboside terpene.

1041

In one of the best known patent remedies for insomnia, Tianwang Buxin Dan (the Ginseng and *Semen ziziphi spinosae* Formula), many herbs with triterpenes are combined together, including zizyphus, ginseng, platycodon, and polygala. The formula also provides herbs that contain steroidal saponins (e.g., ophiopogon and asparagus) that have a similar structure and are likely to have similar effects. The indications and applications of this large formula are almost identical to those of the smaller and more ancient *Semen ziziphi spinosae* Combination. The famous tonic mushroom ganoderma contains triterpenes similar to those found in ginseng and *Semen ziziphi spinosae*, and is classified, like *Semen ziziphi spinosae*, as a tonic sedative. It is sometimes recommended as a single herb remedy for insomnia.

Due to its sweet taste and neutral temperate, it nourishes the heart and liver blood as well as nourishing yin to arrest sweating. The main function of this medicinal seed is to calm and stabilize the Shen. Therefore, *Semen ziziphi spinosae* is most often used in cases of insomnia, irritability, palpitations, forgetfulness, and abnormal sweating due to deficiency. It is also a prophylactic for altitude sickness. Modern research from China shows versatility in the medicinal use of this ancient herb. One study shows improved liver functions as well as improved quality of sleep in chronic hepatitis B patients with insomnia. Other research suggests this medicinal seed has sedative and hypnotic effects on the nervous system, protective ischemic effects on cerebral damage, and antineoplastic effects on cancer cells. Other observed effects are muscle relaxation, reduced pain sensation, increased pain tolerance, and prolonged lowered blood pressure (Zheng *et al.*, 2006). The oil extracts of the seed have also been shown to decrease levels of blood cholesterol, low density lipoprotein, and triglycerides, and inhibit platelet aggregation (Wu, 1991).

ADVERSE EFFECTS AND REACTIONS (ALLERGIES AND TOXICITY)

Semen ziziphi spinosae is reported to have very low toxicity when taken orally. In laboratory animals (mice and rats), a huge single dose of 50 g/kg body weight produced no toxic symptoms, and a daily dose of 20 g/kg for 30 days did not produce toxic reactions. Side effects have not been reported. Modern pharmacology evaluation of *Semen ziziphi spinosae* oil and *Semen ziziphi spinosae* extract suggests that with prolonged feeding they can reduce serum triglycerides and cholesterol (mainly LDL), and reduce fatty degeneration of the liver. These properties have also been attributed to the triterpenes of ginseng and ganoderma. Despite its strong medicinal effects, this mild-natured herb is relatively safe to use long term, and for children. Reported adverse effects from inappropriate use include fever with aversion to cold, cold sweats, and joint pain. Western herbalists are also considering the *Semen ziziphi spinosae* as an alternative to kava kava and valerian, due to the latter two herbs' possible negative impacts on liver function (Zhang *et al.*, 2008).

SUMMARY POINTS

- *Semen ziziphi spinosae* is an excellent natural and safe "sedative" herb. In Chinese medicine, this herb nourishes the two organs responsible for sleep — the liver and heart. When balance is brought back to these organs, then sleep occurs naturally.
- It is common to think that herbs to help you fall asleep should only be taken at night, and this is not always the case. *Semen ziziphi spinosae* can be taken as directed through the day, and will help with sleep naturally.
- Typically, the "insomnia" or sleep problems that this herb helps with will be trouble sleeping due to anxiety, fidgeting, restlessness, night sweating, and such.
- *Semen ziziphi spinosae* helps with the following symptoms, even if they are not compromising sleep: anxiety, overworked mind, restlessness, night sweats, fidgeting, etc.

References

Hsu, H. Y., & Wang, S. Y. (1983). *Chin Kuei Yao Lueh*. Long Beach, CA: Oriental Healing Arts Institute.

Huang, B. S., & Wang, Y. X. (1993). *Thousand formulas and thousand herbs of traditional Chinese medicine Vol. 1*. Harbin, China: Heilongjiang Education Press, Harbin.

Lou, S. N. (1987). Sedative and hypnotic actions of raw and fried suanzaoren. *Journal of Chinese Herbal Research, 21*, 8—19.

Ou, M. (1989). *Chinese—English manual of common-used herbs in traditional Chinese medicine*. Hong Kong, China: Joint Publishing Co.

Qian, C. S., & Chen, H. Y. (2002) *Flora of China, Vol. 12*. Beijing, China: Science Publishing House. 120—121.

Thomas, R., & Christopher, D. (2004). Evolution of insomnia: current status and future direction. *Sleep Medicine, 5*, 23—30.

Wu, S. Y. (1991). Effects of zizyphus seed oil and zizyphus extract on decrease of serum lipoprotein and inhibition of platelet aggregation. *China Journal of Chinese Materia Medica, 16*(7), 435—437.

Xie, Z. W. (1996). *Chinese herbal medicine compilation* (2nd ed.). Beijing, China: The People's Medical Publishing House.

Yang, S. Z. (1998). *The divine farmer's materia medica*. Boulder, CO: Blue Poppy Press.

Yen, K. Y. (1992). *The illustrated materia medica: Crude and prepared*. Taipei, Taiwan: SMC Publishing Inc.

Zhang, X., Ding, C. H., & Li, H. P. (2008). Research progress on the chemical compositions and pharmacological effects of *Semen ziziphi spinosae*. *Journal of Scientific Technology and Food Industry, 3*, 348—350.

Zheng, Y., Qian, S. Y., & You, Z. L. (2006). Advances in study on pharmacological effects of *Semen ziziphi spinosae*. *Sichuan Journal of Physiological Sciences, 28*(1), 35—37.

Zhu, Y. P. (1998). *Chinese materia medica: Chemistry, pharmacology, and applications*. Amsterdam, The Netherlands: Harwood Academic Publishers.

CHAPTER 124

Soursop (*Annona muricata* L.) Seeds, Therapeutic and Possible Food Potential

Julio A. Solís-Fuentes[1], María del Rosario Hernández-Medel[1], María del Carmen Durán-de-Bazúa[2]
[1] Instituto de Ciencias Básicas, Universidad Veracruzana, Xalapa, Ver., México
[2] Chemical Engineering Department, Facultad de Química, Universidad Nacional Autónoma de México, México

CHAPTER OUTLINE

LIST OF ABBREVIATIONS

AACG, annonaceous acetogenin
AMS, *Annona muricata* seed
ATP, adenosine triphosphate
ED_{50}, drug dose effective for 50% of exposed population
FA, fatty acid
Hep G2, human hepatoma cells
Hep 2,2,15, human hepatoma cells infected with hepatitis B virus
IC_{50}, drug concentration required for 50% inhibition *in vitro*
THF, tetrahydrofuran
THP, tetrahydropyran

Nuts & Seeds in Health and Disease Prevention. DOI: 10.1016/B978-0-12-375688-6.10124-0

INTRODUCTION

Soursop (*Annona muricata* L.) is a plant native to South America that is widely distributed in tropical and subtropical regions of the world. It belongs to the Annonaceae family, and its fruits are widely appreciated for their organoleptic characteristics. The family includes several genera characterized by the presence of metabolites with important biological activities. *A. muricata* seeds make up approximately 5% of the ripe fruit weight. In recent years, scientific studies have confirmed the presence of alkaloids, acetogenins, and cyclopeptides, compounds of great importance that have attracted pharmacological interest. The seeds also contain a significant amount of oil, the composition, physical and chemical properties of which make it potentially attractive in the food sector.

BOTANICAL DESCRIPTION

A. muricata is an annonacea belonging to the *Magnoliales* order and the Magnoliopsida class. It is a tree 4—7 m in height, with smooth bark. Its leaves are long and pale green, and its flowers have yellow-green fleshy petals. Its fruits are large, oval, or heart-shaped, with small, green, thin-shelled spines; their size ranges from 10—30 cm long and 15 cm wide, and they can weigh up to 2.5 kg. They have white flesh that is creamy and juicy. The seeds are small and numerous, dark-colored, and of compact consistency. They have a rigid coating that contains a kernel-like almond. At present, there is no information on the number of world varieties. Soursop types have been roughly classified into sweet, subacid, or sour; however, in just one region of Puerto Rico 14 varieties have been distinguished, and it is believed that in other producing regions of the world there are many more (Morton, 1987).

HISTORICAL CULTIVATION AND USAGE

Already being cultivated in Mexico before the Spaniards arrived, the soursop has long been highly appreciated for its edible fruit. It was distributed very early throughout the warm lowlands of eastern and western Africa, South-east Asia and China, where it was (and continues to be) commonly grown on a small scale or as a backyard tree (Janick & Paull, 2008). It was first used as fresh fruit, and later in various food products processed either by hand or on an industrial scale. Historical information on the medicinal use of AMS is scarce, for it was not until the 20th century that its acaricidal property was first reported.

PRESENT-DAY CULTIVATION AND USAGE

The soursop lives in tropical regions with warm weather, at heights ranging from sea level to 500 m. In Mexico, for example, it is cultivated from the state of Sinaloa to the state of Chiapas, and from the state of Veracruz to the Yucatan Peninsula along the Gulf of Mexico, in small plantations of up to 20 hectares. In many other parts of the world, it is still grown as a backyard tree. There are no data available regarding world production, imports and exports. In Surinam, this fruit yields 43 kg/tree and is planted at 278 trees/hectare. The commercial plantations are limited to the Philippines, the Caribbean, and South America.

Currently, the soursop is used mainly for its fresh and processed fruits. The leaves are used commercially for therapeutic applications. Extracts from the seed powder are presently used as an effective insecticide against lice, worms, and aphids, and as a larvicide (Morton, 1987; Johnson *et al.*, 2000).

APPLICATIONS IN HEALTH PROMOTION AND DISEASE PREVENTION

A. muricata seed and its therapeutic potential

In traditional medicine, it is known that the bark, root, leaf, fruit, and seed of *Annona muricata* have various medical applications in the tropics. The crushed seeds are used mainly against internal and external parasites, worms, and lice (Morton, 1987; Johnson *et al.*, 2000). Studies with crude ethanol extracts of AMS have demonstrated their power of inhibition on the growth of larvae of *Spodoptera litura* (Noctuidae) (Leatemia & Isman, 2004) and *Aedes aegypti* (Bobadilla *et al.*, 2005).

The Annonaceae family is characterized, in general, as presenting an exclusive group of secondary metabolites of the family of AACGs, compounds that are also found in large numbers in AMS (Rieser *et al.*, 1996; Zafra-Polo *et al*, 1998; Bermejo *et al.*, 2005). The AACGs are an important group of long-chain FA derivatives, C-32 or C-34, usually with a terminal γ-lactone, saturated or unsaturated. Sometimes the FAs are one to three rings of THF or THP, adjacent or not, which are occasionally replaced with epoxide rings or double bonds near the middle of the aliphatic chain. Accordingly, there can be seven different types of terminal γ-lactones (L-A to L-F), ten skeletons with THF and/or THP (T-A to T-H, with three subtypes of T-G) and mono, di or tri-epoxides (Bermejo *et al.*, 2005). Figure 124.1 shows the general structure of an AACG. The THF ring system can be simple (one ring), adjacent (two rings), or not adjacent, with hydroxyl groups on both sides or not; this creates chiral centers in the molecules, making it very complex to elucidate the structures. Many AACGs isolated as pure compounds are actually a mixture of diastereomers. However, the synthesis of some AACGs makes it possible to assign the chemical configurations of the corresponding natural AACGs unequivocally (Bermejo *et al.*, 2005).

A few more than 400 compounds have been reported in the literature since the discovery of uvaricin in 1982 (Zafra-Polo *et al.*, 1998), all from various genera exclusive to the Annonaceae family, although only some types of structures are present in the AACGs isolated from the AMS (Johnson *et al.*, 2000). To date, the AACGs that have been isolated from the seeds of soursop include a terminal γ-lactone and THF in their structure; a few have epoxides, yet there are none with adjacent THP or THF (Bermejo *et al.*, 2005).

Various pharmacological investigations have shown that these metabolites possess antitumor, antidiarrheal, larvicidal, antimalarial, pesticidal, and fungicidal activities, especially with regard to their antitumor properties *in vitro*. Some are among the most potent inhibitors of complex I (NADH: ubiquinone oxidoreductase) in the system of mitochondrial electron transport between the said complex and the NADH-oxidase in the plasma membrane characteristic of cancer cells; these actions induce apoptosis (programmed cell death), perhaps following deprivation of ATP. The decrease in ATP is especially toxic to multidrug resistant tumor cells, as well as to pesticide-resistant insects that possess ATP-dependent systems of xenobiotic flows, so said metabolites can be regarded as extraordinary antitumor agents and pesticides, especially for inhibiting resistance mechanisms requiring an ATP-dependent flow (Alali *et al.*, 1999).

The AACGs may be among the most potent cytotoxic agents known; for example, trilobacin and asiminocin, AACGs isolated from *Asimina triloba*, have demonstrated values of $DE_{50} < 10^{-12}$ μg/mL in several human tumor cell lines (Alali *et al.*, 1999). A little over 40 AACGs showing cytotoxic activity have been isolated from the AMS, among which *cis*-annonacin may be mentioned as displaying a selective cytotoxicity towards colon adenocarcinoma cells (HT-29) 10,000 times more potent than that of adriamycin ($IC_{50} = 1.0 \times 10^{-8}$ μg/mL,

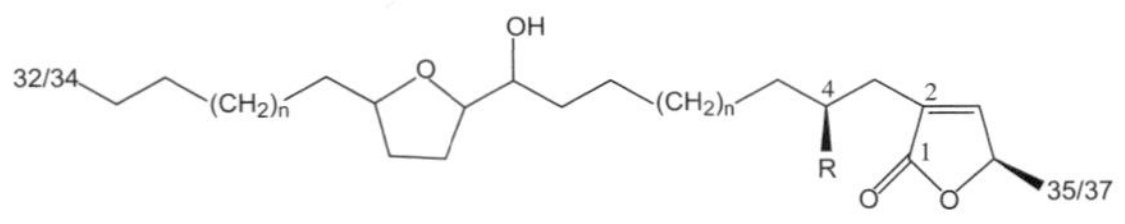

1047

FIGURE 124.1
General structure of an annonaceous acetogenin (AACG). The AACGs are an important group of long-chain FAs derivatives, C-32 or C-34, usually with a terminal γ-lactone, saturated or unsaturated.

TABLE 124.1 Some AACGs with Biological Activity Isolated from AMS

AACGs	x^a	y	z	R^b	R_1	R_2	R_3	Reference
cis-Annonacin	4	2	11	OH	OH	OH	OH	Rieser *et al.*, 1996
Corossolin	4	2	11	H	OH	OH	OH	Wu, 2007
Corossolone	4	2	11	H	=O	OH	OH	Wu, 2007
Longifolicin	4	0	13	H	OH	OH	OH	Wu, 2007
cis-Annomontacin	4	4	11	OH	OH	OH	OH	Liaw *et al.*, 2002
Goniothalamicin	4	2	13	OH	OH	OH	OH	Rieser *et al.*, 1996

^{a}x, y, z: methylen number in Figure 124.2;
^{b}R, R_1, R_2, R_3: substituents in Figure 124.2.

$IC_{50} = 5.1 \times 10^{-4}$ µg/mL, respectively); *cis*-annonacin-10-one showed the same range as adriamycin toward the same type of cell line ($IC_{50} = 9.0 \times 10^{-4}$ µg/mL) (Rieser *et al.*, 1996).

The bioactivity possessed by these compounds has made it possible to obtain a patent for the isolation, identification, and antitumoral use of AACGs from the AMS; these include: muricin A, B, C, D, E, F, and G, among which the most effective were found to be muricin D ($IC_{50} = 6.6 \times 10^{-4}$ µg/mL for Hep G_2 and $IC_{50} = 4.8 \times 10^{-2}$ for Hep 2,2,15); muricin F ($IC_{50} = 4.28 \times 10^{-2}$ µg/mL Hep G_2 and $IC_{50} = 3.86 \times 10^{-3}$ µg/mL for Hep 2,2,15); longifolicin ($IC_{50} = 4.04 \times 10^{-4}$ µg/mL for Hep G_2 and $IC_{50} = 4.9 \times 10^{-3}$ µg/mL for Hep 2,2,15). However, both muricin A and muricin B showed selective activity toward the Hep 2,2,15 cell line, with $IC_{50} = 5.13 \times 10^{-3}$ µg/mL and $IC_{50} = 4.29 \times 10^{-3}$ µg/mL, respectively (Wu, 2007). Table 124.1 shows some of the AACGs isolated from the AMS with cytotoxic activity conforming to the structure shown in Figure 124.2; Table 124.2 presents some of the patented AACGs showing substituents according to Figure 124.3, and Figure 124.4 shows the muricin I structure that exemplifies AACGs found with double bonds.

In addition, the cyclopeptides annomuricatin A, annomuricatin B, and annomuricatin C have been isolated from AMS. These types of compounds have demonstrated activity in other

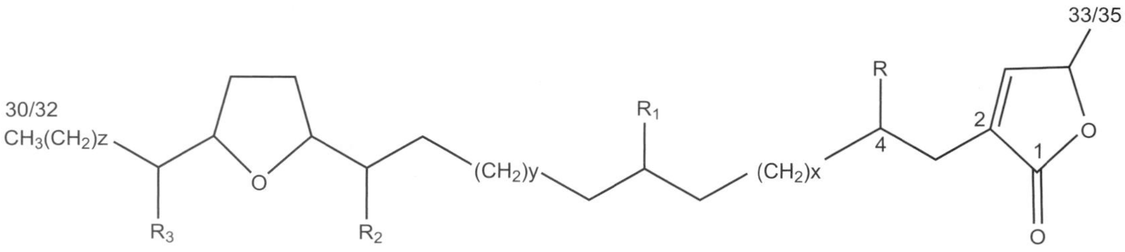

1048

FIGURE 124.2
Basic structure for AACG of Table 124.1. In the figure, when R, R1, R2, and R3 are substituted by OH, and y, x, and z by 4, 2, and 11, respectively, this has the chemical structure of *cis* Annonacin.

TABLE 124.2 Some Patented AACGs with Biological Activity Isolated from AMS (Wu, 2007)

AACGs	x^a	y	z	R^b	R_1	R_2	R_3
Muricatetrocin A	7	2	11	OH	OH	OH	OH
Muricin A	10	6	4	OH	OH	OH	OH
Muricin C	12	2	6	OH	OH	OH	OH
Muricin D	10	2	6	H	OH	OH	H
Muricin E	7	5	6	OH	OH	OH	OH

^{a}x, y, z: methylen number in Figure 124.2;
^{b}R, R_1, R_2, R_3: substituents in Figure 124.3.

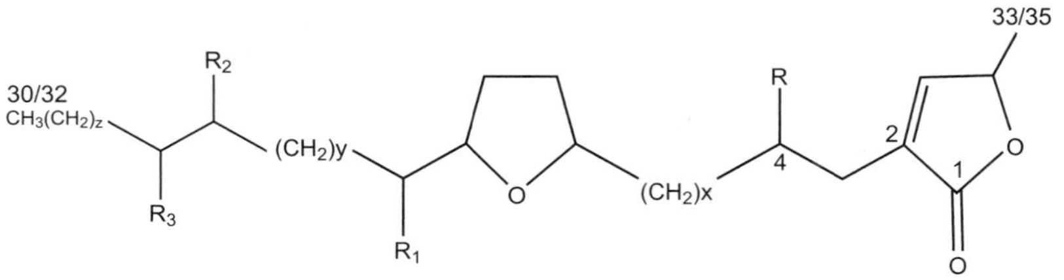

FIGURE 124.3
Basic structure for AACG of Table 124.2. In the figure, when R, R_1, R_2, and R_3 are substituted by OH, and x, y, and z by 7, 2, and 11, respectively, this has the muricatetrocin A chemical structure.

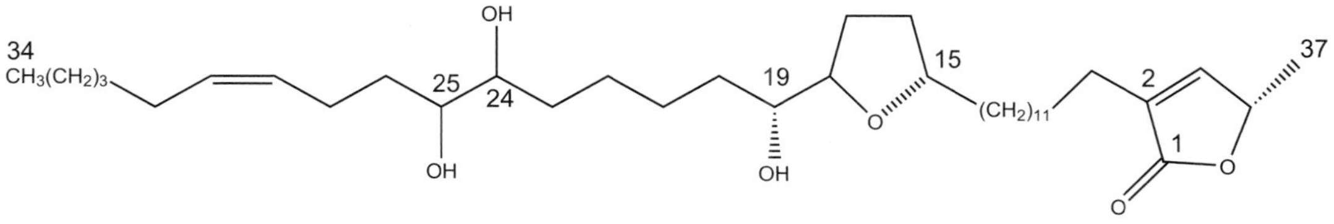

FIGURE 124.4
Muricin I structure. The Muricin I structure exemplifies AACGs found with double bonds.

species of *Annona* (Wéle *et al.*, 2004); so has lectin, which was found to possess inhibitory activity against the fungal pathogens *Fusarium oxysporium*, *Fusarium solana*, and *Colletotrichum musae*, which can produce Fusarium wilt (Panama disease), keratomycosis, and anthracnose, in addition to red cell agglutinating activity (Damico *et al.*, 2003).

Accordingly, AMS is a source of bioactive compounds (AACGs) with immense possibilities for pharmaceutical uses.

A. muricata seed and its possible potential for food use

1049

The use of soursop fruit in the production of foodstuffs results in various so-called waste materials, one of which is the seeds. Studies of AMS in its proximal composition have shown that it contains proteins, carbohydrates, ash, and oil in significant amounts. The reports are scarce, and contain wide ranges of values that reflect the need for further studies regarding varieties and their degree of variability. Table 124.3 shows the composition and property ranges of AMS and its oil as reported in pertinent literature.

There are few studies that have directly explored AMS possibilities as raw material in the food sector, mainly because there is knowledge of the existence of toxic compounds in it. Awan *et al.* (1981), and particularly Fasakin *et al.* (2008), analyzed AMS with the purpose of highlighting its nutritional potential; other authors have studied it for the chemical and physical characteristics and the thermal and phase behavior of its oil (Onimawo, 2002; Ocampo *et al.*, 2007; Solís-Fuentes *et al.*, 2010), with the objective of contributing to knowledge about the re-use of soursop wastes or of characterizing some of its most important components.

The chemical composition of the kernel seed, according to research conducted to date, may be interesting from the standpoint of food. For example, substances that are important for animal feed, like protein and fiber, minerals and oil, can be found in the seed at levels of 2.4–27.3%, 8.0–43.4%, 2.3–13.6%, and 20.5–37.7, respectively.

In recent times, there has been extensive research to locate and expand non-conventional sources of fats and oils for the purpose of finding applications for them in various fields. AMS has an important amount of oil. Composed of oleic (39.5–44.0%), linoleic (27–30%), palmitic (19.0–25.5%), and stearic (5–6%) as its main FAs, it appears similar to the oils from conventional oilseeds. The reported values for its refraction index, saponification index, and iodine value are 1.464–1.468, 100–227, and 87–111, respectively, and studies of its phase and thermal behavior have shown that it has the characteristics of common table or cooking oils;

TABLE 124.3 Reported Composition and Property Ranges of AMS and Oil

Properties	Value Range	Reference
Seeds in fruit[a]	5.4	Solís-Fuentes et al., 2010
Moisture	8.5—34.6	Fasakin et al., 2008; Solís-Fuentes et al., 2010
Protein[b]	2.4—27.3	
Crude Fiber	8.0—43.4	Onimawo, 2002; Fasakin, et al., 2008
Ash	2.3—13.6	
Raw oil	20.5—37.7	Awan et al., 1981; Solís-Fuentes et al., 2010
Oil composition[c]		
Palmitic	19—25.5	Ocampo et al., 2007; Solís-Fuentes et al., 2010
Palmitoleic	1.5—2	
Stearic	5—6	
Oleic	39.5—44	
Linoleic	27.0—30	
SFA	24— 31.5	
UFA	68.5—71	
Oil properties		
Saponification index	100—227	Awan et al., 1981; Onimawo, 2002
Iodine value	87—111	Awan et al., 1981
Refraction index[d]	1.464— 1.468	Onimawo, 2002: Solís-Fuentes et al., 2010
SFC at 10°C	1.3	Solís-Fuentes et al., 2010

[a]%, as-is basis;
[b]% of kernel, dry basis;
[c]% of total FA;
[d]At 40°C.

the yellowish oil remains liquid at warm ambient and refrigeration temperatures, its solid fat content at 10°C being around 1% (see Table 124.3 for references).

Even with all these considerations, it is clear that any application of the AMS or its derived components must take into account the characteristic toxicity of raw AMS. Notwithstanding the bromathological characteristics of the seed and its significant oil yield and physico-chemical properties, its use for human consumption should be viewed with caution. The usability of AMS and its oil undoubtedly requires increased knowledge, particularly with regard to food applications. The reviewed bioactive compounds present in the seed are of toxic and pharmacological importance; consequently, they limit the current use of raw AMS in foods. The processing of AMS to isolate and obtain components of interest to the food industry, such as proteins, oil, or otherwise, also requires further and deeper studies aimed at extraction, purification, toxicological, and nutritional evaluations, etc.

ADVERSE EFFECTS AND REACTIONS, ALLERGIES AND TOXICITY

Soursop seed is not edible, and traditionally has been known to be toxic because of the presence of the described compounds; therefore, its components and derivatives destined for human consumption must be subject to careful research protocols to ensure their toxicological safety. Even when it is used as a drug, in its whole form or from crude extracts, it must be handled with care, because some of its bioactive components have not been fully evaluated for their allergenic or toxicological cumulative effect. One example is annonacin, which has been isolated from most parts of the plant, including the seeds and fruit pulp. It has been shown by clinical studies with rats, and epidemiological studies in human populations (like those from the Caribbean island of Guadeloupe, and those from New Caledonia) with high and prolonged intakes of soursop fruit pulp, that annonacin is associated with the occurrence of a neurodegenerative disease known as atypical parkinsonism (Escobar-Khondiker *et al.*, 2007). This finding, together with the known effects of the AACG on mitochondrial toxicity, makes the consumption of the AMS, its concentrates, and derivates without purification a risky matter (Liaw *et al.*, 2002; Quispe *et al.*, 2006).

SUMMARY POINTS

- Soursop, *Annona muricata*, is a plant native to South America, and is widely cultivated throughout the tropical regions of the world.
- *A. muricata* has shown pharmacological activity in its different parts.
- The *Annona muricata* seeds make up about 5% of the fruit, and are now considered to be industrial processing waste.
- Annonaceous acetogenins are the most studied bioactive metabolites from *Annona muricata* seeds, and have high cytotoxic activity with potential therapeutic applications.
- Other important components of the *Annona muricata* seeds have also been studied from a food perspective.
- *A. muricata* seeds have a significant content of oil, which has a composition and properties similar to those of oils of conventional sources.
- Any component derived from *Annona muricata* seeds for alimentary consumption must be studied regarding possible toxicological risks.

References

Alali, F. Q., Liu, X.-X., & McLaughlin, J. L. (1999). Annonaceous acetogenins: recent progress. *Journal of Natural Products, 62,* 504–540.

Awan, J. A., Kar, A., & Udoudoh, P. J. (1981). Preliminary studies on the seeds of *Annona muricata. Plant Foods for Human Nutrition, 30,* 163–168.

Bermejo, A., Figadère, B., Zafra-Polo, M. C., Barrachina, I., Estornell, E., & Cortes, D. (2005). Acetogenins from Annonaceae recent progress in isolation, synthesis and mechanisms of action. *Natural Product Reports, 22,* 269–303.

Bobadilla, M., Zavala, F., Sisniegas, M., Zavaleta, G., Mostacero, J., & Taramona, L. (2005). Evaluación larvicida de suspensiones acuosas de *Annona muricata* Linnaeus "guanábana" sobre *Aedes aegypti* Linnaeus (Diptera, Culicidae). *Reviews of Physiology Biochemistry, 12,* 145–152.

Damico, D. C. S., Freire, M. G. M., Gomes, V. M., Toyama, M. H., Marangoni, S., Novello, J. C., et al. (2003). Isolation and characterization of a lectin from *Annona muricata* seeds. *Journal of Protein Chemistry, 22,* 655–661.

Escobar-Khondiker, M., Höllerhage, M., Muriel, M. P., Champy, P., Bach, A., Depienne, C., et al. (2007). Annonacin, a natural mitochondrial complex I inhibitor, causes tau pathology in cultured neurons. *Journal of Neuroscience, 27,* 7827–7837.

Fasakin, A. O., Fehintola, E. O., Obijole, O. A., & Oseni, O. A. (2008). Compositional analyses of the seed of sour sop, *Annona muricata* L., as a potential animal feed supplement. *Scientific Research and Essays, 3,* 521–523.

Janick, J., & Paull, R. E. (2008). *The encyclopedia of fruit & nuts.* Cambridge, UK: Cambridge University Press. 42–46.

Johnson, H. A., Oberlies, N. H., Alali, F. Q., & McLaughlin, J. L. (2000). Thwarting resistance: annonaceous acetogenins as new pesticidal and antitumor agents. In S. J. Cutler, & H. G. Cutler (Eds.), *Biologically active natural products: Pharmaceuticals* (pp. 181–191). Boca Raton, FL: CRC Press.

Leatemia, J. A., & Isman, M. B. (2004). Insecticidal activity of crude seed extracts of *Annona* spp., *Lansium domesticum* and *Sandoricum koetjape* against lepidopteran larvae. *Phytoparasitica, 32,* 30–37.

Liaw, Ch.-Ch., Chang, F.-R., Lin, Ch.-Y., Chou, Ch.-J., Chiu, H.-F., Wu, M.-J., et al. (2002). New cytotoxic monotetrahydrofuran annonaceous acetogenins from *Annona muricata. Journal of Natural Products, 65,* 470–475.

Morton, J. (1987). In J. M. Ockerbloom (Ed.), *Fruits of warm climates* (pp. 75–80). Miami, FL: Florida Flair Books.

Ocampo-S., D. M., Betancur-J., L. A., Ortíz, A., & Ocampo-C., R. (2007). Estudio cromatográfico comparativo de los ácidos grasos presentes en semilla de *Annona cherimolioides* y *Annona muricata* L. *Vector, 2,* 103–112.

Onimawo, I. A. (2002). Proximate composition and selected physicochemical properties of the seed, pulp and oil of sour-sop (*Annona muricata*). *Plant Foods for Human Nutrition, 57,* 165–171.

Quispe, A., Zavala, D., Rojas, J., Posso, M., & Vaisberg, A. (2006). Efecto citotóxico selectivo *in vitro* de muricin H (acetogenina de *Annona muricata*) en cultivos celulares de cáncer de pulmón. *Revista Peruana de Medicina Experimental y Salud Pública, 23,* 265–269.

Rieser, M. J., Gu, Z.-M., Fang, X.-P., Zeng, L., Wood, K. V., & McLaughlin, J. L. (1996). Five novel mono-tetrahydrofuran ring acetogenins from the seeds of *Annona muricata*. *Journal of Natural Products, 59*, 100–108.

Solís-Fuentes, J. A., Amador-Hernández, C., Hernández-Medel, M. R., & Durán-de-Bazúa, M. C. (2010). Caracterización fisicoquímica y comportamiento térmico del aceite de "almendra" de guanábana (*Annona muricata*, L). *Grasas y Aceites, 61*, 58–66.

Wélé, A., Zhang, Y., Caux, Ch., Brouard, J.-P., & Bodo, B. (2004). Annomuricatin C, a novel cyclohexapeptide from the seeds of *Annona muricata*. *Comptes Rendus Chimie, 7*, 981–988.

Wu, Y.-Ch (2007). *Cytotoxic annonaceous acetogenins from Annona muricata*. Patent No. US 7,223,792, B2.

Zafra-Polo, M. C., Figadère, B., Gallardo, T., Tormo, J. R., & Cortes, D. (1998). Natural acetogenins from annonaceae, synthesis and mechanisms of action. *Phytochemistry, 48*, 1087–1117.

CHAPTER 125

Effect of Dietary Soybean (*Glycine max* L.) Protein on Lipid Metabolism and Insulin Sensitivity in an Experimental Model of Dyslipidemia and Insulin Resistance

Yolanda B. Lombardo, María Eugenia Oliva, Dante Selenscig, Adriana Chicco
University of Litoral, Department of Biochemistry, School of Biochemistry,
Santa Fe, Argentina

1053

LIST OF ABBREVIATIONS

CPT-1, carnitine palmitoyl transferase 1
DAG, diacylglycerol
FAS, fatty acid synthase
FFA, free fatty acids
GIR, glucose infusion rate

Nuts & Seeds in Health and Disease Prevention. DOI: 10.1016/B978-0-12-375688-6.10125-2

HDLC, high-density lipoprotein cholesterol
IR, insulin resistance
IRS-1, insulin receptor substrate 1
IS, insulin secretion
LCACoA, long-chain acyl-CoA
LDLC, low-density lipoprotein cholesterol
ME, malic enzyme
nPKCθ, novel protein kinase-Cθ
PDH c, pyruvate dehydrogenase complex
PPARα: peroxisome proliferator-activated receptor α
SCD1, stearoyl-CoA desaturase-1
SRD, sucrose-rich diet
SREBP-1, sterol regulatory element-binding protein-1
TG, triglyceride
UCP1, uncoupling protein 1
VLDL-TG, very low density lipoprotein-triglyceride
ZDF fa/fa, Zucker diabetic obese rats

INTRODUCTION

Dietary proteins and amino acids are important modulators of glucose metabolism and insulin sensitivity. It has been reported that the ingestion of vegetable protein instead of animal protein lowers the risk of coronary artery disease. In recent years, there has been a surge of interest in the potential health-beneficial effects of dietary soy bean protein. Several studies have demonstrated the cholesterol-lowering, antilipogenic, and antihypertensive effects of soy protein. Additional health benefits have been suggested to include antidiabetic effects, reduced weight gain, and improved body composition. This chapter focuses on the effect of dietary soy protein upon the prevention and/or improvement of dyslipidemia and IR in an experimental animal model.

BOTANICAL DESCRIPTION

The origins of the soybean plant are obscure, but many botanists believe it to have derived from glycine Ussuriensis, a legume native to East Asia. Soybean, *Glycine max* L., is an annual plant that varies in growth, habit, and height. Its height ranges from 20 cm to 2 m; however, cultivated crops generally reach a height of around 90 cm. Cultivation is successful in climates with hot summers, with optimum growing conditions in mean temperatures of 20–30°C. Soybean grows in a wide range of soils, with optimum growth in moist alluvial soils with good organic content. The mature seed is rich in protein, contains a significant amount of α-linolenic acid (18:3,n-3), and is a concentrated source of isoflavones. Other constituents include fiber, phytosterols, saponins, water- and fat-soluble vitamins, and minerals (Encyclopaedia Britannica, 2009).

HISTORICAL CULTIVATION AND USAGE

Soybean was a crucial crop in eastern Asia long before written records appeared. It remains a major crop in China, Japan, and Korea. Like some others crops of long domestication, the relationship of the modern soybean to wild-growing species can no longer be traced with certainty. It is a cultural variety, with a large number of cultivars. Prior to fermented products such as soy sauce, tempech, natto, and miso, soy was considered sacred for its use in crop rotation as a method of fixing nitrogen. The cultivation of soybean was long confined chiefly to China, but it gradually spread to other countries (Encyclopaedia Britannica, 2009).

PRESENT-DAY CULTIVATION AND USAGE

Despite the fact that soybean is native to East Asia, only about 45% of soy production is located there, with the other 55% being located in the Americas. The leading producers are the United States, Brazil, Argentina, and Paraguay.

Soybean is a very important source of dietary vegetable oil, and an excellent nutritional and inexpensive source of protein for use as either human food or animal feed. Today, soybeans are used in an increasing number of products. Common forms of soy include soy meal, flour, milk, tofu, textured vegetable protein (which is made into a wide variety of vegetarian foods, some of them intended to imitate meat), soy lecithin, and oils. Other products, such as oils, varnishes, etc., are used in industry. Soybean is the primary source of biodiesel in the USA (Encyclopaedia Britannica, 2009).

APPLICATIONS TO HEALTH PROMOTION AND DISEASE PREVENTION

Insulin resistance is a major key factor in the etiology of the plurimetabolic syndrome cluster disease, which includes type 2 diabetes, dyslipidemia, hypertension, cardiovascular disease, and obesity, among other metabolic disturbances. Dietary factors influence insulin action, and play an important role in the prevention or improvement of these metabolic disorders. In recent years, there has been considerable interest in the effect of dietary soy protein on human health. This is based on accumulated evidence that the addition of soy to the diet, or the substitution of soy protein for animal protein, has been shown to reduce plasma total and LDL cholesterol, and TG concentration, in both humans and laboratory animals. In ZDF fa/fa rats, soy protein consumption reduced the accumulation of cholesterol and TG in the liver, preventing the development of fatty livers. Moreover, the reduction of cholesterol was associated with a low expression of liver X receptor α and its target genes (e.g., 7α-hydroxilase and ATP binding cassette A1). Besides this, soy protein decreases lipogenesis through a decrease in the expression of SREBP-1 and several of its target enzymes. The reduction of hepatic lipids could be the result of an increase in fatty acid oxidation, since soy protein increases PPARα and CPT-1 expression (Torres *et al.*, 2006). Soy protein could also improve glucose homeostasis and insulin sensitivity in SHHF (+cp) rats (Davis *et al.*, 2005). In animal models of obesity, soy protein limits or reduces body fat accumulation and improves insulin resistance (Velasquez & Bhathena, 2007).

There are different experimental non-genetic animal models of diet-induced IR, dyslipidemia, overweight, visceral adiposity, and type 2 diabetes, such as high-fat or high-sucrose/fructose fed rats. Numerous studies, including our own (Lombardo & Chicco 2006), have demonstrated that normal rats fed a high-sucrose (SRD) or -fructose diet for a short period of time (3–5 weeks) develop dyslipidemia, enhanced TG accumulation in liver and heart muscles, hyperinsulinemia, IR, and hypertension. In studies of rats fed a SRD for 4 weeks, Lavigne *et al.* (2000) demonstrated that soy protein lowers plasma fasting glucose and insulin concentration as well as TG levels, and improves peripheral insulin sensitivity, compared to rats that had casein as the source of protein. Moreover, Boimvaser *et al.* (2006) showed a decrease of both liver TG content and VLDL-TG secretion rate.

Pfeuffer and Barth (1992) demonstrated that both the plasma cholesterol levels and the rates of VLDL-cholesterol and TG secretion were lower when soy protein replaced casein in rats fed the SRD for 7 weeks. Hurley *et al.* (1995) showed a decrease in plasma total cholesterol, HDL cholesterol, TG, and the insulin/glucagons ratio, as well as liver total cholesterol and TG concentrations, in rats fed a SRD for 4 weeks in which soybean was the source of protein. In brief, these results suggest that when soy protein replaces casein as a source of protein in the rats' diet, it seems to prevent the onset of dyslipidemia and improved insulin sensitivity in rats fed a SRD. A summary of the effects is presented in Table 125.1.

TABLE 125.1 Main Effects of Dietary Soy Protein in the Prevention of Dyslipidemia and Insulin Resistance Induced by a Sucrose-rich Diet

	SRD Casein	SRD-soy Protein
Plasma		
Glucose	↑[a]	⊥[c]
Insulin	↑[a]	↓[c]
Triglyceride	↑[a]	↓[d]
Cholesterol	↑[b]	↓[d]
HDL cholesterol	↑[b]	↓[d]
Peripheral insulin resistance (clamp)	↑[a]	Improved[c]
IVGT	↑[a]	Improved[c]
Insulin/ glucagon ratio	↑[a]	↓[d]
Liver		
Triglyceride	↑[a]	↓[d,e]
Cholesterol	↑[b]	↓[e]
VLDL-triglyceride secretion	↑[a]	↓, ⊥[e,f]
VLDL-cholesterol secretion	↑[a]	⊥[f]
K2 % min^{-1}	↓[a]	↑[e]

↑, Increase; ↓, decrease; ⊥, normal.
[a]*Lombardo & Chicco (2006);*
[b]*Oliva et al. (2009);*
[c]*Lavigne et al. (2000);*
[d]*Hurley et al. (1995);*
[e]*Boimvaser et al. (2006);*
[f]*Pfeuffer & Barth (1992).*

The studies mentioned above focused on the development of the impairment, but only a few examine the effectiveness of dietary nutrients (e.g., varying the quality rather than the quantity of proteins) on modifying diet-induced IR. One possibility is to examine the effect of soy protein in rats chronically fed a SRD (for up to 8 months instead of 5–7 weeks). In this regard, studies carried out in our group (Lombardo & Chicco, 2006) showed that when the feeding period of the SRD is extended up to 4–8 months, the metabolic and hormonal milieu changes and further deteriorates with the length of time on the diet. A steady state of dyslipidemia, and hyperglycemia without changes in insulinemia was observed. At this point, the sucrose-fed rats were moderately overweight with an increase of visceral adiposity. Islets isolated from these rats showed altered insulin secretion patterns in response to glucose. In addition to ectopic fat deposition in liver, heart, and pancreas tissues, an increase of lipid storage accompanied by an impairment of non-oxidative and oxidative pathways of glucose metabolism was observed in the skeletal muscle of the SRD-fed rats. Interestingly, several of these metabolic abnormalities are also present in the plurimetabolic syndrome in humans (Cheal *et al.*, 2004).

Therefore, we investigated the possible beneficial effects of dietary soy protein on both insulin insensitivity and lipid metabolism in rats fed chronically a SRD in which well-established dyslipidemia and IR were present (4 months) before the source of dietary protein casein was isoenergetically replaced by soy protein from 4 to 8 months.

Using this experimental model, Oliva *et al.* (2009) showed that soy protein reduced liver TG, reverting the hepatic steatosis, and normalizing the VLDL-TG secretion (Figure 125.1). These changes could be possible through mechanisms that included opposite changes in SREBP-1 and PPARα expression leading to a significant decrease of hepatic *de novo* lipogenesis. The addition of soy protein reduced the availability of plasma FFA, which in turn contributed to decreasing the synthesis and secretion of liver VLDL-TG, limiting the formation of the LDL particles and therefore reducing plasma TG and cholesterol levels. Moreover, the reduction of serum cholesterol was associated with a stimulation of bile acid synthesis, possible through the reduction of the enzyme 7α-hydroxilase (Torres *et al.*, 2006). Thus, in the SRD-fed rat model, the

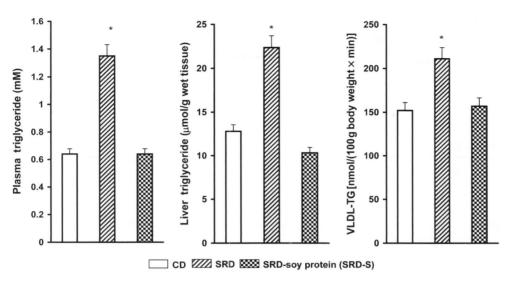

FIGURE 125.1

Plasma and liver TG contents and VLDL-TG secretion rate in rats fed different diets. Mean ± SEM ($n = 6$) , *$P < 0.05$ SRD vs CD and SRD-S.

reversion of dyslipidemia suggests that the principal action of soy protein on hepatic lipid metabolism involves a shift from lipid synthesis and storage to oxidation, and therefore both mechanisms contribute to the observed lipid-lowering effect. However, other mechanisms, such as changes in the insulin to glucagon ratio and the amino acid composition, a rise in thyroid hormone levels, and the presence of lipids, fiber, and isoflavones in the soy protein could also contribute to the hypolipidemic effect (Mezei *et al.*, 2003; Velasquez & Bhathena, 2007).

As shown in Table 125.2, dietary soy protein improved visceral adiposity and decreased body weight gain in rats fed a SRD. The reduction of TG content within the fat cells was accompanied by a decrease of the enzyme activities involved in *de novo* lipogenesis. The mechanisms by which soy protein improves visceral adiposity in the SRD-fed rats are still an open question. However, dietary soy protein reduced the SREBP-1 expression in adipocytes of ZDF fa/fa rats, preventing their hypertrophy (Torres *et al.*, 2006). Besides, Wistar rats fed a soy protein in a high-fat diet gained less weight than those fed casein, partially due to an increase of the thermogenesis capacity of mitochondrial UCP1 (Torre-Villalvazo *et al.*, 2008). Intake of soy protein is associated with increased plasma levels of adiponectin in Wistar rats, suggesting that soy protein may modulate adiponectin production, and therefore enhance insulin sensitivity (Nagasawa *et al.*, 2003).

The administration of soy protein normalized glucose homeostasis without changes in circulating insulin levels, while whole body insulin sensitivity (GIR) was substantially

1057

TABLE 125.2 Body and Adipose Tissue Weight in Rats Fed Different Diets

Diet	Body Weight (g)	Epididymal Fat Pad (g)	Epididymal Fat Pad (g/100 g body weight)
CD	478.6 ± 13.2[a]	8.75 ± 0.78[a]	1.83 ± 0.11[a]
SRD	525.1 ± 10.7[*]	13.94 ± 1.93	2.77 ± 0.14[*]
SRD-soy protein (SRD-S)	492.2 ± 7.2	11.13 ± 0.63	2.39 ± 0.09[**]

[a]Values are mean ± SEM (n = 6),
[*]P < 0.05 SRD vs CD and SRD-S;
[**]P < 0.05 SRD-S vs CD.

TABLE 125.3 Plasma Glucose and Insulin Levels and GIR in Rats Fed Different Diets

Diet	Glucose (mM)	Insulin (μU/ml)	GIR (Euglycemic-Hyperinsulinemic Clamp) (mg glucose/kg per min)
CD	6.00 ± 0.17^{a}	70.2 ± 5.3^{a}	11.48 ± 1.20^{a}
SRD	$8.20 \pm 0.19^{*}$	76.3 ± 7.7	$4.50 \pm 0.50^{*}$
SRD-soy protein(SRD-S)	6.72 ± 0.10	80.4 ± 7.4	$8.48 \pm 0.84^{**}$

[a]Values are mean $\pm$ SEM (n $=$ 6)
[*]$P < 0.05$ SRD vs CD and SRD-S,
[**]$P < 0.05$ SRD-S vs CD.

improved (Table 125.3). Skeletal muscle is the most important tissue for insulin-mediated glucose disposal. Soy protein was able to reverse both the diminished capacity of insulin-stimulated glucose oxidation and disposal in the skeletal muscle of SRD-fed rats (Oliva *et al.*, 2009). Soy protein reduced the spillover of FFA from adipose tissue to non-adipose tissue, decreasing the availability and oxidation of lipid fuel in the muscle, restoring glucose oxidation, and contributing to increased insulin sensitivity in rats fed a SRD. Other studies (Iritani *et al.*, 1997) reported that soybean protein in a high saturated-fat diet increased insulin receptor mRNA in liver and adipose tissue, and decreased IR. Beside, isoflavones, particularly

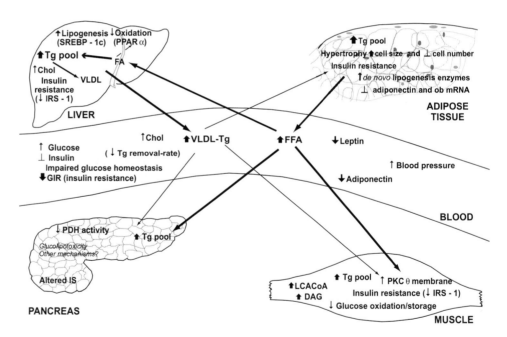

FIGURE 125.2

Effects of long-term sucrose intake on lipid and glucose metabolisms. Chronic intake of a high sucrose diet induces IR in liver, adipose tissue, and skeletal muscle, as well as altered glucose-stimulated IS from isolated perifused β cells. Hepatic TG, cholesterol (Chol), and VLDL-TG secretion are increased. Visceral adiposity is accompanied by increased lipogenic enzyme activities and impaired insulin sensitivity. Fatty acids "spill over" from the hypertrophy fat pad to non-adipose tissue. In liver, the increased flux of plasma fatty acids acts as a substrate for the hepatic TG pool. Increased fatty acid sterification is relatively favored over oxidation, and contributes to the increase of VLDL-TG. The increase of plasma FFA increases TG, LCACoA, and DAG contents in muscle, and enhances the oxidation of lipid fuel as well as stimulating the increase and translocation of the nPKCθ to the cell membrane, which might interfere with the insulin signaling pathway. This contributes to impaired glucose pathways associated with IR. A chronic elevation of both plasma FFA and glucose levels induces an increase of TG pool size while decreasing PDHc activity within the islets and altered IS, which contributes to the demise of β cells, possibly through mechanisms involving glucolipotoxicity. *Adapted from Lombardo & Chicco (2006).*

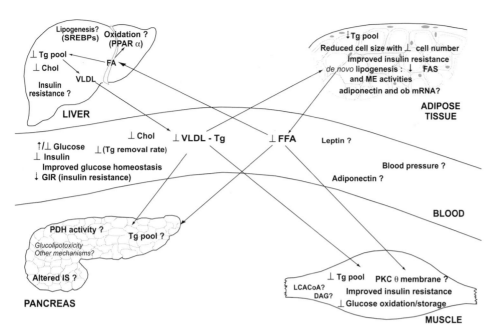

FIGURE 125.3

Possible mechanisms through which soy protein improves or reverses the dyslipidemia and insulin insensitivity induced by a sucrose-rich diet. Soy protein may exert its effects by up-regulating the expression of genes encoding proteins involved in fatty acid oxidation (through PPARs) while simultaneously down-regulating genes encoding proteins of lipid synthesis (through SREBP-1). Another mechanism by which soy protein regulates lipogenesis and Chol homeostasis is through LXRs. The final results of these possible mechanisms in the experimental model are a reduction of hepatic lipogenesis, the TG and Chol pool size, and VLDL-TG secretion, and the normalization of both plasma TG and Chol levels. In adipose tissue, dietary soy protein decreases TG pool size and the activities of enzymes involved in *de novo* lipogenesis possibly through a reduction of the SREBP-1 expression. Soy protein improves the hypertrophy of adipocytes. The smaller adipocytes are more insulin sensitive, and the release of fatty acids is decreased. Therefore, a reduced FFA spillover from adipose tissue to non-adipose tissue is observed. In muscle, soy protein normalizes the TG pool and restores the non-oxidative and oxidative glucose pathway. While lipid content and glucose oxidation in isolated β cells are still an open question, the effects mentioned above could contribute to normalized dyslipidemia and improved glucose homeostasis and IR in this experimental model. *Adapted from Lombardo & Chicco (2006).*

genistein, may favor glucose homeostasis by enhancing IS (Mezei *et al.*, 2003). Differences in the amino acid composition of dietary proteins have also been suggested to mediate protein-dependent changes in glucose and insulin dynamics. A summary of the metabolic changes in the presence of dyslipidemia and IR induced by SRD is depicted in Figures 125.2 and 125.3. Thus, it is possible that all the mechanisms mentioned above, involving the effects of soy protein acting coordinately, might contribute to preventing/improving dyslipidemia, insulin insensitivity, glucose homeostasis, and visceral adiposity experimentally induced in rats fed a SRD, suggesting the role for this macronutrient in the management of these diseases.

ADVERSE EFFECTS AND REACTIONS (ALLERGIES AND TOXICITY)

Foods vary in clinical allergy significance. Although all food proteins have the potential to be allergenic for some people, soybean is cited as one of the eight most frequent human allergenic foods. This group, which accounts for about 90% of food allergies, also includes milk, eggs, fish, shellfish, wheat, peanuts, and three nuts. Some of the major allergen components of soy protein include β-conglycinin, glycinin, soy vacuolar proteins, and Kunitz tripsin inhibitor, and other proteins have been identified in soy-protein sensitive patients.

A substantial body of human clinical and animal model data indicates that soy proteins tend to be less immunologically reactive than many other food proteins. Soy protein allergy occurs only in a minority of children with food allergies, and is relatively uncommon in adults. A meta-analysis of 17 different studies of allergen reactivity in infants and children showed that soy allergy occurs in about 3—4% of subjects, compared to 25% of subjects with an allergy to cow's milk (Cordle, 2004). However, soy protein should be avoided in high risk individuals with food allergies.

SUMMARY POINTS

- Dietary proteins and amino acids modulate glucose metabolism and insulin sensitivity.
- Soybean protein is an excellent vegetable protein which is rich in isoflavones, α-linolenic acid, fiber, and minerals.
- An increasing interest in the health benefits of dietary soybean protein is shown by the many studies that demonstrate, among others things, its antilipogenic and antihypertensive effects.
- Dietary soybean normalizes plasma lipid and glucose levels, reducing visceral adiposity in an experimental model of dyslipidemia and IR.
- Several studies show that soybean is less immunologically reactive than many other food proteins; however, it should be avoided in high risk individuals with food allergies.

ACKNOWLEDGEMENT

The authors acknowledge Lic. S. Boimvaser for her skillful technical assistance.

References

Boimvaser, S., Lombardo, Y. B., & Chicco, A. (2006). Dietary soy protein in the dyslipidemia induced by a sucrose-rich diet. *XVII Reunion Annual SAIB, LI.* 14.

Cheal, K. L., Abbasi, F., Lamendola, C., Mc Laughlin, T., Reaven, G. M., & Ford, E. S. (2004). Relationship to insulin resistance of the adult treatment panel III diagnostic criteria for identification of the metabolic syndrome. *Diabetes, 53,* 1195—1200.

Cordle, C. T. (2004). Soy protein allergy incidence and relative severity. *Journal of Nutrition, 134,* 1213S—1219S.

Davis, J., Steinle, J., Higginbotham, D. A., Oitker, J., Peterson, R. G., & Banz, W. J. (2005). Soy protein influences insulin sensitivity and cardiovascular risk in male lean SHHF rats. *Hormone and Metabolic Research, 37,* 309—315.

Encyclopaedia Britannica. (2009). Soybean. Available at http://www.britannica.com/EBchecked/topic/557184/soybean.

Hurley, C., Galibols, I., & Jacques, H. (1995). Fasting and postprandial lipid and glucose metabolisms are modulated by dietary proteins and carbohydrates: role of plasma insulin concentrations. *Journal of Nutritional Biochemistry, 6,* 540—546.

Iritani, N., Sugimoto, T., Fukuda, H., Komiya, M., & Ikeda, H. (1997). Dietary soy protein increases insulin receptor gene expression in Wistar fatty rats when dietary polyunsaturated fatty acids level is low. *Journal of Nutrition, 127,* 1077—1083.

Lavigne, C., Marette, A., & Jacques, H. (2000). Cod and soy proteins compared with casein improve glucose tolerance and insulin sensitivity in rats. *American Journal of Physiol Endocrinology and Metabolism, 278,* E491—E500.

Lombardo, Y. B., & Chicco, A. G. (2006). Effects of dietary polyunsaturated n-3 fatty acids on dyslipidemia and insulin resistance in rodents and humans. A review. *Journal of Nutritional Biochemistry, 17,* 1—13.

Mezei, O., Banz, W. J., Steger, R. W., Peluso, M. R., Winters, T. A., & Shay, N. (2003). Soy isoflavones exert antidiabetic and hypolipidemic effects through the PPAR pathways in obese Zucker rats and murine raw 264,7 cells. *Journal of Nutrition, 133,* 1238—1243.

Nagasawa, A., Fukui, K., Kojima, M., Kishida, K., Maeda, N., Nagaretani, H., et al. (2003). Divergent effects of soy protein diet on the expression of adipocytokines. *Biochemical and Biophysical Research Communication, 311,* 909—914.

Oliva, M. E., Chicco, A. G., & Lombardo, Y. B. (2009). Soya protein reverses dyslipidemia and the altered capacity of insulin-stimulated glucose utilization in the skeletal muscle of sucrose-rich diet-fed rats. *British Journal of Nutrition, 102,* 60—68.

Pfeuffer, M., & Barth, C. A. (1992). Dietary sucrose but not starch promotes protein-induced differences in rates of VLDL secretion and plasma lipid concentrations in rats. *Journal of Nutrition, 122,* 1582—1586.

Torres, N., Torre-Villalvazo, I., & Tovar, A. R. (2006). Regulation of lipid metabolism by soy protein and its implication in diseases mediated by lipid disorders. *Journal of Nutritional Biochemistry, 17,* 365–373.

Torre-Villalvazo, I., Tovar, A. R., Ramos-Barragán, V. E., Cerbón-Cervantes, M. A., & Torres, N. (2008). Soy protein ameliorates metabolic abnormalities in liver and adipose tissue of rats fed a high fat diet. *Journal of Nutrition, 138,* 462–468.

Velasquez, M. T., & Bhathena, S. J. (2007). Role of dietary soy protein in obesity. *International Journal of Medical Sciences, 4,* 72–82.

CHAPTER 126

Effect of Soybean (*Glycine max* L.) on Hot Flashes, Blood Pressure, and Inflammation

Francine K. Welty, Amna Ali, Nguyen Nguyen, Sunny Jhamnani
Beth Israel Deaconess Medical Center, Division of Cardiology, Boston, Massachusetts, USA

CHAPTER OUTLINE

LIST OF ABBREVIATIONS

apo, apolipoprotein
BP, blood pressure
BMI, body mass index
C, cholesterol
CRP, C-reactive protein
HDL, high density lipoprotein
HRT, hormone replacement therapy
IL-6, interleukin-6

Nuts & Seeds in Health and Disease Prevention. DOI: 10.1016/B978-0-12-375688-6.10126-4

LDLC, low density lipoprotein cholesterol
MMP-9, matrix metalloproteinase-9
NCEP, National Cholesterol Education Program
sICAM-1, soluble intercellular adhesion molecule-1
sVCAM-1, soluble vascular cell adhesion molecule-1
TLC, therapeutic lifestyle changes

INTRODUCTION

Soy foods have been consumed for centuries in Asian countries. Many potential benefits have been linked to the intake of soy products, according to epidemiological investigations. In the present chapter we summarize the effect of soy nuts on hot flashes also known as "hot flushes" and menopausal symptoms, including the effect on blood pressure and inflammation. Note that the terms "soy" and "soya" are used interchangeably; however, the term "soy" is used throughout this chapter.

BOTANICAL DESCRIPTION, AND HISTORICAL CULTIVATION AND USAGE

The soybean, *Glycine max* L., is a subtropical, annual dicot, and member of the pea family (Fabaceae). Soybeans are native to eastern Asia, where they were first cultivated in northeastern China around 11 BC. In the following millennium, soybean cultivation spread to Japan, Korea, Malaysia, and other parts of Asia, accompanied by various processing techniques for human consumption. These traditional techniques produced two classes of soyfoods: fermented soyfoods (such as tempeh, miso, and soysauce) and non-fermented soyfoods (such as fresh green soybeans, dry soybeans, soymilk, and tofu) (Rosengarten, 2004).

PRESENT-DAY CULTIVATION AND USAGE

The adoption of soybean in Western diets, particularly in the United States, has grown steadily since the 1980s. First cultivated as a forage crop, soybean has increased in commercial status as an oilseed crop. With mechanized harvest methods, crop production has expanded, making the United States the world's largest producer of soybean. Processing of soybeans for soy protein has produced a number of products containing soy protein, including meat alternatives, baked goods, soups, meat, poultry, beverages, soups, sauces, confections, infant products, and snacks. Of interest is the whole dry soybean, referred to as a soy nut. With physical properties, taste, and texture similar to those of most tree nuts, soy nuts are a versatile tree nut alternative (Rosengarten, 2004).

APPLICATIONS TO HEALTH PROMOTION AND DISEASE PREVENTION

The soybean is a versatile legume that contains high-quality protein and minimal saturated fat, and is an essential unique dietary source of isoflavones (Messina, 1999) — a group of diphenolic compounds classified as phytoestrogens. Isoflavones bind to both estrogen receptors, although preferentially to estrogen receptor beta (Kuiper *et al.*, 1998), and in people who regularly consume soy foods, serum isoflavone levels reach the low micromolar range. In soybeans, the three isoflavones genistein, diadzein, and glycitein comprise approximately 50, 40, and 10% of the total isoflavone content, respectively (Table 126.1). Soy has received attention as an alternative to conventional hormone replacement therapy (HRT) largely because of these isoflavones. The estrogen-like effects of isoflavones in combination with low reported frequency of hot flashes has prompted the investigation of the effect of soy on menopausal symptoms. The prevalence of hot flashes is much lower in Asian than in Western

TABLE 126.1 Nutritional Composition of One-half Cup of Soy Nuts (56 g)

Ingredient	Content
Calories	240
Total fat, g	8
Saturated fat, g	0
trans fat, g	0
Polyunsaturated fat, g	6
Monounsaturated fat, g	2
Cholesterol, mg	0
Total carbohydrate, g	18
Dietary fiber, g	10
Sugars, g	6
Protein, g	24
Sodium, mg	20
Potassium, mg	780
Calcium, mg	236
Aglycone isoflavones, mg	
Genistein	61
Daidzein	30
Glycitein	10

countries; 10–25% of Chinese women and 10%–20% of Indonesian women have hot flashes, compared with 58%–93% of Western women (Krokenberg, 1999). These findings are thought, at least in part, to be related to consumption of soy phytoestrogen-rich diet (Beaglehole, 1990; Tham *et al.*, 1998). The average daily soybean intake is 17–36 g in Japan, Korea, Taiwan, and Indonesia, compared with 4 g in the United States (Shu *et al.*, 2001).

For all these reasons, Welty *et al.* (2007a) conducted a trial to examine the effect of soy nuts on hot flashes, and cardiovascular risk factors and biomarkers (blood pressure, lipid levels, inflammatory markers and adhesion molecules) of atherosclerosis, in postmenopausal women. The study was a randomized, controlled, crossover trial of the effect of one-half cup of soy nuts daily for 8 weeks on menopausal symptoms and cardiovascular biomarkers (blood pressure, lipid levels) in hypertensive, prehypertensive, and normotensive postmenopausal women. After a 4-week diet run-in, participants adherent to the therapeutic lifestyle changes (TLC) diet (currently recommended by the Adult Treatment Panel of the National Cholesterol Education Program (NCEP, 2001) to lower the risk of coronary heart disease) were randomized in a crossover design between two diet sequences for 8-week periods: the TLC diet without soy, or the TLC diet with prepackaged daily allowances of one-half cup of unsalted soy nuts (Genisoy, Fairfield, California) containing 25 g of soy protein and 101 mg of aglycone isoflavones (Table 126.1) divided into three or four portions spaced throughout the day. A registered dietitian instructed the participants to eat a TLC diet, which consisted of 30% of energy from total fat ($\leq$ 7% saturated fat, 12% monounsaturated fat, and 11% polyunsaturated fat), 15% from protein, and 55% from carbohydrate; less than 200 mg of cholesterol per day (NCEP, 2001); and 1200 mg of calcium, and two fatty fish meals per week. Those ingesting suboptimal dietary calcium were given calcium carbonate supplementation. After a 4-week washout on the TLC diet alone, participants crossed over to the other arm for an additional 8 weeks. Participants were individually advised by a registered dietitian from which sources to decrease their protein intake to compensate for the 25 g of soy protein in the soy diet arm, to keep protein amounts similar on both diet arms. At the end of each 8-week period, fasting blood was collected for measurement of lipid levels, and participants collected a 24-hour urine sample for measurement of isoflavone and creatinine levels. The subjects also completed the menopausal symptom quality of life questionnaire, which included vasomotor, psychosocial, physical, and sexual scores (Hillditch *et al.*, 1996).

Dry-roasted soy beans (soy nuts) were used for the study. The reason for choosing this form of soy was that since the ratio of isoflavones to protein is the highest in unprocessed soybeans and nuts, this reduced the amount of soy protein in the diet necessary to obtain a high level of isoflavones and make it feasible for people unaccustomed to ingesting high amounts of soy to incorporate adequate soy protein and isoflavones in their diet. Secondly, it was done to mimic the Asian diet, where people do not consume isolated soy protein, but rather eat native soy foods and fermented soy foods several times throughout the day. Thirdly, soy nuts do not require refrigeration or any processing before consumption, which makes them portable and ideal for quick consumption.

Participants for this study were recruited with the following inclusion criteria: women with absence of menses for at least 12 months, or irregular periods, and hot flashes. Exclusion criteria were current cigarette smoking or smoking in the previous year, clinical coronary artery disease, peripheral artery disease, or cerebrovascular disease; known diabetes mellitus or a fasting glucose of 126 mg/dl or greater (> 7.0 mmol/l); a history of breast cancer; a fasting triglyceride level greater than 400 mg/dl (> 4.52 mmol/l); systolic BP of 165 mm Hg or greater or diastolic BP of 100 mm Hg or greater; untreated hypothyroidism; systemic or endocrine disease known to affect lipid, mineral, or bone metabolism; and consumption of more than 21 alcoholic drinks per week. Use of lipid-lowering drugs, hormone therapy, medications for osteoporosis, and soy products was discontinued for 2 months before entering the study.

Hot flash and menopausal symptom analysis

Subjects recorded the number of hot flashes daily on monthly calendars. These were averaged over each 8-week diet arm —TLC with soy nuts (soy), and TLC without soy nuts (control). The baseline number of hot flashes was the average number of hot flashes/day during the control diet arm. Women were divided into low (< 4.5 hot flashes/day) and high (> 4.5 flashes/day) hot-flash groups, based on the average number of hot flashes/day during the control diet arm (baseline). At the end of each 8-week period, subjects also completed the menopause-specific quality of life questionnaire, which includes vasomotor, psychosocial, physical, and sexual scores. Table 126.2 shows the baseline characteristics by hot flash frequency. Women with > 4.5 hot flashes/day had a significantly lower body mass index (BMI) than women with ≤ 4.5 hot flashes/day; otherwise, there were no significant differences.

The average number of hot flashes during the first week in the control diet was 22.3 ± 24.9 compared with 23.3 ± 23.5 during the last week of the control diet. Therefore, hot flashes did

TABLE 126.2 Baseline Characteristics by Hot Flash Status

Characteristic	≤4.5 Hot Flashes	>4.5 Hot Flashes	*P* value
Number of subjects	24	15	
Age, Years	51.9 ± 5.5^a	54.6 ± 5.4	0.14
BMI (kg/m^2)	27.9 ± 5.5	23.9 ± 4.0	0.023
Years postmenopause	3.8 ± 4.2	6.1 ± 5.7	0.16
Total C (mg/dL)	227 ± 50	222 ± 30	0.70
LDL-C (mg/dL)	140 ± 41	136 ± 27	0.63
HDL-C (mg/dL)	58 ± 12	57 ± 16	0.94
Triglyceride (mg/dL)	135 ± 78	132 ± 104	0.91

BMI, body mass index (calculated as weight in kilograms divided by the square of height in meters); C, cholesterol; LDL, low-density lipoprotein; HDL, high-density lipoprotein.
From Welty *et al.* (2007b), *J. Women's Health*, 16, 361–369.
^{a}Mean ± SD.

not decrease in the control group. Compared with the control diet arm, soy nut ingestion was associated with a 45% decrease in the hot flashes in women with > 4.5 hot flashes/day at baseline (7.5 ± 3.6 vs 4.1 ± 2.6, respectively, $P < 0.001$) and a 41% decrease in those with ≤ 4.5 flashes/day (2.2 ± 1.2 vs 1.3 ± 1.1, respectively, $P < 0.001$) (Figure 126.1). Both equol producers and non-producers had similar reductions in hot flashes with soy nuts compared with the control diet (44%, $P < 0.017$ for both groups). The reduction in hot flashes was apparent at 2 weeks of treatment in both groups (53% and 54%, respectively) (Figure 126.1). There appeared to be some attenuation in the benefit of soy nuts over time in the low hot-flash group (Figure 126.1). Also, compared to the control arm, soy nut ingestion was associated with significant improvement in scores on the menopausal symptom quality of life questionnaire; 19% decrease in vasomotor score ($P = 0.004$), 12.9% reduction in psychosocial score ($P = 0.001$), and 9.7% decrease in physical score ($P = 0.045$) after 8 weeks of ingestion. There was a trend towards improvement in the sexual score, with a 17.7% reduction in symptoms ($P = 0.129$).

In summary, in this randomized, controlled, crossover trial in postmenopausal women, ingestion of soy nuts containing 25 g soy protein and 101 mg aglycone isoflavones was associated with a decrease in hot flashes and menopausal symptoms in postmenopausal women compared with the TLC diet without soy. Similar reductions were seen in women experiencing > 4.5 hot flashes per day and ≤ 4.5 hot flashes per day. Therefore, the initial baseline frequency of hot flashes had no effect on efficacy in the current study. This benefit may be of interest to symptomatic postmenopausal women, as estrogen/progestin replacement increased the risk of MI, stroke, breast cancer, and dementia, and estrogen alone increased the risk of stroke and dementia, in clinical trials (Welty, 2003; Anderson *et al.*, 2004).

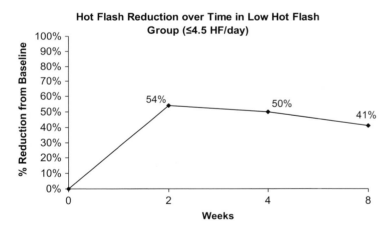

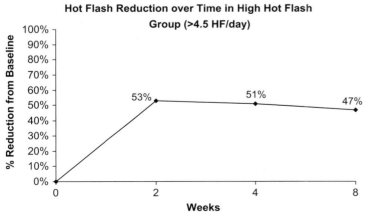

FIGURE 126.1

Percent reduction from baseline in hot flashes after 2, 4, and 8 weeks of treatment in soy diet arm compared with control arm in women with < 4.5 hot flashes/day (top) and > 4.5 hot flashes/day (bottom). *From Welty* et al. *(2007b)*, J. Women's Health, 16, *361–369*.

Prior studies have shown variable reductions in menopausal symptoms with soy preparations (reviewed in Welty *et al.*, 2007b). Ten studies have compared soy foods, soy beverage, or soy powder with placebo or control. Of eight trials examining hot flash frequency, seven observed no improvement compared with control. Nine trials compared soy-derived isoflavones in capsule or tablet form with placebo. Of five published trials examining hot flash frequency, two observed a significant reduction in hot flashes on soy compared with placebo, whereas three found no difference. Soy nut ingestion was also associated with a decrease in menopausal symptoms, as measured by the menopause-specific quality of life questionnaire, in the current study. Seven prior trials have reported symptom outcome, with each trial using a different questionnaire. None of the trials observed a significant improvement in overall score. Only one trial observed improvement in subscale scores on the soy diet compared with placebo: 20 g soy protein with 34 mg isoflavones twice daily improved hot flash severity compared with carbohydrate placebo ($P < 0.001$). We may have achieved greater relief from hot flashes and menopausal symptoms relative to other trials, most of which used isoflavone tablets or isolated soy protein products, because we used a native form of soy, which has all active ingredients and a high ratio of isoflavones/protein. Furthermore, the consumption of native soy at three or four times throughout the day in the current study (done to mimic Asian populations, who consume soy throughout the day) may have kept isoflavone levels more constant, and may be another reason for significant menopausal symptom relief in the current study.

Effect of soy on blood pressure, lipid levels and inflammation in hypertensive, prehypertensive, and normotensive postmenopausal women

This trial also examined the effect of soy on systolic and diastolic BP and lipid levels on all 60 postmenopausal women (Welty *et al.*, 2007a). Women were seated quietly for at least 5 minutes in a chair at 60 to 85°, with feet on the floor and the right arm supported at heart level. Measurements of BP were performed using cycling Dinamaps (GE Medical Systems Information Technologies, Inc., Milwaukee, Wisconsin) at the end of each diet period. Two readings were obtained at the beginning of each visit, at least 30 seconds apart, as in the Treatment of Mild Hypertension Study. The hypertensive group of women was defined by a systolic BP of 140 mm Hg or greater, or diastolic BP of 90 mm Hg or greater. The primary analysis defined normotensive women as all those with a systolic BP less than 140 mm Hg and a diastolic BP less than 90 mm Hg. Normotensive women were then further divided into two groups: one with prehypertension (systolic BP of 120−139 mm Hg and diastolic BP of 80−89 mm Hg) and the second with systolic BP less than 120 mm Hg.

The results showed that soy nut supplementation significantly reduced systolic and diastolic BP in all 12 hypertensive women, and in 40 of the 48 normotensive women. Hypertensive women (mean baseline systolic BP of 152 mm Hg and diastolic BP of 88 mm Hg) had a mean 9.9% decrease in systolic BP ($P = 0.003$) and a 6.8% decrease in diastolic BP ($P = 0.001$) on the soy diet compared with the control diet (Table 126.3). Normotensive women (mean systolic BP of 116 mm Hg) had a 5.2% decrease in systolic BP ($P < 0.001$) and a 2.9% decrease in diastolic BP ($P = 0.02$) on the soy diet compared with the control diet (Table 126.3). There was no change in the BMI (calculated as weight in kilograms divided by the square of height in meters) or exercise on the soy diet vs the control diet for the normotensive and hypertensive women; therefore, neither weight change nor exercise level accounts for the lower BP in the soy diet arm. When normotensive women were subdivided into prehypertensive and normotensive women, both groups had significant reductions in systolic BP, and there was a trend toward a reduction in diastolic BP in the soy diet arm compared with the control diet arm (Table 126.4).

TABLE 126.3 Blood Pressure, BMI and Exercise at the End of Each Diet Period in Normotensive and Hypertensive Women[*]

Variable	Normotensive Women (n=48)				Hypertensive Women (n=12)			
	Control Diet	Soy Diet	Change, %	*P* Value	Control Diet	Soy Diet	Change, %	*P* Value
Systolic BP, mm Hg	116 ± 10	110 ± 11	5.2↓	<0.001	152 ± 12	137 ± 15	9.9↓	0.003
Diastolic BP, mm Hg	69 ± 8	67 ± 7	2.9↓	0.02	88 ± 7	82 ± 8	6.8↓	0.001
BMI	25.7 ± 5.1	25.7 ± 5.1	None	0.26	28.0 ± 4.3	27.7 ± 4.3	1.0↓	0.14
Exercise, min/wk	156 ± 118	143 ± 120	NA	0.09	127 ± 120	152 ± 128	NA	0.08
Exercise, d/wk	4.0 ± 2.3	3.7 ± 2.4	NA	0.14	3.2 ± 2.6	3.1 ± 2.5	NA	0.38

BMI, body mass index (calculated as weight in kilograms divided by the square of height in meters); BP, blood pressure; NA, not applicable; ↓, decrease in percentage.
Reprinted from Welty et al. (2007a), Arch. Int. Med., 167, 1060–1067, with permission. ©2007 American Medical Association. All rights reserved.
[*]Values are given as mean ± SD.

TABLE 126.4 Change in BP at the End of Each Diet Period in Normotensive Women Subdivided into Prehypertensive and Normotensive Groups[*]

Variable	Normotensive Women (n=31)				Hypertensive Women (n=17)			
	Control Diet	Soy Diet	Change, %	*P* Value	Control Diet	Soy Diet	Change, %	*P* Value
Systolic BP, mm Hg	110 ± 7	105 ± 8	4.5↓	0.003	127 ± 5	120 ± 9	5.5↓	0.003
Diastolic BP, mm Hg	67 ± 7	65 ± 7	3.0↓	0.06	73 ± 8	71 ± 8	2.7↓	0.18
BMI	24.9 ± 4.7	25.0 ± 4.6	None	0.21	27.2 ± 5.6	27.2 ± 5.7	None	0.95
Exercise, min/wk	170 ± 153	156 ± 144	8.2↓	0.19	163 ± 115	144 ± 114	11.7↓	0.07
Exercise, d/wk	3.8 ± 2.3	3.5 ± 2.3	7.9↓	0.31	4.5 ± 2.4	4.1 ± 2.5	8.9↓	0.20

BMI, body mass index (calculated as weight in kilograms divided by the square of height in meters); BP, blood pressure; ↓, decrease in percentage.
Reprinted from Welty et al. (2007a), Arch. Int. Med., 167, 1060–1067, with permission. ©2007 American Medical Association. All rights reserved.
[*]Values are given as mean ± SD.

Another major finding in the present study was the reduction of LDLC and apoB with soy in hypertensive women (mean LDLC level of 164 mg/dl [4.25 mmol/l]) in the absence of a change in weight or exercise. Soy nut supplementation lowered low density lipoprotein cholesterol and apolipoprotein B levels 11 and 8% ($P = 0.04$ for both), respectively, in hypertensive women, but had no effect in normotensive women (Table 126.5). There were no changes in glucose level.

To study the mechanisms by which soy nuts mediated reductions in BP, markers of inflammation were measured, including soluble vascular cell adhesion molecule-1 (sVCAM-1), soluble intercellular adhesion molecule-1 (sICAM-1), C-reactive protein (CRP), interleukin-6 (IL-6), and matrix metalloproteinase-9 (MMP-9). Baseline levels of sVCAM-1, sICAM-1, CRP, IL-6, and MMP-9 were not significantly different between women with hypertension and normotensive women (Nasca et al., 2008). Compared with the therapeutic lifestyle change diet alone, levels of sVCAM-1 were significantly lower on the soy diet in women with hypertension (623.6 ± 153.8 vs 553.8 ± 114.4 ng/ml, respectively, $P = 0.003$), whereas no significant differences were observed in normotensive women. This reduction in sVCAM-1 may reflect an overall improvement in the underlying inflammatory process that is now known to underlie the atherosclerotic process. Diets high in saturated fat or glucose increase inflammation and adhesion molecules, and lead to insulin resistance – findings proposed to increase BP (Delichatsios & Welty, 2005). Therefore, the study hypothesized

TABLE 126.5 Lipid and Glucose Results at the End of Each Diet Period in Normotensive and Hypertensive Women[*]

Variable	Normotensive Women				Hypertensive Women			
	Control Diet	Soy Diet	Change, %	P Value	Control Diet	Soy Diet	Change, %	P Value
Total-C, mg/dL	228 ± 39	224 ± 36	1.7↓	0.28	248 ± 62	230 ± 45	7.5↓	0.08
LDL-C, mg/dL	143 ± 32	142 ± 31	0.7↓	0.55	164 ± 57	146 ± 46	11.0↓	0.04
HDL-C, mg/dL	58 ± 15	59 ± 14	1.7↑	0.21	56 ± 10	56 ± 11	None	0.73
Triglycerides, mg/dL	128 ± 97	119 ± 83	7.0↓	0.31	128 ± 74	114 ± 50	10.9↓	0.29
Apolipoprotein B, mg/dL	111 ± 19	108 ± 24	2.7↓	0.25	126 ± 37	116 ± 32	8.0↓	0.04
Glucose, mg/dL	98 ± 10	97 ± 91.0	1.0↓	0.30	97 ± 11	95 ± 12	2.1↓	0.27

C, cholesterol; HDL, high density lipoprotein; LDL, low density lipoprotein; ↓, decrease in percentage. SI conversion factors: to convert cholesterol to millimoles per liter, multiply by 0.0259; triglycerides to millimoles per liter, multiply by 0.0113; glucose to millimoles per liter, multiply by 0.0555.
Reprinted from Welty et al. (2007a), Arch. Int. Med., 167, 1060–1067, with permission. ©2007 American Medical Association. All rights reserved.
[*]Values are given as mean ± SD.

that a whole soy food diet may lower BP by improving inflammation and thereby improving endothelial function.

ADVERSE EFFECTS AND REACTIONS (ALLERGIES AND TOXICITY)
Allergies

Soy is a common food causing allergy (Asthma and Allergy Foundation of America, 2005), with manifestations including urticaria, itching, hives, rash, diarrhea, dyspnea, diaphoresis, and, rarely, anaphylaxis.

In the study by Welty and colleagues (2007b), of the 82 subjects who were enrolled in the study, 22 (25%) dropped out. Reasons for dropouts included flatulence and bloating (three), constipation (one), teeth problems (one), intolerable hot flashes when off HRT or soy (four), not wanting to be off soy in control arm (one), and weight gain (one).

Potential toxicities
WOMEN

As noted above, soy contains isoflavones, which are a class of phytoestrogens that are similar to estradiol. Since isoflavones structurally resemble estrogen and bind to estrogen receptors, probably the most significant potential toxicity in women is the effect of soy isoflavones on the breast and uterus. The fact that Asian women, who consume daily soy, have the lowest rates of breast and uterine cancer in the world suggests that chronic ingestion of soy does not increase the risk of hormonally dependent cancers in women. In fact, epidemiological studies have found that high soy intake during childhood, adolescence, or early adulthood may lessen the risk of breast cancer in later life (Shu et al., 2001; Wu et al., 2002). A study carried out to estimate the relation between isoflavone supplementation and breast biopsies or mammography densities in humans revealed re-assuring results (Maskarinec, et al., 2009; Messina & Wu, 2009). Nevertheless, soy is not recommended for women with breast cancer until further trials have been performed.

INFANTS

The best mode of feeding infants during the first 6 months is breast feeding, but when this is not possible, or the infant does not tolerate human or cow's milk, soy has been found to be a safe feeding option which is also nutritionally complete. This is supported by an ongoing study in children fed with soy formula, which has been shown to have no adverse effects after 5 years of study (Badger et al., 2009).

THYROID GLAND

Soy contains goitrogens, which can potentially lead to goiter and interfere with iodine kinetics in relation to thyroid hormones, possibly to the extent of causing hypothyroidism; however, as long as iodine intake is sufficient, soy ingestion should not cause thyroid function anomalies

SUMMARY POINTS

- The soybean, *Glycine max*, is a subtropical, annual dicot and member of the pea family (Fabaceae), native to eastern Asia.
- The prevalence of hot flashes and hormonally dependent cancers are lower in countries consuming soy than in Western countries.
- Soy nuts (25 g soy protein and 101 mg aglycone isoflavones) consumed three or four times throughout the day lowered hot flashes by approximately 45%, and improved scores on the menopause-specific quality of life questionnaire.
- Soy nuts also decreased systolic and diastolic blood pressures in hypertensive, prehypertensive, and normotensive women.
- Soy nuts lowered levels of LDLC and apoB in hypertensive, hyperlipidemic women.
- Soy nuts also lowered levels of inflammation.
- The main adverse effects of soy include gastrointestinal complaints.
- Soy is not recommended for women with breast cancer until further trials have been completed.

References

Anderson, G. L., Limacher, M., Assaf, A. R., Bassford, T., Beresford, S. A., Black, H., et al. (2004). Women's Health Initiative Steering Committee. Effects of conjugated equine estrogen in postmenopausal women with hysterectomy: the Women's Health Initiative randomized controlled trial. *Journal of American Medical Association, 291*, 1701.

Asthma and Allergy Foundation of America. (2005). *Soy allergy* (http://www.afa.org).

Badger, T. M., Gilchrist, J. M., Pivik, R. T., Andres, A., Shankar, K., Chen, J. R., et al. (2009). The health implications of soy infant formula. *American Journal of Clinical Nutrition, 5*, 1668S–1672S.

Beaglehole, R. (1990). International trends in coronary heart disease mortality, morbidity, and risk factors. *Epidemiology Reviews, 12*, 1.

Delichatsios, H. K., & Welty, F. K. (2005). Influence of the DASH diet and other low-fat, high carbohydrate diets on blood pressure. *Current Atherosclerosis Reports, 7*, 446–454.

Hillditch, J. R., Lewis, J., Peter, A., van Maris, B., Ross, A., Franssen, E., et al. (1996). A menopause-specific quality of life questionnaire: Development and psychometric properties. *Maturitas, 24*, 161.

Krokenberg, F. (1999). Hot flashes. In R. A. Lobo (Ed.), *Treatment of the postmenopausal woman. Basic and clinical aspects*, p. 159. Philadelphia, PA: Lippincott Williams & Wilkins.

Kuiper, G. G., Lemmen, J. G., Carlsson, B., Corton, J. C., Safe, S. H., van der Saag, P. T., et al. (1998). Interaction of estrogenic chemicals and phytoestrogens with estrogen receptor beta. *Endocrinology, 139*, 4252–4263.

Maskarinec, G., Verheus, M., Steinberg, F. M., Amato, P., Cramer, M. K., Lewis, R. D., et al. (2009). Various doses of soy isoflavones do not modify mammographic density in postmenopausal women. *Journal of Nutrition, 5*, 981–986.

Messina, M., & Wu, A. H. (2009). Perspectives on the soy–breast cancer relation. *American Journal of Clinical Nutrition, 5*, 1673S–1679S.

Messina, M. J. (1999). Legumes and soybeans: overview of their nutritional profiles and health effects. *American Journal of Clinical Nutrition, 70*(Suppl), S439–S450.

Nasca, M. M., Zhou, J. R., & Welty, F. K. (2008). Effect of soy nuts on adhesion molecules and markers of inflammation in hypertensive and normotensive postmenopausal women. *American Journal of Cardiology, 102*(1), 84–86.

NCEP. (2001). Expert Panel on Detection, Evaluation and Treatment of High Blood Cholesterol in Adults. Executive Summary of the Third Report of the National Cholesterol Education Program (NCEP) Expert Panel on Detection, Evaluation, and Treatment of High Blood Cholesterol in adults (Adult Treatment Panel III). *Journal of the American Medical Association, 285*, 2486–2497.

Rosengarten, F., Jr. (2004). Soybeans. In *The book of edible nuts, Vol. 1* (pp. 324–326). New York, NY: Dover Publications.

Shu, X. O., Jin, F., Dai, Q., Wen, W., Potter, J. D., Kushi, L. H., et al. (2001). Soyfood intake during adolescence and subsequent risk of breast cancer among Chinese women. *Cancer Epidemiology Biomarkers Prevention, 10*, 483.

Tham, D. M., Gardner, C. D., & Haskell., W. L. (1998). Potential health benefits of dietary phytoestrogens: a review of the clinical, epidemiological, and mechanistic evidence. *Journal of Clinical Endocrinology & Metabolism, 83*, 2223.

Welty, F. K. (2003). Alternative hormone replacement regimens: is there a need for further clinical trials? *Current Opinion in Lipidology, 14*, 585.

Welty, F. K., Lee, K. S., Lew, N. S., & Zhou, J. R. (2007a). Effect of soy nuts on blood pressure and lipid levels in hypertensive, prehypertensive and normotensive postmenopausal women. *Archives of Internal Medicine, 167*, 1060–1067.

Welty, F. K., Lee, K. S., Lew, N. S., Nasca, M., & Zhou, J.-R. (2007b). The association between soy nut consumption and decreased menopausal symptoms. *Journal of Women's Health, 16*, 361–369.

Wu, A. H., Wan, P., Hankin, J., Tseng, C. C., Yu, M. C., & Pike, M. C. (2002). Adolescent and adult soy intake and risk of breast cancer in Asian-Americans. *Carcinogenesis, 9*, 1491–1496.

CHAPTER 127

Antifungal and Antiproliferative Activity of Spotted Bean (*Phaseolus vulgaris* cv.)

Tzi-Bun Ng[1], Jack H. Wong[1], Randy C.F. Cheung[1], Sze-Kwan Lam[1], He-Xiang Wang[2], Xiujuan Ye[1], Patrick H.K. Ngai[3], Evandro F. Fang[1], Yau-Sang Chan[1]

[1] School of Biomedical Sciences, Faculty of Medicine, The Chinese University of Hong Kong, Hong Kong, China
[2] State Key Laboratory of Agrobiotechnology, Department of Microbiology, China Agricultural University, Beijing, China
[3] School of Life Sciences, Faculty of Science, The Chinese University of Hong Kong, Hong Kong, China

INTRODUCTION

Antifungal proteins are produced by a diversity of organisms, including invertebrates (Zhang & Zhu, 2009), vertebrates (Raj & Dentino, 2002), flowering plants (Vigers *et al.*, 1991), fungi (Guo *et al.*, 2005) and bacteria (Wong *et al.*, 2008), to combat fungi.

Leguminous plants produce an array of antifungal proteins comprising chitinase-like proteins, chitinases, chitin-binding proteins, lipid transfer proteins, thaumatin-like proteins, glucanases, ribosome inactivating proteins, nucleases, protease inhibitors, cyclophilin-like proteins, miraculin-like proteins, lectins, and defensins (Selitrennikoff, 2001; Ng, 2004).

Spotted bean, a variety of *Phaseolus vulgaris* that is brownish black with red spots, produces an antifungal protein with antiproliferative activity.

Nuts & Seeds in Health and Disease Prevention. DOI: 10.1016/B978-0-12-375688-6.10127-6

BOTANICAL DESCRIPTION

Phaseolus vulgaris is a herbaceous annual leguminous plant. Bush cultivars produce erect bushes 20–60 cm high. Pole or running cultivars have vines 2–3 m in length. All cultivars bear alternate, green or purple leaves, with three oval leaflets (each 6–15 cm long and 3–11 cm wide). Flowers are white, pink, or purple, about 1 cm long. Pods are 8–20 cm long , 1–1.5 cm wide, black, green, purple or yellow, each with four to six beans that are smooth, plump, kidney-shaped, up to 1.5 cm long, and differ in color, often being mottled with two or more colors.

HISTORICAL CULTIVATION AND USAGE

By 5500 BC, beans had been independently domesticated in Mexico and northern Peru. By the 1400s, beans had become a staple crop in the Americas, and by the mid-1500s, cultivars of *P. vulgaris* had made their appearance in Europe. Beans were probably introduced into Africa by the Portuguese, and spread into the interior faster than European exploration.

Beans are often stored dry, and then rehydrated and cooked as vegetables. Protease inhibitors and lectins, which may adversely affect digestion and intestinal absorption of nutrients, are present in the raw beans. However, these are thermolabile, and inactivated by cooking.

PRESENT-DAY CULTIVATION AND USAGE

Beans need warm, sunny days and well-irrigated, light soil with abundant moisture and a pH of 5.5–6.5 for growth. Beans are cultivated in temperate and subtropical zones, and in mountain valleys in the tropics. Beans grow well with beet, carrots, cauliflowers, celeriac, cucumbers, cabbages, leeks, and strawberries, but are inhibited by alliums and fennel. Beans are symbiotic with nitrogen-fixing bacteria in the root nodules. Part of this nitrogen is utilized by the growing plant, but some can also be used by other plants in the vicinity. At the end of the growing season, the parts of the plant above the ground are removed while the roots are left in the soil to decay and release nitrogen.

APPLICATIONS TO HEALTH PROMOTION AND DISEASE PREVENTION

African and Latin American farmers have adequate dietary carbohydrates (from cassava, maize, rice, wheat, etc.), but are protein-deficient. Dietary proteins can be acquired from eggs, milk, and meat, but are often derived from beans. Legumes are abundant in proteins. Dietary consumption of common beans (*Phaseolus vulgaris* L.) surpasses that of other legume species. About half of the legumes consumed worldwide are common beans in some countries, such as Mexico and Brazil, and those around the Great Lakes in eastern Africa. Beans are the main source of protein in human diets.

Phaseolus vulgaris beans contain polysaccharides, which can undergo fermentation in the colon to form short-chained fatty acids, with effects on health. The beans have been shown to inhibit mammary cancer formation in the rat, and oxidation of low-density lipoprotein induced by copper.

The green pods exert a mildly diuretic action, and have hypoglycemic activity. The seeds display diuretic, hypoglycemic, and hypotensive activities. The seed flour is used in the treatment of ulcers, and cancer of the blood. When boiled with garlic, they can be used to treat intractable coughs. The root has a narcotic effect. A homeopathic remedy prepared from the entire fresh herb is employed in the treatment of arthritis, rheumatism, and urinary tract diseases.

A 7.3-kDa defensin-like antifungal peptide was purified from *Phaseolus vulgaris* cv. "Spotted Bean." The purification protocol entailed anion exchange chromatography on diethyl-aminoethyl (DEAE) cellulose, affinity chromatography on Affi-gel blue gel, cation exchange chromatography on SP-Sepharose, and gel filtration by fast protein liquid chromatography on Superdex 75 (Wang & Ng, 2007). The bean extract was fractionated by anion exchange chromatography on DEAE-cellulose into three fractions: an unbound fraction, D1, and two bound fractions, D2 and D3. Antifungal activity was confined to fraction D1. When fraction D1 was resolved by affinity chromatography on Affi-gel blue gel, antifungal activity appeared in the first bound fraction B2. Cation exchange chromatography of fraction B2 on SP-Sepharose fractionated it into a broad unbound fraction S1 and three sharp bound fractions (S2, S3, and S4). Antifungal activity was found in fraction S2. Gel filtration of S2 on Superdex 75 yielded three fractions, SU1, SU2, and SU3. Antifungal activity resided in the last fraction, which demonstrated a molecular mass of 7.3 kDa in SDS-PAGE, gel filtration, and mass spectrum. Thus, the antifungal peptide could be purified with a protocol previously used for the isolation of antifungal proteins (Wang & Ng, 2000, 2002). In general, antifungal proteins/peptides are unbound on DEAE-cellulose and bound on Affi-gel blue gel (Wang & Ng, 2000, 2002), whereas trypsin inhibitors in leguminous seeds are bound on DEAE-cellulose and unbound on Affi-gel blue gel. Certain leguminous lectins are bound on DEAE-cellulose and Affi-gel blue gel (Ye & Ng, 2001). Hence, the isolation procedure is efficient for separating antifungal protein from other bioactive proteins, such as trypsin inhibitors and lectins. The isolated antifungal peptide had an N-terminal amino acid sequence with close resemblance to defensins. Its molecular mass was also close to those of defensins. In contrast to some defensins (Wong & Ng, 2003, 2005a), the isolated peptide lacked HIV-1 RT inhibitory activity. Like mammalian defensins (Lichtenstein $et\ al.$, 1986), the peptide inhibited mycelial growth in $F.\ oxysporum$ with an IC_{50} value of $1.8 \pm 0.15\ \mu M$ (mean $\pm$ SD, $n = 3$); $M.\ arachidicola$ was also inhibited. The antifungal peptide exerted an antiproliferative action on L1210 cells and MBL2 cells, with an IC_{50} values of 4 ± 0.23 and $9 \pm 0.61\ \mu M$ (mean $\pm$ SD, $n = 3$), respectively. The antifungal peptide did not exhibit RNase, DNase, lectin, and protease inhibitory activities (Table 127.1).

From the haricot bean cultivar of $P.\ vulgaris$, a 7-kDa defensin-like antifungal peptide designated as vulgarinin was isolated (Wong & Ng, 2005b). The peptide had a greater antiproliferative potency toward leukemia L1210 cells ($IC_{50} = 4\ \mu M$ compared with $> 36\ \mu M$ for vulgarinin) and a lower antifungal activity toward $M.\ arachidicola$ ($1.8\ \mu M$ compared with $0.2\ \mu M$ for vulgarinin), and HIV-1 RT inhibitory activity is lacking (compared with $IC_{50} = 30\ \mu M$ for vulgarinin). The variation in activities may be ascribed to by differences in

TABLE 127.1 Characteristic of Antifungal Protein with Antiproliferative Activity from Seeds of *Phaseolus vulgaris* cv. "Spotted Bean" (Wang & Ng, 2007)

Molecular mass	7.3 kDa
N-terminal sequences	KTYENLADTYKGPYFTTGSHDDHYKNKEHLRSGRMRDDFF
Chromatography behavior on:	
DEAE-cellulose	Unadsorbed
Affi-gel blue gel	Adsorbed
SP-Sepharose	Adsorbed
Antifungal activity:	
Fusarium oxysporum	$IC_{50} = 1.8 \pm 0.15\ \mu M$
Mycospharella arachidicola	IC_{50} not determined
Antiproliferative activity:	
L1210 cells	$IC_{50} = 4 \pm 0.23\ \mu M$
MBL2 cells	$IC_{50} = 9 \pm 0.61\ \mu M$
HIV-1 reverse transcriptase inhibitory activity:	Absent
RNase activity	Absent
DNase activity	Absent
Lectin activity	Absent
Protease inhibitory activity	Absent

sequence between the two peptides toward the C-terminal, since the complete sequence is not available. The defensin-like antifungal peptide coccinin, from *P. coccineus*, manifested HIV-1 RT inhibitory activity ($IC_{50} = 40\ \mu M$), and a high antileukemia activity ($IC_{50} = 30\ \mu M$) and a lower antifungal activity against *M. arachidicola* ($IC_{50} = 70\ \mu M$; Ngai & Ng, 2004). In contrast to some ribonucleases (Ng & Wang, 2000), deoxyribonucleases (Wang & Ng, 2001), lectins (Ye & Ng, 2001), and protease inhibitors (Ye & Ng, 2001) that demonstrate antifungal activity, the antifungal peptide had none of these activities. The expression of defensins with antifungal activity in transgenic crops increases protection against fungal pathogens. Tobacco plants expressing radish defensin *Raphanus sativus* antifungal protein exhibited augmented antifungal activity against the tobacco foliar pathogen *Alternaria longipes* (Terras *et al.*, 1993). Gao *et al.* (2000) furnished evidence that expression of alfalfa antifungal peptide imparted resistance against the agronomically important fungus *Verticillium dahliae* in potato under field conditions, indicating that agronomically useful levels of fungal control can be achieved by expressing a single transgene in agricultural crops. Genetically modified legume crops or other plants harboring defensin genes may demonstrate augmented protection against the target pathogens.

ADVERSE EFFECTS AND REACTIONS (ALLERGIES AND TOXICITY)

The toxic lectin phytohemagglutinin is present in many varieties of beans, but is highly concentrated in red kidney beans. Although, using dry beans, it requires 10 minutes at 100°C to degrade the toxin, incidents of poisoning with the use of slow cookers at low temperatures not high enough to degrade the toxin have been reported. The British Public Health Authority, PHLA, has recommended soaking kidney beans for 5 hours before cooking. Sprouts of beans high in hemagglutinins (such as kidney beans) should be avoided.

Consumption of *Phaseolus vulgaris* brought about a decrement of food intake, body weight, and blood glucose level in rats. This is attributed to the suppression of α-amylase by α-amylase inhibitors, interference with the central mechanism controlling appetite, and stimulation of cholecystokinin secretion from intestinal brushborder cells.

SUMMARY POINTS

- The seeds of *Phaseolus vulgaris* cv. "spotted bean" produce a peptide with both antifungal and antiproliferative activities.
- The peptide exhibits a molecular mass of 7.3 kDa and an N-terminal sequence typical of leguminous defensins. It can be obtained from the beans with a procedure used for isolating other defensins, including anion exchange chromatography on DEAE-cellulose, affinity chromatography on Affi-gel blue gel, and cation exchange chromatography on SP-Sepharose.
- The peptide resembles other leguminous defensins in chromatographic media on the aforementioned media.
- It inhibits fungal growth in *Mycospharella arachidicola* and *Fusarium oxysporum* with an IC_{50} of 1.8 μM for the latter fungus.
- It inhibits proliferation of L1210 and MBL2 cancer cells with an IC_{50} values of 0.23 μM and 9 μM, respectively.
- It is devoid of deoxyribonuclease, nuclease, hemagglutinating, and protease inhibitory activities which may be present in some of the antifungal proteins.

References

Gao, A. G., Hakimi, S. M., Mittanck, C. A., Wu, Y., Woerner, B. M., Stark, D. M., et al. (2000). Fungal pathogen protection in potato by expression of a plant defensin peptide. *Nature Biotechnology, 18*, 1307–1310.

Guo, Y., Wang, H., & Ng, T. B. (2005). Isolation of trichogin, an antifungal protein from fresh fruiting bodies of the edible mushroom *Tricholoma giganteum. Peptides, 26*, 575–580.

Lichtenstein, A., Ganz, T., Selsted, M. E., & Lehrer, R. I. (1986). *In vitro* tumor-cell cytolysis mediated by peptide defensins of human and rabbit granulocytes. *Blood, 68*, 1407–1410.

Ng, T. B. (2004). Antifungal proteins and peptides of leguminous and non-leguminous origins. *Peptides, 25*, 1215–1222.

Ng, T. B., & Wang, H. X. (2000). Panaxagin, a new protein from Chinese ginseng possesses antifungal, antiviral, translation-inhibiting and ribosome-inactivating activities. *Life Sciences, 68*, 739–749.

Ngai, P. H., & Ng, T. B. (2004). Coccinin, an antifungal peptide with antiproliferative and HIV-1 reverse transcriptase inhibitory activities from large scarlet runner beans. *Peptides, 25*, 2063–2068.

Raj, P. A., & Dentino, A. R. (2002). Current status of defensins and their role in innate and adaptive immunity. *FEMS Microbiology Letters, 206*, 9–18.

Selitrennikoff, C. P. (2001). Antifungal proteins. *Applied and Environmental Microbiology, 67*, 2883–2894.

Terras, F. R. G., Torrekens, S., VanLeuven, F., Osborn, R. W., Vanderleyden, J., Cammue, B. P. A., et al. (1993). A new family of basic cysteine-rich plant antifungal proteins from Brassicaceae species. *FEBS Letters, 316*, 233–240.

Vigers, A. J., Roberts, W. K., & Selitrennikoff, C. P. (1991). A new family of plant antifungal proteins. *Molecular Plant Microbe Interactions, 4*, 315–323.

Wang, H. X., & Ng, T. B. (2000). Ginkbilobin, a novel antifungal protein from *Ginkgo biloba* seeds with sequence similarity to embryo-abundant protein. *Biochemical and Biophysical Research Communications, 279*, 407–411.

Wang, H. X., & Ng, T. B. (2001). Isolation of a novel deoxyribonuclease with antifungal activity from *Asparagus officinals* seeds. *Biochemical and Biophysical Research Communications, 289*, 120–124.

Wang, H., & Ng, T. B. (2002). Isolation of cicadin, a novel and potent antifungal peptide from juvenile cicadas. *Peptides, 23*, 7–11.

Wang, H. X., & Ng, T. B. (2007). Isolation and characterization of an antifungal peptide with antiproliferative activity from seeds of *Phaseolus vulgaris* cv. "Spotted Bean." *Applied Microbiology and Biotechnology, 74*, 125–130.

Wong, J. H., & Ng, T. B. (2003). Gymnin, a potent defensin-like antifungal peptide from the Yunnan bean (*Gymnocladus chinensis* Baill). *Peptides, 24*, 963–968.

Wong, J. H., & Ng, T. B. (2005a). Sesquin, a potent defensin-like antimicrobial peptide from ground beans with inhibitory activities toward tumor cells and HIV-1 reverse transcriptase. *Peptides, 26*, 1120–1126.

Wong, J. H., & Ng, T. B. (2005b). Vulgarinin, a broad-spectrum antifungal peptide from haricot beans (*Phaseolus vulgaris*. *International Journal of Biochemistry and Cell Biology, 37*, 1626–1632.

Wong, J. H., Hao, J., Cao, Z., Qiao, M., Xu, H., Bai, Y., et al. (2008). An antifungal protein from *Bacillus amyloliquefaciens*. *Journal of Applied Microbiology, 105*, 1888–1898.

Ye, X. Y., & Ng, T. B. (2001). Isolation of unguilin, a cyclophilin-like protein with anti-mitogenic, antiviral, and antifungal activities, from black-eyed pea. *Journal of Protein Chemistry, 20*, 353–359.

Zhang, Z. T., & Zhu, S. Y. (2009). Drosomycin, an essential component of antifungal defence in Drosophila. *Insect Molecular Biology, 18*, 549–556.

CHAPTER 128

Use of Extracts from Squirting Cucumber (*Ecballium elaterium*) Seeds in Health

Everaldo Attard
Institute of Earth Systems, Division of Rural and Food Systems, University of Malta, Msida, Malta

1079

LIST OF ABBREVIATIONS

EEAPI, *Ecballium elaterium* Astacus protease inhibitor
EECI, *Ecballium elaterium* chymotrypsin inhibitor
EEEI, *Ecballium elaterium* elastase inhibitor
EEPI, *Ecballium elaterium* protease inhibitors
EESI, *Ecballium elaterium* subtilisin inhibitor
EETIso, *Ecballium elaterium* trypsin isoinhibitors
PMN, polymorphonuclear

Nuts & Seeds in Health and Disease Prevention. DOI: 10.1016/B978-0-12-375688-6.10128-8

INTRODUCTION

The squirting cucumber, *Ecballium elaterium* (L.) A. Rich., belongs to the Cucurbitaceae family. Although different plant parts have been used to treat several ailments, the fruit was traditionally believed to be an important source of medicine. Nowadays, the fruit is considered to be a source of pharmacologically-active metabolites. Important metabolites include carbohydrates, gum, leukoanthocyanins, tannins, triterpenoids (cucurbitacins), and peptides (Jodral *et al.*, 1990). This chapter deals primarily with the peptides, which are more commonly known as the *Ecballium elaterium* protease inhibitors. Seed extracts obtained from *E. elaterium* contain inhibitors for at least four different serine proteinases (Favel *et al.*, 1989): trypsin inhibitors I, II, and III, also known as the trypsin isoinhibitors (EETIso), chymotrypsin inhibitor (8 kDa), subtilisin inhibitor (9 kDa), and elastase inhibitor, and *Astacus* protease inhibitor.

BOTANICAL DESCRIPTION

Although *Ecballium elaterium* is the official Latin name, it is derived from Greek, due to its various medicinal uses in Greece. The word *Ecballium*, or, more precisely *Ekballein*, means "I expel". This refers to the mode of dispersion of the seeds, which is characteristic to the plant. *Elaterium*, which also means "to cast out," refers to the purgative action of the plant (Bianchini & Corletta, 1985).

E. elaterium is found in a monotypic genus. It is a perennial hispid herb with a tuberous root. The stems are procumbent (15−60 cm), while the leaves are cordate to triangular, and rather fleshy (4−10 cm). The flowers are unisexual, with the male flowers having a yellowish corolla. The fruit is green, hairy, and ovoid in shape (4−5 × 2.5 cm). When mature, the fruit turns from green to yellowish-green, and expels the seed and juice violently. The seeds are ovoid in shape, coffee-brown in color, with a size of 4−5 mm.

E. elaterium is very common in Southern Europe, especially in the Mediterranean region and countries surrounding the Black Sea (Costich, 1997). It is mainly found in Albania, the Balearics, Bulgaria, Corsica, Crete, France, Greece, Italy, the Crimea, Malta, Portugal, Romania, Sardinia, Sicily, Spain, Turkey, and the former Yugoslavia. It probably originated in the Orient, extending as far as India.

HISTORICAL CULTIVATION AND USAGE

Due to its abundance within the Mediterranean and oriental regions, the squirting cucumber was harvested from the wild, and given little importance as a crop for cultivation. In fact, in many countries this plant was considered to be a weed. However, early last century (Grieve & Oswald, 1911) the plant was cultivated on a small scale, for the production of elaterium, at Hitchin and Amptil, and also formerly at Mitcham, in the United Kingdom. Due to the cold weather and lack of sunshine, the proliferative vigor of the plant and the quality of the dried fruit juice (elaterium) were inferior to the characteristics of the same plants growing within the Mediterranean region. In fact, most "elaterium of commerce" on the British market was imported from Malta (Sommier & Caruana Gatto, 1915), which was under British colonial rule at that time.

The squirting cucumber has been used as a remedy for many ailments. The uses of this plant are relatively ancient, and the method used for the preparation of the elaterium today is the same as that described in Greek times, particularly by Dioscorides (Flückiger & Hanbury, 1879). In Arabic folk medicine (Dymock, 1972), Mahometan writers recorded the use of the elaterium as a laxative, and the juice as a treatment for otitis and as a remedy "to purge the brain." The Hindus used the fresh and dried fruit juice in a similar way. In Georgian popular medicine, the plant was used as a remedy for malarial fever (Dymock, 1972). In Turkish folk medicine, the elaterium has been used in treating jaundice and headache. The powdered elaterium (precipitate from the fruit juice) mixed with milk used to be applied in the nostrils to clear icterus and cure persistent headaches. It has also been used in the treatment of sinusitis. *Ecballium elaterium* was popular in Maltese folk medicine, as a cathartic and in the treatment of jaundice (Lanfranco, 1980). The elaterium can also be used to remove edema, which is the accumulation of excessive water in the body tissues. At high doses, the elaterium can cause vomiting; hence, it may have been used in the treatment of poisoning where induced vomiting was necessary.

PRESENT-DAY CULTIVATION AND USAGE

Today, wild stocks of the squirting cucumber have diminished drastically, to the extent that cultivation is necessary for industrial-scale production of drugs from this plant. Several cultivation protocols have been devised (Attard & Scicluna-Spiteri, 2003); however, nowadays the use of *in vitro* propagation methods (Attard & Attard, 2002) complements traditional cultivation practices. In fact, through *in vitro* propagation, suitable wild stocks can be enhanced by starting with seed germination, and then using explant multiplication to yield identical plantlets for cultivation.

Due to its toxic properties, the elaterium is no longer used in medicine. However, the seeds are studied for their pharmacological potential, and therefore clinical applications are still under experimental conditions.

APPLICATIONS TO HEALTH PROMOTION AND DISEASE PREVENTION

Ecballium elaterium protease inhibitors in pharmaceutical preparations: A historical view and recent developments

Peptides possess great potential for the control or treatment of diseases, especially deficiencies. However, in the past they received relatively little attention as regards the development of these peptides into therapeutic agents. It has only been since 1980 that the importance of the development of peptide-based pharmaceuticals has emerged. From the historical point of view, models for peptide–protein interactions were centered on the lock-and-key hypothesis of enzyme–substrate interactions. Here, this model is the most important of all models used to explain peptide–protein interactions.

The relationship between structure and physiochemical properties (e.g., spatial arrangement of functional groups, and ligand–receptor interactions) of peptides and proteins is the first objective of drug design, and it is primarily achieved by computer-aided drug-design packages. The conformational analysis and molecular recognition processes should be coupled with the development of new delivery systems. A concordant theoretical and experimental approach is required.

In 1980, serine protease inhibitors were discovered in Cucurbitaceae plant species by Polanowski (Le-Nguyen *et al.*, 1989). As a result, the individual protease inhibitors were studied in depth, and it was found that *Ecballium elaterium* L. seed extracts were rich in these protease inhibitors. Eventually, the most important protease inhibitor was found to be *Ecballium elaterium* trypsin inhibitor II, leading to further studies performed on its structure and physiochemical properties. The role of the serine protease in several pathological conditions was described. One of the most commonly mentioned examples is elastase, which is the cause of emphysema, resulting from the destruction of the elastic fibers found in the lungs. In 1985 the serine protease inhibitors were extracted from seeds obtained from plants found in the Cucurbitaceae family, and it was discovered that they are:

- the smallest protease inhibitors known (29–32 amino acid residues), and therefore within the reach of chemical synthesis, with a high yield due to correct folding.
- rigid peptides, as they are rich in disulfide bridges.

- considered as lead molecules, leading to potentially therapeutic drugs for inhibiting proteases found in the body, such as plasmin, prothrombin, chymotrypsin, thrombin, elastase, and trypsin.
- highly resistant to denaturation.
- highly specific, and have a high affinity with target enzymes.

Ecballium elaterium trypsin isoinhibitor II (EETI II)

As highlighted earlier, the main protease found in *E. elaterium* seeds is a trypsin inhibitor called *Ecballium elaterium* trypsin inhibitor II (EETI II). It has an association constant of 8×10^{11} M, and contains 28 amino acid residues with a rigid molecular framework (Krätzner *et al.*, 2005). EETI II lacks extensive secondary structure (i.e., α-helix or β-sheet), but it is composed of loops and turns around a core of disulfide bridges (Figure 128.1). As EETI II is a short peptide which is easily synthesized and mutated, it can be modified to become a selective inhibitor against other proteases — for example, modification of EETI II may lead to selective inhibition of porcine pancreatic elastase (Hilpert *et al.*, 2003). To extract a substantial amount of a protease inhibitor, several kilograms of seeds are required. As discussed in the introduction, following chemical and pharmacological experiments, chemical synthesis was attempted; this process was eventually successful with EETI II.

The action of trypsin and elastase: normal physiology

The enzymes required for digestion of proteins in the duodenum (chymotrypsin, carboxypeptidase, trypsin, and elastase) are produced and stored in the pancreas as inactive precursors. Trypsin acts as a common activator of all proteolytic enzymes, and this process occurs when the zymogens enter the duodenum from the pancreas. Elastase is another serine protease that is produced either by human polymorphonuclear (PMN) leukocytes or by the pancreas, as already mentioned. Together with other lysosomal proteases, it plays an important part in phagocytosis, exposure to antigen—antibody complexes, foreign bodies, complement components, and toxins (e.g., endotoxins or uremic toxins) (Sandler & Smith, 1989).

Pathophysiology related to trypsin and elastase

The proteolytic enzymes are stored as zymogens in the pancreas, and their activation is suppressed by the presence of natural protease inhibitors (Stryer, 1988). These prevent the activation of proteases inside the pancreas, and in the ducts of the pancreas. The presence of trypsin in the pancreas leads to the activation of the inactive zymogens, and, eventually, acute inflammation of the organ — i.e., acute pancreatitis. Acute pancreatitis normally presents

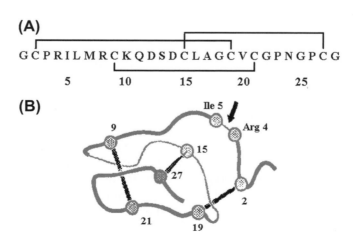

FIGURE 128.1

The (A) primary and (B) tertiary structures of EETI II (Chiche *et al.*, 1989).

with severe pain in the abdomen, shock, and circulatory collapse. Its incidence is higher in females who are usually obese and over the age of 40 years. The mortality in severe cases is 50%, and the patient can die during the first week after the onset, as a result of shock and circulatory failure. In other cases, this condition may result in chronic pancreatitis, suppuration, and gangrene. Pancreatitis is controlled by the alleviation of symptoms and prevention of hypovolemia, shock, and death; gastric lavage and medications are used to relieve pain and decrease stimulation of the pancreatic enzymes (Brunner & Suddarth, 1989; Ohlsson *et al.*, 1989).

The PMN leukocyte elastase is involved in numerous pathological conditions (Sandler & Smith, 1989); in particular, panlubar pulmonary emphysema (Shapiro *et al.*, 2003), rheumatoid arthritis, adult and infantile respiratory distress syndrome, glomerulonephritis, atherosclerosis, and psoriasis. The level of elastase at a site is controlled by an elastase inhibitor known as α_1-antielastase (α_1-AE), formerly known as α_1-antitrypsin. However, some patients are genetically predisposed to or acquire α_1-AE deficiency. This deficiency can be acquired by cigarette smoking, especially if the patient has a genetic predisposition to α_1-AE deficiency. In both cases, there is a disruption in the protease—antiprotease balance. If the patient has an α_1-AE deficiency, then antibiotic prophylactic therapy together with anti-elastase prophylaxis should be considered. In patients on hemodialysis, the nature of the hemodialyzer membrane material is of great importance, particularly if a patient suffers from α_1-AE deficiency. Some dialysis membranes cause the formation of elastase—α_1-AE complexes. In these deficient patients, damage is provoked by the release of elastase, which is not neutralized by α_1-AE.

The mechanism involved in the enzyme—substrate interaction, and the role of inhibitors

For the enzyme—substrate reaction to take place, the substrate requires two functional groups which should correspond to the following two sites on the enzyme (Stryer, 1988). It is worth noting that these two sites are present on most serine proteases, including trypsin, elastase, and chymotrypsin (Figure 128.2):

1. *Binding site.* This site infers the specificity of the enzyme—substrate interaction. The functional group on the substrate determines which enzyme can bind to the substrate. For example, trypsin requires a positively-charged lysine or arginine side chain, while elastase requires smaller, uncharged side chains.
2. *Cleavage site.* A serine—histidine—aspartate triad participates in the cleavage of a susceptible bond — for example, the cleavage on the carboxyl side of arginine or lysine residues, by trypsin. The serine, present in the triad, binds to the substrate at the carbonyl functional group, resulting in a tetrahedral transition state. Also, hydrogen bonds, formed by two NH groups from the enzyme main chain, stabilize the serine—carbonyl complex. This site is called the oxyanion hole.

TRYPSIN INHIBITORS

Several compounds inhibit trypsin, but only a few exhibit selective inhibition. An outstanding example of this is bis-(5-amidino-2-benzimidazolyl) methane, with a dissociation constant of 1.7×10^{-8} M. The dissociation constant of EETI II is 1×10^{-12} M. EETI II has been modified at amino acid 7, replacing methionine with nor-leucine, and the resultant dissociation constant is 4×0^{-10} M. Consequently the latter shows a greater affinity to trypsin than the previous selective inhibitor (Sandler & Smith, 1989).

ELASTASE INHIBITORS

During recent years, elastase inhibitors have been studied more intensively than trypsin inhibitors, mainly due to the fact that elastase is the main cause of most cases of lung disease. For

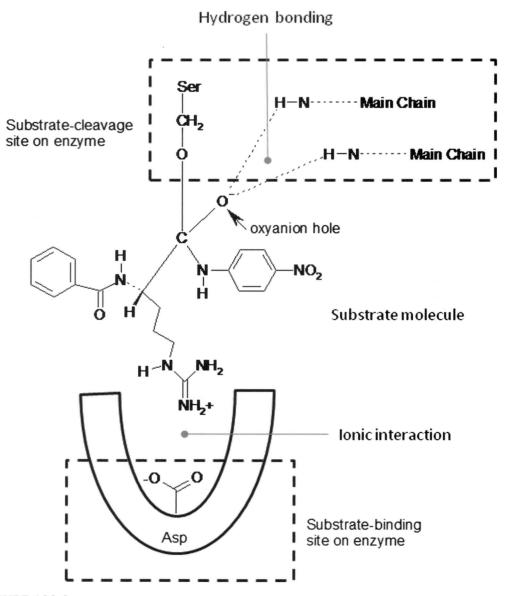

FIGURE 128.2

The interaction of trypsin with a substrate (Nα-benzoyl-L-arginine-4-nitroanilide). The figure shows the two important functional groups on the substrate corresponding to the two sites on the enzyme. This mechanism holds also for other serine proteases.

elastase, both reversible and irreversible inhibitors have been discovered. Reversible inhibitors form a transition state complex which is common to all. In the transition state, the most important moiety is the oxyanion hole, as for the trypsin inhibitors. The irreversible inhibitors include peptide chloromethyl ketones, such as MeO-Suc-Ala-Ala-Pro-Val-CH$_2$C (Sandler & Smith, 1989). These are said to be the smallest inhibitors. They can be easily delivered using human albumin microspheres (HAM), which are non-antigenic, non-toxic, and have a relatively small size. Consequently, they easily become trapped in the pulmonary capillary bed after intravenous injection. HAM derivatives are linked to Ala-Ala-Pro-Val-CH$_2$Cl via covalent bonds. In spite of the importance of elastase inhibitor discovery, the *Ecballium elaterium* elastase inhibitor (EEEI) has not been studied intensively. Primarily, this is because EETI II can be modified to yield inhibitors specific to a particular type of substrate (Hilpert *et al.*, 2003).

ADVERSE EFFECTS AND REACTIONS (ALLERGIES AND TOXICITY)

From ethnobotanical and scientific records, adverse effects and reactions have been mostly related to medicinal sources within the plant other than the seeds. These toxic effects are primarily caused by the juice surrounding the seeds. The juice causes topical and systemic adverse effects, including nausea, diarrhea with colic, vomiting, gastroenteritis, contact dermatitis, diffuse lamellar keratitis (Kocak *et al.*, 2006), and upper airway edema (Kloutsos *et al.*, 2001), amongst others. In pregnancy, the plant induces abortion. In spite of all these adverse effects, no incidence of intoxication from seeds has been reported in the case of the squirting cucumber. This may be due to the fact that seed extracts have not been used in popular folk medicine, and *Ecballium elaterium* protease inhibitor research is still ongoing. To date, no clinical trials have been undertaken to determine the *in vivo* safety, efficacy, and toxicity profiles of these inhibitors. However, owing to their clinical potential, further research will be conducted along these lines.

SUMMARY POINTS

- *Ecballium elaterium* is an herb that grows abundantly in the Mediterranean region.
- In the past, the squirting cucumber was highly esteemed for its medicinal virtues.
- The plant biosynthesizes and stores specific polypeptides, termed protease inhibitors, in the seed extracts.
- The main protease inhibitors act specifically against trypsin, chymotrypsin, elastase, subtilisin, and Astacus protease.
- EETI II is one of the most potent antitrypsin inhibitors; however, this enzyme can be modified to exhibit a different affinity to other proteases (e.g., elastase).
- Although the juice is toxic, no adverse effects have been reported for the ingestion of squirting cucumber seed extracts.

References

Attard, E., & Attard, H. (2002). A micropropagation protocol for *Ecballium elaterium* (L.) A. Rich. *CGC Reports, 25,* 67–70.

Attard, E., & Scicluna-Spiteri, A. (2003). The cultivation and cucurbitacin content of *Ecballium elaterium* (L.) A. Rich. *CGC Reports, 26,* 66–69.

Bianchini, F., & Corletta, F. (1985). *A complete book of health plants.* New York, NY: Crescent Books. pp. 68, 195.

Brunner, L. S., & Suddarth, D. S. (1989). *The Lippincot manual of medical-surgical nursing* (2nd ed.). London, UK: Harper and Row Ltd. pp. 531–532.

Chiche, L., Gaboriaud, C., Heitz, A., Mornon, J. P., Castro, B., & Kollman, P. A. (1989). Use of restrained molecular dynamics in water to determine three-dimensional protein structure: prediction of the three-dimensional structure of *Ecballium elaterium* trypsin inhibitor II. *Proteins, 6,* 405–417.

Costich, D. E. (1997). *Ecballium elaterium:* the squirting cucumber of the Mediterranean. *Plant Genetic Resources Newsletter, 112,* 98–99.

Dymock, W. (1972). *Pharmacographia Indica; A history of the principal drugs of vegetable origin, Vol. XV.* Karachi, Pakistan: The Institute of Health and Tibbi Research. pp. 95–97.

Favel, A., Mattras, H., Coletti-Previero, M. A., Zwilling, R., Robinson, E. A., & Castro, B. (1989). Protease inhibitors from *Ecballium elaterium* seeds. *International Journal of Peptide and Protein Research, 33,* 202–208.

Flückiger, F., & Hanbury, D. (1879). *Pharmacographia: A history of the principle drugs of vegetable origin* (2nd ed.). London, UK: Macmillan & Co. Ltd. pp. 292–295.

Grieve, M., & Oswald, E. (1911). *Squirting cucumber, melons, water melon, pumpkin and colocynth.* Chalfont St Peter, UK: Whin's Cottage.

Hilpert, K., Wessner, K., Schneider-Mergener, J., Welfle, K., Misselwitz, R., Welfle, H., et al. (2003). Design and characterization of a hybrid miniprotein that specifically inhibits porcine pancreatic elastase. *Journal of Biological Chemistry, 278,* 24986–24993.

Jodral, M. M., Martin, J. J., Agil, A. M., Navarro Moll, M. C., & Cabo Cires, M. P. (1990). *Ecballium elaterium* (L.) A. Richard II — Morphological and phytochemical studies. *Boletin da Sociedade Bioteriana, 63,* 213–224.

Kloutsos, G., Balatsouras, D. G., Kaberos, A. C., Kandiloros, D., Ferekidis, E., & Economou, C. (2001). Upper airway edema resulting from use of *Ecballium elaterium*. *Laryngoscope, 111*, 1652–1655.

Kocak, I., Karabela, Y., Karaman, M., & Kaya, F. (2006). Late onset diffuse lamellar keratitis as a result of the toxic effect of *Ecballium elaterium* herb. *Journal of Refractive Surgery, 22*, 826–827.

Krätzner, R., Debreczeni, É., J., Pape, T., Schneider, T. R., Wentzel, A., Kolmar, H., et al. (2005). Structure of *Ecballium elaterium* trypsin inhibitor II (EETI-II): a rigid molecular scaffold. *Acta Crystallographica, D61*, 1255–1262.

Lanfranco, G. (1980). Some recent communications on the folk medicine of Malta. *L-Imnara, 3*, 87.

Le-Nguyen, D., Nalis, D., & Castro, B. (1989). Solid phase synthesis of a trypsin inhibitor isolated from the Cucurbitaceae *Ecballium elaterium*. *International Journal of Peptide and Protein Research, 34*, 492–497.

Ohlsson, K., Olsson, R., Björk, P., Balldin, G., Borgström, A., Lasson, A., et al. (1989). Local administration of human pancreatic secretory trypsin inhibitor prevents the development of experimental acute pancreatitis in rats and dogs. *Scandinavian Journal of Gastroenterology, 24*, 693–704.

Sandler, M., & Smith, H. J. (1989). *Design of enzyme inhibitors as drugs. Inhibitors of proteases.* Oxford, UK: Oxford University Press. pp. 573–649.

Shapiro, S. D., Goldstein, N. M., Houghton, A. M., Kobasyashi, D. K., Kelley, D., & Belaaouaj, A. (2003). Neutrophil elastase contributes to cigarette smoke-induced emphysema in mice. *American Journal of Pathology, 163*, 2329–2335.

Sommier, S., & Caruana Gatto, A. (1915). *Flora Melitensis Nova.* Florence, Italy: Stabilimento Pellas. p. 155.

Stryer, L. (1988). *Biochemistry* (3rd ed.). New York, NY: W.H. Freeman & Co. p. 224.

CHAPTER 129

Biological Properties of Sucupira Branca (*Pterodon emarginatus*) Seeds and Their Potential Usage in Health Treatments

Nádia Rezende Barbosa Raposo, Rafael Cypriano Dutra, Aline Siqueira Ferreira
Department of Food and Toxicology, Faculty of Pharmacy and Biochemistry, Federal University of Juiz de Fora, Juiz de Fora, Brazil

1087

LIST OF ABBREVIATIONS

AI, arthritis index
BF, *Pterodon emarginatus* seed buthanolic fraction
BHT, butylated hydroxytoluene
CIA, collagen II-induced arthritis
DPPH, 2,2-diphenyl-2-picrylhydrazyl hydrate
EO, seed essential oil
HCl, chloridric acid
HF, *Pterodon emarginatus* seed hexanic fraction
IC_{50}, concentration which inhibits 50% of a life system

Nuts & Seeds in Health and Disease Prevention. DOI: 10.1016/B978-0-12-375688-6.10129-X

IL-1α, interleukin-1α

LD$_{50}$, median lethal dose

MF, *Pterodon emarginatus* seed methanolic fraction

MIC, minimum inhibitory concentration

MTT, 3-[4,5-dimethythiazole-2-il]-2,5-dipheniltetrazolium bromide

NO, nitric oxide

p.o., oral route

TNF-α, tumor necrosis factor α

v/v, volume per volume

INTRODUCTION

The Leguminosae family (Fabaceae), the largest of the angiosperms, consists of approximately 650 genera and 18,000 species (Judd *et al.*, 1999). The Faboideae is the largest of the three Leguminosae subfamilies, with approximately 440 genera and 12,000 species. Many species occur in tropical regions, and play a significant part of the life of the Aeschynomeneae, Dalbergieae, Dipteryxeae, and Sophoreae tribes, found in Latin America (Lorenzi & Matos, 2000).

The genus *Pterodon* comprises the following Brazilian native species: *Pterodon abruptus* Benth, *Pterodon appariciori* Pedersoli, *Pterodon emarginatus* Vogel, *Pterodon pubescens* Benth, and *Pterodon polygalaeflorus* Benth. Chemical investigation of these species was first motivated by the discovery that *P. pubescens* seed oil inhibited schistosome cercaria penetration in the skin — a property that was attributed to 14,15-epoxygeranylgeraniol, and later to the accompanying linear diterpenoid 14,15-dihydroxy-14,15-dihydrogeranylgeraniol (Mors *et al.*, 1967). Geranylgeraniol itself also occurs in *P. pubescens* seeds, and is responsible for their characteristic floral odor. Similar biological activity was observed for seed oils of *P. emarginatus*, *P. polygalaeflorus*, and *P. apparicioi*, and the presence of the same or related linear diterpenoids thought probable (Mors *et al.*, 1967; Mahjan & Monteiro, 1970). Fourteen furan diterpenes were described and isolated from the seeds of the *Pterodon* genus, among other substances; eight of these belong to *P. emarginatus* (Fascio *et al.*, 1976).

Chemical studies on the *Pterodon* genus have shown the presence of alkaloid compounds in the bark (Torrenegra *et al.*, 1989), isoflavone and some triterpenes in the wood (Marques *et al.*, 1998), diterpenes (Fascio *et al.*, 1976; Arriaga *et al.*, 2000) and isoflavones (Braz-Filho

TABLE 129.1 Chemical Composition of *Pterodon emarginatus* Seed Essential Oil Determined by Gas Chromatography Coupled to Mass Spectrometer

Chemical Constituent	Retention Time (minutes)	Kovats Index	Percentage
Gamma-elemene	15.64	1327	4.7
Beta-elemene	15.67	1380	15.3
Trans-caryophyllene	18.16	1411	35.9
Beta-gurjunene	18.49	1423	4.6
Alpha-humulene	19.23	1449	6.8
Gamma-muurolene	19.44	1456	2.7
Germacrene-D	19.97	1475	9.8
Bicyclogermacrene	20.44	1492	5.5
Spathulenol	22.53	1570	5.9
Caryophyllene oxide	22.72	1577	3.8
Cis-farnesyl acetate	26.34	1815	4.9
Total			99.9

et al., 1971) in seed oil, and phenolic constituents, such as flavonoids, in the seeds (Dutra *et al.*, 2008a). The composition of seed essential oil (EO) of *P. emarginatus* obtained by steam distillation was analyzed by gas chromatography coupled to a mass spectrophotometer, and the chemical composition is presented in Table 129.1 (Dutra *et al.*, 2009a).

BOTANICAL DESCRIPTION

Pterodon emarginatus Vogel (Leguminosae) is a native aromatic tree that reaches 5–10 m in height and is found throughout the Brazilian Cerrado. It is commonly known as *sucupira branca, faveiro,* or *fava de sucupira.* It is a semideciduous plant, a heliophyte, typically found dispersed irregularly and intermittently on dry land and sandy grasslands, occurring in dense clusters, and often in pure populations. Its wood is considered heavy (0.94 g/cm³ density), with compact, interlocked tissue; it is very hard, difficult to split, and long-lasting, even when in contact with soil and moisture. It is typically used in construction, shipbuilding, bridge pylons, poles, railroad ties, wagons, and wagon floors, and for coal and firewood (Lorenzi & Matos, 2000).

The tree has a large pyramidal crown that reaches 8–16 m in height, and a cylindrical trunk 30–40 cm in diameter, with smooth yellowish-white bark (Figure 129.1). Its roots occasionally form expansions or tubers, called *batata de sucupira,* constituting the organic reserves of the plant. Compound leaves are pinnate, with 20–36 leaflets 3–4 cm in diameter, and rose-colored flowers arranged in terminal inflorescences. The samara fruit pods are rounded, indehiscent, and winged, containing a single seed tightly secured inside a fibrous woody capsule, and held externally in an oily substance in a spongy honeycomb-like structure (Lorenzi & Matos, 2000). Figure 129.2 illustrates the general aspect of *P. emarginatus* seeds.

P. emarginatus ovules are campilotropous, crassinucelate, and bitegmic. At the hilar pole, the characteristic Faboideae seed structure develops, with double palisade layer, subhilar parenchyma, and tracheid bar. The younger nucleus shows thicker pectic cell walls, and is consumed during seed formation. The endosperm is nuclear and, after cellularization, shows peripheral cells with dense lipid content; the seeds are albuminous. The axial embryo shows fleshy cotyledons, which accumulate lipid and protein reserves; starch is rare. Although the seed structure of *P. emarginatus* is characteristic of the Leguminosae family, the inner integument coalesces into the outer integument without being reabsorbed. Large amounts of lipid and protein, and reduced starch reserves, are characteristic of the fleshy cotyledons of *P. emarginatus* seeds (Oliveira & Paiva, 2005).

Recently, our group realized the cytogenetic characterization of *P. emarginatus* seeds. The seeds were treated with 2-mM 8-hydroxyquinoline for 6 hours at room temperature, and 2n = 16 was determined as the chromosome number (unpublished data). The mitotic metaphase is illustrated in Figure 129.3.

1089

FIGURE 129.1
Specimen representative of *Pterodon emarginatus* found in the Horto Florestal do Instituto Brasileiro do Meio Ambiente e dos Recursos Naturais Renováveis (IBAMA) (FLONA), in the city of Paraopeba/Minas Gerais, Brazil. Left, detail of flowers arranged in terminal inflorescences, and fruit pods (samara type); right, detail of the tree trunk. The wood is considered heavy (0.94 g/ch³ density).

FIGURE 129.2
General aspect of *Pterodon emarginatus* Vogel seeds.

FIGURE 129.3
Mitotic metaphase of *Pterodon emarginatus* Vogel. Sixteen pairs of chromosomes were observed. The image was obtained under light microscopy and captured by a video camera attached to the microscope (BX451, Olympus, Japan), digitized and analyzed using the Image-Proplus software (Media Cybernetics). Bar corresponds to 5 μm.

HISTORICAL CULTIVATION AND USAGE

The species *P. emarginatus* produces large quantities of viable seeds, which eventually can be attacked by insects. Flowering occurs during the months September to October, and fruit maturation in June to July, when the plant is almost completely stripped of foliage. Seeds can be obtained when harvested directly from the fruit of the tree, or, when they start to fall spontaneously, be collected from the ground (Lorenzi & Matos, 2000).

PRESENT-DAY CULTIVATION AND USAGE

In Brazilian folk medicine, the seeds have been used to treat rheumatism, sore throats, and respiratory dysfunctions (bronchitis and amygdalitis) (Mahjan & Monteiro, 1970), in addition to having anti-inflammatory (Dutra *et al.*, 2009a), antinociceptive (Dutra *et al.*, 2008b), depurative, and tonic activities (Arriaga *et al.*, 2000). Their seeds are commercially available in the medicinal flora market, being widely used for their pharmacological properties (Arriaga *et al.*, 2000). In popular medicine, seeds of *P. emarginatus* are ground in an hydroalcoholic solution using 50 g of seeds crushed in 250 ml of diluent (concentration of 200 mg/ml) such as ethanol or an alcoholic drink (e.g., brandy), and left under static maceration for 24−48 hours at room temperature. After preparation, the extract is orally consumed at 40 ml/day, divided into two daily doses, over 7 days.

APPLICATIONS TO HEALTH PROMOTION AND DISEASE PREVENTION

The antinociceptive activity of *P. emarginatus* seeds was evaluated by three different assays: acetic acid-induced abdominal writhing, the formalin test, and the hot plate test (Dutra *et al.*, 2008b). *P. emarginatus* EO and the hexanic fraction (HF) from *P. emarginatus* seeds showed a significant diminution of the number of writhings at 500 mg/kg, and their action was similar to that of indomethacin (5 mg/kg, p.o.), used as anti-inflammatory reference drug. In the formalin test, both EO and the seed methanolic fraction (MF) presented a biphasic licking response; this means that they act on the earlier phase, which seems to be caused predominantly by c-fiber activation due to peripheral stimulation. They also act on the later phase, which seems to be caused by tissue and functional changes in the dorsal horn of the spinal cord, and is accompanied by inflammatory mediator release. Conversely, HF and seed buthanolic fraction (BF) were effective only in the first phase, acute neurogenic pain. Morphine (5 mg/kg, subcutaneously), the opioid drug reference used, acted on both phases. Finally, as the hot plate test, HF and BF was found to induce an increase in baseline at 90 and 120 minutes after treatment, in the 100 and 300 mg/kg doses, respectively. *P. emarginatus* EO and MF did not demonstrate antinociceptive activity in this model at tested concentrations.

The action of *P. emarginatus* seed hexanic crude extract (500 mg/kg, p.o., for 6 days) in mouse paw edema induced by different stimuli was investigated (Carvalho *et al.*, 1999). Inhibitions in the formation of granulomatous tissues by 10%, and of approximately 43% on neutrophil migration, were observed when compared to the control group. Similar results were found in a recent study by our group (Dutra *et al.*, 2009a), in which EO (500, 300, and 100 mg/kg) could prevent, in a dose-dependent manner, the number of ulcers and ulcer index in three different models: ethanol, non-steroidal anti-inflammatory drug, and hydrochloric acid (HCl) ethanol-induced ulcer in mouse. *P. emarginatus* seed oil presented a similar protective action when compared to ranitidine (60 mg/kg) in the first and second evaluated models. Figure 129.4 demonstrates the EO effect at the lower tested concentration (100 mg/kg, p.o.) compared with the control (vehicle) and ranitidine (60 mg/kg) in the ethanol-induced gastric ulceration model. In the HCl-ethanol-induced ulcer model a dose-dependent action was observed, and at 500 mg/kg concentration the effect was similar to that of omeprazole (30 mg/kg). Moreover, EO reduced nitric oxide (NO) and interleukin-1α (IL-1α) levels, whereas it did not influence tumor necrosis factor-α (TNF-α) production. These results suggest a powerful anti-inflammatory action of EO. Due to these results, the anti-inflammatory action of *Pterodon* seeds is nowadays a hot topic for investigation, and perhaps a candidate for a new natural phytotherapeutic.

1091

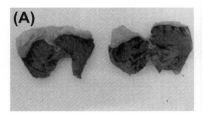  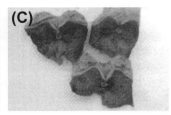

FIGURE 129.4

Therapeutic effect of the essential oil obtained from *Pterodon emarginatus* seeds in the number of ulcers and ulcerogenic index induced by absolute ethanol administration in mice, per oral. Stomachs of each group are presented. Panel (A) represents control (vehicle), panel (B) the drug reference (ranitidine 60 mg/kg), and panel (C) EO (100 mg/kg, p.o.).

P. emarginatus EO and HF were effective on male rabbits in healing cutaneous ulcers, as successfully demonstrated by Dutra *et al.* (2009b). Seed essential oil of *P. emarginatus* and HF, both at 5 and 10% (w/w) incorporated in a cream base, were administrated to shaved and burned skin twice a day, for 10 days. Silver sulfadiazine at 1% was used as drug reference. Both doses of EO and 10% HF presented healing action similar to that observed in silver sulfadiazine-treated areas.

Antibacterial and leishmanicidal activities of *P. emarginatus* seeds were investigated (Dutra *et al.*, 2009c). Antibacterial activity was expressed as the minimum inhibitory concentration (MIC), and leishmanicidal activity as the IC_{50}, which is defined as the concentration which inhibits 50% of a life system. It was found that EO minimal inhibitory concentration was 2.5 mg/ml against *Staphylococcus aureus* ATCC 25923. *P. emarginatus* seed hexanic fration and BF were effective against *Leishmania amazonensis* promastigote forms (IC_{50} of 50 µg/ml and 46.6 µg/ml, respectively). The results indicate that the bioactive molecules present in the seeds of *P. emarginatus* can be used as prototype for the development of antimicrobial/leishmanicidal drugs.

Finally, the antioxidant activity and total phenolic content in *P. emarginatus* seeds were determined (Dutra *et al.*, 2008a). BF and MF fractions could be natural additives to replace synthetic antioxidants, since it was demonstrated that 2,2-diphenyl-2-picrylhydrazyl hydrate (DPPH) showed scavenging activities with IC_{50} values equal to 18.89 and 10.15 µg/ml, respectively, while the ascorbic acid IC_{50} was 2.50 µg/ml and that for butylated hydroxytoluene (BHT) was 7.58 µg/ml. Seed total phenolic contents obtained by reflux were 600 mg/100 g (ethanol/water, 50:50, v/v) and 852 mg/100 g (ethanol/water, 70:30, v/v) when reflux was used as the extraction method.

The *Pterodon emarginatus* seed preparation commonly used in folk medicine is a good source of a variety of active compounds, which suggests a proven therapeutic action. This chapter strongly suggests that *P. emarginatus* seeds could constitute an attractive, safe, and relevant species of interest for development of new analgesic/anti-inflammatory and antimicrobial/antioxidant drugs for the treatment of acute pain and inflammatory diseases (especially those presenting a chronic profile such as gastric ulcers) and for antimicrobial therapy, as well as for its possibilities as an additive in the food industry. In addition, these reports may justify the use of *P. emarginatus* in traditional medicine, and further studies aiming to investigate which compounds in the seeds are responsible for the pharmacological activities and to determine the accurate action mechanisms are under way.

ADVERSE EFFECTS AND REACTIONS (ALLERGIES AND TOXICITY)

There is a lack of information about the adverse effects or toxicological activities of *Pterodon emarginatus* seeds. The *Pterodon* genus has several common constituents in common (Mors *et al.*, 1967; Mahjan & Monteiro, 1970; Fascio *et al.*, 1976), and *P. emarginatus* and *P. pubescens* species were considered by Lorenzi and Matos (2002) as botanical synonyms. With this information, an extrapolation regarding data toxicity could be done.

The median lethal dose (LD_{50}) of *P. emarginatus* seed hexanic crude extract was equal to 4.02 g/kg, and with such doses mice showed stereotypy, convulsions, ataxia, diarrhea, and increased diuresis (Carvalho *et al.*, 1999). In our laboratory, cytotoxicity assays using the brine shrimp *Artemia salina* Leach as a model (Meyer *et al.*, 1982) were performed to determine the LD_{50} of essential oil from *P. emarginatus* seeds. This EO showed potent toxicity in *Artemia salina*, with $LD_{50} = 1.63$ µg/ml. It was about 290-fold more toxic than thymol ($LD_{50} = 480.2$ µg/ml), which was used as reference substance (unpublished data).

At a preliminary stage, *P. pubescens* seed oil clastogenicity *in vivo* on mammalian cells was evaluated, and it showed absence of toxicity (Sabino *et al.*, 1999). Subsequently, other assays were performed (Sabino *et al.*, 1999) using different models. It was found that 0.7 µg/ml was cytotoxic to human peripheral blood mononuclear cells after 24 hours of exposure, quantified by 3-[4,5-dimethythiazole-2-il]-2,5-dipheniltetrazolium bromide (MTT) assay. However, even at 70 µg/ml, *P. pubescens* seed oil did not present mutagenic activity in five different strains of *Salmonella typhimurium* bacteria, with or without metabolic activation. Furthermore, a single oral dose of 8 g/kg administered to 4- to 5-month-old male DBA1/J mice did not cause death, or histological changes in lung, liver, kidneys, stomach, bowel, and brain examined tissues when compared to respective controls. Moreover, clinical toxicity signs, such as vasoconstriction, cyanosis, piloerection and salivation were not observed (Sabino *et al.*, 1999). More recently, it was shown that *P. pubescens* seed crude ethanolic extract shows cytotoxic activity against the human melanoma cell line SKMEL37, which is a useful property for cancer control (Vieira *et al.*, 2008). These authors identified a furan diterpenoid named vouacapan-6α,7β,14β,19-tetraol as the active compound. While evaluating the effect of hydroalcholic *P. pubescens* seed extract on reduced collagen type II-induced arthritis (CIA), monitored by the arthritis index (AI), some toxicological aspects were also evaluated (Coelho *et al.*, 2001). These authors did not observe any difference on hematological or serum clinical biochemical measures, organ weights, and pathological examinations when compared to the control, which suggests the safety of these seeds. Table 129.2 summarizes the toxicological properties of *P. emarginatus*, considering this a synonymy of *P. pubescens*.

In summary, data presented herein show that the seeds of *Pterodon emarginatus* display marked antinociceptive, anti-inflammatory, anti-ulcerogenic, antibacterial, leishmanicidal, and antioxidant effects, as well as healing of cutaneous ulcers, when dosed orally/topically and assessed in several models in rodents. Furthermore, this chapter also provides consistent evidence that *P. emarginatus* seed contains pharmacologically active constituents. These include a predominance of sesquiterpenes in the essential oil, such as trans-caryophyllene, bicyclogermacrene, and α-humulene; phenolic constituents, like flavonoids, in the buthanolic and methanolic fraction; and diterpenoid derivates, such as acid 6α-7β-dihydrox-yvouacapan-17-oic, in the hexanic fraction of seeds from *P. emarginatus*. Different studies also indicate that inhibition of the synthesis release and expression of pro-inflammatory cytokines such as IL-1α, reduction of leukocytes, and neutrophil influx, and inhibition of NO level are certainly implicated in the action mechanisms of the constituents present in *P. emarginatus* seeds.

TABLE 129.2 Evaluation of *Pterodon* Seed Toxicity and Biological Systems

Species	Seed Preparation	Model	Toxic Effect	Reference
Pterodon emarginatus *Pterodon pubencens*	Hexanic fraction Oil Oil	Adult mice *Artemia salina* Leach Mammalian cells Blood mononuclear cells Five different strains of *Salmonella typhimurium* bacteria Adult male mice	Single dose, LD$_{50}$ 4.02 g/kg Cytotoxicity at 1.63 µg/ml Absence of clastogenicity Cytotoxicity at 0.7 µg/m. Absence of mutagenicity up to 70 µg/ml Single oral dose (8 g/kg) was not harmful	Carvalho *et al.*, 1999 Unpublished data Sabino *et al.*, 1999
	Crude ethanolic extract	Human melanoma cell line SKMEL37	Cytotoxicity at 37 µg/ml	Vieira *et al.*, 2008
	Hydroalcoholic extract	Adult male mice	Absence of hematological, clinical, and biochemical parameters or pathological examination differences No difference in organ weights	Coelho *et al.*, 2001

SUMMARY POINTS

- *Pterodon emarginatus* belongs to the Leguminoseae family, the largest of the angiosperms, which consists of approximately 650 genera and 18,000 species.
- *Pterodon emarginatus* produces large quantities of viable seeds, which can be harvested directly from the fruit of the tree, or collected on the floor after the fall of the fruit.
- There is a growing interest in plants as new drug sources, and the genus *Pterodon* is attracting investigation because of its successful use in folk medicine.
- *Pterodon emarginatus* seeds are a promising anti-inflammatory phytotherapic.
- Owing to its antimicrobial and antioxidative properties, *Pterodon emarginatus* could be used as an additive in the food industry.
- In the literature, there are few data regarding *Pterodon* toxicity.
- More studies should be performed to ensure the safety of *Pterodon emarginatus* in scientific applications.

References

Arriaga, A. M. C., Castro, M. A. B., Silveira, E. R., & Braz-Filho, R. (2000). Further diterpenoids isolated from *Pterodon polygalaeflorus*. *Journal of Brazilian Chemical Society, 1*, 187–190.

Braz Filho, R., Gottlieb, O. R., & Assumpção, R. M. V. (1971). The isoflavones of *Pterodon pubescens*. *Phytochemistry, 10*, 2835–2836.

Carvalho, J., Sertié, J., Barbosa, M., Patrício, K., Caputo, L., Sarti, S., et al. (1999). Anti-inflammatory activity of the crude extract from the fruits of *Pterodon emarginatus* Vog. *Journal of Ethnopharmacology, 64*, 127–133.

Coelho, M. G. P., Marques, P. R., Gayer, C. R. M., Vaz, L. C. A., Neto, J. F. N., & Sabino, K. C. D. (2001). Subacute toxicity evaluation of a hydroalcoholic extract of *Pterodon pubescens* seeds in mice with collagen-induced arthritis. *Journal of Ethnopharmacology, 77*, 159–164.

Dutra, R., Leite, M., & Barbosa, N. (2008a). Quantification of phenolic constituents and antioxidant activity of *Pterodon emarginatus* Vogel seeds. *International Journal of Molecular Sciences, 9*, 606–614.

Dutra, R. C., Trevizani, R., Pittella, F., & Barbosa, N. R. (2008b). Antinociceptive activity of the essential oil and fractions of *Pterodon emarginatus* Vogel seeds. *Latin American Journal of Pharmacy, 27*, 865–870.

Dutra, R. C., Fava, M. B., Alves, C. C. S., Ferreira, A. P., & Barbosa, N. R. (2009a). Antiulcerogenic and anti-inflammatory activities of the essential oil from *Pterodon emarginatus* seeds. *Journal of Pharmacy and Pharmacology, 61*, 243–250.

Dutra, R. C., Pitella, F., Ferreira, A. S., Larcher, P., Farias, R. E., & Barbosa, N. R. (2009b). Healing effect of *Pterodon emarginatus* seeds in experimental models of cutaneous ulcers in rabbits. *Latin American Journal of Pharmacy, 28*, 375–382.

Dutra, R. C., Braga, F. G., Coimbra, E. S., Silva, A. D., & Barbosa, N. R. (2009c). Antimicrobial and leishmanicidal activities of seeds of *Pterodon emarginatus*. *Revista Brasileira de Farmacognosia, 19*, 429–435.

Fascio, M., Mors, W. B., Gilbert, B., Mahjan, J. R., Monteiro, M. B., dos Santos Filho, D., et al. (1976). Diterpenoid furans from *Pterodon* species. *Phytochemistry, 15*, 201–203.

Judd, W. S., Campbell, C. S., Kellogg, E. A., & Stevens, P. F. (1999). *Plant systematics: A phylogenetic approach*. Sunderland, MAL: Sinauer.

Lorenzi, H., & Matos, F. J. A. (2002). *Plantas medicinais no Brasil: nativas e exóticas cultivadas*. Nova Odessa, Brazil: Ed. Plantarum.

Mahjan, J. R., & Monteiro, M. B. (1970). New diterpenoids from *Pterodon emarginatus*. *Anais da Academia Brasileira de Ciências, 42*, 103–107.

Marques, D. D., Machado, M. I. L., Carvalho, M. G., Meleira, L. A. C., Braz-Filho, R., & Aughlin, J. L. (1998). Isoflavonoids and triterpenoids isolated from *Pterodon polygalaeflorus*. *Journal of Brazilian Chemical Society, 9*, 295–301.

Meyer, B. N., Ferrigni, N. R., Putnam, J. E., Jacobsen, L. B., Nichols, D. E., & Maclaughlin, J. L. (1982). Brine shrimp: a convenient general bioassay for active constituents. *Planta Medica, 45*, 31–34.

Mors, W. B., Santos Filho, M. F., Monteiro, H. J., Gilbert, B., & Pelegrino, J. (1967). Chemoprophylactic agent in schistosomiasis: 14,15-epoxigeranygeraniol. *Science, 157*, 950–951.

Oliveira, D. M. T., & Paiva, E. A. S. (2005). Anatomy and ontogeny of *Pterodon emarginatus* (Fabaceae: Faboideae) seed. *Brazilian Journal of Biology, 65*, 483–494.

Sabino, K. C. C., Gayer, C. R. M., Vaz, L. C. A., Santos, L. R. L., Felzenszwalb, I., & Coelho, M. G. P. (1999). *In vitro* and *in vivo* toxicological study of the *Pterodon pubescens* seed oil. *Toxicology Letters, 108*, 27–35.

Torrenegra, R., Bauereib, P., & Achenbach, H. (1989). Homoormosanine-type alkaloids from *Bowdichia virgiloides*. *Phytochemistry, 28*, 2219–2221.

Vieira, C., Marques, M., Soares, P., Matuda, L., de Oliveira, C., Kato, L., et al. (2008). Antiproliferative activity of *Pterodon pubescens* Benth. seed oil and its active principle on human melanoma cells. *Phytomedicine, 15*, 528–532.

CHAPTER 130

Sunflower (*Helianthus annuus* L.) Seeds in Health and Nutrition

Dilipkumar Pal
College of Pharmacy, Institute of Foreign Trade & Management, Lodhipur Rajput, Moradabad, Uttar Pradesh, India

CHAPTER OUTLINE

LIST OF ABBREVIATIONS

COX-2, cyclooxygenase-2
HA, *Helianthus annuus* L.
LPS, lipopolysaccharide
NO, nitric oxide
NOS-2, nitric oxide synthase
PGE_2, prostaglandin E_2
SO, sunflower oil
SS, sunflower seed
TNF-α, tumor necrosis factor

INTRODUCTION

In recent years, consumers have become concerned about the consumption of saturated fats in their diets and meals. High levels of saturated fat consumption may lead to increased blood

Nuts & Seeds in Health and Disease Prevention. DOI: 10.1016/B978-0-12-375688-6.10130-6

serum cholesterol, which in turn increases risk of coronary heart disease and other diseases. Prompted by nutritional recommendations, suggestions that they consume fats lower in saturated fatty acids, and food manufacturers' interest in reducing the use of hydrogenated oils, food producers have become interested in specifying the fatty acid profile of sunflower oil (Valtcho et al., 2009).

Vegetable oils are the principal source of fats in many diets and meals. The saturated fatty acid concentration in sunflower (*Helianthus annuus* L.) oil is moderate (120 g/kg) compared to many edible vegetable oils. Its main fatty acid contents are palmitic (65 g/kg) and stearic (65 g/kg) acids. A reduction of saturated fatty acids in sunflower oil to 60—80 g/kg would enhance its acceptability without disturbing its status in health and nutrition.

BOTANICAL DESCRIPTION

Helianthus annuus L. is a member of the Compositeae family, and has many common names, including Suraj mukhi (Hindi, Gujerati, and Bengali), Suria-mukhi Arkakantha (Sansrit); Surya-phul (Mah), and Sunflower (English).

It is an annual herb with an erect, rough, hairy stem, 1—2 m high, and branched at the top. The leaves are mainly alternate, with broadly ovate blades 7—30 cm long, usually slightly acuminate at the apex. The flower heads are usually 8—15 cm in diameter, but can attain 30—60 cm under cultivation (Figure 130.1). The flowers are single or double, situated terminally on the main axis and branches; receptacles are flat, more frequently dilated and convex, and the ray florets are yellow, surrounding a brownish purple center of disc florets (Figure 130.2). The Seeds are cylindrical, obovoid-compressed, 1 cm inch long and 0.6 cm broad, white, black, or striped gray and black, with the pappus falling early. The sunflower is self-sterile, and fertilization is normally effected by insects (The Wealth of India, 2001; Chopra et al., 2006; Kirtikar and Basu 2006). One bioactive annuionone is present in HA leaves (Figure 130.4).

HISTORICAL CULTIVATION AND USAGE

Helianthus annuus (HA) is considered to have originated from *H. lenticularis* Douglas (from the Greek *Helios*, meaning sun, and *Anthos*, meaning a flower), a wild plant indigenous to Mexico. HA is also indigenous to central and eastern North America, and is cultivated worldwide. It has become well established in the Caucasus, Ukraine, the Balkans, Argentina, the USA, Canada, Eastern and Southern Africa, parts of Asia and Australia, Italy, France, Chile, Uruguay, and some other countries.

FIGURE 130.1
Sunflower (*Helianthus annuus* L.) plant, with leaves and flowers. *Source: www.answers.com/sunflower, reproduced with permission.*

FIGURE 130.2
Sunflower (*Helianthus annuus* L.) with seeds. *Source: http:// herbal.medicalonlinemedia.com: sunflower-a-analgesic-herb, reproduced with permission.*

FIGURE 130.3
Different varieties of sunflower seed. *Source: www.mahalo.com/health benefits of sunflower seeds, reproduced with permission.*

For cultivation as oilseed crop, a number of selections of sunflower are known, varying markedly in height; diameter and color of head; shape, size, color, and oil content of seeds; and suitability for different conditions of soil and climate. Seeds can be divided primarily into three distinct types — giant, semi-dwarf, and dwarf — as per their shape, oil content, and place of cultivation. The sunflower requires a warm climate with moderate rainfall, and has a relatively wide range of tolerance to wet and dry conditions.

Though it is cultivated as garden plant, it owes its economic value to its usefulness as an oilseed or fodder crop. The sunflower is special, in that every part of it can be used. The seeds are eaten by people, birds, and livestock, and are an excellent source of healthy unsaturated fats, proteins, fiber, and nutrients like vitamins B1, B5, and B6, phosphorus, copper, manganese, folate, iron, zinc, amino acids, etc. (Figure 130.5). SS may just be nature's perfect food and may be used to enrich any meal.

Sunflower seeds are used medicinally to calm the nerves, muscles, and blood vessels. HA seeds contain vitamin E, which has significant anti-inflammatory effects that result in the reduction of asthma, osteoarthritis, and rheumatoid arthritis. They are rich in magnesium, which helps to reduce asthma attacks and migraines, lower high blood pressure, and reduce the risk of heart attacks (Chopra *et al.*, 1992; The Wealth of India, 2001; Nadkarni, 2007).

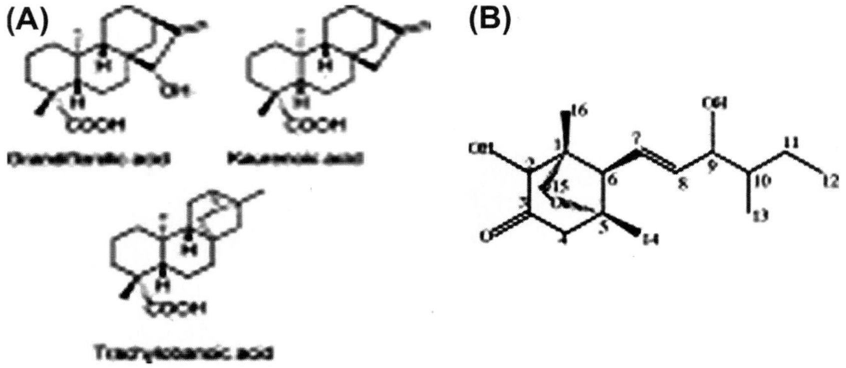

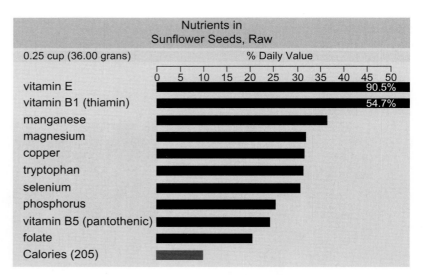

FIGURE 130.4
Structure of diterpenoids (A) and an annuionone (B) isolated from sunflower. *Source: (A) Reprinted from Rodrigo et al. (2008)*, Biochem.Biophysic. Res. Comm., 369, 761–766, *with permission; (B) Reprinted from Anjum & Bajwa (2005)*, Phytochemistry, 66, 1919–1921.

FIGURE 130.5
Nutrients in sunflower seed. *Source: http://www. mahalo.com/health-benefits of sunflower seeds, reproduced with permission.*

PRESENT-DAY CULTIVATION AND USAGE

Sunflower seeds are used for the production of SO, which is a very useful dietary supplement, being enriched in fatty acids and used as an ingredient in many pharmaceutical preparations. Seeds can be sprinkled over cereals, salads, and soups, and mixed with vegetables and snacks. The seeds are diuretic and expectorant, and are now highly valued for the treatment of bronchial, laryngeal and pulmonary infections, coughs, and colds. They are also used to reduce the risk of colon cancer, the severity of hot flushes in women going through menopause, and also diabetic complications. They are also prescribed for snake bite and scorpion sting. Sunflower seed oil is used internally to alleviate constipation, as a lubricant, and is used externally as a massage oil, an oil dressing, and in the treatment of skin lesions, psoriasis, and rheumatism. The oil-cake is a valuable food for cattle and poultry (Bruneton, 1999; The Wealth of India, 2001; LaGow, 2004; Chopra *et al.*, 2006). The oil is used in cooking, soap, lubricants, and candles, and also as biodiesel and biofuel.

APPLICATIONS TO HEALTH PROMOTION AND DISEASE PREVENTION

A variety of sunflower oil (SO) and seed extract preparations are effective against many diseases, like bronchial, laryngeal and pulmonary infections, coughs and cold, etc. A number of studies have been carried out regarding the detailed pharmacology of sunflower, and a few of the reported and established activities are discussed below.

TABLE 130.1 Nutrition through Sunflower Seed

Sunflower oil (high oleic (70% and over))
Nutritional value per 100 g (3.5 oz)

Energy	3699 kJ (884 kcal)
Carbohydrates	0 g
Fat	100 g
saturated	9.748 g
monounsaturated	83.594 g
polyunsaturated	3.798 g
Protein	0 g
Vitamin E	41.08 mg (274%[*])
Vitamin K	5.4 µg (5%[*])

Reproduced from *Sunflower oil, wikipedia, the free encyclopedia*, with permission.
[*]*Percentages are relative to US recommendations for adults.*

Food value and nutritional effects

Sunflower seeds' food and nutritional value is well known. They are excellent sources of healthy unsaturated fats, protein, fiber, and essential nutrients such as vitamin E, the B-complex vitamins, selenium, copper, zinc, folate, iron, and phytochemicals (Table 130.1). SS can be used to complete and enrich any meals, tiffins and snacks (Bruneton, 1999; The Wealth of India, 2001; Dewick, 2003).

Antioxidant activity

Although oxygen is essential for aerobic forms of life, reactive oxygen species (ROS) are known to be implicated in many cellular disorders and abnormalities, and in the development of many diseases, including cardiovascular disease, atherosclerosis, cataracts, chronic inflammation, carcinoma, and neurodegenerative diseases, and premature ageing. Antioxidants, which can inhibit, prevent, or delay the oxidation of an oxidizable substrate in a chain reaction, therefore appear to be very important in the prevention and control of many diseases. However, because of the toxic and carcinogenic effects of synthetic antioxidants, their use is limited and restricted. Therefore, it is of timely interest to search for new antioxidants from plant sources, and attention has been directed towards the bioactivity of flavonoids, ascorbic acid, polyphenols, vitamins, etc., as dietary sources of antioxidants (Pal & Dutta, 2006; Pal & Nimse, 2006; Pal *et al.*, 2008a, 2008b).

The *in vitro* antioxidant activity of stripped SS cotyledon extracts was investigated, along with its different extracts. It was found that aqueous extract at 30 µg/ml showed more potent activity than other SS extracts; the antioxidant activities were concentration-dependent, and compared with the synthetic antioxidant butylated hydroxyl toluene (Maria *et al.*, 2009). The detailed chemical nature of the active principles responsible for such activity is not known; it may be due to the vitamins, particularly vitamin E, present in it. However, the higher antioxidant activity of aqueous extract of this seed suggests its potentiality to prevent several diseases, such as cancer, heart diseases, diabetes, premature ageing, etc. (Mohanta *et al.*, 2007; Pal *et al.*, 2008a, 2008b).

Anticancer activity

In traditional medicine, SS is used for the treatment of tumors and malignant growth. Studies have also been performed to prove the anticancer and antidiabetic activities of HA seed extract.

Sunflower seed is a good source of selenium, a trace mineral that is of fundamental importance to human health. Evidence obtained from intervention trials and studies on animal models of cancer has suggested a strong inverse correlation between selenium intake and cancer incidence. Selenium was found to induce DNA repair and synthesis in damaged cells, and to inhibit the proliferation of cancerous cells and induce their apoptosis — the self-destruct sequence the body uses to eliminate worn out or abnormal cells. Additionally, selenium is incorporated at the active site of many proteins, including glutathione, which is particularly important for cancer protection. Glutathione, the most powerful antioxidant enzyme of the body, is used in the liver to detoxify many potentially harmful molecules. If for any reason glutathione levels are too low, toxic molecules are not disarmed and can damage the cellular DNA of attached cells and promote the development of cancer cells (Simopoulos *et al.*, 1999).

Organic sunflower seeds are an excellent source of vitamin E, which is the body's primary soluble antioxidant. Vitamin E has also been shown to reduce the risk of colon cancer. Recently, promising results also have been found in the treatment of cancers of the stomach and esophagus by SS (Hursting *et al.*, 1990). Research on the guinea pig has led to the discovery that stem marrow and the bottom of sunflower flower (reseptaculum), which contains hemicellulose, is effective in blocking sarcoma and Ehrlich ascites carcinoma (Berquin *et al.*, 2007). The relationship between oil and fat consumption and breast cancer are complex and appear to depend on the specific attributes of the oil or fat in question.

Anti-inflammatory activity

Sunflower seeds have significant anti-inflammatory activity. As per the report of Universidad Complutense de Madrid, Spain, the three diterpene acids (grandiflorolic, kaurenoic, and trachylobanoic acids) (Figure 130.4) isolated from the petroleum ether extract of HA were studied for potential anti-inflammatory activity on the generation of inflammatory mediators in lipopolysaccharide (LPS)-activated RAW 264.7 macrophages. At non-toxic doses, these compounds reduced, in a concentration-dependent manner, NO, PGE2, and TNF-α production, as well as expression of NOS-2 and COX-2. These diterpenoids exhibited significant *in vivo* anti-inflammatory activity, and suppressed 12-O-tetradecanoylphorbol-13-acetate (TPA)-induced mouse ear edema. In addition, inhibition of myeloperoxidase (MPO) action, an index of cellular infiltration, was also found (Rodrigo *et al.*, 2008).

Effect on nerves, muscles, bones, and blood vessels

It has been found that sunflower is a good source of magnesium, which helps to reduce the severity of migraine headaches, as well as the risk of heart attack and stroke. Magnesium is also necessary for healthy bones and for energy production. It helps build the physical structure of bones, and is also stored on the surface of bones for future requirements.

Magnesium counterbalances calcium, thus helping to regulate nerves and muscles. By blocking calcium entry, magnesium keeps nerves, blood vessels, and muscles relaxed. Insufficient dietary magnesium causes free entry of calcium, and thus nerve cells become over-activated, resulting in increased blood pressure, muscle spasm, migraines, tension, soreness, and fatigue. Sunflower seed provides a magnesium source to solve these problems.

Diet and cardiovascular benefits

Sunflower seeds provide sunflower oil, which is enriched with essential vitamin E and comparatively low in saturated fat. The two most common sunflower oils are linoleic and high oleic. Linoleic sunflower oil is a common cooking oil that has high levels of the essential fatty acids known as polyunsaturated fats. It has a clean taste and low levels of trans fats. High-oleic sunflower oils are classified as having monounsaturated levels of 80% and above. Now, newer versions of sunflower oil are available as a hybrid, containing linoleic acid. These have the advantage that their monounsaturated fat levels are lower than in other oleic sunflower oils. The hybrid oil also has lower saturated fat levels than linoleic sunflower oil. Sunflower seeds of all kinds, which contain betaine, are also found to exhibit cardiovascular benefits.

It is well established that diets with a low fat content and high levels of oleic acid lower cholesterol, which in turn lowers the risk of heart disease. A study on adults proved that a balanced diet in which small quantities of saturated fats are replaced with sunflower oil has considerable cholesterol-reducing benefits. Investigation suggests that lowering of cholesterol levels may be caused by the balance of polyunsaturated and monounsaturated fatty acids, and the presence of fiber. SO obtained from SS may contribute to this balance (www.health.am/cholesterol).

Skin protection

Light, non-greasy oil is available in the seeds of the sunflower plant, which also has skin-health benefits. It is loaded with vitamins (A, D, and especially E), calcium, zinc, potassium, iron, and phosphorus. Additionally, SO is enriched with essential fatty acids, including linoleic acids, which make it surprisingly nourishing to the skin (it has been established that the linoleic acid content in skin declines with age, and can be stripped by powerful soaps, shampoos, and cleansers). Sunflower oil has protective, healing effects on the skin, and is especially beneficial for mature, sensitive, dry, damaged, or inflamed skin. It is recommended for all skin types, and is used extensively in skin care products to moisturize, regenerate, protect and condition the skin in part or whole (Rochea *et al.*, 2010).

Sunflower also provides a protective barrier that resists infection. Research using SO on preterm infants, who are often susceptible to infection due to their underdeveloped skin, showed that the infants, especially those of low birth weight, can benefit from SO skin treatments. Infections decreased by 41% in infants who received a daily skin care treatment with SO (www.medscape.com). It is possible that this beneficial skin protection effect may be connected with its antioxidant property.

Other activities

Sunflower seeds possess other properties, including hypocholesterolemic and analgesic activities. They are also used in constipation (as fiber is present in sunflower), dysentery, and urinary problems. It is a fairly efficient foam destroyer, and possesses bactericidal action against many bacteria.

In alternative medicine, medicinal plant preparations have found widespread use, particularly in the case of diseases not amenable to treatment by modern methods. A variety of HA seed extract preparations are effective against a variety of diseases, and are generally non-toxic in nature. However, understanding the mechanism of action of HA seed can be greatly aided by the isolation of active constituents from it and determination of the structure–function relationship. Also, the potent curative effects of HA seed extract against human ailments need to be verified by controlled clinical studies.

ADVERSE EFFECTS AND REACTIONS (ALLERGIES AND TOXICITIES)

Health risks or side effects following the proper administration of designated therapeutic dosages have not been recorded (LaGow, 2004). High consumption of ω6-polyunsaturated fatty acids, which are found in most types of vegetable oils (including sunflower oil), might increase the likelihood that postmenopausal women may develop breast cancer (Sonestedt *et al*, 2008). A similar effect was observed on prostate cancer. Other analyses have suggested an inverse association between total polyunsaturated fatty acids and breast cancer risk.

Sunflower can cause allergic contact dermatitis in those who harvest or prepare plants for the production of SS and SO. The allergens are secreted by trichomes on leaf surfaces. Wind-blown trichomes from dry plants can cause airborne contact dermatitis. The major allergen is known as 1-0-methyl 1-4-5-dihydroniveusin A. The pollen may be described as a minor allergen (Martinez-Force *et al.*, 1999).

Restrictions when taking preparations of HA should be maintained. Sunflower seeds, meal, and oil, taken as part of a well-balanced diet, do not cause any side effects (Kirtikar & Basu, 2006; Nadkarni, 2007).

SUMMARY POINTS

- Since time immemorial, seeds have played an important role in health care and the prevention of disease.
- Sunflower (*Helianthus annuus*) seed preparations have found widespread use in traditional medicine.
- Sunflower seed possesses high food and nutritive value, providing a good source of healthy unsaturated fats, proteins, nutrients, and phytochemicals. It has antioxidant, anticancer, anti-inflammatory, antihypertensive, skin-protective, hypocholesterolemic, analgesic, and antibacterial activities, and calming effects on the nerves, muscles, and blood vessels. It is also used in constipation, dysentery, and urinary problems.
- The detailed chemical nature of the active principle(s) responsible for such activities of sunflower seed require more study. However, chemical and phytochemical analyses of the seed reveal that it is a good source of unsaturated fatty acids, vitamins E, B1, B5, and B6, selenium, magnesium, phosphorus, copper, manganese, folate, fiber, iron, zinc, amino acids, diterpenoids, etc.
- The medicinal potentiality of sunflower seeds needs to be established and verified by controlled clinical studies.

References

Anjum, T., & Bajwa, R. (2005). A bioactive annuionone from sunflower leaves. *Phytochemistry, 66*, 1919–1921.

Berquin, I. M., Min, Y., Wu, R., Wu, J., Perry, D., Cline, J. M., et al. (2007). Modulation of prostate cancer genetic risk by omega-3 and omega-6 fatty acids. *Journal of Clinical Investigation, 117*, 1866–1875.

Bruneton, J. (1999). *Pharmacognosy, phytochemistry of medicinal plants* (2nd ed.). Andover, UK: Intercept Ltd. pp. 151–152.

Chopra, R. N., Nayer, S. L., & Chopra, I. C. (2006). *Glossary of Indian medicinal plants*. New Delhi, India: Publication and Information Directorate, CSIR. p. 131.

Dewick, P. M. (2003). *Medicinal natural products* (2nd ed.). Chichester, UK: John Wiley & Sons Ltd. p. 44.

Hursting, S. D., Thornquist, M., & Henderson, M. M. (1990). Types of dietary fat and the incidence of cancer at five sites. *Preventive Medicine, 19*, 242–253.

Kirtikar, K. R., & Basu, B. D. (2006). Indian medicinal plants (K. S. Mhaskar, & J. F. Caius, eds), Vol. 2, pp. 1370–1371. New Delhi, India: Sri Satguru Publications.

LaGow, B. (2004). *PDR for herbal medicines*. Montvale, NJ: Thomson Healthcare PDR. pp. 803–804.

Maria, D. L. R. G., & Jorge, M. F. (2009). Antioxidant capacity of the striped sunflower (*Helianthus annuus* L.) seed extracts evaluated by three *in vitro* methods. *International Journal of Food Sciences & Nutrition, 60*, 395–401.

Martinez-Force, E., Alvarez-Ortega, R., & Garces, R. (1999). Enzymatic characterization of high palmitic acid sunflower (*Helianthus annuus* L.) mutants. *Planta, 207*, 533–538.

Mohanta, T. K., Patra, J. K., Rath, S. K., Pal, D. K., & Thatoi, H. N. (2007). Evaluation of antimicrobial activity and phytochemical screening of oils and nuts of *Semecarpus anacardium* L.f. *Scientific Research and Essays, 2*, 486–490.

Nadkarni, A. K. (2007). *Indian Materia Medica, Vol 1*. Mumbai, India: Popular Prakashan. p. 614.

Pal, D. K., & Dutta, S. (2006). Evaluation of the antioxidant activity of the roots and rhizomes of *Cyperus rotundus* L. *Indian Jouranl of Pharmaceutical Sciences, 68*, 256–258.

Pal, D. K., & Nimse, S. B. (2006). Screening of the antioxidant activity of *Hydrilla verticillata* plant. *Asian Journal of Chemistry, 18*, 3004–3008.

Pal, D. K., Kumar, M., Chakrabarty, P., & Kumar, S. (2008a). Evaluation of the antioxidant activity of aerial parts of *Cynodon dactylon*. *Asian Journal of Chemistry, 20*, 2479–2481.

Pal, D. K., Kumar, S., Chakrabarty, P., & Kumar, M. (2008b). A study on the antioxidant activity of *Semecarpus anacardium*. L.f. nuts. *Journal of Natural Remedies, 8*, 160–163.

Rochea, J., Alignana, M., Bouniolsa, A., Cernya, M., Moulounguia, Z., Vearc, F., et al. (2010). Sterol content in sunflower seeds (*Helianthus annuus* L.) as affected by genotypes and environmental conditions. *Food Chemistry, 2*(4), 990—995.

Rodrigo, D. V., Sonosoles, H., Natalia, G., Jose, M. M., Benjamin, R., Angel, V., et al. (2008). Modulationn of inflammatory responses by diterpene acids from *Helianthus annuus* L. *Biochemical and Biophysical Research Communications, 369*, 761—766.

The Wealth of India: A dictionary in Indian raw material and industrial products, Vol. 5. (2001). In B. N. Sastri (Ed.). New Delhi, India: Publication and Information Directorate, CSIR. pp. 17—26.

Simopoulos, A. P., Leaf, A., & Salem, N., Jr. (1999). Essentiality of and recommended dietary intakes for omega-6 and omega-3 fatty acids. *Annals of Nutrition and Metabolism, 43*, 127—130.

Sonestedt, E., Ericson, U., Gullberg, B., Skog, K., Olsson, H., & Wirfalt, E. (2008). Do both heterocyclic amines and omega-6-polyunsaturated fatty acids contribute to the incidence of breast cancer in postmenopausal women of the malmo diet and cancer cohort? *International Journal of Cancer, 123*, 1637—1643.

Valtcho, D. Z., Brady, A. V., Brian, S. B., Normie, B., Tess, A., & Johnson, B. (2009). Oil content and saturated fatty acids in sunflower as a function of planting date, nitrogen rate and hybrid. *Journal of Agronomy, 101*, 1003—1011.

CHAPTER 131

Tamarind (*Tamarindus indica*) Seeds: An Overview on Remedial Qualities

Mahadevappa Hemshekhar, Kempaiah Kemparaju, Kesturu Subbaiah Girish
Department of Studies in Biochemistry, University of Mysore, Manasagangothri, Mysore, India

CHAPTER OUTLINE

1107

LIST OF ABBREVIATIONS

ABTS, 2,2-azino-bis-(3-ethyl benzthiozole-6-sulfonic acid)

DTH, delayed type hypersensitivity

fMLP, n-formyl methionyl leucyl phenyl alanine

GOT, glutamate oxaloacetate tranasaminase

GPT, glutamate pyruvate transaminase

H_2O_2, hydrogen peroxide

HDL, high density cholesterol

LDL, low density cholesterol

LPS, lipopolysaccharide

NO, nitric oxide

PAF, platelet activating factor

PLA_2, phospholipase A_2

STZ, streptazotocin

TSE, tamarind seed extract

TSP, tamarind seed polysaccharide

TTI, tamarind trypsin inhibitor

Nuts & Seeds in Health and Disease Prevention. DOI: 10.1016/B978-0-12-375688-6.10131-8

INTRODUCTION

The tamarind is a tree (*Tamarindus indica* L.) comprising multitherapeutic parts, such as the pulp, leaves, and seeds. Though the presence of tannin and other coloring matter in the testa makes the whole seed inedible, the presence of proteins, lipids, poly/oligosaccharides, procyanidins, triterpenes (lupanone and lupeol), xyluloglycans, and, antioxidants such as 2-hydroxy-3,4-dihydroxyacetophenone, methyl 3,4-dihydroxybenzoate, 3,4-dihydroxy-phenylacetate, and epicatechin, makes it therapeutically significant. Tamarind seed extract (TSE) is recommended for the treatment of snakebites, chronic diarrhea, diabetes, dysentery, jaundice, boils, eye diseases, and ulcers, and also as an antiviral, anti-inflammatory, and antirheumatic agent. In addition, TSE is useful in the preparation of facial toners, moisturizers, serums, gels, masks, and anti-ageing formulations.

BOTANICAL DESCRIPTION

Tamarind (*Tamarindus indica* L.) belongs to the dicotyledonous family Fabaceae (Leguminosae), and the genus *Tamarindus* is monotypic. The tree is a long-lived, large evergreen, and generally grows wild. A mature tree may attain a maximum height of 30 m. Leaves are pinnate with opposite leaflets, giving a billowing effect in the wind, and consist of 10–40 leaflets. Flowers are produced in racemes, and are mainly yellow in color. The fruit is a brown pod-like legume, which contains a soft acidic pulp, and is sausage-shaped, curved, or straight, with rounded ends; the pulp is thick and blackish-brown in color, and contains many hard-coated seeds. Tamarind timber consists of hard, dark-red heartwood and softer yellowish sapwood. The seeds are hard, shiny, reddish- or purplish-brown, and can be scarified to assist germination. The tree can grow even in unfertile soil, and has the ability to fix nitrogen and withstand severe drought.

HISTORICAL CULTIVATION AND USAGE

The tamarind tree has long been naturalized in the East Indies and the Pacific islands. According to a survey in 1797, the tree was first planted in Hawaii. It commonly grows in Asian countries, including India, Bangladesh, Sri Lanka, Thailand, and Indonesia, and also in some parts of the African and American continents. The tree was first introduced to India, and then to Persia and Arabia. In 4 BC, the Ancient Egyptians and Greeks are known to have used the fruit as an edible component, the bark for household woodwork, and the leaves as a cure for many ailments.

PRESENT-DAY CULTIVATION AND USAGE

Currently, the tree is cultivated on a large scale throughout tropical and subtropical countries, producing over 275,500 tonnes of tamarind fruit annually. In Thailand, a sweet variety is grown specifically for edible purposes. In Nigeria, the fruit is used to prepare *Kunun Tsamiya*, a traditional dish. The young leaves and flower buds are used as a vegetable in Myanmar, while in India and Mexico the tamarind is popular as a dried and salted snack, in pickles, and in the preparation of many dishes and chutneys. The leaves have traditionally been used in herbal tea for reducing malarial fever, and as an Ayurvedic medicine for gastric or digestion problems, as well as for its cardioprotective activity. Several therapeutic bioactive compounds have been isolated from different parts of the tree; however, TSE has been largely employed in the textile, cosmetic, and pharmaceutical industries (Shankaracharya, 1998).

APPLICATIONS TO HEALTH PROMOTION AND DISEASE PREVENTION

Though the tamarind seed seems to have been overlooked in the past, nowadays TSE and the isolated products of tamarind are receiving a great deal of attention due to the presence of

TABLE 131.1 Food and Non-food Uses of Tamarind Seed

Type of Seed Fraction	Functions
Seed gum	Emulsifier, stabilizer, gelling agent, palatability improver, antimicrobial agent, insulator, odor improver, glazing agent, stickiness preventer, and bodying agent. Converts organic waste into biofertilizer Allows proliferation of gram-negative bacteria
Seed powder	Coating material, gelatinizing agent, emulsifying agent, adsorbent, thickener, binder Used in cosmetic preparations, greaseless ointments, penicillin production, concentration of rubber latex
Seed polysaccharide/ oligosaccharide	Crystallization inhibition, paste material, preservative, mineral disperser, stabilizer, low calorie food
Seed coat	Antioxidative component, anti-acne medicine, antidiabetic activity, tanning material

therapeutic agents like antioxidants, triterpenes, polyphenolic bioactive compounds, procyanidines, polysaccharides, and other unknown factors, which have healing activities for various human pathophysiological disorders (Marangoni *et al.*, 1988; Sudjaroen *et al.*, 2005). Recent studies have shown the vital medicinal applications of TSE and allied products (Table 131.1).

Tamarind seed extract has been shown to exhibit strong antioxidant scavenging activity against hydroxyl radicals and superoxide anions produced by the $ABTS/H_2O_2/FeCl_3$ (Feton reaction) and hypoxanthine-xanthine oxidase (neotetrazolium) systems, respectively. It has also displayed scavenging activity against peroxyl radicals generated by $ABTS/H_2O_2/peroxidase$ and $ABTS/H_2O_2/myoglobin$ systems (Pumthong, 1999). Four antioxidants have been identified: 2-hydroxy-3,4-dihydroxyacetophenone, methyl 3,4-dihydroxybenzoate, 3,4-dihydroxyphenylacetate, and epicatechin (Tsuda *et al.*, 1994; Saowanee *et al.*, 2004). In addition, TSE inhibited the *in vitro* and *in vivo* production of both nitrite and NO. The TSE was found to be non-toxic, as it showed no effect on the viability of mouse peritoneal macrophages (Komutarin *et al.*, 2003).

1109

The antidiabetogenic potential of TSE was evaluated using STZ-induced diabetic rats. Supplementation of TSE resulted in a significant diminution of fasting blood sugar level after 7 days. Continuous supplementation of TSE for 14 days produced a significant elevation in liver and skeletal muscle glycogen content, and the activity of liver glucose-6-phosphate dehydrogenase, compared to the diabetic group. The enzyme activities of liver glucose-6-phosphatase, and liver and kidney GOT and GPT were decreased significantly in the TSE-supplemented diabetic rats, and restored to the control level after 14 days of supplementation. The antidiabetogenic property of TSE could be due to the presence of bioactive molecule(s), which may sensitize the insulin receptor or stimulate the β cells for insulin production, or may regulate the glucose-6-phosphatase enzyme activity. Furthermore, TSE was shown to rectify the dyslipidemia observed in diabetic conditions. TSE supplementation resulted in significant attenuation of serum LDL and HDL levels compared to control, which strengthens the hypolipidemic effect of the TSE, since LDL/HDL and total cholesterol/HDL ratios are the markers of dyslipidemia. Both the ratios increased in STZ-induced diabetic rats, whereas TSE shifted the augmented ratios towards the normal level. The hypolipidemic effect may be due to elevated levels of HDL, which prevents cardiovascular disease. The actual components present in TSE for such correction are not identified from the study (Maiti *et al.*, 2005).

Recently, the anti-snake venom property of TSE has been evaluated. It completely inhibited the PLA_2, protease, hyaluronidase, L-amino acid oxidase, and 5′-nucleotidase enzyme activities of *Vipera russellii* venom dose-dependently. These are the major hydrolytic enzymes responsible for both local and systemic manifestations of snakebites, such as tissue damage, inflammation, hypotension, and hemorrhage in vital organs. TSE neutralized the degradation of the β chain of human fibrinogen (Figure 131.1) and indirect hemolysis caused by the venom. It

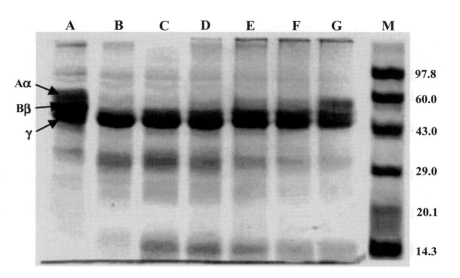

FIGURE 131.1
Effect of tamarind seed extracton on *V. russellii* venom induced fibrinogenolytic activity. Lane A, fibrinogen alone (50 μg), Lane B, fibrinogen + venom (5 μg), and Lanes C—G, venom (5 μg), preincubated with different concentrations (5, 10, 15, 20, and 25 μg) of tamarind seed extract for 15 minutes at RT prior to addition of fibrinogen. Incubation was terminated by adding 20 μl denaturing buffer, and samples were analyzed by 10% SDS-PAGE. M represents molecular weight markers in kDa.

completely neutralized the venom-induced edema, hemorrhage (Table 131.2), and myotoxicity (Figure 131.2), including lethality. The venom-induced lethality ($2LD_{50}$ dose) was antagonized dose-dependently *in vivo*. On the other hand, animals that received TSE 10 minutes after the injection of venom were protected from venom-induced toxicity (Table 131.3). Since it inhibits both hydrolytic enzymes and pharmacological effects, it may be used as an alternative to serum therapy, and, in addition, as a rich source of potential inhibitors of venom hydrolytic enzymes, which are involved in several pathophysiological disorders (Ushanandini *et al.*, 2006). These findings may open up the possibility of characterizing novel inhibitor molecule(s) in the quest for a new therapeutic molecule.

It was also reported that TSE constituted a mucoadhesive polymer TSP, described as a viscosity enhancer exhibiting mucomimetic, mucoadhesive, and bioadhesive activities (Burgalassi *et al.*, 2000). TSP is a high molecular weight, non-ionic, neutral, and branched polysaccharide consisting of a cellulose-like backbone that carries xylose and galactoxylose substituents. The chemical residues of TSP are similar to those of mucin MUC-1 and episialin. These features make TSP as an attractive candidate as a vehicle for ophthalmic medicaments. Being similar to mucins, TSP was able to bind to the cell surface, and intensify the contact between drugs and the adsorbing biological membrane. In the treatment of experimental *Pseudomonas aeruginosa*- and *Staphylococcus aureus*-induced keratitis in rabbits, TSP enhanced transcorneal disposition and intra-aqueous penetration of rufloxacin in both healthy and infected rabbits when administered topically in a drop regimen. The experimental results suggest that the TSP prolongs the precorneal residence of antibiotics, such as rufloxacin and gentamycin, and

TABLE 131.2 Neutralization of Hemorrhagic Activity of *V. russellii* Venom by Tamarind Seed Extract	
Group (*n* = 5)	**Diameter of Hemorrhagic Spot (mm)**
Venom (20 μg) alone	19 ± 3
Venom (20 μg) : TSE (5 μg)	11 ± 2
Venom (20 μg) : TSE (10 μg)	6 ± 1
Venom (20 μg) : TSE (15 μg)	3 ± 2
Venom (20 μg) : TSE (20 μg)	0
Venom (20 μg) : TSE (25 μg)	0
TSE alone (500 μg)	0

Values are mean ± SEM of five animals.

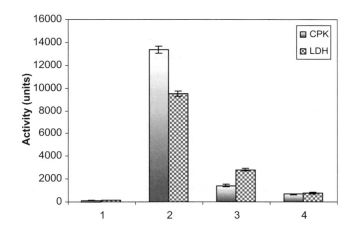

FIGURE 131.2

Serum LDH and CPK activities of mice injected with *V. russellii* venom in the presence and absence of tamarind seed extract. Animals were injected (intramuscularly) at a constant volume of 50 μl with (1) saline; (2) venom alone (1.25 mg/ kg body weight); (3) venom + TSE (venom : TSE, 1 : 1, w/w); and (4) venom + TSE (venom : extract, 1 : 2, w/w). Each denoted value is the mean ± SEM of five independent experiments. After 3 hours, blood was drawn, and serum LDH and CPK levels were determined.

augments drug accumulation in the cornea. A greater amount of mucins may have led to prolonged residence of TSP-rufloxacin on the corneal tissue, allowing increased penetration of rufloxacin into the aqueous humor (Ghelardi *et al.*, 2004). These special qualities of TSP could be exploited for future treatment of intraocular microbial infections or dry eye syndrome (Ronaldo & Valente, 2007).

Seed xyloglucans and pectinic oligogalacturonides also prevented the suppression of *Candida albicans* and alloantigen-induced DTH immune responses in experimental mice. Hence, the ability of TSP to prevent suppression of the T cell-mediated immune response during chronic UV irradiation and to preserve the capacity of UV-irradiated mice to reject a highly antigenic transplanted UV-induced tumor was investigated. TSP progressively lost its ability to protect DTH immune responses to *C. albicans*, but not to alloantigen, during the course of chronic UV irradiation. Despite protection of immunity to alloantigen, the transplanted tumor cells grew equally well in all UV-irradiated animals. This difference might be because the relationship between the tissue damage caused by UV radiation and the host response to the injury is dynamic, and changes with time. The TSP probably does not exert its protective effect directly by stimulating the repair of UV-induced DNA damage, but can preserve DTH immunity by

TABLE 131.3 Survival Time of Mice Injected with *V. russellii* Venom in the Presence and Absence of Tamarind Seed Extract

Group ($n = 7$)	Time of Death (min)
Venom alone (2 LD_{50})	147 ± 26
Venom : TSE (1 : 0.5; w/w)	263 ± 32
Venom : TSE (1 : 0.5; w/w)	216 ± 25
Venom : TSE (1 : 1; w/w)	965 ± 81
Venom : TSE (1 : 1; w/w)	673 ± 52
Venom : TSE (1 : 2; w/w)	Alive
Venom : TSE (1 : 2; w/w)[*]	908 ± 97
TSE alone (10 mg/kg body weight)	Alive

Values are mean ± SEM of seven animals.
[*]*Animals received TSE at 10-min intervals following the injection of venom.*

preventing the production of immunosuppressive cytokines such as IL-10 by UV-irradiated keratinocytes *in vivo* and *in vitro* (Strickland *et al.*, 2001).

Recently, a 14.9-kDa elastase inhibitor has been isolated and characterized from seed. It was found to be highly effective against serine proteases, such as bovine trypsin and human neutrophil elastase, and could not inhibit porcine pancreatic elastase, bovine chymotrypsin, and cysteine proteinases (papain and bromelain). The protein was found to be non-toxic, and preferentially affected exocytosis of elastase of neutrophils by PAF and fMLP. This may indicate selective inhibition of PAF receptors by the protein, and thus it can be considered a potent anti-inflammatory agent (Fook *et al.*, 2004). Further, a 21.4-kDa non-competitive TTI has been purified and evaluated for *in vitro* bioinsecticidal activity, using insect digestive enzymes from different orders. The TTI exerted inhibition to a varied extent against the enzymes from coleopterans *Anthonomus grandis*, *Zabrotes subfasciatus*, *Callosobruchus maculatus*, and *Rhyzopertha dominica*, and lepidopterons *Plodia interpuncptella*, *Alabama argillacea*, and *Spodoptera frugiperda*. The digestive enzymes from Diptera, *Ceratitis capitata* (fruit fly), were also inhibited. *In vivo* assay using *C. capitata* and *C. maculatus* larvae further confirmed the bioinsecticidal activity (Araujo *et al.*, 2005).

Tamarind seed extract was evaluated for the biosorption of aqueous Cr (IV). It exhibited a high sorption capacity, 90 mg/g at equilibrium Cr (IV) concentration of 2.5 mg/l at pH 2. The biosorption capacity significantly varies with pH, ionic strength and temperature. Since it has a high sorptive capacity, it seems to be an economical and worthwhile alternative to conventional methods. TSE is also an efficient flocculant for the treatment of textile industry wastewater (Agarwal *et al.*, 2005). Furthermore, it has been used for the sorptive removal of fluoride from pharmaceutical effluents, as well as from field water samples. Batch sorptive defluoridation was conducted under variable experimental conditions, such as pH, agitation time, initial fluoride concentration, particle size, and sorbent dose. Maximum defluoridation was achieved at pH 7.0. Defluoridation capacity decreases with increase in temperature and particle size; this may be due to coulombic interactions. In addition, TSE increases the excretion rate of fluoride, as suggested by the diet-based study. The augmented level of urinary excretion of fluoride was observed in individuals who received a TSE-based diet compared to normal subjects. In contrast, metal ions (such as Zn^{2+} and Mg^{2+}) and creatinine levels were reduced to a greater extent (Murugan & Subramanian, 2006). Hence, TSE can be used to reduce fluoride toxicity and to prevent dental and skeletal fluorosis in humans.

ADVERSE EFFECTS AND REACTIONS (ALLERGIES AND TOXICITY)

Even though tamarind seeds have medicinal applications, there have been many reports on the toxic or allergic effects of TSE as well as tamarind seed powder. In the textile industries, about 1–4% of workers develop various allergic responses. Many have reported with acute respiratory disease, dyspnea (dry cough to severe bronchospasm), bronchitis, asthma, and intra-dermo-reactions such as skin allergies and lesions, where they use TSE as a sizing agent. Affected individuals exhibited symptoms such as breathlessness, chest tightness, severe cough, headache, an unpleasant and apparently characteristic taste in the mouth accompanied by dryness, eosinophilia, and leukocytosis. In addition, a majority of workers in the rayon and cotton handling industries have been seriously affected with respiratory problems, eye irritation, and purulent sputum where seed extract is involved (Murray *et al.*, 2005). The exact reasons for these toxic and allergic reactions are not known; however, they could be due to the presence of chitin and other polysaccharide fractions in TSE.

SUMMARY POINTS

- Tamarind seeds are a gold mine for bioactive principles such as antioxidants, procyanidines, flavanoids, polysaccharides, and proteins.

- Since these agents play a curative role in many pathophysiological disorders, TSE/isolated components may be recommended to treat oxidative stress, diabetes mellitus, inflammation, cancer, thrombolytic disorders, microbial infections, snakebite, arthritis, ocular lesions, chronic diarrhea, ulcers, serpinopathies, obesity, and dental and skeletal fluorosis.
- The isolated antioxidants from tamarind seeds have drawn the attention of pharmaceutical industries.
- The isolated polysaccharides could be applicable as facial toners, skin moisturizers, biopesticides, and biosorptive agents.
- Tamarind seed polysaccharides could be used as microspeares and nanofiber-based bioscaffolds for drug delivery and tissue-engineering purposes.
- Future studies related to making composite biopolymers of TSP with versatile polysaccharides such as hyaluronan, collagen, PLAGA and chitosan for better mucoadhesive and controlled release properties would be of great interest.

References

Agarwal, G. S., Bhuptawat, H. K., & Chaudhari, S. (2005). Biosorption of aqueous chromium (VI) by *Tamarindus indica* seeds. *Bioresource Technology, 97,* 949−956.

Araujo, C. L., Bezerra, I. W., Oliveira, A. S., Moura, F. T., Macedo, L. L., Gomes, C. E., et al. (2005). *In vivo* bioinsecticidal activity toward *Ceratitis capitata* (fruit fly) and *Callosobruchus maculatus* (cowpea weevil) and *in vitro* bioinsecticidal activity toward bifferent orders of insect pests of a trypsin inhibitor purified from tamarind tree (*Tamarindus indica*) seeds. *Journal of Agricultural and Food Chemistry, 53,* 4381−4387.

Burgalassi, S., Raimondi, L., Pirisino, R., Banchelli, G., Boldrini, E., & Saettone, M. F. (2000). Effect of xyloglucan (tamarind seed polysaccharide) on conjuctival cell adhesion to laminin and corneal epithelium wound healing. *European Journal of Ophthalmology, 10,* 71−76.

Fook, J. M., Macedo, L. L., Moura, G. E., Teixeira, F. M., Oliveira, A. S., Queiroz, A. F., et al. (2004). A serine proteinase inhibitor isolated from *Tamarindus indica* seeds and its effects on the release of human neutrophil elastase. *Life Sciences, 76,* 2881−2891.

Ghelardi, E., Tavanti, A., Davini, P., Celandroni, F., Salvetti, S., Parisio, E., et al. (2004). A mucoadhesive polymer extracted from Tamarind seed improves the intraocular penetration and efficacy of Rufloxacin in topical treatment of experimental bacterial keratitis. *Journal of Antimicrobial Agents and Chemotherapy, 48,* 3396−3401.

Komutarin, T., Azadi, S., Butterworth, L., Keil, D., Chitsomboon, B., Suttajit, M., et al. (2003). Extract of the seed coat of *Tamarindus indica* inhibits nitric oxide production by murine macrophages *in vitro* and *in vivo*. *Food and Chemical Toxicology, 42,* 649−658.

Maiti, R., Das, U. K., & Ghosh, D. (2005). Attenuation of hyperglycemia and hyperlipidemia in streptozotocin-induced diabetic rats by aqueous extract of seed of *Tamarindus indica*. *Biological & Pharmaceutical Bulletin, 28,* 1172−1176.

Marangoni, A., Ali, I., & Kermasha, S. (1988). Composition and properties of seeds of the true legume *Tamarindus indica*. *Journal of Food Science, 53,* 1452−1455.

Murray, R., Dingwall-Fordyce, I., & Lane, R. E. (2005). An outbreak of weaver's cough associated with tamarind seed powder. *British Journal of Industrial Medicine, 14,* 105−110.

Murugan, M., & Subramanian, E. (2006). Studies on defluoridation of water by tamarind seed, an unconventional biosorbent. *Journal of Water Health, 4,* 453−461.

Pumthong, G. (1999). *Antioxidative activity of polyphenolic compounds extracted from seed coat of Tamarindus indicus Linn.* Chiangmai Mai. Thailand: Chiangmai Mai University. IV−V.

Ronaldo, M., & Valente, C. (2007). Establishing the tolerability and performance of tamarind seed polysaccharide (TSP) in treating dry eye syndrome: results of a clinical study. *BMC Ophthalmology, 7,* 5.

Saowanee, L., Dayin, M., Supaporn, D., Peter, L. D., La-Ied, P., & Suwassa, P. (2004). Extraction of antioxidants from sweet Thai tamarind seed coat-preliminary experiments. *Journal of Food Engineering, 63,* 247−252.

Shankaracharya, N. B. (1998). Tamarind — chemistry, technology and uses: a critical appraisal. *Journal of Food Science & Technology, 35,* 193−208.

Strickland, F. M., Sun, Y., Darvill, A., Eberhard, S., Pauly, M., & Albersheim, P. (2001). Preservation of the delayed-type hypersensitivity response to alloantigen by xyloglucans or oligogalacturonide does not correlate with the capacity to reject ultraviolet-induced skin tumors in mice. *Journal of Investigative Dermatology, 116,* 62−68.

Sudjaroen, Y., Haubner, R., Wurtele, G., Hull, W. E., Erben, G., Spiegelnalder, B., et al. (2005). Isolation and structure elucidation of phenolic antioxidants from tamarind (*Tamarindus indica* L.) seeds and pericarp. *Food and Chemical Toxicology, 43,* 1673−1682.

Tsuda, T., Wantanabe, M., Ohshima, K., Yamamoto, A., Kawakishi, S., & Osawa, T. (1994). Antioxidative compo-nents isolated from the seed of tamarind (*Tamarindus indica* L). *Journal of Agricultural and Food Chemistry, 42,* 2671–2674.

Ushanandini, S., Nagaraju, S., Harish Kumar, K., Vedavathi, M., Machiah, D. K., Kemparaju, K., et al. (2006). The anti-snake venom properties of *Tamarindus indica* (Leguminosae) seed extract. *Phytotherapy Research, 20,* 851–858.

CHAPTER 132

Use of Tea (*Camellia oleifera* Abel.) Seeds in Human Health

Chanya Chaicharoenpong[1], Amorn Petsom[2]
[1] Institute of Biotechnology and Genetic Engineering, Chulalongkorn University, Bangkok, Thailand
[2] Department of Chemistry, Faculty of Science, Chulalongkorn University, Bangkok, Thailand

1115

LIST OF ABBREVIATIONS

$\cdot CCl_3$, trichloromethyl free radical
CCl_4, carbon tetrachloride
$[Cl^-]_i$, intracellular chloride ion concentrations
DPPH, α,α-diphenyl-β-picrylhydrazyl
FT-Raman spectroscopy, Fourier Transform Raman spectroscopy
GPx, glutathione peroxidase
GRd, glutathione reductase
GSH, glutathione
GST, glutathione S transferase
H_2O_2, hydrogen peroxide
HUVECs, human umbilical vein endothelial cells
ICAM-1, intercellular adhesion molecule-1
LDL, low density lipoprotein
mICAM-1, membrane-associated intercellular adhesion molecule-1
RBC, red blood cell

Nuts & Seeds in Health and Disease Prevention. DOI: 10.1016/B978-0-12-375688-6.10132-X

ROS, reactive oxygen species
sICAM-1, soluble intercellular adhesion molecule-1
TEAC, Trolox equivalent antioxidant capacity
TNF-α, tumor necrosis factor-α

INTRODUCTION

Camellia oleifera Abel. is a woody shrub that has been widely cultivated in China for its edible oil, commonly known as tea seed oil. The plant has been grown as an ornamental plant in Western countries. Camellia oil or tea seed oil contains a high amount of oleic acid (C18:1) and antioxidants with various biological activities (Shyu *et al.*, 1990), but low saturated fat. Tea seed oil is normally used as a cooking oil because of its high quality and healthful properties. It is also used in Chinese traditional medicine and in cosmetics.

BOTANICAL DESCRIPTION.

C. oleifera is a member of the Theaceae family, and is native to East and South-east Asia. It is an evergreen shrub which is in leaf year-round. It flowers from October to April, and the seeds ripen in September. It can grow to as high as 4.5−6 m, and has brown bark. The leaves are alternately arranged, oblanceolate, serrated, glossy, glabrous, and 3−17 cm long by 2−4 cm wide. The leaf shape is blade-elliptic, with an obtuse or broadly obtuse apex. Its petiole is short (3−7 mm long) and stout, usually giving the leaf an erect position. The flowers are hermaphrodite, fragrant, and white, with 5−10 petals, and usually 5−6 cm in diameter. Fruits are round and 2−4 cm in diameter, and each fruit bears two to five seeds. The seeds are triangular in shape, and 15−20 mm in diameter (Figure 132.1) (Gilman & Watson, 1993).

FIGURE 132.1
Flowers (A), fruits (B), pericarps (C), and seeds (D) of *Camellia oleifera* Abel. *Photographs by C. Chaicharoenpong.*

HISTORICAL CULTIVATION AND USAGE

The *C. oleifera* plant prefers well-drained acidic sandy and loamy soil, and the partial shade of light woodland. It has been widely cultivated for many centuries in central and south China for its seeds. The seeds were put through an extraction process for cooking oil production and folk medicines. Traditional uses of tea seed oil included treatments for stomach ache, burn injuries, and trauma, and as an antitussive, expectorant, and anthelmintic in China (Jiangsu New Medical College, 1977). After the oil has been extracted from the seed, the defatted seed-cake can be used for animal feed, fertilizer, and the development of a natural insecticide, pesticide, and molluscicide (Hu *et al.*, 2005). *C. oleifera* was brought to the West in the 18th century as an ornamental plant.

PRESENT-DAY CULTIVATION AND USAGE

In China, breeding work on *C. oleifera* has been conducted to improve the oil quality and yield, resulting in several cultivars with attractive ornamental features. The hypocotyl grafting technique is a popular cultivation technique used in China to propagate a fine clone of this plant (Zhang *et al.*, 2007). This technique may also have the potential to improve propagation of camellia plants with similar seed characteristics. Nowadays, crop management practices, such as preparing planting sites with organic fertilizer, controlling weeds, thinning, pruning, alternating harvest time, improving harvest techniques, and using new clones, have been implemented. Yields are significantly higher than those of the original stands. The highest yield achieved is 750 kg of fruits per hectare, in the eighth year after implementing new management practices. This result translates to approximately 300 liters of tea oil per acre (Yu, 2005).

APPLICATIONS TO HEALTH PROMOTION AND DISEASE PREVENTION

Tea seed oil has very high-value nutritional and health protection functions because of its many unsaturated fatty acids (Liu *et al.*, 1979). This oil is comprised of 75–80% oleic acid (Lee & Gioia, 2009), and has a low content of saturated fats (Table 132.1). Tea seed oil is bright yellowish and remains liquid even at refrigerated temperatures. High oleic-acid oils are healthy because of their suppressant effect against several degenerative pathologies. Health-promoting effects include lowering blood pressure, cholesterol, and triglycerides, thus preventing cardiovascular diseases and cancer (Chen *et al.*, 1998). Due to its high content of antioxidants, tea seed oil is used as an emollient for skin care, to minimize signs of aging. Additionally, it is rich in vitamins A, B, and E, and contains no natural trans fatty acids (i.e., fatty acids in which there is at least one double bond in the configuration). Trans fatty acids occur in partly hydrogenated fat and in butter fat, and tend to behave, biochemically, more like saturated fatty acids than unsaturated fatty acids. Additionally, tea seed oil is a good source of healthy minerals, such as phosphorus, magnesium, calcium, iron, selenium, and zinc. Tea seed oil can add nutritional value to foods cooked with it. The oil has a high smoke point of 251.67°C, and thus is suitable for use in high-temperature cooking.

TABLE 132.1 Fatty Acid Composition of Tea Seed Oil, Analyzed as Methyl Ester Using the Gas Chromatographic Technique

Fatty Acids	Content (% w/w)
Oleic acid (C18:1)	80
Linoleic acid (C18:2)	8
Palmitic acid (C16:0)	9
Stearic acid (C18:0)	1

Traditional titration was the classic method to determine authenticity of tea seed oil, and for determination of its iodine value (Bangash *et al.*, 2004). Recently, high-efficiency FT-Raman spectroscopy was performed to probe the degree of unsaturated fatty acids, which can be used to monitor adulteration and to evaluate quality of the oil (Weng *et al.*, 2006).

A methanolic extract of tea seed oil exhibited antioxidant activity (Lee & Yen 2006). *In vitro* experiments showed that its α,α-diphenyl-β-picrylhydrazyl (DPPH) scavenging activity was 66.5% at a concentration of 200 μg/ml, and the Trolox equivalent antioxidant capacity (TEAC) assay was equal to 59 μM Trolox at a concentration of 200 μg/ml. Sesamin (1) and 2,5-bis-benzo[1,3]dioxol-5-yl-tetrahydro-furo[3,4-*d*][1,3]dioxine (2) (Figure 132.2) isolated from a methanolic extract of tea seed oil inhibited the oxidative injury of mammalian red blood cells (RBCs) by free radicals by up to 49% and 48%, respectively. Mammalian RBCs are often used as a model system to test the efficacy of antioxidants because they act as an oxygen carrier. Thus, compounds (1) and (2) may bind at cell membranes and protect against damage from free radical attack. Additionally, compound (2) inhibited H_2O_2-stimulated reactive oxygen species (ROS) formation in human erythrocytes by 14% to 86% at increasing concentrations of 10−50 μM, higher than sesamin (1), even up to 16-fold at 50 μM. ROS attack membranes of RBCs and induce cell membrane oxidation. Therefore, compound (2) could potentially reduce the oxidative injury by aerobic metabolism. Moreover, compounds (1) and (2) suppressed low density lipoprotein (LDL) oxidation, which is a marker of arteriosclerosis (Lee & Yen, 2006). Tea seed oil exhibited antioxidant activities to prevent free radical related diseases.

The hepatoprotective activity of tea seed oil on CCl_4-induced acute oxidative damage in rat liver was reported (Lee *et al.*, 2007). Carbon tetrachloride (CCl_4) is a hepatotoxic chemical that is widely used in animal models to investigate chemical toxin-induced liver damage. CCl_4 is used to elucidate the mechanisms of action of hepatotoxic effects, such as fatty degeneration, fibrosis, hepatocellular death, and carcinogenicity (Weber *et al.*, 2003). Tea seed oil reduced serum aspartate aminotransferase, alanine aminotransferase, and lactate dehydrogenase when compared with a CCl_4-treated group in which these serums were elevated. Generally, when there is hepatic injury by chemical agents, aspartate aminotransferase and alanine aminotransferase in plasma are increased. Moreover, treatment of rats with tea seed oil elevated the content of glutathione (GSH), whereas GSH content declined in the group of rats with CCl_4-induced hepatotoxicity. GSH is an intracellular antioxidant, playing a critical role in combating CCl_4-induced injury by covalently binding to trichloromethyl radical ($\cdot CCl_3$). Acute CCl_4 damage significantly decreases the expression of antioxidant enzymes in liver, including glutathione-related enzymes. Glutathione reductase (GRd) is responsible for the regeneration of GSH, and glutathione peroxidase (GPx) works together with GSH in disintegrating hydrogen peroxide and other organic hydroperoxides. To monitor the balance of oxidative stress and chemopreventive ability, glutathione-related enzymes were measured. Tea seed oil increased the expression of GPx and GRd, but had no effect on glutathione *S* transferase (GST) and catalase. Additionally, tea seed oil reduced the content of malondialdehyde in rats. Malondialdehyde is the product of lipid peroxidation, and is a biomarker of the level of oxidative stress in an organism. Thus, tea seed oil protected the liver against CCl_4-induced oxidative damage in rats, and reduced the injury of fatty degeneration and necrosis in rat liver, because of its antioxidant and free-radical scavenger effects.

The defatted tea seed-cake of *C. oleifera* contains a large quantity of saponins. Saponins are amphipathic glycosides that are secondary metabolites found in many plants and some animals. Tea seed saponins are harmless to humans, as they are not absorbed into the bloodstream through the intestinal wall. The defatted tea seed-cake is extensively used in aquaculture to eliminate unwanted fish and harmful insects in prawn ponds (Tang, 1961). Additionally, it is also used as a molluscicide to control golden apple snails,

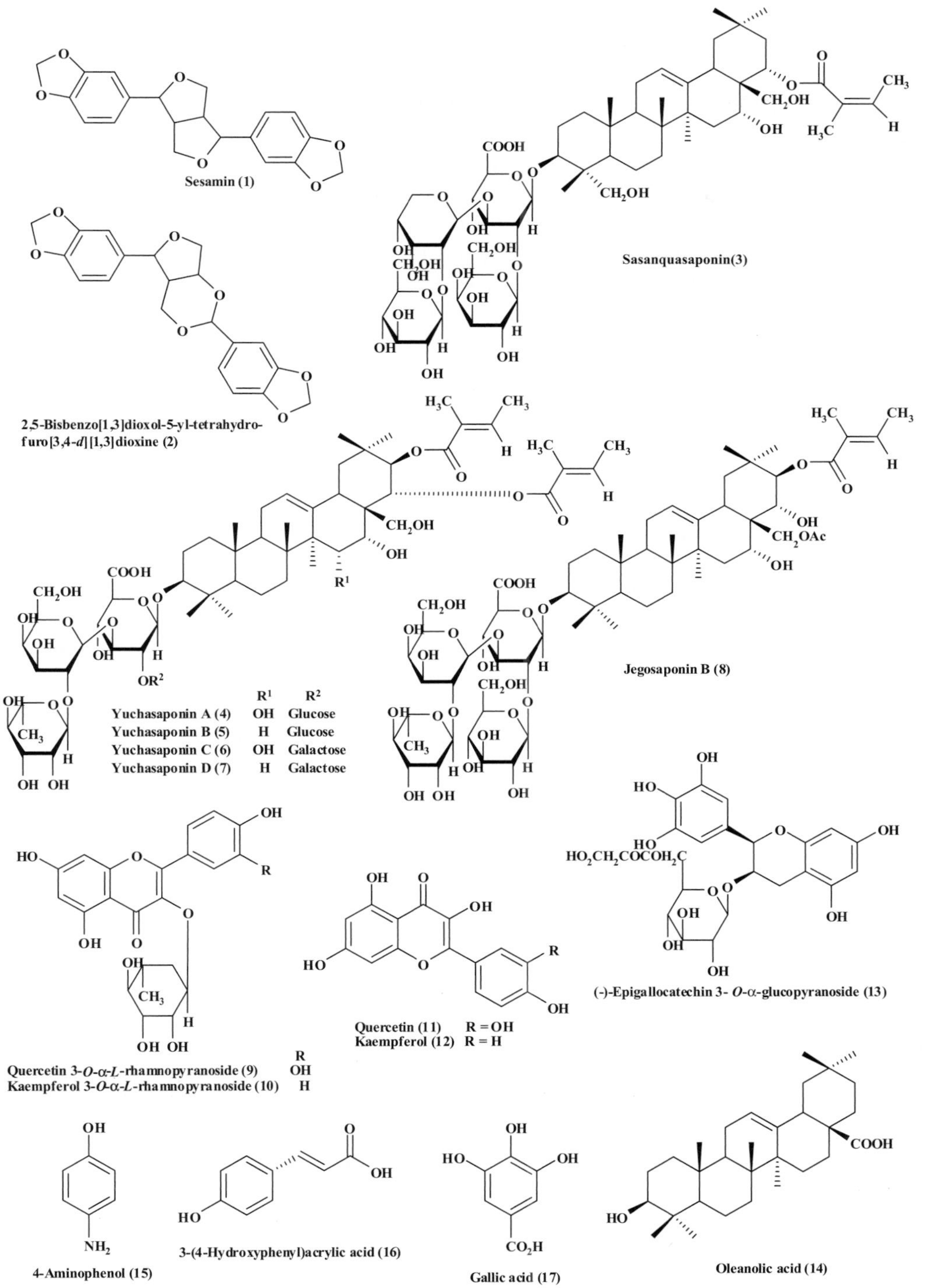

FIGURE 132.2

Structures of active compounds isolated from *Camellia oleifera* Abel.

Pomacea canaliculata Lamarck., one of the major pests in rice fields. Sasanquasaponin (22-O-angleoylcamelliagenin C-3-O-[β-D-glucopyranosyl(1→2)][β-D-glucopyranosyl (1→2)-a-L-arabinopyranosyl (1→3)]-β-D-glucopyranosiduronic acid, (3); Figure 132.2) is a triterpenoid saponin extracted from the defatted tea seed-cake of *C. oleifera* (Lai *et al.*, 2004). It can suppress ischemia-induced elevation of intracellular chloride ion concentrations ($[Cl^-]_i$) in isolated mouse cardiomyocytes. For the *in vitro* experiment, it slightly decreased $[Cl^-]_i$ in non-hypoxic myocytes and delayed the hypoxia/reoxygenation-induced increase in $[Cl^-]_i$ during ischemia and reperfusion. Thus, sasanquasaponin (3) exhibited anti-arrhythmic effects during ischemia and reperfusion in mouse cardiac cells. Moreover, sasanquasaponin (3) affected the expression of the intercellular adhesion molecule-1 (ICAM-1) in burn-induced rats (Huang *et al.*, 2005). ICAM-1 functions in signal transduction, and serum concentration of ICAM-1 can be increased during immune or inflammatory disorders. ICAM-1 has two forms: a soluble intercellular adhesion molecule-1 (sICAM-1), and a membrane-associated intercellular adhesion molecule-1 (mICAM-1). The sICAM-1 is produced by proteolytic cleavage of mICAM-1. Sasanquasaponin (3) inhibited the over-expression of sICAM-1 and mICAM-1 in the rat aorta, and inhibited the transcription of ICAM-1. Thus, sasanquasaponin (3) suppressed an inflammatory reaction induced by burns.

Additionally, sasanquasaponin (3) decreased lactate dehydrogenase activity and malondialdehyde content, and inhibited neutrophil adhesion to human umbilical vein endothelial cells (HUVECs) (Huang *et al.*, 2007). Moreover, sasanquasaponin (3) inhibited nuclear factor kappa B transnuclear activity, and suppressed tumor necrosis factor-α (TNF-α). However, it increased mitochondrial superoxide dismutase and glutathione peroxidase activities. Therefore, sasanquasaponin (3) could protect HUVECs against anoxia and reoxygenation injury, and the protective mechanisms appeared to be related to anti-lipoperoxidation and anti-adhesion. Furthermore, sasanquasaponin (3) exhibited a cardioprotective effect on anoxia/reoxygenation-induced oxidative stress in a neonatal rat cardiomyocyte model (Chen *et al.*, 2007). It suppressed the increase of lactate dehydrogenase activity and decrease of cell viability of anoxia/reoxygenation in a dose-dependent manner up to 10 µM. Sasanquasaponin (3) reduced the concentration of malondialdehyde and improved the activities of various antioxidant enzymes (i.e., superoxide dimutase, catalase, and glutathione peroxidase) in cultured neonatal rat cardiomyocytes undergoing anoxia/reoxygenation. It also suppressed the decrease of glutathione and the increase of oxidized glutathione resulting from anoxia/reoxygenation in a dose-independent manner. In addition, sasanquasaponin (3) decreased intracellular ROS levels and alleviated calcium accumulation in cardiomyocytes undergoing anoxia/reoxygenation. Thus, sasanquasaponin (3) protected cardiomyocytes against oxidative stress induced by anoxia/reoxygenation by way of attenuating ROS generation and increasing activities of endogenous antioxidants (Chen *et al.*, 2007).

Biological activities and chemical constituents in flower buds of *C. oleifera* were reported (Sugimoto *et al.*, 2009). Its *n*-butanol soluble fraction of methanolic extract exhibited a potent protective effect on ethanol- (83.1%) and indomethacin- (76.7%) induced gastric mucosal lesions in rats at 100 mg/kg. Ethyl acetate and *n*-butanol soluble fractions of methanolic extract of flower buds showed inhibitory effects on rat lens aldose reductase with IC_{50} 1.5 and 4.9 µg/ml, respectively. Moreover, the fractions also exhibited scavenging effects on DPPH radical (SC_{50} 4.3 and 2.4 µg/ml, respectively) and superoxide (IC_{50} 2.2 and 1.8 µg/ml, respectively). Yuchasaponins A, B, C, and D (4–7), jegosaponin B (8), quercetin 3-*O*-α-*L*-rhamnopyranoside (9), kaempferol 3-*O*-α-*L*-rhamnopyranoside (10), quercetin (11), kaempferol (12), (−)-epigallocatechin 3-*O*-β-*D*-glucopyranoside (13), oleanolic acid (14), 4-aminophenol (15), 3-(4-hydroxyphenyl)acrylic acid (16), and gallic acid (17) (Figure 132.2) were isolated from the *n*-butanol soluble fraction of methanolic extract of flower buds of *C. oleifera*. Compounds (9) and (10) exhibited moderate inhibitory activities on gastric lesions in rats, induced by 50 mg/kg ethanol, at 34.2% and 38.1%, respectively.

ADVERSE EFFECTS AND REACTIONS (ALLERGIES AND TOXICITY)

No published information has been found on this subject.

SUMMARY POINTS

- Tea-seed oil is suitable for use as a cooking oil because of its high oleic acid content, antioxidants, vitamins, and high smoke point.
- Its potent antioxidant activity confers protection from free-radical related diseases, including the hepatoprotective effect.
- Sasanquasaponin, from defatted tea seed-cake of *C. oleifera*, may be a potent cardioprotective agent against anoxia/reoxygenation injury because it inhibits arrhythmia effects during ischemia and reperfusion in mouse cardiac cells, suppresses an inflammatory reaction induced by burns in rats, and protects against anoxia/reoxygenation injury in HUVECs and the neonatal rat cardiomyocyte model.
- The flower buds of *C. oleifera* also contain potent antioxidants. They exhibit a protective effect on ethanol- and indomethacin-induced gastric mucosal lesions in rats; antioxidant activities as determined by DPPH scavenging effect; and moderate activities on gastric lesions induced by ethanol in rats.

References

Bangash, F. K., Ahmad, T., Atta, S., & Zeb, A. (2004). Effects of irradiation on the storage stability of red palm oil. *Journal of the Chinese Chemical Society, 51*, 991−996.

Chen, H. P., He, M., Huang, Q. R., Liu, D., & Huang, M. (2007). Sasanquasaponin protects rat cardiomyocytes against oxidative stress induced by anoxia-reoxygenation injury. *European Journal of Pharmacology, 575*, 21−27.

Chen, L. F., Qui, S. H., & Peng, Z. H. (1998). Effects of sasanguasaponin on blood lipids and subgroups of high density lipoprotein cholesterol in hyper lipoidemia rat models. *Pharmacology Clinical Chinese Materia Medica, 14*, 13−16.

Gilman, E. F., & Watson, D. G. (1993). *A series of the Environmental Horticulture Department, Florida Cooperative Extension Service.* Institute of Food and Agricultural Sciences, University of Florida. Fact sheet ST−116.

Hu, W. L., Liu, J. X., Ye, J. A., Wu, Y. M., & Guo, Y. Q. (2005). Effect of tea saponin on rumen fermentation *in vitro*. *Animal Feed Science and Technology, 120*, 333−339.

Huang, Q. R., Shao, L. J., He, M., Chen, H. P., Liu, D., Luo, Y. M., et al. (2005). Inhibitory effects of sasanquasaponin on over-expression of ICAM-1 and on enhancement of capillary permeability induced by burns in rats. *Burns, 31*, 637−642.

Huang, Q. R., He, M., Chen, H. P., Shao, L. J., Liu, D., Luo, Y. M., et al. (2007). Protective effects of sasanquasaponin on injury of endothelial cells induced by anoxia and reoxygenation *in vitro*. *Basic & Clinical Pharmacology & Toxicology, 101*, 301−308.

Jiangsu New Medical College. (1977). *Dictionary on Chinese traditional herbs* (2nd ed.) Shanghai, China: People Press. p. 1054.

Lai, Z. F., Shao, Z. Q., Chen, Y. Z., He, M., Huang, Q. R., & Nishi, K. (2004). Effects of sasanquasaponin on ischemia and reperfusion injury in mouse hearts. *Journal of Pharmacological Sciences, 94*, 313−324.

Lee, C. P., & Yen, G. C. (2006). Antioxidant activity and bioactive compounds of tea seed (*Camellia oleifera* Abel.) oil. *Journal of Agricultural and Food Chemistry, 54*, 779−784.

Lee, C. P., Shih, P. H., Hsu, C. L., & Yen, G. C. (2007). Hepatoprotection of tea seed oil (*Camellia oleifera* Abel.) against CCl$_4$-induced oxidative damage in rats. *Food and Chemical Toxicology, 45*, 888−895.

Lee, P. J., & Gioia, A. J. (2009). *Characterization of tea seed oil for quality control and authentication. Application note.* Milford, MA, USA: Water Corporation.

Liu, T. Y., Lee, J. T., & Sun, C. T. (1979). The tea seed composition, oil characteristics, and the correlation between its maturity and oil content. *Food Sciences, 6*, 109−113.

Shyu, S. L., Huang, J. J., Yen, G. C., & Chang, R. L. (1990). Study on properties and oxidative stability of teaseed oil. *Food Sciences, 17*, 114−122.

Sugimoto, S., Chi, G., Kato, Y., Nakamura, S., Matsuda, H., & Yoshikawa, M. (2009). Medicinal flowers. XXVI. Structures of acylated oleanane-type triterpene oligoglycosides, yuchasaponins A, B, C, and D, from the flower buds of *Camellia oleifera* − gastroprotective, aldose reductase inhibitory, and radical scavenging effects. *Chemical Pharmaceutical Bulletin, 57*, 269−275.

Tang, Y. A. (1961). The use of saponin to control predaceous fish in shrimp ponds. *Progressive Fish Culturist, 23*, 43−45.

Weber, L. W., Boll, M., & Stampfl, A. (2003). Hepatotoxicity and mechanism of action of haloalkanes: carbon tetrachlorides as a toxicological model. *Critical Reviews in Toxicology, 33*, 105−136.

Weng, R. H., Weng, Y. M., & Chen, W. L. (2006). Authentication of *Camellia oleifera* Abel. oil by near infrared Fourier Transform Raman spectroscopy. *Journal of the Chinese Chemical Society, 53,* 597—603.

Yu, J. (2005). Management practices to improve yield of *Camellia oleifera* Abel. *HortScience, 40,* 1082—1083.

Zhang, D., Yu, J., Chen, Y., & Zhang, R. (2007). Ornamantal teaoil camellia cultivars and their hypocotyl graft propagation. *Southern Nursery Association Research Conference, 52,* 256—260.

CHAPTER 133

Usage of Tomato (*Lycopersicum esculentum* Mill.) Seeds in Health

Mónica González, M. Carmen Cid, M. Gloria Lobo
Instituto Canario de Investigaciones Agrarias, La Laguna, Spain

1123

LIST OF ABBREVIATIONS

DW, dry weight, composition based on a dry matter basis
FA, fatty acid
FW, fresh weight, composition based on a wet matter basis

INTRODUCTION

World tomato production is approximately 130 million tonnes (FAOSTAT, 2008), 30% of which is used to obtain derived products. Tomato paste manufacturing produces 70–75 kg of solid waste per tonne of fresh tomatoes. Seeds account for approximately 10% of the fruit and 60% of the total waste; the amount of seeds produced by the tomato-processing industries is estimated to be around 1.7 million tonnes worldwide.

It is vital that these seeds be reused, because they constitute an environmental problem. Tomato seeds contain nutrients and healthy phytochemical compounds; therefore, they could be used as sources of ingredients to fortify or functionalize food.

Nuts & Seeds in Health and Disease Prevention. DOI: 10.1016/B978-0-12-375688-6.10133-1

BOTANICAL DESCRIPTION

The tomato, usually referred to as *Lycopersicum esculentum* Mill. (syn. *Solanum lycopersicum* L.), belongs to the Solanaceae family, Solanoideae subfamily, and Solaneae tribe. The correct taxonomic genus is still being debated, although recent studies suggest that Linnaeus was correct in attributing them to *Solanum*.

The *Lycopersicum* genus includes cultivated tomatoes and a small number of closely related wild species native to Central and South America, from Mexico to Peru. Rick (1976) divided the genus in two groups: the *esculentum* complex (six species that are easily crossed with commercial tomatoes), and the *peruvianum* complex (two species that are not easily crossed with *L. esculentum*).

Tomato is a short-lived perennial diploid dicotyledonous ($2n=24$), cultivated as an annual for its fruits, which are savory in flavor (and, accordingly, often considered to be a vegetable). Plants are herbaceous, usually sprawling, and have weak, woody stems (1−3 m long), with pinnate leaves (10−25 cm long) consisting of five to nine serrated leaflets; both stems and leaves are densely glandular-hairy. Inflorescences are cymes with 2−12 yellow flowers (1−2 cm across), having five pointed lobes on the corolla.

HISTORICAL CULTIVATION AND USAGE

The cherry tomato (*L. esculentum* cv. *cerasiforme*) is indigenous to the Peru−Ecuador area, and is most likely an ancestor of modern cultivated varieties. It reached its present status after a long domestication history, probably initiated in Central Mexico, where the Aztecs used it for cooking, calling it *tomatl* or *xtomatl*.

After the colonization of the Americas, the Spanish took the tomato to their colonies in the Caribbean and the Philippines (whence it moved to South-east Asia), and also introduced it to southern Europe in Spain and in Naples, where the plants grew easily. Tomato fruits were definitely used as food in this area of the Mediterranean by the mid-16th century. However, their acceptance in north-western Europe was relatively slow. Initially, fruits were used solely as tabletop decoration, because it was suspected that they could be poisonous (as members of the nightshade family); they were only widely consumed by the end of the 18th century. Throughout the 17th and 18th centuries Europeans took the tomato to Asia and the USA, where its production and consumption expanded rapidly in the 19th century.

PRESENT-DAY CULTIVATION AND USAGE

Tomatoes are one of the most important fruits in the human diet around the world. Global production has increased by about 30% in the past four decades. The leading producers are China and the USA, followed by Turkey, India, Italy, Iran, Egypt, Brazil, and Spain (FAOSTAT, 2008).

Tomatoes can be consumed as fresh fruit or in processed forms — as tomato preserves, dried tomatoes, and tomato-based foods.

Varieties of tomatoes used in processing have a determinate growth, dwarf habit, concentrated and uniform fruit set and ripening, and are harvested mechanically. Fresh market varieties are generally indeterminate, grown in greenhouses, and harvested by hand. There are a large number of cultivars, differing in size, form, color, flavor, taste, and shelf-life, as well as in the content of nutritional and bioactive compounds. Usually they are grouped into five major types: classic round tomatoes, cherry and cocktail tomatoes, plum tomatoes, beefsteak tomatoes, and vine or truss tomatoes.

The tomato-processing industry produces large amounts of bio-wastes: the seeds and skin of the fruit. Tomato by-products are widely used as animal feed (Persia *et al.*, 2003; da Silva *et al.*, 2009). Other attempts at the practical utilization of this fruit waste include the production of enzymes, bioactive compounds, or biomass to produce biofuels (Giannelos *et al.*, 2005) or to generate electricity. Tomato seeds are also used to give an edible oil, and as a protein source that can be employed in the production of food products such as mayonnaise, margarine, tomato paste, or bread.

APPLICATIONS TO HEALTH PROMOTION AND DISEASE PREVENTION

Applications for tomato seeds that promote health and prevent disease are based on the nutrients and phytochemical compounds that it contains. The seeds are rich in dietary fiber (35% DW), fat (20–30%) and polyunsaturated FAs, proteins (25–30% DW) and essential amino acids, and minerals (Persia *et al.*, 2003).

Tomato seeds are made up of about 27% neutral lipids and about 3.5% waxy materials and polar lipids (Al-Wandawi *et al.*, 1985). Linoleic acid is the predominant FA, followed by oleic acid and palmitic acid (Table 133.1). Other FAs present in lower concentrations are stearic and linolenic acid, among others (Giannelos *et al.*, 2005; USDA, 2009). The nutritional value and health benefits of ω-6 and ω-3 polyunsaturated FAs (linoleic and linolenic acid, respectively) are widely known, including the prevention of cardiovascular diseases. However, the ω-6 to ω-3 FA ratio (20 : 1) in tomato seed oil is relatively high. Moreover, consuming 25 g (two tablespoons) of tomato seed oil per day would be an important contribution to meeting the adult dietary reference intake for FAs (Table 133.2). The ω-6 and ω-3 FA contributions are 59–91% and 29–42% (depending on the sex or age of the reference population), respectively. A very small percentage of *trans* FAs has been quantified in tomato seed oil (0.11%), indicating that this oil is practically unaltered during industrial processing. Therefore, tomato seed oil has a very promising FA profile, although further studies should be conducted to establish its suitability for human consumption.

The crude protein (N × 6.25) in full-fat tomato seeds is 32%, and seed flake obtained after lipids extraction indicates a high protein content (around 40%) (Al-Wandawi *et al.*, 1985)

1125

TABLE 133.1 Fatty Acid (FA) Profile of Tomato Seed Oil (% Total FAs)

Fatty Acids	Reference			
	Cámara *et al.*, 2001	Giannelos *et al.*, 2005	Lazos *et al.*, 1998	USDA, 2009
Total saturated	19	18	21	21
12:0 lauric	—	—	—	0
14:0 myristic	0.13	0.10	0.20	0.21
16:0 palmitic	14	12	14	16
17:0 heptadecanoic	0.12	0.10	0.30	—
18:0 stearic	5.2	5.2	6.0	4.6
20:0 eicosanoic	0.37	0.41	0.30	—
22:0 behenic	0.12	0.090	< 0.1	—
Total monounsaturated	22	23	23	24
16:1 palmitoleic	0.55	0.35	0.50	0.52
18:1 oleic	20	22	22	23
20:1 eicosenoic	—	0.12	0.10	0
22:1 erucic	0.043	—	—	0
Total polyunsaturated	59	59	56	55
18:2 linoleic	56	56	54	53
18:3 linolenic	2.5	2.8	2.0	2.4

TABLE 133.2 Fatty Acid (FAs, % Total Weight of the Oil) Profile of Different Edible Oils Compared to Tomato Seed Oil (USDA, 2009); Contribution of Consumption of 25 g of Tomato Seed Oil to Adult Daily Dietary Intake of FAs

Fatty Acids	Corn	Grapeseed	Sunflower[1]	Olive	Palm	Tomato Seed			DRI[a] (g/day)
						Content (%)	Intake (g/day)	% of DRI[a]	
Total saturated	13	9.6	9.9	14	49	20	4.0	—	—
12:0 lauric	0.024	0	0	0	0.10	0	0	—	—
14:0 myristic	0	0.10	0.057	0	1.0	0.20	0.04	—	—
16:0 palmitic	11	6.7	3.7	11	44	15	3.0	—	—
17:0 heptadecanoic	0.067	—	—	0.022	—	—	—	—	—
18:0 stearic	1.9	2.7	4.3	2.0	4.3	4.4	0.88	—	—
20:0 eicosanoic	0.43	—	—	0.41	—	—	—	—	—
22:0 behenic	0	—	1.0	0.13	—	—	—	—	—
Total monounsaturated	28	16	84	73	37	23	4.6	—	—
16:1 palmitoleic	0.11	0.30	0.095	1.3	0.30	0.50	0.10	—	—
18:1 oleic	27	16	83	71	37	22	4.4	—	—
20:1 eicosenoic	0.13	—	0.96	0.31	0.10	0	0	—	—
22:1 erucic	0	—	0	0	0	0	0	—	—
Total polyunsaturated	55	70	3.8	11	9.3	53	11	—	—
18:2 linoleic	54	70	3.6	9.8	9.1	51	10	71−59 (91−83)[b]	14−17 (11−12)[b]
18:3 linolenic	1.2	0.10	0.19	0.76	0.20	2.3	0.46	29 (42)[b]	1.6 (1.1)[b]

The nutritional value, based on the FA profile, of the tomato seed oil is comparable to other edible oils such as corn oil. Moreover, consuming 25 g of tomato seed oil per day would be an important contribution to meeting the adult dietary reference intake (DRI) for proteins and FAs.
[1]Sunflower oil high in oleic;
[a]Based on adequate intakes (in italics); Source: National Academy of Sciences (2005);
[b]When the references are different for males and females, the values for females are indicated in parentheses.

(Table 133.3). Assuming a daily consumption of 20 g (two tablespoons) of tomato seeds per day, for an adult, the contribution to the protein intake in humans is considerably (Table 133.3) higher than that of cereal grains or other seed kernels. With the exception of tryptophan, all essential amino acids are present in tomato seeds (Table 133.4). Glutamic acid and aspartic acid are the most predominant amino acids. The major essential amino acids in seeds are arginine, lysine, valine, and leucine (Persia et al., 2003; da Silva et al., 2009). However, the most limited amino acids in defatted tomato seed flour are methionine and cysteine. Moreover, the contribution to the daily intake of essential amino acids of a 20-g serving of tomato seed is very significant, representing between 10 and 30% (depending on the sex or age of the reference population). It is important to highlight the high content of lysine and threonine in tomato seeds, which could substantially improve the protein quality of cereal products. The functional properties of tomato seed meal and protein concentrates are comparable to other plant proteins. Proteins of de-oiled meal and alkali-extracted concentrate of tomato seeds are classified into albumin, globulin (the major protein), gliadin, and glutenin. The quality of tomato seed protein, evaluated by using biological studies, indicates that tomato seeds contain high-quality plant proteins that could be supplemented into various food products (Sogi et al., 2005).

The mineral content of tomato seed flour is considerable (Table 133.5), being the levels of all elements higher in tomato seed when compared with those of cereal grains and more similar to those of seed kernels (Table 133.3). A 20-g serving of tomato seeds would be a significant contribution to the daily intake of minerals such as copper, iron, magnesium, manganese, and zinc in general, but especially beneficial in respect of copper and manganese. The contribution to the intake of iron and zinc would also be considerable (35−79% and 27−38% of the

TABLE 133.3 Amino Acid [g/100 g Dry Weight (DW)] Profile, Protein Content (g/100 g DW) and Mineral Content (mg/100 g DW)] of Different Cereal Grains and Seed Kernels (USDA, 2009) Compared to Tomato Seed (Al-Wandawi et al., 1985); Contribution of Consumption of 20 g of Tomato Seeds to the Adult Daily Dietary Intake of Amino Acids, Total Proteins, and Minerals

Nutrient	Barley[1]	Corn[2]	Rice[3]	Wheat[4]	Pumpkin	Sesame[5]	Sunflower[6]	Tomato seed Content (%)	Intake (g/day)[a]	% of DRI[b]	DRI[b] (g/day)[a]
Amino acids											
Tryptophan	0.21	0.067	0.10	0.21	0.58	0.39	0.35	—	—	—	0.39 (0.32)c
Threonine	0.42	0.35	0.29	0.40	1.0	0.74	0.93	1.3	0.33	22 (28)c	1.5 (1.2)c
Isoleucine	0.46	0.34	0.34	0.51	1.3	0.76	1.1	0.78	0.20	14 (17)c	1.4 (1.2)c
Leucine	0.85	1.2	0.66	0.93	2.4	1.4	1.7	1.5	0.38	12 (15)c	3.1 (2.5)c
Lysine	0.47	0.27	0.30	0.38	1.2	0.57	0.94	1.7	0.43	15 (19)c	2.9 (2.3)c
Methionine	0.24	0.20	0.18	0.21	0.60	0.59	0.49	0.38	0.10	9.1 (11)c	Met + Cys 1.4 (1.2)c
Cystine	0.28	0.17	0.096	0.32	0.33	0.36	0.45	0.11	0.028	—	—
Phenylalanine	0.70	0.46	0.41	0.65	1.7	0.94	1.2	0.99	0.25	25 (30)c	Phe + Tyr 2.6 (2.2)c
Tyrosine	0.36	0.38	0.30	0.40	1.1	0.74	0.67	1.6	0.40	—	—
Valine	0.61	0.48	0.47	0.62	1.6	0.99	1.3	1.2	0.30	17 (20)c	1.8 (1.5)c
Histidine	0.28	0.29	0.20	0.32	0.78	0.52	0.63	0.52	0.13	13 (16)c	1.0 (0.83)c
Arginine	0.63	0.47	0.60	0.64	5.4	2.6	2.4	1.8	0.45	—	—
Alanine	0.49	0.71	0.46	0.49	1.5	0.93	1.1	1.0	0.25	—	—
Aspartic acid	0.78	0.66	0.74	0.70	3.0	1.6	2.4	2.4	0.60	—	—
Glutamic acid	3.3	1.8	1.6	4.3	6.2	4.0	5.6	5.1	1.3	—	—
Glycine	0.45	0.39	0.39	0.55	1.4	1.2	1.5	0.94	0.24	—	—
Proline	1.5	0.82	0.37	1.4	1.3	0.81	1.2	0.92	0.23	—	—
Serine	0.53	0.45	0.41	0.65	1.7	0.97	1.1	1.0	0.25	—	—

Continued

TABLE 133.3 Amino Acid [g/100 g Dry Weight (DW)] Profile, Protein Content (g/100 g DW) and Mineral Content (mg/100 g DW) of Different Cereal Grains and Seed Kernels (USDA, 2009) Compared to Tomato Seed (Al-Wandawi et al., 1985); Contribution of Consumption of 20 g of Tomato Seeds to the Adult Daily Dietary Intake of Amino Acids, Total Proteins, and Minerals—continued

Nutrient	Barley[1]	Corn[2]	Rice[3]	Wheat[4]	Pumpkin	Sesame[5]	Sunflower[6]	Tomato seed			DRI[b] (g/day)[a]
								Content (%)	Intake (g/day)[a]	% of DRI[b]	
Total proteins	12	9.4	7.9	14	30	18	21	32	8.0	14 (17)[c]	56 (46)[c]
Minerals											
Calcium	33	7	23	34	46	975	78	153	38[a]	4–3	1000–1200[a]
Copper	0.50	0.31	0.28	0.43	1.3	4.1	1.8	5	1.3[a]	144	0.90[a]
Iron	3.6	2.7	1.5	5.4	8.8	15	5.3	25	6.3[a]	79 (79–35)[c]	8 (8–18)[a,c]
Magnesium	133	127	143	90	592	351	325	400	100[a]	25 (32)[c]	400–420 (310–320)[a,c]
Manganese	1.9	0.49	3.7	3.4	4.5	2.5	2.0	13	3.3[a]	143 (183)[c]	2.3 (1.8)[a,c]
Phosphorus	264	210	333	402	1233	629	660	–	–	–	700[a]
Potassium	452	287	223	435	809	468	645	650	163[a]	3	4700[a]
Sodium	12	35	7.0	2.0	7.0	11	9.0	200	50[a]	4–3	1300–1500[a]
Zinc	2.8	2.2	2.0	3.5	7.8	7.8	5.0	12	3.0[a]	27 (38)[c]	11 (8)[a,c]

Tomato seed protein, the essential amino acids scores and the mineral content are similar, or even better, than those of cereal grains and seed kernels. Their contribution to the protein, essential amino acid and mineral intake in humans is considerable.

[1]Hulled;
[2]Yellow;
[3]Brown, long-grain, raw;
[4]Flour, wholegrain;
[5]Whole, dried;
[6]Dried;
[a]For minerals, the data correspond to mg/day;
[b]Dietary reference intake (DRI) based on recommended dietary or adequate intakes (in italics); Source: National Academy of Sciences. (2006);
[c]When the references are different for males and females, the values for females are indicated in parentheses.

TABLE 133.4 Amino Acid (g/100 g Dry Weight) Profile of Tomato Seed

Amino Acids	Reference			
	Al-Wandawi et al., 1985	da Silva et al., 2009[1]	Knoblich et al., 2005[1]	Persia et al., 2003
Tryptophan	—	—	—	—
Threonine	1.3	0.75	0.45	0.82
Isoleucine	0.78	0.78	0.46	0.97
Leucine	1.5	1.3	0.78	1.5
Lysine	1.7	1.1	0.61	1.3
Methionine	0.38	0.33	0.19	0.39
Cystine	0.11	0.30	0.21	0.40
Phenylalanine	0.99	0.93	0.54	1.2
Tyrosine	1.6	—	—	0.90
Valine	1.2	0.90	0.56	1.1
Histidine	0.52	0.43	0.26	0.55
Arginine	1.8	1.6	0.76	2.1
Alanine	1.0	0.94	0.60	1.1
Aspartic acid	2.4	2.2	1.5	2.6
Glutamic acid	5.1	3.1	3.0	4.7
Glycine	0.94	1.1	0.72	—
Proline	0.92	1.1	0.53	1.4
Serine	1.0	0.99	0.56	1.2

With the exception of tryptophan, all essential amino acids are present in tomato seeds.
[1]Tomato seeds and peels.

TABLE 133.5 Mineral (mg/100 g Dry Weight) Content of Tomato Seed

Minerals	Reference			
	Alvarado et al., 2001[1]	Al-Wandawi et al., 1985	da Silva et al., 1999[1]	Knoblich et al., 2005
Calcium	170	153	160	140
Copper	1.3	5.0	—	1.6
Chromium	—	< 1.0	0.13	—
Iron	25	25	49	24
Magnesium	240	400	—	210
Manganese	1.8	13	—	2.5
Nickel	—	0.90	—	—
Phosphorus	300	—	—	400
Potassium	1310	650	1153	1530
Rubidium	—	0.80	—	—
Sodium	—	200	16	280
Strontium	—	1.7	—	—
Sulfur	—	—	—	190
Zinc	17	12	—	3.7

Tomato seeds are rich in minerals such as potassium, magnesium, phosphorus, sodium and calcium and, to a lesser extent, manganese, iron, zinc and copper.
[1]Tomato seeds and peels.

recommended dietary intake, respectively). However, these minerals may be poorly provided by certain plant-based foods, because bioavailability of non-heme iron and dietary zinc is greatly influenced by both dietary inhibitors and enhancers that should be studied more deeply in tomato seeds. Finally, the consumption of tomato seeds contributes relatively little to the intake of calcium and potassium (Table 133.3).

Tomato seed is a good source of antioxidants because it is rich in phytochemical compounds. It is an important reservoir of phenolic compounds. There is growing recognition that many phenolic secondary metabolites present in seeds may be beneficial to human health to some degree, mediated via antioxidant actions. Toor and Savage (2005) evaluated that the hydrophilic phenolic content in the seeds of three commercially grown tomato cultivars (Excell, Tradiro, and Flavourine) is 22 ± 4 mg gallic acid equivalents/100 g FW. Lipophilic phenolics only represent 12–15% of the total phenolic compounds present in the seed fraction of tomatoes. Moreover, the total amount of phenolics (hydrophilic and lipophilic) and flavonoids (12 ± 1 mg rutin equivalents/100 g FW) in seeds of the three cultivars of tomato studied are higher than the mean phenolic content of their pulps. Flavonoids such as quercetin (6.2 ± 0.5 mg/100 g FW), kaempferol (7.1 ± 0.4 mg/100 g FW) and naringenin enantiomers were identified in tomato seeds by Torres *et al.* (2005). *S*-naringenin is the predominant glycosylated enantiomer. Therefore, *S*-naringenin (3.1 ± 0.3 mg/100 g FW) accounts for around 60% of total naringenin in tomato seeds. Moreover, *S*-naringenin has a longer biological half-life than *R*-naringenin, which could be important for bioavailability studies, because chirality may have a significant influence on physiological action and disposition. Peng *et al.* (2008) found that seeds of Anqi, Chunjiao, and Mava tomato cultivars contain naringenin (between 0 and 0.20 mg/100 g FW), rutin (between 0.18 and 1.5 mg/100 g FW), chlorogenic acid (between 2.8 and 3.7 mg/100 g FW), and myricetin (between 0 and 0.64 mg/100 g FW).

Furthermore, carotenoids and different xanthophylls have been identified in tomato seeds (Table 133.6). In general, the carotenoid composition of the seeds reflects the qualitative pattern of the whole fruit (Rymal & Nakayama, 1974). However, the total carotenoid concentration is much lower than in the whole fruit, and the lycopene content is particularly low. β-Carotene, having a comparatively high concentration, emerges as the major carotenoid in the seeds. Lutein is the major xantophyll pigment in the seeds. Lutein, β-carotene, and lycopene are the major pigments of the seeds from the Chico Grande and Rutgers (red fruits) cultivars, and lutein, β-carotene and ζ-carotene (all-*trans* forms) are the major pigments of Golden Jubilee (yellow fruits) seeds (Rymal & Nakayama, 1974). Therefore, β-carotene is more common than lycopene in seeds of red-fruited cultivars, and ζ-carotene in seeds of yellow-fruited cultivars. Evidence has been presented that individuals with low carotenoid intake

TABLE 133.6 Carotenoid Content of Tomato Seed

Carotenoid	Reference					
	Rymal & Nakayama, 1974	Rymal & Nakayama, 1974	Rymal & Nakayama, 1974	Knoblich et al., 2005	Rodríguez et al., 1975	Toor & Savage, 2005
	Content (mg/l oil)			Content (µg/100 g fresh weight)		
β-Carotene	129	11	289	176	1290	—
Lycopene	85	9.0	n.d.	1144	820	1600
α-Carotene	—	—	—	—	60	—
γ-Carotene	—	—	—	—	80	—
ζ-Carotene	—	—	—	—	n.d.	—
Phytofluene	—	—	—	—	40	—
β-Zeacarotene	—	—	—	—	30	—
Lutein	386	180	776	57	50	—
Zeaxanthin	—	—	—	8.8	—	—
Plant material	cv. Chico Grande	cv. Rutgers	cv. Golden Jubilee	By-product	cv. Cerasiforme	Several cultivars

Phytochemical compounds such as carotenoids (lycopene, β-carotene) and xanthophylls (lutein and zeaxanthin) have been identified in tomato seeds. n.d., not detected.

and/or low carotenoid blood levels have an increased risk of degenerative diseases. According to epidemiological studies, carotenoids play an important role in the prevention of cancer, cataracts, and cardiovascular disease.

The following compounds are among the constituents of tomato seed oil unsaponifiables (Lazos *et al.*, 1998; Malecka, 2002): α-, β- and δ-tocopherols, and sterols including β-sitosterol, campesterol, stigmasterol, and Δ-5-avenasterol. In addition, trace to minor amounts of 24-methylenecholesterol, brassicasterol, Δ(7)-campesterol, clerosterol, Δ(7),(24)-stigmastadienol, Δ(7)-stigmastanol, Δ(7)-avenasterol, erythrodiol, and citrostadienol have also been found. Some of these sterols, as well as tocopherols, contribute to the antioxidant properties of tomato seed unsaponifiables. Moreover, plant sterols are effective in lowering total plasma and low density lipoprotein cholesterol by inhibiting the absorption of cholesterol from the small intestine.

Tomato seed also contains other phytochemical compounds, such as ascorbic acid, with a content between 8.2 and 17 mg/100 g FW (Toor & Savage, 2005; Peng *et al.*, 2008).

The antioxidant activity in the hydrophilic extracts ($114 \pm 20 \, \mu M$ Trolox equivalent antioxidant capacity/100 g) of tomato seeds from Excell, Tradiro, and Flavourine tomato cultivars is the major contributor (91–93%) to their total antioxidant activity (hydrophilic and lipophilic) (Toor & Savage, 2005). Furthermore, the seeds are an important contributor to the major antioxidants ($28 \pm 7\%$ of total phenolics, $25 \pm 4\%$ of total flavonoids, $11 \pm 4\%$ of lycopene, and $19 \pm 1\%$ of ascorbic acid) and overall antioxidant activity ($23 \pm 5\%$) of tomatoes. Removing seeds from tomatoes when cooking at home and processing results in a significant loss of all the major antioxidants of tomatoes. Therefore, it is important to consume tomatoes along with their seeds in order to obtain maximum health benefits.

ADVERSE EFFECTS AND REACTIONS (ALLERGIES AND TOXICITY)

No antinutritional factors or harmful constituents have been reported in tomato seeds (Rahma *et al.*, 1986), making them a better source of protein, lipids, or bioactive compounds than other non-conventional sources. Moreover, it is important to highlight that the contribution of a 20-g serving of tomato seeds to the tolerable upper intake level of sodium, defined as the highest level of daily nutrient intake that is likely to pose no risk of adverse health effects to almost all individuals in the general population, is very low (2.2%). This contribution must be considered, because it is known that increased sodium chloride intake increases blood pressure and is associated with an increased risk of cardiovascular disease.

1131

SUMMARY POINTS

- Tomato seed oil is high in unsaturated fatty acids (80%), with a nutritional value comparable to other edible oils, such as corn oil.
- Tomato seed is also high in protein, more than double the amount found in most cereal grains.
- The essential amino acid scores of tomato seed are similar, or even better, than those of different cereal grains and seed kernels.
- Tomato seed is rich in minerals.
- Phytochemicals such as phenolic compounds, carotenoids and xanthophylls, tocopherols, and sterols have been identified in tomato seed.
- Tomato seeds contain a great deal of nutrients and potentially healthy bioactive compounds, could be used as sources of ingredients to fortify or functionalize food.

ACKNOWLEDGEMENTS

M. Gonzalez would like to thank the Instituto Nacional de Investigación y Tecnología Agraria y Alimentaria (INIA) for awarding her contract, financed with the involvement of the European

Social Fund. The authors are grateful to Spain's INIA (Project RTA2006-00187) for financial support.

References

Alvarado, A., Pacheco-Delahaye, E., & Hevia, P. (2001). Value of a tomato byproduct as a source of dietary fiber in rats. *Plant Foods for Human Nutrition, 56*, 335−348.

Al-Wandawi, H., Abdul-Rahman, M., & Al-Shaikhly, K. (1985). Tomato processing wastes as essential raw materials source. *Journal of Agricultural and Food Chemistry, 33*, 804−807.

Cámara, M., del Valle, M., Torija, M. E., & Castillo, C. (2001). Fatty acid composition of tomato pomace. *Acta Horticulturae, 542*, 175−181.

da Silva, J. C., Armelin, M. J. A., & da Silva, A. G. (1999). Determination of the mineral composition in agro-industrial by-products used in animal nutrition, by neuron activation analysis. *Pesquisa Agropecuária Brasileira, 34*, 235−241.

da Silva, E. P., da Silva, D. A. T., Rabello, C. B. V., Lima, R. B., Lima, M. B., & Ludke, J. V. (2009). Physicochemical composition and energy and nutritional characteristics of guava and tomato residues for free range broilers. *Revista Brasileira de Zootecnia, 38*, 1051−1058.

FAOSTAT. (2008). Food and Agriculture Organization of the United Nations Statistical Database. http://faostat.fao.org (accessed January 2010).

Giannelos, P. N., Sxizas, S., Lois, E., Zannikos, F., & Anastopoulos, G. (2005). Physical, chemical and fuel related properties of tomato seed oil for evaluating its direct use in diesel engines. *Industrial Crops and Products, 22*, 193−199.

Knoblich, M., Anderson, B., & Latshaw, D. (2005). Analyses of tomato peel and seed byproducts and their use as a source of carotenoids. *Journal of the Science of Food and Agriculture, 85*, 1166−1170.

Lazos, E. S., Tsaknis, J., & Lalas, S. (1998). Characteristics and composition of tomato seed oil. *Grasas y aceites, 49*, 440−445.

Malecka, M. (2002). Antioxidant properties of the unsaponifiable matter isolated from tomato seeds, oat grains and wheat germ oil. *Food Chemistry, 79*, 327−330.

National Academy of Sciences. (2005). Dietary reference intakes for energy, carbohydrate, fiber, fat, fatty acids, cholesterol, protein and amino acids. http://www.nap.edu (accessed December 2009).

National Academy of Sciences. (2006). Dietary reference intakes: The essential guide to nutrient requirements. In J. J. Otten, J. Pitzi-Hellwig & L. D. Meyers (Eds.). Washington, DC: National Academies Press. http://www.nap.edu (accessed December 2009).

Peng, Y., Zhang, Y., & Ye, J. (2008). Determination of phenolic compounds and ascorbic acid in different factions of tomato by capillary electrophoresis with electrochemical detection. *Journal of Agricultural and Food Chemistry, 56*, 1838−1844.

Persia, M. E., Parsons, C. M., Schang, M., & Azcona, J. (2003). Nutritional value of dried tomato seeds. *Poultry Science, 82*, 141−146.

Rahma, E. H., Moharram, Y. G., & Mostafa, M. M. (1986). Chemical characterization of tomato seed protein (var. Pritchard). *Egyptian Journal of Food Science, 14*, 221−230.

Rick, C. M. (1976). Tomato (family Solanaceae). In N. W. Simmonds (Ed.), *Evolution of crop plants* (pp. 268−273). New York, NY: Longman Publications.

Rodríguez, D. B., Lee, D. C., & Chichester, C. O. (1975). Comparative study of the carotenoid composition of the seeds of ripening *Momordica charantia* and tomatoes. *Plant Physiology, 56*, 626−629.

Rymal, K. S., & Nakayama, T. O. M. (1974). Major carotenoids of the seeds of three cultivars of the tomato, *Lycopersicon esculentum* L. *Journal of Agricultural and Food Chemistry, 22*, 715−717.

Sogi, D. S., Bhatia, R., Garg, S. K., & Bawa, A. S. (2005). Biological evaluation of tomato waste seed meals and protein concentrate. *Food Chemistry, 89*, 53−56.

Toor, R. K., & Savage, G. P. (2005). Antioxidant activity in different fractions of tomatoes. *Food Chemistry, 38*, 487−494.

Torres, C. A., Davies, N. M., Yañez, J. A., & Andrews, P. K. (2005). Disposition of selected flavonoids in fruit tissues of various tomato (*Lycopersicon esculentum* Mill.) genotypes. *Journal of Agricultural and Food Chemistry, 53*, 9536−9543.

USDA. (2009). National nutrient database for standard reference, release 22. http://www.ars.usda.gov/nutrientdata (accessed December 2009).

CHAPTER 134

Antitumor Effects of Extracts from Wallich (*Sophora moorcroftiana*) Seeds

Xingming Ma
Department of Immunology, School of Basic Medical sciences, Lanzhou University, China

1133

LIST OF ABBREVIATIONS

EE$_{95}$, 95% ethanolic extracts
EE$_{75}$, 75% ethanolic extracts
EECCM, Encyclopedia Editorial Committee of Chinese Medicine
HE, hematoxylin-eosin
IL-2, interleukin 2
LC$_{50}$, median lethal concentration
LD$_{50}$, median lethal dose
MTT, 3-(4,5-dimethylthiazol-2-yl)-2,5-diphenyl tetrazolium bromide
SM, *Sophora Moorcroftiana*
TEM, transmission electron microscopy

Nuts & Seeds in Health and Disease Prevention. DOI: 10.1016/B978-0-12-375688-6.10134-3

TNF-α, tumor necrosis factor-α
TNFR, tumor necrosis factor receptor
VP-16, vepeside

INTRODUCTION

Sophora moorcroftiana (*SM*) (Wallich.) Benth. ex Baker seeds have long been in folk medicine. Decoctions of the seeds have been used in Chinese folk medicine for dephlogistication, detoxication, as an emetic, in infectious diseases, and in verminosis (EECCM, 1999; Xingming *et al.*, 2004a, 2005). Prenylflavanones from *Sophora tomentosa* and *SM* showed tumor-specific cytotoxic activity, and the extracts from *SM* seeds showed a protoscolicidal, anti-inflammatory effect (Xingming *et al.*, 2004b) and induced apoptosis of the human stomach cancer cell line SGC-7901 *in vitro* (Xingming *et al.*, 2004c), and have been suggested to have antitumor activity on tumor-bearing mice *in vivo* (Xingming *et al.*, 2009).

BOTANICAL DESCRIPTION

Sophora moorcroftiana, an endemic leguminous dwarf shrub in Tibet, China, which has played an important role in vegetation restoration, is found in the wide valleys and on semi-arid mountain slopes and terraces ranging from 2800 to 4400 m above sea level along the middle reaches of the Yarlung Zangbo River (Nianchu and Lhasa Rivers) (Xingming *et al.*, 2004a; Wenzhi *et al.*, 2007). *SM*, with spinulose in the branch tops, is in leaf all year, in flower in June, and the seeds in the pods ripen from August to October. The beaded pods stay attached to the tree for most of the year, and are up to 11 cm long; they contain one to seven brown or brownish-gray seeds (EECCM, 1999; Wenzhi *et al.*, 2007).

HISTORICAL CULTIVATION AND USAGE

SM originated in the altiplano areas of Tibet, where there is plenty of sunshine. It adapted physically to the drought, cold, and sand. Untreated *SM* seeds germinated and propagated in the altiplano areas (EECCM, 1999; Wenzhi *et al.*, 2007). According to historical records, the ripe seeds were used as medicine. In particular, the decoction of seeds from 3 g to 9 g was used in Chinese folk medicine by oral administration to treat dephlogistication, detoxication, as an emetic, in infectious diseases, and in verminosis (EECCM, 1999; Xingming *et al.*, 2004a).

PRESENT-DAY CULTIVATION AND USAGE

The plant grows in light (sandy), medium (loamy), and heavy (clay) soils, and is as an ideal candidate species for acclimatization to weather conditions (Wenzhi *et al.*, 2007). The decoction of 9 g ripe seeds collected in October showed similar applications to the historical usage (EECCM, 1999 and Xingming *et al.*, 2004a).

APPLICATIONS TO HEALTH PROMOTION AND DISEASE PREVENTION

The crude extracts from *Sophora Moorcroftiana* seeds

The *SM* seeds were collected from the banks of the YaluTsangbo River in the Tibetan altiplano, and air-dried at room temperature. The seeds were powdered using a blender, followed by chloroform (yield 2.6%), 95% ethanol (EE_{95}) (a dark brown solid mass, yield 2.6%), 75% ethanol (EE_{75}) (a dark brown solid mass, yield 3.8%), and distilled-water (a dark brown solid mass, yield 5.2%) extraction (Table 134.1). The dried crude extracts were then dissolved in DMSO to make a final concentration of 0.1% (Xingming *et al.*, 2008).

1134

TABLE 134.1 The Yield of Extracts from *Sophora moorcroftiana* Seeds and their Median Lethal Concentrations (LC_{50}) on the SGC-7901 Cell Line for 24 Hours *In Vitro*

Materials	Yield (%)	LC_{50} (mg/ml)
Chloroform extracts	2.6	4.5
95% ethanol extracts	2.6	1.4
75% ethanol extracts	3.8	1.9
Distilled-water extracts	5.2	41.7
Alkaloids	—	4.5

The alkaloids and all crude extracts from *Sophora moorcroftiana* showed cytotoxicity on the SGC-7901 cell line for 24 hours *in vitro*, but the LC_{50} of 95% ethanol extracts was lower than other material.
Reprinted from Xingming *et al.* (2003), *J. Lanzhou Uni. (Nat. Sci)*, 31, 74–78, and Xingmint *et al.* (2006), *Afr. J. Biotechnol.*, 5, 1669–1674, with permission.

Cytotoxic effects of extracts on tumor cells *in vitro*

The chloroform, ethanol, and distilled-water extracts were tested for cytotoxic activity *in vitro* using an MTT colorimetric assay. Not only did the alkaloids from *SM* show an antitumor effect *in vitro* (Xingming *et al.*, 2003); also, all crude extracts inhibited the proliferation of the human stomach cancer SGC-7901 cell line in a concentration-dependent manner (Xingming *et al.*, 2008). In comparison with the positive cytotoxic control agent, VP-16, which displayed a cytotoxicity level of 16% at a concentration of 0.08 μmol/l, a higher cytotoxicity level in the SGC-7901 cell line was observed after treatment with 0.63 mg/ml of the crude extracts. The median lethal concentrations (LC_{50}) of the chloroform extracts, EE_{95}, EE_{75}, the distilled-water extracts, and alkaloids on the cytotoxicity of the SGC-7901 cell line for 24 hours are shown in Table 134.1. These results suggest that EE_{95} fraction was more effective than other extracts of *SM* seeds against the human stomach cancer cell line SGC-7901 *in vitro* (Xingming *et al.*, 2008).

Induction apoptosis of 95% ethanolic extracts on tumor cells *in vitro*

In order to further investigate the underlying cytotoxicity mechanisms of the EE_{95} on tumor cells, the EE_{95} was added to the culture of the human stomach cancer cell line SGC-7901 *in vitro*. The proliferation and apoptosis of cells were assessed by MTT assay, fluorescence microscopy, transmission electron microscopy, flow cytometry, and agarose gel electrophoresis. The results show that EE_{95} inhibited the proliferation of SGC-7901 cells and its activity in a dose- and time-dependent manner (Figure 134.1). Like alkaloids from *SM*, the crude extracts of 95% ethanol could induce apoptosis on human stomach cancer cell *in vitro*. After treatment with EE_{95}, both morphological study using acridine orange fluorescence staining and TEM revealed the typical morphology of apoptosis in cultured SGC-7901 cells, such as cell pyknosis, chromatin condensation, and nuclear fragmentation (Xingming *et al.*, 2006). Meanwhile, the results of DNA agarose gel electrophoresis showed the typical "DNA ladder," which indicated the internucleosomal DNA fragmentation in apoptotic cells after treatment at concentrations of 1.25 ~ 5.00 mg/ml (Xingming *et al.*, 2006). Additionally, FACScan analysis showed the presence of an apoptotic peak in the sub-G1 phase. SGC-7901 cells treated with EE_{95} displayed a dose-dependent accumulation of apoptosic cells, and the apoptosis percentage of cells at the sub-G1 phase was 14.2% ~ 32.1%, as compared to untreated control (1.4%). These results suggested that EE_{95} could exert its antitumor effects through two fundamental processes: suppression of cell proliferation, and induction of apoptosis on human stomach cancer cells *in vitro* (Table 134.2) (Xingming *et al.*, 2006).

Antitumor effects of 95% ethanolic extracts *in vivo*

The antitumor effects of EE_{95} *in vivo* were studied using murine tumor models. The mice were inoculated subcutaneously in the left flank with 5×10^6 viable S_{180} sarcoma cells (The Medical

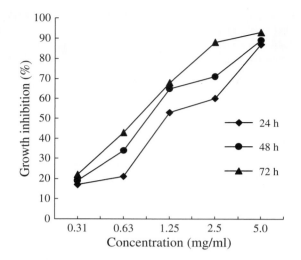

FIGURE 134.1

The inhibitory effect of the 95% ethanolic extracts from *Sophora moorcroftiana* seeds on the SGC-7901 cell line *in vitro*. Results showed that the 95% ethanolic extracts inhibited the proliferation of SGC-7901 cells and their activity in a dose- and time-dependent manner. Results are expressed as mean, $n = 9$. Data from Xingming et al. *(2006), with permission.*

Experiment Center of Lanzhou University, PR China.) in 0.2 ml saline water. After 3 days, the tumor-bearing mice were divided into five groups of eight mice each. Three groups were treated by intragastric gavage (i.g.) with EE_{95} in 0.2 ml distilled water at 200, 400, and 800 mg/kg per day, respectively. The administration of cyclophosphamide singly at a dose of 20 mg/kg/d, as a positive control group, was initiated (day 1) and continued on days 5 and 9. The negative control group was given the same volume of 0.2 ml distilled water by the same route (Xingming *et al.*, 2009).

After 10 days of treatment, all the tumor-bearing mice were anesthetized with pentobarbital sodium according to the institutional guidelines, blood samples were collected from eyeballs, and tumor tissue was stripped from the left flank. The weight of the tumor tissue was measured immediately to determine the S_{180} sarcoma cell growth and the inhibition ratio. Then, tumor tissue was routinely stained with HE. IL-2 and TNF-α in the serum were measured by radioimmunoassay for determination of anticancer mechanisms via immunoregulation efficacy (Xingming *et al.*, 2009).

The weight of tumor tissue decreased gradually upon the treatment with each drug (Table 134.3). The EE_{95}, at a dose of 800 mg/kg per day, inhibited the tumor tissue growth with a rate

TABLE 134.2 The Results of Induction of Tumor Cell Apoptosis with 95% Ethanol Extracts and Alkaloids from *Sophora moorcroftiana* seeds *In Vitro*

Methods	Alkaloids	95% Ethanol Extracts
Concentration	0.80 ~ 1.60 mg/ml	1.25 ~ 5.00 mg/ml
Fluorescence microscope	Morphological change of apoptosic cells	Morphological change of apoptosic cells
Transmission electron microscopy	No data	Morphological change of apoptosic cells
Agarose gel electrophoresis	DNA ladder	DNA ladder
Apoptosis percentage	12.9% ~ 16%	14.2% ~ 32.1%
Induction of apoptosis	Yes	Yes

The 95% ethanolic extracts and alkaloids showed antitumor effects through induction of apoptosis on human stomach cancer cells *in vitro*.
Data from Xingming *et al.* (2003), *J. Lanzhou Uni. (Nat. Sci.), 31,* 74–78, and Xingming et al. (2008), *Nature Prod. Res. Dev.,* 888–891, with permission.

TABLE 134.3 The Weight and Inhibition Rate of Tumor Tissue and Levels of Serum Cytokines in Mice

Group	Tumor Weight (g)	Inhibition Rate (%)	IL-2 (ng/ml)	TNF-α (fmol/ml)
Negative control	1.88 ± 0.39		1.79 ± 0.29	12.01 ± 1.21
Cyclophosphamide control (20 mg/kg per day)	0.42 ± 0.19^b	77.6	1.71 ± 0.42	11.91 ± 2.31
95% Ethanolic extracts (200 mg/kg per day)	1.64 ± 0.44^c	12.0	1.90 ± 0.47	12.21 ± 1.61
95% Ethanolic extracts (400 mg/kg per day)	1.67 ± 0.36^c	11.1	2.07 ± 0.89	12.16 ± 1.49
95% Ethanolic extracts (800 mg/kg per day)	$1.35 \pm 0.21^{a,c}$	28.1	$2.15 \pm 0.47^{a,c}$	12.39 ± 1.69^c

The weight of tumor tissue was significantly decreased upon treatment with 95% ethanolic extracts, and IL-2 and TNF-α in serum from the treated mice significantly increased in ethanolic extract-treated groups, compared with the untreated control animals
Reprinted from Xingming et al. (2009), *Iranian Red Crescent Med. J.*, 11, 18–22.
[a]$P < 0.05$,
[b]$P < 0.01$ vs negative control group;
[c]$P < 0.05$ vs cyclophosphamide control group). Results are expressed as mean $\pm$ SD, n = 8.

of 28.1%, which was slightly lower than the cyclophosphamide at the standard dose rate (20 mg/kg per day). In addition, the appearance of larger areas of necrosis was detected in the tumor tissue treated with EE$_{95}$, and even significantly invaded lymphocytes could be considered as indirect effects of the present treatment. Based on the results stated above, the conclusion was that the EE$_{95}$ treatment at a dose of 800 mg/kg per day showed a weak inhibiting effect on S$_{180}$ sarcoma development *in vivo* (Xingming et al., 2009).

IL-2 and TNF, two of the most important non-specific immune factors, as demonstrated by numerous experiments and clinical applications, have played a positive and estimable role in treating tumors (Xingming et al., 2009). In this study, the therapeutic effects were monitored by testing the IL-2 and TNF levels in mice. An increase of serum IL-2 and TNF-α levels was observed in tumor-bearing mice treated with EE$_{95}$ for 10 days (Table 134.3). In comparison with other experiment groups, a decrease of serum IL-2 and TNF-α levels in tumor-bearing mice was observed with cyclophosphamide treatment. It is possible that after treatment with EE$_{95}$, the growth and structure of tumors may be inhibited or de-structured, and thus the inhibitory action of the tumor on immune cells is attenuated. In comparison with a control group, it is suggested that EE$_{95}$ from *SM* seeds had a weak antitumor role, and in high concentrations in tumor-bearing mice *in vivo*, the EE$_{95}$ was rather effective. One possible mechanism of EE$_{95}$ activity *in vivo* could be because the EE$_{95}$ could enhance the self antitumor response and stimulate the secretion of TNF-α and IL-2 by immune cells (Xingming et al., 2009).

These results suggest that EE$_{95}$ may exert its antitumor effects by four fundamental processes (Table 134.4): (1) cytotoxicity effects and suppression of tumor tissue growth; (2) induction of

TABLE 134.4 The Antitumor Effect Pathway of Extracts from *Sophora moorcroftiana* Seeds

Materials	Pathway
Chloroform extracts, 75% ethanol extracts. distilled-water extracts	Cytotoxicity and suppression effects
95% Ethanol extracts	Cytotoxicity and suppression effects; induction of apoptosis and necrosis Immunoregulatory activity
Alkaloids	Suppression effects and inducted apoptosis

Sophora moorcroftiana seeds could contain antitumor fraction(s) and show antitumor effects through multi-pathways as shown in the table.
Data from the research group of Xingming et al., with permission.

apoptosis and (3) necrosis of the tumor cell; and (4) enhancement of the self antitumor response in tumor-bearing mice. Therefore, it is concluded that EE_{95} from *SM* seeds possesses potent antitumor fraction(s) in animal.

However, what the main biochemical constituents of EE_{95} on antitumor immunity effects in tumor-bearing mice are, and how they inhibit the growth, induce apoptosis and necrosis, and enhance antitumor immunity in tumor-bearing mice remain unknown. Further studies are needed to address those important questions.

ADVERSE EFFECTS AND REACTIONS (ALLERGIES AND TOXICITY)

Acute toxicity tests in mice by abdominal injection were routinely performed for investigating the LD_{50} of alkaloids from *SM* seeds. The LD_{50} on mice with alkaloids was 207.81 ± 20.82 mg/kg (Xingming *et al.*, 2004a), and the LD_{50} on *Pieris rapae* with alkaloids in the food was 326.59 mg/l (Xingming *et al.*, 2008). There are no records regarding allergies and toxicity of the decoction and crude extracts from *SM* seeds.

SUMMARY POINTS

- *Sophora moorcroftiana* is an endemic shrub species in Tibet altiplano, PR China, and there are one to seven seeds in the pods when the seeds ripen, between August and October.
- The ripe seeds are used as medicine, and the decoction of 9 g seeds is used in Chinese folk medicine for dephlogistication, detoxication, as an emetic, in infectious diseases, and in verminosis.
- The EE_{95} LC_{50} of the cytotoxicity in tumor cells is 1.4 mg/ml, which is the lowest among all crude extracts.
- The EE_{95} and alkaloids inhibit the proliferation of and induce apoptosis in tumor cells in a concentration-dependent manner *in vitro*.
- The EE_{95} inhibits the tumor tissue growth, and the inhibition rate when treated with EE_{95} (800 mg/kg per day) was 28.1% *in vivo*.
- The EE_{95} may exert its antitumor effect via inducing tumor cell necrosis and lymphocyte soakage in tumor tissue *in vivo*.
- The EE_{95} could enhance the self antitumor response, and stimulate the secretion of TNF-α and IL-2 by immune cells *in vivo*.
- The EE_{95} from *SM* seeds possesses potential antitumor fraction(s) in animals.

References

Encyclopedia Editorial Committee of Chinese Medicine. (1999). Sophora moorcroftiana seeds. In *China Medical Encyclopedia Tibetan Medicine* (p. 204). Shanghai, China: Shanghai Science and Technology Press.

Wenzhi, Z. H., Zhihui, Z. H., & Qiuyan, L. (2007). Growth and reproduction of *Sophora moorcroftiana* responding to altitude and sand burial in the middle Tibet. *Environmental Geology, 53,* 11–17.

Xingming, M., Hongyu, L., Shaofu, Y., Xueqin, M., Gongke, Z. H., & Xiaoming, L. (2003). The study on bacteriostasis and inhibiting tumor cell proliferative activity of alkaloids from *Sophora moorcroftiana* seeds. *Journal of Lanzhou University (Natural Sciences), 31,* 74–78.

Xingming, M., Hongyu, L., Shaofu, Y., & Bo, W. (2004a). The study on bacteriostasis and anti-inflammatory activity of alkaloids from *sophora moorcroftiana* seeds. *Acta Chinese Medicine and Pharmacology, 32,* 23–25.

Xingming, M., Hongyu, L., Shaofu, Y., & Bo, W. (2004b). The study on killing protoscolex activity and anti-inflammatory of alkaloids extracted from sophora moorcroftiana seeds. *Chinese Journal of Parasitic Diseases Control, 17,* 217–219.

Xingming, M., Hongyu, L., Shaofu, Y., & Bo, W. (2004c). Apoptosis in SGC-7901 cells induced by alkaloids from *sophora moorcroftiana* seeds. *Chinese Traditional Patent Medicine, 26,* 654–657.

Xingming, M., Hongyu, L., Bo, W., & Shaofu, Y. (2005). The determination of bacteriostasis and insecticidal activity of alkaloids from *Sophora moorcroftiana* seeds. *Chinese Journal of Biological Control, 21,* 183–186.

Xingming, M., Yanping, L., Hongjuan, Y., & Yan, C. (2006). Ethanolic extracts of Sophora moorcroftiana seeds induce apoptosis of human stomach cancer cell line SGC-7901 *in vitro*. *African Journal of Biotechnology, 5,* 1669−1674.

Xingming, M., Hongjuan, Y., Yanping, L., & Yan, C. (2008). Antimicrobial and cytotoxic activity of extracts from *Sophora moorcroftiana* seeds *in vitro*. *Natural Product Research and Development, 20,* 888−891.

Xingming, M., Hongjuan, Y., Ying, D., Yanping, L., Weihua, T., Fangyu, F., et al. (2009). Antitumor effects of ethanolic extracts from *Sophora moorcroftiana* seeds in mice. *Iranian Red Crescent Medical Journal, 11,* 18−22.

CHAPTER 135

The Effects of Walnuts (*Juglans regia*) on the Characteristics of the Metabolic Syndrome

Welma Stonehouse
Institute of Food, Nutrition and Human Health, Massey University, North Shore City, Auckland, New Zealand

1141

LIST OF ABBREVIATIONS

ALA, α-linolenic acid
BMI, body mass index
FRAP, ferric ion reducing antioxidant power
HDL, high density lipoprotein
IgE, immunoglobulin E
LDL, low density lipoprotein
LTP, lipid transfer protein
MUFA, monounsaturated fatty acid

Nuts & Seeds in Health and Disease Prevention. DOI: 10.1016/B978-0-12-375688-6.10135-5

ORAC, oxygen radical absorbance capacity
PUFA, polyunsaturated fatty acid
TEAC, Trolox equivalence antioxidant capacity
VCAM-1, vascular adhesion molecule 1

INTRODUCTION

The metabolic syndrome can be defined as a cluster of interrelated metabolic abnormalities, including dyslipidemia (elevated serum triglyceride and reduced high density lipoprotein (HDL) cholesterol concentrations), hyperglycemia/insulin resistance, abdominal obesity, and hypertension. Other conditions associated with the metabolic syndrome include a pro-thrombotic state, increased inflammation, oxidative stress, and endothelial dysfunction (Hansel et al., 2004; Huang, 2009). Because of the clustering of these risk factors, the risk for cardiovascular disease and type 2 diabetes are considerably increased. The first line of treatment of the metabolic syndrome is lifestyle changes, including dietary interventions that address all the components of the metabolic syndrome. The 1-year results from the large PREDIMED randomized trial showed that a Mediterranean diet enhanced with one serving of mixed nuts daily led to a reduction in the overall prevalence of metabolic syndrome compared to a low-fat diet (Salas-Salvado et al., 2008). Various studies have been undertaken on healthy, hypercholesterolemic, and diabetic subjects to investigate the effects of walnut consumption on characteristics of the metabolic syndrome, but only two studies to date have been conducted in subjects with the metabolic syndrome (Davis et al., 2007; Mukuddem-Petersen et al., 2007; Brennan et al., 2010). Walnuts contain various bioactive components that may act independently or synergistically in affecting the multiple abnormalities associated with the metabolic syndrome.

The aim of this chapter is to review the scientific literature regarding the effects of incorporating walnuts into the diet on characteristics of the metabolic syndrome.

BOTANICAL DESCRIPTION

Walnuts belong to the genus *Juglans*, of which the most commonly grown species is the English or Persian walnut (*Juglans regia*). The English walnut is a large, deciduous tree attaining heights of > 20–30 m. The bark is gray, smooth on young trees, and fissuring with age. The leaves are alternate, pinnate (20–45 cm long), and consist of five to seven leaflets. The flowers are monoecious, but male flowers shed their pollen before female flowers are open, resulting in poor fruit-setting. For maximum pollination (which occurs by wind), two complementary cultivars are needed that pollinate each other. Male flowers are borne in catkins on twigs of the previous season's growth, and female flowers in pistillate spikes of two to five flowers at the tips of the current season's shoots in late spring. The fruit may be classed as nuts or dry drupes. The fruit, usually borne in groups of two or three, is 3.7–5 cm in length, with a fleshy husk that covers the brown corrugated seed. When the seed is ripe, the husk opens and the stone falls to the ground. The seed is large, consists of two cotyledons, and is surrounded by a thin shell called the pellicle (Lyle, 2006; Prasad, 2003).

HISTORICAL CULTIVATION AND USAGE

The English walnut originates from Eastern Europe and the foothills of the Himalayas, and probably traveled from ancient Persia to Greece before being distributed by the Romans throughout Southern Europe. The early colonists from the UK brought seeds to North America, and the settlers called the resulting trees "English walnuts" to distinguish them from the Native American black walnut. Walnuts have been used as a food source since at least Neolithic times, 8000 years ago. They were also used in medieval Europe as a herbal medicine, particularly for brain and scalp ailments. In fact, different parts of the walnut tree, including bark, leaves, shells, fruits, and kernels, have been used in folk medicine for various diseases and conditions throughout the world (Lyle, 2006; Prasad, 2003).

PRESENT-DAY CULTIVATION AND USAGE

The English walnut is cultivated commercially in North, Central, and South America, Europe, and Asia. China is the leading producer of walnuts. Other important walnut-growing countries are the USA, Iran, Turkey, India, France, and Romania (Prasad, 2003).

The English walnut needs a long, warm growing season, although too much sun can be damaging. Winter chilling is essential — the English walnut needs 700–1500 hours of winter chilling — and cold acclimation needs to be slow. Preferred soils are deep, well-drained alluvial soils with a pH of 6.0–7.5. Walnut trees like space, and should be planted 6–16 m apart (Lyle, 2006).

Walnuts are cultivated extensively for their high-quality nuts, both eaten fresh and pressed into oil. They are also acknowledged for their various health benefits. The wood is of very high quality, and is used to make furniture and gunstocks. The walnut shells are reduced to flour, and have various applications — for example, as fillers in synthetic resin adhesives, in plastics and industrial tiles, and as insecticide diluents. The outer husk, when crushed, yields an oil that is used in the manufacture of soaps, paints, varnishes, alkyd resins, and styrenated oils (Lyle, 2006; Prasad, 2003).

APPLICATIONS TO HEALTH PROMOTION AND DISEASE PREVENTION

Composition of walnuts

Compared to other nuts, walnuts are unique with regard to their high content of poly-unsaturated fatty acids (PUFAs), including linoleic acid and α-linolenic acid (ALA) (Table 135.1), and their exceptionally high content of phenolics and antioxidants (Table 135.2). Compared to other food sources, nuts, including walnuts, are also rich sources of protein, in particular the amino acid arginine (precursor of the endogenous vasodilator nitric oxide), phytosterols, dietary fiber (including soluble fiber), folate, tocopherols, and magnesium (Tables 135.1, 135.2).

Walnuts and the metabolic syndrome

WALNUTS, ANTIOXIDANT CAPACITY AND INFLAMMATION

The metabolic syndrome is associated with inflammation and elevated oxidative stress (Hansel *et al.*, 2004), which are important features of atherosclerosis and subsequent cardiovascular disease. Walnuts contain the highest amount of phenolics and antioxidants of all nuts, mostly residing in the outer pellicle of the nut (Table 135.2). Compared to berries and dark chocolate, which are well known for their high antioxidant content, walnuts have a higher antioxidant content per 100 g (23.1 mmol vs 8.2 mmol in blueberries, 5.1 mmol in blackberries and 13.4 mmol in dark chocolate (70% cocoa)) (Blomhoff *et al.*, 2006). Intervention studies with walnuts, where markers of antioxidant status were measured in the fasted state after chronic feeding, have, however, been disappointing, showing no effects (Ros *et al.*, 2004; Tapsell *et al.*, 2004; Davis *et al.*, 2007). This has been ascribed to the fact that walnuts are high in PUFAs, which are susceptible to oxidation, and that the high content of antioxidants contained in walnuts may counteract the pro-oxidant effects of the fats (Ros *et al.*, 2004). However, acute feeding of walnuts resulted in a significant increase in plasma polyphenol concentrations, increased *in vivo* antioxidant capacity as measured with oxygen radical absorbance capacity (ORAC) and ferric ion reducing antioxidant power (FRAP), and decreased lipid peroxidation, compared to a control meal (Torabian *et al.*, 2009). Furthermore, walnuts contain melatonin

TABLE 135.1 Selected Macronutrient Composition of Nuts and Tree Nuts (Raw, per 100 g)

	Energy (kJ)	Total Protein (g)	Arginine (g)	Total fat (g)	SFAs (g)	MUFAs (g)	PUFAs (g)	18:2n-6 (g)	18:3n-3 (g)	Total Plant Sterols (mg)	Total Fiber (g)
Almonds	2408	21.2	2.45	49.4	3.73	30.9	12.1	12.1	<0.01	172	12.2
Brazilnuts (dried)	2743	14.3	2.15	66.4	15.1	24.5	20.6	20.5	0.02	NR	7.5
Cashews	2314	18.2	2.12	43.9	7.78	23.8	7.85	7.78	0.06	NR	3.3
Hazelnuts	2629	15.0	2.21	60.8	4.46	45.7	7.92	7.83	0.09	96	9.7
Macadamias	3004	7.91	1.40	75.8	12.1	58.9	1.50	1.30	0.21	116	8.6
Peanuts	2374	25.8	3.09	49.2	6.83	24.4	15.6	15.6	<0.01	220	8.5
Pecans	2889	9.17	1.18	72.0	6.18	40.8	21.6	20.6	1.00	102	9.6
Pine nuts (dried)	2816	13.7	2.41	68.4	4.90	18.8	34.1	33.2	0.11	141	3.7
Pistachios	2352	20.3	2.01	45.4	5.6	23.8	13.7	13.5	0.26	214	10.3
Walnuts	2738	15.2	2.28	65.2	6.13	8.93	47.2	38.1	9.1	72	6.7

NR, not reported; SFAs, saturated fatty acids; MUFAs, monounsaturated fatty acids; PUFAs, polyunsaturated fatty acids; 18:2n-6, linoleic acid; 18:3n-3, α-linolenic acid.
Data from US Department of Agriculture, Agricultural Research Service (2009).

TABLE 135.2 Selected Micronutrient Composition of Nuts and Tree Nuts (Raw, per 100 g), Total Phenolic and Total Antioxidant Content

	Folate (μg DFE)	α-Tocopherol (mg)	γ-Tocopherol (mg)	Magnesium (mg)	Total Phenolics[a] (With Skin) (mg of GAE)	Total Antioxidant Content[b] (With Pellicle) (mmol)
Almonds	50.0	26.2	0.65	268	239	0.41
Brazilnuts (dried)	22.0	5.73	7.87	376	112	0.25
Cashews	25.0	0.90	5.31	292	137	0.39
Hazelnuts	113	15.0	<0.01	163	291	0.70
Macadamias	11	0.54	<0.01	130	46	0.42
Peanuts	240	8.33	NR	168	420	1.97
Pecans	22	1.40	24.4	121	1284	8.33
Pine nuts (dried)	34	9.33	11.2	251	32	0.37
Pistachios	51	2.30	22.6	121	867	1.27
Walnuts	98	0.70	20.8	158	1625	23.1

DFE, dietary folate equivalents; GAE, gallic acid equivalents; NR, not reported.
Data from US Department of Agriculture, Agricultural Research Service (2009).
Sources:
[a]Kornsteiner et al. (2006);
[b]Blomhoff et al. (2006).

(3.5 ± 1.0 ng/g walnuts); this, in a rat feeding trial, resulted in a significant increase in blood melatonin concentrations, which was accompanied by increased antioxidant capacity (FRAP, Trolox equivalence antioxidant capacity (TEAC)) (Reiter *et al.*, 2005). Several walnut intervention trials that explored the effects on markers of inflammation showed inconsistent results (Banel & Hu, 2009).

Thus, while chronic feeding of walnuts and measuring antioxidant capacity in the fasted state does not seem to affect antioxidant status, walnuts may increase antioxidant capacity postprandially.

WALNUTS AND ENDOTHELIAL FUNCTION

Endothelial dysfunction is associated with the metabolic syndrome, and is integral to the pathogenesis of atherosclerosis and cardiovascular disease (Huang, 2009). Two randomized crossover trials, one in hypercholesterolemic subjects (Ros *et al.*, 2004) and one in subjects with type 2 diabetes (Ma *et al.*, 2010), have demonstrated that inclusion of walnuts in the diet compared to a diet without walnuts improved endothelial function. Cortes *et al.* (2006) have furthermore shown that adding walnuts to a high-fat, high saturated fatty acid meal acutely counteracted the detrimental changes in flow-mediated dilation associated with eating a fatty meal, compared with the same meal containing olive oil. According to the systematic review of Banel and Hu (2009), vascular cell adhesion molecule 1 (VCAM-1), a marker of endothelial function, was consistently reduced with the consumption of walnut diets compared to control diets in three studies. These beneficial effects on the endothelium may be ascribed to the various components contained in walnuts, especially ALA, antioxidants, and arginine, which may act independently or synergistically to improve endothelial function. Nitric oxide is one of the most critical substances produced by the vascular endothelium for regulation of vasomotor tone. Endothelial injury reduces nitric oxide production, which results in endothelial abnormalities. Arginine, an essential amino acid contained in walnuts, is required for the production of nitric oxide, and has been shown to improve endothelial function, particularly in subjects with impaired nitric oxide synthesis (summarized by Ros *et al.*, 2004). In addition, the high content of antioxidants and the serum cholesterol lowering effects of walnuts may also contribute to improving endothelial function by protecting the endothelium from injury (Ros *et al.*, 2004; Cortes *et al.*, 2006).

WALNUTS, INSULIN, GLUCOSE, AND BODY WEIGHT

Weight loss is the primary target of intervention for the metabolic syndrome. Nuts are high in fat, which generates concern regarding their potential impact on body weight and insulin resistance. However, Banel and Hu (2009) could not confirm any increases in body mass index (BMI) in their systematic review of studies investigating the effects of walnut-rich diets. Some studies even suggested a trend towards lower BMI. Some of the mechanisms include increased satiety, increased resting metabolic rate, and incomplete absorption of energy from nuts (Sabate, 2003). Brennan *et al.* (2010) showed that walnut consumption over 4 days increased satiety in men and women with metabolic syndrome compared to placebo.

Walnut diets did not affect glycemic control in subjects with metabolic syndrome (Mukuddem-Petersen *et al.*, 2007) or type 2 diabetes (Tapsell *et al.*, 2004) compared to control diets. In both these trials the subjects were overweight or obese, and body weight was maintained throughout the studies. Mukuddem-Petersen *et al.* (2007) speculated that the presence of obesity overshadowed the possible beneficial effects of the nuts, and that maintenance of body weight may have masked the positive metabolic effects of the nut diets, especially if these diets mediate their beneficial effects through central appetite suppression. Tapsell *et al.* (2009) recently showed a significant reduction in fasting insulin concentrations after 3 months in overweight type 2 diabetics who received a low-fat diet including walnuts compared to a control diet. Despite being on a weight maintenance diet, both groups sustained a 1–2 kg weight loss. Walnut diets seem to have the potential for reducing body weight and improving insulin concentrations in subjects with metabolic syndrome, but this needs to be investigated in randomized controlled trials incorporating walnuts in energy-restricted weight-loss diets on subjects with metabolic syndrome or diabetes.

WALNUTS AND THE LIPID PROFILE

A recent meta-analysis of 13 randomized controlled trials on walnut-enriched diets reported decreases in serum total cholesterol and LDL cholesterol concentrations of 10.3 mg/dl (0.27 mmol/l) and 9.2 mg/dl (0.24 mmol/l), respectively, with intakes that ranged from 5% to 25% of total energy intake (30–108 g walnuts/day). HDL-cholesterol and triglyceride concentrations were not significantly affected by the walnut diets (Banel & Hu, 2009). The cholesterol lowering effect of walnuts goes beyond their favorable PUFA content, and is probably due to the different bioactive components in walnuts that act synergistically, particularly dietary fiber (~25% soluble fiber), arginine, phytosterols, tocopherols, and phenolic compounds (Mukuddem-Petersen *et al.*, 2005). However, in the randomized controlled feeding trial on subjects with metabolic syndrome, intakes of 63–108 g walnuts/day did not affect total cholesterol, LDL cholesterol, HDL cholesterol, or triglyceride concentrations compared to a control diet. The authors suggested that this could be due to the high level of obesity in these subjects, which overshadowed the possible beneficial effects of walnuts (discussed above). Improvements in the HDL cholesterol to total cholesterol ratio were seen in subjects with type 2 diabetes who consumed a low-fat diet including 30 g/day of walnuts compared to a control diet (Tapsell *et al.*, 2004).

INCLUDING WALNUTS IN DIETARY GUIDELINES FOR THE METABOLIC SYNDROME

The dietary targets for metabolic syndrome include reduction of body weight, replacing saturated fatty acids with unsaturated fatty acids, and replacing refined carbohydrates with complex carbohydrates like fiber, increased intake of protein, and antioxidants. The incorporation of one serving per day of walnuts in the diet could contribute significantly in achieving these targets.

TABLE 135.3 Allergens in Walnuts Compared to Peanuts (Crespo *et al.*, 2006)

	Allergen	Function/type
Walnut	Jug r 1	Albumin (2S)
	Jug r 2	Vicilin (7S)
	Jug r 3	PR-14 (LTP)
	Jug r 4	Legumin (11S)
Peanut	Ara h 1	Vicilin
	Ara h 2	Conglutin
	Ara h 3	Glicinin
	Ara h 4	Glicinin
	Ara h 5	Profilin
	Ara h 6	Conglutin homologous
	Ara h 7	Conglutin homologous
	Ara h 8	PR-10

LTP, lipid transfer protein; PR, related protein.

In order for people with metabolic syndrome to achieve optimal benefit from incorporating walnuts into the diet, it is important that this should be done in conjunction with an overall healthy lifestyle that includes weight reduction, exercise, and a healthy overall diet.

ADVERSE EFFECTS AND REACTIONS (ALLERGIES AND TOXICITY)

Nuts are well known for their potential to cause food allergy, which affects approximately 1% of the general populations in the UK and the USA. Walnuts are among the different types of nuts and tree nuts known to produce immunoglobulin E (IgE)-mediated allergic reactions following ingestion (Crespo *et al.*, 2006). The allergens identified in walnuts are summarized in Table 135.3. The majority of the allergens in walnuts are seed storage proteins, such as vicilins, legumins, and 2S albumins. Lipid transfer proteins (LTPs) have also been identified in walnuts. Most nut species share antigenic proteins as well as having some unique proteins, which is likely to account for the fact that many tree nut-allergic individuals have a high level of reactivity to a variety of nuts.

SUMMARY POINTS

- Compared to other nuts and plant-based foods, walnuts are exceptionally high in PUFAs, particularly ALA, and antioxidants.
- Walnuts do not appear to affect antioxidant status when measured in the fasted state after chronic feeding, but may increase antioxidant capacity postprandially.
- Inclusion of walnuts in the diet may improve endothelial function compared to a diet without walnuts. Walnuts' high content of ALA, antioxidants, and arginine, and their cholesterol lowering properties, are probably synergistically responsible for the beneficial effects on endothelial function.
- Walnut consumption over 4 days increased satiety by day 3 compared to a placebo in subjects with metabolic syndrome.
- Walnut diets in obese subjects with the metabolic syndrome might only effectively improve lipid profiles, glycemic control, and insulin concentrations when accompanied by a weight loss diet, but this needs to be confirmed in randomized controlled trials on subjects with metabolic syndrome.
- The incorporation of one serving of walnuts into the diets of people with metabolic syndrome may contribute to achieving the dietary targets for metabolic syndrome.

References

Banel, D. K., & Hu, F. B. (2009). Effects of walnut consumption on blood lipids and other cardiovascular risk factors: a meta-analysis and systematic review. *American Journal of Clinical Nutrition, 90*, 56–63.

Blomhoff, R., Carlsen, M. H., Andersen, L. F., & Jacobs, D. R. (2006). Health benefits of nuts: potential role of antioxidants. *British Journal of Nutrition, 96*, S52–S60.

Brennan, A. M., Sweeney, L. L., Liu, X., & Mantzoros, C. S. (2010). Walnut consumption increases satiation but has no effect on insulin resistance or the metabolic profile over a 4-day period. *Obesity, 18*(6), 1176–1182.

Cortes, B., Nunez, I., Cofan, M., Gilabert, R., Perez-Heras, A., Casals, E., et al. (2006). Acute effects of high-fat meals enriched with walnuts or olive oil on postprandial endothelial function. *Journal of the American College of Cardiology, 48*, 1666–1671.

Crespo, J. F., James, J. M., Fernandez-Rodriguez, C., & Rodriguez, J. (2006). Food allergy: nuts and tree nuts. *British Journal of Nutrition, 96*(Suppl. 2), S95–102.

Davis, L., Stonehouse, W., Loots du, T., Mukuddem-Petersen, J., van der Westhuizen, F. H., Hanekom, S. M., et al. (2007). The effects of high walnut and cashew nut diets on the antioxidant status of subjects with metabolic syndrome. *European Journal of Nutrition, 46*, 155–164.

Hansel, B., Giral, P., Nobecourt, E., Chantepie, S., Bruckert, E., Chapman, M. J., et al. (2004). Metabolic syndrome is associated with elevated oxidative stress and dysfunctional dense high-density lipoprotein particles displaying impaired antioxidative activity. *Journal of Clinical Endocrinology & Metabolism, 89*, 4963–4971.

Huang, P. L. (2009). A comprehensive definition for metabolic syndrome. *Disease Models & Mechanisms, 2*, 231–237.

Kornsteiner, M., Wagner, K. H., & Elmadfa, I. (2006). Tocopherols and total phenolics in 10 different nut types. *Food Chemistry, 98*, 381–387.

Lyle, S. (2006). *Discovering fruit & nuts.* Auckland, New Zealand: David Bateman Ltd.

Ma, Y., Njike, V. Y., Millet, J., Dutta, S., Doughty, K., Treu, J. A., et al. (2010). Effects of walnut consumption on endothelial function in type 2 diabetics: a randomized, controlled, cross-over trial. *Diabetes Care, 33*(2), 227–232.

Mukuddem-Petersen, J., Oosthuizen, W., & Jerling, J. C. (2005). A systematic review of the effects of nuts on blood lipid profiles in humans. *Journal of Nutrition, 135*, 2082–2089.

Mukuddem-Petersen, J., Stonehouse Oosthuizen, W., Jerling, J. C., Hanekom, S. M., & White, Z. (2007). Effects of a high walnut and high cashew nut diet on selected markers of the metabolic syndrome: a controlled feeding trial. *British Journal of Nutrition, 97*, 1144–1153.

Prasad, R. (2003). Walnuts and pecans. In B. Caballero, L. Trugo & P. M. Finglas (Eds.), *Encyclopedia of food science and nutrition.* Amsterdam, The Netherlands: Elsevier Science Ltd.

Reiter, R. J., Manchester, L. C., & Tan, D. X. (2005). Melatonin in walnuts: influence on levels of melatonin and total antioxidant capacity of blood. *Nutrition, 21*, 920–924.

Ros, E., Nunez, I., Perez-Heras, A., Serra, M., Gilabert, R., Casals, E., et al. (2004). A walnut diet improves endothelial function in hypercholesterolemic subjects: a randomized crossover trial. *Circulation, 109*, 1609–1614.

Sabate, J. (2003). Nut consumption and body weight. *American Journal of Clinical Nutrition, 78*, 647S–650S.

Salas-Salvado, J., Fernandez-Ballart, J., Ros, E., Martinez-Gonzalez, M. A., Fito, M., Estruch, R., et al. (2008). Effect of a Mediterranean diet supplemented with nuts on metabolic syndrome status: one-year results of the PREDIMED randomized trial. *Archives of Internal Medicine, 168*, 2449–2458.

Tapsell, L. C., Gillen, L. J., Patch, C. S., Batterham, M., Owen, A., Bare, M., et al. (2004). Including walnuts in a low-fat/modified-fat diet improves HDL cholesterol-to-total cholesterol ratios in patients with type 2 diabetes. *Diabetes Care, 27*, 2777–2783.

Torabian, S., Haddad, E., Rajaram, S., Banta, J., & Sabate, J. (2009). Acute effect of nut consumption on plasma total polyphenols, antioxidant capacity and lipid peroxidation. *Journal of Human Nutrition and Dietetics, 22*, 64–71.

US Department of Agriculture, Agricultural Research Service. (2009). USDA National Nutrient Database for Standard Reference, Release 22. Nutrient Data Laboratory Home Page, available at http://www.ars.usda.gov/ba/bhnrc/ndl.

CHAPTER 136

Watermelon (*Citrullus lanatus* (Thunb.) Matsumura and Nakai) Seed Oils and Their Use in Health

Tirupur Venkatachalam Logaraj
Head, Department of Plant Biology & Biotechnology, Government College of Arts and Science, Coimbatore, Tamil Nadu State, India

1149

LIST OF ABBREVIATIONS

   AA, arachidonic acid
   ADHD, attention deficit hyperactivity disorder
   ALA/α-LA, α-linolenic acid
   CHD, coronary heart disease
   DHA, docosahexanoic acid
   EFA, essential fatty acid

Nuts & Seeds in Health and Disease Prevention. DOI: 10.1016/B978-0-12-375688-6.10136-7

en%, energy percentage
EPA, eicosapentanoic acid
HUFA, highly unsaturated fatty acid
LA, linoleic acid
LCPUFA, long-chain polyunsaturated fatty acid
LDL, low density lipoprotein
MRFITRG, Multiple Risk Factors Intervention Trial Research Group
MUFA, monounsaturated fatty acid
NO, nitrogen oxide
OA, oleic acid
PUFA, polyunsaturated fatty acid
SFA, saturated fatty acid
TABP, type-A behavior pattern
VLDL, very low density lipoprotein

INTRODUCTION

One of the pressing needs of today is to utilize crops to the maximum in order to meet the nutritional and health requirements of humankind. Watermelon, a widely used crop, offers calories, water, minerals, and proteins, while its seeds provide oil (therapeutic) and "kernel cake" as filling and fibrous sources. The crop's easy adaptability to dry conditions enables it to be domesticated under cultivable conditions. This chapter concentrates on the rich watermelon seed oils, which are composed chiefly of polyunsaturated fatty acids (i.e., oleic, linoleic, and traces of linolenic acids) along with saturated fatty acids like palmitic and stearic acids, and storage glycerides (El-Adawy & Taha, 2001; Logaraj, 2010).

BOTANICAL DESCRIPTION

The watermelon, *Citrullus lanatus* (Thunb.) Matsumura & Nakai (synonym *Citrullus vulgaris* Schrader), belongs to the Cucurbitaceae family. It is a monoecious trailing herb with softly pubescent leaves, 11–15 ×8–10 cm in size, deeply trifid, with segments further bipinnatifid, obovate-oblong/obtuse-round lobes, and robust, bifid tendrils. The flowers are 2 cm across with a five-lobed calyx, hypantheous, adnate to the ovary; the corolla is yellow and five-lobed; there are basically five stamens but they are partially synandrous; the ovary is inferior and ovoid in shape. The fruit is dark green or white striped, and measures 30–45 cm by 45 cm (Matthew, 1991).

HISTORICAL CULTIVATION AND USAGE

The earliest records on watermelon date back to 3000 BC, appearing in Egyptian hieroglyphics on monument walls. The fruits were left in the tombs of mummified Pharaohs. The plant originates from the Kalahari Desert, in Africa. Probably, the wild species, with its high water content, was an ideal food for the Kalahari aborigines, leading to its selection as a cultivar. With growing experience, three varieties later became popular in northern Africa, and their seeds were used to make cooking oil (known as kalahari or ootanga oil), protein seed-cakes, and condiments. Cultivation spread to Arabian and Mediterranean countries, and later to America and Eastern Asia (Oyenuga & Fetuga, 1975). The seed oils were known to contribute to cardiac health.

PRESENT-DAY CULTIVATION AND USAGE

Currently, 50 varieties of watermelon are known; some of these are seedless. The fruit pulp is usually red, but yellow-orange melons are also available. Owing to its desert origin, it is cultivated throughout the tropics as a long-season plant. The growing literature cites its multifaceted uses. It is a calorific food with a high water content, also containing sugar, minerals, B-complex vitamins, and special amino acids, while its seed oils contain triglycerides and saturated and unsaturated fatty acids with therapeutic values. The seed tea has anti-helminthetic, diuretic and laxative effects, and is anticarcinogenic at the level of the epithelial cells of glands.

In the field of cosmetics, too, watermelon seed oils are popular. Their moisturizing, non-clogging, cleansing, and semi-drying features are useful in preparing skin-care cosmetics and lotions. The abundant ω-PUFAs lead to the above properties, and baby formulations, creams, lotions, soaps, eye creams, facial oils, and skin- and hair-care preparations are also being made and commercialized. Above all, the ω-PUFAs have led to watermelon seed oil's popularity in treating cardiovascular and carcinogenic complications.

Watermelon seeds

There are a few hundred watermelon seeds in total in a mature fruit (Figures 136.1–136.3), although the number may vary from variety to variety and cultivar to cultivar. When the melon

FIGURE 136.1
Morphology of watermelon fruits. A heap of watermelon fruits, of different sizes, shapes, and stripes, for sale at a market in Coimbatore, a southern part of India.

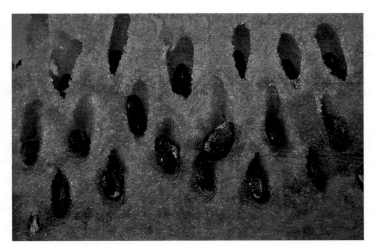

FIGURE 136.2
Longitudinal section of watermelon fruit, showing bright red watery flesh with seeds buried in it.

FIGURE 136.3
Morphology of watermelon seeds. Stored watermelon seeds showing dark reflective and rough surfaces of seed coats.

is freshly cut, the seeds comprise approximately 2–3% of the fruit's total weight. The seeds are slightly pear-shaped, with a strong light brown to brown or black color, sometimes with a mottled surface. The seeds are in parietal placentation, and are covered with mucilage when afresh.

Seed chemistry

Watermelon seeds are procured in tonnes in African and Middle Eastern countries, as delicious cakes can be made from them, or food grade oil pressed from them. The protein composites of the seeds are mostly of amino acids, such as tryptophan, lysine, and glutamic acid. The seeds also contain arginine, which is often used to regulate blood pressure and cure cardiac ailments. The seeds also contain several B vitamins, including folate, thiamin, riboflavin, pantothenate, vitamin B_6, and niacin, totaling 3.8 mg in 31 g of seed — equivalent to 19% of the daily requirement. The prevalent elements are calcium, iron, magnesium, manganese, and zinc. Magnesium regulates blood pressure, and is also involved in controlling the carbohydrate metabolism, thus affecting plasma sugar levels. The fatty acids of watermelon seeds and their health benefits are discussed elsewhere in this chapter.

The seeds can be roasted to make tea or soups. They have diuretic effects, and hence improve kidney and bladder function, particularly during hot weather. They are also anti-inflammatory, chiefly affecting the urinary system. They regulate blood pressure by dilating capillaries, especially in post-menopausal, senile, and obesity-related hypertension.

Water absorption and retention aspects of watermelon seeds

The water absorption and retention aspects of watermelon seeds have been studied, using an average weight of a small cup of about 20 seeds. The fresh seeds weighed 1760 mg and the stored ones 1200 mg. The average weight of the fully dried ones was 1137 mg, and for the fully imbibed ones was 2300 mg (Table 136.1). It was found that water-imbibing capacity is about one-third of the fresh weight. The water-holding capacity improves upon storage (Table 136.2, Figure 136.4). The water retained by the fully dried and stored seeds is double the imbibitional suction power of fresh seeds, suggesting a greater water imbibitional power for watermelon seeds. This evidences the readiness of watermelon seeds for the first and crucial stage of germination. However, germinability tests alone will reveal the successful percentage of germination. Observations of cut stored and totally dried seeds revealed the higher oil content of the kernel (Figure 136.5).

TABLE 136.1 Responses of Watermelon Seeds to Different Storage Conditions*

Seed Type	Average Weight of 20 Seeds (mg)
Fresh	1760
Stored	1200
Fully dried	1137
Fully turgid	2300

*All data are as per the study conducted at the author's laboratory.

TABLE 136.2 Responses of Watermelon Seeds to Soaking, Total Drying, and Storage*

Water loss upon storage	FSW − SSW	1760 − 1200	560
Water loss upon total drying	FSW − TDSW	1760 − 1137	623
Water gained upon soaking	TUSW − FSW	2300 − 1760	540
Water gained upon soaking	TUSW − TDSW	2300 − 1137	1163
Water gained upon soaking	TUSW − SSW	2300 − 1200	1100

Response of watermelon seeds to 12 hours' soaking or total drying over a storage period of 15 days.
FSW, fresh seed weight; TUSW, turgid seed weight; SSW, stored seed weight without drying; TDSW, totally dried seed weight.
The seed behavior is recorded as differences in the average weight of 20 seeds, in mg.
*All the data are as per the study conducted at the author's laboratory.

FIGURE 136.4
Morphometry of turgid (upper row) and dry (lower row) watermelon seeds.

There was not much difference in the seed length between soaked, stored, and fully dried seeds (Table 136.3). There was no difference in the breadth of the kernels between fresh and stored ones, but the totally dried seeds showed a decrease in breadth of 0.25 mm. This suggests that the seeds are mainly composed of oil and other reserves, and the water content is of no great value.

However, the imbibitional capacity of the fully dried seeds suggests that, as well as intra-seed factors, there may be some extra-seed factors responsible for this. It may be that the thick, dried mucilage coat on the dry seed testa absorbs a great amount of water. This must be confirmed by the complete removal of the mucilage coat and recording imbibitional results

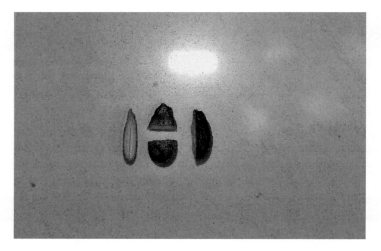

FIGURE 136.5
Longitudinal and transverse sections of watermelon seeds, showing thick testa and heavy oil content.

TABLE 136.3 Morphometric Differences of Watermelon Seeds with Soaking, Total Drying, and Storage*		
Seed Type	**Length (mm)**	**Breadth (mm)**
TUSW	9.5	2.00
SSW	9.5	2.00
TDSW	9.5	1.75

Response of watermelon seeds to 12 hours' soaking or total drying over a storage period of 15 days.
SSW, stored seed weight without drying; TUSW, turgid seed weight; TDSW, totally dried seed weight.
The seed behavior is recorded as differences in the average length and breadth of 20 seeds, in mm.
*All the data are as per the study conducted at the author's laboratory.

thereafter. The data in Tables 136.1—136.3, and the discussion here, are the outcome of the water-relationship studies conducted in the author's laboratory.

APPLICATIONS TO HEALTH PROMOTION AND DISEASE PREVENTION

Watermelon seeds yield 55% oil on a dry weight basis, and mainly contain linoleic acid (50—60%), followed by oleic acid (15%), and considerably lesser amounts of palmitic and stearic acids (Logaraj, 2010). Hence, watermelon seeds offer about 600 calories and 79% of the recommended daily allowance (RDA) of fat. The major composition of the seed oil is ω-6 unsaturated fatty acids, mainly linoleic acid, which helps in decreasing the levels of cholesterol and high blood pressure.

Linoleic acid and α-linolenic acid in moderate concentrations promote good health, but in excess can be harmful. They are involved in energy production, oxygen transfer to blood, hemoglobin synthesis, cell division, growth, and neural functions which is why they are abundant in brain and nervous tissues. Fat deficiency causes chronobiological dysregulation, impaired cognitive sensory responses, and liability to infection, especially in infancy.

In the human body the information traffic is controlled by the nervous, endocrine, and immune systems through electrical and chemical signals either generated or received, which, after modification, activate tissues, muscles, and organs. These three systems therefore involve PUFAs in changing neuronal membrane fluidity, and releasing specific factors, hormones, and cytokines (Yehuda et al., 2000).

The lack of interconversion of ω-6 to ω-3 fatty acids is compensated for by retroconversions in mammals, as in DHA to AA (Sprecher *et al.*, 1995). Contrarily, plants synthesize these precursor PUFAs (LA, ALA), which are abundant in chloroplastic membranes and certain vegetable oils. Hence, the tissues of phytophagous animals, fish, and shellfish are rich with these precursors, while greens and baked beans have a low content (Sinclair, 1993).

LA cascade

The essentiality of LA for growth and reproduction was proved by the resultant scaly-skinned feet and tail, and retarded growth, in rats deprived of ω-6 FAs. The deficient rats mated unsuccessfully, and if fetuses did form they were reabsorbed, or the offspring were born deformed, with histologically abnormal liver and kidneys.

The relative constancy of AA (ω-6) and DHA (ω-3) levels in brain and central nervous system is essential to maintain the brain phospholipid content, which is reduced by prolonged starvation of ω-6 and ω-3 FAs over two generations.

Atopic dermatitis is controlled through steroidal anti-inflammatory drug administration by inhibiting the LA cascade, which can also be achieved by clinical administration of high levels of ω-3 and decreased LA intake, which act synergistically (Kato *et al.*, 2000). Apart from ADHD (attention deficit hyperactivity disorder), certain brain pathological conditions could also be caused by increased LA intake and elevated ω-6 and ω-3 ratios in dietary fats and oils (Yoshida *et al.*, 1998).

Cholesterol and fatty acids

Although SFAs and MUFAs suppress cholesterol synthesis, ω-6 PUFAs (LA and AA) do so to a greater extent (Pischon *et al.*, 2003).While ω-3 fatty acids directly reduce cholesterol levels, ω-6 fatty acids redistribute them, and the preferred ratio for ω-6:ω-3 is 1:4 (Yoshida *et al.*, 1993). Cholesterol synthesis is stimulated in the order of S, M ≥ LA > ALA > EPA, DHA, and AA. The first three possibly elevate the isoprenyl intermediate, suppressing the formation of vasodilatory NO, and expedite atherogenesis through prenylation of the oncogene products (Ras, Rho) leading to cell multiplication.

Overdose effects and prevention of atherosclerosis and related diseases

Although as early as 1982 (MRFITRG) high-LA vegetable oil was recommended to prevent atherosclerosis, hypercholesterolemia, etc., it was found to reduce blood cholesterol only transiently (Okuyama *et al.*, 2000) and to be unable to control a type-A behavior pattern (TABP) heart attack. Thus it was not cholesterol, but high levels of LA and high ω-6:ω-3 PUFA ratios that were the causes of TABP attacks (Okuyama *et al.*, 2000).

Therapeutic advice

There is no general agreement among scientists over the optimal ω-6:ω-3 ratio, which currently ranges from 4.5 to 2; the only common opinion is that the prevailing ratio of ω-6:ω-3 must be brought down; it is recommended that both PUFAs (LA and ALA) and HUFAs (AA, EPA, and DHA) should be almost in equal amounts, and the total free energy must be reduced to < 25%. Though LA and ω-3 FAs are essential, LA should be < 1%.

In conclusion, people with different food habits consume different dosages of essential fatty acids. The EFAs LA and ALA are necessary for healthy living. Regarding watermelon seed oil, its PUFA values are such that it can be recommended and used as a nutraceutical to alter day-to-day cooking oils to give variable EFA proportions to achieve maximum health benefits. A watermelon seed oil emulsion with ω-6 PUFA (LA) can also be used as an effective nutraceutical delivery system for controlled and effective health strategies (Logaraj *et al.*, 2008).

ADVERSE EFFECTS AND REACTIONS (ALLERGIES AND TOXICITY)

Watermelon seeds or nuts contain very few antinutritional and toxic factors compared to the other genera of Cucurbitaceae, and have a very high edible value (Badifu, 2001). Hence, watermelon seed oil has excellent nutritive as well as clinical and therapeutic activities that can be exploited in its use as an effective nutraceutical.

SUMMARY POINTS

- Watermelon (*Citrullus lanatus*) cultivation and the marketing of fruits and seeds are relevant to the nutritional and therapeutic needs of modern lifestyles, as the seeds offer nutraceutically valued oils.
- An additional advantage is the use of watermelon seed oil in cosmetic preparations (cosmoceuticals).
- The abundance of triglycerides (TG), saturated fatty acids (SFA), and ω-6 polyunsaturated fatty acids (PUFAs) meet cooking, cosmetic, and therapeutic oil needs.
- Watermelon seed oil is a popular medicinal and clinical source of ω-PUFAs used in treating coronary heart diseases (CHD), where it is operative at the epithelial and membrane levels.
- The intermediates of linoleic acid and α-linolenic acid cascades could be useful in gene technology.
- Watermelon seed oils rich in linoleic acid (LA) play a vital role in treating nervous, endocrine, and immunological disorders, through PUFA's metabolism.
- Watermelon seed oil, made into emulsions, could also be administered to give an effective and controlled supply or dosage of LA for therapeutic purposes.

ACKNOWLEDGEMENT

The author wishes to acknowledge the valuable help rendered in the preparation of this manuscript by Prof. Joseph Clement A., former Head of the Department where the author presently works.

References

Badifu, G. I. O. (2001). Effect of processing on proximate composition, antinutritional and toxic contents of kernels from Cucurbitaceae species grown in Nigeria. *Journal of Food Composition and Analysis, 14,* 153—161.

El-Adawy, T. A., & Taha, K. M. (2001). Characteristics and composition of different seed oils and flours. *Food Chemistry, 74,* 45—54.

Kato, M., Nagata, Y., Tanabe, A., Ikemoto, A., Watanabe, S., & Kobayashi, T. (2000). Supplementary treatment of atopic dermatitis patients by choosing foods to lower the ω-6/ω-3 ratio of fatty acids. *Journal of Health Science, 46,* 1—10.

Logaraj, T. V. (2010). *Studies on selected plants and microbes with special reference to polyunsaturated fatty acids.* PhD Thesis, University of Mysore, Karnataka, India.

Logaraj, T. V., Bhattacharya, S., Udayasankar, K., & Venkateswaran, G. (2008). Rheological behavior of emulsions of avocado and watermelon oils during storage. *Food Chemistry, 106,* 937—943.

Matthew, K. M. (1991). *An excursion flora of Central Tamilnadu, India.* New Delhi, India: Oxford & IBH Publishing Co. Pvt. Ltd.

Multiple Risk Factors Intervention Trial Research Group. (1982). Multiple Risk Factor Intervention Trial. Risk factor changes and mortality results. *Journal of the American Medical Association, 248,* 1465—1477.

Okuyama, H., Fujii, Y., & Ikemoto, A. (2000). ω-6/ω-3 dietary fatty acids rather than hypercholesterolemia as the major risk factor for atherosclerosis and coronary heart disease. *Journal of Health Science, 46,* 157—177.

Oyenuga, V. A., & Fetuga, B. L. (1975). Some aspects of the biochemistry and nutritive value of the watermelon seed. *Journal of the Science of Food and Agriculture, 26,* 855—860.

Pischon, T., Hankinson, S. E., Hotamisligil, G. S., Rifai, N., Willen, W. C., & Rimm, E. B. (2003). Habitual dietary intake of ω-3 and ω-6 fatty acids in relation to inflammatory markers among US men and women. *Circulation, 108,* 155—160.

Sinclair, A. (1993). The nutritional significance of ω-3 polyunsaturated fatty acids for humans. *Asian Food Journal, 8,* 3—13.

Sprecher, H., Luthria, D. T., Mohammed, B. S., & Baykousharva, S. P. (1995). Reevaluation of the pathways for the biosynthesis of polyunsaturated fatty acids. *Journal of Lipid Research, 36*, 2471–2477.

Yehuda, S., Rabinovitz, S., Carasso, R. L., & Mostofsky, D. I. (2000). Mixture of essential fatty acids rehabilitates stress effects on learning and cortisol and cholesterol level. *International Journal of Neuroscience, 101*, 73–87.

Yoshida, S., Yasuda, A., & Kawasto, H. (1993). Ultrasaturated study of hippocampus synapse in perilla and safflower oil federates. In T. Yasugi, H. Nakamura, & M. Soma (Eds.), *Advances in polyunsaturated fatty acid research.* Amsterdam, The Netherlands: Elsevier.

Yoshida, S., Sato, A., & Okuyama, H. (1998). Pathophysiological effects of dietary essential fatty acids balance on neural systems. *Japanese Journal of Pharmacology, 77*, 11–22.

CHAPTER 137

White Cabbage (*Brassica chinensis*) Seeds and Their Health Promoting Activities

Tzi-Bun Ng[1], Patrick H.K. Ngai[3], Randy C.F. Cheung[1], Jack H. Wong[1], Sze-Kwan Lam[1], He-Xiang Wang[2], Xiujuan Ye[1], Evandro F. Fang[1], Yau-Sang Chan[1]

[1] School of Biomedical Sciences, Faculty of Medicine, The Chinese University of Hong Kong, Hong Kong, China
[2] State Key Laboratory of Agrobiotechnology, Department of Microbiology, China Agricultural University, Beijing, China
[3] School of Life Sciences, Faculty of Science, The Chinese University of Hong Kong, Hong Kong, China

1159

INTRODUCTION

White cabbage, also referred to as snow cabbage, is a leafy vegetable often found in Chinese cuisine. The vegetable is related to the Western cabbage.

BOTANICAL DESCRIPTION

Brassica chinensis, also called Chinese mustard, or bok choy, possesses dark green foliage and thick, white stalks in a loose head with a yellow-flowering center. The most common form of white cabbage has a cylindrical tight head, 10 cm in diameter and up to 45 cm long. The outer leaves are light green with a white midrib, while the inner leaves are creamy yellow. Beijing varieties of white cabbage, also referred to as *Shao* vegetables in Cantonese, have large green outer leaves and pale yellow or white inner leaves. Multiple leaves are wrapped tightly together to form a cylinder, and the majority will form a dense head. The leaves are not exposed to sunlight, and hence are pale yellow in color.

Nuts & Seeds in Health and Disease Prevention. DOI: 10.1016/B978-0-12-375688-6.10137-9

HISTORICAL CULTIVATION AND USAGE

White cabbage is native to China, and was mainly restricted to the Yangtze River Delta area, being very similar to a variant cultivated in Zhejiang around the 14th century. Subsequently, it was cultivated in northern China. Cabbages were exported along the Grand Canal to Zhejiang and the southern provinces, and were taken to Korea during the Ming Dynasty, where they were developed into a staple vegetable for producing kimchi. During the Russo-Japanese War in the early 20th century, Japanese soldiers in northeast China brought it back to Japan.

PRESENT-DAY CULTIVATION AND USAGE

White cabbage is a cool-season crop. A spring crop can be grown from transplants in early spring, and for the fall crop, seeds are sown 60—80 days before the frosts begin. Growth is faster in well-drained, fertile and moist soil.

White cabbages (*Shao* vegetables) are robust, and can be stored for quite some time. At difficult times in northern China, they are the only vegetable available throughout the winter, as families often save a few of them for the cold weather. Chinese cabbage is eaten stewed, fried, salted, or mixed in a variety of ways. The cabbage can be stored outdoors in the winter, and the outer leaves can be dried and insulate the inner leaves. There are several other methods of winter cabbage storage in Korea and northeast China. Cabbages are planted in the autumn, after harvesting the corn, and are then harvested early in the winter. The market price is usually very cheap.

APPLICATIONS TO HEALTH PROMOTION AND DISEASE PREVENTION

The seeds of *Brassica* sp. and the napins they contain constitute an important resource for animal nutrition and industrial oil production. The inclusion of napins, which are rich in cysteine residues, in animal fodder and for human consumption, may make up for the cysteine deficiency in legume grains (Molvig *et al.*, 1997; Barciszewki *et al.*, 2000).

Napins are proteins composed of a 4.5-kDa small subunit and a 10-kDa large subunit (Gehrig & Biemann, 1996). They are derived from proteolytic cleavage of a precursor made up of about 180 amino acid residues (Ericson *et al.*, 1986). Kohlrabi seeds (Neumann *et al.*, 1996a, 1996b), radish seeds (Polya *et al.*, 1993), and *Arabidopsis thaliana* (Krebbers & Vandekerckhove, 1990) produce multiple napins. The most extensively investigated napins are those from oilseed rape *Brassica napus* (Ericson *et al.*, 1991; Neumann *et al.*, 1996a, 1996b). The N-terminal sequences of the large and small subunits of napin are similar in *Sinapis alba* napin (Svendsen *et al.*, 1994) but different in *B. napus* napin (Neumann *et al.*, 1996a, 1996b). The nitrogen storage function of napin, which is synthesized in developing seed embryos, is in conformity with its abundance of amides and arginine residues (Muntz, 1998). Napin serves as a source of nitrogen to germinating seedlings (Lonnerdal & Janson, 1972). Because some napins are allergenic, they are also of clinical interest (Memendez-Arias *et al.*, 1988). Napin demonstrates trypsin-inhibiting activity, but each of the subunits is inactive. Both napin and its subunits are calmodulin antagonists and substrates for plant calcium-dependent protein kinases, since calmodulin and its small subunit exhibit similar α-helix—hinge—α-helix motifs (Neumann *et al.*, 1996a, 1996b). Napin inhibits calmodulin-dependent myosin light-chain kinase (Gehrig & Biemann, 1996; Neumann *et al.*, 1996a, 1996b). It may also elicit an anti-fungal action (Barciszewki et al., 2000).

A heterodimeric 11-kDa napin-like polypeptide composed of a smaller (4-kDa) subunit and a larger (10-kDa) subunit has been purified from Chinese white cabbage (*Brassica chinensis* cv dwarf) seeds. It was tested for trypsin-inhibitory activity, which has been reported for other napins (Neumann *et al.*, 1996a, 1996b), and for other activities, including translation-inhibitory, antibacterial, antifungal, and ribonuclease activities, which are found in other seed proteins, such as antifungal proteins, lectins, ribosome-inactivating proteins, trypsin inhibitors, and ribonucleases (Lam *et al.*, 1998; Fong *et al.*, 2000; Ng *et al.*, 2003).

The isolation procedure is as follows. The seeds were soaked in 10-mM Tris-HCl buffer (pH 7.4), homogenized in the same buffer using a Waring blender, and then centrifuged (12,000×*g*, 30 min, 4°C) The supernatant was then applied to a column of DEAE-cellulose and was fractionated into a large unadsorbed fraction with cell-free translation-inhibiting activity, and a small adsorbed fraction devoid of such activity. When the unadsorbed fraction was applied to an Affi-gel blue gel column, it was resolved into a large unadsorbed fraction (not shown) and three adsorbed fractions, B1, B2, and B3. Activity was enriched in B2. Ion exchange chromatography of B2 on Mono S yielded a large unadsorbed fraction, S1, devoid of activity; a small inactive adsorbed peak, S2; and two large adsorbed peaks, S3 and S4. Translation-inhibiting activity resided in S3. Upon gel filtration on Superdex 75, S3 was separated into a large peak, F1, in which the activity resided, and a small inactive peak, F2. F1, the purified napin-like polypeptide, appeared as two bands in SDS-PAGE, one with a size of 7 kDa and another with a size of 4 kDa.

The 7-kDa subunit markedly resembled napin large chain, albumin, and trypsin inhibitor in N-terminal sequence. The N-terminal sequence of the 4-kDa subunit was similar to that of napin large chain and an antimicrobial peptide. The napin-like polypeptide exhibited translation-inhibitory activity with an IC_{50} of 18.5 nM. It was stable between pH 4 and 11, and between 10°C and 40°C. The polypeptide displayed more potent inhibitory activity toward trypsin ($IC_{50} = 8.5\ \mu M$) than toward chymotrypsin ($IC_{50} = 220\ \mu M$), but was devoid of ribonuclease and antifungal activities. It manifested antibacterial activity against bacteria, including *Bacillus cereus*, *Bacillus megaterium*, *Bacillus subtilis*, and *Pseudomonas aeruginosia* (Ngai & Ng, 2004).

Table 137.1 summarizes of the properties of this napin-like protein.

1161

ADVERSE EFFECTS AND REACTIONS (ALLERGIES AND TOXICITY)

Cabbage may be goitrogenic because of interference with iodine uptake, organification in thyroid cells, and suppression of the formation of thyroxine and triiodothyronine. The

TABLE 137.1 Characteristics of Napin-like Polypeptide with Health-promoting Activity from White Cabbage Seeds (Ngai & Ng, 2004)

Molecular mass	Heterodimeric (one 4-kDa subunit and one 7-kDa subunit)
N-terminal sequence	4-kDa subunit, PAQPFRFPKH
	7-kDa subunit, POGPQTRPPI
Chromatographic behavior on:	
DEAE-cellulose	Unadsorbed
Affi-gel blue gel	Adsorbed
Mono S	Adsorbed
Superdex75	
Antibacterial activity against:	
Bacillus cereus	Present
Bacillus megaterium	Present
Bacillus subtilis	Present
Pseudomonas aeruginosia	Present
Cell-free translation inhibitory activity	$IC_{50} = 18.5$ nM; activity stable at pH 4 and 11, and 10–40°C
Trypsin inhibitory activity	$IC_{50} = 8.5\ \mu M$
Chymotrypsin inhibitory activity	$IC_{50} = 220\ \mu M$

elevated secretion of thyroid stimulating hormone from the pituitary, caused by reduced negative feedback of thyroid hormones, produces thyroidal hypertrophy or goiter. Thus, patients with goiter are advised against consuming cabbage.

Glucosinolate, sulfur-containing organic anions bonded to glucose, exists in plants including cabbages. Glucosinolate is hydrolyzed by myrosinase, and various by-products are produced. One such decomposed product, thiocyanate, exerts harmful effects on thyroidal metabolism, owing to competition with iodine. The Korean diet contains large amounts of cabbage and radishes, which may be correlated with a high incidence of thyroid dysfunction. The average daily intake of thiocyanate through Brassicaceae vegetables in Korea was estimated to be 16.3 μmol SCN(−)/d per person. When this was compared to published animal studies, average thiocyanate intake per person was lower than doses required to produce adverse effects. However, further studies may be warranted to ensure sufficient margins of safety (Han & Kwon, 2009).

Allergic response after consumption of cabbage is rare. Swelling of the face and throat has been reported in a woman after eating cabbage. The patient presented 4+ reactions to *Brassica* plants by skin testing. A RAST employing cabbage extract was positive for specific IgE antibody. Thus, IgE sensitivity to *Brassica* plant can occur (Blaiss *et al.*, 1987).

Digestive discomfort may arise due to gas originating from cabbage. This can be minimized by soaking the entire or chopped cabbage in water for 5 minutes, and replacing the water before cooking.

Although *Brassica* vegetables produce toxic levels of S-methyl-L-cysteine sulfoxide, which can bring about hemolytic anemia in livestock, these vegetables are often safe for human consumption (Benevenga *et al.*, 1989).

SUMMARY POINTS

- White cabbage seeds produce a heterodimeric 11-kDa napin-like polypeptide with health-promoting activities.
- The polypeptide displays antibacterial activity against *Pseudomonas aeruginosia*, *Bacillus subtilis*, *B. megaterium*, and *B. cereus*.
- The polypeptide does not, however, inhibit mycelial growth in several fungal pathogens tested.
- The 7-kDa subunit of the polypeptide closely resembles trypsin inhibitor in sequence. The polypeptide inhibits trypsin with a higher potency than its chymotrypsin inhibitory activity.
- It reduces the incorporation of [^{3}H-methyl]-leucine into protein in the cell-free rabbit reticulocyte system.

References

Barciszewki, J., Szymanski, M., & Haertle, T. (2000). Minireview, analysis of rape seed napin structure and potential roles of the storage protein. *Journal of Protein Chemistry, 19*, 249–254.

Benevenga, N. J., Case, G. L., & Steele, R. D. (1989). Occurrence and metabolism of S-methyl-L-cysteine and S-methyl-L-cysteine sulfoxide in plants and their toxicity and metabolism in animals. In P. R. Cheeke (Ed.), *Proteins and amino acids. Toxicants of plant origin, Vol. III* (pp. 203–228). Boca Raton, FL: CRC Press.

Blaiss, M. S., McCants, M. L., & Lehrer, S. B. (1987). Anaphylaxis to cabbage, detection of allergens. *Annals of Allergy, 58*, 248–250.

Ericson, M. L., Rodin, J., Lenman, M., Glimelius, K., Tosefsson, L. G., & Rask, L. (1986). Structure of the rapeseed 1.7 S storage protein, napin, and its precursor. *Journal of Biological Chemistry, 261*, 14567–14581.

Ericson, L., Muren, E., Gustavvson, H. O., Josefsson, L. G., & Rask, L. (1991). Analysis of the promoter region of napin genes from *Brassica napus* demonstrates binding of nuclear protein *in vitro* to a conserved sequence motif. *European Journal of Biochemistry, 197*, 741–746.

Fong, W. P., Mock, W. Y., & Ng, T. B. (2000). Intrinsic ribonuclease activities in ribonuclease and ribosome-inactivating proteins from the seeds of bitter gourd. *International Journal of Biochemistry & Cell Biology, 32*, 571–577.

Gehrig, P. M., & Biemann, K. (1996). Assignment of the disulfide bonds in napin, a seed storage protein from *Brassica napus*, using matrix-assisted laser description ionization mass spectrometry. *Peptide Research, 9*, 308–314.

Han, H., & Kwon, H. (2009). Estimated dietary intake of thiocyanate from Brassicaceae family in Korean diet. *Journal of Toxicology and Environmental Health A, 72*, 1380–1387.

Krebbers, E., & Vandekerckhove, J. (1990). Production of peptides in plant seeds. *Trends in Biotechnology, 8*, 1–3.

Lam, S. S. L., Wang, H. X., & Ng, T. B. (1998). Purification and characterization of novel ribosome inactivating proteins, alpha- and beta-pisavins, from seeds of the garden pea *Pisum sativum*. *Biochemical and Biophysical Research Communications, 253*, 135–142.

Lonnerdal, B., & Janson, J. C. (1972). Studies on *Brassica* seed proteins. I. The low molecular weight proteins in rapeseed. Isolation and characterization. *Biochimica Biophysica Acta, 78*, 175–184.

Memendez-Arias, L., Moneo, I., Doninguez, J., & Rodriguez, R. (1998). Primary structure of the major allergen of yellow mustard (*Sinapis alba L.*) seeds, Sin a I. *European Journal of Biochemistry, 177*, 159–166.

Molvig, L., Tabe, L. M., Eggum, B. O., Moore, A. E., Craig, S., & Spencer, D. (1997). Enhanced methionine levels and increased nutritive value of seeds of transgenic lupins (*Lupinus angustifolius*) expressing a sunflower seed albumin gene. *Proceedings of the National Academy of Sciences of the United States of America, 94*, 8393–8398.

Muntz, K. (1998). Deposition of storage proteins. *Plant Molecular Biology, 38*, 77–99.

Neumann, G. M., Condron, R., Thomas, I., & Polya, G. M. (1996a). Purification and sequencing of multiple forms of *Brassica napus* seed napin large chains that are calmodulin antagonists and substrates for plant calcium dependent protein kinases. *Biochimica Biophysica Acta, 1295*, 23–33.

Neumann, G. M., Condron, R., Thomas, I., & Polya, G. M. (1996b). Purification and sequencing of multiple forms of *Brassica napus* seed napin small chains that are calmodulin antagonists and substrates for plant calcium dependent protein kinases. *Biochimica Biophysica Acta, 1295*, 34–43.

Ng, T. B., Lam, S. K., & Fong, W. P. (2003). A homodimeric sporamin-type trypsin inhibitor with antiproliferative, HIV reverse transcriptase-inhibitory and antifungal activities from wampee (*Clausena lansium*) seeds. *Biological Chemistry, 384*, 289–293.

Ngai, P. H., & Ng, T. B. (2004). A napin-like polypeptide from dwarf Chinese white cabbage seeds with translation-inhibitory, trypsin-inhibitory, and antibacterial activities. *Peptides, 25*, 171–176.

Polya, G. M., Chandra, S., & Condron, R. (1993). Purification and sequencing of radish seed calmodulin antagonists phosphorylated by calcium-dependent protein kinase. *Plant Physiology, 101*, 545–551.

Svendsen, I. B., Nicolova, D., Goshev, I., & Genov, N. (1994). Primary structure, spectroscopic and inhibitory properties of a two-chain trypsin inhibitor from the seeds of charlock (Sinapis arvensis L.), a member of the napin protein family. *International Journal of Peptide and Protein Research, 43*, 425–430.

CHAPTER 138

Biological Activity of Seeds of Wild Banana (*Ensete superbum* Cheesm, Family Musaceae)

Monica Kachroo[1], Shyam S. Agrawal[2]
[1] Al-Ameen College of Pharmacy, Bangalore, India
[2] HOD Department of Pharmacology, Delhi Institute of Pharmaceutical Sciences and Research, New Delhi, India

1165

LIST OF ABBREVIATIONS

BAP, 6-benzylaminopurine
FTIR, Fourier transform infrared spectroscopy
GA3, gibberellic acid
HPLC, high performance liquid chromatography
HPTLC, high performance thin layer chromatography
NMR, nuclear magnetic resonance
UV, ultra violet spectoscopy

Nuts & Seeds in Health and Disease Prevention. DOI: 10.1016/B978-0-12-375688-6.10138-0

INTRODUCTION

Traditional systems of medicine continue to be widely practiced. Of the 2,500,000 higher plant species on earth, more than 80,000 are medicinal. India is one of the world's 12 biodiversity centers, with the presence of over 45,000 different plant species.

One such plant, a wild banana, *Ensete superbum* (Cheesm), is an ornamental plant indigenous to northern India and the higher elevations of Thailand. Since the seeds of the plant have many traditional uses, the extracts and its constituents have been studied for various pharmacological activities. The present chapter provides a review of the investigations carried out on the seeds of this traditional/folkloric plant.

BOTANICAL DESCRIPTION

Ensete superbum (Cheesman, 1947) is a member of the Musaceae family of the Order Zingiberales, and has the synonym *Musa superba*. Its common names include *banakadli* (Hindi) and wild banana (English). The plant grows to 3—4 m high, and is non-stoloniferous with a stout pseudostem 1.2—1.8 m tall, with an enormous swollen base of 2—2.4 m in circumference at the base, narrowing to 1 m below the leaves (Figure 138.1). The leaves are bright green on sides, 1.5 m long, and 45 cm broad, with a very short, deeply channeled free petiole. The leaf sheath is persistent at the base, and leaves closely set scars on the corm. The inflorescence is globose at first, 30 cm in diameter, later drooping and elongating to one-third the length of the trunk. The bracts are orbicular, dark brownish-red (dull claret-brown), reaching 30 cm in length and breadth, and subtend dense biseriate rows each of 10—15 flowers. The ovary is white, cylindrical, and about 2.5 cm long. The outer perianth is whitish, situated along the ovary, and three-lobed, or formed of three loosely coherent segments. The inner perianth is shorter than the outer, and is tricuspidate with a long, linear central cusp. The rootstock and young inflorescences are eaten, and the fruit is pickled when young. The fruit is subcoriaceous, 7.5 cm long and 3.5 cm in diameter, and more or less triangular in cross-section, has a thick

FIGURE 138.1
Ensete superbum Cheesm (whole plant) with a massive baeband red flower head with broad leaves outlined in deep red and with deep mid-ribs. *Source: www.centralfloridafarms.com/ensete-superbum.htm*

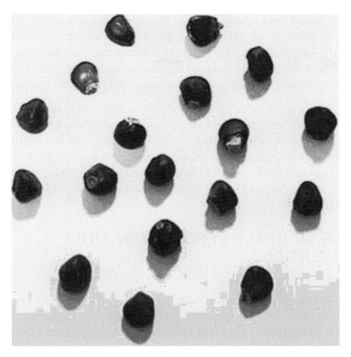

FIGURE 138.2
Ensete superbum Cheesm seeds. *Source: Photographed by author.*

skin, and contains numerous dark brown seeds. The seeds are brownish-black, subglobose but angled by pressure, and 8–12 mm in diameter (Figure 138.2). The plant dies down to an underground corm in the dry season, and forms new leaves at the beginning of the monsoons (Baker 1893, 1894).

The plant has a massive baseband red flower head with broad leaves outlined in deep red, and deep mid-ribs are specific morphological features that add its appeal and esthetic value (http://www.iiitmk.ac.in/~jrnair/ensetepaperfinal.pdf).

The genus *Ensete*, proposed by Horananious in 1862, was further established by Cheesman (1947) as a genus separate from *Musa*. This separation of the two genera was based mainly on the differences in chromosome numbers, and on pseudostem morphology. The basic chromosome number in *Ensete* was reported to be 9, whereas that of *Musa* was observed to be 9, 10, and 11 (Agharakar & Bhaduri, 1933).

HISTORICAL CULTIVATION AND USAGE

As early as 1893, Baker noted that *E. superbum* was in common cultivation, and it was introduced into the Calcutta Botanical Gardens in the year 1800 (Baker, 1893). The plant dies down in the dry season and forms new leaves at the beginning of the monsoon season. This species can be propagated by seeds, but cannot produce vegetative side suckers naturally like other *Ensete* species. It has the capacity to withstand severe drought, and has therefore been used during famines in Ethiopia. It is excellent for outdoor cultivation, where weather permits, and grows well with liberal fertilization when established.

PRESENT-DAY CULTIVATION AND USAGE

Since conventional propagation of *Ensete* is time consuming, there was a need for optimization of tissue culture techniques for its rapid propagation. Research revealed that the encapsulated shoot tips can be handled like a seed, and could be useful in minimizing the cost of production, as 1 ml of medium is sufficient for encapsulation of a single shoot tip, compared to 15—20 ml for conversion of shoot tips into plantlets. By directly sowing the encapsulated shoot tips in soil, the two-stage process (such as rooting and hardening) can be eliminated. As compared to suckers, encapsulated shoot tips are inexpensive, and easier and safer material for germplasm exchange, maintenance, and transportation (Rao *et al.*, 1993).

Recently multiple shoots were induced from *in vitro* cultures of male floral apices of *Ensete superbum* Cheesm. BAP (6-benzylaminopurine) in combination with GA3 (gibberellic acid) was found to be beneficial for the establishment of cultures, and the multiplication and elongation of shoots, while auxins were shown to promote rooting (Kulkarni *et al.*, 1997). Thus, somatic embryogenesis offers an ideal system for the production of somatic embryos on a large scale for use in the preparation of synthetic seeds, propagation, and genetic transformation (Kulkarni, 1997).

APPLICATIONS TO HEALTH PROMOTION AND DISEASE PREVENTION

Traditional uses

Under Ayurvedic practice, the flesh of the fruit is given to diabetics. The ground seeds are also used as Ayurvedic medicine; the powdered seed of this plant is popularly prescribed in the prevention and treatment of smallpox and chickenpox by physicians who practice the indigenous system of therapy.

The seeds are fried, the inner pulp soaked in water, boiled and again soaked in water, and the resultant solution given orally to pregnant women in the third trimester to develop immune resistance in the baby. Processed seeds are consumed for hip pain by Kadars and Malasars. Seed paste is applied to the body for fever with body pains, and for scabies, by the Malasar (Ravishankar *et al.*, 1994).

In southern India, the ripe ground seeds of *Ensete superbum* are eaten to cure dysentery. The seed and the stem are also reported to be given for dog bites.

Seed powder is used for treating kidney stones and painful urination (Yeshodharan & Sujana, 2007). Powdered seeds are also given with milk for diabetes (Sreedharan, 2004), and approximately 5 g seeds are crushed and taken with water for stomachache (Jagtap, 2008). The seeds of *Ensete superbum* are used for the treatment of kidney and vesical calculi, urinary retention, burns and scalds, and menstrual disorders (www.botanical.com/site/column_poudhia/144_kela.html).

Chemistry

The petroleum ether extract of dried seed powder yielded fatty oil and triterpenoid esters. The subsequent aqueous methanol (50 : 50) extract, on concentration and maceration with absolute methanol, yielded a white solid, which tested positive for proanthocyanidin and glucoside. Detailed study showed that it consisted of pro-pelargonidin glucosides of varying degrees of polymerization. A 1-year-old sample of the seeds also gave a violet compound with the properties of the pelargonidin color base, but not identical with it (Mahey *et al.*, 1971).

The three fractions, VIDR-2T, VIDR-2GC, and VIDR-2GD, isolated from the seeds of *E. superbum* Cheesm showed presence of phosphates, and traces of alkaloids, glycoside, and sugars (Dutta *et al.*, 1968).

The extractive values of the aqueous extract, ethanolic extract, and isolated compound of the seeds of *Ensete superbum* Cheesm were found to be 25.0, 1.92, and 0.1%, respectively. The moisture content, total ash, water-soluble ash, acid-insoluble ash, alcohol-soluble extractive value, and water-soluble extractive values were found to be 0.8, 2.8, 1.49, 0.55, 5.8, and 22.92%, respectively (Table 138.1).

TABLE 138.1 Pharmacognostical Evaluation of Powdered Seeds of *Ensete superbum*

S. No.	Foreign Matter	Moisture Content (%)	Total Ash (%)	Water-soluble Ash (%)	Acid-insoluble Ash (%)	Alcohol-soluble Extractive Value (%)	Water-soluble Extractive Value (%)
1	NMT* 2%	0.8	2.8	1.58	0.54	5.8	22.65
2	NMT 2%	0.8	2.8	1.40	0.55	5.8	23.19
Mean	NMT 2%	0.8	2.8	1.49	0.55	5.8	22.92

*NMT, not more than.

Pharmacology

Many studies have been carried out, but the exact detailed pharmacology of *Ensete superbum* is not yet clear. Some of the reported and established activities are discussed below.

1. *Antiviral effect.* The three fractions (VIDR-2T, VIDR-2GC, and VIDR-2GD) isolated from the seeds of *Ensete superbum* Cheesm inhibited the growth of vaccinia and variola viruses on chick chorioallantoic membrane. In variola-infected mice, they prevented the disease from appearing and significantly cured infected mice even when the drug treatment had begun at a late stage (Dutta *et al.*, 1968).

2. *Cardiovascular activity.* VIDR-2GD at the dose of 0.5 −7.5 mg did not produce any effect on the isolated perfused heart of frog. The fraction VIDR-2GD, in dose of 100−150 mg/kg, caused prolonged and sustained hypotension in cats. It caused hypotension in cases of induced hypertension produced by occlusion of the carotid artery. Topical and intravenous administration of the drug produced congestion of the mesenteric blood vessels in rats and mice. Perfusion of frog's vessels with the drug showed a dilatory effect, with an increase in outflow (Roy *et al.*, 1968).

3. *Respiratory activity.* VIDR-2GD did not have any effect on the bronchial musculature. It neither enhanced nor antagonized the effect of histamine on the bronchus, and it did not produce any action on the tracheal chain (Roy *et al.*, 1968).

4. *Cholinergic activity.* A dose of 30 mg/ml caused immediate contraction of the ileum, but after atropine the drug effect was antagonized, showing a cholinergic-like effect. The amplitude of peristaltic movement was increased with 5 μg/ml of the drug, and adrenaline, in large doses, removed the tonicity. Tonicity of the uterus was increased, and adrenaline checked the tonicity. It has also shown a slight diuretic effect (Roy *et al.*, 1968).

5. *Hypnotic and sedative effects.* The drug potentiated phenobarbitone-induced sleep, but had no sedative effect when given alone (Roy *et al.*, 1968).

6. *Coagulant and anticoagulant activity.* In dogs and rabbits, the drug reduced the whole blood clotting time, recalcified plasma clotting time, and bleeding time (Roy *et al.*, 1968).

7. *Hypoglycemic and antiglycemic effects.* VIDR-2GD caused hypoglycemia (15%), at the end of the fifth hour, with a dose of 100 mg/kg. After 24 hours, the blood sugar remained lowered by 8.5%. It did not have any inhibitory effect on the growth of *E. histolytica in vitro* up to the concentration of 50 mg/ml. Toxicity studies showed that doses of up to 5 g/kg orally, 2.5 g/kg intraperitoneally, and 1 g/kg intravenously did not produce any deleterious effects during 7 days observation (Roy *et al.*, 1968).

8. *Antifertility activity.* The antifertility effect of VIDR-2GD, a substance isolated from the seeds of the Musaceae banakadali, was studied in rabbits, mice, rats, hamsters, and guinea pigs. In mice and rats, pregnancy was inhibited if the drug was given during the pre-implantation period; single doses of the drug within 3 days after mating inhibited pregnancy. In rabbits, pregnancy was prevented in 50−60% of the animals. The uteri of all the animals receiving the drug were congested and enlarged. No implantation sites were found. Rabbit endometrium showed hyperplasial cells. Uterine stimulation was noted in ovariectomized rats treated with VIDR-2GD. There was no antagonism to estradiol, and

TABLE 138.2 Comparative Anti-implantation Activity of the Ethanolic Extract and Isolated Compound of *Ensete superbum* Seeds

Treatment	Percentage Anti-implantation, Day 4 (Post-coital)	Percentage Anti-implantation, Days 1–7 (Post-coital)
Ethanolic extract		
Control	28.9	32.8
100 mg	53.65	86.9
125 mg	65.36	94.4
250 mg	82.29	98.77
Isolated compound		
Control	27.6	25.95
2.5 mg	65.9	87.5
5.0 mg	72.6	96.0

The ethanolic extract of the seeds of *Ensete superbum* at the dose of 250 mg/kg on Day 4 and Days 1–7 post-coital possess significant anti-implantation activity, and its isolated fraction (5 mg/kg), a chroman derivative, at the doses of 2.5 mg/kg and 5 mg/kg body weight on Day 4 and Days 1–7 post-coital also possess significant anti-implantation activity.

only a weak antigonadotropic activity was shown (Dutta *et al.*, 1970). The anti-implantation activity of the ethanolic extract of the seeds of *Ensete superbum* Cheesm (collected from the forest area of Waynard district in Kerala, India) at doses of 100, 125, and 250 mg/kg body weight when administered on days 1–7 post-coitally exhibited significant anti-implantation activity, and also very significant anti-estrogenic activity at the dose of 250 mg/kg body weight (Table 138.2).

A compound was also isolated from the ethanolic extract of the seeds of *Ensete superbum* Cheesm (Family Musaceae), from the variety found in Kerala; thin layer chromotography of this reported a single peak with Rf at 0.69 (Figure 138.3). On the basis of physical and spectral parameters (i.e., HPTLC, UV, FTIR, NMR, and mass spectrometry), the isolated compound was identified possibly as a chroman derivative, with the molecular formula $C_{16}O_4H_{22}$ — a non-steroidal phytosterol (Kachroo *et al.*, 2008). This isolated fraction, at doses of 2.5 and 5 mg/kg body weight, was found to exhibit significant anti-implantation activity (Table 138.2), and

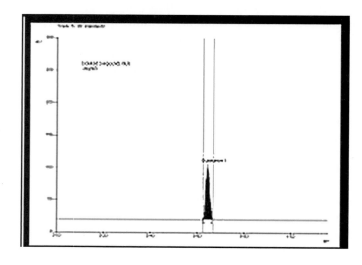

FIGURE 138.3

HPTLC chromatogram of the isolated chroman derivative. The HPTLC profile of the compound isolated from the ethanolic extract of *E. superbum* showed a single spot with Rf of 0.69 using mobile phase (toluene : ethyl acetate : formic acid, 5.0 : 4.5 : 0.5).

TABLE 138.3 Comparative Anti-estrogenic Activity of the Ethanolic Extract and Isolated Compound of *Ensete superbum*

Treatment		Mean Weight of Ovaries	Wet Weight of Uteri
Ethanolic extract	Control	27.02 ± 0.38	78 ± 132
Drug (oral)	125mg/kg	25.03 ± 0.59	65.03 ± 0.80
Oestradiol Benzoate	$(0.05 \ \mu g)$ s.c	36.17 ± 1.05	154.53 ± 194
Oestradiol Benzoate + Drug (125 mg/kg) oral	$(0.05 \ \mu g)$ s.c	30.87 ± 0.66	107.5 ± 0.70
Isolated Compound	Control	26.45 ± 0.82	35.94 ± 1.05
Drug	5mg/kg(oral)	25.16 ± 0.73	34.93 ± 0.98
Oestradiol Benzoate	$(0.05 \ \mu g)$ s.c	41.9 ± 15.7	115.74 ± 4.76
Oestradiol Benzoate + Drug	$(0.05 \ \mu g)$ s.c 5mg/kg(oral)	$35.91 + 0.99$	77.64 ± 4.24

The ethanolic extact of *E.Superbum* and its isolated chroman derivative when administered alone and with estradiol benzoate shows significant anti-estrogenic activity.
source of data personal research work.

also exhibited significant anti-estrogenic activity in immature female rats at the dose of 5 mg/kg (Table 138.3). Histopathological studies revealed significant anti-ovulatory activity in immature female rats at a dose of 5 mg/kg (Kachroo & Agrawal, 2009).

ADVERSE EFFECTS AND REACTIONS (ALLERGIES AND TOXICITY)

No health hazards or side effects are known with the proper administration of designated therapeutic dosages. The seeds of *Ensete superbum* have a long history of use for food and medicinal purposes, and no adverse or side effects have been reported when used within the recommended quantities.

The LD_{50} (intraperitoneally) of VIDR-2GT was 174 mg/kg, while VIDR- 2GD and VIDR-2GC were non-toxic to mice (Dutta *et al.*, 1968). The oral LD_{50} value of the ethanolic extract of *Ensete superbum* was found to be 3235.9 mg/kg, which appears to be safe (Kachroo & Agrawal, 2009).

SUMMARY POINTS

- *Ensete superbum* Cheesm, is essentially a plant of Western Ghats of India. It is known as banakadli in Hindi, and wild banana in English. In Kerala, the plant is known as kal vazhai, or rock banana.
- The ground seeds are used in Ayurvedic medicine, and certain fractions isolated from the seeds have shown to possess anti-variola and anti-vaccinia properties.
- A compound with the molecular formula $C_{16}O_4H_{22}$, appearing to be a chroman derivative, has been isolated from seeds of *Ensete superbum*.
- The ethanolic extract, at doses of 100, 125 and 250 mg/kg body weight, and the isolated chroman, at doses of 2.5 and 5 mg/kg body weight, exhibit significant anti-implantation activity.
- The oral LD_{50} value of the ethanolic extract of *Ensete superbum* was found to be 3235.9 mg/kg, which appears to be safe.
- Although a few promising leads have been identified by studies on seeds of *Ensete superbum*, none of these have been seriously pursued due to the very limited yield of the components isolated.
- The compound isolated and evaluated on the basis of physical and spectral data (i.e., a chroman derivative) can be further be validated by HPLC/HPTLC as a marker compound for elaborate anti-fertility studies.

References

Agharakar, S. P., & Bhaduri, P. N. (1933). Variation of chromosome numbers in Musaceae Agriculture Research Institute Coimbatore. *Indian Journal of Agricultural Sciences, 3*(IV), 1098.

Baker, J. G. (1893). A synopsis of the genera and species of Musaceae. *Annals Botany, 7*, 189–229.

Baker, J. G. (1894). Botany of the hadramaut expedition. *Kew Bulletin, 1*, 241.

Cheesman, E. E. (1947). Classification of the bananas. The genus *Ensete* Horan. *Kew Bulletin, 2*, 97–106.

Dutta, N. K., Dave, K. H., Desai, S. M., & Mhasalkar, M. Y. (1968). Anti-variola and anti-vaccinia principals from the seeds of banakadali (*Ensete superbum*, Cheesm, Musaceae). *Indian Journal of Medical Research, 56*, 735–741.

Dutta, N. K., Mhasalkar, M. Y., & Fernando, G. R. (1970). Studies on the anti-fertility action of VIDR-2GD: a constituent isolated from the seeds of *Ensete superbum* Cheesm, Musaceae (banakadali). *Fertility Sterility, 21*, 247.

Jagtap, S. D. (2008). Ethnomedicobotanical uses of endemic and RET plants utilised by the Korku tribe of Amaravati District, Maharastra. *Indian Journal of Traditional Knowledge, 7*(2), 284–287.

Kachroo, M., Agrawal, S. S., & Sanjay, P. N. (2008). Characterization of a chroman derivative isolated from the seeds of *Ensete superbum* Cheesm. *Pharmacognosy Magazine, 4*, 114–117.

Kachroo, M., & Agrawal, S. S. (2009). Isolation, characterization and anti-fertility activity of the active moiety from the seeds of *Ensete superbum* Cheesm (banakadali). *Journal of Natural Remedies, 9*, 12–20.

Kulkarni, V. M., Ganapati, T. R., Suprasanna, P., Bapat, V. A., & Rao, P. S. (1997). *In vitro* propagation in *Ensete superbum* (Roxb.) Cheesman — a species closely related to Musa. *Indian Journal of Experimental Biology, 35*, 96–98.

Mahey, S., Mukerjee, S. K., Saroja, T., & Seshadri, T. R. (1971). Proanthocyanidin glycosides in Musa accuminata. *Indian Journal of Chemistry, 4*, 381.

Ravishankar, T., Vedavalli, L., Nambi, A. A., & Selvam, V. (1994). *Role of tribal people in the conservation and utilization of plant genetic resource.* Madras, India: MSSRF.

Rao, P. S., Ganapathi, T. R., Suprasanna, P., & Bapat, V. A. (1993). Encapsulated shoot tips of banana: a new propagation and delivery system. *InfoMusa, 2*(2), 4–5.

Roy, R. N., Bhagwager, S., & Dutta, N. K. (1968). Pharmacological observations on the fractions isolated from seed of banakadali. *Indian Journal of Pharmaceutical Sciences, 30*, 285.

Sreedharan, T. P. (2004). *Biological diversity of Kerala: A survey of Kalliasseri panchayat, Kannur district, Kerala.* Research Programme on Local Level Development Centre for Development Studies Thiruvananthapuram. Discussion Paper No. 62.

Yeshodharan, K., & Sujana, K. A. (2007). Ethno-medicinal knowledge among the Malamalsar tribe of Parambikulam wildlife sanctuary, Kerala. *Indian Journal of Traditional Knowledge, 6*(3), 481–485.

1182

1184

1187